Essential Biochemistry

INTERNATIONAL ADAPTATION

Essential Biochemistry

Fifth Edition

INTERNATIONAL ADAPTATION

Charlotte W. Pratt
Kathleen Cornely

WILEY

Essential Biochemistry

Fifth Edition

INTERNATIONAL ADAPTATION

Contributing Subject Matter Experts: Dr. Arijit Bhattacharya, Associate Professor, Adamas University, West Bengal, India; Dr. Jana Balaji Chandra Mouli, Assistant Professor, Adikavi Nannaya University, Andhra Pradesh, India; Dr. Nasir Salam, Associate Professor, Jamia Millia Islamia, New Delhi, India; Dr. Savita Bansal, Assistant Professor, Institute of Home Economics, University of Delhi, New Delhi, India; Dr. Shiv Kumar Dubey, Assistant Professor, Govind Ballabh Pant University of Agriculture and Technology, Uttarakhand, India; Dr. Sorokhaibam Jibankumar Singh, Assistant Professor, Manipur University, Manipur, India.

Founded in 1807, John Wiley & Sons, Inc. has been a valued source of knowledge and understanding for more than 200 years, helping people around the world meet their needs and fulfill their aspirations. Our company is built on a foundation of principles that include responsibility to the communities we serve and where we live and work. In 2008, we launched a Corporate Citizenship Initiative, a global effort to address the environmental, social, economic, and ethical challenges we face in our business. Among the issues we are addressing are carbon impact, paper specifications and procurement, ethical conduct within our business and among our vendors, and community and charitable support. For more information, please visit our website: www.wiley.com/go/citizenship.

ISBN: 978-1-394-22327-5

ISBN: 978-1-394-22328-2 (ePub)

ISBN: 978-1-394-22329-9 (ePdf)

Printed and bound by CPI Group (UK) Ltd, Croydon, CR0 4YY

C9781394223275_150324

CHARLOTTE PRATT received a B.S. in biology from the University of Notre Dame and a Ph.D. in biochemistry from Duke University. She is a protein chemist who has conducted research in blood coagulation and inflammation at the University of North Carolina at Chapel Hill. She is currently Associate Professor in the Biology Department at Seattle Pacific University. Her interests include molecular evolution, enzyme action, and the relationship between metabolic processes and disease. She has written numerous research and review articles, has worked as a textbook editor, and is a co-author, with Donald Voet and Judith G. Voet, of *Fundamentals of Biochemistry*, published by John Wiley & Sons, Inc.

KATHLEEN CORNELY holds a B.S. in chemistry from Bowling Green (Ohio) State University, an M.S. in biochemistry from Indiana University, and a Ph.D. in nutritional biochemistry from Cornell University. She is currently the Robert H. Walsh '39 Endowed Professor in Chemistry and Biochemistry at Providence College, where she has focused on expanding the use of case studies and guided inquiry across a broad spectrum of classes. Her interest in active pedagogy has led to her involvement in national programs including Project Kaleidoscope, the POGIL Project, and the Howard Hughes Medical Institute SEA PHAGES program, which has also fueled her current experimental research in phage genomics. She has been a member of the editorial board of *Biochemistry and Molecular Biology Education* and has served for several years as coordinator of the undergraduate poster competition at the annual meeting of the American Society for Biochemistry and Molecular Biology.

*Available Online

Our goal in writing *Essential Biochemistry* has been to provide students with a succinct guide to modern biochemistry that includes plenty of chemistry, biological context, and problem-solving opportunities. We want our book to be readable and practical, presenting the basic concepts of molecular structure and function, metabolism, and molecular biology along with some insights into the biochemistry of health and disease.

This International Edition of *Essential Biochemistry, fifth edition* is our most ambitious revision—almost every page has at least minor changes. We looked for better ways to focus on the chemistry behind the biology, and we added some new topics to help students make connections between biochemistry and other subjects. We worked hard to integrate the updates so that students would find this book—like the previous editions—clear, concise, and manageable.

We kept the organization mostly the same, and the chapters include all the helpful pedagogical features you're used to. And because we believe that students learn by doing, at the end of each chapter you'll find even more problems for students to test their understanding. Of the 1867 problems—averaging about 85 per chapter—20% are new to this edition.

Among the major differences in the new edition is a streamlined Chapter 3, now called Nucleic Acid Structure and Function, Chapter 4, now called Amino Acid and Protein Structure to provide a solid foundation for understanding protein structure and function. Chapter 6 now called **Enzymes: Classification and Catalysis**, which gives understanding from the name itself for enzymes uses. Chapter 8 now called Lipids and Biological Membrane, Chapter 9 now called Biochemical Signaling. Chapter 13 now called Carbohydrate Metabolism which gives general briefing of all carbohydrates metabolism. New material also helps students relate biochemistry to previously encountered biological concepts such as chromosome division and the laws of inheritance. Material related to manipulating DNA has been moved to Chapter 20 (DNA Replication and Repair), which provides the appropriate context for appreciating techniques for copying, pasting, and sequencing DNA.

In this edition, Chapter 3 includes a brief introduction on Human Genome Project. Chapter 4 includes an updated section on Ramachandran plot, protein folding, and stability, with information about metamorphic proteins and intrinsically disordered proteins. We added a new section on antibody structure and function to Chapter 5 to present key features of immunoglobulins and some current applications. This chapter now have all pressure units to bar. We used a new approach to define uncompetitive and noncompetitive enzyme inhibition, breif discussion on SARS-CoV-2 main protease, and lysozyme which should help students better visualize these situations (Chapter 7). A new subsection on steroid hormones has been added, which gives students a knowledge of steroids hormones and their function in the body (Chapter 8). A new section on Clinical Connection: Glucose and Diabetes Mellitus/Lectin has been added to Chapter 11.

Additional diagrams now present all the intermediates of the pentose phosphate pathway (Chapter 13) and the Calvin cycle (Chapter 16).

Other notable changes are 17 new boxes covering topics such as archaeal cells, virus structures and viral replication, thioesters, types of G-proteins, α and ω-oxidation of fatty acids, enzymes in nucleotide biosynthesis, telomers in cancer opioids, the Maillard reaction, iron metabolism, the nuclear pore complex reverse transcription and cDNA. Significant updates in the text itself address practical application of genomics (Chapter 3), uses of enzyme in clinical diagnosis (Chapter 6) autophagy and ionophores (Chapter 9), the structures and mechanisms of sensory and other signaling proteins (Chapter 10), restructuring within chapter (Chapter 12 and Chapter 15), supermolecular complexes in respiration and photosynthesis (Chapters 15 and 16), the regulation of metabolism by sterols and oxygen (Chapters 17 and 19), and next-generation DNA sequencing (Chapter 20).

In addition to many small adjustments to improve clarity and readability, we have refreshed some diagrams and have moved information from figure captions to the text so that students won't overlook key details. Finally, the list of selected readings for each chapter has been revised with the goal of providing current and accessible reviews that students can follow.

We hope you'll find the fifth edition a valuable resource for your students. As always, we welcome your feedback and suggestions,

Charlotte Pratt
Kathleen Cornely

Pedagogical Elements

We hope that students will not simply memorize what they read but will use the textbook as a guide and learn to think and explore on their own. To that end, we have built into the chapters a variety of features to facilitate learning outside the lecture hall.

- **Do You Remember** review questions at the start of each chapter help students tie new topics to previously encountered material.

- **Key Concepts** for each section are based on verbs, giving students an indication of what they need to be able to *do*, not just *know*.

- **Concept check** study hints at the end of each section suggest activities that support learning.

- **Questions** following selected tables and figures prompt students to inspect information more closely and make comparisons.

- **Key sentences** summarizing main points are printed in italics to help students focus and review.

- **Metabolism overview figures** introduced in Chapter 12 and revisited in subsequent chapters help students place individual metabolic pathways into a broader context.
- **Chapter Summaries**, organized by section headings, highlight the most important concepts in each section.
- **Key terms** are in boldface and are listed at the end of the chapter, with complete definitions in the **Glossary**.
- **Tools and Techniques** sections appear at the end of Chapters 2, 4, and 20 to showcase practical aspects of biochemistry and provide an overview of experimental techniques that students will encounter in their reading or laboratory experience.
- **Clinical Connection** sections in Chapters 2, 4, 5, 6, 7, 11, 13, 19, and 20 provide more in-depth exploration of the biochemical basis of diseases.
- **Selected Readings** are listed at the end of each chapter, with short descriptions, for students to consult as sources of additional information and context.
- **Problems** for each chapter—20% more than in the last edition—are grouped by section and are offered in pairs, with the answers to odd-numbered problems provided in an appendix.

Organization

We have chosen to focus on aspects of biochemistry that tend to receive little coverage in other courses or present a challenge to many students. Thus, in this textbook, we devote proportionately more space to topics such as acid–base chemistry, enzyme mechanisms, enzyme kinetics, oxidation–reduction reactions, oxidative phosphorylation, photosynthesis, and the enzymology of DNA replication, transcription, and translation. At the same time, we appreciate that students can become overwhelmed with information. To counteract this tendency, we have intentionally left out some details, particularly in the chapters on metabolic pathways, in order to emphasize some general themes, such as the stepwise nature of pathways, their evolution, and their regulation.

The 22 chapters of *Essential Biochemistry* are relatively short, so that students can spend less time reading and more time extending their learning through active problem-solving. Most of the problems require some analysis rather than simple recall of facts. Many problems based on research data provide students a glimpse of the "real world" of science and medicine.

Although each chapter of *Essential Biochemistry, International Adaptation* is designed to be self-contained so that it can be covered at any point in the syllabus, the 22 chapters are organized into four parts that span the major themes of biochemistry, including some chemistry background, structure–function relationships, the transformation of matter and energy, and how genetic information is stored and made accessible.

Part 1 of the textbook includes an introductory chapter and a chapter on water. Students with extensive exposure to chemistry can use this material for review. For students with little previous experience, these two chapters provide the chemistry background they will need to appreciate the molecular structures and metabolic reactions they will encounter later.

Part 2 begins with a chapter on the genetic basis of macromolecular structure and function (Chapter 3, Nucleic Acid Structure and Function). This is followed by chapters on protein structure (Chapter 4, Amino Acid and Protein Structure) and protein function (Chapter 5), with coverage of myoglobin and hemoglobin, cytoskeletal and motor proteins, and antibodies. An explanation of how enzymes work (Chapter 6, Enzymes: Classifications and Catalysis) precedes a discussion of enzyme kinetics (Chapter 7), an arrangement that allows students to grasp the importance of enzymes and to focus on the chemistry of enzyme-catalyzed reactions before delving into the more quantitative aspects of enzyme kinetics. A chapter on lipid chemistry (Chapter 8, Lipids and Biological Membranes) is followed by two chapters that discuss critical biological functions of membranes (Chapter 9, Transport Through Membrane, and Chapter 10, Biochemical Signaling). The section ends with a chapter on carbohydrate chemistry (Chapter 11), completing the survey of molecular structure and function.

Part 3 begins with an introduction to metabolism that provides an overview of fuel acquisition, storage, and mobilization as well as the thermodynamics of metabolic reactions (Chapter 12). This is followed, in traditional fashion, by chapters on glucose and glycogen metabolism (Chapter 13, Carbohydrates Metabolism); the citric acid cycle (Chapter 14); electron transport and oxidative phosphorylation (Chapter 15); the light and dark reactions of photosynthesis (Chapter 16); lipid catabolism and biosynthesis (Chapter 17); and pathways involving nitrogen-containing compounds, including the synthesis and degradation of amino acids, the synthesis and degradation of nucleotides, and the nitrogen cycle (Chapter 18). The final chapter of Part 3 explores the integration of mammalian metabolism, with extensive discussions of hormonal control of metabolic pathways, disorders of fuel metabolism, and cancer (Chapter 19).

Part 4, the management of genetic information, includes three chapters, covering DNA replication and repair (Chapter 20), transcription (Chapter 21, Transcription and RNA Processing), and protein synthesis (Chapter 22). Because these topics are typically also covered in other courses, Chapters 20–22 emphasize the relevant biochemical details, such as topoisomerase action, nucleosome structure, mechanisms of polymerases and other enzymes, structures of accessory proteins, proofreading strategies, and chaperone-assisted protein folding.

Instructor Resources

- PowerPoint Art Slides.
- **Brief Bioinformatics Exercises** crafted by Rakesh Mogul at California State Polytechnic University, Pomona, provide detailed instructions for novices to access and use bioinformatics databases and software tools. Each of the 57 exercises includes multiple-choice questions to help students gauge their success in learning from these resources.
- Test Bank Questions by Scott Lefler, Arizona State University and Anne Grippo, Arkansas State University.

- **Bioinformatics Projects**, written by Paul Craig at Rochester Institute of Technology, provide guidance for 12 extended explorations of online databases, with questions, many openended, for students to learn on their own.
- PowerPoint Lecture Slides by Mary Peek, Georgia Institute of Technology.

- Personal Response System ("Clicker") Questions by Gail Grabner, University of Texas at Austin, and Mary Peek, Georgia Institute of Technology.
- Solutions to Odd-Numbered Problems.

ACKNOWLEDGMENTS

Unless otherwise noted, molecular graphics images were created using data from the Protein Data Bank (www.rcsb.org) with the PyMol Molecular Graphics System Version 2.1, Schrödinger LLC, originally developed by Warren DeLano, or with the Swiss-Pdb Viewer [Guex, N., Peitsch, M.C., SWISS-MODEL and the Swiss-Pdb Viewer: an environment for comparative protein modeling, *Electrophoresis* 18, 2714–2723 (1997).

We would like to thank everyone who helped develop *Essential Biochemistry, Fifth Edition,* including Associate Publisher Sladjana Bruno, Senior Editor Jennifer Yee, Senior Course Production Operations Specialist Patricia Gutierrez, Course Content Developers Corrina Santos and Andrew Moore, Senior Designer Thomas Nery, Editorial Assistant Samantha Hart, and the Lumina Datamatics team.

We also thank all the reviewers who provided essential feedback on manuscript and media, corrected errors, and made valuable suggestions for improvements that have been so important in the writing and development of *Essential Biochemistry, Fifth Edition*

Fifth Edition Reviewers:

Arkansas
Cindy L. White, *Harding University*
California
M. Nidanie Henderson-Stull, *Soka University of America*
Rakesh Mogul, *California State Polytechnic University, Pomona*
Florida
Vijaya Narayanan, *Florida International University*
Massachusetts
Emily Westover, *Brandeis University*
Michigan
Allison C. Lamanna, *University of Michigan*
Nebraska
Jing Zhang, *University of Nebraska, Lincoln*
North Dakota
Dennis R. Viernes, *University of Mary*
New York
Youngjoo Kim, *SUNY College at Old Westbury*
Texas
Autumn L. Sutherlin, *Abilene Christian University*
Virginia
Michael Klemba, *Virginia Tech*
Canada
Isabelle Barrette-Ng, *University of Windsor*
Sian T. Patterson, *University of Toronto*

Previous Edition Reviewers:

Arkansas
Anne Grippo, *Arkansas State University*
Arizona
Allan Bieber, *Arizona State University*
Matthew Gage, *Northern Arizona University*
Scott Lefler, *Arizona State University, Tempe*
Allan Scruggs, *Arizona State University, Tempe*
Richard Posner, *Northern Arizona State University*
California
Elaine Carter, *Los Angeles City College*
Daniel Edwards, *California State University, Chico*
Gregg Jongeward, *University of the Pacific*
Pavan Kadandale, *University of California, Irvine*
Paul Larsen, *University of California, Riverside*
Rakesh Mogul, *California State Polytechnic University, Pomona*
Brian Sato, *University of California, Irvine*
Colorado
Paul Azari, *Colorado State University*
Andrew Bonham, *Metropolitan State University of Denver*
Johannes Rudolph, *University of Colorado*
Connecticut
Matthew Fisher, *Saint Vincent's College*
Florida
David Brown, *Florida Gulf Coast University*
Georgia
Chavonda Mills, *Georgia College*
Mary E. Peek, *Georgia Tech University*
Rich Singiser, *Clayton State University*
Hawaii
Jon-Paul Bingham, *University of Hawaii-Manoa, College of Tropical Agriculture and Human Resources*
Illinois
Lisa Wen, *Western Illinois University*
Gary Roby, *College of DuPage*
Jon Friesen, *Illinois State University*
Constance Jeffrey, *University of Illinois, Chicago*
Stanley Lo, *Northwestern University*
Kristi McQuade, *Bradley University*
Indiana
Brenda Blacklock, *Indiana University-Purdue University Indianapolis*
Todd Hrubey, *Butler University*
Christine Hrycyna, *Purdue University*
Mohammad Qasim, *Indiana University-Purdue University*

Iowa

Don Heck, *Iowa State University*

Kansas

Peter Gegenheimer, *The University of Kansas*

Ramaswamy Krishnamoorthi, *Kansas State University*

Louisiana

James Moroney, *Louisiana State University*

Jeffrey Temple, *Southeastern Louisiana University*

Maine

Robert Gundersen, *University of Maine, Orono*

Massachusetts

Jeffry Nichols, *Worcester State University*

Michigan

Marilee Benore, *University of Michigan*

Kim Colvert, *Ferris State University*

Kathleen Foley, *Michigan State University*

Deborah Heyl-Clegg, *Eastern Michigan University*

Melvin Schindler, *Michigan State University*

Jon Stoltzfus, *Michigan State University*

Mark Thomson, *Ferris State University*

Minnesota

Sandra Olmsted, *Augsburg College*

Tammy Stobb, *St. Cloud State University*

Mississippi

Jeffrey Evans, *University of Southern Mississippi*

James R. Heitz, *Mississippi State University*

Arthur Chu, *Delta State University*

Missouri

Karen Bame, *University of Missouri, Kansas City*

Nuran Ercal, *Missouri University of Science & Technology*

Nebraska

Jodi Kreiling, *University of Nebraska, Omaha*

Madhavan Soundararajan, *University of Nebraska*

Russell Rasmussen, *Wayne State College*

New Jersey

Yufeng Wei, *Seton Hall University*

Bryan Spiegelberg, *Rider University*

New Mexico

Beulah Woodfin, *University of New Mexico*

New York

Wendy Pogozelski, *SUNY Geneseo*

Susan Rotenberg, *Queens College of CUNY*

Sergio Abreu, *Fordham University*

Ohio

Edward Merino, *University of Cincinnati*

Heeyoung Tai, *Miami University*

Lai-Chu Wu, *The Ohio State University*

Oklahoma

Charles Crittell, *East Central University*

Oregon

Jeannine Chan, *Pacific University*

Steven Sylvester, *Oregon State University*

Pennsylvania

Mahrukh Azam, *West Chester University of Pennsylvania*

Jeffrey Brodsky, *University of Pittsburgh*

David Edwards, *University of Pittsburgh School of Pharmacy*

Robin Ertl, *Marywood University*

Amy Hark, *Muhlenberg College*

Justin Huffman, *Pennsylvania State University, Altoona*

Michael Sypes, *Pennsylvania State University*

Sandra Turchi-Dooley, *Millersville University*

Laura Zapanta, *University of Pittsburgh*

Rhode Island

Lenore Martin, *University of Rhode Island*

Erica Oduaran, *Roger Williams University*

South Carolina

Carolyn S. Brown, *Clemson University*

Weiguo Cao, *Clemson University*

Ramin Radfar, *Wofford College*

Paul Richardson, *Coastal Carolina University*

Kerry Smith, *Clemson University*

Tennessee

Meagan Mann, *Austin Peay State University*

Texas

Johannes Bauer, *Southern Methodist University*

David W. Eldridge, *Baylor University*

Edward Funkhouser, *Texas A&M University*

Gail Grabner, *University of Texas, Austin*

Barrie Kitto, *University of Texas at Austin*

Marcos Oliveira, *Feik School of Pharmacy, University of the Incarnate Word*

Richard Sheardy, *Texas Woman's University*

Linette Watkins, *Southwest Texas State University*

Utah

Craig Thulin, *Utah Valley University*

Wisconsin

Sandy Grunwald, *University of Wisconsin La Crosse*

Canada

Isabelle Barrette-Ng, *University of Windsor*

The Chemical Basis of Life

DSP/deepseaphotography

The chemical reactions of living systems take place across a wide range of conditions. Although many microbial species can tolerate extreme heat, multicellular organisms require much more temperate habitats. One exception is *Alvinella pompejana*, the Pompeii worm, which lives near deep-sea hydrothermal vents and thrives at 42°C (107°F). Hair-like colonies of symbiotic bacteria may help insulate its body.

This first chapter offers a preview of the study of biochemistry, broken down into three sections that reflect how topics in this book are organized. First come brief descriptions of the four major types of small biological molecules and their polymeric forms. Next is a summary of the thermodynamics that apply to metabolic reactions. Finally, there is a discussion of the origin of self-replicating life-forms and their evolution into modern cells. These short discussions introduce some of the key players and major themes of biochemistry and provide a foundation for the topics that will be encountered in subsequent chapters.

1.1 What Is Biochemistry?

KEY CONCEPTS

Recognize the main themes of biochemistry.

Biochemistry is the scientific discipline that seeks to explain life at the molecular level. It uses the tools and terminology of chemistry to describe the various attributes of living organisms. Biochemistry offers answers to such fundamental questions as "What are we made of?" and "How do we work?" Biochemistry is also a practical science: It generates powerful techniques that underlie advances in other fields, such as genetics, cell biology, and immunology; it offers insights into the treatment of diseases such as cancer and diabetes; and it improves the efficiency of industries such as wastewater treatment, food production, and drug manufacturing.

Some aspects of biochemistry can be approached by studying individual molecules isolated from cells. A thorough understanding of each molecule's physical structure and chemical reactivity helps lead to an understanding of how molecules cooperate and combine to form larger functional units and, ultimately, the intact organism (**Fig. 1.1**). However, just as a clock completely disassembled no longer resembles a clock, information about a multitude of biological molecules does not necessarily reveal how an organism lives. Biochemists therefore investigate how organisms behave under different conditions or when a particular molecule is modified or absent. In addition, they collect vast amounts of information about molecular structures and functions—information that is stored and analyzed by computer, a field of study known as **bioinformatics.** A biochemist's laboratory is as likely to hold racks of test tubes as flasks of bacteria or computers.

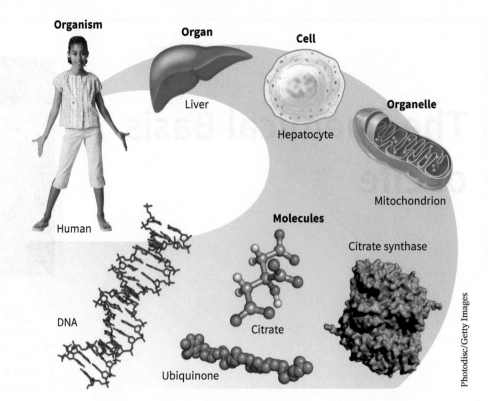

Figure 1.1 Levels of organization in a living organism. Biochemistry focuses on the structures and functions of molecules. Interactions between molecules give rise to higher-order structures (for example, organelles), which may themselves be components of larger entities, leading ultimately to the entire organism.

Chapters 3 through 22 of this book are divided into three groups that roughly correspond to three major themes of biochemistry:

1. *Living organisms are made of macromolecules.* Some molecules are responsible for the physical shapes of cells. Others carry out various activities in the cell. (For convenience, we often use *cell* interchangeably with *organism* since the simplest living entity is a single cell.) In all cases, the structure of a molecule is intimately linked to its function. Understanding a molecule's structural characteristics is therefore an important key to understanding its functional significance.

2. *Organisms acquire, transform, store, and use energy.* The ability of a cell to carry out metabolic reactions—to synthesize its constituents and to move, grow, and reproduce—requires the input of energy. A cell must extract this energy from the environment and spend it or store it in a manageable form.

3. *Biological information is transmitted from generation to generation.* Modern human beings look much like they did 100,000 years ago. Certain bacteria have persisted for millions, if not billions, of years. In all organisms, the genetic information that specifies a cell's structural composition and functional capacity must be safely maintained and transmitted each time the cell divides.

Several other themes run throughout biochemistry and we will highlight these where appropriate.

4. *Cells maintain a state of homeostasis.* Even within its own lifetime, a cell may dramatically alter its shape or metabolic activities, but it does so within certain limits. In order to remain in a steady, nonequilibrium state—**homeostasis**—the cell must recognize changing internal and external conditions and regulate its activities.

5. *Organisms evolve.* Over long periods of time, the genetic composition of a population of organisms changes. Examining the molecular makeup of living organisms allows biochemists to identify the genetic features that distinguish groups of organisms and to trace their evolutionary history.

6. *Diseases can be explained at the biochemical level.* Identifying the molecular defects that underlie human diseases, or investigating the pathways that allow one organism to infect another, is the first step in diagnosing, treating, preventing, or curing a host of ailments.

1.2 The Origin of Cells

KEY CONCEPTS

Summarize the evolutionary history of cells.

- List the events that must have occurred during prebiotic evolution.
- Name the three domains of life.
- Distinguish prokaryotic and eukaryotic cells.
- Summarize the importance of the human microbiota.

Every living cell originates from the division of a parental cell. Thus, the ability to **replicate** (make a replica or copy of itself) is one of the universal characteristics of living organisms. *In order to leave descendants that closely resemble itself, a cell must contain a set of instructions—and the means for carrying them out—that can be transmitted from generation to generation.* Over time, the instructions change gradually, so that species also change, or **evolve.** By carefully examining an organism's genetic information and the cellular machinery that supports it, biochemists can draw some conclusions about the organism's relationship to more ancient life-forms. The history of evolution is therefore contained not just within the fossil record but also in the molecular makeup of all living cells. For example, nucleic acids participate in the storage and transmission of genetic information in all organisms, and the oxidation of glucose is an almost universal means for generating metabolic free energy. Consequently, DNA, RNA, and glucose must have been present in the ancestor of all cells.

Prebiotic evolution led to cells

Experimental evidence strongly supports the hypothesis that early in the earth's history, small biological molecules such as amino acids and monosaccharides—and even the more elaborate nucleotides—arose spontaneously from inorganic (prebiotic) materials like H_2O, CH_4, and HCN. Such reactions, which need an energy source, might have occurred near hydrothermal vents, where water as hot as 350°C emerges from the sea floor (**Fig. 1.2**), or in terrestrial ponds exposed to lighting and ultraviolet radiation. A less-popular hypothesis is that organic molecules were delivered to the earth by meteorites, although the origin of those molecules also needs an explanation. Researchers

Figure 1.2 A hydrothermal vent.   Life may have originated at these "black smokers," where high temperatures, H_2S, and metal sulfides might have stimulated the formation of biological molecules.

have recovered amino acids and other biological molecules from hydrothermal vents and from laboratory experiments designed to mimic conditions on the early earth. Even with some uncertainty about the exact temperatures, reactant concentrations, and the involvement of catalysts like iron or nickel, organic molecules could have formed through hypothetical processes such as these:

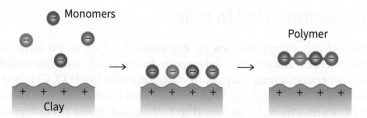

Glyceraldehyde
(carbohydrate)

Adenine
(nucleotide base)

Glycine
(amino acid)

Regardless of how they formed, the first biological building blocks must have accumulated, acquired reactive phosphate or thioester groups, and reached concentrations that allowed the formation of polymers. Polymerization might have been stimulated when the organic molecules—often bearing anionic (negatively charged) groups—aligned themselves on a cationic (positively charged) mineral surface:

Monomers

Polymer

Clay

In fact, in the laboratory, common clay promotes the polymerization of nucleotides into RNA. *Primitive polymers would have had to gain the capacity for self-replication.* Otherwise, no matter how stable or chemically versatile, such molecules would never have given rise to anything larger or more complicated: The probability of assembling a fully functional cell from a solution of thousands of separate small molecules is practically nil. Because RNA in modern cells represents a form of genetic information and participates in all aspects of expressing that information, it may be similar to the first self-replicating biopolymer. It might have made a copy of itself by first making a **complement,** a sort of mirror image, that could then make a complement of *itself*, which would be identical to the original molecule (**Fig. 1.3**).

A replicating molecule's chances of increasing in number depend on **natural selection,** the phenomenon whereby the entities best suited to the prevailing conditions are the likeliest to survive and multiply (**Box 1.A**). This would have favored a replicator that was chemically stable and had a ready supply of building blocks and free energy for making copies of itself. Accordingly, it would have been advantageous to become enclosed in some sort of membrane that could prevent valuable small molecules from diffusing away. Natural selection would also have favored replicating systems that developed the means for synthesizing their own building blocks and for more efficiently harnessing sources of free energy. Fossil evidence for microscopic life dates to about 3.5 billion years ago, about a billion years after the earth formed from interstellar dust.

The first cells were probably able to "fix" CO_2—that is, convert it to reduced organic compounds—using the free energy released in the oxidation of readily available inorganic compounds such as H_2S or Fe^{2+}. Vestiges of these processes can be seen in modern metabolic reactions that involve sulfur and iron.

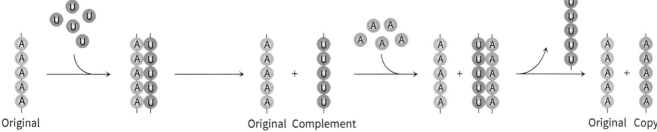

1. *The polyA molecule serves as a template for the synthesis of a polymer containing uracil nucleotides, U, which are complementary to adenine nucleotides (in modern RNA, A pairs with U).*

2. *The two polymer chains separate.*

3. *The polyU molecule serves as a template for the synthesis of a new complementary polyA chain.*

4. *The chains again separate and the polyU polymer is discarded, leaving the original polyA molecule and its exact copy.*

Original   Original Complement   Original Copy

Figure 1.3 Possible mechanism for the self-replication of a primitive RNA molecule. For simplicity, the RNA molecule is shown as a polymer of adenine nucleotides, A.

Question Draw a diagram showing how polyU would be replicated.

Box 1.A How Does Evolution Work?

Documenting evolutionary change is relatively straightforward, but the mechanisms whereby evolution occurs are prone to misunderstanding. Populations change over time, and new species arise as a result of natural selection. Selection operates on individuals, but its effects can be seen in a population only over a period of time. Most populations are collections of individuals that share an overall genetic makeup but also exhibit small variations due to random alterations (mutations) in their genetic material as it is passed from parent to offspring. In general, the survival of an individual depends on how well suited it is to the particular conditions under which it lives.

Individuals whose genetic makeup grants them the greatest rate of survival have more opportunities to leave offspring with the same genetic makeup. Consequently, their characteristics become widespread in a population and, over time, the population appears to adapt to its environment. A species that is well suited to its environment tends to persist; a poorly adapted species fails to reproduce and therefore dies out.

Because evolution is the result of random variations and changing probabilities for successful reproduction, it is inherently random and unpredictable. Furthermore, natural selection acts on the raw materials at hand. It cannot create something out of nothing but must operate in increments. For example, the insect wing did not suddenly appear in the offspring of a wingless parent but most likely developed bit by bit, over many generations, by modification of a heat-exchange appendage. Each step of the wing's development

would have been subject to natural selection, eventually making an individual that bore the appendage more likely to survive, perhaps by being able to first glide and then actually fly in pursuit of food or to evade predators.

Although we tend to think of evolution as an imperceptibly slow process, occurring on a geological time scale, it can be observed. For example, under optimal conditions, the bacterium *Escherichia coli* requires only about 20 minutes to produce a new generation. In the laboratory, a culture of *E. coli* cells can progress through about 2500 generations in a year (in contrast, 2500 human generations would require about 60,000 years). Hence, it is possible to subject a population of cultured bacterial cells to some "artificial" selection—for example, by making an essential nutrient scarce—and observe how the genetic composition of the population changes over time.

Experiments such as these have revealed that even after an initial period of rapid adaptation to the new conditions, the population continues to change, suggesting that evolution is an ongoing process with no fixed end point. In addition, so-called neutral mutations, which do not significantly affect an organism's fitness, randomly accumulate. For this reason, it is impossible to explain all genetic changes in terms of adaptations to specific conditions.

Question Why can't acquired (rather than genetic) characteristics serve as the raw material for evolution?

Later, photosynthetic organisms similar to present-day cyanobacteria (also called blue-green algae) used the sun's energy to fix CO_2:

$$CO_2 + H_2O \rightarrow (CH_2O) + O_2$$

The concomitant oxidation of H_2O to O_2 dramatically increased the concentration of atmospheric O_2, about 2.5 billion years ago, and made it possible for **aerobic** (oxygen-using) organisms to take advantage of this powerful oxidizing agent. The **anaerobic** origins of life are still visible in the most basic metabolic reactions of modern organisms; these reactions proceed in the absence of oxygen. Now that the earth's atmosphere contains about 20% oxygen, anaerobic organisms have not disappeared, but they have been restricted to microenvironments where O_2 is scarce, such as the digestive systems of animals or underwater sediments.

Eukaryotes are more complex than prokaryotes

The earth's present-day life-forms are of two types, which are distinguished by their cellular architecture:

1. ***Prokaryotes*** *are small unicellular organisms that lack a discrete nucleus and usually contain no internal membrane systems.* This group comprises two subgroups that are remarkably different metabolically, although they are similar in appearance: the eubacteria (usually just called **bacteria**), exemplified by *E. coli,* and the **archaea** (or archaebacteria), best known as organisms that inhabit extreme environments, although they are actually found almost everywhere (**Fig. 1.4**).

2. ***Eukaryotic*** *cells are usually larger than prokaryotic cells and contain a nucleus and other membrane-bounded cellular compartments* (such as mitochondria, chloroplasts, and endoplasmic reticulum). Eukaryotes may be unicellular or multicellular. This group (also called the **eukarya**) includes microscopic organisms as well as familiar macroscopic plants and animals (**Fig. 1.5**).

By analyzing the sequences of nucleotides in certain genes that are present in all species, it is possible to construct a diagram that indicates how the bacteria, archaea, and eukarya are related. *The number of sequence differences between two groups of organisms indicates how long ago they diverged from a common ancestor.* Species with similar sequences have a longer shared evolutionary history than species with dissimilar sequences. This sort of analysis has produced the evolutionary tree shown in **Figure 1.6**.

Eukaryotic cells are likely to have evolved from a mixed population of prokaryotic cells about 1.8 billion years ago. Over many generations of living in close proximity and sharing each other's metabolic products, some of the bacterial cells became stably incorporated inside archaeal cells, which accounts for the mosaic character of modern eukaryotic cells (**Fig. 1.7**).

The descendants of the once free-living bacteria became mitochondria, which carry out most of the eukaryote's oxidative metabolism, or chloroplasts, which carry out photosynthesis in plants and closely resemble the photosynthetic cyanobacteria. In fact, both mitochondria and chloroplasts contain their own genetic material and protein-synthesizing machinery and can grow and divide independently of the rest of the cell. Remnants of the archaeal ancestor of eukaryotes are found in the eukaryotic cytoskeleton and DNA-replication enzymes (**Box 1.B**).

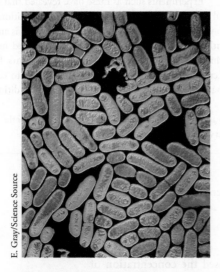

Figure 1.4 Prokaryotic cells.
These single-celled *Escherichia coli* bacteria lack a nucleus and internal membrane systems.

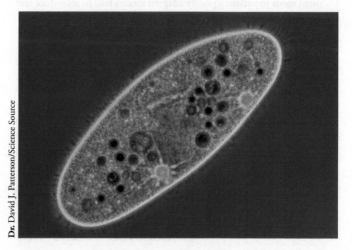

Figure 1.5 A eukaryotic cell. The paramecium, a one-celled organism, contains a nucleus and other membrane-bounded compartments.

Question **Describe the visible differences between prokaryotic and eukaryotic cells (Figures 1.4 and 1.5).**

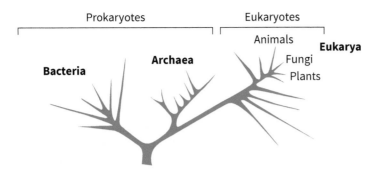

Figure 1.6 Evolutionary tree based on nucleotide sequences. This diagram reveals that the ancestors of archaea and bacteria separated before the eukarya emerged from an archaea-like ancestor. Note that the closely spaced fungi, plants, and animals are actually more similar to each other than are many groups of prokaryotes.

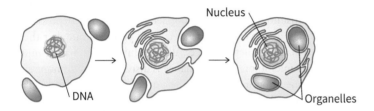

Figure 1.7 Possible origin of eukaryotic cells. The close association of different kinds of free-living cells gradually led to the modern eukaryotic cell, which appears to be a mosaic of bacterial and archaeal features and contains organelles that resemble whole bacterial cells.

The evolution of eukaryotes included the development of extensive intracellular membranes, many of which enclose discrete compartments, or **organelles,** with specialized functions, such as lysosomes (for the degradation of macromolecules), peroxisomes (for some oxidative reactions), and vacuoles (for storage). Like mitochondria and chloroplasts, the nucleus is surrounded by a double membrane; the outer nuclear membrane bends and folds extensively to form the endoplasmic reticulum. Some key features of eukaryotic cell structure are shown in **Figure 1.8**. A few types of prokaryotic cells also contain membrane-enclosed compartments devoted to storage or to chemical processes that could potentially damage the rest of the cell, but these cells lack a nucleus and the variety of organelles that characterize eukaryotic cells.

Many types of prokaryotic and eukaryotic cells live in colonies, which may increase metabolic efficiency through the cooperation of individual cells. However, a true multicellular lifestyle, which entails the division of labor and specialization among different types of cells, occurs only in eukaryotes. Multicellular organisms first appeared in the fossil record about 600 million years ago.

The earth currently sustains about 10 million different species (although estimates vary widely). Perhaps some 500 million species have appeared and vanished over the course of evolutionary history. It is unlikely that the earth harbors more than a few mammals that have yet to be discovered, but new microbial species are routinely described. Although the number of known prokaryotes (about 10,000) is much less than the number of known eukaryotes (for example, there are about 900,000 known species of insect), prokaryotic metabolic lifestyles are amazingly varied.

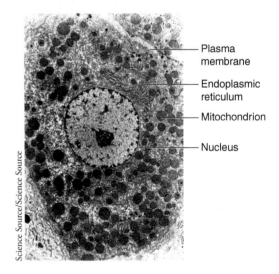

Science Source/Science Source

Figure 1.8 Eukaryotic cell structure.
Key components of an animal cell are labeled in this electron micrograph. A typical plant cell also contains chloroplasts and is surrounded by a cell wall.

The human body includes microorganisms

The human body consists of an estimated 10 trillion (10^{13}) cells, and there may be as many as 40 to 100 trillion microorganisms living in and on the body, including bacteria, archaea, and fungi (viruses are also present but are molecular parasites rather than true cells). The "foreign" cells, mostly living in the intestine, form an integrated community called the **microbiota.** Very few of these cells are pathogens that cause disease; in fact, a diverse and stable microbiota actually helps to prevent the growth of harmful species.

Analysis of microbial DNA has allowed characterization of the **microbiome,** the genetic information contributed by the organisms that make up the microbiota. The microbiome reveals that the human species hosts several thousand different species of microorganisms. Many have not yet been identified, and relatively few have been cultured. However, an individual person typically harbors only a few hundred species of microorganisms. Establishing this community of cells begins at birth and is mostly complete after about a year. The mix of species remains fairly constant throughout the person's lifetime, although the proportions may fluctuate somewhat. The microbiota can vary markedly among individuals, even in the same household. However, the overall metabolic capabilities of the microbial community seem to matter more than which species are actually present. Contrary to expectations, there is no core microbiota that is common to all individuals.

The gut microbiota in particular plays a large role in digestion, providing nutrients to the host and regulating metabolic functions. However, byproducts of microbial metabolism can act as hormones and neurotransmitters with possible links to anxiety and depression. In turn, antibiotics, other types of drugs, and even some personal care products can alter the community of microorganisms and the chemicals they produce. All the while, the human immune system must ignore—or "tolerate"—the microbiota. If this balance is disrupted, the immune system may react to the microorganisms, generating inflammatory responses that can lead to diabetes or inflammatory bowel disease. Untangling these complicated relationships is the goal of the Integrated Human Microbiome Project (https://hmpdacc.org/ihmp/), which organizes data describing the species that inhabit the human body and contribute to human health and disease.

Box 1.B   Archaeal cells

Archaeal cells represent one of the largest single-celled prokaryotic microorganisms. In 1980, Carl Woes recognized archaea as a distinct domain living in extreme environments such as salt lakes, hot springs, highly acidic bogs, and ocean depth. Archaea were initially isolated from extreme environments; however, over the past few years, they have been identified in an enormous range of environments. The archaea constitute one of the great domains of living creatures showing similarities with bacteria and eukaryotes. They have no nuclear compartment, no complex endomembrane system, circular genome lacking introns and are often co-located in operons and co-expressed as in the case of bacteria. However, they show similarities with eukaryotic systems in types of machinery that participate in gene replication and their expression. They also possess a set of common cellular features that separate them from both bacteria and eukaryotes. The most striking features that differentiate archaea from other living things include their cell membrane components having chirality of glycerol, ether linkage, isoprenoid chains, and branching of side chains. They also have a unique motility apparatus called the archaellum, and a unique type of metabolism called methanogenesis. The ability of Archaea to adapt to extreme environments is due to the presence of unique cell wall components. Archaeal cells have a wide variety of shapes and sizes. Many of them have a morphology similar to bacteria, such as cocci and rods. Others have geometric morphologies having flat squares and triangle shapes. These shapes arise from the cellular control of local growth and the division under the control of cytoskeleton filaments and an overlying coat of glycosylated proteins called S-layer. The cell division of archaeal cells requires the rapid re-organization of cellular space, mid-cell constriction, and membrane scission. These steps are predominantly mediated by two main families of proteins namely tubulin FtsZ and the second one is ESCRT-III. Archaeal RNA polymerase of transcriptional machinery is complex, consisting of approximately 14 subunits. Archaea cell poses a single RNA polymerase like bacteria but resembles those of eukaryotes in multiple subunit complexity and sequence homology. Also, archaeal RNA polymerase is unable to initiate transcription *in vitro* like the feature of eukaryotic transcription initiation where transcriptional factors are required.

Question   Question   **What are the unique features in cell wall component of archaebacteria that adapt them to survive under extreme environments?**

Concept Check

1. Describe how simple prebiotic compounds could give rise to biological monomers and polymers.
2. Explain why anaerobic organisms arose before aerobic organisms, and why are they relatively scarce now.
3. Describe the differences between prokaryotes and eukaryotes.
4. Explain why eukaryotic cells appear to be mosaics.
5. List some functions of the human microbiota.

1.3 Biological Molecules

KEY CONCEPTS

Identify the major classes of biological molecules.

- List the elements found in biological molecules.
- Draw and name the common functional groups in biological molecules.
- Draw and name the common linkages in biological molecules.
- Distinguish the main structural features of amino acids, carbohydrates, nucleotides, and lipids.
- Identify the monomers and linkages in polypeptides, polysaccharides, and nucleic acids.
- Summarize the biological functions of the major classes of biological molecules.

Even the simplest organisms contain a staggering number of different molecules, yet this number represents only an infinitesimal portion of all the molecules that are chemically possible. For one thing, *only a small subset of the known elements are found in living systems* (**Fig. 1.9**). The most abundant of these are C, N, O, and H, followed by Ca, P, K, S, Cl, Na, and Mg. Certain **trace elements** are also present in very small quantities.

Virtually all the molecules in a living organism contain carbon, so biochemistry can be considered to be a branch of organic chemistry. In addition, nearly all biological molecules are constructed from H, N, O, P, and S. Most of these molecules belong to one of a few structural classes, which are described below.

Similarly, *the chemical reactivity of biomolecules is limited relative to the reactivity of all chemical compounds.* A few of the functional groups and intramolecular linkages that are common in biochemistry are listed in **Table 1.1**. Familiarity with these functional groups is essential for understanding the behavior of the different types of biological molecules we will encounter throughout this book.

Cells contain four major types of biomolecules

Most of the cell's small molecules can be divided into four classes. Although each class contains many members, *they are united under a single structural or functional definition.* Identifying a particular molecule's class may help predict its chemical properties and possibly its role in the cell.

1. Amino Acids
Among the simplest compounds are the **amino acids,** so named because they contain an amino group (—NH_2) and a carboxylic acid group (—COOH). Under

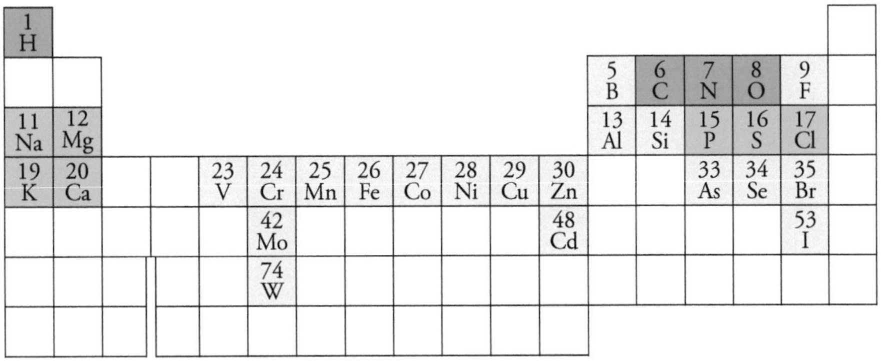

Figure 1.9 Elements found in biological systems. The most abundant elements are most darkly shaded; trace elements are most lightly shaded. Not every organism contains every trace element. Biological molecules primarily contain H, C, N, O, P, and S.

Table 1.1 Common Functional Groups and Linkages in Biochemistry

Compound name	Structure[a]	Functional group
Amine[b]	RNH_2 or RNH_3^+ R_2NH or $R_2NH_2^+$ R_3N or R_3NH^+	$-N\langle$ or $\overset{+}{-}N-$ (amino group)
Alcohol	ROH	—OH (hydroxyl group)
Thiol	RSH	—SH (sulfhydryl group)
Ether	ROR	—O— (ether linkage)
Aldehyde	$R-\overset{\displaystyle O}{\overset{\|}{C}}-H$	$-\overset{\displaystyle O}{\overset{\|}{C}}-$ (carbonyl group), $R-\overset{\displaystyle O}{\overset{\|}{C}}-$ (acyl group)
Ketone	$R-\overset{\displaystyle O}{\overset{\|}{C}}-R$	$-\overset{\displaystyle O}{\overset{\|}{C}}-$ (carbonyl group), $R-\overset{\displaystyle O}{\overset{\|}{C}}-$ (acyl group)
Carboxylic acid[b] (Carboxylate)	$R-\overset{\displaystyle O}{\overset{\|}{C}}-OH$ or $R-\overset{\displaystyle O}{\overset{\|}{C}}-O^-$	$-\overset{\displaystyle O}{\overset{\|}{C}}-OH$ (carboxyl group) or $-\overset{\displaystyle O}{\overset{\|}{C}}-O^-$ (carboxylate group)
Ester	$R-\overset{\displaystyle O}{\overset{\|}{C}}-OR$	$-\overset{\displaystyle O}{\overset{\|}{C}}-O-$ (ester linkage)
Thioester	$R-\overset{\displaystyle S}{\overset{\|}{C}}-OR$	$-\overset{\displaystyle S}{\overset{\|}{C}}-O-$ (thioester linkage)
Amide	$R-\overset{\displaystyle O}{\overset{\|}{C}}-NH_2$ $R-\overset{\displaystyle O}{\overset{\|}{C}}-NHR$ $R-\overset{\displaystyle O}{\overset{\|}{C}}-NR_2$	$-\overset{\displaystyle O}{\overset{\|}{C}}-N\langle$ (amido group)
Imine[b]	$R{=}NH$ or $R{=}NH_2^+$ $R{=}NR$ or $R{=}NHR^+$	$\rangle C{=}N-$ or $\rangle C{=}\overset{+}{N}\langle^{H}$ (imino group)
Phosphoric acid ester[b, c]	$R-O-\overset{\displaystyle O}{\underset{\displaystyle OH}{\overset{\|}{P}}}-OH$ or $R-O-\overset{\displaystyle O}{\underset{\displaystyle O^-}{\overset{\|}{P}}}-O^-$	$-O-\overset{\displaystyle O}{\underset{\displaystyle OH}{\overset{\|}{P}}}-O-$ (phosphoester linkage) $-\overset{\displaystyle O}{\underset{\displaystyle OH}{\overset{\|}{P}}}-OH$ or $-\overset{\displaystyle O}{\underset{\displaystyle O^-}{\overset{\|}{P}}}-O^-$ (phosphoryl group)
Diphosphoric acid ester[b, d]	$R-O-\overset{\displaystyle O}{\underset{\displaystyle OH}{\overset{\|}{P}}}-O-\overset{\displaystyle O}{\underset{\displaystyle OH}{\overset{\|}{P}}}-OH$ or $R-O-\overset{\displaystyle O}{\underset{\displaystyle O^-}{\overset{\|}{P}}}-O-\overset{\displaystyle O}{\underset{\displaystyle O^-}{\overset{\|}{P}}}-O^-$	$-O-\overset{\displaystyle O}{\underset{\displaystyle OH}{\overset{\|}{P}}}-O-\overset{\displaystyle O}{\underset{\displaystyle OH}{\overset{\|}{P}}}-O-$ (phosphoanhydride linkage) $-\overset{\displaystyle O}{\underset{\displaystyle OH}{\overset{\|}{P}}}-O-\overset{\displaystyle O}{\underset{\displaystyle OH}{\overset{\|}{P}}}-OH$ or $-\overset{\displaystyle O}{\underset{\displaystyle O^-}{\overset{\|}{P}}}-O-\overset{\displaystyle O}{\underset{\displaystyle O^-}{\overset{\|}{P}}}-O^-$ (diphosphoryl group, pyrophosphoryl group)

[a] R represents any carbon-containing group. In a molecule with more than one R group, the groups may be the same or different.

[b] Under physiological conditions, these groups are ionized and hence bear a positive or negative charge.

[c] When R = H, the molecule is inorganic phosphate (abbreviated P_i), usually $H_2PO_4^-$ or HPO_4^{2-}.

[d] When R = H, the molecule is pyrophosphate (abbreviated PP_i).

Question Cover the structure column and draw the structure for each compound listed on the left. Do the same for each functional group.

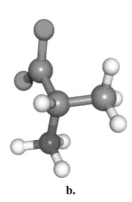

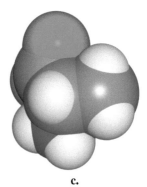

$$COO^-$$
$$H—\overset{\displaystyle |}{\underset{\displaystyle |}{C}}—CH_3$$
$$NH_3^+$$
a.

A structural formula includes all the atoms and the major bonds. Some bonds, such as the C—O and N—H bonds, are implied. The central carbon has tetrahedral geometry. The horizontal bonds extend slightly above the plane of the page and the vertical bonds extend slightly behind it.

b.

A ball-and-stick representation more accurately reveals the arrangement of atoms in space, although it does not show their relative sizes or electrical charges. The atoms are color-coded by convention: C gray, N blue, O red, and H white.

c.

The space-filling model best represents the actual shape of the molecule but may obscure some of its atoms and linkages. Each atom is shown as a sphere whose radius (the van der Waals radius) corresponds to the distance of closest approach by another atom.

Figure 1.10 Representations of alanine.   a. Structural formula, **b.** ball-and-stick model, and **c.** space-filling model.

physiological conditions, these groups are actually ionized to —NH$_3^+$ and —COO$^-$. The common amino acid alanine—like other small molecules—can be depicted in different ways, for example, by a structural formula, a ball-and-stick model, or a space-filling model (**Fig. 1.10**). Other amino acids resemble alanine in basic structure, but instead of a methyl group (—CH$_3$), they have another group—called a side chain or R group—that may also contain N, O, or S; for example,

Asparagine   Cysteine

2. Carbohydrates

Simple **carbohydrates** (also called **monosaccharides** or just sugars) have the formula $(CH_2O)_n$, where n is ≥ 3. Glucose, a monosaccharide with six carbon atoms, has the formula $C_6H_{12}O_6$. It is sometimes convenient to draw it as a ladder-like chain (*left*); however, glucose forms a cyclic structure in solution (*right*):

Glucose

In the representation of the cyclic structure, the darker bonds project in front of the page and the lighter bonds project behind it. In many monosaccharides, one or more hydroxyl groups are replaced by other groups, but the ring structure and multiple —OH groups of these molecules allow them to be easily recognized as carbohydrates.

3. Nucleotides

A five-carbon sugar, a nitrogen-containing ring, and one or more phosphate groups are the components of **nucleotides.** For example, adenosine triphosphate (ATP)

contains the nitrogenous group adenine linked to the monosaccharide ribose, to which a triphosphate group is also attached:

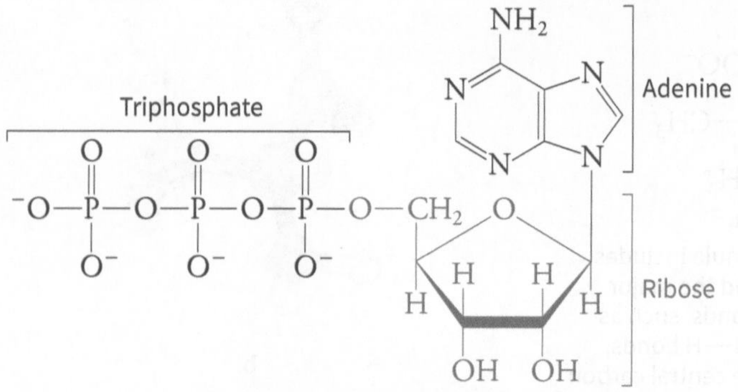

Adenosine triphosphate (ATP)

The most common nucleotides are mono-, di-, and triphosphates containing the nitrogenous ring compounds (or "bases") adenine, cytosine, guanine, thymine, or uracil (abbreviated A, C, G, T, and U).

4. Lipids   The fourth major group of biomolecules consists of the **lipids.** These compounds cannot be described by a single structural formula since they are a diverse collection of molecules. However, they all tend to be poorly soluble in water because the bulk of their structure is hydrocarbon-like. For example, palmitic acid consists of a highly insoluble chain of 15 carbons attached to a carboxylic acid group, which is ionized under physiological conditions. The anionic lipid is therefore called palmitate.

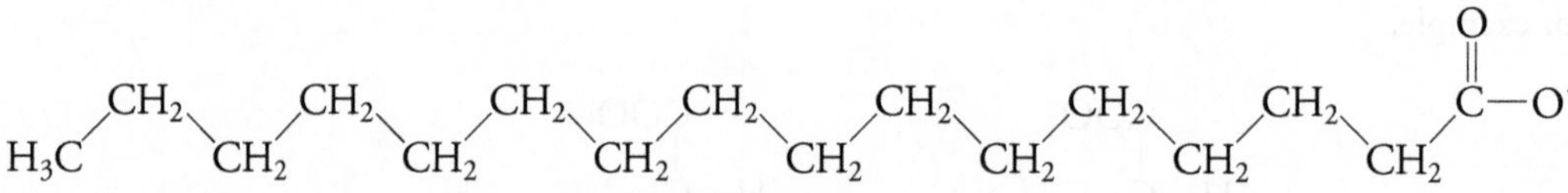

Palmitate

Cholesterol, although it differs significantly in structure from palmitate, is also poorly soluble in water because of its hydrocarbon-like composition.

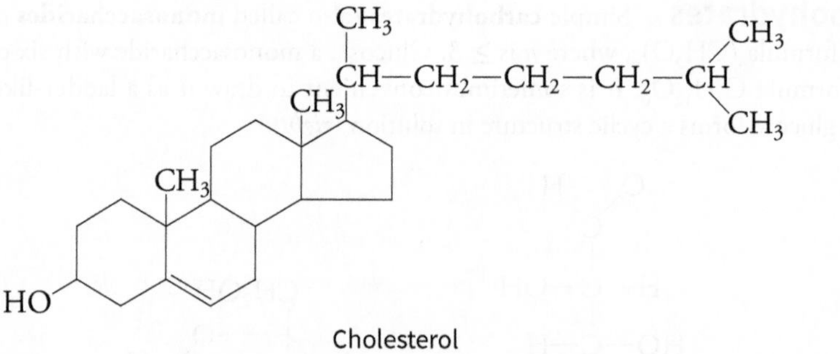

Cholesterol

Cells also contain a few other small molecules that cannot be easily classified into the groups above or that are constructed from molecules belonging to more than one group.

There are three major kinds of biological polymers

In addition to small molecules consisting of relatively few atoms, organisms contain macromolecules that may consist of thousands of atoms. Such huge molecules are not synthesized in one piece but are built from smaller units. This is a universal feature of nature: *A few kinds of building blocks can be combined in different ways to produce a wide variety of larger structures.* This is advantageous for a cell, which can get by with a limited array of raw materials. In addition, the very act of chemically linking individual units (**monomers**) into longer strings (**polymers**) is a way of encoding information (the sequence of the monomeric units) in a stable form. Biochemists use certain units of measure to describe both large and small molecules (**Box 1.C**).

Box 1.C   A Units Used in Biochemistry

Biochemists follow certain conventions when quantifying objects on a molecular scale. For example, the mass of a molecule can be expressed in atomic mass units; however, the masses of biological molecules—especially very large ones—are typically given without units. Here it is understood that the mass is expressed relative to one-twelfth the mass of an atom of the common carbon isotope ^{12}C (12.011 atomic mass units). Occasionally, units of daltons (D) are used (1 dalton = 1 atomic mass unit), often with the prefix kilo, k (kD). This is useful for macromolecules such as proteins, many of which have masses in the range from 20,000 (20 kD) to over 1,000,000 (1000 kD).

The standard metric prefixes are also necessary for expressing the minute concentrations of biomolecules in living cells. Concentrations are usually given as moles per liter (mol · L^{-1} or M), with the appropriate prefix, such as m, μ, or n:

mega (M)	10^6	nano (n)	10^{-9}
kilo (k)	10^3	pico (p)	10^{-12}
milli (m)	10^{-3}	femto (f)	10^{-15}
micro (μ)	10^{-6}		

For example, the concentration of the sugar glucose in human blood is about 5 mM, but many intracellular molecules are present at concentrations of μM or less.

Distances are customarily expressed in angstroms, Å (1 Å = 10^{-10} m) or in nanometers, nm (1 nm = 10^{-9} m). For example, the distance between the centers of carbon atoms in a C—C bond is about 1.5 Å and the diameter of a DNA molecule is about 20 Å.

Question The diameter of a typical spherical bacterial cell is about 1 μm. What is the cell's volume?

Amino acids, monosaccharides, and nucleotides each form polymeric structures with widely varying properties. In most cases, the individual monomers become covalently linked in head-to-tail fashion:

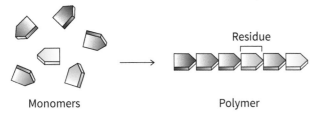

The linkage between monomeric units is characteristic of each type of polymer. The monomers are called **residues** after they have been incorporated into the polymer. Strictly speaking, lipids do not form polymers, although they do tend to aggregate to form larger structures such as cell membranes. Most of the mass of a cell consists of polymers, with proteins accounting for the greatest share (**Fig. 1.11**).

1. Proteins

Polymers of amino acids are called **polypeptides** or **proteins.** Twenty different amino acids serve as building blocks for proteins, which may contain many hundreds of amino acid residues. The amino acid residues are linked to each other by amide bonds called **peptide bonds.** A peptide bond (*arrow*) links the two residues in a dipeptide (the side chains of the amino acids are represented by R$_1$ and R$_2$).

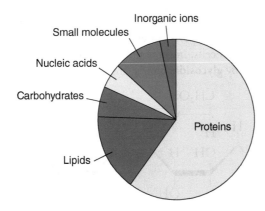

Figure 1.11 Mass of a mammalian cell. Proteins and lipids account for about 75% of the dry mass of a typical mammalian cell.

Because the side chains of the 20 amino acids have different sizes, shapes, and chemical properties, the exact **conformation** (three-dimensional shape) of the polypeptide chain depends on its amino acid composition and sequence. For example, the small polypeptide endothelin, with 21 residues, assumes a compact shape in which the polymer bends and folds to accommodate the functional groups of its amino acid residues (**Fig. 1.12**).

The 20 different amino acids can be combined in almost any order and in almost any proportion to produce myriad polypeptides, all of which have unique three-dimensional shapes. This property makes proteins as a class the most structurally variable and therefore the most functionally versatile of all the biopolymers. Accordingly, *proteins perform a wide variety of tasks in the cell, such as mediating chemical reactions and providing structural support.*

2. Nucleic Acids

Polymers of nucleotides are termed **polynucleotides** or **nucleic acids,** better known as DNA and RNA. Unlike polypeptides, with 20 different amino acids available for polymerization, each nucleic acid is made from just four different nucleotides. For example, the residues in RNA contain the bases adenine, cytosine, guanine, and uracil, whereas the residues in DNA contain adenine, cytosine, guanine, and thymine. Polymerization involves the phosphate and sugar groups of the nucleotides, which become linked by **phosphodiester bonds.**

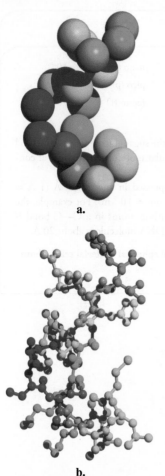

Figure 1.12 Structure of human endothelin. The 21 amino acid residues of this polypeptide, shaded from blue to red, form a compact structure. In **a,** each amino acid residue is represented by a sphere. The ball-and-stick model **b** shows all the atoms except hydrogen.

In part because nucleotides are much less variable in structure and chemistry than amino acids, nucleic acids tend to have more regular structures than proteins. *This is in keeping with their primary role as carriers of genetic information, which is contained in their sequence of nucleotide residues rather than in their three-dimensional shape* (**Fig. 1.13**). Nevertheless, many nucleic acids do bend and fold into compact globular shapes, as proteins do.

3. Polysaccharides

Polysaccharides usually contain only one or a few different types of monosaccharide residues, so even though a cell may synthesize dozens of different kinds of monosaccharides, most of its polysaccharides are homogeneous polymers. This tends to limit their potential for carrying genetic information in the sequence of their residues (as nucleic acids do) or for adopting a large variety of shapes and mediating chemical reactions (as proteins do). On the other hand, *polysaccharides perform essential cell functions by serving as fuel-storage molecules and by providing structural support.* For example, plants link the monosaccharide glucose, which is a fuel for virtually all cells, into the polysaccharide starch for long-term storage. The glucose residues are linked by **glycosidic bonds** (the bond is shown in red in this disaccharide):

Glucose monomers are also the building blocks for cellulose, the extended polymer that helps make plant cell walls rigid (**Fig. 1.14**). The starch and cellulose polymers differ in the arrangement of the glycosidic bonds between glucose residues.

The brief descriptions of biological polymers given above are generalizations, meant to convey some appreciation for the possible structures and functions of these macromolecules. *Exceptions to the generalizations abound.* For example, some small polysaccharides encode information that allows cells bearing the molecules on their surfaces to recognize each other. Likewise, some nucleic acids perform structural roles, for example, by serving as scaffolding in ribosomes, the small machines where protein synthesis takes place. Under certain conditions, proteins are called on as fuel-storage molecules. A summary of the major and minor functions of proteins, polysaccharides, and nucleic acids is presented in **Table 1.2**.

CGUACG
a.

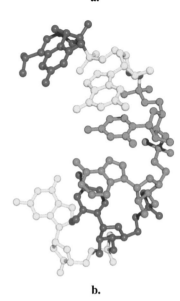

b.

Figure 1.13 Structure of a nucleic acid. **a.** Sequence of nucleotide residues, using one-letter abbreviations. **b.** Ball-and-stick model of the polynucleotide, showing all atoms except hydrogen (this structure is a six-residue segment of RNA).

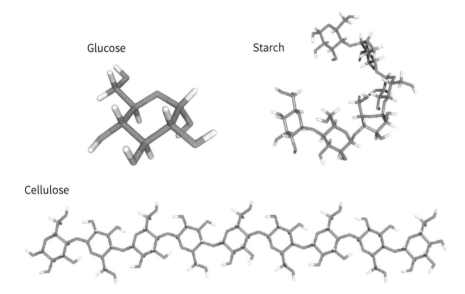

Glucose   Starch

Cellulose

Figure 1.14 Glucose and its polymers.   Both starch and cellulose are polysaccharides containing glucose residues. They differ in the type of chemical linkage between the monosaccharide units. Starch molecules have a loose helical conformation, whereas cellulose molecules are extended and relatively stiff.

Table 1.2 Functions of Biopolymers

Biopolymer	Encode information	Carry out metabolic reactions	Store energy	Support cellular structures
Proteins	—	✔	✓	✔
Nucleic acids	✔	✓	—	✓
Polysaccharides	✓	—	✔	✔

✔ major function
✓ minor function

Concept Check

1. List the six most abundant elements in biological molecules.
2. Name the common functional groups and linkages shown in Table 1.1.
3. Give the structural or functional definitions for amino acids, monosaccharides, nucleotides, and lipids.
4. Describe the advantage of building a polymer from monomers.
5. Give the structural definitions and major functions of proteins, polysaccharides, and nucleic acids.
6. Why is it efficient for macromolecules to be polymers?
7. What is the relationship between a monomer and a residue?

1.4 Energy and Metabolism

KEY CONCEPTS

Explain how enthalpy, entropy, and free energy apply to biological systems.

- Define enthalpy, entropy, and free energy.
- Write the equation that links changes in enthalpy, entropy, and free energy.
- Relate changes in enthalpy and entropy to the spontaneity of a process.
- Describe the energy flow that makes living systems thermodynamically possible.

Assembling small molecules into polymeric macromolecules requires energy. And unless the monomeric units are readily available, a cell must synthesize the monomers, which also requires energy. In fact, *cells require energy for all the functions of living, growing, and reproducing.*

It is useful to describe the energy in biological systems using the terminology of thermodynamics (the study of heat and power). An organism, like any chemical system, is subject to the laws of thermodynamics. According to the first law of thermodynamics, energy cannot be created or destroyed. However, it can be transformed. For example, the energy of a river flowing over a dam can be harnessed as electricity, which can then be used to produce heat or perform mechanical work. Cells can be considered to be very small machines that use chemical energy to drive metabolic reactions, which may also produce heat or carry out mechanical work.

Enthalpy and entropy are components of free energy

The energy relevant to biochemical systems is called the Gibbs free energy (after the scientist who defined it) or just **free energy.** It is abbreviated **G** and has units of joules per mol (J · mol⁻¹). Free energy has two components: enthalpy and entropy. **Enthalpy** (abbreviated **H**, with units of J · mol⁻¹) *is taken to be equivalent to the heat content of the system.* **Entropy** (abbreviated **S**, with units of J · K⁻¹ · mol⁻¹) *is a measure of how the energy is dispersed within that system.* Entropy can therefore be considered to be a measure of the system's disorder or randomness, because the more ways a system's components can be arranged, the more dispersed its energy. For example, consider a pool table at the start of a game when all 15 balls are arranged in one neat triangle (a state of high order or low entropy). After play has begun, the balls are scattered across the table, which is now in a state of disorder and high entropy (**Fig. 1.15**).

Free energy, enthalpy, and entropy are related by the equation

$$G = H - TS \tag{1.1}$$

Figure 1.15 Illustration of entropy. Entropy is a measure of the dispersal of energy in a system, so it reflects the system's randomness or disorder. **a.** Entropy is low when all the balls are arranged in a single area of the pool table. **b.** Entropy is high after the balls have been scattered, because there are now a large number of different possible arrangements of the balls on the table.

Question **Compare the entropy of a ball of yarn before and after a cat has played with it.**

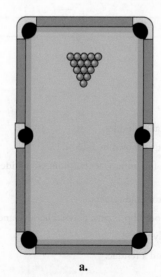
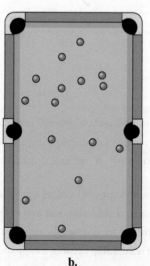

a.   b.

SAMPLE CALCULATION 1.1

Problem   Use the information below to calculate the change in enthalpy and the change in entropy for the reaction A → B.

	Enthalpy (kJ · mol^{-1})	Entropy (J · K^{-1} · mol^{-1})
A	60	22
B	75	97

Solution

$$\Delta H = H_B - H_A$$
$$= 75 \text{ kJ} \cdot \text{mol}^{-1} - 60 \text{ kJ} \cdot \text{mol}^{-1}$$
$$= 15 \text{ kJ} \cdot \text{mol}^{-1}$$
$$= 15{,}000 \text{ J} \cdot \text{mol}^{-1}$$

$$\Delta S = S_B - S_A$$
$$= 97 \text{ J} \cdot \text{K}^{-1} \cdot \text{mol}^{-1}$$
$$- 22 \text{ J} \cdot \text{K}^{-1} \cdot \text{mol}^{-1}$$
$$= 75 \text{ J} \cdot \text{K}^{-1} \cdot \text{mol}^{-1}$$

where T represents temperature in Kelvin (equivalent to degrees Celsius plus 273). Temperature is a coefficient of the entropy term because entropy varies with temperature; the entropy of a substance increases when it is warmed because more thermal energy has been dispersed within it. The enthalpy of a chemical system can be measured, although with some difficulty, but it is next to impossible to measure a system's entropy because this would require counting all the possible arrangements of its components or all the ways its energy could be spread out among them. Therefore, it is more practical to deal with *changes* in these quantities (change is indicated by the Greek letter delta, Δ) so that

$$\Delta G = \Delta H - T\Delta S \tag{1.2}$$

Biochemists can measure how the free energy, enthalpy, and entropy of a system differ before and after a chemical reaction. For example, **exothermic reactions** are accompanied by the release of heat to the surroundings ($H_{final} - H_{initial} = \Delta H < 0$), whereas **endothermic reactions** absorb heat from the surroundings ($\Delta H > 0$). Similarly, the entropy change, $S_{final} - S_{initial} = \Delta S$, can be positive or negative. When ΔH and ΔS for a process are known, Equation 1.2 can be used to calculate the value of ΔG at a given temperature (see Sample Calculation 1.1).

ΔG is less than zero for a spontaneous process

A china cup dropped from a great height will break, but the pieces will never reassemble themselves to restore the cup. The thermodynamic explanation is that the broken pieces have less free energy than the intact cup. *In order for a process to occur, the overall change in free energy (ΔG) must be negative.* For a chemical reaction, this means that the free energy of the products must be less than the free energy of the reactants:

$$\Delta G = G_{products} - G_{reactants} < 0 \tag{1.3}$$

When ΔG is less than zero, the reaction is said to be **spontaneous** or **exergonic**. A **nonspontaneous** or **endergonic** reaction has a free energy change greater than zero; in this case, the reverse reaction is spontaneous:

A → B	B → A
$\Delta G > 0$	$\Delta G < 0$
Nonspontaneous	Spontaneous

Note that thermodynamic spontaneity does not indicate how *fast* a reaction occurs, only whether it will occur as written. (The rate of a reaction depends on other factors, such as the concentrations of the reacting molecules, the temperature, and the presence of a catalyst.) When a reaction, such as A → B, is at equilibrium, the rate of the forward reaction is equal to the rate of the reverse reaction, so there is no net change in the system. In this situation, $\Delta G = 0$.

A quick examination of Equation 1.2 reveals that *a reaction that occurs with a decrease in enthalpy and an increase in entropy is spontaneous at all temperatures because ΔG is always less than zero.* These results are consistent with everyday experience. For example, heat moves spontaneously from a hot object to a cool object, and items that are neatly arranged tend to become disordered, never the other way around. (This is a manifestation of the second law of thermodynamics, which

SAMPLE CALCULATION 1.2

Problem   Use the information given in Sample Calculation 1.1 to determine whether the reaction A → B is spontaneous at 25°C.

Solution   Substitute the values for ΔH and ΔS, calculated in Sample Calculation 1.1, into Equation 1.2. To express the temperature in Kelvin, add 273 to the temperature in degrees Celsius: $273 + 25 = 298$ K.

$$\Delta G = \Delta H - T\Delta S$$
$$= 15,000 \text{ J} \cdot \text{mol}^{-1} - 298 \text{ K} (75 \text{ J} \cdot \text{K}^{-1} \cdot \text{mol}^{-1})$$

$$= 15,000 - 22,400 \text{ J} \cdot \text{mol}^{-1}$$
$$= -7400 \text{ J} \cdot \text{mol}^{-1}$$
$$= -7.4 \text{ kJ} \cdot \text{mol}^{-1}$$

Because ΔG is less than zero, the reaction is spontaneous. Even though the change in enthalpy is unfavorable, the large increase in entropy makes ΔG favorable.

states that energy tends to spread out.) Accordingly, reactions in which the enthalpy increases and entropy decreases do not occur. If enthalpy and entropy both increase or both decrease during a reaction, the value of ΔG then depends on the temperature, which governs whether the $T\Delta S$ term of Equation 1.2 is greater than or less than the ΔH term. This means that a large increase in entropy can offset an unfavorable (positive) change in enthalpy. Conversely, the release of a large amount of heat ($\Delta H < 0$) during a reaction can offset an unfavorable decrease in entropy (see Sample Calculation 1.2).

Life is thermodynamically possible

In order to exist, life must be thermodynamically spontaneous. Does this hold at the molecular level? When analyzed in a test tube (***in vitro,*** literally "in glass"), many of a cell's metabolic reactions have free energy changes that are less than zero, but some reactions do not. Nevertheless, the nonspontaneous reactions are able to proceed ***in vivo*** (in a living organism) because they occur in concert with other reactions that are thermodynamically favorable. Consider two reactions *in vitro,* one nonspontaneous ($\Delta G > 0$) and one spontaneous ($\Delta G < 0$):

$$A \rightarrow B \qquad \Delta G = +15 \text{ kJ} \cdot \text{mol}^{-1} \quad \text{(nonspontaneous)}$$
$$B \rightarrow C \qquad \Delta G = -20 \text{ kJ} \cdot \text{mol}^{-1} \quad \text{(spontaneous)}$$

When the reactions are combined, their ΔG values are added, so the overall process has a negative change in free energy:

$$A + B \rightarrow B + C \qquad \Delta G = (15 \text{ kJ} \cdot \text{mol}^{-1}) + (-20 \text{ kJ} \cdot \text{mol}^{-1})$$
$$A \rightarrow C \qquad \Delta G = -5 \text{ kJ} \cdot \text{mol}^{-1}$$

This phenomenon is shown graphically in **Figure 1.16**. In effect, the unfavorable "uphill" reaction A → B is pulled along by the more favorable "downhill" reaction B → C.

Cells couple unfavorable metabolic processes with favorable ones so that the net change in free energy is negative. Note that it is permissible to add ΔG values because the free energy, G, depends only on the initial and final states of the system, without regard to the specific chemical or mechanical work that occurred in going from one state to the other.

Most macroscopic life on earth today is sustained by the energy of the sun (this was not always the case, nor is it true of all organisms). In photosynthetic organisms, such as green plants, light energy excites certain molecules so that their subsequent chemical reactions occur with a net negative change in free energy. These thermodynamically favorable (spontaneous) reactions are coupled to the unfavorable synthesis of monosaccharides from atmospheric CO_2 (**Fig. 1.17**). In this process, the carbon is **reduced.** Reduction, the gain of electrons, is accomplished by the addition of hydrogen or the removal of oxygen (the oxidation states of carbon are reviewed in **Table 1.3**). The plant—or an animal that eats the plant—can then break down the monosaccharide to use it as a fuel to power other metabolic activities. In the process, the carbon

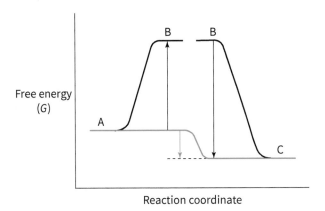

Figure 1.16 Free energy changes in coupled reactions. A nonspontaneous reaction, such as A → B, which has a positive value of ΔG, can be coupled to another reaction, B → C, which has a negative value of ΔG and is therefore spontaneous. The reactions are coupled because the product of the first reaction, B, is a reactant for the second reaction.

Question Which reaction occurs spontaneously in reverse: **C → B, B → A, or C → A?**

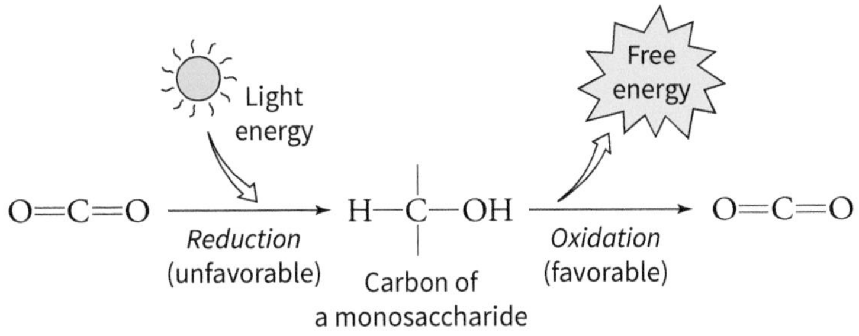

Figure 1.17 Reduction and reoxidation of carbon compounds. The sun provides the free energy to convert CO_2 to reduced compounds such as monosaccharides. The reoxidation of these compounds to CO_2 is thermodynamically spontaneous, so free energy can be made available for other metabolic processes. Note that free energy is not actually a substance that is physically released from a molecule.

Table 1.3 Oxidation States of Carbon

Compound[a]	Formula
Carbon dioxide *most oxidized* (*least reduced*)	O=C=O
Acetic acid	(structure)
Carbon monoxide	C≡O
Formic acid	(structure)
Acetone	(structure)
Acetaldehyde	(structure)
Formaldehyde	(structure)
Acetylene	H—C≡C—H
Ethanol	(structure)
Ethene	(structure)
Ethane	(structure)
Methane *least oxidized* (*most reduced*)	(structure)

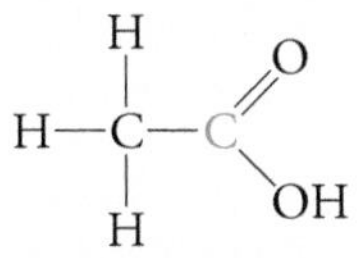

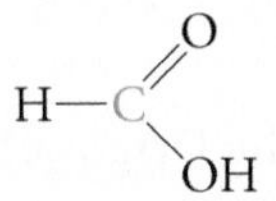

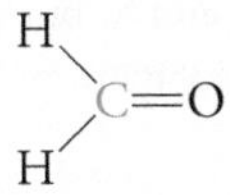

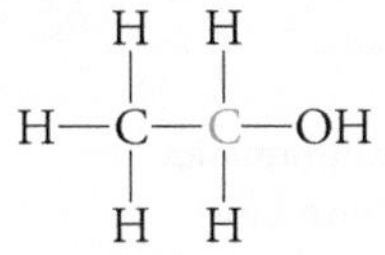

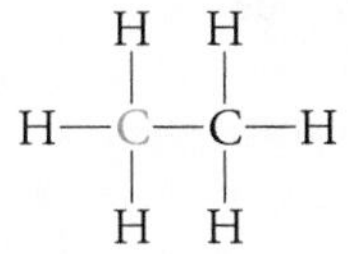

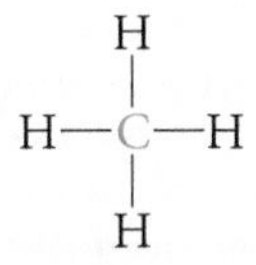

[a]Compounds are listed in order of decreasing oxidation state of the red carbon atom.

is **oxidized**—it loses electrons through the addition of oxygen or the removal of hydrogen—and ultimately becomes CO_2. The oxidation of carbon is thermodynamically favorable, so it can be coupled to energy-requiring processes such as the synthesis of building blocks and their polymerization to form macromolecules.

Virtually all metabolic processes occur with the aid of catalysts called **enzymes,** most of which are proteins (a catalyst greatly increases the rate of a reaction without itself undergoing any net change). For example, specific enzymes catalyze the formation of peptide, phosphodiester, and glycosidic linkages during polymer synthesis. Other enzymes catalyze cleavage of these bonds to break the polymers into their monomeric units.

A living organism—with its high level of organization of atoms, molecules, and larger structures—represents a state of low entropy relative to its surroundings. Yet the organism can maintain this thermodynamically unfavorable state as long as it continually obtains free energy from its food. Thus, living organisms do indeed obey the laws of thermodynamics. When the organism ceases to obtain a source of free energy from its surroundings or exhausts its stored food, the chemical reactions in its cells reach equilibrium ($\Delta G = 0$), which results in death.

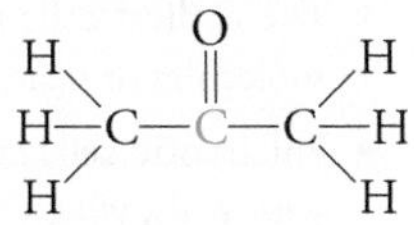

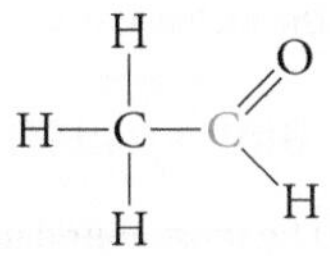

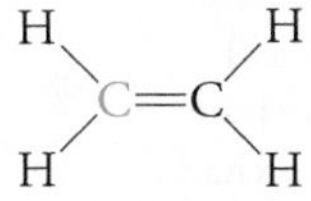

Concept Check

1. What do enthalpy and entropy mean and how are they related to free energy?
2. Show how increasing temperature affects ΔG when ΔH and ΔS are constant.
3. Explain how thermodynamically unfavorable reactions proceed *in vivo*.
4. Explain why an organism must have a steady supply of food.
5. Describe the cycle of carbon reduction and oxidation in photosynthesis and in the breakdown of a compound such as a monosaccharide.

SUMMARY

1.2 The Origin of Cells

- The earliest cells may have evolved in concentrated solutions of molecules or near hydrothermal vents.
- Eukaryotic cells contain membrane-bounded organelles. Prokaryotic cells, which are smaller and simpler, include the bacteria and the archaea.

1.3 Biological Molecules

- The most abundant elements in biological molecules are H, C, N, O, P, and S, but a variety of other elements are also present in living systems.

- The major classes of small molecules in cells are amino acids, monosaccharides, nucleotides, and lipids. The major types of biological polymers are proteins, nucleic acids, and polysaccharides.

1.4 Energy and Metabolism

- Free energy has two components: enthalpy (heat content) and entropy (disorder). Free energy decreases in a spontaneous process.
- Life is thermodynamically possible because unfavorable endergonic processes are coupled to favorable exergonic processes.

KEY TERMS

bioinformatics	polymer	polysaccharide	exergonic reaction	replication	eukaryote
homeostasis	residue	glycosidic bond	nonspontaneous	evolution	eukarya
trace element	polypeptide	free energy (*G*)	process	complement	organelle
amino acid	protein	enthalpy (*H*)	endergonic reaction	natural selection	microbiota
carbohydrate	peptide bond	entropy (*S*)	*in vitro*	aerobic	microbiome
monosaccharide	conformation	exothermic reaction	*in vivo*	anaerobic	
nucleotide	polynucleotide	endothermic reaction	reduction	prokaryote	
lipid	nucleic acid	ΔG	oxidation	bacteria	
monomer	phosphodiester bond	spontaneous process	enzyme	archaea	

BIOINFORMATICS

Brief Bioinformatics Exercises

1.1 The Periodic Table of the Elements and Domains of Life

1.2 Organic Functional Groups and the Three-Dimensional Structure of Vitamin C

PROBLEMS

1.2 The Origin of Cells

1. Why is molecular information so important for classifying and tracing the evolutionary relatedness of bacterial species but less important for vertebrate species?

2. The first theories to explain the similarities between bacteria and mitochondria or chloroplasts suggested that an early eukaryotic cell engulfed but failed to fully digest a free-living prokaryotic cell. Why is such an event unlikely to account for the origin of mitochondria or chloroplasts?

3. Draw a simple evolutionary tree that shows the relationships between species A, B, and C based on the DNA sequences given here.

Species A	T C G T C G A G T C
Species B	T G G A C T A G C C
Species C	T G G A C C A G C C

4. A portion of the evolutionary tree for a flu virus is shown here. Different strains are identified by an H followed by a number. **a.** Identify two pairs of closely related flu strains. **b.** Which strain(s) is(are) most closely related to strain H3?

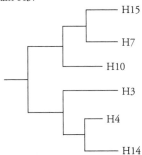

5. Propose an explanation why taking antibiotics sometimes leads to illness caused by the intestinal bacterium *Clostridium difficile*.

6. Clinicians know that a drug may be ineffective as an antibiotic when tested with pure cultures of bacteria in the laboratory. Yet when the drug is orally administered to a patient, it has the desired antibacterial effect against the same bacteria. Explain.

1.3 Biological Molecules

7. Use Table 1.1 to assign the appropriate compound name to each molecule.

a. $H_3C-(CH_2)_{14}-\overset{\overset{\displaystyle O}{\|}}{C}-OH$

b. H_3C-CH_2-CHO

c. $H_3C-\overset{\overset{\displaystyle O}{\|}}{C}-O-CH_2-CH_3$

d. H_3C-CH_2-OH

8. Use Table 1.1 to assign the appropriate compound name to each molecule.

a. $H_3C-O-CH_3$

b. $H_3C-O-\overset{\overset{\displaystyle OH}{|}}{\underset{\underset{\displaystyle O}{\|}}{P}}-O-CH_3$

c. H_3C-CH_2-SH

d. $H_3C-\overset{\overset{\displaystyle O}{\|}}{C}-NH_2$

e.

f.

g.

h.

9. The structures of three forms of vitamin A are shown. Use Table 1.1 to assign the appropriate compound name to each molecule.

Retinol

Retinal

Retinoic acid

10. Investigators synthesized a series of compounds that showed promise as drugs for the treatment of Alzheimer's disease. The structure of one of the compounds is shown below. Use Table 1.1 to identify the functional groups in this compound.

11. The structures of several molecules are shown below. Use Table 1.1 to identify the functional groups in each structure.

Phylloquinone
(Vitamin K)

Nicotinamide

Coenzyme Q

12. Coenzyme A is an important carrier of acetyl groups in metabolism. Its structure is shown in Figure 3.2. Use Table 1.1 to identify the functional groups in this molecule.

13. If the carboxylic group of one amino acid is condensed with an amino group of another amino acid, what new functional group is formed?

14. Name the four types of small biological molecules. Which three are capable of forming polymeric structures? What are the names of the polymeric structures that are formed?

15. To which of the four classes of biomolecules do the following compounds belong?

a.

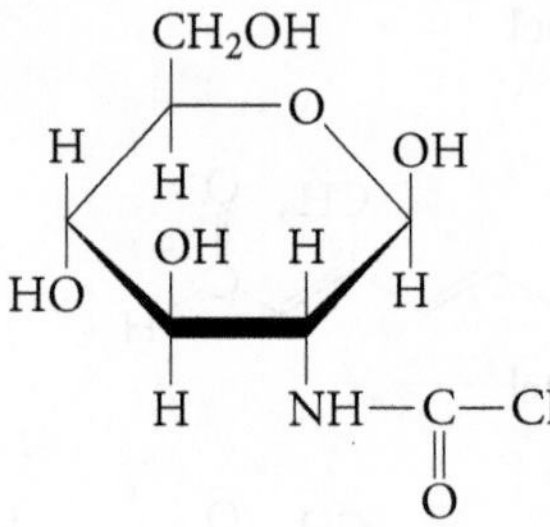

b.

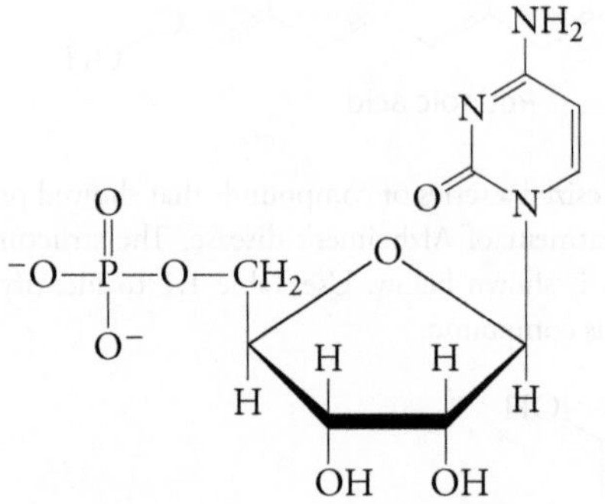

c. HS—CH₂—CH₂—CH—COO⁻
            |
           NH₃⁺

$$\text{HS—CH}_2\text{—CH}_2\text{—CH—COO}^-$$
$$|$$
$$\text{NH}_3^+$$

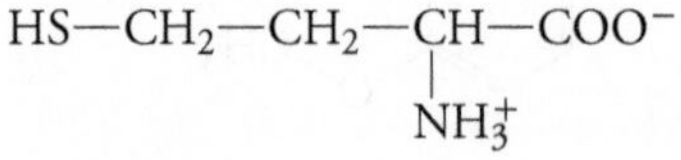

d.

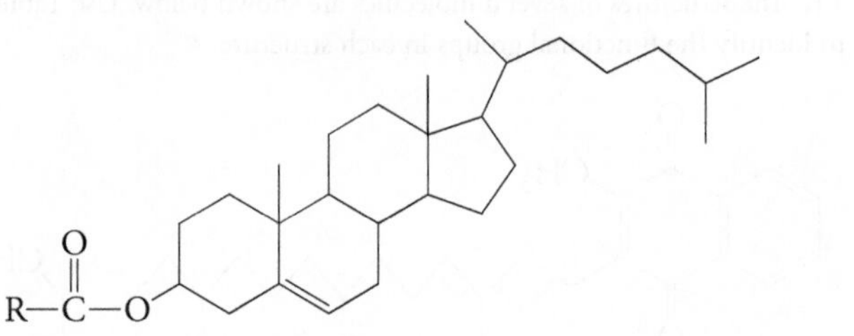

16. The compound shown below was isolated from the starfish *Anthenea aspera*. What two types of biomolecules are found in this compound?

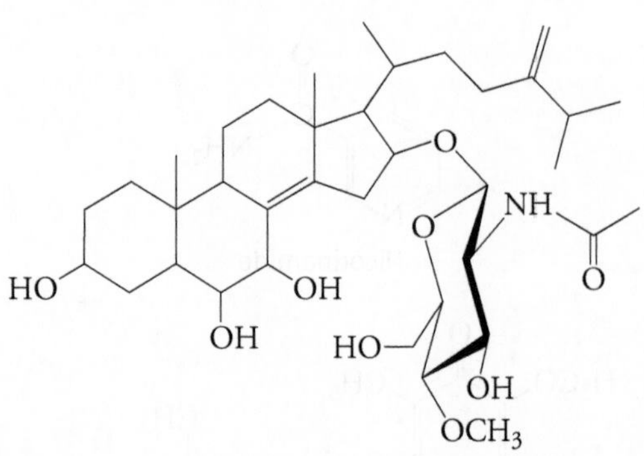

17. The nutritive quality of food can be analyzed by measuring the amounts of the chemical elements it contains. Most foods are mixtures of the three major types of molecules: **a.** fats (lipids), **b.** carbohydrates, and **c.** proteins. What elements are present in each of these types of molecules?

18. A compound present in animals, plants and microorganisms has the formula $CH_3(CH_2)_{14}COOH$. To which class of molecules does this compound belong? Explain your answer.

19. A healthy diet must include some protein. Assuming you had a way to measure the amount of each element in a sample of food, which element would you measure in order to tell whether the food contained protein?

20. The structures of three compounds are shown below. Based on your answer to Problem 15, which of the three compounds would you add to a food sample so that it would appear to contain more protein? Which of the three compounds would already be present in a food sample that actually did contain protein? Explain.

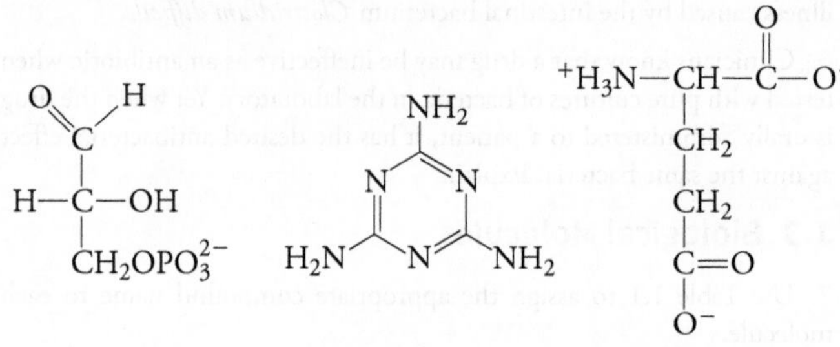

A   B   C

21. The structure of the compound urea is shown. Urea is a waste product of metabolism excreted by the kidneys into the urine. Why do doctors tell patients with kidney damage that they should consume a low-protein diet?

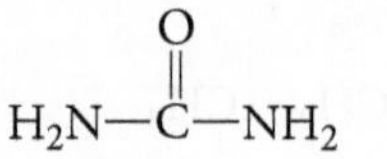

Urea

22. The structures of the amino acids glutamine (Gln) and cysteine (Cys) are shown in Section 1.2. What functional group does Gln have that Cys does not? What functional group does Cys have that Gln does not?

23. Consult Table 4.1 for the structures of the amino acids cysteine and lysine. What functional group does cysteine have that lysine does not?

24. Many amino acid side chains are modified after translation. Draw the structures of the following modified amino acids: **a.** phosphothreonine, **b.** hydroxylyproline, and **c.** acetyllysine.

25. The "straight-chain" structure of glucose is shown in Section 1.2. What functional groups are present in the glucose molecule?

26. Consider the monosaccharide glucose. **a.** How does its molecular formula differ from that of fructose? **b.** How does its structure differ from the structure of fructose?

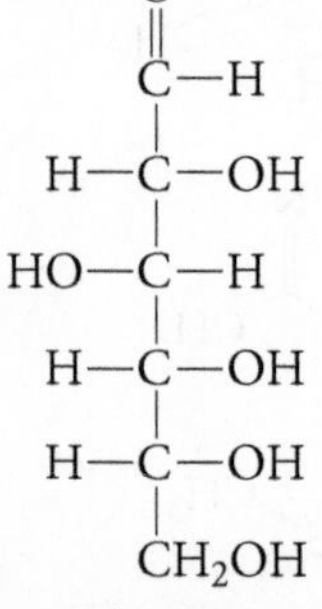

Glucose

27. The structures of the nitrogenous bases uracil and adenine are shown below. How do their functional groups differ?

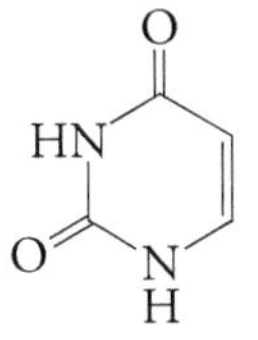

Uracil 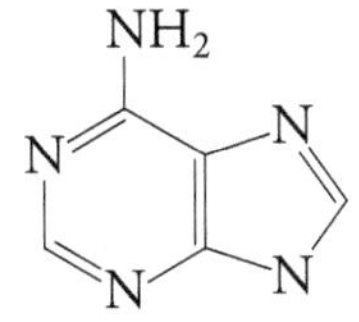   Adenine

28. What types of linkages are found in the following compounds?

a.

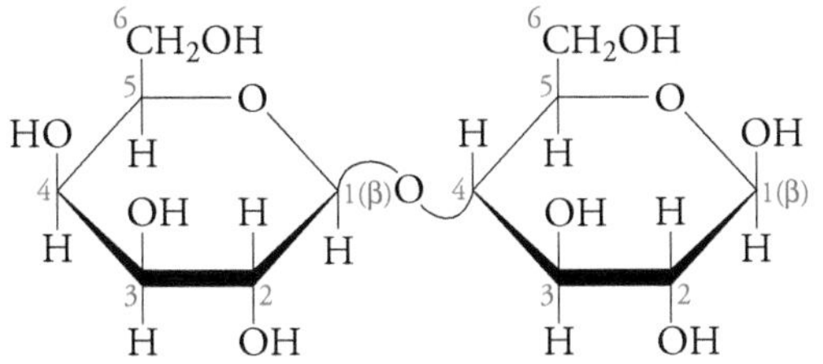

Galactose   Glucose

Lactose

b.

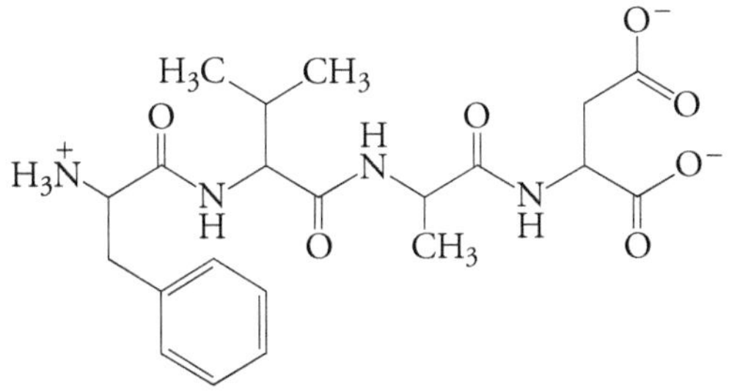

c.

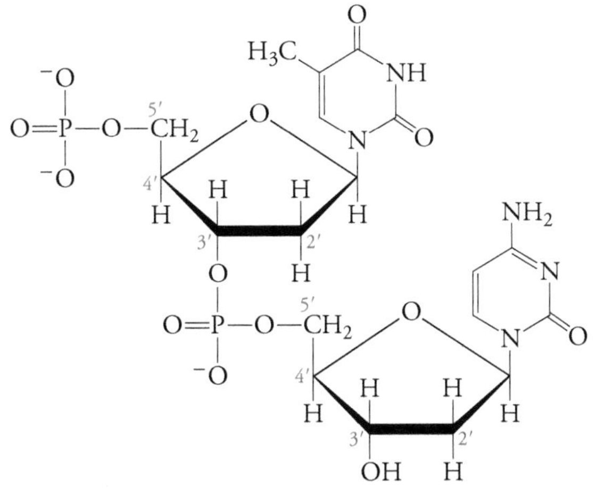

29. The structure of the artificial sweetener aspartame is shown below. What type of linkage is found in this compound?

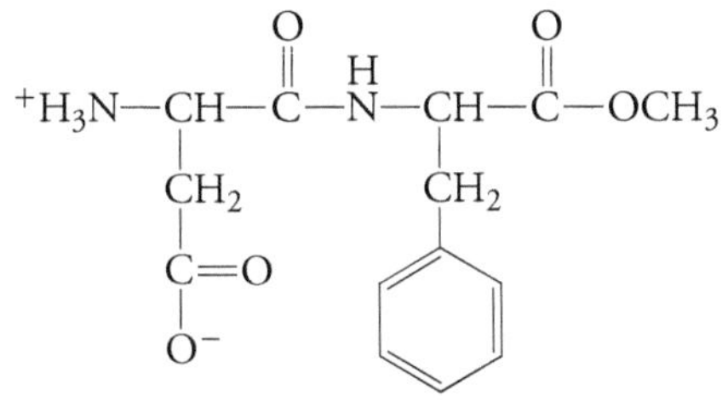

Aspartame

30. Compare the solubilities in water of serine, fructose, stearate, and cholesterol, and explain your reasoning.

31. Cell membranes are largely hydrophobic structures. Which compound will pass through a membrane more easily, glucose or 2,4-dinitrophenol? Explain.

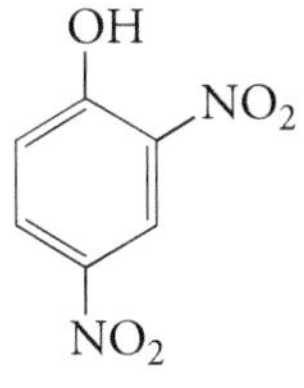

2,4-Dinitrophenol

32. What polymeric molecule forms a more regular structure, DNA or protein? Explain this observation in terms of the cellular roles of the two different molecules.

33. What are the two major biological roles of polysaccharides?

34. Pancreatic amylase digests the glycosidic bonds that link glucose residues together in starch. Would you expect this enzyme to digest the glycosidic bonds in cellulose as well? Explain why or why not.

35. Mammals cannot digest the cellulose but can easily digest the starch. However, both starch and cellulose are polymers of glucose. Explain.

1.4 Energy and Metabolism

36. What is the sign of the entropy change for each of the following processes? **a.** Water freezes. **b.** Water evaporates. **c.** Dry ice sublimes. **d.** Sodium chloride dissolves in water. **e.** Several different types of lipid molecules assemble to form a membrane.

37. When a stretched rubber band is allowed to relax, it feels cooler. **a.** Is the enthalpy change for this process positive or negative? **b.** Since a stretched rubber band spontaneously relaxes, what can you conclude about the entropy change during relaxation?

38. Does entropy increase or decrease in the following reactions in aqueous solution?

a.

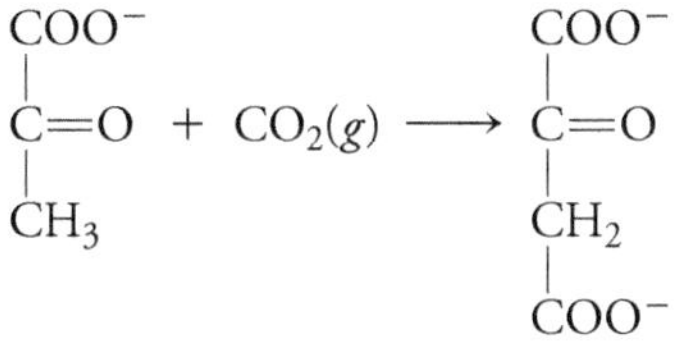

b.

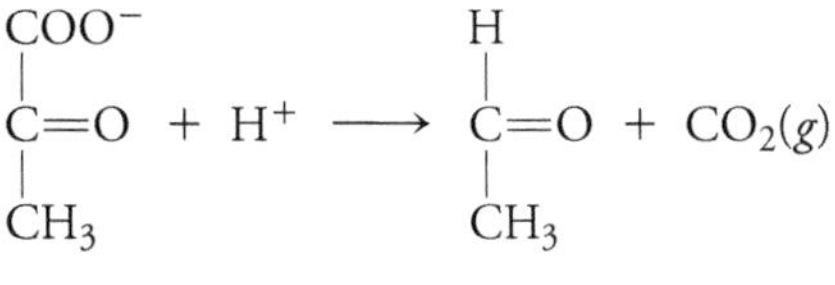

39. How does the entropy change when glucose undergoes combustion?

$$C_6H_{12}O_6 + 6\,O_2 \rightarrow 6\,CO_2 + 6\,H_2O$$

40. A soccer coach keeps a couple of instant cold packs in her bag in case one of her players suffers a muscle injury. Instant cold packs are composed of a plastic bag containing a smaller water bag and solid ammonium nitrate. In order to activate the cold pack, the bag is kneaded until the smaller water bag breaks, which allows the released water to dissolve the ammonium nitrate. The equation for the dissolution of ammonium nitrate in water is shown below. How does the cold pack work?

$$NH_4NO_3\,(s) \xrightarrow{\ H_2O\ } NH_4^+\,(aq) + NO_3^-\,(aq)$$
$$\Delta H = 26.4\ kJ \cdot mol^{-1}$$

41. Campers carry hot packs with them, especially when camping during the winter months or at high altitudes. The design is similar to that described in Problem 37, except that calcium chloride is used in place of the ammonium nitrate. The equation for the dissolution of calcium chloride in water is shown below. How does the hot pack work?

$$CaCl_2\,(s) \xrightarrow{\ H_2O\ } Ca^{2+}(aq) + 2\,Cl^-(aq) \qquad \Delta H = -81\ kJ \cdot mol^{-1}$$

42. Urea (NH_2CONH_2) dissolves readily in water; i.e., this is a spontaneous process. A beaker containing the dissolved compound is cold to the touch. What conclusions can you make about the sign of the **a.** enthalpy change and **b.** entropy change for this process?

43. Would you expect the reaction shown in Problem 36 to be endergonic or exergonic? Explain.

44. For the reaction in which reactant A is converted to product B, tell whether this process is favorable at **a.** 4°C and **b.** 37°C.

	H (kJ · mol^{-1})	S (J · K^{-1} · mol^{-1})
A	43	11
B	50	37

45. For a given reaction, the value of ΔH is 25 kJ · mol^{-1} and the value of ΔS is 85 J · K^{-1} · mol^{-1}. Above what temperature will this reaction be spontaneous?

46. The ΔG value at 37°C of the arginine kinase reaction was recently determined. The ΔH for the reaction is –8.19 kJ · mol^{-1} and the value of ΔS is 2.20 J · K^{-1} · mol^{-1}. **a.** Is the reaction exothermic or endothermic? **b.** What is the value of ΔG for the reaction? Is the reaction spontaneous? **c.** Which component makes a greater contribution to the free energy value—the ΔH or ΔS value? Comment on the significance of this observation.

47. Which of the following processes are spontaneous? **a.** A reaction that occurs with any size decrease in enthalpy and any size increase in entropy. **b.** A reaction that occurs with a small increase in enthalpy and a large increase in entropy. **c.** A reaction that occurs with a large decrease in enthalpy and a small decrease in entropy. **d.** A reaction that occurs with any size increase in enthalpy and any size decrease in entropy.

48. Given that the process described in Problem 38 is spontaneous, what are the signs of ΔS and ΔG? Confirm your answer by using the enthalpy change provided to calculate the sign and magnitude of ΔS. Assume a temperature of 25°C.

49. The hydrolysis of pyrophosphate at 25°C is spontaneous. The enthalpy change for this reaction is –18 kJ · mol^{-1}. What is the sign and the magnitude of ΔS for this reaction?

50. Phosphoenolpyruvate donates a phosphate group to ADP to produce pyruvate and ATP. The ΔG value for this reaction at 25°C is –63 kJ · mol^{-1} and the value of ΔS is 190 J · K^{-1} · mol^{-1}. What is the value of ΔH? Is heat absorbed from or released to the surroundings?

51. A monoclonal antibody binds to the protein cytochrome c. The ΔH value for binding at 25°C is –87.9 kJ · mol^{-1} and the ΔS is –118 J · K^{-1} · mol^{-1}. **a.** Does entropy increase or decrease when the antibody binds to the protein? **b.** Calculate ΔG for the formation of the antibody–protein complex. Does the complex form spontaneously? **c.** The ΔG value for the binding of a second monoclonal antibody to cytochrome c is –58.2 kJ · mol^{-1}. Which antibody binds more readily to the protein?

52. Phosphofructokinase catalyzes the transfer of a phosphate group (from ATP) to fructose-6-phosphate to produce fructose-1,6-bisphosphate. The ΔH value for this reaction is –9.5 kJ · mol^{-1} and the ΔG is –17.2 kJ · mol^{-1} at 37°C. **a.** Is heat absorbed from or released to the surroundings? **b.** What is the value of ΔS for the reaction? Does this reaction proceed with an increase or decrease in entropy? **c.** Which component makes a greater contribution to the free energy change: the ΔH or ΔS value? Comment on the significance of this observation.

53. Glucose can be converted to glucose-6-phosphate:

$$glucose + phosphate \rightarrow glucose\text{-}6\text{-}phosphate + H_2O$$
$$\Delta G \text{ is } 13.8\ kJ \cdot mol^{-1}$$

a. Is this reaction favorable? Explain.

b. Suppose the synthesis of glucose-6-phosphate is coupled with the hydrolysis of ATP. Write the overall equation for the coupled process and calculate the ΔG for the coupled reaction. Is the conversion of glucose to glucose-6-phosphate favorable under these conditions? Explain.

$$ATP + H_2O \rightarrow ADP + phosphate \qquad \Delta G = -30.5\ kJ \cdot mol^{-1}$$

54. Glyceraldehyde-3-phosphate (GAP) is converted to 1,3-bisphosphoglycerate (1,3BPG) as shown.

$$GAP + P_i + NAD^+ \rightarrow 1,3BPG + NADH \qquad \Delta G = 6.7\ kJ \cdot mol^{-1}$$

a. Is this reaction spontaneous?

b. The reaction shown above is coupled to the following reaction in which 1,3BPG is converted to 3-phosphoglycerate (3PG):

$$1,3BPG + ADP \rightarrow 3PG + ATP \qquad \Delta G = -18.8\ kJ \cdot mol^{-1}$$

Write the equation for the overall conversion of GAP to 3PG. Is the coupled reaction favorable?

55. Place these molecules in order from the most oxidized to the most reduced.

A   B   C

56. Identify the process described in the following statements as an oxidation or reduction process. **a.** Monosaccharides are synthesized from carbon dioxide by plants during photosynthesis. **b.** An animal eats the plant and breaks down the monosaccharide in order to obtain energy for cellular processes.

57. Given the following reactions, **a.** tell whether the reactant is being oxidized or reduced. Reactions may not be balanced. **b.** tell whether an oxidizing agent or a reducing agent is needed to accomplish the reaction.

a.
$$CH_3-(CH_2)_{14}-\overset{O}{\overset{\|}{C}}-O^- \longrightarrow 8\,CH_3-\overset{O}{\overset{\|}{C}}-S-CoA$$

b.

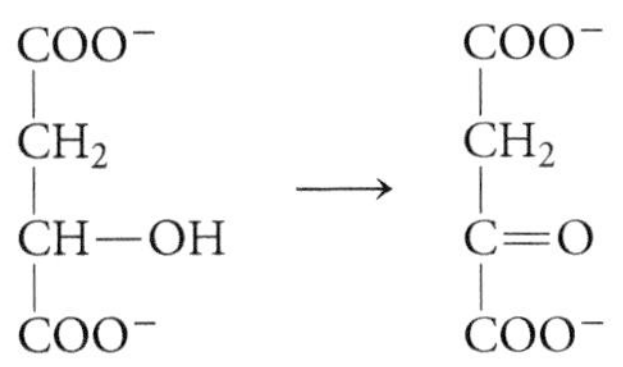

$$\begin{array}{ccc}
\text{COO}^- & & \text{COO}^- \\
| & & | \\
\text{CH}_2 & & \text{CH}_2 \\
| & \longrightarrow & | \\
\text{CH}-\text{OH} & & \text{C}=\text{O} \\
| & & | \\
\text{COO}^- & & \text{COO}^-
\end{array}$$

c.

$$\begin{array}{ccc}
\text{COO}^- & & \text{COO}^- \\
| & & | \\
\text{CH} & & \text{CH}_2 \\
\| & \longrightarrow & | \\
\text{CH} & & \text{CH}-\text{OH} \\
| & & | \\
\text{COO}^- & & \text{COO}^-
\end{array}$$

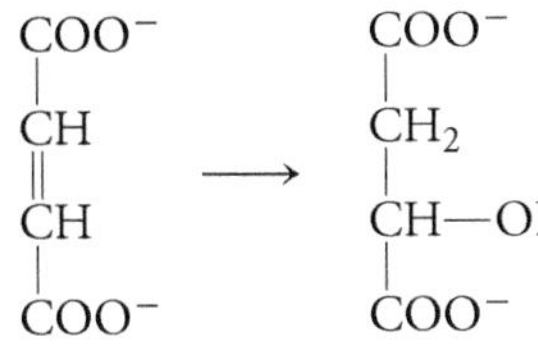

d.

$$^+\text{H}_3\text{N}-\underset{\underset{\text{S}}{\overset{\text{CH}_2}{|}}}{\overset{\overset{\text{O}}{\|}}{\text{CH}-\text{C}-\text{O}^-}}$$

$$^-\text{O}-\underset{\overset{\text{O}}{\|}}{\text{C}}-\underset{\text{CH}_2}{|}$$

+ H$_2$ $\longrightarrow$ 2 $^+\text{H}_3\text{N}-\underset{\underset{\text{SH}}{\overset{\text{CH}_2}{|}}}{\overset{\overset{\text{O}}{\|}}{\text{CH}-\text{C}-\text{O}^-}}$

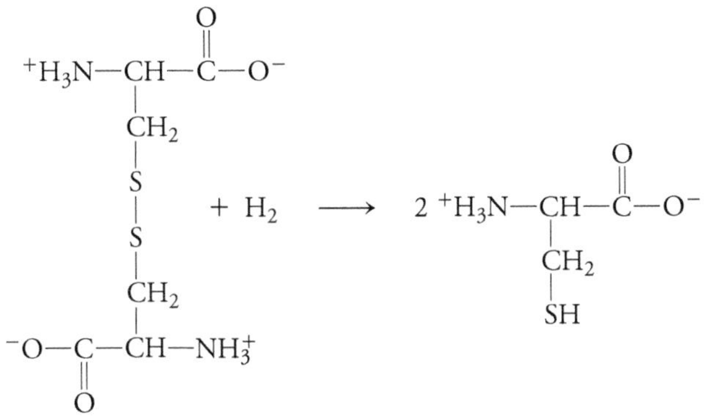

58. Rank the forms of vitamin A (see Problem 9) in order from most oxidized to most reduced.

59. The reaction shown in Problem 54 requires the coenzyme NAD$^+$, which is reduced to NADH during the reaction. Which is more oxidized, GAP or 1,3BPG?

60. In some cells, lipids such as palmitate (shown in Section 1.2), rather than monosaccharides, serve as the primary metabolic fuel. **a.** Consider the oxidation state of palmitate's carbon atoms and explain how it fits into a scheme such as the one shown in Figure 1.17. **b.** On a per-carbon basis, which would make more energy available for metabolic reactions: palmitate or glucose?

61. Which yields more free energy when completely oxidized, stearate or α-linolenate?

$$\text{H}_3\text{C}-(\text{CH}_2)_{16}-\text{COO}^-$$
Stearate

$$\text{H}_3\text{C}-\text{CH}_2-(\text{CH}=\text{CHCH}_2)_3-(\text{CH}_2)_6-\text{COO}^-$$
α-Linolenate

62. A protein folds from a random coil into a highly ordered structure. How does this process, which proceeds with a loss of entropy, obey the laws of thermodynamics?

63. An enzyme binds to a substrate, forming an enzyme–substrate complex. Explain how this process obeys the laws of thermodynamics.

SELECTED READINGS

Archibald, J.M., Endosymbiosis and eukaryotic cell evolution, *Current Biol.* 25, R911–R921, doi: 10.1016/j.cub.2015.07.055 (2015). [An extensive review of hypotheses about the origins of mitochondria and chloroplasts in eukaryotic cells.]

Baum, B. and Baum, D.A., The merger that made us, *BMC Biology* 18, 72, doi: 10.1186/s12915-020-00806-3 (2020). [Briefly summarizes the outside-in and inside-out models for the fusion of prokaryotic cells that generated the first eukaryotes.]

Cullen, C.M., Aneja, K.K., Beyhan, S., Cho, C.E., Woloszynek, S., Convertino, M., McCoy, S.J., Zhang, Y., Anderson, M.Z., Alvarez-Ponce, D., Smirnova, E., Karstens, L., Dorrestein, P.C., Li, H., Sen Gupta, A., Cheung, K., Powers, J.G., Zhao, Z., and Rosen, G.L., Emerging priorities for microbiome research, *Front. Microbiol.* 11, doi: 10.3389/fmicb.2020.00136 (2020). [Summarizes current understanding and outlines some approaches to studying microbiota.]

Koonin, E.V., Splendor and misery of adaptation, or the importance of neutral null for understanding evolution, *BMC Biology* 14, 114, doi: 10.1186/s12915-016-0338-2 (2016). [Includes a caution about assigning an explanatory story to every genetic change.]

Pagel, M., Natural selection 150 years on, *Nature* 457, 808–811 (2009). [A short summary of the key points of evolution by natural selection, including the branching pattern of descent and the process of speciation.]

Szostak, J.W., The narrow road to the deep past: In search of the chemistry of the origin of life, *Agnew Chem. Int. Ed. Engl.* 56, 11037–11043, doi: 10.1002/anie.201704048 (2017). [A review of prebiotic evolution and the origin of cells.]

CHAPTER 1 CREDITS

Figure 1.11 Adapted from Palm, W. and Thompson, C.B., *Nature* 546, 234–242 (2017).

Figure 1.12 Image based on 1EDN. Janes, R.W., Peapus, D.H., Wallace, B.A., The crystal structure of human endothelin, *Nat. Struct. Biol.* 1, 311–319 (1994).

Figure 1.13 Image based on Nucleic Acids Database ARF0108. Biswas, R., Mitra, S.N., Sundaralingam, M., 1.76 A structure of a pyrimidine start alternating A-RNA hexamer r(CGUAC)dG, *Acta Crystallogr., Sect.* D 54, 570–576 (1998).

Figure 1.16 Adapted from Wheelis, M.L., Kandler, O., and Woese, C.R., *Proc. Natl. Acad. Sci.* USA 89, 2930–2934 (1992).

CHAPTER 2

Aqueous Chemistry

H.Helene/Shutterstock.com

Wood ants defend themselves by squirting enemies with formic acid from a gland on their abdomen. They also use formic acid to disinfect their nests. One species, *Formica paralugubris*, takes this antimicrobial defense even further. When these ants drag particles of tree resin—which is already naturally antimicrobial—back to their nest, they intentionally coat it with formic acid to further enhance its antifungal activity.

Do You Remember?

- Organisms maintain a state of homeostasis (Section 1.1).
- Biological molecules are composed of a subset of all possible elements and functional groups (Section 1.3).
- The free energy of a system is determined by its enthalpy and entropy (Section 1.4).

Water is a fundamental requirement for life, so it is important to understand the structural and chemical properties of water. Not only are most biological molecules surrounded by water, but their molecular structure is in part governed by how their component groups interact with water. In addition, water plays a role in how these molecules assemble to form larger structures or undergo chemical transformation. In fact, water itself—or its H^+ and OH^- constituents—participates directly in many biochemical processes. Therefore, an examination of water is a logical prelude to exploring the structures and functions of biomolecules in the following chapters.

2.1 Water Molecules and Hydrogen Bonds

KEY CONCEPTS

Explain water's properties in term of its ability to form hydrogen bonds.

- Describe the electronic structure of a water molecule.
- Identify hydrogen bond donor and acceptor groups.
- List the other types of weak noncovalent forces that affect biological molecules.
- Describe how water interacts with polar and charged solutes.

What is the nature of the substance that accounts for about 70% of the mass of most organisms? The human body, for example, is about 60% water by weight, most of it in the extracellular fluid (the fluid surrounding cells) and inside cells:

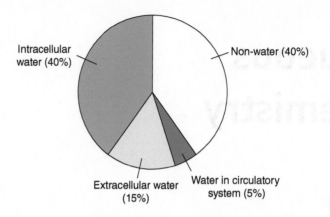

Intracellular water (40%)

Non-water (40%)

Extracellular water (15%)

Water in circulatory system (5%)

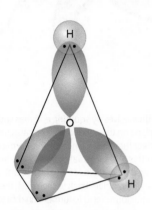

Figure 2.1 Electronic structure of the water molecule. Four electron orbitals, in an approximately tetrahedral arrangement, surround the central oxygen. Two orbitals participate in bonding to hydrogen (gray), and two contain unshared electron pairs.

In an individual H_2O molecule, the central oxygen atom forms covalent bonds with two hydrogen atoms, leaving two unshared pairs of electrons. The molecule therefore has approximately tetrahedral geometry, with the oxygen atom at the center of the tetrahedron, the hydrogen atoms at two of the four corners, and electrons at the other two corners (**Fig. 2.1**).

As a result of this electronic arrangement, *the water molecule is **polar***; that is, it has an uneven distribution of charge. The oxygen atom bears a partial negative charge (indicated by the symbol δ^-), and each hydrogen atom bears a partial positive charge (indicated by the symbol δ^+):

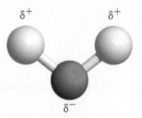

δ^+ δ^+

δ^-

This polarity is the key to many of water's unique physical properties.

Neighboring water molecules tend to orient themselves so that each partially positive hydrogen is aligned with a partially negative oxygen:

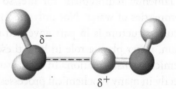

δ^-

δ^+

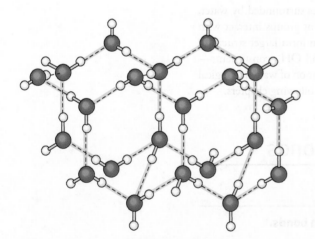

Figure 2.2 Structure of ice. Each water molecule acts as a donor for two hydrogen bonds and an acceptor for two hydrogen bonds, thereby interacting with four other water molecules in the crystal. (Only two layers of water molecules are shown here.)

Question Identify a hydrogen bond donor and acceptor in this structure.

This interaction, shaded yellow here, is known as a **hydrogen bond.** This weak electrostatic attraction between oppositely charged particles actually has some covalent character. In addition, the bond has directionality, or a preferred orientation.

Each water molecule can potentially participate in four hydrogen bonds, since it has two hydrogen atoms to "donate" to a hydrogen bond and two pairs of unshared electrons that can "accept" a hydrogen bond. In ice, a crystalline form of water, each water molecule does indeed form hydrogen bonds with four other water molecules (**Fig. 2.2**). This regular, lattice-like structure breaks down when the ice melts.

In liquid water, each molecule can potentially form hydrogen bonds with up to four other water molecules, but each bond has a life-time of only about 10^{-12} s. As a result, *the structure of water is continually flickering as water molecules rotate, bend, and reorient themselves.*

Theoretical calculations and spectroscopic data suggest that water molecules participate in only two strong hydrogen bonds, one as a donor and one as an acceptor, generating transient hydrogen-bonded clusters such as the prism structure shown here:

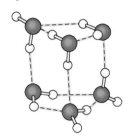

Because of its ability to form hydrogen bonds, water is highly cohesive. This accounts for its high surface tension, which allows certain insects to walk on water (**Fig. 2.3**). The cohesiveness of water molecules also explains why water remains a liquid, whereas molecules of similar size, such as CH_4 and H_2S, are gases at room temperature (25°C). At the same time, water is less dense than other liquids because hydrogen bonding demands that individual molecules not just approach each other but interact with a certain orientation. This geometrical constraint also explains why ice floats; for other materials, the solid is denser than the liquid.

Figure 2.3 A water strider supported by the surface tension of water.

Hydrogen bonds are one type of electrostatic force

Powerful covalent bonds define basic molecular constitutions, but much weaker noncovalent bonds, including hydrogen bonds, govern the final three-dimensional shapes of molecules and how they interact with each other. For example, about 460 kJ · mol⁻¹ (110 kcal · mol⁻¹) of energy is required to break a covalent O—H bond. However, a hydrogen bond in water has a strength of only about 20 kJ · mol⁻¹ (4.8 kcal · mol⁻¹). Other noncovalent interactions are weaker still.

Hydrogen bonds, with a length of about 1.8 Å, are longer and hence weaker than a covalent O—H bond (which is about 1 Å long). However, a completely noninteracting O and H would approach no closer than about 2.7 Å, which is the sum of their **van der Waals radii** (the van der Waals radius of an isolated atom is the distance from its nucleus to its effective electronic surface).

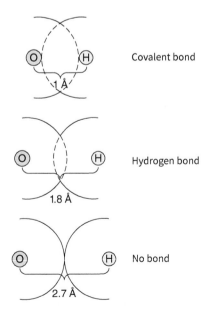

Hydrogen bonds usually involve N—H and O—H groups as hydrogen donors and the electronegative N and O and occasionally S atoms as hydrogen acceptors (**electronegativity** is a measure of an atom's affinity for electrons; **Table 2.1**). *Water, therefore, can form hydrogen bonds not just*

Table 2.1 Electronegativities of Some Elements

Element	Electronegativity
H	2.20
C	2.55
S	2.58
N	3.04
O	3.44
F	3.98

with other water molecules but with a wide variety of other compounds that bear N- and O-containing functional groups:

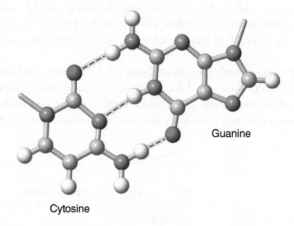

Water–alcohol   Water–amine

Likewise, these functional groups can form hydrogen bonds among themselves. For example, the complementarity of bases in DNA and RNA is determined by their ability to form hydrogen bonds with each other. Here, three N—H groups are hydrogen bond donors, and N and O atoms are acceptors:

Guanine

Cytosine

Among the noncovalent interactions that occur in biological molecules are electrostatic interactions between charged groups such as carboxylate (—COO⁻) and amino (—NH₃⁺) groups. These **ionic interactions** are intermediate in strength to covalent bonds and hydrogen bonds (**Fig. 2.4**).

Other electrostatic interactions occur between particles that are polar but not actually charged, for example, two carbonyl groups:

$$\text{C=O} \cdots \cdots \text{C=O}$$

These forces, called **van der Waals interactions,** are usually weaker than hydrogen bonds. The interaction between two strongly polar groups is known as a **dipole–dipole interaction** and has a strength of about 9 kJ · mol⁻¹. Very weak van der Waals interactions, called **London dispersion forces,** occur between nonpolar molecules as a result of small fluctuations in their distribution of electrons that create a temporary

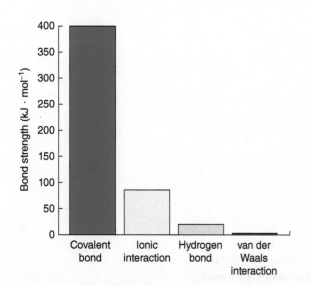

Figure 2.4 Relative strengths of bonds in biological molecules.

separation of charge. Nonpolar groups such as methyl groups can therefore experience a small attractive force, in this case about $0.3 \ kJ \cdot mol^{-1}$:

$$\overset{H}{\underset{H}{\overset{|}{\underset{|}{C}}}} \overset{\delta^-}{} \underset{\delta^+}{H} - - - - - \underset{\delta^-}{H} \overset{\delta^+}{\underset{H}{\overset{H}{\underset{|}{\overset{|}{C}}}}}$$

Not surprisingly, these forces act only when the groups are very close, and their strength quickly falls off as the groups draw apart. If the groups approach too closely, however, their van der Waals radii collide and a strong repulsive force overcomes the attractive force.

Although hydrogen bonds and van der Waals interactions are individually weak, biological molecules usually contain multiple groups capable of participating in these intermolecular interactions, so their cumulative effect can be significant (**Fig. 2.5**). These weak forces also determine how biological moleucles can "recognize" or bind noncovalently to each other. Drug molecules are typically designed to optimize the weak interactions that govern their therapeutic activity (**Box 2.A**).

Figure 2.5 The cumulative effect of small forces. Just as the fictional giant Gulliver was restrained by many small tethers at the hands of the tiny Lilliputians, the structures of macromolecules are constrained by the effects of many weak noncovalent interactions.

Water dissolves many compounds

Unlike most other solvent molecules, water molecules are able to form hydrogen bonds and participate in other electrostatic interactions with a wide variety of compounds. Water has a relatively high **dielectric constant,** which is a measure of a solvent's ability to diminish the electrostatic attractions between dissolved ions (**Table 2.2**). The higher the dielectric constant of the solvent, the less able the ions are to associate with each other. The polar water molecules surround ions

---

| **Box 2.A** | **Why Do Some Drugs Contain Fluorine?** |

As mentioned in Section 1.3, the most abundant elements in biological molecules are H, C, N, O, P, and S. Fluorine, the 13th most abundant element on earth, appears in animal bodies mainly in bones and teeth. Fluoride ions occasionally take the place of hydroxyl groups in crystals of hydroxyapatite, $Ca_{10}(PO_4)_6(OH)_2$, the mineral that accounts for most of the mass of bones and teeth. Although fluoride's role in bone structure is controversial, it strengthens dental enamel and makes teeth more resistant to the demineralization that occurs during the formation of dental caries (cavities). For this reason, small amounts of fluoride are often added to toothpastes and to municipal water supplies, particularly in the U.S.

Despite its participation in tooth structure, fluorine almost never appears in naturally occurring organic compounds. Why, then, do about one-quarter of all drug molecules, including the widely prescribed Prozac (fluoxetine, an antidepressant; Box 9.B), fluorouracil (an anticancer agent; Section 7.3), and Ciprofloxacin (an antibacterial agent; Section 20.1), contain F?

In designing an effective drug, pharmaceutical scientists often intentionally introduce F in order to alter the drug's chemical or biological properties without significantly altering its shape. The small fluorine can take the place of hydrogen in a chemical structure, but with its high electronegativity (see Table 2.1), F behaves much more

like O than H. Consequently, transforming a relatively inert C—H group into an electron-withdrawing C—F group can decrease the basicity of nearby amino groups (see Section 2.3). Fewer positive charges in a drug allow it to more easily pass through membranes to enter cells and exert its biological effect.

Prozac (Fluoxetine)

In addition, the polar C—F bond can participate in hydrogen bonding (C—F $\cdots$ H—C) or other dipole–dipole interactions (such as C—F $\cdots$ C=O), potentially augmenting the intermolecular attraction between a drug and its target molecule in the body. Better binding usually means that the drug will be effective at lower concentrations and will have fewer side effects.

Question **Identify the hydrogen-bonding groups in Prozac.**

Table 2.2 Dielectric Constants for Some Solvents at Room Temperature

Solvent	Dielectric constant
Formamide ($HCONH_2$)	109
Water	80
Methanol (CH_3OH)	33
Ethanol (CH_3CH_2OH)	25
1-Propanol ($CH_3CH_2CH_2OH$)	20
1-Butanol ($CH_3CH_2CH_2CH_2OH$)	18
Benzene (C_6H_6)	2

Question Compare the hydrogen-bonding ability of these solvents.

(for example, the Na^+ and Cl^- ions from the salt NaCl) by aligning their partial charges with the oppositely charged ions.

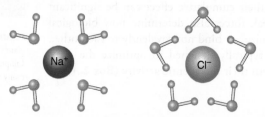

Because the interactions between the polar water molecules and the ions are stronger than the attractive forces between the Na^+ and Cl^- ions, the salt dissolves (the dissolved particle is called a **solute**). Each solute ion surrounded by water molecules is said to be **solvated** (or **hydrated,** to indicate that the solvent is water).

Biological molecules that bear polar or ionic functional groups are also readily solubilized, in this case because the groups can form hydrogen bonds with the solvent water molecules. Glucose, for example, with its six hydrogen-bonding oxygens, is highly soluble in water:

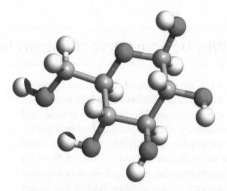

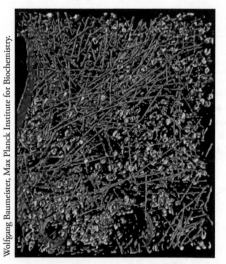

Figure 2.6 Portion of a *Dictyostelium* cell visualized by cryoelectron tomography. In this technique, the cells are rapidly frozen so that they retain their fine structure, and two-dimensional electron micrographs taken from different angles are merged to re-create a three-dimensional image. The red structures are filaments of the protein actin, ribosomes and other macromolecular complexes are colored green, and membranes are blue. Small molecules (not visible) fill the spaces between these larger cell components.

Note that when we describe the behavior of a single molecule, such as glucose in this example, we are really describing the average behavior of a huge number of molecules. (Most biochemical techniques cannot assess the activity of an individual molecule.)

The concentration of glucose in human blood is about 5 mM. In a solution of 5 mM glucose in water, there are about 10,000 water molecules for every glucose molecule (the water molecules are present at a concentration of about 55.5 M). However, biological molecules are never found alone in such dilute conditions *in vivo*, because a large number of small molecules, large polymers, and macromolecular aggregates collectively form a solution that is more like a hearty stew than a thin, watery soup (**Fig. 2.6**).

Inside a cell, the spaces between molecules may be only a few Å wide, enough room for only two water molecules to fit. This allows solute molecules, each with a coating of properly oriented water molecules, to slide past each other. This thin coating, or shell, of water may be enough to keep molecules from coming into van der Waals contact (van der Waals interactions are weak but attractive), thereby helping maintain the cell's contents in a crowded but fluid state.

Consequently, although there are a lot of water molecules in biological systems, those molecules may not be free to behave like the water molecules in bulk solution. In fact, the water molecules that associate closely with biological molecules, especially through hydrogen bonding, can be considered to be part of the macromolecules' structure. For example, every gram of DNA has about 0.6 g of tightly bound water molecules that affect its overall stability.

Concept Check

1. Explain why a water molecule is polar.
2. Draw three hydrogen-bonded water molecules.
3. Describe the structure of liquid water.
4. Compare the strengths of covalent bonds, hydrogen bonds, ionic interactions, and van der Waals interactions.
5. Describe what happens when an ionic substance dissolves in water.
6. Explain why water is a more effective solvent than ammonia or methanol.

2.2 The Hydrophobic Effect

KEY CONCEPTS

Relate the solubility of substances to the hydrophobic effect.

- Explain the hydrophobic effect in terms of water's entropy.
- Predict the water solubility of hydrophobic and hydrophilic substances.
- Describe how amphiphilic substances behave in water.
- Explain why a lipid bilayer is a barrier to diffusion.

Glucose and other readily hydrated substances are said to be **hydrophilic** (water-loving). In contrast, a compound such as dodecane (a C_{12} alkane),

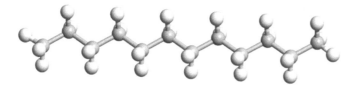

which lacks polar groups, is relatively insoluble in water and is said to be **hydrophobic** (water-fearing). Although pure hydrocarbons are rare in biological systems, many biological molecules contain hydrocarbon-like portions that are insoluble in water.

When a nonpolar substance such as vegetable oil (which consists of hydrocarbon-like molecules) is added to water, it does not dissolve but forms a separate phase. In order for the water and oil to mix, free energy must be added to the system (for example, by stirring vigorously or applying heat). Why is it thermodynamically unfavorable to dissolve a hydrophobic substance in water? One possibility is that enthalpy is required to break the hydrogen bonds among solvent water molecules in order to create a "hole" into which a nonpolar molecule can fit. Experimental measurements, however, show that the free energy barrier (ΔG) to the solvation process depends much more on the entropy term (ΔS) than on the enthalpy term (ΔH; recall from Chapter 1 that $\Delta G = \Delta H - T\Delta S$; Equation 1.2). This is because when a hydrophobic molecule is hydrated, it becomes surrounded by a layer of water molecules that cannot participate in normal hydrogen bonding with each other but instead must align themselves so that their polar ends are not oriented toward the nonpolar solute. *This constraint on the structure of water represents a loss of entropy*

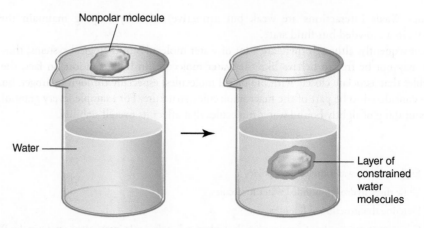

Figure 2.7 Hydration of a nonpolar molecule.   When a nonpolar substance (green) is added to water, the system loses entropy because the water molecules surrounding the nonpolar solute (orange) lose their freedom to form hydrogen bonds. The loss of entropy is a property of the entire system, not just the water molecules nearest the solute, because these molecules are continually changing places with water molecules from the rest of the solution. The loss of entropy presents a thermodynamic barrier to the hydration of a nonpolar solute.

in the system, because now the highly mobile water molecules have lost some of their freedom to rapidly form, break, and re-form hydrogen bonds with other water molecules (**Fig. 2.7**). Note that the loss of entropy is not due to the formation of a frozen "cage" of water molecules around the nonpolar solute, as commonly pictured, because in liquid water the solvent molecules are in constant motion.

When a large number of nonpolar molecules are introduced into a sample of water, they do not disperse and become individually hydrated, each surrounded by a layer of water molecules. Instead, the nonpolar molecules are forced together, away from contact with water molecules. (This explains why small oil droplets coalesce into one large oily phase.) Although the entropy of the nonpolar molecules is thereby reduced, this thermodynamically unfavorable event is more than offset by the increase in the entropy of the surrounding water molecules, which regain their ability to interact freely with other water molecules (**Fig. 2.8**).

The exclusion of nonpolar substances from an aqueous solution is known as the **hydrophobic effect.** It is a powerful force in biochemical systems, even though it is not a bond or an attractive interaction in the conventional sense. The nonpolar molecules do not experience any additional attractive force among themselves; they aggregate only because they are driven out of the aqueous phase by the unfavorable entropy cost of individually hydrating them. The hydrophobic effect governs the structures and functions of many biological molecules. For example, each polypeptide chain of a

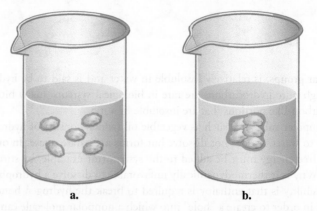

Figure 2.8 Aggregation of nonpolar molecules in water.   a. The individual hydration of dispersed nonpolar molecules (green) decreases the entropy of the system because the hydrating water molecules (orange) are not as free to form hydrogen bonds. **b.** Aggregation of the nonpolar molecules increases the entropy of the system, since the number of water molecules required to hydrate the aggregated solutes is less than the number of water molecules required to hydrate the dispersed solute molecules. This increase in entropy accounts for the spontaneous aggregation of nonpolar substances in water.

Question Explain why it is incorrect to describe the behavior shown in part b in terms of "hydrophobic bonds."

protein folds into a globular mass so that its hydrophobic groups are in the interior, away from the solvent, and its polar groups are on the exterior, where they can interact with water. Similarly, the structure of the lipid membrane that surrounds all cells is maintained by the hydrophobic effect acting on the lipids.

Amphiphilic molecules experience both hydrophilic interactions and the hydrophobic effect

Consider a molecule such as the fatty acid palmitate:

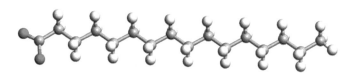

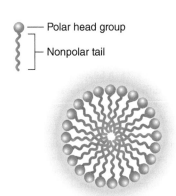

Polar head group

Nonpolar tail

The hydrocarbon "tail" of the molecule (on the right) is nonpolar, while its carboxylate "head" (on the left) is strongly polar. Molecules such as this one, which have both hydrophobic and hydrophilic portions, are said to be **amphiphilic** or **amphipathic.** What happens when amphiphilic molecules are added to water? In general, *the polar groups of amphiphiles orient themselves toward the solvent molecules and are therefore hydrated, while the nonpolar groups tend to aggregate due to the hydrophobic effect.* As a result, the amphiphiles may form a spherical **micelle,** a particle with a solvated surface and a hydrophobic core (**Fig. 2.9**).

Depending in part on the relative sizes of the hydrophilic and hydrophobic portions of the amphiphiles, the molecules may form a sheet rather than a spherical micelle. The amphiphilic lipids that provide the structural basis of biological membranes form two-layered sheets called **bilayers,** in which a hydrophobic layer is sandwiched between hydrated polar surfaces (**Fig. 2.10**). The structures of biological membranes are discussed in more detail in Chapter 8. The formation of micelles or bilayers is thermodynamically favored because the polar head groups can interact with solvent water molecules and the nonpolar tails are sequestered from the solvent.

Figure 2.9 Cross-section of a micelle formed by amphiphilic molecules. The hydrophobic tails of the molecules aggregate, out of contact with water, due to the hydrophobic effect. The polar head groups are exposed to and can interact with the solvent water molecules.

The hydrophobic core of a lipid bilayer is a barrier to diffusion

To eliminate its solvent-exposed edges, a lipid bilayer tends to close up to form a **vesicle,** shown cut in half:

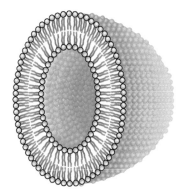

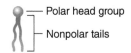

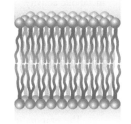

Polar head group

Nonpolar tails

Figure 2.10 A lipid bilayer. The amphiphilic lipid molecules form two layers so that their polar head groups are exposed to the solvent while their hydrophobic tails are sequestered in the interior of the bilayer, away from water. The likelihood of amphiphilic molecules forming a bilayer rather than a micelle depends in part on the sizes and nature of the hydrophobic and hydrophilic groups. One-tailed lipids tend to form micelles (see Fig. 2.9), and two-tailed lipids tend to form bilayers.

Question Indicate where a sodium ion and a benzene molecule would be located.

Many of the subcellular compartments (organelles) in eukaryotic cells have a similar structure.

When the vesicle forms, it traps a volume of the aqueous solution. Polar solutes in the enclosed compartment tend to remain there because they cannot easily pass through the hydrophobic interior of the bilayer. The energetic cost of transferring a hydrated polar group through the nonpolar lipid tails is too great. (In contrast, small nonpolar molecules such as O_2 can pass through the bilayer relatively easily.)

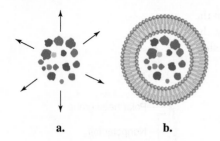

Figure 2.11 A bilayer prevents the diffusion of polar substances.
a. Solutes spontaneously diffuse from a region of high concentration to a region of low concentration. **b.** A lipid bilayer, which presents a thermodynamic barrier to the passage of polar substances, prevents the diffusion of polar substances out of the inner compartment (it also prevents the inward diffusion of polar substances from the external solution).

Normally, substances that are present at high concentrations tend to diffuse to regions of lower concentration. (This movement "down" a concentration gradient is a spontaneous process driven by the increase in entropy of the solute molecules.) *A barrier such as a bilayer can prevent this diffusion* (**Fig. 2.11**). This helps explain why cells, which are universally enclosed by a membrane, can maintain their specific concentrations of ions, small molecules, and biopolymers even when the external concentrations of these substances are quite different (**Fig. 2.12**). The solute composition of intracellular compartments and other biological fluids is carefully regulated. Not surprisingly, organisms spend a considerable amount of metabolic energy to maintain the proper concentrations of water and salts, and losses of one or the other must be compensated (**Box 2.B**).

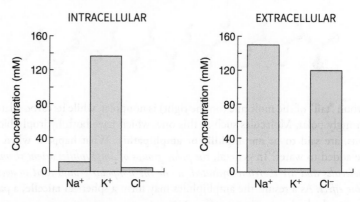

Figure 2.12 Ionic composition of intracellular and extracellular fluid. Human cells contain much higher concentrations of potassium than of sodium or chloride; the opposite is true of the fluid outside the cell. The cell membrane helps maintain the concentration differences.

Box 2.B Sweat, Exercise, and Sports Drinks

Animals, including humans, generate heat, even at rest, due to their metabolic activity. Some of this heat is lost to the environment by radiation, convection, conduction, and—in terrestrial animals—the vaporization of water. Evaporation has a significant cooling effect because about 2.5 kJ of heat is given up for every gram (mL) of water lost. In humans and certain other animals, an increase in skin temperature triggers the activity of sweat glands, which secrete a solution containing (in humans) about 50 mM Na$^+$, 5 mM K$^+$, and 45 mM Cl$^-$. The body is cooled as the sweat evaporates from its surface.

The evaporation of water accounts for a small portion of a resting body's heat loss, but sweating is the main mechanism for dissipating heat generated when the body is highly active. During vigorous exercise or exertion at high ambient temperatures, the body may experience a fluid loss of up to 2 L per hour. Athletic training not only improves the performance of the muscles and cardiopulmonary system, it also increases the capacity for sweating so that the athlete begins to sweat at a lower skin temperature and loses less salt in the secretions of the sweat glands. However, regardless of training, a fluid loss representing more than 2% of the body's weight may impair cardiovascular function. In fact, "heat exhaustion" in humans is usually due to dehydration rather than an actual increase in body temperature.

Numerous studies have concluded that athletes seldom drink enough before or during exercise. Ideally, fluid intake should match the losses due to sweat and the rate of intake should keep pace with the rate of sweating. So what should the conscientious athlete drink? For activities lasting less than about 90 minutes, especially when periods of high intensity alternate with brief periods of rest, water alone is sufficient. Commercial sports drinks containing carbohydrates can replace the water lost as sweat and also provide a source of energy. However, this carbohydrate boost may be an advantage only during prolonged sustained activity, such as during a marathon, when the body's own carbohydrate stores are depleted. A marathon runner or a manual laborer in the hot sun might benefit from the salt found in sports drinks, but most athletes do not need the supplemental salt (although it does make the carbohydrate solution more palatable). A normal diet usually contains enough Na$^+$ and Cl$^-$ to offset the losses in sweat.

Question **Compare the ion concentrations of sweat and extracellular fluid.**

Concept Check

1. What does entropy have to do with the solubility of nonpolar substances in water?
2. Explain how you can distinguish hydrophobic and hydrophilic substances.
3. Explain why polar molecules dissolve more easily than nonpolar substances in water.
4. Explain how a molecule can be both hydrophilic and hydrophobic. Give an example.
5. Explain why a lipid bilayer is a barrier to the diffusion of polar molecules.

2.3 Acid–Base Chemistry

KEY CONCEPTS

Determine the effect of acids and bases on a solution's pH.

* Recognize the relationship between the concentrations of H^+ and OH^-.
* Predict how the pH changes when acid or base is added to water.
* Relate an acid's pK value to its tendency to ionize.
* Perform calculations using the Henderson–Hasselbalch equation.
* Predict the ionization states of acid–base groups at a given pH.

Water is not merely an inert medium for biochemical processes; it is an active participant. Its chemical reactivity in biological systems is in part a result of its ability to ionize. This can be expressed in terms of a chemical equilibrium:

$$H_2O \rightleftharpoons H^+ + OH^-$$

The products of water's dissociation are a hydrogen ion or proton (H^+) and a hydroxide ion (OH^-).

Aqueous solutions do not actually contain lone protons. Instead, the H^+ can be visualized as combining with a water molecule to produce a **hydronium ion** (H_3O^+):

However, the H^+ is somewhat delocalized, so it probably exists as part of a larger, fleeting structure such as

Because a proton does not remain associated with a single water molecule, it appears to be relayed through a hydrogen-bonded network of water molecules (**Fig. 2.13**). This rapid **proton jumping** means that the effective mobility of H^+ in water is much greater than the mobility of other ions that must physically diffuse among water molecules. Consequently, acid–base reactions are among the fastest biochemical reactions.

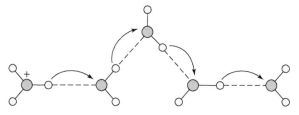

Figure 2.13 Proton jumping.   A proton associated with one water molecule (as a hydronium ion, at left) appears to jump rapidly through a network of hydrogen-bonded water molecules.

[H⁺] and [OH⁻] are inversely related

Pure water exhibits only a slight tendency to ionize, so the resulting concentrations of H^+ and OH^- are actually quite small. According to the law of mass action, the ionization of water can be described by a dissociation constant, K, which is equivalent to the concentrations of the reaction products divided by the concentration of un-ionized water:

$$K = \frac{[H^+][OH^-]}{[H_2O]} \tag{2.1}$$

The square brackets represent the molar concentrations of the indicated species.

Because the concentration of H_2O (55.5 M) is so much greater than $[H^+]$ or $[OH^-]$, it is considered to be constant, and the dissociation of water is described by K_w, the **ionization constant of water:**

$$K_w = K[H_2O] = [H^+][OH^-] \tag{2.2}$$

K_w is 10^{-14} at 25°C. In a sample of pure water, $[H^+] = [OH^-]$, so $[H^+]$ and $[OH^-]$ must both be equal to 10^{-7} M:

$$K_w = 10^{-14} = [H^+][OH^-] = (10^{-7}\,M)(10^{-7}\,M) \tag{2.3}$$

Since *the product of [H⁺] and [OH⁻] in any solution must be equal to 10^{-14}*, a hydrogen ion concentration greater than 10^{-7} M is balanced by a hydroxide ion concentration less than 10^{-7} M (**Fig. 2.14**).

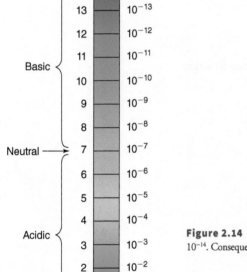

Figure 2.15 Relationship between pH and [H⁺].
Because pH is equal to $-\log[H^+]$, the greater the $[H^+]$, the lower the pH. A solution with a pH of 7 is neutral, a solution with a pH < 7 is acidic, and a solution with a pH > 7 is basic.

Question What is the difference in H⁺ concentration between a solution at pH 4 and a solution at pH 8?

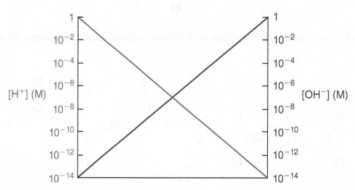

Figure 2.14 Relationship between [H⁺] and [OH⁻]. The product of $[H^+]$ and $[OH^-]$ is K_w, which is equal to 10^{-14}. Consequently, when $[H^+]$ is greater than 10^{-7} M, $[OH^-]$ is less than 10^{-7} M, and vice versa.

A solution in which $[H^+] = [OH^-] = 10^{-7}$ M is said to be **neutral;** a solution with $[H^+] > 10^{-7}$ M ($[OH^-] < 10^{-7}$ M) is **acidic,** and a solution with $[H^+] < 10^{-7}$ M ($[OH^-] > 10^{-7}$ M) is **basic.** To more easily describe such solutions, the hydrogen ion concentration is expressed as a **pH:**

$$pH = -\log[H^+] \tag{2.4}$$

Accordingly, a neutral solution has a pH of 7, an acidic solution has a pH < 7, and a basic solution has a pH > 7 (**Fig. 2.15**). Note that because the pH scale is logarithmic, *a difference of one pH unit is equivalent to a 10-fold difference in [H⁺]*. The so-called physiological pH, the normal pH of human blood, is a near-neutral 7.4. The pH values of some other body fluids are listed in **Table 2.3**. The pH of the environment is also a concern (**Box 2.C**).

Table 2.3 pH Values of Some Biological Fluids

Fluid	pH
Pancreatic juice	7.8–8.0
Blood	7.4
Saliva	6.4–7.0
Urine	5.0–8.0
Gastric juice	1.5–3.0

Box 2.C   **Atmospheric CO_2 and Ocean Acidification**

The human-generated increase in atmospheric carbon dioxide that is contributing to global warming is also impacting the chemistry of the world's oceans. Atmospheric CO_2 dissolves in water and reacts with it to generate carbonic acid. The acid immediately dissociates to form protons (H^+) and bicarbonate (HCO_3^-):

$$CO_2 + H_2O \rightleftharpoons H_2CO_3 \rightleftharpoons H^+ + HCO_3^-$$

The addition of hydrogen ions from CO_2-derived carbonic acid therefore leads to a decrease in the pH. Currently, the earth's oceans are slightly basic, with a pH of approximately 8.0. It has been estimated that over the next 100 years, the ocean pH will drop to about 7.8. Although the oceans act as a CO_2 "sink" that helps mitigate the increase in atmospheric CO_2, the increase in acidity in the marine environment represents an enormous challenge to organisms that must adapt to the new conditions.

Many marine organisms, including mollusks, many corals, and some plankton, use dissolved carbonate ions (CO_3^{2-}) to construct protective shells of calcium carbonate ($CaCO_3$). However, carbonate ions can combine with H^+ to form bicarbonate:

$$CO_3^{2-} + H^+ \rightleftharpoons HCO_3^-$$

Consequently, the increase in ocean acidity could decrease the availability of carbonate and thereby slow the growth of shell-building organisms. This not only would affect the availability of shellfish for human consumption but also would impact huge numbers of unicellular organisms at the base of the marine food chain. It is possible that acidification of the oceans could also dissolve existing calcium carbonate–based materials, such as coral reefs:

$$CaCO_3 + H^+ \rightleftharpoons HCO_3^- + Ca^{2+}$$

This could have disastrous consequences for these species-rich ecosystems.

Question   **Paradoxically, some marine organisms appear to benefit from increased atmospheric CO_2. Write an equation that describes how increased bicarbonate concentrations in seawater could promote shell growth.**

The pH of a solution can be altered

The pH of a sample of water can be changed by adding a substance that affects the existing balance between [H^+] and [OH^-]. Adding an acid increases the concentration of [H^+] and decreases the pH; adding a base has the opposite effect. *Biochemists define an **acid** as a substance that can donate a proton and a **base** as a substance that can accept a proton.* For example, adding hydrochloric acid (HCl) to a sample of water increases the hydrogen ion concentration ([H^+] or [H_3O^+]) because the HCl donates a proton to water:

$$HCl + H_2O \rightarrow H_3O^+ + Cl^-$$

Note that in this reaction, H_2O acts as a base that accepts a proton from the added acid.

Similarly, adding the base sodium hydroxide (NaOH) increases the pH (decreases [H^+]) by introducing hydroxide ions that can recombine with existing hydrogen ions:

$$NaOH + H_3O^+ \rightarrow Na^+ + 2\,H_2O$$

In this reaction, H_3O^+ is the acid that donates a proton to the added base. Note that acids and bases must operate in pairs: An acid can function as an acid (proton donor) only if a base (proton acceptor) is present, and vice versa. Water molecules can serve as both acid and base.

The final pH of the solution depends on how much H^+ (for example, from HCl) has been introduced or how much H^+ has been removed from the solution by its reaction with a base (for example, the OH^- ion of NaOH). Substances such as HCl and NaOH are known as "strong" acids and bases because they ionize completely in water. The Na^+ and Cl^- ions are called spectator ions and do not affect the pH. Keep in mind that pH is a property of a solution, not a characteristic of any individual reactant in the mixture. Calculating the pH of a solution of strong acid or base is straightforward (see Sample Calculation 2.1).

SAMPLE CALCULATION 2.1

Problem   Calculate the pH of 1 L of water to which is added **a.** 10 mL of 5.0 M HCl or **b.** 10 mL of 5.0 M NaOH.

Solution

a. The final concentration of HCl is $\dfrac{(0.01\ \text{L})(5.0\ \text{M})}{(1.01\ \text{L})} = 0.050\ \text{M}$

Since HCl dissociates completely, the added $[H^+]$ is equal to [HCl], or 0.050 M (the existing hydrogen ion concentration, 10^{-7} M, can be ignored because it is much smaller).

$$pH = -\log[H^+]$$
$$= -\log 0.050$$
$$= 1.3$$

b. The final concentration of NaOH is 0.050 M. Since NaOH dissociates completely, the added $[OH^-]$ is 0.050 M. Use Equation 2.2 to calculate $[H^+]$.

$$K_w = 10^{-14} = [H^+][OH^-]$$
$$[H^+] = 10^{-14}/[OH^-]$$
$$= 10^{-14}/(0.050\ \text{M})$$
$$= 2.0 \times 10^{-13}\ \text{M}$$
$$pH = -\log[H^+]$$
$$= -\log(2.0 \times 10^{-13})$$
$$= 12.7$$

A pK value describes an acid's tendency to ionize

Most biologically relevant acids and bases, unlike HCl and NaOH, do not dissociate completely when added to water. In other words, proton transfer to or from water is not complete. Therefore, the final concentrations of the acidic and basic species (including water itself) must be expressed in terms of an equilibrium. For example, acetic acid partially ionizes, or donates only some of its protons to water:

$$CH_3COOH + H_2O \rightleftharpoons CH_3COO^- + H_3O^+$$

The equilibrium constant for this reaction takes the form

$$K = \frac{[CH_3COO^-][H_3O^+]}{[CH_3COOH][H_2O]} \tag{2.5}$$

Because the concentration of H_2O is much higher than the other concentrations, it is considered constant and is incorporated into the value of K, which is then formally known as K_a, the **acid dissociation constant:**

$$K_a = K[H_2O] = \frac{[CH_3COO^-][H^+]}{[CH_3COOH]} \tag{2.6}$$

The acid dissociation constant for acetic acid is 1.74×10^{-5}. The larger the value of K_a, the more likely the acid is to ionize; that is, the greater its tendency to donate a proton to water. The smaller the value of K_a, the less likely the compound is to donate a proton.

Acid dissociation constants, like hydrogen ion concentrations, are often very small numbers. Therefore, it is convenient to transform the K_a to a **pK** value as follows:

$$pK = -\log K_a \tag{2.7}$$

The term pK_a is also used, but for simplicity we will use pK. For acetic acid,

$$pK = -\log(1.74 \times 10^{-5}) = 4.76 \tag{2.8}$$

The larger an acid's K_a, the smaller its pK and the greater its "strength" as an acid. In a solution of an acid with a low pK value, most of the molecules are in the unprotonated form. For an acid with a high pK value, most of the molecules remain protonated.

Consider an acid such as the ammonium ion, NH_4^+:

$$NH_4^+ \rightleftharpoons NH_3 + H^+$$

Its K_a is 5.62×10^{-10}, which corresponds to a pK of 9.25. This indicates that the ammonium ion is a relatively weak acid, a compound that tends not to donate a proton. On the other hand, ammonia (NH_3), which is the **conjugate base** of the acid NH_4^+, readily accepts a proton. The pK values of some compounds are listed in **Table 2.4**. A **polyprotic acid,** a compound with more than

Table 2.4 pK Values of Some Acids

Name	Formula[a]	pK
Trifluoroacetic acid	CF_3COOH	0.18
Phosphoric acid	H_3PO_4	2.15[b]
Formic acid	$HCOOH$	3.75
Succinic acid	$HOOCCH_2CH_2COOH$	4.21[b]
Acetic acid	CH_3COOH	4.76
Succinate	$HOOCCH_2CH_2COO^-$	5.64[c]
Thiophenol	C_6H_5SH	6.60
Phosphate	$H_2PO_4^-$	6.82[c]
N-(2-acetamido)-2-amino-ethanesulfonic acid (ACES)	$H_2NCOCH_2\overset{+}{N}H_2CH_2CH_2SO_3^-$	6.90
Imidazolium ion		7.00
p-Nitrophenol		7.24
N-2-hydroxyethylpiperazine-N'-2-ethanesulfonic acid (HEPES)		7.55
Glycinamide	$^+H_3NCH_2CONH_2$	8.20
Tris(hydroxymethyl)-aminomethane (Tris)	$(HOCH_2)_3C\overset{+}{N}H_3$	8.30
Boric acid	H_3BO_3	9.24
Ammonium ion	NH_4^+	9.25[c]
Phenol	C_6H_5OH	9.90
Methylammonium ion	$CH_3NH_3^+$	10.60
Phosphate	HPO_4^{2-}	12.38[d]

[a]The acidic hydrogen is highlighted in red; [b]pK_1; [c]pK_2; [d]pK_3.

Question Determine the charge of the conjugate base of each acid.

one acidic hydrogen, has a pK value for each dissociation (called pK_1, pK_2, etc.). The first proton dissociates with the lowest pK value. Subsequent protons are less likely to dissociate and so have higher pK values.

The pH of a solution of acid is related to the pK

When an acid (represented as the proton donor HA) is added to water, the final hydrogen ion concentration of the solution depends on the acid's tendency to ionize:

$$HA \rightleftharpoons A^- + H^+$$

In other words, the final pH depends on the equilibrium between HA and A^-,

$$K_a = \frac{[A^-][H^+]}{[HA]} \tag{2.9}$$

so that

$$[H^+] = K_a\frac{[HA]}{[A^-]} \tag{2.10}$$

We can express $[H^+]$ as a pH, and K_a as a pK, which yields

$$-\log [H^+] = -\log K_a - \log \frac{[HA]}{[A^-]} \tag{2.11}$$

or

$$pH = pK + \log \frac{[A^-]}{[HA]} \tag{2.12}$$

Equation 2.12 is known as the **Henderson–Hasselbalch equation.** It relates the pH of a solution to the pK of an acid and the concentration of the acid (HA) and its conjugate base (A⁻). This equation makes it possible to perform practical calculations to predict the pH of a solution (see Sample Calculation 2.2) or the concentrations of an acid and its conjugate base at a given pH (see Sample Calculation 2.3).

SAMPLE CALCULATION 2.2

Problem Calculate the pH of a 1-L solution to which has been added 6.0 mL of 1.5 M acetic acid and 5.0 mL of 0.4 M sodium acetate.

Solution First, calculate the final concentrations of acetic acid (HA) and acetate (A⁻). The final volume of the solution is $1 L + 6 mL + 5 mL = 1.011 L$.

$$[HA] = \frac{(0.006\ L)(1.5\ M)}{1.011\ L} = 0.0089\ M$$

$$[A^-] = \frac{(0.005\ L)(0.4\ M)}{1.011\ L} = 0.0020\ M$$

Next, substitute these values into the Henderson–Hasselbalch equation using the pK for acetic acid given in Table 2.4:

$$pH = pK + \log \frac{[A^-]}{[HA]}$$

$$pH = 4.76 + \log \frac{0.0020}{0.0089}$$

$$= 4.76 - 0.65$$

$$= 4.11$$

SAMPLE CALCULATION 2.3

Problem Calculate the concentration of formate in a 10-mM solution of formic acid at pH 4.15.

Solution The solution of formic acid contains both the acid species (formic acid) and its conjugate base (formate). Use the Henderson–Hasselbalch equation to determine the ratio of formate (A⁻) to formic acid (HA) at pH 4.15, using the pK value given in Table 2.4.

$$pH = pK + \log \frac{[A^-]}{[HA]}$$

$$\log \frac{[A^-]}{[HA]} = pH - pK = 4.15 - 3.75 = 0.40$$

$$\frac{[A^-]}{[HA]} = 2.51 \text{ or } [A^-] = 2.51[HA]$$

Since the total concentration of formate and formic acid is 0.01 M, $[A^-] + [HA] = 0.01$ M, and $[HA] = 0.01$ M $- [A^-]$. Therefore,

$$[A^-] = 2.51[HA]$$

$$[A^-] = 2.51(0.01\ M - [A^-])$$

$$[A^-] = 0.0251\ M - 2.51[A^-]$$

$$3.51[A^-] = 0.0251\ M$$

$$[A^-] = 0.0072\ M \text{ or } 7.2\ mM$$

The Henderson–Hasselbalch equation indicates that *when the pH of a solution of acid is equal to the pK of that acid, then the acid is half dissociated; that is, exactly half of the molecules are in the protonated HA form and half are in the unprotonated A⁻ form.* You can prove to yourself that when $[A^-] = [HA]$, the log term of the Henderson–Hasselbalch equation becomes zero (log 1 = 0) and pH = pK. When the pH is far below the pK, the acid exists mostly in the HA form; when the pH is far above the pK, the acid exists mostly in the A⁻ form. Note that A⁻ signifies a deprotonated acid; if the acid (HA) bears a positive charge to begin with, the dissociation of a proton yields a neutral species, still designated A⁻.

Knowing the ionization state of an acidic substance at a given pH can be critical. For example, a drug that has no net charge at pH 7.4 may readily enter cells, whereas a drug that bears a net positive or negative charge at that pH may remain in the bloodstream and be therapeutically useless (see Sample Calculation 2.4).

SAMPLE CALCULATION 2.4

Problem Determine which molecular species of phosphoric acid predominates at pH values of **a.** 1.5, **b.** 4, **c.** 9, and **d.** 13.

Solution From the pK values in Table 2.4, we know that:

Below pH 2.15, the fully protonated H_3PO_4 species predominates.

At pH 2.15, $[H_3PO_4] = [H_2PO_4^-]$.

Between pH 2.15 and 6.82, the $H_2PO_4^-$ species predominates.

At pH 6.82, $[H_2PO_4^-] = [HPO_4^{2-}]$.

Between pH 6.82 and 12.38, the HPO_4^{2-} species predominates.

At pH 12.38, $[HPO_4^{2-}] = [PO_4^{3-}]$.

Above pH 12.38, the fully deprotonated PO_4^{3-} species predominates.

Therefore, the predominant species at the indicated pH values are **a.** H_3PO_4, **b.** $H_2PO_4^-$, **c.** HPO_4^{2-}, and **d.** PO_4^{3-}.

The pH-dependent ionization of biological molecules is also key to understanding their structures and functions. *Under physiological conditions, some of the functional groups on biological molecules act as acids and bases.* Their ionization states depend on their respective pK values and on the pH ($[H^+]$) of their environment. For example, at physiological pH, a polypeptide bears multiple ionic charges because its carboxylic acid (—COOH) groups are ionized to carboxylate (—COO⁻) groups and its amino (—NH₂) groups are protonated (—NH₃⁺). This is because the pK values for the carboxylic acid groups are about 4, and the pK values for the amino groups are above 10. Consequently, below pH 4, both the carboxylic acid and amino groups are mostly protonated; above pH 10, both groups are mostly deprotonated.

pH < 4	4 < pH < 10	pH > 10
—COOH	—COO⁻	—COO⁻
—NH₃⁺	—NH₃⁺	—NH₂

Note that a compound containing a —COO⁻ group is sometimes still called an "acid," even though it has already given up its proton. Similarly, a "basic" compound may already have accepted a proton.

In addition to amino and carboxyl groups, pH-sensitive phosphoryl and imino groups (see Table 1.1) influence the chemistry of biological molecules. For example, phosphoryl groups, with pK values near neutral, tend to exist as mixtures of ionic forms under cellular conditions:

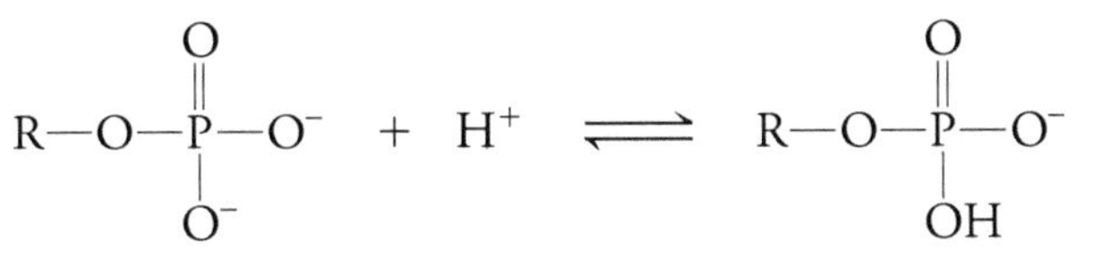

Likewise, imino groups are sensitive to the ambient pH. An imino group, also known as a Schiff base (except when R = H), has a pK value near neutral, so it often exists as a mixture of acidic and basic forms:

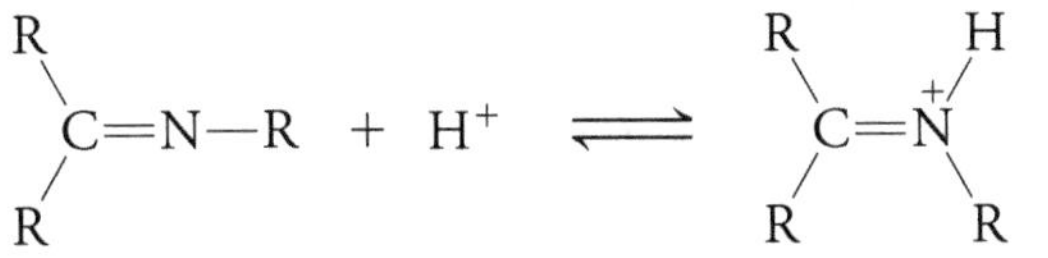

Imine
(Schiff base)

Iminium ion
(protonated Schiff base)

Under physiological conditions, amido, hydroxyl, and sulfhydryl groups (see Table 1.1) are only rarely found in ionized forms in biological molecules.

Concept Check

1. Draw the structure of a water molecule, including all its electrons, before and after ionization.
2. What is pH and how is it related to $[H^+]$ and $[OH^-]$?
3. For each fluid in Table 2.3, determine whether the addition of acid or base would bring the pH to neutral.
4. Rearrange the Henderson–Hasselbalch equation to isolate each term.
5. Predict the net charge of each molecule in Table 2.4 at pH 7.0.
6. What is the relationship between pH and the pK of an acid?
7. Draw the ionic forms of phosphoryl and imino groups that would predominate at very low and very high pH.

2.4 Tools and Techniques: Buffers

KEY CONCEPTS

Describe how buffer solutions resist changes in pH.

- Recognize the acidic and basic species in a buffer solution.
- Use the Henderson–Hasselbalch equation to devise a recipe for a buffer solution.
- Determine the useful pH range of a buffer solution.

When a strong acid such as HCl is added to pure water, all the added acid contributes directly to a decrease in pH. However, when HCl is added to a solution containing a weak acid (HA) in equilibrium with its conjugate base (A^-), the pH does not change so dramatically, because some of the added protons combine with the conjugate base to re-form the acid and therefore do not contribute to an increase in $[H^+]$.

$$HCl \rightarrow H^+ + Cl^- \quad \text{large increase in } [H^+]$$
$$HCl + A^- \rightarrow HA + H^+ + Cl^- \quad \text{small increase in } [H^+]$$

Conversely, when a strong base (such as NaOH) is added to the solution of weak acid/conjugate base, some of the added hydroxide ions accept protons from the acid to form H_2O and therefore do not contribute to a decrease in $[H^+]$.

$$NaOH \rightarrow Na^+ + OH^- \quad \text{large decrease in } [H^+]$$
$$NaOH + HA \rightarrow Na^+ + A^- + OH^- + H_2O \quad \text{small decrease in } [H^+]$$

The weak acid/conjugate base system (HA/A^-) acts as a **buffer** against the added acid or base by preventing the dramatic changes in pH that would otherwise occur.

The buffering activity of a weak acid, such as acetic acid, can be traced by titrating the acid with a strong base (**Fig. 2.16**). At the start of the titration, all the acid is present in its protonated (HA) form. As base (for example, NaOH) is added, protons begin to dissociate from the acid, producing A^-. The continued addition of base eventually causes all the protons to dissociate, leaving all the acid in its conjugate base (A^-) form. At the midpoint of the titration, exactly half the protons have dissociated, so [HA] = [A^-] and pH = pK (Equation 2.12). The broad, flat shape of the titration curve shown in Figure 2.16 indicates that the pH does not change drastically with added acid or base when the pH is near the pK. *The effective buffering capacity of an acid is generally taken to be within one pH unit of its pK.* For acetic acid (pK = 4.76), this would be pH 3.76–5.76.

Biochemists nearly always perform experiments in buffered solutions in order to maintain a constant pH when acidic or basic substances are added or when chemical reactions produce or consume protons. Without buffering, fluctuations in pH would alter the ionization state of the

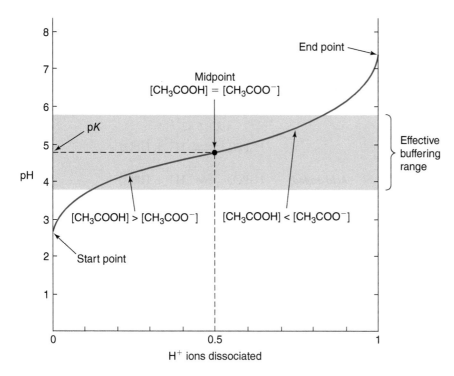

Figure 2.16 Titration of acetic acid. At the start point (before base is added), the acid is present mainly in its CH_3COOH form. As small amounts of base are added, protons dissociate until, at the midpoint of the titration (where $pH = pK$), $[CH_3COOH] = [CH_3COO^-]$. The addition of more base causes more protons to dissociate until nearly all the acid is in the CH_3COO^- form (the end point). The shaded area indicates the effective buffering range of acetic acid. Within one pH unit of the pK, additions of acid or base do not greatly perturb the pH of the solution.

Question **Sketch the titration curve for ammonia.**

molecules under study, which might then behave differently. Before biochemists appreciated the importance of pH and buffering, experimental results were often poorly reproducible, even within the same laboratory.

A buffer solution is typically prepared from a weak acid and the salt of its conjugate base (see Sample Calculation 2.5). The two are mixed together in the appropriate ratio, according to the Henderson–Hasselbalch equation, and the final pH is adjusted if necessary by adding a small amount of concentrated HCl or NaOH. In addition to choosing a buffering compound with a pK value near the desired pH, a biochemist must consider other factors, including the compound's solubility, stability, toxicity to cells, reactivity with other molecules, and cost.

SAMPLE CALCULATION 2.5

Problem How many mL of a 2.0-M solution of boric acid must be added to 600 mL of a solution of 10 mM sodium borate in order for the pH to be 9.45?

Solution Rearrange the Henderson–Hasselbalch equation to isolate the $[A^-]/[HA]$ term:

$$pH = pK + \log\frac{[A^-]}{[HA]}$$

$$\log\frac{[A^-]}{[HA]} = pH - pK$$

$$\frac{[A^-]}{[HA]} = 10^{(pH-pK)}$$

Substitute the known pK (from Table 2.4) and the desired pH:

$$\frac{[A^-]}{[HA]} = 10^{(9.45-9.24)} = 10^{0.21} = 1.62$$

The starting solution contains $(0.6 \text{ L})(0.01 \text{ mol} \cdot \text{L}^{-1}) = 0.006$ moles of borate (A^-). The amount of boric acid (HA) needed is $0.006 \text{ mol}/1.62 = 0.0037 \text{ mol}$.

Since the stock boric acid is 2.0 M, the volume of boric acid to be added is $(0.0037 \text{ mol})/(2.0 \text{ mol} \cdot \text{L}^{-1}) = 0.0019 \text{ L}$ or 1.9 mL.

One commonly used laboratory buffer system that mimics physiological conditions contains a mixture of dissolved NaH_2PO_4 and Na_2HPO_4 for a total phosphate concentration of 10 mM. The Na^+ ions are spectator ions and are usually not significant because the buffer solution usually also contains about 150 mM NaCl (see Fig. 2.12). In this "phosphate-buffered saline," the equilibrium between the two species of phosphate ions can "soak up" added acid (producing more $H_2PO_4^-$ or added base (producing more HPO_4^{2-}).

$$pK = 6.82$$
$$H_2PO_4^- \rightleftharpoons H^+ + HPO_4^{2-}$$

Acid added $H_2PO_4^- \quad\longleftarrow\quad H^+ + HPO_4^{2-}$

Base added $H_2PO_4^- \quad\longrightarrow\quad H^+ + HPO_4^{2-}$

This phenomenon illustrates **Le Châtelier's principle,** which states that a change in concentration of one reactant will shift the concentrations of other reactants in order to restore equilibrium. In the human body, the major buffering systems involve bicarbonate, phosphate, and other ions.

Concept Check

1. Write an equation that describes the equilibrium in a solution containing an acetate/acetic acid buffer and describe what happens when HCl or NaOH is added.
2. What is the useful pH range of a buffer?

2.5 Clinical Connection: Physiological Buffers

KEY CONCEPTS

Explain how the human body maintains a constant pH.

- Write the equations that describe operation of the bicarbonate buffer system in the human body.
- Describe the roles of the lungs and kidneys in maintaining blood pH homeostasis.
- Summarize the causes and treatments of acidosis and alkalosis.

The cells of the human body typically maintain an internal pH of 6.9–7.4. The body does not normally have to defend itself against strong inorganic acids, but many metabolic processes generate acids, which must be buffered so that they do not cause the pH of blood to drop below its normal value of 7.4 as almost all life processes are pH dependent. The functional groups of proteins and phosphate groups can serve as biological buffers; however, the most important buffering system involves CO_2 (itself a product of metabolism) in the blood plasma (plasma is the fluid component of blood).

CO_2 reacts with water to form carbonic acid, H_2CO_3:

$$CO_2 + H_2O \rightleftharpoons H_2CO_3$$

This freely reversible reaction is accelerated *in vivo* by the enzyme carbonic anhydrase, which is present in most tissues and is particularly abundant in red blood cells. Carbonic acid ionizes to bicarbonate, HCO_3^- (see Box 2.C):

$$H_2CO_3 \rightleftharpoons H^+ + HCO_3^-$$

so that the overall reaction is

$$CO_2 + H_2O \rightleftharpoons H^+ + HCO_3^-$$

The pK for this process is 6.1 (the ionization of HCO_3^- to CO_3^{2-} occurs with a pK of 10.3 and is therefore not significant at physiological pH).

Normal conditions

$$H^+ + HCO_3^- \rightleftharpoons H_2CO_3 \rightleftharpoons H_2O + CO_2$$

Excess acid

$$H^+ + HCO_3^- \longrightarrow H_2CO_3 \rightleftharpoons H_2O + CO_2$$
$$H^+ + HCO_3^- \rightleftharpoons H_2CO_3 \longrightarrow H_2O + CO_2$$
$$H^+ + HCO_3^- \rightleftharpoons H_2CO_3 \rightleftharpoons H_2O + CO_2$$

Insufficient acid

$$H^+ + HCO_3^- \rightleftharpoons H_2CO_3 \longleftarrow H_2O + CO_2$$
$$H^+ + HCO_3^- \longleftarrow H_2CO_3 \rightleftharpoons H_2O + CO_2$$
$$H^+ + HCO_3^- \rightleftharpoons H_2CO_3 \rightleftharpoons H_2O + CO_2$$

Figure 2.17 The bicarbonate buffer system. Elimination or retention of CO_2 can shift the equilibrium in order to promote or prevent the loss of H^+ from the body.

Although a pK of 6.1 appears to be just outside the range of a useful physiological buffer (which would be within one pH unit of 7.4), the effectiveness of the bicarbonate buffer system is augmented by the fact that excess hydrogen ions can not only be buffered but can also be eliminated from the body. This is possible because after the H^+ combines with HCO_3^- to re-form H_2CO_3, which rapidly equilibrates with $CO_2 + H_2O$, some of the CO_2 can be given off as a gas in the lungs. If the body needs to retain more H^+ to maintain a constant pH, breathing can be adjusted so that less gaseous CO_2 is lost during exhalation (**Fig. 2.17**).

Changes in pulmonary function can adequately adjust blood pH on the order of minutes to hours; however, longer-term adjustments of hours to days are made by the kidneys, which use a variety of mechanisms to excrete or retain H^+, bicarbonate, and other ions. In fact, the kidneys play a major role in the buffering of metabolic acids. Normal metabolic activity generates acids as the result of the degradation of amino acids, the incomplete oxidation of glucose and fatty acids, and the ingestion of acidic groups in the form of phosphoproteins and phospholipids. The HCO_3^- required to buffer these acids is initially filtered out of the bloodstream in the kidneys, but the kidneys actively reclaim most of this bicarbonate before it is lost in the urine (**Fig. 2.18**).

In addition to reabsorbing filtered HCO_3^-, the kidneys also generate additional HCO_3^- to offset losses due to the buffering of metabolic acids and the exhalation of CO_2. Metabolic activity in the kidney cells produces CO_2, which is converted to $H^+ + HCO_3^-$. The cells actively secrete

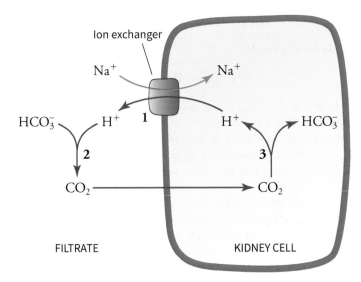

Figure 2.18 Bicarbonate reabsorption. H^+ leaves the kidney cells in exchange for Na^+ (step 1). The expelled H^+ combines with HCO_3^- in the filtrate, forming CO_2 (step 2). Because it is nonpolar, the CO_2 can diffuse into the kidney cell, where it is converted back to $H^+ + HCO_3^-$ (step 3).

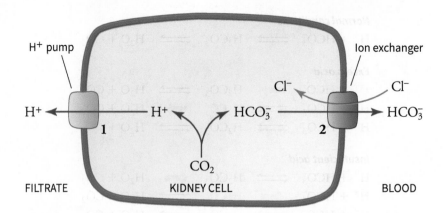

Figure 2.19 Bicarbonate production. Carbonic acid–derived protons are pumped out of the kidney cell (step 1), and the bicarbonate remaining in the cell is returned to the bloodstream in exchange for Cl⁻ (step 2).

the H⁺, which is lost via the urine, accounting for the mildly acidic pH of normal urine. The bicarbonate remaining in the cell is returned to the bloodstream in exchange for Cl⁻ (**Fig. 2.19**).

Some HCO_3^- also accumulates as a result of the metabolism of the amino acid glutamine in the kidneys.

$$H-\overset{\overset{\displaystyle COO^-}{|}}{\underset{\underset{\displaystyle {}^+NH_3}{|}}{C}}-CH_2-CH_2-\overset{\overset{\displaystyle O}{\|}}{\underset{\underset{\displaystyle NH}{}}{C}}$$

Glutamine

The two amino groups are removed as ammonia (NH_3), which is ultimately excreted in the urine. Because the ammonium ion (NH_4^+) has a pK value of 9.25, nearly all the ammonia molecules become protonated at physiological pH. The consumption of protons from carbonic acid (ultimately, CO_2) leaves an excess of HCO_3^-.

The protein buffer system also acts as a good buffering system because of the presence of amino acids. The amino acids have two ionizable groups, i.e., a protonated amino group (NH_4^+) which is effective in the basic pH zone, and a carboxylate group (COO^-) which is effective in the acidic pH zone. Buffering by proteins accounts for two-thirds of the buffering power of the blood and most of the buffering done within cells.

The phosphate buffer is also an important intracellular buffer in mammals and consists of $H_2PO_4^-$ and HPO_4^{2-} ions.

$$H_2PO_4 + H_2O \rightleftharpoons H_3O^+ + HPO_4^{2-}$$

or it may be simply described as

$$H_2PO_4 \rightleftharpoons H^+ + HPO_4^{2-}$$

The pK_a of H_2PO_4 is 6.86, this system resists changes in pH from 6.1 to 7.7 whereas the intracellular fluid has a pH range of 6.9–7.4.

Certain medical conditions can disrupt normal acid–base balance, leading to **acidosis** (blood pH less than 7.35) or **alkalosis** (blood pH greater than 7.45). The activities of the lungs and kidneys, as well as other organs, may contribute to the imbalance or respond to help correct the imbalance.

The most common disorder of acid–base chemistry is **metabolic acidosis,** which is caused by the accumulation of the acidic products of metabolism and can develop during shock, starvation, severe diarrhea (in which bicarbonate-rich digestive fluid is lost), certain genetic diseases, and renal failure (in which damaged kidneys eliminate too little acid). Metabolic acidosis can interfere with cardiac function and oxygen delivery and contribute to central nervous system depression. Despite its many different causes, one common symptom of metabolic acidosis is rapid, deep breathing. The increased ventilation helps compensate for the acidosis by "blowing off" more acid in the form of CO_2 derived from H_2CO_3. However, this mechanism also impairs O_2 uptake by the lungs.

Metabolic acidosis can be treated by administering sodium bicarbonate ($NaHCO_3$). In chronic metabolic acidosis, the mineral component of bone serves as a buffer, which leads to the loss of calcium, magnesium, and phosphate, and ultimately to osteoporosis and fractures.

Metabolic alkalosis, a less common condition, can result from prolonged vomiting, which represents a loss of HCl in gastric fluid. This specific disorder can be treated by infusing NaCl. Metabolic alkalosis is also caused by overproduction of mineralocorticoids (hormones produced by the adrenal glands), which leads to abnormally high levels of H^+ excretion and Na^+ retention. In this case, the disorder does not respond to saline infusion. Individuals with metabolic alkalosis may experience apnea (cessation of breathing) for 15 s or more and cyanosis (blue coloration) due to inadequate oxygen uptake. Both these symptoms reflect the body's effort to compensate for the high blood pH by minimizing the loss of CO_2 from the lungs.

What happens when abnormal lung function is the *cause* of an acid–base imbalance? **Respiratory acidosis** can result from impaired pulmonary function caused by airway blockage, asthma (constriction of the airways), and emphysema (loss of alveolar tissue). In all cases, the kidneys respond by adjusting their activity, mainly by increasing the synthesis of the enzymes that break down glutamine in order to produce NH_3. Excretion of NH_4^+ helps correct the acidosis. However, renal compensation of respiratory acidosis takes hours to days (the time course for adjusting enzyme levels), so this condition is best treated by restoring pulmonary function through bronchodilation, supplemental oxygen, or mechanical ventilation (assisted breathing).

Some of the diseases that impair lung function (for example, asthma) can also contribute to **respiratory alkalosis,** but this relatively rare condition is more often caused by hyperventilation brought on by fear or anxiety. Unlike the other acid–base disorders, this form of respiratory alkalosis is seldom life-threatening.

Concept Check

1. Review the equations that describe the interconversion of carbon dioxide and bicarbonate.
2. Compare the metabolic adjustments made by the lungs and kidneys during acid–base homeostasis.
3. Discuss how impaired lung or kidney function could lead to acidosis or alkalosis.

SUMMARY

2.1 Water Molecules and Hydrogen Bonds

- Water molecules are polar; they form hydrogen bonds with each other and with other polar molecules bearing hydrogen bond donor or acceptor groups.
- The electrostatic forces acting on biological molecules also include ionic interactions and van der Waals interactions.
- Water dissolves polar and ionic substances.

2.2 The Hydrophobic Effect

- Nonpolar (hydrophobic) substances tend to aggregate rather than disperse in water in order to minimize the decrease in entropy that would be required for water molecules to surround each nonpolar molecule. This is the hydrophobic effect.
- Amphiphilic molecules, which contain both polar and nonpolar groups, may aggregate to form micelles or bilayers.

2.3 Acid–Base Chemistry

- The dissociation of water produces hydroxide ions (OH^-) and protons (H^+) whose concentration can be expressed as a pH value.

- The pH of a solution can be altered by adding an acid (which donates protons) or a base (which accepts protons).
- The tendency for a proton to dissociate from an acid is expressed as a pK value.
- The Henderson–Hasselbalch equation relates the pH of a solution of a weak acid and its conjugate base to the pK and the concentrations of the acid and base.

2.4 Tools and Techniques: Buffers

- A buffered solution, which contains an acid and its conjugate base, resists changes in pH when more acid or base is added.

2.5 Clinical Connection: Physiological Buffers

- The body uses the bicarbonate buffer system to maintain a constant internal pH. Homeostatic adjustments are made by the lungs, where CO_2 is released, and by the kidneys, which excrete H^+ and ammonia.

KEY TERMS

polarity	London dispersion	amphiphilic	neutral solution	conjugate base	alkalosis
hydrogen bond	forces	amphipathic	acidic solution	polyprotic acid	metabolic acidosis
ionic interaction	dielectric constant	micelle	basic solution	Henderson–	metabolic alkalosis
van der Waals radius	solute	bilayer	pH	Hasselbalch	respiratory acidosis
electronegativity	solvation	vesicle	acid	equation	respiratory alkalosis
van der Waals	hydration	hydronium ion	base	buffer	
interaction	hydrophilic	proton jumping	acid dissociation	Le Châtelier's	
dipole–dipole	hydrophobic	ionization constant	constant (K_a)	principle	
interaction	hydrophobic effect	of water (K_w)	pK	acidosis	

BIOINFORMATICS

Brief Bioinformatics Exercises

2.1 Structure and Solubility

2.2 Amino Acids, Ionization, and pK Values

PROBLEMS

2.1 Water Molecules and Hydrogen Bonds

1. The H—C—H bond angle in the perfectly tetrahedral CH_4 molecule is 109°. Explain why the H—O—H bond angle in water is only about 104.5°.

2. Each C=O bond in CO_2 is polar, yet the whole molecule is nonpolar. Explain.

3. Which compound has a higher boiling point, H_2O or H_2S? Explain.

4. Consider the following molecules and their melting points listed below. How can you account for the differences in melting points among these molecules of similar size?

	Molecular weight (g · mol⁻¹)	Melting point (°C)
Water, H_2O	18.0	0
Ammonia, NH_3	17.0	−77
Methane, CH_4	16.0	−182

5. An exhaustive study with over 6000 fluorinated compounds was carried out in order to determine the ability of these compounds to participate in hydrogen bonding (see Box 2.A). The structure of one of these compounds is shown below. Identify the hydrogen bond acceptor and donor groups. Use an arrow to point toward each acceptor and away from each donor.

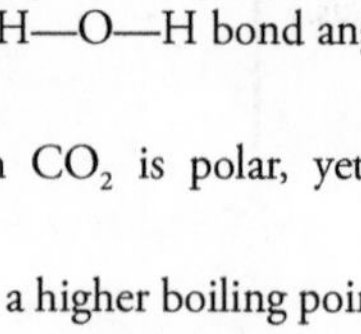

6. In 2007, pets consuming a brand of pet food containing melamine (Compound B in Problem 1.16) and cyanuric acid (both nontoxic when consumed alone) suffered from renal failure when the two compounds combined to form crystalline melamine cyanurate in the kidney. Draw the structure of melamine cyanurate, which forms when melamine and cyanuric acid (see below) form hydrogen bonds with each other.

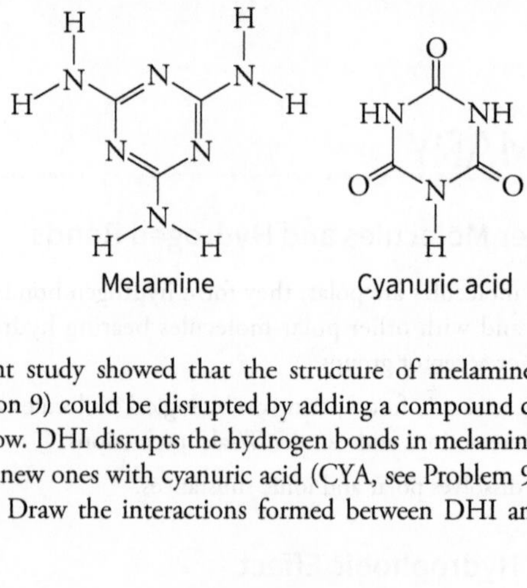

Melamine   Cyanuric acid

7. A recent study showed that the structure of melamine cyanurate (see Solution 9) could be disrupted by adding a compound called DHI, shown below. DHI disrupts the hydrogen bonds in melamine cyanurate and forms new ones with cyanuric acid (CYA, see Problem 9), releasing melamine. Draw the interactions formed between DHI and cyanuric acid.

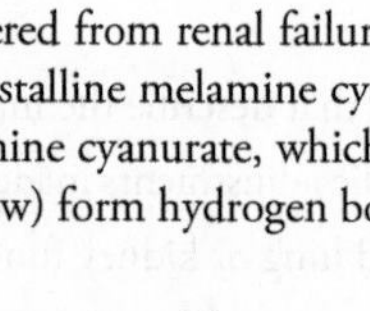

8. Examine Table 2.1. What is the relationship between an atom's electronegativity and its ability to participate in hydrogen bonding?

9. Do intermolecular hydrogen bonds form in the compounds below? Draw the hydrogen bonds where appropriate.

A $H_3C-\overset{\overset{\displaystyle O}{\|}}{C}-H$

B (imidazole structure with N–H)

C H_3C-CH_2-OH

D H_3C-CH_2-Cl and $H-\overset{\displaystyle O}{}-H$

E $H_3C-CH_2-O-CH_2-CH_3$ and $H-\overset{\overset{\displaystyle H}{|}}{N}-H$ with H below

10. Occasionally, when a carbon atom is polarized by an adjacent electronegative atom, a C—H group can form a weak hydrogen bond with a carbonyl group. Draw a structure to represent this situation and add δ^+ and δ^- symbols where appropriate.

11. Two hydrogen bonds form between the DNA bases adenine and thymine. Draw these bonds as dashed lines.

(adenine structure)   (thymine structure with CH₃)

Adenine   Thymine

12. Rank the following compounds in order of melting points. Classify the compounds as soluble, slightly soluble, or insoluble in water and explain your answer.

a. $H_3C-CH_2-O-CH_3$

b. $H_3C-\overset{\overset{\displaystyle O}{\|}}{C}-NH_2$

c. $H_2N-\overset{\overset{\displaystyle O}{\|}}{C}-NH_2$

d. $H_3C-(CH_2)_3-CH_3$

e. $H_3C-CH_2-\overset{\overset{\displaystyle O}{\|}}{C}H$

13. What intermolecular interactions are likely to form between the two amino acid side chains shown in the protein diagrams below?

a.
(Asn side chain with C=O, CH₂, NH₂ and Ser HO—CH₂)
Asn   Ser

b.

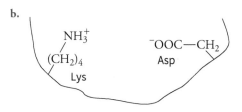

c.

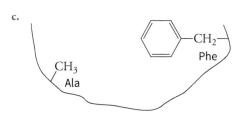

14. Practitioners of homeopathic medicine believe that good health can be restored by administering a small amount of a harmful substance that stimulates the body's natural healing processes. The remedy is typically prepared by serially diluting an animal or plant extract. In a so-called 30× dilution, the active substance is diluted 10-fold, 30 times in succession. **a.** If the substance is initially present at a concentration of 1 M, what is its concentration after a 30× dilution? **b.** A typical homeopathic dose is a few drops. How many molecules of the active substance would be present in 1 mL? **c.** When confronted with the results of an exercise such as the one in part (b), proponents of homeopathy claim that a molecule can leave an imprint, or memory of itself, on the surrounding water molecules. What information presented in this chapter might be used to support this claim? To refute it?

15. The solubilities of several alcohols in water are shown below (their structures are shown in Table 2.2). Note that propanol is miscible in water (i.e., any amount will dissolve). However, as the size of the alcohol increases, solubility decreases. Explain.

Alcohol	Solubility in water (mmol · 100 g⁻¹)
Propanol	Miscible
Butanol	0.11
Pentanol	0.05

16. Water is unusual in that its solid form is less dense than its liquid form. This means that when a pond freezes in the winter, ice is found as a layer on top of the pond, not on the bottom. What are the biological advantages of this?

17. Ice skaters skate not on the solid ice surface itself but on a thin film of water that forms between the skater's blade and the ice. What unique property of water makes ice skating possible?

18. Ammonium sulfate, $(NH_4)_2SO_4$, is a water-soluble salt. Draw the structures of the hydrated ions that form when ammonium sulfate dissolves in water.

19. Which of the four primary alcohols listed in Table 2.2 would be the best solvent for ammonium ions? What can you conclude about the polarity of these solvents, which all contain an —OH group that can form hydrogen bonds?

20. The amino acid glycine is sometimes drawn as structure A below. However, the structure of glycine is more accurately represented by structure B. Glycine has the following properties: white crystalline solid, high melting point, and water solubility. Why does structure B more accurately represent the structure of glycine than structure A?

$$H_2N—CH_2—COOH \qquad {}^+H_3N—CH_2—COO^-$$

Structure A   Structure B

21. A typical spherical bacterial cell with a diameter of 1 μm contains 1000 molecules of a certain protein. **a.** What is the protein's concentration in units of mM? (*Hint:* The volume of a sphere is $4\pi r^3/3$.) **b.** How many molecules of glucose does the cell contain if the internal glucose concentration is 5 mM?

22. a. Water has a surface tension that is nearly three times greater than that of ethanol. Explain. **b.** The surface tension of water decreases with increasing temperature. Explain.

23. Explain why water forms nearly spherical droplets on the surface of a freshly waxed car. Why doesn't water bead on a clean windshield?

2.2 The Hydrophobic Effect

24. The structures of **a.** hexadecyltrimethylammonium (a cationic detergent) **b.** cholate (an anionic detergent) **c.** dimethyldecylphosphine oxide (a nonionic detergent), are shown here. Identify the polar and the nonpolar regions of these amphipathic molecules.

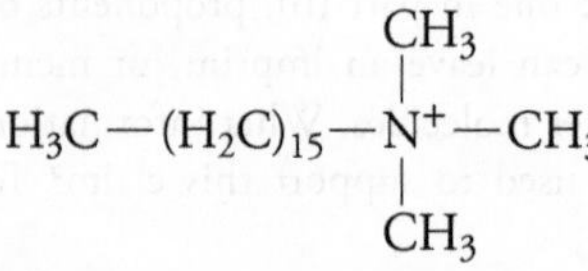

$$H_3C—(H_2C)_{15}—\overset{\overset{\displaystyle CH_3}{|}}{\underset{\underset{\displaystyle CH_3}{|}}{N^+}}—CH_3$$

Hexadecyltrimethylammonium

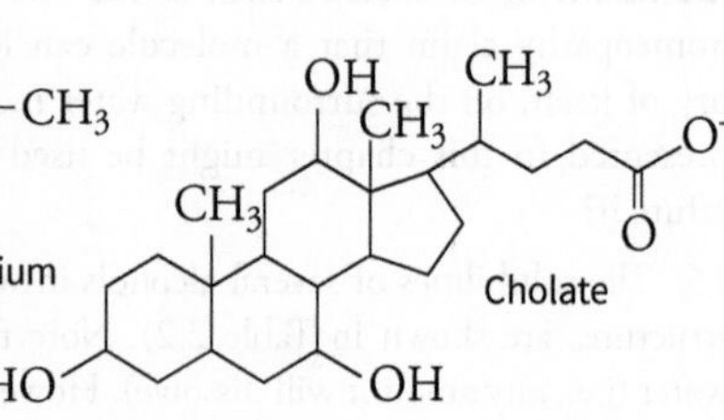

Cholate

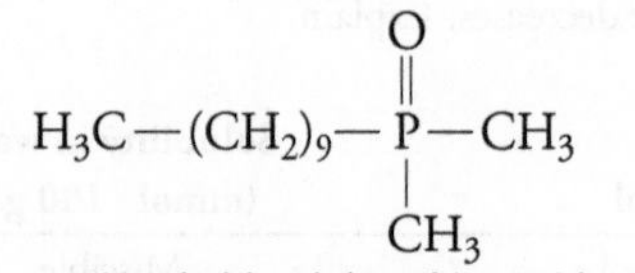

$$H_3C—(CH_2)_9—\overset{\overset{\displaystyle O}{\|}}{\underset{\underset{\displaystyle CH_3}{|}}{P}}—CH_3$$

Dimethyldecylphosphine oxide

25. Identify the regions of the detergents shown in Problem 24 that are able to form favorable interactions with water. Identify these interactions.

26. The structures of dimethyldecylphosphine oxide and *n*-octyl glucoside, both nonionic detergents, are shown here. Identify the polar and the nonpolar regions of these amphipathic molecules.

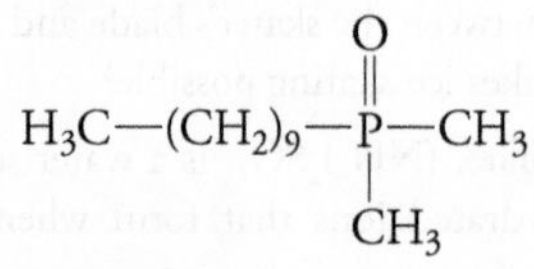

$$H_3C—(CH_2)_9—\overset{\overset{\displaystyle O}{\|}}{\underset{\underset{\displaystyle CH_3}{|}}{P}}—CH_3$$

Dimethyldecylphosphine oxide

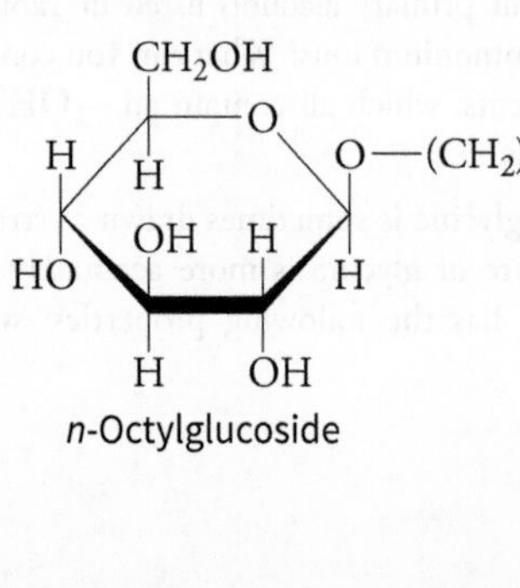

n-Octylglucoside

27. Identify the regions of the detergents shown in Problem 29 that are able to form favorable interactions with water. Identify these interactions.

28. Consider the structures of the molecules below. **a.** Are these molecules polar, nonpolar, or amphiphilic? **b.** Which are capable of forming micelles? **c.** Which are capable of forming bilayers?

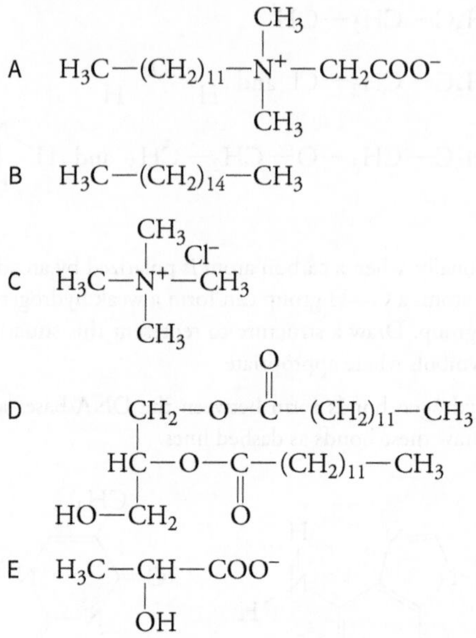

A $H_3C—(CH_2)_{11}—\overset{\overset{\displaystyle CH_3}{|}}{\underset{\underset{\displaystyle CH_3}{|}}{N^+}}—CH_2COO^-$

B $H_3C—(CH_2)_{14}—CH_3$

C $H_3C—\overset{\overset{\displaystyle CH_3}{|}}{\underset{\underset{\displaystyle CH_3}{|}}{N^+}}—CH_3 \quad Cl^-$

D

$$CH_2—O—\overset{\overset{\displaystyle O}{\|}}{C}—(CH_2)_{11}—CH_3$$
$$HC—O—\overset{}{C}—(CH_2)_{11}—CH_3$$
$$HO—CH_2 \qquad O$$

E $H_3C—\overset{}{\underset{\underset{\displaystyle OH}{|}}{CH}}—COO^-$

29. The compound bis-(2-ethylhexyl)sulfosuccinate (abbreviated AOT) is capable of forming "reverse" micelles in the hydrocarbon solvent isooctane (2,2,4-trimethylpentane). Scientists have investigated the use of reverse micelles for extracting water-soluble proteins. A two-phase system is formed: the hydrocarbon phase containing the reverse micelles and the water phase containing the protein. After a certain period of time, the protein is transferred to the reverse micelle. **a.** Draw the structure of the reverse micelle that AOT would form in isooctane. **b.** Where would the protein be located in the reverse micelle?

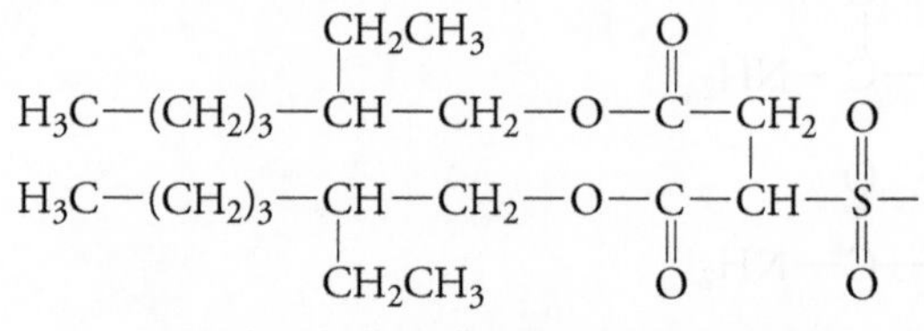

Bis-(2-ethylhexyl)sulfosuccinate (AOT)

30. Many household soaps are amphiphilic substances, often the salts of long-chain fatty acids, that form water-soluble micelles. An example is sodium dodecyl sulfate (SDS), an anionic detergent. **a.** Identify the polar and nonpolar portions of the SDS molecule. **b.** Draw the structure of the micelle formed by SDS. **c.** Explain how the SDS micelles "wash away" water-insoluble substances such as cooking grease.

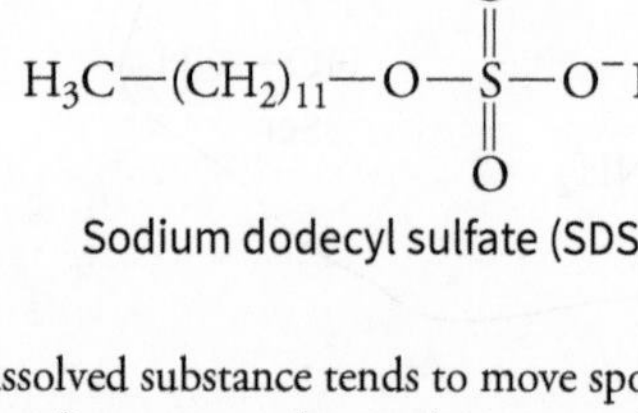

$$H_3C—(CH_2)_{11}—O—\overset{\overset{\displaystyle O}{\|}}{\underset{\underset{\displaystyle O}{\|}}{S}}—O^-Na^+$$

Sodium dodecyl sulfate (SDS)

31. Just as a dissolved substance tends to move spontaneously down its concentration gradient, water also tends to move from an area of high concentration (low solute concentration) to an area of low concentration (high solute concentration), a process known as osmosis. **a.** Explain why a lipid bilayer would be a barrier to osmosis. **b.** Why are isolated human cells placed in a solution that typically contains about 150 mM NaCl? What would happen if the cells were placed in pure water?

32. Fresh water is obtained from seawater in a desalination process using reverse osmosis. In reverse osmosis, seawater is forced through a membrane; the salt remains on one side of the membrane and fresh water is produced on the other side. In what ways does reverse osmosis differ from osmosis described in Problem 31?

33. Drug delivery is a challenging problem because the drug molecules must be sufficiently water-soluble to dissolve in the blood, but also sufficiently nonpolar to pass through the cell membrane for delivery. A medicinal chemist proposes to encapsulate a water-soluble drug into a vesicle (see Section 2.2). How does this strategy facilitate the delivery of the drug to its target cell?

34. A specialized protein pump in the red blood cell membrane exports Na^+ ions and imports K^+ ions in order to maintain the sodium and potassium ion concentrations shown in Figure 2.12. Does the movement of these ions occur spontaneously or is this an energy-requiring process? Explain.

35. Estimate the amount of Na^+ lost in sweat during 15 minutes of vigorous exercise. What is the mass of potato chips (300 mg Na^+ per ounce) you would have to consume in order to replace the lost sodium?

36. The bacterium *E. coli* can adapt to changes in the solute concentration of its growth medium. The cell consists of a cytoplasmic compartment bounded by a cell membrane surrounded by a porous cell wall; both the membrane and the cell wall allow the passage of water and ions. Under nongrowing conditions, only cytoplasmic water content is regulated. What happens to the cytoplasmic volume if *E. coli* is grown in a growth medium with **a.** a high salt concentration or **b.** a low salt concentration?

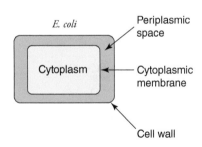

E. coli

Periplasmic space

Cytoplasm

Cytoplasmic membrane

Cell wall

37. Under growth conditions, *E. coli* regulates cytoplasmic K^+ content (in addition to water; see Problem 41) in order to prevent large changes in cell volume. How might *E. coli* regulate both the cytoplasmic concentrations of K^+ and water when grown in a low-salt medium? What happens when *E. coli* is grown in a high-salt medium?

38. Which of the following substances might be able to cross a bilayer? Which substances could not? Explain your answer.

a. CO_2

b.

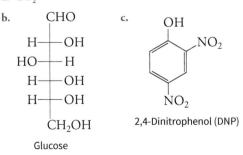

Glucose

c. 2,4-Dinitrophenol (DNP)

d. Na^+

2.3 Acid–Base Chemistry

39. Compare the concentrations of H_2O and H^+ in a sample of pure water at pH 7.0 at 25°C.

40. What is the pH of a neutral solution at a temperature where $K_w = 8 \times 10^{-14}$?

41. What is the pH of a solution of 1.0×10^{-7} M HCl?

42. What is the pH of a solution of 1.0×10^{-8} M NaOH?

43. Draw a diagram similar to Figure 2.13 showing how a hydroxide ion appears to jump through an aqueous solution.

44. Fill in the blanks of the following table:

	Acid, base, or neutral?	pH	$[H^+]$ (M)	$[OH^-]$ (M)
Pancreatic juice	____	____	1.3×10^{-8}	____
Blood	____	7.42	____	____
Saliva	neutral	____	____	____
Urine	____	____	____	6.3×10^{-8}
Gastric juice	____	____	7.9×10^{-3}	____

45. Several hours after a meal, partially digested food leaves the stomach and enters the small intestine, where pancreatic juice is added. How does the pH of the partially digested mixture change as it passes from the stomach into the intestine (see Table 2.3)?

46. The pH of urine has been found to be correlated with diet. Acidic urine results when meat and dairy products are consumed, because the oxidation of sulfur-containing amino acids in the proteins produces protons. Consumption of fruits and vegetables leads to alkaline urine, because these foods contain plentiful amounts of potassium and magnesium carbonate salts. Why would the presence of these salts result in alkaline urine?

47. Give the conjugate acid and base of the following acids: **a.** $HC_2O_4^-$; **b.** HSO_3^-; **c.** $H_2PO_4^-$; **d.** HCO_3^-; **e.** $HAsO_4^{2-}$; **f.** HPO_4^{2-}; **g.** HO_2^-.

48. Calculate the pH of 200 mL of water to which **a.** 10 mL of 1.0 M HNO_3 or **b.** 5 mL of 1.0 M KOH has been added.

49. What is the pH of 2.0 L of water to which (a) 1.5 mL of 2.0 M HCl or (b) 1.5 mL of 2.0 M NaOH has been added?

50. Calculate the pH of 500 mL of water to which **a.** 10 mL of 0.50 M HBr or **b.** 25 mL of 0.25 M NaOH has been added.

51. Calculate the pH of 1.0 L of water to which **a.** 15 mL of 2.0 M HI or **b.** 25 mL of 2.0 M $Ba(OH)_2$ has been added.

52. Identify the acidic hydrogens in the following compounds, using Table 2.4 as a guide:•

CH_2-COOH

$HO-C-COOH$

CH_2-COOH

Citric acid

Piperidine

$H_3\overset{+}{N}-CH-\overset{O}{\overset{\|}{C}}-OH$

CH_2

CH_2

CH_2

CH_2

$\overset{+}{N}H_3$

Lysine

Oxalic acid

$HO-\overset{O}{\overset{\|}{C}}-\overset{O}{\overset{\|}{C}}-OH$

Barbituric acid

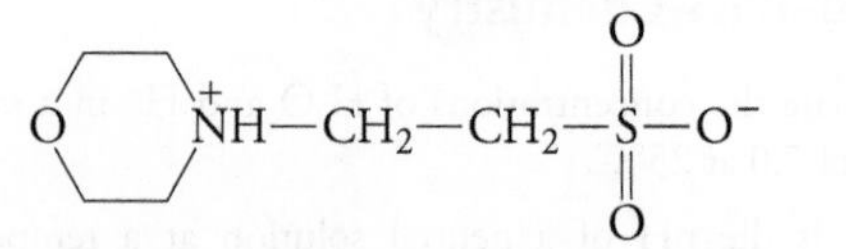

4-Morphine ethanesulfonic acid (MES)

53. Rank the following according to their strength as acids:

	Acid	K_a	pK
A	Citrate		4.76
B	Succinic acid	6.17×10^{-5}	
C	Succinate	2.29×10^{-6}	
D	Formic acid	1.78×10^{-4}	
E	Citric acid		3.13

54. Imidazole, shown here in its unprotonated form, has a pK value near 7.0 (see Table 2.4). Draw the structure of imidazole that predominates at the pH of blood.

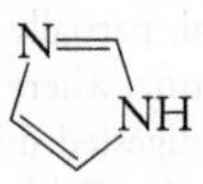

Imidazole

55. The blood of individuals with uncontrolled diabetes contains ketone bodies, such as the one shown here. Identify the acidic functional group(s) in acetoacetic acid and draw the structure of the molecule as it would appear at the pH of blood.

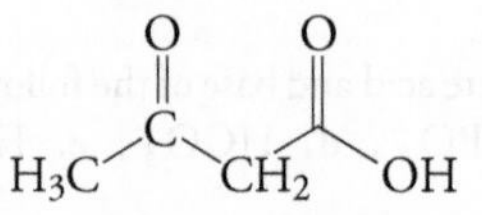

Acetoacetic acid

56. Calculate the pH of a 200-mL solution to which has been added 15 mL of 50 mM boric acid and 25 mL of 20 mM sodium borate.

57. Calculate the pH of a 500-mL solution to which has been added 20 mL of 100 mM glycinamide hydrochloride (Cl$^-$ $^+$H$_3$NCH$_2$CONH$_2$) and 35 mL of 50 mM glycinamide (H$_2$NCH$_2$CONH$_2$).

58. A solution is made by mixing 100 mL of a stock solution of 2.0 M K$_2$HPO$_4$ and 50 mL of a stock solution of 2.0 M KH$_2$PO$_4$. The solution is diluted to a final volume of 500 mL. What is the final pH of the solution?

59. Many insect larvae, like the ant shown at the start of the chapter, secrete acid for defense and to facilitate feeding on plant leaves. The fluid secreted by the *Theroa zethus* caterpillar contains approximately 6.53 M formic acid. Calculate the pH of this solution, assuming that only 1% of the acid is in the dissociated (A$^-$) form.

60. The fluid secreted by the *Theroa zethus* caterpillar (see Problem 65) also contains approximately 50 mM butyric acid ($pK = 4.82$). If the secretion contained only the butyric acid and 90% was in the undissociated (HA) form, what would be the solution's pH?

61. Calculate the concentration of acetate in a 50 mM solution of acetic acid buffer at pH 5.0.

62. Calculate the concentration of HPO$_4^{2-}$ in a 100 mM solution of phosphate at pH 8.0.

63. The pK values for succinic acid and succinate are provided in Table 2.4. Draw the structure of the species that predominates at pH 2, 5, and 7.

64. The amino acid glycine (H$_2$N—CH$_2$—COOH) has pK values of 2.35 and 9.78. Indicate the structure and net charge of the molecular species that predominates at pH 2, 7, and 10.

65. The pK of CH$_3$CH$_2$NH$_3^+$ is 10.7. Would the pK of FCH$_2$CH$_2$NH$_3^+$ be higher or lower?

66. The structure of pyruvic acid is shown below. **a.** Draw the structure of pyruvate. **b.** Using what you have learned about acidic functional groups, which form of this compound is likely to predominate in the cell at pH 7.4? Explain.

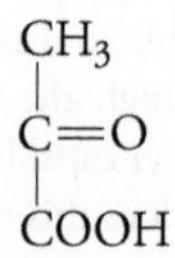

Pyruvic acid

2.4 Tools and Techniques: Buffers

67. Which would be a more effective buffer at pH 8.0 (see Table 2.4)? **a.** 10 mM HEPES buffer or 10 mM glycinamide buffer; **b.** 10 mM Tris buffer or 20 mM Tris buffer; **c.** 10 mM boric acid or 10 mM sodium borate.

68. Phosphoric acid, H$_3$PO$_4$, has three ionizable protons. **a.** Sketch the titration curve. Indicate the pK values and the species that predominate in each area of the curve. **b.** Write the equations for the dissociation of the three ionizable protons. **c.** Which two phosphate species are present in the blood at pH 7.4? **d.** Which two phosphate species would be used to prepare a buffer solution of pH 11?

69. The structure of acetylsalicylic acid (aspirin) is shown. Is aspirin more likely to be absorbed (pass through a lipid membrane) in the stomach (pH ~2) or in the small intestine (pH ~8)? Explain.

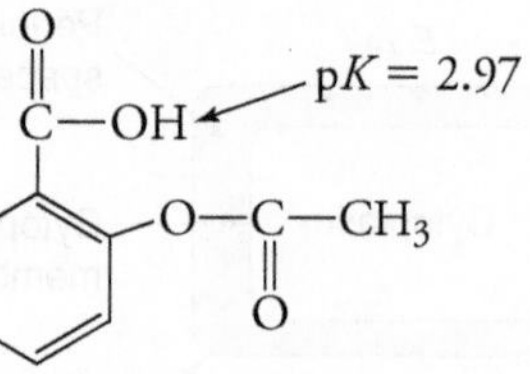

Acetylsalicylic acid (aspirin)

70. Calculate the ratio of imidazole to the imidazolium ion in a solution at pH 7.8.

71. What volume (in mL) of a 2.0 M solution of imidazolium chloride must be added to a 500-mL solution of 10 mM imidazole in order to achieve a pH of 6.5?

72. Calculate the ratio of methylamine to the methylammonium ion in a solution at pH 10.5.

73. Methylamine is sold by a manufacturer as a 40% solution by weight. The molar mass of methylamine is 31.1 g · mol^{-1}. What volume of this solution would need to be added to 1.0 L of a 50-mM solution of methylammonium bromide in order to achieve a pH of 10.5?

74. What is the volume (in mL) of glacial acetic acid (17.4 M) that would need to be added to 500 mL of a solution of 0.20 M sodium acetate in order to achieve a pH of 5.0?

75. What is the mass of NaOH that would need to be added to 500 mL of a solution of 0.20 M acetic acid in order to achieve a pH of 5.0?

76. An experiment requires the buffer HEPES, pH = 8.0 (see Table 2.4). **a.** Write an equation for the dissociation of HEPES in water. Identify the weak acid and the conjugate base. **b.** What is the effective buffering range for HEPES? **c.** The buffer will be prepared by making 1.0 L of a 0.10 M solution of HEPES. Hydrochloric acid will be added to achieve the desired pH. Describe how you will make 1.0 L of 0.10 M HEPES (A^-). (HEPES is supplied by the chemical company as a sodium salt with a molar mass of 260.3 g · mol^{-1}.) **d.** What is the volume (in mL) of 6.0 M HCl that must be added to the 0.10 M HEPES to achieve the desired pH of 8.0? Describe how you will make the buffer.

77. An alternative way of preparing the buffer described in Problem 83 involves preparing solutions of the sodium salt form of HEPES (A^-) and the weak acid form of HEPES (HA) and then combining the solutions to obtain a buffer with the desired pH and concentration. **a.** Describe how you would prepare 250 mL of 1.0-M stock solutions of the A^- and HA forms of HEPES (supplied as solids by the manufacturer with molar masses of 260.3 g · mol^{-1} and 238.3 g · mol^{-1}, respectively). **b.** Calculate the volume of each stock solution needed to prepare 1.0 L of the 0.10 M HEPES buffer described in Problem 83 and describe how you would prepare the buffer.

78. One liter of a 0.10 M Tris buffer (see Table 2.4) is prepared and adjusted to a pH of 8.2.

$$CH_2OH$$
$$HOH_2C-C-NH_2$$
$$CH_2OH$$

Tris(hydroxymethyl)aminomethane

a. Write the equation for the dissociation of Tris in water. Identify the weak acid and the conjugate base. **b.** What is the effective buffering range for Tris? **c.** What are the concentrations of the conjugate acid and weak base at pH 8.2? **d.** What is the ratio of conjugate base to weak acid if 1.5 mL of 3.0 M HCl is added to 1.0 L of the buffer? What is the new pH? Has the buffer functioned effectively? Compare the pH change to that of Problem 54a in which the same amount of acid was added to the same volume of pure water. **e.** What is the ratio of conjugate base to weak acid if 1.5 mL of 3.0 M NaOH is added to 1.0 L of the buffer? What is the new pH? Has the buffer functioned effectively? Compare the pH change to that of Problem 54b in which the same amount of base was added to the same volume of pure water.

79. One liter of a 0.1 M Tris buffer (see Table 2.4) is prepared and adjusted to a pH of 2.0. **a.** What are the concentrations of the conjugate base and weak acid at this pH? **b.** What is the pH when 1.5 mL of 3.0 M HCl is added to 1.0 L of the buffer? Has the buffer functioned effectively? Explain. **c.** What is the pH when 1.5 mL of 3.0 M NaOH is added to 1.0 L of the buffer? Has the buffer functioned effectively? Explain.

80. An experiment requires an imidazole buffer at pH 7.0 (see Problem 77). The buffer is prepared by making a solution of imidazole, then adding hydrochloric acid to achieve the desired pH. **a.** Describe how to prepare 1.0 L of a 0.10-M solution of imidazole (supplied by the manufacturer as a solid with a molar mass of 68.1 g · mol^{-1}). **b.** What is the volume (in mL) of 6.0 M HCl that must be added to this solution to achieve the desired pH? Describe how you will make the buffer.

81. An alternative way of preparing the buffer described in Problem 87 is described in Problem 84 and involves preparing solutions of imidazole (A^-) and imidazolium (HA; supplied by the manufacturer as a hydrobromide salt with a molar mass of 149 g · mol^{-1}) and then combining the solutions to obtain a buffer with the desired pH and concentration. **a.** Describe how you would prepare 100 mL of 1.0-M stock solutions of imidazole and imidazolium hydrobromide. **b.** Calculate the

volume of each stock solution needed to prepare 1.0 L of the 0.10 M imidazole buffer described in Problem 87 and describe how you would prepare the buffer.

2.5 Clinical Connection: Physiological Buffers

82. The pH of blood is maintained within a narrow range (7.35–7.45). Carbonic acid, H_2CO_3, participates in blood buffering. **a.** Write the equations for the dissociation of the two ionizable protons. **b.** The pK for the first ionizable proton is 6.1; the pK for the second ionizable proton is 10.3. Use this information to identify the weak acid and the conjugate base present in the blood. **c.** Calculate the concentration of carbonic acid in a sample of blood with a bicarbonate concentration of 24 mM and a pH of 7.4.

83. Impaired pulmonary function can contribute to respiratory acidosis. Using the appropriate equations, explain how the failure to eliminate sufficient CO_2 through the lungs leads to acidosis.

84. Metabolic acidosis often occurs in patients with impaired circulation from cardiac arrest. Mechanical hyperventilation can be used to alleviate acidosis. Explain why this strategy is effective.

85. Mechanical hyperventilation (a standard treatment for acidosis as described in Problem 91) cannot be used in patients with impaired pulmonary function because these patients often have acute lung injury (ALI). **a.** Sodium bicarbonate is an effective treatment for metabolic acidosis. Would this be an acceptable treatment for ALI patients? **b.** A group of physicians at San Francisco General Hospital advocates using Tris (see Problem 85) to treat metabolic acidosis in these patients. Explain why this procedure is effective and why it is an acceptable treatment for ALI patients.

86. An individual who develops alkalosis by hyperventilating is encouraged to breathe into a paper bag for several minutes. Why does this treatment correct the alkalosis?

87. A patient who took a large overdose of aspirin is brought into the emergency room for treatment. She suffers from respiratory alkalosis and the pH of her blood is 7.5. Determine the ratio of HCO_3^- to H_2CO_3 in the patient's blood at this pH. How does this compare to the ratio of HCO_3^- to H_2CO_3 in normal blood? Can the H_2CO_3/HCO_3^- system work effectively as a buffer in this patient under these conditions?

88. Metabolic acidosis is a general term that describes a number of disorders in metabolism in the body that result in a lowering of the blood pH from 7.4 to 7.35 or below. The kidney plays a vital role in regulating blood pH. The kidney can either excrete or reabsorb various ions, including phosphate, $H_2PO_4^-$; ammonium, NH_4^+; or bicarbonate, HCO_3^-. Which ions are excreted and which ions are reabsorbed in metabolic acidosis? Explain, using relevant chemical equations.

89. Metabolic alkalosis occurs when the blood pH rises to 7.45 or greater. Which ions are excreted and which ions are reabsorbed in metabolic alkalosis (see Problem 95)?

90. Kidney cells excrete H^+ and reabsorb HCO_3^- from the filtrate. The movement of each of these ions is thermodynamically unfavorable. However, the movement of each becomes possible when it is coupled to another, thermodynamically favorable ion transport process. Explain how the movement of other ions drives the transport of H^+ and HCO_3^-.

91. In uncontrolled diabetes, the body converts fats to ketone bodies. One of these ketone bodies is acetoacetic acid (see Problem 60), which accumulates in the bloodstream. Do ketone bodies contribute to acidosis or alkalosis? How might the body compensate for this acid–base imbalance?

92. Kidney cells have a carbonic anhydrase on their external surface as well as an intracellular carbonic anhydrase. What are the functions of these two enzymes?

93. Explain why the lungs can rapidly compensate for metabolic acidosis, whereas the kidneys are slow to compensate for respiratory acidosis.

94. Explain why metabolically generated CO_2 does not accumulate in tissues but is quickly converted to carbonic acid by the action of carbonic anhydrase in red blood cells.

95. Ammonia produced in kidney cells can diffuse out of the cell, but only in its unprotonated (NH_3) form. **a.** Explain. **b.** The NH_4^+ ion, however, can exit the cell in exchange for another ion. What ion is most likely involved?

SELECTED READINGS

Ball, P., Water, water, everywhere? *Nature* 427, 19–20 (2004). [A brief discussion of water as the matrix of life.]

Garde, S. and Patel, A.J., Unraveling the hydrophobic effect, one molecule at a time, *Proc. Natl. Acad. Sci.* 108, 16491–16492, doi: 10.1073/pnas.1113256108 (2011). [A brief commentary that summarizes some key features of the hydrophobic effect.]

Privalov, P.L. and Crane-Robinson, C., Role of water in the formation of macromolecular structures, *Eur. Biophys. J.* 46, 203–224, doi: 10.1007/s00249-016-1161-y (2017). [An extensive review that includes descriptions of the contribution of water to the structures and binding properties of proteins and DNA.]

Richardson, J.O., Pérez, C., Lobsiger, S., Reid, A.A., Temelso, B., Shields, G.C., Kisiel, Z., Wales, D.J., Pate, B.H., and Althorpe, S.C., Concerted hydrogen-bond breaking by quantum tunneling in the water hexamer prism, *Science* 351, 1310–1313, doi: 10.1126/science.aae0012 (2016). [Describes the structure and dynamics of a hexamer of water molecules.]

Seifter, J.L. and Chang, H.-Y., Extracellular acid–base balance and ion transport between body fluid compartments, *Physiology* 32, 367–379, doi: 10.1152/physiol.00007 (2017). [Focuses on plasma membrane transporters involved in maintaining intra- and extracellular pH.]

Nucleic Acid Structure and Function

The inherited characteristics of every organism are determined by its DNA, but the amount of DNA varies considerably. Each cell of the pine tree that produced this cone contains about 20 billion base pairs of DNA, about seven times more DNA than in a human cell.

Do You Remember?

- Cells contain four major types of biological molecules and three major types of polymers (Section 1.3).
- Modern prokaryotic and eukaryotic cells apparently evolved from simpler nonliving systems (Section 1.2).
- Noncovalent forces, including hydrogen bonds, ionic interactions, and van der Waals forces, act on biological molecules (Section 2.1).

All the structural components of a cell and the machinery that carries out the cell's activities are ultimately specified by the cell's genetic material—DNA. Therefore, before examining other types of biological molecules and their metabolic transformations, we must consider the nature of DNA, including its chemical structure and how its biological information is organized and expressed.

3.1 Nitrogenous Bases

KEY CONCEPTS

Recognize the structures of nucleosides and nucleotides.

- Identify the base, sugar, and phosphate groups of nucleosides and nucleotides.
- Recognize nucleotide derivatives.

Nucleosides

Nucleosides consist of a purine or pyrimidine nitrogenous base connected to ribose or deoxyribose pentose sugar by glycosidic linkage. Linking atom N9 in a purine or atom N1 in a pyrimidine to a five-carbon sugar forms a **nucleoside.** *In DNA, the sugar is 2′-deoxyribose; in RNA, the sugar is ribose* (the sugar atoms are numbered with primes to distinguish them from the atoms of the attached bases).

Ribose   2′-Deoxyribose

Depending on the type of nitrogenous bases ribonucleoside or deoxyribonucleoside may be of different types as shown in **Table 3.1**. Nucleoside analogs are used as therapeutic targets to prevent the growth of pathogenic viruses inside the host cells. Nucleoside analogs may also be used as anticancer agents.

Table 3.1 Nucleic Acid Bases, Nucleosides, and Nucleotides

Base	Nucleoside[a]	Nucleotides[a]
Adenine (A)	Adenosine	Adenylate; adenosine monophosphate (AMP)
		adenosine diphosphate (ADP)
		adenosine triphosphate (ATP)
Cytosine (C)	Cytidine	Cytidylate; cytidine monophosphate (CMP)
		cytidine diphosphate (CDP)
		cytidine triphosphate (CTP)
Guanine (G)	Guanosine	Guanylate; guanosine monophosphate (GMP)
		guanosine diphosphate (GDP)
		guanosine triphosphate (GTP)
Thymine (T)[b]	Thymidine	Thymidylate; thymidine monophosphate (TMP)
		thymidine diphosphate (TDP)
		thymidine triphosphate (TTP)
Uracil (U)[c]	Uridine	Uridylate; uridine monophosphate (UMP)
		uridine diphosphate (UDP)
		uridine triphosphate (UTP)

[a]Nucleosides and nucleotides containing 2′-deoxyribose rather than ribose may be called deoxynucleosides and deoxynucleotides. The nucleotide abbreviation is then preceded by "d."

[b]Thymine is found in DNA but not in RNA.

[c]Uracil is found in RNA but not in DNA.

Nucleotides

Gregor Mendel was certainly not the first to notice that an organism's characteristics (for example, flower color or seed shape in pea plants) were passed to its progeny, but in 1865 he was the first to describe their predictable patterns of inheritance. By 1903, Mendel's inherited factors (now called genes) were recognized as belonging to **chromosomes** (a word that means "colored bodies"), which are visible by light microscopy (**Fig. 3.1**).

Eventually, chromosomes were shown to be composed of proteins, which had first been described in 1838 by Gerardus Johannes Mulder, and **nucleic acids,** which had been discovered in 1869 by Friedrich Miescher. Proteins, with their 20 different types of amino acids and great diversity in size and shape, were the obvious candidates to be carriers of genetic information in chromosomes. Nucleic acids, in contrast, seemed uninteresting and contained only four different types of structural units, called **nucleotides.** In **DNA (deoxyribonucleic acid),** these components—abbreviated A, C, G, and T—were thought to occur as simple repeating tetranucleotides, for example,

—ACGT-ACGT-ACGT-ACGT—

In 1950, when Erwin Chargaff showed that the nucleotides in DNA were not all present in equal numbers and that the nucleotide composition varied among species, it became apparent that DNA might be complex enough to be the genetic material after all. Several other lines of research also pointed to the importance of DNA, and the race was on to decipher its molecular structure.

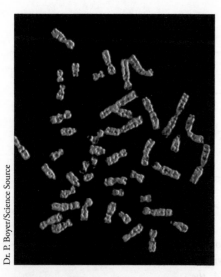

Figure 3.1 Human chromosomes from amniocentesis.   In this image, the chromosomes have been stained with fluorescent dyes.

A nucleotide is a nucleoside to which one or more phosphate groups are linked, usually at C5′ of the sugar. Depending on whether there are one, two, or three phosphate groups, the nucleotide is known as a nucleoside monophosphate, nucleoside diphosphate, or nucleoside triphosphate and is represented by a three-letter abbreviation, for example,

Adenosine monophosphate
(AMP)

Guanosine diphosphate
(GDP)

Cytidine triphosphate
(CTP)

Deoxynucleotides are named in a similar fashion, and their abbreviations are preceded by "d." The deoxy counterparts of the compounds shown above would therefore be deoxyadenosine monophosphate (dAMP), deoxyguanosine diphosphate (dGDP), and deoxycytidine triphosphate (dCTP). The names and abbreviations of the common bases, nucleosides, and nucleotides are summarized in **Table 3.1**.

After they have been incorporated into DNA or RNA, the bases of some nucleotides may be chemically modified. This occurs extensively in some classes of RNA (described more fully in Section 21.3). The most common DNA modification is methylation, generating residues containing 5-methylcytosine and N^6-methyladenine.

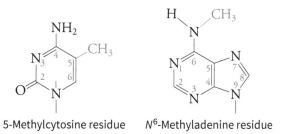

5-Methylcytosine residue N^6-Methyladenine residue

Nucleic acids are polymers of nucleotides

Each nucleotide of DNA includes a nitrogen-containing **base.** The bases adenine (A) and guanine (G) are known as **purines** because they resemble the organic compound purine:

Adenine

Guanine

Purine

The bases cytosine (C) and thymine (T) are known as **pyrimidines** because they resemble the organic compound pyrimidine:

Cytosine

Thymine

Pyrimidine

Ribonucleic acid (RNA) contains the pyrimidine uracil (U) rather than thymine:

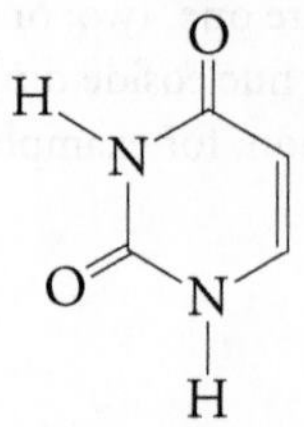

Uracil

so that DNA contains the bases A, C, G, and T, whereas RNA contains A, C, G, and U. The purines and pyrimidines are known as *bases* because they can participate in acid–base reactions. However, they donate or accept protons only at extremely low or high pH, so this behavior is not relevant to their function inside cells.

Some nucleotides have other functions

In addition to serving as the building blocks for DNA and RNA, nucleotides perform a variety of functions in the cell. They are involved in energy transduction, intracellular signaling, and regulation of enzyme activity. Some nucleotide derivatives are essential players in the metabolic pathways that synthesize biomolecules or degrade them in order to "capture" free energy. For example, coenzyme A (CoA; **Fig. 3.2a**) is a carrier of other molecules during their synthesis and degradation. Two nucleotides are linked in the compounds nicotinamide adenine dinucleotide (NAD; Fig. 3.2b) and flavin adenine dinucleotide (FAD; Fig. 3.2c), which undergo reversible oxidation and reduction during a number of metabolic reactions. Interestingly, a portion of the structures of each of these molecules is derived from a **vitamin,** a compound that must be obtained from the diet.

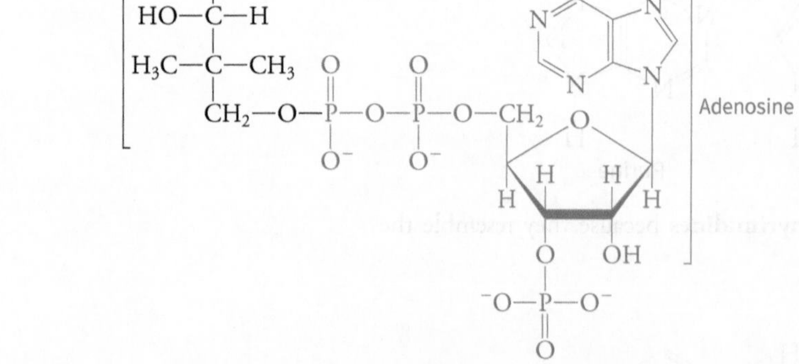

Coenzyme A (CoA)

Coenzyme A (CoA) contains a residue of pantothenic acid (pantothenate), also known as vitamin B_5. The sulfhydryl group is the site of attachment of other groups.

a.

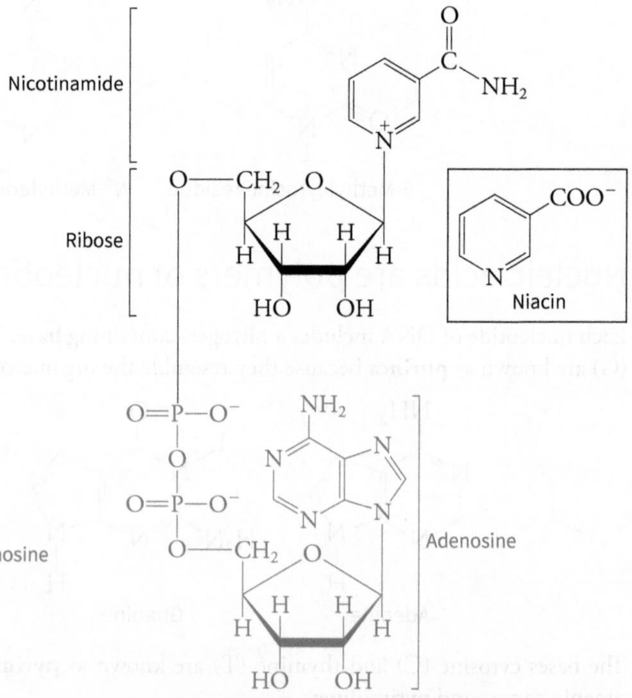

Nicotinamide adenine dinucleotide (NAD)

The nicotinamide group of nicotinamide adenine dinucleotide (NAD) is a derivative of the vitamin niacin (also called nicotinic acid or vitamin B_3; see inset) and undergoes oxidation and reduction. The related compound nicotinamide adenine dinucleotide phosphate (NADP) contains a phosphoryl group at the adenosine C2′ position.

b.

Figure 3.2 (Continued)

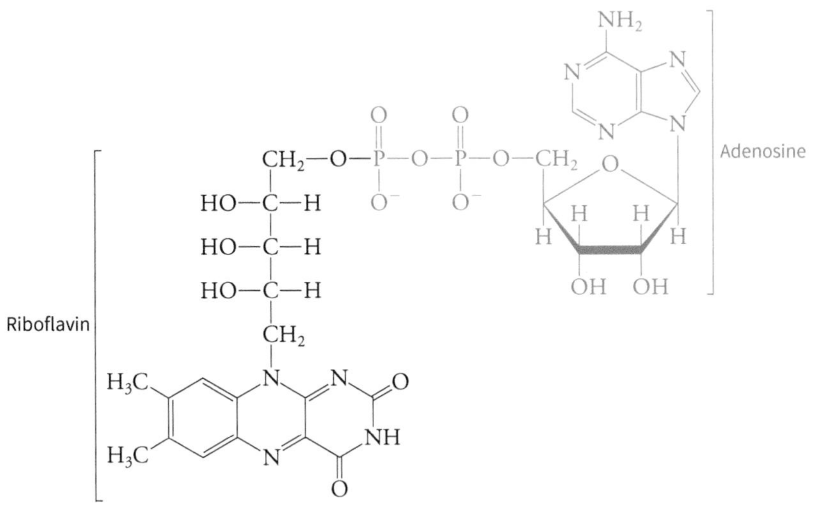

Oxidation and reduction of flavin adenine dinucleotide (FAD) occurs at the riboflavin group (also known as vitamin B$_2$).

c.

Figure 3.2 Some nucleotide derivatives.   The adenosine group of each of these compounds is shown in red. Note that each also contains a vitamin derivative.

Question Locate the nitrogenous base(s) and sugar(s) in each structure.

Concept Check

1. Practice drawing the structures of the two purine bases, the three pyrimidine bases, ribose, and deoxyribose.
2. What are the relationships among purines, pyrimidines, nucleosides, nucleotides, and nucleic acids?
3. Sketch the overall structure of a nucleoside and a nucleotide.

3.2 Nucleic Acid Structure

KEY CONCEPTS

Describe the structure and stabilizing forces in DNA.

- Summarize the physical features of the DNA double helix.
- Distinguish the structures of RNA and DNA.
- Recount the events in nucleic acid denaturation and renaturation.

In a nucleic acid, the linkage between nucleotides is called a **phosphodiester bond** because a single phosphate group forms ester bonds to both C5′ and C3′. During DNA synthesis in a cell, when a nucleoside triphosphate is added to the **polynucleotide** chain, a diphosphate group is eliminated. Once incorporated into a polynucleotide, the nucleotide is formally known as a nucleotide **residue.** Nucleotides consecutively linked by phosphodiester bonds form a polymer in which the bases project out from a backbone of repeating sugar–phosphate groups.

The end of the polymer that bears a phosphate group attached to C5′ is known as the **5′ end,** and the end that bears a free OH group at C3′ is the **3′ end.** By convention, the base sequence in a polynucleotide is read from the 5′ end (on the left) to the 3′ end (on the right).

DNA is a double helix

A DNA molecule consists of two polynucleotide strands linked by hydrogen bonds (hydrogen bonding is discussed in Section 2.1). The structure of this molecule, elucidated by James Watson and Francis Crick in 1953, incorporated Chargaff's earlier observations about DNA's base composition. Specifically, Chargaff noted that the amount of A is equal to the amount of T, the amount of C is equal to the amount of G, and the total amount of A + G is equal to the total amount of C + T. *Chargaff's "rules" could be satisfied by a molecule with two polynucleotide strands in which A and C in one strand pair with T and G in the other.* Two hydrogen bonds link adenine and thymine, and three hydrogen bonds link guanine and cytosine:

*All the **base pairs,** which consist of a purine and a pyrimidine, have the same molecular dimensions* (about 11 Å wide). Consequently, the **sugar–phosphate backbones** of the two strands of DNA are separated by a constant distance, regardless of whether the base pair is A:T, G:C, T:A, or C:G.

Although DNA can be shown as a ladder-like structure (*left*), with the two sugar–phosphate backbones as the vertical supports and the base pairs as the rungs, the two strands of DNA twist around each other to generate the familiar double helix (*right*).

Sugar–phosphate backbones

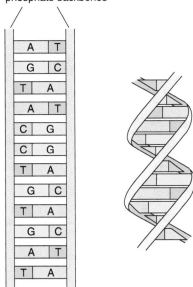

This conformation allows successive base pairs, which are essentially planar, to stack on top of each other with a center-to-center distance of only 3.4 Å. In fact, Watson and Crick derived this model for DNA not just from Chargaff's rules but also from Rosalind Franklin's studies of the diffraction (scattering) of an X-ray beam by a DNA fiber, which suggested a helix with a repeating spacing of 3.4 Å.

The major features of the DNA molecule include the following (**Fig. 3.3**):

1. The two polynucleotide strands are **antiparallel;** that is, their phosphodiester bonds run in opposite directions. One strand has a $5' \rightarrow 3'$ orientation, and the other has a $3' \rightarrow 5'$ orientation.

2. The DNA "ladder" is twisted in a right-handed fashion. (If you climbed the DNA helix as if it were a spiral staircase, you would hold the outer railing—the sugar–phosphate backbone—with your right hand.)

3. The diameter of the helix is about 20 Å and it completes a turn about every 10 base pairs, which corresponds to an axial distance of about 34 Å.

4. The twisting of the DNA "ladder" into a helix creates two grooves of unequal width, the **major** and **minor grooves.**

5. The sugar–phosphate backbones define the exterior of the helix and are exposed to the solvent. The negatively charged phosphate groups bind Mg^{2+} cations *in vivo*, which helps minimize electrostatic repulsion between these groups.

6. The base pairs are located in the center of the helix, approximately perpendicular to the helix axis.

7. The base pairs stack on top of each other, so the core of the helix is solid (see Fig. 3.3b). Although the planar faces of the base pairs are not accessible to the solvent, their edges are exposed in the major and minor grooves (this allows certain DNA-binding proteins to recognize specific bases).

In nature, DNA seldom assumes a perfectly regular conformation because of small sequence-dependent irregularities. For example, base pairs can roll or twist like propeller blades, and the

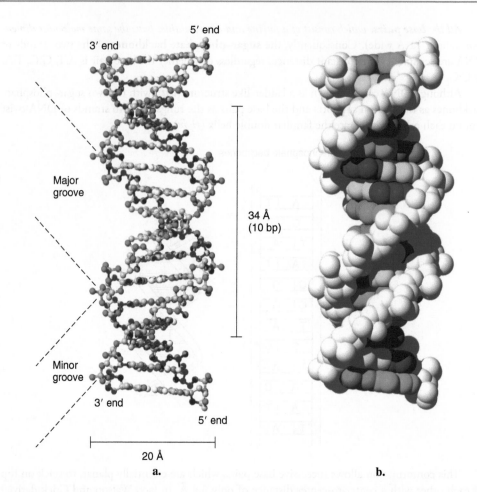

Figure 3.3 Model of DNA.
a. Ball-and-stick model with atoms colored: C gray, O red, N blue, and P gold (H atoms are not shown).
b. Space-filling model with the sugar–phosphate backbone in gray and the bases color-coded: A green, C blue, G yellow, and T red.

Question How many nucleotides are shown in this double helix?

helix may wind more tightly or loosely at certain nucleotide sequences. DNA-binding proteins may take advantage of these small variations to locate their specific binding sites, and they in turn may further distort the DNA helix by causing it to bend or partially unwind.

The size of a DNA segment is expressed in units of base pairs (**bp**) or kilobase pairs (1000 bp, abbreviated **kb**). Most naturally occurring DNA molecules comprise thousands to millions of base pairs. A short single-stranded polymer of nucleotides is usually called an **oligonucleotide** (*oligo* is Greek for "few"). In a cell, nucleotides are polymerized by the action of enzymes known as **polymerases.** The phosphodiester bonds linking nucleotide residues can be hydrolyzed by the action of **nucleases.** An **exonuclease** removes a residue from the end of a polynucleotide chain, whereas an **endonuclease** cleaves at some other point along the chain. Polymerases and nucleases are usually specific for either DNA or RNA. In the absence of these enzymes, the structures of nucleic acids are remarkably stable. The hydrogen bonds between polynucleotide strands, however, are relatively weak and break to allow the strands to separate during replication and transcription, described below.

RNA is single-stranded

RNA, which is a single-stranded polynucleotide, has greater conformational freedom than DNA, whose structure is constrained by the requirements of regular base-pairing between its two strands. *An RNA strand can fold back on itself so that base pairs form between complementary segments of the same strand.* **Complementarity** refers to the ability of bases to form hydrogen bonds with their standard partners: A is complementary to T and U, and G is complementary to C. Consequently, RNA molecules tend to assume intricate three-dimensional shapes (**Fig. 3.4**). Unlike DNA, whose regular structure is suited for the long-term storage of genetic information, RNA can assume more active roles in expressing that information. For example, the molecule shown in Figure 3.4, which carries the amino acid phenylalanine, interacts with a number of proteins and other RNA molecules during protein synthesis.

Figure 3.4 A transfer RNA molecule.   This 76-nucleotide single-stranded RNA molecule folds back on itself so that base pairs form between complementary segments.

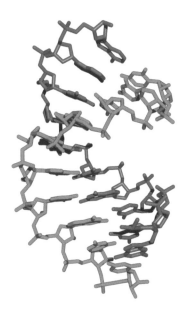

Figure 3.5 An RNA–DNA hybrid helix.   In a double helix formed by one strand of RNA (red) and one strand of DNA (blue), the planar base pairs are tilted and the helix does not wind as steeply as in a standard DNA double helix (compare with Fig. 3.3).

The residues of RNA are also capable of base-pairing with a complementary single strand of DNA to produce an RNA–DNA hybrid double helix (**Fig. 3.5**). A double helix involving RNA is wider and flatter than the standard DNA helix (its diameter is about 26 Å, and it makes one helical turn every 11 residues). In addition, its base pairs are inclined to the helix axis by about 20°. These structural differences relative to the standard DNA helix primarily reflect the presence of the 2′ OH groups in RNA.

A double-stranded DNA helix can adopt this same helical conformation; it is known as **A-DNA.** The standard DNA helix shown in Figure 3.3 is known as **B-DNA.** Other conformations of DNA have been described, and there is evidence that they exist *in vivo*, at least for certain nucleotide sequences, but their functional significance is not completely understood.

Nucleic acids can be denatured and renatured

The pairing of polynucleotide strands in a double-stranded nucleic acid is possible because bases in each strand form hydrogen bonds with complementary bases in the other strand: A is the complement of T (or U) and G is the complement of C. However, the structural stability of a double helix does not depend significantly on hydrogen bonding between complementary bases. (If the strands were separated, the bases could still satisfy their hydrogen-bonding requirements by forming hydrogen bonds with solvent water molecules.) Instead, *stability depends mostly on* **stacking interactions,** *which are a form of van der Waals interaction, between adjacent base pairs.* A view down the helix axis shows that stacked base pairs do not overlap exactly, due to the winding of the helix (**Fig. 3.6**). Although individual stacking interactions are weak, they are additive along the length of a DNA molecule.

In addition to stacking interactions, tightly associated water molecules help stabilize the DNA helix. The major groove is wide enough to accommodate several somewhat disordered water molecules, but in the minor groove, water molecules fit in a more rigid single file (**Fig. 3.7**). This water must be displaced when the helix structure is disrupted. Interestingly, *strands containing mostly A and T separate more easily than strands containing mostly G and C.* This could be due to the slightly greater stacking energy of G:C base pairs compared to A:T base pairs, but it could also reflect the greater constriction of water molecules in AT-rich DNA, which has

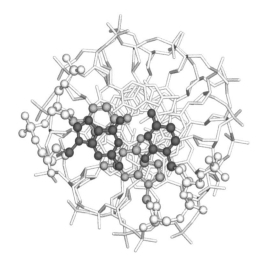

Figure 3.6 Axial view of DNA base pairs.   A view down the central axis of the DNA helix shows the overlap of neighboring base pairs (only the first two nucleotide pairs are highlighted).

Question Locate the base and sugar in the nucleotides with the blue bases. Identify the bases.

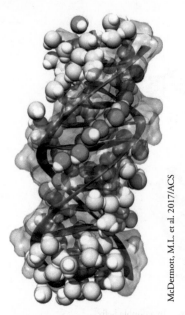

Figure 3.7 A model of DNA with bound water molecules. Water molecules (red) in the minor groove form an ordered "spine" along the double helix, whereas water molecules (blue) in the major groove are less ordered.

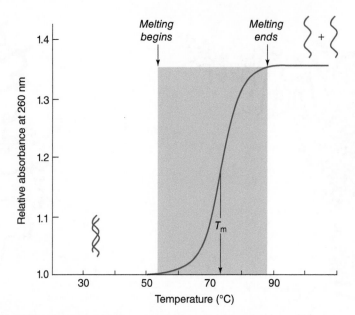

Figure 3.8 A DNA melting curve. Thermal denaturation (melting, or strand separation) of DNA results in an increase in ultraviolet absorbance relative to the absorbance at 25°C. The melting point, T_m, of the DNA sample is defined as the midpoint of the melting curve.

a narrower minor groove, such that disrupting the helical structure results in a larger increase in the entropy of the released water molecules. Although G:C base pairs have one more hydrogen bond than A:T base pairs, this feature does *not* account for the easier separation of AT-rich DNA strands.

The loss of helical structure in a sample of DNA molecules can be quantified in the **melting temperature (T_m).** To determine the melting point of the DNA, the temperature is slowly increased. At a sufficiently high temperature, the base pairs begin to unstack, hydrogen bonds break, and the two strands begin to separate. This process continues as the temperature rises, until the two strands come completely apart. The melting, or **denaturation,** of the DNA can be recorded as a melting curve (**Fig. 3.8**) by monitoring an increase in the absorbance of ultraviolet (260 nm) light (the aromatic bases absorb more light when unstacked). The midpoint of the melting curve (that is, the temperature at which half the DNA has separated into single strands) is the T_m. **Table 3.2** lists the GC content and the melting point of the DNA from different species. Since manipulating DNA in the laboratory frequently requires the thermal separation of paired DNA strands, it is sometimes helpful to know the DNA's GC content.

When the temperature is lowered slowly, denatured DNA can **renature;** that is, *the separated strands can re-form a double helix by reestablishing hydrogen bonds between the complementary strands and by restacking the base pairs.* The maximum rate of renaturation occurs at about 20–25°C below the melting temperature. If the DNA is cooled too rapidly, it may not fully renature because base pairs may form randomly between short complementary segments. At low temperatures, the improperly paired segments are frozen in place since they do not have enough thermal energy to melt apart and find their correct complements (**Fig. 3.9**). The rate of renaturation of denatured DNA depends on the length of the double-stranded molecule: Short segments come together (**anneal**) faster than longer segments because the bases in each strand must locate their partners along the length of the complementary strand.

Table 3.2 GC Content and Melting Points of DNA

Source of DNA	GC content (%)	T_m (°C)
Dictyostelium discoideum (fungus)	23.0	79.5
Clostridium butyricum (bacterium)	37.4	82.1
Homo sapiens	40.3	86.5
Streptomyces albus (bacterium)	72.3	100.5

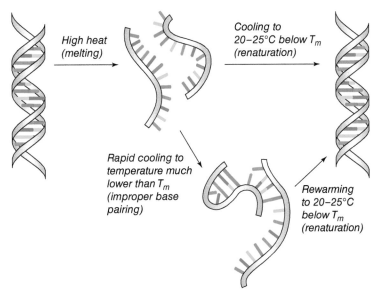

Figure 3.9 Renaturation of DNA.   DNA strands that have been melted apart can renature at a temperature of 20–25°C below the T_m. At much lower temperatures, base pairs may form between short complementary segments within and between the single strands. Correct renaturation is possible only if the sample is rewarmed so that the improperly paired strands can separate and reanneal.

The ability of short single-stranded nucleic acids (either DNA or RNA) to hybridize with longer polynucleotide chains is the basis for a number of useful laboratory techniques (described in detail in Section 20.6). For example, an oligonucleotide **probe** that has been tagged with a fluorescent group can be used to detect the presence of a complementary nucleic acid sequence in a complex mixture.

Concept Check

1. Explain how Chargaff's rules helped reveal the structure of DNA.
2. Describe the arrangement of the base pairs and sugar–phosphate backbones in DNA.
3. List the ways that RNA differs from DNA.
4. Describe the molecular events in DNA denaturation and renaturation.
5. How can a single-stranded oligonucleotide be used to reveal the presence of a particular segment of DNA?

3.3 The Central Dogma

KEY CONCEPTS

Summarize the biological roles of DNA and RNA.

- Distinguish replication, transcription, and translation.
- Decode a nucleotide sequence to an amino acid sequence.
- Describe how a mutation can cause a disease.
- Summarize the goals and challenges of gene therapy.

The complementarity of the two strands of DNA is essential for its function as the storehouse of genetic information, since this information must be **replicated** (copied) for each new generation. As first suggested by Watson and Crick, the separated strands of DNA direct the synthesis of complementary strands, thereby generating two identical double-stranded

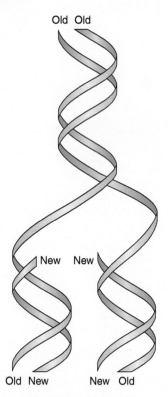

Old   Old

New   New

Old New   New Old

Figure 3.10 DNA replication.
The double helix unwinds so that each parental strand can serve as a template for the synthesis of a new complementary strand. The result is two identical double-helical DNA molecules.

Question Label the 5′ and 3′ ends of each strand.

molecules (**Fig. 3.10**). The parental strands are said to act as templates for the assembly of the new strands because their sequence of nucleotides determines the sequence of nucleotides in the new strands. When a cell divides, each daughter cell receives genetic information—sequences of nucleotides—in the form of DNA molecules containing one old strand and one new strand (**Box 3.A**).

A similar phenomenon is responsible for the **expression** of that genetic information, a process in which nucleotide sequences are used to direct the synthesis of proteins that carry out the cell's activities. First, *a portion of the DNA, a **gene**, is **transcribed** to produce a complementary strand of RNA; then the RNA is **translated** into a polypeptide chain (protein).* This paradigm, known as the **central dogma of molecular biology,** was formulated by Francis Crick. It can be shown schematically as

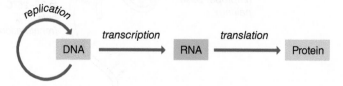

Although it is tempting to think of a cell's DNA as its "brain," the DNA does not issue commands to the rest of the cell. Instead, DNA simply holds genetic information—the instructions for synthesizing proteins.

DNA must be decoded

Even in the simplest organisms, DNA is an enormous molecule, and many organisms contain several different DNA molecules as separate chromosomes. An organism's complete set of genetic information is called its **genome.** A genome may comprise several hundred to perhaps 35,000 genes.

To transcribe a gene, one of the two strands of DNA serves as a template for an enzyme called RNA polymerase to synthesize a complementary strand of RNA. The RNA therefore has

Box 3.A	**Replication, Mitosis, and Meiosis**

Each chromosome contains one long double-stranded DNA molecule that must be replicated at some point before cell division begins. Prokaryotic cells typically harbor one chromosome in the form of a circular DNA molecule, and replication and cell division are relatively straightforward. However, eukaryotic cells generally have multiple linear chromosomes that must each be replicated and then

allocated equally to the two daughter cells during cell division—a more complicated process.

Linking the chemistry of DNA to the biology of cell division demands a closer look at a typical eukaryotic chromosome. The two DNA strands are shown in blue. Replication generates two identical DNA molecules, each with one old (blue) and one new (purple) strand.

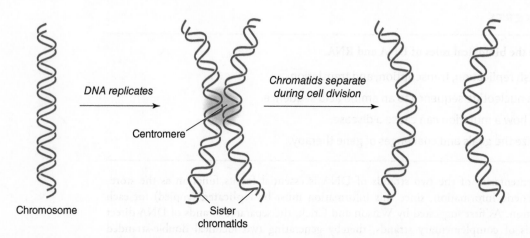

The two DNA molecules, called **sister chromatids,** remain attached at a more-or-less central point known as the **centromere.** The two chromatids are clearly apparent early in cell division, when the chromosomes are highly condensed.

Biophoto Associates/Science Source

Each daughter cell receives one of the sister chromatids when the centromere attachment is broken. The process of **mitosis** describes the allocation of eukaryotic chromosomes during cell division. For a cell with six chromosomes, the process can be diagrammed as follows:

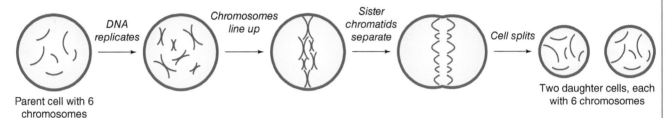

The final result is two daughter cells that are genetically identical to the parent cell (in this example, the parent cell and the daughter cells all have six chromosomes). This mode of cell division is standard for unicellular organisms and for the cells within multicellular organisms.

In animals that reproduce sexually, virtually all the cells are **diploid.** They have two sets of chromosomes, one inherited from each parent. However, in order to reproduce, these animals have specialized cells in the ovaries and testes to generate **gametes,** which are **haploid,** having just one set of genetic information. At fertilization, two haploid gametes—one from each parent—combine to produce a new diploid organism.

Meiosis is the variation of mitosis that occurs in the reproductive tissues of animals as well as plants in order to produce haploid cells. For example, diploid human cells with 46 chromosomes undergo meiosis to produce egg or sperm cells with just 23 chromosomes. Before meiosis, the DNA of each chromosome is replicated. Meiosis begins with the pairing of **homologous chromosomes,** which have the same genes but slightly different DNA sequences. In humans, the two versions of Chromosome 1 form a pair, the two versions of Chromosome 2 form a pair, and so on for a total of 23 pairs. The model cell above, with six chromosomes, forms three pairs at the start of meiosis. Here, the sets of homologous chromosomes are colored blue and red. When the cell splits, the chromosome pairs also split up, so that the two daughter cells are haploid. However, because each chromosome still consists of two sister chromatids, a second division, exactly like mitosis, must then occur to produce four daughter cells, the gametes, containing only one DNA molecule to represent each chromosome.

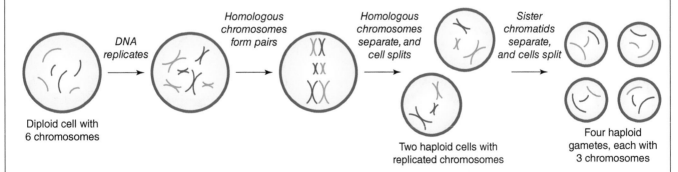

At fertilization, a haploid gamete from one parent fuses with a haploid gamete from a different parent to generate almost unlimited new diploid combinations of genetic information in the next generation. Consequently, meiosis leads to the genetic diversity that is essential for natural selection. Additional diversity is generated early in meiosis, when homologous pairs exchange segments of DNA.

Meiosis also explains why parental traits may or may not appear in the offspring. A diploid parent has two sets of homologous chromosomes, so there are two possible versions, or **alleles,** for each gene. During meiosis, when pairs of homologous chromosomes separate, the two alleles end up in different daughter cells. Thus, only one allele is passed from the parent to an individual in the next generation. For example, in a certain animal, a gene that codes for an enzyme called "B" has two alleles: the B allele codes for a functional version of the enzyme and the b allele codes for a nonfunctional version of the enzyme. A parent bearing two B alleles passes a B allele to each of its offspring, and a parent bearing two b alleles passes on a b allele.

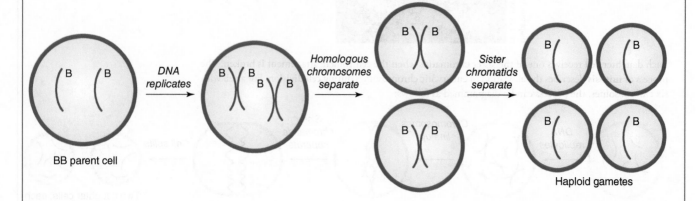

BB parent cell

DNA replicates

Homologous chromosomes separate

Sister chromatids separate

Haploid gametes

When a BB individual mates with a bb individual, all the offspring will receive one B allele and one b allele. Since the B allele codes for a functional enzyme, all the Bb offspring will have that functionality.

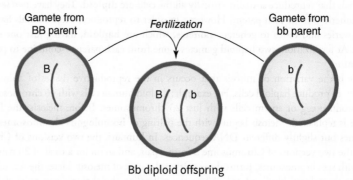

Gamete from BB parent

Fertilization

Gamete from bb parent

Bb diploid offspring

When a Bb individual mates with another Bb individual, each parent produces gametes containing either a B allele or a b allele.

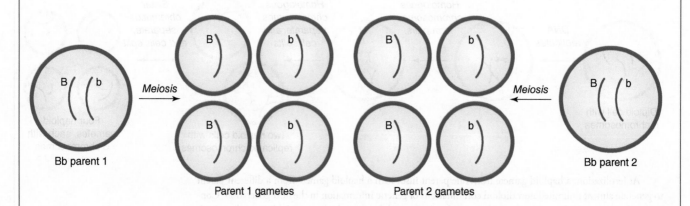

Bb parent 1

Meiosis

Parent 1 gametes

Parent 2 gametes

Meiosis

Bb parent 2

All the possible combinations of gametes during fertilization can be predicted by setting up a simple grid:

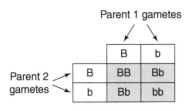

Parent 1 gametes

	B	b
B	BB	Bb
b	Bb	bb

Parent 2 gametes

There is a 1-in-4 (25%) chance that a given offspring individual will have two B alleles, a 2-in-4 (50%) chance that it will have one B and one b allele, and a 1-in-4 (25%) chance that it will have two b alleles. In a large litter, approximately three-quarters of the offspring will have a functional enzyme (they are either BB or Bb), and one-quarter of the offspring will have the nonfunctional enzyme (they are bb). This three-to-one ratio was one of the patterns that Mendel documented in his classic studies of inheritance in pea plants. Although Mendel's work preceded the discoveries of DNA, mitosis, and meiosis, the laws of inheritance follow solidly from the structure of DNA, its mode of replication, and the way chromosomes are passed to daughter cells.

the same sequence (except for the substitution of U for T) and the same $5' \rightarrow 3'$ orientation as the nontemplate strand of DNA. This strand of DNA is often called the **coding strand** or sense strand (the template strand is called the **noncoding strand**).

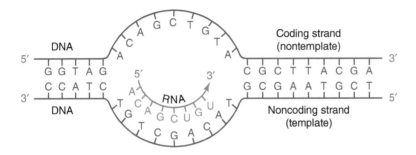

The transcribed RNA is known as **messenger RNA (mRNA)** because it carries the same genetic message as the gene.

The mRNA is translated by a **ribosome,** a cellular particle consisting of protein and **ribosomal RNA (rRNA).** At the ribosome, small molecules called **transfer RNA (tRNA),** which carry amino acids, recognize sequential sets of three bases (known as **codons**) in the mRNA through complementary base-pairing (a tRNA molecule is shown in Fig. 3.4). The ribosome covalently links the amino acids carried by successive tRNAs to form a polypeptide. *The protein's amino acid sequence therefore ultimately depends on the nucleotide sequence of the DNA.*

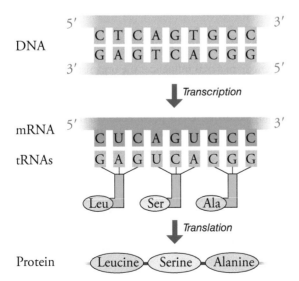

The correspondence between amino acids and mRNA codons is known as the **genetic code.** There are a total of 64 codons: 3 of these are "stop" signals that terminate translation and the

Table 3.3 The Standard Genetic Code[a]

First position (5' end)	Second position				Third position (3' end)
	U	C	A	G	
U	UUU Phe	UCU Ser	UAU Tyr	UGU Cys	U
	UUC Phe	UCC Ser	UAC Tyr	UGC Cys	C
	UUA Leu	UCA Ser	UAA Stop	UGA Stop	A
	UUG Leu	UCG Ser	UAG Stop	UGG Trp	G
C	CUU Leu	CCU Pro	CAU His	CGU Arg	U
	CUC Leu	CCC Pro	CAC His	CGC Arg	C
	CUA Leu	CCA Pro	CAA Gln	CGA Arg	A
	CUG Leu	CCG Pro	CAG Gln	CGG Arg	G
A	AUU Ile	ACU Thr	AAU Asn	AGU Ser	U
	AUC Ile	ACC Thr	AAC Asn	AGC Ser	C
	AUA Ile	ACA Thr	AAA Lys	AGA Arg	A
	AUG Met	ACG Thr	AAG Lys	AGG Arg	G
G	GUU Val	GCU Ala	GAU Asp	GGU Gly	U
	GUC Val	GCC Ala	GAC Asp	GGC Gly	C
	GUA Val	GCA Ala	GAA Glu	GGA Gly	A
	GUG Val	GCG Ala	GAG Glu	GGG Gly	G

[a]The 20 amino acids are abbreviated; Ala, alanine; Arg, arginine; Asn, asparagine; Asp, aspartate; Cys, cysteine; Gly, glycine; Gln, glutamine; Glu, glutamate; His, histidine; Ile, isoleucine; Leu, leucine; Lys, lysine; Met, methionine; Phe, phenylalanine; Pro, proline; Ser, serine; Thr, threonine; Trp, tryptophan; Tyr, tyrosine; and Val, valine.

Question How many amino acids would be uniquely specified by a genetic code that consisted of just the first two nucleotides in each codon?

remaining 61 represent, with some redundancy, the 20 standard amino acids found in proteins. **Table 3.3** shows which codons specify which amino acids. In theory, knowing a gene's nucleotide sequence should be equivalent to knowing the amino acid sequence of the protein encoded by the gene. However, as we will see, genetic information is often "processed" at several points before the protein reaches its mature form. Keep in mind that the rRNA and tRNA required for protein synthesis, as well as other types of RNA, are also encoded by genes. The "products" of these genes are the result of transcription without translation.

A mutated gene can cause disease

Because an organism's genetic material influences the organism's entire repertoire of activities, it is vitally important to unravel the sequence of nucleotides in that organism's DNA. Over the past 60 years, researchers have developed a variety of methods to sequence DNA. These techniques take advantage of naturally occurring DNA polymerase enzymes that construct complementary copies of a given template in a way that allows the identification of each new nucleotide incorporated (Section 20.6 provides more details). Huge amounts of nucleic acid sequence data—representing individual genes as well as entire genomes—are available for analysis.

The process of DNA replication is highly accurate, but mistakes do happen. In addition, DNA can be physically and chemically damaged, and not all cellular repair mechanisms are able to perfectly restore it. Consequently, the information in DNA may change over time. A **mutation** is simply a permanent change in DNA and may take the form of a single-nucleotide substitution, an insertion or deletion of nucleotides, or a rearrangements of chromosomal segments. Some of these sequence variations cause disease.

In traditional approaches to understanding human genetic diseases, researchers used the defective protein associated with a particular disease to track down the corresponding gene. (Current methods for discovering disease mutations rely on directly sequencing the DNA.)

One classic example is the genetic change responsible for sickle cell disease. In the gene for one hemoglobin protein chain, the codon GAG normally codes for the amino acid glutamate (Glu). In the defective gene, the sequence has mutated to GTG, which codes for valine (Val). This amino acid substitution affects the protein's structure and function.

Normal gene	$\cdots$ ACT CCT GAG GAG AAG $\cdots$
Protein	$\cdots$ Thr – Pro – Glu – Glu – Lys $\cdots$
Mutated gene	$\cdots$ ACT CCT GTG GAG AAG $\cdots$
Protein	$\cdots$ Thr – Pro – Val – Glu – Lys $\cdots$

As many as 10,000 human disorders are **monogenetic diseases,** such as sickle cell disease and cystic fibrosis, where a defect in a specific gene explains the molecular basis of the disease. In many cases, a variety of different mutations have been catalogued for a particular disease gene, which explains in part why symptoms of the disease vary between individuals. The database known as OMIM (Online Mendelian Inheritance in Man, omim.org) contains information on thousands of genetic variants, including the clinical features of the resulting disorder and its biochemical basis. The Genetic Testing Registry (www.ncbi.nlm.nih.gov/gtr/) is a database of the diseases that can be detected through analysis of DNA, carried out by either clinical or research laboratories.

Unlike monogenetic diseases, **polygenic diseases** are linked to variations in a number of different genes (many common human disorders such as cardiovascular disease and cancer also have environmental components). Researchers use **genome-wide association studies (GWAS)** to correlate the locations of sequence differences with particular diseases. For example, as many as 350 genetic variations have been associated with type 2 diabetes, but some variations are quite rare, and probably only 40–60 actually increase a person's risk of developing the disorder. The risk tied to any particular genetic variant is low, and the entire set of variations cannot entirely explain the heritability of the disease. In any case, DNA sequence variations can be difficult to interpret, as they may affect protein-coding genes as well as noncoding regulatory DNA segments. Several commercial enterprises offer individual genome-sequencing services (at a cost of only a few hundred dollars), but until genetic information can be reliably translated into effective plans to prevent or treat diseases, the practical value of "personal genomics" is somewhat limited.

On average, *the DNA of any two humans differs at 3 million sites*. These **single-nucleotide polymorphisms (SNPs)** are compiled in databases. A person begins life with an average of 60 new genetic changes that were not present in either parent. Of course, not all genetic variations have negative consequences. The vast majority of variations have no discernible effect on an individual's ability to survive and reproduce. Some variations may be beneficial, depending on the circumstances, and natural selection acts on this inherent genetic variation, favoring the persistence of certain genes in subsequent generations.

Genes can be altered

Understanding genetic information and how it is expressed makes genetic engineering possible. Using a variety of laboratory techniques (described more fully in Section 20.6), researchers can extract a gene from one organism and combine it with other DNA to produce **recombinant DNA.** The new DNA can then be further manipulated in the laboratory or introduced into another organism to direct the synthesis of a gene product. Sometimes a gene is intentionally modified by "site-directed mutagenesis." The near-universal interpretation of the genetic code means that almost any host cell can "read" the codons in a foreign gene to construct the appropriate protein. Introducing human genes into cultured bacterial, fungal, or mammalian cells is an economical way to produce certain proteins that are useful as drugs (**Table 3.4**). The same technology is used to genetically modify whole organisms (**Box 3.B**).

Making intentional changes to an individual human's genetic makeup is the goal of **gene therapy.** In traditional gene therapy, an extra gene is delivered to a patient's cells in order to compensate for a defective gene. The first successful gene therapy trials treated children with severe combined immunodeficiency (SCID), a normally fatal condition caused by a single-gene defect.

Table 3.4 Some Recombinant Protein Products

Protein	Purpose
Insulin	Treat insulin-dependent diabetes
Growth hormone	Treat certain growth disorders in children
Erythropoietin	Stimulate production of red blood cells; useful in kidney dialysis
Tissue plasminogen activator	Promote clot lysis following myocardial infarction or stroke
Colony stimulating factor	Promote white blood cell production after bone marrow transplant

Box 3.B   Genetically Modified Organisms

Introducing a foreign gene into a single host cell alters the genetic makeup of that cell and all its descendants. However, if the cell is part of a multicellular organism, such as an animal or plant, more work is required to generate a **transgenic organism** whose cells all contain the foreign gene. In mammals, modified DNA can be introduced into fertilized eggs, which are then implanted in a foster mother. Some of the resulting embryos' cells (possibly including their reproductive cells) will contain the foreign gene. When the animals mature, they must be bred in order to yield offspring whose cells are all transgenic. Transgenic plants are easier to develop: a few cells containing the modified DNA can sometimes be coaxed into developing into an entire plant.

Most of the transgenic animals that have been developed are used for research or are still being developed for commercial purposes. Some examples are virus-resistant pigs and chickens, cows that resist bovine tuberculosis, and cows that produce milk with a healthier assortment of fats. Transgenic salmon that grow faster than their wild relatives are already on the market. Mosquitoes that have been engineered for infertility (to prevent the spread of malaria) are being field-tested.

Transgenic crops have been commercially available for over 20 years and in many areas are more popular than conventional crops. Approximately 80–90% of the U.S. corn (maize), soybean, and cotton harvests are genetically modified. Some genetically modified produce is destined for animal feed or industrial uses, but transgenic fruits and vegetables have also been developed for direct human consumption. Among the desirable traits that have been engineered into various plant species is protection from insect pests, leading to higher yields and less reliance on pesticides (which are costly and potentially toxic to animals). Plants engineered for resistance to weed-killing herbicides, such as the *Brassica* species that produce canola oil (see photo), also give higher yields.

Daniel/123RF

Researchers have developed transgenic crops with improved nutritional value, such as a strain of rice containing high levels of iron and vitamin A, and crops that can flourish under suboptimal conditions such as drought, high temperatures, and high soil salinity.

Concerns about the safety of foods containing foreign genes have limited their acceptance by consumers in some places. Transgenic organisms also present some biological risks. For example, genes that code for insecticides (such as the bacterial toxin known as Bt, which is intended to kill the insect larvae that would otherwise feast on the plants) can make their way into wild plants that support beneficial insects. Similarly, herbicide-resistance genes can jump to weed species, making them even more difficult to control. Nevertheless, crops that can more efficiently sustain the world's growing population, particularly as the global climate changes, are necessary, and it is difficult to image how to develop these plants without taking advantage of the tools of genetic engineers.

Question   What molecular components should be analyzed in order to detect the presence of transgenic material in food items?

Bone marrow cells were removed from each patient and cultured in the presence of an engineered virus capable of transferring a normal version of the defective gene. When the modified cells were infused into the patient, they differentiated into functional immune system cells that produced the missing protein.

Table 3.5 Some Diseases Treated by Gene Therapy

Disease	Symptoms
Adrenoleukodystrophy	Neurodegeneration
Hemophilia	Excessive bleeding
Junctional epidermolysis bullosa	Severe skin blistering
Leber's congenital amaurosis	Blindness
Severe combined immunodeficiency	Immunodeficiency
β-Thalassemia	Anemia

Some of the diseases treated by gene therapy are listed in **Table 3.5**. Disorders related to blood cells have been attractive targets for gene therapy, since a small number of treated stem cells from bone marrow can regenerate a healthy population of blood cells. Introducing engineered genes to other tissues with lower rates of cellular turnover, such as muscle or brain, is more difficult. In any case, using viruses to deliver genes is inherently risky. For one thing, the engineered viruses can behave unpredictably, sometimes triggering a fatal immune response or inserting themselves into the host cell's chromosomes at random, which may interrupt the functions of other genes and cause cancer. Other challenges involve delivering the therapeutic gene to enough host cells to "correct" the disease, ensuring that the gene's expression is regulated properly so that the appropriate amount of protein is produced, and making the genetic change permanent.

Some of these potential problems can be minimized by newer gene-therapy protocols that use CRISPR gene-editing technology (described in Box 20.B) to fix—rather than just supplement—a defective gene. Hundreds of potential therapies are being tested, but as in traditional gene therapy, none are entirely risk-free. Off-target effects (altering gene sequences in addition to the defective gene) and the irreversible nature of gene editing present technical and ethical issues that require careful consideration.

Concept Check

1. Draw a diagram to illustrate each step of the central dogma.
2. Practice locating the codons for each of the 20 amino acids.
3. Explain the relationship between mutations and disease.
4. What is the relationship between the nucleotide sequence in a gene and the amino acid sequence of a protein?
5. List some positive and negative outcomes of gene therapy.

3.4 Genomics

KEY CONCEPTS

Identify the types of information provided by genomic analysis.

- Explain how genes are identified.
- Compare the genomes of different species.

Analyzing one gene at a time provides limited information, but thanks to DNA sequencing technology, entire genomes can be explored and compared. By tallying the nucleotide sequence differences between two similar groups of organisms, researchers can estimate how long the groups have been independently evolving and accumulating changes since splitting from a common ancestor. This sort of analysis allows the construction of branching trees (such as the one in Fig. 1.15) to depict the course of evolution. However, the work is far from finished.

Human Genome Project

The Human Genome Project (HGP) was a large-scale, and highly collaborative international effort to generate the sequence of a complete human genome of 3×10^9 base pairs along the 24 chromosomes. The project was launched in 1990 and completed in 2003. This project was funded by the United States government through NIH and also by different research groups around the world. A similar project was also conducted by a non-government organization called Celera Corporation. To sequence the entire genome of organisms, numerous efforts were made to improve the methods of DNA sequencing. The important techniques utilized to sequence the entire DNA were based on gene mapping, and DNA sequencers (Sanger's method). It took ten years to complete, and the first draft of the human genome sequence was announced in June 2000 by the International Human Genome Sequencing Consortium which accounted for 90% of the human genome with many gaps where the DNA sequence was unknown. An improved version of the first draft was announced by the Consortium in April 2003 which accounted for 92% of the human genome and was more accurate with fewer gaps. The first truly complete human genome sequence was announced on March 31, 2022, by the Telomere-to-Telomere (T2T) consortium. A major outcome of HGP was to provide information on the human blueprint and to estimate the number of protein-coding genes. These data offer many opportunities to study human genetic variability, and their association with diseases, and phenotype variability. Also, HGP provides a way to study human origin and evolutionary relationships.

For many organisms, including humans, the exact number of genes has not yet been determined, and different methods for identifying genes yield different estimates. For example, a computer can scan a long DNA sequence for an **open reading frame (ORF),** that is, a stretch of nucleotides that can potentially be transcribed and translated to a polypeptide. The ORF begins with a "start" codon: ATG in the coding strand of DNA, which corresponds to AUG in RNA (see Table 3.3). This codon specifies methionine, the initial residue of all newly synthesized polypeptides. The ORF ends with one of the three "stop" codons: DNA coding sequences of TAA, TAG, or TGA, which correspond to the three mRNA stop codons (see Table 3.3). For example,

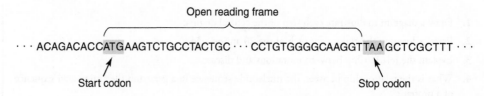

An alternative approach is to examine the population of RNA molecules in a single cell or organism. While this kind of survey captures many protein-coding segments of DNA (which correspond to messenger RNA), it also sometimes reveals that a significant portion of an organism's genome is transcribed but not translated. The importance of all this **noncoding RNA (ncRNA)** is not fully understood.

Genes can also be identified by comparing a DNA sequence with the sequences of known genes in the same species or in other species. *Genes with similar functions tend to have similar sequences*; such genes are said to be **homologous.** Even an inexact match can still indicate a protein's overall functional category, such as enzyme or hormone receptor, although its exact role in an organism may not be obvious.

How many genes are in the human genome? The best estimates put the tally at about 21,000 protein-coding genes. The human genome probably also contains a comparable number of noncoding genes as well as thousands of **pseudogenes,** which are nonfunctional copies of true genes and appear to be evolutionary leftovers. An analysis of 17,000 protein-coding genes in 44 different human tissues has revealed that only about half of these genes are expressed in all locations. Other studies in which protein-coding genes are inactivated, one at a time, suggest that even fewer human genes—roughly 2000 or about 10%—are absolutely essential for most cells. This core set of genes encodes proteins that are produced in relatively large amounts and carry out the most basic cellular activities.

When searching for genetic variations that could be tied to diseases, researchers can avoid labor-intensive genome-wide association studies (Section 3.3) and instead focus only on the **exome,** the set of DNA sequences that are actually used to direct protein synthesis in most cells.

The human exome represents only a tiny fraction of the genome but includes at least 180,000 distinct segments of DNA, far more than the number of genes. This is because virtually all protein-coding genes in complex eukaryotes include multiple short expressed sequences (called **exons**). The exons in each gene are separated by intervening sequences (called **introns**) that are cut out of the messenger RNA before it is translated. The mRNA splicing process is described in Section 21.3.

Genome size varies

The genomes of almost 100,000 different organisms have been sequenced—from the miniscule circular DNAs of parasitic bacteria to the enormous multi-chromosome genomes of plants and mammals. Some organisms whose genomes have been fully sequenced are listed in **Table 3.6**. The list includes species that are widely used as model organisms for different types of biochemical studies (**Fig. 3.11**). Note that for plants and animals, which are diploid (see Box 3.A), genomic information usually refers to the haploid state.

Not surprisingly, *organisms with the simplest lifestyles tend to have the smallest amount of DNA and the fewest genes*. For example, *M. genitalium* (see Table 3.6) is a human pathogen that relies on its host to provide nutrients; this organism contains fewer genes than free-living bacteria such as *E. coli*. Multicellular organisms generally have even more DNA and more genes, presumably to support the activities of their many specialized cell types. Exceptions abound; for example, the largest known genome belongs to the plant *Paris japonica*, which has about 50 times more DNA than humans but probably has a similar number of genes.

In prokaryotic genomes, all but a few percent of the DNA represents genes for proteins and RNA. The proportion of noncoding DNA generally increases with the complexity of the organism. For example, about 30% of the yeast genome, about half of the *Arabidopsis* genome, and over 98% of the human genome is noncoding DNA. Although up to 80% of the human genome may actually be transcribed to RNA, *the protein-coding segments account for only about 1.4% of the total* (**Fig. 3.12**).

In large genomes, much of the noncoding DNA consists of repeating sequences with no known function. Some repeating sequences are the remnants of virus-like **transposable**

Table 3.6 Genome Size and Gene Number

Organism	Genome size (Mb)[a]	Protein-coding genes[b]
Bacteria		
Mycoplasma genitalium	0.58	515
Haemophilus influenzae	1.85	1,620
Escherichia coli	4.64	4,240
Archaea		
Methanocaldococcus jannaschii	1.74	1,760
Fungi		
Saccharomyces cerevisiae (yeast)	12.16	6,000
Plants		
Arabidopsis thaliana	119.70	25,300
Oryza sativa (rice)	382.40	21,900
Animals		
Caenorhabditis elegans (nematode)	102.00	18,800
Drosophila melanogaster (fruit fly)	137.70	13,000
Homo sapiens	3279.00	21,000

[a] 1 Mb is 1000 kb or 1 million base pairs of DNA.

[b] RNA genes are not included.

Data from NCBI https://www.ncbi.nlm.nih.gov/genome/.

Question What is the relationship between genome size and gene number in prokaryotes? How does this differ in eukaryotes?

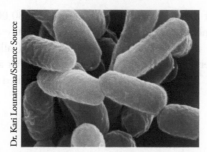

Escherichia coli, a normal inhabitant of the mammalian digestive tract, is a metabolically versatile bacterium that tolerates both aerobic and anaerobic conditions.

Baker's yeast, *Saccharomyces cerevisiae,* is one of the simplest eukaryotic organisms, with around 6000 genes.

Caenorhabditis elegans is a small (1 mm) and transparent roundworm. As a multicellular organism, it bears genes not found in unicellular organisms.

The plant kingdom is represented by *Arabidopsis thaliana,* which has a short generation time and readily takes up foreign DNA.

Figure 3.11 Some model organisms.

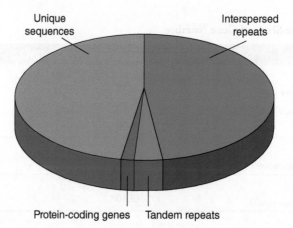

Figure 3.12 Coding and noncoding portions of the human genome.   Approximately 1.4% of the genome codes for proteins. Interspersed repetitive sequences make up 48% of the genome, and highly repetitive tandem repeat sequences account for about 3%.

Question **What percentage of the human genome consists of unique (nonrepeating) DNA sequences?**

elements, short segments of DNA that are copied many times and inserted randomly into the chromosomes. Not all viruses behave in this way, but all viruses can be considered to be molecular parasites (**Box 3.C**).

The human genome contains two types of repetitive DNA sequences. About 48% of human DNA consists of **interspersed repetitive sequences,** which are blocks of hundreds or

Box 3.C   Viruses

Although **viruses** are sometimes lumped with bacteria as "germs," the two groups have little in common. Viruses are not cells, they do not carry out metabolic activities, and they cannot replicate on their own. Instead, they must infect a host cell, either a prokaryote or eukaryote, and take advantage of that cell's resources. Bacterial viruses are sometimes called **bacteriophages** (literally, *bacteria-eaters*) because they kill their host cells.

Most viruses are much smaller than the cells they infect and have a simple structure: a nucleic acid genome surrounded by a protein **capsid** (shell), which may be surrounded by an outer membrane envelope with proteins embedded in it. Not all viruses have an outer membrane, and some virus capsids also enclose proteins that assist with infection or viral replication.

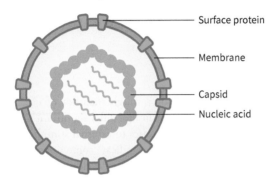

Surface protein

Membrane

Capsid

Nucleic acid

In all cases, the virus first attaches to a protein or another molecule on the surface of a host cell. This attachment may be quite specific, which limits the range of hosts that the virus can infect. In general, bacteriophages inject their nucleic acids into a bacterial cell, leaving the emptied capsid on the surface, whereas eukaryotic viruses are internalized before the capsid breaks apart and releases the nucleic acids.

Viral genomes exhibit huge variety. They may be RNA or DNA, single- or double-stranded, single molecules or multiple chains or even circles. Viral genomes range in size from about 4000 nucleotides in a single RNA chain to 2,500,000 base pairs of double-stranded DNA, with anywhere from 4 to 1000 genes. Not surprisingly, some viruses have a simple and quick life cycle, whereas others follow a much more elaborate sequence of infection, replication, and assembly of new **virions** (individual virus particles).

All viruses require some of the host cell's enzymes to help replicate their genomes and produce viral proteins. Double-stranded viral DNA genomes are replicated and transcribed in the usual way, but single-stranded viral DNA is usually converted by host enzymes to double-stranded DNA that can guide the synthesis of new viral DNA genomes and viral mRNAs. RNA viruses use a variety of strategies. So-called positive-sense RNA viruses have single-stranded genomes that can be directly translated into viral proteins, but their replication requires the synthesis of a complementary chain to act as a template for synthesis of more single-stranded RNA genomes.

The virus known as SARS-CoV-2, the cause of the COVID-19 pandemic, is among the largest RNA viruses, with a positive-sense RNA genome of about 30,000 nucleotides. Four of its genes code for structural proteins, including the large spike protein that gives the virus its crownlike appearance. Several other genes code for proteins that replicate the RNA genome and interfere with the host cell's antiviral defenses.

Negative-sense RNA viruses must first be transcribed, often by a viral RNA-directed RNA polymerase, to complementary RNA that can be translated. The retroviruses use their own reverse transcriptase enzyme to make a DNA copy of their RNA genome, and the DNA may then be integrated into the host cell's chromosome. The subsequent excision of viral DNA may leave characteristic "footprint" sequences (the source of some repetitive DNA sequences). The excised virus may also bring along a host gene or two; viruses are one way for genes to move around the genome or be transferred between species.

While the host cell's DNA or RNA polymerases and ribosomes are occupied with producing viral components, normal cellular activities may be neglected. In addition, some viruses encode proteins that specifically inhibit the host cell's normal functions and block antiviral responses, such as the degradation of double-stranded RNA, which almost always represents a replicating virus.

New viruses assemble inside the host cell or at its surface as viral proteins and nucleic acids accumulate. The capsid proteins tend to self-assemble, typically forming hollow 20-sided icosahedral shapes or helical tubes.

Icosahedral capsid   Helical capsid

Some bacteriophage genomes are actively threaded into preassembled capsids, whereas other virus capsids assemble around the viral nucleic acids. Additional steps of viral genome replication and protein maturation may occur inside the newly assembled virus before it becomes a complete virion capable of infecting a new host cell. The size of the virion "burst" ranges from perhaps 50 (for certain bacteriophage-infected prokaryotes) to over 50,000 (for some virus-infected eukaryotic cells).

thousands of nucleotides scattered throughout the genome. The most numerous of these are present in hundreds of thousands of copies. **Tandemly repeated DNA** accounts for another 3% of the human genome. This type of DNA consists of short sequences (usually 2 to 10 bp) that are repeated, side by side, many times. The purpose—if any—of all this repetitive DNA is not understood, but it does help to explain why certain very large genomes actually include

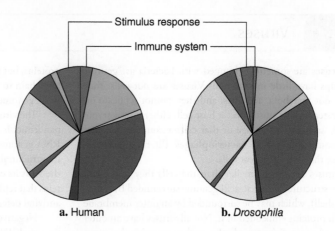

Figure 3.13 Functional classification of genes. This diagram groups **a.** human genes and **b.** *Drosophila* genes according to the biochemical function of the gene product. Humans devote a larger proportion of genes to the immune response (3.1% versus 1.0% in *Drosophila*) and to responses to external stimuli (6.0% versus 3.2% in *Drosophila*).

only a modest number of genes, and it explains why large genomes are able to accumulate sequence variations that have little effect on an individual's fitness (see Section 1.4).

Genomics has practical applications

Genomics, the study of genomes, can sometimes provide a rough snapshot of the metabolic capabilities of a given organism. For example, humans and fruit flies differ significantly in the proportion of genes involved in the immune response and responses to external stimuli (**Fig. 3.13**). An unusual number of genes belonging to one category might indicate some unusual biological properties in an organism. This sort of information can be useful for developing drugs to inhibit the growth of a pathogenic organism according to its unique metabolism. For example, as many as 2000 of the 6000 genes in the malaria parasite *Plasmodium* may code for proteins that are potential drug targets.

Curiously, many prokaryotes and small eukaryotes can acquire genes from other species through **horizontal gene transfer.** (Vertical gene transfer is the conventional pathway from parent to offspring, described in Box 3.A.) Horizontal gene transfer is one mechanism responsible for the spread of antibiotic-resistance genes among different bacterial species. In general, swapping genes with another organism or picking up stray DNA from the environment is a way to increase genetic diversity and potentially increase fitness. Although bacteria can readily acquire mammalian genes, the opposite does not appear to occur. The human microbiota, however, adds millions of microbial genes to the overall genetic makeup of the human body.

Genomic analysis can also provide insights into the functionality of microbial communities in almost any sample, such as a spoonful of soil or a jar of seawater. In an approach known as **metagenomics,** all the bits of DNA in a sample are sequenced, and the genomes of the component species are reconstructed from the fragments. Although this method does not always reveal the presence of rare species, it is a practical way to assess the overall diversity of the community. In fact, it is often the only way, because most prokaryotes cannot be grown in the laboratory for close study.

For larger organisms, **DNA barcoding** is a technique for identifying the species present in a sample of environmental DNA. In this case, researchers look for the sequence (the "barcode") of a specific gene that is known to vary between species but not between members of that species. DNA barcoding can be used to detect the presence of endangered animals from traces they leave behind (fur, feces, and so on) without having to capture or harm the animal.

Concept Check

1. Describe the approaches used to identify genes.
2. Outline the correlation between gene number and organismal lifestyle.
3. List some ways in which the human genome differs from a bacterial genome.
4. Explain the value and limitations of genome-wide association studies.

SUMMARY

3.1 Nitrogenous Bases

- The genetic material in virtually all organisms consists of DNA, a polymer of nucleotides. A nucleotide contains a purine or pyrimidine base linked to a ribose group (in RNA) or a deoxyribose group (in DNA) that also bears one or more phosphate groups.

3.2 Nucleic Acid Structure

- DNA contains two antiparallel helical strands of nucleotides linked by phosphodiester bonds. Each base pairs with a complementary base in the opposite strand: A with T and G with C. The structure of RNA, which is single-stranded and contains U rather than T, is more variable.

- Nucleic acid structures are stabilized primarily by stacking interactions between bases. The separated strands of DNA can reanneal.

3.3 The Central Dogma

- The central dogma summarizes how the sequence of nucleotides in DNA is transcribed into RNA, which is then translated into protein according to the genetic code.

- The sequence of nucleotides in a segment of DNA can reveal mutations that cause disease. Genetic variations across the genome may contribute to polygenic disease.

- In gene therapy, a normal gene is introduced or a gene is edited in order to cure a genetic disease.

3.4 Genomics

- Even with multiple techniques for identifying genes, the total number of genes in the human genome is not known.

- In general, the size of the genome and the number of protein-coding genes increases with organismal complexity. Large genomes may contain repetitive DNA sequences.

- Genomic analysis may reveal the functionality of organisms or the species diversity of communities.

KEY TERMS

nucleoside
chromosome
nucleic acid
nucleotide
DNA (deoxyribonucleic acid)
deoxynucleotide
base
purine
pyrimidine
RNA (ribonucleic acid)
vitamin
phosphodiester bond
polynucleotide
residue
5′ end
3′ end

base pair
sugar–phosphate backbone
antiparallel
major groove
minor groove
bp
kb
oligonucleotide
polymerase
nuclease
exonuclease
endonuclease
complement
A-DNA
B-DNA
stacking interactions

melting temperature (T_m)
denaturation
renaturation
anneal
probe
replication
gene expression
sister chromatids
centromere
mitosis
diploid
haploid
gametes
meiosis
homologous chromosomes
allele

gene
transcription
translation
central dogma of molecular biology
genome
coding strand
noncoding strand
messenger RNA (mRNA)
ribosome
ribosomal RNA (rRNA)
transfer RNA (tRNA)
codon
genetic code
mutation

monogenetic disease
polygenic disease
genome-wide association study (GWAS)
single-nucleotide polymorphism (SNP)
recombinant DNA
transgenic organism
gene therapy
open reading frame (ORF)
noncoding RNA (ncRNA)
homologous genes
pseudogene
exome

exon
intron
transposable element
virus
bacteriophage
capsid
virion
interspersed repetitive DNA
tandemly repeated DNA
genomics
horizontal gene transfer
metagenomics
DNA barcoding

BIOINFORMATICS

Brief Bioinformatics Exercises

3.1 Drawing and Visualizing Nucleotides

3.2 The DNA Double Helix

3.3 Melting Temperature and the GC Content of Duplex DNA

3.4 Analysis of Genomic DNA

Bioinformatics Project

Databases for the Storage and "Mining" of Genome Sequences

PROBLEMS

3.1 Nitrogenous Bases

1. The identification of DNA as the genetic material began with Griffith's "transformation" experiment conducted in 1928. Griffith worked with *Pneumococcus*, an encapsulated bacterium that forms smooth colonies when plated on agar and causes death when injected into mice. A mutant *Pneumococcus* lacking the enzymes needed to synthesize the polysaccharide capsule (required for virulence) forms rough colonies when plated on agar and does not cause death when injected into mice. Griffith found that heat-treated wild-type *Pneumococcus* did not cause death when injected into the mice because the heat treatment destroyed the polysaccharide capsule. However, if Griffith mixed heat-treated wild-type *Pneumococcus* and the mutant unencapsulated *Pneumococcus* together and injected this mixture, the mice died. Even more surprisingly, upon autopsy, Griffith found live, encapsulated *Pneumococcus* bacteria in the mouse tissue. Griffith concluded that the mutant *Pneumococcus* had been "transformed" into disease-causing *Pneumococcus*, but he could not explain how this occurred. Using your current knowledge of how DNA works, explain how the mutant *Pneumococcus* became transformed.

2. In 1944, Avery, MacLeod, and McCarty set out to identify the chemical agent capable of transforming mutant unencapsulated *Pneumococcus* to the deadly encapsulated form (see Problem 1). They isolated a viscous substance with the chemical and physical properties of DNA that was capable of transformation. If proteases (enzymes that degrade proteins) or ribonucleases (enzymes that degrade RNA) were added prior to the experiment, transformation could still occur. What did these treatments tell the investigators about the molecular identity of the transforming factor?

3. In 1952, Alfred Hershey and Martha Chase carried out experiments using bacteriophages, which consist of nucleic acid enclosed by a protein capsid (coat). They first labeled the bacteriophages with the radioactive isotopes ^{35}S and ^{32}P. Because proteins contain sulfur but not phosphorus, and DNA contains phosphorus but not sulfur, each type of molecule was separately labeled. The radiolabeled bacteriophages were allowed to infect the bacteria and then the preparation was treated to separate the empty capsids (ghosts) from the bacterial cells. The ghosts were found to contain most of the ^{35}S label, whereas 30% of the ^{32}P was found in the new bacteriophages produced by the infected cells. What does this experiment reveal about the roles of bacteriophage DNA and protein?

4. In February 1953 (two months before Watson and Crick published their paper describing DNA as a double helix), Linus Pauling and Robert Corey published a paper in which they proposed that DNA adopts a triple-helical structure. In their model, the three chains were tightly packed together, with the phosphates on the inside of the triple helix and the nitrogenous bases on the outside. They proposed that the triple helix was stabilized by hydrogen bonds between the interior phosphate groups. What are the flaws in this model?

5. Classify the following as purines or pyrimidines: **a.** hypoxanthine; **b.** xanthine; **c.** the nitrogenous base shown in Problem 9; and **d.** the compound described in Problem 12.

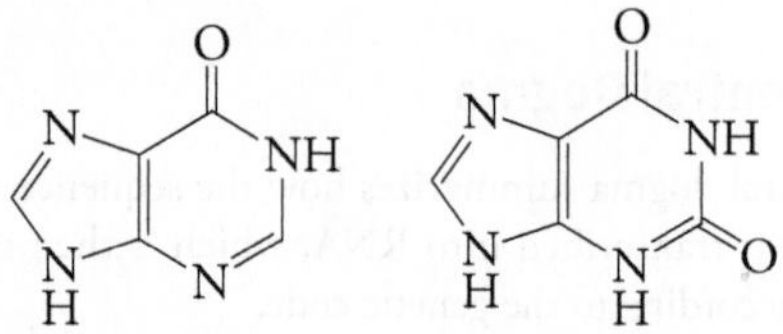

Hypoxanthine   Xanthine

6. Examine the structure of GDP in Section 3.1. Identify the functional groups and linkages in this molecule (see Table 1.1).

7. In some organisms, DNA is modified by addition of methyl groups. Draw the structure of 3-methylguanine.

8. Describe the chemical difference between uracil and thymine and draw their structures.

9. a. How do the adenosine groups in FAD and CoA differ (see Fig. 3.2)? **b.** What kind of linkage joins the two nucleotides in NAD and FAD (see Fig. 3.2)? **c.** Locate the base and monosaccharide in each residue of the dinucleotide NAD and the dinucleotide FAD (see Fig. 3.2).

10. Certain strains of *E. coli* incorporate the nitrogenous base shown below into nucleotides. For which base is this one a substitute?

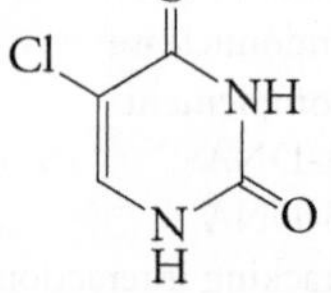

11. An *E. coli* culture is grown in the presence of the base shown in Problem 9. A control culture is grown in the absence of this modified base. Compare the masses of the DNA isolated from *E. coli* in these two cultures.

12. The simple synthesis of the antiviral compound 5-bromo-2′-deoxyuridine was recently reported. Draw the structure of this compound. For what base is this compound a substitute?

13. In some RNA molecules, a cytosine base is first methylated at N3 and then oxidatively deaminated in a process that replaces the amino group at position 4 with a carbonyl group. Draw the modified base.

14. DNA damage occurs when adenine, guanine, and cytosine undergo oxidative deamination. Draw the structures of the modified bases.

15. In certain pathogenic bacteria, the methylation of certain adenines in DNA is required in order for the bacteria to cause disease. **a.** Draw the structure of N^6-methyladenine. **b.** Why might scientists be interested in studying the bacterial enzyme N^6-DNA methyltransferase, which catalyzes the transfer of methyl groups to adenine?

16. Many cellular signaling pathways involve the conversion of ATP to cyclic AMP, in which a single phosphate group is esterified to both C3′ and C5′. Draw the structure of cyclic AMP.

17. Investigators recently identified a bacterial enzyme that acts on 5-hydroxyuridine. Draw the structure of this nucleoside.

18. Nucleosides are used as therapeutic drugs against certain viruses and are also used as anticancer agents. Explain.

19. The compound bredinin has antiviral activity. **a.** Is bredinin a base, a nucleoside, or a nucleotide? **b.** One of its functional groups has a pK lower than 7 and donates a proton to another functional group with a pK greater than 7. Using Table 2.4 as a guide to the types of functional groups that can donate and accept protons, draw the structure of ionized bredinin.

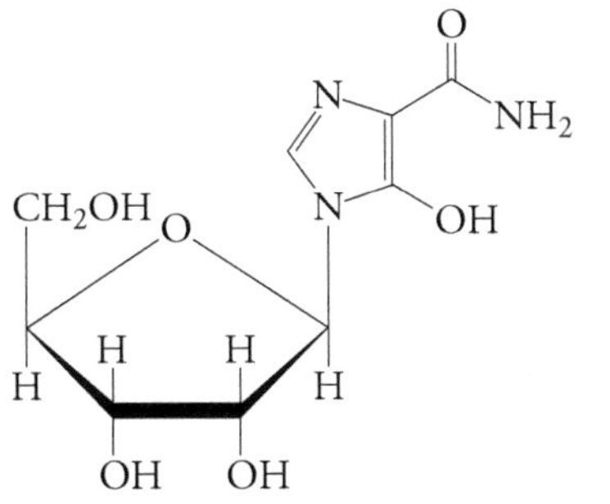

Bredinin

3.2 Nucleic Acid Structure

20. Draw a CA (ribo)dinucleotide and label the phosphodiester bond. How would the structure differ if it were DNA?

21. Draw the structure of the signaling molecule cyclic triadenylate (a cyclic trinucleotide with three adenosine monophosphate residues).

22. The dinucleotide cyclic guanosine monophosphate–adenosine monophosphate (cGAMP) is an intracellular signaling molecule involved in antiviral defense. The molecule has two phosphodiester linkages, one between the 2′-OH of GMP and the 5′-phosphate of AMP and the other between the 3′-OH of AMP and the 5′-phosphate of GMP. Draw this dinucleotide.

23. A diploid organism with a 30,000-kb haploid genome contains 30% T residues. Calculate the number of A, C, G, and T residues in the DNA of each cell in this organism.

24. Explain the structural basis for Chargaff 's observation that the amount of A + G in DNA is equal to the total amount of C + T. Does this rule hold for RNA?

25. Do Chargaff's rules hold true for RNA? Explain why or why not.

26. A well-studied bacteriophage has 98,534 base pairs in its complete genome. **a.** There are 22,356 G residues in the genome. Calculate the number of C, A, and T residues. **b.** Why does GenBank report a total of 49,267 bases for this bacteriophage genome?

27. Explain why RNA gets hydrolyzed on exposure to alkali treatment.

28. The SARS-CoV-2 virus was responsible for a global pandemic in 2020. Its genome contains 29,811 nucleotides, with 8903 A residues, 5482 C residues, 5852 G residues and 9574 U residues. **a.** Calculate the percentages of A, C, G, and U in the genome. **b.** What type of genome does this virus have (see Problem 23)?

29. Rotavirus infection is the leading cause of childhood diarrhea worldwide. The genome of the rotavirus A strain consists of 2265 A residues, 1037 C residues, 1037 G residues, and 2265 U residues. **a.** Calculate the percentages of each base in the genome. **b.** What type of genome does this virus have (see Problem 23)?

30. The complete genome of a virus contains 1328 T residues, 1250 G residues, 1536 A residues, and 1212 C residues. What can you conclude about the structure of the viral genome, given this information?

31. The adenine derivative hypoxanthine (see Problem 5) can base pair with cytosine, adenine, and uracil. Draw the structures of these base pairs.

32. The nucleotide bases can undergo tautomerization, a form of isomerization that involves the rearrangement of hydrogen atoms. For example, guanine can form the enol tautomer shown here, which can pair with thymine. Draw the structure of the resulting base pair, which has three hydrogen bonds.

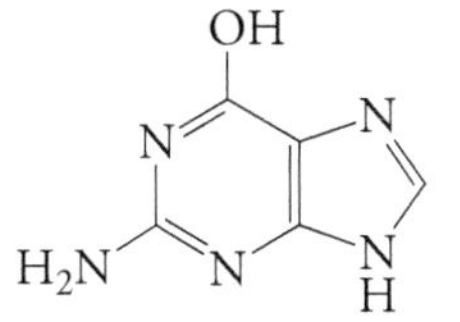

Guanine (enol tautomer)

33. Explain whether the following statement is true or false: Because a G:C base pair is stabilized by three hydrogen bonds, whereas an A:T base pair is stabilized by only two hydrogen bonds, GC-rich DNA is harder to melt than AT-rich DNA.

34. DNA having a high GC content requires more energy to melt as compared to DNA having a low GC content. Explain.

35. How can the hydrophobic effect (Section 2.2) explain why DNA adopts a helical structure?

36. a. Would you expect proteins to bind more favorably to the major groove or the minor groove of DNA? Explain. **b.** Eukaryotic DNA is packaged with histones, small proteins with a high lysine and arginine content (see Figure 4.2 for the amino acid structures). Why do histones have a high affinity for DNA?

37. a. What is the T_m of the DNA sample whose melting curve is shown in Figure 3.8? **b.** Draw melting curves that would be obtained from the DNA of *Dictyostelium discoideum* and *Streptomyces albus* (see Table 3.2).

38. What might you find in comparing the GC content of DNA from *Thermus aquaticus* (in hot springs) or *Pyrococcus furiosus* (in hydrothermal vents) and DNA from bacteria in a typical backyard pond?

39. Explain why the melting temperature of a sample of double-helical DNA increases when the K^+ concentration increases.

40. a. You have a short piece of synthetic RNA that you want to use as a probe to identify a gene in a sample of DNA. The RNA probe has a tendency to hybridize with sequences that are only weakly complementary. Should you increase or decrease the temperature to improve your chances of tagging the correct sequence? **b.** You have a short piece of single-stranded DNA that you want to hybridize to another strand of DNA with one mismatched base pair between the two strands. Should you increase or decrease the temperature to improve your chances of annealing the two strands?

41. Explain why the absorption of double-stranded DNA increases in the UV region at 260 nm on its heating.

3.3 The Central Dogma

42. Discuss the shortcomings of the following definitions for *gene*: **a.** A gene is the information that determines an inherited characteristic such as flower color. **b.** A gene is a segment of DNA that encodes a protein. **c.** A gene is a segment of DNA that is transcribed in all cells.

43. The semiconservative nature of DNA replication (as shown in Fig. 3.10) was proposed by Watson and Crick in 1953, but it was not experimentally verified until 1958. Meselson and Stahl grew bacteria on the "heavy" nitrogen isotope ^{15}N, producing DNA that was denser than normal. The food source was then abruptly switched to one containing only ^{14}N. Bacteria were harvested and the DNA isolated by density gradient centrifugation. **a.** What is the density of the DNA of the first-generation daughter DNA molecules? Explain. **b.** What is the density of the DNA isolated after two generations? Explain. **c.** What results would Meselson and Stahl have obtained had DNA replicated conservatively?

44. A portion of a gene is shown below.

5′-ATCTAACCGGTCCGCACTCCTTAGTCGCTCCTGCG-

3′-TAGATTGGCCAGGCGTGAGGAATCAGCGAGGACGCGCT-
　　　　　　　　　　　　　GACGCTGCTCGCCA-3′
　　　　　　　　　　　　　CGACTGCGACGAGCGGT-5′

The sequence of the mRNA transcribed from this gene has the following sequence:

5′-AUCUAACCGGUCCGCACUCCUUAGUCGCUCCUGCG-
　　　　　　　　　　　　GUGACGCUGCUCGCCA-3′

a. Identify the coding and noncoding strands of the DNA. **b.** Explain why only the coding strands of DNA are commonly published in databanks.

45. A segment of the coding strand of a gene is shown below.

ATCACATGGTGCATCTGACT

a. Write the sequence of the complementary strand that DNA polymerase would make. **b.** Write the sequence of the mRNA that RNA polymerase would make from the gene segment.

46. In the early 1960s, Marshall Nirenberg deciphered the genetic code by designing an experiment in which he synthesized a polynucleotide strand consisting solely of U residues, then added this strand to a test tube containing all of the components needed for protein synthesis. **a.** What polypeptide was produced by this "cell-free" system? **b.** What polypeptides were produced when poly A, poly C, and poly G were added to the cell-free system?

47. How many different codons are possible in nucleic acids containing four different nucleotides if a codon consists of **a.** three nucleotides, or consecutive sequences of **b.** five nucleotides, **c.** six nucleotides, or **d.** seven nucleotides? Does your answer help explain why codons consist of three nucleotides?

48. Synthetic biologists at the Scripps Institute expanded the genetic repertoire by adding two new bases into living bacterial cells. The two bases are named d5SICS and dNaM, and they base-pair with one another. How many different codons are possible in nucleic acids containing six different nucleotides if a codon consists of a consecutive sequence of three nucleotides?

49. An open reading frame (ORF) is a portion of the genome that potentially codes for a protein. A given mRNA sequence potentially has three different reading frames, only one of which is correct (the selection of the correct ORF will be discussed more fully in Section 22.3). A portion of the gene for a type II human collagen is shown. **a.** What are the amino acid sequences that can potentially be translated from each of the three possible reading frames? **b.** Collagen's amino acid sequence consists of repeating triplets in which every third amino acid is glycine. Does this information help you identify the correct reading frame?

AGGTCTTCAGGGAATGCCTGGCGAGAGGGGAGCAGC

50. Shown below is the coding strand of a portion of the gene that codes for the β-globin chain of hemoglobin. **a.** What are the amino acid sequences that could potentially be translated from each of the three reading frames? **b.** Examine your results and decide which of the three reading frames is the most likely to be correct.

GTGCATCTGACTCCTGAGGAGAAGTC

51. A portion of a sequence from human chromosome 22 is shown below. **a.** What are the amino acid sequences that could potentially be translated from this sequence if it is from the coding strand? **b.** What are the amino acid sequences that could potentially be translated if this sequence is from the noncoding strand? **c.** If you are told that the sequence is in the middle of a gene, does this help you to identify the correct reading frame?

TTCCAATGACTGAGTTCTCTCTCTAGAGAG

52. Is it possible for the same segment of DNA to encode two different proteins? Explain.

53. Examine the following nucleotide sequence from the coding strand of DNA. What is the amino acid sequence of the encoded protein?

GTAATTCAAAT ATG CCT TAC GCC

CCT GGA GAC GAA AAG AAG GGT

GCT ATT ACG TAT TTG AAG AAG

GCC ACC TCT GAG TAA ATGTGA

54. a. One form of the disease adrenoleukodystrophy (ALD) is caused by the substitution of serine for asparagine in the ALD protein. List the possible single-nucleotide alterations in the DNA of the ALD gene that could cause this genetic disease. **b.** In another form of ALD, a CGA codon is converted to a UGA codon. Explain how this mutation affects the ALD protein.

55. A mutation occurs when there is a base change in the DNA sequence. Some base changes do not lead to changes in the amino acid sequence of the resulting protein. Explain why.

56. The disease cystic fibrosis is the result of a mutation in the gene that encodes the cystic fibrosis transmembrane regulator (CFTR), a channel protein that allows chloride to exit the cell. A partial sequence of the CFTR gene is shown below, with the correct reading frame indicated. The most serious form of the disease results from the deletion of three consecutive nucleotides, as shown.

normal gene　504　505　506　507　508　509　510　511　512
sequence　⋯GAA AAT ATC ATC TTT GGT GTT TCC TAT⋯
mutated gene
sequence　⋯GAA AAT ATC AT–　– –T GGT GTT TCC TAT⋯

a. What are the sequences of the normal and mutated proteins? **b.** Explain why this form of the disease is referred to as ΔF508.

57. Another patient with a less severe form of cystic fibrosis (see Problem 53) has a mutation in a different region of the CFTR gene. The patient's DNA sequence is ⋯AAT AGA TAC AG⋯ (the normal sequence of the CFTR gene is ⋯AAT ATA GAT ACA G⋯). How has the DNA sequence changed and how does this affect the encoded protein?

58. Describe the result when DNA containing a normal G:C base pair undergoes replication during which G tautomerizes (see Problem 28) and pairs with T in the new DNA strand. After cell division, what will the DNA look like in the two daughter cells? Assuming that the G remains in its tautomeric form, what will the DNA look like after those cells divide again?

3.4 Genomics

59. The genome of the bacterium *Carsonella ruddii* contains 159 kb of DNA with 182 ORFs. What can you conclude about the habitat or lifestyle of this bacterium?

60. For many years, biologists and others have claimed that humans and chimpanzees are 98% identical at the level of DNA. Both the human and chimp genomes, which are roughly the same size, have now been sequenced, and the data reveal approximately 35 million nucleotide differences between the two species. How does this number compare to the original claim?

61. When genomes of various organisms were sequenced, biologists expected that the DNA content (the C-value) would always be positively correlated with organismal complexity. However, no such correlation has been demonstrated. In fact, some plant and algae genomes are many times the size of the human genome. The *C-value paradox* is the term that refers to this puzzling lack of correlation between DNA content and organismal complexity. What questions do biologists need to ask as they attempt to solve the paradox?

62. A portion of the sequence from the DNA coding strand of the myoglobin gene from a species of mollusk is shown below. Identify the correct open reading frame and give the amino acid sequence for the first few encoded amino acids.

ACCACCCCAACTGAAAATGTCTCTCTCTGATGC

63. A portion of the sequence from the DNA coding strand of the chick ovalbumin gene is shown. Determine the partial amino acid sequence of the encoded protein.

CTCAGAGTTCACCATGGGAGACATCGGTAAAGCAAGCATGGAA-
(1104 bp)-TTCTTTGGCAGATGTGTTTCCCGCTAAAAAGAA

64. A partial sequence of a newly discovered bacteriophage is shown below. **a.** Identify the longest open reading frame (ORF). **b.** Assuming that the ORF has been correctly identified, where is the most likely start site?

TATGGGATGGCTGAGTACAGCACGTTGAATGAGGCGATGG-
CCGCTGGTGATG

65. The genome of the bacteriophage described in Problem 63 contains 59 kb and 105 ORFs. None of the ORFs code for tRNAs. How does the bacteriophage replicate its DNA and synthesize the structural proteins necessary to replicate itself?

66. What is The Human Genome Project (HGP) and when it was launched?

67. If each person's genome contains an SNP every 300 nucleotides or so, how many SNPs are in that person's genome?

68. A genome-wide association study was carried out to identify the SNPs located on chromosome 1 that were correlated with an intestinal disease. The locations of three genes on chromosome 1 (between positions 6.3×10^7 and 6.6×10^7) are shown in the figure below. A $-\log_{10}$ P-value of 7 or greater is assumed to be associated with the disease. **a.** Which chromosomal locations show the strongest correlation with the disease? **b.** Which genes contain SNPs associated with the disease and which do not?

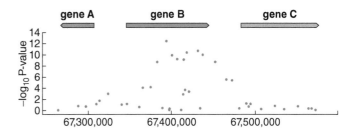

69. Use the information in the figure below to determine the chromosomal locations for the SNPs that are most closely associated with a colon disease. In this study, a $-\log_{10}$ P-value of 5 was used as the cutoff.

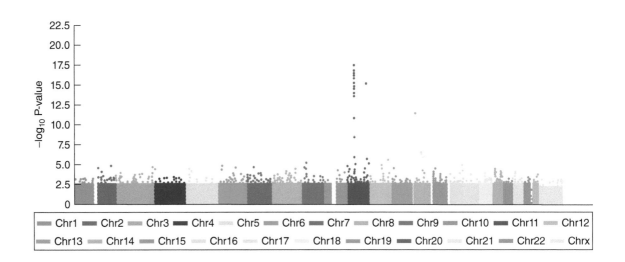

SELECTED READINGS

Cannon, M.E. and Mohlke, K.L., Deciphering the emerging complexities of molecular mechanisms at GWAS loci, *Am. J. Hum. Genet.* 103, 637–653, doi: 10.1016/j.ajhg.2018.10.001 (2018). [Reviews the methods and limitations of genome-wide association studies.]

High, K.A. and Roncarolo, M.G., Gene therapy, *N. Engl. J. Med.* 381, 455–464, doi: 10.1056/NEJMra1706910 (2019). [Summarizes the general approach of gene therapy and describes some proven applications.]

Lander, E.S., Initial impact of the sequencing of the human genome, *Nature* 470, 187–197 (2011). [Reviews genome structure and provides insights into its variation and links to human diseases.]

McDermott, M.L., Vanselous, H., Corcelli, S.A., and Petersen, P.B., DNA's chiral spine of hydration, *ACS Cent. Sci.* 3, 708–714, doi: 10.1021/acscentsci.7b00100 (2017). [Presents spectroscopic evidence of water binding in DNA's major and minor grooves.]

Rigden, D.J. and Fernández, X.M., The 28th annual Nucleic Acids Research database issue and molecular biology database collection, *Nucleic Acids Res.* 49. D1–D9. doi: 10.1093/nar/gkaa1216 (2021). [Introduces the annual database issue, which includes a summary of bioinformatics resources and some additions to the list of 1641 molecular biology databases. The entire issue is available at https://academic.oup.com/nar.]

Salzberg, S.L., Open questions: How many genes do we have? *BMC Biology* 16, 94, doi: 10.1186/s12915-018-0564-x (2018). [A brief summary of differences between sequence databases.]

Travers, A. and Muskhelishvili, G., DNA structure and function, *FEBS J.* 282, 2279–2295, doi: 10.1111/febs.13307 (2015). [Argues that the physicochemical properties of DNA are ideal for information storage and retrieval.]

CHAPTER 3 CREDITS

Figure 3.4 Image based on 4TRA. Westhof, E., Dumas, P., Moras, D., Restrained refinement of two crystalline forms of yeast aspartic acid and phenylalanine transfer RNA crystals, *Acta Crystallogr.* A 44, 112–123 (1988).

Figure 3.5 Image based on 1FIX. Horton, N.C., Finzel, B.C., The structure of an RNA/DNA hybrid: a substrate of the ribonuclease activity of HIV-1 reverse transcriptase, *J. Mol. Biol.* 264, 521–533 (1996).

Figure 3.7 Image from ACS Cent. Sci. 2017, 3, 7, 708–714 Publication Date: May 24, 2017. Copyright © 2017 American Chemical Society.

Table 3.2 Data from T.A. Brown (ed.), *Molecular Biology LabFax*, VI., Academic Press (1998), pp. 233–237.

Figure 3.13 Data from the Protein Analysis through Evolutionary Relationships classification system, www.pantherdb.org/.

Amino Acid and Protein Structure

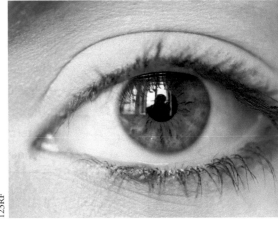

123RF

Behind the colored iris lies the lens of the eye, a 1-cm-diameter flattened sphere of cells that contain little besides proteins known as crystallins. Despite their name, these proteins do not actually form crystals, which would scatter light and defeat the purpose of the lens as a tool to focus light. Instead, the crystallins, which are very soluble, form a highly concentrated solution—almost like glass—that allows light to pass through. The conformations of the crystallins must be maintained for many decades to preserve the transparency of the lens.

Do You Remember?

- Cells contain four major types of biological molecules and three major types of polymers (Section 1.3).

- Noncovalent forces, including hydrogen bonds, ionic interactions, and van der Waals forces, act on biological molecules (Section 2.1).

- The hydrophobic effect, which is driven by entropy, excludes nonpolar substances from water (Section 2.2).

- An acid's pK value describes its tendency to ionize (Section 2.3).

- The biological information encoded by a sequence of DNA is transcribed to RNA and then translated into the amino acid sequence of a protein (Section 3.3).

Proteins account for about half of the solid mass of a typical cell, and they carry out a wide variety of functions. They provide structural stability and motors for movement; they form the molecular machinery for harvesting free energy and using it to carry out other metabolic activities; they participate in the expression of genetic information; and they mediate communication between the cell and its environment. In subsequent chapters, we will describe in more detail these protein-driven phenomena, but for now we will focus on protein structure.

We will look first at the amino acid components of proteins. Next comes a discussion of how protein chains fold into three-dimensional shapes. Finally, the Tools and Techniques section of this chapter examines some of the procedures for purifying and sequencing proteins and determining their structures.

4.1 Amino Acids, the Building Blocks of Proteins

KEY CONCEPTS

Identify the 20 amino acids that occur in proteins.

- Locate the functional groups in amino acids.
- Classify amino acid side chains as hydrophobic, polar, or charged.
- Draw a simple peptide and label its parts.
- Determine the net charge of peptides.
- Define the four levels of protein structure.

Amino acids are classified by R

Proteins come in a huge variety of shapes and sizes (**Fig. 4.1**) but they are all built in the same way. Each **protein** consists of one or more **polypeptides,** which are chains of polymerized amino acids. A cell may contain dozens of different amino acids, but only 20 of these—called the "standard" amino acids—are commonly found in proteins. As introduced in Section 1.2, an **amino acid** is a small molecule containing an amino group (—NH_3^+) and a carboxylate group (—COO^-) as well as a side chain of variable structure, called an **R group:**

$$H-\underset{\underset{NH_3^+}{|}}{\overset{\overset{COO^-}{|}}{C}}-R$$

Note that at physiological pH, the carboxyl group is unprotonated and the amino group is protonated, so an isolated amino acid bears both a negative and a positive charge.

The 20 amino acids have different chemical properties

The identities of the R groups distinguish the 20 standard amino acids. The R groups can be classified by their overall chemical characteristics as hydrophobic, polar, or charged, as shown in **Figure 4.2**, which also includes the one- and three-letter codes for each amino acid. These compounds are formally called **α-amino acids** because the amino and carboxylate (acid) groups are both attached to a central carbon atom known as the α carbon (abbreviated **Cα**).

Figure 4.2 also includes the one- and three-letter codes for each amino acid. The three-letter code is usually the first three letters of the amino acid's name. The one-letter code is derived as follows: If only one amino acid begins with a particular letter, that letter is used: C = cysteine, H = histidine, I = isoleucine, M = methionine, S = serine, and V = valine. If more than one amino acid begins with a particular letter, the letter is assigned to the most abundant amino acid: A = alanine, G = glycine, L = leucine, P = proline, and T = threonine. Most of the others are phonetically suggestive: D = aspartate ("asparDate"), F = phenylalanine ("Fenylalanine"), N = asparagine ("asparagiNe"), R = arginine ("aRginine"), W = tryptophan ("tWyptophan"), and Y = tyrosine ("tYrosine"). The rest are assigned as follows: E = glutamate (near D, aspartate), K = lysine, and Q = glutamine (near N, asparagine). The carbon atoms of amino acids are sometimes assigned Greek letters, beginning with Cα, the carbon to which the R group is attached. Thus, glutamate has a γ-carboxylate group, and lysine has an ε-amino group.

Nineteen of the 20 standard amino acids are asymmetric, or chiral, molecules. Their **chirality,** or handedness (from the Greek *cheir,* "hand"), results from the asymmetry of the alpha carbon. The four different substituents of Cα can be arranged in two ways. For alanine, a small amino acid with a methyl R group, the possibilities are

$$H_3N^+\!\!-\!\!\underset{\underset{CH_3}{|}}{\overset{\overset{COO^-}{|}}{C_\alpha}}\!\!-\!\!H \qquad H\!\!-\!\!\underset{\underset{CH_3}{|}}{\overset{\overset{COO^-}{|}}{C_\alpha}}\!\!-\!\!NH_3^+$$

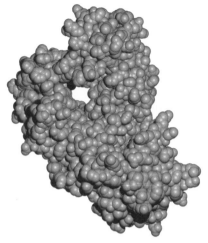

DNA polymerase (*E. coli* Klenow fragment)
Synthesizes a new DNA chain using an existing
DNA strand as a template (more in Section 20.1)

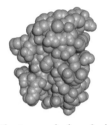

Plastocyanin (poplar)
Shuttles electrons as part of the
apparatus for converting light energy to
chemical energy (more in Section 16.2)

Insulin (pig)
Released from the pancreas to signal
the availability of the metabolic
fuel glucose (more in Section 19.2)

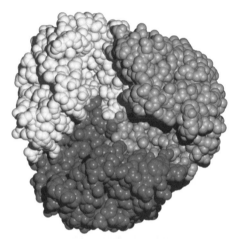

Maltoporin (*E. coli*)
Permits sugars to cross the bacterial cell
membrane (more in Section 9.2)

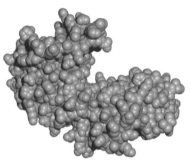

Phosphoglycerate kinase (yeast)
Catalyzes one of the central reactions in
metabolism (more in Section 13.1)

Figure 4.1 A gallery of protein structures.   These space-filling models are all shown at approximately the same scale. In proteins that consist of more than one chain of amino acids, the chains are shaded differently.

You can use a simple model-building kit to satisfy yourself that the two structures are not identical. They are nonsuperimposable mirror images, like right and left hands.

The amino acids found in proteins all have the form on the left. For historical reasons, these are designated L amino acids (from the Greek *levo,* "left"). Their mirror images, which rarely occur in proteins, are the D amino acids (from *dextro,* "right").

Molecules related by mirror symmetry are physically indistinguishable and are usually present in equal amounts in synthetic preparations. However, the two forms behave differently in biological systems (**Box 4.A**).

It is advisable to become familiar with the structures of the standard amino acids, since their side chains ultimately help determine the three-dimensional shape of a protein, its solubility, its ability to interact with other molecules, and its chemical reactivity.

Amino Acids with Hydrophobic Side Chains   *Some amino acids have nonpolar (hydrophobic) side chains that interact very weakly or not at all with water.* The aliphatic (hydrocarbonlike) side chains of alanine (Ala), valine (Val), leucine (Leu), isoleucine (Ile), and phenylalanine (Phe) obviously fit into this group. Although the side chains of methionine (Met) and tryptophan (Trp) include atoms with unshared electron pairs, the bulk of their side chains is nonpolar. Proline (Pro) is unique among the amino acids because its aliphatic side chain is also

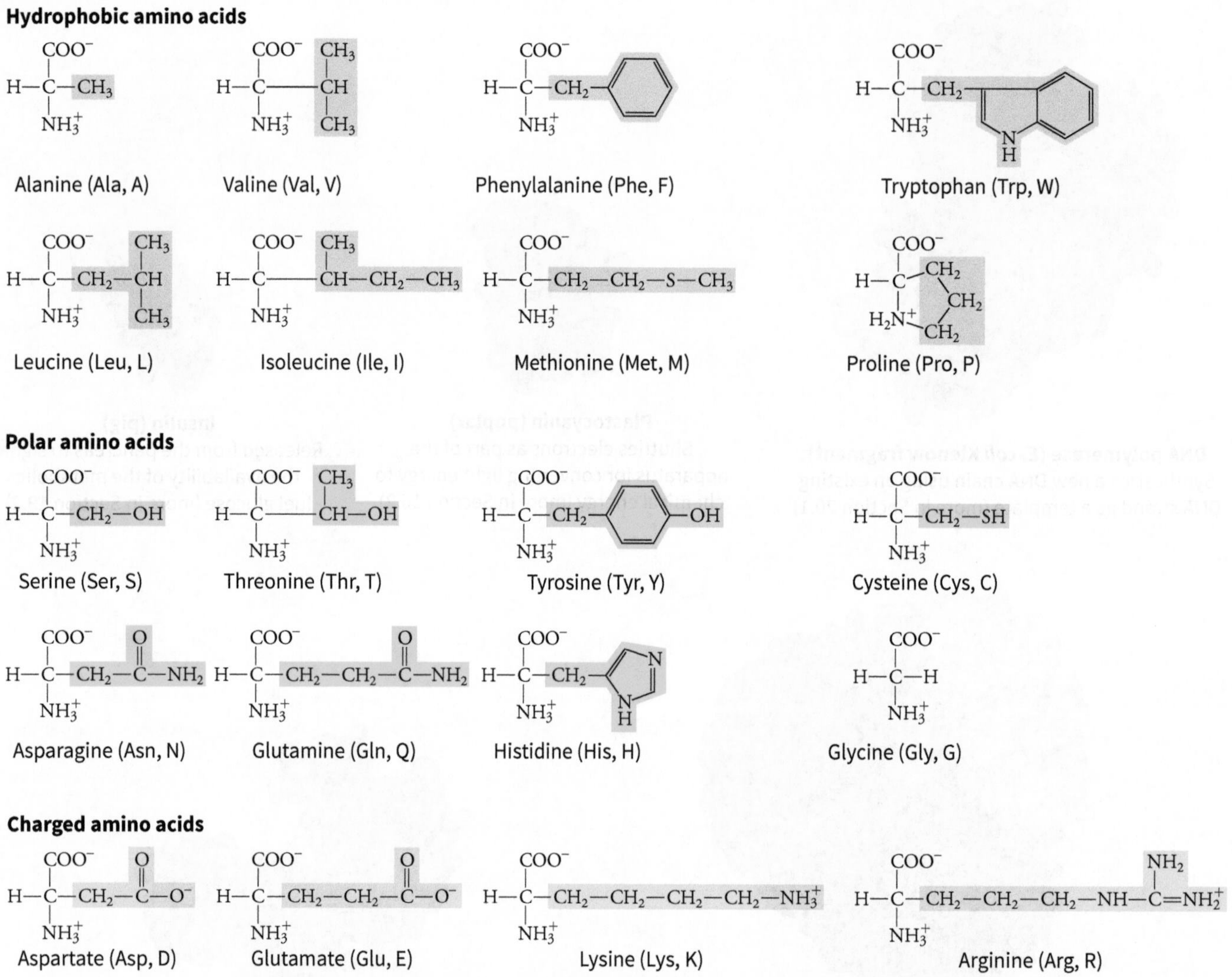

Hydrophobic amino acids

Alanine (Ala, A)   Valine (Val, V)   Phenylalanine (Phe, F)   Tryptophan (Trp, W)

Leucine (Leu, L)   Isoleucine (Ile, I)   Methionine (Met, M)   Proline (Pro, P)

Polar amino acids

Serine (Ser, S)   Threonine (Thr, T)   Tyrosine (Tyr, Y)   Cysteine (Cys, C)

Asparagine (Asn, N)   Glutamine (Gln, Q)   Histidine (His, H)   Glycine (Gly, G)

Charged amino acids

Aspartate (Asp, D)   Glutamate (Glu, E)   Lysine (Lys, K)   Arginine (Arg, R)

Figure 4.2 Structures and abbreviations of the 20 standard amino acids. The amino acids can be classified according to the chemical properties of their R groups as hydrophobic, polar, or charged. The side chain (R group) of each amino acid is shaded.

Question Identify the functional groups in each amino acid. Refer to Table 1.1.

| **Box 4.A** | **Does Chirality Matter?** |

The importance of chirality in biological systems was brought home in the 1960s when pregnant women with morning sickness were given the sedative thalidomide, which was a mixture of right- and left-handed forms. The active form of the drug has the structure shown here.

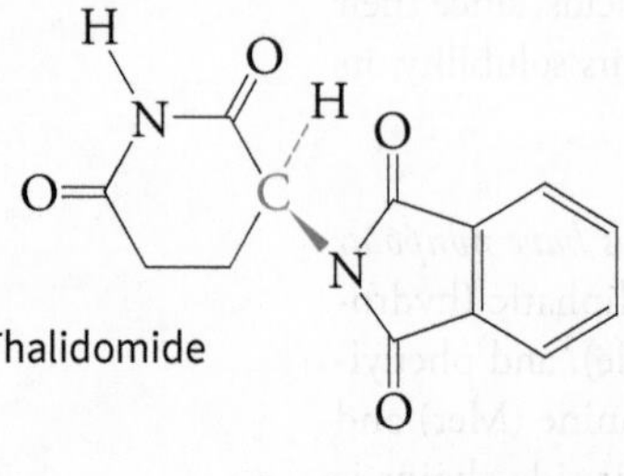

Thalidomide

Tragically, its mirror image, which was also present, caused severe birth defects, including abnormally short or absent limbs.

Although the mechanisms of action of the two forms of thalidomide are not well understood, different responses to the two forms can be rationalized. An organism's ability to distinguish chiral molecules results from the handedness of its molecular constituents. For example, proteins contain all l amino acids, and polynucleotides coil in a right-handed helix (see Fig. 3.3). The lessons learned from thalidomide have made drugs more costly to develop and test but have also made them safer.

Question Which of the 20 amino acids is not chiral?

covalently linked to its amino group. Glycine (Gly) is sometimes included with the hydrophobic amino acids.

In proteins, the hydrophobic amino acids are almost always located in the interior of the molecule, among other hydrophobic groups, where they do not interact with water. And because they lack reactive functional groups, the hydrophobic side chains do not directly participate in mediating chemical reactions.

Amino Acids with Polar Side Chains   *The side chains of the polar amino acids can interact with water because they contain hydrogen-bonding groups.* Serine (Ser), threonine (Thr), and tyrosine (Tyr) have hydroxyl groups; cysteine (Cys) has a thiol group; and asparagine (Asn) and glutamine (Gln) have amide groups. All these amino acids, along with histidine (His, which bears a polar imidazole ring), can be found on the solvent-exposed surface of a protein, although they also occur in the protein interior, provided that their hydrogen-bonding requirements are satisfied by their proximity to other hydrogen bond donor or acceptor groups. Glycine (Gly), whose side chain consists of only an H atom, cannot form hydrogen bonds but is included with the polar amino acids because it is neither hydrophobic nor charged.

Depending on the presence of nearby groups that increase their polarity, *some of the polar side chains can ionize at physiological pH values.* For example, the neutral (basic) form of histidine can accept a proton to form an imidazolium ion (an acid):

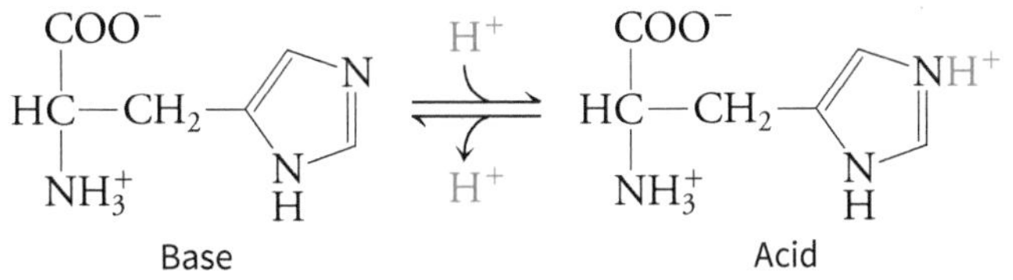

Base   Acid

As we will see, the ability of histidine to act as an acid or a base gives it great versatility in catalyzing chemical reactions.

Similarly, the thiol group of cysteine can be deprotonated, yielding a thiolate anion:

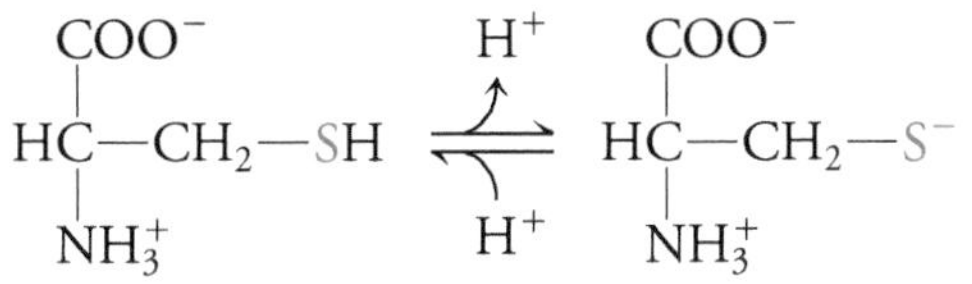

Occasionally, cysteine's thiol group undergoes oxidation with another thiol group, such as another Cys side chain, to form a **disulfide bond:**

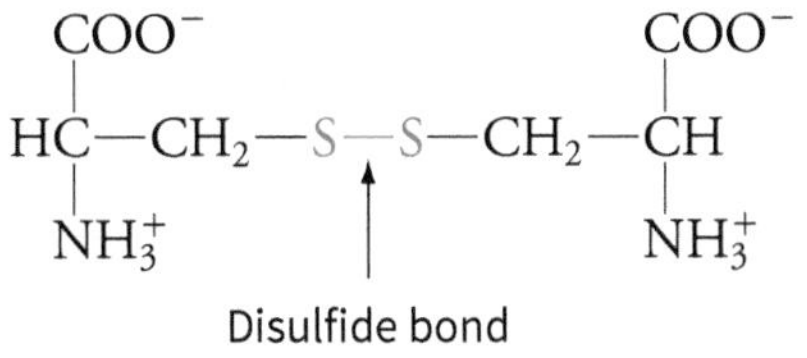

Disulfide bond

In certain situations, the hydroxyl groups of serine, threonine, and tyrosine undergo chemical reactions in which the O—H bond is cleaved.

Amino Acids with Charged Side Chains   *Four amino acids have side chains that are virtually always charged under physiological conditions.* Aspartate (Asp) and glutamate (Glu), which bear carboxylate groups, are negatively charged. Lysine (Lys) and arginine (Arg) are positively charged. Histidine, described above, can also bear a positive charge. Amino acids with charged side chains are usually located on the protein's surface, where the ionic groups can be surrounded by water molecules or interact with other polar or ionic substances. Note that the charges of these side chains depend on their ionization state, which is sensitive to the local pH. The amino acids with acidic or basic side chains can also participate in acid–base reactions.

Monosodium Glutamate

A number of amino acids and compounds derived from them function as signaling molecules in the nervous system (we will look at some of these in more detail in Section 18.2). Among the amino acids with signaling activity is glutamate, which most often operates as an excitatory signal and is necessary for learning and memory. Because glutamate is abundant in dietary proteins and because the human body can manufacture it, glutamate deficiency is rare. However, is there any danger in eating too much glutamate?

Glutamate binds to receptors on the tongue that register the taste of umami—one of the five human tastes, along with sweet, salty, sour, and bitter. By itself, the umami taste is not particularly pleasing, but when combined with other tastes, it imparts a sense of savoriness and induces salivation. For this reason, glutamate in the

form of monosodium glutamate (MSG) is sometimes added to processed foods as a flavor enhancer. For example, a low-salt food item can be made more appealing by adding MSG to it.

At one time, it was thought that consuming excess MSG from restaurant meals or processed foods led to symptoms such as muscle tingling, headaches, and drowsiness—all of which could potentially reflect the role of glutamate in the nervous system. However, MSG is also naturally present in many foods, including cheese, tomatoes, and soy sauce. Furthermore, scientific studies have been unable to show a consistent link between MSG intake and neurological symptoms in humans.

Question **Which group of glutamate is ionized to pair with a single sodium ion in MSG?**

Although it is convenient to view amino acids merely as the building blocks of proteins, many amino acids play key roles in regulating physiological processes (**Box 4.B**).

Peptide bonds link amino acids in proteins

The polymerization of amino acids to form a polypeptide chain involves the condensation of the carboxylate group of one amino acid with the amino group of another (a **condensation reaction** is one in which a water molecule is eliminated):

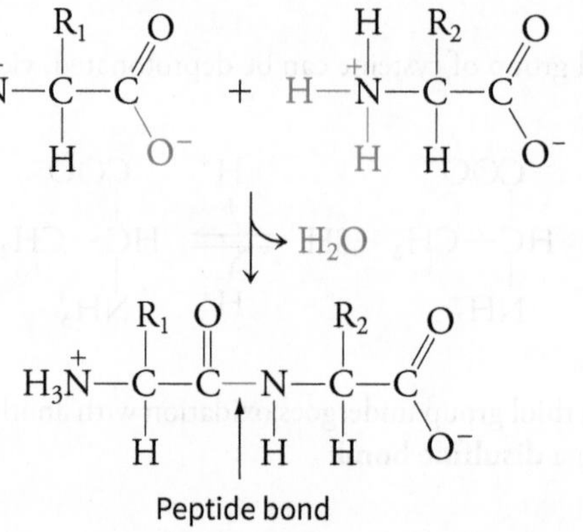

$$\overset{+}{H_3}N-\underset{\underset{H}{|}}{\overset{\overset{R_1}{|}}{C}}-\overset{O}{\underset{O^-}{\overset{\|}{C}}} \quad + \quad H-\overset{+}{\underset{\underset{H}{|}}{N}}-\underset{\underset{H}{|}}{\overset{\overset{R_2}{|}}{C}}-\overset{O}{\underset{O^-}{\overset{\|}{C}}}$$

$$\searrow H_2O$$

$$\overset{+}{H_3}N-\underset{\underset{H}{|}}{\overset{\overset{R_1}{|}}{C}}-\overset{O}{\overset{\|}{C}}-\underset{\underset{H}{|}}{N}-\underset{\underset{H}{|}}{\overset{\overset{R_2}{|}}{C}}-\overset{O}{\underset{O^-}{\overset{\|}{C}}}$$

Peptide bond

The resulting amide bond linking the two amino acids is called a **peptide bond.** The remaining portions of the amino acids are called amino acid **residues.** In a cell, peptide bond formation is carried out in several steps involving the ribosome and additional RNA and protein factors (Section 22.3). Peptide bonds can be broken, or **hydrolyzed,** by the action of **exopeptidases** or **endopeptidases** (enzymes that act from the end or the middle of the chain, respectively). A hydrolysis reaction is the reverse of a condensation reaction.

By convention, a chain of amino acid residues linked by peptide bonds is written or drawn so that the residue with a free amino group is on the left (this end of the polypeptide is called the **N-terminus**) and the residue with a free carboxylate group is on the right (this end is called the **C-terminus**):

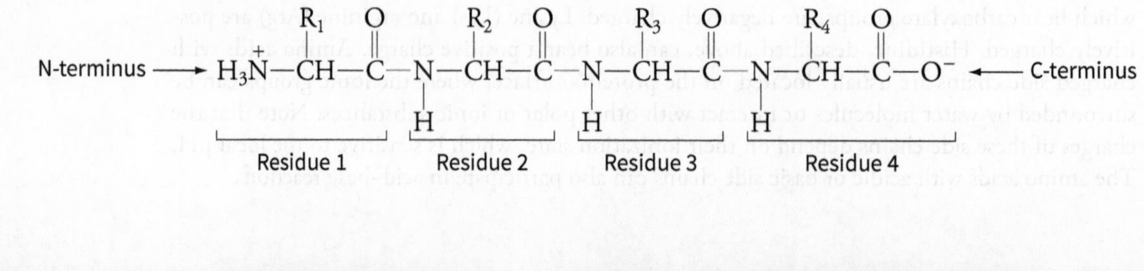

N-terminus ⟶ $\overset{+}{H_3}N-CH-\overset{O}{\overset{\|}{C}}-N-CH-\overset{O}{\overset{\|}{C}}-N-CH-\overset{O}{\overset{\|}{C}}-N-CH-\overset{O}{\overset{\|}{C}}-O^-$ ⟵ C-terminus

Residue 1 · Residue 2 · Residue 3 · Residue 4

Table 4.1 pK Values of Ionizable Groups in Amino Acids

Group[a]		pK
C-terminus	—COOH	3.5
Asp	—CH$_2$—C(=O)—OH	3.9
Glu	—CH$_2$—CH$_2$—C(=O)—OH	4.1
His	—CH$_2$—(imidazole) NH$^+$ / N—H	6.0
Cys	—CH$_2$—SH	8.4
N-terminus	—NH$_3^+$	9.0
Tyr	—CH$_2$—(C$_6$H$_4$)—OH	10.5
Lys	—CH$_2$—CH$_2$—CH$_2$—CH$_2$—NH$_3^+$	10.5
Arg	—CH$_2$—CH$_2$—CH$_2$—NH—C(NH$_2$)=NH$_2^+$	12.5

[a]The ionizable proton is indicated in red.

Note that, except for the two terminal groups, the charged amino and carboxylate groups of each amino acid are eliminated in forming peptide bonds. *The electrostatic properties of the polypeptide therefore depend primarily on the identities of the side chains (R groups) that project out from the polypeptide* **backbone.**

The pK values of all the charged and ionizable groups in amino acids are given in **Table 4.1** (recall from Section 2.3 that a pK value is a measure of a group's tendency to ionize). Thus, it is possible to calculate the net charge of a protein at a given pH (see Sample Calculation 4.1). At best, this value is only an estimate, since the side chains of polymerized amino acids do not behave as they do in free amino acids. This is because of the electronic effects of the peptide bond and other functional groups that may be brought into proximity when the polypeptide chain folds into a three-dimensional shape. The chemical properties of a side chain's immediate neighbors, its **microenvironment,** may alter its polarity, thereby altering its tendency to lose or accept a proton.

Nevertheless, *the chemical and physical properties of proteins depend on their constituent amino acids,* so proteins exhibit different behaviors under given laboratory conditions. These differences can be exploited to purify a protein, that is, to isolate it from a mixture containing other molecules (see Section 4.6). Most proteins contain all 20 amino acids, with some tending to appear more often than others (**Fig. 4.3**).

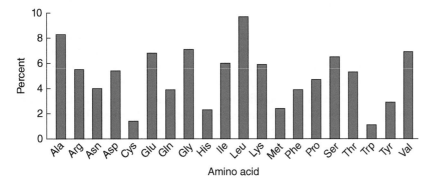

Figure 4.3 Percent occurrence of amino acids in proteins.

SAMPLE CALCULATION 4.1

Problem   Estimate the net charge of the polypeptide chain below at physiological pH (7.4) and at pH 5.0.

Ala–Arg–Val–His–Asp–Gln

Solution   The polypeptide contains the following ionizable groups, whose pK values are listed in Table 4.1: the N-terminus (pK = 9.0), Arg (pK = 12.5), His (pK = 6.0), Asp (pK = 3.9), and the C-terminus (pK = 3.5).

At pH 7.4, the groups whose pK values are less than 7.4 are mostly deprotonated and the groups with pK values greater than 7.4 are mostly protonated. The polypeptide therefore has a net charge of 0:

Group	Charge
N-terminus	+1
Arg	+1
His	0
Asp	−1

Group	Charge
C-terminus	−1
net charge	0

At pH 5.0, His is likely to be protonated, giving the polypeptide a net charge of +1:

Group	Charge
N-terminus	+1
Arg	+1
His	+1
Asp	−1
C-terminus	−1
net charge	+1

Most polypeptides contain between 100 and 1000 amino acid residues, although some contain thousands of amino acids (**Table 4.2**). Very short polypeptides are often called **oligopeptides** (*oligo* is Greek for "few") or just **peptides.** Some examples of peptides are the 9-residue hormones oxytocin (involved in childbirth and social bonding in mammals) and vasopressin (which regulates water homeostasis). Conotoxins are 10- to 30-residue peptides produced by *Conus* snails to paralyze their prey. Like many polypeptides that are secreted from cells, these peptides include Cys residues that form intramolecular disulfide bonds.

Cys–Tyr–Ile–Gln–Asn–Cys–Pro–Leu–Gly
Oxytocin

Cys–Tyr–Phe–Gln–Asn–Cys–Pro–Arg–Gly
Vasopressin

Glu–Cys–Cys–Asn–Pro–Ala–Cys–Gly–Arg–His–Tyr–Ser–Cys
A conotoxin

Since there are 20 different amino acids that can be polymerized to form polypeptides, even peptides of similar size can differ dramatically from each other, depending on their complement of amino acids. The potential for sequence variation is enormous. For a modest-sized polypeptide of 100 residues, there are 20^{100} or 1.27×10^{130} possible amino acid sequences. This number is clearly unattainable in nature, since there are only about 10^{79} atoms in the universe, but it illustrates the tremendous structural variability of proteins.

Unraveling the amino acid sequence of a protein may be relatively straightforward if its gene has been sequenced (see Section 3.4). In this case, it is just a matter of reading successive sets of three nucleotides in the DNA as a sequence of amino acids in the protein. However, this exercise may not be accurate if the gene's mRNA is spliced before being translated or if the protein is hydrolyzed or otherwise covalently modified immediately after it is synthesized. Of course, nucleic acid sequencing is of no use if the protein's gene has not been identified. The alternative is to use a technique such as mass spectrometry to directly determine the protein's amino acid sequence (Section 4.6).

Table 4.2 Composition of Some Proteins

Protein	Number of amino acid residues	Number of polypeptide chains	Molecular mass (D)
Insulin (bovine)	51	2	5733
Rubredoxin (*Pyrococcus*)	53	1	5878
Myoglobin (human)	153	1	17,053
Phosphorylase kinase (yeast)	416	1	44,552
Hemoglobin (human)	574	4	61,972
Reverse transcriptase (HIV)	986	2	114,097
Nitrite reductase (*Alcaligenes*)	1029	3	111,027
C-reactive protein (human)	1030	5	115,160
Pyruvate decarboxylase (yeast)	1112	2	121,600
Immunoglobulin (mouse)	1316	4	145,228
Ribulose bisphosphate carboxylase (spinach)	5048	16	567,960
Glutamine synthetase (*Salmonella*)	5628	12	621,600
Carbamoyl phosphate synthetase (*E. coli*)	5820	8	637,020

The amino acid sequence is the first level of protein structure

The sequence of amino acids in a polypeptide is called the protein's **primary structure.** There are as many as four levels of structure in a protein (**Fig. 4.4**). Under physiological conditions, a polypeptide very seldom assumes a linear extended conformation but instead usually folds up to form a more compact shape, typically consisting of several layers. The conformation of the polypeptide backbone (exclusive of the side chains) is known as **secondary structure.** The complete three-dimensional conformation of the polypeptide, including its backbone atoms and all its side chains, is the polypeptide's **tertiary structure.** In a protein that consists of more than one polypeptide chain, the **quaternary structure** refers to the spatial arrangement of all the chains. In the following sections we will consider the second, third, and fourth levels of protein structure.

Primary structure
The sequence of amino acid residues

Secondary structure
The spatial arrangement of the polypeptide backbone

Tertiary structure
The three-dimensional structure of an entire polypeptide, including all its side chains

Quaternary structure
The spatial arrangement of polypeptide chains in a protein with multiple subunits

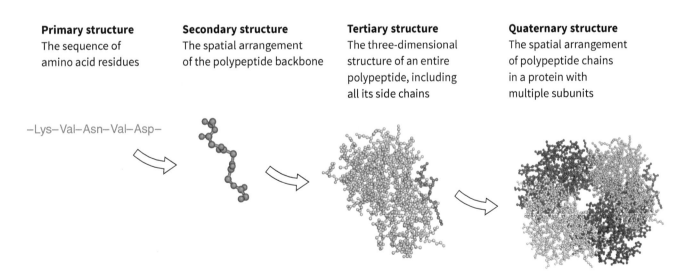

–Lys–Val–Asn–Val–Asp–

Figure 4.4 Levels of protein structure in hemoglobin.

Concept Check

1. Draw the structures and give the one- and three-letter abbreviations for the 20 standard amino acids.
2. Divide the 20 amino acids into groups that are hydrophobic, polar, and charged.
3. Identify the polar amino acids that are sometimes charged.
4. Draw a tripeptide and identify its peptide bonds, backbone, side chains, N-terminus, C-terminus, and net charge.
5. Describe the four levels of protein structure.

4.2 Secondary Structure: The Conformation of the Peptide Group

KEY CONCEPTS

Recognize the common types of regular secondary structure.

- Explain the limited flexibility of the peptide chain.
- Describe the features of an α helix.
- Describe the features of parallel and antiparallel β sheets.
- Define irregular secondary structure.

In the peptide bond that links successive amino acids in a polypeptide chain, the electrons are somewhat delocalized so that the peptide bond has two resonance forms:

Due to this partial (about 40%) double-bond character, there is no rotation around the C—N bond. In a polypeptide backbone, the repeating N—Cα—C units of the amino acid residues can therefore be considered to be a series of planar peptide groups (where each plane contains the atoms involved in the peptide bond):

Here the H atom and R group attached to Cα are not shown.

The polypeptide backbone can still rotate around the alpha carbon. The degree of rotation is described by **torsion angles** known as ϕ (for the N—Cα bond) and ψ (for the Cα—C bond). However, the values of ϕ and ψ are somewhat limited because rotation of the two peptide groups around Cα can bring atoms from neighboring residues too close, as shown here:

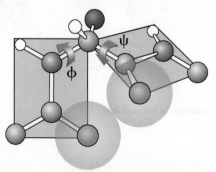

The atoms are color-coded: C gray, O red, N blue, and H white, and the van der Waals surfaces of the two carbonyl O atoms are shown.

As the resonance structures indicate, the groups involved in the peptide bond are strongly polar, with a tendency to form hydrogen bonds. The backbone amino groups are hydrogen bond donors and the carbonyl oxygens are hydrogen bond acceptors. *Under physiological conditions, the polypeptide chain tends to fold so that it can satisfy its hydrogen-bonding requirements.* At the same time, *the polypeptide backbone must adopt a conformation (a secondary structure) that minimizes steric strain.* In addition, side chains must be positioned in a way that minimizes their steric interference. To meet these criteria, the polypeptide backbone often assumes a repeating conformation, known as **regular secondary structure,** such as an α helix or a β sheet.

The α helix exhibits a twisted backbone conformation

The **α helix** was first identified through model-building studies carried out by Linus Pauling. In this type of secondary structure, the polypeptide backbone twists in a right-handed helix (the DNA helix is also right-handed; see Section 3.2 for an explanation). There are 3.6 residues per turn of the helix and, for every turn, the helix rises 5.4 Å along its axis. In the α helix, the carbonyl oxygen of each residue forms a hydrogen bond with the backbone NH group four residues ahead. The backbone hydrogen-bonding tendencies are thereby met, except for the four residues at each end of the helix (**Fig. 4.5**). The average α helix is about 10 residues long.

Like the DNA helix, whose side chains fill the helix interior (see Figure 3.3b), the α helix is solid—the atoms of the polypeptide backbone are in van der Waals contact. However, in the α helix, the side chains extend outward from the helix (**Fig. 4.6**).

The β sheet contains multiple polypeptide strands

Pauling, along with Robert Corey, also built models of the **β sheet.** This type of secondary structure consists of aligned strands of polypeptide whose hydrogen-bonding requirements are met by bonding between neighboring strands. The strands of a β sheet can be arranged in two ways (**Fig. 4.7**): In a **parallel β sheet,** neighboring chains run in the same direction; in an **antiparallel β sheet,** neighboring chains run in opposite directions. Each residue forms two hydrogen bonds with a neighboring strand, so all hydrogen-bonding requirements are met. The first and last strands of the sheet must find other hydrogen-bonding partners for their outside edges.

A single β sheet may contain from 2 to more than 12 polypeptide strands, with an average of 6 strands, and each strand has an average length of 6 residues. In a β sheet, the amino acid side chains extend from both faces (**Fig. 4.8**).

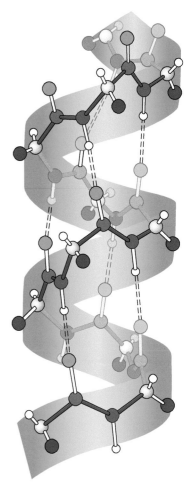

Figure 4.5 The α helix.
In this conformation, the polypeptide backbone twists in a right-handed fashion so that hydrogen bonds (dashed lines) form between C=O and N—H groups four residues farther along. Atoms are color-coded: Cα light gray, carbonyl C dark gray, O red, N blue, side chain purple, H white. (Based on a drawing by Irving Geis.)

Question How many amino acid residues are shown here? How many hydrogen bonds?

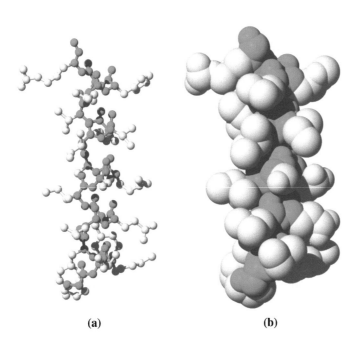

(a)   (b)

Figure 4.6 An α helix from myoglobin.   a. Ball-and-stick model of residues 100–118 of myoglobin. **b.** Space-filling model. Backbone atoms are green and side chains are gray.

Figure 4.7 β sheets.   In a parallel β sheet and an antiparallel β sheet, the polypeptide backbone is extended. In both types of β sheet, hydrogen bonds form between the amino and carbonyl groups of adjacent strands. The H and R attached to Cα are not shown. Note that the strands are not necessarily separate polypeptides but may be segments of a single chain that loops back on itself.

Question How many amino acid residues are shown in each chain?

Ramachandran plot and dihedral angles

Torsion Angles/Dihedral Angles/Rotation Angles   In the context of peptide structure, the Cα—C—N—Cα bond angles serve to separate adjacent carbon atoms in amino acid residues. Notably, due to the partial double bond character in the bond between C—N within peptides, the peptide C—N is marginally shorter than the bond between C—N in a simple amine. The X-ray diffraction investigations of amino acids, dipeptides, and tripeptides reveal that the atoms connected to the peptide bond lie in a coplanar configuration, which is indicative of electron resonance or partial sharing between the amide nitrogen and carbonyl oxygen. This arrangement results in the creation of a minor electric dipole due to the partial negative

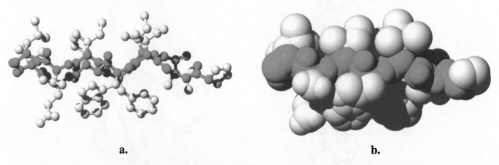

a.   **b.**

Figure 4.8 Side view of two parallel strands of a β sheet.   **a.** Ball-and-stick model of a β sheet from carboxypeptidase A. **b.** Space-filling model. Backbone atoms are green and side chains (gray) point alternately to each side of the β sheet.

charge and the net partial positive charge on the oxygen and hydrogen respectively, bonded to the nitrogen.

These findings led Pauling and Corey to propose that the partial double-bond nature of the peptide C—N bonds denoted by ω (omega) restricts their free rotation which can have either *cis* or *trans* conformation. In contrast, rotation is allowed for the N—Cα and Cα—C bonds. Consequently, the polypeptide chain's backbone can be envisioned as a collection of planes having a singular point of rotation. This results in a protein's backbone consisting of interconnected planar peptide units, and the conformation of a polypeptide's backbone can be described by the torsion angles (or dihedral angles) about the Cα—N bond (φ) and Cα—C bond (ψ) of each amino acid residue.

These three dihedral angles, φ (phi), ψ (psi), and ω (omega) are crucial for characterizing peptide conformation. A dihedral angle is the angle formed when two planes intersect, and peptide planes are determined by the bond vectors in the peptide backbone. Each plane is described by two consecutive bond vectors, and the angle between these planes is used to determine the conformation of the peptide.

In the fully extended, planar (all-trans) conformation of a polypeptide chain, the angles φ and ψ are both set to 180°, and they increase with a clockwise rotation when viewed from Cα. However, the conformation of a polypeptide backbone is subject to various steric constraints on the torsion angles φ and ψ. It is worth noting that a single (σ) bond's electronic structure exhibits cylindrical symmetry around its bond axis, suggesting free rotation. For example, a C—C bond allows for equally probable torsion angles in ethane if such were the case. Nonetheless, due to quantum mechanical effects arising from molecular orbital interactions in ethane, certain conformations are favored. Ethane exists in two distinct configurations, with the **staggered conformation** (torsion angle = 180°) being the most stable and the **eclipsed conformation** (torsion angle = 0°) the least stable.

Ramachandran Plot The Ramachandran plot visually illustrates the permissible or allowed combinations of φ and ψ angles in the peptide backbone, as well as those which are disallowed or unfavourable due to steric constraints. By calculating the interatomic distances between the atoms of a tripeptide across all φ and ψ values for the central peptide unit, one can establish the angles that are sterically allowed. G. N. Ramachandran's creation, known as the **Ramachandran diagram**, provides a condensed representation of these rotations involving two dihedral angles of dipeptide.

A polypeptide chain is restricted from accessing 77% of the Ramachandran diagram, representing the majority of φ and ψ combinations, due to conformational limitations. The van der Waals radii used to construct the Ramachandran diagram define which regions of the diagram indicate permissible conformations. Notably, all common forms of regular secondary structures found in proteins are located within the permitted zones of the Ramachandran diagram. In fact, most residues other than glycine in proteins with known X-ray structures exhibit conformational angles that fall within these allowed ranges.

Most points in the prohibited zones are situated between the two completely permissible regions, particularly near ψ = 0. These conformations, resulting from interactions between successive amide groups, become permissible when slight twists of a few degrees around the peptide bond are permitted. This is reasonable given the peptide bond's limited resistance to minor deviations from planarity. Notably, Gly stands out to some extent as sterically unhindered compared to remaining amino acid residues because it lacks a carbon atom.

Proteins also contain irregular secondary structure

α Helices and β sheets are classified as regular secondary structures, because *their component residues exhibit backbone conformations that are the same from one residue to the next.* In fact, these elements of secondary structure are easily recognized in the three-dimensional structures of a huge variety of proteins, regardless of their amino acid composition. Of course, depending on the identities of the side chains and other groups that might be present, α helices and β sheets may be slightly distorted from their ideal conformations. For example, the final turn of some α helices becomes "stretched out" (longer and thinner than the rest of the helix).

In every protein, elements of secondary structure (individual α helices or strands in a β sheet) are linked together by polypeptide loops of various sizes. A loop may be a relatively simple hairpin turn, as in the connection of two antiparallel β strands (which are shown below as flat arrows; *left*), or it may be quite long, especially if it joins successive strands in a parallel β sheet (*right*):

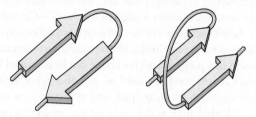

These connecting loops, as well as other segments of the polypeptide chain, are usually described as **irregular secondary structure;** that is, the polypeptide does not adopt a defined secondary structure in which successive residues have the same backbone conformation.

Most proteins contain a combination of regular and irregular secondary structure. But note that "irregular" does not necessarily mean "disordered." In many proteins, the peptide backbone adopts a single, unique conformation. However, as described below, some proteins include segments that are truly disordered and can adopt many different conformations.

Concept Check

1. Draw a polypeptide backbone and indicate which bonds can rotate freely.
2. Identify all the hydrogen bond donor and acceptor groups in the backbone.
3. What limits the rotation of the backbone?
4. Explain how an α helix and a β sheet satisfy a polypeptide's hydrogen-bonding needs.
5. Compare parallel and antiparallel β sheets.
6. Summarize the difference between regular and irregular secondary structure.
7. Explain why does virtually every protein contain loops.

4.3 Tertiary Structure and Protein Stability

KEY CONCEPTS

Describe the forces that stabilize protein structure.

- Analyze protein structures presented in different styles.
- Explain how the hydrophobic effect stabilizes globular protein structures.
- Describe the intramolecular interactions that can stabilize protein structures.
- Summarize the process of protein folding.
- Identify different types of disorder in protein structures.
- List some functions of disordered protein regions.

According to the traditional view, the three-dimensional shape of a polypeptide is its tertiary structure, which includes the chain's regular and irregular secondary structure (that is, the overall folding of its peptide backbone) as well as the spatial arrangement of all its side chains. However, the idea that every atom has a single designated place mainly reflects historical approaches for evaluating protein structures and does not apply to all proteins. The current understanding is that protein structures lie along a spectrum of possibilities from highly ordered, fixed shapes to highly disordered and dynamic shapes. However, all proteins are subject to the same thermodynamic forces, so, for convenience, we will focus first on the principles that govern proteins with unique shapes and then investigate proteins that do not conform as closely to these traditional "rules."

Proteins can be described in different ways

Most of the ~160,000 proteins whose structures are known in atomic detail were studied using X-ray crystallography (Section 4.6), a technique that captures the conformation of a protein fixed in a crystalline array. This was the technique used by John Kendrew to determine the first protein structure—of myoglobin—in 1958. After painstakingly determining the position of every backbone and side-chain group, Kendrew was mildly disappointed to find that, unlike the elegant, symmetric DNA structure that had been published just a few years earlier, the myoglobin structure was irregular and complex (myoglobin structure is examined in detail in Section 5.1).

Many early studies of protein structure examined compact **globular proteins** that formed crystals suitable for analysis. **Fibrous proteins,** in contrast, are usually highly elongated (examples of these are presented in Section 5.3). The proteins shown in Figure 4.1 are all globular proteins. To make sense of structures like these, researchers use different approaches. For example, the tertiary structure of the well-studied triose phosphate isomerase can be represented as atoms in space-filling form, as lines that connect Cα of each residue, or as ribbons that highlight secondary structures (**Fig. 4.9**). Different rendering styles convey different types of information, but keep in mind that *proteins, like all molecules, are solid objects with no empty space inside.*

Well-ordered globular proteins can be classified by their common features, such as the arrangement of helices and sheets. For example, four classes are recognized by the CATH system (the name refers to a hierarchy of organizational levels: Class, Architecture, Topology, and Homology). Ordered proteins may contain mostly α structure, mostly β structure, a combination of α and β, or very few regular secondary structural elements. Examples of each class are shown in **Figure 4.10**.

Globular proteins have a hydrophobic core

Globular protein shapes typically contain at least two layers of a secondary structure. This means that the protein has definite surface and core regions. On the protein's surface, some backbone and side-chain groups are exposed to the aqueous environment; in the core, these groups are

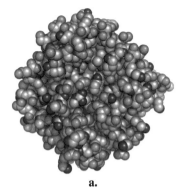

a.

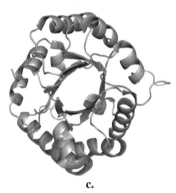

b.

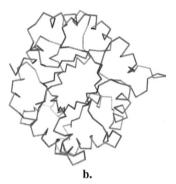

c.

Figure 4.9 Triose phosphate isomerase. a. Space-filling model. All atoms (except H) are shown (C gray, O red, N blue). **b.** Polypeptide backbone. The trace connects the α carbons of successive amino acid residues. **c.** Ribbon diagram. The ribbon represents the overall conformation of the backbone.

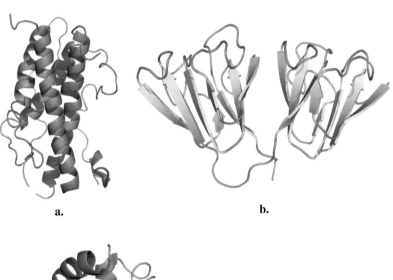

a.   b.

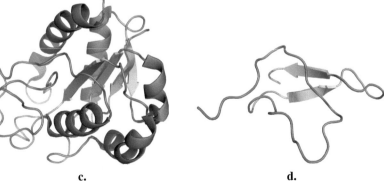

c.   d.

Figure 4.10 Classes of protein structure.   In each protein, α helices are colored red, and β strands are colored yellow. **a.** Growth hormone, an all-α protein. **b.** β/γ-Crystallin, an all-β protein. **c.** Flavodoxin, an α/β protein. **d.** Tachystatin, a protein with little secondary structure.

Figure 4.11 A two-domain protein.   In this model of glyceraldehyde-3-phosphate dehydrogenase, the small domain is red and the large domain is green.

Question Which of the proteins in Figure 4.10 consists of two domains?

Figure 4.12 Hydrophobic and hydrophilic residues in myoglobin.   Hydrophobic side chains belonging to Ala, Ile, Leu, Met, Phe, Pro, Trp, and Val (green) mostly cluster in the protein interior and polar and charged side chains (purple) predominate on the protein surface. Backbone atoms are gray.

sequestered from water. In other words, *the protein comprises a hydrophilic surface and a hydrophobic core.*

A polypeptide segment that has folded into a single structural unit with a hydrophobic interior is often called a **domain.** Some small proteins consist of a single domain. Larger proteins may contain several domains, which may be structurally similar or dissimilar (**Fig. 4.11**).

The core of a domain or a small protein is typically rich in a regular secondary structure. This is because the formation of α helices and β sheets, which are internally hydrogen-bonded, minimizes the hydrophilicity of the polar backbone groups. Irregular secondary structures (loops) are more often found on the solvent-exposed surface of the domain or protein, where the polar backbone groups can form hydrogen bonds with water molecules.

The requirement for a hydrophobic core and a hydrophilic surface also places constraints on amino acid sequences. The location of a particular side chain in a protein's tertiary structure is related to its hydrophobicity: *the more hydrophobic the residue, the more likely it is to be located in the protein interior.* For example, highly hydrophobic residues such as Phe and Met are almost always buried. In the protein interior, side chains pack together, leaving essentially no empty space or space that could be occupied by a water molecule. (**Fig. 4.12**). Polar and charged groups that are positioned in the nonpolar interior must be "neutralized" by interacting with other polar or charged groups.

Protein structures are stabilized mainly by the hydrophobic effect

Surprisingly, the fully folded conformation of a typical globular protein is only marginally more stable than its unfolded form. The difference in thermodynamic stability amounts to about $0.4 \text{ kJ} \cdot \text{mol}^{-1}$ per amino acid, or about $40 \text{ kJ} \cdot \text{mol}^{-1}$ for a 100-residue polypeptide. This is equivalent to the amount of free energy required to break just two hydrogen bonds (about $20 \text{ kJ} \cdot \text{mol}^{-1}$ each). This quantity seems incredibly small—considering the number of potential interactions among all of a protein's backbone and side-chain atoms—yet many proteins do fold into a stable three-dimensional arrangement of atoms.

The largest force governing protein structure is the hydrophobic effect (introduced in Section 2.2), which causes nonpolar groups to aggregate in order to minimize their contact with water. This partitioning is not due to any strong attractive force between hydrophobic groups. Rather, the hydrophobic effect is driven by the increase in entropy of the solvent water molecules, which would otherwise have to order themselves around each hydrophobic group. As we have seen, hydrophobic side chains are located predominantly in the interior of a protein. This arrangement stabilizes the folded polypeptide backbone, since unfolding it or extending it would expose the hydrophobic side chains to the solvent (**Fig. 4.13**). In the core of the protein, the hydrophobic groups experience only very weak attractive forces (van der Waals interactions); they stay in place mainly due to the hydrophobic effect.

Hydrogen bonding by itself is not a major determinant of protein stability, because in an unfolded protein, polar groups could just as easily form energetically equivalent hydrogen bonds with water molecules. Instead, hydrogen bonding—such as occurs in the formation of α helices or β sheets—may help the protein to fine-tune a folded conformation that is already largely stabilized by the hydrophobic effect.

Folded   Unfolded

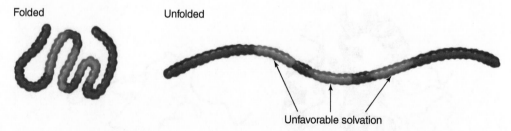

Unfavorable solvation

Figure 4.13 The hydrophobic effect in protein folding.   In a folded protein, hydrophobic regions (represented by green segments of the polypeptide chain) are sequestered in the protein interior. Unfolding the protein exposes these segments to water. This arrangement is energetically unfavorable, because the presence of the hydrophobic groups interrupts the hydrogen-bonded network of water molecules.

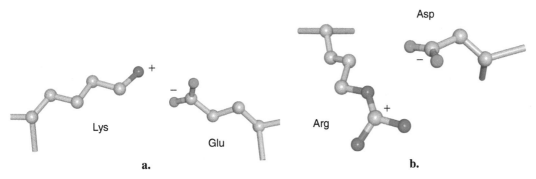

Figure 4.14 Examples of ion pairs in myoglobin. **a.** The ε-amino group of Lys 77 interacts with the carboxylate group of Glu 18. **b.** The carboxylate group of Asp 60 interacts with Arg 45. The atoms are color-coded: C gray, N blue, and O red. Note that these intramolecular interactions occur between side chains that are near each other in the protein's tertiary structure but are far apart in the primary structure.

Question **Identify amino acid side chains that could form two other types of ion pairs.**

Once folded, many polypeptides appear to maintain their shapes through various other types of interactions, the most common being ion pairs, interactions with zinc ions, and covalent cross-links such as disulfide bonds.

Ion pairs are electrostatic interactions between charged groups, which may be amino acid side chains or the N- and C-terminal groups of a polypeptide (**Fig. 4.14**). Although the resulting interaction is strong, it does not contribute much to protein stability. This is because the favorable free energy of the electrostatic interaction is offset by the loss of entropy when the charged groups become fixed in the ion pair. For a buried ion pair, there is the additional energetic cost of dehydrating the charged groups in order for them to enter the hydrophobic core.

Domains known as **zinc fingers** are common in DNA-binding proteins. These structures consist of 20–60 residues with one or two Zn^{2+} ions. The Zn^{2+} ions are tetrahedrally coordinated by the side chains of cysteine and/or histidine and sometimes aspartate or glutamate (**Fig. 4.15**). Protein domains of this size are too small to assume a stable tertiary structure without a metal ion cross-link. Zinc is an ideal ion for stabilizing proteins: It can interact with ligands (S, N, or O) provided by several amino acids, and it has only one oxidation state (unlike Cu or Fe ions, which readily undergo oxidation–reduction reactions under cellular conditions).

Disulfide bonds (a type of covalent bond, shown in Section 4.1) can form within and between polypeptide chains. Experiments show that even when the cysteine residues of certain proteins are chemically blocked, the proteins may still fold and function normally. This suggests that disulfide bonds are not essential for stabilizing these proteins. In fact, disulfides are rare in intracellular proteins, since the cytoplasm is a reducing environment. They are more plentiful in proteins that are secreted to an extracellular (oxidizing) environment (**Fig. 4.16**). Here, the bonds may help prevent protein unfolding under relatively harsh extracellular conditions.

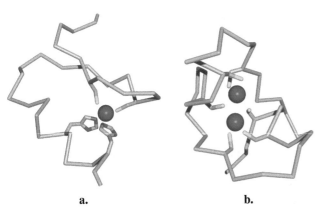

Figure 4.15 Zinc fingers. **a.** A zinc finger with one Zn^{2+} (purple sphere) coordinated by two Cys and two His residues, from *Xenopus* transcription factor IIIA. **b.** A zinc finger with two Zn^{2+} coordinated by six Cys residues, from the yeast transcription factor GAL4.

Figure 4.16 Disulfide bonds in lysozyme, an extracellular protein. This 129-residue enzyme from hen egg white contains eight Cys residues (yellow), which form four disulfide bonds that link different sites on the polypeptide backbone.

Box 4.C   Thioester Bonds as Spring-Loaded Traps

Thioester bonds (Table 1.1) appear in some small biological molecules such as acetyl-CoA (Section 12.3) as well as in some proteins. For example, some enzymes briefly form thioesters during the transformation of reactants to products. The small protein ubiquitin temporarily attaches to a protein called E2 via a thioester linkage before making a covalent cross-link with the lysine side chain of another protein (Section 12.1). Thioesters also appear as internal cross-links in several specialized proteins that snap open and grab on to other molecules.

These internal thioester bonds appear to form spontaneously when polypeptide folding brings a Cys side chain close to a Gln side chain.

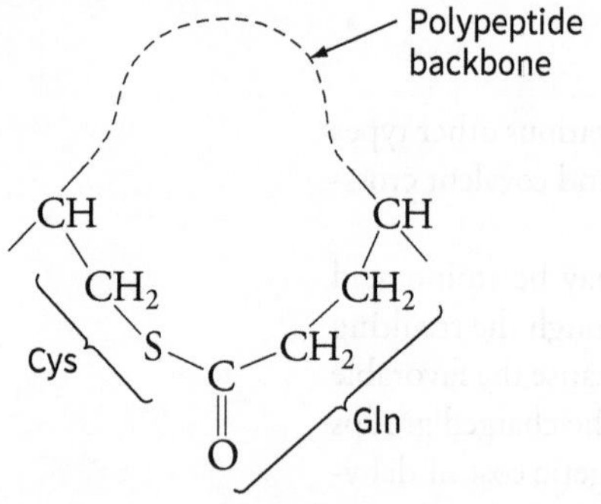

Polypeptide backbone

The buried thioester groups are inaccessible to water and other nucleophiles that readily attack the electron-poor carbon atom. Breaking a thioester bond is a highly favorable reaction; it releases

about the same amount of free energy as removing a phosphate group from ATP. Thioesters are more reactive than conventional esters because the electron delocalization (resonance) involving the large S atom is not as efficient as the C—O resonance that stabilizes oxygen esters.

Proteins that contain internal thioesters must undergo a conformational change that exposes the reactive group to hydroxyl or amino groups, which rapidly react to form covalent adducts:

In this way, the thioester-containing protein can latch on to something else. For example, the complement proteins C3 and C4, which help the immune system tag and eliminate pathogens (disease-causing organisms), become activated, undergo a conformational change, and quickly react with the amino groups of proteins and the hydroxyl groups of cell-wall carbohydrates on the pathogen's surface. In C3 and C4, the Cys and Gln residues that form the thioester are separated by only two other amino acids. Consequently, springing the thioester "trap" may also be facilitated by physical strain in the polypeptide when the protein changes conformation. Some pathogens also position proteins with internal thioesters at the end of long surface extensions; when they contact a prospective host cell, the protein conformation changes, allowing the reactive thioester to form a covalent bond with the target, facilitating infection.

Other covalent cross-links include **thioester bonds,** which appear in a few proteins (**Box 4.C**), and **isopeptide bonds,** in which a carboxyl group on a side chain or at the C-terminus condenses with an amino group on a side chain or at the N-terminus. A Lys–Glu isopeptide bond is shown here:

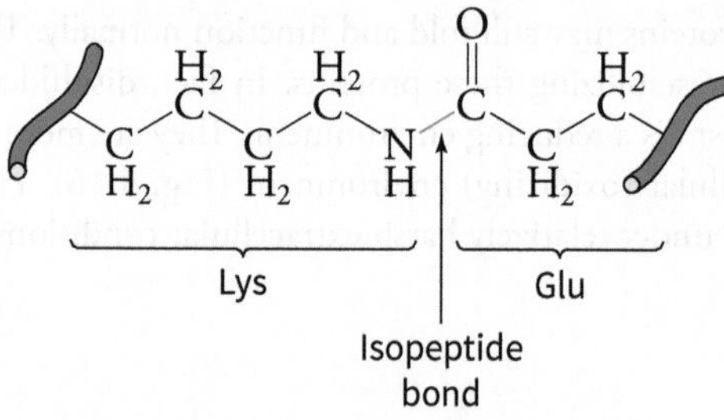

Protein folding is a dynamic process

How do newly made proteins assume their proper conformation in the crowded environment of the cell? A polypeptide begins to fold as soon as it emerges from the ribosome, so part of the chain may adopt its mature tertiary structure before the entire chain has been synthesized.

It is difficult to monitor this process in the cell, so studies of protein folding *in vitro* usually rely on small globular proteins that have been chemically unfolded (**denatured**) and then allowed to refold (**renature**). In the laboratory, proteins can be denatured by adding highly soluble substances such as salts or urea (NH_2—CO—NH_2). Large amounts of these solutes interfere with the structure of the solvent water, thereby attenuating the hydrophobic effect and causing the proteins to unfold. When the solutes are removed, the proteins renature. Irreversible protein denaturation is a key part of cooking all sorts of foods, including baked goods (**Box 4.D**).

Although denaturation–renaturation experiments are difficult or impossible for large proteins, biochemists assume that a newly synthesized polypeptide follows some sort of folding pathway

Box 4.D   Baking and Gluten Denaturation

A familiar example of protein denaturation is the structural transformation of an egg white—from a clear viscous liquid to an opaque solid—when heated. Ovalbumin and other proteins in the egg white unfold during cooking and aggregate, irreversibly in this case. Thermal denaturation also explains what happens to bread dough in the oven.

Contrary to popular belief, raw wheat flour does not contain any protein called gluten. Rather, it contains proteins called glutenins and gliadins. Glutenins form extended coiled shapes and are not highly soluble, due to an abundance of uncharged side chains. When added to water to make bread dough and then kneaded, the glutenins stretch into sheets and trap the smaller globular gliadins. This network of proteins is known as gluten. The glutenins make gluten elastic, while the interspersed gliadins prevent it from becoming too solid. During the dough-rising phase, the sheets of gluten capture bubbles of CO_2 gas generated by yeast or by chemical agents such as sodium bicarbonate (baking soda). When the dough is cooked, the gas bubbles expand further. The heat also drives out water (and any ethanol produced by the yeast) and fully denatures the gluten, causing it to stiffen. The result is a spongy and chewy loaf of bread.

Cake batter also contains flour but differs from bread dough by the presence of additional fats (lipids), often introduced in the form of butter, vegetable oil, or eggs (although egg whites are largely fat-free, the yolks are not). The hydrophobic lipids bind to nonpolar segments of glutenins and gliadins to inhibit their intermolecular interactions. Moreover, after the flour is added, cake batter is only minimally mixed and not kneaded, so gluten formation is less extensive than in bread dough. Baking denatures the proteins and solidifies the limited gluten network, yielding a fluffy, crumbly cake rather than a chewy loaf.

---

in order to reach its most stable conformation or **native structure.** Contrary to expectations, secondary structures such as α helices and β sheets do not form immediately. Instead, the first stages of protein folding can be described as a "hydrophobic collapse," when nonpolar groups are forced out of contact with water. Once the hydrophobic core of the protein begins to take shape, the hydrogen bonds responsible for regular secondary structure can form, and other structures and side chains jostle themselves into their final positions in the tertiary structure (**Fig. 4.17**). Different proteins follow different folding pathways, and even a single polypeptide chain may achieve its native structure in slightly different ways.

In a cell, the protein folding process may be aided by additional proteins known as **molecular chaperones.** Some of these associate with the ribosome to assist with the initial stages of folding (described in more detail in Section 22.4). Other chaperones escort newly made proteins as they move to other parts of the cell or undergo **post-translational processing.** Depending on the protein, this might mean removal of some amino acid residues or the covalent attachment of another group such as a lipid, carbohydrate, or phosphate group (**Fig. 4.18**). The attached groups usually have a discrete biological function and may also help to stabilize the folded conformation of the protein. Metal ions or small organic molecules may also associate with the protein. Some proteins contain several polypeptide chains, which fold individually before assembling.

All the information required for a protein to fold is contained in its amino acid sequence. Unfortunately, there are no completely reliable methods for predicting how a polypeptide chain will fold. In fact, it is difficult to determine whether a relatively short amino acid sequence will form an α helix, β sheet, or no particular structure at all. This uncertainty presents a formidable obstacle to assigning three-dimensional structures—and possible functions—to the burgeoning number of proteins identified through genome sequencing (see Section 3.4).

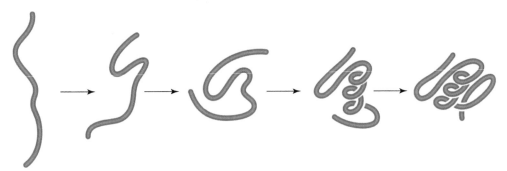

Figure 4.17 A protein folding pathway.   In this diagram for a hypothetical globular protein, the polypeptide assumes a progressively more ordered conformation.

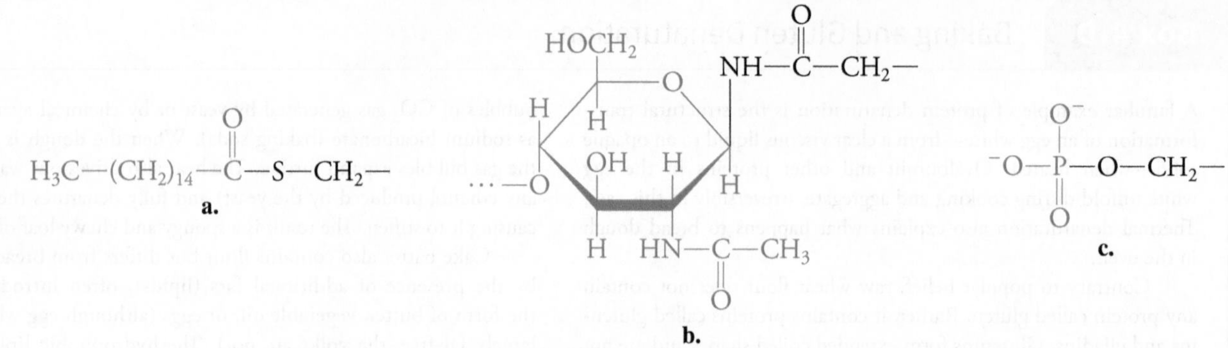

$$H_3C-(CH_2)_{14}-\overset{\overset{\displaystyle O}{\|}}{C}-S-CH_2-$$

a.

b.

$$^-O-\overset{\overset{\displaystyle O}{\|}}{\underset{\underset{\displaystyle O}{\|}}{P}}-O-CH_2-$$

c.

Figure 4.18 Some covalent modifications of proteins. a. A 16-carbon fatty acid (palmitate, in red) is linked by a thioester bond to a Cys residue. **b.** A chain of several carbohydrate units (here only one sugar residue is shown, in red) is linked to the amide N of an Asn side chain. **c.** A phosphoryl group (red) is esterified to a Ser side chain.

Disorder is a feature of many proteins

Traditional views of protein structure—largely built on experiments with small globular proteins—hold that each protein has just one stable, low-energy conformation. However, a few proteins have been captured in different conformations in crystals, and many crystallized proteins include short mobile segments. A few proteins appear to entirely lack any secondary or tertiary structure yet are able to perform distinct functions. Most tellingly, analyses of thousands of sequences obtained from genomic data suggest that many polypeptide chains do not include enough non-polar residues to fold into a conventional globular shape with a hydrophobic core. In the modern view of protein structure, globular proteins with only one conformation are simply one end of a continuum of the structural options for polypeptide chains.

Of course, no protein exists as a completely rigid block. *All proteins are inherently flexible,* because individual bonds in the polypeptide chain can rotate, bend, and stretch, and secondary and tertiary structures are stabilized by relatively weak noncovalent forces. In addition to the small, unavoidable, and random fluctuations in their conformations, many proteins must undergo other types of structural changes in order to carry out their biological functions. For example, to catalyze a chemical reaction, an enzyme must closely contact the reacting substances while they are being transformed to reaction products. When an extracellular signaling molecule binds to a cell-surface receptor protein, the protein's conformation changes in a way that triggers additional events inside the cell. Such conformational adjustments might involve repositioning a few side chains or a loop of polypeptide chain, or they might require more dramatic shifts in the overall arrangement of secondary structures or entire domains.

The absence of conformational order in proteins can take different forms, with a range of functional implications. For example, some proteins are prone to denaturation, requiring periodic visits to chaperones to restore their functional conformations. So-called chameleonic sequences of up to 20 residues may form an α helix in one protein but a β sheet in another. Although all proteins fold in a way that allows them to proceed from an unstable, high-energy state to a lower-energy state, the energetic landscape may not always resemble a simple funnel (**Fig. 4.19**). Conventional or **monomorphic proteins** have a single stable tertiary structure, whereas **metamorphic proteins** may have two or more possible conformations. **Intrinsically disordered proteins** have no fixed structure at all.

Figure 4.19 Energy diagrams for proteins.   Monomorphic proteins have a single low-energy conformation; metamorphic proteins can adopt multiple shapes that have roughly equivalent energy. Intrinsically disordered proteins are not energetically constrained to occupy any particular structural space. These are idealized diagrams, and proteins may include multiple segments with different degrees of disorder.

Question **Which sides of this diagram correspond to low entropy and high entropy?**

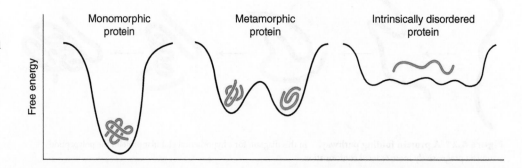

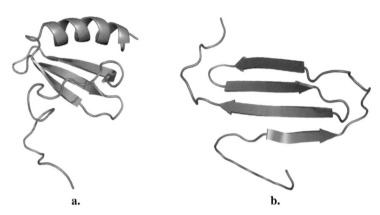

a.   b.

Figure 4.20 A metamorphic protein with two stable conformations.
a. One form of the protein lymphotactin (also called XCL1) consists of a three-stranded β sheet and an α helix. **b.** An alternate form with an all-β structure interconverts rapidly with the first form.

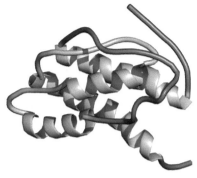

Figure 4.21 A protein with intrinsically disordered regions. The signaling molecule interleukin-21 includes four ordered helices (light blue) and disordered segments of polypeptide (magenta) at the N-terminus, C-terminus, and an internal loop. The disordered segments can adopt different positions; just one possible protein structure is depicted here.

Metamorphic proteins switch easily between energetically equivalent conformations, typically about once per second (**Fig. 4.20**). Different forms of these "transformer" proteins are in dynamic equilibrium, but the balance can shift in response to changes in cellular pH, oxidation or reduction of disulfide groups, the addition of phosphoryl groups, the presence of ions such as Mg^{2+}, or binding to other molecules. The alternate structures may exhibit different degrees of functionality (for example, an enzyme that is more active or less active in a certain conformation) or they may have entirely different functions (some dual-function proteins are known as "moonlighting" proteins).

Many proteins include disordered segments interspersed with more defined globular regions. An estimated 44% of the human proteome contains disordered segments of over 30 residues (the **proteome** is the set of all proteins made by an organism). These intrinsically disordered regions, which may contain as many as several hundred amino acids, have no fixed secondary or tertiary structure (**Fig. 4.21**). As expected, they are rich in amino acids bearing polar and charged side chains.

Protein functions may depend on disordered regions

What can disordered protein segments do? Some function as linkers or spacers, such as the "hinge" regions that connect domains in antibody proteins (Section 5.5). A few disordered proteins actually perform like molecular springs. Disordered segments commonly appear in proteins that participate in the **biomineralization** of hard structures such as bones and teeth. Here, the flexible proteins wrap around and solubilize calcium phosphate, which prevents the crystals from precipitating or forming large brittle aggregates. However, the functions of most disordered protein regions are related to binding other proteins.

Although it might be tempting to picture intrinsically disordered protein regions as long, extended polypeptide segments, the chains may also transiently adopt more compact structures with well-defined secondary structures, as if "trying out" different conformations. One of these conformations may be suitable for interacting with another molecule, and that conformation may then become locked in place when the two molecules bind (in this respect, the proteins resemble metamorphic proteins). In some cases, a disordered region allows a protein to interact with many different potential binding partners, each of which requires a slightly different conformation for a productive binding interaction (**Fig. 4.22**). Hence, the disorder itself is key to the protein's function.

An intrinsically disordered protein region may retain some flexibility, even after binding another molecule. If the protein functions as scaffolding for assembling a large multiprotein complex, it can adjust its conformation to accommodate each new addition. Intrinsic disorder also means that the complex can adopt more than one structural state. Rather than existing in a fully bound or fully free state,

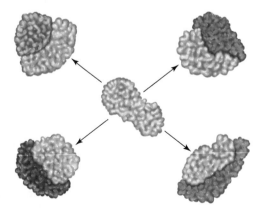

Figure 4.22 Schematic view of an intrinsically disordered protein interacting with other proteins. In this diagram, the protein at the center can bind only one other protein at a time; some intrinsically disordered proteins can bind multiple molecules simultaneously.

one of these "fuzzy complexes" can exhibit multiple behaviors consistent with different degrees of partial binding. Presumably, such options give cells more ways to fine-tune the functions of proteins.

Finally, one newly recognized function of disordered proteins is to aggregate in a tangled network that undergoes liquid-liquid phase separation, generating a droplet with a more gel-like consistency than the surrounding fluid. The protein-rich droplet is sometimes called a **membraneless organelle** (because conventional organelles are surrounded by a lipid bilayer; Section 8.4). The droplet can harbor additional proteins and other cellular components such as RNA. Membraneless organelles, which may occur in the cytosol or the nucleus, appear to function primarily in organizing and concentrating molecules in order to maximize the efficiency of certain cellular activities, such as processing RNA.

Concept Check

1. Explain why a globular protein has a hydrophilic surface and a hydrophobic core.
2. Name some residues that are likely to be located on the protein surface and some residues that are likely to be located in its core.
3. List the covalent and noncovalent forces that can stabilize protein structures.
4. Describe what happens as a polypeptide chain folds.
5. Explain why proteins are not rigid.
6. Distinguish monomorphic, metamorphic, and intrinsically disordered proteins and propose some possible functions for each type.
7. Explain why biochemists know more about globular proteins than disordered proteins.

4.4 Quaternary Structure

KEY CONCEPTS

List the advantages of having a protein with quaternary structure.

- Recognize quaternary structure in proteins.
- Explain why large proteins almost always have quaternary structure.

Most small proteins consist of a single polypeptide chain, but most proteins with masses greater than 100 kD contain multiple chains. The individual chains, called **subunits,** may all be identical, in which case the protein is known as a **homodimer, homotrimer, homotetramer,** and so on (*homo-* means "the same"). If the chains are not all identical, the prefix *hetero-* ("different") is used. The spatial arrangement of these polypeptides is known as the protein's quaternary structure.

The forces that hold subunits together are similar to those that determine the tertiary structures of the individual polypeptides. That is, the hydrophobic effect is mostly responsible for maintaining quaternary structure. Accordingly, the interface (the area of contact) between two subunits is mostly nonpolar, with closely packed side chains. Hydrogen bonds, ion pairs, and disulfide bonds contribute to a lesser extent but help dictate the exact geometry of the interacting subunits.

Among the most common quaternary structures in proteins are symmetrical arrangements of two or more identical subunits (**Fig. 4.23**). Even in proteins with nonidentical subunits, the symmetry is based on groups of subunits. For example, hemoglobin, a heterotetramer with two αβ units, can be considered to be a dimer of dimers (see Fig. 4.23c). Some metamorphic proteins can form different quaternary structures. The subunits apparently dissociate, change their tertiary structure, then reassociate, sometime forming a complex with a different number of subunits.

The advantages of a multisubunit protein structure are many. For starters, extremely large proteins can be constructed by the incremental addition of small building blocks that are encoded by a single gene. For example, the capsid that serves as a protective shell for a virus (see Box 3.C) self-assembles from multiple copies of a few proteins. Herpesviruses are an extreme case, where about 3000 individual proteins snap together to form an almost spherical capsid. Modular construction also makes sense for certain structural proteins that—due to their enormous

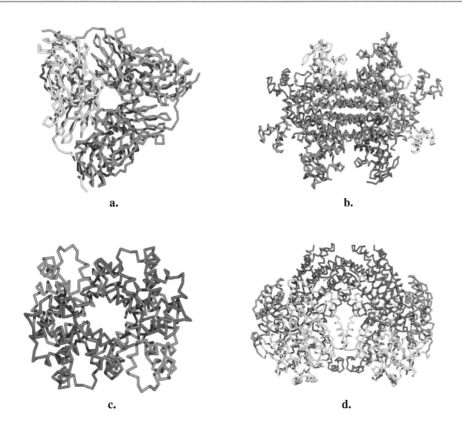

Figure 4.23 Some proteins with quaternary structure. The alpha carbon backbone of each polypeptide is shown. **a.** Nitrite reductase, an enzyme with three identical subunits, from *Alcaligenes*. **b.** *E. coli* fumarase, a homotetrameric enzyme. **c.** Human hemoglobin, a heterotetramer with two α subunits (blue) and two β subunits (red). **d.** Bacterial methane hydroxylase, whose two halves (*right* and *left* in this image) each contain three kinds of subunits.

Question **Which of the proteins shown in Figure 4.1 has quaternary structure?**

size—cannot be synthesized all in one piece or must be assembled outside the cell. Moreover, the impact of the inevitable errors in transcription and translation can be minimized if the affected polypeptide is small and readily replaced.

Finally, the interaction between subunits in a multisubunit protein affords an opportunity for the subunits to influence each other's behavior or work cooperatively. The result is a way of regulating function that is not possible in single-subunit proteins or in multisubunit proteins whose subunits each operate independently. In Chapter 5, we will examine the cooperative behavior of hemoglobin, which has four interacting oxygen-binding sites.

Concept Check

1. Explain how to tell whether a protein has quaternary structure.
2. Explain why large proteins almost always have quaternary structure.

4.5 Clinical Connection: Protein Misfolding and Disease

KEY CONCEPTS

Identify the common features of amyloid diseases.

- List the possible fates of a misfolded protein.
- Describe the overall structure of an amyloid fibril.

In a cell, a protein may occasionally fail to assume its proper tertiary or quaternary structure. Sometimes this is due to a genetic mutation, such as the substitution of one amino acid for another, that makes it difficult or impossible for the protein to fold. Regardless of the cause, the result may be disastrous for the cell. In fact, a variety of human diseases have been linked to the presence of misfolded proteins.

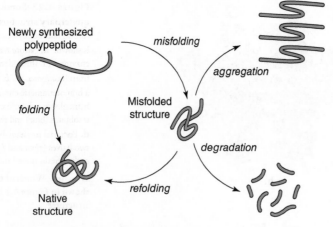

Newly synthesized polypeptide

misfolding

aggregation

folding

Misfolded structure

degradation

Native structure

refolding

Figure 4.24 The fates of a misfolded protein.

Figure 4.25 Brain tissue from Alzheimer's disease. Amyloid deposits (large red areas) and intracellular tangles (smaller dark shapes) form in parts of the brain responsible for memory and cognition.

Dennis Selkoe and Marcia Podlisny, Harvard University Medical School

Normally, chaperones help a misfolded protein recover its native conformation. If the protein cannot be salvaged in this way, it is usually degraded to its component amino acids (**Fig. 4.24**). The operation of this quality control system could explain why some mutated proteins, which are synthesized at a normal rate but fold incorrectly, never reach their intended cellular destinations.

In some diseases, misfolded proteins aggregate to form long insoluble fibers. Although the fibers can occur throughout the body, their appearance in the brain is a feature of the deadliest disorders, including Alzheimer's disease, Parkinson's disease, and the transmissible spongiform encephalopathies. The aggregated proteins—a different type in each disease—are commonly called **amyloid deposits** (a name originally referring to their starch-like appearance).

At one time, it seemed obvious that fibrous deposits within and between neurons (nerve cells) caused neurological abnormalities and eventually cell death. This made sense because neurons are long-lived cells and the damage could accumulate slowly over time (the diseases mentioned above develop over many years). However, neurodegeneration and symptoms such as memory loss seem to begin before protein aggregates are detectable. Furthermore, almost any protein can be coaxed into forming fibers under the right conditions. These observations seem to rule out a straightforward cause-and-effect scenario. Some evidence suggests that the fibers themselves might not cause disease but instead provide a way to safely detoxify and tuck away misbehaving proteins, particularly when the cell's normal protein-disposal system is overwhelmed. The real mystery may not be what the fibers do but why they form in the first place.

Alzheimer's disease, the most common neurodegenerative disease, is accompanied by both extracellular "plaques" and intracellular "tangles" in brain tissue (**Fig. 4.25**). The extracellular amyloid material consists primarily of a protein called amyloid-β, which is a 40- or 42-residue fragment of the much larger amyloid precursor protein located in the cell membrane. Normal brain tissue contains some extracellular amyloid-β, but neither its function nor the function of its precursor protein is completely understood.

Curiously, amyloid-β exists as monomers, dimers, and larger insoluble aggregates in equilibrium. Each form behaves differently, but the most toxic form appears to be the dimer. A variety of evidence suggests that amyloid-β might interfere with the operation of membrane transport proteins that take up glutamate (which acts as a signaling molecule between neurons), or it might directly affect membrane integrity in the neurons or the cells that support them. At some point, amyloid-β chains, which are relatively hydrophobic, form extended β strands that stack next to each to generate long fibrils (fine fibers; **Fig. 4.26**).

The intracellular protein known as Tau also forms amyloid fibers in Alzheimer's disease. Normally, Tau is involved in the assembly of microtubules, a component of the cytoskeleton (Section 5.3). On its own, Tau is an intrinsically disordered protein and for this reason may be prone to aggregate. However, Tau fibrils are not merely an alternate protein conformation, because their formation is irreversible. (Once formed, amyloid fibers tend to be resistant to degradation by cellular enzymes.) Like amyloid-β, a segment of the Tau protein adopts a mostly β secondary structure so that strands align perpendicularly to the fibril axis (**Fig. 4.27**). Tau fibers accompany

Gremer et al. 2017/AAAS

Figure 4.26 Model of an amyloid-β fibril containing two filaments. Each 7-nm-diameter filament consists of stacked β strands (flat arrows), with each 42-residue amyloid-β peptide corresponding to one "rung" of the filament.

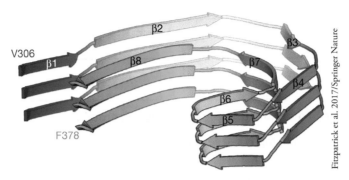

Figure 4.27 Model of the filament-forming segments of three Tau peptides.   The eight β strands of each 72-residue segment stack next to the β strands of adjacent Tau polypeptides.

a number of neurogenerative diseases other than Alzheimer's disease, including chronic traumatic encephalopathy from repeated head impact injuries sustained by soldiers in combat or by players of high-contact sports. Different disorders feature Tau fibers with slightly different structures, consistent with chemical modification of the protein, such as phosphorylation. The toxicity of Tau or its fibrils is not understood, but as in other amyloid diseases, symptoms may appear before the fibers are detectable.

In Parkinson's disease, cells in one portion of the brain accumulate fragments of a protein known as α-synuclein, which appears to play a role in neurotransmission. α-Synuclein is a small (140-residue) intrinsically disordered protein that seems to adopt relatively compact structures under cellular conditions. Like other amyloid proteins, it shifts to a β conformation—possibly in response to some intra- or extracellular event—when it aggregates. The accumulation of α-synuclein deposits (called Lewy bodies) is associated with the death of neurons, leading to the tremor, muscular rigidly, and slow movements that are symptoms of Parkinson's disease.

The **transmissible spongiform encephalopathies (TSEs)** are a group of rare fatal disorders in which the brain develops a spongy appearance. Once thought to be caused by a virus, the TSEs are actually triggered by an infectious protein called a **prion.** Interestingly, normal human brain tissue contains the same 253–amino acid protein, named PrP^C (C for cellular), which occurs on neural cell membranes and seems to be required for normal brain function. PrP^C has mostly α secondary structure. Introduction of the prion version of the protein, which has mostly β structure, causes PrP^C to switch to the same β conformation and aggregate. This generates a stable fiber of hydrogen-bonded β strands that resembles other amyloid fibers. Neurodegeneration may eventually result from the loss of functional PrP^C as it misfolds, or from the toxic effects of the amyloid fibers.

Prion diseases can be transmitted by ingesting foods derived from infected animals. Somehow, the prion protein must be absorbed, without being digested, and transported to the central nervous system. Intriguingly, there is evidence that Parkinson's disease may be triggered by changes in the intestinal microbiota (see Section 1.4), suggesting that signs of amyloid diseases might appear in the digestive system before the brain is affected. Early diagnosis would be helpful, because by the time neurological symptoms are obvious, there is no effective treatment.

Concept Check

1. List some diseases that are associated with the presence of amyloid fibers.
2. Summarize the evidence suggesting that amyloid fibers cause disease and the evidence suggesting that amyloid fibers are a by-product of disease.
3. Describe the common structural features of the proteins that form fibrils.

4.6 Tools and Techniques: Analyzing Protein Structure

KEY CONCEPTS

Describe the techniques for purifying and analyzing proteins.

- Explain how chromatography can separate molecules on the basis of size, charge, or specific binding behavior.
- Determine the isoelectric points of amino acids.
- Summarize how the sequence of a polypeptide can be determined by mass spectrometry.
- Describe how the three-dimensional arrangement of atoms in a protein can be deduced by analyzing nuclear magnetic resonance or by measuring the diffraction of X-rays or electrons.

Like nucleic acids (Section 3.5), proteins can be purified and analyzed in the laboratory. In this section, we examine some commonly used methods for isolating proteins, determining their sequence of amino acids, and visualizing their three-dimensional structures.

Chromatography takes advantage of a polypeptide's unique properties

As described in Section 4.1, a protein's amino acid sequence determines its overall chemical characteristics, including its size, shape, charge, and ability to interact with other substances. A variety of laboratory techniques have been devised to exploit these features in order to separate proteins from other cellular components. Keep in mind that the exact charge and conformation of individual molecules vary slightly in a population of otherwise identical molecules, because acid–base groups exist in equilibrium between protonated and unprotonated forms and bonded atoms have some rotational freedom. Consequently, laboratory techniques assess the average chemical and physical properties of the population of molecules.

One of the most powerful techniques is **chromatography.** Originally performed with solvents moving across paper, chromatography now typically uses a column packed with a porous matrix (the stationary phase) and a buffered solution (the mobile phase) that percolates through the column. *Proteins or other solutes pass through the column at different rates, depending on how they interact with the stationary phase.*

In **size-exclusion chromatography** (also called gel filtration chromatography), the stationary phase consists of tiny beads with pores of a characteristic size. If a solution containing proteins of different sizes is applied to the top of the column, the proteins will move through the column as fluid drips out the bottom. Larger proteins will be excluded from the spaces inside the beads and will pass through the column faster than smaller proteins, which will spend time inside the beads. The proteins gradually become separated and can be recovered by collecting the solution that exits the column (**Fig. 4.28**).

A protein's net charge at a particular pH can be exploited for its purification by **ion exchange chromatography.** In this technique, the solid phase typically consists of beads derivatized with positively charged diethylaminoethyl (DEAE) groups or negatively charged carboxymethyl (CM) groups:

DEAE ◯$-CH_2-CH_2-NH^+\begin{smallmatrix}CH_2-CH_3\\\\CH_2-CH_3\end{smallmatrix}$ **CM** ◯$-CH_2-COO^-$

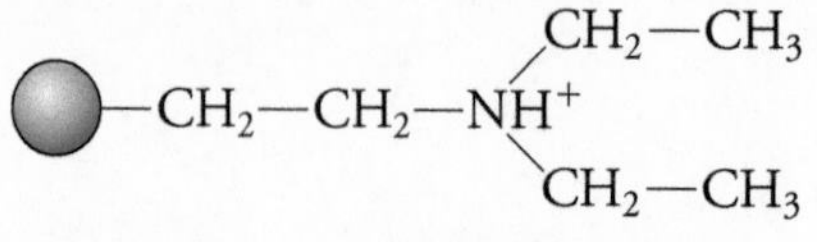

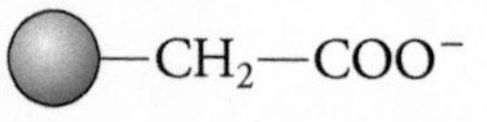

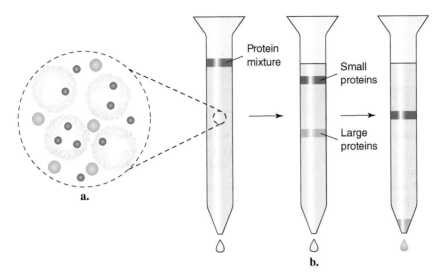

Figure 4.28 Size-exclusion chromatography.   **a.** Small molecules (blue) can enter the spaces inside the porous beads of the stationary phase, while larger molecules (gold) are excluded. **b.** When a mixture of proteins (green) is applied to the top of a size-exclusion column, the large proteins (gold) migrate more quickly than small proteins (blue) through the column and are recovered by collecting the material that flows out of the bottom of the column. In this way, a mixture of proteins can be separated according to size.

Negatively charged proteins will bind tightly to the DEAE groups, while uncharged and positively charged proteins pass through the column. The bound proteins can then be dislodged by passing a high-salt solution through the column so that the dissolved ions can compete with the protein molecules for binding to DEAE groups (**Fig. 4.29**). Alternatively, the pH of the solvent can be decreased so that the bound protein's anionic groups become protonated, loosening their hold on the DEAE matrix. Similarly, positively charged proteins will bind to CM groups (while uncharged and anionic proteins flow through the column) and can subsequently be dislodged by solutions with a higher salt concentration or a higher pH.

The success of ion exchange can be enhanced by knowing something about the protein's net charge (see Sample Calculation 4.1) or its **isoelectric point, pI,** the pH at which it carries no net charge. For a molecule with two ionizable groups, the pI lies between the pK values of those two groups:

$$pI = {}^1\!/_2\,(pK_1 + pK_2) \qquad\qquad (4.1)$$

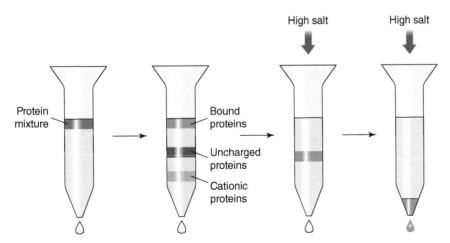

Figure 4.29 Ion exchange chromatography.   When a mixture of proteins is applied to the top of a positively charged anion exchange column (e.g., a DEAE matrix), negatively charged proteins bind to the matrix, while uncharged and cationic proteins flow through the column. The desired protein can be dislodged by applying a high-salt solution (whose anions compete with the protein for binding to the DEAE groups).

Calculating the pI of an amino acid is relatively straightforward (Sample Calculation 4.2). However, a protein may contain many ionizable groups, so although its pI can be estimated from its amino acid composition, its pI is more accurately determined experimentally.

SAMPLE CALCULATION 4.2

Problem   Estimate the isoelectric point of arginine.

Solution   In order for arginine to have no net charge, its α-carboxyl group must be unprotonated (negatively charged), its α-amino group must be unprotonated (neutral), and its side chain must be protonated (positively charged). Because protonation of the

α-amino group or deprotonation of the side chain would change the amino acid's net charge, the pK values of these groups (9.0 and 12.5) should be used with Equation 4.1:

$$pI = \frac{1}{2}(9.0 + 12.5) = 10.75$$

Other binding behaviors can be adapted for chromatographic separations. For example, a small molecule can be immobilized on the chromatographic matrix, and proteins that can specifically bind to that molecule will stick to the column while other substances exit the column without binding. This technique, called **affinity chromatography,** is a particularly powerful separation method because it takes advantage of a protein's unique ability to interact with another molecule, rather than one of its general features such as size or charge. **High-performance liquid chromatography (HPLC)** is the name given to chromatographic separations, often analytical in nature rather than preparative, that are carried out in closed columns under high pressure, with precisely controlled flow rates and automated sample application.

Electrophoresis is a powerful technique to separate charge biomolecules

Proteins are sometimes analyzed or isolated by **electrophoresis,** in which charged molecules move through a gel-like matrix such as polyacrylamide under the influence of an electric field. In sodium dodecyl sulfate polyacrylamide gel electrophoresis **(SDS-PAGE),** both the sample and the gel contain the detergent SDS, which binds to proteins to give them a uniform density of negative charges. When the electric field is applied, the proteins all move toward the positive electrode at a rate depending on their size, with smaller proteins migrating faster than larger ones. After staining, the proteins are visible as bands in the gel **(Fig. 4.30).**

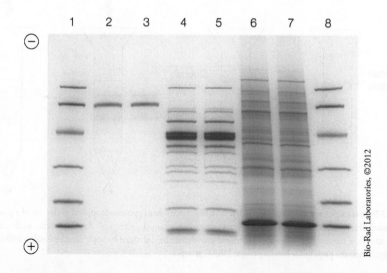

Figure 4.30 SDS-PAGE.
Samples were applied to the top of the gel. Following electrophoresis, the gel was stained with Coomassie blue dye so that each protein is visible as a blue band. Lanes 1 and 8 contain proteins of known molecular mass that serve as standards for estimating the masses of other proteins.

Bio-Rad Laboratories, ©2012

Mass spectrometry reveals amino acid sequences

A standard approach to sequencing a protein has several steps:

1. *The sample of protein to be sequenced is purified* (for example, by chromatographic or other methods) so that it is free of other proteins.

2. If the protein contains more than one kind of polypeptide chain, *the chains are separated* so that each can be individually sequenced. In some cases, this requires breaking (reducing) disulfide bonds.

3. *Large polypeptides must be broken into smaller pieces (<100 residues) that can be individually sequenced.* Cleavage can be accomplished chemically, for example, by treating the polypeptide with cyanogen bromide (CNBr), which cleaves the peptide bond on the C-terminal side of Met residues. Cleavage can also be accomplished with **proteases** (another term for peptidases) that hydrolyze specific peptide bonds. For example, the protease trypsin cleaves the peptide bond on the C-terminal side of the positively charged residues Arg and Lys:

Some commonly used proteases and their preferred cleavage sites are listed in **Table 4.3**.

4. The sequence of each peptide is determined. In a classic procedure known as **Edman degradation**, the N-terminal residue of a peptide is chemically derivatized, cleaved off, and identified; the process is then repeated over and over again so that the peptide's sequence can be deduced, one residue at a time. Alternatively, each peptide can be sequenced by mass spectrometry.

5. To reconstruct the sequence of the intact polypeptide, *a different set of fragments that overlaps the first is generated* so that the two sets of sequenced fragments can be aligned.

Table 4.3 Specificities of Some Proteases

Protease	Residue preceding cleaved peptide bond[a]
Chymotrypsin	Leu, Met, Phe, Trp, Tyr
Elastase	Ala, Gly, Ser, Val
Thermolysin	Ile, Met, Phe, Trp, Tyr, Val
Trypsin	Arg, Lys

[a]Cleavage does not occur if the following residue is Pro.

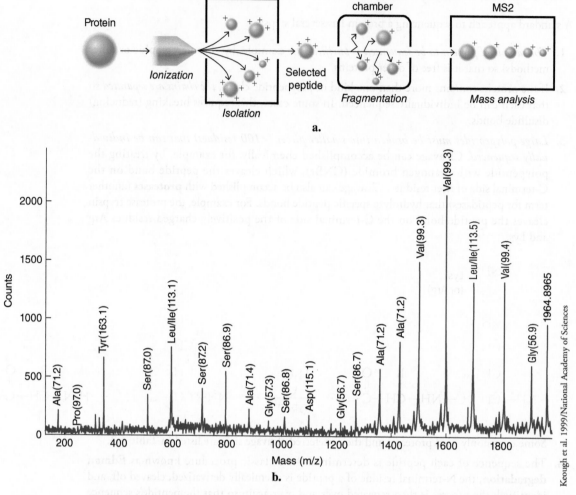

Figure 4.31 Peptide sequencing by mass spectrometry. a. A solution of charged peptides is sprayed into the first mass spectrometer (MS1). One peptide ion is selected to enter the collision chamber to be fragmented. The second mass spectrometer (MS2) then measures the mass-to-charge ratio of the ionic peptide fragments. The peptide sequence is determined by comparing the masses of increasingly larger fragments. **b.** An example of peptide sequencing by mass spectrometry. The difference in mass of each successive peak identifies each residue, allowing the amino acid sequence to be read from right to left.

A more efficient approach for analyzing protein structure measures the sizes of peptide fragments by **mass spectrometry.** In standard mass spectrometry, a solution of the protein is sprayed from a tiny nozzle at high voltage. This yields droplets of positively charged molecular fragments, from which the solvent quickly evaporates. Each gas-phase ion then passes through an electric field. The ions are deflected, with smaller ions deflected more than larger ions, so that they are separated by their mass-to-charge ratios. In this way, the masses of the fragments can be measured and the mass of the intact molecule can be deduced.

Two mass spectrometers in series can be used to determine the sequence of amino acids in a polypeptide. The first instrument sorts the peptide ions so that only one emerges. This species is then allowed to collide with an inert gas, such as helium, which breaks the peptide, usually at a peptide bond. The second mass spectrometer then measures the mass-to-charge ratios of the peptide fragments (**Fig. 4.31**). *Because successively larger fragments all bear the same charge but differ by one amino acid, and because the mass of each of the 20 amino acids is known, the sequence of amino acids in a set of fragments can be deduced.* Although mass spectrometry is not practical for sequencing large polypeptides, even a partial sequence may be valuable (**Box 4.E**). Mass spectrometry can also reveal additional details of protein structure, such as covalently modified amino acid residues.

Box 4.E   Mass Spectrometry Applications

Mass spectrometry has been used for several decades in clinical and forensics laboratories to identify normal metabolic compounds as well as toxins and drugs (both therapeutic and illicit). Instruments that detect traces of explosives at airport security checkpoints also use mass spectrometry, which is rapid, sensitive, and reliable. The analysis of small compounds by mass spectrometry is relatively straightforward. It is much more challenging, however, to analyze large molecules in complex mixtures for the purpose of diagnosing a disease such as cancer or tracking its effects on human tissues.

Fluids such as blood contain so many different proteins, with concentrations ranging over many orders of magnitude, that it is difficult to detect rare proteins against a backdrop of very abundant ones such as serum albumin and immunoglobulins, which together account for about 75% of the plasma proteins. Urinalysis by mass spectrometry holds more promise, as urine normally contains relatively few proteins and none with a mass greater than about 15,000 D. Even so, over 2000 different proteins are detectable in urine.

One common approach for identifying proteins in biological samples is to fractionate the mixture by electrophoresis, then extract the separated proteins from the gel, partially digest them with a protease, and subject them to mass spectrometry. The resulting pattern of peaks, a "fingerprint," can be compared to a database to identify the protein. Even when tandem mass spectrometers are used, it may not be necessary to completely sequence each polypeptide. A partial sequence of just a few amino acids is often sufficient to identify the parent protein. Of course, the availability of complete genomic sequences makes this approach possible, and a number of software programs have been developed to translate mass spectral data into query sequences for searching sequence databases.

Question   **Explain why it is possible that a 5-residue sequence can uniquely identify a protein whose total length is 200 residues.** (*Hint:* **compare the total number of possible pentapeptide sequences to the actual number of pentapeptide fragments in the polypeptide.**)

Protein structures are determined by NMR spectroscopy, X-ray crystallography, and cryo-electron microscopy

Protein structures can be probed by several methods that reveal details at the atomic level. For example, relatively small soluble proteins can be analyzed by **nuclear magnetic resonance (NMR) spectroscopy,** which takes advantage of the ability of atomic nuclei (most commonly, hydrogen) to resonate in an applied magnetic field according to their interactions with nearby atoms. An NMR spectrum consists of numerous peaks that can be analyzed to reveal the distances between two H atoms that are close together in space or are covalently connected through one or two other atoms. These measurements, along with information about the protein's amino acid sequence, are used to construct a three-dimensional model of the protein. Due to the inherent imprecision of the measurements, NMR spectroscopy typically yields a set of closely related structures, which may convey a sense of the protein's natural conformational flexibility (**Fig. 4.32**).

Figure 4.32 The NMR structure of glutaredoxin. Twenty structures for this 85-residue protein, all compatible with the NMR data, are shown as α-carbon traces, colored from the N-terminus (blue) to the C-terminus (red).

About 90% of known protein structures have been determined by **X-ray crystallography.** This technique uses samples of protein that have been induced to form crystals. A protein preparation must be exceptionally pure in order to crystallize without imperfections. A protein crystal, often no more than 0.5 mm in diameter, usually contains 40–70% water by volume and is therefore more gel-like than solid (**Fig. 4.33**).

When bombarded with a narrow beam of X-rays, the electrons of the atoms in the crystal scatter the X-rays, which reinforce and cancel each other to produce a **diffraction pattern** of light and dark spots that can be captured electronically or on a piece of X-ray film (**Fig. 4.34**). Mathematical analysis of the intensities and positions of the diffracted X-rays yields a three-dimensional map of electron density in the crystallized molecule. The level of detail of the image depends in part on the quality of the crystal. Slight conformational variations among the crystallized protein molecules often limit resolution to about 2 Å. However, this is usually sufficient to trace the polypeptide backbone and discern the general shapes of the side chains.

Various lines of evidence indicate that the structure of a protein in a crystal closely resembles the structure of the protein in solution. In fact, crystallized proteins retain some ability to move and even bind small molecules that diffuse into the crystal. Newer techniques that take advantage of high-energy synchrotron radiation can monitor protein movements on the order of microseconds, which is useful for studying enzyme mechanisms. However, X-ray crystallography generally cannot capture the conformations of disordered protein segments and is useless for large protein complexes that do not crystallize.

Figure 4.33 Crystals of the protein streptavidin.

I. Le Trong and R. E. Stenkamp, University of Washington

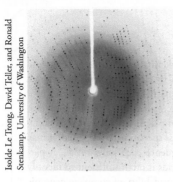

Figure 4.34 An X-ray diffraction pattern.

Large proteins and protein complexes can be analyzed by **cryo-electron microscopy (cryo-EM)**. An electron microscope generates images by aiming a beam of electrons to either pass through an object or interact with its surface. In cryo-EM, samples are cooled to the temperature of liquid N_2 (−196°C) or less. This treatment freezes water molecules in place, preserving macromolecular structure and minimizing radiation damage (electrons interact strongly with atoms, damaging specimens even more than X-rays). Originally, conventional two-dimensional film images, captured at different angles, were used to reconstruct three-dimensional models of proteins. Advances in camera technology and computer analysis have simplified data acquisition and processing to the point where cryo-EM is becoming a popular tool among structural biologists. The structures of amyloid fibers (see Figures 4.26 and 4.27) and the ribosome (Section 22.2) have been visualized using cryo-electron microscopy.

Structural data—consisting of the three-dimensional coordinates for each atom—for proteins as well as other macromolecules are available in online databases. The use of software for visualizing and manipulating such structures provides valuable insight into molecular structure and function and is a key part of the field of bioinformatics.

Concept Check

1. Choose a type of chromatography to separate proteins of different size or of different charge.
2. Explain what happens in affinity chromatography.
3. Choose several amino acids with and without ionizable side chains and calculate their pI values.
4. Summarize the strategy used to determine the sequence of a large protein.
5. Compare the physical state of a protein to be analyzed by NMR spectroscopy, X-ray crystallography, and cryo-electron microscopy.

SUMMARY

4.1 Amino Acids, the Building Blocks of Proteins

- The 20 amino acid constituents of proteins are differentiated by the chemical properties of their side chains, which can be roughly classified as hydrophobic, polar, or charged.
- Amino acids are linked by peptide bonds to form a polypeptide.

4.2 Secondary Structure: The Conformation of the Peptide Group

- The protein secondary structure includes the α helix and β sheet, in which hydrogen bonds form between backbone carbonyl and amino groups. Irregular secondary structure has no regularly repeating conformation.

4.3 Tertiary Structure and Protein Stability

- The three-dimensional shape (tertiary structure) of a protein includes its backbone and all side chains. A protein may contain all α, all β, or a mix of α and β structures.
- A globular protein has a hydrophobic core and is stabilized primarily by the hydrophobic effect. Ion pairing, disulfide bonds, and other cross-links may also help stabilize proteins.
- A denatured protein may refold to achieve its native structure. In a cell, chaperones assist protein folding.

- Some proteins include intrinsically disordered segments, and the additional conformational freedom may be essential for their biological functions.

4.4 Quaternary Structure

- Proteins with quaternary structure have multiple subunits.

4.5 Clinical Connection: Protein Misfolding and Disease

- Proteins that do not fold properly and are not degraded may adopt β-rich conformations that aggregate as amyloid fibers and damage cells, particularly neurons.

4.6 Tools and Techniques: Analyzing Protein Structure

- In the laboratory, proteins can be purified by chromatographic techniques that take advantage of the proteins' size, charge, and ability to bind other molecules.
- The sequence of amino acids in a polypeptide can be determined by chemically derivatizing and removing them in Edman degradation or by measuring the mass-to-charge ratio of peptide fragments in mass spectrometry.
- NMR spectroscopy, X-ray crystallography, and cryo-electron microscopy provide information about the three-dimensional structures of proteins at the atomic level.

KEY TERMS

protein	C-terminus	parallel β sheet	thioester bond	subunit	pI
polypeptide	backbone	antiparallel β sheet	denaturation	dimer	affinity
amino acid	microenvironment	staggered	renaturation	trimer	chromatography
R group	oligopeptide	conformation	native structure	tetramer	HPLC
α-amino acid	peptide	eclipsed	molecular chaperone	homo-	electrophoresis
Cα	primary structure	conformation	post-translational	hetero-	SDS-PAGE
chirality	secondary structure	Ramachandran	processing	amyloid deposit	protease
disulfide bond	tertiary structure	diagram	monomorphic	transmissible spongi-	Edman degradation
condensation	quaternary structure	irregular secondary	protein	form encephalopa-	mass spectrometry
reaction	torsion angle	structure	metamorphic protein	thy (TSE)	NMR spectroscopy
peptide bond	φ	globular protein	intrinsically	prion	X-ray crystallography
residue	ψ	fibrous protein	disordered protein	chromatography	diffraction pattern
hydrolysis	regular secondary	domain	proteome	size-exclusion	cryo-EM
exopeptidase	structure	ion pair	biomineralization	chromatography	
endopeptidase	α helix	zinc finger	membraneless	ion exchange	
N-terminus	β sheet	isopeptide bond	organelle	chromatography	

BIOINFORMATICS

Brief Bioinformatics Exercises

4.1 Drawing and Analyzing Amino Acids and Peptides

4.2 Chemical Properties and Amino Acid Composition of Green Fluorescent Protein

4.3 Protein Properties and Purification Methods

4.4 Visualizing and Analyzing α Helices

4.5 Visualizing and Analyzing β Sheets

4.6 Visualizing and Analyzing Cytochrome *c*

Bioinformatics Projects

Visualizing Three-Dimensional Protein Structures Using the Molecular Visualization Programs JSmol and PyMOL

Structural Alignment and Protein Folding

PROBLEMS

4.1 Amino Acids, The Building Blocks of Proteins

1. What are the four groups attached to the central (α) carbon of a proteinogenic amino acid?

2. Which amino acids fit into each of these categories? (Note: The number in parentheses indicates the number of amino acids in that category.) **a.** Nonpolar, nonaromatic (6) **b.** Aromatic amino acid.

3. For a generic amino acid, $NH_2CRHCOOH$, with an uncharged side chain, what would be the predominant form at pH 7?

4. Give examples of the following: **a.** optically inactive amino acid. **b.** sulphur containing amino acid. **c.** aromatic amino acid. **d.** basic and acidic amino acids.

5. What is the difference between an amino acid residue, a dipeptide, a tripeptide, an oligopeptide, and a polypeptide?

6. If chymotrypsin cleaves at the carboxyl end of phenylalanine, tryptophan, and tyrosine, how many oligopeptides would be formed in enzymatic cleavage of the following molecule with chymotrypsin? Val – Phe – Glu – Lys – Tyr – Phe – Trp – Ile – Met – Tyr – Gly – Ala

7. β-Alanine is the only naturally occurring β-amino acid. It is formed as a breakdown product in metabolism and is also a component of vitamin B₅. Draw the structures of α-Ala and β-Ala.

8. At what pH would an amino acid bear both a COOH and an NH_2 group? Is your answer consistent with the observation that most amino acids have high melting points (~300°C) and are generally soluble in water?

9. **a.** Draw the structure of magnesium glutamate (see Box 4.B). **b.** Would magnesium glutamate have the same physiological effects as MSG? Note that the average pH of saliva is about 7.

10. L-2-amino-4-phosphonobutyrate, an analogue of glutamate, can bind to the same umami taste receptors as glutamate (see Box 4.B). Draw this glutamate analog's structure. What structural differences exist between this and glutamate, and why may the analogue bind to the same glutamate receptor?

11. Answer the following: **a.** What are aquaporins? **b.** explain the stability of solenoid and zig-zag model of DNA packaging.

12. Cite the examples of serine protease and aspartate protease.

13. The protein known as green fluorescent protein (GFP) was first purified from bioluminescent jellyfish and can be linked to other proteins in order to study them using fluorescent techniques. The fluorophore in GFP (shown below) is a derivative of three consecutive amino acids that undergo cyclization of the polypeptide chain and an oxidation. Identify **a.** the three residues and the bonds that result from the **b.** cyclization and **c.** oxidation reactions.

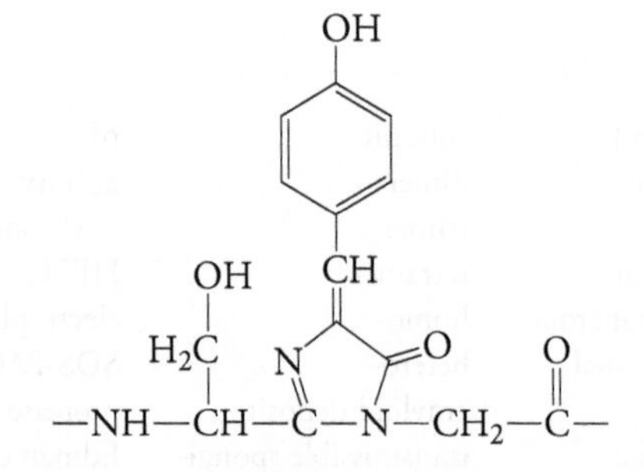

OH

(structure of a peptide drawn above problem 14, containing a tyrosine-derived aromatic ring with OH, and the backbone labeled)

OH

CH

H₂C   N   O   O

—NH—CH—C—N—CH₂—C—

14. Draw the structure of the tripeptide His–Lys–Glu at pH 8.0. Label the following: peptide bonds, N-terminus, C-terminus, an α-amino group and an ε-amino group, and an α-carboxylate group and a γ-carboxylate group.

15. The artificial non-nutritive sweetener aspartame is a dipeptide with the sequence Asp–Phe. The C-terminus is methylated. Draw the structure of aspartame at pH 7.0.

16. The zebrafish is a good animal model to test the actions of various drugs. The zebrafish produces a novel heptapeptide that binds to opiate receptors in the brain. The sequence of the peptide is Tyr–Gly–Gly–Phe–Met–Gly–Tyr. Draw the structure of the heptapeptide at pH 7.

17. Glutathione (abbreviated GSH) removes reactive oxygen species that can irreversibly damage hemoglobin and cell membranes in the red blood cell. The sequence of glutathione is γ-Glu–Cys–Gly. The γ-Glu indicates that the first peptide bond forms between the γ-carboxylate group of the Glu side chain and the α-amino group of Cys. Draw the structure of glutathione at pH 7.

18. Two molecules of reduced glutathione (see Problem 17) react with an organic peroxide (abbreviated ROOH). The glutathione is oxidized to GSSG and the organic peroxide is reduced to a less harmful alcohol. Use the abbreviations provided, the equation below, and the structure that you drew for GSH to deduce the structure of GSSG.

$$2\ GSH + R—O—O—H \xrightarrow{\text{glutathione peroxidase}} GSSG + ROH + H_2O$$

19. Which of the amino acid has been shown as one of the active site in the following enzyme: a. Phosphoglucomutase b. Aldolase c. Enolase

20. Estimate the net charge of the following peptides: a. Glu–Tyr at pH 6.0, b. Asp–Asp–Asp at pH 7.0, and c. His–Lys–Glu at pH 8.0 (see Solution 14).

21. Estimate the charge on oxytocin at pH 7 (see Section 4.1).

22. Estimate the charge on vasopressin at pH 7 (see Section 4.1).

23. Certain bacteria synthesize cyclic tetrapeptides. a. Estimate the net charge at pH 7.0 of such a compound that consists of two Pro and two Tyr residues. b. If the peptide were linear rather than cyclic, what would be its net charge?

24. Cyclic peptides are showing promise as anticancer drugs. One cyclic peptide contains one Lys residue, one Glu residue, and two Cys residues; the remaining residues are uncharged. The N- and C-termini form a covalent bond, and a disulfide bond forms between the Cys side chains. What is the charge of this cyclic peptide?

25. Poly-aspartate peptides have a wide variety of applications in industry and medicine. Is a poly-aspartate polypeptide more soluble at neutral pH or low pH? Explain.

26. Estimate the net charge of a His–His–His–His tetrapeptide at pH 6.0. Which is more soluble: a solution of the histidine tetrapeptide or a solution of free histidine?

27. In the conotoxin shown in Section 4.1, the first and third Cys residues form a disulfide bond and the second and fourth Cys residues form a disulfide bond. Using one-letter abbreviations for the amino acid residues, sketch a diagram of the disulfide-bonded peptide.

28. Peptide drugs cannot be administered orally because their peptide bonds are easily hydrolyzed by digestive peptidases. One possible solution is to synthesize peptides with thioamide linkages rather than peptide bonds. Draw two amino acids joined by a thioamide bond.

29. The pK values of the amino and carboxylate groups in free amino acids differ from the pK values of the N- and C-termini of polypeptides. Explain.

30. In structural studies of a staphylococcal nuclease enzyme, a valine residue buried in the protein's interior was experimentally changed, first to an Asp residue and then, in a second experiment, to a Lys residue. How do the pK values of the a. Asp and b. Lys side chains in the mutant nucleases compare to those listed for the free amino acids in Table 4.1?

31. In how many combinations do the tripeptide containing Lys, Arg, and Pro can exist?

32. You have isolated a tetrapeptide with an unknown sequence and, after hydrolyzing its peptide bonds, you recover Ala, Glu, Lys, and Thr. How many different sequences are possible for this tetrapeptide?

33. Identify each of the following sentences as describing the primary, secondary, tertiary, or quaternary structure of the protein: a. The shape of myoglobin is roughly spherical. b. Hemoglobin consists of four polypeptide chains. c. About one-third of the amino acid residues in collagen are glycines. d. The lysozyme molecule contains regions of helical structure.

4.2 Secondary Structure: The Conformation of the Peptide Group

34. The structure of the peptide bond is drawn in Section 4.1 in the *trans* configuration in which the carbonyl group and the amino hydrogen are on opposite sides of the peptide bond. Draw the structure of the peptide bond in the *cis* configuration. Which configuration is more likely to be found in proteins and why?

35. a. Would you expect the peptide group's C—N bond to be longer or shorter than the N—Cα single bond? b. Would you expect the peptide group's C=O bond to be longer or shorter than the C=O bond commonly found in aldehydes and ketones? Explain.

36. Would you expect an antiparallel β sheet or a parallel β sheet to be more stable? Explain.

37. Compare and contrast the structures of the DNA helix and the α helix.

38. The α helix is stabilized by hydrogen bonds that are directed along the length of the helix. The peptide carbonyl group of the nth residue forms a hydrogen bond with the peptide —NH group of the $(n + 4)$th residue. Draw the structure of this hydrogen bond.

39. Proline is known as a helix disrupter; it sometimes appears at the beginning or the end of an α helix but never in the middle. Explain.

40. Comment on the stability of types of beta-pleated secondary protein structure.

41. A 24-residue peptide called Pandinin 2, isolated from scorpion venom, was found to have both antimicrobial and hemolytic properties. The sequence of the first 18 residues of this peptide is shown below. The peptide forms an amphipathic helix in which nonpolar amino acids are on one side of helix and polar or charged amino acids are on the other

side of the helix. **a.** Examine the sequence below and predict which amino acid side chains are likely to be found on the nonpolar side of the amphipathic helix. **b.** Propose a hypothesis that explains why the peptide is able to lyse bacteria and red blood cells.

FWGALAKGALKLIPSLFS

42. The G protein ARF1 is involved in cell signaling. It has a small 14-residue segment that forms an amphipathic α helix (see Problem 41). What happens to ARF1's ability to associate with a membrane when several of its Leu residues are mutated to Ala residues?

43. The red blood cell protein glycophorin is a transmembrane protein that crosses the membrane once with an α-helical structure. A portion of the amino acid sequence is shown below. Identify the transmembrane domain of glycophorin. (*Hint:* If the average thickness of a membrane is 3 nm and if the average length of an amino acid in an α helix is 0.15 nm, how many amino acids are required to span the membrane?)

HHFSEPEITLIIFGVMAGVIGTILLISYGIRRLIKKSPSD

4.3 Tertiary Structure and Protein Stability

44. Discuss the key interactions involved in the hierarchy of protein structures.

45. In site-directed mutagenesis experiments, Gly is often successfully substituted for Val, but Val cannot substitute for Gly. Explain.

46. a. The guanidinium group on the Arg side chain is stabilized by resonance. Draw the contributing resonance structures for the Arg side chain. **b.** A study of 60 proteins showed that Arg is about 50% more likely to be buried than Lys. Provide an explanation for this observation.

47. a. The carboxylate group on the Asp and Glu side chains is stabilized by resonance. Draw the contributing resonance structures for the carboxylate side chain. **b.** A study of 60 proteins showed that of the 20 amino acids, Glu and Asp residues are the least likely to be found in a protein's hydrophobic core. Provide an explanation for this observation.

48. In three separate site-directed mutagenesis experiments, scientists altered a Lys residue in corn PEP carboxylase to an Asn, Glu, or Arg. Which substitution would you expect to have the least effect on enzymatic activity? Which substitution would have the greatest effect? Explain.

49. Draw two amino acid side chains that can interact with each other via the following intermolecular interactions: **a.** ion pair, **b.** hydrogen bond, and **c.** van der Waals interaction (London dispersion forces).

50. Exposing a protein to either low or high pH conditions can denature the protein. What types of intermolecular forces in the native protein could be disrupted when the pH is altered?

51. Laboratory techniques for randomly linking together amino acids typically generate an insoluble polypeptide, yet a naturally occurring polypeptide of the same length is usually soluble. Explain.

52. A type of muscular dystrophy results from a mutation in the gene for a 50-kD muscle protein. The defective protein leads to muscle necrosis. Detailed studies of this protein reveal that an arginine residue at position 98 is mutated to a histidine. Why might replacing an arginine with histidine result in a defective protein?

53. In 1957, Christian Anfinsen carried out a denaturation experiment with ribonuclease (a pancreatic enzyme that catalyzes the digestion of RNA), which consists of a single chain of 124 amino acids cross-linked by four disulfide bonds. Urea and 2-mercaptoethanol (a reducing agent that breaks disulfide bonds by converting them to the —SH form) were added to a solution of ribonuclease, which caused it to unfold, or denature. The loss of tertiary structure resulted in a loss of biological

activity. When the denaturing agent (urea) and the reducing agent (2-mercaptoethanol) were simultaneously removed, the ribonuclease spontaneously folded back up to its native conformation and regained full enzymatic activity in a process called renaturation. What is the significance of this experiment?

54. The pancreatic hormone insulin consists of an A chain and a B chain held together with disulfide bonds. *In vivo*, insulin is processed from proinsulin, a single polypeptide chain. The C chain is removed from proinsulin to form insulin.

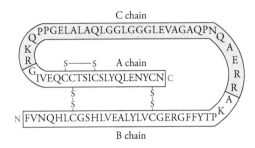

a. A denaturation/renaturation experiment similar to the one carried out by Anfinsen with ribonuclease (see Problem 53) was carried out using insulin. However, only 2–4% of the activity of the native protein was recovered when the urea and 2-mercaptoethanol were removed by dialysis (this is the level of activity to be expected if the disulfide bridges form randomly). When the experiment was repeated with proinsulin, 60% of the activity was restored upon renaturation. Explain these observations.

b. The investigators noted that the refolding of proinsulin depended on pH. For example, when the proinsulin was incubated in pH 7.5 buffer, only 10% of the proinsulin was renatured. However, at pH 10.5, 60% of the proinsulin was renatured. Explain these observations. (*Hint:* The pK value of the cysteine side chain in proinsulin is similar to the pK value of the sulfhydryl group in free cysteine.)

55. In the mid-1980s, scientists noted that if cells were incubated at 42°C instead of the normal 37°C, the synthesis of a group of proteins dramatically increased. For lack of a better name, the scientists called these heat-shock proteins. It was later determined that the heat-shock proteins were chaperones. Why do you think that the cell increases its synthesis of chaperones when the temperature increases?

56. A protein engineering laboratory studying monoclonal antibody proteins characterized the thermal stability of these proteins by measuring their melting temperature (T_m), defined as the temperature at which the proteins are half unfolded. The investigators found a positive correlation between T_m and the proteins' —SH content. In other words, proteins with more —SH groups were more thermally stable. Explain this observation.

57. Spectrin is a protein found in the cytoskeleton of the red blood cell. The cytoskeleton consists of proteins anchored to the cytosolic surface of the membrane and gives the cell the strength and flexibility to squeeze through capillaries. The spectrin protein chains consist of repeating segments that form α-helical bundles. Recently, a mutant spectrin was isolated in which a glutamine was replaced with proline. What would be the effect of this mutation on the spectrin protein? What are the consequences for the red blood cell?

58. During evolution, why do insertions, deletions, and substitutions of amino acids occur more often in loops than in elements of regular secondary structure?

59. The lens of the eye contains two families of crystallins known as α and β/γ (see the chapter-opening page). α-Crystallins function as chaperones for other crystallins. Why is this chaperone activity important for the lens?

60. Intrinsically disordered segments of proteins appear to evolve more rapidly than segments that form distinct folded domains. Explain.

61. Intrinsically disordered protein segments, particularly those that form gel-like aggregates, tend to be rich in serine, glutamine, asparagine, and glycine residues. Explain.

62. A membraneless organelle contains a mixture of protein and RNA. What residues would you expect to find in abundance in the proteins of a membraneless organelle?

4.4 Quaternary Structure

63. Restriction nucleases recognize and then hydrolyze palindromic DNA sequences (sequences that read the same on both strands, such as –GAATTC–). Many restriction enzymes are dimeric (two-subunit) enzymes. Based on how these proteins interact with DNA, do you expect them to be homo- or heterodimeric?

64. A protein with two identical subunits can often be rotated 180° (halfway) around its axis so as to generate an identical structure; such a protein is said to have rotational symmetry. Why is it not possible for a protein to have mirror symmetry (that is, its halves would be related as if reflected in a mirror)?

65. Glutathione transferase consists of a homodimer in equilibrium with its constituent monomers. In site-directed mutagenesis studies, two arginine residues were mutated to glutamines and two aspartates were mutated to asparagines. These substitutions caused the equilibrium to shift in favor of the monomeric form of the enzyme. Where are the arginines and aspartates likely to be found on the protein and what is their role in stabilizing the dimeric form of the enzyme?

66. A tetrameric protein dissociates into dimers when the detergent sodium dodecyl sulfate (SDS) is added to a solution of the protein. The dimers are termed SDS-resistant because they do not further dissociate into monomers in the presence of the detergent. What intermolecular forces might be acting at the dimer–dimer interface? Are the intermolecular forces acting at the monomer–monomer interface different? Explain.

67. Malic enzyme consists of a dimer of dimers. A site-directed mutagenesis experiment was carried out in order to determine the intermolecular interactions important in tetramer formation. **a.** When Asp 568 was mutated to Ala, the enzyme assumed the tetrameric form, but when both His 142 and Asp 568 were mutated to Ala, the enzyme formed only dimers. What can you conclude from this experiment? **b.** In a second experiment, the investigators found that mutation of Tyr 570, Trp 572, and Pro 673 resulted in an enzyme that formed dimers only. What can you conclude from this experiment?

68. Site-directed mutagenesis experiments were performed on a hydroxysteroid dehydrogenase in order to study its quaternary structure. The enzyme is a tetramer of identical monomers. In five different experiments, the investigators mutated His 93, Gln 148, Lys 158, Tyr 253, and Cys 255 to Ala residues. **a.** Why do investigators typically mutate selected amino acid residues to Ala residues? **b.** What types of interactions were being investigated in constructing the mutant enzymes?

4.5 Clinical Connection: Protein Misfolding and Disease

69. The gene for the amyloid precursor protein is located on chromosome 21. Individuals with Down syndrome (trisomy 21) have three rather than two copies of chromosome 21. Explain why individuals with Down syndrome tend to exhibit Alzheimer-like symptoms by middle age.

70. Mice that have been genetically engineered to lack the PrP gene (but are normal in all other ways) do not develop spongiform encephalopathy after being inoculated with the prion form of PrP, which causes disease in normal mice. What do these results reveal about the relationship between PrP and TSE? What do they reveal about the cellular function of PrP?

71. Myoglobin is a protein that contains mostly α helices and no β sheets (see Fig. 5.1). **a.** Would you expect myoglobin to form amyloid fibers? **b.** Under certain laboratory conditions, myoglobin can be induced to form amyloid fibers. What does this suggest about a polypeptide's ability to adopt β secondary structure?

72. Discuss the role of amino acid side chains in the formation of an amyloid fiber.

73. Transthyretin, a protein composed of four identical subunits, transports thyroid hormone in the plasma and cerebrospinal fluid. Several mutations in the gene that codes for this protein disrupt the protein's tetrameric structure, favoring the monomer, which is more likely to form amyloid fibers. Why might a Leu → Pro transthyretin mutant be amyloidogenic?

4.6 Tools and Techniques: Analyzing Protein Structure

74. Estimate the pI values of **a.** alanine **b.** glutamate, and **c.** lysine.

75. What types of amino acids are likely to be relatively abundant in a protein with a pI of **a.** 4.3 and **b.** 11.0?

76. What can you conclude about the net charge of any protein at a pH **a.** less than 3.5 or **b.** greater than 11.0?

77. In your laboratory, you plan to use ion exchange chromatography to separate the peptide shown below from a mixture of different peptides at pH 7.0. Should you choose a matrix containing DEAE or CM groups?

Peptide: GLEKSLVRLGDVQPSLGKESRAKKFQRQ

78. The Ara h8 protein allergen from peanuts was recently purified, a challenging task because the peanut Ara h6 protein has a similar size and pI. Separation of the two proteins was finally achieved when it was noted that the Ara h6 protein contains 10 disulfide-bonded Cys residues whereas Ara h8 contains no Cys. The peanut protein mixture was treated with dithiothreitol (DTT, to reduce the disulfide bonds) and then iodoacetic acid (ICH₂COOH, a reagent that alkylates — SH groups). The mixture was then loaded on to an anion exchange column and the two proteins were successfully separated. **a.** Show the structural changes that occur when Cys residues are exposed to DTT followed by iodoacetic acid. **b.** Draw a plausible elution profile (protein concentration versus solvent volume) for the separation of the proteins by anion exchange chromatography. **c.** Explain why this strategy resulted in the successful separation of the two proteins.

79. The sequence of crinia-angiotensin, an angiotensin II-like undecapeptide from the skin of the Australian frog, is determined. A single round of Edman degradation releases DNP-Ala. A second sample of the peptide is then treated with chymotrypsin. Two fragments are released with the following amino acid compositions: Fragment I (His, Pro, Phe, Val) and Fragment II (Ala, Asp, Arg, Gly, Pro, Ile, Tyr). Next, a third sample of peptide is treated with trypsin, which results in two fragments with the following amino acid compositions: Fragment III (Ala, Asp, Arg, Gly, Pro) and Fragment IV (His, Ile, Pro, Phe, Tyr, Val). Treatment of another sample with elastase yields three fragments, two of which are sequenced: Fragment V (His–Pro–Phe) and Fragment VI (Ala–Pro–Gly). What is the sequence of the undecapeptide?

80. Before sequencing, a protein whose two identical polypeptide chains are linked by a disulfide bond must be reduced and alkylated (to chemically block them). Why should reduction and alkylation also be performed for a single polypeptide chain that includes an intramolecular disulfide bond?

81. Proteins post-translationally modified by phosphorylation can be isolated by affinity chromatography using a matrix of immobilized iron ions. **a.** Explain how this works. **b.** What class of cellular molecules are likely to be major contaminants of the purified phosphoproteins?

82. A group of investigators carried out a theoretical study of the behavior of a dimeric protein during gel filtration chromatography. A dimer may exist in a dynamic equilibrium with its monomeric units as described by the following equation:

$$\text{dimer} \rightleftharpoons \text{monomer}$$

The investigators determined that when the dissociation (forward) and association (reverse) rates were slow, two peaks appeared on the chromatogram, one corresponding to the dimer and one corresponding to the monomer. **a.** Which species would elute first? **b.** What are the expected results if the association rate is much faster than the dissociation rate? **c.** What are the expected results if the association rate is much slower than the dissociation rate?

83. Four major proteins from egg white were separated and purified for use in the pharmaceutical and food industries. Some characteristics of these proteins are listed in the table below. What technique could you use to isolate lysozyme, an enzyme used as a food preservative, from the other egg white proteins?

Protein	Molecular mass (D)	pI
Transferrin	79,900	6.0
Mucoid	30,000	4.1
Albumin	45,000	4.5
Lysozyme	14,500	11.0

84. Would it be possible to separate the remaining three proteins from egg white (see Problem 83) using anion exchange chromatography?

85. Avidin is a minor (0.05%) protein component of egg white. It binds very tightly to the B vitamin biotin, a small organic molecule. None of the egg white proteins shown in Problem 91 bind to biotin. What strategy could you use to separate avidin from other egg white proteins?

86. Two forms of lactoglobulin, a milk protein, have slightly different amino acid sequences that give them slightly different pI values. Lactoglobulin A (LgA) has a pI of 5.1 and lactoglobulin B (LgB) has a pI of 5.2.

If separation of these two proteins is attempted using DEAE anion exchange chromatography, what is the order of elution of the proteins if the mobile phase buffer is adjusted to a pH of 5.2?

87. Determining the sequence of the 284-residue protein bacteriorhodopsin was quite an undertaking when it was accomplished in the 1970s. The protein was cleaved into shorter peptides using cyanogen bromide (CNBr); then each peptide was sequenced by cleaving it into smaller fragments. Determine the sequence of one of these peptides using the data in the table below.

Sequence of fragments obtained by digestion with chymotrypsin	Sequence of fragments obtained by digestion with trypsin
W	VYSYR
SY	IGTGLVGALTK
RF	FVWWAISTAAM
VW	
AISTAAM	
IGTGLVGALTKVY	

88. Explain why sequencing using traditional methods (e.g., enzymatic cleavage) would be difficult to accomplish with this peptide:

METDTLLLWVLLLWVPGSTG

89. In sequencing by mass spectrometry, not every peptide bond may break. **a.** If cleavage between two Gly residues does not occur, which amino acid would be identified in place of the two glycines? **b.** What amino acid would be identified if a bond between Ser and Val did not break?

90. The peptide shown here is subjected to mass spectrometry to determine its sequence. **a.** If all its peptide bonds (but no other bonds) are broken, what is the mass of the smallest fragment? Assume that the only charged group is the N-terminus. **b.** What is the difference in mass between the smallest and the next-smallest fragment?

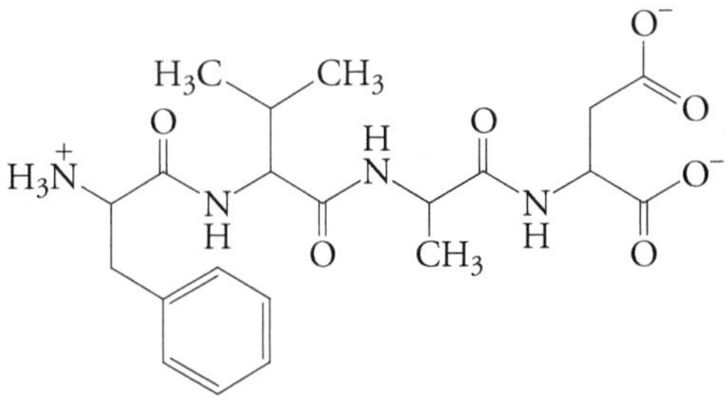

SELECTED READINGS

Cheng, Y., Single-particle cyro-EM—how did it get here and where will it go, *Science* 361, 876–880, doi: 10.1126/science.aat4346 (2018). [Summarizes some of the challenges and advances in cryo-EM determination of protein structures.]

Davey, N.E., The functional importance of structure in unstructured protein regions, *Curr. Opin. Struct. Biol.* 56, 155–163, doi: 10.1016/j.sbi.2019.03.009 (2019). [Discusses the binding properties of intrinsically disordered protein regions.]

Dishman, A.C., and Volkman, B.F., Unfolding the mysteries of protein metamorphosis, *ACS Chem. Biol.* 13, 1438–1446, doi: 10.1021/acschembio.8b00276 (2018). [Recounts the development of ideas about disordered protein structure, with a focus on metamorphic proteins.]

Goodsell, D.S., Burley, S.K., and Berman, H.M., Revealing structural views of biology, *Biopolymers* 99, 817–824, doi: 10.1002/bip.22338 (2013). [Summarizes the history of protein structure determination, including the application of new techniques.]

Iadanza, M.G., Jackson, M.P., Hewitt, E.W., Ranson, N.A., and Radford, S.E., A new era for understanding amyloid structures and disease, *Nat. Rev. Mol. Cell Biol.* 19, 755–773, doi: 10.1038/

s41580-018-0060-8 (2018). [Summarizes what is known about the structures of amyloid fibrils and includes a list of amyloid diseases.]

Proteopedia, www.proteopedia.org [This interactive resource showcases tutorials on the three-dimensional structures of proteins, nucleic acids, and other molecules.]

Scheraga, H.A., My 65 years in protein chemistry, *Q. Rev. Biochem.* 48, 117–177, doi: 10.1017/S0033583514000134 (2015). [A personal account that touches all aspects of proteins, from hydrogen bonding to protein misfolding.]

Sillitoe, I., Lewis, T.E., Cuff, A., Das, S., Ashford, P., Dawson, N.L., Furnham, N., Laskowski, R.A., Lee, D., Lees, J.G., Lehtinen, S., Studer, R.A., Thornton, J., and Orengo, C.A., CATH: comprehensive structural and functional annotations for genome sequences, *Nuc. Acids Res.* 43, D376–381 (2015). [Summarizes a widely used approach to classifying protein structures.]

CHAPTER 4 CREDITS

Figure 4.1 Image of DNA polymerase based on 1KFS. Brautigam, C.A., Steitz, T.A., DNA Polymerase I Klenow fragment (E.C.2.7.7.7) mutant/DNA complex, *J. Mol. Biol.* 277, 363–377 (1998). Image of plastocyanin based on 1PND. Fields, B.A., Guss, J.M., Freeman, H.C., Accuracy and precision in protein crystal structure analysis, two independent refinements of the structure of poplar plastocyanin at 173K, *Acta Crystallogr. D Biol. Crystallogr.* 50, 709–730 (1994). Image of insulin based on 1ZNI. Bentley, G., Dodson, E., Dodson, G., Hodgkin, D., Mercola, D., Structure of insulin in 4-zinc insulin, *Nature* 261, 166–168 (1976). Image of maltoporin based on 1MPM. Dutzler, R., Schirmer, T., Maltoporin maltose complex, *Structure* 4, 127–134 (1996). Image of phosphoglycerate kinase based on 3PGK. Shaw, P.J., Walker, N.P., Watson, H.C., The structure of yeast phosphoglycerate kinase at 0.25 nm resolution, *EMBO J.* 1, 1635–1640 (1982).

Figure 4.4 Image based on 2HHB. Fermi, G., Perutz, M.F., The crystal structure of human deoxyhaemoglobin at 1.74 Angstroms resolution, *J. Mol. Biol.* 175, 159–174 (1984).

Figure 4.6 Image based on 1MBD. Phillips, S.E.V. X-Ray structure of sperm whale deoxymoglobin refined at 1.4A resolution [unpublished].

Figure 4.8 Image based on 3CPA. Christianson, D.W., Lipscomb, W.N., X-Ray crystallographic investigation of substrate binding to carboxypeptidase A at subzero temperature, *Proc. Natl. Acad. Sci. USA* 83, 7568–7572 (1986).

Figure 4.9 Image based on 1YPI. Lolis, E., Alber, T., Davenport, R.C., Rose, D., Hartman, F.C., Petsko, G.A., Structure of yeast triosephosphate isomerase at 1.9-Angstroms resolution, *Biochemistry* 29, 6609–6618 (1990).

Figure 4.10a Image based on 1HGU. Chantalat, L., Jones, N., Korber, F., Navaza, J., Pavlovsky, A.G., Human growth hormone, *Protein Pept. Lett.* 2, 333–340 (1995).

Figure 4.10b Image based on 1GCS. Lindley, P.F., Najmudin, S., Bateman, O., Slingsby, C., Myles, D., Kumaraswamy, S., Glover, I., Structure of the bovine gamma-B crystallin at 150K, *J. Am. Chem. Soc.* 89, 2677 (1993).

Figure 4.10c Image based on 1CZN. Drennan, C.L., Pattridge, K.A., Weber, C.H., Metzger, A.L., Hoover, D.M., Ludwig, M.L., Refined structures of oxidized flavodoxin from Anacystis nidulans, *J. Mol. Biol.* 294, 711–724 (1999).

Figure 4.10d Image based on 1CIX. Fujitani, N., Kawabata, S., Osaki, T., Kumaki, Y., Demura, M., Nitta, K., Kawano, K., Structure of the antimicrobial peptide tachystatin A, *J. Biol. Chem.* 277, 23651–23657 (2002).

Figure 4.11 Image based on 1GPD. Moras, D., Olsen, K. W., Sabesan, M.N., Buehner, M., Ford, G.C., Rossmann, M.G., Studies of asymmetry in the three-dimensional structure of lobster D-glyceraldehyde-3-phosphate dehydrogenase, *J. Biol. Chem.* 250, 9137–9162 (1975).

Figure 4.12 Image based on 1MBD. Phillips, S.E.V. X-Ray structure of sperm whale deoxymoglobin refined at 1.4A resolution [unpublished].

Figure 4.15a Image based on 1TF6. Nolte, R.T., Conlin, R.M., Harrison, S.C., Brown, R.S., Differing roles for zinc fingers in DNA recognition, structure of a six-finger transcription factor IIIA complex, *Proc. Natl. Acad. Sci. USA* 95, 2938–2943 (1998).

Figure 4.15b image based on 1D66. Marmorstein, R., Carey, M., Ptashne, M., Harrison, S.C., DNA recognition by GAL4, structure of a protein/DNA complex, *Nature* 356, 408–414 (1992).

Figure 4.16 Image based on 1E8L. Schwalbe, H., Grimshaw, S.B., Spencer, A., Buck, M., Boyd, J., Dobson, C.M., Redfield, C., Smith, L.J., A refined solution structure of hen lysozyme determined using residual dipolar coupling data, *Protein Sci.* 10, 677–688 (2001).

Figure 4.20a Image based on 1J9O. Kuloglu, E.S., McCaslin, D.R., Kitabwalla, M., Pauza, C.D., Markley, J.L., Volkman, B.F., Monomeric solution structure of the prototypical 'C' chemokine lymphotactin, *Biochemistry* 40, 12486–12496 (2001).

Figure 4.20b image based on 2JP1. Tuinstra, R.L., Peterson, F.C., Kutlesa, S., Elgin, E.S., Kron, M.A., Volkman, B.F., Interconversion between two unrelated protein folds in the lymphotactin native state, *Proc. Natl. Acad. Sci. USA* 105, 5057–5062 (2008).

Figure 4.21 Image based on 2OQP. Bondensgaard, K., Breinholt, J., Solution structure of human interleukin-21, *J. Biol. Chem.* 282, 23326–23336 (2007).

Figure 4.23a Image based on 1AS8. Murphy, M.E., Turley, S., Adman, E.T., Structure of nitrite bound to copper-containing nitrite reductase from Alcaligenes faecalis. Mechanistic implications, *J. Biol. Chem.* 272, 28455–28460 (1997).

Figure 4.23b Image based on 1FUQ. Weaver, T., Banaszak, L., Fumarase with bound 3-trimethylsilylsuccinic acid, *Biochemistry* 35, 13955–13965 (1996).

Figure 4.23c Image based on 2HHB. Fermi, G., Perutz, M.F., The crystal structure of human deoxyhaemoglobin at 1.74 Angstroms resolution, *J. Mol. Biol.* 175, 159–174 (1984).

Figure 4.23d Image based on 1MMO. Rosenzweig, A.C., Frederick, C.A., Lippard, S.J., Nordlund, P., Crystal structure of a bacterial non-haem iron hydroxylase that catalyses the biological oxidation of methane, *Nature* 366, 537–543 (1993).

Figure 4.26 Image from Gremer, L., Schölzel, D., Schenk, C., Reinartz, E., Labahn, J., Ravelli, R.B.G., Tusche, M., Lopez-Iglesias, C., Hoyer, W., Heise, H., Willbold, D., Schröder, G.F. Fibril structure of amyloid-β(1–42) by cryo–electron microscopy. *Science,* 358, 116–119. © 2017 AAAS. Reproduced with permission of AAAS.

Figure 4.27 Image from Fitzpatrick, A.W.P., Falcon, B., He, S., Murzin, A.G., Murshudov, G., Garringer, H.J., Crowther, R.A., Ghetti, B., Goedert, M., Scheres, S.H.W. Cryo-EM structures of tau filaments from Alzheimer's disease. *Nature,* 547, 185–190. © 2017. Reproduced with permission of Springer Nature.

Figure 4.31b Image from Keough, T., Youngquist, R.S., Lacey, M.P.A method for high-sensitivity peptide sequencing using postsource decay matrix-assisted laser desorption ionization mass spectrometry. *Proc. Natl. Acad. Sci. USA* 96, 7131–7136. ©Copyright (1999) National Academy of Sciences, U.S.A. Reproduced with permission of National Academy of Sciences.

Figure 4.32 Image based on 1EGR. Sodano, P., Xia, T., Bushweller, J.H., Björnberg, O., Holmgren, A., Billeter, M., & Wüthrich, K. Sequence-specific 1H n.m.r. assignments and determination of the three-dimensional structure of reduced Escherichia coli glutaredoxin. *J. Mol. Biol.* 221, 1311–1324 (1991).

Protein Function

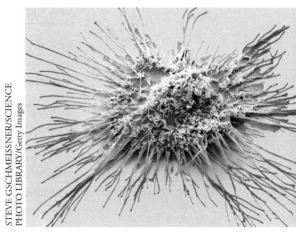

This dendritic cell, a type of white blood cell, patrols the body in search of pathogens. It is able to move through tight spaces by rearranging its cytoskeleton. Actin filaments are most active in the slender cell extensions, but the cell's network of microtubules ultimately determines which direction the cell will travel.

Do You Remember?

- Noncovalent forces, including hydrogen bonds, ionic interactions, and van der Waals forces, act on biological molecules (Section 2.1).
- A protein's structure may be described at four levels, from primary to quaternary (Section 4.1).
- Some proteins can adopt more than one conformation (Section 4.3).
- Proteins containing more than one polypeptide chain have quaternary structure (Section 4.4).

Every protein, with its unique three-dimensional structure, can perform some unique function for the organism that produces it. This chapter begins by examining myoglobin, an oxygen-binding protein that gives vertebrate muscle a reddish color, and hemoglobin, a major protein of red blood cells that transports O_2 from the lungs to other tissues. These two proteins have been studied for many decades and provide a wealth of information about protein function.

Unlike myoglobin and hemoglobin, which are globular proteins, many of the most abundant proteins are fibrous proteins that form extended structures that determine the shape and other physical attributes of cells and entire organisms. These structural proteins include the extracellular matrix protein collagen and a variety of intracellular proteins. Other than forming fibrous networks, these proteins have little in common, exhibiting a variety of secondary, tertiary, and quaternary structures related to their distinct physiological functions.

The supportive role of fibrous proteins in cellular architecture may seem obvious, but it turns out that many of the dynamic functions of cells are also intricately tied to these proteins. The movements of cells and the movements of organelles within cells reflect the action of motor proteins that operate along fibrous protein tracks; they provide some additional lessons in protein structure and function.

Finally, antibody proteins provide an example of naturally occurring proteins with seemingly impossible sequence variations. Such proteins are essential components of the vertebrate immune response to pathogens of all kinds. In the laboratory and clinic, modified versions of antibodies have become valuable tools that bind specifically to other molecules to identify them or block their activity.

5.1 Myoglobin and Hemoglobin: Oxygen-Binding Proteins

KEY CONCEPTS

Compare the structures and functions of myoglobin and hemoglobin.

- Recognize the role of the heme prosthetic group.
- Describe oxygen binding in quantitative terms.
- Identify conserved and variable residues in protein sequences.
- Explain cooperative oxygen binding to hemoglobin.
- Describe how oxygen is efficiently delivered to tissues.

Myoglobin is a relatively small protein with a compact shape about $44 \times 44 \times 25$ Å (**Fig. 5.1a**). Myoglobin lacks β structure entirely, and all but 32 of its 153 amino acids are part of eight α helices, which range in length from 7 to 26 residues and are labeled A through H (Fig. 5.1b). Hemoglobin is a tetrameric protein whose four subunits each resemble myoglobin.

The fully functional myoglobin molecule contains a polypeptide chain plus the iron-containing porphyrin derivative known as **heme** (shown below). The heme is a type of **prosthetic group,** an organic compound that allows a protein to carry out some function that the polypeptide alone cannot perform—in this case, binding oxygen.

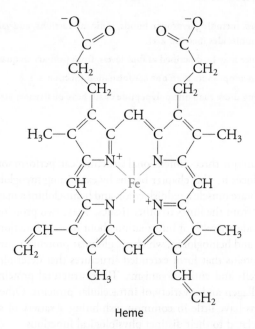

Heme

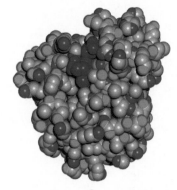

a.

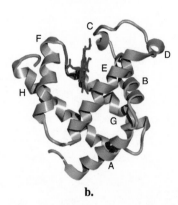

b.

Figure 5.1 Models of myoglobin structure. a. Space-filling model. All atoms (except H) are shown (C gray, O red, N blue). The heme group, where oxygen binds, is purple. **b.** Ribbon diagram with the eight α helices labeled A–H.

Question To which class of the CATH system (Section 4.3) does myoglobin belong

The planar heme is tightly wedged into a hydrophobic pocket between helices E and F of myoglobin. It is oriented so that its two nonpolar vinyl ($-CH{=}CH_2$) groups are buried and its two polar propionate ($-CH_2-CH_2-COO^-$) groups are exposed to the solvent. The central iron atom, with six possible coordination bonds, is liganded by four N atoms of the porphyrin ring system. A fifth ligand is provided by a histidine residue of myoglobin known as His F8 (the eighth residue of helix F). Molecular oxygen (O_2) can bind reversibly to form the sixth coordination bond. (This is what allows certain heme-containing proteins, such as myoglobin and hemoglobin, to function physiologically as oxygen carriers.) Residue His E7 (the seventh residue of helix E) forms a hydrogen bond to the O_2 molecule (**Fig. 5.2**). By itself, heme is not an effective oxygen carrier because the central ferrous Fe(II) (or Fe^{2+}) atom is easily oxidized to the ferric Fe(III) (or Fe^{3+}) state, which cannot bind O_2. Oxidation does not readily take place when the heme is part of a protein such as myoglobin or hemoglobin.

Oxygen binding to myoglobin depends on the oxygen concentration

The muscles of diving mammals are especially rich in myoglobin. In these animals, myoglobin probably serves as an oxygen-storage protein to provide additional oxygen during long dives. The concentration of myoglobin in the muscles of terrestrial mammals—about 100–300 µM—is not enough to supply more than a few seconds' worth of O_2. However, the presence of myoglobin boosts the rate of oxygen diffusion from the blood capillaries to the cells' mitochondria, where most of the O_2 is consumed in the aerobic production of ATP. Myoglobin also binds nitric oxide (·NO), a free radical that acts as a signal to help regulate blood flow and the activity of the mitochondrial oxidative machinery. Neurons (nerve cells) contain a globin known as neuroglobin, which might operate like myoglobin to facilitate the diffusion of O_2 into these metabolically active cells. Curiously, many other types of cells contain a protein called cytoglobin, which might function as an O_2 or ·NO sensor to regulate cellular activities.

Myoglobin's O_2-binding behavior can be quantified. To begin, the reversible binding of O_2 to myoglobin (Mb) is described by a simple equilibrium

$$Mb + O_2 \rightleftharpoons Mb \cdot O_2$$

with a dissociation constant, K:

$$K = \frac{[Mb][O_2]}{[Mb \cdot O_2]} \tag{5.1}$$

where the square brackets indicate molar concentrations. (Note that biochemists tend to describe binding phenomena in terms of dissociation constants, sometimes given as K_d, which are the reciprocals of the association constants, K_a, used by chemists.) The proportion of the total myoglobin molecules that have bound O_2 is called the **fractional saturation** and is abbreviated **Y:**

$$Y = \frac{[Mb \cdot O_2]}{[Mb] + [Mb \cdot O_2]} \tag{5.2}$$

Since $[Mb \cdot O_2]$ is equal to $[Mb][O_2]/K$ (Equation 5.1, rearranged),

$$Y = \frac{[O_2]}{K + [O_2]} \tag{5.3}$$

O_2 is a gas, so its concentration is expressed as pO_2, the **partial pressure of oxygen,** in units of millibar (or mbar) (where 1013 mbar = 1 atm). Thus,

$$Y = \frac{pO_2}{K + pO_2} \tag{5.4}$$

In other words, *the amount of O_2 bound to myoglobin (Y) is a function of the oxygen concentration (pO_2) and the affinity of myoglobin for O_2 (K).*

A plot of fractional saturation (Y) versus pO_2 yields a hyperbola (**Fig. 5.3**). As the O_2 concentration increases, more and more O_2 molecules bind to the heme groups of myoglobin molecules until, at very high O_2 concentrations, virtually all the myoglobin molecules have bound O_2. Myoglobin is then said to be **saturated** with oxygen. The oxygen concentration at which myoglobin is half-saturated—that is, the concentration of O_2 at which Y is half-maximal—is equivalent to K. For convenience, K is usually called p_{50}, the oxygen pressure at 50% saturation. For human myoglobin, p_{50} is 3.72 mbar (see Sample Calculation 5.1).

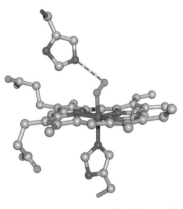

Figure 5.2 Oxygen binding to the heme group of myoglobin. The central Fe(II) atom of the heme group (purple) is liganded to four porphyrin N atoms and to the N of His F8 below the porphyrin plane. O_2 (red) binds reversibly to the sixth coordination site, above the porphyrin plane. Residue His E7 forms a hydrogen bond to O_2.

SAMPLE CALCULATION 5.1

Problem   Calculate the fractional saturation of myoglobin when $pO_2 = 1$ mbar, 10 mbar, and 100 mbar.

Solution   Use Equation 5.4 and let $K = 3.72$ mbar.

When $pO_2 = 1$ mbar, $Y = \dfrac{1}{3.72 + 1} = 0.212$

When $pO_2 = 10$ mbar, $Y = \dfrac{10}{3.72 + 10} = 0.728$

When $pO_2 = 100$ mbar, $Y = \dfrac{100}{3.72 + 100} = 0.964$

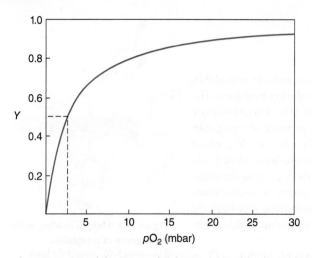

Figure 5.3 Myoglobin oxygen-binding curve. The relationship between the fractional saturation of myoglobin (Y) and the oxygen concentration (pO_2) is hyperbolic. When $pO_2 = K = 3.72$ mbar, myoglobin is half-saturated ($Y = 0.5$). Note that this curve describes a population of myoglobin molecules, not the behavior of a single protein.

Figure 5.4 Tertiary structures of myoglobin and the α and β chains of hemoglobin. Backbone traces of α globin (blue) and β globin (red) are aligned with myoglobin (green) to show their structural similarity. The heme group of myoglobin is shown in gray.

Question **Which portion of globin structure shows the most variability? The least? Explain.**

Myoglobin and hemoglobin are related by evolution

Hemoglobin is a heterotetramer containing two α chains and two β chains. Each of these sub-units, called a **globin,** looks a lot like myoglobin. The hemoglobin α chain, the hemoglobin β chain, and myoglobin have remarkably similar tertiary structures (**Fig. 5.4**). All have a heme group in a hydrophobic pocket, a His F8 that ligands the Fe(II) ion, and a His E7 that forms a hydrogen bond to O_2.

Somewhat surprisingly, the amino acid sequences of the three globin polypeptides are only 18% identical. **Figure 5.5** shows the aligned sequences, with the necessary gaps (for example, the hemoglobin α chain has no D helix). The lack of striking sequence similarities among these proteins highlights an important principle of protein three-dimensional structure: *Certain tertiary structures—for example, the backbone folding pattern of a globin polypeptide—can accommodate a variety of amino acid sequences. In fact, many proteins with completely unrelated sequences adopt similar structures.*

Clearly, the globins are **homologous proteins** that have evolved from a common ancestor through genetic mutation (see Section 3.3). The α and β chains of human hemoglobin share a number of residues; some of these are also identical in human myoglobin. A few residues are found in all vertebrates hemoglobin and myoglobin chains. The **invariant residues,** those that are identical in all the globins, are essential for the structure and/or function of the proteins and cannot be replaced by other residues. Some positions are under less selective pressure to maintain a particular amino acid match and can be **conservatively substituted** by a similar amino acid (for example, isoleucine for leucine or serine for threonine). Still other positions are **variable,** meaning that they can accommodate a variety of residues, none of which is critical for the protein's structure or function. *By looking at the similarities and differences in sequences among evolutionarily related proteins such as the globins, it is possible to deduce considerable information about elements of protein structure that are central to protein function.*

Sequence analysis also provides a window on the course of globin evolution, since *the number of sequence differences roughly corresponds to the time since the genes diverged.* An estimated 1100 million years ago (1100 mya), a single globin gene was duplicated, possibly by aberrant genetic recombination, leaving two globin genes that then could evolve independently (**Fig. 5.6**). Over time, the gene sequences diverged by mutation. One gene became the myoglobin gene. The other coded for a monomeric hemoglobin, which is still found in some primitive vertebrates such as the lamprey (an organism that originated about 425 mya). Subsequent duplication of the hemoglobin gene around 165 mya and additional sequence changes yielded the α and β globins, which made possible the evolution of a tetrameric hemoglobin (whose structure is abbreviated $\alpha_2\beta_2$).

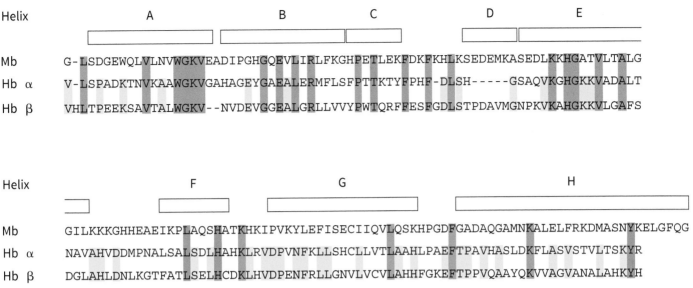

Helix	A	B	C	D	E

Mb   G-LSDGEWQLVLNVWGKVEADIPGHGQEVLIRLFKGHPETLEKFDKFKHLKSEDEMKASEDLKKHGATVLTALG

Hb α   V-LSPADKTNVKAAWGKVGAHAGEYGAEALERMFLSFPTTKTYFPHF-DLSH-----GSAQVKGHGKKVADALT

Hb β   VHLTPEEKSAVTALWGKV--NVDEVGGEALGRLLVVYPWTQRFFESFGDLSTPDAVMGNPKVKAHGKKVLGAFS

Helix	F	G	H

Mb   GILKKKGHHEAEIKPLAQSHATKHKIPVKYLEFISECIIQVLQSKHPGDFGADAQGAMNKALELFRKDMASNYKELGFQG

Hb α   NAVAHVDDMPNALSALSDLHAHKLRVDPVNFKLLSHCLLVTLAAHLPAEFTPAVHASLDKFLASVSTVLTSKYR

Hb β   DGLAHLDNLKGTFATLSELHCDKLHVDPENFRLLGNVLVCVLAHHFGKEFTPPVQAAYQKVVAGVANALAHKYH

Figure 5.5 The amino acid sequences of myoglobin and the hemoglobin α and β chains. The sequence of human myoglobin (Mb) and the human hemoglobin (Hb) chains are written so that their helical segments (bars labeled A through H) are aligned. Residues that are identical in the α and β globins are shaded yellow; residues identical in myoglobin and the α and β globins are shaded blue, and residues that are invariant in all vertebrate myoglobin and hemoglobin chains are shaded purple. The one-letter abbreviations for amino acids are given in Figure 4.2.

Question **Can you identify positions occupied by structurally similar amino acids in all three globins?**

Other gene duplications and mutations produced the ζ chain (from the α chain) and the γ and ε chains (from the β chain). In fetal mammals, hemoglobin has the composition $\alpha_2\gamma_2$, and early human embryos synthesize a $\zeta_2\varepsilon_2$ hemoglobin. In primates, a recent duplication of the β chain gene has yielded the δ chain. An $\alpha_2\delta_2$ hemoglobin occurs as a minor component (about 2%) of adult human hemoglobin. At present, the δ chain appears to have no unique biological function, but it may eventually evolve one.

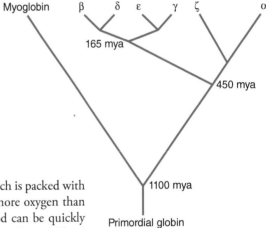

Figure 5.6 Evolution of the globins. Duplication of a primordial globin gene allowed the separate evolution of myoglobin and a monomeric hemoglobin. Additional duplications among the hemoglobin genes gave rise to six different globin chains that combine to form tetrameric hemoglobin variants at various times during development.

Oxygen binds cooperatively to hemoglobin

A milliliter of human blood contains about 5 billion red blood cells, each of which is packed with about 300 million hemoglobin molecules. Consequently, blood can carry far more oxygen than a comparable volume of pure water. The oxygen-carrying capacity of the blood can be quickly assessed by measuring the hematocrit (the percentage of the blood volume occupied by red blood cells, which ranges from about 40% (in women) to 45% (in men). Individuals with **anemia,** too few red blood cells, can sometimes be treated with the hormone erythropoietin to increase red blood cell (erythrocyte) production by bone marrow.

The hemoglobin in red blood cells, like myoglobin, binds O_2 reversibly, but it does not exhibit the simple behavior of myoglobin. A plot of fractional saturation (Y) versus pO_2 for hemoglobin is sigmoidal (S-shaped) rather than hyperbolic (**Fig. 5.7**). Furthermore, hemoglobin's overall oxygen affinity is lower than that of myoglobin: Hemoglobin is half-saturated at an oxygen pressure of 34.58 mbar (p_{50} = 34.58 mbar), whereas myoglobin is half-saturated at 3.74 mbar.

Why is hemoglobin's binding curve sigmoidal? At low O_2 concentrations, hemoglobin appears to be reluctant to bind the first O_2, but as the pO_2 increases, O_2 binding increases sharply, until hemoglobin is almost fully saturated. A look at the binding curve in reverse shows that at high O_2 concentrations, oxygenated hemoglobin is reluctant to give up its first O_2, but as the pO_2 decreases, all the O_2 molecules are easily given up. This behavior suggests that the binding of the first O_2 increases the affinity of the remaining O_2-binding sites. Apparently, *hemoglobin's four heme groups are not independent but communicate with each other in order to work in a unified fashion.* This is known as **cooperative binding** behavior. In fact, the fourth O_2 taken up by hemoglobin binds with about 100 times greater affinity than the first.

Figure 5.7 Oxygen binding to hemoglobin. The relationship between fractional saturation (Y) and oxygen concentration (pO_2) is sigmoidal. The pO_2 at which hemoglobin is half-saturated ($p50$) is 34.58 mbar. For comparison, myoglobin's O_2-binding curve is indicated by the dashed line. The difference in oxygen affinity between hemoglobin and myoglobin ensures that O_2 bound to hemoglobin in the lungs is released to myoglobin in the muscles. This oxygen-delivery system is efficient because the tissue pO_2 corresponds to the part of the hemoglobin binding curve where the O_2 affinity falls off most sharply.

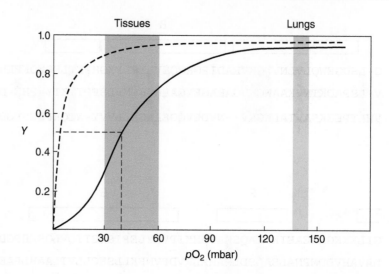

Question Could this O_2-delivery system still operate if the $p50$ for myoglobin was twice as high?

Hemoglobin's relatively low oxygen affinity and its cooperative binding behavior are the keys to its physiological function (see Fig. 5.7). In the lungs, where the pO_2 is about 133 mbar, hemoglobin is about 95% saturated with O_2. In the tissues, where the pO_2 is only about 26.6 to 53.2 mbar, hemoglobin's oxygen affinity drops off rapidly (it is only about 55% saturated when the pO_2 is 40 mbar). Under these conditions, *the O_2 released from hemoglobin is readily taken up by myoglobin in muscle cells, since myoglobin's affinity for O_2 is much higher*, even at the lower oxygen pressure. Myoglobin can therefore relay O_2 from the blood to the muscle cells' mitochondria, where it is consumed in the oxidative reactions that sustain muscle activity. Agents such as carbon monoxide, which interferes with O_2 binding to hemoglobin, prevent the efficient delivery of O_2 to cells (**Box 5.A**).

A conformational shift explains hemoglobin's cooperative behavior

The four heme groups of hemoglobin must be able to sense one another's oxygen-binding status so that they can bind or release their O_2 in a coordinated manner. However, the four heme groups are 25 to 37 Å apart, too far for them to communicate via an electronic signal. Therefore, the signal

---

| **Box 5.A** | **Carbon Monoxide Poisoning** |

The affinity of hemoglobin for carbon monoxide is about 250 times higher than its affinity for oxygen. However, the concentration of CO in the atmosphere is only about 0.1 ppm (parts per million by volume), compared to an O_2 concentration of about 200,000 ppm. Normally, only about 1% of the hemoglobin molecules in an individual are in the carboxyhemoglobin (Hb·CO) form, probably as a result of endogenous production of CO in the body (CO acts as a signaling molecule, although its physiological role is not well understood).

Danger arises when the fraction of carboxyhemoglobin rises, which can occur when individuals are exposed to high levels of environmental CO. For example, the incomplete combustion of fuels, as occurs in gas-burning appliances and vehicle engines, releases CO. The concentration of CO can rise to about 10 ppm in these situations and to as high as 100 ppm in highly polluted urban areas. The concentration of carboxyhemoglobin may reach 15% in some heavy smokers, although the symptoms of CO poisoning are usually not apparent.

CO toxicity, which occurs when the concentration of carboxyhemoglobin rises above about 25%, causes neurological impairment, usually dizziness and confusion. High doses of CO, which cause carboxyhemoglobin levels to rise above 50%, can trigger coma and death. When CO is bound to some of the heme groups of hemoglobin, O_2 is not able to bind to those sites because its low affinity means that it cannot displace the bound CO. In addition, the carboxyhemoglobin molecule remains in a high-affinity conformation, so that even if O_2 does bind to some of the hemoglobin heme groups in the lungs, O_2 release to the tissues is impaired. The effects of mild CO poisoning are largely reversible through the administration of O_2, but because the CO remains bound to hemoglobin with a half-life of several hours, recovery is slow.

Question Sketch the oxygen-binding curves of hemoglobin and carboxyhemoglobin [Hb·(CO)$_2$].

must be mechanical. In a mechanism first proposed by Max Perutz, the four globin subunits undergo conformational changes when they bind O_2.

In **deoxyhemoglobin** (hemoglobin without any bound O_2), the heme Fe ion has five ligands, so the porphyrin ring is somewhat dome-shaped and the Fe lies about 0.6 Å out of the plane of the porphyrin ring. As a result, the heme group is bowed slightly toward His F8 (**Fig. 5.8**). When O_2 binds to produce **oxyhemoglobin,** the Fe—now with six ligands—moves into the center of the porphyrin plane. This movement of the Fe ion pulls His F8 farther toward the heme group, and this in turn drags the entire F helix so that it moves as much as 1 Å. The F helix cannot move in this manner unless the entire protein alters its conformation, culminating in the rotation of one αβ unit relative to the other. Consequently, *hemoglobin can alternate between two quaternary structures, corresponding to the oxy and deoxy states.*

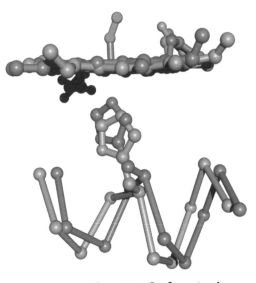

Figure 5.8 Conformational changes in hemoglobin upon O_2 binding. In deoxyhemoglobin (blue), the porphyrin ring is slightly bowed down toward His F8 (shown in ball-and-stick form). The remainder of the F helix is represented by its alpha carbon atoms. In oxyhemoglobin (purple), the heme group becomes planar, pulling His F8 and its attached F helix upward. The bound O_2 is shown in red.

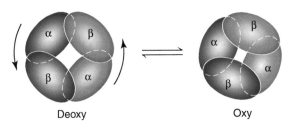

Deoxy   Oxy

The shift in conformation between the oxy and deoxy states primarily involves rotation of one αβ unit relative to the other. Oxygen binding decreases the size of the central cavity between the four subunits and alters some of the contacts between subunits. The two conformational states of hemoglobin are formally known as **T** (for "tense") and **R** (for "relaxed"). The T state corresponds to deoxyhemoglobin, and the R state corresponds to oxyhemoglobin.

Deoxyhemoglobin is reluctant to bind the first O_2 molecule because the protein is in the deoxy (T) conformation, which is unfavorable for O_2 binding (the Fe atom lies out of the heme plane). However, once O_2 has bound, probably to the α chain in each αβ pair, the entire tetramer switches to the oxy (R) conformation as the Fe atom and the F helix move. An intermediate conformation would be unstable, because the contacts between the αβ units would be unfavorable (**Fig. 5.9**).

Subsequent O_2 molecules bind with higher affinity because the protein is already in the oxy (R) conformation, which is favorable for O_2 binding. Similarly, oxyhemoglobin tends to retain its bound O_2 molecules until the oxygen pressure drops significantly. Then some O_2 is released, triggering

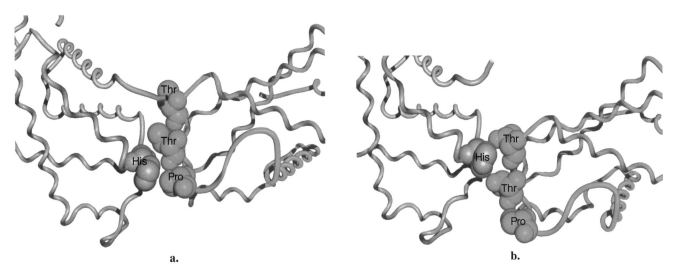

a.   b.

Figure 5.9 Some of the subunit interactions in hemoglobin.   The interactions between the αβ units of hemoglobin include contacts between side chains. The relevant residues are shown in space-filling form. **a.** In deoxyhemoglobin, a histidine residue on the β chain (blue, left) fits between a proline and a threonine residue on the α chain (green, right). **b.** Upon oxygenation, the His residue moves between two Thr residues. An intermediate conformation (between the deoxy and oxy conformations) is unlikely in part because the highlighted side chains would experience strain.

the change to the deoxy (T) conformation. This decreases the affinity of the remaining bound O_2 molecules, making it easier for hemoglobin to unload its bound oxygen. Because measurements of O_2 binding reflect the average behavior of many individual hemoglobin molecules, the result is a smooth curve (as shown in Fig. 5.7).

The first models for hemoglobin's cooperative binding behavior required that all four protein subunits instantaneously snap from one conformation to the other, in an all-or-none fashion. The model was subsequently refined, based on more detailed structural data, to allow each subunit to alternate between two tertiary structures that sequentially contribute to the overall change in quaternary structure:

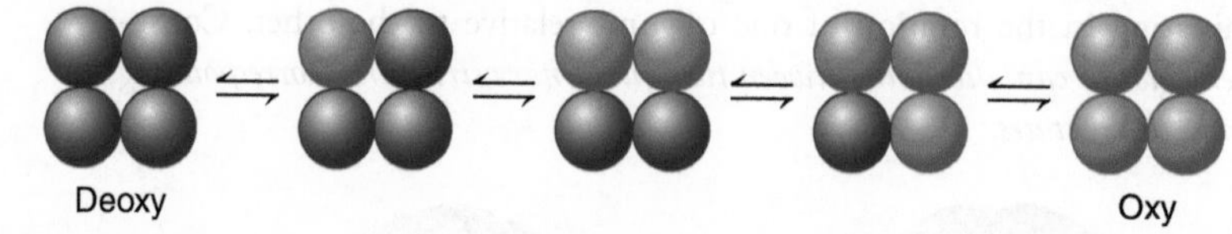

This model seemed to be more mechanistically plausible than the first. However, methods other than X-ray crystallography (Section 4.6) have provided evidence that hemoglobin—like many globular proteins (Section 4.3)—is more dynamic than originally thought, and the tetramer probably fluctuates among a large number of slightly different conformations. The precise structural changes that couple O_2 binding to the deoxy-to-oxy quaternary shift are still not well understood

Hemoglobin and many other proteins with multiple binding sites are known as **allosteric proteins** (from the Greek *allos,* meaning "other," and *stereos,* meaning "space"). In these proteins, *the binding of a small molecule (called a **ligand**) to one site alters the ligand-binding affinity of the other sites.* In principle, the ligands need not be identical, and their binding may either increase or decrease the binding activity of the other sites. In hemoglobin, the ligands are all oxygen molecules, and O_2 binding to one subunit of the protein increases the O_2 affinity of the other subunits.

H⁺ ions and bisphosphoglycerate regulate oxygen binding to hemoglobin *in vivo*

Decades of study have revealed the detailed chemistry behind hemoglobin's activity (and have also revealed how molecular defects can lead to disease). The conformational change that transforms deoxyhemoglobin to oxyhemoglobin alters the microenvironments of several ionizable groups in the protein, including the two N-terminal amino groups of the α subunits and the two histidine residues near the C-terminus of the β subunits. As a result, these groups become more acidic and release H^+ when O_2 binds to the protein:

$$Hb \cdot H^+ + O_2 \rightleftharpoons Hb \cdot O_2 + H^+$$

Therefore, increasing the pH of a solution of hemoglobin (decreasing $[H^+]$) favors O_2 binding by "pushing" the reaction to the right, as written above. Decreasing the pH (increasing $[H^+]$) favors O_2 dissociation by "pushing" the reaction to the left. *The reduction of hemoglobin's oxygen-binding affinity when the pH decreases is known as the **Bohr effect**.*

The Bohr effect plays an important role in O_2 transport *in vivo*. Tissues release CO_2 as they consume O_2 in respiration. The dissolved CO_2 enters red blood cells, where it is rapidly converted to bicarbonate (HCO_3^-) by the action of the enzyme carbonic anhydrase (see Section 2.5):

$$CO_2 + H_2O \rightleftharpoons HCO_3^- + H^+$$

The H^+ released in this reaction induces hemoglobin to unload its O_2 (**Fig. 5.10**). In the lungs, the high concentration of oxygen promotes O_2 binding to hemoglobin. This causes the release of protons that can then combine with bicarbonate to re-form CO_2, which is breathed out.

Red blood cells use one additional mechanism to fine-tune hemoglobin function. These cells contain a three-carbon compound, 2,3-bisphosphoglycerate (BPG):

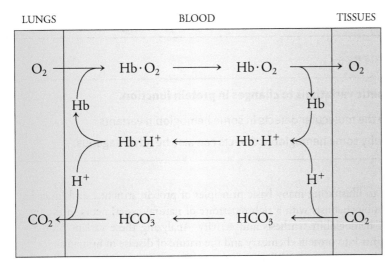

LUNGS   BLOOD   TISSUES

Figure 5.10 Oxygen transport and the Bohr effect.   Hemoglobin picks up O_2 in the lungs. In the tissues, H^+ derived from the metabolic production of CO_2 decreases hemoglobin's affinity for O_2, thereby promoting O_2 release to the tissues. Back in the lungs, hemoglobin binds more O_2, releasing the protons, which recombine with bicarbonate to re-form CO_2.

Question **Write the net equation for the process shown in the diagram.**

2,3-Bisphosphoglycerate (BPG)

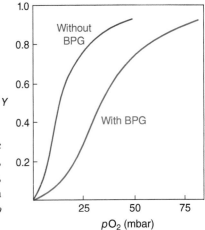

Figure 5.11 Effect of BPG on hemoglobin.   BPG binds to deoxyhemoglobin but not to oxyhemoglobin. It therefore reduces hemoglobin's O_2 affinity by stabilizing the deoxy conformation.

BPG binds in the central cavity of hemoglobin, but only in the T (deoxy) state. The five negative charges in BPG interact with positively charged groups in deoxyhemoglobin; in oxyhemoglobin, these cationic groups have moved and the central cavity is too narrow to accommodate BPG. Thus, *the presence of BPG stabilizes the deoxy conformation of hemoglobin.* Without BPG, hemoglobin would bind O_2 too tightly to release it to cells. In fact, hemoglobin stripped of its BPG *in vitro* exhibits very strong O_2 affinity, even at low pO_2 (**Fig. 5.11**).

The fetus takes advantage of this chemistry to obtain O_2 from its mother's hemoglobin. Fetal hemoglobin has the subunit composition $\alpha_2\gamma_2$. In the γ chains, position H21 is not histidine (as it is in the mother's β chain) but serine. His H21 bears one of the positive charges important for binding BPG in adult hemoglobin. The absence of this interaction in fetal hemoglobin reduces BPG binding. Consequently, *hemoglobin in fetal red blood cells has a higher O_2 affinity than adult hemoglobin, which helps transfer O_2 from the maternal circulation across the placenta to the fetus.*

Concept Check

1. Prove that $pO_2 = p_{50}$ when $Y = 0.5$.
2. Why does myoglobin require a prosthetic group?
3. Identify His F8 and His E7 in Figure 5.5.
4. Explain how the sequences of homologous proteins provide information about residues that are essential and nonessential for a protein's function.
5. Sketch the oxygen-binding curves for myoglobin and hemoglobin.
6. Link hemoglobin's structural features to its ability to bind O_2 cooperatively.
7. Describe how myoglobin and hemoglobin together make an efficient O_2-delivery system.
8. Explain how the Bohr effect and BPG regulate O_2 transport *in vivo*.

5.2 Clinical Connection: Hemoglobin Variants

KEY CONCEPTS

Relate genetic variations to changes in protein function.

- Describe the molecular defects in some hemoglobin variants.
- Explain why some hemoglobin defects can also be advantageous.

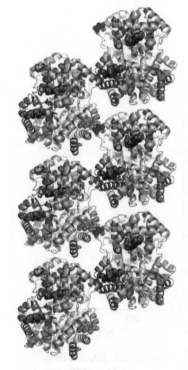

Figure 5.12 Polymerized hemoglobin S. In this model, the heme groups are red and the mutant valine residues are blue.

In addition to illustrating many basic principles of protein structure and function, hemoglobin also supplies biochemists with a rich repertoire of natural experiments in the form of inherited disorders of hemoglobin synthesis and activity. Analyzing these variant proteins has provided further insights into protein chemistry and the nature of disease in humans.

As much as 7% of the world's population carries a variant of the genes that encode the α and β chain of hemoglobin. These mutations produce hemoglobin proteins with altered amino acid sequences. In most cases, the mutation is benign and the hemoglobin molecules function more or less normally. However, in other cases, the mutation results in serious physical complications for the individual, as the ability of the mutant hemoglobin to deliver oxygen to cells is compromised. Mutant hemoglobins are often unstable, which may also result in destruction of the red blood cell, leading to anemia. Over 1200 hemoglobin variants have been discovered, but one of the best-known is sickle cell hemoglobin (known as hemoglobin S or Hb S; the normal hemoglobin is called hemoglobin A). Individuals with two copies of the defective gene develop sickle cell disease, a debilitating condition that predominantly affects populations of African descent.

The discovery of the molecular defect that causes sickle cell disease was a groundbreaking event in biochemistry. The disease was first described in 1910, but for many years there was no direct evidence that sickle cell disease—or any genetic disease—was the result of an alteration in the molecular structure of a protein. Then in 1949, Linus Pauling showed that hemoglobin from patients with sickle cell disease had a different electrical charge than hemoglobin from healthy individuals. The difference turned out to be due to a single amino acid difference: the sixth residue of the β chain, which is normally glutamate, is replaced by valine in hemoglobin S. This was the first evidence that an alteration in a gene caused an alteration in the amino acid sequence of the corresponding polypeptide. The mutation is described in Section 3.3.

In normal hemoglobin, the switch from the oxy to the deoxy conformation exposes a hydrophobic patch on the protein surface between the E and F helices. The hydrophobic valine residues on hemoglobin S are optimally positioned to bind to this patch. This intermolecular association leads to the rapid aggregation of hemoglobin S molecules to form long, rigid fibers (**Fig. 5.12**).

These fibers physically distort the red blood cell, producing the familiar sickle shape. Because hemoglobin S aggregation occurs only among deoxyhemoglobin S molecules, sickling tends to occur when the red blood cells pass through oxygen-poor capillaries. The misshapen cells can obstruct blood flow and rupture, leading to the intense pain, organ damage, and loss of red blood cells that characterize the disease (**Fig. 5.13**).

Figure 5.13 Normal and sickled red blood cells. Normal cells (at center) are rounded and can flex slightly when they pass through capillaries. Sickled cells have an elongated C shape and are more likely to rupture when passing through a small capillary.

The high frequency of the allele for sickle cell disease (that is, the mutated β globin gene) was at first puzzling: Alleles that lead to disabling conditions tend to be rare because individuals with two copies of the allele usually die before they can pass the allele to their offspring (see Box 3.A). However, carriers of the sickle cell variant appear to have a selective advantage. They are protected against malaria, an illness caused primarily by the intracellular parasite *Plasmodium falciparum.* Malaria afflicts about 225 million people and kills about 1 million each year, mostly children. In fact, the sickle cell hemoglobin variant is common in regions of the world where malaria is endemic.

In heterozygotes (individuals with one normal and one defective β globin allele), only about 2% of red blood cells undergo sickling. The sickling and loss of a small proportion of parasite-infected red blood cells probably has no significant impact on the overall number of parasites circulating in the body, as was once thought. Instead, the lysis (breakage) of the sickled cells releases hemoglobin, whose free heme groups are toxic. In response to the heme groups, the body increases production of an enzyme, called heme oxygenase, to degrade the heme. A byproduct of this reaction is carbon monoxide, which in small amounts is a signal for cells to minimize their response to inflammatory signals. Consequently, the inflammation and tissue damage that would normally occur during a *Plasmodium* infection are less severe. Experiments in mice indicate that hemoglobin S does not interfere with the life cycle of the *Plasmodium* parasite, but it makes a typical bout of malaria less likely to be fatal.

Like hemoglobin S, hemoglobin C confers protection against malaria. Coincidentally, the same amino acid of the hemoglobin β chain is affected, but in this case the glutamate is replaced by lysine. The variant hemoglobin C does not aggregate as hemoglobin S does, but it makes the red blood cells a bit more rigid than normal, leading to mild anemia. The antimalarial effect of hemoglobin C is not well understood; one possibility is that the variant hemoglobin interferes with the placement of a parasite protein on the surface of the red blood cell. As a result, parasite-infected cells are less able to stick to the walls of blood vessels, a feature of *Plasmodium* infection, and instead are more likely to be destroyed by the spleen, where worn-out red cells are normally taken out of circulation.

The **thalassemias** result from genetic defects that reduce the rate of synthesis of the α or β globin chains. Over 300 forms of thalassemia have been identified. These disorders are prevalent in the Mediterranean area (the name comes from the Greek word *thalassa,* meaning "sea") and in South Asia. Depending on the nature of the mutation, individuals with thalassemia may experience mild to severe anemia and their red blood cells may be smaller than normal. However, like heterozygotes for hemoglobin S and hemoglobin C, individuals with thalassemia are more likely to survive malaria.

Table 5.1 lists some hemoglobin residues that are critical for normal function; their mutation produces clinical symptoms. Because hemoglobin is a transport protein that must efficiently bind O_2 in the lungs but give it up in the tissues, variations that interfere with O_2 binding or allow it to bind too tightly can both lead to impaired O_2 delivery.

Table 5.1 Some Hemoglobin Variants

Variant[a]	Chain	Position of mutation	Amino acid change	Role of normal residue
Milledgeville	α	44	Pro → Leu	Participates in the formation of the α–β interface in the deoxy form but not the oxy form.
Chesapeake	α	92	Arg → Leu	Participates in α–β contacts.
Singapore	α	141 (C-terminus)	Arg → Pro	Its COO⁻ forms an ion pair with Lys 127 and its side chain forms an ion pair with Asp 126 in the deoxy form.
Providence	β	82	Lys → Asn	Forms an ion pair to BPG in the central cavity.
Kansas	β	102	Asn → Thr	Participates in subunit–subunit contacts.
Syracuse	β	146	His → Pro	The side-chain imidazole ring forms an ion pair with Asp 94. It also forms an ion pair with BPG in the central cavity.

[a]Hemoglobin variants are usually named for the place where they were first observed or characterized.

Concept Check

1. Predict the physiological effects of hemoglobin genetic changes that lead to higher O_2 affinity, lower O_2 affinity, loss of cooperativity, unstable protein structure, or red blood cell lysis.

5.3 Structural Proteins

KEY CONCEPTS

Compare the structures and functions of structural proteins.

- Describe the cellular functions of actin filaments, microtubules, and intermediate filaments.
- Contrast the assembly of fibers from globular and fibrous protein subunits.
- Relate fiber structures to their ability to assemble and disassemble.
- Describe the amino acid sequence constraints in intermediate filaments and collagen.

The shapes of eukaryotic cells, particularly those without an external cell wall, are determined by an intracellular network of proteins known as the **cytoskeleton.** Typically, three types of cytoskeletal proteins form fibers that extend throughout the cell (**Fig. 5.14**). These are **actin filaments** (with a diameter of about 70 Å), **intermediate filaments** (with a diameter of about 100 Å), and **microtubules** (with a diameter of about 240 Å). In large multicellular organisms, fibers of the protein collagen provide structural support extracellularly. Bacterial cells also contain proteins that form structures similar to actin filaments and microtubules. In the following discussion, note how the structure of each protein influences the overall structure and flexibility of the fiber as well as the fiber's ability to disassemble and reassemble.

Actin filaments are most abundant

A major portion of the eukaryotic cytoskeleton consists of actin filaments, also known as microfilaments, which are polymers of the protein actin. In many cells, a network of actin filaments supports the plasma membrane and therefore determines cell shape (see Fig. 2.6 and Fig. 5.14).

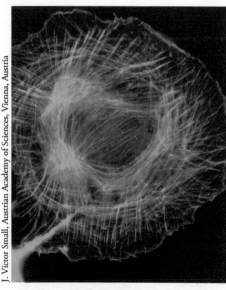

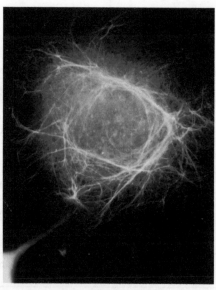

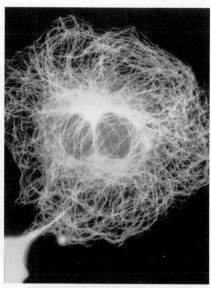

J. Victor Small, Austrian Academy of Sciences, Vienna, Austria

Actin filaments   **Intermediate filaments**   **Microtubules**

Figure 5.14 Distribution of cytoskeletal fibers in a single cell.   To make these micrographs, each type of fiber was labeled with a fluorescent probe that binds specifically to one type of cytoskeletal protein. Note how the distribution of actin filaments differs somewhat from that of intermediate filaments and microtubules.

Question What type of fiber best defines the cell's nucleus?

Monomeric actin is a globular protein with about 375 amino acids (**Fig. 5.15**). On its surface is a cleft in which adenosine triphosphate (ATP) binds. The adenosine group slips into a pocket on the protein, and the ribose hydroxyl groups and the phosphate groups form hydrogen bonds with the protein.

Polymerized actin is sometimes referred to as **F-actin** (for filamentous actin, to distinguish it from **G-actin,** the globular monomeric form). *The actin polymer is actually a double chain of subunits in which each subunit contacts four neighboring subunits* (**Fig. 5.16**). Each actin subunit has the same orientation (for example, all the nucleotide-binding sites point up in Fig. 5.16), so the assembled fiber has a distinct polarity. The end with the ATP site is known as the **(–) end,** and the opposite end is the **(+) end.**

Initially, polymerization of actin monomers is slow because actin dimers and trimers are unstable. However, once a longer polymer has formed, subunits add to both ends. Addition is usually much more rapid at the (+) end (hence its name) than at the (–) end (**Fig. 5.17**).

Actin polymerization is driven by the hydrolysis of ATP (splitting ATP by the addition of water) to produce ADP + inorganic phosphate (**P$_i$**):

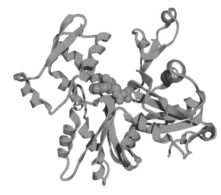

Figure 5.15 Actin monomer.
This protein assumes a globular shape with a cleft where ATP (green) binds.

Adenosine triphosphate (ATP)   + H$_2$O $\longrightarrow$

Adenosine diphosphate (ADP)   Inorganic phosphate (P$_i$)

This reaction is catalyzed by F-actin but not by G-actin. Consequently, *most of the actin subunits in a filament contain bound ADP.* Only the most recently added subunits still contain ATP. Because ATP-actin and ADP-actin assume slightly different conformations, proteins that interact with actin filaments may be able to distinguish rapidly polymerizing (ATP-rich) actin filaments from longer-established (ADP-rich) actin filaments.

Actin filaments continuously extend and retract

Actin filaments are dynamic structures. *Polymerization of actin monomers is a reversible process, so the polymer undergoes constant shrinking and growing as subunits add to and dissociate from one or both ends of the microfilament* (see Fig. 5.17). When the rate of addition of subunits to one end of an actin filament matches the rate of removal of subunits at the other end, the polymer is said to be **treadmilling** (**Fig. 5.18**).

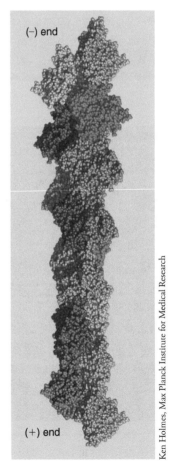

(–) end

(+) end

Ken Holmes, Max Planck Institute for Medical Research

Figure 5.16 Model of an actin filament.   The structure of F-actin was determined from X-ray diffraction data and computer model-building. Fourteen actin subunits are shown (all are different colors except the central actin subunit, whose two halves are blue and gray).

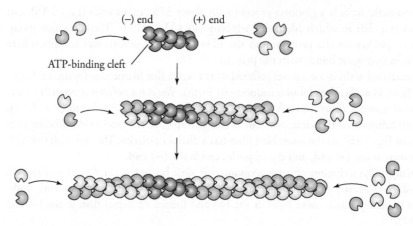

(−) end   (+) end

ATP-binding cleft

Figure 5.17 Actin filament assembly.   An actin filament grows as subunits add to its ends. Subunits usually add more rapidly to the (+) end, which therefore grows faster than the (−) end. The original segment is shaded more darkly. Actual actin filaments are much longer than depicted here.

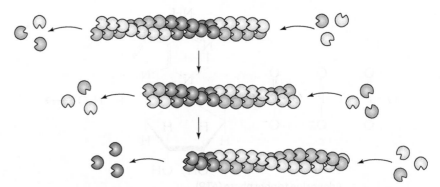

Figure 5.18 Actin filament treadmilling.   Assembly at one end balances dissociation at the other end so that there is no net change in length. The original segment (darker color) appears to travel along the filament during treadmilling.

Question   Explain why the original segment usually moves toward the (−) end.

a.

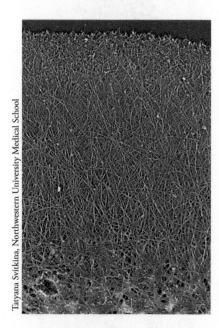

b.

Figure 5.19 Actin filament dynamics in cell crawling. **a.** Scanning electron micrograph of crawling cells. The leading edges of the cells are at the lower left, and the trailing edges are at the upper right. **b.** Organization of actin filaments in a fish epithelial cell. The leading edge of the cell is at the top.

Calculations suggest that under cellular conditions, the equilibrium between monomeric actin and polymeric actin favors the polymer. However, the growth of actin filaments *in vivo* is limited by capping proteins that bind to and block further polymerization at the (+) or (−) ends. A process that removes an actin filament cap will target growth to the uncapped end. New actin filaments can also grow as branches that emerge from an exiting filament at an angle of about 70°.

A supply of actin monomers to support actin filament growth in one area must come at the expense of actin filament disassembly elsewhere. In a cell, certain proteins sever *actin filaments* by binding to a polymerized actin subunit and inducing a small structural change that weakens actin–actin interactions and thereby increases the likelihood that the filament will break at that point. Actin subunits can then dissociate from the newly exposed ends unless they are subsequently capped.

Capping, branching, and severing proteins, along with other proteins whose activity is sensitive to extracellular signals, regulate the assembly and disassembly of actin filaments. A cell containing a network of actin filaments can therefore change its shape as the filaments lengthen in one area and regress in another. Certain cells use this system to move. When a cell crawls along a surface, *actin polymerization extends its "leading" edge, while depolymerization helps retract its "trailing" edge.* The leading edges, called lamellipodia (if broad) or filopodia (if narrow), appear ruffled where they become detached from the surface and are in the process of extending (**Fig. 5.19a**). The trailing edges of the cells, still attached to the surface, are gradually pulled toward the leading edge. The rate of actin polymerization is greatest at the leading edge, and this area shows a high density of branched filaments (Fig. 5.19b). Deeper within the cell, the filaments are sparser.

In an animal body, the rapid formation and outward extension of actin filaments allow certain cells to travel long distances during embryonic development, immune surveillance, wound healing, and **metastasis** (the spread of cancerous cells). Even in animal cells that do not move much, actin is the most abundant intracellular protein, accounting for as much as 5% of the cell's protein mass. A network of actin filaments, known as cortical actin, lies beneath the cell membrane in most cells, lending some support to the membrane and helping to define the cell's shape.

Among the dozens of proteins that interact with actin are cross-linking proteins that bundle 10–30 actin filaments, in parallel or antiparallel fashion, to form even stronger fibers deep within the cell. These so-called stress fibers are part of the cell's apparatus for maintaining its shape and generating force. This system is particularly well developed in muscle cells, where actin filaments are an essential part of the contractile machinery (see Section 5.4).

Tubulin forms hollow microtubules

Like actin filaments, microtubules are cytoskeletal fibers built from small globular protein subunits. Consequently, they share with actin filaments the ability to assemble and disassemble on a time scale that allows the cell to rapidly change shape in response to external or internal stimuli. Compared to a microtubule, however, an actin filament is a thin and flexible rod. *A microtubule is about three times thicker and much more rigid because it is constructed as a hollow tube.* Consider the following analogy: A metal rod with the dimensions of a pencil is easily bent. The same quantity of metal, fashioned into a hollow tube with a larger diameter but the same length, is much more resistant to bending.

Bicycle frames, plant stems, and bones are built on this same principle. Eukaryotic cells use hollow microtubules to reinforce other elements of the cytoskeleton (see Fig. 5.14), to construct cilia and flagella, and to align and separate pairs of chromosomes during mitosis.

The basic structural unit of a microtubule is the protein tubulin. Two monomers, known as α-tubulin and β-tubulin, form a dimer, and a microtubule grows by the addition of tubulin dimers. Each tubulin monomer contains about 450 amino acids, 40% of them identical in α- and β-tubulin. The core of tubulin consists of a four-stranded and a six-stranded β sheet surrounded by 12 α helices (**Fig. 5.20**).

Each tubulin subunit includes a nucleotide-binding site. Unlike actin, tubulin binds a guanine nucleotide, either guanosine triphosphate (GTP) or its hydrolysis product, guanosine diphosphate (GDP). When the dimer forms, the α-tubulin GTP-binding site becomes buried in the interface between the monomers. The nucleotide-binding site in β-tubulin remains exposed to the solvent (**Fig. 5.21**). After the tubulin dimer is incorporated into a microtubule and another dimer binds on top of it, the β-tubulin nucleotide-binding site is also sequestered from solvent. The GTP is then

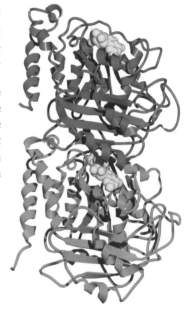

Figure 5.21 The tubulin dimer.   The guanine nucleotide (gold) in the α-tubulin subunit (*bottom*) is inaccessible in the dimer, whereas the nucleotide in the β-tubulin subunit (*top*) is more exposed to the solvent.

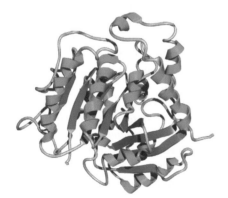

Figure 5.20 Structure of β-tubulin.   The strands of the two β sheets are shown in blue, and the 12 α helices that surround them are green.

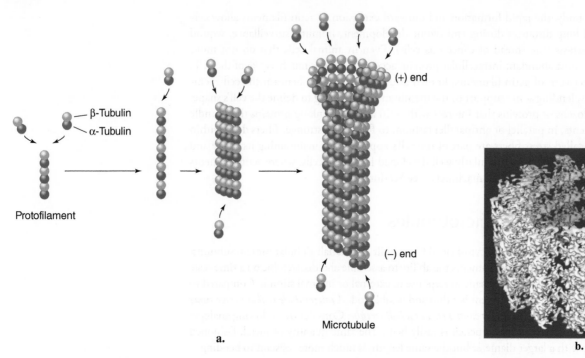

Kenneth Downing, Lawrence Berkeley National Laboratory

β-Tubulin
α-Tubulin

Protofilament

(+) end

(−) end

Microtubule

a.

b.

Figure 5.22 Assembly of a microtubule. **a.** αβ Dimers of tubulin initially form a linear protofilament. Protofilaments associate side by side, ultimately forming a tube. Tubulin dimers can add to either end of the microtubule, but growth is about twice as fast at the (+) end. **b.** Cryo-electron microscopy view of a microtubule.

Question Compare a microtubule and an actin filament (Fig. 5.16) in terms of strength and speed of assembly.

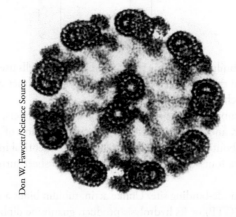

Don W. Fawcett/Science Source

Figure 5.23 Cross-section of a eukaryotic flagellum. This electron micrograph shows the nine microtubule doublets linked by "spoke" proteins to the central pair. Motor proteins (Section 5.4) connect adjacent microtubule doublets.

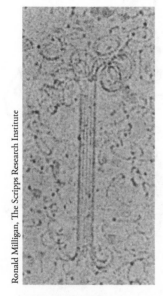

Ronald Milligan, The Scripps Research Institute

Figure 5.24 Electron micrograph of a depolymerizing microtubule. The ends of protofilaments apparently curve away from the microtubule and separate before tubulin dimers dissociate.

hydrolyzed, but the resulting GDP remains bound to β-tubulin because it cannot diffuse away (the GTP in the α-tubulin subunit remains where it is and is not hydrolyzed).

Assembly of a microtubule begins with the end-to-end association of tubulin dimers to form a short linear **protofilament.** Protofilaments then align side-to-side in a curved sheet, which wraps around on itself to form a hollow tube that usually contains 13 protofilaments (**Fig. 5.22**). However, variations are common, and microtubules may contain anywhere from 9 to 16 protofilaments, depending on the species and the cell type. In addition, microtubule "doublets" run the length of eukaryotic flagella, which must be strong as well as flexible. The most common microtubule arrangement consists of nine doublets surrounding two central microtubules (**Fig. 5.23**). Cilia have a similar structure but are generally shorter than flagella.

A single microtubule can grow as tubulin dimers add to both ends. Like an actin filament, the microtubule is polar and one end grows more rapidly. *The (+) end, terminating in β-tubulin, grows about twice as fast as the (−) or α-tubulin end because tubulin dimers bind preferentially to the (+) end.* Disassembly of a microtubule also takes place at both ends but occurs more rapidly at the (+) end. Under conditions that favor depolymerization, the ends of the microtubule appear to fray (**Fig. 5.24**). This suggests that tubulin dimers do not simply dissociate individually from the microtubule ends but that the interactions between protofilaments weaken before the tubulin dimers come loose.

Microtubule assembly and disassembly are regulated by proteins that cross-link microtubules and promote or prevent depolymerization. Tubulin is also subject to a range of chemical modifications. Under certain conditions, microtubule treadmilling can occur when tubulin subunits add to the (+) end as fast as they leave the (–) end. In a cell, however, most microtubule growth and regression occur at the (+) ends, because the (–) ends are often anchored to some sort of organizing center.

During mitosis and meiosis (see Box 3.A), microtubules emanating from two organizing centers at opposite ends of the cell attach to the replicated chromosomes. When the microtubules shorten, the two sets of chromosomes are pulled apart (**Fig. 5.25**). Each chromosome is linked to the (+) end of a microtubule by a complex of proteins called the **kinetochore,** which slides along the microtubule as the microtubule frays (**Fig. 5.26**). In fact, this fraying action appears to generate all the force necessary to pull the chromosomes to opposite sides of the cell before the cell splits in half.

Drugs that interfere with microtubule dynamics can block cell division. For example, colchicine, a product of the meadow saffron plant, causes microtubules to depolymerize.

Jennifer Waters & Adrian Salic/Science Source

Figure 5.25 Microtubules in a dividing cell.   During mitosis, microtubules (green fluorescence) connect replicated chromosomes (blue fluorescence) to two organizing centers at opposite sides of the cell.

Colchicine

Colchicine binds at the interface between α- and β-tubulin in a dimer, facing the inside of the microtubule cylinder. The bound drug may induce a slight conformational change that weakens the lateral contacts between protofilaments. If enough colchicine is present, microtubules shorten and eventually disappear. Colchicine was first used over 2000 years ago to treat **gout** (inflammation stemming from the precipitation of uric acid in the joints) because it inhibits the action of the white blood cells that mediate inflammation.

Paclitaxel binds to β-tubulin subunits in a microtubule, but not to free tubulin, so it stabilizes the microtubule, preventing its depolymerization.

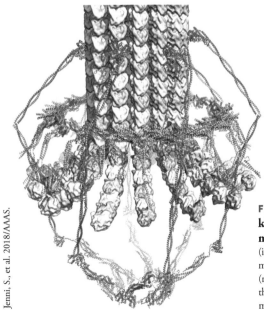

Jenni, S., et al. 2018/AAAS.

Figure 5.26 Model of a yeast kinetochore attached to a microtubule.   The kinetochore proteins (in color) encircle the microtubule (gray) and move upward as it frays. The chromosome (not shown here) is attached to the bottom of the kinetochore and also moves upward as the microtubule shortens.

Paclitaxel

The paclitaxel–tubulin interaction appears to include close contacts between paclitaxel's phenyl groups and hydrophobic residues such as phenylalanine, valine, and leucine. Paclitaxel was originally extracted from the slow-growing and endangered Pacific yew tree, but it can also be purified from more renewable sources or be chemically synthesized. *Paclitaxel is used as an anticancer agent because it blocks cell division and is therefore toxic to rapidly dividing cells such as tumor cells.*

Keratin is an intermediate filament

In addition to actin filaments and microtubules, eukaryotic cells—particularly those in multicellular organisms—contain intermediate filaments. With a diameter of about 100 Å, these fibers are intermediate in thickness to actin filaments and microtubules. *Intermediate filaments are exclusively structural proteins.* They play no part in cell motility and, unlike actin filaments and microtubules, they have no associated motor proteins. However, they do interact with actin filaments and microtubules via cross-linking proteins.

Intermediate filament proteins as a group are much more heterogeneous than the highly conserved actin and tubulin. For example, humans have about 65 intermediate filament genes. The lamins are the intermediate filaments that help form the nuclear lamina in animal cells, a 30–100-Å-thick network inside the nuclear membrane that helps define the nuclear shape and may play a role in DNA replication and transcription. In certain cells, intermediate filaments are much more abundant than actin filaments or microtubules and are most prominent in the dead remnants of epidermal cells—that is, in the hard outer layers of the skin—where they may account for 85% of the total protein (**Fig. 5.27**). The best-known intermediate filament proteins are the keratins, a large group of proteins that include the "soft" keratins, which help define internal body structures, and the "hard" keratins of skin, hair, and claws.

The basic structural unit of an intermediate filament is a dimer of α helices that wind around each other—that is, a **coiled coil.** The amino acid sequence in such a structure consists of seven-residue repeating units in which the first and fourth residues are predominantly nonpolar. In an α helix, these nonpolar residues line up along one side (**Fig. 5.28**). Because a nonpolar group appears on average every 3.5 residues, but there are 3.6 residues per α-helical turn, the strip of nonpolar residues actually winds slightly around the surface

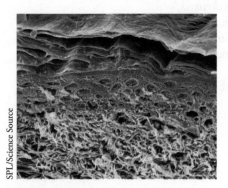

SPL/Science Source

Figure 5.27 Scanning electron micrograph of sectioned human skin. The layers of dead epidermal cells at the top consist mostly of keratin.

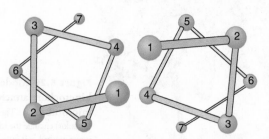

Figure 5.28 Arrangement of residues in a coiled coil. This view down the axis of two seven-residue α helices shows that amino acids at positions 1 and 4 line up on one side of each helix. Nonpolar residues occupying these positions form a hydrophobic strip along the sides of the helices.

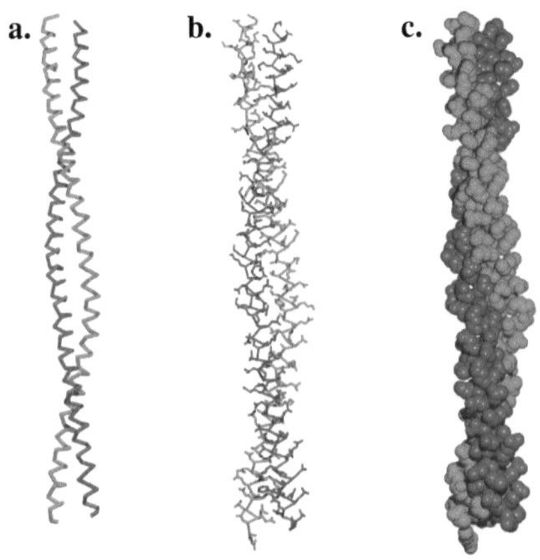

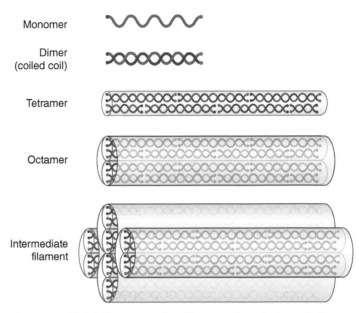

Figure 5.29 Three views of a coiled coil.
These models show a segment of the coiled coil from the protein tropomyosin. **a.** Backbone model. **b.** Stick model. **c.** Space-filling model. Each α-helical chain contains 100 residues. The nonpolar strips along each helix contact each other, so the two helices wind around each other in a gentle left-handed coil.

Question Which nonpolar residues would be most likely to appear at positions 1 and 4? Which would be least likely?

Figure 5.30 Model of an intermediate filament. Pairs of polypeptides form coiled coils. These dimers associate to form tetramers and so on, ultimately producing an intermediate filament composed of 16 to 32 polypeptides in cross-section. Although drawn as straight rods here, the intermediate filament and its component structures are probably all twisted around each other in some way, much like a man-made rope or cable.

of the helix. Two helices whose nonpolar strips contact each other therefore adopt a coiled structure with a left-handed twist (**Fig. 5.29**).

Each intermediate filament subunit contains a stretch of α helix flanked by nonhelical regions at the N- and C-termini. Two of these polypeptides interact in register (parallel and with ends aligned) to form a coiled coil. The dimers then associate in a staggered antiparallel arrangement to form higher-order fibrous structures (**Fig. 5.30**). The fully assembled intermediate filament may consist of 16 to 32 polypeptides in cross-section. The subunits may be identical or different. Note that no nucleotides are required for intermediate filament assembly. The N- and C-terminal domains may help align subunits during polymerization and interact with proteins that cross-link intermediate filaments to other cell components. Keratin fibers themselves are cross-linked through disulfide bonds between cysteine residues on adjacent chains.

An animal hair—for example, sheep's wool or the hair on your head—consists almost entirely of keratin filaments (**Fig. 5.31**). Hair resists deformation but can be stretched. Tensile stress breaks the hydrogen bonds between carbonyl and amino groups four residues apart in the keratin α helix. The helices can then be pulled until the polypeptides are fully extended. Additional force causes the polypeptide chain to break. If unbroken, the protein can spring back—at least partially—to its original α-helical conformation when the force is removed. This is why a wool sweater stretched out of shape gradually reverts to its former style.

Perhaps due to their cable-like construction, intermediate filaments undergo less remodeling than cellular fibers constructed from globular subunits such as actin and tubulin. For example, nuclear lamins disassemble and reassemble only once during the cell cycle, when the cell divides. Keratins, as part of dead cells, remain intact for years. In the innermost layers of animal skin, epidermal cells synthesize large amounts of keratin. As layers of cells move outward and die, their keratin molecules are pushed together to form a strong waterproof coating. Because keratin is so important for the integrity of the epidermis, mutations in keratin genes are linked to certain skin disorders. In diseases such as epidermolysis bullosa simplex (EBS), cells rupture when subject to normal mechanical stress. The result is separation of epidermal layers, which leads to blistering. The most severe cases of the disease arise from mutations in the most highly conserved regions of the keratin molecules, near the ends of the α-helical regions.

Figure 5.31 Scanning electron micrograph of a human hair. This nonliving structure is built from the remnants of keratin-packed cells, with a covering of flattened scale-like cells.

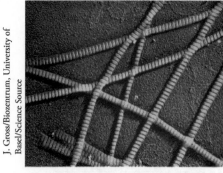

J. Gross/Biozentrum, University of Basel/Science Source

Figure 5.32 Electron micrograph of collagen fibers. Thousands of aligned collagen proteins yield structures with a diameter of 500 to 2000 Å, which are visible here.

Collagen is a triple helix

Unicellular organisms can get by with just a cytoskeleton, but multicellular animals must have a way to hold their cells together according to some characteristic body plan. Large animals—especially nonaquatic ones—must also support the body's weight. This support is provided largely by collagen, which is the most abundant animal protein. It plays a major structural role in the **extracellular matrix** (the material that helps hold cells together), in connective tissue within and between organs, and in bone. Its name was derived from the French word for *glue* (at a time when glue was derived from animal connective tissue).

There are at least 28 types of collagen with slightly different three-dimensional structures and physiological functions. The most familiar is the collagen from animal bones and tendons, which forms thick, ropelike fibers (**Fig. 5.32**). This type of collagen is a trimeric molecule about 3000 Å long but only 15 Å wide. As in all forms of collagen, the polypeptide chains have an unusual amino acid composition and an unusual conformation. Except in the extreme N- and C-terminal regions of the polypeptides (which are cleaved off once the protein exits the cell), every third amino acid is glycine, and about 30% of the remaining residues are proline and hydroxyproline (Hyp). Hyp residues result from the hydroxylation of Pro residues after the polypeptide has been synthesized, in a reaction that requires ascorbate (vitamin C; **Box 5.B**).

Proline   Hydroxyproline (Hyp)

For a stretch of about 1000 residues, each collagen chain consists of repeating triplets, the most common of which is Gly–Pro–Hyp. Glycine residues, which have only a hydrogen atom for a side chain, can normally adopt a wide range of secondary structures. However, the **imino groups** of proline and hydroxyproline residues (that is, their connected side chains and amino groups) constrain the geometry of the peptide group. *The most stable conformation for a polypeptide sequence containing repeating units of Gly–Pro–Hyp is a narrow left-handed helix* (**Fig. 5.33a**).

Box 5.B   Vitamin C Deficiency Causes Scurvy

Ascorbate (vitamin C)

In the absence of ascorbate (vitamin C), collagen contains too few hydroxyproline residues and hydroxylated lysine residues, so the resulting collagen fibers are relatively weak. Ascorbate also participates in enzymatic reactions involved in fatty acid breakdown and production of certain hormones. The symptoms of ascorbate deficiency include poor wound healing, loss of teeth, and easy bleeding—all of which can be attributed to abnormal collagen synthesis—as well as lethargy and depression.

Historically, ascorbate deficiency, known as scurvy, was common in sailors on long voyages where fresh fruit was unavailable. A remedy, in the form of a daily ration of limes, was discovered in the mid-eighteenth century. Unfortunately, citrus juice, which was also widely administered, proved much less effective in preventing scurvy because ascorbate is destroyed by heating and by prolonged exposure to air. For this reason, factors such as exercise and good hygiene were also believed to prevent scurvy.

Fruit is not the only source of ascorbate. Most animals—with the exception of bats, guinea pigs, and primates—produce ascorbate, so a diet containing fresh meat can supply sufficient ascorbate, which is vital in locations such as the far North, where fruit is not available.

Given the presence of ascorbate in many foods, a true deficiency is rare in adults. Scurvy does still occur, however, as a side effect of general malnutrition or in individuals consuming odd diets. Fortunately, the symptoms of scurvy, which is otherwise fatal, can be easily reversed by administering ascorbate or consuming fresh food.

Question Refer to the structure of ascorbate to explain why it **does not accumulate in the body but instead is readily lost via the kidneys.**

*In collagen, three polypeptides wind around each other to form a right-handed **triple helix*** (Fig. 5.33b–d). The chains are parallel but staggered by one residue so that glycine appears at every position along the axis of the triple helix. The glycine residues are all located in the center of the helix, whereas all other residues are on the periphery. A look down the axis of the triple helix shows why glycine—but no other residue—occurs in the center of the helix (**Fig. 5.34**). The side chain of any other residue would be too large to fit. In fact, replacing glycine with alanine, the next-smallest amino acid, greatly perturbs the structure of the triple helix.

The collagen triple helix is stabilized through hydrogen bonding. One set of interactions links the backbone N—H group of each glycine residue to a backbone C=O group in another chain. The geometry of the triple helix prevents the other backbone N—H and C=O groups from forming hydrogen bonds with each other, but they are able to interact with a highly ordered

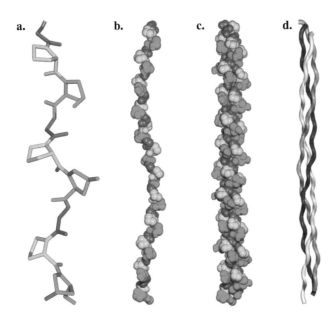

Figure 5.33 Collagen structure. **a.** A sequence of repeating Gly–Pro–Hyp residues adopts a secondary structure in which the polypeptide forms a narrow left-handed helix. The residues in this stick model are color-coded: Gly gray, Pro orange, Hyp red. H atoms are not shown. **b.** Space-filling model of a single collagen polypeptide. **c.** Space-filling model of the triple helix. **d.** Backbone trace showing the three polypeptides in different shades of gray. Each polypeptide has a left-handed twist, but the triple helix has a right-handed twist.

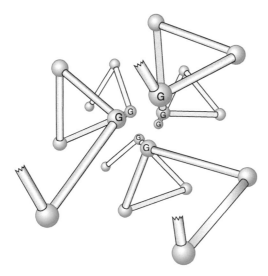

Figure 5.34 Cross-section of the collagen triple helix. In this view, looking down the axis of the three-chain molecule, each ball represents an amino acid and the bars represent peptide bonds. Glycine residues (which lack side chains and are marked by "G") are located in the center of the triple helix, whereas the side chains of other residues point outward from the triple helix.

Question Compare this axial view of a collagen triple helix to the axial view of a coiled coil (Fig. 5.28).

network of water molecules surrounding the triple helix like a sheath. Altogether, the collagen triple helix is strong and nonelastic—it resists stretching—but still slightly bendable.

Collagen molecules are covalently cross-linked

Trimeric collagen molecules assemble in the endoplasmic reticulum. After they are secreted from the cell, they are trimmed by proteases and align side-to-side and end-to-end to form the enormous fibers visible by electron microscopy (see Fig. 5.32). The fibers are strengthened by several kinds of cross-links. Because collagen polypeptides contain almost no cysteine, these links are not disulfide bonds. Instead, *the cross-links are covalent bonds between side chains that have been chemically modified following polypeptide synthesis.* For example, one kind of cross-link requires the enzyme-catalyzed oxidation of two lysine side chains, which then react to form a covalent bond. The number of these and other types of cross-links tends to increase with age, which explains why meat from older animals is tougher than meat from younger animals. Once in place, collagen fibers are resistant to degradation by proteases, which helps to account for their longevity in animal bodies.

Collagen fibers have tremendous tensile strength. On a per-weight basis, collagen is stronger than steel. Not all types of collagen form thick linear fibers, however. Many nonfibrillar collagens form sheetlike networks of fibers that support layers of cells in tissues. Often, several types of collagen are found together. Not surprisingly, defects in collagen affect a variety of organ systems (**Box 5.C**).

Box 5.C   Bone and Collagen Defects

Connective tissue, such as cartilage and bone, consists of cells embedded in a matrix containing proteins (mainly collagen) and a space-filling "ground substance" (mostly polysaccharides; see Section 11.3). The polysaccharides, which are highly hydrated, are resilient and return to their original shape after being compressed. The collagen fibers are strong and relatively rigid, resisting tensile (stretching) forces. Together, the polysaccharides and collagen give ligaments (which attach bone to bone) and tendons (which join muscles to bones) the appropriate degree of resistance and flexibility. The regular arrays of collagen fibers in a tendon are visible as horizontal bands in the photo below. Fibroblasts, which are collagen-producing cells, occupy the spaces between the bands. The connective tissues that surround muscles and organs contain collagen fibers arranged in sheetlike networks.

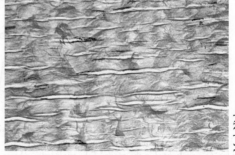

Mark Nielsen

In bone, the extracellular matrix is supplemented by minerals, mainly hydroxyapatite, $Ca(PO_4)_3(OH)$. This form of calcium phosphate forms extensive white crystals that account for up to 50% of the mass of bone. By itself, calcium phosphate is brittle. Yet bone is thousands of times stronger than the crystals because it is a composite structure in which the mineral is interspersed with collagen fibers. This layered arrangement dissipates stresses so that the bone remains strong but can "give" a little under pressure without shattering. Bone tissue also includes passageways for small blood vessels.

During development, most of the skeleton takes shape as cartilaginous tissue becomes mineralized, forming hard bone. The repair of a fractured bone follows a similar process in which fibroblasts moving to the injured site synthesize large amounts of collagen, chondroblasts produce cartilage, and osteoblasts gradually replace the cartilage with bony tissue. Mature bone continues to undergo remodeling, which begins with the release of enzymes and acid from cells known as osteoclasts. The enzymes digest collagen and other extracellular matrix components, while the low pH helps dissolve calcium phosphate. Osteoblasts then fill the void with new bone material. Some bones are remodeled faster than others, and the system responds to physical demands so that the bone becomes stronger and thicker when subjected to heavy loads. This is the basis for orthodontics: Teeth are realigned by placing stress on the bone in the tooth sockets for a period long enough for remodeling to take place. Bone remodeling also plays a role in cancer, as microscopic sites of bone destruction may provide access to metastatic (spreading) cancer cells; bone is a common site for metastatic tumor growth, especially in breast cancer.

The importance of collagen for the structure and function of connective tissues means that irregularities in the collagen protein itself or in the enzymes that process collagen molecules can lead to serious physical abnormalities. Hundreds of collagen-related mutations have been identified as the cause of at least 40 distinct diseases in humans. Because most tissues contain more than one type of collagen, the physiological manifestations of collagen mutations are highly variable.

Defects in collagen type I (the major form in bones and tendons) cause the congenital disease **osteogenesis imperfecta.** The primary symptoms of the disease include bone fragility, leading to easy fracture, long-bone deformation, and abnormalities of the skin and teeth. The photo below is an X-ray of a child with a moderately severe case of osteogenesis imperfecta.

Collagen type I, a trimeric molecule, contains two different types of polypeptide chains. Therefore, the severity of the disease

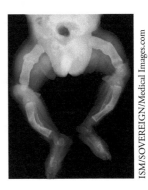

depends in part on whether one or two chains in a collagen molecule are affected. Furthermore, *the location and nature of the mutation determine whether the abnormal collagen retains some normal function.* For example, in one severe form of osteogenesis imperfecta, a 599-base deletion in a collagen gene represents the loss of a large portion of triple helix. The resulting protein is unstable and is degraded intracellularly. Milder cases of osteogenesis imperfecta result from amino acid substitutions, for example, the replacement of glycine by a bulkier residue. Other amino acid changes may slow intracellular processing and excretion of collagen polypeptides, which affects the assembly of collagen fibers. Osteogenesis imperfecta affects about one in 10,000 people.

Mutations in collagen type II, a form found in cartilage, lead to osteoarthritis. This genetic disease, which becomes apparent in childhood, is distinct from the osteoarthritis that can develop later in life, often after years of wear and tear on the joints. Defects in the proteins that process collagen extracellularly and help assemble collagen fibers lead to disorders such as dermatosparaxis, which is characterized by extreme skin fragility.

Ehlers–Danlos syndrome results from abnormalities in collagen type III, a molecule that is abundant in most tissues but is scarce in skin and bone. Symptoms of this phenotypically variable disorder include easy bruising, thin or elastic skin, and joint hyperextensibility. In one form of the disease, which is accompanied by a high risk for arterial rupture, the molecular defect is a mutation in a collagen type III gene. In another form of the disease, in which individuals often suffer from scoliosis (curvature of the spine), the collagen genes are normal. In these cases, the disease results from a deficiency of lysyl oxidase, the enzyme that modifies lysine residues so that they can participate in collagen cross-links. Ehlers–Danlos syndrome is both rarer and less severe than osteogenesis imperfecta, with many affected individuals surviving to adulthood.

Concept Check

1. Make a table to compare the subunit structure, overall size, dynamic character, and biological functions of actin filaments, microtubules, intermediate filaments, and collagen.
2. Discuss why some fibers are more easily disassembled and reassembled.
3. What is the basis of actin filament polarity?
4. Explain why not all filamentous structures have polarity.
5. List some factors that control the growth of structural fibers.
6. List some ways that cells can increase the strength of structural fibers.

5.4 Motor Proteins

KEY CONCEPTS

Explain how motor proteins operate.

- Compare the overall structures of myosin and kinesin.
- Describe how the energy of the ATP hydrolysis reaction is used to perform work.
- Contrast the independent action of myosin and the processive action of kinesin.

In eukaryotic cells, **motor proteins** act on structural elements such as actin filaments and microtubules to generate the movements that allow the cells to reorganize their contents, change their shape, and move. For example, muscle contraction is accomplished by the motor protein myosin pulling on actin filaments. Eukaryotic cells that move via the wavelike motion of cilia or flagella rely on the motor protein dynein, which acts by bending the array of microtubules first on one side and then on the other. Intracellular transport is carried out by motor proteins that move along actin filament and microtubule tracks. In all cases, the molecular machinery uses the chemical energy of ATP hydrolysis to carry out mechanical work. In this section, we focus on two well-characterized motor proteins, myosin and kinesin.

Myosin has two heads and a long tail

There are at least 35 different types of myosin, which is present in nearly all eukaryotic cells. Myosin works with actin to produce movement by transducing the free energy of the ATP hydrolysis reaction to mechanical energy. Muscle myosin, with a total molecular mass of about 540 kD, consists mostly of two large polypeptides that form two globular heads attached to a long tail (**Fig. 5.35**). Each head includes a binding site for actin and a binding site for an adenine nucleotide. In the tail region, the two polypeptides twist around each other to form a single rodlike coiled coil (the same structural motif that occurs in intermediate filaments). The "neck" that joins each myosin head to the tail region consists of an α helix about 100 Å long, around which are wrapped two small polypeptides (**Fig. 5.36**). These so-called light chains help stiffen the neck helix so that it can act as a lever.

Each myosin head can interact noncovalently with a subunit in an actin filament, but the two heads act independently and only one head binds to the actin filament at a given time. *In a series of steps that include protein conformational changes and the hydrolysis of ATP, the myosin head releases its bound actin subunit and rebinds another subunit closer to the (+) end of the actin filament.* Repetition of this reaction cycle allows myosin to progressively walk along the length of the actin filament.

In a muscle cell, hundreds of myosin tails associate to form a **thick filament** with the head domains sticking out (**Fig. 5.37**). These heads act as cross-bridges to **thin filaments,** which each consist of an actin filament and actin-binding proteins that regulate the accessibility of the actin subunits to myosin heads. When a muscle contracts, the multitude of myosin heads individually bind and release actin, like rowers working asynchronously, which causes the thin and thick filaments to slide past each other (**Fig. 5.38**). Because of the arrangement of filaments in the muscle cell, the action of myosin on actin results in an overall shortening of the muscle. This phenomenon is commonly called contraction, but the muscle does not undergo any compression and its volume remains constant—it actually becomes thicker around the middle. A shortening on the order of 20% of a muscle's length is typical; 40% is extreme.

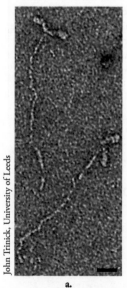

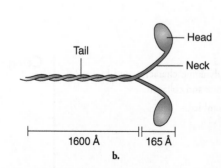

Tail
Head
Neck
1600 Å
165 Å

a.

b.

Figure 5.35 Structure of muscle myosin. **a.** Electron micrograph. **b.** Drawing of a myosin molecule. Myosin's two globular heads are connected via necks to myosin's tail, where the polypeptide chains form a coiled coil.

Question In what part of the protein are large hydrophobic residues most likely to occur?

John Trinick, University of Leeds

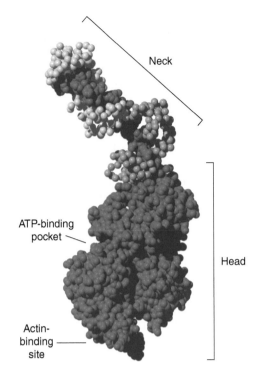

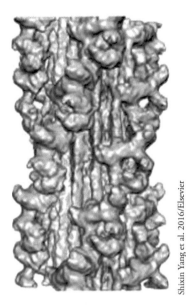

Figure 5.36 Myosin head and neck region. The myosin neck forms a molecular lever between the head domain and the tail. Two light chains (lighter shades of purple) help stabilize the α-helical neck. The actin-binding site is at the far end of the myosin head. ATP binds in a cleft near the middle of the head. Only the alpha carbons of the light chains are visible in this model.

Figure 5.37 Cryo-electron micrograph of a thick filament. The heads of many myosin molecules project laterally from the thick rod formed by the aligned myosin tails.

Shixin Yang et al. 2016/Elsevier

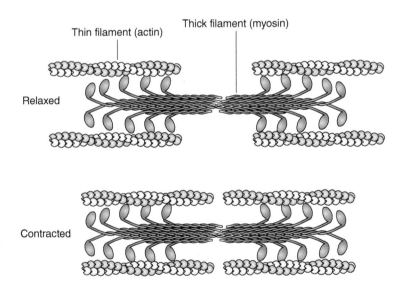

Figure 5.38 Movement of thin and thick filaments during muscle contraction.

Myosin operates through a lever mechanism

How does myosin work? The key to its mechanism is the hydrolysis of the ATP that is bound to the myosin head. Although the ATP-binding site is about 35 Å away from the actin-binding site, *the conversion of ATP to ADP + P_i (phosphate) triggers conformational changes in the myosin head that are communicated to the actin-binding site as well as to the lever (the neck region).*

The chemical reaction of ATP hydrolysis thereby drives the physical movement of myosin along an actin filament. In other words, the free energy of the ATP hydrolysis reaction is transformed into mechanical work.

The four steps of the myosin–actin reaction sequence are shown in **Figure 5.39**. Note how each event at the nucleotide-binding site correlates with a conformational change related to either actin binding or bending of the lever. An α helix makes an ideal lever because it can be quite long. It is also relatively incompressible, so it can pull the coiled-coil myosin tail along with it. Altogether, the lever swings by about 70° relative to the myosin head. The return of the lever to its original conformation (step 4 of the reaction cycle) is the force-generating step. The shortening and bulging of a contracting muscle actually augment its power by optimizing the angle at which the myosin heads tug on actin filaments. When adjusted for the difference in mass, the myosin–actin system has a power output comparable to a typical automobile.

Each cycle of ATP hydrolysis moves the myosin head by an estimated 50–100 Å. Since individual actin subunits are spaced about 55 Å apart along the thin filament, the myosin head advances by at least one actin subunit per reaction cycle. Each myosin head can hydrolyze 20 ATP per second. Because the reaction cycle involves several steps, some of which are essentially irreversible (such as ATP → ADP + P_i), the entire cycle is unidirectional.

Various types of myosin operate in many cells, not just in muscles. For example, myosin works with actin during **cytokinesis** (the splitting of the cell into two halves following mitosis), and some myosin proteins use their motor activity to transport certain cell components along

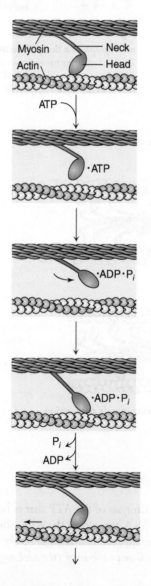

Myosin — Neck
Actin — Head

ATP

·ATP

·ADP·P_i

·ADP·P_i

P_i
ADP

Figure 5.39 The myosin–actin reaction cycle. For simplicity, only one myosin head is shown.

Question Write an equation that describes the reaction catalyzed by myosin.

1. *The reaction sequence begins with a myosin head bound to an actin subunit of the thin filament. ATP binding alters the conformation of the myosin head so that it releases actin.*

2. *The rapid hydrolysis of ATP to ADP + P_i triggers a conformational change that rotates the myosin lever and increases the affinity of myosin for actin.*

3. *Myosin binds to an actin subunit farther along the thin filament.*

4. *Binding to actin causes P_i and then ADP to be released. As these reaction products exit, the myosin lever returns to its original position. This causes the thin filament to move relative to the thick filament (the power stroke).*

ATP replaces the lost ADP to repeat the reaction cycle.

actin filament tracks. Myosin molecules also act as tension rods, pulling on the actin filaments that constitute a cell's stress fibers. In the sensory cells of the ear, myosin regulates the sensitivity of the sound-detecting structures (**Box 5.D**). The dimpled shape of red blood cells (see Fig. 5.13) is maintained by a type of myosin interacting with actin filaments positioned just beneath the cell membrane. All forms of myosin have an actin-activated motor domain, but they often differ in the proportion of the reaction cycle spent attached to actin. Muscle myosin spends most of its time detached from actin filaments, while myosin varieties involved in transport spend relatively more time anchored to actin.

Kinesin is a microtubule-associated motor protein

Many cells also contain motor proteins, such as kinesin, that move along microtubule tracks. There are 14 classes of kinesins; we will describe the prototypical one, also known as conventional kinesin.

Kinesin, like myosin, is a relatively large protein (with a molecular mass of 380 kD) and has two large globular heads and a coiled-coil tail domain (**Fig. 5.40**). Each 100-Å-long head consists of an eight-stranded β sheet flanked by three α helices on each side and includes a tubulin-binding site and a nucleotide-binding site. The light chains, situated at the opposite end of the protein, bind to proteins in the membrane shell of a **vesicle.** The vesicle and its contents become kinesin's cargo. *Kinesin moves its cargo toward the (+) end of a microtubule by stepping along the length of a single protofilament.* Other microtubule-associated

---

Box 5.D	Myosin Mutations and Deafness

Inside the cochlea, the spiral-shaped organ of the inner ear, are thousands of hair cells, each of which is topped with a bundle of bristles known as **stereocilia:**

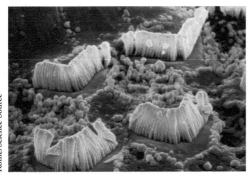

Prof. P. M. Motta/Univ. "La Sapienza", Rome/Science Source

Each stereocilium contains several hundred cross-linked actin filaments and is therefore extremely rigid, except at its base, where there are fewer actin filaments. Sound waves deflect the stereocilia at the base, initiating an electrical signal that is transmitted to the brain.

Myosin molecules probably help control the tension inside each stereocilium, so the ratcheting activity of the myosin motors along the actin filaments may adjust the sensitivity of the hair cells to different degrees of stimulus. Other myosin molecules whose tails bind certain cell constituents may use their motor activity to redistribute these substances along the length of the actin filaments. Abnormalities in any of these proteins could interfere with normal hearing.

About half of all cases of deafness have a genetic basis and over 100 different genes have been linked to deafness. One of

these codes for myosin type VIIa (muscle myosin is type II). Gene-sequencing studies indicate that myosin VIIa has 2215 residues and forms a dimer with two heads and a long tail. Its head domains convert the chemical energy of ATP into mechanical energy for movement relative to an actin filament. Over a hundred mutations in the myosin VIIa gene have been identified, including premature stop codons, amino acid substitutions, and deletions—all of which compromise the protein's function. Such mutations are responsible for many cases of **Usher syndrome,** the most common form of deaf-blindness in the United States. Usher syndrome is characterized by profound hearing loss, retinitis pigmentosa (which leads to blindness), and sometimes vestibular (balance) problems.

The congenital deafness of Usher syndrome results from the failure of the cochlear hair cells to develop properly. The unresponsiveness of the stereocilia to sound waves probably also accounts for their inability to respond normally to the movement of fluid in the inner ear, which is necessary for maintaining balance. Abnormal myosin also plays a role in the blindness that often develops in individuals with Usher syndrome, usually by the second or third decade. The intracellular transport function of myosin VIIa is responsible for distributing bundles of pigment in the retina. In retinitis pigmentosa, retinal neurons gradually lose their ability to transmit signals in response to light; in advanced stages of the disease, pigment actually becomes clumped on the retina.

Question **Not all mutations are associated with a loss of function. Explain how a myosin mutation that *increases* the rate of ADP release could lead to deafness.**

a.

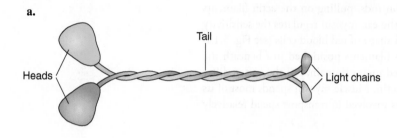

Heads

Tail

Light chains

b.

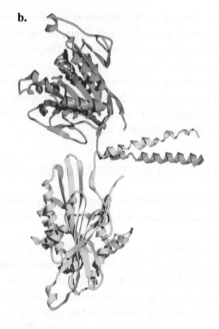

Figure 5.40 Structure of kinesin.   a. Diagram of the molecule. **b.** Model of the head and neck region. Each globular head, which contains α helices and β sheets, connects to an α helix that winds around its counterpart to form a coiled coil. Two light chains at the end of the coiled-coil tail can interact with a membranous vesicle, the "cargo."

Question **Compare the structure of kinesin to the structure of myosin (Fig. 5.35).**

motor proteins appear to use a similar mechanism but move toward the (–) end of the microtubule.

The motor activity of kinesin requires the free energy of ATP hydrolysis. However, *kinesin cannot follow the myosin lever mechanism because its head domains are not rigidly fixed to its neck regions.* In kinesin, a relatively flexible polypeptide segment joins each head to an α helix that eventually becomes part of the coiled-coil tail (see Fig. 5.40b). (Recall that in myosin, the lever is a long α helix that extends from the head to the coiled-coil region and is stiffened by the two light chains; see Fig. 5.36.) Nevertheless, the relative flexibility of kinesin's neck is critical for its function.

Kinesin's two heads are not independent but work in a coordinated fashion so that the two heads alternately bind to successive β-tubulin subunits along a protofilament, as if walking. Conformational changes elicited by ATP binding and hydrolysis are relayed to other regions of the molecule (**Fig. 5.41**). *This transforms the free energy of ATP hydrolysis into the mechanical movement of kinesin.* Each ATP-binding event yanks the trailing head forward by about 160 Å, so the net movement of the attached cargo is about 80 Å, or the length of a tubulin dimer.

Kinesin is a processive motor

As in the myosin–actin system (see Fig. 5.39), the kinesin–tubulin reaction cycle proceeds in only one direction. Although most molecular movement is associated with ATP binding, ATP hydrolysis is a necessary part of the reaction cycle. The slowest step of the reaction cycle shown in Figure 5.41 is the dissociation of the trailing kinesin head from the microtubule. ATP binding to the leading head may help promote trailing-head release as part of the forward-swing step. Because the free head quickly rebinds tubulin, the kinesin heads spend most of their time bound to the microtubule track.

One consequence of kinesin's almost constant hold on a microtubule is that many—perhaps 100 or more—cycles of ATP hydrolysis and kinesin advancement can occur before the motor dissociates from its microtubule track. *Kinesin is therefore said to have high* **processivity.** A motor protein such as myosin, which dissociates from an actin filament after a single stroke, is not processive.

In a muscle cell, low processivity is permitted because the many myosin–actin interactions occur more or less simultaneously to cause the thin and thick filaments to slide past each other (see Fig. 5.38). *High processivity is advantageous for a transport engine such*

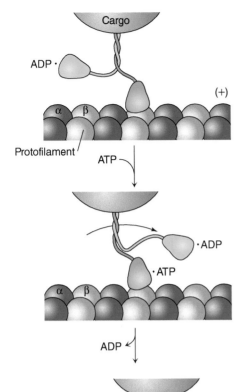

1. *ATP binding to the leading head (orange) induces a conformational change in which the neck docks against the head. This movement swings the trailing head (yellow) forward by 180° toward the (+) end of the microtubule. This is the force-generating step.*

2. *The new leading head (yellow) quickly binds to a tubulin subunit and releases its ADP. This step moves kinesin's cargo forward along the protofilament.*

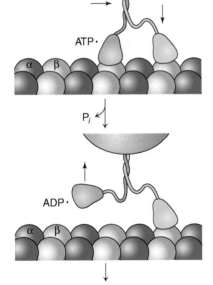

3. *In the trailing head (orange), ATP is hydrolyzed to ADP + P_i. The P_i diffuses away, and the trailing head begins to detach from the microtubule.*

4. *ATP binds to the leading head to repeat the reaction cycle.*

Figure 5.41 The kinesin reaction cycle. The cycle begins with one kinesin head bound to a tubulin subunit in a protofilament of a microtubule. The trailing head has ADP in its nucleotide-binding site. For clarity, the relative size of the neck region is exaggerated.

Question **What prevents kinesin from moving backward?**

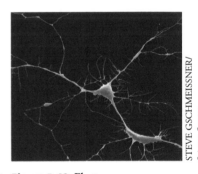

Figure 5.42 Electron micrograph of neurons. Microtubule-associated motor proteins move cargo between the cell body and the ends of the axon and other cell processes.

STEVE GSCHMEISSNER/ Science Source

as kinesin, because its cargo (which is relatively large and bulky) can be moved long distances without being lost. Consider the need for efficient transport in a neuron. Neurotransmitters and membrane components are synthesized in the cell body, where ribosomes are located, but must be moved to the end of the axon, which may be several meters long in some cases (**Fig. 5.42**).

Concept Check

1. List the ways that myosin resembles and differs from kinesin.
2. Without looking at the text, describe the myosin and kinesin reaction cycles (Figures 5.39 and 5.41).
3. Explain the role of nucleotide binding and hydrolysis in molecular movement.
4. List some cellular activities that require processive motors.

5.5 Antibodies

KEY CONCEPTS

Describe the structure and function of immunoglobulins.

- Identify the parts of an immunoglobulin.
- Explain the sources of diversity in an individual's antibodies.
- List some applications of antibodies in the laboratory and clinic.

All organisms are susceptible to infection, even the simplest bacterial cells that are invaded by bacteriophages (literally, *bacteria-eaters*). In eukaryotes, **pathogens** include viruses, bacteria, and other eukaryotes, both single-celled and multicellular. Not surprisingly, potential host organisms have evolved an array of defenses against pathogens, such as the CRISPR system in bacteria (described in Box 20.B), the signal-based system in plants, and the cellular immune system in animals.

In addition to the innate defenses that have been shaped by natural selection to recognize generic features of pathogens (such as viral RNAs or bacterial cell walls), *vertebrates have an "adaptive" immune system that can potentially recognize, respond to, and remember millions or billions of antigens.* An antigen may be a foreign molecule or a molecule already present in the body (a self-antigen). Although many types of molecules are antigenic, the vast majority of antigens that the immune system deals with are proteins or peptide fragments. The ability to recognize pathogens, allergens (antigens that are not intrinsically harmful), and self-antigens (such as those that appear on cancer cells) is a property of cells known as **T lymphocytes** and **B lymphocytes** (or T cells and B cells). As they develop, these white blood cells express unique receptors, one type per cell, collectively generating as many as 10^{12} different antigen receptors. A circulating **antibody** is the secreted form of a B cell's receptor protein. These proteins are the focus of this section.

Immunoglobulin G includes two antigen-binding sites

Antibodies, or **immunoglobulins**, are the second-most abundant type of protein (after albumin) circulating in the plasma, the fluid component of blood. Immunoglobulin G (IgG) is the most plentiful of the five types of immunoglobulins, and its four subclasses all have the same overall structure: two identical heavy chains and two identical light chains linked by a set of disulfide bonds (**Fig. 5.43**). Proteolytic cleavage of the flexible hinge regions of the Y-shaped molecule separates the two "arms" or **Fab fragments** (ab for antigen-binding) from the "stem" or **Fc fragment** (c because it readily crystallizes; **Fig. 5.44a**).

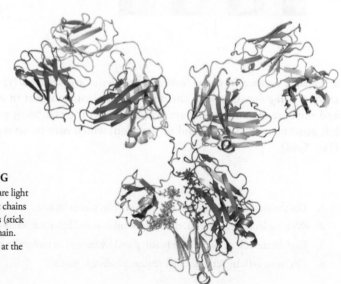

Figure 5.43 Immunoglobulin G structure. The two heavy chains are light blue and dark blue, and the two light chains are orange and yellow. Carbohydrates (stick models) are attached to each heavy chain. The antigen-binding sites are located at the ends of the upper arms.

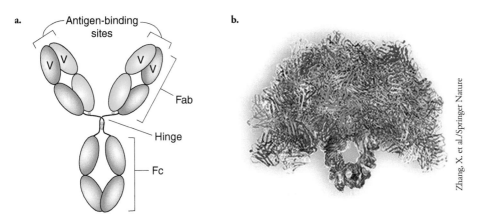

a.

Antigen-binding sites

b.

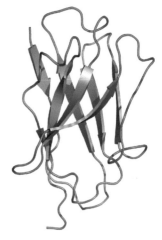

Figure 5.44 Immunoglobulin anatomy. **a.** The antigen-binding Fab fragments can be separated from the Fc fragment by proteolysis. The variable (V) domains of each chain bind to antigens. Orange dots represent disulfide bonds. **b.** The flexible hinge segments bend and twist so that the Fab fragments can occupy a range of positions in space.

Question Why are the two antigen-binding sites identical?

Cryo-electron microscopy and other structure-determining techniques reveal the dynamic properties of an intact immunoglobulin: The Fab arms can swing around, presumably to allow each antigen-binding site to interact independently with an antigen (Fig. 5.44b). Bivalent binding is an advantage when a pathogen bears multiple copies of an antigen on its surface. Each IgG has two binding sites (other types of immunoglobulins have up to 10 Fab arms), so that even if one binding site releases the antigen, the other binding site holds the immunoglobulin in place, and the first site can quickly rebind the antigen:

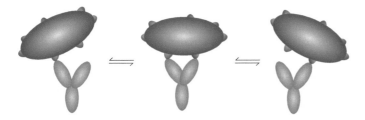

Figure 5.45 The immunoglobulin-fold domain. The outermost (variable) domain of the heavy chain of the immunoglobulin in Figure 5.43 is shown here. The three polypeptide loops that are involved in antigen binding are colored magenta.

Question How many β strands make up each β sheet in this immunoglobulin-fold domain?

As a result, the immunoglobulin may have an overall antigen-binding **avidity** that is stronger than the binding **affinity** measured for a single Fab–antigen interaction.

As Figures 5.43 and 5.44 indicate, immunoglobulins have a modular structure in which the heavy and light chains are constructed from multiple iterations of a small protein domain known as an immunoglobulin-fold domain. In IgG molecules, the heavy chain contains four of these domains, and the light chains two. *An immunoglobulin-fold domain consists of about 110 residues that form two small antiparallel β sheets held together in part by a disulfide bond* (**Fig. 5.45**). Interestingly, immunoglobulin-fold domains appear in the structures of hundreds of proteins involved in immune responses, including the T cell receptor, which is built from four such domains. Apparently, gene duplication was a common event in the evolution of the vertebrate immune system, and the sturdy β-sandwich structure seems to be a versatile protein building block.

The outermost domains of the heavy and light chains in IgG (that is, the ends of the Fab arms) form the antigen-binding site. The specificity of antigen binding is defined by three polypeptide loops from each domain. The loops are called **hypervariable loops** and are highlighted in Figure 5.45. *Some or all of the six loops at each binding site (three from the heavy chain and three from the light chain) interact with the antigen through weak noncovalent forces, including ionic interactions, hydrogen bonds, and van der Waals interactions* (**Fig. 5.46**). The complementarity of the antibody and antigen surfaces yields typical binding affinities in the nanomolar to picomolar range.

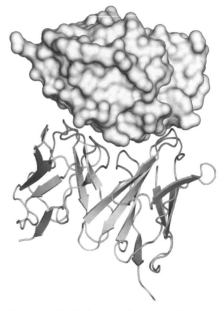

Figure 5.46 Antigen–antibody complex. The surface of the antigen (hen egg white lysozyme) is gray and the antigen-binding domains from the antibody heavy and light chains are green and blue, with their hypervariable loops in magenta.

Because the amino acid sequences of the antigen-biding domains—particularly the three loops—differ markedly among immunoglobulins, the domains are known as **variable domains** (note the "V" marking the variable domains in Fig. 5.44a). In contrast, the sequences of the other heavy and light chain domains are largely the same from one IgG to another and are therefore known as **constant domains.** Thus, *each IgG molecule, with its unique set of variable domains, binds a different antigen* (even if two different IgGs happen to bind to the same antigen, the binding affinity is likely to differ).

B lymphocytes produce diverse antibodies

How can immunoglobulin chains contain regions that are constant as well as regions that exhibit high variability, especially when the number of different variable domains far exceeds the apparent number of genes in the genome? The answer, initially worked out by Susumu Tonegawa, is that developing B cells build a gene for the B cell receptor (the membrane-anchored version of an antibody) by cutting and pasting shorter gene segments in a mostly random fashion. For example, to build a heavy chain gene, the cell must assemble so-called V, D, J, and C segments, recombining them into one continuous stretch of DNA that can be transcribed to messenger RNA and then translated into a polypeptide chain. Because the genome includes multiple sets of V, D, J, and C segments to choose from, the combinatorial possibilities are enormous (**Fig. 5.47**). (The situation is a bit like preparing a pizza when given multiple choices for crust, sauce, cheese, and toppings.)

Recombination alone could theoretically generate about 6200 different heavy chains and 370 light chains, yielding a total of about 2.3 million (6200 × 370) different sets. Even so, the cellular machinery that recombines the immunoglobulin gene segments introduces *additional* sequence variation at the joints between segments, adding or deleting nucleotides and thereby changing the identities of the amino acids encoded in that stretch of DNA. *The V-D-J joint region, the site of greatest sequence diversity, corresponds to one of the antigen-binding hypervariable loops in the variable domain of the heavy chain.* A similar program of gene splicing generates diversity in a hypervariable loop in the variable domain of the light chain. In Figure 5.46, the loops that result from gene recombination are the two at the center of the antigen-binding site.

Although an individual could theoretically make billions of different immunoglobulins, *only a small percentage of them are actually synthesized at any time.* This seems to be sufficient to protect most people from most pathogens. However, the random nature of generating a diverse repertoire of immunoglobulins means that the risk of accidentally generating antibodies to self-antigens (and triggering an autoimmune disease) is always present. Note that the capacity to recognize a wide variety of potential antigens is present even before those antigens are encountered.

When a B cell bearing an appropriate receptor comes across an antigen it can recognize, the cell becomes activated. This leads to several rounds of cell division and a shift in production from membrane-anchored receptors to soluble versions of the same protein (antibodies). The result is an expanded population of cells producing huge amounts of antigen-specific antibodies. While the B cells proliferate, two additional changes typically occur. First, *the cells allow their immunoglobulin genes to mutate at a high rate* (an activity that is suppressed in other cells of the body).

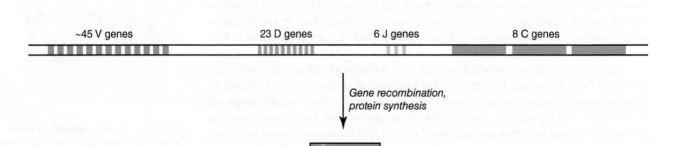

Figure 5.47 Human heavy chain gene segments. Developing B cells recombine a V, D, J, and C gene segment to construct a heavy chain gene. The gene segments and intervening DNA are not drawn to scale.

At the same time, the cells undergo a selection process so that only the cells with the tightest-binding receptors survive. The second development involves carrying out another round of DNA recombination, so that the cell can swap one set of heavy-chain constant domains for another. *These alternatives allow the lymphocytes to produce other classes of immunoglobulins that have the same antigen-binding site.* The high rate of mutation along with class switching means that over the few weeks of a typical immune response, the body produces increasingly more effective antibodies.

Different classes of antibodies exhibit different **effector functions** that depend on the Fc region of the molecule, which includes two constant domains from each heavy chain (see Fig. 5.44a). The antigen-binding (Fab) portions of antibodies can coat and thereby prevent pathogens (such as viruses or bacteria) from attacking or entering host cells. However, fully eliminating the antigen (and the pathogen that displays it) often requires a phagocytic cell to engulf and destroy the antigen–antibody complex (**Fig. 5.48**). Phagocytic cells bear Fc receptors, which recognize the Fc portion of IgG after the Fab portions have latched on to antigens. In effect, *antibodies specifically tag antigens, which can then be cleaned up nonspecifically by macrophages.* The Fc regions of other immunoglobulin classes have additional activities. For example, the Fc region of immunoglobulin A (IgA) permits it to be transported through certain cells to reach areas outside the body, including the intestinal lumen and secretions such as saliva, tears, and milk. The Fc region of immunoglobulin E (IgE) causes specialized white blood cells to release signals that lead to mucus production and itching; these responses are useful for eliminating worm parasites but are also triggered by IgE-binding allergens like dust mites and pollen.

During an immune response, some B lymphocytes differentiate into **memory cells** that can respond more quickly—and with higher-affinity antibodies of the proper immunoglobulin class—the next time the antigen is encountered. This immunological memory is the key to **immunization** (vaccination): *Prior exposure to harmless antigens from a pathogen ensures a robust and effective response if the actual pathogen later enters the individual's body.* T lymphocytes are essential for developing an immune response to all kinds of pathogens, but depending on the type of threat, antibody production by B lymphocytes may or may not be a decisive factor in eliminating the pathogen.

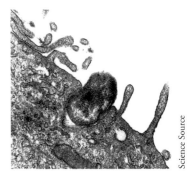

Figure 5.48 A phagocyte engulfing a bacterium.
Pathogens that have been tagged with antibodies are easily recognized by the phagocyte's Fc receptors.

Science Source

Researchers take advantage of antibodies' affinity and specificity

Outside the body, immunoglobulins have no effector functions, but their exquisite antigen-binding specificity makes them attractive as laboratory reagents and therapeutic drugs. The earliest attempts to harness the power of antibodies involved injecting an antigen into a horse or rabbit and collecting the antibodies from the animal's blood. Because a number of B lymphocytes with different receptors can respond to the same antigen, the resulting antibodies are heterogeneous, or polyclonal. The development of clever cell-culture techniques allowed researchers to produce mouse **monoclonal antibodies,** which are produced by one cell and its genetically identical progeny (a clone is simply a copy). Monoclonal antibodies, which all have the same structure and antigen-binding properties, can be linked to reporter enzymes so that when an antibody binds to a specific antigen, the enzyme generates a colored product. Variations of this technique are the basis for simple and fast laboratory tests and home pregnancy tests. For immunofluorescence microscopy, a monoclonal antibody bearing a fluorescent group binds tightly and specifically to its antigen to light up certain cellular structures (**Fig. 5.49**).

More than 60 monoclonal antibodies have been approved for use as drugs; many more are being developed. Because mouse proteins are recognized as foreign and tend to trigger an immune response when injected into humans, the usefulness of mouse monoclonal antibodies is limited in some recipients. To get around this problem, the mouse immunoglobulins are "humanized." Genetic engineers paste the mouse DNA sequences for the antigen-binding hypervariable loops into human genes for the immunoglobulin heavy and light chains. The reconstructed genes are then introduced into cultured cells that can produce large amounts of proteins that exhibit the antigen-binding properties of the original mouse

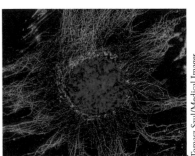

Figure 5.49 Immunofluorescence microscopy. In this example, green fluorescent antibodies mark the Golgi apparatus, and red fluorescent antibodies mark microtubules. The DNA is stained with a blue fluorescent dye.

Tomasz Szul/Medical Images

antibodies but behave like human antibodies. Adalimumab is one such drug; it is used to reduce inflammation in some autoimmune diseases.

Antibody drugs are known as "biologics," because they cannot be synthesized *in vitro* and must be produced by living cells. For this reason, their production is difficult and expensive. Still, the high affinity and high specificity of antibodies are hard to beat. Some highly effective cancer treatments use T lymphocytes that have been engineered to express cancer-recognizing immunoglobulin variable domains. These so-called CAR-T cells, an application of gene therapy (Section 3.3), might also be effective for treating other conditions. Researchers are also exploring the use of smaller pieces of immunoglobulins, such as single variable domains, that could still bind antigens but be less costly to produce.

Concept Check

1. Without checking the text, draw a simple diagram of an IgG molecule and label its parts.
2. Describe how an individual's B lymphocytes generate a population of antibodies with different antigen-binding properties.
3. Explain why the effectiveness of antibodies improves over time.
4. List some effector functions of antibodies.
5. Summarize the advantages and disadvantages of using immunoglobulins as drugs.

SUMMARY

5.1 Myoglobin and Hemoglobin: Oxygen-Binding Proteins

- Myoglobin contains a heme prosthetic group that reversibly binds oxygen. The amount of O_2 bound depends on the O_2 concentration and on myoglobin's affinity for oxygen.

- Hemoglobin's α and β chains are homologous to myoglobin, indicating a common evolutionary origin.

- O_2 binds cooperatively to hemoglobin with low affinity, so hemoglobin can efficiently bind O_2 in the lungs and deliver it to myoglobin in the tissues.

- Hemoglobin is an allosteric protein whose four subunits alternate between the T (deoxy) and R (oxy) conformations in response to O_2 binding to the heme groups. The deoxy conformation is favored by low pH (the Bohr effect) and by the presence of BPG.

5.2 Clinical Connection: Hemoglobin Variants

- Hemoglobin variants provide information about protein structure and function. The substitution of glutamate by valine in the β chain of hemoglobin S leads to polymerization that causes red blood cell sickling and anemia. Carriers of the genetic variant are more likely to survive malaria, as are carriers of the variant that produces hemoglobin C.

- Genetic variations that affect hemoglobin production, oxygen binding affinity, and intersubunit interactions may have mild to severe clinical effects.

5.3 Structural Proteins

- Actin filaments in the eukaryotic cytoskeleton are built from ATP-binding globular actin subunits that polymerize as a

double chain. Polymerization is reversible, so actin filaments undergo growth and regression. Their dynamics may be modified by proteins that mediate filament capping, branching, and severing.

- GTP-binding tubulin dimers polymerize to form a hollow microtubule. Polymerization is more rapid at one end, and the microtubule can disassemble rapidly by fraying. Drugs that affect microtubule dynamics interfere with cell division.

- The intermediate filament keratin contains two long α-helical chains that coil around each other so that their hydrophobic residues are in contact. Keratin filaments associate and are cross-linked to form semipermanent structures.

- Collagen polypeptides contain a large amount of proline and hydroxyproline and include a glycine residue at every third position. Each chain forms a narrow left-handed helix, and three chains coil around each other to form a right-handed triple helix with glycine residues at its center. Covalent cross-links strengthen collagen fibers.

5.4 Motor Proteins

- Myosin, a protein with a coiled-coil tail and two globular heads, interacts with actin filaments to perform mechanical work. ATP-driven conformational changes allow the myosin head to bind, release, and rebind actin. This mechanism, in which myosin acts as a lever, is the basis of muscle contraction.

- The motor protein kinesin has two globular heads connected by flexible necks to a coiled-coil tail. Kinesin transports vesicular cargo along the length of a microtubule by a processive stepping mechanism that is driven by conformational changes triggered by ATP binding and hydrolysis.

5.5 Antibodies

- By combining individual gene segments, B lymphocytes produce immunoglobulin G proteins that differ in the variable domains of their heavy and light chains. Mutations and class switching add further diversity to the antibody population.

- The high specificity and affinity for antigens makes antibodies effective at recognizing pathogens and useful as reagents and drugs.

KEY TERMS

heme	anemia	intermediate	gout	cytokinesis	Fc fragment
prosthetic group	cooperative binding	filament	coiled coil	stereocilia	avidity
Y	deoxyhemoglobin	microtubule	extracellular matrix	Usher syndrome	affinity
pO_2	oxyhemoglobin	F-actin	imino group	vesicle	hypervariable loop
saturation	T state	G-actin	triple helix	processivity	variable domain
p_{50}	R state	(−) end	osteogenesis	pathogen	constant domain
globin	allosteric protein	(+) end	imperfecta	antigen	effector function
homologous proteins	ligand	P_i	Ehlers–Danlos	T lymphocyte	memory cell
invariant residue	Bohr effect	treadmilling	syndrome	B lymphocyte	immunization
conservative	thalassemia	metastasis	motor protein	antibody	monoclonal antibody
substitution	cytoskeleton	protofilament	thick filament	immunoglobulin	
variable residue	actin filament	kinetochore	thin filament	Fab fragment	

BIOINFORMATICS

Brief Bioinformatics Exercises

5.1 Visualizing and Analyzing Hemoglobin

5.2 Globin Sequence Alignment and Evolution

5.3 Deoxyhemoglobin and Oxyhemoglobin

Bioinformatics Projects

Using Databases to Compare and Identify Related Protein Sequences

The Pfam Database

PROBLEMS

5.1 Myoglobin and Hemoglobin: Oxygen-Binding Proteins

1. Explain why neither globin alone nor heme alone can function effectively as an oxygen carrier.

2. Hexacoordinate Fe(II) in heme is bright red. Pentacoordinate Fe(II) is blue. Explain how these electronic changes account for the different colors of arterial (scarlet) and venous (purple) blood.

3. Myoglobin transfers oxygen obtained from hemoglobin to mitochondrial proteins involved in the catabolism of metabolic fuels to produce energy for the muscle cell. Using what you have learned in this chapter about myoglobin and hemoglobin, what can you conclude about the structures of these mitochondrial proteins?

4. Myoglobin most effectively facilitates oxygen diffusion through muscle cells when the intracellular oxygen partial pressure is comparable to the p_{50} value of myoglobin. Explain why.

5. Calculate the fractional saturation for myoglobin when p_{50} is **a.** 3 mbar (millibar) and **b.** 100 mbar.

6. Calculate the fractional saturation for emperor penguin myoglobin, which has a p_{50} value of 3.325 mbar at **a.** 39.9 mbar **b.** 133 mbar.

7. At what pO_2 value is myoglobin **a.** 30% saturated and **b.** 80% saturated?

8. At what pO_2 value is emperor penguin myoglobin (see Problem 6) **a.** 25% saturated and **b.** 90% saturated?

9. Intracellular pO_2 in heart and muscle cells is around 3.3 mbar. Explain how an increase and decrease of 1 mbar in pO_2 might affect O_2 binding by myoglobin ($p_{50myoglobin} = 3.64$ mbar).

10. When oxygen levels are low, cells in the walls of blood vessels produce nitric oxide (·NO, a free radical), which signals the vessel's smooth muscle cells to relax. As a result, the vessel dilates and more blood (and oxygen) can flow to the tissues. However, ·NO inhibits the mitochondrial protein complexes that carry out the O_2-dependent process that generates most of a cell's ATP. Oxygenated myoglobin participates in this reaction:

$$MbO_2 (Fe^{2+}) + \cdot NO \rightarrow Mb (Fe^{3+}) + NO_3^-$$

Deoxymyoglobin participates in this reaction:

$$Mb (Fe^{2+}) + NO_2^- + H^+ \rightarrow Mb (Fe^{3+}) + \cdot NO + OH^-$$

Explain how oxymyoglobin and deoxymyoglobin help regulate the use of O_2 in cells.

11. Why is it necessary for cells to have a mechanism to convert ferric (Fe^{3+}) myoglobin to ferrous (Fe^{2+}) myoglobin (see Problem 10)?

12. Myoglobin in which the Fe^{2+} is oxidized to Fe^{3+} is referred to as met-myoglobin and appears brown. The sixth ligand in met-myoglobin is water rather than O_2. **a.** When a fresh, uncooked beef roast is sliced in two, the surface of the meat first appears purplish but then turns red. Explain why. **b.** Why does meat turn brown when it is cooked? **c.** Why might a retailer choose to package meat using oxygen-permeable plastic wrap instead of a vacuum-sealed package?

13. Below is a list of the first 10 residues of the B helix in myoglobin from different organisms. Based on this information, which positions **a.** cannot tolerate substitution, **b.** can tolerate conservative substitution, and **c.** are highly variable?

Position	Human	Chicken	Alligator	Turtle	Tuna	Carp
1	D	D	K	D	D	D
2	I	I	L	L	Y	F
3	P	A	P	S	T	E
4	G	G	E	A	T	G
5	H	H	H	H	M	T
6	G	G	G	G	G	G
7	Q	H	H	Q	G	G
8	E	E	E	E	L	E
9	V	V	V	V	V	V
10	L	L	I	I	L	L

14. The oxygen binding curves for three species of fish—tuna, bonito, and mackerel—are shown below. **a.** Use the graph to estimate a p_{50} value for each species. **b.** Which species of fish has the highest oxygen binding affinity? The least?

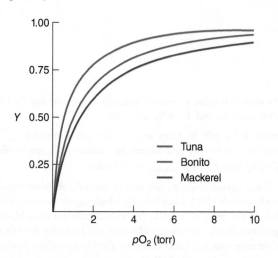

15. Compare the p_{50} values and the oxygen binding affinities of the fish described in Problem 14 with the emperor penguin (see Problem 6).

16. The heme group in myoglobin, shown in Figure 5.2, is tightly wedged into a pocket, where it is held in place by the ValE11 and PheCD1 (the first residue in the segment between the C and D helices) side chains, which presumably swing aside to allow oxygen to bind. Consult Figure 5.5 to determine whether these two residues are highly conserved and, if so, provide a plausible explanation.

17. How might oxygen binding be affected when the ValE11 residue of myoglobin (see Problem 16) is mutated to **a.** Ile? or **b.** Ser?

18. The sequence of the E and F helices in harbor seal myoglobin is shown below. Carry out an exercise similar to that illustrated in Figure 5.5 to compare the sequences of this region in harbor seal myoglobin and in human myoglobin. Shade the invariant residues one color and the structurally similar residues a different color.

SEDLRKHGKTVLTALGGILKKKGHHDAELKPLAQSHA

19. Invariant residues are those that are essential for the structure and/or function of the protein and cannot be replaced by other residues. Which amino acid residue in the globin chain is most likely to be invariant? Is your answer consistent with your answer to Problem 18? Explain.

20. Hemoglobin is 50% saturated with oxygen when $pO_2 = 35$ mbar. If hemoglobin exhibited hyperbolic binding (as myoglobin does), what would be the fractional saturation when $pO_2 = 130$ mbar? What does this tell you about the physiological importance of hemoglobin's sigmoidal oxygen-binding curve?

21. Hemoglobin's oxygen binding curve is sigmoidal (see Fig. 5.7) rather than hyperbolic (see Problem 20). The equation for hemoglobin's oxygen binding curve is modified from Equation 5.4 and can be used to calculate the fractional saturation (Y) for hemoglobin:

$$Y = \frac{(pO_2)^n}{(p_{50})^n + (pO_2)^n}$$

The quantity n is the Hill coefficient and its value is correlated with the degree of cooperative O_2 binding. For hemoglobin, $n \approx 3$. (The value of n is 1 in the absence of cooperativity.) What is Y when $pO_2 = 33.25$ mbar, a typical venous pO_2? What is Y when $pO_2 = 160$ mbar, a typical oxygen partial pressure in the lungs? (*Note:* At $pCO_2 = 8.71$ mbar, $p_{50} = 20$ mbar.)

22. Carbon monoxide, CO, binds to hemoglobin about 250 times as tightly as oxygen (see Box 5.A). The resulting complex is a red color that is much brighter than that observed when oxygen is bound. Why might a researcher studying the properties of oxygenated Hb decide to use Hb · CO instead?

23. What are some potential intermolecular interactions (see Fig. 2.4) that occur at the α–β interface when Hb is in the T form (see Fig. 5.9a)? How might these interactions change upon conversion to the R form (see Fig. 5.9b)?

24. Highly active muscle generates lactic acid by respiration so fast that the blood passing through the muscle actually experiences a drop in pH from about 7.4 to about 7.2. Under these conditions, hemoglobin releases about 10% more O_2 than it does at pH 7.4. Explain.

25. About two dozen histidine residues in hemoglobin are involved in binding the protons produced by cellular metabolism. In this manner, hemoglobin contributes to buffering in the blood, and the imidazole groups able to bind and release protons contribute to the Bohr effect. One important contributor to the Bohr effect is the C-terminal His 146 on the β chain of hemoglobin, whose side chain is in close proximity to the side chain of Asp 94 in the deoxy form of hemoglobin but not the oxy form. **a.** What kind of interaction occurs between Asp 94 and His 146 in deoxyhemoglobin? **b.** The proximity of Asp 94 alters the pK value of the imidazole ring of His. In what way?

26. The C-terminal His 146 on the β chain of hemoglobin (see Problem 25) also forms an interaction with a Lys residue on the neighboring α chain in the T form. What is the nature of this interaction?

27. While most humans are able to hold their breath for only a minute or two, crocodiles can stay submerged under water for an hour or longer. This adaptation allows the crocodiles to kill their prey by drowning. Investigators hypothesized that bicarbonate (HCO_3^-) might act as an allosteric effector in crocodiles in a manner similar to BPG in humans. **a.** What is the source of HCO_3^- in crocodile tissues? **b.** Draw oxygen-binding curves for crocodile hemoglobin in the presence and absence of HCO_3^-.

Which conditions increase the p_{50} value for crocodile hemoglobin? What does this tell you about the oxygen-binding affinity of hemoglobin under those conditions? **c.** The investigators found that the HCO_3^- binding site on crocodile hemoglobin was located at the α–β subunit interface, where the two subunits slide with respect to each other during the oxy ↔ deoxy transition. Based on their results, the researchers modeled a plausible binding site that included the phenolate anions of Tyr 41β and Tyr 42α and the ε-amino group of Lys 38β. What interactions might form between these amino acid side chains and HCO_3^-?

28. Fish use ATP or GTP as allosteric effectors to encourage hemoglobin to release its bound O_2. What structural characteristics do these allosteric effectors have in common with the BPG allosteric effector used by mammals and the HCO_3^- effector used by crocodiles (see Problem 27)?

29. Lampreys are among the world's most primitive vertebrates and are therefore interesting organisms to study. It has been shown that lamprey hemoglobin forms a tetramer when deoxygenated, but when the hemoglobin becomes oxygenated, the tetramers dissociate into monomers. The binding of oxygen to lamprey hemoglobin is influenced by pH; as in human hemoglobin, the deoxygenated form is favored when the pH decreases. Glutamate residues on the surface of the monomers play an important role. How might glutamate residues influence the equilibrium between the monomer and tetramer form with changes in pH?

30. The grelag goose lives year-round in the Indian plains, while its close relative, the bar-headed goose, spends the summer months in the Tibetan lake region. The p_{50} value for bar-headed goose hemoglobin is less than the p_{50} value for grelag goose hemoglobin. What is the significance of this adaptation?

31. In the developing fetus, fetal hemoglobin (Hb F) is synthesized beginning at the third month of gestation and continuing until birth. After the baby is born, the concentration of Hb F declines and is replaced, by the age of six months, with adult hemoglobin (Hb A) as synthesis of the γ chain declines and synthesis of the β chain increases. **a.** In the graph below, which curve represents fetal hemoglobin? **b.** Why does Hb F have a higher oxygen affinity than Hb A? Why does this provide an advantage to the developing fetus?

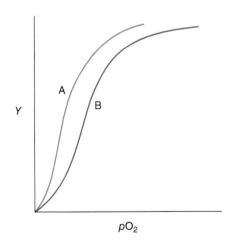

32. People who live at or near sea level undergo *high altitude acclimatization* when they go up to moderately high altitudes of about 5000 m. **a.** The low pO_2 at these altitudes initially induces hypoxia, which results in hyperventilation. Explain why this occurs. What happens to blood pH during hyperventilation? **b.** After a period of weeks, the alveolar pCO_2 decreases and the 2,3-BPG concentration increases. Explain these observations.

33. Drinking a few drops of a preparation called "vitamin O," which consists of oxygen and sodium chloride dissolved in water, purportedly

increases the concentration of oxygen in the body. **a.** Use your knowledge of oxygen transport to evaluate this claim. **b.** Would vitamin O be more or less effective if it was infused directly into the bloodstream?

34. Animals indigenous to high altitudes, such as yaks, llamas, and alpacas, have adapted to the low pO_2 at high altitudes because they have high-affinity hemoglobins, achieved through a combination of globin chain mutations and the continued synthesis of fetal hemoglobin in the mature animal. This is a true evolutionary adaptation and not the high-altitude acclimatization described in Problem 32. **a.** Compare the p_{50} values of the high-altitude animals with the p_{50} values of their low-altitude counterparts. **b.** How does the continued synthesis of fetal hemoglobin help these animals to survive at high altitudes?

5.2 Clinical Connection: Hemoglobin Variants

35. *Plasmodium falciparum*, the protozoan that causes malaria, slightly decreases the pH of the red blood cells it infects. Invoke the Bohr effect to explain why *Plasmodium*-infected cells are more likely to undergo sickling in individuals with the Hb S variant.

36. The drug hydroxyurea can be used to treat sickle cell disease, although it is not used often because of undesirable side effects. Hydroxyurea is thought to function by stimulating the afflicted person's synthesis of fetal hemoglobin. In a clinical study, patients who took hydroxyurea showed a 50% reduction in frequency of hospital admissions for severe pain. Why would this drug alleviate the symptoms of sickle cell disease?

37. Hb Milledgeville (α44Pro → Leu) results in a mutated hemoglobin with altered oxygen affinity. Explain how the oxygen affinity is altered (see Table 5.1).

38. The oxygen-binding curves for normal hemoglobin (Hb A) and a mutant hemoglobin (Hb Great Lakes) are shown in the figure. **a.** Compare the shapes of the curves for the two hemoglobins and comment on their significance. **b.** Which hemoglobin has a higher affinity for oxygen when $pO_2 = 26.66$ mbar? **c.** Which hemoglobin has a higher affinity for oxygen when $pO_2 = 99.99$ mbar? **d.** Which hemoglobin is most efficient at delivering oxygen from arterial blood ($pO_2 = 99.99$ mbar) to active muscle ($pO_2 = 26.66$ mbar)?

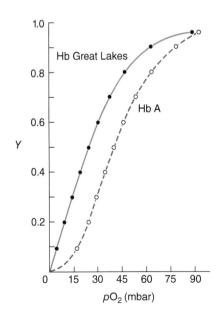

39. Investigators of mutant human hemoglobins often subject the proteins to cellulose acetate electrophoresis at pH 8.5, a pH at which most hemoglobins are negatively charged. The proteins are applied at the negative pole and migrate toward the positive pole when the current is turned on. The more negatively charged the hemoglobin, the faster it migrates to the positive pole. Results for normal (Hb A) and sickle-cell hemoglobin (Hb S) are shown. Draw a diagram that shows the results of electrophoresis of Hb A and Hb Ohio (β142Ala → Asp).

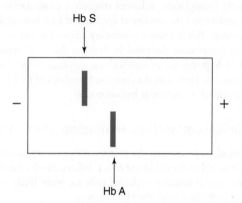

40. Individuals with Hb Hornchurch, in which a Glu residue in the β chain is replaced by Lys, are often asymptomatic. **a.** In general, would you expect a Glu → Lys substitution to disrupt a protein's structure? **b.** The change occurs at position 43 in the β chain. Use Figures 5.5 and 5.1b to identify the location of the amino acid substitution. Does this affect your answer to part **a**?

41. Hb Yakima (β99Asp → His) has a p_{50} value of 13.3 mbar and an n value of 1 (see Problem 21). Calculate the saturation Y of Hb Yakima at **a.** $pO_2 = 39$ mbar and **b.** $pO_2 = 133$ mbar. Perform the same calculations for normal Hb ($p_{50} = 35$ mbar and $n = 3$). **c.** Given the information that β99Asp makes important contacts with the α chain in the T form, how does the mutation affect the ability of Hb Yakima to bind and deliver oxygen?

42. Hb Santa Giusta Sardegna (β93Cys → Trp) was recently identified in a patient in Sardinia, Italy. The mutant hemoglobin has a p_{50} value of 16 torr and an n value of 2.2 (see Problem 21). The β93Cys residue is adjacent to the His F8 residue and is in close proximity to the Asp 94 side chain (see Problems 25 and 26). How does the amino acid substitution affect the mutant hemoglobin?

43. The hemoglobin α–β interface is important in the reversible transition from the T to the R form. Hydrogen bonding between βAsp99 and αTyr42 and αAsn97, which are important interactions in the T form, are disrupted upon conversion to the R form. **a.** Draw a plausible hydrogen bonding pattern between the side chains of these three amino acids, using dotted lines between hydrogen bond donors and acceptors. **b.** Hb Coimbra (β99Asp → Glu) has a lower p_{50} value and a lower n value (see Problem 21) than normal hemoglobin. Would you expect this amino acid substitution to produce a hemoglobin mutant with these characteristics? Consult the diagram that you drew in part **a** and provide a hypothesis that might explain these results.

44. Hb Providence (β82Lys → Asn) results from a single point mutation of the DNA (see Table 5.1). **a.** Compare the oxygen affinities of Hb Providence and Hb A. **b.** There are actually two forms of Hb Providence in affected individuals. Hb Providence in which the β82Lys has been replaced with Asn is referred to as Hb Providence Asn. This hemoglobin variant can undergo deamidation to produce Hb Providence Asp. Draw the reaction that converts Hb Providence Asn to Hb Providence Asp. **c.** Compare the oxygen affinities of Hb Providence Asp and Hb Providence Asn.

45. In a high-affinity hemoglobin isolated from an individual, Lys at β144 was replaced with Asn. (The normal C-terminal sequence of the β chain is –Lys[144]–Tyr[145]–His[146]–COO⁻.) The C-terminal portion of the β chain, which resides in hemoglobin's central cavity, normally forms interactions that affect the position of the F helix that contains His F8. How might the Lys → Asn substitution result in a higher-affinity hemoglobin?

46. Patients with the mutation described in Problem 45 are able to undergo high-altitude acclimatization (described in Problem 32) much more readily than "lowlanders" with normal hemoglobin. Explain why.

5.3 Structural Proteins

47. Compare and contrast globular and fibrous proteins.

48. Of the various proteins highlighted in this chapter (actin, tubulin, keratin, collagen, myosin, and kinesin), **a.** which are exclusively structural (not involved in cell shape changes); **b.** which are considered to be motor proteins; **c.** which are not motor proteins but can undergo structural changes; and **d.** which contain nucleotide-binding sites?

49. Explain why actin filaments and microtubules are polar whereas intermediate filaments are not.

50. In order to obtain crystals suitable for X-ray crystallography, G-actin was first mixed with another protein. **a.** Why was this step necessary? **b.** What additional information was required to solve the structure of G-actin?

51. Phalloidin, a peptide isolated from a poisonous mushroom, binds to F-actin but not G-actin. How does the addition of phalloidin affect cell motility?

52. Phalloidin (see Problem 51) is used as an imaging tool by covalently attaching a fluorescent tag to it. What structures can be visualized in cells treated with fluorescently labeled phalloidin? What is not visible?

53. How could a microtubule-binding protein distinguish a rapidly growing microtubule from one that was growing more slowly?

54. Why would fraying be a faster mechanism for microtubule disassembly than dissociation of tubulin dimers?

55. Microtubule ends with GTP bound tend to be straight, whereas microtubule ends with GDP bound are curved and tend to fray. An experiment was carried out in which monomers of β-tubulin were allowed to polymerize in the presence of a nonhydrolyzable analog of GTP called guanylyl-($\alpha\beta$)-methylenediphosphonate (GMP·CPP). Compare the stability of this polymer with one polymerized in the presence of GTP.

GMP·CPP

56. Explain why colchicine (which promotes microtubule depolymerization) and paclitaxel (taxol, which prevents depolymerization) both prevent cell division.

57. A recent review listed dozens of compounds that are being investigated as possible agents for treating cancer. Tubulin was the target for all of the compounds. Why is tubulin a good target for anticancer drugs?

58. The three-dimensional structure of paclitaxel bound to the tubulin dimer has been solved. Paclitaxel binds to a pocket in β-tubulin on the inner surface of the microtubule and causes a conformational change that counteracts the effects of GTP hydrolysis. How does the binding of paclitaxel affect the stability of the microtubule?

59. Gout is a disease characterized by the deposition of sodium urate crystals in the synovial (joint) fluid. The crystals are ingested by neutrophils, which are white blood cells that circulate in the bloodstream before they crawl through tissues to reach sites of injury. Upon ingestion of the crystals, a series of biochemical reactions leads to a release of mediators from the neutrophils that cause inflammation, resulting in joint pain. Colchicine is used as a drug to treat gout because colchicine can inhibit neutrophil mobility. How does colchicine accomplish this?

60. Vinblastine, a compound in the periwinkle, has been shown to affect microtubule assembly by stabilizing the (+) ends and destabilizing the (–) ends of microtubules. **a.** How does vinblastine affect the formation of the mitotic spindle during mitosis? **b.** Why would you expect vinblastine to have a more dramatic effect on rapidly dividing cells such as cancer cells?

61. Why would you expect vinblastine (Problem 60) to have a more dramatic effect on rapidly dividing cells such as cancer cells?

62. Hydrophobic residues usually appear at the first and fourth positions in the seven-residue repeats of polypeptides that form coiled coils. **a.** Why do polar or charged residues usually appear in the remaining five positions? **b.** Why is the sequence Ile–Gln–Glu–Val–Glu–Arg–Asp more likely than the sequence Trp–Gln–Glu–Tyr–Glu–Arg–Asp to appear in a coiled coil?

63. Globular proteins are typically constructed from several layers of secondary structure, with a hydrophobic core and a hydrophilic surface. Is this true for a fibrous protein such as keratin?

64. Straight hair can be curled and curly hair can be straightened by exposing wet hair to a reducing agent, repositioning the hair with rollers, then exposing the hair to an oxidizing agent. Explain how this procedure alters the shape of the hair.

65. "Hard" keratins such as those found in hair, horn, and nails have a high sulfur content whereas "soft" keratins found in the skin have a lower sulfur content. Explain the structural basis for this observation.

66. Describe the primary, secondary, tertiary, and quaternary structures of **a.** actin and **b.** collagen. Why is it difficult to assign the four structural categories to these proteins?

67. Radioactively labeled [^{14}C]-proline is incorporated into collagen in cultured fibroblasts. The radioactivity is detected in the collagen protein. However, collagen synthesized in the presence of [^{14}C]-hydroxyproline is not radiolabeled. Explain why.

68. In a collagen polypeptide consisting mostly of repeating Gly–Pro–Hyp sequences, which residue(s) is(are) most likely to be substituted by 5-hydroxylysine (see Solution 73)?

69. The triple helix of collagen is stabilized through hydrogen bonding. Draw an example of one of these hydrogen bonds, as described in the text.

70. A retrospective study conducted over 25 years examined a group of patients with similar symptoms: bruising, joint swelling, fatigue, and gum disease. Examination of patient records showed that some patients suffered from gastrointestinal diseases, poor dentition, or alcoholism, while others followed various fad diets. **a.** What disease afflicts this diverse group of patients? **b.** Explain why the patients suffer from the symptoms listed. **c.** Explain why your diagnosis is consistent with the patients' medical histories.

71. The effect of the drug minoxidil on collagen synthesis in skin fibroblasts was investigated by measuring the activities of the prolyl hydroxylase and lysyl hydroxylase enzymes in the cells (see the table). **a.** What is the effect of minoxidil on the cultured fibroblasts? **b.** Why would this drug be effective in treating fibrosis (a skin condition associated with an accumulation of collagen)? **c.** What are the potential dangers in the long-term use of minoxidil as a topical ointment to treat hair loss?

	Prolyl hydroxylase activity units	Lysyl hydroxylase activity units
Control cells	13,100	12,400
Minoxidil-treated cells	13,200	1900

72. The highly pathogenic bacterium *Clostridium perfringens* causes gangrene, a disease that results in the destruction of animal tissue. The bacterium secretes two types of collagenases that cleave the peptide bonds linking Gly and Pro residues. A biochemical company sells a preparation of these collagenases to investigators interested in culturing cells derived from bone, cartilage, muscle, or endothelial tissue. How does this enzyme preparation assist the investigator in obtaining cells for culturing?

73. A pharmaceutical company sells a collagenase ointment that is applied to the surface of skin wounds. How does the ointment promote the healing process? What precaution should patients take when using the ointment?

74. The T_m (melting temperature) is defined as that temperature at which collagen is half-denatured and is used as a criterion for collagen stability. T_m measurements for collagen from rats and sea urchins are presented in the table. **a.** Assign each collagen to the proper organism and **b.** explain the correlation between imino acid content (defined as the combined content of Pro and Hyp) and T_m.

Collagen	Number of Hyp per 1000 residues	Number of Pro per 1000 residues	T_m (°C)
A	48.5	81.3	27.0
B	68.5	111	38.5

75. Because collagen molecules are difficult to isolate from the connective tissue of animals, investigators who study collagen structure often use synthetic peptides (see Problems 83–84). Would it be practical to try to purify collagen molecules from cultures of bacterial cells that have been engineered to express collagen genes?

76. The deep-sea hydrothermal vent worm has a thick collagen-containing cuticle that protects it from the drastic temperature changes and low oxygen content of its habitat. Sequence analyses indicated that the worm collagen has the customary Gly–X–Y triplet but Hyp occurs only in the X position and Y is often a glycosylated threonine (a threonine side chain with a covalently attached galactose sugar residue). The melting temperatures (see Problem 74) of a series of synthetic peptides were measured to assess their stabilities. The T_m values are shown in the table. **a.** Compare the melting temperatures of (Pro–Pro–Gly)$_{10}$ and (Pro–Hyp–Gly)$_{10}$. What is the structural basis for the difference? **b.** Compare the melting temperatures of (Pro–Pro–Gly)$_{10}$ and (Gly–Pro–Thr(Gal))$_{10}$ and provide an explanation. **c.** Why was (Gly–Pro–Thr)$_{10}$ included by the investigators?

Gal-Thr

Synthetic peptide	Forms a triple helix?	T_m (°C)
(Pro–Pro–Gly)$_{10}$	Yes	41
(Pro–Hyp–Gly)$_{10}$	Yes	60
(Gly–Pro–Thr)$_{10}$	No	N/A
(Gly–Pro–Thr(Gal))$_{10}$	Yes	41

77. A small portion of the middle of the gene for the α1(II) chain of type II collagen (the entire gene contains 3421 base pairs) is shown below as the sequence of the nontemplate, or coding, strand of the DNA. Also shown is the gene isolated from the skin fibroblasts of a patient with a mutant α1(II) gene. The patient was a neonate delivered at 38 weeks who died shortly after birth. The neonate had shortened limbs and a large head, with a flat nasal bridge and short bones and ribs. The symptoms are consistent with the disease hypochondrogenesis. Analysis of tissue taken from the neonate revealed a single point mutation in the α1(II) collagen gene.

Normal α1(II) collagen gene:

…TAACGGCGAGAAGGGAGAAGTTGGACCTCCT…

Mutant α1(II) collagen gene:

…TAACGGCGAGAAGGCAGAAGTTGGACCTCCT…

a. What is the sequence of the protein encoded by both the normal and mutant genes? (Note that you must first determine the correct open reading frame.) What is the change in the protein sequence? **b.** How would the Tm values for normal collagen and mutant collagen compare? **c.** Why did the patient die shortly after birth?

78. Why do Hyp-containing collagen molecules have greater stability? To investigate this question, a group of investigators synthesized a collagen containing 4-fluoroproline (Flp), which resembles hydroxyproline except that fluorine substitutes for the hydroxyl group. The T_m values (see Problem 74) of three synthetic collagen molecules are shown in the table. **a.** Compare the structures of hydroxyproline and fluoroproline. Why did the investigators choose fluorine? **b.** Compare the T_m values of the three synthetic collagens. What factor contributes the most to the stability of the molecule?

Synthetic peptide	T_m (°C)
(Pro–Pro–Gly)$_{10}$	41
(Pro–Hyp–Gly)$_{10}$	60
(Pro–Flp–Gly)$_{10}$	91

79. Gelatin is a food product obtained from the thermal degradation of collagen. Powdered gelatin is soluble in hot water and forms a gel upon cooling to room temperature. Why is gelatin nutritionally inferior to other types of protein?

80. Papaya and pineapple fruits contain proteolytic enzymes that can degrade collagen. **a.** Cooking meat with these fruits helps to tenderize the meat. How does this work? **b.** A cook decides to flavor a gelatin dessert with fresh papaya and pineapple. The fruits are added to a solution of gelatin in hot water and then the mixture is allowed to cool (see Problem 79). What is the result and how could the cook have avoided this?

81. Osteogenesis imperfecta is a genetic disease, but most cases result from new mutations rather than ones that are passed from parent to child. Explain.

82. Because networks of collagen fibers lend resilience to the deep (living) layers of skin, some "anti-aging" cosmetics contain purified collagen. **a.** Explain why this exogenous (externally supplied) collagen is unlikely to attenuate the skin wrinkling that accompanies aging. **b.** Another anti-aging strategy involves taking collagen supplements orally—in some cases, literally ground-up bone—to improve skin appearance. Despite a few industry-funded studies showing promise, most nutrition experts remain skeptical. Explain why.

83. A series of collagen-like peptides, each containing 30 amino acid residues, were synthesized in order to study the important interactions among the three chains in the triple helix. Peptide 1 is (Pro–Hyp–Gly)$_{10}$, peptide 2 is (Pro–Hyp–Gly)$_4$–Glu–Lys–Gly–(Pro–Hyp–Gly)$_5$, and peptide 3 is Gly–Lys–Hyp–Gly–Glu–Hyp–Gly–Pro–Lys–Gly–Asp–Ala–(Gly–Ala–Hyp)$_2$–(Gly–Pro–Hyp)$_4$. Their structural properties and melting temperatures (see Problem 53) are summarized in the table below. Note that "imino acid content" refers to the combined content of Pro and Hyp.

Peptide	Forms trimer	Imino acid content	T_m (°C)	
1	Yes	67%	pH = 1	61
			pH = 7	58
			pH = 11	60
2	Yes	60%	pH = 1	44
			pH = 7	46
			pH = 1	49
3	Yes	30%	pH = 7	18
			pH = 7	26.5
			pH = 13	19

a. Rank the stability of the three collagen-like peptides. What is the reason for the observed stability? **b.** Compare the T_m values of peptide 3 at the various pH values. Why does peptide 3 have a maximum T_m value at pH = 7? What interactions are primarily responsible? **c.** Compare the T_m values of peptide 1 at the various pH values. Why is there less of a difference of T_m values at different pH values for peptide 1 than for peptide 3? **d.** Consider the answers to the above questions. Which factor is more important in stabilizing the structure of the collagen-like peptides: ion pairing or imino acid content?

84. An enzyme found in tadpoles cleaves the bond indicated in the figure below. What kind of enzyme is this, and what is its role in the metamorphosis of the tadpole to the adult frog?

—Gly—Pro—Gln—Gly—↓—Ile—Ala

5.4 Motor Proteins

85. Is myosin a fibrous protein or a globular protein? Explain.

86. Myosin type V is a two-headed myosin that operates as a transport motor to move its attached cargo along actin filaments. Its mechanism is similar to that of muscle myosin, but it acts processively, like kinesin. The reaction cycle diagrammed here begins with both myosin V heads bound to an actin filament. ADP is bound to the leading head, and the nucleotide-binding site in the trailing head is empty.

Based on your knowledge of the muscle myosin reaction cycle (Fig. 5.39), propose a reaction mechanism for myosin V, starting with the entry of ATP. How does each step of the ATP hydrolysis reaction correspond to a conformational change in myosin V?

87. Early cell biologists, examining living cells under the microscope, observed that the movement of certain cell constituents was rapid, linear, and targeted (that is, directed toward a particular point). **a.** Why are these qualities inconsistent with diffusion as a mechanism for redistributing cell components? **b.** List the minimum requirements for an intracellular transport system that is rapid, linear, and targeted.

88. Explain why the movement of myosin along an actin filament is "hopping," whereas the movement of kinesin along a microtubule is "walking."

89. Rigor mortis is the stiffening of a corpse that occurs in the hours following death. **a.** Using what you know about the mechanism of motor proteins, explain what causes this stiffness. **b.** Why is rigor mortis of concern to those in the meat industry? **c.** How could a forensic pathologist use the concept of rigor mortis to investigate the circumstances of an individual's death?

90. In individuals with muscular dystrophy, a mutation in a gene for a structural protein in muscle fibers leads to progressive muscle weakening beginning in early childhood. Explain why muscular dystrophy patients typically also exhibit abnormal bone development.

91. Explain why a one-headed kinesin motor would be ineffective.

5.5 Antibodies

92. Enlist the number of hypervariable loops in IgM, IgG, and IgA.

93. Outline how F(ab)$_2$ and F(ab) fragments can be generated from matured IgG.

94. In immunoglobulin M (IgM), five immunoglobulins resembling IgG molecules are assembled in a ring structure. How many hypervariable loops are in IgM?

95. Some proteases cleave the hinge region of IgG below the level of the interchain disulfide bonds (as depicted in Fig. 5.44a), which generates a so-called F(ab)$_2$ fragment. **a.** How many immunoglobulin-fold domains are present in an F(ab)$_2$ fragment? **b.** How many hypervariable loops are in one F(ab)$_2$ fragment?

96. Compare the antigen-binding affinity and avidity of an Fab fragment and an F(ab)$_2$ fragment derived from the same immunoglobulin G.

97. Some mammals produce a type of immunoglobulin that consists of two identical heavy chains and no light chains. Each heavy chain includes one variable domain, a hinge segment, and two constant domains. Sketch a diagram of the structure of this immunoglobulin molecule. Label the antigen-binding site(s) and the part of the molecule that would be recognized by a macrophage that is responsible for removing antigen–antibody complexes.

98. Researchers have attempted to develop monoclonal antibodies that recognize antigens on cancer cells as a way to tag the cells for subsequent destruction by macrophages. Unfortunately, such monoclonal antibodies tend to bind to noncancerous cells that happen to express similar antigens. Could this problem be solved by engineering a bi-specific antibody with binding sites for two different antigens? Explain.

99. Some bacterial antigens are present in multiple copies on the surface of a bacterial cell. Immunoglobulin G can cross-link these bacteria because each IgG can bind more than one antigen, and each bacterium can bind more than one IgG. Draw a simple diagram to illustrate this phenomenon.

100. Some bacterial proteases can cleave the hinge region of immunoglobulin molecules without affecting the antigen binding sites. Given your answer to Problem 99, explain why these proteases would give the bacteria a better chance of starting an infection.

101. Explain why immunoglobulins that are used as drugs must be infused or injected rather than administered orally to the patient.

102. Not all engineered immunoglobulins use the antigen-binding portion of the molecule. For example, one research team has constructed a protein that consists of the Fc portion of a human IgG covalently linked to a viral protein that binds tightly to the cell wall of the pathogenic bacterium *Staphylococcus aureus*. Explain why this engineered protein might be an effective treatment for *S. aureus* infections.

103. Dengue virus, which can cause hemorrhagic fever, has four strains. A person's first exposure to any strain of dengue virus triggers a strong immune response with antibody production. If the person is later exposed to a different strain of dengue virus, the existing anti-dengue antibodies can bind to the new strain. However, rather than neutralizing the virus, antibody binding actually promotes viral infection of cells such as macrophages, which can lead to serious illness. **a.** Explain why exposure to the second viral strain results in a more severe infection. **b.** Does your answer to part **a** explain why a dengue vaccine (an injection of viral antigens) given to young children actually cause more cases of illness than it theoretically prevents?

104. At some point after their activation by antigens, B lymphocytes express an Fc receptor. Explain how this receptor could help the immune system regulate the strength of an immune response to a pathogen.

SELECTED READINGS

Chaaban, S. and Brouhard, G.J., A microtubule bestiary: structural diversity in tubulin polymers, *Mol. Biol. Cell* 28, 2924–2931, doi: 10.1091/mbc.E16-05-0271 (2017). [Describes the structure of conventional 13-protofilament microtubules as well as variations in different phylogenetic groups.]

Fidler, A.L., Boudko, S.P., Rokas, A., and Hudson, B.G., The triple helix of collagens—an ancient protein structure that enabled animal multicellularity and tissue evolution, *J. Cell Sci.* 131, jcs203950. doi: 10.1242/jcs.203950 (2018). [Reviews the structure and probable evolution of the triple helix and lists diseases caused by collagen defects.]

Hohmann, T. and Dehghani, F., The cytoskeleton—a complex interacting meshwork, *Cells* 8, 362, doi: 10.3390/cells8040362 (2019). [A detailed review of actin filaments, intermediate filaments, and microtubules and their roles in cell motility, using glioma (brain cancer) cells as an example.]

Jay, J.W., Bray, B., Qi, Y., Igbinigie, E., Wu, H., Li, J., and Ren, G., IgG antibody 3D structures and dynamics, *Antibodies* 7, 18, doi: 10.3390/antib7020018 (2018). [Provides basic details of immunoglobulin structure in summarizing techniques for visualizing the three-dimensional shapes of IgG.]

Storz, J.F., Gene duplication and evolutionary innovations in hemoglobin-oxygen transport, *Physiology* 31, 223–232, doi: 10.1152/physiol.00060 (2016). [Summarizes the evolutionary history of globin proteins and the functional implications of multiple forms of hemoglobin in different organisms.]

Svitkina, T.M., Ultrastructure of the actin cytoskeleton, *Curr. Opin. Cell Biol.* 54, 1–8, doi: 10.1016/j.ceb.2018.02.007 (2018). [A brief review of the dynamics of actin filaments.]

Thom, C.S., Dickson, C.F., Gell, D.A., and Weiss, M.J., Hemoglobin variants: Biochemical properties and clinical correlates, *Cold Spring Harb. Perspect. Med.* 2013; 3: a011858. doi: 10.1101/cshperspect.a011858 (2013). [Summarizes hemoglobin structure and function and how these are disrupted by different types of genetic changes.]

Walklate, J., Ujfalusi, Z., and Geeves, M.A., Myosin isoforms and the mechanochemical cross-bridge cycle, *J. Exp. Biol.* 219, 168–174, doi: 10.1242/jeb.124594 (2016). [Describes the myosin reaction cycle and summarizes the differences among myosin isoforms.]

Wang, W., Cao, L., Wang, C., Gigant, B., and Knossow, M., Kinesin, 30 years later: Recent insights from structural studies, *Prot. Sci.* 24, 1047–1056, doi: 10.1002/pro.2697 (2015). [Includes a detailed description of the kinesin reaction mechanism.]

Yuan, Y., Tam, M.F., Simplaceanu, V., and Ho, C., A new look at hemoglobin allostery, *Chem. Rev.* 115, 1702–1724, doi: 10.1021/cr500495x (2105). [Reviews various models of allostery and structural studies of hemoglobin, showing that the molecule is more dynamic than depicted in early models.]

CHAPTER 5 CREDITS

Figure 5.1 Image based on 1MBO. Phillips, S.E., Structure and refinement of oxymyoglobin at 1.6 A resolution, *J. Mol. Biol.* 142, 531–554 (1980).

Figures 5.4, 5.9a, and 5.12 Images based on 2HHB. Fermi, G., Perutz, M.F., The crystal structure of human deoxyhaemoglobin at 1.74 Angstroms resolution, *J. Mol. Biol.* 175, 159–174 (1984).

Figure 5.5 After Dickerson, R.E., and Geis, I., Hemoglobin, pp. 68–69, Benjamin/Cummings (1983).

Figure 5.9b Image based on 1HHO. Shaanan, B., Structure of human oxyhaemoglobin at 2.1 Angstroms resolution, *J. Mol. Biol.* 171, 31–59 (1983).

Figure 5.15 Image based on 1J6Z. Otterbein, L.R., Graceffa, P., Dominguez, R., The crystal structure of uncomplexed actin in the ADP state, *Science* 293, 708–711 (2001).

Figures 5.20 and 5.21 Images based on 1TUB. Nogales, E., Wolf, S.G., Downing, K.H., Structure of the alpha beta tubulin dimer by electron crystallography, *Nature* 391, 199–203 (1998).

Figure 5.26 Image from Jenni, S., Harrison, S.C., "Structure of the DASH/Dam1 complex shows its role at the yeast kinetochore-microtubule interface", 2018. Science, AAAS.

Figure 5.29 Image based on 1C1G. Whitby, F.G., Phillips Jr., G.N., Crystal structure of tropomyosin at 7 Angstroms resolution in the spermine-induced crystal form, *Proteins* 38, 49–59 (2000).

Figure 5.33 Image based on 2CLG. Chen, J.M., Kung, C.E., Feairheller, S.H., Brown, E.M., An energetic evaluation of a "Smith" collagen microfibril model, *J. Protein Chem.* 10, 535–552 (1991).

Figure 5.36 Image based on 2MYS. Rayment, I., Rypniewski, W.R., Schmidt-Base, K., Smith, R., Tomchick, D.R., Benning, M.M., Winkelmann, D.A., Wesenberg, G., Holden, H.M., Three-dimensional structure of myosin subfragment-1: a molecular motor, *Science* 261, 50–58 (1993).

Figure 5.37 Image from Yang, S., Woodhead, L.J., Zhao, Q.F., Sulbarán, G., Craig, R., An approach to improve the resolution of helical filaments with a large axial rise and flexible subunits, *J Struct Biol.* 2016 January; 193(1): 45–54.

Figure 5.40b Image based on 3KIN. Kozielski, F., Sack, S., Marx, A., Thormahlen, M., Schonbrunn, E., Biou, V., Thompson, A., Mandelkow, E.M., Mandelkow, E., The crystal structure of dimeric kinesin and implications for microtubule-dependent motility, *Cell* 91, 985–994 (1997).

Figure 5.43 Image based on 1IGY. Harris, L.J., Skaletsky, E., McPherson, A., Crystallographic structure of an intact IgG1 monoclonal antibody, *J. Mol. Biol.* 275, 861–872 (1998).

Figure 5.44b Image from Zhang, X. et al. 3D Structural Fluctuation of IgG1 Antibody Revealed by Individual Particle Electron Tomography. *Sci. Rep.* 5, 09803 (2015). Reproduced with permission of Springer Nature.

Figures 5.45 and 5.46 Images based on 1C08. Kondo, H., Shiroishi, M., Matsushima, M., Tsumoto, K., Kumagai, I., Crystal structure of anti-hen egg white lysozyme antibody (HyHEL-10) Fv-antigen complex. Local structural changes in the protein antigen and water-mediated interactions of Fv-antigen and light chain-heavy chain interfaces, *J. Biol. Chem.* 274, 27623–27631 (1999).

Enzymes: Classification and Catalysis

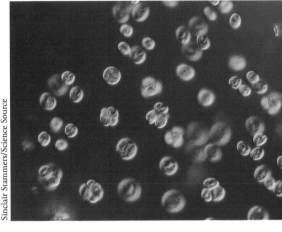

Sinclair Stammers/Science Source

Photosynthetic organisms like *Chlorella* absorb light energy to drive carbohydrate production, but one species of these algae also uses a unique light-powered enzyme to convert fatty acids into hydrocarbons. The algal photoenzyme could give biotechnologists a new way to tap solar energy for synthetic reactions.

Do You Remember?

- Living organisms obey the laws of thermodynamics (Section 1.4).
- Noncovalent forces, including hydrogen bonds, ionic interactions, and van der Waals forces, act on biological molecules (Section 2.1).
- An acid's pK value describes its tendency to ionize (Section 2.3).
- The 20 amino acids differ in the chemical characteristics of their R groups (Section 4.1).
- Some proteins can adopt more than one stable conformation (Section 4.3).

We have already seen how protein structures relate to their physiological functions. We are now ready to examine enzymes, which directly participate in the chemical reactions by which matter and energy are transformed by living cells. In this chapter, we examine the fundamental features of enzymes, including the thermodynamic underpinnings of their activity. We describe various mechanisms by which enzymes accelerate chemical reactions, focusing primarily on the digestive enzyme chymotrypsin to illustrate how different structural features influence catalytic activity. The following chapter continues the discussion of enzymes by describing how enzymatic activity is quantified and how it can be regulated.

6.1 What Is an Enzyme?

KEY CONCEPTS

Describe how enzymes differ from other catalysts.

- Explain why enzymes exhibit specificity for their substrates and products.
- Describe how enzymes are classified.

In vitro, under physiological conditions, the peptide bond that links amino acids in peptides and proteins has a half-life of about 20 years. That is, after 20 years, about half of the peptide bonds in a given sample of a peptide will have been broken down through **hydrolysis** (cleavage by water):

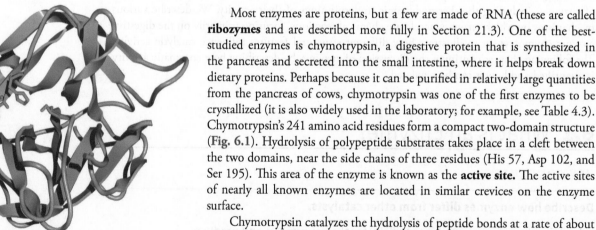

Obviously, the long half-life of the peptide bond is advantageous for living organisms, since many of their structural and functional characteristics depend on the integrity of proteins. On the other hand, many proteins—some regulatory proteins, for example—must be broken down very rapidly so that their biological effects can be limited. Clearly, an organism must be able to accelerate the rate of peptide bond hydrolysis.

In general, there are three ways to increase the rate of hydrolysis or any other chemical reaction:

1. *Increasing the temperature (adding energy in the form of heat).* Unfortunately, this is not very practical, since the vast majority of organisms cannot regulate their internal temperature and thrive only within relatively narrow temperature ranges. Furthermore, an increase in temperature accelerates all chemical reactions, not just the desired reaction.

2. *Increasing the concentrations of the reacting substances.* Higher concentrations of **reactants** increase the likelihood that they will encounter each other in order to react. However, a cell may contain tens of thousands of different types of molecules, space is limited, and many essential reactants are scarce inside as well as outside the cell.

3. *Adding a **catalyst**.* A catalyst is a substance that participates in the reaction yet emerges at the end in its original form. A huge variety of chemical catalysts are known. For example, the catalytic converter in an automobile engine contains a mixture of platinum and palladium that accelerates the conversion of carbon monoxide and unburned hydrocarbons to the relatively harmless carbon dioxide. *Living systems use catalysts called **enzymes** to increase the rates of chemical reactions.*

Most enzymes are proteins, but a few are made of RNA (these are called **ribozymes** and are described more fully in Section 21.3). One of the best-studied enzymes is chymotrypsin, a digestive protein that is synthesized in the pancreas and secreted into the small intestine, where it helps break down dietary proteins. Perhaps because it can be purified in relatively large quantities from the pancreas of cows, chymotrypsin was one of the first enzymes to be crystallized (it is also widely used in the laboratory; for example, see Table 4.3). Chymotrypsin's 241 amino acid residues form a compact two-domain structure (**Fig. 6.1**). Hydrolysis of polypeptide substrates takes place in a cleft between the two domains, near the side chains of three residues (His 57, Asp 102, and Ser 195). This area of the enzyme is known as the **active site.** The active sites of nearly all known enzymes are located in similar crevices on the enzyme surface.

Chymotrypsin catalyzes the hydrolysis of peptide bonds at a rate of about 190 per second, which is about 1.7×10^{11} times faster than in the absence of a catalyst. This is also orders of magnitude faster than would be possible with a simple chemical catalyst. In addition, *chymotrypsin and other enzymes act under mild conditions* (atmospheric pressure and physiological temperature), whereas many chemical catalysts require extremely high temperatures and pressures for optimal performance.

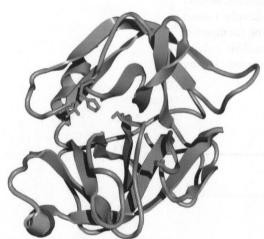

Figure 6.1 Ribbon model of chymotrypsin. The polypeptide chain (gray) folds into two domains. Three residues essential for the enzyme's activity are shown in red.

Table 6.1 Rate Enhancements of Enzymes

Enzyme	Half-time (uncatalyzed)[a]	Uncatalyzed rate (s⁻¹)	Catalyzed rate (s⁻¹)	Rate enhancement (catalyzed rate/ uncatalyzed rate)
Orotidine-5′-monophosphate decarboxylase	78,000,000 years	2.8×10^{-16}	39	1.4×10^{17}
Staphylococcal nuclease	130,000 years	1.7×10^{-13}	95	5.6×10^{14}
Adenosine deaminase	120 years	1.8×10^{-10}	370	2.1×10^{12}
Chymotrypsin	20 years	1.0×10^{-9}	190	1.7×10^{11}
Triose phosphate isomerase	1.9 years	4.3×10^{-6}	4,300	1.0×10^{9}
Chorismate mutase	7.4 hours	2.6×10^{-5}	50	1.9×10^{6}
Carbonic anhydrase	5 seconds	1.3×10^{-1}	1,000,000	7.7×10^{6}

[a]The half-times of very slow reactions were estimated by extrapolating from measurements made at very high temperatures.

Chymotrypsin's catalytic power is not unusual: *Rate enhancements of 10^8 to 10^{12} are typical of enzymes* (**Table 6.1** gives the rates of some enzyme-catalyzed reactions). Of course, the slower the rate of the uncatalyzed reaction, the greater the opportunity for rate enhancement by an enzyme (see, for example, orotidine-5′-monophosphate decarboxylase in Table 6.1). Interestingly, even relatively fast reactions are subject to enzymatic catalysis in biological systems. For example, the conversion of CO_2 to carbonic acid in water

$$CO_2 + H_2O \rightleftharpoons H_2CO_3$$

has a half-time of 5 seconds (half the molecules will have reacted within 5 seconds). This reaction is accelerated over a millionfold by the enzyme carbonic anhydrase (see Table 6.1).

Another feature that sets enzymes apart from nonbiological catalysts is their **reaction specificity.** Most enzymes are highly specific for their reactants (called **substrates**) and products. The functional groups in the active site of an enzyme are so carefully arranged that the enzyme can distinguish its substrates from among many other molecules that are similar in size and shape and can then mediate a single chemical reaction involving those substrates. This reaction specificity stands in marked contrast to the permissiveness of most organic catalysts, which can act on many different kinds of reactants and, for a given reactant, sometimes yield more than one product.

Chymotrypsin and some other digestive enzymes are somewhat unusual in acting on a relatively broad range of substrates and, at least in the laboratory, catalyzing several types of reactions. For instance, chymotrypsin catalyzes the hydrolysis of the peptide bond following almost any large nonpolar residue such as phenylalanine, tryptophan, or tyrosine. It can also catalyze the hydrolysis of other amide bonds and ester bonds. This behavior has proved to be convenient for quantifying the activity of purified chymotrypsin. An artificial substrate such as *p*-nitrophenylacetate (an ester) is readily hydrolyzed by the action of chymotrypsin (the name of the enzyme appears next to the reaction arrow to indicate that it participates as a catalyst):

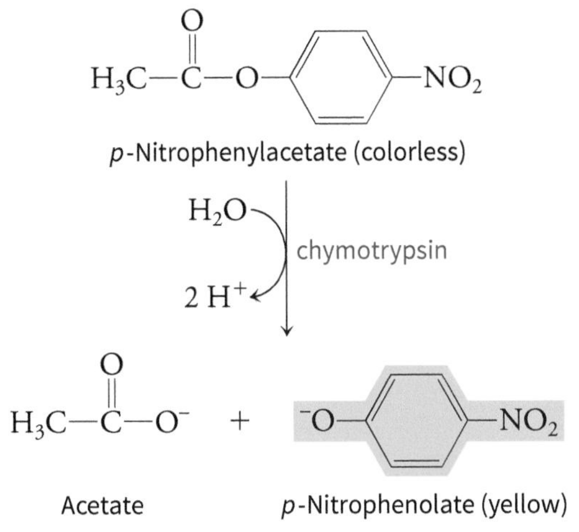

p-Nitrophenylacetate (colorless)

Acetate   *p*-Nitrophenolate (yellow)

Table 6.2 Enzyme Classification

Class of enzyme	Type of reaction catalyzed
1. Oxidoreductases	Oxidation–reduction reactions
2. Transferases	Transfer of functional groups
3. Hydrolases	Hydrolysis reactions
4. Lyases	Group elimination to form double bonds
5. Isomerases	Isomerization reactions
6. Ligases	Bond formation coupled with ATP hydrolysis
7. Translocases	Solute transport through membranes

The *p*-nitrophenolate reaction product is bright yellow, so the progress of the reaction can be easily monitored by a spectrophotometer.

Finally, enzymes differ from nonbiological catalysts in that *the activities of many enzymes are regulated so that the organism can respond to changing conditions or follow genetically determined developmental programs.* For this reason, biochemists seek to understand *how* enzymes work as well as *when* and *why*. These aspects of enzyme behavior are fairly well understood for chymotrypsin, which makes it an ideal subject to showcase the fundamentals of enzyme activity.

Enzymes are usually named after the reaction they catalyze

The enzymes that catalyze biochemical reactions have been formally classified into seven groups according to the type of reaction carried out (**Table 6.2**). Basically, *most biochemical reactions involve either the addition of some substance to another, or its removal, or the rearrangement of that substance.* One exception is the translocase group; these enzymes catalyze reactions in which ions or molecules move across membranes (Section 9.3). Keep in mind that although the substrates of many biochemical reactions appear to be quite large (for example, proteins or nucleic acids), the action really involves just a few chemical bonds and a few small groups (sometimes H_2O or even just an electron).

The name of an enzyme frequently provides a clue to its function. In some cases, an enzyme is named by incorporating the suffix *-ase* into the name of its substrate. For example, fumarase is an enzyme that acts on fumarate (Reaction 7 in the citric acid cycle; see Section 14.2). Chymotrypsin can similarly be called a proteinase, a protease, or a peptidase. Most enzyme names contain more descriptive words (also ending in *-ase*) to indicate the nature of the reaction catalyzed by that enzyme. For example, pyruvate decarboxylase catalyzes the removal of a CO_2 group from pyruvate:

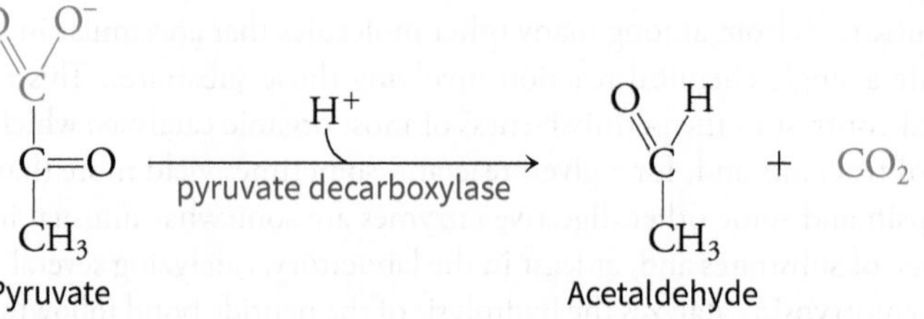

Alanine aminotransferase catalyzes the transfer of an amino group from alanine to an α-keto acid:

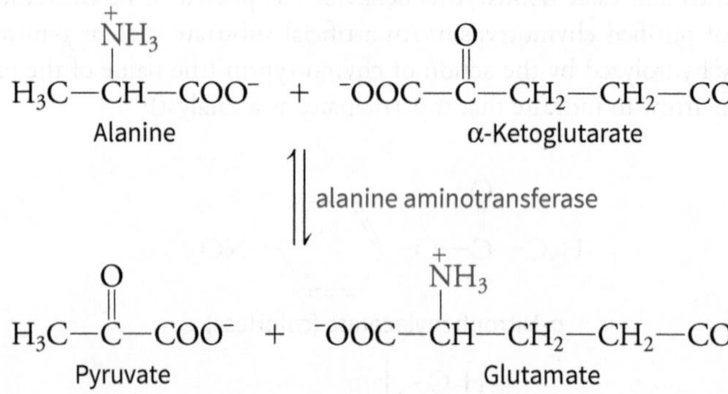

Such a descriptive naming system tends to break down in the face of the many thousands of known enzyme-catalyzed reactions, but it is adequate for the small number of well-known reactions that are included in this book. A more precise classification scheme systematically groups enzymes in a four-level hierarchy and assigns each enzyme a unique number. For example, chymotrypsin is known as EC 3.4.21.1 (*EC* stands for Enzyme Commission, part of the nomenclature committee of the International Union of Biochemistry and Molecular Biology; the EC database can be accessed at enzyme.expasy.org).

Keep in mind that even within an organism, more than one protein may catalyze a given chemical reaction. Multiple enzymes catalyzing the same reaction are called **isozymes.** Although they usually share a common evolutionary origin, isozymes differ in their catalytic properties. Consequently, the various isozymes that are expressed in different tissues or at different developmental stages can perform slightly different metabolic functions.

Concept Check

1. Make a list of things that enzymes do but purely chemical systems cannot do.
2. Identify the substrates of the enzymes listed in Table 6.1.
3. Explain why an enzyme's common name may not reveal its biological role.

6.2 Chemical Catalytic Mechanisms

KEY CONCEPTS

Describe the chemical mechanisms enzymes use to accelerate reactions.

- Relate a reaction's activation energy to its rate.
- Recognize that an enzyme accelerates a reaction by allowing it to proceed with a lower activation energy.
- Identify the three types of chemical catalytic mechanisms.
- Assign roles to specific amino acid side chains during catalysis.
- Trace the events of chymotrypsin catalysis.

In a biochemical reaction, the reacting species must come together and undergo electronic rearrangements that result in the formation of products. In other words, some old bonds break and new bonds form. Let us consider an idealized transfer reaction in which compound A—B reacts with compound C to form two new compounds, A and B—C:

$$\text{A—B} + \text{C} \rightarrow \text{A} + \text{B—C}$$

In order for the first two molecules to react, they must approach closely enough for their constituent atoms to interact. Normally, atoms that approach too closely repel each other. However, if the groups have sufficient free energy, they can pass this point and react with each other to form products. The progress of the reaction can be depicted on a diagram (**Fig. 6.2**) in which the

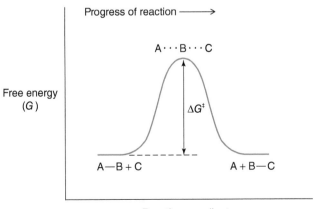

Figure 6.2 Reaction coordinate diagram for the reaction A—B + C → A + B—C.   The progress of the reaction is shown on the horizontal axis, and free energy is shown on the vertical axis. The transition state of the reaction, represented as A · · · B · · · C, is the point of highest free energy. The free energy difference between the reactants and the transition state is the free energy of activation ($\Delta G^{\ddagger}$).

horizontal axis represents the progress of the reaction (the **reaction coordinate**) and the vertical axis represents the free energy (G) of the system. The energy-requiring step of the reaction is shown as an energy barrier, called the **free energy of activation** or **activation energy** and symbolized $\Delta G^{\ddagger}$. The point of highest energy is known as the **transition state** and can be considered as something midway between the reactants and products.

The lifetimes of transition states are extremely short, on the order of 10^{-14} to 10^{-13} seconds. Because they are too short-lived to be accessible to most analytical techniques, the transition states of many reactions cannot be identified with absolute certainty. However, it is useful to visualize the transition state as a molecular species in the process of breaking old bonds and forming new bonds. For the reaction above, we can represent this as A···B···C. The reactants require free energy ($\Delta G^{\ddagger}$) to reach this point (for example, some energy is required to break existing bonds and bring other atoms together to begin forming new bonds). The analogy of going uphill in order to undergo a reaction is appropriate.

The height of the activation energy barrier determines the rate of a reaction (the amount of product formed per unit time). The higher the activation energy barrier, the less likely the reaction is to occur (the slower it is). Although the reactant molecules have varying free energies, very few of them have enough free energy to reach the transition state during a given time interval. The lower the energy barrier, the more likely the reaction is to occur (the faster it is), because more reactant molecules happen to have enough free energy to achieve the transition state during the same time interval. Note that the transition state, at the peak, can potentially roll down either side of the free energy hill. Therefore, not all the reactants that get together to form a transition state actually proceed all the way to products; they may return to their original state. Similarly, the products (A and B—C in this case) can react, pass through the same transition state (A···B···C), and yield the original reactants (A—B and C).

In nature, the free energies of the reactants and products of a chemical reaction are seldom identical, so the reaction coordinate diagram looks more like **Figure 6.3a**. When the products have a lower free energy than the reactants, then the overall free energy change of the reaction ($\Delta G_{\text{reaction}}$, or $G_{\text{products}} - G_{\text{reactants}}$) is less than zero. A negative free energy change indicates that a reaction proceeds spontaneously as written. Note that "spontaneously" does not mean "quickly." A reaction with a negative free energy change is thermodynamically favorable, but the height of the activation energy barrier ($\Delta G^{\ddagger}$) determines how fast the reaction actually occurs. If the products have greater free energy than the reactants (Fig. 6.3b), then the overall free energy change for the reaction ($\Delta G_{\text{reaction}}$) is greater than zero. This reaction does not proceed as written (because it cannot go "uphill"), but it does proceed in the reverse direction.

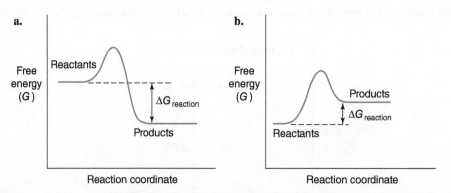

Figure 6.3 Reaction coordinate diagram for a reaction in which reactants and products have different free energies. The free energy change for the reaction (ΔG) is equivalent to $G_{\text{products}} - G_{\text{reactants}}$. **a.** When the free energy of the reactants is greater than that of the products, the free energy change for the reaction is negative, so the reaction proceeds spontaneously. **b.** When the free energy of the products is greater than that of the reactants, the free energy change for the reaction is positive, so the reaction does not proceed spontaneously (however, the reverse reaction does proceed).

Question In a cell, some enzyme-catalyzed reactions proceed in both the forward and reverse directions. Sketch a reaction coordinate diagram for such a reaction.

A catalyst provides a reaction pathway with a lower activation energy barrier

A catalyst, whether inorganic or enzymatic, allows a reaction to proceed with a lower activation energy barrier ($\Delta G^{\ddagger}$; Fig. 6.4). It does so by interacting with the reacting molecules such that they are more likely to assume the transition state. A catalyst speeds up the reaction because as more reacting molecules achieve the transition state per unit time, more molecules of product can form per unit time. (An increase in temperature increases the rate of a reaction for a similar reason: The input of thermal energy allows more molecules to achieve the transition state per unit time.) Thermodynamic calculations indicate that lowering $\Delta G^{\ddagger}$ by about 5.7 kJ·mol^{-1} accelerates the reaction 10-fold. A rate increase of 10^6 requires lowering $\Delta G^{\ddagger}$ by six times this amount, or about 34 kJ·mol^{-1}.

An enzyme catalyst does not alter the net free energy change for a reaction; it merely provides a pathway from reactants to products that passes through a transition state that has lower free energy than the transition state of the uncatalyzed reaction. *An enzyme, therefore, lowers the height of the activation energy barrier ($\Delta G^{\ddagger}$) by lowering the energy of the transition state.* The hydrolysis of a peptide bond is always thermodynamically favorable, but the reaction occurs quickly only when a catalyst (such as the protease chymotrypsin) is available to provide a lower-energy route to the transition state. Note that enzymes do not work at a distance; they must closely interact with their substrates to catalyze reactions.

Enzymes use chemical catalytic mechanisms

The idea that living organisms contain agents that can promote the change of one substance into another has been around at least since the early nineteenth century, when scientists began to analyze the chemical transformations carried out by organisms such as yeast. However, it took some time to appreciate that these catalytic agents were not part of some "vital force" present only in intact, living organisms. In 1878, the word *enzyme* was coined to indicate that there was something *in* yeast (Greek *en* = "in," *zyme* = "yeast"), rather than the yeast itself, that was responsible for breaking down (fermenting) sugar. In fact, the action of enzymes can be explained in purely chemical terms. What we currently know about enzyme mechanisms rests solidly on a foundation of knowledge about simple chemical catalysts.

In an enzyme, certain functional groups in the enzyme's active site perform the same catalytic function as small chemical catalysts. In some cases, the amino acid side chains of an enzyme cannot provide the required catalytic groups, so a tightly bound **cofactor** participates in catalysis. For example, many oxidation–reduction reactions require a metal ion cofactor, since a metal ion can exist in multiple oxidation states, unlike an amino acid side chain. Some enzyme cofactors are organic molecules known as **coenzymes,** which may be derived from vitamins.

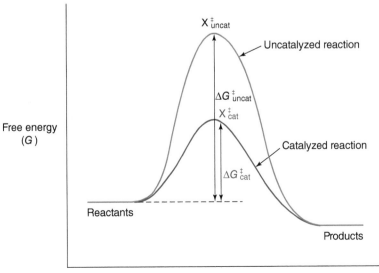

Figure 6.4 Effect of a catalyst on a chemical reaction. Here, the reactants proceed through a transition state symbolized $X^{\ddagger}$ during their conversion to products. In the presence of a catalyst, the free energy of activation ($\Delta G^{\ddagger}$) for the reaction is lower, so that $\Delta G^{\ddagger}_{cat} < \Delta G^{\ddagger}_{uncat}$. Lowering the free energy of the transition state ($X^{\ddagger}$) accelerates the reaction because more reactants are able to achieve the transition state per unit time.

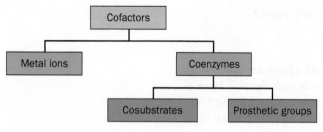

Figure 6.5 Types of enzyme cofactors.

Enzymatic activity still requires the protein portion of the enzyme, which helps position the cofactor and reactants for the reaction (this situation is reminiscent of myoglobin and hemoglobin, where the globin and heme group together function to bind oxygen; see Section 5.1). Some coenzymes, termed cosubstrates, enter and exit the active site as substrates do; a tightly bound coenzyme that remains in the active site between reactions is called a prosthetic group (Fig. 6.5).

There are three basic kinds of chemical catalytic mechanisms used by enzymes: acid–base catalysis, covalent catalysis, and metal ion catalysis. We will examine each of these, using model reactions to illustrate some of their fundamental features.

1. Acid–Base Catalysis Many enzyme mechanisms include **acid–base catalysis,** in which a proton is transferred between the enzyme and the substrate. This mechanism of catalysis can be further divided into **acid catalysis** and **base catalysis.** Some enzymes use one or the other; many use both. Consider the following model reaction, the tautomerization of a ketone to an enol (**tautomers** are interconvertible isomers that differ in the placement of a hydrogen and a double bond):

$$
\begin{array}{ccc}
\text{R} & \text{R} & \text{R} \\
| & | & | \\
\text{C}=\text{O} \rightleftharpoons & \left[\text{C}\cdots\text{O}^- \right] \rightleftharpoons & \text{C}-\text{O}-\text{H} \\
| & \vdots & || \\
\text{CH}_2 & \text{CH}_2 & \text{CH}_2 \\
| & \vdots & \\
\text{H} & \text{H}^+ & \\
\text{Ketone} & \text{Transition state} & \text{Enol}
\end{array}
$$

Here the transition state is shown in square brackets to indicate that it is an unstable, transient species. The dotted lines represent bonds in the process of breaking or forming. The uncatalyzed reaction occurs slowly because formation of the carbanion-like transition state has a high activation energy barrier (a **carbanion** is a compound in which the carbon atom bears a negative charge).

If a catalyst (symbolized H—A) donates a proton to the ketone's oxygen atom, it reduces the unfavorable carbanion character of the transition state, thereby lowering its energy and hence lowering the activation energy barrier for the reaction:

$$
\begin{array}{ccc}
\text{R} & \text{R} & \text{R} \\
| & | & | \\
\text{C}=\text{O} + \text{H}-\text{A} \rightleftharpoons & \left[\text{C}\cdots\text{O}^-\cdots\text{H}^+\cdots\text{A}^- \right] \rightleftharpoons & \text{C}-\text{O}-\text{H} + \text{H}-\text{A} \\
| & \vdots & || \\
\text{CH}_2 & \text{CH}_2 & \text{CH}_2 \\
| & \vdots & \\
\text{H} & \text{H}^+ & \\
\end{array}
$$

This is an example of acid catalysis, since the catalyst acts as an acid by donating a proton. Note that *the catalyst is returned to its original form at the end of the reaction.*

The same keto-enol tautomerization reaction shown above can be accelerated by a catalyst that can accept a proton, that is, by a base catalyst. Here, the catalyst is shown as :B, where the dots represent unpaired electrons:

$$
\begin{array}{ccc}
\text{R} & \text{R} & \text{R} \\
| & | & \text{H}^+ \quad | \\
\text{C}=\text{O} + :\text{B} \rightleftharpoons & \left[\text{C}\cdots\text{O}^- \right] \rightleftharpoons & \text{C}-\text{O}-\text{H} + :\text{B} + \text{H}^+ \\
| & \vdots & || \\
\text{CH}_2 & \text{CH}_2 & \text{CH}_2 \\
| & \vdots & \\
\text{H} & \text{H}^+ & \\
& \vdots & \\
& \text{B} & \\
\end{array}
$$

Base catalysis lowers the energy of the transition state and thereby accelerates the reaction.

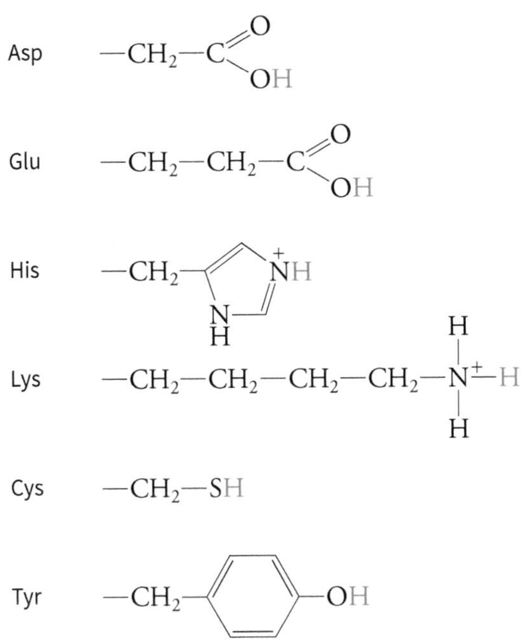

Asp $-CH_2-C\begin{smallmatrix}O\\OH\end{smallmatrix}$

Glu $-CH_2-CH_2-C\begin{smallmatrix}O\\OH\end{smallmatrix}$

His   —CH₂—[imidazole ring with $\overset{+}{N}H$ and NH]

Lys $-CH_2-CH_2-CH_2-CH_2-\overset{H}{\underset{H}{\overset{+}{N}}}-H$

Cys $-CH_2-SH$

Tyr $-CH_2-$[benzene ring]$-OH$

Figure 6.6 Amino acid side chains that can act as acid–base catalysts. These groups can act as acid or base catalysts, depending on their state of protonation in the enzyme's active site. The side chains are shown in their protonated forms, with the acidic proton highlighted.

Question **At neutral pH, which of these side chains most likely function as acid catalysts? Which most likely function as base catalysts?** (*Hint:* See Table 4.1.)

In enzyme active sites, several amino acid side chains can potentially act as acid or base catalysts. These are the groups whose pK values are in or near the physiological pH range. The residues most commonly identified as acid–base catalysts are shown in **Figure 6.6**. Because the catalytic functions of these residues depend on their state of protonation or deprotonation, the catalytic activity of the enzyme may be sensitive to changes in pH.

2. Covalent Catalysis   In **covalent catalysis,** the second major chemical reaction mechanism used by enzymes, a covalent bond forms between the catalyst and the substrate during formation of the transition state. Consider as a model reaction the decarboxylation of acetoacetate. In this reaction, the movement of electron pairs among atoms is indicated by red curved arrows (**Box 6.A**).

Box 6.A   Depicting Reaction Mechanisms

While it is often sufficient to draw the structures of a reaction's substrates and products, a full understanding of the reaction mechanism requires knowing what the electrons are doing. Recall that a covalent bond forms when two atoms share a pair of electrons, and a vast number of biochemical reactions involve breaking and forming covalent bonds. Although single-electron reactions also occur in biochemistry, we will focus on the more common two-electron reactions.

The curved arrow convention shows how electrons are rearranged during a reaction. The arrow emanates from the original location of an electron pair. This can be either an unshared electron pair, on an atom such as N or O, or the electrons of a covalent bond. The curved arrow points to the final location of the electron pair. For example, to show a bond breaking:

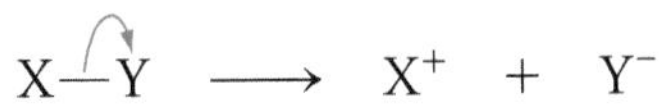

$$X-Y \longrightarrow X^+ + Y^-$$

and to show a bond forming:

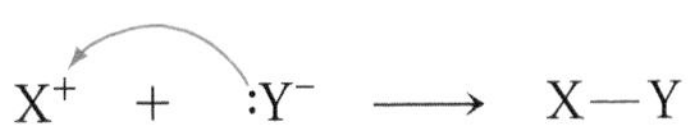

$$X^+ + :Y^- \longrightarrow X-Y$$

A familiarity with Lewis dot structures and an understanding of electronegativity (see Section 2.1) is helpful for identifying electron-rich groups (often the source of electrons during a reaction) and electron-poor groups (where the electrons often end up).

A typical biochemical reaction requires several curved arrows, for example,

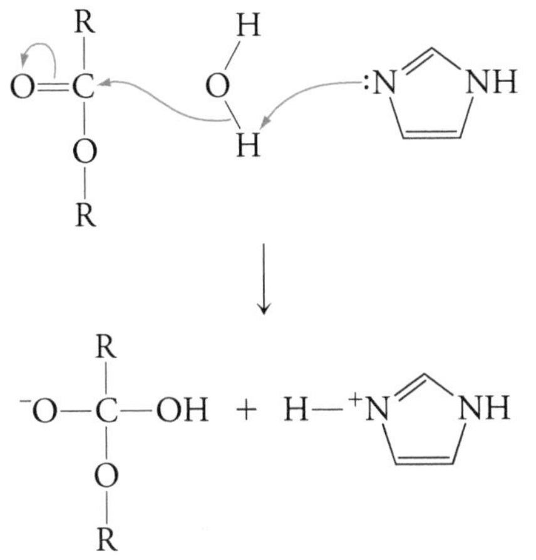

[reaction scheme showing carbonyl compound reacting with water and imidazole, yielding product below]

Question   **Draw arrows to show the electron movements in the reaction 2 H₂O → OH⁻ + H₃O⁺.**

The transition state, an enolate, has a high free energy of activation. This reaction can be catalyzed by a primary amine (RNH_2), which reacts with the carbonyl group of acetoacetate to form an **iminium ion** or protonated Schiff base (an **imine** or **Schiff base** is a compound containing a $C{=}N$ bond:

In this covalent intermediate, the protonated nitrogen atom acts as an electron sink to reduce the enolate character of the transition state in the decarboxylation reaction:

Finally, the protonated Schiff base decomposes, which regenerates the amine catalyst and releases the product, acetone:

In enzymes that use covalent catalysis, an electron-rich group in the enzyme active site forms a covalent adduct with a substrate. This covalent complex can sometimes be isolated; it is much more stable than a transition state. *Enzymes that use covalent catalysis undergo a two-part reaction process, so the reaction coordinate diagram contains two energy barriers with the reaction intermediate between them* (**Fig. 6.7**).

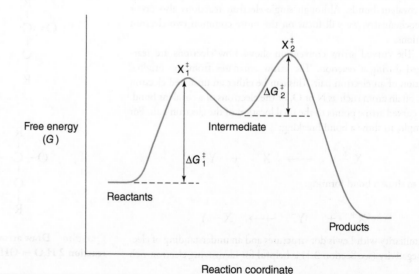

Figure 6.7 Diagram for a reaction accelerated by covalent catalysis. Two transition states flank the covalent intermediate. The relative heights of the activation energy barriers (to achieve the two transition states, $X_1^{\ddagger}$ and $X_2^{\ddagger}$) vary depending on the reaction.

Question Explain why the free energy of the reaction intermediate must be greater than the free energy of either the reactants or products.

Many of the same groups that make good acid–base catalysts (see Fig. 6.6) also make good covalent catalysts because they contain unshared electron pairs (**Fig. 6.8**). Covalent catalysis is often called nucleophilic catalysis because the catalyst is a **nucleophile,** that is, an electron-rich group in search of an electron-poor center (a compound with an electron deficiency is known as an **electrophile**).

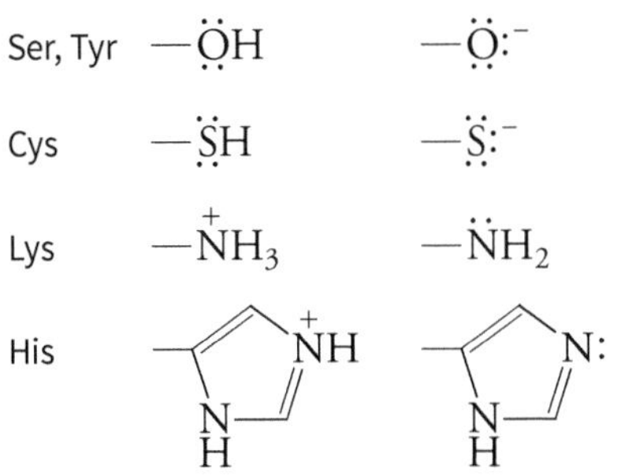

3. Metal Ion Catalysis

Metal ion catalysis occurs when metal ions participate in enzymatic reactions by mediating oxidation–reduction reactions, as mentioned earlier, or by promoting the reactivity of other groups in the enzyme's active site through electrostatic effects. A protein-bound metal ion can also interact directly with the reacting substrate. For example, during the conversion of acetaldehyde to ethanol as catalyzed by the liver enzyme alcohol dehydrogenase, a zinc ion stabilizes the developing negative charge on the oxygen atom during formation of the transition state:

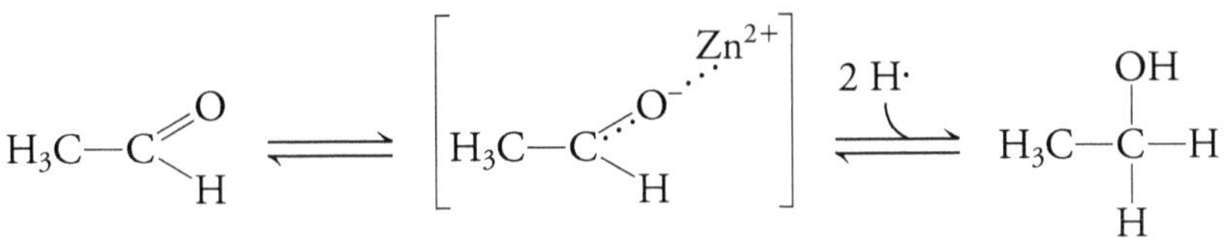

Figure 6.8 Protein groups that can act as covalent catalysts. In their deprotonated forms (*right*), these groups act as nucleophiles. They attack electron-deficient centers to form covalent intermediates.

Acetaldehyde   Ethanol

The catalytic triad of chymotrypsin promotes peptide bond hydrolysis

Chymotrypsin uses both acid–base catalysis and covalent catalysis to accelerate peptide bond hydrolysis. These activities depend on three active-site residues whose identities and catalytic importance have been the object of intense study since the 1960s. Two of chymotrypsin's catalytic residues were identified using a technique called **chemical labeling.** When chymotrypsin is incubated with the compound diisopropylphosphofluoridate (DIPF), one of its 27 serine residues (Ser 195) becomes covalently tagged with the diisopropylphospho (DIP) group, and the enzyme loses activity.

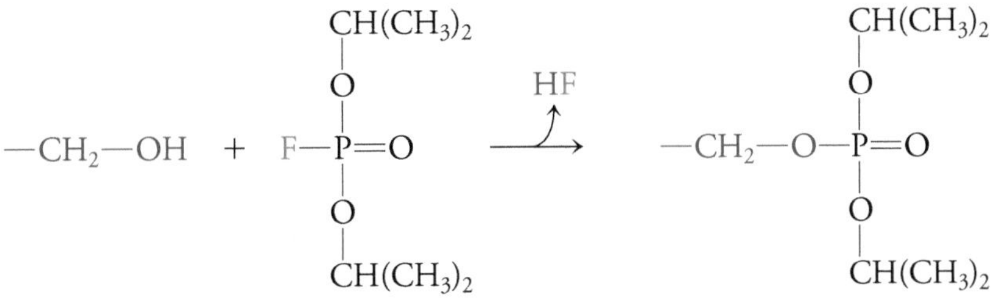

Ser 195   Diisopropylphosphofluoridate   DIP-Ser 195
(active)   (DIPF)   (inactive)

This observation provided strong evidence that Ser 195 is essential for catalysis. Chymotrypsin is therefore known as a **serine protease.** It is one of a large family of enzymes that use the same Ser-dependent catalytic mechanism. A similar labeling technique was used to identify the catalytic importance of His 57. The third residue involved in catalysis by chymotrypsin—Asp 102—was identified only after the fine structure of chymotrypsin was visualized through X-ray crystallography.

The hydrogen-bonded arrangement of the Asp, His, and Ser residues of chymotrypsin and other serine proteases is called the **catalytic triad** (**Fig. 6.9**). The substrate's **scissile bond** (the bond to be cleaved by hydrolysis) is positioned near Ser 195 when the substrate binds to the enzyme. The side chain of serine is not normally a strong enough nucleophile to attack an amide bond. However, His 57, acting as a base catalyst, abstracts a proton from Ser 195 so that the oxygen can act as a covalent catalyst. Asp 102 promotes catalysis by stabilizing the resulting positively charged imidazole group of His 57.

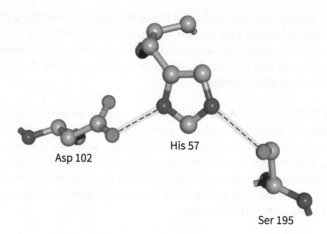

Figure 6.9 The catalytic triad of chymotrypsin. Asp 102, His 57, and Ser 195 are arrayed in a hydrogen-bonded network. Atoms are color-coded (C gray, N blue, O red), and the hydrogen bonds are shaded yellow.

Question Add hydrogen atoms to the three side chains.

Chymotrypsin-catalyzed peptide bond hydrolysis actually occurs in two phases that correspond to the formation and breakdown of a covalent reaction intermediate. The steps of catalysis are detailed in **Figure 6.10**. Nucleophilic attack of Ser 195 on the substrate's carbonyl carbon leads to a transition state in which the carbonyl carbon assumes tetrahedral geometry. This structure then collapses to an intermediate in which the N-terminal portion of the substrate remains covalently attached to the enzyme. The second part of the reaction, during which the oxygen of a water molecule attacks the carbonyl carbon, also includes a tetrahedral transition state. Although the enzyme-catalyzed reaction requires multiple steps, the net reaction is the same as the uncatalyzed reaction shown in Section 6.1.

The roles of Asp, His, and Ser in peptide bond hydrolysis, as catalyzed by chymotrypsin and other members of the serine protease family, have been tested through site-directed mutagenesis (Section 3.3). Replacing the catalytic aspartate with another residue decreases the rate of substrate hydrolysis about 5000-fold. Adding a methyl group to histidine by chemical labeling (so that it cannot accept or donate a proton) has a similar effect. Replacing the catalytic serine with another residue decreases enzyme activity about a millionfold. Surprisingly, replacing all three catalytic residues—Asp, His, and Ser—through site-directed mutagenesis does not completely abolish protease activity: The modified enzyme still catalyzes peptide bond hydrolysis at a rate about 50,000 times greater than the rate of the uncatalyzed reaction. Clearly, chymotrypsin and its relatives rely on the acid–base catalysis and covalent catalysis carried out by the Asp–His–Ser catalytic triad, but these enzymes must have additional catalytic mechanisms that allow them to achieve reaction rates 10^{11} times greater than the rate of the uncatalyzed reaction.

Concept Check

1. Draw a free energy diagram for a reaction with and without a catalyst, and label reactants, products, transition state, activation energy, and free energy change for the reaction.

2. Explain why must a reaction that proceeds via covalent catalysis have two transition states.

3. Make a list of amino acid side chains that can function as acid–base catalysts or covalent catalysts.

4. Explain why some enzymes require a metal ion or organic group in the active site.

5. Draw the groups that make up chymotrypsin's catalytic triad and explain their roles.

6. Describe each step of the chymotrypsin reaction without looking at the text.

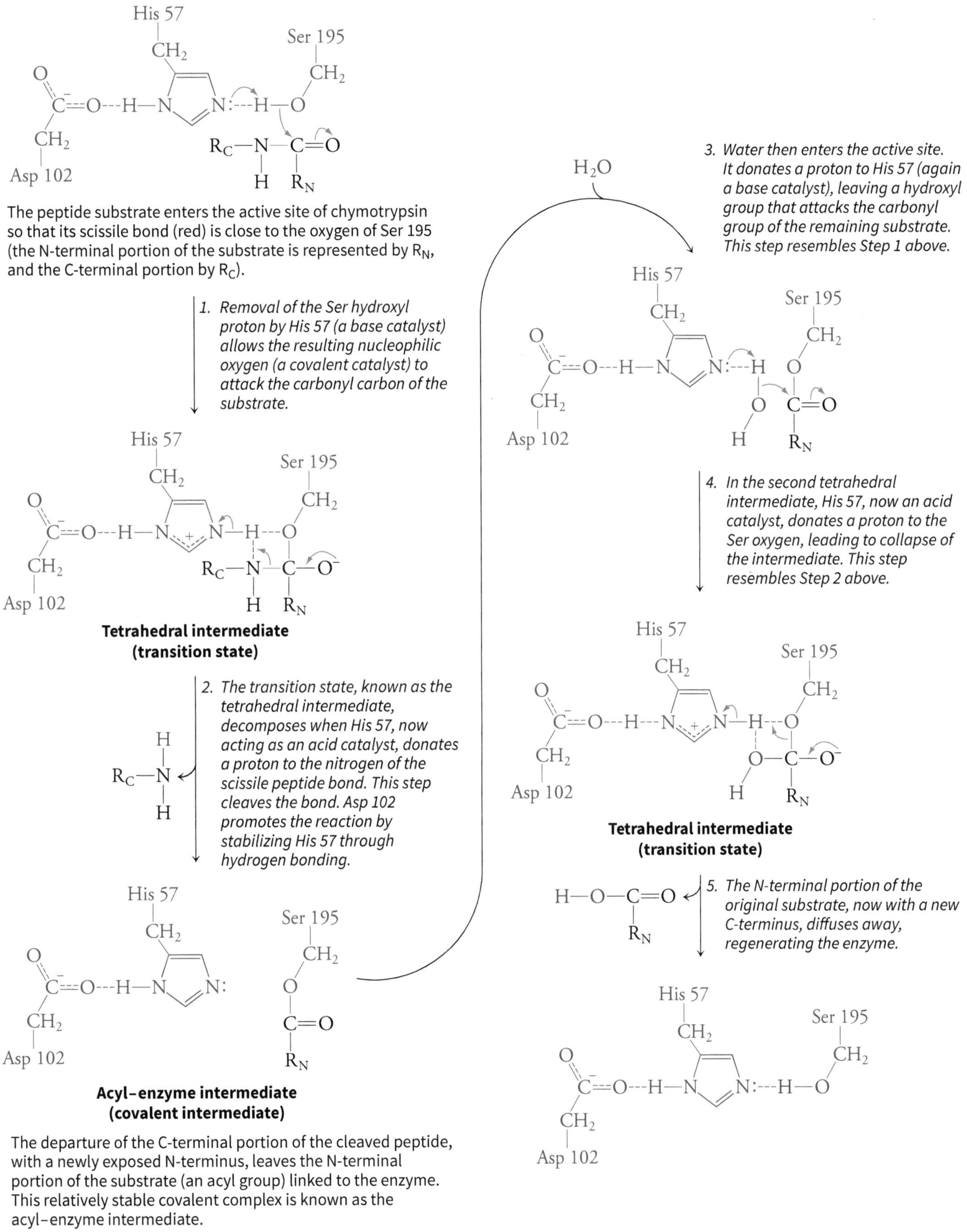

The peptide substrate enters the active site of chymotrypsin so that its scissile bond (red) is close to the oxygen of Ser 195 (the N-terminal portion of the substrate is represented by R_N, and the C-terminal portion by R_C).

1. Removal of the Ser hydroxyl proton by His 57 (a base catalyst) allows the resulting nucleophilic oxygen (a covalent catalyst) to attack the carbonyl carbon of the substrate.

Tetrahedral intermediate (transition state)

2. The transition state, known as the tetrahedral intermediate, decomposes when His 57, now acting as an acid catalyst, donates a proton to the nitrogen of the scissile peptide bond. This step cleaves the bond. Asp 102 promotes the reaction by stabilizing His 57 through hydrogen bonding.

Acyl–enzyme intermediate (covalent intermediate)

The departure of the C-terminal portion of the cleaved peptide, with a newly exposed N-terminus, leaves the N-terminal portion of the substrate (an acyl group) linked to the enzyme. This relatively stable covalent complex is known as the acyl–enzyme intermediate.

3. Water then enters the active site. It donates a proton to His 57 (again a base catalyst), leaving a hydroxyl group that attacks the carbonyl group of the remaining substrate. This step resembles Step 1 above.

4. In the second tetrahedral intermediate, His 57, now an acid catalyst, donates a proton to the Ser oxygen, leading to collapse of the intermediate. This step resembles Step 2 above.

Tetrahedral intermediate (transition state)

5. The N-terminal portion of the original substrate, now with a new C-terminus, diffuses away, regenerating the enzyme.

Figure 6.10 The catalytic mechanism of chymotrypsin and other serine proteases. Two tetrahedral transition states lead to and from the acyl–enzyme intermediate.

Question For each step of the reaction, identify all the bonds that break and form in that step.

6.3 Unique Properties of Enzyme Catalysts

KEY CONCEPTS

Explain how active site structure contributes to catalysis.

- Discuss the limitations of the lock-and-key model.
- Describe the importance of transition state stabilization, proximity and orientation effects, induced fit, and electrostatic catalysis.

If only a few residues in an enzyme directly participate in catalysis (for example, Asp, His, and Ser in chymotrypsin), why are enzymes so large? One obvious answer is that the catalytic residues must be precisely aligned in the active site, so a certain amount of surrounding structure is required to hold them in place. In 1894, long before the first enzyme structure had been determined (and several decades before it had been shown that enzymes are proteins), Emil Fischer noted the exquisite substrate specificity of enzymes and proposed that the substrate fit the enzyme like a key in a lock. However, the **lock-and-key model** for enzyme action does not explain how an active site that perfectly accommodates substrates could also accommodate reaction products before they are released from the enzyme. Moreover, the lock-and-key model cannot explain how an enzyme inhibitor can bind tightly in the active site but not react. The answer is that enzymes, like other proteins (see Section 4.3), are not rigid molecules but instead can—and must—flex while binding substrates. In other words, the enzyme–substrate interaction must be more dynamic than a key in a lock, more like a hand in a glove. In some cases, the enzyme can physically distort the substrate as it binds, in effect pushing it toward a higher-energy conformation closer to the reaction's transition state. Current theories attribute much of the catalytic power of enzymes to these and other specific interactions between enzymes and their substrates.

Enzymes stabilize the transition state

The lock-and-key model does contain a grain of truth, a principle first formulated by Linus Pauling in 1946. He proposed that *an enzyme increases the reaction rate not by binding tightly to the substrates but by binding tightly to the reaction's transition state* (that is, substrates that have been strained toward the structures of the products). In other words, the tightly bound key of the lock-and-key model is the transition state, not the substrate. In an enzyme, tight binding (stabilization) of the transition state occurs in addition to acid–base, covalent, or metal ion catalysis. In general, transition state stabilization is accomplished through the close complementarity in shape and charge between the active site and the transition state. A nonreactive substance that mimics the transition state can therefore bind tightly to the enzyme and block its catalytic activity (enzyme inhibition is discussed further in Section 7.3).

Transition state stabilization appears to be an important part of the chymotrypsin reaction. In this case, the two tetrahedral transition states (see Fig. 6.10), are stabilized through interactions that do not occur at any other point in the reaction. Rate acceleration is believed to result from an increase in both the number and strength of the bonds that form between active site groups and the substrate in the transition state.

1. During formation of the tetrahedral intermediate, the planar peptide group of the substrate changes its geometry, and the carbonyl oxygen, now an anion, moves into a previously unoccupied cavity near the Ser 195 side chain. In this cavity, called the **oxyanion hole,** the substrate oxygen ion can form two new hydrogen bonds with the backbone NH groups of Ser 195 and Gly 193 (**Fig. 6.11**). The backbone NH group of the substrate residue preceding the scissile bond forms another hydrogen bond to Gly 193 (not shown in Fig. 6.11). Thus, the transition state is stabilized (its energy lowered) by three hydrogen bonds that cannot form when the enzyme first binds its substrate. The stabilizing effect of these three new hydrogen bonds could account for a significant portion of chymotrypsin's catalytic power, since the energy of a standard hydrogen bond is about 20 kJ · mol^{-1} and the reaction rate increases 10-fold for every decrease in $\Delta G^{\ddagger}$ of 5.7 kJ · mol^{-1}.

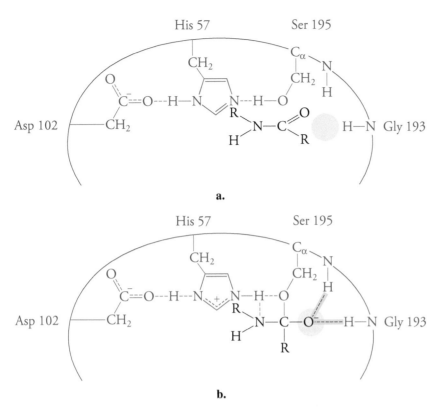

Figure 6.11 Transition state stabilization in the oxyanion hole. **a.** The chymotrypsin active site is shown with the oxyanion hole shaded in pink. The carbonyl carbon of the peptide substrate has trigonal geometry, so the carbonyl oxygen cannot occupy the oxyanion hole. **b.** Nucleophilic attack by the oxygen of Ser 195 on the substrate carbonyl group leads to a transition state, in which the carbonyl carbon assumes tetrahedral geometry. At this point, the substrate's anionic oxygen (the oxyanion) can move into the oxyanion hole, where it forms hydrogen bonds (shaded yellow) with two enzyme backbone groups.

2. NMR studies, which can identify individual hydrogen-bonding interactions, suggest that the hydrogen bond between Asp 102 and His 57 becomes shorter during formation of the two transition states (**Fig. 6.12**). Such a bond is called a **low-barrier hydrogen bond** because the hydrogen is shared equally between the original donor and acceptor atoms (in a standard hydrogen bond, the hydrogen still "belongs" to the donor atom and there is an energy barrier for its full transfer to the acceptor atom). A decrease in bond length from ~2.8 Å to ~2.5 Å in forming the low-barrier hydrogen bond is accompanied by a three- to fourfold increase in bond strength. The low-barrier hydrogen bond that forms during catalysis in chymotrypsin helps stabilize the transition state and thereby accelerate the reaction.

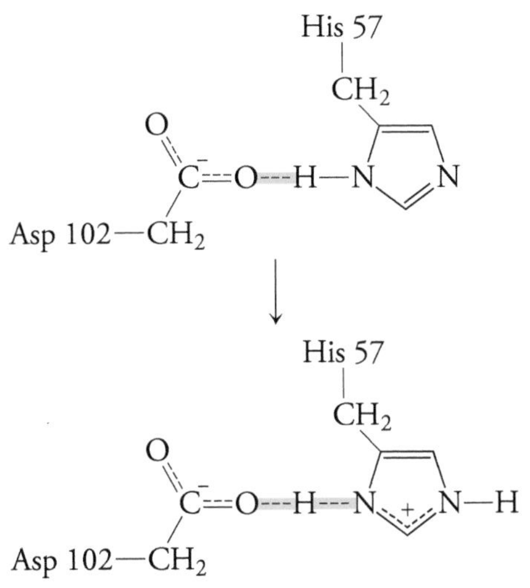

Figure 6.12 Formation of a low-barrier hydrogen bond during catalysis in chymotrypsin. The Asp 102—His 57 hydrogen bond becomes shorter and stronger, so the imidazole proton comes to be shared equally between the O of aspartate and the N of histidine in a low-barrier hydrogen bond.

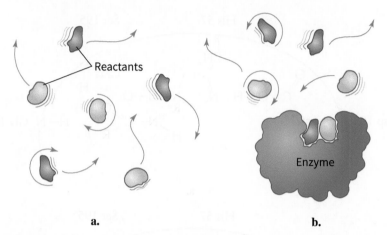

Figure 6.13 Proximity and orientation effects in catalysis. In order to react, two groups must come together and collide with the correct orientation. **a.** Reactants in solution are separated in space and have translational and rotational motions that must be overcome. **b.** When the reactants bind to an enzyme, their motion is limited, and they are held in close proximity and with the correct alignment for a productive reaction.

Efficient catalysis depends on proximity and orientation effects

Enzymes increase reaction rates by bringing reacting groups into close proximity so as to increase the frequency of collisions that can lead to a reaction. Furthermore, when substrates bind to an enzyme, their translational and rotational motions are frozen out so that they can be oriented properly for reaction (**Fig. 6.13**). These **proximity and orientation effects** likely explain some of the residual activity of chymotrypsin whose catalytic residues have been altered. Nevertheless, *an enzyme must be more than a template for assembling and aligning reacting groups.*

The active-site microenvironment promotes catalysis

Working out the step-by-step mechanism of an enzyme-catalyzed reaction may rely on classic approaches such as chemical modification of reacting groups, interference by transition-state analogues, tracing the fates of isotope-containing reactants, and X-ray crystallography. More recently, high-energy synchrotron radiation and free-electron laser beams have been used to probe protein structures on time scales of milliseconds or less. Collectively, these techniques can yield an almost frame-by-frame movie of enzyme action. Not surprisingly, conformational flexibility is a key feature of enzyme-catalyzed reactions, from binding substrates to repositioning them in the active site, forming the transition state, and releasing products.

In nearly all cases, an enzyme's active site is somewhat removed from the solvent, with its catalytic residues in some sort of cleft or pocket on the enzyme surface. Upon binding substrates, some enzymes undergo a pronounced conformational change so that they almost fully enclose the substrates. Daniel Koshland has called this phenomenon **induced fit.** A classic case of induced fit occurs in hexokinase, which catalyzes the phosphorylation of glucose by ATP (Reaction 1 of glycolysis; Section 13.1):

Glucose + ATP $\xrightarrow{\text{hexokinase}}$ Glucose-6-phosphate + ADP

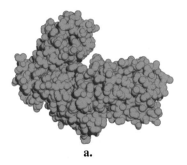
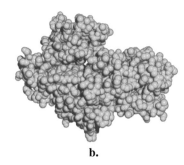

a.   b.

Figure 6.14 Conformational changes in hexokinase. **a.** The enzyme consists of two lobes connected by a hinge region. The active site is located in a cleft between the lobes. **b.** When glucose (not shown) binds to the active site, the enzyme lobes swing together, enclosing glucose and preventing the entry of water.

The enzyme consists of two hinged lobes with the active site located between them (**Fig. 6.14a**). When glucose binds to hexokinase, the lobes swing together, engulfing the sugar (Fig. 6.14b). The result of hinge bending is that the substrate glucose is positioned near the substrate ATP such that a phosphoryl group can be easily transferred from the ATP to a hydroxyl group of the sugar. Not even a water molecule can enter the closed active site. This is beneficial, since water in the active site could lead to wasteful hydrolysis of ATP:

$$ATP + H_2O \rightarrow ADP + P_i$$

A glucose molecule in solution is surrounded by ordered water molecules in a hydration shell (see Section 2.1). These water molecules must be shed in order for glucose to fit into the active site of an enzyme, such as hexokinase. However, once the desolvated substrate is in the enzyme active site, the reaction can proceed quickly because there are no solvent molecules to interfere. In solution, rearranging the hydrogen bonds of surrounding water molecules as the reactants approach each other and pass through the transition state is energetically costly. *By sequestering substrates in the active site, an enzyme can eliminate the energy barrier imposed by the ordered solvent molecules, thereby accelerating the reaction.*

This phenomenon is sometimes described as **electrostatic catalysis** since the nonaqueous active site allows more powerful electrostatic interactions between the enzyme and substrate than could occur in aqueous solution (for example, a low-barrier hydrogen bond can form in an active site but not in the presence of solvent molecules that would form ordinary hydrogen bonds).

Concept Check

1. Discuss why the lock-and-key model does not fully describe enzyme action.
2. Explain why excluding water from the active site promotes catalysis.
3. Describe the roles of the oxyanion hole and low-barrier hydrogen bonds in chymotrypsin.
4. Contrast chemical catalysts and enzymes with respect to transition state stabilization, proximity and orientation effects, induced fit, and electrostatic catalysis.

6.4 Chymotrypsin in Context

KEY CONCEPTS

Recognize that chymotrypsin illustrates general features of enzyme evolution and physiology.

- Distinguish divergent and convergent evolution.
- Identify determinants of substrate specificity.
- Explain why some enzymes undergo activation and inhibition.

Figure 6.15 Structures of chymotrypsin, trypsin, and elastase. The superimposed backbone traces of bovine chymotrypsin (blue), bovine trypsin (green), and porcine elastase (red) are shown along with the side chains of the active site Asp, His, and Ser residues.

Chymotrypsin serves as a model for the structures and functions of a large family of serine proteases. As with myoglobin and hemoglobin (Section 5.1), a close look at chymotrypsin reveals some general features of enzyme function, including evolution, substrate specificity, and inhibition.

Not all serine proteases are related by evolution

The first three proteases to be examined in detail were the digestive enzymes chymotrypsin, trypsin, and elastase, which have strikingly similar three-dimensional structures (**Fig. 6.15**). This was not expected on the basis of their limited sequence similarity. For example, bovine (cow) chymotrypsin is only 53% identical to bovine trypsin, and porcine (pig) elastase is only 48% identical. However, careful examination revealed that most of the sequence variation is on the enzyme surface, and the positions of the catalytic residues in the three active sites are virtually identical. It is believed that *these proteins are the result of **divergent evolution;** that is, they evolved from a common ancestor and have retained their overall structure and catalytic mechanism.*

Some bacterial proteases with a catalytically essential serine are structurally related to the mammalian digestive serine proteases. However, the bacterial serine protease subtilisin (**Fig. 6.16**) shows no sequence similarity to chymotrypsin and no overall structural similarity, although it has the same Asp–His–Ser catalytic triad and an oxyanion hole in its active site. Subtilisin is an example of **convergent evolution,** a phenomenon whereby unrelated proteins evolve similar characteristics.

As many as five groups of serine proteases, each with a different overall backbone conformation, have undergone convergent evolution to arrive at the same Asp, His, and Ser catalytic groups. In some other hydrolases, the substrate is attacked by a nucleophilic serine or threonine residue that is located in a catalytic triad such as His–His–Ser or Asp–Lys–Thr. It would appear that natural selection favors this sort of arrangement of catalytic residues.

Enzymes with similar mechanisms exhibit different substrate specificity

Despite similarities in their catalytic mechanisms, chymotrypsin, trypsin, and elastase differ significantly from one another in their substrate specificity. Chymotrypsin preferentially cleaves peptide bonds following large hydrophobic residues. Trypsin prefers the basic residues arginine and lysine, and elastase cleaves the peptide bonds following small hydrophobic residues such as alanine, glycine, and valine (these residues predominate in elastin, an animal protein responsible for the elasticity of some tissues). *The varying specificities of these enzymes are largely*

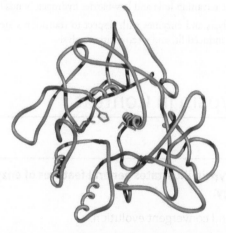

Figure 6.16 Structure of subtilisin from *Bacillus amyloliquefaciens*. The residues of the catalytic triad are highlighted in red.

Question Compare the structure of this enzyme with those of the three serine proteases shown in Figure 6.15.

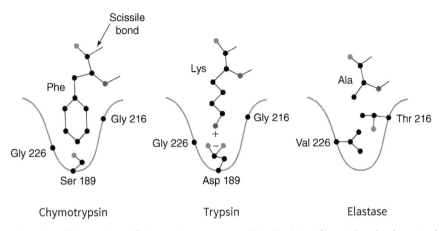

Figure 6.17 Specificity pockets of three serine proteases. The side chains of key residues that determine the size and nature of the specificity pocket are shown along with a representative substrate for each enzyme. Chymotrypsin prefers large hydrophobic side chains; trypsin prefers Lys or Arg; and elastase prefers Ala, Gly, or Val. For convenience, the residues of all three enzymes are numbered to correspond to the sequence of residues in chymotrypsin.

Question **What would the specificity pocket look like in a protease that cleaved bonds following Asp or Glu residues?**

explained by the chemical character of the so-called **specificity pocket,** *a cavity on the enzyme surface at the active site that accommodates the residue on the N-terminal side of the scissile peptide bond* (**Fig. 6.17**).

In chymotrypsin, the specificity pocket is about 10 Å deep and 5 Å wide, which offers a snug fit for an aromatic ring (whose dimensions are 6 Å × 3.5 Å). The specificity pocket in trypsin is similarly sized but has an aspartate residue rather than serine at the bottom. Consequently, the trypsin specificity pocket readily binds the side chain of arginine or lysine, which has a diameter of about 4 Å and a cationic group at the end. In elastase, the specificity pocket is only a small depression due to the replacement of two glycine residues on the walls of the specificity pocket (residues 216 and 226 in chymotrypsin) with the bulkier valine and threonine. Elastase therefore preferentially binds small nonpolar side chains. Although these same side chains could easily enter the chymotrypsin and trypsin specificity pockets, they do not fit well enough to immobilize the substrate at the active site, as required for efficient catalysis. Other serine proteases, such as those that participate in blood coagulation, exhibit exquisite substrate specificity, in keeping with their highly specialized physiological functions.

Chymotrypsin is activated by proteolysis

Relatively nonspecific proteases could do considerable damage to the cells where they are synthesized unless their activity is carefully controlled. In many organisms, *the activity of proteases is limited by the action of protease inhibitors* (some of which are discussed further below) *and by synthesizing the proteases as inactive precursors (called* **zymogens**) *that are later activated when and where they are needed.*

The inactive precursor of chymotrypsin is called chymotrypsinogen and is synthesized by the pancreas along with the zymogens of trypsin (trypsinogen), elastase (proelastase), and other hydrolytic enzymes. All these zymogens are activated by proteolysis after they are secreted into the small intestine. An intestinal protease called enteropeptidase activates trypsinogen by catalyzing the hydrolysis of its Lys 6—Ile 7 bond. Enteropeptidase catalyzes a highly specific reaction; it appears to recognize a string of aspartate residues near the N-terminus of its substrate:

$$H_3\overset{+}{N}—Val—Asp—Asp—Asp—Asp—Lys—Ile—\cdots$$

$$H_2O \searrow \text{enteropeptidase}$$

$$H_3\overset{+}{N}—Val—Asp—Asp—Asp—Asp—Lys—COO^- \ + \ H_3\overset{+}{N}—Ile—\cdots$$

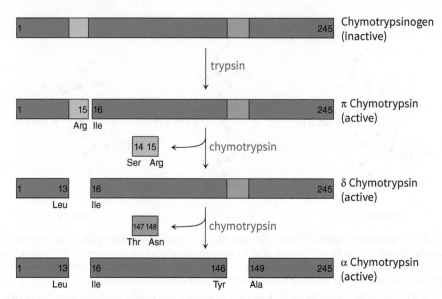

Figure 6.18 Activation of chymotrypsinogen. Trypsin activates chymotrypsinogen by catalyzing hydrolysis of the Arg 15—Ile 16 bond of the zymogen. The resulting active chymotrypsin then excises the Ser 14–Arg 15 dipeptide (by cleaving the Leu 13—Ser 14 bond) and the Thr 147–Asn 148 dipeptide (by cleaving the Tyr 146—Thr 147 and Asn 148—Ala 149 bonds). All three species of chymotrypsin (π, δ, and α) have proteolytic activity.

Trypsin, now itself active, cleaves the N-terminal peptide of the other pancreatic zymogens, including trypsinogen. The activation of trypsinogen by trypsin is an example of **autoactivation.**

The Arg 15—Ile 16 bond of chymotrypsinogen is susceptible to trypsin-catalyzed hydrolysis. Cleavage of this bond generates a species of active chymotrypsin (called π chymotrypsin), which then undergoes two autoactivation steps to generate fully active chymotrypsin (also called α chymotrypsin; **Fig. 6.18**). A similar process, in which zymogens are sequentially activated through proteolysis, occurs during blood coagulation (Section 6.5).

The two dipeptides that are excised during chymotrypsinogen activation are far removed from the active site (**Fig. 6.19**). How does their removal boost catalytic activity? A comparison of the X-ray structures of chymotrypsin and chymotrypsinogen reveals that the conformations of their active site Asp, His, and Ser residues are virtually identical (in fact, the zymogen can catalyze hydrolysis extremely slowly). However, the substrate specificity pocket and the oxyanion hole are incompletely formed in the zymogen. Proteolysis of the zymogen elicits small conformational changes that open up the substrate specificity pocket and oxyanion hole. *Thus, the enzyme becomes maximally active only when it can efficiently bind its substrates and stabilize the transition state.*

Protease inhibitors limit protease activity

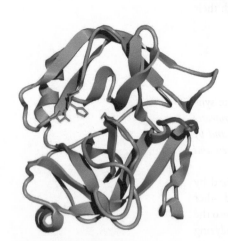

Figure 6.19 Location of the dipeptides removed during the activation of chymotrypsinogen. The Ser 14–Arg 15 dipeptide (*lower right*, green) and the Thr 147–Asn 148 dipeptide (*right*, blue) are located at some distance from the active site residues (red) in chymotrypsinogen.

The pancreas, in addition to synthesizing the zymogens of digestive proteases, synthesizes small proteins that act as **protease inhibitors.** The liver also produces a variety of protease inhibitor proteins that circulate in the bloodstream. If the pancreatic enzymes were prematurely activated or escaped from the pancreas through trauma, they would be rapidly inactivated by protease inhibitors. *The inhibitors pose as protease substrates but are not completely hydrolyzed.* For example, when trypsin attacks a lysine residue of bovine pancreatic trypsin inhibitor, the reaction halts during formation of the first transition state. The inhibitor remains in the active site, preventing any further catalytic activity (**Fig. 6.20**). The complex between trypsin and bovine pancreatic trypsin inhibitor is one of the strongest noncovalent protein–protein interactions known, with a dissociation constant of 10^{-14} M. An imbalance between the activities of proteases and the activities of protease inhibitors may contribute to disease (Section 6.5).

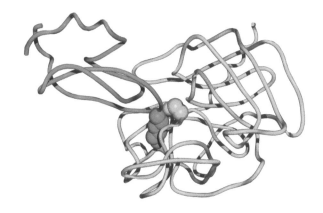

Figure 6.20 The complex of trypsin with bovine pancreatic trypsin inhibitor. Ser 195 of trypsin (gold) attacks the peptide bond of Lys 15 (green) in the inhibitor, but the reaction is arrested on the way to the tetrahedral intermediate.

Concept Check

1. Explain why chymotrypsin, trypsin, and elastase are similar.
2. Explain the structural basis for the different substrate specificities of the three enzymes.
3. Describe how to determine whether two proteins are related by convergent or divergent evolution.
4. Explain how chymotrypsin is activated and why this step is necessary.
5. Explain why protease inhibitors are necessary.

6.5 Clinical Connection: Blood Coagulation

KEY CONCEPTS

Describe blood coagulation as a protease cascade.

When a blood vessel is injured by mechanical force, infection, or some other pathological process, red and white blood cells and the plasma (fluid) that surrounds them can leak out. Except in the most severe trauma, the loss of blood can be halted through formation of a clot at the site of injury. The clot consists of aggregated platelets (tiny cell fragments that rapidly adhere to the damaged vessel wall and to each other) and a mesh of the protein fibrin, which reinforces the platelet plug and traps larger particles such as red blood cells (**Fig. 6.21**).

Fibrin polymers can form rapidly because they are generated at the site of injury from the soluble protein fibrinogen, which circulates in the blood plasma. Fibrinogen is an elongated molecule with a molecular mass of 340,000 and consists of three pairs of polypeptide chains. The proteolytic removal of short (14- or 16-residue) peptides from the N-termini of four of the six chains causes the protein to polymerize in end-to-end and side-to-side fashion to produce a thick fiber:

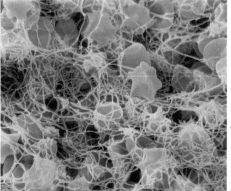

Figure 6.21 A blood clot. Red blood cells are trapped in a mesh of fibrin and platelets.

P. Motta/Dept. of Anatomy, University La Sapienza, Rome/Science Photo Library/ Photo Researchers

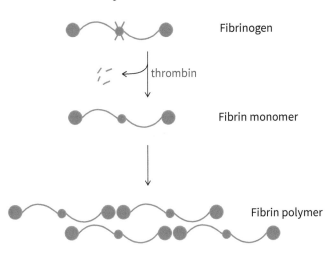

Fibrinogen

thrombin

Fibrin monomer

Fibrin polymer

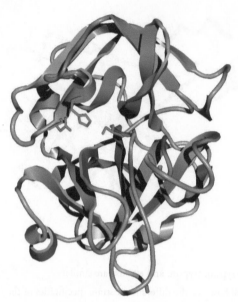

Figure 6.22 Thrombin.   The catalytic residues (aspartate, histidine, and serine) are highlighted in red.

Question **Compare the structures of thrombin and chymotrypsin (Fig. 6.1).**

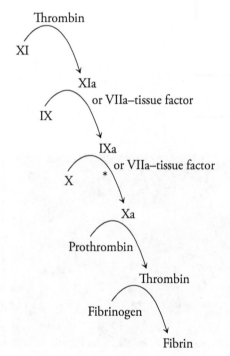

Figure 6.23 The coagulation cascade.
The triggering step is marked with an asterisk.

Table 6.3 Plasma Concentrations of Some Human Coagulation Factors

Factor	Concentration (µM)
XI	0.06
IX	0.09
VII	0.01
X	0.18
Prothrombin	1.39
Fibrinogen	8.82

The conversion of fibrinogen to fibrin is the final step of **coagulation,** a series of proteolytic reactions involving a number of proteins and additional factors from platelets and damaged tissue. The enzyme responsible for cleaving fibrinogen to fibrin is known as thrombin (**Fig. 6.22**). It is similar to trypsin in its sequence (38% of their residues are identical) and in its structure and catalytic mechanism.

Like trypsin, thrombin cleaves peptide bonds following arginine residues, but it is highly specific for the two cleavage sites in the fibrinogen sequence. Thrombin, like fibrinogen, circulates as an inactive precursor. Its zymogen, called prothrombin, contains a serine protease domain along with several other structural motifs. These elements interact with other coagulation factors to help ensure that thrombin—and therefore fibrin—is produced only when needed.

A serine protease known as factor Xa catalyzes the specific hydrolysis of prothrombin to generate thrombin. Factor Xa (the *a* stands for *active*) is the protease form of the zymogen factor X. To initiate coagulation, factor X is activated by a protease known as factor VIIa, working in association with an accessory protein called tissue factor, which is exposed when a blood vessel is broken. During the later stages of coagulation, factor Xa is generated by the activity of factor IXa, which is generated from its zymogen by the activity of factor XIa or factor VIIa–tissue factor. Factor XIa is in turn generated by the proteolysis of its zymogen by trace amounts of thrombin produced earlier in the coagulation process. The cascade of activation reactions is depicted in **Figure 6.23**. Many of these enzymatic reactions require accessory factors that are not shown in this simplified diagram.

Note that the coagulation proteases are named according to their order of discovery, not their order of action (thrombin is also known as factor II). All the coagulation proteases appear to have evolved from a trypsin-like enzyme but have acquired narrow substrate specificity and a correspondingly narrow range of physiological activities.

The coagulation reactions have an amplifying effect because each protease is a catalyst for the activation of another catalyst. Thus, a very small amount of factor IXa can activate a larger amount of factor Xa, which can then activate an even larger amount of thrombin. This amplification effect is reflected in the plasma concentrations of the coagulation factors (**Table 6.3**).

A complex process such as coagulation is subject to regulation at a variety of points, including the activation and inhibition of the various proteases. Protease inhibitors account for about 10% of the protein that circulates in plasma. One inhibitor, known as antithrombin, blocks the proteolytic activities of factor IXa, factor Xa, and thrombin, thereby limiting the extent and duration of clot formation. An arginine residue serves as a "bait" for the arginine-specific coagulation proteases (**Fig. 6.24**). The protease recognizes the inhibitor as a substrate but is unable to complete the hydrolysis reaction. The protease and inhibitor form a stable acyl–enzyme intermediate that is removed from the circulation within a few minutes.

Heparin, a sulfated polysaccharide (Section 11.3) first purified from liver, enhances the activity of antithrombin by two mechanisms: A short segment (5 monosaccharide residues) acts as an allosteric activator of antithrombin (allostery is introduced in Section 5.2), and a longer heparin polymer (containing at least 18 residues) can bind simultaneously to both antithrombin and its target protease such that their co-localization dramatically increases the rate of their reaction. Heparin or a synthetic version of it is used clinically as an anticoagulant following surgery.

Defects in many of the proteins involved in blood coagulation or its regulation have been linked to bleeding (or clotting) disorders. For example, one form of hemophilia, a tendency to bleed following minor trauma, results from a genetic deficiency of factor IX. A deficiency of antithrombin results in an increased risk of clot formation in the veins. If dislodged, the clots may end up blocking an artery in the lungs or brain, with dire consequences.

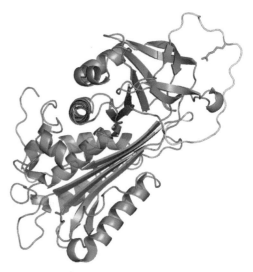

Figure 6.24 Antithrombin. A loop of the protein (yellow, residues 377–400 of the 432-residue protein) offers arginine 393 (red) as a "substrate" for several proteases.

Uses of enzymes in clinical diagnosis

Enzymes are widely used in clinical diagnosis due to their specific catalytic activities and presence in various tissues and body fluids. Enzyme-based assays provide valuable information about the functioning of organs, tissues, and cells, aiding in diagnosing, monitoring, and managing various diseases. Here are a few examples of enzymes used in clinical diagnosis.

1. Enzymes as **Biomarkers**: Certain enzymes are released into the bloodstream or other body fluids in response to specific diseases or conditions. Measuring the levels of these enzymes can provide diagnostic information. For instance: a. Creatine kinase (CK): Elevated levels of CK in the blood indicate muscle damage, such as myocardial infarction (heart attack) or muscular dystrophy. b. Alanine aminotransferase (ALT) and aspartate aminotransferase (AST): Increased levels of these liver enzymes can indicate liver damage or disease, such as hepatitis or cirrhosis. c. Amylase and lipase: Elevated levels of these enzymes in the blood can suggest pancreatitis.

2. Other Examples of Enzymes as **Biomarkers** are: a. Acid phosphatase: Increased acid phosphatase activity in the blood can indicate prostate cancer or bone diseases. b. Glucose-6-phosphate dehydrogenase (G6PD): Assaying G6PD activity is used to diagnose G6PD deficiency, an inherited enzyme disorder that can lead to hemolytic anemia. c. Cholinesterase: Measurement of cholinesterase activity is used in the diagnosis of pesticide or organophosphate poisoning.

3. Enzyme **Immunoassays** (EIA): Enzyme immunoassays utilize enzyme-labeled antibodies or antigens to detect the presence or quantity of specific substances in body fluids. This technique is used for various diagnostic purposes, including a. Enzyme-linked immunosorbent assay (ELISA): ELISA is commonly used to detect antibodies or antigens associated with infectious diseases, such as HIV, hepatitis, or COVID-19. b. Prostate-specific antigen (PSA): Measurement of PSA levels using enzyme immunoassays aids in detecting and monitoring prostate cancer.

4. Enzymes in **Genetic Testing**: Enzymes are utilized in genetic testing to detect specific genetic mutations or variations associated with inherited disorders. For example, a. Polymerase chain reaction (PCR): PCR, which relies on the heat-resistant DNA polymerase enzyme, is used to amplify specific DNA sequences to detect genetic mutations or pathogens. b. Restriction enzyme analysis: Restriction enzymes are used to cleave DNA at specific recognition sites, enabling the identification of genetic variations or mutations.

Concept Check

1. Describe the coagulation process beginning with the step marked by an asterisk in Fig. 6.23.
2. Discuss the advantages of a cascade system involving zymogen activation.
3. Explain why antithrombin can inhibit more than one coagulation protease.
4. Enzymes used in clinical diagnosis. Explain.

SUMMARY

6.1 What Is an Enzyme?

- Enzymes accelerate chemical reactions with high specificity under mild conditions.

6.2 Chemical Catalytic Mechanisms

- A reaction coordinate diagram illustrates the change in free energy between the reactants and products as well as the activation energy required to reach the transition state. The higher the activation energy, the fewer the reactant molecules that can reach the transition state and the slower the reaction.

- An enzyme provides a route from reactants (substrates) to products, which has a lower activation energy than the uncatalyzed reaction. Enzymes, sometimes with the assistance of a cofactor, use chemical catalytic mechanisms such as acid–base catalysis, covalent catalysis, and metal ion catalysis.

- In chymotrypsin, an Asp–His–Ser triad catalyzes peptide bond hydrolysis through acid–base and covalent catalysis and by stabilizing the transition state via the oxyanion hole and low-barrier hydrogen bonds.

6.3 Unique Properties of Enzyme Catalysts

- In addition to transition state stabilization, enzymes use proximity and orientation effects, induced fit, and electrostatic catalysis to facilitate reactions.

6.4 Chymotrypsin in Context

- Serine proteases that have evolved from a common ancestor share their overall structure and catalytic mechanism but differ in their substrate specificity.

- The activities of some proteases are limited by their synthesis as zymogens that are later activated and by their interaction with protease inhibitors.

6.5 Clinical Connection: Blood Coagulation

- Formation of a blood clot involves the sequential activation of serine protease zymogens, culminating in the generation of active thrombin, which converts fibrinogen to fibrin.

- Antithrombin inhibits several proteases to limit clot formation.

- Enzyme-based assays provide valuable information about the functioning of organs, tissues, and cells, aiding in diagnosing, monitoring, and managing various diseases.

KEY TERMS

hydrolysis	isozymes	acid catalysis	electrophile	low-barrier hydrogen	specificity pocket
reactant	reaction coordinate	base catalysis	metal ion catalysis	bond	zymogen
catalyst	$\Delta G^{\ddagger}$	tautomer	chemical labeling	proximity and	autoactivation
enzyme	transition state	carbanion	serine protease	orientation effects	protease inhibitor
ribozyme	$\Delta G_{reaction}$	covalent catalysis	catalytic triad	induced fit	coagulation
active site	cofactor	imine	scissile bond	electrostatic catalysis	enzyme
reaction specificity	coenzyme	Schiff base	lock-and-key model	divergent evolution	immunoassays
substrate	acid–base catalysis	nucleophile	oxyanion hole	convergent evolution	

BIOINFORMATICS

Brief Bioinformatics Exercises

6.1 Introduction to Enzyme Informatics

6.2 Interactions in the Serine Protease Active Site

Bioinformatics Projects

Enzyme Commission Classes and Catalytic Site Alignments with PyMOL

Enzyme Evolution

PROBLEMS

6.1 What Is an Enzyme?

1. Enzymes can accelerate the rate of reactions but cannot initiate them. What is the reason behind enzymes being unable to initiate chemical reactions?

2. Why do most enzymes adopt a globular rather than a fibrous shape?

3. Explain why the motor proteins myosin and kinesin (described in Section 5.4) are enzymes and write the reaction that each catalyzes.

4. An aminoglycoside complexed with copper has been shown to cleave DNA and can potentially be used as an antitumor agent. The glycoside cleaves DNA at a rate of 3.57 h^{-1} whereas the uncatalyzed rate of DNA hydrolysis is $3.6 \times 10^{-8} \text{ h}^{-1}$. What is the rate enhancement for DNA cleavage by the aminoglycoside?

5. Calculate the rate enhancement for DNA cleavage by the deoxyribonuclease. The deoxyribonuclease cleaves DNA at a rate of 3.57 h^{-1} whereas the uncatalyzed rate of DNA hydrolysis is $3.6 \times 10^{-8} \text{ h}^{-1}$.

6. Carboxypeptidase A is a protease that hydrolyzes peptide bonds sequentially from the C-terminus of proteins. The rate of uncatalyzed hydrolysis is $3.0 \times 10^{-9} \text{ s}^{-1}$ whereas the rate of the catalyzed reaction is 580 s^{-1}. What is the rate enhancement of carboxypeptidase A?

7. What is the rate constant for the uncatalyzed hydrolysis of a C-terminal peptide bond, as described in Problem 6? (*Hint*: For a first-order reaction, the rate constant, k, is equal to $0.693/t_{1/2}$.) The $t_{1/2}$ for this reaction is 7.3 years.

8. The half-life for the hydrolysis of the glycosidic bond in the sugar trehalose is 6.6×10^{6} years. **a.** What is the rate constant for the uncatalyzed hydrolysis of this bond if the reaction is first order (see Problem 7)? **b.** What is the rate enhancement for glycosidic bond hydrolysis catalyzed by trehalase if the rate constant for the catalyzed reaction is $2.6 \times 10^{3} \text{ s}^{-1}$?

9. The rate constant for the reaction catalyzed by cytidine deaminase, in the absence of enzyme, is $3.2 \times 10^{-10} \text{ s}^{-1}$. **a.** What is the half-life for this reaction (see Problem 7) in years? **b.** What is the rate enhancement for this reaction if the enzyme-catalyzed rate is 300 s^{-1}?

10. Compare the rate enhancements of cytidine deaminase (see Solution 9b) and adenosine deaminase (see Table 6.1). Would you expect these values to be similar?

11. The reactions catalysed by the enzymes listed here are presented in this chapter. To which class does each enzyme belong? Explain your answers. **a.** Lactase; **b.** Ribonuclease; **c.** Catalase; **d.** Amylase; **e.** DNA polymerase.

12. What is the relationship between the rate of an enzyme-catalyzed reaction and the rate of the corresponding uncatalyzed reaction? Do enzymes enhance the rates of slow uncatalyzed reactions as much as they enhance the rates of fast uncatalyzed reactions?

13. a. Draw an arrow to identify the bonds in the peptide below that would be hydrolyzed in the presence of the appropriate peptidase. **b.** To what class of enzymes does the peptidase belong?

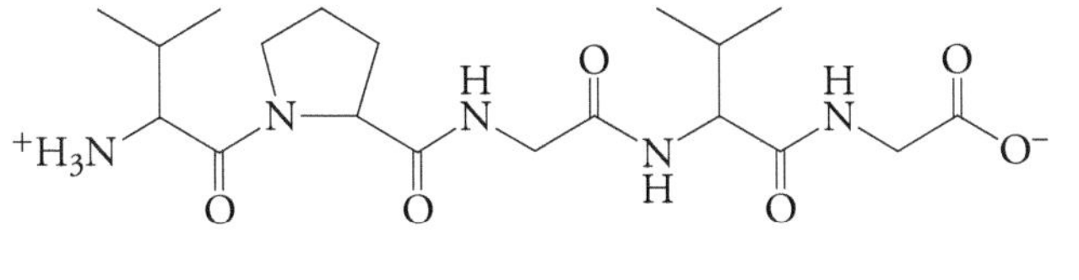

14. a. The molecule shown here is a substrate for trypsin. Draw the reaction products. **b.** Design an assay that would allow you to measure the rate of trypsin-catalyzed hydrolysis of the molecule.

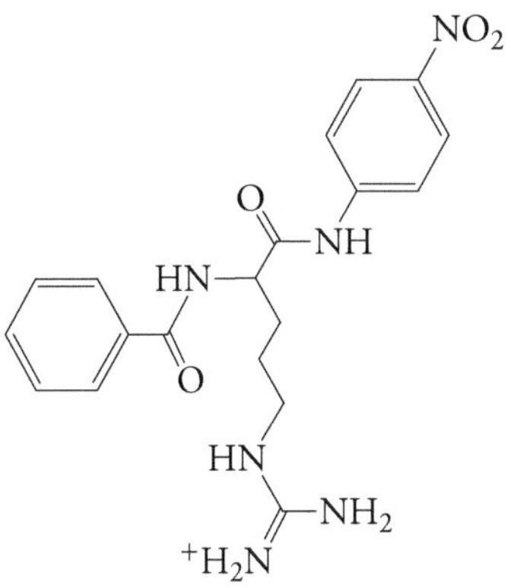

15. The reactions catalyzed by these enzymes are presented in this chapter. To which class does each enzyme belong? Explain your answers. **a.** pyruvate decarboxylase, **b.** alanine aminotransferase, **c.** alcohol dehydrogenase, **d.** hexokinase, **e.** chymotrypsin.

16. To which class do the enzymes that catalyze the following reactions belong?

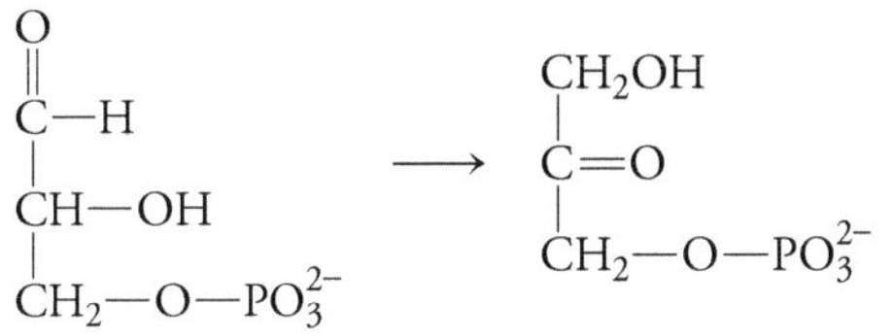

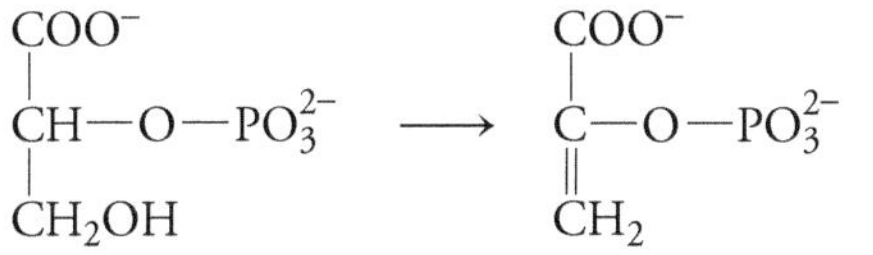

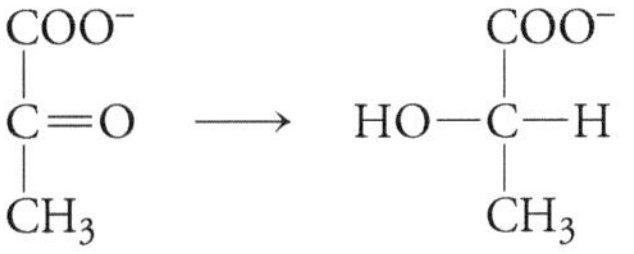

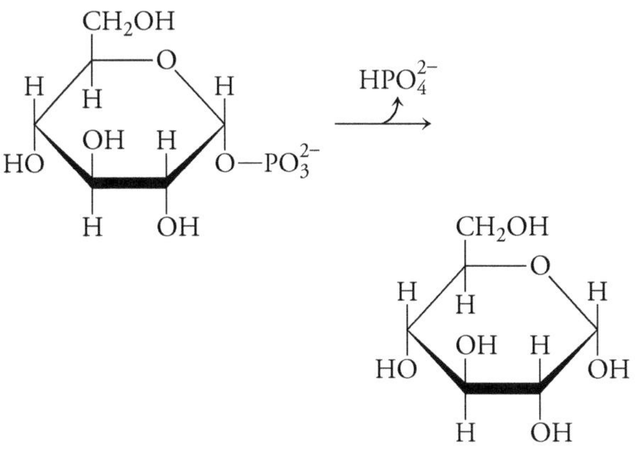

17. Use what you have learned about the reaction catalyzed by pyruvate decarboxylase, presented in this chapter, to predict the products of the reaction catalyzed by the aromatic L-amino acid decarboxylase using phenylalanine (see Table 4.1) as a substrate.

18. The ALT and AST tests measure the plasma activities of the liver enzymes alanine aminotransferase (presented in this chapter) and aspartate aminotransferase, respectively, which are released from the liver into the blood when the liver is damaged. Draw the reaction catalyzed by aspartate aminotransferase.

19. a. Succinate is oxidized by succinate dehydrogenase. Draw the structure of the product. **b.** Malate dehydrogenase catalyzes a reaction in which C2 of malate is oxidized. Draw the structure of the product. **c.** To which class of enzymes do succinate dehydrogenase and malate dehydrogenase belong?

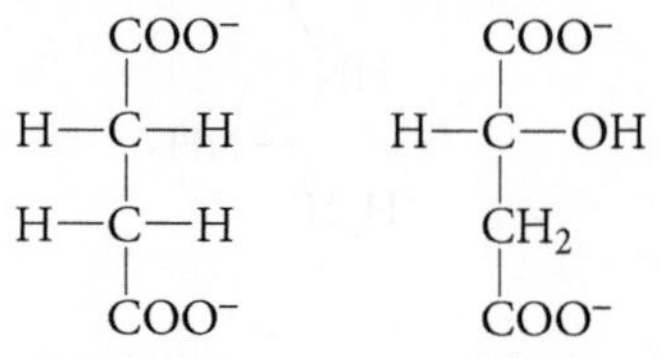

Succinate   Malate

20. a. Examine the reaction catalyzed by hexokinase in Section 6.3. Draw the product of the reaction catalyzed by creatine kinase, which acts on creatine in a similar manner. What would you predict to be the usual function of a kinase? **b.** The reaction catalyzed by hexokinase can be reversed by the enzyme glucose-6-phosphatase. To what class of enzymes do phosphatases belong? What would you predict to be the usual function of a phosphatase?

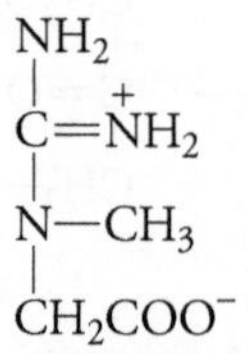

Creatine

21. Use Table 6.1 to name the enzymes that could act on the following substrates: **a.** protein, **b.** glyceraldehyde-3-phosphate (see Problem 16a), **c.** deoxyadenosine, **d.** DNA.

22. Propose a name for the enzymes that catalyze the following reactions (reactions may not be balanced):

a. 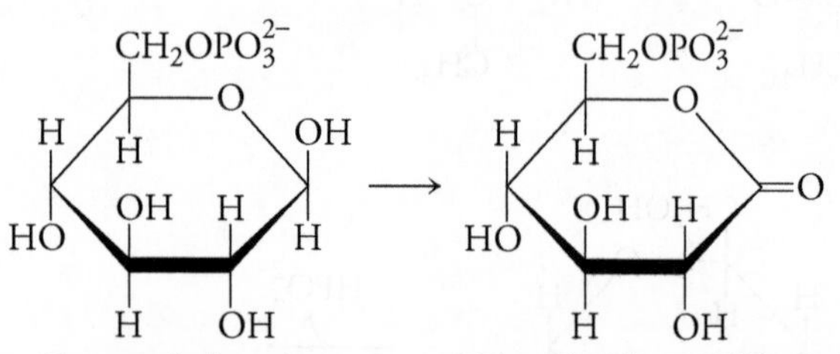
Glucose-6-phosphate   6-Phosphoglucono-δ-lactone

b. 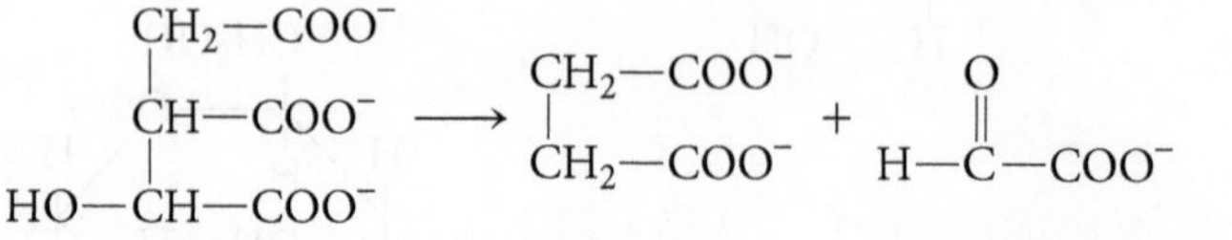
Isocitrate   Succinate   Glyoxylate

c. 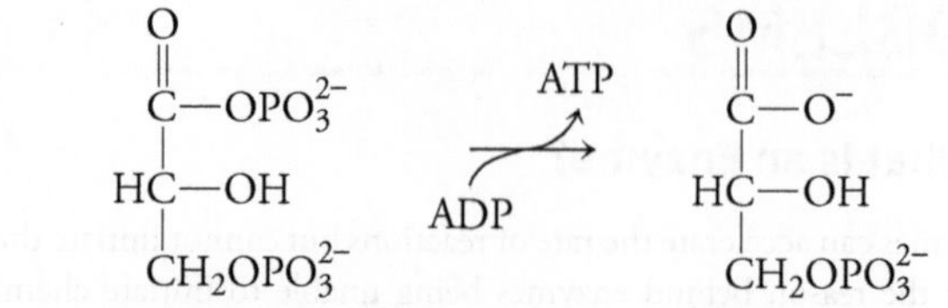
1,3-Bisphosphoglycerate   3-Phosphoglycerate

d. 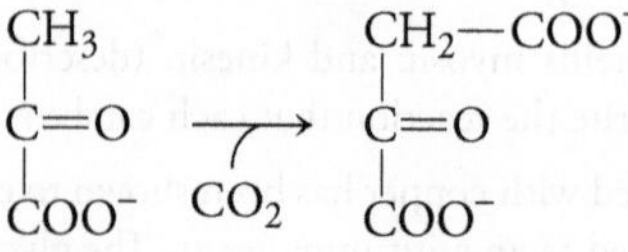
Pyruvate   Oxaloacetate

23. The amino acid tryptophan is converted to the hormone melatonin, as shown below. Indicate which reaction is catalyzed by each of the following types of enzymes: **a.** methyltransferase, **b.** hydroxylase, **c.** acetyltransferase, **d.** decarboxylase.

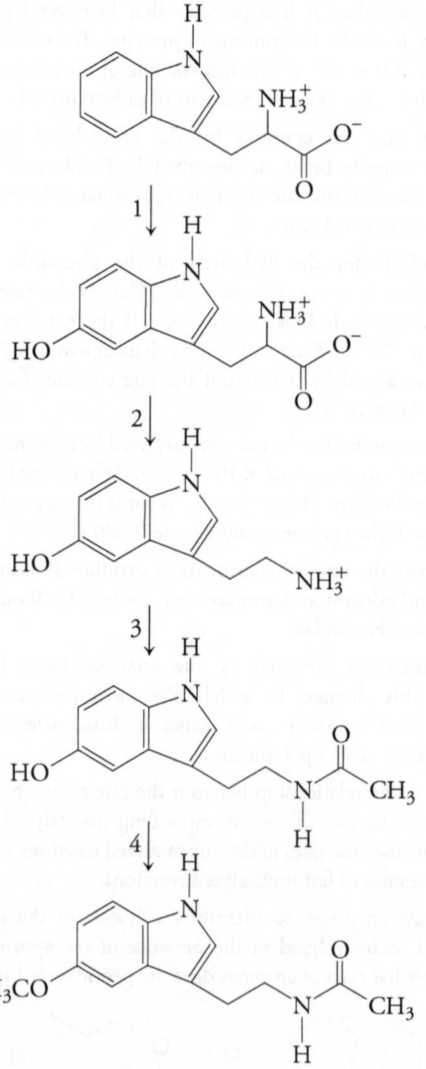

24. Use the enzyme nomenclature database at enzyme.expasy.org to determine what reaction is catalyzed by the enzyme **a.** catalase and **b.** glutathione peroxidase.

25. Given the following EC numbers, use the enzyme nomenclature database (see Problem 24) to provide the common name of the enzyme **a.** 4.3.2.1 and **b.** 1.7.2.2.

26. The EC number for chymotrypsin is 3.4.21.1. Use the enzyme nomenclature database (see Problem 24) to determine which amino acid residues precede the peptide bonds preferentially hydrolyzed by chymotrypsin.

27. **a.** Use the enzyme nomenclature database (see Problem 24) to determine the EC number for orotidine-5′-monophosphate decarboxylase (Table 6.1). (*Hint*: Enter "OMP decarboxylase" in the search engine.) **a.** According to the database, what are the products of this reaction? **b.** Is this enzyme similar to any other enzymes presented in Section 6.1?

6.2 Chemical Catalytic Mechanisms

28. Approximately how much do the enzymes **a.** adenosine deaminase and **b.** triose phosphate isomerase (see Table 6.1) decrease the activation energy ($\Delta G^{\ddagger}$) of the reactions they catalyze?

29. Approximately how much do the following enzymes decrease the activation energy ($\Delta G^{\ddagger}$) of the reactions they catalyze? **a.** The aminoglycoside complex described in Problem 3 and **b.** the carboxypeptidase described in Problem 4.

30. The reaction coordinate diagram shown in Figure 6.7 illustrates a two-step reaction. Which step is slower, according to the diagram?

31. Draw pairs of free energy diagrams that show the differences between the following reactions: **a.** a fast reaction vs. a slow reaction, **b.** a one-step reaction vs. a two-step reaction, **c.** a reaction with a positive free energy change vs. a reaction with a negative free energy change, and **d.** a two-step reaction with an initial slow step vs. a two-step reaction with an initial fast step.

32. Under certain conditions, peptide bond formation is more thermodynamically favorable than peptide bond hydrolysis. Would you expect chymotrypsin to catalyze peptide bond formation?

33. What is the relationship between the nucleophilicity and the acidity of an amino acid side chain?

34. **a.** Why do amino acids such as valine, leucine, and isoleucine generally not participate directly in acid-base or covalent catalysis? **b.** Despite their limited catalytic properties, why can mutating a valine, leucine, or isoleucine residue in an enzyme's active site still significantly affect catalysis?

35. Using what you know about the structures of the amino acid side chains and the mechanisms presented in this chapter, choose an amino acid side chain to play the following roles in an enzymatic mechanism: **a.** participate in proton transfer, **b.** act as a nucleophile.

36. Ribozymes are RNA molecules that catalyze chemical reactions. **a.** What features of a nucleic acid would be important for it to act as an enzyme? **b.** What part of the RNA structure might fulfill the roles listed in Problem 35? **c.** Why can RNA but not DNA act as a catalyst?

37. RNA is susceptible to base catalysis in which the hydroxide ion abstracts a proton from the 2′ OH group and then the resulting 2′ O⁻ nucleophilically attacks the 5′ phosphate. **a.** Is DNA susceptible to base hydrolysis? **b.** Does this observation help you to explain why DNA rather than RNA evolved as the genetic material?

38. The aldolase reaction, a reverse aldol condensation in which fructose-1,6-bisphosphate is converted to glyceraldehyde-3-phosphate and dihydroxyacetone phosphate in glycolysis, is shown in Figure 13.4. Which types of catalysis occur during the aldolase reaction: acid–base catalysis, covalent catalysis, and/or metal ion catalysis?

39. What is the role of the cysteine side chain in the glyceraldehyde-3-phosphate dehydrogenase reaction (see Fig. 13.6)?

40. Use what you know about the mechanism of chymotrypsin to explain why DIPF inactivates the enzyme.

41. Sarin is an organophosphorus compound similar to DIPF. In 1995, terrorists released it on a Japanese subway. Injuries resulted from the reaction of sarin with a serine esterase involved in nerve transmission, called acetylcholinesterase. Draw the structure of the enzyme's catalytic residue modified by sarin.

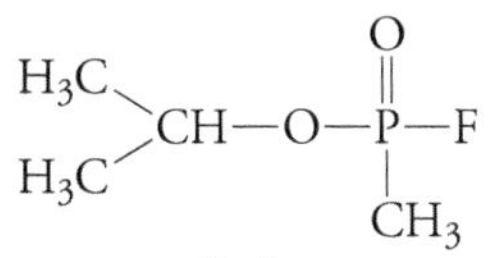

Sarin

42. Angiogenin, an enzyme involved in the formation of blood vessels, was treated with bromoacetate ($BrCH_2COO^-$). An essential histidine was modified and the activity of the enzyme was reduced by 95%. Show the chemical reaction for the modification of the histidine side chain with bromoacetate. Why would such a modification inactivate an enzyme that requires a His side chain for activity?

43. Treatment of the enzyme d-amino acid oxidase with 1-fluoro-2, 4-dinitrophenol (FDNP, below) inactivates the enzyme. Analysis of the derivatized enzyme revealed that one of the tyrosine side chains in the enzyme was unusually reactive. In a second experiment, when benzoate (a compound that structurally resembles the enzyme's substrate) was added prior to the addition of the FDNP, the tyrosine side chain did not react. **a.** Show the chemical reaction between FDNP and the tyrosyl residue. **b.** Why does benzoate prevent the reaction? What does this tell you about the role of the tyrosine side chains in this enzyme?

FDNP

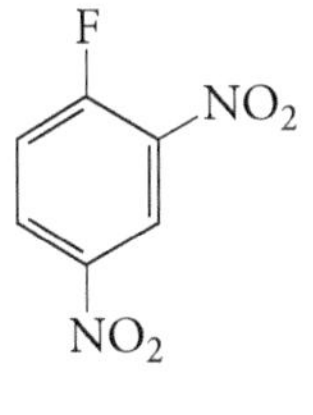

44. Sulfhydryl groups can react with the alkylating reagent *N*-ethylmaleimide (NEM). When NEM is added to a solution of creatine kinase, Cys 278 is alkylated, but no other Cys residues in the protein are modified. What can you infer about the role of the Cys 278 residue based on this information?

45. Cysteine proteases have a catalytically active Cys residue and can often be inactivated by iodoacetate (ICH_2COO^-). Show the chemical reaction for the modification of the Cys side chain by iodoacetate. Why would such a modification inactivate an enzyme that requires a Cys side chain for activity?

46. Consult Figure 6.10 and draw a reaction coordinate diagram for the chymotrypsin-catalyzed hydrolysis of a peptide bond.

47. Chymotrypsin readily hydrolyzes *p*-nitrophenylacetate using a mechanism similar to peptide bond hydrolysis (see Section 6.1). When *p*-nitrophenylacetate is mixed with chymotrypsin, *p*-nitrophenolate initially forms extremely rapidly. This is followed by a steady-state phase in which the product forms at a uniform rate. **a.** Using what you know about the mechanism of chymotrypsin, explain these observations. **b.** Draw a reaction coordinate diagram for the reaction. **c.** Would *p*-nitrophenylacetate be useful for monitoring the activity of trypsin *in vitro*?

48. How would chymotrypsin's catalytic triad be affected by extremely low and extremely high pH values (assuming that the rest of the protein structure remained intact)?

49. A bioinformatics study revealed that vasoinhibin enzymes have a Cys–His–Ser catalytic triad. These enzymes regulate angiogenesis (the formation of new blood vessels) and are thus potential drug targets for cancer treatment. If the Cys residue serves as the nucleophile, what are the possible roles of His and Ser in enzyme catalysis?

50. The crystal structure of a protease from an avian virus was recently solved. The enzyme has a Cys–His–Asp catalytic triad. Draw the structure of a plausible tetrahedral intermediate for this enzyme.

51. The mechanism of a steroid hydrolase was recently determined. The enzyme has potential industrial applications in the large-scale production of cholesterol-lowering drugs. The first step of the mechanism, very similar to that of chymotrypsin, is shown below. What is the role of the following active site amino acid side chains? **a.** Tyr 170, **b.** Lys 60, and **c.** Ser 57.

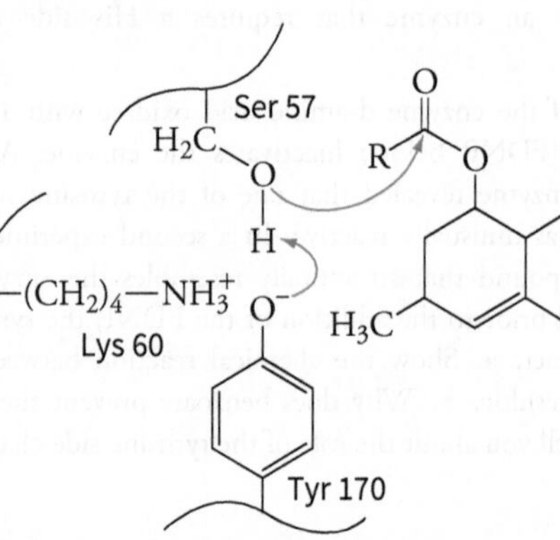

52. *E. coli* ribonuclease H1 catalyzes the hydrolysis of phosphodiester bonds in RNA. Its proposed mechanism involves a "carboxylate relay." **a.** Using the structures below as a guide, draw arrows that indicate how the hydrolysis reaction might be initiated. **b.** The pK values for all of the histidines in ribonuclease H1 were determined and include values of 7.1, 5.5, and < 5.0. Which value is most likely to correspond to His 124? Explain. **c.** Substituting an alanine residue for His 124 results in a dramatic decrease in enzyme activity. Explain.

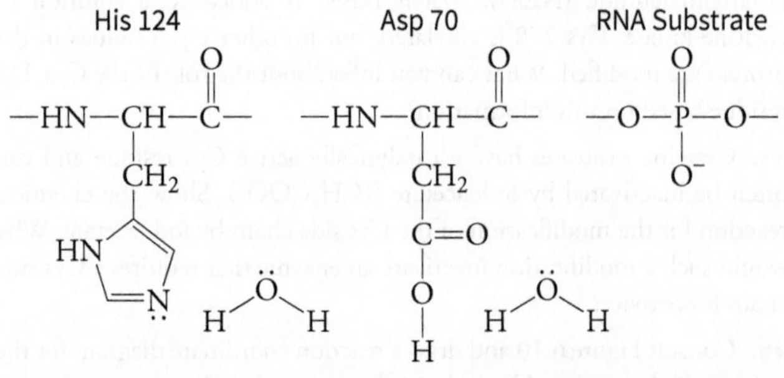

53. Glucosaminidase enzymes are essential for synthesis of the bacterial cell wall. Understanding the mechanism of these enzymes will aid in the development of drugs that could inhibit the enzyme in bacterial

pathogens. The mechanism of a glucosaminidase is shown. The enzyme active site has a unique Asp–His "catalytic dyad" in which Asp 232 positions the correct tautomer of His 234 in the active site. **a.** What are the roles of the Asp 318 and His 234 side chains in the reaction mechanism? **b.** Does the enzyme proceed via acid–base catalysis, covalent catalysis, metal ion catalysis, or some combination of these? Explain. **c.** The pK values of the side chains involved in the reaction were determined to be 5.0 and 7.0. Examine the mechanism and assign these pK values to the His 234 and Asp 318 side chains. Predict the pH optimum for the reaction. **d.** Replacing either His 234 or Asp 318 with a Gly residue abolishes enzyme activity. How does this result help to clarify the role of the Asp–His catalytic dyad in catalysis?

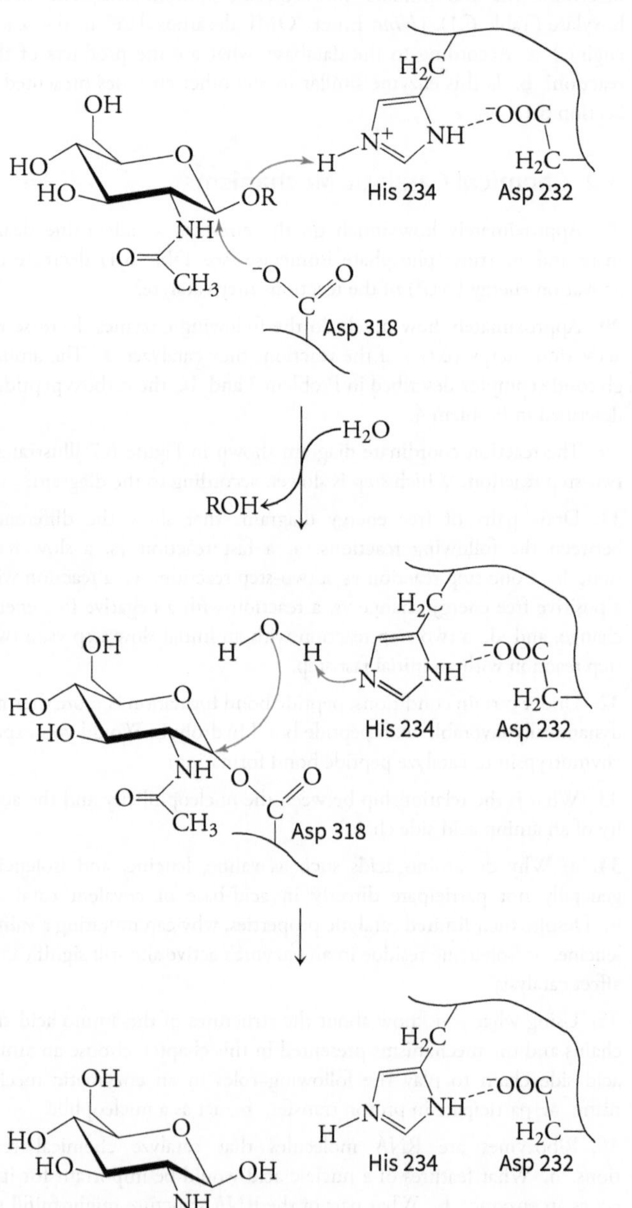

54. Renin belongs to a class of enzymes called aspartyl proteases in which aspartate plays a catalytic role similar to serine in chymotrypsin. Renin inhibitors could potentially be used to treat high blood pressure. Two aspartate residues in the renin active site, Asp 32 and Asp 215, constitute a catalytic dyad, rather than the catalytic triad found in chymotrypsin. **a.** Begin with the figure provided and draw the reaction mechanism for peptide hydrolysis catalyzed by renin. **b.** What type of catalytic strategy is utilized by renin? **c.** Compare the pK values of the Asp 32 and Asp 215 residues.

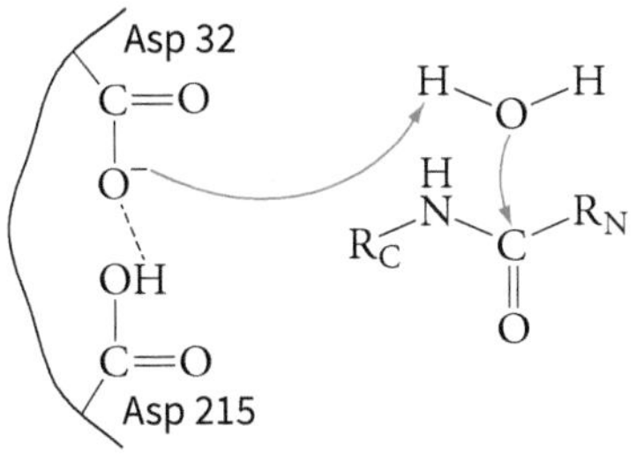

55. The enzyme bromelain (found in pineapple) is a cysteine protease in which the Cys residue plays a catalytic role similar to serine in chymotrypsin. The active site also contains a His residue but lacks Asp. Bromelain relieves inflammation and may have antitumor activity. **a.** Using chymotrypsin as a model, draw a reaction mechanism for the bromelain protease. **b.** Does the mechanism you have drawn employ acid–base catalysis, covalent catalysis, or both? **c.** Draw a reaction coordinate diagram consistent with the mechanism of bromelain. **d.** Why did natives of tropical countries use pineapple as a meat tenderizer? **e.** The pK values for the active site residues in bromelain are ~3 and ~8. The pH optimum for the reaction is 6.0. Assign pK values to the appropriate amino acid side chains and explain your reasoning.

56. The neurotoxin secreted by *Clostridium botulinum* contains a protease whose active site includes a Zn^{2+} ion coordinated by two histidine residues and a glutamate residue. The active site contains a second glutamate residue (Glu 224). Use the figure provided as a starting point to draw the mechanism of peptide bond hydrolysis catalyzed by this enzyme.

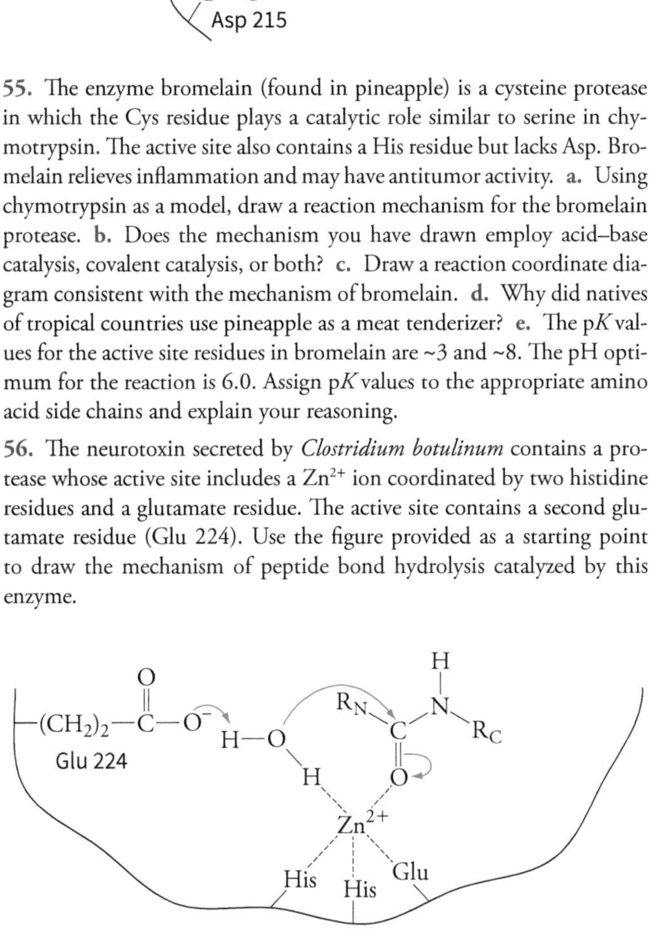

57. The enzyme carbonic anhydrase catalyzes the hydration of carbon dioxide to form carbonic acid (see Section 6.1), which dissociates to form a bicarbonate ion and a hydrogen ion. The enzyme's active site contains a zinc ion coordinated to the imidazole rings of three histidine residues. (A fourth histidine residue is located nearby and participates in catalysis.) A fourth coordination position is occupied by a water molecule. Draw the reaction mechanism (the first step is shown) of the hydration of carbon dioxide and show the regeneration of the enzyme.

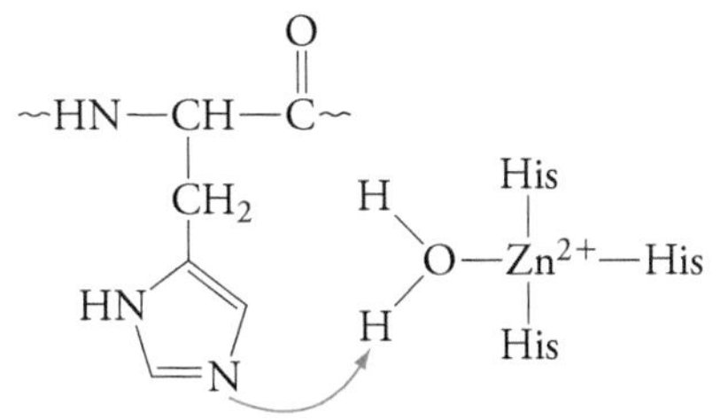

58. Asn and Gln residues in proteins can be nonenzymatically hydrolytically deamidated to Asp and Glu, respectively, a process that might function as a molecular timer for protein turnover. Studies show that Asn residues susceptible to deamidation are more likely to be preceded by Ser, Thr, or Lys and to be followed by Gly, Ser, or Thr. **a.** Write the balanced chemical equation for the hydrolytic deamidation of Asn to Asp. **b.** The first step of the mechanism of the deamidation of Gln is shown below, along with a required acid catalyst (HA). Draw the rest of the reaction mechanism.

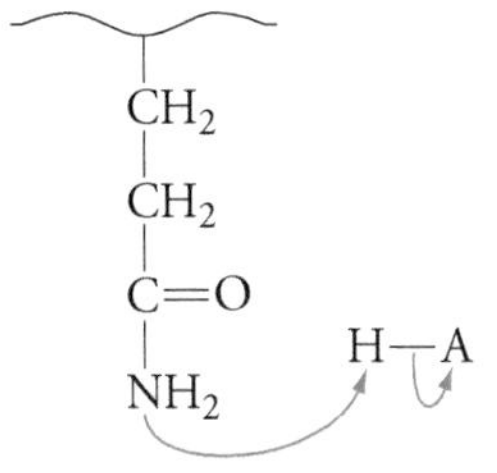

c. The HA catalyst in nonenzymatic deamidation is believed to be provided by neighboring amino acids in the deamidated protein. Describe how amino acid side chains that either precede or follow the labile Asn could serve as catalytic groups in the deamidation process. **d.** Amino terminal Gln residues undergo deamidation much more rapidly than internal Gln residues. Write a mechanism for this deamidation process, which includes the formation of a five-membered pyrrolidone ring. Amino terminal Asn residues do not undergo deamidation. Explain why.

59. Proteases in the amidase family have a Ser–Ser–Lys catalytic triad, as shown below. Propose a mechanism for the amidase enzyme.

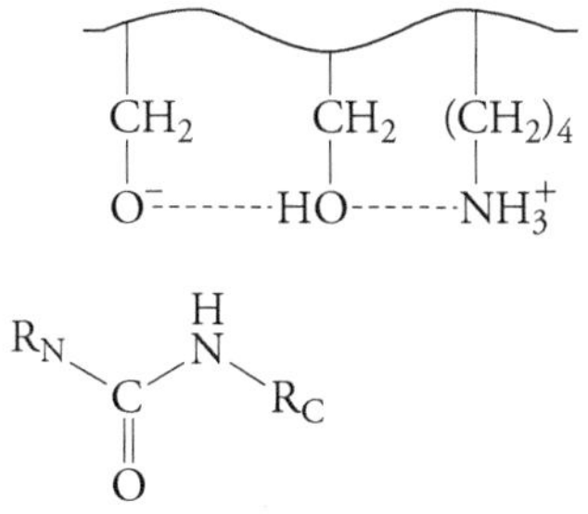

60. Lysozyme catalyzes the hydrolysis of a polysaccharide component of bacterial cell walls. The damaged bacteria subsequently lyse (rupture). Part of lysozyme's mechanism is shown below. The enzyme catalyzes cleavage of a bond between two sugar residues (represented by hexagons). Catalysis involves the side chains of Glu 35 and Asp 52. One of the residues has a pK of 4.5; the other has a pK of 5.9. **a.** Assign the given pK values to Glu 35 and Asp 52. **b.** Lysozyme is inactive at pH 2.0 and at pH 8.0. Explain. **c.** The lysozyme mechanism proceeds via a covalent intermediate. Use this information to complete the drawing of the lysozyme mechanism.

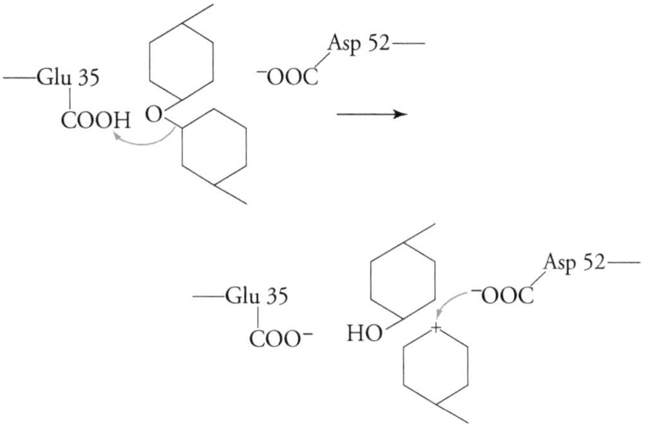

61. RNase A is a digestive enzyme secreted by the pancreas into the small intestine, where it hydrolyzes RNA into its component nucleotides. The optimum pH for RNAse A is about 6, and the pK values of the two histidines that serve as catalytic residues are 5.4 and 6.4. The first step of the mechanism is shown. **a.** Does ribonuclease proceed via acid–base catalysis, covalent catalysis, metal ion catalysis or some combination of these strategies? Explain. **b.** Assign the appropriate pK values to His 12 and His 119. **c.** Explain why the pH optimum of ribonuclease is pH 6. **d.** Ribonuclease catalyzes the hydrolysis of RNA but not DNA. Explain why.

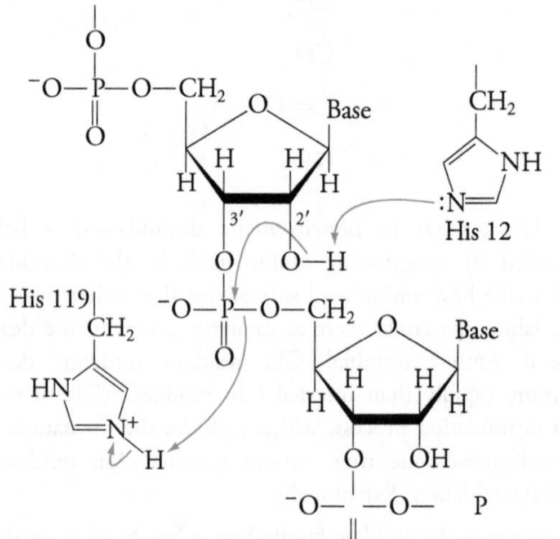

62. Chymotrypsin hydrolyzes *p*-nitrophenylacetate using a similar mechanism to that of peptide bond hydrolysis. When *p*-nitrophenylacetate is mixed with chymotrypsin, *p*-nitrophenolate initially forms extremely rapidly. This is followed by a steady-state phase in which the product forms at a uniform rate. **a.** Using what you know about the mechanism of chymotrypsin, explain these observations. **b.** Draw a reaction coordinate diagram for the reaction. **c.** Would *p*-nitrophenylacetate be useful for monitoring the activity of trypsin *in vitro*?

6.3 Unique Properties of Enzyme Catalysts

63. When substrates bind to an enzyme, their free energy may be lowered. Why doesn't this binding defeat catalysis?

64. Substrates and reactive groups in an enzyme's active site must be precisely aligned in order for a productive reaction to occur. Why, then, is some conformational flexibility also a requirement for catalysis?

65. A protease from a pathogenic bacterium uses a transition-state stabilization strategy similar to that used by chymotrypsin, in which Ser and Gly residues in the active site of the bacterial protease donate hydrogens to the oxyanion hole. When the Gly residue of the bacterial protease is mutated to Asp, the activity of the enzyme decreases to 55% compared to the wild-type enzyme. Explain.

66. The decarboxylation of orotidine-5′-monophosphate (OMP) to UMP is catalyzed by OMP decarboxylase (see Problem 27 and Table 6.1). The reaction intermediate is the carbanion shown below. Several amino acid side chains are important for stabilizing the transition state. **a.** How might an Arg residue and a Tyr residue stabilize the transition state by interacting with the phosphate group? **b.** How might a Gln residue and a Ser residue stabilize the transition state by interacting with the nitrogenous base?

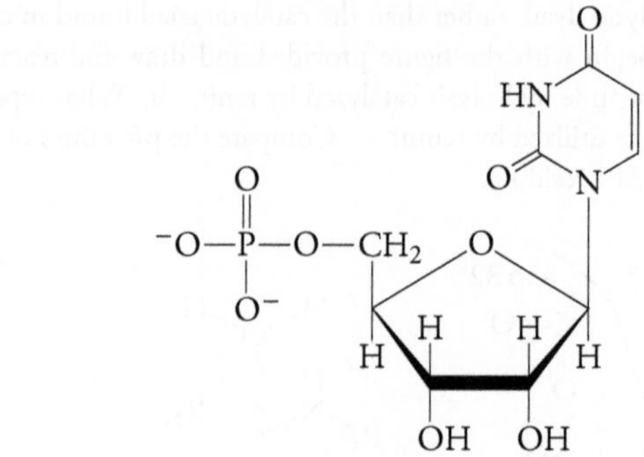

67. a. Daniel Koshland has noted that hexokinase undergoes a large conformational change on substrate binding, which apparently prevents water from entering the active site and participating in hydrolysis. However, the serine proteases do not undergo large conformational changes on substrate binding. Explain. **b.** When ATP and the sugar xylose (which resembles glucose but has only 5 carbons) are added to hexokinase, the enzyme produces a small amount of xylose-5-phosphate along with a large amount of ADP. Does this observation support Koshland's induced fit hypothesis?

68. Refer to Problem 55. What is the role of the zinc ion in catalysis? In transition state stabilization?

69. Refer to Problem 58. How might the amidase stabilize the transition state?

70. The enzyme adenosine deaminase catalyzes the conversion of adenosine to inosine. The compound 1,6-dihydropurine ribonucleoside binds to the enzyme with a much greater affinity than the adenosine substrate. What does this tell you about the mechanism of adenosine deaminase?

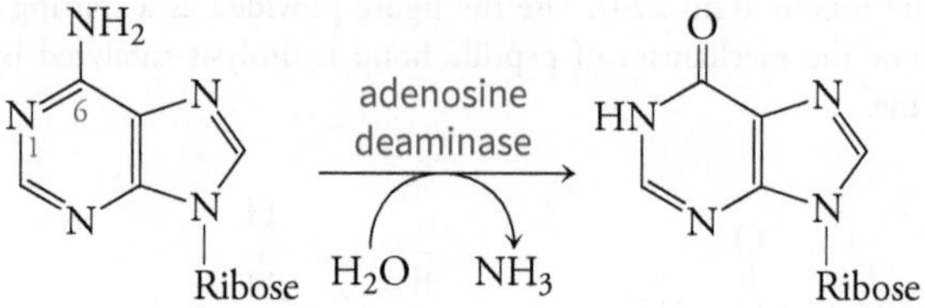

Adenosine H_2O NH_3 Inosine

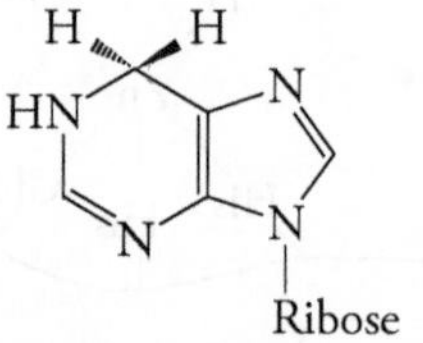

1,6-Dihydropurine ribonucleoside

71. Flurofamide is a potent inhibitor of urease, an enzyme that catalyzes the conversion of urea to CO_2 and NH_3 (shown below). What does this tell you about the geometry of the reaction's transition state?

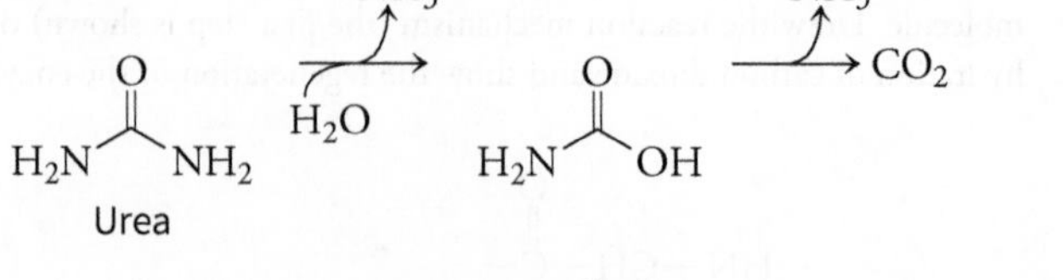

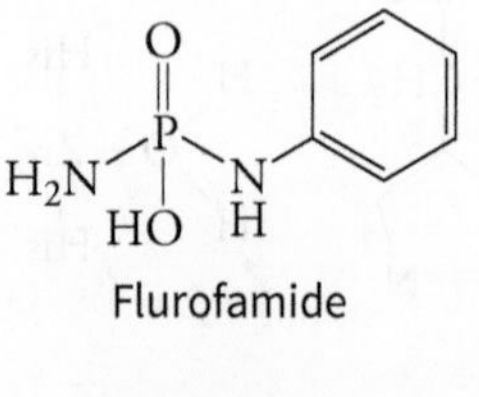

Flurofamide

72. Why are transition state analogs effective as drugs?

73. Imidazole reacts with *p*-nitrophenylacetate to produce the *N*-acetylimidazolium and *p*-nitrophenolate ions.

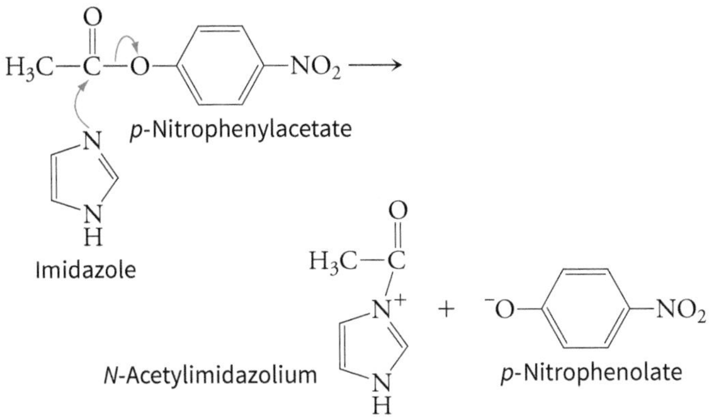

p-Nitrophenylacetate

Imidazole

N-Acetylimidazolium + *p*-Nitrophenolate

a. The compound shown below reacts to form *p*-nitrophenolate. Draw the reaction mechanism for this reaction.

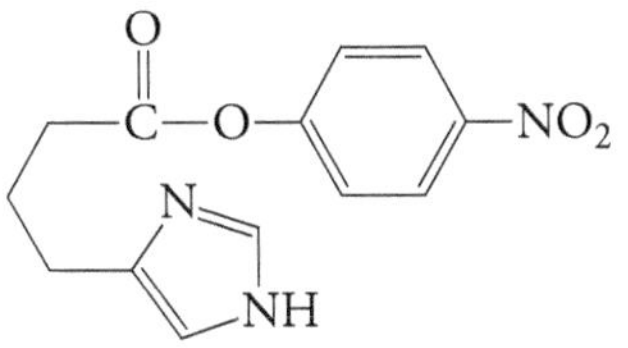

b. The compound described in part **a** reacts 24 times faster than imidazole. Explain why.

c. What do the results of this experiment tell you about how enzymes speed up reaction rates?

6.4 Chymotrypsin in Context

74. Many genetic mutations prevent the synthesis of a protein or give rise to an enzyme with diminished catalytic activity. Is it possible for a mutation to increase the catalytic activity of an enzyme?

75. The protease from the human immunodeficiency virus (HIV) is a target of the drugs used to treat HIV/AIDS. The protease has a mechanism similar to that of renin, described in Problem 54, except that renin consists of a single protein with two catalytic Asp residues, whereas the HIV protease consists of two identical subunits, each of which contributes an Asp residue. Is this an example of convergent or divergent evolution? Explain.

76. Carboxypeptidases catalyze the hydrolysis of peptide bonds from the C-terminal end of proteins (see Problem 6), as shown in the figure. **a.** Carboxypeptidase A (CBP A) preferentially hydrolyzes peptide bonds where R is the side chain of Phe, Trp, Tyr, Leu, or Ile. What does this tell you about the specificity pocket of CBP A? **b.** The crystal structure of carboxypeptidase B (CBP B) was recently determined. CBP B preferentially catalyzes the hydrolysis of peptide bonds where R is the side chain of Lys or Arg. The specificity pocket of CBP B contains, among others, Leu, Ile, and Asp side chains that interact with the substrate. What types of interactions are formed in the enzyme–substrate complex?

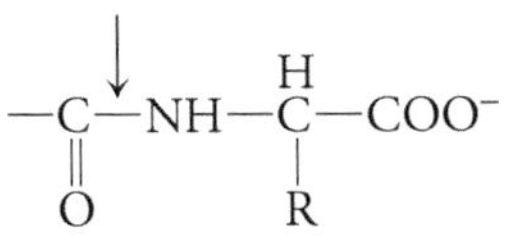

77. The amino acid Asp 189 lies at the base of the substrate specificity pocket in the enzyme trypsin (see Fig. 6.17). **a.** How is this related to trypsin's substrate specificity? What kinds of interactions take place between Asp 189 and the amino acid side chain on the carboxyl side of the scissile bond? **b.** In site-directed mutagenesis studies, Asp 189 was replaced with lysine. How do you think this would affect substrate specificity? **c.** The investigators who carried out the experiment described in part **b** analyzed the three-dimensional structure of the mutant enzyme and found that Lys 189 is actually not located in the substrate specificity pocket. Instead, the Lys side chain reaches out of the base of the pocket, rendering the specificity pocket nonpolar. With this additional information, how would the substrate specificity differ in the Lys 189 mutant enzyme?

78. Describe how the activation of the digestive enzyme chymotrypsin follows a cascade mechanism. Draw a diagram of your cascade.

79. The aspartyl aminopeptidase enzyme (involved in regulation of blood pressure) hydrolyzes peptide bonds on the carboxyl side of aspartate and is involved in the regulation of blood pressure. In order to study the enzyme's substrate binding site, a number of tetrapeptides were synthesized and the ability of the enzyme to bind to these peptides was measured. It is likely that both the P1′ residue and the P2′ residue fit in adjacent "pockets" on the aspartyl aminopeptidase enzyme, with hydrolysis occurring between residues P1 and P1′. Describe the characteristics of these pockets on the enzyme, using the data in the table.

Peptide (P1–P1′–P2′–P3′)	Catalyzed rate (s^{-1})
Asp–Phe–Ala–Leu	9.9
Asp–Lys–Ala–Leu	2.8
Asp–Ala–Phe–Leu	17.2
Asp–Ala–Lys–Leu	5.0
Asp–Ala–Asp–Leu	2.3

80. Chymotrypsin is usually described as an enzyme that catalyzes hydrolysis of peptide bonds following Phe, Trp, or Tyr residues. Is this information consistent with the description of the bonds cleaved during chymotrypsin activation (Fig. 6.18)? What does this tell you about chymotrypsin's substrate specificity?

81. Would chymotrypsin catalyze hydrolysis of the substrate shown in Problem 14? Explain why or why not.

82. Does DIPF inactivate trypsin or elastase (see Problem 40)? Explain.

83. In an Enzyme Immunoassay (EIA) test, a patient's sample's optical density (OD) was measured as 0.85, while the OD of the positive control was 2.10. Calculate the sample's percent positivity using the formula: Percent Positivity = (Sample OD/Positive Control OD) × 100.

84. One type of hereditary pancreatitis is caused by a mutation in the autocatalytic domain of trypsinogen that results in persistent activity of the enzyme. **a.** What are the physiological consequences of this disease? **b.** Describe a strategy to treat this disease.

85. Another type of hereditary pancreatitis (see Problem 84) results from a mutation in the pancreatic trypsin inhibitor. Would the treatment strategy that you designed in Problem 84 also be effective in treating this type of pancreatitis?

86. Not all proteases are synthesized as zymogens or have inhibitors that block their activity. What limits the potentially destructive power of these proteases?

87. Some plants contain compounds that inhibit serine proteases. It has been hypothesized that these compounds protect the plant from proteolytic enzymes of insects and microorganisms that would damage the plant. Tofu, or bean curd, possesses these compounds. Manufacturers of tofu treat it to eliminate serine protease inhibitors. Why is this treatment necessary?

88. A chymotrypsin inhibitor isolated from snake venom resembles the bovine pancreatic trypsin inhibitor (Fig. 6.20) but has an Asn residue in place of Lys 15. Why is this finding unexpected?

6.5 Clinical Connection: Blood Coagulation

89. Hemophiliacs who are deficient in factor IX can be treated with injections of pure factor IX. Occasionally, a patient develops antibodies to this material and it becomes ineffective. In such cases, the patient can be given factor VII instead. Explain why factor VII injections would be a useful treatment for a factor IX deficiency.

90. A genetic defect in factor IX causes hemophilia, a serious disease. However, a genetic defect in factor XI may have no clinical symptoms. Explain this discrepancy in terms of the cascade mechanism for activation of coagulation proteases.

91. Factor IX deficiency has been successfully treated by gene therapy, in which a normal copy of the gene is introduced into the body (see Section 3.3). Explain why restoration of factor IX levels to just a few percent of normal is enough to cure the abnormal bleeding.

92. Of all the proteases included in the cascade diagram of coagulation, which one is the oldest in evolutionary terms?

93. A vitamin K–dependent enzyme catalyzes the modification of Glu residues in the amino-terminal domain of prothrombin to produce γ-carboxyglutamate. **a.** Draw the structure of a γ-carboxyglutamate residue. **b.** The γ-carboxyglutamate side chain binds Ca^{2+} (which is essential for the blood-clotting process) more effectively than a glutamate side chain. Explain why.

94. Following severe trauma or infection, a patient may develop disseminated intravascular coagulation (DIC) in which numerous small clots form throughout the circulatory system. Explain why patients with DIC subsequently exhibit excessive bleeding.

95. In one thrombin variant, a point mutation produces a protease with normal activity toward fibrinogen but decreased activity toward antithrombin. Would this genetic defect increase the risk of bleeding or of clotting?

96. Why can't heparin be administered orally?

97. Researchers have developed drugs based on hirudin, a thrombin inhibitor from the medicinal leech *Hirudo medicinalis*. Why would the leech be a good source for an anticoagulant?

98. What challenges are associated with using enzymes as biomarkers in clinical diagnosis, and how can these challenges be addressed?

99. How does measuring enzyme activity in the blood help diagnose specific medical conditions?

100. Compare and contrast the diagnostic applications of enzyme immunoassays and enzyme activity assays. Highlight situations where one method might be more suitable than the other in clinical practice.

101. In a patient's blood sample, Creatine Kinase (CK) activity was measured as 250 U/L. Is this within the normal range, and what does this measurement indicate?

102. An Enzyme Immunoassay (EIA) test detected the presence of Hepatitis B surface antigen (HBsAg) in a patient's blood sample. What does this result indicate, and why is it significant?

103. Why are the activities of many enzymes, which are specific to various tissues/organs, usually detected in the blood/body fluids even when there is no damage, injury, or disease affecting the tissues/organs?

104. Can any factors or medications interfere with enzyme test results, potentially leading to false positives or negatives?

SELECTED READINGS

Di Cera, E., Serine proteases, *IUBMB Life* 61, 510–515 (2009). [Summarizes some of the biological roles and mechanistic features of proteases.]

Gutteridge, A. and Thornton, J.M., Understanding nature's catalytic toolkit, *Trends Biochem. Sci.* 30, 622–629 (2005). [Describes how certain sets of amino acid side chains form catalytic units that appear in many different enzymes.]

Martínez Cuesta, S., Rahman, S.A., Furnham, N., and Thornton, J.M., The classification and evolution of enzyme function, *Biophys. J.* 109, 1082–1086, doi: 10.1016/j.bpj.2015.04.020 (2015). [A brief review of enzyme classes (six at the time) and the evolution of new enzymes with the same or different classification.]

Radisky, E.S., Lee, J.M., Lu, C.-J.K., and Koshland, D.E., Jr., Insights into the serine protease mechanism from atomic resolution structures of trypsin reaction intermediates, *Proc. Nat. Acad. Sci.* 103, 6835–6840 (2006).

Ringe, D. and Petsko, G.A., How enzymes work, *Science* 320, 1428–1429 (2008). [Briefly summarizes some general features of enzyme function.]

CHAPTER 6 CREDITS

Figure 6.1 Image based on 4CHA. Tsukada, H., Blow, D.M., Structure of alpha-chymotrypsin refined at 1.68 A resolution, *J. Mol. Biol.* 184, 703–711 (1985).

Table 6.1 Data mostly from Radzicka, R. and Wolfenden, R., *Science* 267, 90–93 (1995).

Figure 6.14a Image based on 2YHX. Anderson, C.M., Stenkamp, R.E., Steitz, T.A., Sequencing a protein by X-ray crystallography. II. Refinement of yeast hexokinase B co-ordinates and sequence at 2.1 A resolution, *J. Mol. Biol.* 123, 15–33 (1978).

Figure 6.14b Image based on 1HKG. Steitz, T.A., Shoham, M., Bennett Jr., W.S., Structural dynamics of yeast hexokinase during catalysis, *Phil. Trans. R. Soc. London B* 293, 43–52 (1981).

Figure 6.15 Image of chymotrypsin based on 4CHA. Tsukada, H., Blow, D.M. Structure of alpha-chymotrypsin refined at 1.68 A resolution, *J. Mol. Biol.* 184, 703–711 (1985).

Image of trypsin based on 3PTN. Walter, J., Steigemann, W., Singh, T.P., Bartunik, H., Bode, W., Huber, R., On the disordered activation domain in trypsinogen. Chemical labelling and low-temperature crystallography, *Acta Crystallogr. B* 38, 1462–1472 (1982).

Image of elastase based on 3EST. Meyer, E.F., Cole, G., Radhakrishnan, R., Epp, O., Structure of native porcine pancreatic elastase at 1.65 Angstroms resolution, *Acta Crystallogr. B* 44, 26–38 (1988).

Figure 6.16 Image based on 1CSE. Bode, W., The high-resolution X-ray crystal structure of the complex formed between subtilisin Carlsberg and eglin C, an elastase inhibitor from the leech Hirudo medicinalis. Structural analysis, subtilisin structure and interface geometry, *Eur. J. Biochem.* 166, 673–692 (1987).

Figure 6.19 Image based on 2CG. Wang, D., Bode, W., Huber, R., Bovine chymotrypsinogen A X-ray crystal structure analysis and refinement of a new crystal form at 1.8 A resolution, *J. Mol. Biol.* 185, 595–624 (1985).

Figure 6.20 Image based on 2PTC. Huber, R., Deisenhofer, J., The geometry of the reactive site and of the peptide groups in trypsin, trypsinogen and its complexes with inhibitors, *Acta Crystallogr. B* 39, 480 (1983).

Figure 6.22 Image based on 1PPB. Bode, W., The refined 1.9 Angstroms crystal structure of human alpha-thrombin: interaction with d-Phe-Pro-Arg chloromethylketone and significance of the Tyr-Pro-Pro-Trp insertion segment, *EMBO J.* 8, 3467–3475 (1989).

Table 6.3 Concentrations calculated from data in High, K.A. and Roberts, H.R., eds., *Molecular Basis of Thrombosis and Hemostasis, Marcel Dekker* (1995).

Figure 6.24 Image based on 2ANT. Skinner, R., Abrahams, J.P., Whisstock, J.C., Lesk, A.M., Carrell, R.W., Wardell, M.R., The 2.6 A structure of antithrombin indicates a conformational change at the heparin binding site, *J. Mol. Biol.* 266, 601–609 (1997).

Enzyme Kinetics and Inhibition

The cell walls of woody plants account for a huge proportion of biomass that can potentially be converted to green fuels and other chemical resources. Enzymologists are exploring ways to optimize the activity of fungal hydrolases and oxidases in order to efficiently break down the complex polysaccharides and other polymers that generally resist degradation.

Do You Remember?

- O_2 binds to myoglobin such that binding is half-maximal when the oxygen concentration is equal to the dissociation constant (Section 5.1).
- O_2 can bind cooperatively to hemoglobin as the protein shifts conformation (Section 5.1).
- Enzymes differ from simple chemical catalysts in their efficiency and specificity (Section 6.1).
- The height of the activation energy barrier determines the rate of a reaction (Section 6.2).
- The catalytic activity of enzymes depends on transition state stabilization (Section 6.3).

In the preceding chapter, we examined the basic features of enzyme catalytic activity, primarily by exploring the various catalytic mechanisms used by chymotrypsin. This chapter extends the discussion of enzymes by introducing enzyme kinetics, the mathematical analysis of enzyme activity. Here we describe how an enzyme's reaction speed and specificity can be quantified and how this information can be used to evaluate the enzyme's physiological function. We also look at the regulation of enzyme activity by inhibitors, including drugs, that bind to the enzyme and alter its activity. The discussion also focuses on allosteric regulation, a mechanism for inhibiting as well as activating enzymes.

7.1 Introduction to Enzyme Kinetics

KEY CONCEPTS

Explain why an enzyme's activity varies with the substrate concentration.

- Describe how an enzyme's activity is measured.
- Depict enzyme saturation in graphical form.

The structure and chemical mechanism of an enzyme (for example, chymotrypsin, as discussed in Chapter 6) often reveal a great deal about how that enzyme functions *in vivo*. However, structural information alone does not provide a full accounting of an enzyme's physiological role. For example, one might need to know exactly how fast the enzyme catalyzes a reaction or how well it recognizes different substrates or how its activity is affected by other substances. These questions are not trivial—consider that a single cell contains thousands of different enzymes, all operating simultaneously and in the presence of one anothers' substrates and products. To fully describe enzyme activity, enzymologists apply mathematical tools to quantify an enzyme's catalytic power and its substrate affinity as well as its response to inhibitors. This analysis is part of the area of study known as enzyme **kinetics** (from the Greek *kinetos,* which means "moving").

Some of the earliest biochemical studies examined the reactions of crude preparations of yeast cells and other organisms. Even without isolating any enzymes, researchers could mathematically analyze their activity by measuring the concentrations of substrates and reaction products and observing how these quantities changed over time. For example, consider the simple reaction catalyzed by triose phosphate isomerase, which interconverts two three-carbon sugars (trioses):

Glyceraldehyde-3-phosphate → (triose phosphate isomerase) → Dihydroxyacetone phosphate

Over the course of the reaction, the concentration of the substrate glyceraldehyde-3-phosphate falls as the concentration of the product dihydroxyacetone phosphate rises (**Fig. 7.1**). *The progress of this or any reaction can be expressed as a* **velocity (v)**, *either the rate of disappearance of the substrate (S) or the rate of appearance of the product (P):*

$$v = -\frac{d[S]}{dt} = \frac{d[P]}{dt} \tag{7.1}$$

where [S] and [P] represent the molar concentrations of the substrate and product, respectively. Not surprisingly, *the more catalyst (enzyme) present, the faster the reaction* (**Fig. 7.2**).

Enzymologists typically study the early stages of a reaction, when the substrate concentration is just starting to decrease and the product concentration is just starting to increase. It wouldn't make sense to wait until the reaction goes to completion (that is, reaches equilibrium), since the

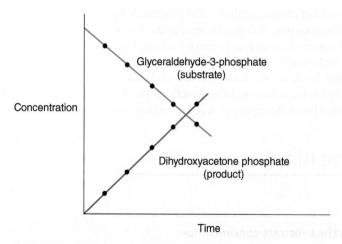

Figure 7.1   Progress of the triose phosphate isomerase reaction.   Over time, the concentration of the substrate glyceraldehyde-3-phosphate decreases and the concentration of the product dihydroxyacetone phosphate increases.

Question **Extend the lines in the graph to show that the reaction eventually reaches equilibrium, when the ratio of product concentration to reactant concentration is about 20.**

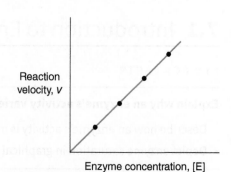

Figure 7.2   Progress of the triose phosphate isomerase reaction.   The more enzyme present, the faster the reaction.

substrate and product concentrations don't change at that point. In fact, in a cell, many reactions never reach equilibrium, since new substrates are constantly arriving and products are used up as soon as they are made.

How does the rate of an enzyme-catalyzed reaction change when the substrate concentration changes? When the enzyme concentration is held constant, the reaction velocity varies with the substrate concentration, but in a nonlinear fashion (**Fig. 7.3**). The shape of this velocity versus substrate curve is an important key to understanding how enzymes interact with their substrates. *The hyperbolic, rather than linear, shape of the curve suggests that an enzyme physically combines with its substrate to form an **enzyme–substrate (ES) complex.*** Therefore, the enzyme-catalyzed conversion of S to P

$$S \xrightarrow{\text{E}} P$$

can be more accurately written as

$$E + S \rightarrow ES \rightarrow E + P$$

As small amounts of substrate are added to the enzyme preparation, enzyme activity (measured as the reaction velocity) appears to increase almost linearly. However, the enzyme's activity increases less dramatically as more substrate is added. At very high substrate concentrations, enzyme activity appears to level off as it approaches a maximum value. This behavior shows that at low substrate concentrations, the enzyme quickly converts all the substrate to product, but as more substrate is added, the enzyme becomes **saturated** with substrate—that is, there are many more substrate molecules than enzyme molecules, so not all the substrate can be converted to product in a given time. These so-called saturation kinetics are a feature of many binding phenomena, including the binding of O_2 to myoglobin (see Section 5.1).

The curve shown in Figure 7.3 reveals considerable information about a given enzyme and substrate under a chosen set of reaction conditions. All simple enzyme-catalyzed reactions yield a hyperbolic velocity versus substrate curve, but the exact shape of the curve depends on the enzyme, its concentration, the concentrations of enzyme inhibitors, the pH, the temperature, and so on. By analyzing such curves, it is possible to address some basic questions, for example,

- How fast does the enzyme operate?
- How efficiently does the enzyme convert different substrates to products?

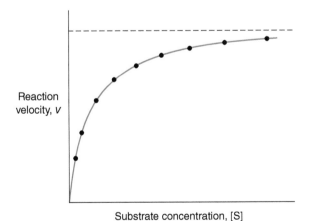

Figure 7.3   A plot of reaction velocity versus substrate concentration.   Varying amounts of substrate are added to a fixed amount of enzyme. The reaction velocity is measured for each substrate concentration and plotted. The resulting curve takes the form of a hyperbola, a mathematical function in which the values first increase steeply but eventually approach a maximum.

Question Compare this diagram to Figure 5.3, which shows oxygen binding to myoglobin.

- How susceptible is the enzyme to various inhibitors and how do these inhibitors affect enzyme activity?

The answers to these questions, in turn, may reveal

- Whether an enzyme is likely to catalyze a particular reaction *in vivo*
- What substances are likely to serve as physiological regulators of the enzyme's activity
- Which enzyme inhibitors might be effective drugs

Concept Check

1. Describe two ways you could measure the rate of an enzyme-catalyzed reaction.
2. Sketch a velocity versus substrate curve like the one in Fig. 7.3.
3. Explain why the velocity differs at low and high substrate concentrations.
4. Make a list of the questions that can be answered by studying an enzyme's kinetics.

7.2 The Michaelis–Menten Equation for an Enzyme-Catalyzed Reaction

KEY CONCEPTS

Use the Michaelis–Menten equation to describe enzyme behavior.

- Distinguish first-order and second-order reactions.
- Describe the changes in the concentrations of S, P, E_T, and ES during the course of an enzyme-catalyzed reaction.
- Define K_M and k_{cat}.
- Derive the values of K_M and V_{max} from graphical data.
- List the limitations of the Michaelis–Menten model.

The mathematical analysis of enzyme behavior centers on the equation that describes the hyperbolic shape of the velocity versus substrate plot (see Fig. 7.3). We can analyze an enzyme-catalyzed reaction, such as the one catalyzed by triose phosphate isomerase, by conceptually breaking it down into smaller steps and using the terms that apply to simple chemical processes.

Rate equations describe chemical processes

Consider a **unimolecular reaction** (one that involves a single reactant) such as the conversion of compound A to compound B:

$$A \rightarrow B$$

The progress of this reaction can be mathematically described by a **rate equation** in which the reaction rate (the velocity) is expressed in terms of a constant (the **rate constant**) and the reactant concentration [A]:

$$v = -\frac{d[A]}{dt} = k[A] \tag{7.2}$$

Here, k is the rate constant and has units of reciprocal seconds (s^{-1}). This equation shows that the reaction velocity is directly proportional to the concentration of reactant A. Such a reaction is said to be **first-order** because its rate depends on the concentration of one substance.

A **bimolecular** or **second-order reaction,** which involves two reactants, can be written

$$A + B \rightarrow C$$

Its rate equation is

$$v = -\frac{d[A]}{dt} = -\frac{d[B]}{dt} = k[A][B] \tag{7.3}$$

Here, k is a second-order rate constant and has units of $M^{-1} \cdot s^{-1}$. The velocity of a second-order reaction is therefore proportional to the product of the two reactant concentrations (see Sample Calculation 7.1).

SAMPLE CALCULATION 7.1

Problem   Determine the velocity of the reaction $X + Y \rightarrow Z$ when the sample contains 3 μM X and 5 μM Y and k for the reaction is $400\ M^{-1} \cdot s^{-1}$.

Solution   Use Equation 7.3 and make sure that all units are consistent:

$$\begin{aligned}
v &= k[X][Y] \\
&= (400\,M^{-1} \cdot s^{-1})(3\,\mu M)(5\,\mu M) \\
&= (400\,M^{-1} \cdot s^{-1})(3 \times 10^{-6}\,M)(5 \times 10^{-6}\,M) \\
&= 6 \times 10^{-9}\,M \cdot s^{-1} = 6\ nM \cdot s^{-1}
\end{aligned}$$

The Michaelis–Menten equation is a rate equation for an enzyme-catalyzed reaction

In the simplest case, an enzyme binds its substrate (in an enzyme–substrate complex) before converting it to product, so *the overall reaction actually consists of first-order and second-order processes, each with a characteristic rate constant:*

$$E + S \underset{k_{-1}}{\overset{k_1}{\rightleftharpoons}} ES \overset{k_2}{\longrightarrow} E + P \tag{7.4}$$

The initial collision of E and S is a bimolecular reaction with the second-order rate constant k_1. The ES complex can then undergo one of two possible unimolecular reactions: k_2 is the first-order rate constant for the conversion of ES to E and P, and k_{-1} is the first-order rate constant for the conversion of ES back to E and S. The bimolecular reaction that would represent the formation of ES from E and P is not shown because we assume that this step is not significant at the start of the reaction.

The rate equation for product formation is

$$v = \frac{d[P]}{dt} = k_2[ES] \tag{7.5}$$

To calculate the rate constant, k_2, for the reaction, we would need to know the reaction velocity and the concentration of ES. The velocity can be measured relatively easily, for example, by using a synthetic substrate that is converted to a light-absorbing or fluorescent product. (The velocity of the reaction is just the rate of appearance of the product as monitored by a spectrophotometer or fluorometer.) However, measuring [ES] is more difficult because the concentration of the enzyme–substrate complex depends on its rate of formation from E and S and its rate of decomposition to E + S and E + P:

$$\frac{d[ES]}{dt} = k_1[E][S] - k_{-1}[ES] - k_2[ES] \tag{7.6}$$

To simplify our analysis, we choose experimental conditions such that the substrate concentration is much greater than the enzyme concentration ($[S] \gg [E]$). Under these conditions, after E and S have been mixed together, *the concentration of ES remains constant until nearly all the substrate molecules have been converted to product.* This is shown graphically in **Figure 7.4**. [ES] is said to maintain a **steady state** (it has a constant value) and

$$\frac{d[ES]}{dt} = 0 \tag{7.7}$$

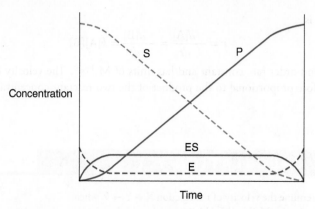

Figure 7.4 Changes in concentration for a simple enzyme-catalyzed reaction. For most of the duration of the reaction, [ES] remains constant while S is converted to P. In this idealized reaction, all the substrate is converted to product.

According to the steady-state assumption, the rate of ES formation must therefore balance the rate of ES consumption:

$$k_1[E][S] = k_{-1}[ES] + k_2[ES] \tag{7.8}$$

At any point during the reaction, [E]—like [ES]—is difficult to determine, but the total enzyme concentration, $[E]_T$, is usually known:

$$[E]_T = [E] + [ES] \tag{7.9}$$

Thus, $[E] = [E]_T - [ES]$. This expression for [E] can be substituted into Equation 7.8 to give

$$k_1([E]_T - [ES])[S] = k_{-1}[ES] + k_2[ES] \tag{7.10}$$

Rearranging (by dividing both sides by [ES] and k_1) gives an expression in which all three rate constants are together:

$$\frac{([E]_T - [ES])[S]}{[ES]} = \frac{k_{-1} + k_2}{k_1} \tag{7.11}$$

At this point, we can define the **Michaelis constant, K_M,** as a collection of rate constants:

$$K_M = \frac{k_{-1} + k_2}{k_1} \tag{7.12}$$

Consequently, Equation 7.11 becomes

$$\frac{([E]_T - [ES])[S]}{[ES]} = K_M \tag{7.13}$$

or

$$K_M[ES] = ([E]_T - [ES])[S] \tag{7.14}$$

Dividing both sides by [ES] gives

$$K_M = \frac{[E]_T[S]}{[ES]} - [S] \tag{7.15}$$

or

$$\frac{[E]_T[S]}{[ES]} = K_M + [S] \tag{7.16}$$

Solving for [ES] yields

$$[ES] = \frac{[E]_T[S]}{K_M + [S]} \tag{7.17}$$

The rate equation for the formation of product (Equation 7.5) is $v = k_2[ES]$, so we can express the reaction velocity as

$$v = k_2[ES] = \frac{k_2[E]_T\,[S]}{K_M + [S]} \tag{7.18}$$

Now we have an equation containing known quantities: $[E]_T$ and $[S]$. Although some S is consumed in forming the ES complex, we can ignore it because $[S]_T \gg [E]_T$.

Typically, kinetic measurements are made soon after the enzyme and substrate are mixed together, before more than about 10% of the substrate molecules have been converted to product molecules (this is also the reason why we can ignore the reverse reaction, $E + P \rightarrow ES$). Therefore, the velocity at the start of the reaction (at time zero) is expressed as v_0 (the **initial velocity**):

$$v_0 = \frac{k_2[E]_T\,[S]}{K_M + [S]} \tag{7.19}$$

We can make one additional simplification: When $[S]$ is very high, virtually all the enzyme is in its ES form (it is saturated with substrate) and therefore approaches its point of maximum activity (see Fig. 7.3). The maximum reaction velocity, designated V_{max}, can be expressed as

$$V_{max} = k_2[E]_T \tag{7.20}$$

which is similar to Equation 7.5. By substituting Equation 7.20 into Equation 7.19, we obtain

$$v_0 = \frac{V_{max}[S]}{K_M + [S]} \tag{7.21}$$

This relationship is called the **Michaelis–Menten equation** after Leonor Michaelis and Maude Menten, who derived it in 1913. *It is the rate equation for an enzyme-catalyzed reaction and is the mathematical description of the hyperbolic curve shown in Figure 7.3.* (See Sample Calculation 7.2.)

SAMPLE CALCULATION 7.2

Problem   An enzyme-catalyzed reaction has a K_M of 1 mM and a V_{max} of 5 nM·s^{-1}. What is the reaction velocity when the substrate concentration is 0.25 mM?

Solution   Use the Michaelis–Menten equation (Equation 7.21):

$$v_0 = \frac{(5\text{ nM}\cdot\text{s}^{-1})\,(0.25\text{ mM})}{(1\text{ mM}) + (0.25\text{ mM})}$$
$$= \frac{1.25}{1.25}\text{ nM}\cdot\text{s}^{-1}$$
$$= 1\text{ nM}\cdot\text{s}^{-1}$$

K_M is the substrate concentration at which velocity is half-maximal

We have seen that the Michaelis constant, K_M, is a combination of three rate constants (Equation 7.12), but it can be fairly easily determined from experimental data. Kinetic measurements are usually made over a range of substrate concentrations. When $[S] = K_M$, the reaction velocity (v_0) is equal to half its maximum value ($v_0 = V_{max}/2$), as shown in **Figure 7.5**. You can prove that this is true by substituting K_M for $[S]$ in the Michaelis–Menten equation (Equation 7.21). *Since K_M is the substrate concentration at which the reaction velocity is half-maximal, it indicates how efficiently an enzyme selects its substrate and converts it to product.* The lower the value of K_M, the more effective the enzyme is at low substrate concentrations; the higher the value of K_M, the less effective the enzyme is. The K_M is unique for each enzyme–substrate pair. Consequently, K_M values are useful for comparing the activities of two enzymes that act on the same substance or for assessing the ability of different substrates to be recognized by a single enzyme.

In practice, K_M *is often used as a measure of an enzyme's affinity for a substrate.* In other words, it approximates the dissociation constant of the ES complex:

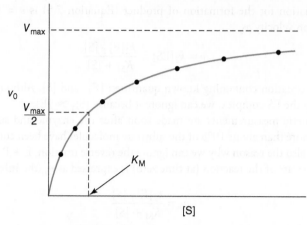

Figure 7.5   Graphical determination of K_M. K_M corresponds to the substrate concentration at which the reaction velocity is half-maximal. It can be visually estimated from a plot of v_0 versus [S].

Question   **Explain why doubling the substrate concentration does not necessarily double the reaction rate.**

$$K_M \approx \frac{[E][S]}{[ES]} \tag{7.22}$$

Note that this relationship is strictly true only when the rate of the ES → E + P reaction is much slower than the rate of the ES → E + S reaction (that is, when $k_2 \ll k_{-1}$). Technically, K_M cannot be an equilibrium constant because it is based on a steady state situation, not an equilibrium.

The catalytic constant describes how quickly an enzyme can act

It is also useful to know how fast an enzyme operates after it has bound its substrate. In other words, how fast does the ES complex proceed to E + P? This parameter is termed the **catalytic constant** and is symbolized as k_{cat}. For any enzyme-catalyzed reaction,

$$k_{cat} = \frac{V_{max}}{[E]_T} \tag{7.23}$$

For a simple reaction, such as the one diagrammed in Equation 7.4,

$$k_{cat} = k_2 \tag{7.24}$$

Thus, k_{cat} *is the rate constant of the reaction when the enzyme is saturated with substrate* (when $[ES] \approx [E]_T$ and $v_0 \approx V_{max}$). We have already seen this relationship in Equation 7.20. k_{cat} is also known as the enzyme's **turnover number** because it is the number of catalytic cycles that each active site undergoes per unit time, or the number of substrate molecules transformed to product molecules by a single enzyme in a given period of time. The turnover number is a first-order rate constant and therefore has units of s^{-1}. As shown in **Table 7.1**, the catalytic constants of enzymes vary over many orders of magnitude.

Keep in mind that the rate of an enzymatic reaction is a function of the number of reactant molecules that can achieve the high-energy transition state per unit time (as explained in Section 6.2). While an enzyme can accelerate a chemical reaction by providing

Table 7.1 Catalytic Constants of Some Enzymes

Enzyme	k_{cat} (s^{-1})
Staphylococcal nuclease	95
Cytidine deaminase	299
Triose phosphate isomerase	4300
Cyclophilin	13,000
Ketosteroid isomerase	66,000
Carbonic anhydrase	1,000,000

a mechanistic pathway with a lower activation energy barrier, the enzyme cannot alter the free energies of the reactants and products. This means that the enzyme can make a reaction happen faster, but only when the overall change in free energy is less than zero (that is, the products have lower free energy than the reactants).

k_{cat}/K_M indicates catalytic efficiency

An enzyme's effectiveness as a catalyst depends on how avidly it binds its substrates and how rapidly it converts them to products. Thus, a measure of catalytic efficiency must reflect both binding and catalytic events. The quantity k_{cat}/K_M satisfies this requirement. At low concentrations of substrate ([S] < K_M), very little ES forms and [E] ≈ [E]$_T$. Equation 7.18 can then be simplified (the [S] term in the denominator becomes insignificant):

$$v_0 = \frac{k_2 [E]_T [S]}{K_M + [S]} \tag{7.25}$$

$$v_0 \approx \frac{k_2}{K_M} [E][S] \tag{7.26}$$

Equation 7.26 is the rate equation for the second-order reaction of E and S. k_{cat}/K_M, which has units of $M^{-1} \cdot s^{-1}$, is the apparent second-order rate constant. As such, it indicates how the reaction velocity varies according to how often the enzyme and substrate combine with each other. *The value of k_{cat}/K_M, more than either K_M or k_{cat} alone, represents the enzyme's overall ability to convert substrate to product.*

What limits the catalytic power of enzymes? Electronic rearrangements during formation of the transition state occur on the order of 10^{-13} s, about the lifetime of a bond vibration. However, enzyme turnover numbers are much slower than this (see Table 7.1). An enzyme's overall speed is further limited by how often it collides productively with its substrate. The upper limit for the rate of this second-order reaction (a bimolecular reaction) is about 10^8 to 10^9 $M^{-1} \cdot s^{-1}$, which is the maximum rate at which two freely diffusing molecules can collide with each other in aqueous solution.

This so-called **diffusion-controlled limit** for the second-order reaction between an enzyme and a substrate is achieved by several enzymes, including triose phosphate isomerase, whose value of k_{cat}/K_M is 2.4×10^8 $M^{-1} \cdot s^{-1}$. This enzyme is therefore said to have reached **catalytic perfection** because *its overall rate is diffusion-controlled: It catalyzes a reaction as rapidly as it encounters its substrate.* However, many enzymes perform their physiological roles with more modest k_{cat}/K_M values.

K_M and V_{max} are experimentally determined

Kinetic data are usually collected by adding a small amount of an enzyme to varying amounts of substrate and then monitoring the reaction mixture for the appearance of product over a period of time. In order to meet the assumptions of the Michaelis–Menten model, the concentration of the substrate must be much greater than the concentration of the enzyme (so that the concentration of the ES complex will be constant and the formation of the ES complex will be limited by the affinity of E for S, not the amount of S available), and measurements must be taken of the initial velocity, before product begins to accumulate and the reverse reaction becomes significant.

Velocity versus substrate plots such as Figure 7.5 can be useful for visually estimating the kinetic parameters K_M and V_{max} (from which k_{cat} can be derived by Equation 7.23). However, in practice, hyperbolic curves are prone to misinterpretation because it is difficult to estimate the upper limit of the curve (V_{max}). In order to more accurately determine V_{max} and K_M (the substrate concentration at $V_{max}/2$), it is necessary to perform one of the following steps:

1. Analyze the data by a curve-fitting computer program that mathematically calculates the upper limit for the reaction velocity.

2. Transform the data to a form that can be plotted as a line. The best-known linear transformation of the velocity versus substrate curve is known as a **Lineweaver–Burk plot,** whose equation is

$$\frac{1}{v_0} = \left(\frac{K_M}{V_{max}}\right)\frac{1}{[S]} + \frac{1}{V_{max}} \tag{7.27}$$

Equation 7.27 has the familiar form $y = mx + b$. A plot of $1/v_0$ versus $1/[S]$ gives a straight line whose slope is K_M/V_{max} and whose intercept on the $1/v_0$ axis is $1/V_{max}$. The extrapolated intercept on the $1/[S]$ axis is $-1/K_M$ (**Fig. 7.6**). A comparison of Figures 7.5 and 7.6, made from the same data, illustrates how evenly spaced points on a velocity versus substrate plot become compressed in a Lineweaver–Burk plot (see Sample Calculation 7.3).

Ideally, experimental conditions are chosen so that velocity measurements can be made for substrate concentrations that are both higher and lower than K_M. This yields the most accurate values for K_M and V_{max}. A Lineweaver–Burk plot, whether constructed manually or by computer, offers the advantage that K_M and V_{max} can be quickly estimated by eye. Linear plots are also more convenient than curves for comparing multiple data sets, such as different enzyme preparations, or a single enzyme in the presence of different concentrations of an inhibitor.

How closely do kinetic parameters measured *in vitro* match the values *in vivo*? Given that laboratory methods rely on assumptions about how enzymes behave and on artificial conditions (saturating the enzyme with substrate, ignoring the reverse reaction, and so on), is it safe to extrapolate laboratory values to living systems? Computer models based on the cellular concentrations of enzymes and the rate that materials (substrates and products) flow through a series of enzyme-catalyzed reactions yield estimates for k_{cat} values that—for most enzymes—are reasonably close to the values measured in laboratory settings.

Not all enzymes fit the simple Michaelis–Menten model

So far, the discussion has focused on the very simplest of enzyme-catalyzed reactions, namely, a reaction with one substrate and one product. Such reactions represent only a small portion of known enzymatic reactions, which often involve multiple substrates and products, proceed via multiple steps, or do not meet the assumptions of the Michaelis–Menten kinetic model for other reasons. Nevertheless, the kinetics of these reactions can still be evaluated.

1. Multisubstrate Reactions   More than half of all known biochemical reactions involve two substrates. Most of these **bisubstrate reactions** are either oxidation–reduction reactions or transfer reactions. In an oxidation–reduction reaction, electrons are transferred between substrates:

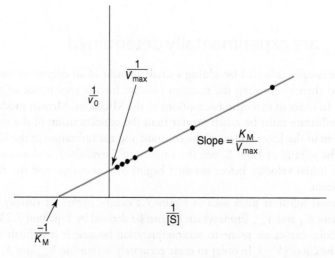

Figure 7.6   A Lineweaver–Burk plot.   Plotting the reciprocals of [S] and v_0 yields a line whose slope and intercepts yield values of K_M and V_{max}. The plotted points correspond to the points in Figure 7.5.

Question Sketch a line to represent a reaction with a **larger** value of K_M and the **same** value of V_{max}.

SAMPLE CALCULATION 7.3

Problem   The velocity of an enzyme-catalyzed reaction was measured at several substrate concentrations. Calculate K_M and V_{max} for
the reaction.

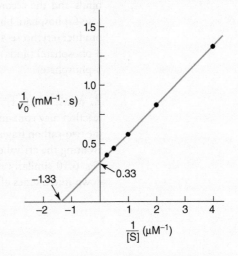

[S] (μM)	v_0 (mM·s^{-1})
0.25	0.75
0.5	1.20
1.0	1.71
2.0	2.18
4.0	2.53

Solution   Calculate the reciprocals of the substrate concentration
and velocity, and then make a plot of $1/v_0$ versus $1/[S]$ (a Lineweaver–
Burk plot).

$1/[S]$ (μM^{-1})	$1/v_0$ (mM^{-1}·s)
4.0	1.33
2.0	0.83
1.0	0.58
0.5	0.46
0.25	0.40

The intercept on the $1/[S]$ axis (which is equal to $-1/K_M$) is -1.33
μM^{-1}. Therefore,

$$K_M = -\left(\frac{1}{-1.33 \ \mu M^{-1}}\right) = 0.75 \ \mu M$$

The intercept on the $1/v_0$ axis (which is equal to $1/V_{max}$) is
0.33 mM^{-1}·s. Therefore,

$$V_{max} = \frac{1}{0.33 \ mM^{-1} \cdot s} = 3.0 \ mM \cdot s^{-1}$$

$$X_{oxidized} + Y_{reduced} \rightarrow X_{reduced} + Y_{oxidized}$$

In a transfer reaction, such as the one catalyzed by transketolase, a group is transferred between
two molecules:

CH$_2$OH
|
C=O
|
HO—C—H
|
H—C—OH   +
|
H—C—OH
|
CH$_2$OPO$_3^{2-}$

O   H
\\ //
C
|
H—C—OH
|
CH$_2$OPO$_3^{2-}$

$\xrightarrow{\text{transketolase}}$

O   H
\\ //
C
|
H—C—OH
|
H—C—OH
|
CH$_2$OPO$_3^{2-}$ +

CH$_2$OH
|
C=O
|
HO—C—H
|
H—C—OH
|
CH$_2$OPO$_3^{2-}$

Fructose-6-phosphate   Glyceraldehyde-3-phosphate   Erythrose-4-phosphate   Xylulose-5-phosphate

The transketolase reaction is ubiquitous in nature; it functions in the synthesis and degradation
of carbohydrates. As written here, it transforms a six-carbon sugar and a three-carbon sugar to
a four-carbon sugar and a five-carbon sugar. *Each of the substrates interacts with the enzyme with
a characteristic* K_M. To experimentally determine each K_M, the reaction velocity is measured at
different concentrations of one substrate while the other substrate is present at a saturating concentration. V_{max} *is the maximum reaction velocity when both substrates are present at concentrations
that saturate their binding sites on the enzyme.*

   In some bisubstrate reactions, the substrates can bind in any order, as long as they both end
up in the active site at the same time. These reactions are said to follow a **random mechanism.**

Enzymes in which one substrate must bind before the other follow an **ordered mechanism.** In a **ping pong mechanism,** one substrate binds and one product is released before the other substrate binds and the second product is released. Transketolase catalyzes a ping pong reaction: Fructose-6-phosphate binds first and surrenders a two-carbon fragment to the enzyme, and the first product (erythrose-4-phosphate) leaves the active site before the second substrate (glyceraldehyde-3-phosphate) binds and receives the two-carbon fragment to yield the second product (xylulose-5-phosphate).

2. Multistep Reactions

As the transketolase reaction illustrates, an enzyme-catalyzed reaction may contain many steps. In this example, the reaction includes an intermediate in which the two-carbon fragment removed from fructose-6-phosphate remains bound to the enzyme while awaiting the arrival of the second substrate. (The chymotrypsin reaction mechanism outlined in Fig. 6.10 similarly requires several steps.) The multistep transketolase reaction can be broken down into a series of simple mechanistic steps and diagrammed as follows:

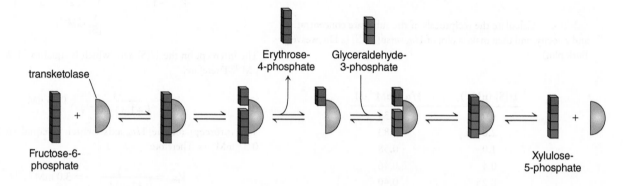

Each step of this process has characteristic forward and reverse rate constants. Consequently, k_{cat} for the overall reaction

$$\text{fructose-6-phosphate} + \text{glyceraldehyde-3-phosphate} \rightleftharpoons$$
$$\text{erythrose-4-phosphate} + \text{xylulose-5-phosphate}$$

is a complicated function of many individual rate constants (only for a very simple reaction, for example, Equation 7.4, does $k_{cat} = k_2$). Nevertheless, *the meaning of k_{cat}—the enzyme's turnover number—is the same as for a single-step reaction.*

The rate constants of individual steps in a multistep reaction can sometimes be measured during the initial stages of the reaction, that is, *before* a steady state is established. This requires instruments that can rapidly mix the reactants and then monitor the mixture on a time scale from 1 s to 10^{-7} s.

3. Nonhyperbolic Reactions

Many enzymes, particularly oligomeric enzymes with multiple active sites, do not obey the Michaelis–Menten rate equation and therefore do not yield hyperbolic velocity versus substrate curves. In these **allosteric enzymes,** *the presence of a substrate at one active site can affect the catalytic activity of the other active sites.* This **cooperative** behavior occurs when the enzyme subunits are structurally linked to each other so that a substrate-induced conformational change in one subunit elicits conformational changes in the remaining subunits. (Cooperative behavior also occurs in hemoglobin, when O_2 binding to the heme group in one subunit alters the O_2 affinity of the other subunits; see Section 5.1.) Like hemoglobin, an allosteric enzyme has two possible quaternary structures. The T (or "tense") state has lower catalytic activity, and the R (or "relaxed") state has higher activity. Because of the interactions between individual subunits, the entire enzyme can switch between the T and R conformations. The result of allosteric behavior is a sigmoidal (S-shaped) velocity versus substrate curve (**Fig. 7.7**). Although the standard Michaelis–Menten equation does not apply here, K_M and V_{max} can be estimated and used to characterize enzyme activity.

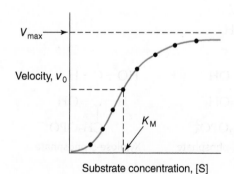

Figure 7.7   Effect of cooperative substrate binding. The velocity versus substrate curve is sigmoidal rather than hyperbolic when substrate binding to one active site in an oligomeric enzyme alters the catalytic activity of the other active sites. The maximum reaction velocity is V_{max}, and K_M is the substrate concentration when the velocity is half-maximal.

Question Compare this diagram to Figure 5.7, which shows oxygen binding to hemoglobin.

1. Write the rate equations for first-order and second-order reactions.
2. Describe the two possible fates of an ES complex.
3. Plot the concentrations of S, P, E_T, and ES during the course of an enzyme-catalyzed reaction.
4. Write the rate equation for an enzyme-catalyzed reaction.
5. Write definitions for K_M, v_0, V_{max}, k_{cat}, and k_{cat}/K_M.
6. Explain why k_{cat}/K_M is a better indicator of enzyme efficiency than either k_{cat} or K_M alone.
7. Define catalytic perfection.
8. Why is triose phosphate considered to be a catalytically perfect enzyme?
9. Sketch a velocity versus substrate plot and a Lineweaver–Burk plot and indicate which parameters reveal K_M and V_{max}.
10. Give the number of K_M values and V_{max} values that pertain to a bisubstrate reaction.
11. Describe the shape of the velocity versus substrate curve when an enzyme exhibits cooperative behavior.

7.3 Enzyme Inhibition

KEY CONCEPTS

Distinguish the effects of different types of enzyme inhibitors.

- Compare the action of reversible and irreversible inhibitors.
- Describe the effects of competitive, noncompetitive, mixed, and uncompetitive inhibitors on a reaction's apparent K_M and V_{max}.
- Express inhibitor strength in terms of a K_I value.
- Explain why transition state analogs often act as competitive inhibitors.
- Explain why allosteric enzymes can be activated or inhibited.
- Summarize the ways that cells regulate enzyme activity.

Inside a cell, an enzyme is subject to a variety of factors that can influence its behavior. Substances that interact with the enzyme can interfere with substrate binding and/or catalysis. Many naturally occurring antibiotics, pesticides, and other poisons are substances that inhibit the activity of essential enzymes. From a strictly scientific point of view, these inhibitors are useful probes of an enzyme's active-site structure and catalytic mechanism. Enzyme inhibitors are also used therapeutically as drugs. The ongoing pursuit of more effective drugs requires knowledge of how enzyme inhibitors work and how they can be altered to better inhibit their target enzymes.

Some inhibitors act irreversibly

Certain compounds interact with enzymes so tightly that their effects are essentially irreversible. For example, diisopropylphosphofluoridate (DIPF), the reagent used to identify the active-site Ser residue of chymotrypsin (see Section 6.2), is an **irreversible inhibitor** of the enzyme. When DIPF reacts with chymotrypsin, leaving the DIP group covalently attached to the Ser hydroxyl group, the enzyme becomes catalytically inactive. In general, *any reagent that covalently modifies an amino acid side chain in a protein can potentially act as an irreversible enzyme inhibitor.*

Some irreversible enzyme inhibitors are called **suicide substrates** because they enter the enzyme's active site and begin to react, just as a normal substrate would. However, they are unable to undergo the complete reaction and hence become "stuck" in the active site. For example, thymidylate synthase is the enzyme that converts the nucleotide deoxyuridylate (dUMP) to deoxythymidylate (dTMP) by adding a methyl group to C5:

thymidylate synthase

Deoxyribose-5-phosphate
dUMP

Deoxyribose-5-phosphate
dTMP

5-Fluorouracil

When the synthetic compound 5-fluorouracil (at left) is taken up by cells, it is readily converted to the nucleotide 5-fluorodeoxyuridylate. This compound, like dUMP, enters the active site of thymidylate synthase, where a Cys —SH group adds to C6. Normally, this enhances the nucleophilicity (electron richness) of C5 so that it can accept an electron-poor methyl group. However, the presence of the electron-withdrawing F atom prevents methylation. The inhibitor therefore remains in the active site, bound to the cysteine side chain, rendering thymidylate synthase inactive. For this reason, 5-fluorouracil is used to disrupt DNA synthesis in rapidly dividing cancer cells.

Competitive inhibition is the most common form of reversible enzyme inhibition

As its name implies, reversible enzyme inhibition results when a substance binds reversibly (that is, noncovalently) to an enzyme so as to alter its catalytic properties. A reversible inhibitor may affect the enzyme's K_M, k_{cat}, or both. The most common form of reversible enzyme inhibition is known as **competitive inhibition.** In this situation, *the inhibitor is a substance that directly competes with a substrate for binding to the enzyme's active site* (**Fig. 7.8**). In all cases, binding of the inhibitor and the substrate is mutually exclusive. As expected, the inhibitor usually resembles the substrate in overall size and chemical properties so that it can bind to the enzyme, but it lacks the exact electronic structure that allows it to react.

One well-known competitive inhibitor affects the activity of succinate dehydrogenase, which catalyzes the oxidation (dehydrogenation) of succinate to produce fumarate:

succinate dehydrogenase

Succinate

Fumarate

The compound malonate

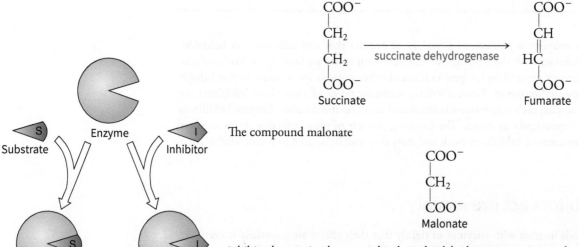

Malonate

inhibits the reaction because it binds to the dehydrogenase active site but cannot be dehydrogenated. Apparently, the enzyme active site can accommodate either the substrate succinate or the competitive inhibitor malonate, although they differ slightly in size.

A plot of an enzyme's reaction velocity in the presence of a competitive inhibitor, as a function of the substrate concentration, is shown in **Figure 7.9**. Because the inhibitor prevents some of the substrate from reaching the active site, the K_M appears to increase (the enzyme's affinity for the substrate appears to decrease). However, because the inhibitor binds reversibly, it constantly dissociates from and reassociates with the enzyme, which allows a substrate molecule to occasionally enter the active site. High concentrations of substrate can overcome the effect of the inhibitor (I) because, when [S] ≫ [I],

Figure 7.8 Competitive enzyme inhibition. In its simplest form, competitive inhibition of an enzyme occurs when the inhibitor and substrate compete for binding in the enzyme active site. A competitive inhibitor often resembles the substrate in size and shape but cannot undergo a reaction.

the enzyme is more likely to bind S than I. The presence of a competitive inhibitor does not affect the enzyme's k_{cat}, so as [S] approaches infinity, v_0 approaches V_{max}. To summarize, *a competitive inhibitor increases the apparent K_M of the enzyme but does not affect k_{cat} or V_{max}.*

The Michaelis–Menten equation for a competitively inhibited reaction has the form

$$v_0 = \frac{V_{max}[S]}{\alpha K_M + [S]} \tag{7.28}$$

where α is a factor that makes K_M appear larger. The value of α—the degree of inhibition—depends on the inhibitor's concentration and its affinity for the enzyme:

$$\alpha = 1 + \frac{[I]}{K_I} \tag{7.29}$$

K_I is the **inhibition constant;** it is the dissociation constant for the **enzyme–inhibitor (EI) complex:**

$$K_I = \frac{[E][I]}{[EI]} \tag{7.30}$$

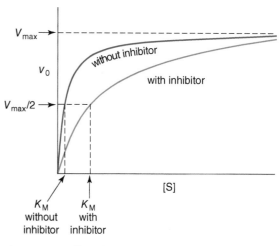

Figure 7.9   Effect of a competitive inhibitor on reaction velocity. In a plot of velocity versus substrate concentration, the inhibitor increases the apparent K_M because it competes with the substrate for binding to the enzyme. The inhibitor does not affect k_{cat}, so at high [S], v_0 approaches V_{max}.

The lower the value of K_I, the tighter the inhibitor binds to the enzyme. It is possible to derive α (and therefore K_I) by plotting the reaction velocity as a function of substrate concentration in the presence of a known concentration of inhibitor. When the data are replotted in Lineweaver–Burk form, the intercept on the 1/[S] axis is $-1/\alpha K_M$ (**Fig. 7.10**; also see Sample Calculation 7.4).

K_I values are useful for assessing the inhibitory power of different substances, such as a series of compounds being tested for usefulness as drugs. For example, atorvastatin, a widely used cholesterol-lowering drug, binds to the enzyme HMG-CoA reductase with a K_I value of about 8 nM (the enzyme's substrate has a K_M of about 4 µM). Keep in mind that an effective drug is not necessarily the compound with the lowest K_I (the tightest binding), since other factors, such as the drug's solubility or its stability, must also be considered. Some researchers assess an inhibitor in terms of its **IC$_{50}$** value, which is the inhibitor concentration at which the enzyme is 50% active. However, activity measurements depend on a host of factors, such as the enzyme and substrate concentrations used in a particular experiment, and are therefore variable. Despite its apparent usefulness as a description of an inhibitor's effect, an IC$_{50}$ value has no relationship to the biochemical quantity K_I.

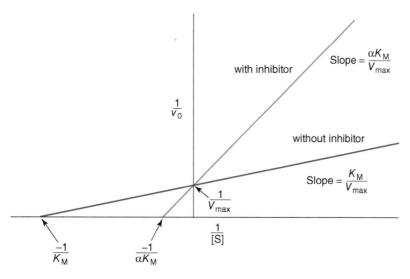

Figure 7.10   A Lineweaver–Burk plot for competitive inhibition. The presence of a competitive inhibitor alters the apparent value of $-1/K_M$, the intercept on the 1/[S] axis, by the factor α. Note that the inhibitor does not affect the value of $1/V_{max}$, the intercept on the $1/v_0$ axis.

SAMPLE CALCULATION 7.4

Problem   An enzyme has a K_M of 8 μM in the absence of a competitive inhibitor and an apparent K_M of 12 μM in the presence of 3 μM of the inhibitor. Calculate K_I.

Solution   The inhibitor increases K_M by a factor α (Equation 7.28). Since the value of K_M with the inhibitor is 1.5 times greater than the value of K_M without the inhibitor (12 μM ÷ 8 μM), α = 1.5. Equation 7.29, which gives the relationship between α, [I], and K_I, can be rearranged to solve for K_I:

$$K_I = \frac{[I]}{\alpha - 1}$$
$$= \frac{3 \ \mu M}{1.5 - 1}$$
$$= \frac{3 \ \mu M}{0.5} = 6 \ \mu M$$

Product inhibition occurs when the product of a reaction occupies the enzyme's active site, thereby preventing the binding of additional substrate molecules. This is another reason why measurements of enzyme activity are made early in the reaction, before product has significantly accumulated.

Transition state analogs inhibit enzymes

Studies of enzyme inhibitors can reveal information about the chemistry of the reaction and the enzyme's active site. For example, the inhibition of succinate dehydrogenase by malonate, shown earlier, suggests that the dehydrogenase active site recognizes and binds substances with two carboxylate groups. Similarly, the ability of an inhibitor to bind to an enzyme's active site may confirm a proposed reaction mechanism. The triose phosphate isomerase reaction (introduced in Section 7.1) is believed to proceed through an enediolate transition state (shown inside brackets):

Glyceraldehyde-3-phosphate   Enediolate   Dihydroxyacetone phosphate

Recall from Section 6.2 that the transition state corresponds to a high-energy structure in which bonds are in the process of breaking and forming. The compound phosphoglycohydroxamate

Phosphoglycohydroxamate

resembles the proposed transition state and, in fact, binds to triose phosphate isomerase about 300 times more tightly than glyceraldehyde-3-phosphate or dihydroxyacetone phosphate binds to the enzyme.

Numerous studies demonstrate that *whereas substrate analogs make good competitive inhibitors, **transition state analogs** make even better inhibitors*. This is because in order to catalyze a reaction, the enzyme must bind to (stabilize, or lower the energy of) the reaction's transition state. A compound that mimics the transition state can take advantage of features in the active site in a way that a substrate analog cannot. For example, the nucleoside adenosine is converted to inosine as follows:

The K_M of the enzyme for the substrate adenosine is 3×10^{-5} M. The product inosine acts as an inhibitor of the reaction, with a K_I of 3×10^{-4} M. The transition state analog 1,6-dihydroinosine (at right) inhibits the reaction with a K_I of 1.5×10^{-13} M.

1,6-Dihydroinosine

Not only do such inhibitors shed light on the probable structure of the reaction's transition state, they may provide a starting point for the design of even better inhibitors. Some of the drugs used to treat infection by HIV (the human immunodeficiency virus) were first designed by considering how transition state analogs inhibit viral enzymes (**Box 7.A**).

Other types of inhibitors affect V_{max}

A competitive inhibitor interferes with substrate binding but does not block its conversion to product. In other cases, a substance may interact with an enzyme *after* the substrate has bound, triggering a conformational change that affects the chemical properties of the active site. In this way, the substance, known as an **uncompetitive inhibitor**, *does not interfere with substrate binding but prevents the reaction from continuing and yielding product* (**Fig. 7.11**). The inhibitor lowers k_{cat}, leading to a decrease in the observed V_{max} value. The observed K_M value also decreases by the same amount, because the loss of functional ES complexes (as ESI complexes) lowers the apparent value of k_2 in the expression of K_M (which is a collection of rate constants; Equation 7.12). Uncompetitive inhibition is rare, but it can occur in multisubstrate reactions.

Box 7.A	Inhibitors of HIV Protease and SARS-CoV-2 Main Protease

Human immunodeficiency virus (HIV), the causative agent of acquired immune deficiency syndrome (AIDS), has an RNA genome that codes for 15 different proteins. Six of these are structural proteins (derived from the green genes in the diagram below), three are enzymes (from the purple genes), and six are accessory proteins (blue genes) that are required for viral gene expression and assembly of new viral particles. Several genes overlap, and two genes are composed of noncontiguous segments of RNA.

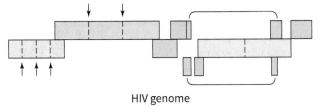

HIV genome

HIV's structural proteins and its three enzymes are initially synthesized as **polyproteins** whose individual members are later separated by proteolysis (at the sites indicated by arrows). The enzyme responsible for this activity is HIV protease, one of the three viral enzymes. A small amount of the protease is present in the virus particle when it first infects a cell; more is generated as the viral genome is transcribed and translated. HIV protease catalyzes the hydrolysis of Tyr—Pro or Phe—Pro peptide bonds in the viral polyproteins. Catalytic activity is centered on two Asp residues (shown in green in the model below), each contributed by one subunit of the homodimeric enzyme. The gold structure represents a peptide substrate analog. The side chains of peptide substrates bind in hydrophobic pockets near the active site.

HIV protease

HIV protease inhibitors are the result of rational drug design (see Section 7.4) based on detailed knowledge of the enzyme's structure. For example, studies of inhibitor–protease complexes revealed that strong inhibitors must be at least the size of a tetrapeptide but need not be symmetrical (although the enzyme itself is symmetrical). Saquinavir was the first widely used HIV protease inhibitor. It is a transition state analog with bulky side chains that mimics the protease's natural Phe—Pro substrates (the scissile bonds are shown in red in the structures at left).

Saquinavir acts as a competitive inhibitor with a K_I of 0.15 nM (for comparison, synthetic peptide substrates have K_M values of about 35 μM). Efforts to develop other drugs have focused on improving the solubility (and therefore the bioavailability) of protease inhibitors by adding polar groups without diminishing binding to the protease. Such efforts have yielded, for example, ritonavir, with a K_I of 0.17 nM.

One of the challenges of developing antiviral drugs is to *select a target that is unique to the virus so that the drugs will not disrupt the host's normal metabolic reactions*. The HIV protease inhibitors

—Phe—Pro—

Saquinavir

Ritonavir

are effective antiviral agents because mammalian proteases do not recognize compounds containing amide bonds to proline or proline analogs. Nevertheless, the drugs that target HIV protease do have side effects. In addition, the high rate of mutation in HIV increases the risk of the virus developing resistance to a drug. For this reason, HIV infection is typically treated with a combination of several drugs, including protease inhibitors and inhibitors of reverse transcriptase and integrase, the virus's other two enzymes (see Box 20.A).

The Severe Acute Respiratory Syndrome Coronavirus 2 (SARS-CoV-2) is a novel coronavirus that is responsible for the pandemic of global coronavirus disease in 2019 (COVID-19). The virus was first identified in December 2019 in Wuhan, Hubei province, China. It belongs to the same family of viruses as the original SARS-CoV (Severe Acute Respiratory Syndrome Coronavirus) and MERS-CoV (Middle East Respiratory Syndrome Coronavirus). SARS-CoV-2 is an enveloped, positive-sense, single-stranded RNA virus. Its genetic material encodes for various proteins, including the spike (S) protein, which is responsible for viral entry into the human cells. The S protein binds to the angiotensin-converting enzyme 2 (ACE2) receptor on the surface of host cells, facilitating viral attachment, and entry. SARS-CoV-2 primarily spreads through respiratory droplets produced when an infected person coughs, sneezes, talks, or breathes. It can also spread by touching surfaces or objects contaminated with the virus and then touching the face, especially the mouth, nose, or eyes. The virus can be highly contagious, leading to community transmission and outbreaks. COVID-19 can range from asymptomatic or mild symptoms to severe

respiratory illness and death. Common symptoms include fever, cough, shortness of breath, fatigue, body aches, loss of taste or smell, and sore throat. In severe cases, patients may develop acute respiratory distress syndrome (ARDS), pneumonia, and other complications.

Several treatment strategies and vaccines for COVID-19 are in use or development. Vaccination has been a critical tool in controlling the pandemic, with various vaccines developed and authorized for emergency use to induce immunity against the virus. Vaccines widely administered include mRNA vaccines of Pfizer-BioNTech and Moderna vaccines that provide instructions for the body to produce a viral spike protein, stimulating an immune response; viral vector vaccines of Oxford-AstraZeneca, Johnson & Johnson, and Sputnik V vaccines use harmless adenoviruses to deliver the spike protein genetic material; protein subunit vaccines of Novavax employs a protein fragment of the virus's spike protein to trigger an immune response; and inactivated Virus Vaccines of Sinopharm, Sinovac, and Bharat Biotech vaccines.

Among drugs in development and research, those targeting the main protease of SARS-CoV-2, also known as the 3C-like protease (3CLpro) or the non-structural protein 5 protease (NSP5), are of immense interest. The structure of the SARS-CoV-2 main protease has been extensively studied, and its three-dimensional structure has been determined through techniques like X-ray crystallography and cryo-electron microscopy. It is a homodimeric cysteine protease having three structural domains. A catalytic dyad of Cys145–His41 residues in the cleft between structural domains I and II is responsible for cleaving the two large viral proteins pp1a and pp1ab, producing 12 functional proteins (nsp4–16) that execute functions such as the replication and transcription of the viral genome. Therefore, the catalytic pocket of 3CLpro is a potential site for developing inhibitors against SARS-CoV-2. Importantly, no equivalent human protein with a similar cleavage site as 3CLpro has been identified, minimizing off-target. Because of its vital role in viral replication and maturation, the main protease is an attractive target for antiviral drug development. Inhibiting the activity of the main protease could block viral replication and reduce the viral load in infected individuals. Researchers have been designing small molecule inhibitors that can bind to the catalytic site of the main protease and prevent its activity. Several antiviral compounds have been

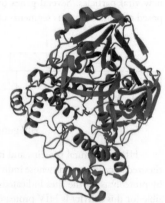

SARS-CoV-2 Main protease (3CLpro) [Structure (pdb 6XB0) determined by Kneller, D.W., Kovalevsky, A., and Coates, L.].
Source: https://www.rcsb.org/structure/6XHU

developed or repurposed to target the main protease of SARS-CoV-2. These compounds aim to inhibit the protease's enzymatic activity, thereby preventing the cleavage of viral polyproteins and impeding viral replication. Drugs like lopinavir and ritonavir, initially developed to treat HIV, have been investigated for their potential to inhibit the main protease. It is worth noting that drug development is a complex and iterative process. While inhibitors targeting the main protease have shown promise in vitro and preclinical studies, their efficacy and safety in clinical settings require thorough testing and validation.

Question Can you identify the amino acid residues in saquinavir and ritonavir?

Competitive and uncompetitive inhibition can be considered as two extremes: One affects substrate binding exclusively, and the other affects catalytic activity exclusively. More commonly, a substance inhibits an enzyme by interfering with *both* substrate binding and catalysis. This mode of inhibition, which includes elements of both competitive and uncompetitive inhibition, is known as **mixed inhibition** or **noncompetitive inhibition.** A noncompetitive inhibitor binds to a site on the enzyme other than the active site, either before or after the substrate binds (**Fig. 7.12**).

Metal ions may act as noncompetitive enzyme inhibitors. For example, trivalent ions such as aluminum (Al^{3+}) inhibit the activity of acetylcholinesterase, which catalyzes the hydrolysis of the neurotransmitter acetylcholine:

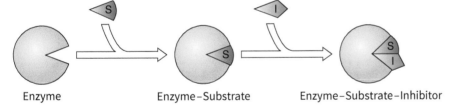

This reaction limits the duration of certain nerve impulses (see Section 9.4). Al^{3+} inhibits acetylcholinesterase noncompetitively by binding to the enzyme at a site distinct from the active site. Consequently, Al^{3+} can bind to the free enzyme or to the enzyme–substrate complex.

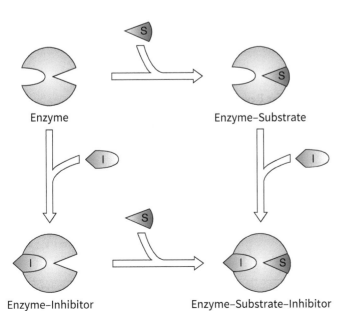

Enzyme   Enzyme–Substrate   Enzyme–Substrate–Inhibitor

Figure 7.11   Uncompetitive enzyme inhibition. The inhibitor binds to the enzyme after the substrate binds. As a result, V_{max} and K_M appear to be reduced by the same amount.

Question Sketch a velocity versus substrate curve for an enzymatic reaction in the absence and presence of an uncompetitive inhibitor.

Enzyme   Enzyme–Substrate

Enzyme–Inhibitor   Enzyme–Substrate–Inhibitor

Figure 7.12   Noncompetitive enzyme inhibition. Inhibitor binding does not prevent substrate binding but alters the catalytic activity of the enzyme (thereby decreasing the apparent V_{max}). The inhibitor can also affect substrate binding, so the K_M appears to increase or decrease.

Although a noncompetitive inhibitor does not completely prevent substrate binding (as in competitive inhibition), it can alter the K_M, yielding either a higher apparent K_M value (signifying less efficient formation of the ES complex) or a lower apparent K_M (as in uncompetitive inhibition). *Because a noncompetitive inhibitor also interferes with catalysis, the apparent V_{max} for the reaction decreases.* A Lineweaver–Burk plot for noncompetitive inhibition is shown in **Figure 7.13**. In very rare cases, the apparent V_{max} decreases with no change in K_M; this is known as "pure" noncompetitive inhibition.

Table 7.2 summarizes the Michaelis–Menten equations for reversible enzyme inhibition. The different forms of inhibition—competitive, uncompetitive, and noncompetitive—can be distinguished by the enzyme's kinetic behavior. For example, increasing the concentration of substrate relieves inhibition by a competitive inhibitor. Increasing the substrate concentration does not alleviate uncompetitive or noncompetitive inhibition because the inhibitor does not prevent substrate binding to the active site. However, enzyme behavior can be difficult to predict. In practice, many substances that resemble substrates and are expected to act as simple competitive inhibitors turn out to also affect the enzyme's catalytic activity and decrease the enzyme's apparent V_{max}.

Lysozymes

Lysozyme is a widely distributed enzyme found in various organisms, including humans, animals, plants, and bacteria. It primarily hydrolyzes the β-1,4-glycosidic bonds in peptidoglycan, a major component of bacterial cell walls, breaking down the peptidoglycan (**Fig.11.11**) network and causing the bacterial cell to lose its structural integrity. The antimicrobial activity of lysozyme is

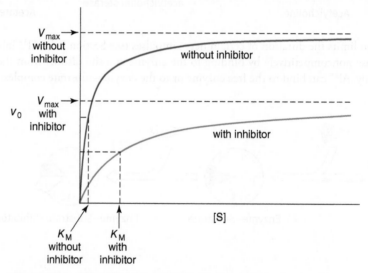

Figure 7.13   Effect of a noncompetitive inhibitor on reaction velocity. As shown here, the inhibitor affects both substrate binding (represented as K_M) and k_{cat} so that the apparent K_M increases and the apparent V_{max} decreases. For some noncompetitive inhibitors, the apparent K_M may decrease or remain unchanged.

Table 7.2 Michaelis–Menten Equations for Reversible Inhibition

Type of inhibition	Michaelis-Menten equation[a]	Effect of inhibitor
None	$v_0 = \dfrac{V_{max}[S]}{K_M + [S]}$	None
Competitive	$v_0 = \dfrac{V_{max}[S]}{\alpha K_M + [S]}$	Increases apparent K_M
Uncompetitive	$v_0 = \dfrac{V_{max}[S]}{K_M + \alpha'[S]}$	Decreases apparent V_{max} and apparent K_M
Noncompetitive (mixed)	$v_0 = \dfrac{V_{max}[S]}{\alpha K_M + \alpha'[S]}$	Decreases apparent V_{max} and may increase or decrease apparent K_M

[a] $\alpha = 1 + \dfrac{[I]}{K_I}$

not limited to its enzymatic action on peptidoglycan. It also contributes to bacterial clearance by promoting opsonization, which enhances immune cells' recognition and phagocytosis of bacteria.

The structure of lysozyme has been extensively studied, with numerous X-ray crystallography and nuclear magnetic resonance (NMR) studies revealing its three-dimensional conformation (**Fig. 4.16**). It is a small protein of 129 amino acids with a molecular weight of around 14.3 kDa. It adopts a globular structure consisting of a single polypeptide chain with four distinct helices (α-helices) and a central β-sheet. The enzyme's active site is situated within a cleft between the α-helices and the β-sheet, forming a catalytic groove that accommodates the substrate binding. The active site contains several vital residue, including glutamic acid (Glu) and aspartic acid (Asp), which play essential roles in catalysis.

Understanding lysozyme inhibitors is crucial for exploring potential therapeutic applications and enhancing our understanding of the enzyme's role in biological systems. Various natural and synthetic compounds have been identified as lysozyme inhibitors. Some natural inhibitors include chitotriose and chitopentaose, which compete with the active site substrate and reduce the enzyme's catalytic activity. Allosamidin is a natural inhibitor that binds to a site distinct from the active site on lysozyme. Its binding induces a conformational change in the enzyme's structure, altering the active site's geometry and reducing its catalytic efficiency. Synthetic inhibitors have been designed based on the enzyme's active site structure and interactions with substrate analogs.

Allosteric enzyme regulation includes inhibition and activation

Oligomeric enzymes—those with multiple active sites in one multisubunit protein—are commonly subject to **allosteric regulation,** which includes inhibition as well as activation. Just as ligand binding to one subunit of an oligomeric enzyme may alter the activity of the other active sites, inhibitor (or activator) binding to one subunit of an enzyme may decrease (or increase) the catalytic activity of all the subunits.

Allosteric effects are part of the physiological regulation of the enzyme phosphofructokinase, which catalyzes the reaction

Fructose-6-phosphate → Fructose-1,6-bisphosphate

This phosphorylation reaction is step 3 of glycolysis, the glucose degradation pathway that is an important source of ATP in virtually all cells (see Section 13.1). The phosphofructokinase reaction is inhibited by phosphoenolpyruvate, the product of Reaction 9 of glycolysis:

Phosphoenolpyruvate

Phosphoenolpyruvate is an example of a **feedback inhibitor:** When its concentration in the cell is sufficiently high, *it shuts down its own synthesis by blocking an earlier step in its biosynthetic pathway:*

Glucose → → ——————→ → → → → → → Phosphoenolpyruvate
phosphofructokinase

Phosphofructokinase from the bacterium *Bacillus stearothermophilus* is a tetramer with four active sites. The subunits are arranged as a dimer of dimers (**Fig. 7.14**). Each of the four fructose-6-phosphate binding sites is made up of residues from both dimers. *B. stearothermophilus* phosphofructokinase binds fructose-6-phosphate with hyperbolic kinetics and a K_M of 23 μM.

In the presence of 300 µM of the inhibitor phosphoenolpyruvate, fructose-6-phosphate binding becomes sigmoidal and the K_M increases to about 200 µM (**Fig. 7.15**). The inhibitor does not affect V_{max}, but phosphofructokinase becomes less active because its apparent affinity for fructose-6-phosphate decreases.

How does phosphoenolpyruvate exert its inhibitory effects? The sigmoidal velocity versus substrate curve (see Fig. 7.15) indicates that the phosphofructokinase active sites behave cooperatively in the presence of phosphoenolpyruvate. In each subunit, the inhibitor binds in a pocket that is separated from the fructose-6-phosphate binding site of the neighboring dimer by a loop of protein. When phosphoenolpyruvate occupies its binding site, the protein closes in around it. This causes a conformational change in which two residues in the loop switch positions: Arg 162 moves away from the fructose-6-phosphate binding site of the neighboring subunit and is replaced by Glu 161 (**Fig. 7.16**).

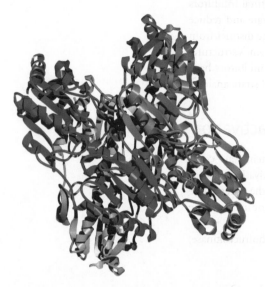

Figure 7.14 Structure of phosphofructokinase from *B. stearothermophilus*. The four identical subunits are arranged as a dimer of dimers (one dimer is shown with blue subunits, the other with purple subunits).

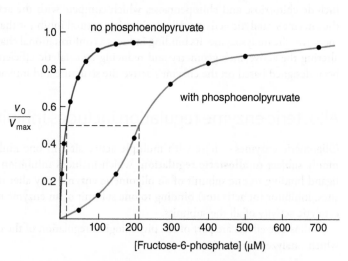

Figure 7.15 Effect of phosphoenolpyruvate on phosphofructokinase activity. In the absence of the inhibitor (green line), *B. stearothermophilus* phosphofructokinase binds the substrate fructose-6-phosphate with a K_M of 23 µM. In the presence of 300 µM phosphoenolpyruvate (red line), the K_M increase to about 200 µM.

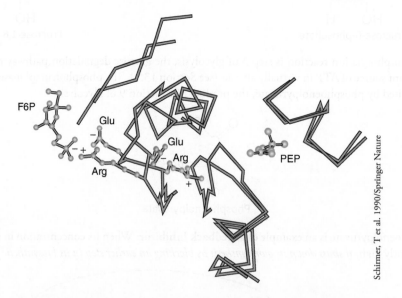

Schirmer, T et al. 1990/Springer Nature

Figure 7.16 Conformational change upon phosphoenolpyruvate binding to phosphofructokinase. The green structure represents the conformation of the enzyme that readily binds the substrate fructose-6-phosphate (labeled F6P). The red structure represents the enzyme with a bound allosteric inhibitor (a phosphoenolpyruvate analog labeled PEP). Phosphoenolpyruvate binding to the enzyme causes a conformational change in which Arg 162 (which forms part of the fructose-6-phosphate binding site of the neighboring subunit) changes places with Glu 161. Because the subunits act cooperatively, the inhibitor diminishes substrate binding to the entire enzyme, causing the K_M to increase.

This conformational switch diminishes fructose-6-phosphate binding because the positively charged side chain of Arg 162, which helps stabilize the negatively charged phosphate group of fructose-6-phosphate, is replaced by the negatively charged side chain of Glu 161, which repels the phosphate group. *The effect of phosphoenolpyruvate is communicated to the entire protein (thereby explaining the cooperative effect) because phosphoenolpyruvate binding to one subunit of phosphofructokinase affects fructose-6-phosphate binding to the neighboring subunit in the other dimer.* Using the terminology for **allosteric proteins,** phosphoenolpyruvate binding causes the entire tetramer to switch to the T (low-activity) conformation, as measured by fructose-6-phosphate binding affinity. Phosphoenolpyruvate is therefore known as a **negative effector** of the enzyme.

Phosphofructokinase is allosterically inhibited by phosphoenolpyruvate, but it can also be allosterically activated by ADP, a **positive effector** of the enzyme. Although ADP is a product of the phosphofructokinase reaction, it is also a general signal of the cell's need for more ATP since the metabolic consumption of ATP yields ADP:

$$ATP + H_2O \rightarrow ADP + P_i$$

Because phosphofructokinase catalyzes step 3 of the 10-step glycolytic pathway (one of whose ultimate products is ATP), increasing phosphofructokinase activity can increase the rate of ATP produced by the pathway as a whole.

Interestingly, the activator ADP binds not to the active site (which accommodates the substrate ATP and the reaction product ADP) but to the same site where the inhibitor phosphoenolpyruvate binds. However, because ADP is much larger than phosphoenolpyruvate, the enzyme cannot close around it, and the conformational change to the low-activity T state cannot occur. Instead, ADP binding forces Arg 162 to remain where it can stabilize fructose-6-phosphate binding (in other words, it helps keep the enzyme in the high-activity R state). The overall result is that ADP counteracts the inhibitory effect of phosphoenolpyruvate and boosts phosphofructokinase activity.

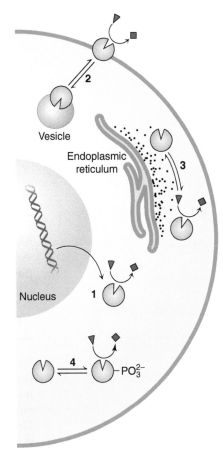

Figure 7.17   Some mechanisms for regulating enzyme activity.   In this diagram, enzymes are shown as circular shapes, substrates as small triangles, and products as small squares. The amount of enzyme may depend on its rates of synthesis and degradation (1). The reaction velocity may depend on the enzyme's location (2). A signal such as a burst of Ca^{2+} ions released from the endoplasmic reticulum may affect the enzyme's activity (3). Covalent modification, such as phosphorylation, may activate an enzyme; the enzyme may be inactive when dephosphorylated (4).

Question Which of the mechanisms shown in the diagram would be fastest (or slowest) to alter an enzyme's activity?

Several factors may influence enzyme activity

So far, we have examined how small molecules that bind to an enzyme inhibit (or sometimes activate) that enzyme. These relatively simple phenomena are not the only means for regulating enzyme activity *in vivo*. Listed below and in **Figure 7.17** are some additional mechanisms. Keep in mind that several of these mechanisms, along with enzyme inhibition or activation, may operate in concert to precisely adjust the activity of a given enzyme.

1. A change in the rate of an enzyme's synthesis or degradation can alter the amount of enzyme available to catalyze a reaction (since $V_{max} = k_{cat}[E]_T$; Equation 7.23).

2. A change in subcellular location, for example, from an intracellular membrane to the cell surface, can bring an enzyme into proximity to its substrate and thereby increase reaction velocity. The opposite effect—sequestering an enzyme away from its substrate—dampens the reaction velocity.

3. An ionic "signal" such as a change in pH or the release of stored Ca^{2+} ions can activate or deactivate an enzyme by altering its conformation.

4. Covalent modification of an enzyme can affect the enzyme's K_M or k_{cat}, just as with an allosteric activator or inhibitor. Most commonly, a phosphoryl ($—PO_3^{2-}$) group or a fatty acyl (lipid) group is added to an enzyme so as to alter its catalytic activity. The effects of covalent modification are actually considered to be reversible, since cells contain enzymes that catalyze the removal of the modifying group as well as its addition. As we will see in Chapter 10 on signaling, phosphorylation and dephosphorylation can dramatically alter the activities of certain proteins.

Concept Check

1. Draw shapes to represent the binding of competitive, uncompetitive, and noncompetitive inhibitors to an enzyme.
2. Why do some reaction products and transition state analogs act as competitive enzyme inhibitors?
3. Sketch the velocity versus substrate curve and the Lineweaver–Burk plot for each type of inhibitor.
4. List some general features of competitive enzyme inhibitors.
5. Explain why quaternary structure is required for allosteric regulation.
6. List all the ways that a cell could increase or decrease the activity of an enzyme.

7.4 Clinical Connection: Drug Development

KEY CONCEPTS

Describe the process of drug development.

- List some factors that influence a drug's usefulness.
- Explain the purpose of clinical trials.

The development of a drug from an enzyme inhibitor in a biochemist's laboratory to a pill in a patient's medicine cabinet is typically long, arduous, and expensive. Whether a new drug results from a surprise discovery or dedicated efforts by pharmaceutical scientists, a newly identified drug candidate is almost never exactly the same compound that makes it to the pharmacist's shelves. Drug development is a process of refinement and testing that can take years and cost billions of dollars, but it usually yields a substance that is therapeutically useful and safe.

The majority of drugs currently in use block the activity of proteins that participate in cell-signaling pathways, but they have much in common with drugs that act as enzyme inhibitors. Every potential drug starts out as a synthetic compound or a natural product that is then altered and tested for the desired biological effect. For enzyme inhibitors, drug developers take advantage of knowledge about an enzyme's active-site structure and mechanism to design a compound that will precisely block catalysis. This process is sometimes called **rational drug design.** A drug candidate, or lead compound, may be systematically altered by adding or deleting various chemical groups (including fluorine, Box 2.A) and then retested for inhibitory activity. Robotic procedures for chemical synthesis and analysis can handle thousands of substances at a time. Alternatively, computer simulations, based on the enzyme's structure, can predict whether a modified structure might be a better inhibitor. In a related approach, virtually screening hundreds of millions of different chemical structures can identify previously unrecognized drug candidates that can then be synthesized and tested.

The goal throughout the drug development process is to create a compound with the following properties: The drug must be a tight-binding inhibitor, it must be highly selective for its target enzyme (so that it doesn't interfere with the activity of other enzymes), and it must not be toxic. In addition, the drug's **pharmacokinetics,** or behavior in the body, must be assessed. For example, the drug must be water-soluble so that it can be transported via the bloodstream; at the same time, it must be lipid-soluble so that it can pass through the walls of the intestine to be absorbed and through the walls of blood vessels so that it can reach the tissues. In many cases, the drug is inactive until this point; once inside a cell, it is converted to its bioactive form by cellular enzymes.

Acyclovir, which is used to treat herpes virus infections, provides an interesting example of this phenomenon. The drug mimics guanosine with an incomplete ribose group. Infected cells contain a viral kinase that converts the drug to a nucleotide analog that then interferes with DNA synthesis.

Acyclovir

Uninfected cells lack the kinase and are therefore unaffected by acyclovir.

A large part of the research relevant to drug development focuses on how the body metabolizes foreign compounds. The cytochrome P450 class of enzymes is responsible for detoxifying a variety of naturally occurring compounds that make their way into the body; drugs are also potential substrates. Cytochrome P450 contains a heme prosthetic group (see Section 5.1) that participates in an oxidation–reduction reaction that adds a hydroxyl group to the substrate to make it more water-soluble and therefore more easily excreted. Consequently, the activity of a cytochrome P450 enzyme can decrease the effectiveness of a drug. In some cases, the hydroxylation reaction can convert a drug to a toxin. This occurs when acetaminophen, a commonly used fever-reducer (also known as paracetamol), is consumed in large doses:

Acetaminophen

cytochrome P450
O_2 H_2O

spontaneous
H_2O

Acetimidoquinone
(toxic)

Individuals vary in the amount and type of cytochrome P450 enzymes they express, so it is difficult to predict how a drug might be metabolized.

In addition to accounting for variations in human metabolism, drug developers must contend with the possibility of drug metabolism by the microbiota. Laboratory studies indicate that a common selection of gut microorganisms can oxidize, reduce, or chemically modify a wide assortment of orally administered drug molecules, reducing their effective concentration in the body. Consequently, the best laboratory findings—even testing in animals—cannot guarantee that a drug will be therapeutically useful in a particular individual.

Ultimately, a drug's effectiveness and safety must be assessed in **clinical trials.** This type of testing is organized into three consecutive phases, each involving a larger number of subjects. In Phase I clinical trials, the drug candidate is administered to a small group of healthy volunteers at levels up to the expected therapeutic dose. The goal of Phase I trials is to ensure that the drug is safe and is well tolerated by humans. Researchers can also examine the drug's pharmacokinetics and select a version of the drug and a dosing regimen suitable for larger-scale testing.

After the safety of the drug has been verified, its effectiveness is tested in Phase II trials, which typically involve several hundred subjects who have the disease targeted by the drug. Assessments of safety and optimum dosing continue during Phase II. Patients are usually randomly assigned to test or control groups. Since many clinical trials test whether a new drug works better than an existing drug, the control group receives the old drug rather than no drug at all. To avoid the placebo effect, in which a patient's or physician's expectations can alter the outcome, the trial is "blinded" so that the patients do not know which treatment they are receiving (a single-blind trial) or, better yet, neither the patients nor their physicians know (a double-blind trial). Drug trials are usually easy to blind, since the test and control substances can be made to appear identical, but it is difficult if not impossible to conduct blind trials in other situations, such as comparing a drug to a surgical intervention. In these cases, objectivity can be maximized by not telling the statisticians who analyze the results which patient group received which treatment.

Phase III clinical trials are conducted with large numbers of patients (hundreds to thousands). Large numbers are needed to provide robust statistical evidence that the new drug works as hoped and is safe for widespread use. Like Phase II trials, Phase III trials are randomized and blinded if possible. Successful completion of Phase III, the longest phase of testing, usually leads to regulatory approval of the drug by the Food and Drug Administration (in the United States), although the drug may be marketed to a limited extent before full approval has been granted.

Even after a drug has been approved, marketed, and widely adopted, the patient population is continually scrutinized for rare side effects or side effects that might not have developed during the short time period of Phase II or Phase III trials. This surveillance period is sometimes called Phase IV.

Its importance is highlighted by the case of the widely used pain reliever rofecoxib, which had been approved but was withdrawn when it was discovered that it increased the risk of heart attacks. Similarly, the use of the antidiabetic drug rosiglitazone has been limited since analysis of large numbers of patients revealed an increase in heart attacks.

Ongoing challenges for the drug industry include the development of new drugs that are more effective than existing drugs, so that a patient can use less of a drug and therefore experience fewer side effects. However, bringing the new drug to market is expensive, and drug developers must recover their investment. This can be difficult to accomplish without pricing the drug out of patients' reach. A consequence of this trade-off is that some of the most widely used (and most profitable) drugs target common disorders that are not immediately life-threatening (**Table 7.3**), while rarer and deadlier diseases have fewer drug treatment options.

Table 7.3 Some Commonly Prescribed Drugs

Drug[a]	Drug action	Target disease or condition
Albuterol	binds to adrenergic receptors to trigger bronchodilation	asthma, chronic obstructive pulmonary disease
Amoxicillin	antibiotic, inhibits enzyme required for synthesis of bacterial cell walls	bacterial infections
Atorvastatin	inhibits HMG-CoA reductase, a key enzyme of cholesterol synthesis	hypercholesterolemia, atherosclerosis
Duloxetine	inhibits reuptake of neurotransmitters serotonin and norepinephrine in central nervous system	depression, anxiety disorder
Esomeprazole	inhibits a proton pump to block stomach acid production	peptic ulcers, gastroesophageal reflux disease
Levothyroxine	mimics thyroid hormone, binds to thyroid hormone receptor	hypothyroidism
Lisinopril	inhibits angiotensin converting enzyme to block vasoconstriction	hypertension (high blood pressure), congestive heart failure
Metformin	increases cellular [AMP] to lower liver glucose production and increase insulin sensitivity	type 2 diabetes

Question Which of these drugs are enzyme inhibitors?

Concept Check

1. Compare rational drug design and design by trial and error.
2. Explain why developers study a drug's pharmacokinetics.
3. Summarize the purpose of each phase of a clinical trial.

SUMMARY

7.1 Introduction to Enzyme Kinetics

- Rate equations describe the velocity of simple unimolecular (first-order) or bimolecular (second-order) reactions in terms of a rate constant.

7.2 The Michaelis–Menten Equation for an Enzyme-Catalyzed Reaction

- An enzyme-catalyzed reaction can be described by the Michaelis–Menten equation. The overall rate of the reaction is a function

of the rates of formation and breakdown of an enzyme–substrate (ES) complex.

- The Michaelis constant, K_M, is a combination of the three rate constants relevant to the ES complex. It is also equivalent to the substrate concentration at which the enzyme is operating at half-maximal velocity. The maximum velocity is achieved when the enzyme is fully saturated with substrate.

- The catalytic constant, k_{cat}, for a reaction is the first-order rate constant for the conversion of the enzyme–substrate complex to product. The quotient k_{cat}/K_M, a second-order rate constant for the overall conversion of substrate to product, indicates an enzyme's catalytic efficiency because it accounts for both the binding and catalytic activities of the enzyme.

- Values for K_M and V_{max} (from which k_{cat} can be calculated) are often derived from Lineweaver–Burk, or double-reciprocal, plots. Not all enzymatic reactions obey the simple Michaelis–Menten model, but their kinetic parameters can still be estimated.

7.3 Enzyme Inhibition

- Some substances react irreversibly with enzymes to permanently block catalytic activity.
- The most common reversible enzyme inhibitors, which may be transition state analogs, compete with substrate for binding to the active site, thereby increasing the apparent K_M.

- Reversible enzyme inhibitors that affect an enzyme's V_{max} may be uncompetitive or noncompetitive (mixed) inhibitors.
- Oligomeric enzymes such as bacterial phosphofructokinase are regulated by allosteric inhibitors and activators.
- The activities of enzymes may also be regulated by changes in enzyme concentration, location, ion concentrations, and covalent modification.

7.4 Clinical Connection: Drug Development

- Some drugs act as enzyme inhibitors; these substances are modified to maximize their binding to target proteins and to optimize their pharmacokinetics.
- The effectiveness and safety of a prospective drug are tested in clinical trials, starting with a small group of subjects and including blinded assessments in larger groups.

KEY TERMS

kinetics
v
ES complex
saturation
unimolecular
 reaction
rate equation
rate constant (k)
first-order reaction
bimolecular reaction

second-order
 reaction
steady state
Michaelis
 constant (K_M)
v_0
V_{max}
Michaelis–Menten
 equation
catalytic constant (k_{cat})

turnover number
k_{cat}/K_M
diffusion-controlled
 limit
catalytic perfection
Lineweaver–Burk
 plot
bisubstrate reaction
random mechanism
ordered mechanism

ping pong
 mechanism
allosteric enzyme
cooperative binding
irreversible inhibitor
suicide substrate
competitive
 inhibition
inhibition constant
 (K_I)

EI complex
IC_{50}
product inhibition
transition state
 analog
polyprotein
uncompetitive
 inhibition
mixed inhibition

noncompetitive
 inhibition
lysozyme
allosteric regulation
feedback inhibitor
negative effector
positive effector
rational drug design
pharmacokinetics
clinical trial

BIOINFORMATICS

Brief Bioinformatics Exercises

7.1 Enzyme Cofactors and Kinetics

7.2 Survey of Enzyme–Inhibitor Complexes

Bioinformatics Projects

Enzyme Inhibitors and Rational Drug Design

PROBLEMS

7.1 Introduction to Enzyme Kinetics

1. At about the time scientists began analyzing the hyperbolic velocity versus substrate concentration curve (Fig. 7.3), Emil Fischer was formulating his lock-and-key hypothesis of enzyme action (Section 6.3), which described the enzyme as a lock and the substrate as a key. Later experiments showed that the enzyme–substrate relationship is more dynamic, but kinetic data are generally consistent with his model. Explain.

2. Explain why it is usually easier to calculate an enzyme's reaction velocity from the rate of appearance of the product rather than the rate of disappearance of the substrate.

3. The rate of hydrolysis of sucrose to glucose and fructose is quite slow in the absence of a catalyst.

$$\text{sucrose} + H_2O \rightarrow \text{glucose} + \text{fructose}$$

If the initial concentration of sucrose is 0.10 M, it takes 440 years for the concentration of the sucrose to decrease by half to 0.05 M.

What is the rate of disappearance of sucrose in the absence of a catalyst?

4. If a catalyst is present, the hydrolysis of sucrose (see Problem 3) is much more rapid. If the initial concentration of sucrose is 0.10 M, it takes 6.9×10^{-5} s for the concentration to decrease by half to 0.05 M. What is the rate of disappearance of sucrose in the presence of a catalyst?

5. The rate of hydrolysis of trehalose to its constituent glucose monomers in the absence of a catalyst is even slower than hydrolysis of sucrose (see Problem 3) and may indicate that the two sugars are hydrolyzed by different mechanisms. If the initial concentration of trehalose is 0.10 M, it takes 6.6×10^6 years for the concentration to decrease by half. What is the rate of disappearance of trehalose in the absence of a catalyst?

6. If a catalyst is present, the hydrolysis of trehalose (see Problem 5) is much more rapid. It takes 2.7×10^{-4} s for the concentration of trehalose to decrease by half from 0.10 M to 0.05 M. **a.** What is the rate of disappearance of trehalose in the presence of a catalyst? **b.** Is the rate enhancement greater for sucrose (see Problem 4) or trehalose?

7. In the absence of a catalyst, 1-methylorotic acid undergoes decarboxylation to 1-methyluracil. If the initial concentration of 1-methylorotic acid is 1.0 M, it takes 2.5×10^{15} s for the concentration of 1-methylorotic acid to decrease by half to 0.5 M. What is the rate of conversion of 1-methyorotic acid to product in the absence of a catalyst?

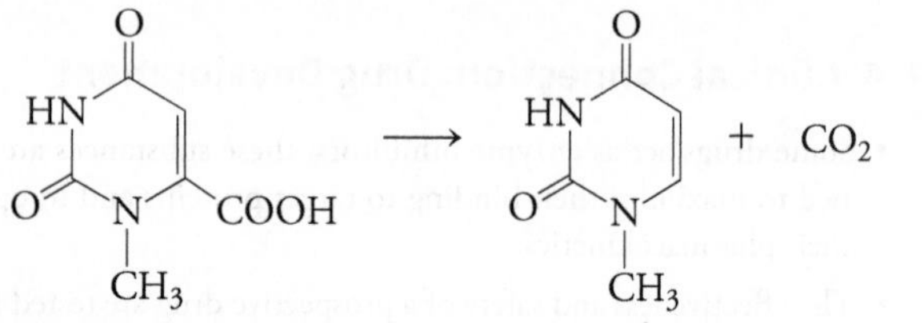

1-Methylorotic acid → 1-Methyluracil + CO_2

8. The reaction shown in Problem 7 is carried out in the presence of OMP decarboxylase. **a.** Consult Table 6.1 and then calculate the rate of formation of product when the reactant concentration decreases by half from 1.0 M to 0.5 M in the presence of OMP decarboxylase. (*Hint*: For a first-order reaction, the rate constant, k, is equal to $0.693/t_{1/2}$.) **b.** Use your answers for part **a** and Solution 7 to calculate the rate enhancement in the presence of the enzyme. Does your calculated value agree with the value presented in Table 6.1?

9. The uncatalyzed rate of amide bond hydrolysis in hippurylphenylalanine was measured by monitoring the formation of the product, phenylalanine. A 60 mM solution of hippurylphenylalanine was monitored for amide bond hydrolysis for 50 days. At the end of this time, 50 μM phenylalanine was detected in the solution. **a.** What is the rate of formation of phenylalanine (in units of $M \cdot s^{-1}$) in this reaction? **b.** Draw a graph of the reaction to show the velocity of the reaction (measured as the rate of the appearance of phenylalanine per unit time) as a function of the reactant concentration.

10. The amide bond in hippurylphenylalanine (see Problem 9) is hydrolyzed 4.7×10^{11} times faster in the presence of an enzyme. What is the rate of formation of phenylalanine in the presence of an enzyme?

11. Draw a graph of the reaction described in Problem 10 to show the rate of appearance of phenylalanine as a function of substrate concentration. Compare this graph to the one you drew in Problem 9. How is this graph different, and why?

12. A bacterial enzyme catalyzes hydrolysis of the disaccharide maltose to produce two glucose monosaccharides. During an interval of 1 minute, the concentration of maltose decreases by 65 mM. What is the rate of disappearance of maltose in the enzyme-catalyzed reaction?

13. For the reaction in Problem 12, what is the rate of appearance of glucose?

14. The rate of hydrolysis of lactose to glucose and galactose is relatively slow without a catalyst. If the initial lactose concentration is 0.040 M, it takes 600 years for the lactose concentration to decrease by half to 0.020 M. What is the lactose disappearance rate without a catalyst?

15. The hydrolysis of cellulose to its constituent glucose molecules without a catalyst is even slower than hydrolysis of lactose (see Problem 14). If the initial cellulose concentration is 0.030 M, it takes 8.5×10^7 years to decrease by half. What is the cellulose disappearance rate in the absence of a catalyst?

7.2 The Michaelis–Menten Equation for an Enzyme-Catalyzed Reaction

16. a. Complete the table for the following one-step reactions:

Reaction	Molecularity	Rate equation	Units of k	Reaction velocity is proportional to...	Order
$A \rightarrow B + C$					
$A + B \rightarrow C$					
$2 A \rightarrow B$					
$2 A \rightarrow B + C$					

b. The formation of an enzyme–substrate complex for a one-substrate reaction can be expressed as $E + S \rightarrow ES$. For an enzymatic reaction with two substrates, A and B, the expression would be $E + A + B \rightarrow EAB$. Explain why the second process is approached as two bimolecular reactions rather than as a single termolecular (three-reactant) reaction.

17. The rates of the reactions described in Problems 3 and 5 double when the concentrations of sucrose and trehalose double. The reaction rates are not affected by the concentration of water. Write rate equations for the reactions that are consistent with these observations. What is the order of the reactions?

18. Refer to Sample Calculation 7.1 and to the rate equation you wrote for the hydrolysis of sucrose in Problem 17 to determine the velocity of the uncatalyzed reaction when the concentration of sucrose is 0.050 M. The rate constant k for the uncatalyzed reaction is 5.0×10^{-11} s^{-1}.

19. Repeat the calculation you performed in Problem 18 for the enzyme-catalyzed hydrolysis of sucrose. The rate constant k for the catalyzed reaction is 1.0×10^4 s^{-1}.

20. Refer to Sample Calculation 7.1 and to the rate equation you wrote for Problem 17 to determine the velocity of the uncatalyzed reaction when the concentration of trehalose is 0.050 M. The rate constant k for the uncatalyzed reaction is 3.3×10^{-15} s^{-1}.

21. Repeat the calculation you performed in Problem 20 for the enzyme-catalyzed hydrolysis of trehalose. The rate constant k for the catalyzed reaction is 2.6×10^3 s^{-1}.

22. At what substrate concentration is $v = 5.0$ $mM \cdot s^{-1}$ for the enzyme-catalyzed hydrolysis of sucrose (see Problem 19)?

23. At what substrate concentration is $v = 5.0$ $mM \cdot s^{-1}$ for the enzyme-catalyzed hydrolysis of trehalose (see Problem 21)?

24. Sugars entering bacteria are phosphorylated during transport via a complex series of reactions. In one of these reactions, an enzyme is phosphorylated; the phosphate group is subsequently transferred from the enzyme to the sugar. **a.** Write a rate equation for the one-step reaction: Enzyme + $P_i \rightarrow$ Enzyme–P_i. **b.** What is the velocity of this reaction when the concentration of phosphate is 50 mM and the concentration of the enzyme is 15 pM? The rate constant k is 3.9×10^6 $M^{-1} \cdot s^{-1}$.

25. The Michaelis-Menten equation assumes that the enzyme-substrate complex (ES) rapidly reaches equilibrium. What factors might influence the kinetics of ES formation and breakdown, and how could these factors affect the validity of the equation?

26. An unprotonated primary amino group in a blood protein can react with carbon dioxide to form a carbamate as shown here:

$$R-NH_2 + CO_2 \rightarrow R-NH-COO^- + H^+$$
Carbamate

The rate constant k for this reaction is 4950 $M^{-1} \cdot s^{-1}$. **a.** What is the order of this reaction? **b.** Calculate the velocity of the reaction of an α-amino group in a blood protein at 37°C if its concentration is 0.5 mM and the partial pressure of carbon dioxide is 45 torr. (*Hint*: Convert the units of partial pressure to molar concentration using the ideal gas law. The value of R is 0.0821 L·atm·K^{-1}·mol^{-1}.) **c.** How would the rate constant for this reaction vary with pH? Explain. **d.** What CO_2 partial pressure is required to yield a velocity of 0.045 M·s^{-1} for the reaction?

27. The spontaneous hydrolysis of ATP has a rate constant of 1.8×10^{-5} s^{-1}. In the presence of ATPase enzymes, the rate constant increases by 200. Calculate the rate constant of the catalyzed reaction.

28. What portions of the velocity versus substrate concentration curve (Fig. 7.3) correspond to zero-order and first-order processes?

29. Draw curves that show the appropriate relationship between the variables of each plot. **a.** [P] vs. time. **b.** [S] vs. time. **c.** [ES] vs. time. **d.** [E] vs. time. **e.** v_0 vs. [E]. **f.** v_0 vs. [S].

30. The enzyme-catalyzed reaction described in Problem 13 has a K_M of 0.135 μM and a V_{max} of 65 μmol·min^{-1}. What is the reaction velocity when the concentration of maltose is 1.0 μM?

31. The enzyme phenylalanine hydroxylase (PAH) catalyzes the hydroxylation of phenylalanine to tyrosine and is deficient in patients with the disease phenylketonuria (PKU). The K_M for PAH is 0.5 mM and the V_{max} is 7.5 μmol·min^{-1}·mg^{-1}. What is the velocity of the reaction when the phenylalanine concentration is 0.25 mM?

32. The enzyme polyphenol oxidase acts on several substrates, one of which is dopamine. What is the K_M for the enzyme if the velocity for the reaction is 0.23 U·min^{-1}·mL^{-1} when the concentration of substrate is 10 mM? The V_{max} for the reaction is 0.36 U·min^{-1}·mL^{-1} (U stands for enzyme units).

33. A phosphatidylinositol-4-kinase (PI-4-kinase) catalyzes the phosphorylation of membrane lipids as part of a signaling pathway in yeast. What is the K_M for the enzyme if the reaction velocity is 7 nmol·min^{-1} when the concentration of substrate is 140 μM? The V_{max} for the reaction is 12 nmol·mL^{-1}.

34. In the absence of allosteric effectors, the enzyme phosphofructokinase displays Michaelis–Menten kinetics (see Fig. 7.15). The v_0/V_{max} ratio is 0.9 when the concentration of the substrate, fructose-6-phosphate, is 0.10 mM. Calculate the K_M for phosphofructokinase under these conditions.

35. A mutant phosphofructokinase has a v_0/V_{max} ratio of 0.9 when the concentration of the substrate, fructose-6-phosphate, is 0.28 mM. The K_M for the mutant enzyme is 0.1 mM. Does the mutant enzyme display Michaelis–Menten kinetics?

36. Brain glutaminase has a V_{max} of 1.1 μmol·min^{-1}·mL^{-1} and a K_M of 0.6 mM. What is the substrate concentration when the velocity is 0.3 μmol·min^{-1}·mL^{-1}?

37. Use the plot to estimate values of K_M and V_{max} for an enzyme-catalyzed reaction.

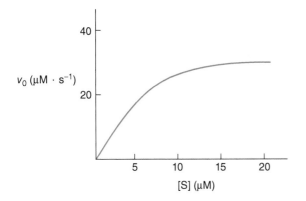

38. Estimate the K_M and the V_{max} for each enzyme from the plot. **a.** Which enzyme generates product more rapidly when [S] = 1 mM? What is the relationship between [S] and K_M for each enzyme at this substrate concentration? **b.** Answer part **a** when [S] = 10 mM.

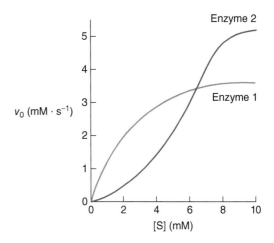

39. An enzyme catalyzes a reaction with a K_M of 50 μM. If the substrate concentration is 500 μM and the reaction velocity is 75% of V_{max}, then calculate the enzyme concentration.

40. When [S] = 5 K_M, how close is v_0 to V_{max}? When [S] = 20K_M, how close is v_0 to V_{max}? What do these results tell you about the accuracy of estimating V_{max} from a plot of v_0 versus [S]?

41. You are attempting to determine K_M by measuring the reaction velocity at different substrate concentrations, but you do not realize that the substrate tends to precipitate under the experimental conditions you have chosen. How would this affect your measurement of K_M?

42. You are constructing a velocity versus substrate concentration curve for an enzyme whose K_M is believed to be about 2 μM. The enzyme concentration is 200 nM and the substrate concentrations range from 0.1 μM to 10 μM. What is wrong with this experimental setup and how could you fix it?

43. The K_M values for the reaction of chymotrypsin with two different substrates are given. **a.** Which substrate has the higher apparent affinity for the enzyme? Use what you learned about chymotrypsin in Chapter 6 to explain why. **b.** Which substrate is likely to give a higher value for V_{max}?

Substrate	K_M (M)
N-Acetylvaline ethyl ester	8.8×10^{-2}
N-Acetyltyrosine ethyl ester	6.6×10^{-4}

44. The enzyme hexokinase acts on both glucose and fructose. Using the K_M and V_{max} values in the table, compare and contrast the interaction of hexokinase with each substrate.

Substrate	K_M (M)	V_{max} (relative)
Glucose	1.0×10^{-4}	1.0
Fructose	7.0×10^{-4}	1.8

45. a. Calculate k_{cat} for the polyphenol oxidase described in Problem 32 for the dopamine substrate. The enzyme concentration is 2.0×10^{-2} U·mL^{-1}. **b.** The polyphenol oxidase also acts on catechol as a substrate, with a V_{max} of 0.39 U·min^{-1}·mL^{-1}. Assuming the same enzyme concentration, what is the k_{cat} for the oxidase-catalyzed catechol reaction?

46. Using data from Problems 32 and 45, calculate and compare the catalytic efficiencies of the polyphenol oxidase enzyme using **a.** dopamine or **b.** catechol as a substrate. The K_M for catechol is 7.9×10^{-4} M.

47. An enzyme has a K_m of 4.5×10^{-5} M. If the V_{max} of the enzyme preparation is 25 μmoles·liter^{-1}·min^{-1}, what velocity will be observed in the presence of 2.2×10^{-4} M substrate and 5.2×10^{-4} M of **a.** a competitive inhibitor, **b.** a noncompetitive inhibitor, and **c.** an uncompetitive inhibitor. The K_I in all three cases is 3.2×10^{-4} M.

48. a. Use the data provided to determine whether the reactions catalyzed by enzymes A, B, and C are diffusion-controlled. **b.** A reaction is carried out in which 5 nM substrate S is added to a reaction mixture containing equivalent amounts of enzymes A, B, and C. After 30 seconds, which product will be more abundant, P, Q, or R?

Enzyme	Reaction	K_M	k_{cat}
A	S → P	0.3 mM	5000 s^{-1}
B	S → Q	1 nM	2 s^{-1}
C	S → R	2 μM	850 s^{-1}

49. Protein engineers constructed a mutant subtilisin enzyme, a serine protease used in the detergent industry, in an attempt to improve its catalytic activity. The isoleucine at position 31 (adjacent to the Asp residue in the Asp–His–Ser catalytic triad) was replaced with leucine. Both the mutant and wild-type enzymes were assayed for catalytic activity using the artificial substrate N-succinyl-Ala-Ala-Pro-Phe-p-nitroanilide (AAPF). The results are shown in the table. **a.** Explain how reaction velocity is measured using the AAPF substrate. **b.** Calculate the catalytic efficiencies for both enzymes. What effect did the replacement of Leu for Ile at position 31 have on the ability of subtilisin to catalyze hydrolysis of AAPF? Explain. **c.** Would subtilisin as a detergent additive be effective in removing protein-based stains (such as milk or blood) from clothing?

Enzyme	K_M (mM)	k_{cat} (s^{-1})
Ile 31 wild-type subtilisin	1.9	21
Leu 31 mutant subtilisin	2.0	120

50. Phosphatases catalyze the hydrolysis of phosphate groups from amino acid side chains (typically Ser, Thr, or Tyr) in target proteins. The activity of a phosphatase enzyme isolated from the bacterium that causes Legionnaire's disease was measured in the presence of various divalent cations. In addition, researchers constructed versions of the enzyme in which active site residues were mutated to alanine residues. The results are shown in the table. Calculate the catalytic efficiencies of each of the enzymes. What can you conclude about the roles of the amino acids and cofactors in catalysis?

Enzyme	K_M (mM)	k_{cat} (s^{-1})
Wild-type + Mn^{2+}	0.011	0.040
Wild-type + Ni^{2+}	0.0244	0.0023
His 213 → Ala mutant	305	2.53
Arg 185 → Ala mutant	0.086	0.13
Arg 369 → Ala mutant	0.046	0.053

51. The transferase enzyme β3-GalT catalyzes a reaction in which a galactose residue is transferred from the substrate UDP–galactose to a serine side chain on a protein important in immune recognition. Use the Lineweaver–Burk plot provided to determine the K_M and the V_{max} for β3-GalT.

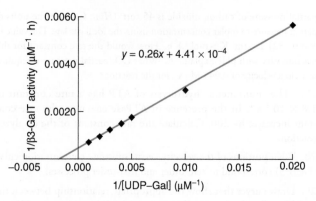

$y = 0.26x + 4.41 \times 10^{-4}$

52. Tryptase, a trypsin-like enzyme, catalyzes the hydrolysis of a single peptide bond in plasminogen activator (PA), which activates the enzyme and allows it to convert the inactive plasminogen to plasmin, an important reaction in the breakdown of fibrin polymers formed in blood clots. Use the Lineweaver–Burk plot provided to determine the K_M and the V_{max} for tryptase.

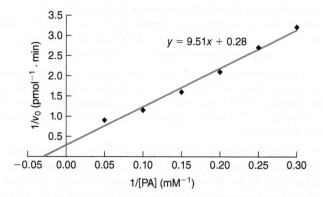

$y = 9.51x + 0.28$

53. Determine whether the following enzymes are likely to obey Michaelis–Menten kinetics: **a.** A tadpole collagenase hydrolyzes collagen in the tadpole tail during metamorphosis. **b.** When the enzyme hexokinase binds the glucose substrate, a conformational change permits the subsequent binding of the ATP substrate. **c.** Threonine dehydratase isolated from spinach catalyzes the deamination of threonine to ammonia and α-ketobutyrate; it can also act on serine, but threonine is the preferred substrate. **d.** ATCase, the enzyme that catalyzes the first step in pyrimidine biosynthesis, is subject to regulation by both ATP and CTP.

54. Three enzymatic reactions carried out under different conditions are shown in the figure above right. Estimate the K_M and V_{max} values from the plot. Which reaction has the lowest K_M? Which reaction has the highest V_{max}?

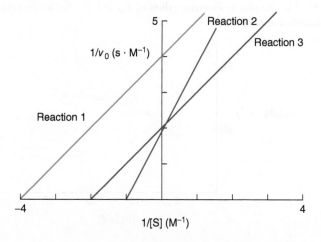

55. The enzyme γ-glutamylcysteine synthetase (γ-GCS) from the protozoan *Trypanosoma brucei*, the parasite that causes African sleeping sickness, catalyzes the first step of the biosynthesis of trypanothione, a compound the parasite requires to maintain proper redox balance. The K_M values for each substrate were determined. **a.** Does this reaction obey Michaelis–Menten kinetics? **b.** Describe how the K_M values for each of the three substrates were determined. **c.** How would V_{max} be achieved for the γ-GCS reaction?

$$\text{aminobutyric acid} + \text{glutamate} + \text{ATP} \xrightarrow{\text{γ-GCS}} \rightarrow \rightarrow \text{trypanothione}$$

Substrate	K_M (mM)
Glutamate	5.9
Aminobutyric acid	6.1
ATP	1.4

56. Calculate K_M and V_{max} from the following data:

[S] (mM)	v_0 (mM · s^{-1})
1	1.82
2	3.33
4	5.71
8	8.89
18	12.31

57. The kinetics of a bacterial dehalogenase were investigated. Calculate K_M and V_{max} from the following data:

[S] (mM)	v_0 (nmol · min^{-1})
0.04	0.229
0.13	0.493
0.40	0.755
0.90	0.880
1.30	0.917

7.3 Enzyme Inhibition

58. Based on some preliminary measurements, you suspect that a sample of enzyme contains an irreversible inhibitor. You decide to dilute the sample 100-fold and remeasure the enzyme's activity. What would your results show if the inhibitor in the sample is **a.** irreversible or **b.** reversible?

59. How would diisopropylphosphofluoridate (DIPF) affect the apparent K_M and V_{max} of a sample of chymotrypsin?

60. Shown below are five Lineweaver–Burk plots. The reaction without inhibitor is shown in blue; the reaction with inhibitor is shown in red. Identify the type of inhibition in each plot.

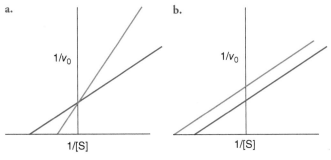

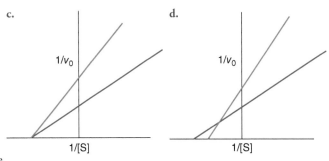

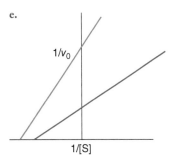

61. For each of the plots in Problem 60, describe the changes that occur (if any) in the values of K_M and V_{max} in the presence of the inhibitor.

62. Complete the following table:

Type of inhibition	I binds to E, ES, or both?	I binds to active site?
Competitive		
Uncompetitive		
Noncompetitive		

63. Complete the following table and determine the type of inhibition. Note that V_{max}^I is the V_{max} in the presence of the inhibitor and K_M^{app} is the "apparent" K_M in the presence of the inhibitor.

	K_M (mM)	K_M^{app} (mM)	$\dfrac{K_M^{app}}{K_M}$	V_{max} (nmol · min^{-1})	V_{max}^I (nmol · min^{-1})	$\dfrac{V_{max}}{V_{max}^I}$
a.	6	2		36	12	
b.	6	18		36	18	
c.	6	12		36	36	
d.	6	1		36	12	
e.	6	6		36	18	

64. Estimate the K_M and V_{max} values using the Michaelis–Menten plot provided. What type of inhibition is represented here?

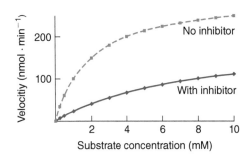

65. Estimate the K_M and V_{max} values using the Michaelis–Menten plot provided. What type of inhibition is represented here?

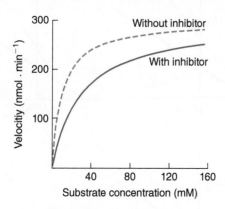

66. Phosphatase enzymes catalyze the hydrolysis of phosphate groups from phosphotyrosine residues in target proteins in signaling pathways (see Chapter 10), as shown below.

Phosphatase inhibitors, such as the compound shown below, abbreviated OBA, could be used to treat inflammation, diabetes, and cancer. **a.** What kind of inhibitor is OBA? Explain. **b.** Can inhibition by OBA be overcome *in vitro* by increasing the substrate concentration? Explain. **c.** Does this inhibitor bind reversibly or irreversibly to the enzyme? Explain.

OBA

67. Would indole (see structure below) be a more effective competitive inhibitor of chymotrypsin, trypsin, or elastase? Explain.

Indole

68. Calculate the K_I value for the inhibitor shown in Problem 64 in the presence of 5.0 μM inhibitor.

69. Calculate the K_I value for the inhibitor shown in Problem 65 in the presence of 5.0 μM inhibitor.

70. The K_I value for an inhibitor is 2 μM. When no inhibitor is present, the K_M value is 10 μM. Calculate the apparent K_M when 4 μM inhibitor is present.

71. Inhibitor A at a concentration of 3 μM doubles the apparent K_M for an enzymatic reaction, whereas inhibitor B at a concentration of 12 μM quadruples the apparent K_M. What is the ratio of the K_I for inhibitor B to the K_I for inhibitor A?

72. Glucose-6-phosphate dehydrogenase catalyzes the reaction shown below (see Section 13.4). Kinetic data for the enzyme isolated from the thermophilic bacterium *T. maritima* are presented in the table. **a.** NADPH inhibits glucose-6-phosphate dehydrogenase. What kind of inhibitor is NADPH? **b.** How are the values of K_M and V_{max} likely to differ from those listed in the table above if NADPH is present? **c.** Does glucose-6-phosphate dehydrogenase prefer NAD^+ or $NADP^+$ as a cofactor?

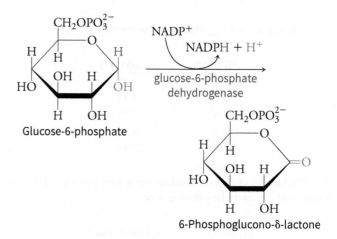

Substrate	K_M (mM)	V_{max} (U·mg^{-1})
Glucose-6-phosphate	0.15	20
$NADP^+$	0.03	20
NAD^+	12	6

73. Glucose-6-phosphate dehydrogenase catalyzes the same reaction shown in Problem 72 in yeast. The K_M for yeast glucose-6-phosphate is 2.0×10^{-5} M and the K_M for $NADP^+$ is 2.0×10^{-6} M. The yeast enzyme can be inhibited by a number of cellular agents whose K_I values are listed in the table. **a.** Which is the most effective inhibitor? Explain. **b.** Which inhibitor(s) is (are) likely to be completely ineffective under normal cellular conditions? Explain.

Inhibitor	K_I (M)
Inorganic phosphate	1.0×10^{-1}
Glucosamine-6-phosphate	7.2×10^{-4}
NADPH	2.7×10^{-5}

74. A phospholipase hydrolyzes its lipid substrate with a K_M of 10 μM and a V_{max} of 7 μmol·mg^{-1}·min^{-1}. In the presence of 30 μM palmitoylcarnitine inhibitor, the K_M increases to 40 μM and the V_{max} remains unchanged. **a.** What is the K_I of the inhibitor? **b.** What type of inhibitor is palmitoylcarnitine?

75. A compound named P56 was found to inhibit a malarial parasite cysteine protease and has the potential to be used as a drug to treat malaria. Data are plotted below. **a.** What kind of inhibitor is P56? **b.** Use the equations of the lines provided to calculate K_M and V_{max} for the enzyme with and without 0.22 mM inhibitor. **c.** Calculate the value of K_I in the presence of 0.22 mM inhibitor.

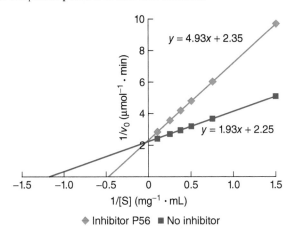

$y = 4.93x + 2.35$

$y = 1.93x + 2.25$

◆ Inhibitor P56   ■ No inhibitor

76. The HIV protease, an important enzyme in the life cycle of the human immunodeficiency virus, is a good drug target for the treatment of HIV and AIDS (see Box 7.A). A protein produced by the virus, p6*, inhibits the protease. The activity of the protease was measured in the presence (10 μM) and absence of p6* using an assay with a modified peptide that is hydrolyzed by the protease to release a derivative of *p*-nitrophenolate that absorbs light at 295 nm (this assay is similar to the one described for chymotrypsin in Section 6.1). **a.** Construct a Lineweaver–Burk plot and determine the K_M and V_{max} in the presence and in the absence of inhibitor. **b.** What type of inhibitor is p6*? Explain. **c.** Calculate the K_I for the inhibitor.

[S] (μM)	v_0 without p6* (nmol · min⁻¹)	v_0 with p6* (nmol · min⁻¹)
10	4.63	2.70
15	5.88	3.46
20	6.94	4.74
25	9.26	6.06
30	10.78	6.49
40	12.14	8.06
50	14.93	9.71

77. The drug troglitazone was used to treat diabetes but was withdrawn from the market when patients who took the drug suffered from severe side effects. The activity of a hydroxylase enzyme, which acts on progesterone as a substrate in a steroid synthetic pathway, was measured in the presence and in the absence of 10 μM troglitazone. **a.** Use the data in the table to construct a Lineweaver–Burk plot. What type of inhibitor is troglitazone? **b.** Calculate the K_I for the inhibitor.

[Progesterone] (μM)	v_0 without I (pmol · min⁻¹)	v_0 with I (pmol · min⁻¹)
0.50	0.00272	0.0015
1.00	0.00457	0.0026
2.00	0.00672	0.0044
4.00	0.00769	0.0067

78. The phosphatase enzyme PTP1B catalyzes the removal of phosphate groups from specific proteins and is involved in insulin signaling. Phosphatase inhibitors such as vanadate may be useful in the treatment of diabetes. The activity of PTP1B was measured in the presence and absence of vanadate using an artificial substrate, fluorescein diphosphate (FDP), which produces a light-absorbing product, along with free phosphate. **a.** Construct a Lineweaver–Burk plot using the data provided. Calculate K_M and V_{max} for PTP1B in the absence and in the presence of vanadate. **b.** What kind of inhibitor is vanadate? Explain.

[FDP] (μM)	v_0 without vanadate (nM · s⁻¹)	v_0 with vanadate (nM · s⁻¹)
6.67	5.7	0.71
10	8.3	1.06
20	12.5	2.04
40	16.7	3.70
100	22.2	8.00
200	25.4	12.5

79. An alternative way to calculate K_I for the vanadate inhibitor in the PTP1B reaction (see Problem 78) is to measure the velocity of the enzyme-catalyzed reaction in the presence of increasing amounts of inhibitor and a constant amount of substrate. These data are shown in the table for a substrate concentration of 6.67 μM. To calculate K_I, rearrange Equation 7.28 and solve for α. Then construct a graph plotting α versus [I]. Since $\alpha = 1 + [I]/K_I$, the slope of the line is equal to $1/K_I$. Determine K_I for vanadate using this method.

[Vanadate] (μM)	v_0 (nM · s⁻¹)
0.0	5.70
0.2	3.83
0.4	3.07
0.7	2.35
1.0	2.04
2.0	1.18
4.0	0.71

80. The bacterial enzyme proline racemase catalyzes the interconversion of two isomers of proline:

The compound shown below is an inhibitor of proline racemase. Explain why.

81. Cytidine deaminase catalyzes the following reaction:

Cytidine → → Uridine

Both of the compounds shown below inhibit the reaction. The compounds have K_I values of 3×10^{-5} M and 1.2×10^{-12} M. Assign the appropriate K_I

value to each inhibitor. Which compound is the more effective inhibitor? Give a structural basis for your answer.

A

B

82. The observation that adenosine deaminase is inhibited by 1,6-dihydroinosine allowed scientists to propose a structure for the transition state of this enzyme. The compound coformycin also inhibits adenosine deaminase; its K_I value is about 0.25 µM. Does this observation support or refute the proposed transition state for adenosine deaminase?

Coformycin

83. Why is the SARS-CoV-2 protease, specifically the 3CLpro, a crucial target for developing antiviral drugs? How does inhibiting this protease affect the virus's lifecycle and replication?

84. Explain the significance of the catalytic dyad Cys145–His41 in the SARS-CoV-2 protease's cleavage activity. How does the interaction between these residues contribute to the protease's function?

85. How does the three-dimensional structure of the SARS-CoV-2 protease, as revealed by techniques like X-ray crystallography and cryo-electron microscopy, influence its function and potential as a drug target?

86. Why is the absence of a homologous human protein with a similar cleavage site to the SARS-CoV-2 protease crucial when considering the development of protease inhibitors? How does this factor impact the specificity and safety of potential drugs?

87. The flu virus enzyme neuraminidase hydrolyzes sialic acid residues from cell-surface glycoproteins, allowing newly made viruses to escape from the host cell. The drugs oseltamivir and zanamivir are transition state analogs that inhibit neuraminidase to block viral infection. Wild-type and mutant strains of the flu virus, in which neuraminidase residue 274 is changed from His to Tyr, exhibit the kinetic parameters shown below. **a.** For the wild-type virus, which drug would work better? **b.** Does the mutation appear to affect substrate binding or turnover? **c.** How does the mutation affect inhibition by oseltamivir and zanamivir? Which drug would be more effective against the mutant flu strain?

	K_M for sialic acid (µM)	V_{max} (relative units)
Wild-type enzyme	6.3	1.0
His 274 → Tyr mutant	27.0	0.8

	K_I for oseltamivir (nM)	K_I for zanamivir (nM)
Wild-type enzyme	0.32	0.10
His 274 → Tyr mutant	85	0.19

88. The drug troglitazone was used to treat diabetes but was withdrawn from the market when patients who took the drug suffered from severe side effects. The activity of a hydroxylase enzyme, which acts on progesterone as a substrate in a steroid synthetic pathway, was measured in the presence and in the absence of 10 µM troglitazone. **a.** Use the data in the table to construct a Lineweaver–Burk plot. What type of inhibitor is troglitazone? **b.** Calculate the K_I for the inhibitor.

[Pregnenolone] (µM)	v_0 without I (pmol · min^{-1})	v_0 with I (pmol · min^{-1})
0.50	0.00272	0.0015
1.00	0.00457	0.0026
2.00	0.00672	0.0044
4.00	0.00769	0.0067

89. Protein phosphatase 1 (PP1) helps regulate cell division and is a possible drug target to treat certain types of cancer. The enzyme catalyzes the hydrolysis of a phosphate group from myelin basic protein (MBP). The activity of PP1 was measured in the presence and absence of the inhibitor phosphatidic acid (PA). **a.** Use the data to construct a Lineweaver–Burk plot for PP1 in the presence and absence of PA. What kind of inhibitor is PA? **b.** Report the K_M and V_{max} values for PP1 in the presence and absence of the inhibitor.

[MBP] (mg · mL^{-1})	v_0 without PA (nmol · mL^{-1} · min^{-1})	v_0 with PA (nmol · mL^{-1} · min^{-1})
0.010	0.0209	0.00381
0.015	0.0355	0.00620
0.025	0.0419	0.00931
0.050	0.0838	0.01400

90. The tyrosinase enzyme catalyzes reactions that produce brown-colored products. In mammals, it is responsible for the production of melanin in the skin; in plants, it is responsible for the browning that occurs when foods such as apples or potatoes are sliced open and exposed to air. Food scientists interested in preventing the browning process in plant-based foods have shown that dodecyl gallate inhibits tyrosinase. Using the Lineweaver–Burk plot provided, calculate the V_{max} and K_M for the enzyme in the presence and absence of the inhibitor. What type of inhibitor is dodecyl gallate?

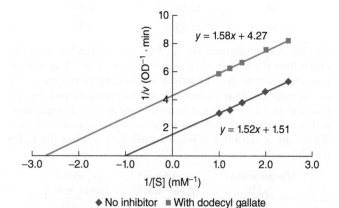

◆ No inhibitor ■ With dodecyl gallate

91. The phosphofructokinase (PFK) enzyme of parasites is an attractive drug target to treat parasitic infections in cattle. PFK catalyzes a reaction in glycolysis in which fructose-6-phosphate (F6P) is phosphorylated (see Section 13.1). The activity of the parasitic enzyme was measured using an assay in which the PFK reaction was coupled to a second reaction that converts NADH (which absorbs ultraviolet light at 340 nm) to NAD$^+$ (which does not). The activity of the enzyme in the presence and absence of 5 nM suramin was measured. The data are shown in the table. **a.** Use these data to construct a Lineweaver–Burk plot. **b.** Calculate K_M and V_{max} for PFK in the presence and in the absence of suramin. **b.** What kind of inhibitor is suramin? Explain. **c.** Calculate the K_I for suramin. **d.** The K_I for suramin for mammalian PFK is 40 nM. Compare this value to your answer to part c and assess the effectiveness of suramin as a drug to treat parasitic infections in cattle.

[F6P] (mM)	v_0 without inhibitor ($\Delta A \cdot min^{-1}$)	v_0 with inhibitor ($\Delta A \cdot min^{-1}$)
0.5	0.0202	0.0051
1.0	0.0310	0.0080
2.0	0.0555	0.0129
4.0	0.0590	0.0145
8.0	0.0625	0.0167
15.0	0.0670	0.0172

92. The γ-glutamyl transpeptidase (GGT) enzyme is an attractive drug target, because it is over-expressed in cancer cells, making them resistant to traditional cancer treatments. A high-throughput screening procedure identified a benzenesulfonamide compound that was able to inhibit GGT. GGT activity was measured in the presence and in the absence of 125 μM inhibitor using a modified glutamic acid substrate (GpNA) that, when hydrolyzed, yields a product that absorbs light at 405 nm. The data are shown in the table. **a.** Use these data to construct a Lineweaver–Burk plot. Calculate K_M and V_{max} for GGT in the absence and in the presence of inhibitor. **b.** What kind of inhibitor is the compound? Explain. **c.** Calculate the K_I for the inhibitor. Compare this value to the K_M value for the substrate in the absence of the inhibitor. **d.** Evaluate the use of benzenesulfonamide as a possible cancer drug.

[GpNA] (mM)	v_0 without inhibitor ($nmol \cdot min^{-1}$)	v_0 with inhibitor ($nmol \cdot min^{-1}$)
0.25	0.0765	0.054
0.5	0.125	0.079
1.0	0.189	0.101
2.0	0.325	0.113

93. Explain why there are so few examples of inhibitors that decrease V_{max} but do not affect K_M.

94. Homoarginine has been shown to inhibit the activity of an alkaline phosphatase found in bone. Alkaline phosphatase enzymes are widely distributed in tissues, and serum levels of these enzymes are often used as a diagnostic tool for various diseases. The artificial substrate phenyl phosphate was incubated with bone alkaline phosphatase in the presence of 4 mM homoarginine. The data are shown in the table.

[Phenyl phosphate] (mM)	v_0, no inhibitor ($mM \cdot min^{-1}$)	v_0, with inhibitor ($mM \cdot min^{-1}$)
4.00	1.176	0.476
2.00	0.909	0.436
1.00	0.667	0.385
0.67	0.556	0.333
0.50	0.455	0.286
0.33	0.345	0.244

a. Construct a Lineweaver–Burk plot using the data provided. Calculate K_M and V_{max} for the alkaline phosphatase enzyme in the absence and in the presence of homoarginine. **b.** What kind of inhibitor is homoarginine? Explain. **c.** Homoarginine does not inhibit intestinal alkaline phosphatase. Why might homoarginine inhibit bone alkaline phosphatase but not the enzyme found in the intestine?

95. Computer-modeling studies have shown that uncompetitive and noncompetitive inhibitors of enzymes are more effective than competitive inhibitors. These studies have important implications in drug design. Propose a hypothesis that explains these results.

96. Aspartate transcarbamoylase (ATCase) catalyzes the formation of N-carbamoyl aspartate from carbamoyl phosphate and aspartate, a step in the multienzyme process that synthesizes cytidine triphosphate (CTP). Measurements of ATCase activity as a function of aspartate concentration yield the results shown in the graph. **a.** Is ATCase an allosteric enzyme? How do you know? **b.** What kind of an effector is CTP? Explain. What is the biological significance of CTP's effect on ATCase? **c.** What kind of an effector is ATP? Explain. What is the biological significance of ATP's effect on ATCase?

carbamoyl phosphate + aspartate $\xrightarrow{\text{ATCase}}$

N-carbamoylaspartate →→→ UMP→→UTP→→CTP

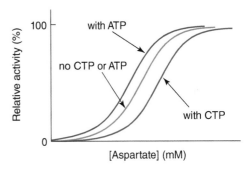

97. The activity of ATCase (see Problem 95) was measured in the presence of several nucleotide combinations. The results are shown in the table. A value greater than 1 indicates that the enzyme was activated; a value less than 1 indicates inhibition. **a.** What combination gives the most effective inhibition? **b.** What is the physiological significance of this combination? **c.** Redraw the graph shown in Problem 96 to include a curve representing ATCase activity in the presence of the most inhibitory nucleotide combination. How do the K_M and V_{max} values of this combination compare with the values for ATCase with ATP and CTP alone?

Nucleotide effectors	Activity in the presence of 5 mM aspartate
ATP	1.35
CTP	0.43
UTP	0.95
ATP + CTP	0.85
ATP + UTP	1.52
CTP + UTP	0.06

98. The redox state of the cell (the likelihood that certain groups will be oxidized or reduced) is believed to regulate the activities of some enzymes. Explain how the reversible formation of an intramolecular disulfide bond (—S—S—) from two Cys —SH groups could affect the activity of an enzyme.

99. Pyruvate kinase catalyzes the last reaction in the glycolytic pathway:

Phosphoenolpyruvate (PEP) + ADP → pyruvate + ATP

There are four different mammalian forms of pyruvate kinase. All catalyze the same reaction but differ in their response to the glycolytic metabolite fructose-1,6-bisphosphate (F16BP). The activity of the M_1 form of pyruvate kinase was measured at various concentrations of PEP, both in the presence (blue) and absence (red) of F16BP. The results are shown in the graph. **a.** Is the M_1 form of pyruvate kinase an allosteric enzyme? Does F16BP affect enzyme activity, and if so, how?

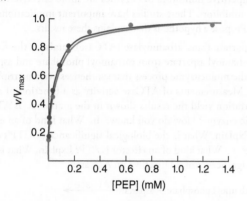

b. The investigators carried out site-directed mutagenesis to replace Ala 398 with Arg. The mutated residue is located at one of the intersubunit contact positions. The activity of the mutant enzyme in the presence and absence of F16BP was measured as in part **a.** The results are shown in the graph. How did the mutation affect the enzyme?

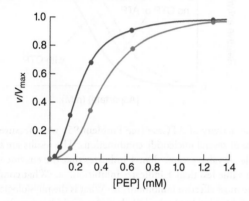

7.4 Clinical Connection: Drug Development

100. Medicinal chemists apply certain rules to assess the suitability of drug candidates. For example, they prefer to work with compounds that have a molecular mass less than 500 g·mol⁻¹ and contain fewer than five hydrogen bond donor groups and fewer than 10 hydrogen bond acceptor groups. Explain why these properties would be desirable in a drug.

101. The structure of rosiglitazone is shown. Does it meet the "rules" described in Problem 100?

102. Explain why some drugs must be administered intravenously rather than swallowed in the form of a pill.

103. The toxicity of the acetaminophen derivative acetamidoquinone results from its ability to react with the Cys groups of proteins. Explain why the toxic effects of acetamidoquinone are localized to the liver.

104. The drug warfarin is widely prescribed as a "blood thinner" because it inhibits the post-translational modification of some of the proteins involved in blood coagulation (see Section 6.5). In order to choose an effective dose of warfarin, physicians may test patients to identify which cytochrome P450 enzymes they express. Explain the purpose of this genetic testing.

105. Enalapril inhibits an enzyme that converts the peptide angiotensin I to angiotensin II, a potent vasoconstrictor. **a.** What type of disorders might be treated by enalapril? **b.** Enalapril is inactive until acted upon by an esterase. Draw the structure of the resulting bioactive derivative.

Enalapril

SELECTED READINGS

Corey, E.J., Czakó, B., and Kürti, L., *Molecules and Medicine,* Wiley (2007). [Describes the discovery, structure, and action of over a hundred drugs affecting a variety of medical conditions.]

Cornish-Bowden, A., The origins of enzyme kinetics, *FEBS Lett.* 587, 2725–2730, doi: 10.1016/j.febslet.2013.06.009 (2013). [Illuminates some key features of the Michaelis–Menten equation through its history.]

Davidi, D., Noor, E., Liebermeister, W., Bar-Even, A., Flamholz, A., Tummler, K., Barenholz, U., Goldenfeld, M., Shlomi, T., and Milo, R., Global characterization of *in vivo* enzyme catalytic rates and their correspondence to *in vitro* k_{cat} measurements, *Proc. Natl. Acad. Sci. USA* 113, 3401–3406, doi: 10.1073/pnas.1514240113 (2106). [Compares predicted k_{cat} values to k_{cat} measurements for over a hundred *E. coli* enzymes.]

Lv, Z., Chu, Y., and Wang, Y., HIV protease inhibitors: a review of molecular selectivity and toxicity, *HIV AIDS (Auckl.)* 7, 95–104 (2015). [A short review of the structures and action of HIV protease inhibitors.]

Schramm, V.L., Enzymatic transition state theory and transition state analogue design, *J. Biol. Chem.* 282, 28297–28300 (2007). [Describes how transition state analog inhibitors work, with examples.]

CHAPTER 7 CREDITS

Table 7.1 Data from Radzicka, A. and Wolfenden, R., *Science* 267, 90–93 (1995).

Box 7.A Image based on 1HXW. Kempf, D.J., Marsh, K.C., Denissen, J.F., McDonald, E., Vasavanonda, S., Flentge, C.A., Green, B.E., Fino, L., Park, C.H., Kong, X.P., ABT-538 is a potent inhibitor of human immunodeficiency virus protease and has high oral bioavailability in humans, *Proc. Natl. Acad. Sci. USA* 92, 2484–2488 (1995).

Figure 7.14 Image based on 6PFK. Schirmer, T., Evans, P.R., Structural basis of the allosteric behaviour of phosphofructokinase, *Nature* 343, 140–145 (1990).

Figure 7.15. Data from Zhu, X., Byrnes, N., Nelson, J.W., and Chang, S.H., *Biochemistry* 34, 2560–2565 (1995).

Figure 7.16 Image from Schirmer, T. and Evans, P.R., Structural basis of the allosteric behaviour of phosphofructokinase, *Nature* 343, 142 (1990). Reproduced with permission of Springer Nature.

Lipids and Biological Membranes

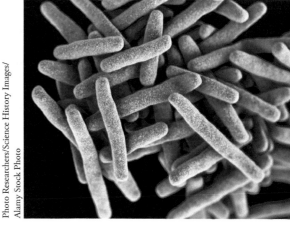

One of the challenges of fighting *Mycobacterium tuberculosis* is the pathogen's ability to live inside its host cells and feed on intracellular lipid droplets. Lipid-soluble antibiotics that preferentially localize to the droplets can more efficiently enter the host cells and kill the bacteria.

Do You Remember?

- Cells contain four major types of biological molecules (Section 1.3).
- Noncovalent forces, including hydrogen bonds, ionic interactions, and van der Waals forces, act on biological molecules (Section 2.1).
- The hydrophobic effect, which is driven by entropy, excludes nonpolar substances from water (Section 2.2).
- Amphipathic molecules form micelles or bilayers (Section 2.2).
- A folded polypeptide tends to assume a shape with a hydrophilic surface and a hydrophobic core (Section 4.3).

All cells—and the various compartments inside eukaryotic cells—are surrounded by membranes. In fact, the formation of a membrane is believed to be a defining event in the evolutionary history of the cell (Section 1.2); without membranes, a cell would be unable to retain essential resources. To begin to understand how membranes work, we will examine them as composite structures containing both lipids and proteins.

The lipids that occur in biological membranes aggregate to form sheets that are relatively impermeable to ions and other solutes. The key to this behavior is the hydrophobicity of the lipid molecules. Hydrophobicity is also a useful feature of lipids that perform other roles, such as energy storage. Although lipids exhibit enormous variety in shape and size and carry out all sorts of biological tasks, they are united by their hydrophobicity.

8.1 Lipids

KEY CONCEPTS

Recognize the physical characteristics of lipids.

- Explain how fatty acids are esterified to make larger structures.
- Describe the structures of amphipathic lipids.
- List some functions of lipids.

The molecules that fit the label of *lipid* do not follow a single structural template or share a common set of functional groups, as nucleotides and amino acids do. In fact, *lipids are defined primarily by the absence of functional groups.* Because they consist mainly of C and H atoms and have few if any N- or O-containing functional groups, they lack the ability to form hydrogen bonds and are therefore largely insoluble in water (most lipids are soluble in nonpolar organic solvents). Although some lipids do contain polar or charged groups, the bulk of their structure is hydrocarbon-like.

Fatty acids contain long hydrocarbon chains

The simplest lipids are the **fatty acids,** which are long-chain carboxylic acids (at physiological pH, they are ionized to the carboxylate form). These molecules may contain up to 24 carbon atoms, but the most common fatty acids in plants and animals are the even-numbered C_{16} and C_{18} species such as palmitate and stearate:

Palmitate   Stearate   Oleate   Linoleate

Such molecules are called **saturated fatty acids** because all their tail carbons are "saturated" with hydrogen. **Unsaturated fatty acids** (which contain one or more double bonds) such as oleate and linoleate are also common in biological systems. In these molecules, the double bond usually has the *cis* configuration (in which the two hydrogens are on the same side). Some common saturated and unsaturated fatty acids are listed in **Table 8.1**. Although human cells can synthesize a variety of unsaturated fatty acids, they are unable to make any with double bonds past carbon 9 (counting from the carboxylate end). Some organisms can do so, however, producing what are known as omega-3 and omega-6 fatty acids (**Box 8.A**).

Free fatty acids are relatively scarce in biological systems. Instead, they are usually esterified, for example, to glycerol.

Glycerol

The fats and oils found in animals and plants are **triacylglycerols** (sometimes called **triglycerides**) in which the **acyl groups** (the R—CO— groups) of three fatty acids are esterified to the three hydroxyl groups of glycerol.

$$CH_2 \text{—} CH \text{—} CH_2$$

(Glycerol / 3 fatty acyl groups structure)

The ester bond linking each acyl group is the result of a condensation reaction. This is as close as lipids come to forming polymers. They cannot be linked end-to-end to form long chains, as the other types of biological molecules can. The three fatty acids of a given triacylglycerol may be the same or different.

For reasons that are outlined below, triacylglycerols do not form bilayers and so are not important components of biological membranes. However, they do aggregate in large globules, acting as a storage depot for fatty acids that serve as metabolic fuels (these reactions are described in Section 17.1). The hydrophobic nature of triacylglycerols and their tendency to aggregate means that cells can store a large amount of this material without it interfering with other activities that take place in an aqueous environment.

Table 8.1 Some Common Fatty Acids

Number of carbon atoms	Common name	Systematic name[a]	Structure
Saturated fatty acids			
12	Lauric acid	Dodecanoic acid	$CH_3(CH_2)_{10}COOH$
14	Myristic acid	Tetradecanoic acid	$CH_3(CH_2)_{12}COOH$
16	Palmitic acid	Hexadecanoic acid	$CH_3(CH_2)_{14}COOH$
18	Stearic acid	Octadecanoic acid	$CH_3(CH_2)_{16}COOH$
20	Arachidic acid	Eicosanoic acid	$CH_3(CH_2)_{18}COOH$
22	Behenic acid	Docosanoic acid	$CH_3(CH_2)_{20}COOH$
24	Lignoceric acid	Tetracosanoic acid	$CH_3(CH_2)_{22}COOH$
Unsaturated fatty acids			
16	Palmitoleic acid	9-Hexadecenoic acid	$CH_3(CH_2)_5CH{=}CH(CH_2)_7COOH$
18	Oleic acid	9-Octadecenoic acid	$CH_3(CH_2)_7CH{=}CH(CH_2)_7COOH$
18	Linoleic acid	9,12-Octadecadienoic acid	$CH_3(CH_2)_4(CH{=}CHCH_2)_2(CH_2)_6COOH$
18	α-Linolenic acid	9,12,15-Octadecatrienoic acid	$CH_3CH_2(CH{=}CHCH_2)_3(CH_2)_6COOH$
18	γ-Linolenic acid	6,9,12-Octadecatrienoic acid	$CH_3(CH_2)_4(CH{=}CHCH_2)_3(CH_2)_3COOH$
20	Arachidonic acid	5,8,11,14-Eicosatetraenoic acid	$CH_3(CH_2)_4(CH{=}CHCH_2)_4(CH_2)_2COOH$
20	EPA	5,8,11,14,17-Eicosapentaenoic acid	$CH_3CH_2(CH{=}CHCH_2)_5(CH_2)_2COOH$
22	DHA	4,7,10,13,16,19-Docosahexaenoic acid	$CH_3CH_2(CH{=}CHCH_2)_6CH_2COOH$

[a]Numbers indicate the starting position of the double bond; the carboxylate carbon is in position 1.

Question Which of the unsaturated fatty acids listed here are omega-3 fatty acids?

Box 8.A   Omega-3 Fatty Acids

An **omega-3 fatty acid** has a double bond starting three carbons from its methyl end (the last carbon in the fatty acid chain is called the omega carbon). Marine algae are notable producers of the long-chain omega-3 fatty acids EPA and DHA (see Table 8.1); these lipids tend to accumulate in the fatty tissues of cold-water fish. Consequently, fish oil has been identified as a convenient source of omega-3 fatty acids, which have purported health benefits. Somewhat shorter omega-3 fatty acids such as α-linolenic acid are manufactured by plants. Humans whose diets don't include fish oil acquire α-linolenic acid from plant sources and convert it to the longer varieties of omega-3 fatty acids by lengthening the fatty acid chain from the carboxyl end (Section 17.3).

Omega-3 fatty acids were identified as essential for normal human growth in the 1930s, but it was not until the 1970s that consumption of omega-3 fatty acids such as EPA was linked to decreased risk of cardiovascular disease. The correlation came to light from the observation that native Arctic populations, who ate fish and meat but few vegetables, had a surprisingly low incidence of heart disease. One possible biochemical explanation is that the omega-3 fatty acids compete with omega-6 fatty acids for the enzymes that convert the fatty acids to certain signaling molecules. The omega-6 derivatives are stronger triggers of inflammation, which underlies conditions such as atherosclerosis. The relative amounts of omega-3 and omega-6 fatty acids might therefore matter more than the absolute amount of omega-3 fatty acids consumed.

The 22-carbon DHA is abundant in the brain and retina, and its concentration decreases with age. DHA is converted *in vivo* to substances that are believed to protect neural tissues from damage following a stroke. However, DHA supplements do not appear to reverse the cognitive decline associated with disorders such as Alzheimer's disease (see Section 4.5).

Numerous studies have attempted to demonstrate the ability of omega-3 fatty acids to prevent or treat other conditions such as arthritis and cancer, but the findings have been mixed, with the omega-3 fatty acid supplements in some cases exacerbating the disease. For this reason, the true role of omega-3 fatty acids in human health demands additional investigation.

Question   **Explain why vegans exhibit DHA levels only slightly lower than the levels in individuals who consume large amounts of fish. What does this information reveal about the usefulness of consuming DHA supplements?**

Classification of lipids basis structural and physiological functions

Among the major lipids of biological membranes in bacteria and eukaryotes are the **glycerophospholipids,** which contain a glycerol backbone with fatty acyl groups esterified at positions 1 and 2 and a phosphate derivative, called a head group, esterified at position 3. As in triacylglycerols, the fatty acyl components of glycerophospholipids vary from molecule to molecule. These lipids are usually named according to their head group, for example,

Phosphatidylcholine

Phosphatidylethanolamine

Phosphatidylglycerol

Phosphatidylserine

Note that the glycerophospholipids are not completely hydrophobic: *They are **amphipathic**, with hydrophobic tails attached to polar or charged head groups.* As we will see, their structure is ideal for forming bilayers.

The bonds that link the various components of a glycerophospholipid can be hydrolyzed by **phospholipases** to release the acyl chains or portions of the head group (**Fig. 8.1**). These enzymatic reactions are not just for degrading lipids. Some products derived from membrane lipids act as signaling molecules inside cells or between neighboring cells.

Eukaryotic membranes also contain amphipathic lipids known as **sphingolipids.** The **sphingomyelins,** with phosphocholine or phosphoethanolamine head groups, are sterically similar to their glycerophospholipid counterparts. The major difference is that sphingomyelins are not built on a glycerol backbone. Instead, their basic component is sphingosine, a derivative of serine and the fatty acid palmitate. Attaching a fatty acyl group via an amide bond to the serine amino group of sphingosine produces a ceramide (**Fig. 8.2**). A sphingomyelin consists of a ceramide plus a head

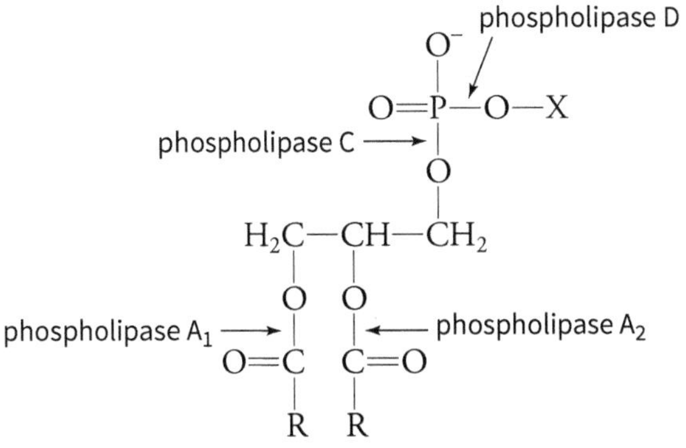

Figure 8.1 Sites of action of phospholipases.

Question **Which phospholipase-catalyzed reactions release charged portions of the phospholipid?**

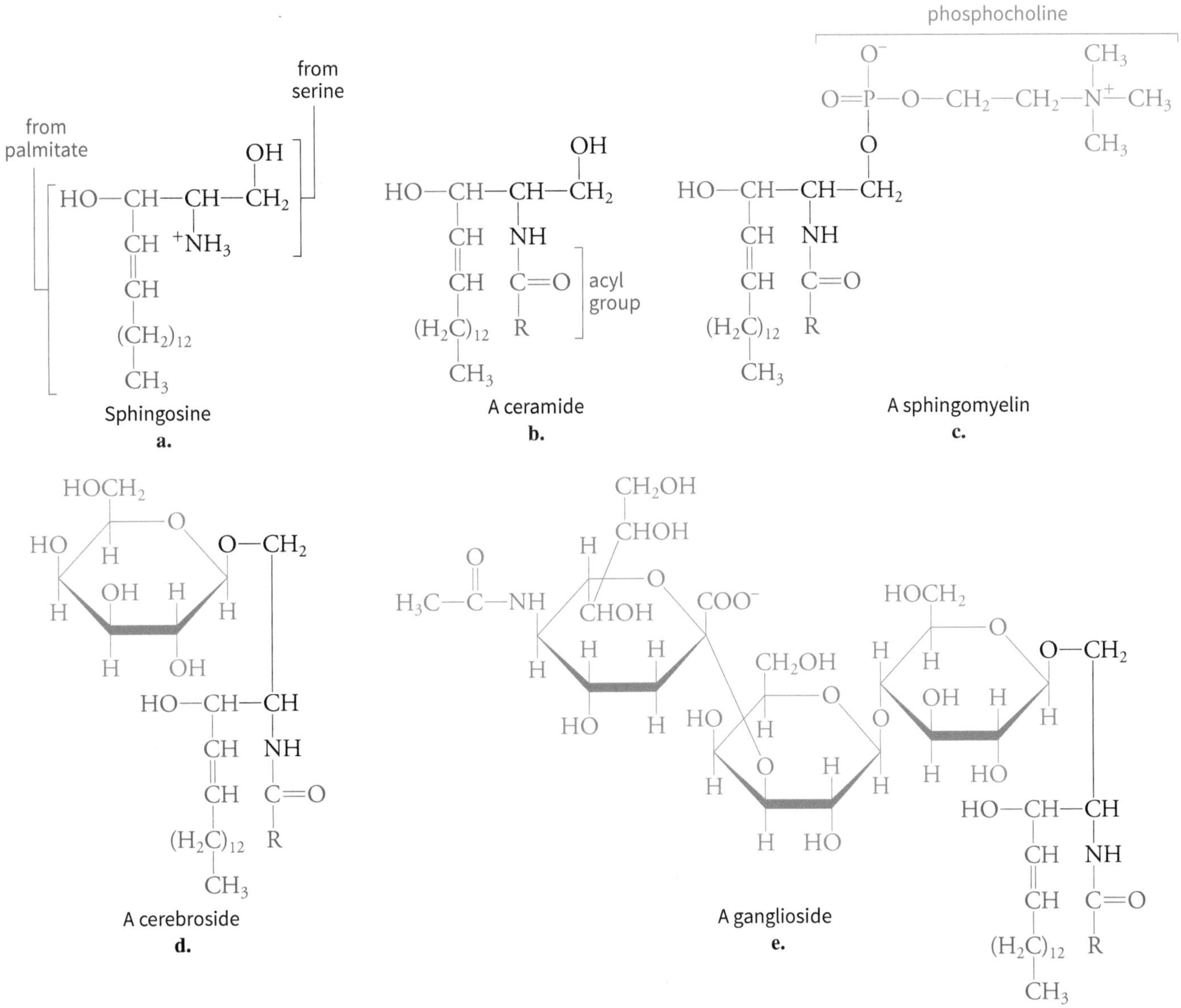

Figure 8.2 Sphingolipids. **a.** The sphingosine backbone is derived from serine and palmitate. **b.** Attaching a second acyl group yields a ceramide. **c.** Adding a phosphocholine (or phosphoethanolamine) head group yields a sphingomyelin. **d.** A cerebroside has a monosaccharide as a head group rather than a phosphate derivative. **e.** A ganglioside includes an oligosaccharide head group.

Question **Compare the structure of a sphingomyelin to the structure of a glycerophospholipid.**

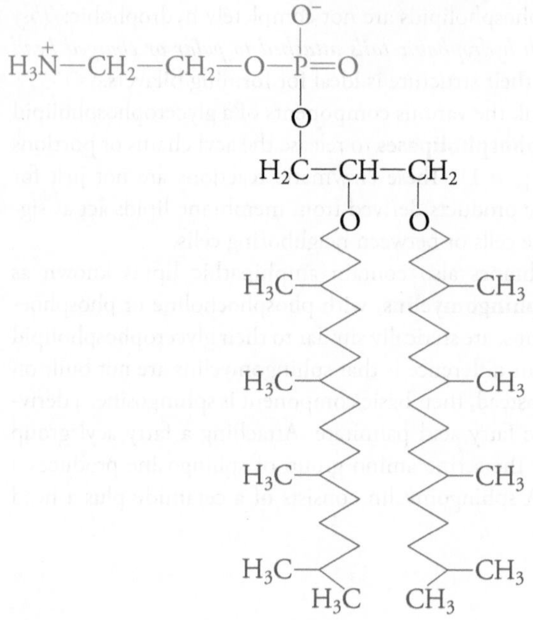

$$H_3\overset{+}{N}-CH_2-CH_2-O-\overset{\overset{\textstyle O^-}{|}}{\underset{\underset{\textstyle O}{||}}{P}}=O$$

Figure 8.3 An archaeal membrane lipid with an ethanolamine head group.

Question Compare this structure to that of the phosphatidylethanolamine lipid shown earlier.

group. Some sphingolipids have large head groups containing one or more carbohydrate groups. These **glycolipids** are known as cerebrosides and gangliosides.

Archaebacterial membranes contain a different sort of amphipathic lipid that has the same overall shape and function as a bacterial or eukaryotic glycerophospholipid. However, in the archaeal version, the hydrocarbon tails are linked to the glycerol backbone by ether rather than ester linkages, and the glycerol group has the opposite chirality (handedness, see Box 4.A). In addition, the tails are typically 20-carbon branched structures rather than straight chains (**Fig. 8.3**). The polar and charged head groups of the archaeal membrane lipids are similar to those in bacteria and eukaryotes.

In addition to glycerophospholipids and sphingolipids, many other types of lipids occur in membranes and elsewhere in the cell. One of these is cholesterol, a 27-carbon, four-ring molecule:

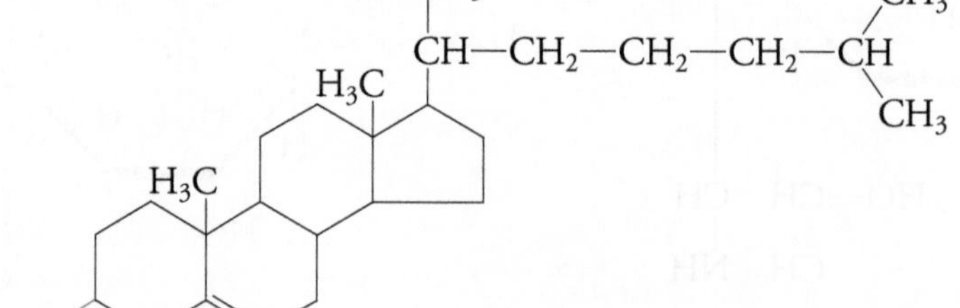

Cholesterol

Cholesterol is an important component of membranes and is also a metabolic precursor of steroid hormones such as estrogen and testosterone.

Cholesterol is one of many types of terpenoids, or **isoprenoids,** lipids that are constructed from 5-carbon units with the same carbon skeleton as isoprene.

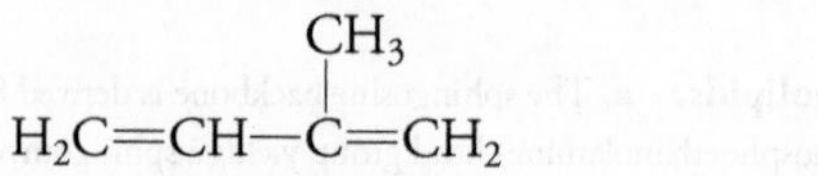

$$H_2C=CH-\overset{\overset{\textstyle CH_3}{|}}{C}=CH_2$$

Isoprene

For example, the isoprenoid ubiquinone is a compound that is reversibly reduced and oxidized in the mitochondrial membrane (it is described further in Section 12.2):

Ubiquinone

The tails of archaeal membrane lipids (Fig. 8.3) are also isoprenoids, as are the molecules known as vitamins A, D, E, and K, which perform a variety of physiological roles not related to membrane structure (**Box 8.B**).

Box 8.B   The Lipid Vitamins A, D, E, and K

The plant kingdom is rich in isoprenoid compounds, which serve as pigments, molecular signals (hormones and pheromones), and defensive agents. During the course of evolution, vertebrate metabolism has co-opted several of these compounds for other purposes. The compounds have become **vitamins,** which are substances that an animal cannot synthesize but must obtain from its food. Vitamins A, D, E, and K are lipids, but many other vitamins are water-soluble. Other than being lipids, vitamins A, D, E, and K have little in common.

The first vitamin to be discovered was vitamin A, or retinol.

Retinol
(vitamin A)

It is derived mainly from plant pigments such as β-carotene (an orange pigment that is present in green vegetables as well as carrots and tomatoes). Retinol is oxidized to retinal, an aldehyde, which functions as a light receptor in the eye. Light causes the retinal to isomerize, triggering an impulse through the optic nerve. A severe deficiency of vitamin A can lead to blindness. The retinol derivative retinoic acid behaves like a hormone by stimulating tissue repair. It is sometimes used to treat severe acne and skin ulcers.

The steroid derivative vitamin D is actually two similar compounds—one (vitamin D$_2$) derived from plants and the other (vitamin D$_3$) from endogenously produced cholesterol.

Vitamin D$_2$

Vitamin D$_3$

Ultraviolet light is required for the formation of vitamins D$_2$ and D$_3$, giving rise to the saying that sunlight makes vitamin D. Two hydroxylation reactions carried out by enzymes in the liver and kidney convert vitamin D to its active form, which stimulates calcium absorption in the intestine. The resulting high concentration of Ca^{2+} in the bloodstream promotes Ca^{2+} deposition in the bones and teeth. Rickets, a vitamin D–deficiency disease characterized by stunted growth and deformed bones, is easily prevented by good nutrition and exposure to sunlight.

α-Tocopherol, or vitamin E,

α-Tocopherol
(vitamin E)

is a highly hydrophobic molecule that is incorporated into cell membranes. The traditional view is that it reacts with free radicals generated during oxidative reactions. Vitamin E activity would thereby help prevent the peroxidation of polyunsaturated fatty acids in membrane lipids. However, compounds that are closely related to α-tocopherol in structure do not exhibit this free radical–scavenging activity, and it has been proposed that the observed antioxidant effect of vitamin E instead stems from its activity as a regulatory molecule that may suppress free radical formation by inhibiting the production or activation of oxidative enzymes. In this respect, vitamin E resembles other lipids that function as signaling molecules.

Vitamin K is named for the Danish word *koagulation*. It participates in the enzymatic carboxylation of glutamate residues in some of the proteins involved in blood coagulation (see Section 6.5). A vitamin K deficiency prevents Glu carboxylation, which inhibits the normal function of the proteins, leading to excessive bleeding. Vitamin K can be obtained from green plants, as phylloquinone,

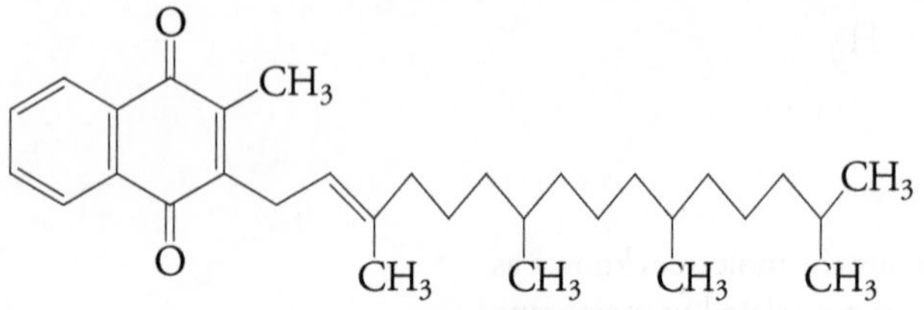

However, about half the daily uptake of the vitamin is supplied by intestinal bacteria.

Because vitamins A, D, E, and K are water-insoluble, they can accumulate in fatty tissues over time. Excessive vitamin D accumulation can lead to kidney stones and abnormal calcification of soft tissues. High levels of vitamin K have few adverse effects, but extremely high levels of vitamin A can produce a host of nonspecific symptoms as well as birth defects. In general, vitamin toxicities are rare and usually result from overconsumption of commercial vitamin supplements rather than from natural causes.

Phylloquinone
(vitamin K)

What are some other functions of lipids that are not used to construct bilayers? Due to their hydrophobicity, some lipids function as waterproofing agents. For example, waxes produced by plants protect the surfaces of leaves and fruits against water loss. Beeswax contains an ester of palmitate and a 30-carbon alcohol that makes this substance extremely water-insoluble.

$$H_3C-(CH_2)_{14}-\overset{\displaystyle O}{\overset{\displaystyle \|}{C}}-O-(CH_2)_{29}-CH_3$$

In humans, derivatives of the C_{20} fatty acid arachidonate are signaling molecules that help regulate blood pressure, pain, and inflammation (Section 10.4). Many plant lipids function as attractants for pollinators or repellants for herbivores. For example, geraniol is produced by many flowering plants (it has a roselike smell). Limonene gives citrus fruits their characteristic odor.

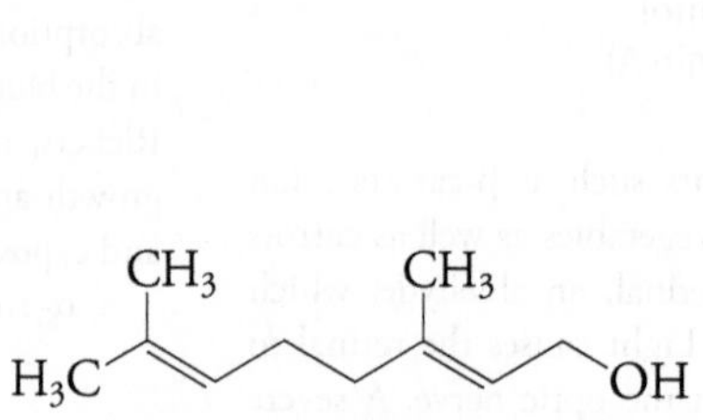

Geraniol

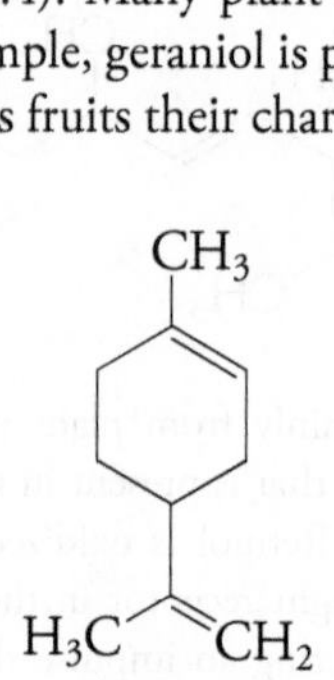

Limonene

Capsaicin, the compound that gives chili peppers their "hot" taste, is an irritant to the digestive tracts of many animals, but it is consumed by humans worldwide.

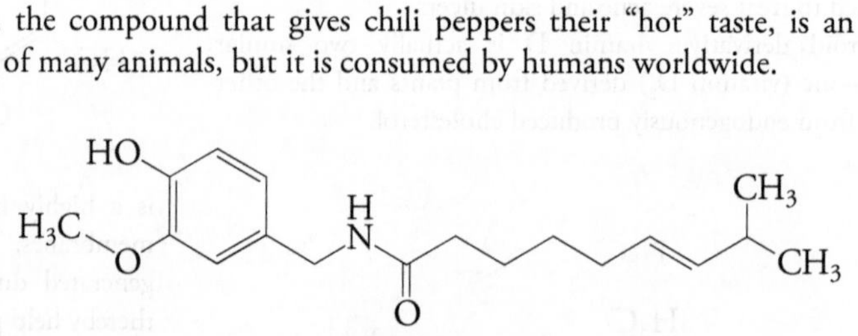

Capsaicin

Its hydrophobicity explains why it cannot be washed away with water. Capsaicin has been used therapeutically as a pain reliever. It appears to activate receptors on neurons that sense both pain and heat; by overwhelming the receptors with a "hot" signal, capsaicin prevents the neurons from receiving pain signals.

Steroid Hormones

Steroid Hormones: Orchestrators of Physiological Signaling and Regulation

Steroid hormones, a subset of bioactive lipids, occupy a crucial position in the intricate web of physiological signaling and regulation within organisms. Derived from cholesterol through a series of enzymatic reactions, these molecules exhibit a characteristic structure comprising four interconnected rings. Their remarkable versatility stems from their ability to engage with specific intracellular receptors, triggering a cascade of events that span the realms of growth, development, metabolism, and reproduction.

Biosynthesis and Diversity of Steroid Hormones

The biosynthesis of steroid hormones is a finely tuned process, characterized by distinct enzymatic steps occurring in specific cellular compartments. Cholesterol, the precursor molecule, serves as the foundation for the creation of diverse steroid hormones. Through enzymatic conversions, cholesterol is transformed into pregnenolone, a precursor that branches into various pathways leading to the synthesis of glucocorticoids, mineralocorticoids, and sex hormones. This orchestrated process ensures the production of an array of steroid hormones, each with specialized functions.

Glucocorticoids: Balancing Metabolism and Immunity

Glucocorticoids, exemplified by cortisol, emerge as essential players in modulating metabolic processes and immune responses. Their molecular structure involves a hydroxyl group at the C17 position and a ketone at C3. Cortisol's interactions with its intracellular receptor modulate gene expression, orchestrating processes vital for glucose homeostasis, lipid metabolism, and anti-inflammatory responses. These hormones empower organisms to withstand stressors by mobilizing energy reserves. Cortisol aids in the mobilization of energy stores during periods of stress or fasting, while its anti-inflammatory actions mitigate immune responses, preventing excessive inflammation that could harm tissues.

Mineralocorticoids: Guardians of Electrolyte Balance

Mineralocorticoids, notably aldosterone, exhibit a molecular framework characterized by a ketone at C3 and a hydrogen at C17. It exerts precise control over electrolyte and water balance, particularly within the renal system. Aldosterone acts on the distal tubules and collecting ducts of the kidneys, promoting sodium reabsorption and potassium excretion. This orchestrated process ensures the maintenance of blood pressure and electrolyte equilibrium, underscoring their pivotal role in cardiovascular and renal homeostasis.

Sex Hormones: Architects of Reproduction and Development

The realm of steroid hormones expands to encompass sex hormones, comprising androgens (e.g., testosterone), estrogens (e.g., estradiol), and progestogens (e.g., progesterone). These hormones exert profound influences on sexual development, secondary sexual characteristics, and the intricate intricacies of the reproductive cycle. Androgens exhibit a 19-carbon structure with a ketone at C3, while estrogens possess an aromatic A ring and a hydroxyl group at C3. Progestogens showcase a ketone at C3 and a methyl at C10. Androgens contribute to the development of male reproductive tissues and behaviors, while estrogens govern female reproductive function and contribute to bone health. Progestogens play a critical role in preparing the uterus for pregnancy and maintaining pregnancy.

Mechanisms of Action and Regulation

Steroid hormones execute their actions through intracellular receptors that function as transcription factors. Upon binding to their ligand, the hormone-receptor complex translocates to the nucleus, where it binds to specific DNA sequences, modulating gene transcription. This transcriptional regulation leads to a cascade of cellular responses that culminate in physiological outcomes. Intriguingly, steroid hormones exhibit negative feedback loops that ensure homeostasis, involving complex interactions between the hypothalamus, pituitary gland, and target organs.

Disruptions and Clinical Implications   Imbalances in steroid hormone levels are implicated in various disorders. Hypersecretion or hyposecretion of glucocorticoids, for instance, can lead to metabolic disorders, immune dysfunction, and stress-related ailments. Disturbances in sex hormone levels are associated with infertility, menstrual irregularities, and even the development of certain cancers. Notably, anabolic steroid misuse has garnered attention for its impact on physical and psychological health.

Structures: A Glimpse into Molecular Diversity

1. Cortisol:

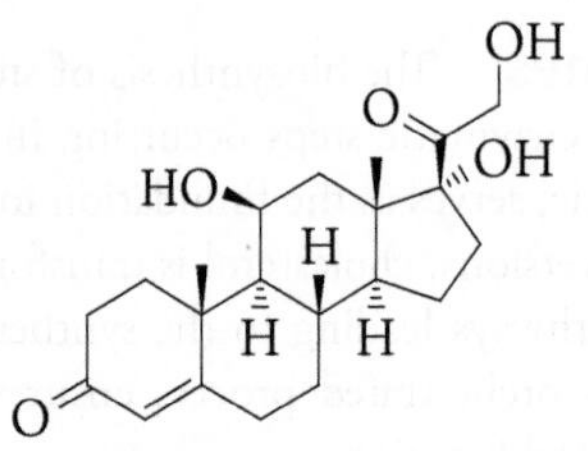

2. Aldosterone:

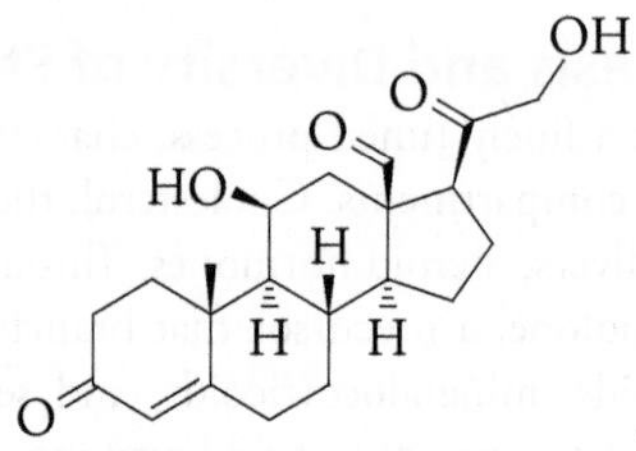

3. Testosterone:

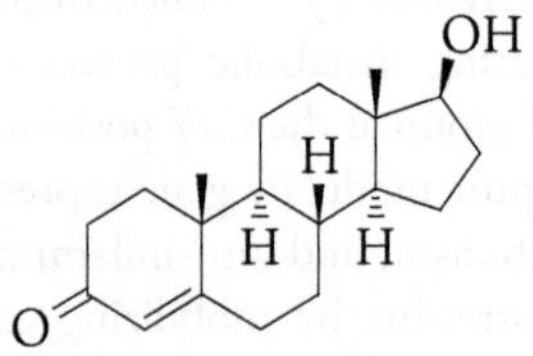

4. Estradiol:

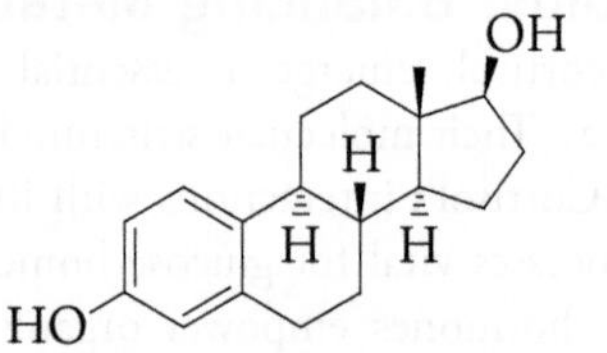

5. Progesterone:

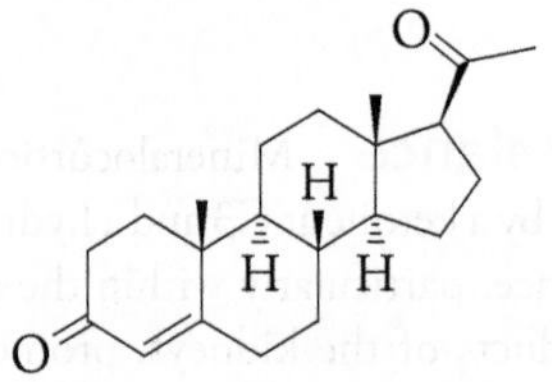

Steroid hormones constitute a dynamic class of molecules that function as master regulators in the orchestra of physiological processes. Their diverse functions, intricacies of regulation, and far-reaching effects underscore their paramount importance in maintaining health and homeostasis. Beyond their foundational roles in biology, steroid hormones continue to unveil novel avenues of research and therapeutic interventions, shining light on the intricate balance that governs life's intricate symphony. The molecular architects of physiological orchestration, exhibit remarkable molecular diversity mirrored in their intricate functions. Their pivotal roles in metabolic regulation, electrolyte balance, sexual development, and reproduction underscore their indispensability. As the intricate nexus of molecular structures, regulatory mechanisms, and clinical implications unfolds, the awe-inspiring symphony of steroid hormones continues to captivate scientists and clinicians alike, revealing the intricacies of life's molecular tapestry.

Concept Check

1. What are the structural features that distinguish fatty acids, triacylglycerols, glycerophospholipids, sphingomyelins, and cholesterol?

2. Explain why lipids cannot form true polymers.

3. For each lipid mentioned in this section, describe how its hydrophobic character determines its location or function.

4. Explain, what are the steroid hormones.

5. Explain the disruptions and clinical implications of steroid hormone.

8.2 The Lipid Bilayer

Describe the physical properties of the lipid bilayer.

- List the ways that membrane lipids can move.
- Relate lipid composition to bilayer fluidity.
- Compare transverse and lateral lipid diffusion.

The fundamental component of a biological membrane is the **lipid bilayer,** *a two-dimensional array of amphipathic molecules whose tails associate with each other, out of contact with water, and whose head groups interact with the aqueous solvent* (**Fig. 8.4**). The beauty of the bilayer as a barrier for biological systems is that it forms spontaneously (without the input of free energy) due to the influence of the hydrophobic effect, which favors the aggregation of nonpolar groups that are energetically costly to individually hydrate. In other words, the lipids do not aggregate because they are attracted to each other; instead they are forced together because this maximizes the entropy of the surrounding water molecules (Section 2.2). In addition, a bilayer is self-sealing and, despite its thinness, it can enclose a relatively vast compartment or an entire cell. Once it has formed, a bilayer is quite stable.

Glycerophospholipids and sphingolipids—with two tails and a large head group—have the appropriate geometry to form bilayers. Although they are amphipathic, fatty acids form spherical particles (micelles, Fig. 2.9) rather than bilayers. Triacylglycerols are almost completely nonpolar and therefore cannot form bilayers. Likewise, pure cholesterol, with only a single polar hydroxyl group, cannot form a bilayer on its own. Nevertheless, molecules of these non-bilayer-forming lipids can be found buried in the hydrophobic region of a membrane, among the acyl chains of other lipids.

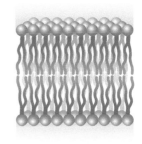

Figure 8.4 A lipid bilayer.

The bilayer is a fluid structure

Naturally occurring bilayers are mixtures of many different lipids. This is one reason why *a lipid bilayer has no clearly defined geometry.* Most lipid bilayers have a total thickness of between 30 and 40 Å (3–4 nm), with a hydrophobic core about 25 to 30 Å (2.5–3 nm) thick. The exact thickness varies according to the lengths of the acyl chains and how they bend and interdigitate. In addition, the head groups of membrane lipids are not of uniform size, and their distance from the membrane center depends on how they nestle in with neighboring head groups. Finally, the lipid bilayer is impossible to describe in precise terms because it is not a static structure. Rather, it is a dynamic assembly: The head groups bob up and down and the hydrocarbon tails of the lipids are in constant rapid motion. At its very center, the bilayer is as fluid as a sample of pure hydrocarbon (**Fig. 8.5**). The bilayer is flexible enough to accommodate changes in cell shape, but its hydrocarbon interior remains intact, as an oily layer between two aqueous compartments.

It is useful to describe the fluidity of a given membrane lipid in terms of its **melting point,** the temperature of transition from an ordered crystalline state to a more fluid state. Note that the melting point does not refer to the conformational shift of just one individual molecule but instead describes the collective behavior of a population of molecules. In the crystalline phase, all the acyl chains in the sample pack together tightly in van der Waals contact. In the fluid phase, the methylene (—CH_2—) groups of the acyl chains in the sample can rotate freely. The melting point of a particular acyl chain depends on its length and degree of saturation. For a saturated acyl chain, the melting point increases with increasing chain length. This is because more free energy (a higher temperature) is required to disrupt the more extensive van der Waals interactions between longer chains. A shorter acyl chain melts at a lower temperature because it has less surface area to make van der Waals contacts. A double bond introduces

Figure 8.5 Simulation of a lipid bilayer. In this model of a dipalmitoyl phosphatidylcholine bilayer, C atoms are gray (except for the terminal carbon of each lipid tail, which is yellow), ester O atoms are red, phosphate groups are green, and choline head groups are magenta. Water molecules on each side of the bilayer are shown as blue spheres.

Question Indicate where the molecules shown in Box 8.B would be located.

a kink into the acyl chain, so an unsaturated acyl chain is less able to pack efficiently against its neighbors:

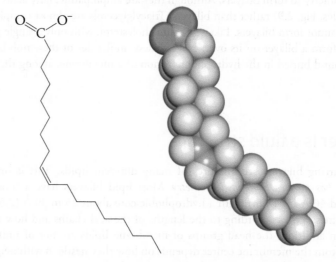

Consequently, the melting point of an acyl chain decreases as the degree of unsaturation increases.

Of course, biological membranes contain more than one type of acyl chain and do not typically experience sudden changes in temperature. However, in a mixed bilayer at constant temperature, *longer acyl chains tend to be less mobile (more crystalline) than shorter acyl chains, and saturated acyl chains are less mobile than unsaturated chains.* Because a fluid membrane is essential for many metabolic processes, organisms endeavor to maintain constant membrane fluidity by adjusting the lipid composition of the bilayer. For example, during adaptation to lower temperatures, an organism may increase its production of lipids with shorter and less saturated acyl chains.

The membranes of most organisms remain fluid over a range of temperatures. This is partly because biological membranes include a variety of different lipids (with different melting points) and do not undergo a sharp transition between liquid and crystalline phases, as a sample of pure lipid would. In eukaryotic membranes, a planar sterol such as cholesterol slips in between the acyl tails of membrane lipids and helps maintain constant membrane fluidity over a range of temperatures through two opposing mechanisms:

1. In a bilayer of mixed lipids, cholesterol's rigid ring system restricts the movements of nearby acyl chains, which prevents the bilayer from becoming too fluid.

2. By inserting between membrane lipids, cholesterol prevents their close packing (that is, their crystallization), which prevents the bilayer from becoming too solid.

The isoprenoid tails of the lipids in archaeal membranes, which lack cholesterol, sometimes include rings of five or six carbons, which appear to modulate bilayer fluidity in the same manner as cholesterol.

Different areas of a naturally occurring membrane may be characterized by different degrees of fluidity. For example, regions known as membrane **rafts** contain tightly packed cholesterol and sphingolipids and have a near-crystalline consistency. Certain proteins appear to associate with rafts, so these structures have functional importance for processes such as transport and signaling. However, the physical characteristics of lipid rafts have been difficult to pin down, and it is possible that such structures may sometimes have only a fleeting existence.

Natural bilayers are asymmetric

The two leaflets of the bilayer in a biological membrane seldom have identical compositions. For example, sphingolipids with carbohydrate head groups occur almost exclusively on the outer leaflet of the animal plasma membrane, facing the extracellular space. The polar head groups of phosphatidylcholine also usually face the cell exterior, whereas phosphatidylserine is usually found in the inner leaflet. In fact, anionic lipids give the inner face of the plasma membrane a distinct negative charge in all eukaryotes.

The distinct compositions of the inner and outer leaflets are preserved by the extremely slow rate at which most membrane lipids undergo **transverse diffusion,** or **flip-flop** (the movement from one leaflet to the other):

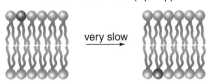

This movement is thermodynamically unfavorable since it would require the passage of a solvated polar head group through the hydrophobic interior of the bilayer. Lipid molecules undergo rapid **lateral diffusion,** that is, movement within one leaflet:

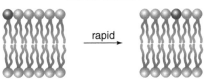

In a membrane bilayer, a lipid changes places with its neighbors as often as 10^7 times per second. Thus, the image of the bilayer in Figure 8.5 represents a bilayer frozen for an instant in time.

Concept Check

1. Explain why glycerophospholipids and sphingolipids form bilayers while fatty acids, triacylglycerols, and cholesterol do not.
2. Draw a diagram of a lipid bilayer and add arrows to indicate all the ways that membrane lipids can move.
3. How do chain length and saturation influence the melting point of a fatty acid?
4. Describe the effects of cholesterol on membrane fluidity.
5. What are the differences between transverse and lateral diffusion?

8.3 Membrane Proteins

KEY CONCEPTS

Explain how proteins associate with membranes.

- Distinguish integral, peripheral, and lipid-linked membrane proteins.
- Describe how an α helix or β barrel spans the membrane.

Biological membranes consist of proteins as well as lipids. On average, a membrane is about 50% protein by weight, but this value varies widely, depending on the source of the membrane. Some bacterial plasma membranes and organelle membranes are as much as three quarters protein. By itself, a lipid bilayer serves mainly as a barrier to the diffusion of polar substances, and virtually all the additional functions of a biological membrane depend on membrane proteins. For example, some membrane proteins sense exterior conditions and communicate them to the cell interior. Other membrane proteins carry out specific metabolic reactions or function as transporters to move substances from one side of the membrane to the other. Membrane proteins fall into different groups, depending on how they are specialized for interaction with the hydrophobic interior of the lipid bilayer (**Fig. 8.6**).

Integral membrane proteins span the bilayer

In the membrane proteins known as **integral** or **intrinsic membrane proteins,** a portion of the structure is fully buried in the lipid bilayer. These proteins are customarily contrasted with **peripheral** or **extrinsic membrane proteins,** which are more loosely associated with the membrane via interactions with lipid head groups or integral membrane proteins (see Fig. 8.6). Except for their weak affiliation with the membrane, peripheral proteins are not notably different from ordinary water-soluble proteins.

Most integral membrane proteins completely span the lipid bilayer, so they are exposed to the hydrophobic interior as well as the aqueous environment on each side of the membrane. The solvent-exposed portions of an integral membrane protein are typical of other proteins: a polar surface surrounding a hydrophobic core. However, the portion of the protein that penetrates the lipid bilayer must have a hydrophobic surface, since the energetic cost of burying a polar protein group (and its solvating water molecules) is too great. Some membrane proteins are embedded only part way in the membrane. No proteins are known to be entirely situated within the lipid bilayer.

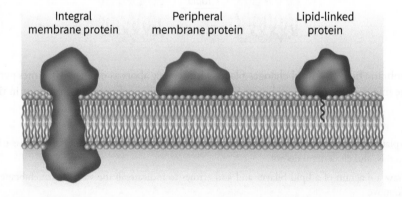

Figure 8.6 Types of membrane proteins. This schematic cross-section of a membrane shows an integral protein spanning the width of the membrane, a peripheral membrane protein associated with the membrane surface, and a lipid-linked protein whose attached hydrophobic tail is incorporated into the lipid bilayer.

Question **Which type of membrane protein would be easiest to separate from the lipid bilayer?**

Integral membrane proteins in signaling and cellular processes

Integral membrane proteins, also known as **transmembrane proteins**, are a diverse group of proteins that are firmly embedded within the lipid bilayer of biological membranes. They constitute a significant portion of the proteome and are essential for various cellular functions. Integral membrane proteins play a pivotal role in cellular signaling and various cellular processes due to their strategic position within the lipid bilayer. These proteins are embedded within the cell membrane, where they mediate critical functions such as signal transduction, transport of ions and molecules, cell adhesion, and structural support.

Structural Features of Integral Membrane Proteins Integral membrane proteins exhibit a variety of structural features, but a common characteristic is their transmembrane domains (TMDs) or hydrophobic regions that span the lipid bilayer. These TMDs typically consist of alpha helices or beta sheets with hydrophobic amino acid residues, allowing them to interact with the hydrophobic core of the lipid bilayer. Integral membrane proteins also possess extracellular and intracellular domains that mediate interactions with other biomolecules, such as ligands, receptors, or cytoskeletal elements.

Roles of Integral Membrane Proteins in Signaling

1. **Receptor Proteins:**

   Many integral membrane proteins function as receptors for extracellular ligands, including hormones, neurotransmitters, and growth factors. The binding of ligands to these receptors initiates a cascade of intracellular events, leading to a cellular response. For example, G protein-coupled receptors (GPCRs) are a class of integral membrane proteins involved in signal transduction.

2. **Ion Channels:**

   Ion channels are integral membrane proteins that form pores in the membrane, allowing the selective passage of ions across the membrane. These channels are crucial for electrical signaling in excitable cells (e.g., neurons and muscle cells) and the regulation of ion homeostasis.

3. **Transporters and Pumps:**

   Integral membrane proteins also serve as transporters and pumps that facilitate the movement of ions, nutrients, and other molecules across the membrane. Examples include the sodium-potassium pump (Na^+/K^+ pump) and glucose transporters.

4. **Cell-Cell Communication:**

   Cell adhesion molecules, such as cadherins and integrins, are integral membrane proteins that play a pivotal role in cell-cell communication and tissue organization. They are involved in processes like cell adhesion, migration, and tissue morphogenesis.

5. **Enzyme Activity:**

   Some integral membrane proteins possess enzymatic activities, participating in intracellular signal transduction pathways. For instance, receptor tyrosine kinases (RTKs) phosphorylate target proteins in response to ligand binding, thereby initiating signal transduction cascades.

Integral Membrane Proteins in Other Cellular Processes Apart from their roles in signaling, integral membrane proteins are integral to several other cellular processes, including *membrane trafficking:* Vesicle transport and fusion processes depend on membrane-associated proteins, and *cell structure:* Proteins like spectrin and dystrophin provide structural support to cell membranes, and *immune response:* Major histocompatibility complex (MHC) proteins present antigens to immune cells.

Conclusion Integral membrane proteins are indispensable components of cellular membranes, with diverse structural and functional roles. Their involvement in cellular signaling, ion transport, adhesion, and other processes underscores their significance in maintaining cellular homeostasis and coordinating complex physiological responses. A thorough understanding of

a. Pro–Glu–Trp–Ile–Trp–Leu–Ala–Leu–
Gly–Thr–Ala–Leu–Met–Gly–Leu–
Gly–Thr–Leu–Tyr–Phe–Leu–Val–
Lys–Gly

b.

Figure 8.7 A membrane-spanning α helix. **a.** A portion of the amino acid sequence from the protein bacteriorhodopsin. **b.** Three-dimensional structure of the same sequence. Polar residues are purple and nonpolar residues are green.

these proteins is crucial for unraveling the intricate mechanisms that govern cellular function and for the development of targeted therapies for various diseases.

An α helix can cross the bilayer

One way for a polypeptide chain to cross a lipid bilayer is by forming an α helix whose side chains are all hydrophobic. The hydrogen-bonding tendencies of the functional groups of the backbone are satisfied through hydrogen bonding in the α helix (see Fig. 4.5). The hydrophobic side chains project outward from the helix to mingle with the acyl chains of the lipids.

To span a 30-Å hydrophobic bilayer core, an α helix must contain at least 20 amino acids. A transmembrane helix is often easy to spot by its sequence: It is rich in highly hydrophobic amino acids such as isoleucine, leucine, valine, and phenylalanine. Polar aromatic groups (tryptophan and tyrosine) and asparagine and glutamine often occur where the helix approaches the more polar lipid head groups. Charged residues such as aspartate, glutamate, lysine, and arginine often mark the point where the polypeptide leaves the membrane and is exposed to the solvent (**Fig. 8.7**).

Many integral membrane proteins contain several membrane-spanning α helices bundled together (**Fig. 8.8**). These α helices interact much like the left-handed coiled coils in keratin (Section 5.3). Some of the helix–helix interactions involve the electrostatic pairing of polar residues, but the surface of the helix bundle—where it contacts the lipid tails—is predominantly hydrophobic.

A transmembrane β sheet forms a barrel

A polypeptide that crossed the membrane as a β strand would leave its hydrogen-bonding backbone groups unsatisfied. However, if several β strands together form a fully hydrogen-bonded β sheet, they can cross the membrane in an energetically favorable way. In order to maximize hydrogen bonding, the β sheet must close up on itself to form a **β barrel.**

The smallest possible β barrel contains eight strands. The exterior surface of the barrel includes a band of hydrophobic side chains. This band is flanked on each side by aromatic side chains, which are more polar and form an interface with the lipid head groups (**Fig. 8.9**). Larger β barrels, containing up to 24 strands, sometimes include a central water-filled passageway that allows small molecules to diffuse from one side of the membrane to the other (Section 9.2).

Figure 8.9 A membrane-spanning β barrel. The eight strands of this *E. coli* protein, known as OmpX, are fully hydrogen-bonded where they span the width of the bilayer. Hydrophobic side chains (green) on the barrel exterior face the bilayer core. Aromatic residues (gold) are located mostly near the lipid head groups.

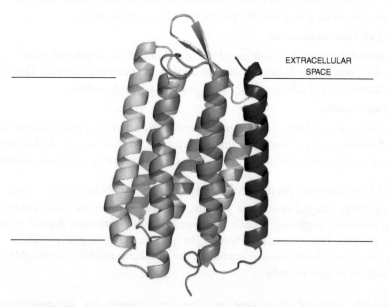

EXTRACELLULAR SPACE

Figure 8.8 Bacteriorhodopsin. This integral membrane protein consists of a bundle of seven membrane-spanning α helices connected by loops that project into the solution on each side of the membrane. The helices are colored in rainbow order from blue (N-terminus) to red (C-terminus). The horizontal lines approximate the outer surfaces of the membrane.

Because the side chains in a β sheet point alternately to each face, some side chains in a β barrel point into the barrel interior, and others face the lipid bilayer. The absence of a discrete stretch of hydrophobic residues, as in a membrane-spanning α helix, makes it difficult to detect membrane-spanning β strands by examining a protein's sequence.

Lipid-linked proteins are anchored in the membrane

A second group of membrane proteins consists of **lipid-linked proteins.** Many of these are otherwise soluble proteins that are anchored in the lipid bilayer by a covalently attached lipid group. A few lipid-linked proteins also contain membrane-spanning polypeptide segments. In some lipid-linked proteins, a fatty acyl group such as a myristoyl residue (from the 14-carbon saturated fatty acid myristate) is attached via an amide bond to the N-terminal glycine residue of a protein (**Fig. 8.10a**). Other proteins contain a palmitoyl group (from the 16-carbon palmitate) attached to the sulfur of a cysteine side chain via a thioester bond (Fig. 8.10b). Palmitoylation, in contrast to myristoylation, is reversible *in vivo*. Consequently, proteins with a myristoyl group are permanently anchored to the membrane, but proteins with a palmitoyl group may become soluble if the acyl group is removed.

Other lipid-linked proteins in eukaryotes are prenylated; that is, they contain a 15- or 20-carbon isoprenoid group linked to a C-terminal cysteine residue via a thioether bond (Fig. 8.10c). The C-terminus is also usually carboxymethylated.

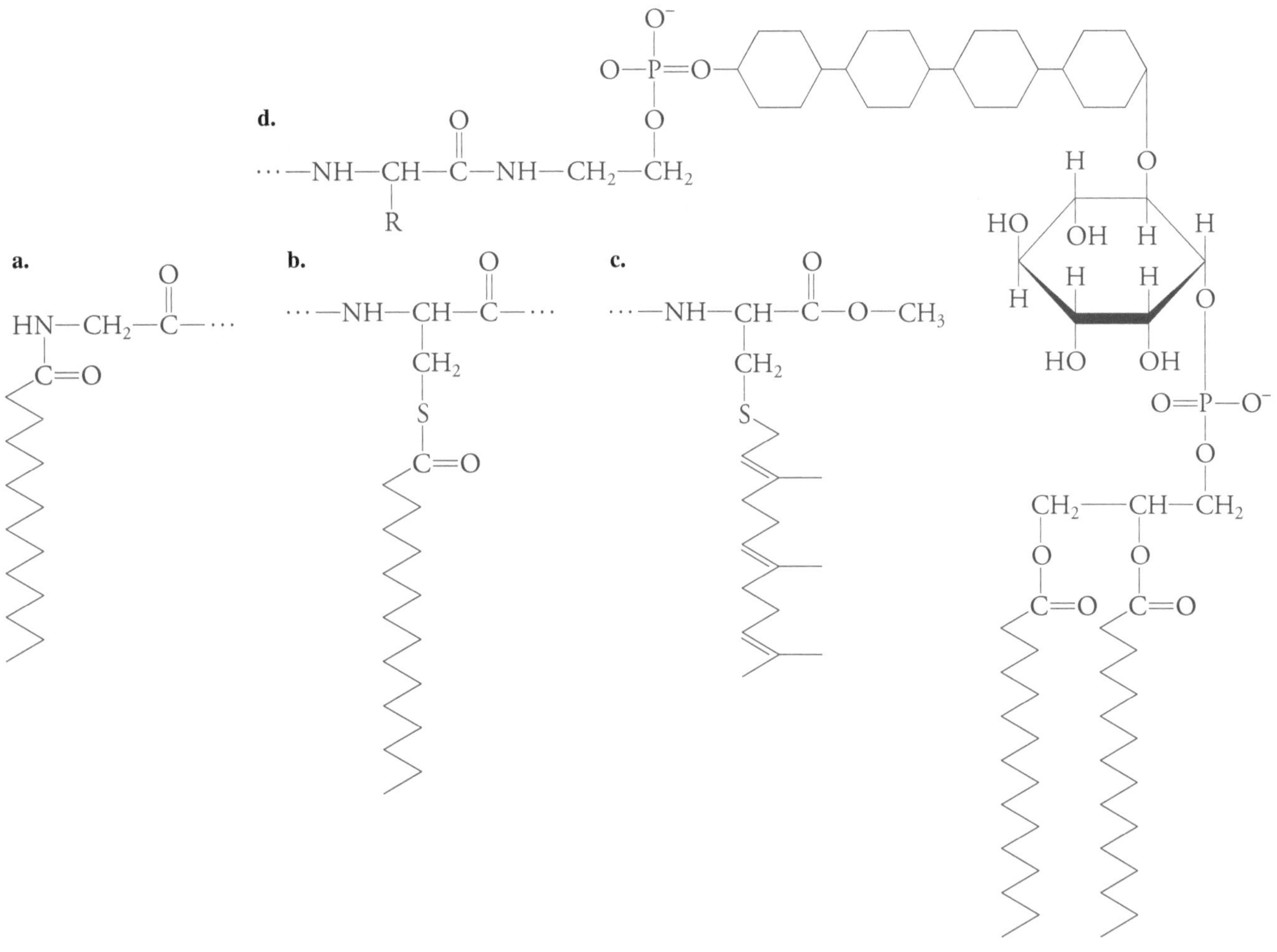

Figure 8.10 Attachment of lipids in some lipid-linked proteins. Proteins are shown in blue and the lipid anchors are in green. Other groups are purple. **a.** Myristoylation. **b.** Palmitoylation. **c.** Prenylation. The lipid anchor is the 15-carbon farnesyl group. **d.** Linkage to a glycosylphosphatidylinositol group. The hexagons represent different monosaccharide residues.

Finally, many eukaryotes, particularly protozoans, contain proteins linked to a lipid–carbohydrate group, known as glycosylphosphatidylinositol, at the C-terminus (Fig. 8.10d). These lipid-linked proteins almost always face the external surface of the cell and are often found in sphingolipid–cholesterol rafts.

Concept Check

1. Describe the sequence requirements for a membrane-spanning α helix and β barrel.
2. List the similarities and differences between integral membrane proteins, peripheral membrane proteins, and lipid-linked proteins.
3. What are some examples of integral membrane proteins that function as transporters and pumps?
4. What is the enzymatic activity of some integral membrane proteins?

8.4 The Fluid Mosaic Model

KEY CONCEPTS

Summarize the features of the fluid mosaic model.

- Explain the limitations on diffusion of membrane proteins.
- Describe how lipids are distributed in a membrane.

Biological membranes consist of both proteins and lipids, although the mixture may not be entirely random. For example, *certain lipids appear to associate specifically with certain proteins,* possibly to stabilize the protein's structure or modulate its function. Because the associations are noncovalent, an integral membrane protein or a lipid-linked protein can diffuse within the plane of the bilayer, albeit more slowly than a membrane lipid. This sort of movement is a key feature of the **fluid mosaic model** of membrane structure described in 1972 by S. Jonathan Singer and Garth Nicolson. According to their model, membrane proteins are like icebergs floating in a lipid sea (**Fig. 8.11**).

Although a membrane is thin and not particularly strong, its hydrophobic core makes it a powerful barrier to the passage of large and polar or charged substances. All cells are enclosed by a plasma membrane, and in eukaryotic cells, organelles are separated from the cytosol by at least one membrane (Section 1.2). So-called membraneless organelles in the cytosol and nucleus are actually separate liquid phases, often containing intrinsically disordered proteins (Section 4.3).

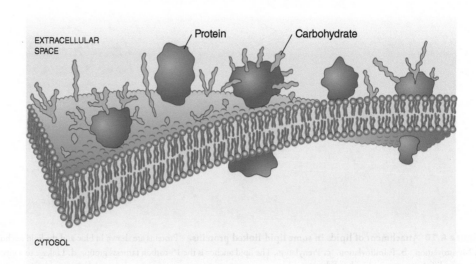

Figure 8.11 The fluid mosaic model of membrane structure. Integral membrane proteins (blue) float in a sea of lipid and can move laterally but cannot undergo transverse movement. The gold structures are the carbohydrate chains of glycolipids and glycoproteins.

Question Identify integral and peripheral membrane proteins in the model.

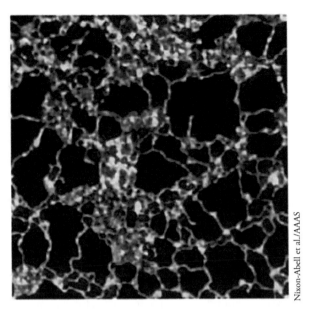

Figure 8.12 The endoplasmic reticulum. High-resolution microscopy of living cells reveals that the endoplasmic reticulum is a dense meshwork of fine tubules.

Lipids and membrane proteins are manufactured in the endoplasmic reticulum, a system of membranes extending throughout the cell. The endoplasmic reticulum is much more dynamic than traditional microscopic methods have revealed; it is a constantly changing network of tubules (**Fig. 8.12**). The tubules make temporary contacts with the plasma membrane and with most organelles so that lipids and proteins can flow between them. The membranes do not actually touch at these contact sites, so proteins are involved in ferrying materials across the gap. Vesicles also bud off from the endoplasmic reticulum and travel through the cytosol to deliver membrane materials to other sites. The ability to form tubes and spherical shapes means that cellular membranes are not necessarily sheetlike but must be able to curve dramatically.

Membrane proteins have a fixed orientation

A given membrane protein has a characteristic orientation; that is, it faces one side of the membrane or the other. For many eukaryotic proteins, membrane orientation is established while the protein is being synthesized by a ribosome attached to the surface of the endoplasmic reticulum (a process described in more detail in Section 22.4). Some proteins are embedded in a lipid bilayer later on. In either case, once a protein has reached its mature configuration, it does not undergo flip-flop because this would require the passage of large polar polypeptide regions through the hydrophobic bilayer core. However, lateral movement is still possible.

Over the years, the fluid mosaic model has remained generally valid, although it has been refined. For example, *many membrane proteins do not diffuse as freely as first imagined*. Their movements are hindered to some degree according to whether they interact with other membrane proteins or with cytoskeletal elements that lie just beneath the membrane. Thus, a given membrane protein may be virtually immobile (if it is firmly attached to the cytoskeleton), mobile within a small area (if it is confined within a space defined by other membrane and cytoskeletal proteins), or fully free to diffuse (**Fig. 8.13**). The presence of lipid rafts, another feature not described in the original fluid mosaic model, may further define the boundaries for a membrane protein.

The lipid bilayer is a viscous environment for a protein: A transmembrane protein moves about a hundred times more slowly than it would in solution. Studies of proteins in artificial bilayers containing phospholipids with different hydrocarbon chain lengths indicate that proteins diffuse more rapidly through thinner bilayers.

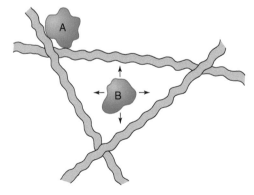

Figure 8.13 Limitations on the mobility of membrane proteins. A protein (labeled A) that interacts tightly with the underlying cytoskeleton appears to be immobile. Another protein (B) can diffuse within a space defined by cytoskeletal proteins. Some proteins appear to diffuse throughout the membrane with no constraints.

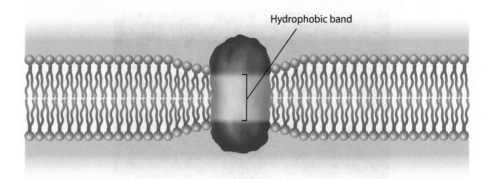

Hydrophobic band

Figure 8.14 Accelerated lateral diffusion by hydrophobic mismatch   A protein whose hydrophobic surface is shorter than the typical height of the lipid bilayer distorts nearby lipids, facilitating its diffusion.

This makes sense, because the shorter lipids form fewer van der Waals contacts with the protein surface. However, cells do not appear to adjust bilayer thickness just so that some proteins can move around more quickly. Curiously, at least some integral membrane proteins have a lipid-exposed hydrophobic band that is shorter than the 25–30 Å (2.5–3 nm) height that is typical for membrane-spanning proteins. Proteins with a shorter band cause neighboring lipids to reorient. This "hydrophobic mismatch" creates local thinning of the membrane, boosting the protein's rate of lateral diffusion (**Fig. 8.14**).

Like membrane lipids, membrane proteins are distributed asymmetrically between the two leaflets. For example, most lipid-linked proteins face the cell interior (glycosylphosphatidylinositol-linked proteins are an exception). The exterior face of the membrane in vertebrate cells is rich in carbohydrate-bearing glycolipids (such as cerebrosides and gangliosides) and **glycoproteins.** The oligosaccharide chains (polymers of monosaccharide residues) that are covalently attached to membrane lipids and proteins shroud the cell in a fuzzy coat (see Fig. 8.11). When fully solvated, the highly hydrophilic carbohydrates tend to occupy a large volume.

As we will see in Chapter 11, monosaccharide residues can be linked to each other in different ways and in potentially unlimited sequences. This diversity, present in both glycolipids and glycoproteins, is a form of biological information. For example, the well-known ABO blood group system is based on differences in the composition of the carbohydrate components of glycolipids and glycoproteins on red blood cells (discussed in Box 11.C). Many other cells appear to recognize each other through mutual interactions between membrane proteins.

Lipid asymmetry is maintained by enzymes

In eukaryotes, glycerophospholipids are synthesized on the cytoplasmic side of the endoplasmic reticulum. Left on their own, the lipids would only slowly distribute themselves evenly between the two leaflets of the bilayer, since the rate of transverse diffusion is low (Section 8.2). In cells, a type of **translocase**, nicknamed **scramblase**, catalyzes the bidirectional movement of the lipids between the two leaflets (**Fig. 8.15**). This prevents the cytoplasmic leaflet from expanding more rapidly than the other leaflet during membrane biogenesis. Scramblase-catalyzed equilibration is a process that maximizes the entropy of the system, so it does not require the input of free energy.

Once new membrane material has been synthesized, cells generate and maintain non-equilibrium distributions of certain lipids with the assistance of other translocases. A **flippase** catalyzes the movement of a lipid from the noncytoplasmic leaflet (the leaflet facing the endoplasmic reticulum lumen) to the cytoplasmic side. A **floppase** catalyzes movement in the opposite direction. Translocating a membrane lipid to the leaflet where it is more abundant is a free energy–requiring process, so the action of flippases and floppases involves ATP.

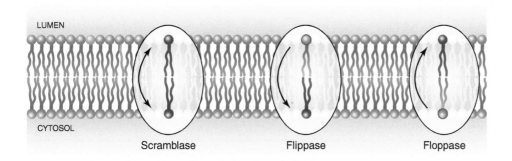

Figure 8.15 Action of translocases. A scramblase equilibrates the distribution of lipids. A flippase moves lipids to the cytoplasmic leaflet, and a floppase moves lipids to the other leaflet; both of these enzymes require ATP.

Concept Check

1. Describe the fluid mosaic model of membrane structure.
2. List some factors that limit membrane protein mobility.
3. Explain why glycoproteins and glycolipids face the cell exterior.
4. Distinguish the three types of lipid translocases.

SUMMARY

8.1 Lipids

- Lipids are largely hydrophobic molecules. Fatty acids can be esterified to form triacylglycerols.

- Glycerophospholipids contain two fatty acyl groups attached to a glycerol backbone that bears a phosphate derivative head group. Sphingomyelins are functionally similar but lack a glycerol backbone. Cholesterol, archaeal membrane lipids, and some other lipids are isoprenoids.

8.2 The Lipid Bilayer

- Lipid bilayers are dynamic structures. Their fluidity depends on the length and degree of saturation of their fatty acyl groups. Shorter and less saturated chains are more fluid. Cholesterol helps maintain membrane fluidity over a range of temperatures.

- Membrane lipids can freely diffuse laterally but undergo transverse diffusion very slowly. Membranes may contain crystalline rafts composed of cholesterol and sphingolipids.

8.3 Membrane Proteins

- An integral membrane protein spans the lipid bilayer as one or a bundle of α helices or as a β barrel. Some membrane proteins are anchored in the bilayer by a covalently linked lipid group.

8.4 The Fluid Mosaic Model

- According to the fluid mosaic model, membrane proteins diffuse within the plane of the bilayer. The mobility of proteins may be limited by their interaction with cytoskeletal proteins. Translocases move lipids from one leaflet to the other.

KEY TERMS

lipid
fatty acid
saturated fatty acid
unsaturated fatty acid
omega-3 fatty acid
triacylglycerol (triglyceride)

acyl group
glycerophospholipid
amphipathic
phospholipase
sphingolipid
sphingomyelin
glycolipid

isoprenoid
steroid hormone
vitamin
lipid bilayer
melting point
raft

transverse diffusion (flip-flop)
lateral diffusion
integral (intrinsic) protein
peripheral (extrinsic) protein

β barrel
lipid-linked protein
fluid mosaic model
glycoprotein
translocase
scramblase

flippase
floppase

BIOINFORMATICS

Brief Bioinformatics Exercises

8.1 Drawing and Naming Fatty Acids

8.2 Properties of Membrane Proteins

PROBLEMS

8.1 Lipids

1. The structures of fatty acids can be represented in a shorthand form consisting of two numbers separated by a colon. The first number is the number of carbons; the second number is the number of double bonds. For example, palmitate is represented by the shorthand 16:0. For unsaturated fatty acids, the quantity *n-x* is used, where *n* is the total number of carbons and *x* is the last double-bonded carbon from the methyl end. Unless indicated otherwise, it is assumed that double bonds are *cis* and that one methylene group separates each double bond. For example, oleate would be represented by the shorthand 18:1*n*-9 (see Table 8.1). Using the shorthand form as a guide, draw the structures of the following fatty acids: **a.** Arachidic acid, 20:0, **b.** Oleic acid, 18:1*n*-9, **c.** γ-Linolenic acid, 18:3*n*-6, **d.** Nervonate, 24:1*n*-9.

2. Fish oils contain the fatty acids EPA and DHA. People who eat at least two fish meals a week have a lower incidence of cardiovascular disease because of the positive physiological effects of these lipids (see Box 8.A). Use the shorthand form described in Problem 1 to draw the structures of **a.** EPA (eicosapentaenoate, 20:5*n*-3) and **b.** DHA (docosahexaenoate 22:6*n*-3).

3. Some species of plants contain desaturase enzymes capable of producing fatty acids not found in animals. An example of an unusual fatty acid is sciadonate, which is designated as all-*cis*-$\Delta^{5,11,14}$-eicosatrienoate. In this shorthand nomenclature, the superscripts refer to the positions of the double bonds beginning at the carboxyl end. This nomenclature is less common than the scheme described in Problem 1 but is sometimes used when the positions of the double bonds do not conform to the pattern described in Problem 1. Using this form of shorthand as a guide, draw the structure of sciadonate.

4. Several hundred unusual fatty acids have been found in plants. Some of these fatty acids are unusually short or long; some have double bonds in unexpected positions (see Problem 3); others have additional functional groups. These fatty acids are important as starting materials in the industrial synthesis of lubricants and polymers. Using the shorthand notation described in Problem 3, draw the structures of the following plant fatty acids: **a.** erucic acid (*cis*-Δ^{13}-docosenoic acid), **b.** calendic acid (*trans, trans, cis*-$\Delta^{8,10,12}$-octadecatrienoic acid), **c.** ricinoleic acid (12-hydroxy-*cis*-Δ^9-octadecenoic acid).

5. A class of fatty acids called demospongic fatty acids was so named because of their occurrence in *Demospongia* sponges; however, it has since been discovered that these fatty acids have a wider distribution. Use the shorthand method described in Problem 3 as a guide to draw the structures of the following lipids found in marine mollusks: **a.** *cis, cis*-$\Delta^{5,9}$-tetracosadienoate, **b.** all *cis*-$\Delta^{5,9,15,18}$-tetracosatetraenoate.

6. *Trans* fatty acids occur naturally in beef and milk products but are also produced when oils undergo partial hydrogenation to convert the liquid oil into a semi-solid fat. Draw the structure (see Problem 3) of elaidic acid (*trans*-Δ^9-octadecenoic acid).

7. Platinum-induced fatty acids are produced by stem cells in response to the administration of cis-platin, a platinum-containing drug used to treat cancer. One of these fatty acids, 16:4*n*-3, contributes to the cells' resistance to the chemotherapy regimen. Draw the structure of this fatty acid.

8. Enynoic fatty acids containing both double and triple bonds have been isolated from seed oils. The name is derived from ethene (double bond) and ethyne (triple bond). The systematic name (see Table 8.1) of one enynoic fatty acid is *cis*-9-octadecen-12-ynoic acid. Draw the structure of this fatty acid.

9. Triacylglycerols are named by specifying the identities of the fatty acids esterified to each of the three carbons on the glycerol backbone. Draw the structure of another common lipid found in the *Cuphea* seed (see Problem 12): 1,3-di-lauroyl-2-myristoyl-glycerol (see Table 8.1).

10. The nonfood oilseed plant *Camelina sativa* was genetically modified to produce acetyl triacylglycerols, lipids that can be used to produce biofuels. Draw the structure of 1,2-dioleoyl-3-acetyl-glycerol.

11. Human milk differs from cow's milk in that 70% of the fatty acids on the C2 position of the triacylglycerols consist of palmitate (16:0). Draw the structure of 1-oleoyl-2-palmitoyl-3-linoleoyl-glycerol (see Table 8.1).

12. The surface of the eye is wet by tears, and to minimize evaporation, the tear layer is coated with lipids. One of these lipids is an ester-containing a 50-carbon *O*-acyl-ω-hydroxy fatty acid abbreviated as 18:1(*n*-9, *cis*)/ω-*O*-32:1(*n*-9, *cis*). This complex lipid consists of two fatty acids that form an ester linkage via the carboxylate group of an oleic acid and a hydroxyl group on the ω carbon of the 32:1*n*-9 fatty acid. Draw the structure of this lipid.

13. During digestion, pancreatic lipase catalyzes the hydrolysis of the fatty acids from the 1 and 3 positions of triacylglycerols. **a.** Draw the products of lipase-catalyzed digestion of tripalmitin (see Problem 11). **b.** What type of structure is formed by the products? (*Hint*: See Section 2.2).

14. After ingestion of fat, the vertebrate intestine produces an appetite-regulating molecule that contains oleic acid and ethanolamine linked by an amide bond. Draw its structure.

15. Soap (a sodium salt of a fatty acid) is made by treating fat (triacylglycerols) with lye (an aqueous solution of sodium hydroxide) in a process called saponification. Draw the products of the saponification of trilaurin (see Problem 12).

16. The label on a can of oven cleaner lists sodium hydroxide as one of its ingredients. Consult your answer to Problem 19 and explain how oven cleaners work.

17. Marine organisms are good sources of unusual fatty acids that have potential as therapeutic agents. A monoacylglycerol isolated from a sponge contains 10-methyl-9-*cis*-octadecenoic acid esterified to C1 of glycerol. Draw its structure.

18. When certain nutrients are limiting, some marine phytoplankton can change their membrane lipid composition, producing substitute lipids such as sulfoquinovosyldiacylglycerol (SQDG, below). **a.** Is SQDG more likely to substitute for phosphatidylethanolamine or phosphatidylglycerol? **b.** What element must be in short supply to induce the organism to increase its synthesis of SQDG?

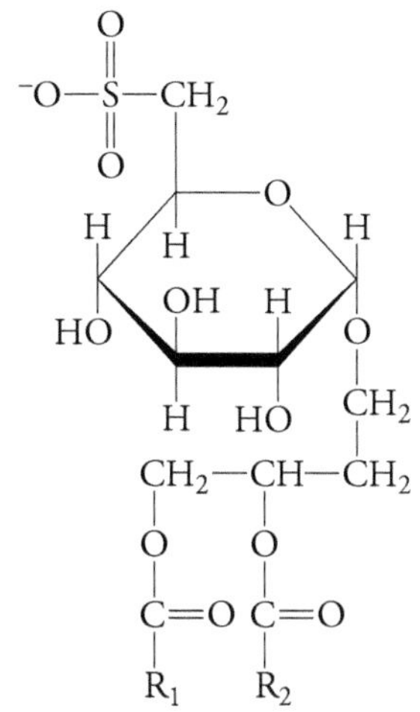

Sulfoquinovosyldiacylglycerol
(SQDG)

19. Which of the glycerophospholipids shown in Section 8.1 have hydrogen-bonding head groups?

20. Which of the glycerophospholipids shown in Section 8.1 are charged? Which are neutral?

21. What are some of the physiological effects of eicosanoids in the human body?

22. Phosphatidylinositols (PI) are glycerophospholipids important in cell signaling. **a.** Draw the structure of a PI, given the structure of *myo*-inositol below. The hydroxyl group involved in the bond between the inositol and the phosphate group is circled. **b.** Indicate the polar and nonpolar domains of the molecule.

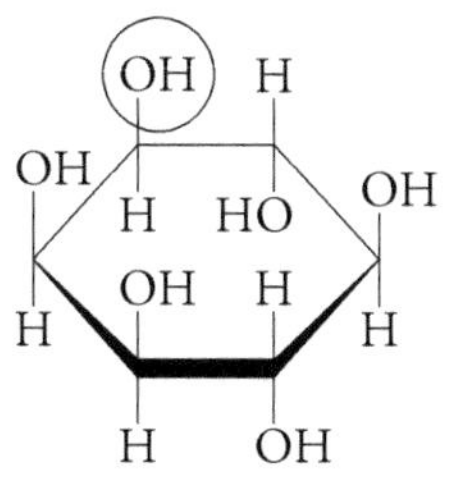

myo-Inositol

23. Some signaling pathways generate signaling molecules derived from phosphatidylinositol that has been phosphorylated at multiple sites. How many additional phosphate groups can potentially be attached to phosphatidylinositol?

24. Dipalmitoylphosphatidylcholine (DPPC) is the major lipid of lung surfactant, a protein–lipid mixture essential for pulmonary function. Surfactant production in the developing fetus is low until just before birth, so infants may develop respiratory difficulties if born prematurely. Draw the structure of DPPC.

25. An unusual sphingosine variant has recently been isolated from the nerve of the squid *Loligo pealeii*. Its systematic name is 2-amino-9-methyl-4,8,10-octadecatriene-1,3-diol. Draw the structure of this sphingosine variant.

26. Complex lipids in mammalian skin serve as a waterproof layer. One of these lipids is a glucocerebroside in which the amide-linked acyl group has 28 carbons and an ω-hydroxyl group to which linoleate is esterified. **a.** Draw the structure of this lipid. **b.** The lipid undergoes hydrolysis to remove the glucose and linoleate groups, followed by linkage of the ω-hydroxyl group to the side chain of a protein Glu residue. Draw the structure of the protein–lipid complex.

27. Very long chain acyl ceramides, in addition to the glucocerebrosides described in Problem 31, help prevent dehydration of mammalian skin. Draw the structure that results from the addition of an amide-linked saturated C_{30} fatty acid to sphingosine, followed by ω-hydroxylation of the fatty acid and its esterification with linoleate.

28. The points of attack of several phospholipases are indicated in Figure 8.1. Draw the products of the following reactions: **a.** phosphatidylserine + phospholipase A_1, **b.** phosphatidylcholine + phospholipase C, **c.** phosphatidylglycerol + phospholipase D.

29. Archaeal lipids (see Fig. 8.3) are found in thermophiles, bacteria that thrive at extremely high temperatures. Why are archaeal lipids more stable than glycerophospholipids at high temperatures?

30. What are the three major classes of eicosanoids and which enzymes are responsible for their synthesis?

31. In a nutrition study, volunteers consumed salads with and without added avocado (a rich source of monounsaturated lipids). Blood samples drawn from the volunteers showed a dramatic increase in β-carotene (an orange pigment found in plants that is processed by cells to yield vitamin A) following the consumption of the salad containing avocado. Explain these findings.

32. Why does the consumption of excessive amounts of vitamins D and A cause adverse health effects whereas consumption of vitamin C (see Box 5.B) well in excess of the recommended daily allowance generally does not lead to toxicity?

33. Bacterial infection stimulates nearby host cells to increase production of an enzyme that oxidizes the aldehyde group of retinal (see Box 8.B). Identify the product of the reaction.

34. Because it stimulates the activity of immune system cells, retinoic acid has been proposed as a treatment for infections. However, administering retinoic acid can also exacerbate conditions such as arthritis and other inflammatory diseases. Explain.

35. a. Does vitamin D fit the definition of a vitamin as a substance that an organism requires but cannot synthesize? **b.** Vitamin D deficiency is hypothesized to contribute to the development of multiple sclerosis because the prevalence of this autoimmune disease increases with increasing latitude. Explain.

36. Inhaling vitamin E acetate, a more stable artificial form of α-tocopherol (see Box 8.B), may lead to lung injury in some e-cigarette users. Draw the structure of this molecule.

37. Explain the following: **a.** Long-term use of antibiotics may lead to a vitamin K deficiency. **b.** Obese individuals require larger amounts of vitamins A, D, E, and K in their diets.

8.2 The Lipid Bilayer

38. Classify the following molecules as polar, nonpolar, or amphipathic:

a. $CH_3CH_2(CH=CHCH_2)_3(CH_2)_7COO^-$

b.

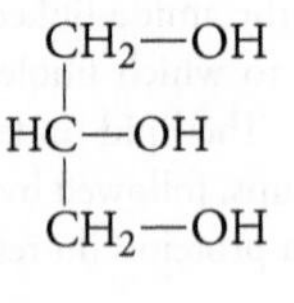

$$CH_2-OH$$
$$HC-OH$$
$$CH_2-OH$$

c.

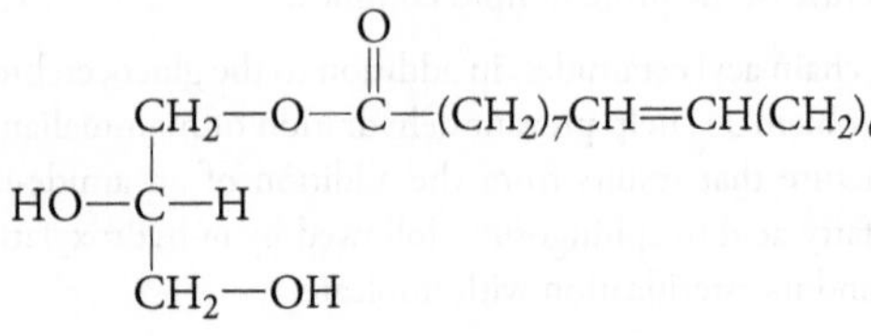

$$CH_2-O-\overset{O}{\overset{\|}{C}}-(CH_2)_7CH=CH(CH_2)_6CH_3$$
$$HO-\overset{|}{C}-H$$
$$CH_2-OH$$

d.

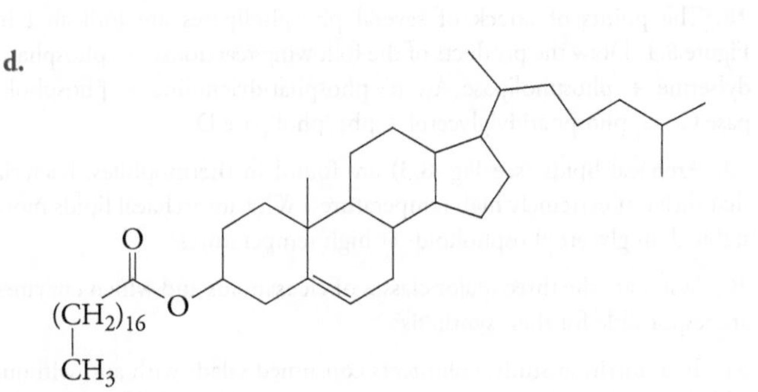

e.

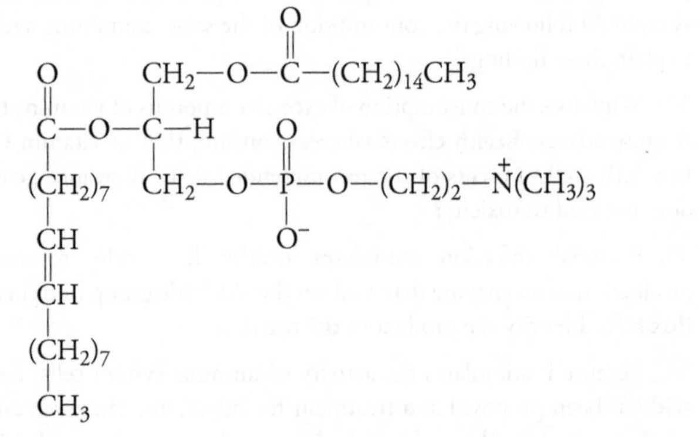

39. Which molecules in Problem 47 can form bilayers? For the molecules that do not form bilayers, explain why not.

40. The lipid shown below is a plasmalogen. **a.** How does it differ from a glycerophospholipid? **b.** Would the presence of this lipid have a dramatic effect on a bilayer that contains only phosphatidylcholine?

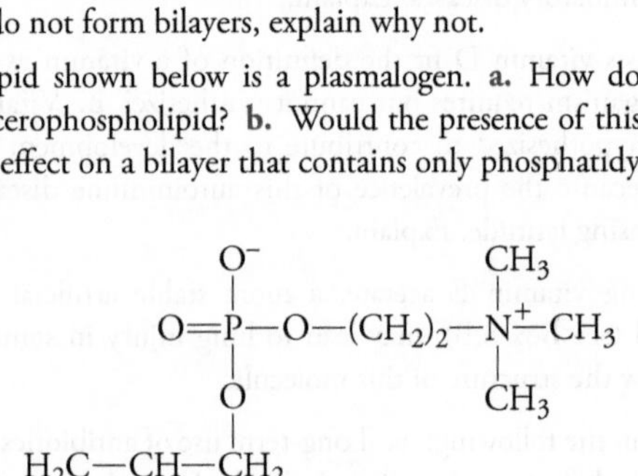

A plasmalogen

41. Use a simple diagram to show why bilayer curvature would be affected by replacing glycerophospholipids bearing two saturated acyl chains with ones bearing two highly unsaturated acyl chains.

42. Why can't triacylglycerols form a lipid bilayer?

43. Red blood cells lyse (break apart) when treated with phospholipase A₁, an enzyme found in the venom of many poisonous insects (see Fig. 8.1 and Solution 34a). Why does treatment with the enzyme result in the destruction of the red blood cell membrane?

44. The melting points of some common saturated and unsaturated fatty acids are shown in the table. What factors influence a fatty acid's melting point?

Fatty acid	Melting point (°C)
Laurate (12:0)	44.2
Linoleate (18:2)	–9
Linolenate (18:3)	–17
Myristate (14:0)	52
Oleate (18:1)	13.4
Palmitate (16:0)	63.1
Stearate (18:0)	69.1

45. a. How do the melting temperatures of triacylglycerols from *Cuphea* seeds (see Problems 12 and 13) compare with seed triacylglycerols composed of saturated long-chain fatty acids? **b.** Compare the melting points of the *Cuphea* seed triacylglycerols to triacylglycerols composed of unsaturated long-chain fatty acids. **c.** Plant scientists noted that *Cuphea* seeds lose viability when stored frozen in seed banks and require a heat pulse at 45°C prior to planting. Explain why this is the case.

46. How does the melting point of elaidic acid (see Problem 7) compare with the melting point of oleic acid (see Table 8.1 and Problem 53)?

47. Rank the melting points of the following fatty acids: **a.** *cis*-oleate (18:1), **b.** *trans*-oleate (18:1), **c.** linoleate (18:2).

48. Would you expect the melting point of 9-octadecynoic acid (an 18-carbon fatty acid containing a triple bond, see Problem 10) to be closer to that of stearic acid or oleic acid (see Table 8.1)? Explain.

49. A trade publication states that acetyl triacylglycerols (see Problem 14) have lower viscosity and better cold-temperature properties than most vegetable oils, and as such, perform well as biofuels and biodegradable lubricants. Explain why.

50. The triacylglycerols of animals tend to be solids (fats), whereas the triacylglycerols of plants tend to be liquids (oils) at room temperature. What can you conclude about the nature of the fatty acyl chains in animal and plant triacylglycerols?

51. Peanut oil contains a high percentage of monounsaturated triacylglycerols (having acyl chains with only one double bond), whereas vegetable oil contains a higher percentage of polyunsaturated triacylglycerols (having acyl chains with more than one double bond). A bottle of peanut oil and a bottle of vegetable oil are stored in a pantry with an outside wall. During a cold spell, the peanut oil freezes but the vegetable oil remains liquid. Explain why.

52. Membrane lipids in tissue samples obtained from different parts of the leg of a reindeer show different fatty acid compositions. The proportion of unsaturated fatty acyl chains increases from the top of the leg to the hoof. Provide an explanation for this observation.

53. Phytol is an alcohol produced from chlorophyll that becomes part of the diet of mammals consuming plants. Phytol is converted to phytanic acid (shown below) in a three-step process, then oxidized to obtain metabolic energy. In individuals with a defect in one of the enzymes of the oxidative pathway, phytanic acid accumulates in the membranes of nerve cells and impairs neurological functions. How does the presence of phytanic acid affect nerve cell membrane fluidity?

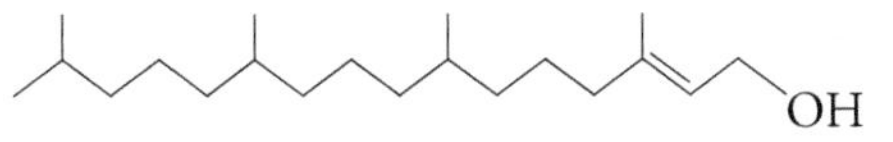

Phytol (3,7,11,15-Tetramethyl-2-hexadecen-1-ol)

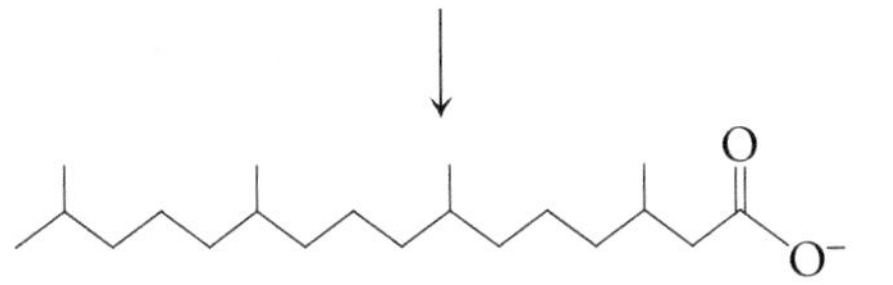

Phytanic acid (3,7,11,15-Tetramethylhexadecanoic acid)

54. Bacteria of the genus *Lactobacillus* colonize the human digestive tract and are considered "friendly" bacteria that are often used to treat digestive disorders. These bacteria produce lactobacillic acid, a 19-carbon fatty acid containing a cyclopropane ring. Is the melting point of this fatty acid closer to the melting point of stearate (18:0) or oleate (18:1)? Rank the melting points of these three fatty acids.

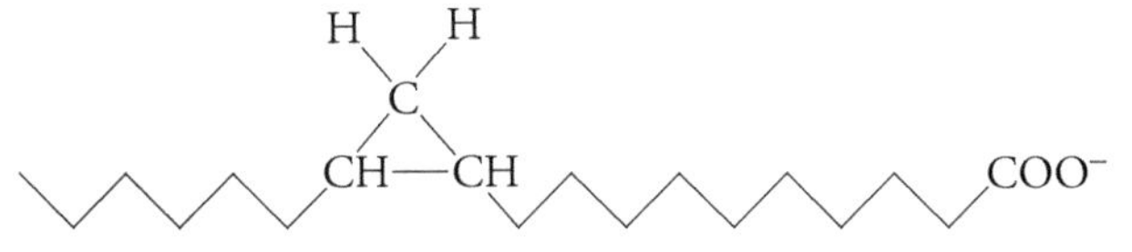

55. Bacteria are typically grown in the laboratory at a temperature of 37°C. What happens to the membrane lipid composition if the temperature is increased to 42°C?

56. Plants can synthesize trienoic acids (fatty acids with three double bonds) by introducing another double bond into a dienoic acid. Would you expect plants growing at higher temperatures to convert more of their dienoic acids into trienoic acids?

57. A membrane consisting only of phospholipids undergoes a sharp transition from the crystalline form to the fluid form as it is heated. However, a membrane containing 80% phospholipid and 20% cholesterol undergoes a more gradual change from crystalline to fluid form when heated over the same temperature range. Explain why.

58. *Chorispora bungeana* is a plant that is well adapted to growth at freezing temperatures. Plants grown at –4°C have a greater percentage of 18:3 fatty acids compared to control plants grown at 25°C. The increase in 18:3 fatty acids was accompanied by a decrease in 18:0, 18:1, and 18:2 fatty acids. Propose a hypothesis consistent with these data.

59. Plants are able to change their lipid compositions to adapt to seasonal changes in temperature, but a study confirmed that alpine and desert plants, which are exposed to wide fluctuations in temperature over a 24-hour period, do not vary their lipid compositions with changes in temperatures. Provide an explanation for this observation.

8.3 Membrane Proteins

60. Purification of transmembrane proteins requires the addition of detergents to buffers in order to solubilize the proteins. **a.** Why would transmembrane proteins be insoluble in the absence of detergent? **b.** Draw a schematic diagram that shows how the detergent sodium dodecyl sulfate interacts with a transmembrane protein.

61. Cytochrome *c*, a protein of the electron transport chain in the inner mitochondrial membrane, can be removed by relatively mild means, such as extraction with salt solution. In contrast, cytochrome oxidase from the same source can be removed only by extraction into detergent

solutions or organic solvents. What kind of membrane proteins are cytochrome *c* and cytochrome oxidase? Explain. Draw a schematic diagram that shows how each protein is positioned in the membrane.

62. Draw the structures of the lipid-protein linkages described: **a.** A protein involved in the Wnt signaling pathway is modified by the attachment of palmitoleate (see Table 8.1) to a Ser residue. Blocking this modification may be an effective cancer treatment. **b.** Ghrelin, an appetite-stimulating factor, requires the attachment of an octanoate residue to a Ser residue for full activity. Blocking this attachment may be an effective strategy for treating obesity.

63. The glycosylphosphatidylinositol (GPI) group shown in Figure 8.10d is susceptible to hydrolysis by phospholipase C (see Fig. 8.1) *in vitro*. If this same reaction were to occur *in vivo*, what would be the result?

64. Glycophorin A is a 131-residue integral membrane protein that includes one bilayer-spanning segment. Identify that segment in the glycophorin A amino acid sequence (which uses one-letter abbreviations):

LSTTEVAMHTTTSSSVSKSYISSQTNDTHKRDTYAATPRAHEV-SEISVRTVYPPEEETGERVQLAHHFSEPEITLIIFGVMAGVIG-TILLISYGIRRLIKKSPSDVKPLPSPDTDVPLSSVEIENPETSDQ

65. Proteins that form a transmembrane β barrel always have an even number of β strands. **a.** Explain why. **b.** Why are the strands antiparallel? **c.** Could some of them possibly be parallel?

66. Peptide hormones must bind to receptors on the extracellular surface of their target cells before their effects are communicated to the cell interior. In contrast, receptors for steroid hormones such as estrogen are intracellular proteins. Why is this possible?

67. Melittin, a 26-amino acid peptide, is known to associate with membranes. The presence of a tryptophan residue in the peptide allows fluorescence spectroscopy to be used to monitor the structure assumed by melittin when it is membrane-associated. Artificial membranes were prepared using phosphatidylcholine (PC) esterified with palmitate (16:0) at position 1. Lipids at position 2 were varied. Melittin was conformationally restricted when associated with PC containing oleate (18:1) at position 2. However, when the lipids contained arachidonate (20:4) at position 2, the melittin peptide was less conformationally restricted. Propose a hypothesis consistent with the observed results.

68. In an experiment, intact cells are exposed to trypsin. The cells are then lysed using a detergent, and the preparation is analyzed by SDS-PAGE (see Section 4.6). Control cells are not exposed to trypsin. Describe how trypsin treatment affects the electrophoretic behavior of **a.** an integral membrane protein with an extracellular domain, **b.** a cytosolic peripheral membrane protein, and **c.** an integral membrane protein with a cytosolic domain.

69. Cholesterol is transported through the blood in association with phospholipids and proteins that form complexes called lipoproteins. The lipoprotein known as LDL consists of an inner core of cholesteryl ester molecules (see Problem 47d) covered with a layer of phospholipids and cholesterol. A single molecule of apolipoprotein B (apoB) winds around the particle. High levels of LDL are associated with an increased risk of cardiovascular disease. **a.** Why is it necessary for cholesterol and cholesteryl esters to be packaged into LDL for transport through the blood? **b.** How is the structure of LDL similar to the structure of a membrane? How is it different? **c.** The protein apolipoprotein B was purified in the mid-1980s. Why was this protein so difficult to purify?

8.4 The Fluid Mosaic Model

70. Around the turn of the twentieth century, Charles Overton noted that low-molecular-weight aliphatic alcohols, ether, chloroform, and acetone could pass through membranes easily, while sugars, amino acids, and salts could not. This was a radical notion at the time, since most scientists believed that membranes were impermeable to all compounds but water. **a.** Using what you know about membrane structure, explain Charles Overton's results. **b.** Propose a hypothesis to explain how water could be transported across a membrane.

71. In 1935, Davson and Danielli described a "sandwich model" for membrane structure, which proposed that the membrane consisted of outer and inner layers of protein (the sandwich bread) with a "filling" of lipid (see diagram). This model is no longer accepted because of inconsistencies between the model and experimental data. Using what is currently known about membrane structure, explain some of the shortcomings of the sandwich model.

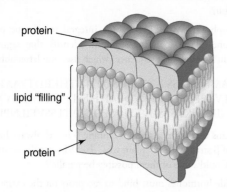

72. Identify the membrane proteins shown in the figure below as integral, peripheral, lipid-linked, or GPI-anchored.

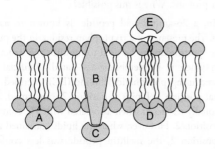

73. In fluorescence photobleaching recovery studies, fluorescent groups are attached to membrane components in a cell. An intense laser beam pulse focused on a very small area destroys (bleaches) the fluorophores in that area. Use the fluid mosaic model of the membrane to explain what happens to the fluorescence in the bleached area over time.

74. In a famous experiment, Michael Edidin labeled the proteins on the surface of mouse and human cells with green and red fluorescent markers, respectively. The two types of cells were induced to fuse, forming hybrid cells. Immediately after fusion, green markers could be seen on the surface of one half of the hybrid cell and red markers on the other half. After a 40-minute incubation at 37°C, the green and red markers became intermingled over the entire surface of the hybrid cell. If the hybrid cells were instead incubated at 15°C, this mixing did not occur. Explain these observations and why they support the fluid mosaic model of membrane structure.

75. The proteins labeled by Edidin in his cell fusion experiment (see Problem 83) were able to move freely in the membrane without constraints. What would he have observed if the labeled proteins resembled **a.** protein A or **b.** protein B in Figure 8.13?

76. There is some evidence that membranes contain structured domains called lipid rafts. These structures are thought to be composed of loosely packed glycosphingolipids, with the gaps filled in with cholesterol. The fatty acyl chains of the phospholipids associated with lipid rafts tend to be saturated. **a.** Why do glycosphingolipids pack together loosely? **b.** Given the description provided here, do you expect a lipid raft to be more or less fluid than the surrounding membrane?

77. The lipid distribution in membranes is asymmetric. Phosphatidylserine (PS) is exclusively found in the cytosol-facing leaflet of the membrane bilayer. Phosphatidylethanolamine (PE) is also more likely to be found in this leaflet. In contrast, phosphatidylcholine (PC) and sphingomyelin (SM) are more likely to be found in the extracellular leaflet of the membrane bilayer. **a.** What functional group do PS and PE have in common? **b.** What functional group do PC and SM have in common? **c.** Is one side of a membrane more likely to carry a charge than the other side, or do both sides of the membrane have the same charge?

78. Experiments with a phospholipid "flippase" enzyme from red blood cells show that the flippase translocates phospholipids from the extracellular leaflet to the cytosolic leaflet of the membrane. The flippase prefers phosphatidylserine and translocates phosphatidylethanolamine more slowly. Phosphatidylcholine and sphingomyelin are not translocated. Translocation does not occur if cells are deprived of ATP or magnesium ions or if red blood cells are treated with a reagent that alkylates sulfhydryl groups. Write a paragraph that describes the essential features of the flippase. Are these observations consistent with the data presented in Problem 87?

SELECTED READINGS

Edidin, M., Lipids on the frontier: A century of cell-membrane bilayers, *Nat. Rev. Mol. Cell Biol.* 4, 414–418 (2003). [Briefly reviews the history of the study of membranes.]

Nicolson, G.L., The fluid-mosaic model of membrane structure: Still relevant to understanding the structure, function and dynamics of biological membranes after more than 40 years. *Biochim. Biophys. Acta* 1838, 1451–1466 (2014). [Describes the model and how it has been updated.]

Puthenveetil, R. and Vinogradova, O., Solution NMR: A powerful tool for structural and functional studies of membrane proteins in reconstituted environments, *J. Biol. Chem.* 294, 15914–15931, doi: 10.1074/jbc.REV119.009178 (2019). [Discusses the challenges of determining membrane protein structures and includes numerous examples.]

van Meer, G., Voelker, D.R., and Feigenson, G.W., Membrane lipids: Where they are and how they behave, *Nat. Rev. Mol. Cell Biol.* 9, 112–124 (2008). [Reviews the structures and functions of membrane lipids, including their asymmetry and their liquid and gel phases.]

Watson, H., Biological membranes, *Essays Biochem.* 59, 43–69 (2015). [An overview of membrane structure, membrane proteins, and membrane functions.]

Yang, Y., Lee, M., and Fairn, G.D., Phospholipid subcellular localization and dynamics, *J. Biol. Chem.* 293, 6230–6240, doi: 10.1074/jbc.R117.000582 (2018). [Describes the distribution of lipids in different types of membranes as well as mechanisms to translocate membrane lipids.]

CHAPTER 8 CREDITS

Figures 8.7 and 8.8 Images based on 1QHJ. Belrhali, H., Nollert, P., Royant, A., Menzel, C., Rosenbusch, J.P., Landau, E.M., Pebay-Peyroula, E., Protein, lipid and water organization in bacteriorhodopsin crystals: a molecular view of the purple membrane at 1.9 A resolution, *Structure* 7, 909–917 (1999).

Figure 8.9 Image based on 1QJ9. Vogt, J., Schulz, G.E., The structure of the outer membrane protein OmpX from Escherichia coli reveals possible mechanisms of virulence, *Structure* 7, 1301–1309 (1999).

Figure 8.14. Based on information in Kreutzberger, A.J.B., Ji, M., Aaron, J., Mihaljević, L., Urban, S., Rhomboid distorts lipids to break the viscosity-imposed speed limit of membrane diffusion, *Science* 363, 497 (2019).

Membrane Transport

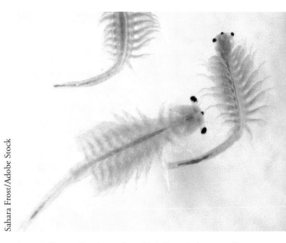

One challenge for *Artemia*, which lives in hypersaline lakes many times saltier than seawater, is the elimination of excess sodium. Specialized cells with a highly folded plasma membrane use ATP-powered pumps to eject Na⁺ ions from the tiny shrimp.

Do You Remember?

- Living organisms obey the laws of thermodynamics (Section 1.4).
- Amphiphilic molecules form micelles or bilayers (Section 2.2).
- An enzyme provides a lower-energy pathway from reactants to products (Section 6.2).
- Integral membrane proteins completely span the bilayer by forming one or more α helices or a β barrel (Section 8.3).

Some of the best-understood membrane-related events occur during nerve signaling. The ability of neurons to transmit signals from cell to cell depends on electrical changes that result from the regulated flow of charged particles through the cells' plasma membranes. However, because the plasma membrane presents a barrier to the free diffusion of ions and other substances, membrane proteins are needed to mediate their movement across membranes. In this chapter, we will examine some of these proteins as well as the membrane shape changes involved in nerve signaling and other processes.

9.1 The Thermodynamics of Membrane Transport

KEY CONCEPTS

Explain how ion movements affect membrane potential.

- Calculate the membrane potential from ion concentrations.
- Describe the ion movements of an action potential.
- Analyze the thermodynamics of ion movement across membranes.
- Distinguish active and passive transport.

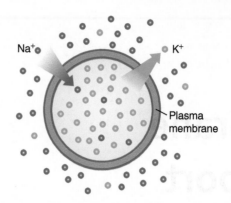

Figure 9.1 Distribution of Na⁺ and K⁺ ions in an animal cell. The extracellular Na⁺ concentration (about 150 mM) is much greater than the intracellular concentration (about 12 mM), whereas the extracellular K⁺ concentration (about 4 mM) is much less than the intracellular concentration (about 140 mM). If the plasma membrane were completely permeable to ions, Na⁺ would flow into the cell down its concentration gradient (blue arrow) and K⁺ would flow out of the cell down its concentration gradient (orange arrow).

Question Would ion movement ever stop? What would be the final ion concentrations?

All animal cells—including neurons—maintain intracellular ion concentrations that differ from those outside the cell (see Fig. 2.12). For example, intracellular sodium ion concentrations are much lower than extracellular sodium ion concentrations, and the opposite is true for potassium ions. Neither ion is at equilibrium. To reach equilibrium, Na⁺ would need to enter the cell, by spontaneously moving down its concentration gradient. Likewise, K⁺ would need to exit the cell, also moving down its concentration gradient (**Fig. 9.1**). This occurs extremely slowly, because biological membranes are largely impermeable to ions.

Even so, a small percentage of K⁺ ions do leak out of the cell. The movement of K⁺ and other ions places relatively more positive charges outside the cell and leaves relatively more negative charges inside the cell. The resulting charge imbalance, though small, generates a voltage across the membrane, which is called the **membrane potential** and is symbolized $\Delta\psi$ ($\Delta\psi = \psi_{inside} - \psi_{outside}$). In the simplest case, $\Delta\psi$ is a function of the ion concentration on each side of a membrane:

$$\Delta\psi = \frac{RT}{Z\mathcal{F}} \ln \frac{[\text{ion}]_{in}}{[\text{ion}]_{out}} \tag{9.1}$$

where R is the **gas constant** (8.3145 J · K⁻¹ · mol⁻¹), T is temperature in Kelvin (20°C = 293 K), Z is the net charge per ion, $\mathcal{F}$ is the **Faraday constant,** the charge of one mole of electrons (96,485 coulombs · mol⁻¹ or 96,485 J · V⁻¹ · mol⁻¹), and the cytosol is *in*. $\Delta\psi$ is expressed in units of volts (V) or millivolts (mV). For a monovalent ion ($Z = 1$) at 20°C, the equation reduces to

$$\Delta\psi = 0.058 \text{ V} \log_{10} \frac{[\text{ion}]_{in}}{[\text{ion}]_{out}} \tag{9.2}$$

See Sample Calculation 9.1. In a neuron, the membrane potential is actually a more complicated function of the concentrations and membrane permeabilities of several different ions, although K⁺ is the most important.

SAMPLE CALCULATION 9.1

Problem   Calculate the intracellular concentration of Na⁺ when the extracellular concentration is 160 mM. Assume that the membrane potential, –50 mV at 20°C, is due entirely to Na⁺.

Solution   Use Equation 9.2 and solve for $[\text{Na}^+]_{in}$:

$$\Delta\psi = 0.058 \text{ V} \log \frac{[\text{Na}^+]_{in}}{[\text{Na}^+]_{out}}$$

$$\frac{\Delta\psi}{0.058 \text{ V}} = \log[\text{Na}^+]_{in} - \log[\text{Na}^+]_{out}$$

$$\log[\text{Na}^+]_{in} = \frac{\Delta\psi}{0.058 \text{ V}} + \log[\text{Na}^+]_{out}$$

$$\log[\text{Na}^+]_{in} = \frac{-0.050 \text{ V}}{0.058 \text{ V}} + \log(0.160)$$

$$\log[\text{Na}^+]_{in} = -0.862 - 0.796$$

$$\log[\text{Na}^+]_{in} = -1.66$$

$$[\text{Na}^+]_{in} = 0.022 \text{ M} = 22\text{m M}$$

Ion movements alter membrane potential

Most animal cells maintain a membrane potential of about −70 mV. The negative sign indicates that the inside (the cytosol) is more negative than the outside (the extracellular fluid). A sudden flux of ions across the cell membrane can dramatically alter the membrane potential, and this is exactly what happens when a neuron fires.

When a nerve is stimulated, either mechanically or by a signal ultimately derived from one of the sensory organs, Na⁺ channels in the plasma membrane open. Sodium ions immediately move into the cell, since their concentration inside is much less than outside. The inward movement of Na⁺ makes the membrane potential more positive, increasing it from its resting value of −70 mV to as much as +50 mV. This reversal of membrane potential, or depolarization, is called the **action potential.**

The Na⁺ channels remain open for less than a millisecond. However, the action potential has already been generated, and it has two effects. First, *it triggers the opening of nearby voltage-gated K⁺ channels* (these channels open only in response to the change in membrane potential). The open K⁺ channels allow K⁺ ions to diffuse out of the cell, following their concentration gradient. This action restores the membrane potential to about −70 mV (**Fig. 9.2**).

The action potential also stimulates the opening of additional Na⁺ channels farther along the axon (the elongated portion of the cell). This induces another round of depolarization and repolarization, and then another. In this way, the action potential travels down the axon. The signal cannot travel backward because once the ion channels have shut, they remain closed for a few milliseconds. These events are summarized in **Figure 9.3**.

In mammals, action potentials propagate extremely rapidly because the axons are insulated by a so-called **myelin sheath.** This structure consists of several layers—up to 15—of membrane, derived from another cell, coiled around the axon (**Fig. 9.4**). The myelin sheath is rich in sphingomyelins and contains little protein (about 18%; a typical membrane contains about 50% protein). Because the myelin sheath prevents ion movements except at the points, or nodes, in between myelinated segments of the axon, the action potential appears to jump from node to node, propagating about 20 times faster than it would in an unwrapped axon. Deterioration of the myelin sheath in diseases such as multiple sclerosis results in the progressive loss of motor control.

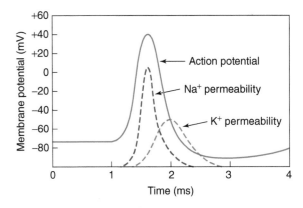

Figure 9.2 An action potential. The neuron membrane undergoes depolarization as Na⁺ channels are opened (green dashed line), then repolarizes as K⁺ channels are opened (red dashed line). Following the action potential, the membrane may be hyperpolarized (Δψ < −70 mV) but returns to normal within a few milliseconds.

Membrane proteins mediate transmembrane ion movement

The Na⁺ and K⁺ channels that participate in the propagation of an action potential are just two members of a large group of transport proteins that occur in the plasma membranes of all cells and in the internal membranes of eukaryotes. Transport proteins go by many different names, depending somewhat arbitrarily on their mode of action: transporters, translocases, permeases, pores, channels, and pumps, to list a few. These proteins can also be classified by the type of substance they transport across the membrane and by whether they are always open or gated (open only when stimulated). About 20,000 transporters have been grouped into five mechanistic classes in the Transporter Classification Database (www.tcdb.org). The most important distinction among transport proteins, however, is whether they require a source of free energy to operate. The neuronal Na⁺ and K⁺ channels are considered **passive transporters** because they provide a means for ions to move down a concentration gradient, a thermodynamically favorable event.

For any transport protein operating independently of the effects of membrane potential, the free energy change for the transmembrane movement of a substance X from the outside to the inside is

$$\Delta G = RT \ln \frac{[\text{X}]_{in}}{[\text{X}]_{out}} \tag{9.3}$$

Consequently, *the free energy change is negative (the process is spontaneous) only when X moves from an area of high concentration on the outer side of the membrane to an area of low concentration on the inner side of the membrane* (see Sample Calculation 9.2).

Figure 9.3 Propagation of a nerve impulse.

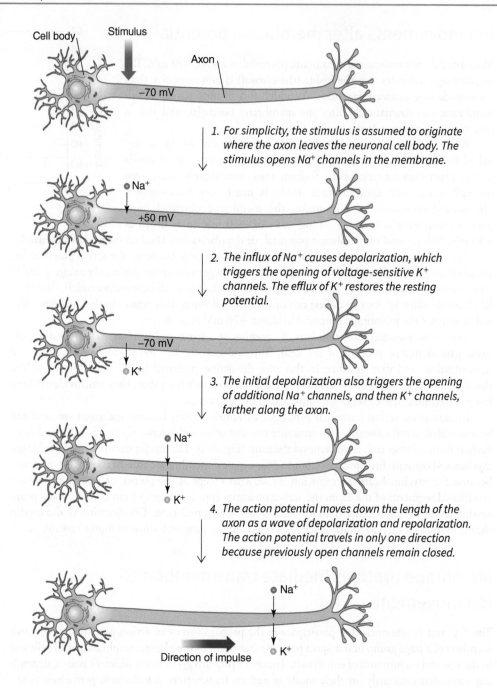

1. For simplicity, the stimulus is assumed to originate where the axon leaves the neuronal cell body. The stimulus opens Na⁺ channels in the membrane.

2. The influx of Na⁺ causes depolarization, which triggers the opening of voltage-sensitive K⁺ channels. The efflux of K⁺ restores the resting potential.

3. The initial depolarization also triggers the opening of additional Na⁺ channels, and then K⁺ channels, farther along the axon.

4. The action potential moves down the length of the axon as a wave of depolarization and repolarization. The action potential travels in only one direction because previously open channels remain closed.

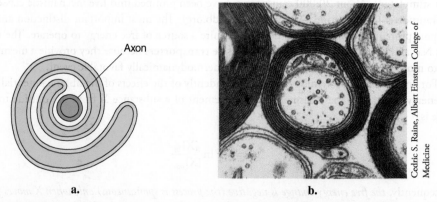

Cedric S. Raine, Albert Einstein College of Medicine

a.   **b.**

Figure 9.4 Myelination of an axon. **a.** Cross-sectional diagram showing how an accessory cell coils around the axon so that multiple layers of its plasma membrane coat the axon. **b.** Electron micrograph of myelinated axons.

SAMPLE CALCULATION 9.2

Problem   Show that $\Delta G < 0$ when glucose moves from outside the cell (where its concentration is 10 mM) to the cytosol (where its concentration is 0.1 mM).

Solution   The cytosol is *in* and the extracellular space is *out*.

$$\Delta G = RT \ln \frac{[glucose]_{in}}{[glucose]_{out}}$$

$$= RT \ln \frac{[10^{-4}]}{[10^{-2}]} = RT(-4.6)$$

Because the logarithm of $(10^{-4}/10^{-2})$ is a negative quantity, ΔG is also negative.

If the transported substance is an ion, there will be a charge difference across the membrane, so a term containing the membrane potential must be added to Equation 9.3:

$$\Delta G = RT \ln \frac{[X]_{in}}{[X]_{out}} + Z\mathcal{F}\Delta\psi \qquad (9.4)$$

See Sample Calculation 9.3. To determine the free energy change for transporting an ion when *out* is the ion's initial location and *in* is the final location, use Equation 9.4 but reverse the sign of the free energy value (this is equivalent to switching the concentration terms $[X]_{in}$ and $[X]_{out}$ and reversing the sign of $\Delta\psi$). Note that for an anionic substance with charge Z, transport may not be thermodynamically favored, depending on the membrane potential $\Delta\psi$, even if the concentration gradient alone favors transport.

SAMPLE CALCULATION 9.3

Problem   Calculate the free energy change for the movement of Na^+ into a cell when its concentration outside is 150 mM and its cytosolic concentration is 10 mM. Assume that $T = 20°C$ and $\Delta\psi = -50$ mV (inside negative).

Solution   Use Equation 9.4:

$$\Delta G = RT \ln \frac{[X]_{in}}{[X]_{out}} + Z\mathcal{F}\Delta\psi$$

$$= (8.3145 \ J \cdot K^{-1} \cdot mol^{-1})(293 \ K) \ln \frac{(0.010)}{(0.150)}$$

$$+ (1)(96{,}485 \ J \cdot V^{-1} \cdot mol^{-1})(-0.05 \ V)$$

$$= -6600 \ J \cdot mol^{-1} - 4820 \ J \cdot mol^{-1}$$

$$= -11{,}600 \ J \cdot mol^{-1} = -11.6 \ kJ \cdot mol^{-1}$$

In contrast to the passive ion channels in neurons, the protein that initially establishes and maintains the cell's Na^+ and K^+ gradients is an **active transporter** that needs the free energy of the ATP hydrolysis reaction to move ions against their concentration gradients. In the following sections we will examine various types of transport proteins. Keep in mind that small nonpolar substances can cross a membrane without the aid of any transport protein; they simply diffuse through the lipid bilayer.

Ionophores

The word **ionophore** was coined by Pressman in 1964. Ionophore comprises diversified molecules that are able to transport specific ions across the membrane by forming a lipid ion-soluble complex. They can be synthetic or natural molecules that possess a unique ability to change the ion gradient and electric potential across the membrane, thereby affecting the permeability of biological membranes to certain ions and antibiotics. An ionophore typically has hydrophobic amino acids at their exterior surfaces to interact with the lipid bilayer, and hydrophilic amino acids are present in their interior to interact with ions to facilitate ion movement across the membrane (**Figure 9.5**). Ionophores can be classified into two categories based on their properties and functions: Carriers and Channels. A list of ionophores according to their structure and ion specificity is given in **Table 9.1**.

The carrier works by shielding the specific ions from the hydrophobic environment of the membrane. During this process, carriers shift the ion by diffusing across the membrane and itself come back to its original place. The electrical potential generated across the membrane due to the

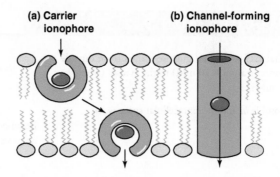

Figure 9.5 Ionophores: The Carrier, and the Channel forming Ionophores to transport certain ions across the membrane.

unequal distribution of ions facilitates the binding of ions to ionophores. The ions movement by ionophores through the membrane may decrease or balance the membrane potential and form the basis of their antibiotic properties. Sometimes these ionophores may act as uncouplers in mitochondrial respiration and affect the synthesis of ATP. Structural studies of carrier ionophores have revealed the binding of ions to oxygen atoms at the center of the ionophore. They can form a complex with monovalent- or divalent-cations. The carrier ionophores are further classified into different categories based on their interactions with cations, mode of action, and structural properties. The well studied class of carrier ionophores is polyether including more than 130 types having antibiotic and anti-tumor activities.

Carrier ionophores are not found to be effective against gram-negative bacteria but effective against gram-positive bacteria. Some of the important ionophores are valinomycin, calcimycin, salinomycin, and monesin. Valinomycin is produced from various strains of Streptomyces and is highly selective for potassium and rubidium ions and doesn't allow sodium and lithium transport. The other examples of ionophores include calcium ionophores such as calcimycin and ionomycin having the ability to hydrolyze the phosphatidylinositides in the T-cell membrane resulting in activation of phosphokinase C (PKC) and cause the autophosphorylation of CD4 and CD8 T-cells. Calcium ionophores also promote the cytosolic calcium movement in low native calcium permeability cells by certain mechanisms such as activation of receptor-mediated calcium transport by Ca^{2+}-channels into the cytosol, or calcium mobilization from intracellular stores through Phospholipase C-dependent manner. Hemisodium is a synthetic ionophore having selectivity for sodium ions. Beauvericin, an ionophore has phenylalanine and hydroxyl-iso-valine alternatively forming a cyclic hexapeptide structure and having antibiotic and insecticidal properties. They transport the alkaline molecules across the membrane. Nigericin is responsible for exchange of K^+ for protons and has been used to study mitochondrial bioenergetics to alter the electrical and chemical gradient for protons.

Ionophores are also responsible for transport across the membrane by forming channels across the membrane. They have non-standard amino acids in their structure. Gramicidin channel forming ionophore having 15 amino acids arranged in D and L alternate manner with amino-terminal formaldehyde and carboxyl-terminal ethanol amine. The diffusion rate of ions through the channel

Table 9.1 Type of Ionophores and their ion-selectivity.

	Ion Specificity
Carriers	
Valinomycin	K^+
Calcium ionophores	Ca^{2+}/Mg^{2+}; $Ca^{2+}/2H^+$
Hemisodoium	Na^+
Nigericin	K^+/H^+
Channels	
Gramicidin	$Cs^+ > Rb^+ > K^+ > Na^+ > Li^+$
Nystatin	Monovalent cation and anions

is high as compared to the carrier ionophore. Gramicidin channels allow the movement of protons and cations but don't allow the movement of divalent cations. Affinity for various cations through the gramicidin channel occurred in the following order $Cs^+> Rb^+> K^+ > Na^+ > Li^+$. Gramicidin as an antibiotic was found to be effective against gram-positive, gram-negative bacteria, and pathogenic fungi. Nystatin, a channel forming ionophore is able to alter the cation composition of cells by permitting the movement of monovalent cations, and anions across the membrane.

Concept Check

1. Explain the role of the membrane in maintaining membrane potential.
2. Practice using Equations 9.1 and 9.4.
3. Describe how an action potential is generated and propagated.
4. Explain how concentration differences and membrane potential influence the free energy change for ion movements across a membrane.
5. Summarize the differences between active and passive transport.
6. Explain how ionophores affect the ion's gradient across the membrane.

9.2 Passive Transport

KEY CONCEPTS

Describe the operation of passive transport systems.

- Compare the structures of porins, channels, and transporters.
- Explain the mechanisms of solute selectivity in the different types of transporters.
- Describe the role of conformational changes in the GLUT proteins.
- Compare transport proteins to enzymes.

Transporters of one kind or another have been described for virtually every substance that cannot easily diffuse through a bilayer on its own. Some transporters create a straightforward opening through the membrane, while more complicated transporters function much like enzymes. We begin our discussion with porins, the simplest transporters.

Porins are β barrel proteins

Porins are located in the outer membranes of bacteria, mitochondria, and chloroplasts (some bacteria and the organelles descended from them have a second outer membrane in addition to the membrane that encloses the cytosol). *Most porins are trimers in which each subunit forms a 16- or 18-stranded membrane-spanning β barrel* (**Fig. 9.6**). A β barrel of this size has a water-filled core lined with hydrophilic side chains, which forms a passageway for the transmembrane movement of ions or molecules with a molecular mass up to about 1000 D. In the eight-stranded β barrel shown in Figure 8.9, the protein core is too tightly packed with amino acid side chains for it to function as a pore.

In the 16-stranded OmpF barrel, long loops connect the β strands (Fig. 9.6c). One of these loops in each monomer folds down into the β barrel and constricts its diameter to about 7 Å at one point, thereby preventing the passage of substances larger than 600 D. The loop bears several carboxylate side chains, which make this porin weakly selective for cationic substances. Other porins exhibit a greater degree of solute selectivity, depending on the geometry of the barrel interior and the nature of the side chains that project into it. For example, some porins are specific for anions or small carbohydrates. A porin is considered to be always open, and a solute can travel through it in either direction, depending on which side of the membrane has a lower solute concentration.

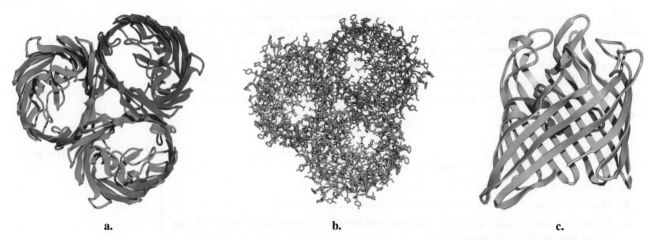

a.   b.   c.

Figure 9.6 The *E. coli* OmpF porin.   Each subunit of the trimeric protein forms a transmembrane β barrel that permits the passage of ions or small molecules. **a.** Ribbon model, viewed from the extracellular side of the membrane. **b.** Stick model. **c.** In each subunit, the 16 β strands are connected by loops, one of which (blue) constricts the barrel core and makes the porin specific for small cationic solutes.

Ion channels are highly selective

The ion channels in neurons and in other eukaryotic and prokaryotic cells are more complicated proteins than the porins. Many are multimers of identical or similar subunits with α-helical membrane-spanning segments. The ion passageway itself lies along the central axis of the protein, where the subunits meet. One of the best known of these proteins is the K+ channel from the bacterium *Streptomyces lividans*. Each subunit of this tetrameric protein includes two long α helices. One helix forms part of the wall of the transmembrane pore, and the other helix faces the hydrophobic membrane interior (**Fig. 9.7**). A third, smaller helix is located on the extracellular side of the protein.

The K+ channel is about 10,000 times more permeant to K+ than to Na+, even though Na+ is smaller and should easily pass through the central pore. *The high selectivity for K+ reflects the geometry of the selectivity filter,* an arrangement of protein groups that define the extracellular mouth of the pore. At one point, the pore narrows to ~3 Å, and the four polypeptide backbones fold so that their carbonyl groups project into the pore. The carbonyl oxygen atoms are arranged

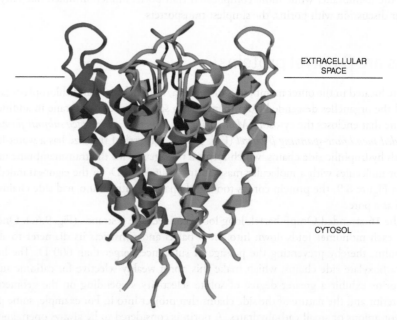

EXTRACELLULAR
SPACE

CYTOSOL

Figure 9.7 Structure of the K+ channel from *S. lividans*.   The four subunits are shown in different colors. Each subunit consists mostly of an inner helix that forms part of the central pore and an outer helix that contacts the membrane interior.

with a geometry suitable for coordinating desolvated K$^+$ ions (diameter 2.67 Å) as they move through the pore. A desolvated Na$^+$ ion (diameter 1.90 Å) is too small to coordinate with the carbonyl groups and is therefore excluded from the pore (**Fig. 9.8**).

The voltage-gated K$^+$ channel in neurons is larger than the bacterial channel, with six helices in each of its four subunits, and it associates with other proteins to form a large complex. However, like most K$^+$ channels, it contains the same type of selectivity filter. Other ion channels have similar overall structures but different filtering mechanisms. For example, the mitochondrial calcium uniporter, which allows Ca^{2+} ions to enter the mitochondria in all eukaryotes, is a tetramer with a central cone-shaped pore. The narrow end of the pore is lined with four aspartate and four glutamate residues whose carboxyl groups have the ideal geometry to coordinate a Ca^{2+} ion. K$^+$ ions are too large to fit inside this selectivity filter, and Na$^+$ ions—although the same size as Ca^{2+} ions—do not coordinate as well with the glutamate side chains.

Chloride channel proteins belonging to the CLC family of proteins have multiple transmembrane helices surrounding a hydrated channel that is lined with small hydrogen-bonding side chains. An anion-specific selectivity filter is located midway between the channel entrance and exit. Here, the channel is constricted, with the N-terminal ends of three α helices generating a positive electrostatic force that is oriented toward the transiting Cl$^-$ ions.

Gated channels undergo conformational changes

If K$^+$ or Na$^+$ channels were always open, nerve cells would not experience action potentials, and the intracellular and extracellular concentrations of ions would quickly reach equilibrium, thereby killing the cell (**Box 9.A**). Consequently, these channels—and many others—are **gated**; that is, they open or close in response to a specific signal. Some ion channels respond to changes in pH or

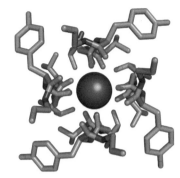

Figure 9.8 The K$^+$ channel selectivity filter. This model shows a portion of the pore looking from the extracellular space into the selectivity filter. The pore is lined by backbone carbonyl groups with a geometry suitable for coordinating a K$^+$ ion (purple sphere). A rigid protein network that includes tyrosine residues prevents the pore from contracting to accommodate the smaller Na$^+$ ion. Atoms are color-coded: C green, N blue, and O red.

Question Explain why a small anion would not pass through the channel.

Box 9.A   Pores Can Kill

The antifungal agent amphotericin B kills a variety of pathogenic fungi. Amphotericin B is a relatively small cyclic compound with decidedly hydrophobic and hydrophilic faces. Atoms are color-coded: C green, N blue, O red, and H white.

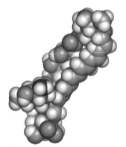

An estimated four to six amphotericin molecules insert into the fungal cell membrane, where the hydrophobic portions interact strongly with ergosterol (a fungal counterpart of the cholesterol found in mammalian cell membranes), and the hydrophilic portions define a passageway from one side of the membrane to the other. The amphotericin molecule, with a length of only 23 Å, barely spans the hydrophobic core of the lipid bilayer, and the opening it forms is too small to allow the mass exit of the cell's contents. However, the pore is apparently sufficient to permit the flow of Na$^+$, K$^+$, and other ions. The resulting disruption of ion concentration gradients and the loss of membrane potential are lethal to the cell, which otherwise remains intact.

Amphotericin, produced by soil bacteria, is just one example of a wide array of microbial products that are intended to disrupt the membrane integrity of competitors or predators. Typically, these

chemical weapons are excreted individually and, in pairs or small multimers, assemble in the target cell's membrane to form a passageway for ions. Some of these compounds, often in chemically modified form, have been adopted for use as antibiotics.

The mammalian immune system also relies on pore-forming mechanisms to combat bacterial and fungal infections. A wide variety of antimicrobial peptides are synthesized by cells that line the body's surfaces, including the skin and digestive tract. These peptides tend to have a net positive charge so that they are attracted to microbial membranes, which are usually negatively charged, and not to the host's own mostly neutral membranes. Although some antimicrobial peptides must pass through the pathogen's membrane in order to reach an intracellular target, most antimicrobial peptides insert into the cell membrane and either form a pore or locally dissolve the lipid bilayer.

In a more elaborate defense system, certain microbial cell-wall components trigger the activation of **complement,** a set of circulating proteins that sequentially activate each other and lead to the formation of a doughnut-shaped structure, the so-called membrane attack complex, that creates a large pore in the target cell's membrane. The resulting loss of ions kills the cell. The inappropriate assembly of the membrane attack complex on the surface of human cells contributes to the pathology of some diseases that are caused by the immune system mistakenly responding to the body's own components.

Question What prevents pore formation in the membrane of the cell that produces amphotericin or a similar compound? Explain why bacteria with elaborate outer membranes are more resistant to pore-forming antibiotics.

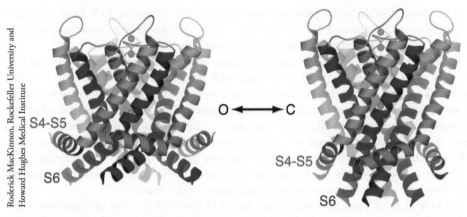

Roderick MacKinnon, Rockefeller University and Howard Hughes Medical Institute

Figure 9.9 Operation of the voltage-gated K⁺ channel.   In the closed conformation (right), the so-called S4-S5 linker helix (shown in red) pushes down on the S6 helix, pinching off the intracellular end of the pore. When depolarization occurs, the linker helix swings upward, and the S6 helix bends, opening up the pore (left).

to the binding of a specific ligand such as Ca^{2+} or a small molecule. Some channel proteins open when phosphorylated (have a phosphate group covalently attached). In the hair cells of the inner ear, mechanical displacement of the stereocilium (see Box 5.D) opens a cation channel to initiate a nerve impulse.

Analysis of gated channels reveals a variety of mechanisms for opening and closing off a pore. The neuronal K⁺ channel is voltage-gated; it opens in response to depolarization. The gating mechanism involves the motion of helices near the intracellular side of the membrane, which move enough to expose the entrance of the pore without significantly disrupting the structure of the rest of the protein (**Fig. 9.9**).

In addition to voltage gating, the K⁺ channel in neurons is subject to inactivation by a process in which an N-terminal segment of the protein (not shown in Fig. 9.9) is repositioned to block the cytoplasmic opening of the pore. This inactivation occurs a few milliseconds after the K⁺ channel first opens and explains why the channel cannot immediately reopen. As a result, the action potential can only travel forward.

In bacterial mechanosensitive channels, which open in response to membrane tension, a set of α helices slide past each other to alter their packing arrangement (**Fig. 9.10**). Interestingly, in the closed state, the pore is not 100% occluded. However, neither water nor ions can pass through because the opening is lined with bulky hydrophobic residues. Although a single water molecule or desolvated ion might fit geometrically, the high energetic cost of passing a polar solute past this hydrophobic barrier effectively closes off the pore.

Figure 9.10 Closed and open conformations of mechanosensitive channels.   In the bacterial proteins MscS and MscL, a set of α helices surround the pore (the rest of the proteins are not shown). The helices slide past each other, opening and closing the pore, much like the iris of the eye. Hydrophobic residues that block the pore in the closed state are shown in magenta.

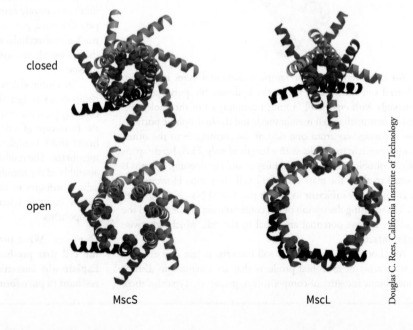

Douglas C. Rees, California Institute of Technology

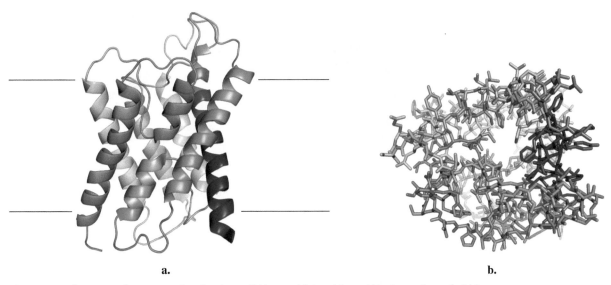

Figure 9.11 Structure of an aquaporin subunit. **a.** Ribbon model viewed from within the membrane. **b.** Stick model viewed from one end.

Aquaporins are water-specific pores

For many years, water molecules were assumed to cross membranes by simple diffusion (technically **osmosis,** the movement of water from regions of low solute concentration to regions of high solute concentration). Because water is present in large amounts in biological systems, this premise seemed reasonable. However, certain cells, such as in the kidney, can sustain unexpectedly rapid rates of water transport, which suggested the existence of a previously unrecognized pore for water. The elusive protein was discovered in 1992 by Peter Agre, who coined the term **aquaporin.**

Aquaporins are widely distributed in nature; plants may have as many as 50 different aquaporins. The 13 mammalian aquaporins are expressed at high levels in tissues where fluid transport is important, including the kidney, salivary glands, and lacrimal glands (which produce tears). Most aquaporins are extremely specific for water molecules and do not permit the transmembrane passage of other small polar molecules such as glycerol or urea.

$$CH_2-CH-CH_2$$
$$\;|\quad\;|\quad\;|$$
$$OH\quad OH\quad OH$$
Glycerol

$$O$$
$$\|$$
$$H_2N-C-NH_2$$
Urea

The best-defined member of the aquaporin family (aquaporin 1, or AQP1) is a homotetramer with carbohydrate chains on its extracellular surface. Each subunit consists mostly of six membrane-spanning α helices plus two shorter helices that lie within the boundaries of the bilayer (**Fig. 9.11**). The subunits associate with each other via hydrogen bonding between helices and through interactions among the loops outside the membrane.

Unlike the K^+ channel, whose pore lies in the center of the four protein subunits, each aquaporin subunit contains a pore. At its narrowest, the pore is about 3 Å in diameter (the diameter of a water molecule is 2.8 Å). The dimensions of the pore clearly restrict the passage of larger molecules. The pore is lined with hydrophobic residues except for two asparagine side chains, which have an important function (**Fig. 9.12**).

If water were to pass through aquaporin as a chain of hydrogen-bonded molecules, then protons could also easily pass through (recall from Section 2.3 that a proton is equivalent to H_3O^+ and that a proton can appear to jump rapidly through a network of hydrogen-bonded water molecules). However, aquaporin does not transport protons (other proteins that do transport protons play important roles in energy metabolism). To prevent proton transport, aquaporin interrupts the hydrogen-bonded chain of water molecules in its pores, which occurs when the asparagine side chains transiently form hydrogen bonds to a water molecule passing by.

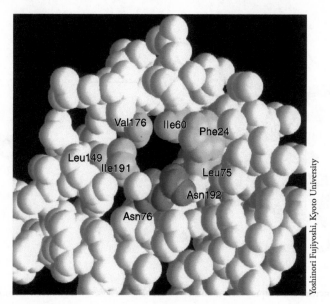

Figure 9.12 View of the aquaporin pore. Hydrophobic residues are colored yellow and the two asparagine residues are red.

Some transport proteins alternate between conformations

Not all proteins that mediate transmembrane traffic have an obvious membrane-spanning pore, as in porins and ion channels. Many transport proteins undergo conformational changes in order to move a solute from one side of the membrane to the other. Some of the best known transporters are the glucose-transporting family of GLUT proteins. There are 14 forms of GLUT in humans, and similar proteins occur in all types of cells.

The GLUT protein has a glucose-binding site that alternately faces the cell exterior and interior. In a typical transport cycle, extracellular glucose binds to the outward-facing conformation of the protein. Ligand binding triggers conformational changes that involve a sort of rocking motion in the protein, which exposes the bound glucose to the intracellular side of the membrane (**Fig. 9.13**). Because the glucose concentration is typically lower inside the cell, the bound glucose dissociates. The GLUT protein is not perfectly symmetrical, however, so the inward-facing conformation switches back to the outward-facing conformation, ready to bind another glucose molecule. The two conformational states of GLUT are in equilibrium, so this passive transporter can move glucose in either direction across the cell membrane, depending on the relative concentrations of glucose inside and outside the cell.

The GLUT transporters consist of 12 membrane-spanning α helices arranged in two domains (**Fig. 9.14**). Modest shifts in the orientations of a few helices convert the proteins from outward-facing to inward-facing and back again. *Many other transport proteins structurally resemble the GLUT proteins and use the same alternating-access mechanism to bind and release a ligand on opposite sides of the membrane.* These transporters function like enzymes by accelerating the rate at which a substance crosses the membrane. And like enzymes, they can be saturated by high concentrations of their "substrate," and they are susceptible to competitive and other types of inhibition. For obvious reasons, transport proteins tend to be more solute-selective than porins or ion channels. Their great variety reflects the need to transport many different kinds of metabolic fuels and building blocks into and out of cells and organelles. An estimated 10% of the genes in microorganisms encode transport proteins.

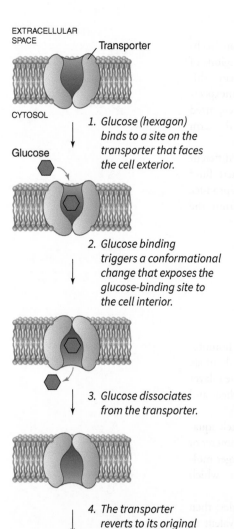

EXTRACELLULAR
SPACE

Transporter

CYTOSOL

Glucose

1. *Glucose (hexagon) binds to a site on the transporter that faces the cell exterior.*

2. *Glucose binding triggers a conformational change that exposes the glucose-binding site to the cell interior.*

3. *Glucose dissociates from the transporter.*

4. *The transporter reverts to its original conformation.*

Figure 9.13 Operation of the red blood cell glucose transporter.

Question Would this transport protein allow water or ions to move across the membrane?

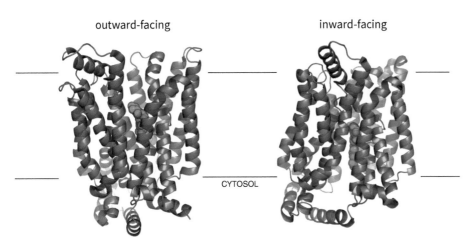

outward-facing   inward-facing

CYTOSOL

Figure 9.14 Conformations of GLUT proteins.   Two GLUT proteins are shown as ribbon models. Analogs of glucose (red) occupy the binding sites of the proteins in the outward-facing (left) and inward-facing (right) conformations.

Some transport proteins can bind more than one type of ligand, so it is useful to classify them according to how they operate (**Fig. 9.15**):

1. A **uniporter** such as the glucose transporter moves a single substance at a time.
2. A **symporter** transports two different substances across the membrane.
3. An **antiporter** moves two different substances in opposite directions across the membrane.

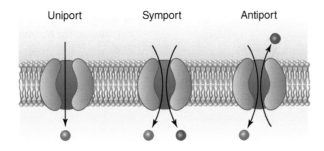

Uniport   Symport   Antiport

Figure 9.15 Some types of membrane transport systems.

Concept Check

1. Compare the overall structure, solute selectivity, and general mechanism of porins, ion channels, aquaporins, and the GLUT transporters.
2. Which of these transporters shown in Figure 9.15 is continually open?
3. Explain why these transport systems allow solute movement in either direction.
4. Recount the reaction sequence of a GLUT transporter.

9.3 Active Transport

KEY CONCEPTS

Describe the operation of active transport systems.

- Distinguish primary and secondary active transport.
- Describe the reaction sequence of the Na,K-ATPase.
- Explain why active transport is unidirectional.

The differing Na^+ and K^+ concentrations inside and outside of eukaryotic cells are maintained largely by an antiport protein known as the Na,K-ATPase. This active transporter pumps Na^+ out

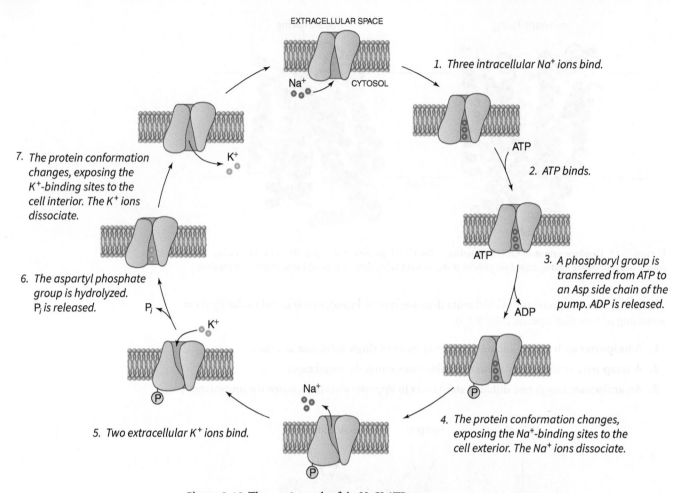

EXTRACELLULAR SPACE

Na⁺   CYTOSOL

1. *Three intracellular Na⁺ ions bind.*

ATP

2. *ATP binds.*

ATP

3. *A phosphoryl group is transferred from ATP to an Asp side chain of the pump. ADP is released.*

ADP

K⁺

7. *The protein conformation changes, exposing the K⁺-binding sites to the cell interior. The K⁺ ions dissociate.*

6. *The aspartyl phosphate group is hydrolyzed. Pᵢ is released.*

Pᵢ

K⁺

P

5. *Two extracellular K⁺ ions bind.*

Na⁺

P

4. *The protein conformation changes, exposing the Na⁺-binding sites to the cell exterior. The Na⁺ ions dissociate.*

Figure 9.16 The reaction cycle of the Na,K-ATPase.

Question What does the ion transport mechanism have in common with the mechanism of motor proteins (see Figs. 5.39 and 5.41)?

of and K⁺ into the cell, working against the ion concentration gradients. As the name **ATPase** implies, ATP hydrolysis is its source of free energy. Other ATP-requiring transport proteins pump a variety of substances against their concentration gradients.

The Na,K-ATPase changes conformation as it pumps ions across the membrane

With each reaction cycle, the Na,K-ATPase hydrolyzes 1 ATP, pumps 3 Na⁺ ions out, and pumps 2 K⁺ ions in:

$$3 \text{ Na}^+_{in} + 2 \text{ K}^+_{out} + \text{ATP} + \text{H}_2\text{O} \rightarrow 3 \text{ Na}^+_{out} + 2 \text{ K}^+_{in} + \text{ADP} + \text{P}_i$$

Like other membrane transport proteins, the Na,K-ATPase has two conformations that alternately expose the Na⁺ and K⁺ binding sites to each side of the membrane. As diagrammed in **Figure 9.16**, the protein pumps out 3 Na⁺ ions at a time, then transports in 2 K⁺ ions at a time as it hydrolyzes ATP. *The energetically favorable reaction of converting ATP to ADP + Pᵢ drives the energetically unfavorable transport of Na⁺ and K⁺.* The ATP hydrolysis reaction is coupled to ion transport so that phosphoryl-group transfer from ATP to the protein triggers one conformational change (steps 3 and 4) and the subsequent release of the phosphoryl group as Pᵢ triggers another conformational change (steps 5 and 6). This multistep process, which involves a phosphorylated protein intermediate, ensures that the transporter operates in only one direction and prevents Na⁺ and K⁺ from diffusing back down their concentration gradients. A similar mechanism operates in motor proteins (Section 5.4), where ADP and Pᵢ are released in separate steps and so cannot recombine to re-form ATP and drive the reaction cycle in reverse.

The Na,K-ATPase consists of a large α subunit with 10 transmembrane helices plus smaller β and γ subunits containing one transmembrane helix each. The structure of the pump in its outward-facing form is shown in **Figure 9.17**. The ATP binding site and the aspartate residue that becomes phosphorylated during the reaction cycle are located in cytoplasmic domains, indicating that ATP-binding and phosphate-transfer events must be communicated over a considerable distance to the membrane-spanning region where cations are bound and released.

In the Na,K-ATPase and in the structurally related Ca^{2+}-ATPase, the perpendicular movement of transmembrane α helices appears to be a key part of the pumping mechanism. Although such rearrangements could potentially expose bilayer-embedded residues to the cytoplasm, this does not occur because the entire protein tilts relative to the membrane, keeping hydrophobic groups buried. The protein rights itself again, because bulky tryptophan side chains act as "floats," and arginine and lysine side chains maintain an electrostatic grip on the head groups of phospholipids that are closely associated with the protein.

Both the Na,K-ATPase and the Ca^{2+}-ATPase are known as P-type ATPases (P stands for phosphorylation). Other types of ATP-dependent pumps are the V-type ATPases, which operate in plant vacuoles and other organelles, and the F-type ATPases, which actually operate in reverse to *synthesize* ATP in mitochondria (Section 15.3) and chloroplasts (Section 16.2).

ABC transporters mediate drug resistance

All cells have some ability to protect themselves from toxic substances that insert themselves into the lipid bilayer and alter membrane structure and function. This defense depends on the action of membrane proteins known as **ABC transporters** (ABC refers to ATP-binding cassette, a common structural motif in these proteins). Unfortunately, many antibiotics and other drugs are lipid-soluble and are therefore substrates for these same transporters. Drug resistance in cancer chemotherapy and antibiotic resistance in bacteria have been linked to the expression or overexpression of ABC transporters. In humans, this transporter is also known as P-glycoprotein or the multidrug-resistance transporter.

ABC transporters function much like other transport proteins and ATPase pumps: *ATP-dependent conformational changes in cytoplasmic portions of the protein are coupled to conformational changes in the membrane-embedded portion of the protein.* As expected, the transporter consists of two halves that presumably reorient relative to each other to expose the ligand-binding site to each side of the membrane in turn (as in the GLUT transporter mechanism shown in Fig. 9.13). Each half of the transporter includes a bundle of membrane-spanning α helices linked to a globular nucleotide-binding domain where the ATP reaction takes place (**Fig. 9.18**).

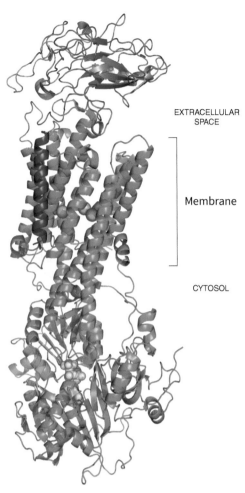

EXTRACELLULAR SPACE

Membrane

CYTOSOL

Figure 9.17 Structure of the Na,K-ATPase. The α subunit is green and the β and γ subunits are blue. Three Na^+ ions (orange) occupy the ion-binding sites, and an ATP analog (yellow) marks the site of phosphorylation.

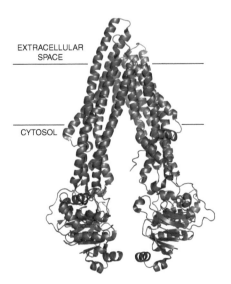

EXTRACELLULAR SPACE

CYTOSOL

Figure 9.18 Structure of mouse P-glycoprotein. The transporter, which is built from a single polypeptide chain, is shown as a ribbon model. Note that the internal cavity is open to both the cytoplasm and the inner leaflet of the membrane.

Some ABC transporters are specific for ions, sugars, amino acids, or other polar substances. P-glycoprotein and other drug-resistance transporters prefer nonpolar substrates. In this case, the transmembrane domain of the protein allows the entry of substances from within the lipid bilayer. The substance may then be entirely expelled from the cell, as occurs in drug resistance. Alternatively, the substance may simply move from one leaflet to the other. Some lipid flippases and floppases (Section 8.4) are ABC transporters that transport lipids between leaflets, generating nonequilibrium distributions of certain lipids in membranes.

Secondary active transport exploits existing gradients

In some cases, the "uphill" transmembrane movement of a substance is not directly coupled to the conversion of ATP to ADP + P_i. Instead, *the transporter takes advantage of a gradient already established by another pump,* which is often an ATPase. This indirect use of the free energy of ATP is known as **secondary active transport.** For example, the high Na^+ concentration outside of intestinal cells (a gradient established by the Na,K-ATPase) helps drive glucose into the cells via a symport protein (**Fig. 9.19**). The free energy released by the movement of Na^+ into the cells (down its concentration gradient) drives the inward movement of glucose (against its gradient). This mechanism allows the intestine to collect glucose from digested food and then release it into the bloodstream. Other types of secondary active transporters take advantage of the free energy of a proton gradient, transporting solutes across a membrane along with a proton.

Concept Check

1. Compare the mechanisms of active and passive transporters.
2. List the ways that the ATP reaction is used to move solutes across membranes.
3. Recount the steps of the Na,K-ATPase mechanism.
4. How do secondary active transporters compare to other transporters in terms of mechanism and energy requirement?

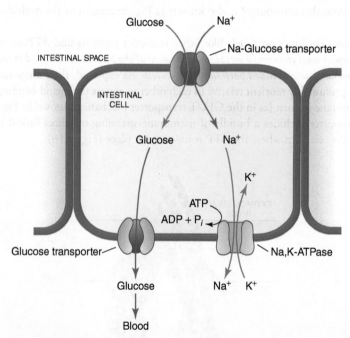

Figure 9.19 Glucose transport into intestinal cells. The Na,K-ATPase establishes a concentration gradient in which $[Na^+]_{out} > [Na^+]_{in}$. Sodium ions move into the cell, down their concentration gradient, along with glucose molecules via a symport protein that transports Na^+ and glucose simultaneously. Glucose thereby becomes more concentrated inside the cell, which it then exits, down its concentration gradient, via a passive uniport GLUT transporter. Energetically favorable movements are indicated by green arrows; energy-requiring movements are indicated by red arrows.

Question Identify the uniport, symport, and antiport proteins in this diagram.

9.4 Membrane Fusion

KEY CONCEPTS

Describe the process of membrane fusion.

- Summarize the events that occur at a nerve–muscle synapse.
- Describe the role of SNARES and membrane curvature in vesicle fusion.
- Compare exocytosis and endocytosis.
- Describe the role of membranes in autophagy.

The final steps in the transmission of a signal from one neuron to the next, or to a gland or muscle cell, culminate in the release of substances known as **neurotransmitters** from the end of the axon. Common neurotransmitters include amino acids and compounds derived from them. In the case of a synapse linking a motor neuron and its target muscle cell, the neurotransmitter is acetylcholine:

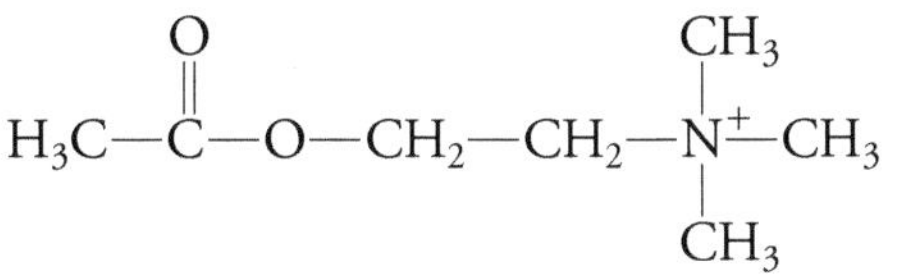

$$H_3C-\overset{\displaystyle O}{\overset{\|}{C}}-O-CH_2-CH_2-\overset{\displaystyle CH_3}{\underset{\displaystyle CH_3}{\overset{|}{\underset{|}{N^+}}}}-CH_3$$

Acetylcholine

Acetylcholine is stored in membrane-bounded compartments, called **synaptic vesicles,** about 40 nm in diameter. When the action potential reaches the axon terminus, voltage-gated Ca^{2+} channels open and allow the influx of extracellular Ca^{2+} ions. The increase in the local intracellular Ca^{2+} concentration, from less than 1 µM to as much as 100 µM, triggers **exocytosis** of the vesicles (exocytosis is the fusion of the vesicle with the plasma membrane such that the vesicle contents are released into the extracellular space). The acetylcholine diffuses across the synaptic cleft, the narrow space between the axon terminus and the muscle cell, and binds to integral membrane protein receptors on the muscle cell surface. This binding initiates a sequence of events that result in muscle contraction (**Fig. 9.20**). In general, the response of the cell that receives a neurotransmitter depends on the nature of the neurotransmitter and the cellular proteins that are activated when the neurotransmitter binds to its receptor. We will explore other receptor systems in Chapter 10.

The events at the nerve–muscle synapse occur rapidly, within about one millisecond, but the effects of a single action potential are limited. First, the Ca^{2+} that triggers neurotransmitter release is quickly pumped back out of the cell by the Ca^{2+}-ATPase. Second, acetylcholine in the synaptic cleft is degraded within a few milliseconds by a lipid-linked or soluble acetylcholinesterase, which catalyzes the reaction

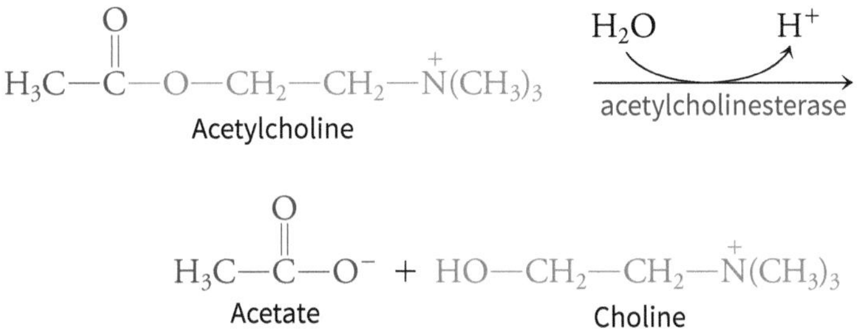

$$H_3C-\overset{\displaystyle O}{\overset{\|}{C}}-O-CH_2-CH_2-\overset{+}{N}(CH_3)_3 \quad \xrightarrow[\text{acetylcholinesterase}]{H_2O \quad\quad H^+}$$

Acetylcholine

$$H_3C-\overset{\displaystyle O}{\overset{\|}{C}}-O^- \ + \ HO-CH_2-CH_2-\overset{+}{N}(CH_3)_3$$

Acetate   Choline

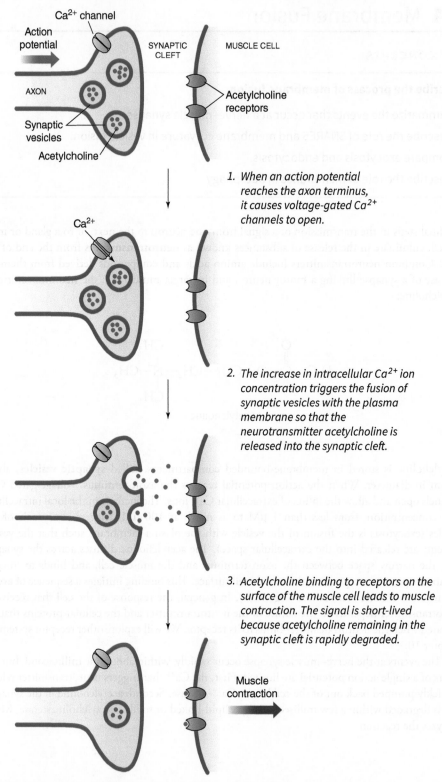

Figure 9.20 Events at the nerve–muscle synapse.

Question Describe the events that must occur to restore the neuron and muscle cell to their original states.

Other types of neurotransmitters are recycled rather than destroyed. They are transported back into the cell that released them by the action of secondary active transporters (**Box 9.B**). A neuron may contain hundreds of synaptic vesicles, only a small percentage of which undergo exocytosis at a time, so the cell can release neurotransmitters repeatedly (as often as 50 times per second).

Box 9.B   Antidepressants Block Serotonin Transport

The neurotransmitter serotonin, a derivative of tryptophan, is released by cells in the central nervous system.

Serotonin

Fluoxetine

Sertraline

Serotonin signaling leads to feelings of well-being, suppression of appetite, and wakefulness, among other things. Seven different families of receptor proteins respond to serotonin signals, sometimes in opposing ways, so the pathways by which this neurotransmitter affects mood and behavior have not been completely defined.

Unlike acetylcholine, serotonin is not broken down in the synapse, but instead about 90% of it is transported back into the cell that released it and is reused. Because the extracellular concentration of serotonin is relatively low, it reenters cells via symport with Na^+, whose extracellular concentration is greater than its intracellular concentration. The rate at which the serotonin transporter takes up the neurotransmitter helps regulate the extent of signaling. Research suggests that genetic variation in the transporter protein may explain an individual's susceptibility to conditions such as depression and post-traumatic stress disorder, but such correlations are difficult to prove, since the level of expression of the transporter appears to vary between individuals and even within an individual.

Drugs known as selective serotonin reuptake inhibitors (SSRIs) block the transporter and thereby enhance serotonin signaling. Some of the most widely prescribed drugs are SSRIs, including fluoxetine and sertraline.

These drugs are used primarily as antidepressants, although they are also prescribed for anxiety disorders and obsessive-compulsive disorder. Despite decades of research, the interactions between the serotonin transporter and the drugs are not completely understood at the molecular level, and some studies indicate that different inhibitors may bind to different sites on the transporter.

The results of rigorous clinical tests suggest that SSRIs are most effective at treating severe disorders, whereas in mild cases of depression, the SSRIs are about as effective as a placebo. One of the challenges in assessing the clinical effectiveness of drugs such as fluoxetine and sertraline is that depression is difficult to define biochemically. Furthermore, serotonin signaling pathways are complex, and the body responds to SSRIs with changes in gene expression and adjustments in other signaling pathways so that antidepressive effects may not be apparent for several weeks. The list of SSRI side effects is long and highly variable among individuals, and, disturbingly, includes a small increase in risk of suicidal behavior.

Question   **What types of food could contribute to serotonin production in the body?**

SNAREs link vesicle and plasma membranes

Membrane fusion is a multistep process that begins with the targeting of one membrane (for example, the vesicle) to another (for example, the plasma membrane). A number of proteins participate in tethering the two membranes and readying them for fusion. However, many of these proteins may be only accessory factors for the SNAREs, the proteins that physically pair the two membranes and induce them to fuse.

SNAREs are integral membrane proteins (their name is coined from the term "soluble *N*-ethylmaleimide-sensitive-factor attachment protein receptor"). Two SNAREs from the plasma membrane and one from the synaptic vesicle form a complex that includes a 120-Å-long coiled-coil structure containing four helices (two of the SNAREs contribute one helix each, and one SNARE contributes two helices). The four helices, each with about 70 residues, line up in parallel fashion (**Fig. 9.21**). Unlike other coiled-coil proteins such as keratin (Section 5.3), the four-helix bundle is not a perfectly geometric structure but varies irregularly in diameter.

Figure 9.21 Structure of the four-helix bundle of the SNARE complex.   The three proteins (one includes two helices) are in different colors. Portions of the SNAREs that do not form the helix bundle were cleaved off before X-ray crystallography.

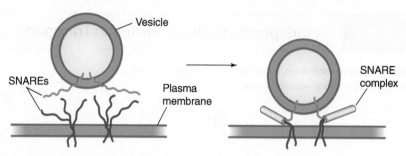

Figure 9.22 Model for SNARE-mediated membrane fusion. Formation of the complex of SNAREs from the vesicle and plasma membranes brings the membranes close together so that they can fuse.

Question Why does disassembly of the SNARE complex require ATP?

The mutual interactions between SNAREs in the vesicle and plasma membranes serve as an addressing system so that the proper membranes fuse with each other. Initially, the individual SNARE proteins are unfolded, and they spontaneously zip up to form the four-helix complex. This action necessarily brings the two membranes close together (**Fig. 9.22**). *Although the formation of the SNARE complex is thermodynamically favorable, membrane fusion may require the presence of additional proteins.* The rapid rate of acetylcholine release *in vivo* indicates that at least some synaptic vesicles are already docked at the plasma membrane, awaiting the Ca^{2+} signal for fusion to proceed.

Experiments with pure lipid vesicles demonstrate that SNAREs are not essential for membrane fusion to occur *in vitro,* but the rate of fusion does depend on the membranes' lipid compositions. The explanation for this observation is that the lipid bilayers of the fusing membranes undergo rearrangement: The lipids in the contacting leaflets must mix before a pore forms (**Fig. 9.23**). Certain types of lipids appear to promote the required changes in membrane curvature.

In living cells, bilayer shape changes could be facilitated by the tension exerted by the SNARE complex. In addition, membrane lipids may undergo active remodeling. For example, the enzymatic removal of an acyl chain would convert a cylindrical lipid to a cone-shaped lipid. Clustering of such lipids would cause the bilayer to bow outward.

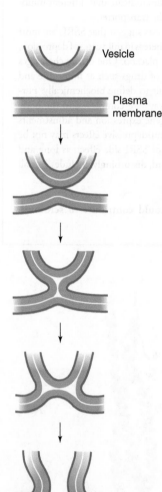

Vesicle

Plasma membrane

Figure 9.23 Schematic view of membrane fusion. For simplicity, the vesicle and plasma membranes are depicted as bilayers.

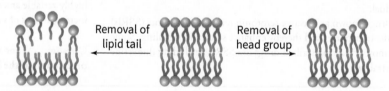

Removal of lipid tail

Removal of head group

Conversely, removing large lipid head groups would cause the bilayer to bow inward.

Endocytosis is the reverse of exocytosis

When neurotransmitter vesicles fuse with the plasma membrane during exocytosis, the plasma membrane gains membrane proteins and lipids. These materials must be removed and recycled in order to maintain the shape of the neuron and to generate new neurotransmitter vesicles for the next round of neurotransmission. One possibility is that soon after the vesicle and plasma membranes fuse, the "fusion pore" (shown in Fig. 9.23) pinches shut and the vesicle re-forms. The empty vesicle can then be refilled with neurotransmitters.

Another mechanism involves **endocytosis,** in which a new vesicle forms by budding inward from the plasma membrane. This process also requires the membrane to pinch off (the reverse of the process shown in Fig. 9.23) but could occur anywhere on the cell surface. Cells are capable of several kinds of endocytosis (**Fig. 9.24**). **Pinocytosis** is the formation of small intracellular vesicles that contain extracellular fluid plus whatever small molecules happen to be present. In **receptor-mediated endocytosis,** the materials brought into the cell must first bind specifically to a protein receptor on the cell surface. This binding event triggers the membrane shape changes that lead to formation of an intracellular vesicle called an **endosome.** The endosome may subsequently fuse

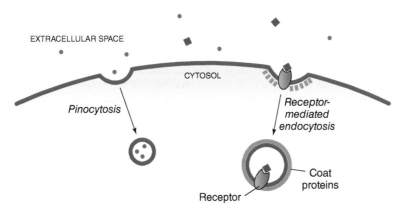

Figure 9.24 Endocytosis. Pinocytosis captures extracellular materials nonspecifically. In receptor-mediated endocytosis, a molecule binds to a specific receptor, which is then internalized along with its bound ligand. Coat proteins just beneath the membrane interact with the receptors and help mediate the membrane shape changes required to pinch off a portion of the cell membrane and form a vesicle.

with a lysosome (or lysosomal enzymes may be delivered to the vesicle) so that its contents can be digested by the lysosomal enzymes.

Formation of membranous vesicles—either at the cell surface or involving other membrane systems—often involves "coated" vesicles. So-called coat proteins form a lattice or cage on the cytosolic side of the membrane, forcing the membrane to bud into a spherical shape and then maintaining that structure as the vesicle makes its way to another site in the cell. Vesicle transport often involves a motor protein such as kinesin (Section 5.4), which follows microtubule tracks.

Vesicle trafficking between the endoplasmic reticulum and Golgi apparatus, and between the Golgi apparatus and the plasma membrane, involves coated vesicles. One well known coat protein is clathrin, whose spindly subunits assemble into a regular geometric structure around a vesicle (**Fig. 9.25**). Clathrin and other coat proteins must form strong yet flexible networks that change shape as the vesicle matures and separates from its parent membrane. An interesting type of vesicle formation occurs in the production of exosomes (**Box 9.C**).

Autophagosomes enclose cell materials for degradation

Vesicle formation within the cytoplasm is the basis for **autophagy,** literally, *self-eating*. Autophagy, which occurs in all eukaryotes, maintains cellular homeostasis by recycling intracellular components

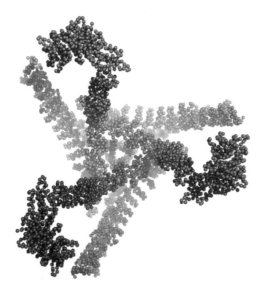

Figure 9.25 Clathrin assembly. Three clathrin subunits are shown in this model derived from electron micrographs. Multiple subunits assemble to form a lattice that surrounds and defines the shape of an intracellular vesicle.

Box 9.C   Exosomes

Exosomes, also known as **microvesicles,** are small extracellular membrane-enclosed particles. Although they were once believed to function as a sort of microscopic garbage can containing unwanted cellular components, they are now believed to participate in informal cell–cell communication. Virtually all eukaryotic cells release exosomes, which are typically 30 to 100 nm in diameter and contain an assortment of the cell's proteins, lipids, and nucleic acids.

Rather than budding directly from the plasma membrane, exosomes are formed within an intracellular compartment, which then fuses with the plasma membrane to release the exosomes (see diagram below).

Some cells produce exosomes constitutively (constantly), but others seem to release them in response to stress or some other stimulation. In animals, exosomes may circulate throughout the body, so this transport system may be a mechanism for sharing information

about what's going on inside a certain type of cell. Recipient cells appear to take up exosomes and their cargo by receptor-mediated endocytosis or by direct membrane fusion.

The RNA molecules delivered by exosomes may be particularly useful in cell–cell communication, because mRNAs indicate which genes are being expressed and small RNAs help regulate gene expression. Cancer cells release exosomes containing components that could make neighboring cells more accommodating of tumor expansion. Exosomes also appear to play a role in coordinating defenses during an immune response.

Because exosomes are present in all body fluids, including blood and urine, they may provide diagnostic information that previously required more invasive procedures such as tissue biopsies. Exosomes have also attracted attention as systems for delivering drugs, gene-edited DNA, or materials that promote tissue healing after injury.

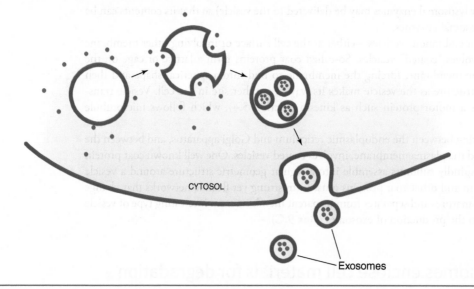

CYTOSOL

Exosomes

that are no longer needed or are potentially harmful. The undesirable items include normal cytoplasmic contents as well as damaged organelles, aggregated proteins, or certain intracellular pathogens. The rate of autophagy increases in response to certain stressors, such as starvation, hypoxia, and viral infection, and poorly regulated autophagy may underlie some disease states.

The process of autophagy begins with membrane budding from the endoplasmic reticulum (the main source of a cell's lipids; Section 8.4). The resulting vesicle, the **phagophore,** assumes a cuplike shape as it begins to surround material to be degraded (**Fig. 9.26**). In some instances, the phagophore randomly encloses a portion of cytoplasm, but in other cases, autophagy specifically targets some unwanted cellular component. The phagophore expands as vesicles deliver additional membrane material, until membrane fusion produces the **autophagosome,** a compartment surrounded by a double membrane that fully encloses the items to be recycled. Finally, the delivery of lysosomal enzymes generates the **autolysosome,** in which the captured macromolecules are degraded to monomeric units that can be re-used by the cell. Proteins belonging to the ATG (autophagy-related gene) family help orchestrate autophagosome maturation and fusion with a lysosome, as shown by abnormal autophagy in organisms with ATG mutations. However, the factors that select targets for autophagy and regulate the overall process are not completely understood.

Inadequate autophagy is believed to contribute to the accumulation of misfolded proteins that characterize a number of neurodegenerative diseases (Section 4.5). Similarly, a low rate of

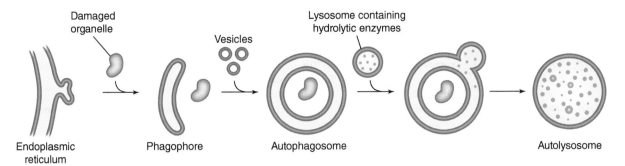

Figure 9.26 Autophagy. The cuplike phagophore originates from the endoplasmic reticulum and surrounds unneeded cellular material, such as a damaged organelle. Expansion and membrane fusion produce the autophagosome, a compartment with a double membrane. Fusion with a lysosome introduces hydrolytic enzymes, generating an autolysosome.

Question **Explain why lysosomal enzymes must be able to degrade a lipid bilayer in addition to macromolecules.**

autophagy may lead to the pathology of metabolic syndrome (Section 19.3); in this so-called lifestyle disease, cells saturated with sugar and fat may not engage in the starvation-induced housecleaning that autophagy normally accomplishes. Conversely, high rates of autophagy have been documented in some types of cancer cells, which apparently use autophagy to maximize the supply of small molecules that support rapid cell growth. Inhibiting autophagy is therefore an attractive option for treating some forms of cancer, provided that normal rates of autophagy can be maintained in noncancerous cells.

Concept Check

1. Explain how acetylcholine inside a nerve cell reaches a muscle cell.
2. Relate the structures of SNARES and clathrin to their functions.
3. Explain why membrane fusion and changes in bilayer curvature are essential for endocytosis, exocytosis, and autophagy.

SUMMARY

9.1 The Thermodynamics of Membrane Transport

- The transmembrane movements of ions generate changes in membrane potential, $\Delta\psi$, during neuronal signaling.
- The free energy change for the transmembrane movement of a substance depends on the concentrations on each side of the membrane and, if the substance is charged, on the membrane potential.

9.2 Passive Transport

- Passive transport proteins such as porins allow the transmembrane movement of substances according to their concentration gradients. Aquaporins mediate the transport of water molecules.
- Ion channels have a selectivity filter that allows passage of one type of ion. Gated channels open and close in response to some other event.
- Membrane proteins such as the passive glucose transporter undergo conformational changes that alternately expose ligand-binding sites on each side of the membrane.

9.3 Active Transport

- Active transporters such as the Na,K-ATPase and ABC transporters require the free energy of ATP to drive the transmembrane movement of substances against their concentration gradients.
- Secondary active transport allows the favorable movement of one substance to drive the unfavorable transport of another substance.

9.4 Membrane Fusion

- During the release of neurotransmitters, intracellular vesicles fuse with the cell membrane. SNARE proteins in the vesicle and target membranes form a four-helix structure that brings the fusing membranes close together. Changes in bilayer curvature are also necessary for fusion to occur.
- In endocytosis, a vesicle buds off from an existing membrane to form an intracellular vesicle.
- In autophagy, a double-membrane organelle forms around unneeded cellular materials and fuses with a lysosome.

KEY TERMS

membrane potential $(\Delta\psi)$	axon	gated channel	antiporter	synaptic vesicle	endosome
gas constant (R)	myelin sheath	complement	ATPase	exocytosis	exosome (microvesicle)
Z	passive transport	osmosis	ABC transporter	endocytosis	autophagy
Faraday constant $(\mathcal{F})$	active transport	aquaporin	secondary active transport	pinocytosis	phagophore
action potential	ionophore	uniporter	neurotransmitter	receptor-mediated endocytosis	autophagosome
	porin	symporter			autolysosome

BIOINFORMATICS

Brief Bioinformatics Exercises

9.1 Maltoporin and OmpF

PROBLEMS

9.1 The Thermodynamics of Membrane Transport

1. Calculate the membrane potential at 20°C when **a.** $[Na^+]_{in}$ = 20 mM and $[Na^+]_{out}$ = 200 mM and **b.** $[Na^+]_{in}$ = 50 mM and $[Na^+]_{out}$ = 25 mM.

2. Calculate the membrane potential at 20°C when **a.** $[K^+]_{in}$ = 155 mM and $[K^+]_{out}$ = 4 mM and **b.** $[K^+]_{in}$ = 410 mM and $[K^+]_{out}$ = 12 mM.

3. Calculate the intracellular concentration of Na^+ in the giant squid axon when the extracellular concentration is 550 mM. Assume a membrane potential of −60 mV at 20°C.

4. Calculate the extracellular concentration of K^+ in frog muscle when the intracellular concentration is 124 mM. Assume a membrane potential of 100 mV at 20°C.

5. The resting membrane potential maintained in most nerve cells is about −70 mV. Use Equation 9.2 to calculate the ratio of $[Na^+]_{in}/[Na^+]_{out}$ at the resting potential.

6. Use Equation 9.4 to calculate the free energy change for the movement of Na^+ into the depolarized nerve cell described in Problem 6. Assume the temperature is 37°C. How does this compare with the value you calculated in Problem 7 and what is the significance of the difference?

7. In typical marine organisms, the intracellular concentrations of Na^+ and Ca^{2+} are 20 mM and 0.2 μM, respectively. Extracellular concentrations of Na^+ and Ca^{2+} are 500 mM and 5 μM, respectively. Use Equation 9.4 to calculate the free energy changes at 20°C for the transmembrane movement of these ions. In which direction do the ions move? Assume the membrane potential is −70 mV.

8. Calculate the free energy changes at 20°C for the transmembrane movement of Na^+ and K^+ ions using the conditions presented in Figure 9.1. Assume the membrane potential is −70 mV. In which direction do the ions move?

9. The concentration of Ca^{2+} in the cytosol (*inside*) is 0.1 μM and the concentration of Ca^{2+} in the extracellular medium (*outside*) is 2 mM.

Calculate ΔG at 37°C, assuming the membrane potential is −50 mV. In which direction is Ca^{2+} movement thermodynamically favored?

10. Calculate the free energy change for the movement of Cl^- from the extracellular medium, where $[Cl^-]$ = 120 mM, into the cytosol, where $[Cl^-]$ = 5 mM. Assume that T = 37°C and $\Delta\psi$ = −50 mV. Is this process spontaneous?

11. The concentration of H^+ in the lysosomes (*outside*) is 32 μM, and the concentration of H^+ in the cytosol (*inside*) is 63 nM. Calculate ΔG at 37°C when the membrane potential is −30 mV (cytosol negative). In which direction is H^+ movement thermodynamically favorable?

12. A high fever can interfere with normal neuronal activity. Since temperature is one of the terms in Equation 9.1, which defines membrane potential, a fever could potentially alter a neuron's resting membrane potential. **a.** Calculate the effect of a change in temperature from 98°F to 104°F (37°C to 40°C) on a neuron's membrane potential. Assume that the normal resting potential is −70 mV and that the distribution of ions does not change. **b.** How else might an elevated temperature affect neuronal activity?

13. During mitochondrial electron transport (Chapter 15), protons are transported across the inner mitochondrial membrane from inside the mitochondrial matrix to the intermembrane space. The pH of the mitochondrial matrix is 7.78. The pH of the intermembrane space is 6.88. **a.** What is the membrane potential under these conditions? **b.** Use Equation 9.4 to determine the free energy change for proton transport at 37°C.

14. Calculate the value of ΔG for the movement of glucose from outside to inside at 20°C when the extracellular concentration is 5 mM and the cytosolic concentration is 0.5 mM.

15. Glucose absorbed by the epithelial cells lining the intestine leaves these cells and travels to the liver via the portal vein. After a high-carbohydrate meal, the concentration of glucose in the portal vein can reach 15 mM. **a.** What is the ΔG for transport of glucose from portal vein blood to the interior of the liver cell where the concentration of glucose is 0.5 mM? **b.** What is the ΔG for transport under fasting conditions when the blood glucose level falls to 4 mM? Assume the temperature is 37°C.

16. Rank the rate of transmembrane diffusion of the following compounds:

$$H_3C-\overset{\overset{\displaystyle O}{\|}}{C}-NH_2$$

A. Acetamide

$$H_3C-CH_2-CH_2-\overset{\overset{\displaystyle O}{\|}}{C}-NH_2$$

B. Butyramide

$$H_2N-\overset{\overset{\displaystyle O}{\|}}{C}-NH_2$$

C. Urea

17. The permeability coefficient indicates a solute's tendency to move from an aqueous solvent to a nonpolar lipid membrane. The permeability coefficients for the compounds shown in Problem 23 are listed in the table. How do the permeability coefficients assist you in ranking the rate of the transmembrane diffusion of these compounds?

	Permeability coefficient ($cm \cdot s^{-1}$)
Acetamide	9×10^{-6}
Butyramide	1×10^{-5}
Urea	1×10^{-7}

18. The permeability coefficients (see Problem 24) of glucose and mannitol for both natural and artificial membranes are shown in the table below. **a.** Compare the permeability coefficients for the two solutes (structures shown below). Which solute moves more easily through the synthetic bilayer and why? **b.** Compare the permeability coefficients of each solute for the two types of membranes. For which solute is the difference more dramatic and why?

	Permeability coefficient ($cm \cdot s^{-1}$)	
	Glucose	**Mannitol**
Synthetic lipid bilayer	2.4×10^{-10}	4.4×10^{-11}
Red blood cell membrane	2.0×10^{-4}	5.0×10^{-9}

$$
\begin{array}{ccc}
\text{CHO} & & \text{CH}_2\text{OH} \\
\text{H}-\text{C}-\text{OH} & & \text{HO}-\text{C}-\text{H} \\
\text{HO}-\text{C}-\text{H} & & \text{HO}-\text{C}-\text{H} \\
\text{H}-\text{C}-\text{OH} & & \text{H}-\text{C}-\text{OH} \\
\text{H}-\text{C}-\text{OH} & & \text{H}-\text{C}-\text{OH} \\
\text{CH}_2\text{OH} & & \text{CH}_2\text{OH} \\
\text{Glucose} & & \text{Mannitol}
\end{array}
$$

19. Explain how the treatment of valinomycin- and nigericin-ionophores affects the electron transport chain and ATP production in actively respiring mitochondria.

20. Why does valinomycin have a high affinity for K^+ as compared to Na^+?

21. What is the effect of valinomycin on K^+ transport when the membrane is cooled below its transition temperature?

9.2 Passive Transport

22. The bacterium *Pseudomonas aeruginosa* expresses a phosphate-specific porin when phosphate in its growth medium is limiting. Noting that there are three lysines clustered in the surface-exposed amino terminal region of the protein, investigators constructed mutants in which the lysine residues were replaced with glutamate residues. **a.** Why did the investigators hypothesize that lysine residues might play an important role in phosphate transport in the bacterium? **b.** Predict the effect of the Lys → Glu substitution on the transport activity of the porin.

23. As noted in the text, the OmpF porin in *E. coli* has a constricted diameter that prevents the passage of large substances. The protein loops in this constricted site contain a D-E-K-A sequence, which makes the porin weakly selective for cations. If you wanted to construct a mutant porin that was highly selective for calcium ions, what changes would you make to the amino acid sequence at the constricted site?

24. As a result of the activity of the acetylcholine receptor, muscle cells undergo depolarization, but the change in membrane potential is less dramatic and slower than in neurons. **a.** The acetylcholine receptor is also a gated ion channel. What triggers the gate to open? **b.** The acetylcholine receptor/ion channel is specific for Na^+ ions. Do Na^+ ions flow in or out? **c.** How does the Na^+ flow through the ion channel change the membrane potential?

25. When a bacterial cell is transferred from a solute-rich environment to pure water, mechanosensitive channels open and allow the efflux (outflow) of cytoplasmic contents. Use osmotic effects to explain how this prevents cell lysis.

26. Ammonia, like water, was long believed to diffuse across membranes without the aid of a channel protein. Researchers deleted the gene *Rhcg* from mice and examined the NH_3 permeability of their kidney cells. Cells from wild-type mice exhibited an NH_3 flux about three times higher than the mutant cells. **a.** What do these results suggest about the role of the protein encoded by *Rhcg*? **b.** In the light of what you know about aquaporin, does this surprise you?

27. The selectivity filter in the bacterial chloride channel ClC is formed in part by the OH groups of Ser and Tyr side chains and main chain NH groups. **a.** Draw a figure showing plausible interactions between these groups and Cl^-. **b.** Explain how these groups would allow Cl^- but not cations to pass through.

28. A channel protein specific for calcium ions was found to have a pore lined with six glutamate residues that participate in coordinating the calcium ion. What would happen to the selectivity of the pore if the Glu residues were mutated to Asp residues?

29. The kinetics of transport through protein transporters can be described using Michaelis–Menten terminology. The transported substance is analogous to the substrate, and the protein transporter is analogous to the enzyme. K_M and V_{max} values can be determined to describe the binding and transport process where K_M is a measure of the affinity of the transported substance for its transporter and V_{max} is a measure of the rate of the transport process. **a.** A plot of the glucose transport rate versus glucose concentration for the passive glucose transporter of red blood cells is shown below. Explain why the plot yields a hyperbolic curve. **b.** Estimate the V_{max} and the K_M values for this glucose transporter.

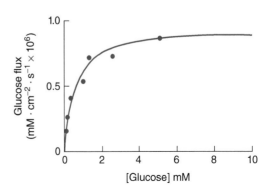

30. The compound 6-*O*-benzyl-D-galactose competes with glucose for binding to the glucose transporter. **a.** Sketch a curve similar to the one shown in Problem 35 that illustrates the kinetics of glucose transport in the presence of this inhibitor. **b.** How would the K_M and V_{max} values of 6-*O*-benzyl-D-galactose compare to those estimated for glucose in Problem 35? Explain.

31. Experiments with erythrocyte "ghosts" were carried out to learn more about the glucose transporter in red blood cells. "Ghosts" are prepared by lysing red blood cells in a hypotonic medium and washing away the cytoplasmic contents. Suspension in an isotonic buffer allows the ghost membranes to reseal. If ghosts are prepared so that the enzyme trypsin is incorporated into the ghost interior, glucose transport does not occur. However, glucose transport is not affected if trypsin is located in the extracellular ghost medium. What can you conclude about the glucose transporter, given these observations?

32. If a propyl group is added to the hydroxyl group on C1 of glucose, the modified glucose is unable to bind to its transporter on the extracellular surface. If the propyl group is added on C6 of glucose, the modified glucose is unable to bind to its transporter on the cytosolic surface of the membrane. What do these observations reveal about the mechanism of passive glucose transport?

33. Glutamate acts as a neurotransmitter in the brain and is reabsorbed and recycled by glial cells associated with the neurons. The glutamate transporter also transports 3 Na^+ and 1 H^+ along with glutamate, and it transports 1 K^+ in the opposite direction. What is the net charge imbalance across the cell membrane for each glutamate transported?

34. The bacterium *Oxalobacter formigenes* is found in the intestine, where it plays a role in digesting the oxalic acid found in some fruits and vegetables (spinach is a particularly rich source). The bacterium takes up oxalic acid from the extracellular medium in the form of oxalate. Once inside the cell, the oxalate undergoes decarboxylation to produce formate, which leaves the cell in antiport with the oxalate. **a.** Draw a diagram of this process. **b.** What is the net charge imbalance generated by this system? **c.** Why do you suppose the investigators who elucidated this mechanism referred to the process as a "hydrogen pump"?

$$^-OOC-COO^- \xrightarrow[\text{decarboxylase}]{} H-COO^-$$
Oxalate H^+ CO_2 Formate

35. As discussed in Section 5.1, tissues produce CO_2 as a waste product during respiration. The CO_2 enters the red blood cell and combines with water to form carbonic acid in a reaction catalyzed by carbonic anhydrase. A red blood cell protein called Band 3 transports HCO_3^- ions in exchange for Cl^- ions. **a.** What role does Band 3 play in transporting CO_2 to the lungs where it can be exhaled? **b.** Band 3 is unusual in that it can transport HCO_3^- in exchange for Cl^- ions, in either direction, via passive transport. Why is it important that Band 3 operates in both directions?

36. A group of investigators was interested in the mechanism of lactate transport in *Plasmodium vivax*, which causes malaria. They carried out experiments in which they isolated parasites and suspended them in a buffer containing [^{14}C]-lactate at two different pH values, as shown in the figure. Propose a mechanism for lactate transport consistent with the data.

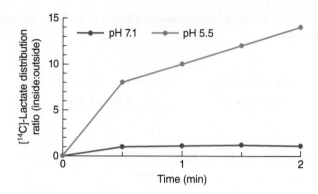

37. Glucose uptake by cancer cells via the GLUT1 transport protein occurs far more rapidly than in normal cells. The cancer cell metabolizes the glucose to lactic acid, which exits the cell via the MCT4 transport protein. In response, the activity of carbonic anhydrase (see Section 2.5 and Problem 41), whose catalytic domain is on the outside of the cancer cell, increases. The carbonic anhydrase is associated with a bicarbonate transporter, which assists in maintaining the slightly alkaline pH that cancer cells prefer. **a.** Why is it advantageous for the cancer cell to up-regulate the activity of carbonic anhydrase? **b.** List several drug targets that have the potential to interfere with this process and kill the cancer cell.

38. Liver cells use a choline transport protein to import choline from the portal circulation. Choline transport was measured in mouse cells transfected with the gene for the hepatic transporter. The uptake of radioactively labeled choline by the transfected cells was measured at increasing choline concentrations.

$$HO-(CH_2)_2-\overset{+}{N}\overset{\displaystyle CH_3}{\underset{\displaystyle CH_3}{|}}-CH_3$$
Choline

a. Use Michaelis–Menten terminology as described in Problem 35 to estimate K_M and V_{max} from the curve shown.

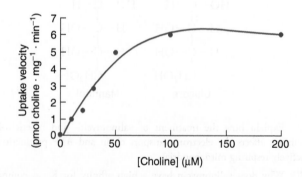

b. Plasma concentrations of choline range from 10 to 80 μM, although the concentration may be higher in the portal circulation after ingestion of choline. Does the transporter operate effectively at these concentrations? **c.** The choline transport protein is inhibited by low external pH and stimulated by high external pH. What role might protons play in the transport of choline? **d.** Propose a mechanism that explains how tetramethylammonium ions (TEA, see structure below) inhibit choline transport.

$$H_3C—CH_2—\overset{\overset{\displaystyle CH_3}{\overset{|}{\displaystyle CH_2}}}{\underset{\underset{\displaystyle CH_3}{\underset{|}{\displaystyle CH_2}}}{\overset{+}{N}}}—CH_2—CH_3$$

Tetraethylammonium (TEA)

39. Unidirectional glucose transport into the brain was measured in the presence and absence of phlorizin. The velocity of transport was determined at various glucose concentrations and the data are displayed as a Lineweaver–Burk plot. **a.** Calculate the K_M and the V_{max} (see Problem 35) in the presence and absence of phlorizin. **b.** What kind of inhibitor is phlorizin? Explain.

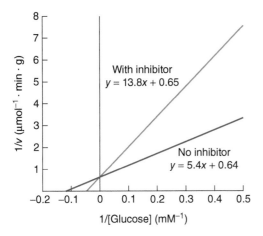

40. Bacterial fermentation of fiber-containing foods in the intestine generates butyrate, a short-chain fatty acid. A carboxylate transporter in the intestinal epithelial cells takes up butyrate, which is used by the cells as an energy source. **a.** Estimate the K_M and V_{max} (see Problem 35) for the carboxylate transporter from the graph below. **b.** The carboxylate transporter is regulated by the hormone leptin, which is secreted by adipose tissue and acts as an appetite suppressant. What is the effect of leptin on the K_M and V_{max} values of the transporter? Explain the significance of these values.

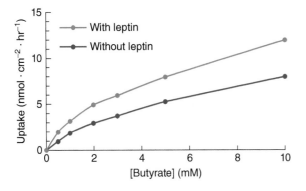

9.3 Active Transport

41. Ouabain, an extract of the East African ouabio tree, has been used as an arrow poison. Ouabain binds to an extracellular site on the Na,K-ATPase and prevents hydrolysis of the phosphorylated intermediate. Why is ouabain a lethal poison?

42. The parasite *Trypanosoma brucei*, which causes sleeping sickness, uses proline as an energy source during one stage of its life cycle. The properties of its proline-specific transporter were investigated in a series of experiments. It was discovered that L-hydroxyproline (see Section 5.3) is a potent inhibitor of the transporter whereas D-proline is not, and that proline transport is not affected by pH, Na⁺, K⁺, or ouabain (see Problem 48). Incubation of the cells with iodoacetate (which decreases intracellular ATP) decreases the rate of proline transport into the parasite. What do these experiments reveal about the nature of the proline transporter in trypanosomes?

43. In eukaryotes, ribosomes (approximate mass 4×10^6 D) are synthesized inside the nucleus, which is enclosed by a double membrane. Protein synthesis occurs in the cytosol. **a.** Would a glucose transporter-type protein be capable of transporting ribosomes into the cytosol? Would a porin-type transporter be able to do so? Explain why or why not. **b.** Do you think that free energy would be required to move ribosomes from the nucleus to the cytoplasm? Why or why not?

44. The retina of the eye contains equal amounts of endothelial and pericyte cells. Basement membrane thickening in pericytes occurs during the early stages of diabetic retinopathy, eventually leading to blindness. Glucose uptake was measured in both types of cells in culture in the presence of increasing amounts of sodium. The results are shown in the graph. **a.** How does glucose uptake vary in the two cell types? **b.** What information is conveyed by the shapes of the curves? **c.** By what mechanism might the pericytes use sodium ions to assist with glucose import?

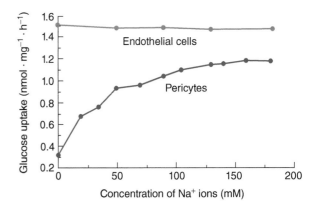

45. Many ABC transporters are inhibited by vanadate, a phosphate analog. Why is vanadate an effective inhibitor of these transporters?

46. The ABC multidrug transporter LmrA from *Lactococcus* exports cytotoxic compounds like vinblastine. There are two binding sites for vinblastine on the transporter; one vinblastine binds with an association constant (equivalent to $1/K_d$) of 150 nM; the association constant of the second vinblastine is 30 nM. What do these values tell you about the mode of vinblastine binding to LmrA?

47. Many cells have a mechanism for exporting ammonium ions. Describe how this could occur through secondary active transport.

48. The PEPT1 transporter aids in digestion by transporting di- and tripeptides into the cells lining the small intestine. There are three components of this system: (a) a symport transporter that ferries di- and tripeptides across the membrane, along with an H^+ ion, (b) an Na^+/H^+ antiporter, and (c) an Na,K-ATPase. Draw a diagram that illustrates how these three transporters work together to transport peptides into the cell.

49. The H,K-ATPase in gastric parietal cells is responsible in part for the secretion of hydrochloric acid into the stomach to aid in protein digestion. Parietal cells have two membrane domains: the apical membrane, which faces the stomach lumen, and the basolateral membrane, which has access to the circulatory system. There are five components of the secretion system: (a) the H,K-ATPase antiporter, (b) a Cl^-/HCO_3^- antiporter similar to Band 3 in blood cells (see Problem 41), (c) the Na,K-ATPase, (d) a K^+/Cl^- symporter, and (e) cytosolic carbonic anhydrase. Draw a diagram that illustrates how these transporters work to secrete HCl into the stomach lumen.

50. The X-ray structure of a Ca^{2+}-ATPase was determined by crystallizing the enzyme along with adenosine-5′-(βγ-methylene) triphosphate (AMPPCP), an ATP analog (structure shown below). How did the co-crystallization strategy assist the crystallographers in imaging this protein?

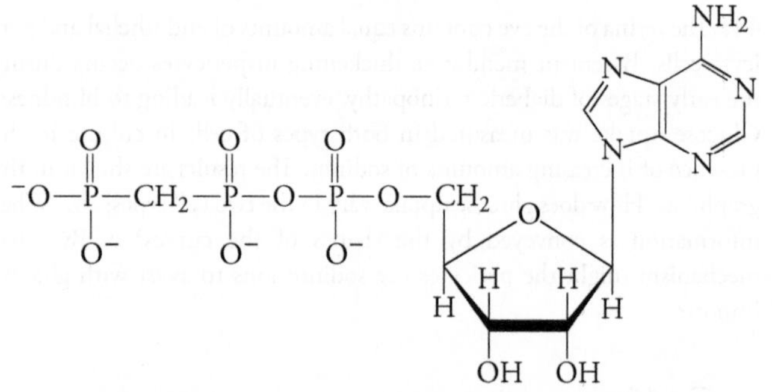

Adenosine-5′-(β,γ-methylene) triphosphate (AMPPCP)

9.4 Membrane Fusion

51. Myasthenia gravis is an autoimmune disorder characterized by muscle weakness and fatigue. Patients with the disease generate antibodies against the acetylcholine receptor in the postsynaptic cell; this results in a decrease in the number of receptors. The disease can be treated by administering drugs that inhibit acetylcholinesterase. Why is this an effective strategy for treating the disease?

52. Lambert–Eaton syndrome is another autoimmune muscle disorder (see Problem 59), but in this disease, antibodies against the voltage-gated calcium channels in the presynaptic cell prevent the channels from opening. Patients with this disease suffer from muscle weakness. Explain why.

53. Parathion and malathion are organophosphorus compounds similar to DIPF (see Problem 61). These compounds are sometimes used as insecticides. Why are these compounds deadly poisons?

54. The Na,K-ATPase is essential for establishing the ion gradients that make neuronal signaling possible. Explain why the pump is also required to recycle serotonin.

55. The illicit drug 3,4-methylenedioxymethylamphetamine (MDMA, also known as Ecstasy) decreases the expression of the serotonin transporter in the brain. Explain how this would alter the MDMA user's mood.

56. In addition to being an SSRI, fluoxetine (see Box 9.B) may also bind to and block signaling by one type of serotonin receptor. Would you expect this activity to be consistent with fluoxetine's ability to treat the symptoms of depression?

57. Serotonin and other so-called monoamine signaling molecules are degraded in the liver, beginning with a reaction catalyzed by a monoamine oxidase (MAO). Explain why doctors avoid prescribing an MAO inhibitor along with an SSRI.

58. The toxin produced by *Clostridium tetani*, which causes tetanus, is a protease that cleaves and destroys SNAREs. Explain why this activity would lead to muscle paralysis.

59. The drug known as Botox is a preparation of botulinum toxin, which is similar to the tetanus toxin (Problem 67). Describe the biochemical basis for its use by plastic surgeons, who inject small amounts of it to alleviate wrinkling in areas such as around the eyes.

60. Phosphatidylinositol is a membrane glycerophospholipid whose head group includes a monosaccharide (inositol) group. A kinase adds another phosphate group to a phosphorylated phosphatidylinositol. Why might this activity be required during the production of a new vesicle, which forms by budding from an existing membrane?

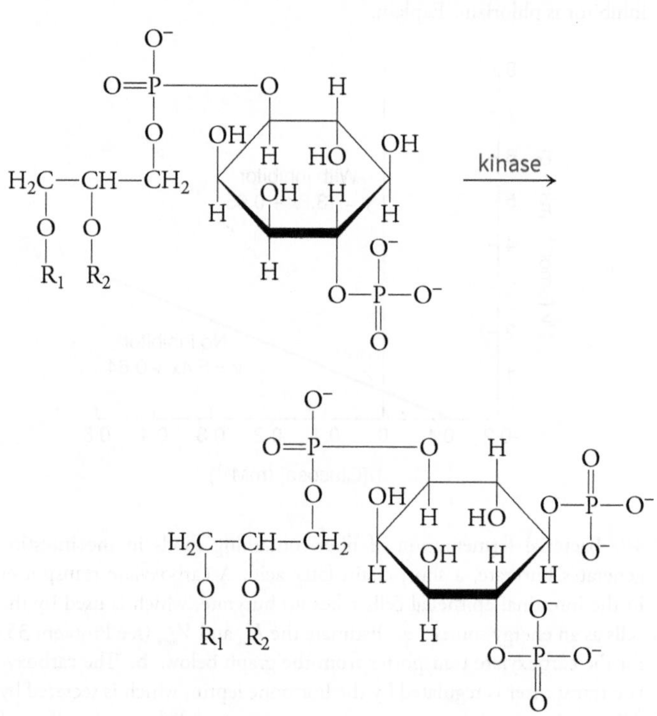

61. Some studies show that prior to membrane fusion, the proportion of diacylglycerol in the bilayer increases. Explain how the presence of this lipid would aid the fusion process.

62. During starvation, yeast cells increase the rate of autophagy. Explain why this makes sense.

63. An early event in autophagy is the phosphorylation of phosphatidylinositol in the endoplasmic reticulum membrane to generate a lipid with an additional phosphate group in its head group (phosphatidylinositol-3-phosphate). Explain how this lipid modification could promote the formation of the phagophore.

64. Low-density lipoprotein (LDL, see Problem 8.78 for a description) enters the cell via receptor-mediated endocytosis by binding to a specific cell-surface receptor that localizes to an invaginated area on the membrane surface called a clathrin-coated pit (see Fig. 9.23). Once internalized, the LDL particle is degraded; cholesterol is available to the cell for biosynthetic reactions, and the rate-limiting enzyme that catalyzes the cell's own synthesis of cholesterol is inactivated. The experimental data used to elucidate this process in a normal person are shown below. JD, a patient with severe familial hypercholesterolemia, inherited defective alleles from each parent: an allele from his mother that prevented production of LDL receptors and an allele from his father that produced defective LDL receptors. Redraw these graphs to show how the normal process would be altered in this patient and explain your reasoning.

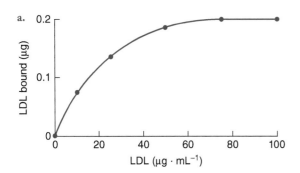

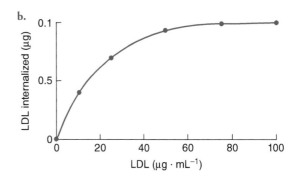

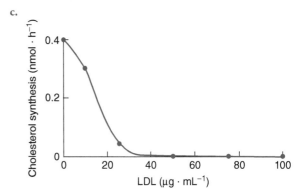

SELECTED READINGS

de la Ballina, L.R., Munson, M.J., and Simonsen, A., Lipids and lipid-binding proteins in selective autophagy. *J. Mol. Biol.* 432, 135–159, doi: 10.1016/j.jmb.2019.05.051 (2020). [A thorough review of current knowledge of some of the molecules that direct autophagy, with helpful diagrams.]

Deng, D. and Yan, N., GLUT, SGLT, and SWEET: Structural and mechanistic investigations of the glucose transporters, *Prot. Sci.* 25, 546–558, doi: 10.1002/pro.2858 (2015). [Provides a brief overview of the different types of glucose transport proteins and their physiological significance.]

Jahn, R. and Fasshauer, D., Molecular machines governing exocytosis of synaptic vesicles, *Nature* 490, 201–207 (2012). [Reviews many of the events involved in nerve cell signaling, including vesicle fusion.]

Morth, J.P., Pederson, B.P., Buch-Pederson, M.J., Andersen, J.P., Vilsen, B., Palmgren, M.G., and Nissen, P., A structural overview of the plasma membrane Na$^+$, K$^+$-ATPase and H$^+$-ATPase ion pumps, *Nat. Rev. Mol. Cell Biol.* 12, 60–70 (2011). [Summarizes the structures and physiological importance of two active transporters.]

Ozu, M., Galizia, L., Acuña, C., and Amodeo, G., Aquaporins: More than functional monomers in a tetrameric arrangement, *Cells* 7, 209, doi: 10.3390/cells7110209 (2018). [Reviews aquaporin structure and discusses how the flow of water and other substances is regulated.]

Roux, B., Ion channels and ion selectivity, *Essays Biochem.* 61, 201–209, doi: 10.1042/EBC20160074 (2017). [Discusses the operation of ion channel proteins, focusing mostly on potassium transport.]

Zhang, Y., Liu, Y., Liu, H., and Tang, W.H., Exosomes: Biogenesis, biologic function and clinical potential, *Cell Biosci.* 9, 19, doi: 10.1186/s13578-019-0282-2 (2019). [Reviews the formation, activities, and practical applications of exosomes.]

CHAPTER 9 CREDITS

Figure 9.6 Image based on 1OPF. Cowan, S.W., Schirmer, T., Pauptit, R.A., Jansonius, J.N., The structure of OmpF porin in a tetragonal crystal form, *Structure* 3, 1041–1050 (1995).

Figures 9.7 and 9.8 Images based on 1BL8. Doyle, D.A., Cabral, J.M., Pfuetzner, R.A., Kuo, A., Gulbis, J.M., Cohen, S.L., Chait, B.T., Mackinnon, R., Potassium channel (KcsA) from Streptomyces lividans, *Science* 280, 69–77 (1998).

Figure 9.11 Image based on 1FQY. Murata, K., Mitsuoka, K., Hirai, T., Walz, T., Agre, P., Heymann, J.B., Engel, A., Fujiyoshi, Y., Structural determinants of water permeation through aquaporin-1, *Nature* 407, 599–605 (2000).

Figure 9.14 Image of GLUT3 (left) based on 4ZWC. Deng, D., Sun, P.C., Yan, C.Y., Yan, N., Crystal structure of maltose-bound human GLUT3 in the outward-open conformation at 2.6 Angstrom, *Nature* 526, 391–396 (2015). Image of GLUT1 (right) based on 4PYP. Deng, D., Yan, C.Y., Xu, C., Wu, J.P., Sun, P.C., Hu, M.X., Yan, N., Crystal structure of the human glucose transporter GLUT1, *Nature* 510, 121–125 (2014).

Figure 9.17 Image based on 4HQJ. Nyblom, M., Poulsen, H., Gourdon, P., Reinhard, L., Andersson, M., Lindahl, E., Fedosova, N., Nissen, P., Crystal structure of Na⁺, K⁺-ATPase in the Na⁺-bound state, *Science* 342, 123–127 (2013).

Figure 9.18 Image based on 3G5U. Aller, S.G., Yu, J., Ward, A., Weng, Y., Chittaboina, S., Zhuo, R., Harrell, P.M., Trinh, Y.T., Zhang, Q., Urbatsch, I.L., Chang, G., Structure of P-glycoprotein reveals a molecular basis for poly-specific drug binding, *Science* 323, 1718–1722 (2009).

Figure 9.21 Image based on 1SFC. Sutton, R.B., Brunger, A.T., Neuronal synaptic fusion complex, *Nature* 395, 347–353 (1998).

Figure 9.25 Image based on 1XI4. Fotin, A., Cheng, Y., Sliz, P., Grigorieff, N., Harrison, S.C., Kirchhausen, T., Walz, T., Molecular model for a complete clathrin lattice from electron cryomicroscopy, *Nature* 432, 573–579 (2004).

Signaling

Shutterbug75/907 images/Pixabay

The taste of a piece of fruit—sweet or bitter—depends on whether its component molecules can bind to certain receptors on the surface of cells in the taste buds. By themselves, the molecules and the receptors have no effect, but the appropriate combination of signal plus receptor generates a significant intracellular response.

Do You Remember?

- Some proteins can adopt more than one stable conformation (Section 4.3).
- Allosteric regulators can inhibit or activate enzymes (Section 7.3).
- Cholesterol and other lipids that do not form bilayers have a variety of other functions (Section 8.1).
- Integral membrane proteins completely span the bilayer by forming one or more α helices or a β barrel (Section 8.3).
- Conformational changes resulting from ATP hydrolysis drive Na⁺ and K⁺ transport in the Na,K-ATPase (Section 9.3).

All cells, including prokaryotes, must have mechanisms for sensing external conditions and responding to them. Because the cell membrane creates a barrier between outside and inside, communication typically involves an extracellular molecule binding to a cell-surface receptor. The receptor then changes its conformation to transmit information to the cell interior. A signaling pathway requires many proteins, from the receptor itself to the intracellular proteins that ultimately respond to the signal by changing their behavior. We begin this chapter by describing some characteristics of signaling pathways and then examine some well-known systems that involve G proteins and receptor tyrosine kinases.

10.1 General Features of Signaling Pathways

KEY CONCEPTS

Summarize the properties of a receptor.

- Quantify ligand binding in terms of a dissociation constant.
- Recount the events in the two main types of signal transduction.
- Describe the factors that limit signaling.

Table 10.1 Examples of Extracellular Signals

Hormone	Chemical class	Source	Physiological function
Auxin	Amino acid derivative	Most plant tissues	Promotes cell elongation and flowering in plants
Cortisol	Steroid	Adrenal gland	Suppresses inflammation
Epinephrine	Amino acid derivative	Adrenal gland	Prepares the body for action
Erythropoietin	Polypeptide (165 residues)	Kidneys	Stimulates red blood cell production
Growth hormone	Polypeptide (19 residues)	Pituitary gland	Stimulates growth and metabolism
Nitric oxide	Gas	Vascular endothelial cells	Triggers vasodilation
Thromboxane	Eicosanoid	Platelets	Activates platelets and triggers vasoconstriction

Every signaling pathway requires a **receptor,** most commonly an integral membrane protein, that specifically binds a small molecule called a **ligand.** A receptor does not merely bind its ligand in the way that hemoglobin binds oxygen; rather, *a receptor interacts with its ligand in such a way that some kind of response occurs inside the cell.* This is **signal transduction:** the signal itself does not enter the cell, but information is transmitted.

A ligand binds to a receptor with a characteristic affinity

Extracellular signals in animals and plants can take many forms, including amino acids and their derivatives, peptides, lipids, and other small molecules (**Table 10.1**), as well as larger molecules such as glycoproteins. Some signal molecules are formally called **hormones,** which are substances produced in one tissue that affect the functions of other tissues, but many signal molecules go by other names. For example, interleukins are involved in communication between white blood cells (leukocytes, from the Greek *leukos,* white). Signal molecules may travel around the entire body or reach only a local area; some signals act on the same cells that produce them. Bacteria often produce small molecules for a form of communication known as **quorum sensing:** The cells can assess their population density before committing to produce a protective biofilm (described in Section 11.2) or release a toxin.

Signaling molecules behave much like enzyme substrates: *They bind to their receptors with high affinity, reflecting the structural and electronic complementarity between each ligand and its binding site.* Receptor–ligand binding can be written as a reversible reaction, where R represents the receptor and L the ligand:

$$R + L \rightleftharpoons R \cdot L$$

Biochemists express the strength of receptor–ligand binding as a **dissociation constant,** K_d, which is the reciprocal of the association constant. For this reaction,

$$K_d = \frac{[R][L]}{[R \cdot L]} \qquad (10.1)$$

(See Sample Calculation 10.1.) In keeping with other binding phenomena, such as oxygen binding to myoglobin (Section 5.1) or substrate binding to an enzyme (Section 7.2), K_d *is the ligand concentration at which the receptor is half-saturated with ligand;* in other words, half the receptor molecules have bound ligand (**Fig. 10.1**).

A ligand that binds to a receptor and elicits a biological effect is known as an **agonist.** For example, adenosine is the natural agonist of the adenosine receptor. Adenosine signaling in cardiac muscle slows the heart, and adenosine signaling in the brain leads to a decrease in neurotransmitter release, producing a sedative effect.

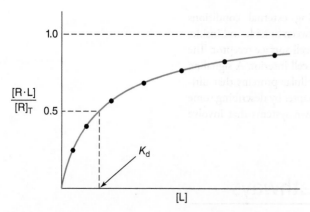

Figure 10.1 Receptor–ligand binding. As the ligand concentration [L] increases, more receptor molecules bind ligand. Consequently, the fraction of receptors that have bound ligand [R·L] approaches 1.0. [R]$_T$ is the total concentration of receptors.

Question Compare this graph to Figures 5.3 and 7.5.

SAMPLE CALCULATION 10.1

Problem   A sample of cells has a total receptor concentration of 10 mM. Twenty-five percent of the receptors are occupied with ligand and the concentration of free ligand is 15 mM. Calculate K_d for the receptor–ligand interaction.

Solution   Because 25% of the receptors are occupied, $[R \cdot L] = 2.5$ mM and $[R] = 7.5$ mM. Use Equation 10.1 to calculate K_d:

$$K_d = \frac{[R][L]}{[R \cdot L]}$$

$$= \frac{(0.0075)(0.015)}{(0.0025)}$$

$$= 0.045 \text{ M} = 45 \text{ mM}$$

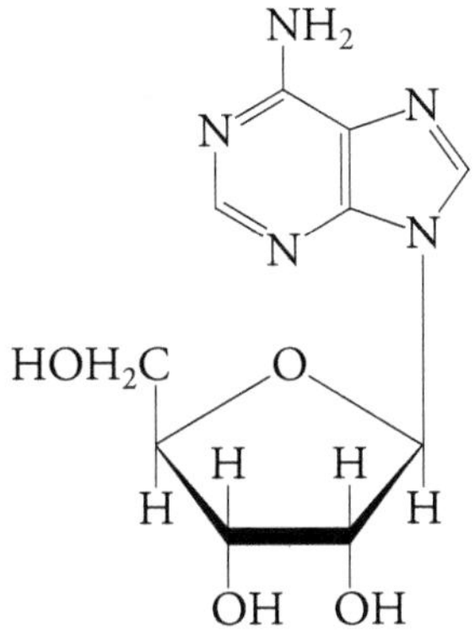

Adenosine

Caffeine is an **antagonist** of the adenosine receptor because it binds to the receptor but does not trigger a response.

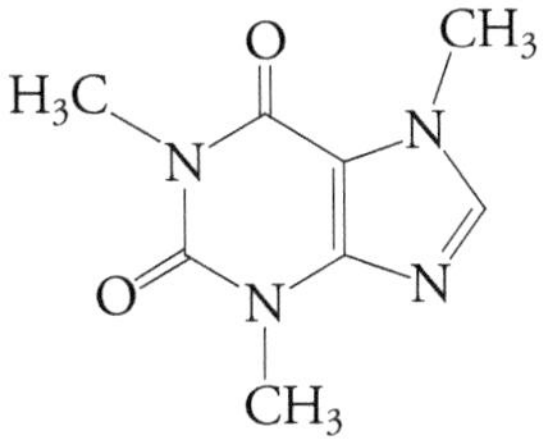

Caffeine

It functions like a competitive enzyme inhibitor (Section 7.3). As a result, caffeine keeps the heart rate high and produces a sense of wakefulness. Like caffeine, the majority of drugs currently in clinical use act as agonists or antagonists for various receptors involved in regulating such things as blood pressure, reproduction, and inflammation.

Ligands typically bind to a receptor with high affinity and high specificity, but because the ligand–receptor interactions are noncovalent, binding is reversible. Consequently, a cell responds to a signal molecule—or a drug—only while that molecule is associated with its receptor.

Most signaling occurs through two types of receptors

When an agonist binds to the adenosine receptor, which is a transmembrane protein, the receptor undergoes a conformational change so that it can interact with an intracellular protein called a **G protein.** Such receptors are therefore called **G protein–coupled receptors (GPCRs).** G proteins are named for their ability to bind guanine nucleotides (GTP and GDP). In response to receptor–ligand binding, the G protein becomes activated and in turn interacts with and thereby activates additional intracellular proteins. Often, one of these is an enzyme that

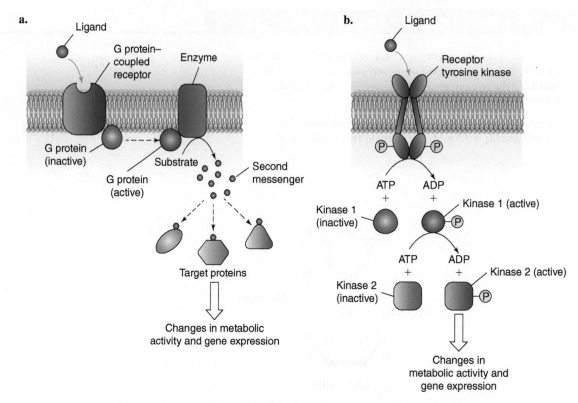

Figure 10.2 Overview of signal transduction pathways. Ligand binding to a cell-surface receptor causes a signal to be transduced to the cell interior, leading eventually to changes in the cell's behavior. **a.** Ligand binding to a G protein–coupled receptor triggers the activation of a G protein, which then activates an enzyme that produces a second messenger. Second messenger molecules diffuse away to activate or inhibit the activity of target proteins in the cell. **b.** Ligand binding to a receptor tyrosine kinase activates the kinase activity of the receptor so that intracellular proteins become phosphorylated. A series of kinase reactions activates or inhibits target proteins by adding phosphoryl groups to them.

Question **Which pathway components are enzymes?**

generates a small molecule product that diffuses throughout the cell. These small molecules are called **second messengers** because they represent the intracellular response to the extracellular, or first, message that binds to the GPCR. A variety of substances serve as second messengers in cells, including nucleotides, nucleotide derivatives, and the polar and nonpolar portions of membrane lipids. The presence of a second messenger can alter the activities of cellular proteins, leading ultimately to changes in metabolic activity and gene expression. These events are summarized in **Figure 10.2a.**

A second type of receptor, also a transmembrane protein, becomes activated as a kinase as a result of ligand binding. A **kinase** is an enzyme that transfers a phosphoryl group from ATP to another molecule. In this case, the phosphoryl group is condensed with the hydroxyl group of a tyrosine side chain on a target protein, so these receptors are termed **receptor tyrosine kinases.** In some signal transduction pathways involving receptor tyrosine kinases, the target protein is also a kinase that becomes catalytically active when phosphorylated. The result may be a series of kinase-activation events that eventually lead to changes in metabolism and gene expression (Fig. 10.2b).

Some receptor systems include both G proteins and tyrosine kinases, and others operate by entirely different mechanisms. For example, the acetylcholine receptor on muscle cells (Fig. 9.20) is a ligand-gated ion channel. When acetylcholine is released into the neuromuscular synapse and binds to the receptor, Na^+ ions flow into the muscle cell, causing depolarization that leads to an influx of Ca^{2+} ions to trigger muscle contraction.

The effects of signaling are limited

The multistep nature of signaling pathways and the participation of enzyme catalysts ensure that the signal represented by an extracellular ligand will be amplified as it is transduced inside the cell (see Fig. 10.2). Consequently, *a relatively small extracellular signal can have a dramatic effect on a cell's behavior.* The cell's responses to the signal, however, are regulated in various ways.

The speed, strength, and duration of a signaling event may depend on the cellular location of the components of the signaling pathway. There is evidence that components for some pathways are preassembled in multiprotein complexes in or near the plasma membrane so that they can be quickly activated when the ligand docks with its receptor. Components that must diffuse long distances to reach their targets, or that move from the cytoplasm to the nucleus, may need more time to trigger cellular responses.

Because signaling pathways tend to be branched rather than completely linear, the same intracellular components may participate in more than one signal transduction pathway, so two different extracellular signals could ultimately achieve the same intracellular results. Alternatively, two signals could cancel each other's effects. The response of a given cell, which expresses many different types of receptors, therefore depends on how various signals are integrated. Similarly, different types of cells may include different intracellular components and therefore respond to the same ligand in different ways.

In a biological system that obeys the law of homeostasis, *any process that is turned on must eventually be turned off.* Such control applies to signaling pathways. For example, shortly after a G protein has been activated by its associated receptor, it becomes inactive again. The action of kinases is undone by the action of enzymes that remove phosphoryl groups from target proteins. These and other reactions restore the signaling components to their resting state so that they can be ready to respond again when another ligand binds to its receptor.

Finally, most people are aware that a strong odor loses its pungency after a few minutes. This occurs because olfactory receptors, like other types of receptors, become **desensitized.** In other words, the receptors become less able to transmit a signal even when continually exposed to ligand. Desensitization may allow the signaling machinery to reset itself at a certain level of stimulation so that it can better respond to subsequent changes in ligand concentration.

Concept Check

1. Compare hormone receptors to enzymes and simple binding proteins.
2. Explain why an extracellular signal needs a second messenger.
3. Describe how extracellular signals are amplified inside a cell.
4. Draw simple diagrams of signaling pathways involving G proteins and receptor tyrosine kinases.
5. Explain why a receptor would need to be turned off or desensitized.

10.2 G Protein Signaling Pathways

KEY CONCEPTS

Describe signaling via G protein–coupled receptors.

- Summarize the roles of receptors, G proteins, nucleotides, and lipids in signaling.
- Describe how a kinase is activated.
- List the mechanisms that limit, compete with, and terminate the G protein signaling pathway.
- Explain how the same hormone can elicit different responses in different cells and how different hormones can elicit the same response in a cell.
- List some examples of G protein–coupled receptors and their ligands.

Over 800 genes in the human genome encode G protein–coupled receptors, and these proteins are responsible for transducing the majority of extracellular signals. In this section we will describe some features of these receptors, their associated G proteins, and various second messengers and their intracellular targets.

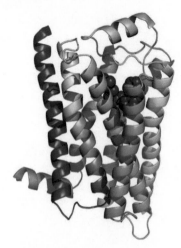

Figure 10.3 The β₂-adrenergic receptor. The backbone structure of the protein is colored in rainbow order from the N-terminus (blue) to the C-terminus (red). A ligand is shown in space-filling form in blue.

Question Explain why the receptor for a polar hormone must be a transmembrane protein.

G protein–coupled receptors include seven transmembrane helices

The GPCRs are known as 7-transmembrane (7TM) receptors because they include seven α helices, which are arranged much like those of the membrane protein bacteriorhodopsin (Fig. 8.8). Many G protein–coupled receptors are palmitoylated at a cysteine residue, so they are also lipid-linked proteins (Section 8.3). In the GPCR family, the helical segments are more strongly conserved than the loops that join them on the intracellular and extracellular sides of the membrane.

The structure of one of these proteins, the β₂-adrenergic receptor, is shown in **Figure 10.3**. The ligand-binding site of this GPCR is defined by a portion of the helical core of the protein as well as its extracellular loops. Other GPCRs bind signal molecules in this same general location, with the exact size and polarity of the binding pocket depending on the nature of the signal, which is unique for each receptor. Some GPCRs are equipped with an extracellular protein domain that contributes to the binding site for a ligand, particularly if it is a large peptide. Still other receptors have extensive extracellular domains even though they bind relatively small ligands.

In some G protein–coupled receptors, domains that extend beyond the transmembrane portion of the protein mediate the formation of receptor dimers that are essential for signal transduction (**Fig. 10.4**). In fact, experiments with receptor proteins in living cells or reconstituted in artificial membranes indicate that many GPCRs may form dimers or oligomers, either with the same or different GPCRs. Although the association in some cases may be transient, lasting up to a few seconds, it may affect the kinetics of ligand binding or signal transduction.

The physiological ligands for the β₂-adrenergic receptor are the hormones epinephrine and norepinephrine, which are synthesized by the adrenal glands from the amino acid tyrosine.

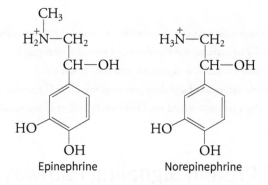

Epinephrine   Norepinephrine

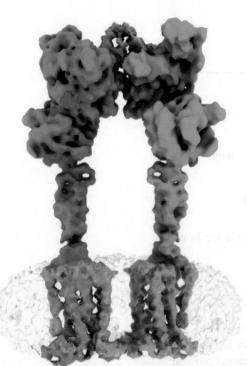

Koehl A et al 2019/Springer Nature

These same substances, sometimes called adrenaline and noradrenaline, also function as neurotransmitters. They are responsible for the fight-or-flight response, which is characterized by fuel mobilization, dilation of blood vessels and bronchi (airways), and increased cardiac action. Antagonists that prevent signaling via the β₂-adrenergic receptor, known as β-blockers, are used to treat high blood pressure.

How does the receptor transmit the extracellular hormonal signal to the cell interior? *Signal transduction depends on conformational changes involving the receptor's transmembrane helices.* Two of the helices shift slightly to accommodate the ligand, which repositions one of the cytoplasmic protein loops. Studies with a variety of different ligands indicate that the receptor can actually adopt a range of conformations, suggesting that the receptor is not merely an on–off switch but can mediate the effects of strong as well as weak agonists.

Figure 10.4 Model of the glutamate receptor mGlu5. The functional receptor is a dimer with two 7TM domains embedded in the membrane (at the bottom). The glutamate binding sites are in the "Venus flytrap" domains (at top), which come together after ligand binding.

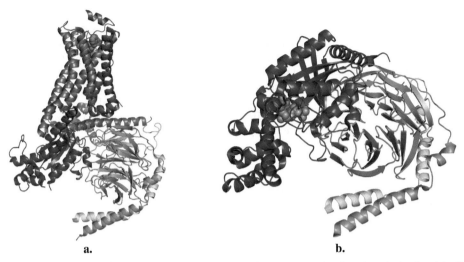

Figure 10.5 A GPCR–G protein complex. **a.** Side view. The GPCR is purple, a bound agonist is red, and the G protein is yellow, green, and blue. **b.** The G protein β subunit (green) has a propeller-like structure. The small γ subunit (yellow) associates tightly with the β subunit. The α subunit (blue) binds a guanine nucleotide (GDP, orange) in a cleft between two domains.

The receptor activates a G protein

Ligand-induced conformational changes in a G protein–coupled receptor open up a pocket on its cytoplasmic side to create a site where another protein—either a G protein or a protein called arrestin—can bind. The G protein is presumably already close to the receptor, since it has covalently attached lipids that keep it tethered to the cytoplasmic leaflet of the plasma membrane. The trimeric G proteins associated with GPCRs consist of three subunits, designated α, β, and γ (**Fig. 10.5**); other types of G proteins do not have this three-part structure (**Box 10.A**).

Box 10.A	Types of G proteins

G proteins, or guanine nucleotide-binding proteins, are a group of proteins that help cells communicate with the extracellular environment. They participate in various cellular processes, including cell growth, differentiation, and response to environmental signals. G proteins act as molecular switches, cycling between an inactive (GDP-bound) and an active (GTP-bound) state.

The following are the two main types of G proteins.

Heterotrimeric G proteins are the classical G proteins composed of three subunits—α, β, and γ. They transmit G protein-coupled receptors (GPCRs) signals to intracellular effectors such as enzymes or ion channels. GPCRs are a large family of cell surface receptors activated by various ligands, such as hormones, neurotransmitters, and sensory stimuli. The subunits of the heterotrimeric G proteins are attached to the cell membrane. The alpha subunit is the primary effector in signal transduction, and it has intrinsic GTPase activity, allowing it to hydrolyze GTP to GDP and thereby switch off the signaling cascade. Gα subunits fall into four families: $G\alpha_s$, $G\alpha_i$, $G\alpha_q$, and $G\alpha_{12}$. The $G\alpha_s$ family includes $G\alpha_s$ and $G\alpha_{olf}$, with $G\alpha_s$ expressed widely, while $G\alpha_{olf}$ is specific to olfactory sensory neurons. The diverse $G\alpha_i$ family comprises $G\alpha_{i1}$, $G\alpha_{i2}$, $G\alpha_{i3}$, $G\alpha_o$, $G\alpha_t$, $G\alpha_g$, and $G\alpha_z$, with $G\alpha_i$ proteins found in various cell types. $G\alpha_o$ is prevalent in neurons with two spliced variants; $G\alpha_t$ has isoforms in eye cells, $G\alpha_g$ is in taste receptor cells, and $G\alpha_z$ is present in neuronal tissues and platelets. The $G\alpha_q$ family ($G\alpha_q$, $G\alpha_{11}$, $G\alpha_{14}$, $G\alpha_{16}$) is ubiquitously expressed, while $G\alpha_{14}$ is mainly in the kidney, lung, and liver, and $G\alpha_{15/16}$ is specific to hematopoietic cells. The $G\alpha_{12}$ family encompasses $G\alpha_{12}$ and $G\alpha_{13}$, expressed widely.

Gβγ subunits of heterotrimeric G proteins usually function as a complex and participate in signaling events independently of the Gα subunit. The human and mouse genomes have five Gβ and 12 Gγ genes. $G\beta_{1-4}$ share high sequence similarities, while $G\beta_5$ is ~50% similar. $G\beta_5$ is mainly found in the brain, while other Gβ subunits are widely distributed. Gγ subunits, more diverse, share sequence similarities from 20% to 80%. Despite similar biochemical activities in vitro, gene-deletion experiments in mice suggest different physiological functions for Gβ and Gγ genes, likely due to their varied distributions and expression levels.

Monomeric small G proteins are small, single-subunit proteins involved in a variety of cellular processes, including cell proliferation, cytoskeletal organization, and vesicular trafficking. The most well-known family of monomeric G proteins is the Ras superfamily, which includes Ras, Rho, Rab, and Ran proteins. Monomeric G proteins, like Ras, have intrinsic GTPase activity, allowing them to hydrolyze GTP to GDP and vice versa. This ability to switch between the GDP-bound (inactive) and GTP-bound (active) states is crucial for their function. Monomeric G proteins typically have conserved domains, such as the G domain (GTPase domain) and switch regions. The G domain is responsible for binding and hydrolyzing GTP, while the switch regions undergo conformational changes during the GTPase cycle. They undergo post-translational modifications, such as prenylation, to anchor them to cellular membranes.

Prenylation involves adding lipid groups (farnesyl or geranylgeranyl) to the protein, facilitating their association with the cell membrane.

G proteins are implicated in several diseases, and their dysregulation can contribute to pathological conditions. For example: Mutations in G proteins, especially Ras proteins, are common in various cancers. Ras mutations can lead to uncontrolled cell growth and survival. G proteins are involved in neurotransmitter signaling, and abnormalities in their function can contribute to neurological diseases like Parkinson's disease and schizophrenia. G proteins regulate heart rate and vascular tone, and dysregulation can contribute to cardiovascular diseases.

Because of their involvement in diseases, G proteins and their associated pathways are potential targets for therapeutic interventions. Drugs modulating G protein activity or targeting downstream effectors in the signaling cascade are being developed for various diseases. For example, beta-blockers target $G\alpha_s$ signaling and treat conditions like hypertension and heart failure. Drugs targeting specific GPCRs or their associated G proteins are also used to treat conditions such as asthma, pain, and certain neurological disorders. The development of therapies targeting G proteins is an active area of research in pharmacology.

Question: **Discuss the advantages and disadvantages of having multiple G protein subtypes (e.g., G_s, G_i, G_q) in a cell. How might this contribute to cellular versatility, and under what circumstances could it lead to complexity or potential issues?**

In the resting state, GDP is bound to the G protein's α subunit, but association with the hormone–receptor complex causes the G protein to release the GDP and bind GTP in its place. The third phosphate group of GTP is not easily accommodated in the αβγ trimer, so the α subunit dissociates from the β and γ subunits, which remain together. Once they dissociate, the α subunit and the βγ dimer are both active; that is, *they interact with additional cellular components in the signal transduction pathway.* However, since both the α and β subunits include lipid anchors, the G protein subunits do not diffuse far from the receptor that activated them.

The signaling activity of the G protein is limited by the intrinsic GTPase activity of the α subunit, which slowly converts the bound GTP to GDP:

$$GTP + H_2O \rightarrow GDP + P_i$$

Hydrolysis of the GTP allows the α and βγ units to reassociate as an inactive trimer (**Fig. 10.6**). The cost for the cell to switch a G protein on and then off again is the free energy of the GTP hydrolysis reaction (GTP is energetically equivalent to ATP).

The second messenger cyclic AMP activates protein kinase A

Although humans have over 800 GPCRs, the number of G proteins is more modest: There are 16 genes for the α subunit, 5 for β, and 12 for γ. Together, these genes code for a small set of trimeric G proteins that interact with various targets in the cell and activate or inhibit them. A single receptor may interact with more than one G protein, so the effects of ligand binding are amplified at this point. One of the major targets of activated G proteins is an integral membrane

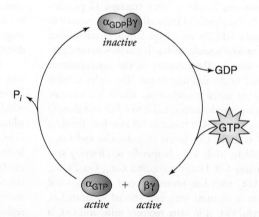

Figure 10.6 The G protein cycle. The αβγ trimer, with GDP bound to the α subunit, is inactive. Ligand binding to a receptor associated with the G protein triggers a conformational change that causes GTP to replace GDP and the α subunit to dissociate from the βγ dimer. Both portions of the G protein are active in the signaling pathway. The GTPase activity of the α subunit returns the G protein to its inactive trimeric state.

enzyme called adenylate cyclase. When the α subunit of the G protein binds, the enzyme's catalytic domains convert ATP to a molecule known as cyclic AMP (**cAMP**). cAMP is a second messenger that can freely diffuse through the cytosol.

ATP

PP$_i$

adenylate cyclase

Cyclic AMP (cAMP)

Among the targets of cAMP is an enzyme called protein kinase A or PKA. In the absence of cAMP, this kinase is an inactive tetramer of two regulatory (R) and two catalytic (C) subunits (**Fig. 10.7**). A segment of each R subunit occupies the active site in a C subunit so that the kinase is unable to phosphorylate any substrates. cAMP binding to the regulatory subunits relieves the inhibition, causing the tetramer to release the two active catalytic subunits.

R C / C R *inactive*   4 cAMP →   R R   + C *active* + C *active*

In a cell, the kinase subunits may become active even without full dissociation of the tetramer. cAMP functions as an allosteric activator of the kinase, and *the level of cAMP determines the level of activity of protein kinase A.*

Protein kinase A is known as a Ser/Thr kinase because it transfers a phosphoryl group from ATP to the serine or threonine side chain of a target protein.

$-CH_2-O-PO_3^{2-}$ Phospho-Ser

$-CH-O-PO_3^{2-}$ with CH_3 Phospho-Thr

The substrates for the reaction bind in a cleft defined by the two lobes of the protein (**Fig. 10.8a**). Other kinases share this same core structure but often have additional domains that determine their subcellular location or provide additional regulatory functions.

In addition to regulation by cAMP binding to the R subunit, *protein kinase A itself is regulated by phosphorylation.* The protein's so-called activation loop, a segment of polypeptide near the entrance to the active site, includes a phosphorylatable threonine residue. When the loop is not phosphorylated, the kinase's active site is blocked. When phosphorylated, the loop swings aside and the kinase's catalytic activity increases; for some protein kinases, activity increases by several orders of magnitude. This activation effect is not merely a matter of improving substrate access to the active site but also appears to involve conformational changes that affect catalysis. For example, the negatively charged phospho-Thr interacts with a positively charged arginine residue in the active site. Efficient catalysis requires that this arginine residue and an adjacent aspartate residue be re-positioned for phosphoryl-group transfer from ATP to a protein substrate (Fig. 10.8b).

The targets of protein kinase A include enzymes involved in glycogen metabolism (Section 19.2). One result of signaling via the β$_2$-adrenergic receptor, which leads to protein kinase A activation by cAMP, is the phosphorylation and activation of an enzyme called glycogen phosphorylase, which catalyzes the removal of glucose residues, the cell's primary metabolic fuel, from glycogen, the cell's glucose-storage depot. Consequently, a signal

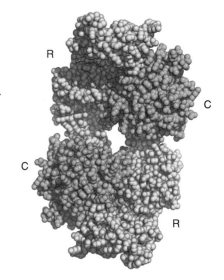

Figure 10.7 Inactive protein kinase A. In the inactive complex, the two regulatory subunits (R) block the active sites of the two catalytic subunits (C).

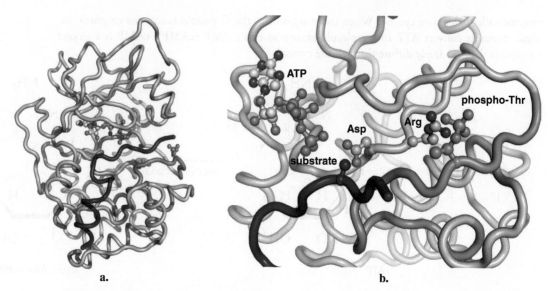

Figure 10.8 Protein kinase A.   a. The backbone of the catalytic subunit is light green, with its activation loop dark green. The phospho-Thr residue (right side) and ATP (left side) are shown in ball-and-stick form. A peptide that mimics a target protein is blue. **b.** Close-up view of the active site region. When the activation loop is phosphorylated, the phospho-Thr residue interacts with an Arg residue, and the adjacent Asp residue is positioned near the third phosphate group of ATP and the peptide substrate. Atoms are color-coded: C gray or green, O red, N blue, and P gold.

such as epinephrine can mobilize the metabolic fuel needed to power the body's fight-or-flight response.

The enzymes that phosphorylate the activation loops of protein kinase A and other cell-signaling kinases apparently operate when the kinase is first synthesized, so the kinase is already "primed" and needs only to be allosterically activated by the presence of a second messenger. However, this regulatory mechanism begs the question of what activates the kinase that phosphorylates the kinase. As we will see, kinases that act in series are common in biological signaling pathways, and many of these pathways are interconnected, making it difficult to trace simple cause-and-effect relationships.

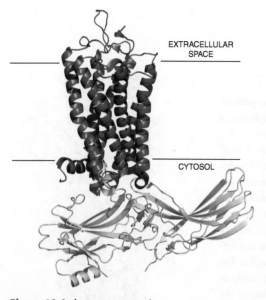

Figure 10.9 A receptor–arrestin complex. Rhodopsin, a model G protein–coupled receptor, is shown as a purple ribbon, and the intracellular arrestin protein as a pink ribbon.

Question Compare this figure and Figure 10.5a.

Arrestin competes with G proteins

Interacting with a G protein is just one option for the GPCR–ligand complex. An arrestin protein can occupy the same binding site the G protein uses on the cytoplasmic side of the receptor (**Fig. 10.9**). *Because arrestin competes with the G protein, it hinders or stops the receptor's ability to signal via the G protein pathway* (hence the name *arrestin*). Arrestins recognize the GPCR–ligand complex and undergo a conformational change that exposes Arg and Lys side chains that are particularly effective at binding GPCRs that have been phosphorylated at some point in the signaling process. Consequently, an arrestin can interact with the receptor in two ways, either of which may activate the arrestin.

The activated arrestin can interact with additional intracellular components, including an assortment of kinases. In a way, the arrestin serves as a scaffold or bridge between the enzymes and the receptor, thereby expanding the intracellular effects of the initial hormone signal that docked with the receptor. In fact, the arrestin may remain active even after it dissociates from the G protein–coupled receptor, prolonging signaling while the receptor interacts with additional arrestins or G proteins.

Arrestin binding to a G protein–coupled receptor also promotes the receptor's internalization by endocytosis (Section 9.4). This diminishes the cell's ability to receive subsequent hormone signals and accounts for the early hypothesis that arrestin functioned simply as an off switch for G protein–coupled receptors. Internalized receptors may retain some signal transduction activity inside the cell. The significance of this intracellular signaling is not known, although GPCRs have been identified on a variety of intracellular membranes, including the nuclear membrane and mitochondrial membrane, well out of reach of extracellular signals.

Signaling pathways must be switched off

What happens after a ligand binds to a receptor, a G protein or an arrestin responds, a second messenger is produced, an effector enzyme such as a protein kinase is activated, and target proteins are phosphorylated? *To restore the cell to a resting state, any or all of the events of the signal transduction pathway can be blocked or reversed.* First, the dissociation of an extracellular ligand from its receptor can halt a signal transduction process, or the receptor can become desensitized. Desensitization of a G protein–coupled receptor begins with phosphorylation of the ligand-bound receptor by a GPCR kinase, followed by arrestin binding and endocytosis.

Even without arrestin, G protein signaling is limited. We have already seen that the intrinsic GTPase activity of G proteins stops their activity. Moreover, second messengers often have short lifetimes due to their rapid degradation in the cell. For example, cAMP is hydrolyzed by the enzyme cAMP phosphodiesterase:

cAMP   AMP

Caffeine, for example, in addition to being an adenosine receptor antagonist, can diffuse inside cells and inhibit cAMP phosphodiesterase. As a result, the cAMP concentration remains high, the action of protein kinase A is sustained, and stored metabolic fuels are mobilized, readying the body for action rather than sleep.

Some of the cell's G proteins may inhibit rather than activate adenylate cyclase and therefore decrease the cellular cAMP level. Some G proteins activate cAMP phosphodiesterase, with similar effects on cAMP-dependent processes. *A cell's response to a hormone signal depends in part on which G proteins respond.* Because a single type of hormone may stimulate several types of G proteins, the signaling system may be active for only a brief time before it is turned off.

The phosphorylations catalyzed by protein kinase A and other kinases can be reversed by the work of protein **phosphatases,** which catalyze a hydrolytic reaction to remove phosphoryl groups from protein side chains. Like kinases, phosphatases are generally specific for serine or threonine, or tyrosine, although some "dual specificity" phosphatases remove phosphoryl groups from all three side chains. The active-site pocket of the Tyr phosphatases is deeper than the pocket of Ser/Thr phosphatases in order to accommodate the larger phospho-Tyr side chain. Some protein phosphatases are transmembrane proteins; others are entirely intracellular. They tend to have multiple domains or multiple subunits, consistent with their ability to form numerous protein–protein interactions and participate in complex regulatory networks.

The phosphoinositide signaling pathway generates two second messengers

The diversity of G protein–coupled receptors, together with the diversity of G proteins, creates almost unlimited possibilities for adjusting the levels of second messengers and altering the activities of cellular enzymes. Epinephrine, the hormone that activates the β_2-adrenergic receptor, also binds to a receptor known as the α-adrenergic receptor, which uses the **phosphoinositide signaling system.**

The α- and β-adrenergic receptors are situated in different tissues and mediate different physiological effects, even though they bind the same hormone. The G protein associated with the α-adrenergic receptor activates the cellular enzyme phospholipase C, which acts on the membrane lipid phosphatidylinositol bisphosphate. Phosphatidylinositol is a minor component of the plasma membrane (4–5% of the total phospholipids), and the bisphosphorylated form (with a total of three phosphate groups) is rarer still. Phospholipase C converts this lipid to inositol trisphosphate and diacylglycerol.

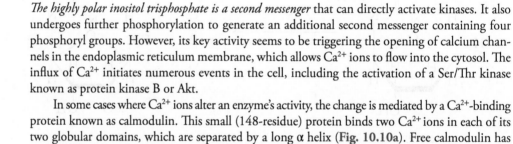

Phosphatidylinositol bisphosphate → (H₂O, phospholipase C) → Inositol trisphosphate + Diacylglycerol

The highly polar inositol trisphosphate is a second messenger that can directly activate kinases. It also undergoes further phosphorylation to generate an additional second messenger containing four phosphoryl groups. However, its key activity seems to be triggering the opening of calcium channels in the endoplasmic reticulum membrane, which allows Ca^{2+} ions to flow into the cytosol. The influx of Ca^{2+} initiates numerous events in the cell, including the activation of a Ser/Thr kinase known as protein kinase B or Akt.

In some cases where Ca^{2+} ions alter an enzyme's activity, the change is mediated by a Ca^{2+}-binding protein known as calmodulin. This small (148-residue) protein binds two Ca^{2+} ions in each of its two globular domains, which are separated by a long α helix (**Fig. 10.10a**). Free calmodulin has an extended shape, but in the presence of Ca^{2+} and a target protein, the helix partially unwinds and calmodulin bends in half to grasp the target protein to activate or inhibit it (Fig. 10.10b).

The hydrophobic diacylglycerol product of the phospholipase C reaction is also a second messenger. Although it remains in the cell membrane, it can diffuse laterally to activate protein kinase C, which phosphorylates its target proteins at serine or threonine residues. In its resting state, protein kinase C is a cytosolic protein with an activation loop blocking its active site. Noncovalent binding to diacylglycerol docks the enzyme at the membrane surface so that it changes its conformation, repositions the activation loop, and becomes catalytically active. As in protein kinase A (whose sequence is about 40% identical), phosphorylation of a threonine residue in the activation loop, a requirement for catalytic activity, has already occurred. Full activation of some forms of protein kinase C also requires Ca^{2+}, which is presumably available as a result of the activity of the inositol trisphosphate second messenger. Among the targets of protein kinase C are proteins involved in the regulation of gene expression and cell division.

Phospholipase C can be activated not only by a G protein–coupled receptor such as the α-adrenergic receptor but also by other signaling systems involving receptor tyrosine kinases. This is an example of **cross-talk,** the interconnections between signaling pathways that share some intracellular components. The phosphoinositide signaling pathway is regulated in part by the action of lipid phosphatases that remove phosphoryl groups from the phosphatidylinositol bisphosphate substrate that gives rise to the second messengers.

Another example of the overlap between signaling pathways involves sphingolipids such as sphingomyelin (Fig. 8.2), which is a normal component of membranes. Ligand binding to certain receptor tyrosine kinases leads to activation of sphingomyelinases that release sphingosine and ceramide (ceramide is sphingomyelin without its phosphocholine head group). Ceramide is a second messenger that activates kinases, phosphatases, and other cellular enzymes. Sphingosine, which undergoes phosphorylation (by a receptor tyrosine kinase–dependent mechanism) to sphingosine-1-phosphate, is both an intracellular and extracellular signaling molecule. It inhibits phospholipase inside the cell, apparently exits the cell via an ABC transport protein (Section 9.3), and then binds to a G protein–coupled receptor to trigger additional cellular responses.

a.

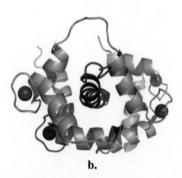

b.

Figure 10.10 Calmodulin.
a. Isolated calmodulin has an extended shape. The four bound Ca^{2+} ions are shown as blue spheres. **b.** When bound to a target protein (blue helix), calmodulin's long central helix unwinds and bends so that the protein can wrap around its target.

Many sensory receptors are GPCRs

As estimated one-third of therapeutic drugs—including antihistamines, antipsychotics, and opioid painkillers—target G protein–coupled receptors and their associated proteins (**Box 10.B**). GPCRs also respond to light, odors, and tastes. In humans, **olfaction,** the sense of smell, depends on roughly 400 GPCRs (some mammals have over 1000). Each sensory neuron in the nose expresses just one type of receptor, but a given ligand, called an **odorant,** can bind to several different receptors, so the perception of an odor depends on the signaling activity in multiple neurons. This combinatorial system allows humans to distinguish hundreds of thousands of different odorants despite a somewhat limited set of receptors.

When an odorant binds to an olfactory receptor, its associated G protein activates adenylate cyclase, and the resulting cAMP second messenger opens a nonspecific ion channel, which facilitates the influx of Na^+ and Ca^{2+} ions. This, in turn, triggers the opening of a Ca^{2+}-dependent chloride channel, and the ensuing exit of Cl^- ions further depolarizes the cell. These changes generate the action potential that travels the length of the neuron toward the brain.

In animals, **gustation,** the sense of taste, works a bit differently. First, neurons themselves do not have taste receptors but instead receive input indirectly from epithelial cells located throughout the mouth but mostly in taste buds on the tongue. Of the five taste qualities—sweet, salty, sour, bitter, and umami—two are not actually detected by receptors. The perception of sourness (acidity) depends mainly on the ability of protons to affect ion channels, opening Na^+ channels and closing K^+ channels. The cell becomes depolarized, Ca^{2+} channels open, and neurotransmitters are released. Nearby neurons then convey the signal to the brain. In a similar fashion, saltiness is detected by the opening of sodium channels. Curiously, low concentrations of Na^+ are perceived as appetizing, whereas very high concentrations are distasteful. The "hot" substances in peppers, wasabi, and mustard bind to receptors that perceive pain or heat.

The G protein–coupled receptors that respond to sweet, bitter, and umami tastes, although not as numerous as olfactory receptors, can respond to a few thousand different ligands, called **tastants.** Ligand binding to these GPCRs initiates the phosphoinositide pathway. Inositol trisphosphate opens Ca^{2+} channels; this leads to the opening of Na^+ channels and cell depolarization. Voltage-gated Ca^{2+} channels then open, and the influx of Ca^{2+} triggers neurotransmitter release that activates a sensory neuron.

Box 10.B | Opioids

Morphine is arguably the world's most powerful and most abused drug. Produced by the opium poppy *Papaver somniferum,* its usefulness lies in its ability to relieve pain without loss of consciousness. Morphine has been used for millennia; it was the major ingredient of laudanum, popular in the eighteenth and nineteenth centuries. The poppy metabolite codeine, which is readily converted to morphine by cytochrome P450 in the liver (Section 7.4), was until recently a common ingredient in cough suppressants. Morphine along with its natural and synthetic derivatives, such as heroin, oxycodone, methadone, and fentanyl, are collectively known as **opioids.**

Morphine

Morphine and other opioids bind to three different types of G protein–coupled receptors that are distributed in an overlapping fashion in many parts of the brain as well as in other tissues, notably the intestine (which helps explain the gastrointestinal symptoms associated with opioid use). The receptors interface with arrestin and with an inhibitory G protein that diminishes cAMP production within the cell. G protein–mediated signaling also increases K^+ channel function that hyperpolarizes the cell, blocks Ca^{2+} transport, and inhibits neurotransmitter release. This set of responses effectively reduces the transmission of pain signals. Opioid binding to the receptor also activates arrestin, which leads to internalization of the receptor and counters the G protein effects.

The human body naturally produces about 30 endogenous opioids, including the enkephalins and endorphins, peptides that have morphinelike effects and are presumably essential for regulating normal pain perception, mood, and other functions. Administering exogenous opioids replicates the pain-relieving and mood-elevating effects of the endogenous signals but leads to drug dependence because the body stops manufacturing its own opioids. When an opioid drug is discontinued, the individual experiences severe withdrawal symptoms that include very real physical distress, not just a sensation of craving.

As with any drug, chronic use leads to tolerance, an adaptation that requires ever-increasing doses to achieve the same effect.

Unfortunately, in addition to blocking pain, opioids suppress the respiratory center in the brain, which controls breathing. Because tolerance to a drug's pain-relieving and euphoric effects develops faster than tolerance to respiratory depression, opioid-addicted individuals are at risk of ingesting a dose that causes them to stop breathing and die.

Understanding ligand–receptor systems suggests some possible solutions to opioid addiction and overdosing. For example, administering an opioid receptor antagonist such as naloxone, which quickly enters the brain and competes with agonists for binding to the opioid receptors, can quickly reverse the effects of an opioid overdose. Newly developed opioid drugs are designed to exploit the subtle conformational differences in the ligand–receptor complex in order to bias the system toward the G protein pathway for analgesia (pain relief) rather than the arrestin pathway that favors drug tolerance and increases the risk of respiratory side effects.

Sweet substances bind to a dimeric GPCR that resemble the brain's glutamate receptor (Fig. 10.4), with subunits known as TAS1R2 and TAS1R3 (**Fig. 10.11**). Umami receptors, which bind substances such as glutamate, aspartate, and the purine nucleotides IMP and GMP, are dimers of TAS1R1 and TAS1R3. Humans also have 25 types of bitter taste receptor subunits, the TAS2Rs, which collectively appear able to recognize a large number of agonists, although with relatively low affinity. Somewhat nonspecific binding could be advantageous for detecting a wide array of potentially toxic substances, which tend to have a bitter taste. Surprisingly, olfactory receptors and taste receptors have been identified in many nonsensory organs in the body, but these GPCRs do not necessarily respond to conventional odorants or tastants or transmit information to the brain. Their function is still mostly a puzzle.

Concept Check

1. Draw a simple diagram that includes all the components of the β2-adrenergic receptor signaling pathway. Draw a similar diagram for the α-adrenergic receptor signaling pathway.
2. Describe the structure of a G protein–coupled receptor.
3. Describe the activity cycle of a G protein.
4. Draw a diagram to illustrate the steps of signaling via a G protein, from hormone–receptor binding to second messenger degradation. Include arrestin.
5. Draw a diagram to illustrate the phosphoinositide signaling pathway.
6. For each pathway, identify points where signaling activity can be switched off.
7. Explain why it is an advantage for receptors and G proteins to be lipid-linked.
8. Compare the relative concentrations of extracellular signal molecules and second messengers.
9. Discuss the advantages and disadvantages of cross-talk.
10. Propose an evolutionary explanation for the ability to sense sweet, salty, sour, bitter, and umami substances.

Kim S-K et al. 2017/PNAS

Figure 10.11 Model of the TAS1R2–TAS1R3 sweet receptor. This structure has a more "closed" conformation than the dimeric glutamate receptor shown in Figure 10.4.

Question Identify the ligand-binding and transmembrane domains.

10.3 Receptor Tyrosine Kinases

KEY CONCEPTS

Describe the receptor tyrosine kinase signaling pathway.

- Compare G protein–coupled receptors and receptor tyrosine kinases.
- Distinguish the two mechanisms by which receptor tyrosine kinases activate target proteins.
- Explain how kinases and transcription factors mediate cellular responses over different time scales.

A number of hormones and other signaling molecules that regulate cell growth, division, and immune responses bind to cell-surface glycoproteins that operate as tyrosine kinases. These receptors are typically built from protein subunits containing an extracellular ligand-binding site, a single membrane-spanning helix, and an intracellular tyrosine kinase domain. At one time, ligand binding was believed to bring two separate monomeric receptors together to form a catalytically active dimer, but the current understanding is that the receptor subunits are often already associated but inactive. Ligand binding induces conformational changes that bring the receptor subunits closer together, in a scissor-like rearrangement. The intracellular tyrosine kinase domains, which are far apart in the absence of ligand, are repositioned close enough so that each can phosphorylate the other, a key step in signal transduction.

Receptor tyrosine kinases function as homodimers or heterodimers, depending on the type of ligand. Each monomer has a ligand-binding site, and many structural studies reveal symmetric complexes, typically with two ligands per receptor dimer. However, much remains unknown about the exact sequence of events that link the binding of the first ligand, the binding of the second, changes in quaternary structure, and activation of the tyrosine kinase domains. Like G protein–coupled receptors, receptor tyrosine kinases are likely to adopt a series of intermediate structures—including asymmetric complexes—in the transition from inactive to fully active.

The insulin receptor dimer changes conformation

The insulin receptor is one of approximately 60 receptor tyrosine kinases in humans. Cells throughout the body express the receptor, which recognizes insulin, a 51-residue polypeptide hormone that regulates many aspects of fuel metabolism. The receptor is constructed from two long polypeptides that are cleaved after their synthesis, so the mature receptor has an $(\alpha\beta)_2$ structure in which all four polypeptide segments are held together by disulfide bonds (**Fig. 10.12a**). Each $\alpha\beta$ unit functions as one monomer. The extracellular ligand-binding portion of the insulin receptor has an inverted V shape with multiple structural domains (Fig. 10.12b).

Once the first insulin molecule makes contact, the receptor may "explore" additional conformations before the second contact is established. This process requires segments of the α chain to dramatically reorient. When fully bound, the insulin interacts with portions of each $\alpha\beta$ monomer. According to some models, the second insulin molecule binds with a lower affinity (an example of negative cooperative behavior; Section 5.1). By this point, *the receptor–insulin complex is T-shaped, a conformation that brings the transmembrane helices and their attached tyrosine kinase domains close together* (**Fig. 10.13**). Two more insulin molecules appear to bind cooperatively to two additional

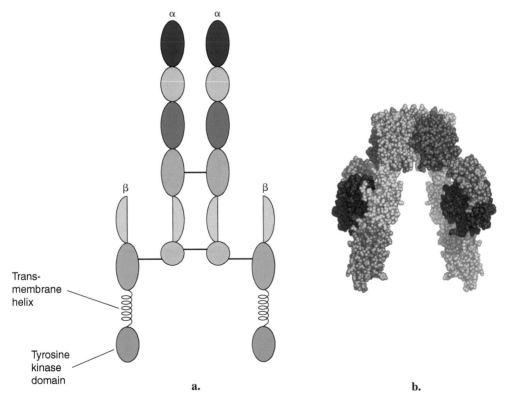

a.   **b.**

Figure 10.12 Structure of the insulin receptor.   a. Schematic diagram showing the two α and two β subunits joined by disulfide bonds (horizontal lines). **b.** Model of the extracellular portion of the receptor, with the structural domains shown in the same colors as in part **a.** The cell surface is at the bottom.

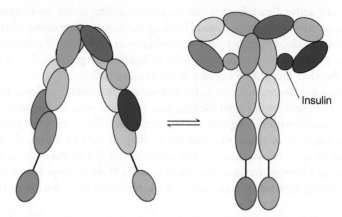

Figure 10.13 Conformational changes in the insulin receptor. This diagram is based on X-ray and cryo-EM models of the insulin receptor. Two bound insulin molecules are shown. The structural domains are colored as in Figure 10.12a.

low-affinity sites to stabilize the receptor in a fully active conformation. The existence of four discrete insulin-binding sites on one receptor raises the possibility that different signals could be transduced across a range of physiological insulin concentrations.

The receptor undergoes autophosphorylation

Inside the cell, *the two tyrosine kinase domains phosphorylate each other,* using ATP as a source of the phosphoryl groups. Because the receptor subunits appear to phosphorylate themselves, this process is termed **autophosphorylation.** Each tyrosine kinase domain has the typical kinase core structure, including an activation loop that lies across the active site to prevent substrate binding. In the insulin receptor, phosphorylation of three tyrosine residues in the kinase activation loop causes the loop to swing aside to better expose the active site (**Fig. 10.14**). This conformational change allows the enzyme to interact with additional protein substrates and to transfer a phosphoryl group from ATP to a tyrosine side chain on these target proteins.

Not all receptor tyrosine kinases phosphorylate other proteins; some activate their intracellular target by other mechanisms. For example, to initiate the responses that promote cell growth and division, many growth factor receptors phosphorylate various intracellular target proteins. They also switch on other pathways that involve small monomeric G proteins (which are also GTPases) such as Ras. The tyrosine kinase domain of the receptor does not directly interact with Ras but instead relies on one or more adapter proteins that form a bridge between Ras and the phospho-Tyr residues of the receptor (**Fig. 10.15**). These proteins also stimulate Ras to release GDP and bind GTP.

Like other G proteins, *Ras is active as long as it has GTP bound to it.* The Ras · GTP complex allosterically activates a Ser/Thr kinase, which becomes active and phosphorylates another kinase, activating it, and so on. Cascades of several kinases can therefore amplify the initial growth factor signal.

The ultimate targets of Ras-dependent signaling cascades are nuclear proteins, which, when phosphorylated, bind to specific sequences of DNA to induce (turn on) or repress (turn off) gene expression. The altered activity of these **transcription factors** means that the original hormonal signal not only alters the activities of cellular enzymes on a short time scale (seconds to minutes) via phosphorylation but also affects protein synthesis, a process that may require several hours or more.

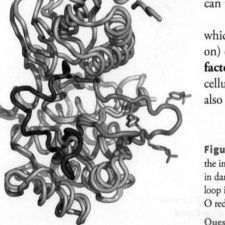

Figure 10.14 Activation of the insulin receptor tyrosine kinase. The backbone structure of the inactive tyrosine kinase domain of the insulin receptor is shown in light blue, with the activation loop in dark blue. The structure of the active tyrosine kinase domain is shown in light green, with the activation loop in dark green. The phospho-Tyr side chains are shown in stick form with atoms color coded: C green, O red, and P orange.

Question How would a phosphatase affect the activity of the insulin receptor?

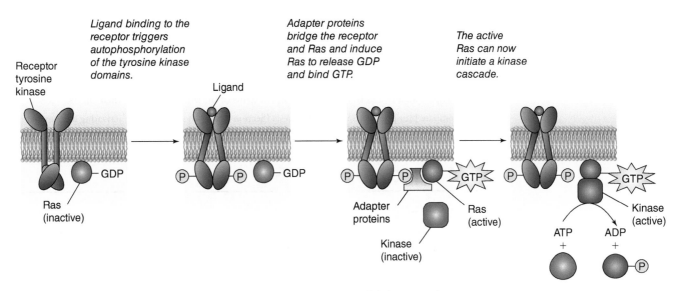

Figure 10.15 The Ras pathway.   Ras links receptor–hormone binding to an intracellular kinase cascade.

Question   **Describe the role of protein conformational changes in each step of the pathway.**

Ras signaling activity is shut down by the action of proteins that enhance the GTPase activity of Ras so that it returns to its inactive GDP-bound form. In addition, phosphatases reverse the effects of the various kinases. The hormone receptor itself may eventually be inactivated by phosphorylation or dephosphorylation or removed from the cell surface by endocytosis. Like the other signaling pathways we have examined, the receptor tyrosine kinase pathways are not linear and they are capable of cross-talk. For example, some receptor tyrosine kinases directly or indirectly (via Ras) activate the kinase that phosphorylates phosphatidylinositol lipids, thereby promoting signaling through the phospho-inositide pathway. Abnormalities in these signaling pathways can promote tumor growth (**Box 10.C**).

Box 10.C   Cell Signaling and Cancer

The progress of a cell through the cell cycle, from DNA replication through the phases of mitosis, depends on the orderly activity of signaling pathways. Cancer, which is the uncontrolled growth of cells, can result from a variety of factors, including overactivation of the signaling pathways that stimulate cell growth. In fact, *the majority of cancers include mutations in the genes for proteins that participate in signaling via Ras and the phosphoinositide pathways.* These altered genes are termed **oncogenes,** from the Greek *onkos,* meaning "tumor."

Oncogenes were first discovered in certain cancer-causing viruses. The viruses presumably picked up the normal gene from a host cell, then mutated. In some cases, an oncogene encodes a growth factor receptor that has lost its ligand-binding domain but retains its tyrosine kinase domain. As a result, the kinase is constitutively (constantly) active, promoting cell growth and division even in the absence of the growth factor. Some *RAS* oncogenes generate mutant forms of Ras that hydrolyze GTP extremely slowly, thus maintaining the signaling pathway in the "on" state. Note that oncogenic mutations can strengthen an activating event or weaken an inhibitory event; in either case, the outcome is excessive signaling activity.

The importance of various kinases in triggering or sustaining tumor growth has made these enzymes attractive targets for anti-cancer drugs. Some forms of leukemia (a cancer of white blood cells) are triggered by a chromosomal rearrangement that generates a kinase with constitutive signaling activity, called Bcr-Abl. The drug

imatinib specifically inhibits this kinase without affecting any of the cell's numerous other kinases. The result is an effective anticancer treatment with few side effects.

Imatinib

The engineered antibody known as trastuzumab binds as an antagonist to a growth factor receptor that is overexpressed in many breast cancers. Other antibody-based drugs target similar receptors in other types of cancers. Clearly, understanding the operation of growth-signaling pathways—both normal and mutated—is essential for the ongoing development of effective anticancer treatments.

Question   **Using Figures 10.2 and 10.15 as a guide, identify other types of signaling proteins that are potential anticancer drug targets.**

Concept Check

1. Draw a diagram to illustrate receptor tyrosine kinase signaling, including tyrosine phosphorylation and Ras activation.

2. Explain how the free energy of ATP and GTP hydrolysis is used to turn on cellular responses to extracellular signals.

3. Compare kinases and transcription factors in terms of the time required for their effects.

4. Compare phosphorylation and proteolysis (Section 6.4) as mechanisms for activating an enzyme.

10.4 Lipid Hormone Signaling

KEY CONCEPTS

Compare lipid signaling to other signal transduction pathways.

- Recognize lipid hormones.
- Describe how lipid hormones regulate gene expression.
- Explain how eicosanoids differ from other signaling molecules.

Some hormones do not need to bind to cell-surface receptors because they are lipids and can cross the membrane to interact with intracellular receptors. For example, retinoic acid and the thyroid hormones thyroxine (T_4) and triiodothyronine (T_3) belong to this class of hormones (**Fig. 10.16**). Retinoic acid (retinoate), a compound that regulates cell growth and differentiation, particularly in the immune system, is synthesized from retinol, a derivative of β-carotene (Box 8.B). The thyroid hormones, which generally stimulate metabolism, are derived from a large precursor protein called thyroglobulin: Tyrosine side chains are enzymatically iodinated, then two of these residues undergo condensation, and the hormones are liberated from thyroglobulin by proteolysis.

The 27-carbon cholesterol, introduced in Section 8.1, is the precursor of a large number of hormones that regulate metabolism, salt and water balance, and reproductive functions. Androgens (which are primarily male hormones) have 19 carbons, and estrogens (which are primarily female hormones) contain 18 carbons. Cortisol, a C_{21} glucocorticoid hormone, affects the

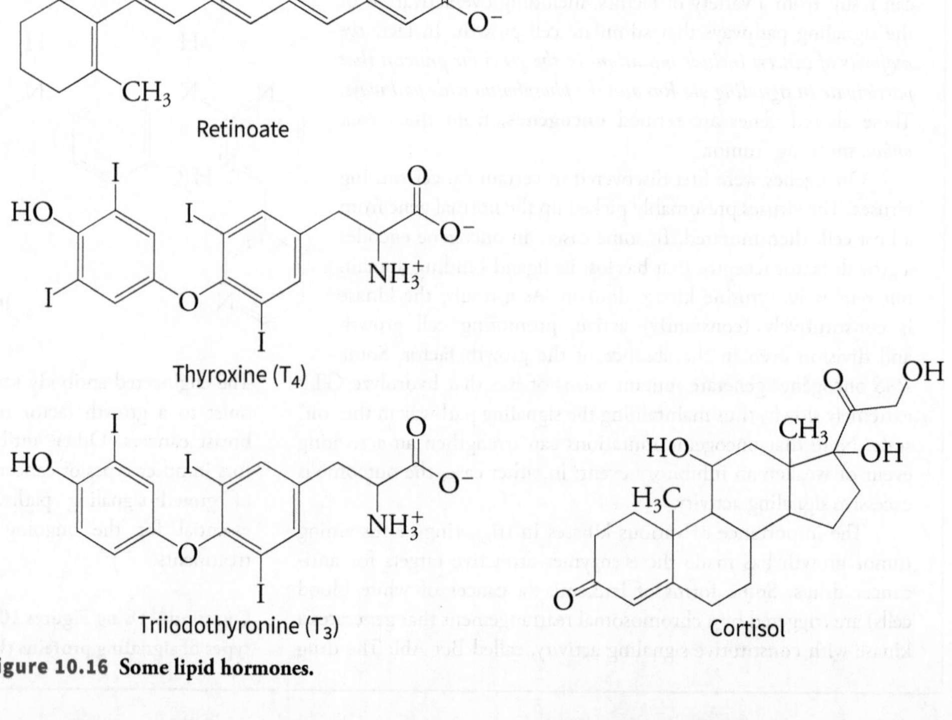

Figure 10.16 Some lipid hormones.

metabolic activities of a wide variety of tissues. Retinoic acid, thyroid hormones, and steroids are all hydrophobic molecules that are carried in the bloodstream either by specific carrier proteins or by albumin, a sort of all-purpose binding protein.

The receptors to which the lipid hormones bind are located inside the appropriate target cell, either in the cytoplasm or the nucleus. Ligand binding often—but not always—causes the receptors to form dimers. Each receptor subunit is constructed from several modules, which include a ligand-binding domain and a DNA-binding domain. The ligand-binding domains are as varied as their hormone ligands, but the DNA-binding domains exhibit a common structure that includes two zinc fingers, which are cross-links formed by the interaction of four cysteine side chains with a Zn^{2+} ion (Section 4.3). In the absence of a ligand, the receptor cannot bind to DNA.

Following ligand binding and dimerization, the receptor moves to the nucleus (if it is not already there) and binds to specific DNA sequences called **hormone response elements.** Although the nucleotide sequences of the hormone response elements vary for each receptor–ligand complex, they are all composed of two identical 6-bp sequences separated by a few base pairs. Simultaneous binding of the two hormone response element sequences explains why many of the lipid hormone receptors are dimers (**Fig. 10.17**).

The receptors function as transcription factors so that the genes near the hormone response elements may experience higher or lower levels of expression. For example, glucocorticoids such as cortisol stimulate the production of phosphatases, which dampen the stimulating effects of kinases. This property makes cortisol and its derivatives useful as drugs to treat conditions such as chronic inflammation or asthma. However, because so many tissues respond to glucocorticoids, the side effects of these drugs can be significant and tend to limit their long-term use.

The changes in gene expression triggered by steroids and other lipid hormones require many hours to take effect. However, cellular responses to some lipid hormones are evident within seconds or minutes, indicating that the hormones also participate in signaling pathways with shorter time courses, such as those centered on G proteins and/or kinases. In these instances, the receptors must be located on the cell surface.

Eicosanoids are short-range signals

Many of the hormones discussed in this chapter are synthesized and stockpiled to some extent before they are released, but some lipid hormones are synthesized as a response to other signaling events (sphingosine-1-phosphate is one example; Section 10.2). The lipid hormones called **eicosanoids** are produced when the enzyme phospholipase A_2 is activated by phosphorylation and by the presence of Ca^{2+}. One substrate of the phospholipase is the membrane lipid phosphatidylinositol. In this lipid, cleavage of the acyl chain attached to the second glycerol carbon often releases arachidonate, a C_{20} fatty acid (the term *eicosanoid* comes from the Greek *eikosi,* meaning "twenty").

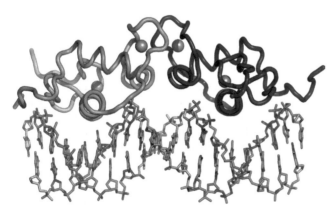

Figure 10.17 The glucocorticoid receptor–DNA complex. The two DNA-binding zinc finger domains of the glucocorticoid receptor are blue and green. The Zn^{2+} ions are shown as gray spheres. The hormone response element sequences of the DNA (bottom) are colored red. Two protein helices make sequence-specific contacts with nucleotides.

Question **What is the surface charge of the receptor's DNA-binding domain?**

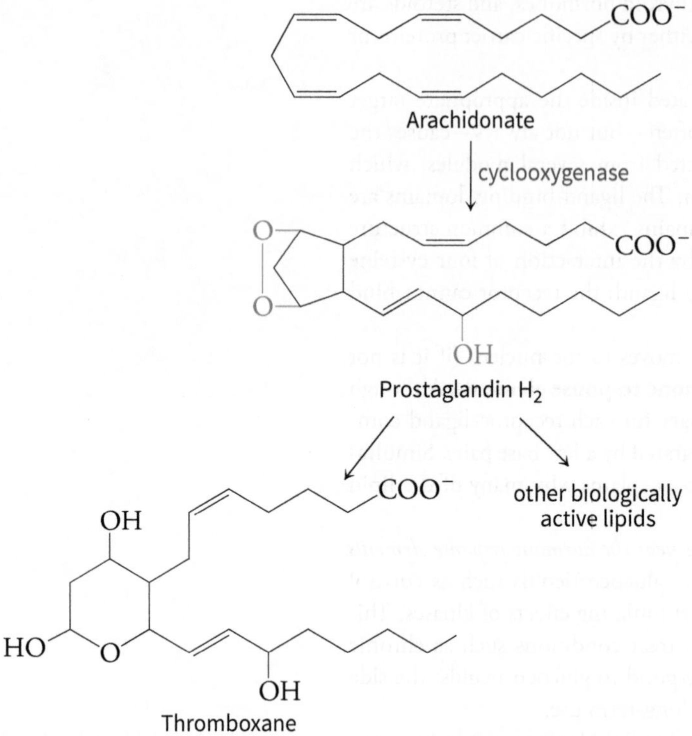

Figure 10.18 Arachidonate conversion to eicosanoid signal molecules. The first step is catalyzed by cyclooxygenase. Only two of the many dozens of eicosanoids are shown here.

Arachidonate, a polyunsaturated fatty acid with four double bonds, is further modified by the action of enzymes that catalyze cyclization and oxidation reactions (**Fig. 10.18**). A wide variety of eicosanoids can be produced in a tissue-dependent fashion, and their functions are similarly varied. Eicosanoids regulate such things as blood pressure, blood coagulation, inflammation, pain, and fever. The eicosanoid thromboxane, for example, helps activate platelets (cell fragments that participate in blood coagulation) and induces vasoconstriction. Other eicosanoids have the opposite effects: They prevent platelet activation and promote vasodilation. The use of aspirin as a "blood thinner" stems from its ability to inhibit the enzyme that initiates the conversion of arachidonate to thromboxane (see Fig. 10.18). A number of other drugs interfere with the production of eicosanoids by blocking the same enzymatic step (**Box 10.D**).

The receptors for eicosanoids are G protein–coupled receptors that trigger cAMP-dependent and phosphoinositide-dependent responses. However, eicosanoids are degraded relatively quickly. This instability, along with their hydrophobicity, means that *their effects are relatively limited in time and space.* Eicosanoids tend to elicit responses only in the cells that produce them and in nearby cells. In contrast, many other hormones travel throughout the body, eliciting effects in any tissue that exhibits the appropriate receptors. For this reason, eicosanoids are sometimes called local mediators rather than hormones.

Similar signaling molecules operate in plants. For example, the lipid hormone jasmonate is synthesized as a result of localized herbivore damage and quickly spreads to other parts of the plant to trigger defensive responses.

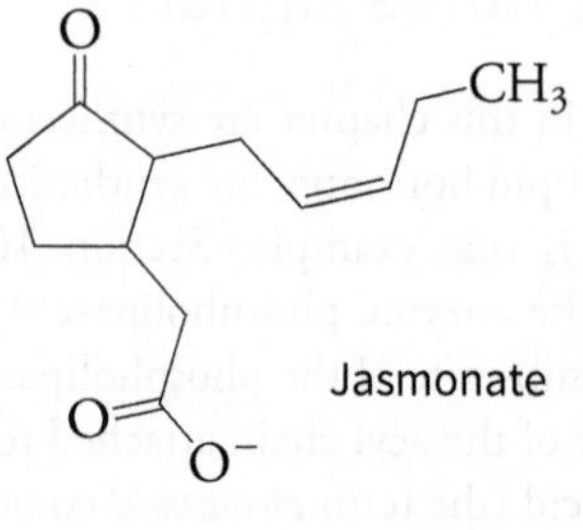

Volatile derivatives of jasmonate can travel through the air to spread the alarm to neighboring plants.

Box 10.D   Inhibitors of Cyclooxygenase

The bark of the willow *Salix alba* has been used since ancient times to relieve pain and fever. The active ingredient is acetylsalicylate, or aspirin.

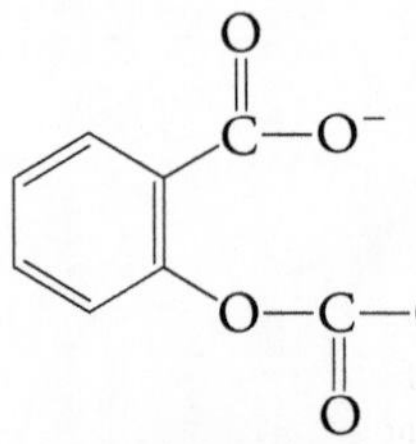

Acetylsalicylate
(aspirin)

Aspirin was first prepared in 1853, but it was not used clinically for another 50 years or so. Effective promotion of aspirin by the Bayer chemical company at the start of the twentieth century marked the beginning of the modern pharmaceutical industry.

Despite its popularity, aspirin's mode of action was not discovered until 1971. It inhibits the production of prostaglandins (which induce pain and fever, among other things) by inhibiting the activity of cyclooxygenase (also known as COX), the enzyme that acts on arachidonate (see Fig. 10.18). COX inhibition results from acetylation of a serine residue located near the active site in a cavity that accommodates the arachidonate substrate. Other pain-relieving substances such as ibuprofen also bind to COX to prevent the synthesis of prostaglandins, although these drugs do not acetylate the enzyme.

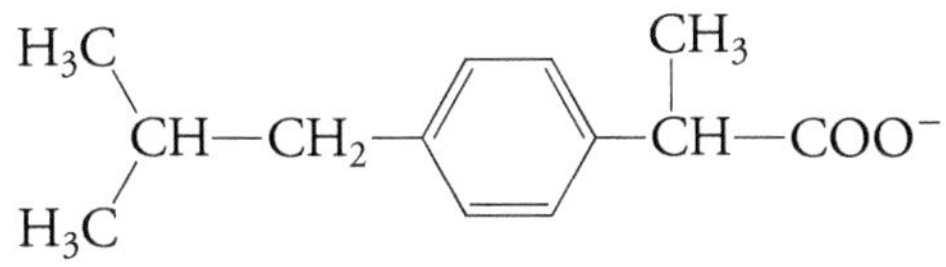

H₃C
    CH—CH₂—⟨ ⟩—CH—COO⁻
H₃C   CH₃

Ibuprofen

One shortcoming of aspirin is that it inhibits more than one COX isozyme. COX-1 is a constitutively expressed enzyme that is responsible for generating various eicosanoids, including those that maintain the stomach's protective layer of mucus. COX-2 expression increases during tissue injury or infection and generates eicosanoids involved in inflammation. Long-term aspirin use suppresses the activity of both isozymes, which can lead to side effects such as gastric ulcers.

Rational drug design (Section 7.4), based on the slightly different structures of COX-1 and COX-2, led to the development of drugs that bind only to the active site of COX-2 because they are too large to fit into the COX-1 active site. The drugs therefore can selectively block the production of pro-inflammatory eicosanoids without damaging gastric tissue. Unfortunately, the side effects of the drugs include an increased risk of heart attacks, through a mechanism that is not fully understood. As a result, only one drug—celecoxib—is currently in use. If nothing else, this story illustrates the complexity of biological signaling pathways and the difficulty of understanding how to manipulate them for therapeutic reasons.

A third COX isozyme, COX-3, is expressed at high levels in the central nervous system. It is the target of the widely used drug acetaminophen (Section 7.4), which reduces pain and fever and does not appear to incur the side effects of the COX-2–specific inhibitors.

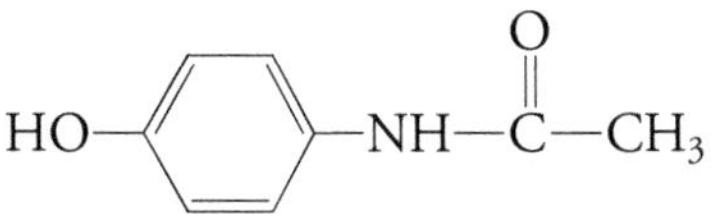

                    O
HO—⟨ ⟩—NH—C—CH₃

Acetaminophen

Question   Which of the drugs shown here is chiral (see Section 4.1), with two different configurations?

Concept Check

1. List some types of lipid hormones and their physiological effects.

2. Explain why lipid hormones have intracellular receptors.

3. Compare the timing of the responses to steroid hormones and eicosanoids.

4. Draw a model of a cell and add shapes to represent all the types of receptors, target enzymes, and other signal-transduction machinery mentioned in the chapter, placing each component in the membrane, cytosol, or nucleus as appropriate.

5. Make a list of the drugs mentioned in this chapter and indicate how they interfere with signaling.

SUMMARY

10.1 General Features of Signaling Pathways

- Agonist or antagonist binding to a receptor can be quantified by a dissociation constant.

- G protein–coupled receptors and receptor tyrosine kinases are the most common types of receptors.

- While signaling systems amplify extracellular signals, they are also regulated so that signaling can be turned off, and the receptor may become desensitized.

10.2 G Protein Signaling Pathways

- A ligand such as epinephrine binds to a G protein–coupled receptor. A G protein responds to the receptor–ligand complex by releasing GDP, binding GTP, and splitting into an α subunit and a βγ dimer. Alternatively, arrestin may bind to the complex and activate additional proteins.

- The α subunit of the G protein activates adenylate cyclase, which converts ATP to cAMP. cAMP is a second messenger that triggers a conformational change in protein kinase A that repositions its activation loop to achieve full catalytic activity.

- cAMP-dependent signaling activity is limited by the reduction of second messenger production through the GTPase activity of G proteins and the action of phosphodiesterases and by the activity of phosphatases that reverse the effects of protein kinase A. Ligand dissociation and receptor desensitization through phosphorylation and arrestin binding also limit signaling via G protein–coupled receptors.

- G protein–coupled receptors that lead to activation of phospholipase C generate inositol trisphosphate and diacylglycerol second messengers, which activate protein kinase B and protein kinase C, respectively.

- Signaling pathways originating with different G protein–coupled receptors and receptor tyrosine kinases overlap through activation or inhibition of the same intracellular components, such as kinases, phosphatases, and phospholipases.

- Odorant binding to olfactory receptors leads to cAMP production that triggers cell depolarization. Tastants activate the phosphoinositide pathway that leads to neurotransmitter release.

10.3 Receptor Tyrosine Kinases

- Ligand binding to a receptor tyrosine kinase causes conformational changes in the receptor dimer that bring the cytoplasmic tyrosine kinase domains close enough to phosphorylate each other.

- In addition to acting as kinases, the receptor tyrosine kinases initiate other kinase cascades by activating the small monomeric G protein Ras.

10.4 Lipid Hormone Signaling

- Steroids and other lipid hormones bind primarily to intracellular receptors that dimerize and bind to hormone response elements on DNA to induce or repress the expression of nearby genes.

- Eicosanoids, which are synthesized from membrane lipids, function as signals over short ranges and for a limited time.

KEY TERMS

receptor
ligand
signal transduction
hormone
quorum sensing

dissociation constant
(K_d)
agonist
antagonist
G protein
GPCR

second messenger
kinase
receptor tyrosine
kinase
desensitization
cAMP

phosphatase
phosphoinositide sig-
naling system
cross-talk
olfaction
odorant

opiod
gustation
tastant
autophosphorylation
transcription factor
oncogene

hormone response
element
eicosanoid

BIOINFORMATICS

Brief Bioinformatics Exercises

10.1. G Protein–Coupled Receptors and Receptor Tyrosine Kinases

10.2. Biosignaling and the KEGG Database

PROBLEMS

10.1 General Features of Signaling Pathways

1. Which of the signaling molecules listed in Table 10.1 would not require a cell-surface receptor?

2. Animals "taste" the carbonation (fizziness) of sodas when the dissolved CO_2 reacts with water to generate bicarbonate and protons in a reaction catalyzed by carbonic anhydrase expressed on the surface of cells in the tongue. Does this enzyme fit the definition of a CO_2 receptor?

3. Discuss the significance of signaling molecules behaving as enzyme substrates in their binding affinity to receptors, drawing parallels between receptor–ligand, and enzyme–substrate interactions.

4. A sample of cells has a total receptor concentration of 25 mM. Eighty percent of the receptors have bound ligand and the concentration of free ligand is 125 µM. What is the K_d for the receptor–ligand interaction?

5. A sample of cells has a total receptor concentration of 100 mM. Fifty percent of the receptors have bound ligand and the concentration of free ligand is 10 mM. **a.** What is the K_d for the receptor–ligand interaction? **b.** What is the relationship between K_d and [L]?

6. The K_d for a receptor–ligand interaction is 5 mM. When the concentration of free ligand is 18 mM and the concentration of free receptor is 8 mM, what is the concentration of receptor that is occupied by ligand?

7. The total concentration of receptors in a sample is 40 mM. The concentration of free ligand is 5 mM and the K_d is 20 mM. Calculate the percentage of receptors that are occupied by ligand.

8. The total concentration of receptors in a sample is 12 mM. The concentration of free ligand is 3 mM and the K_d is 1.5 mM. Calculate the percentage of receptors that are occupied by ligand.

9. The total concentration of receptors in a sample is 10 mM. The concentration of free ligand is 2.5 mM and the K_d is 0.3 mM. Calculate the percentage of receptors that are occupied by ligand. Compare the answer to this problem with the answer you obtained in Problem 7 and explain the difference.

10. Use the plot below to estimate a value for K_d.

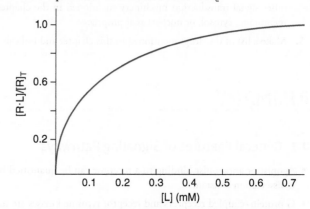

11. **a.** In an experiment, the ligand adenosine is added to heart cells in culture. The number of receptors with ligand bound is measured and the data yield a curve like the one shown in Figure 10.1. What would the results look like if the experiment were repeated in the presence of caffeine? **b.** How does the apparent K_d in the presence of caffeine compare to the K_d in its absence?

12. K_d is defined in Equation 10.1, which shows the relationship between the free receptor concentration [R], the ligand concentration [L], and the concentration of receptor–ligand complexes [R · L]. The value of [R], like [R · L], is difficult to evaluate, but various experimental techniques can be used to determine $[R]_T$, the total number of receptors. $[R]_T$ is the sum of [R] and [R · L]. Using this information, begin with Equation 10.1 and derive an expression for the [R · L]/$[R]_T$ ratio. (Note that your derived expression will be similar to the Michaelis–Menten equation and that Equations 7.9 through 7.17 may give you an idea of how to proceed.)

13. The [R · L]:$[R]_T$ ratio gives the fraction of receptors that have bound ligand. Use the expression you derived in Problem 12 to express [R · L] as a fraction of $[R]_T$ for the following situations: **a.** K_d = 3[L], **b.** K_d = [L], and **c.** $3K_d$ = [L].

14. If a cell has 1000 surface receptors for erythropoietin, and if only 15% of those receptors need to bind the ligand to achieve a maximal response, what ligand concentration is required to achieve a maximal response? Use the equation you derived in Problem 12. The K_d for erythropoietin is 1.0×10^{-10} M.

15. Suppose the number of surface receptors on the cell described in Problem 14 decreases to 200. What ligand concentration is required to achieve a maximal response?

16. ADP binds to platelets in order to initiate the activation process. Two binding sites were identified on platelets, one with a K_d of 0.35 μM and one with a K_d of 7.9 μM. **a.** Which site is a low-affinity binding site and which is a high-affinity binding site? **b.** The ADP concentration required to activate a platelet is in the range of 0.1–0.5 μM. Which receptor will be more effective at activating the platelet? **c.** Two ADP agonists were also found to bind to platelets: 2-methythio-ADP bound with a K_d of 7 μM and 2-(3-aminopropylthio)-ADP bound with a K_d of 200 μM. Can these agonists effectively compete with ADP for binding to platelets? **d.** In the study, 160,000 high-affinity binding sites were identified on each platelet. What is the concentration of ADP required to achieve 85% binding to the high-affinity site? Use the equation you derived in Problem 11.

17. The peptide hormone glucagon, released during fasting, binds to a high-affinity receptor with a K_d value of 0.8 nM. **a.** What percentage of glucagon receptors are occupied when the blood glucagon concentration is 70 pM after one day of fasting? Use the equation you derived in Problem 11. **b.** A glucagon derivative missing the N-terminal histidine residue has a K_d value of 50 nM. What does this reveal about the structural requirements for glucagon binding to its receptor?

18. Like the Michaelis–Menten equation, the equation derived in Solution 11 can be converted to an equation for a straight line. A double-reciprocal plot for a ligand binding to its receptor is shown below. Use the information in the plot to estimate a value for K_d.

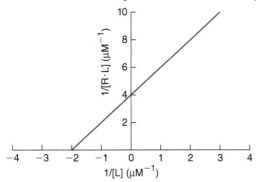

19. A double-reciprocal plot (see Problem 18) for a ligand binding to its receptor is shown below. Use the information in the plot to calculate a value for K_d.

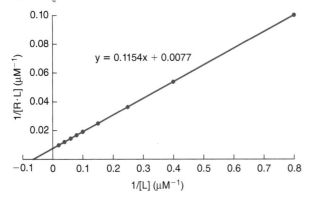

20. A Scatchard plot is another method of representing ligand binding data using a straight line (see Problem 18). In a Scatchard plot, $[R \cdot L]/[L]$ is plotted versus $[R \cdot L]$. The slope is equal to $-1/K_d$. Use the Scatchard plot provided to calculate a value for K_d for calmodulin binding to calcineurin.

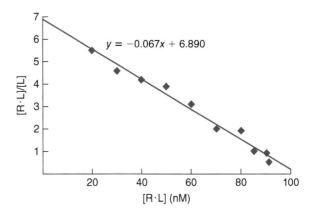

21. A nucleoside derivative binds to lymphocytes and elicits a wide variety of cellular responses. The binding of the nucleoside to its receptor was analyzed using a Scatchard plot (see Problem 20). Use the plot provided to calculate the K_d for nucleoside–receptor binding.

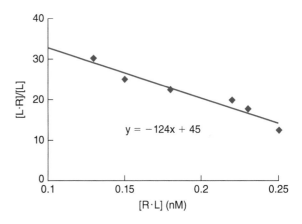

22. Why might it be difficult to purify cell-surface receptors using the techniques described in Section 4.6?

23. Affinity chromatography is often used as a technique to purify cell-surface receptors. Describe the steps you would take to purify a cell-surface receptor using this technique.

24. Epinephrine can bind to several different types of G protein–linked receptors. Each of these receptors triggers a different cellular response. Explain how this is possible.

25. In the liver, glucagon and epinephrine bind to different members of the G protein–coupled receptor family, yet binding of each of these ligands results in the same response—glycogen breakdown. How is it possible that two different ligands can trigger the same cellular response?

26. Many receptors become desensitized in the presence of high concentrations of signaling ligand. This can occur in a variety of ways, such as removal of the receptors from the cell surface by endocytosis. Why is this an effective desensitization strategy?

27. What is the advantage of activating D using the strategy shown in the figure? Why is this strategy more effective than a simple activation of D that occurs in one step?

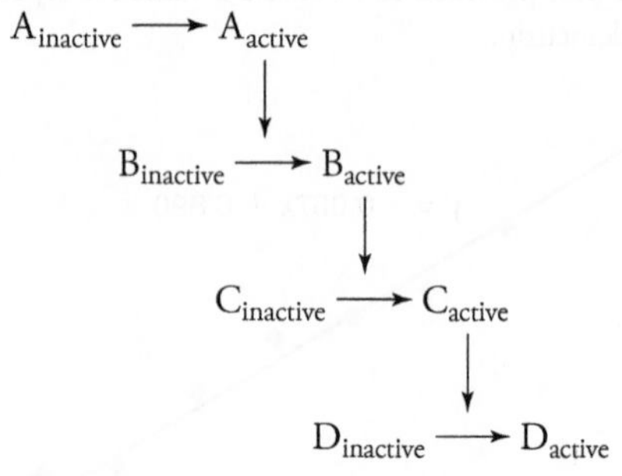

$$A_{inactive} \longrightarrow A_{active}$$
$$B_{inactive} \longrightarrow B_{active}$$
$$C_{inactive} \longrightarrow C_{active}$$
$$D_{inactive} \longrightarrow D_{active}$$

10.2 G Protein Signaling Pathways

28. Explain the evolutionary advantage of having a signaling pathway involving G proteins and G protein-coupled receptors (GPCRs) rather than relying solely on direct cell-cell communication.

29. As described in the text, G protein–coupled receptors are often palmitoylated at a Cys residue. **a.** Draw the structure of the palmitate (16:0) residue covalently linked to a Cys residue. **b.** What is the role of this fatty acyl chain? **c.** What is the expected result if the Cys residue is mutated to a Gly?

30. The cell contains many types of guanine nucleotide exchange factors (GEFs) that regulate the activity of monomeric GTPases by promoting the exchange of GDP for GTP. Explain how G protein–coupled receptors also function as GEFs.

31. a. Naturally occurring mutations in the genes that code for GPCRs have provided insights into GPCR function and the diseases that often result from these mutations. List the types of mutations in a GPCR that would result in loss of the receptor's function. **b.** Mutations in a GPCR gene occasionally result in increased receptor activity. List the types of mutations in a GPCR that would result in gain of function of the receptor.

32. Some G protein–linked receptors are associated with a protein called RGS (regulator of G protein signaling). RGS stimulates the GTPase activity of the G protein associated with the receptor. What effect does RGS have on the signaling process?

33. An Asp residue in the third transmembrane helix of the epinephrine receptor (a GPCR) interacts with the ligand epinephrine. **a.** What type of interaction is likely to form between the receptor and epinephrine? **b.** How might an Asp → Glu mutation affect ligand binding? **c.** How might an Asp → Asn mutation affect ligand binding?

34. How do epinephrine and norepinephrine differ from tyrosine, their parent amino acid?

35. β_2-Adrenergic receptor antagonists known as β-blockers prevent epinephrine and norepinephrine binding to their receptors. Why are β-blockers effective at treating high blood pressure?

36. If a mutation occurred in the GTP-binding site of a G protein, how might this affect the overall signaling process, and what cellular consequences could arise?

37. A toxin secreted by the bacterium *Vibrio cholerae* catalyzes the covalent attachment of an ADP-ribose group to the α subunit of a G protein. This results in the inhibition of the intrinsic GTPase activity of the G protein. How does this affect the activity of adenylate cyclase? How are intracellular levels of cAMP affected?

38. Some G protein–linked receptors are associated with G proteins that inhibit rather than stimulate the activity of adenylate cyclase.

A toxin secreted by the bacterium *Bordetella pertussis* (the causative agent of whooping cough) catalyzes the covalent attachment of an ADP-ribose group to the α subunit of the inhibitory G protein, preventing it from carrying out its normal function. How does this affect the activity of adenylate cyclase? How are intracellular levels of cAMP affected?

39. Addition of the nonhydrolyzable analog GTPγS to cultured cells is a common practice in signal transduction experiments. What effect does GTPγS have on cellular cAMP levels for **a.** a stimulatory G protein (see Problem 37) and **b.** an inhibitory G protein (see Problem 38)?

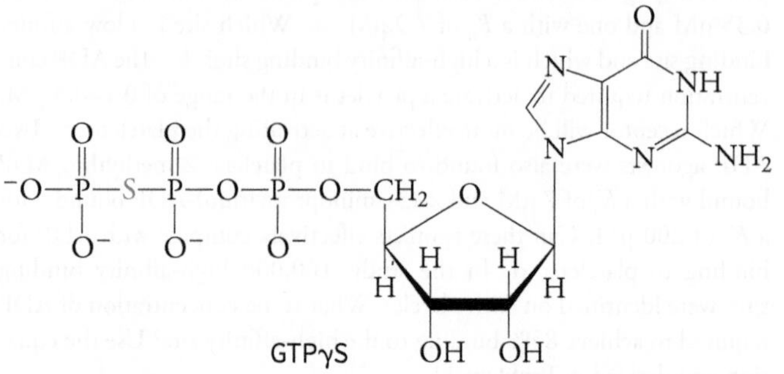

GTPγS

40. Consider a scenario where a cell selectively loses its ability to produce the β-subunit of a trimeric G protein. How might this impact the cell's responsiveness to extracellular signals, and what cellular processes could be affected?

41. a. Draw the reaction that shows the protein kinase A–catalyzed phosphorylation of a threonine residue on a target protein. **b.** Draw the reaction that shows the phosphatase-catalyzed hydrolysis of the phosphorylated threonine. **c.** Some bacterial signaling systems involve kinases that transfer a phosphoryl group to a His side chain. Draw the structure of the phospho-His side chain.

42. Phorbol esters, which are compounds isolated from plants, are structurally similar to diacylglycerol. How does the addition of phorbol esters affect the cellular signaling pathways of cells in culture?

43. As described in the text, ligand binding to certain receptor tyrosine kinases results in the activation of a sphingomyelinase enzyme. Draw the reaction that shows the sphingomyelinase-catalyzed hydrolysis of sphingomyelin to ceramide.

44. In unstimulated T cells, a transcription factor called NFAT (nuclear factor of activated T cells) resides in the cytosol in a phosphorylated form. When the cell is stimulated, the cytosolic Ca^{2+} concentration increases and activates a phosphatase called calcineurin. The activated calcineurin catalyzes the hydrolysis of the phosphate group from NFAT, exposing a nuclear localization signal that allows the NFAT to enter the nucleus and stimulate the expression of genes essential for T cell activation. Describe the cell-signaling events that result in the activation of NFAT.

45. The immunosuppressive drug cyclosporine A is an inhibitor of calcineurin (see Problem 44). Why is cyclosporine A an effective immunosuppressant?

46. Pathways that lead to the activation of protein kinase B (Akt) are considered to be anti-apoptotic (apoptosis is programmed cell death). In other words, protein kinase B stimulates a cell to grow and proliferate. Like all biological events, signaling pathways that are turned on must also be turned off. A phosphatase called PTEN plays a role in removing phosphate groups from proteins, but it is highly specific for removing a phosphate group from inositol trisphosphate. **a.** If PTEN is overexpressed in mammalian cells, do these cells grow or do they undergo apoptosis? **b.** Would you expect to find mutations in the gene for PTEN in human cancers? Explain why or why not.

47. Nitric oxide (NO) is a naturally occurring signaling molecule (see Table 10.1) that is produced from the decomposition of arginine to NO and citrulline in endothelial cells. The enzyme that catalyzes this reaction, NO synthase, is stimulated by cytosolic Ca^{2+}, which increases when acetylcholine binds to endothelial cells. **a.** What is the source of the acetylcholine ligand? **b.** Propose a mechanism that describes how acetylcholine binding leads to the activation of NO synthase. **c.** NO formed in endothelial cells quickly diffuses into neighboring smooth muscle cells and binds to a cytosolic protein that catalyzes the formation of the second messenger cyclic GMP. Cyclic GMP then activates protein kinase G (see Problem 38), resulting in smooth muscle cell relaxation. How might protein kinase G bring about smooth muscle relaxation?

48. As discussed in the text, any signal transduction event that is turned on must subsequently be turned off. Refer to your answer to Problem 47 and describe the events that would lead to the cessation of each step of the signaling pathway you described.

49. NO synthase knockout mice (animals missing the NO synthase enzyme) have elevated blood pressure, an increased heart rate, and enlarged left ventricle chambers. Explain the reasons for these symptoms.

50. Clotrimazole is a calmodulin antagonist (see Solution 47b). How does the addition of clotrimazole affect endothelial cells in culture?

51. Nitroglycerin placed under the tongue has been used since the late nineteenth century to treat angina pectoris (chest pains resulting from reduced blood flow to the heart). Only recently have scientists elucidated its mechanism of action. Propose a hypothesis that explains why nitroglycerin placed under the tongue relieves the pain of angina.

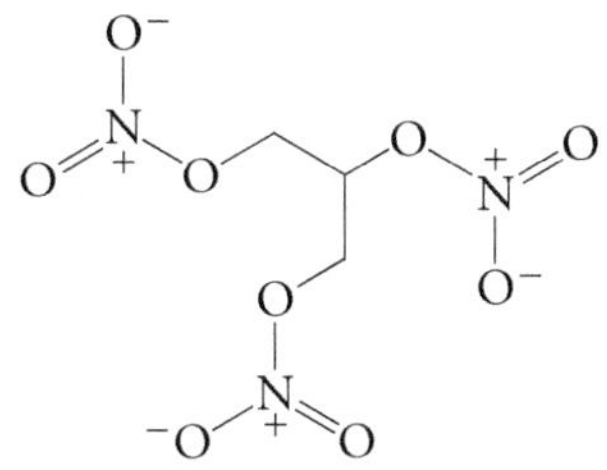

Nitroglycerin

52. Viagra, a drug used to treat erectile dysfunction, is a cGMP phosphodiesterase inhibitor. Propose a mechanism that explains why the drug is effective in treating this condition.

53. Propose an explanation why arrestin can be activated by the molecule shown here.

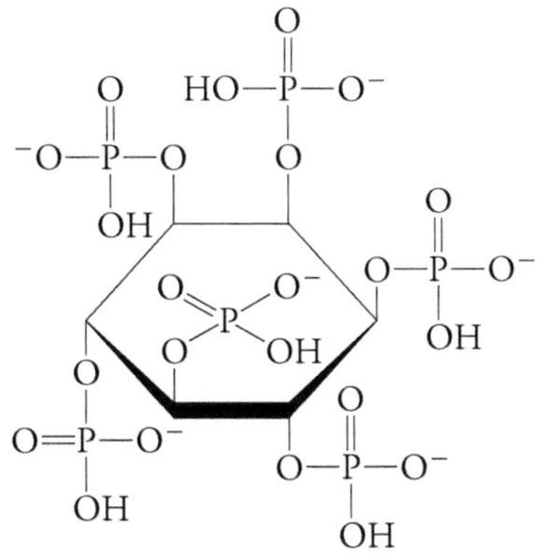

54. Arrestin has been shown to bind to clathrin (see Fig. 9.24). What is the significance of this observation?

55. Thrombin generated at the site of a blood vessel injury (see Section 6.5) binds to the thrombin receptor, a GPCR, on the surface of platelets, which activates the platelets, leading to their aggregation and the formation of a plug to help seal the broken blood vessel. Draw a diagram of the thrombin signaling pathway, given the following information: phospholipase C is activated; Ca^{2+} ions activate myosin-light chain kinase; the activated myosin light chain is responsible for platelet shape change; and protein kinase C stimulates the release of granules (substances essential in blood coagulation) when activated.

56. A series of experiments showed that thrombin signaling (see Problem 55) is terminated when the GPCR is inactivated by phosphorylation catalyzed by a specific kinase called G protein coupled receptor kinase (GRK). **a.** Why does the receptor become radioactively labeled when $[\gamma^{32}P]$-ATP is added to cells in culture? **b.** Would this happen if Ser and Thr residues in the receptor were mutated? **c.** What is the expected result if GRK were overexpressed in cultured cells? **d.** Why does GRK activity decrease by 90% when a lysine residue in the kinase is mutated to an alanine? **e.** Predict what would happen if high concentrations of thrombin were added to cultured cells.

57. *Bacillus anthracis,* the causative agent of anthrax, produces a three-part toxin. One part facilitates the entry of the two other toxins into the cytoplasm of a mammalian cell. The toxin known as edema factor (EF) is an adenylate cyclase. **a.** Explain how EF could disrupt normal cell signaling. **b.** EF must first be activated by Ca^{2+}-calmodulin binding to it. Explain how this requirement could also disrupt cell signaling.

58. *Bacillus anthracis* (see Problem 57) also makes a toxin called lethal factor (LF). LF is a protease that specifically cleaves and inactivates a protein kinase that is part of a pathway for stimulating cell proliferation. Explain why the entry of LF into white blood cells promotes the spread of *B. anthracis* in the body.

59. Ligand binding to some growth-factor receptors triggers kinase cascades and also leads to activation of enzymes that convert O_2 to hydrogen peroxide (H_2O_2), which acts as a second messenger. Describe the likely effect of H_2O_2 on the activity of cellular phosphatases.

60. Hydrogen peroxide acts as a second messenger, as described in Problem 59, and affects PTEN (see Problem 46) as well as other cellular phosphatases. Does H_2O_2 activate or inhibit PTEN?

10.3 Receptor Tyrosine Kinases

61. Stimulation of the insulin receptor by ligand binding and autophosphorylation eventually leads to the activation of both protein kinase B (Akt) and protein kinase C. Protein kinase B phosphorylates glycogen synthase kinase 3 (GSK3) and inactivates it. (Active GSK3 inactivates glycogen synthase by phosphorylating it.) Glycogen synthase catalyzes synthesis of glycogen from glucose. In the presence of insulin, GSK3 is inactivated, so glycogen synthase is not phosphorylated and is active. Protein kinase C stimulates the translocation of glucose transporters to the plasma membrane by a mechanism not currently understood. One strategy for treating diabetes is to develop drugs that act as inhibitors of the phosphatases that remove phosphate groups from the phosphorylated tyrosines on the insulin receptor. Why might this be an effective treatment for diabetes?

62. When insulin binds to its receptor, a conformational change occurs that results in autophosphorylation of the receptor on specific Tyr residues. In the next step of the signaling pathway, an adaptor protein called IRS-1 (insulin receptor substrate-1) docks with the phosphorylated receptor (the involvement of adaptor proteins in cell signaling is shown in Fig. 10.15). This step is essential for the downstream activation of protein kinases B and C (see Problem 61). If IRS-1 is overexpressed in muscle cells in culture, what effects, if any, would you expect to see on glucose transporter translocation and glycogen synthesis?

63. The activity of Ras is regulated in part by two proteins, a guanine nucleotide exchange factor (GEF) and a GTPase activating protein (GAP). The GEF protein binds to Ras · GDP and promotes dissociation of bound GDP. The GAP protein binds to Ras · GTP and stimulates the intrinsic GTPase activity of Ras. **a.** How is downstream activity of a signaling pathway affected by the presence of GEF? **b.** By the presence of GAP?

64. Mutant Ras proteins have been found in various types of cancers. What is the effect on a cell if the mutant Ras is able to bind GTP but is unable to hydrolyze it?

65. As shown in Figure 10.15, Ras can activate a kinase cascade. The most common cascade is the MAP kinase pathway, which is activated when growth factors bind to cell surface receptors and activate Ras. This leads to the activation of transcription factors and other gene regulatory proteins and results in growth, proliferation, and differentiation. Use this information to explain why phorbol esters (see Problem 42) promote tumor development.

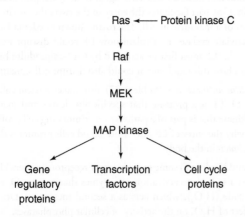

66. How might a signaling molecule activate the MAP kinase cascade (see Problem 65) via a G protein–coupled receptor rather than a receptor tyrosine kinase?

67. Mutations in proteins involved in the receptor tyrosine kinase pathway (see Fig. 10.15 and Problem 65) are found in many different types of cancers. How might such mutations alter signaling activity in a cancerous cell? Give several examples.

68. Hsp90, a molecular chaperone that assists with protein folding (see Fig. 4.24) is overexpressed in cancer cells and is often found associated with proteins involved in signaling pathways. Why would cancer cells require additional assistance from chaperones?

69. Ras can activate a cascade of reactions (see Fig. 10.12) involving membrane phosphatidylinositol (PI) lipids (see Solutions 8.22 and 8.23). Ras activates PI 3-kinase (PI3K), which phosphorylates PI (4,5) bisphosphate (PIP$_2$) to PI (3,4,5) trisphosphate (PIP$_3$). PIP3 then activates Akt (see Problem 46). **a.** Why would PI3K inhibitors be effective at treating cancer? **b.** PTEN (see Problem 46) can catalyze the hydrolysis of PIP$_3$ to PIP$_2$. Does PTEN promote cell survival or cell death?

70. PKR is a protein kinase that recognizes double-stranded RNA molecules such as those that form during the intracellular growth of certain viruses. The structure of PKR includes the standard kinase domains as well as an RNA-binding module. In the presence of viral RNA, PKR undergoes autophosphorylation and is then able to phosphorylate cellular target proteins that initiate antiviral responses. Short (<30 bp) RNAs inhibit activation of PKR, but RNAs longer than 33 bp are strong activators of PKR. Explain the role of RNA in PKR activation.

71. The bacterium *Yersinia pestis*, the pathogen responsible for bubonic plague, caused the deaths of about a third of the population of Europe in the fourteenth century. The bacterium produces a phosphatase called YopH, which hydrolyzes phosphorylated tyrosines and is much more catalytically active than mammalian phosphatases. **a.** What happens when the *Yersinia* bacterium injects YopH into a mammalian cell? **b.** Why is the bacterium itself not affected by YopH? **c.** Scientists are interested in developing YopH inhibitors in order to treat *Yersinia* infection, a re-emerging disease. What are some important considerations in the development of a YopH inhibitor?

10.4 Lipid Hormone Signaling

72. Steroid hormone receptors have different cellular locations. The progesterone receptor is located in the nucleus and interacts with DNA once progesterone has bound. However, the glucocorticoid receptor is located in the cytosol and does not move into the nucleus until its ligand has bound. What structural feature must be different in these two receptor molecules?

73. Abnormal changes in steroid hormone levels in the breast, uterus, ovaries, prostate, and testes are observed in cancers of these steroid-responsive tissues. **a.** Using what you know about the mechanism of steroid hormone–induced stimulation in these tissues, what strategies could you use to design drugs to treat cancers in these tissues? **b.** Given what you know about cell signaling and cancer (see Box 10.C), is it reasonable to suspect that a "nonclassical" pathway may be involved in the development of cancers in these tissues?

74. The production of sphingosine-1-phosphate and ceramide 1-phosphate is shown in the diagram. These signaling molecules can act in the cell where they are produced (intracellularly) or they can exit the cell and act on neighboring cells, as described in Section 10.2. There is quite a bit of cross-talk between the pathway shown in the diagram and other pathways discussed in this chapter. Ceramide-1-phosphate (C1P) promotes the release of arachidonic acid from the membrane. Sphingosine-1-phosphate (S1P) stimulates the activity of COX-2. Are these observations consistent with the inflammatory properties attributed to C1P and S1P?

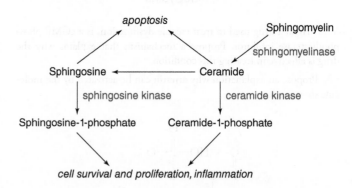

75. Sphingosine-1-phosphate (see Problem 74) stimulates the activation of protein kinase B (Akt) (see Problem 46). What effect would this have on the cell?

76. Sphingosine-1-phosphate's pro-survival effects probably result from the interaction of S1P with multiple cellular pathways. In addition to S1P's ability to activate Akt (see Problem 75), how else might S1P act to promote cell survival?

77. There is a strong association between inflammation and cancer. Using the information presented in Problem 74, identify possible targets for anticancer drugs.

78. Aspirin inhibits COX by acetylating a Ser residue on the enzyme (see Box 10.D). **a.** Draw structures to show how aspirin acetylates the serine side chain. **b.** Propose a hypothesis that explains why Ser acetylation inhibits the enzyme. **c.** What type of inhibitor is aspirin (see Section 7.3)?

79. Cyclooxygenase uses arachidonic acid as a substrate for the synthesis of prostaglandins (see Box 10.D). In platelets, a similar pathway yields thromboxanes, compounds that stimulate vasoconstriction and platelet aggregation (see Fig. 10.18). Why do some people take a daily aspirin as protection against heart attacks?

80. Analogs of cortisol, such as prednisone, are used as anti-inflammatory drugs, although their mechanism of action is not entirely understood. Explain how inhibitors of phospholipase A_2 by prednisone could decrease inflammation.

81. Tetrahydrocannabinol (THC), the active ingredient in marijuana, binds to a receptor in the brain. The natural ligand for the receptor is anandamide. Anandamide has a short half-life because it is rapidly broken down by a hydrolase. One product is ethanolamine. Name the other product of anandamide breakdown.

Anandamide

82. A complex signaling pathway in yeast allows the cells to accumulate high concentrations of glycerol if they are exposed to high extracellular concentration of salt or glucose. The increased osmolarity of the extracellular medium activates Ras, which in turn activates adenylate cyclase. A second pathway, the HOG (high osmolarity glycerol) pathway activates the MAP kinase pathway (see Problem 65). The target protein is the enzyme PFK2, which is activated by phosphorylation. (PFK2 produces an allosteric regulator that activates glycolysis, which ultimately produces glycerol.) Draw a diagram that shows how the Ras and MAP kinase pathways converge to result in the phosphorylation and activation of PFK2.

83. Yeast mutants lacking components of the HOG pathway (see Problem 82) were first exposed to high concentrations of glucose and then the PFK2 activity was measured. How does the PFK2 activity in the mutants compare with the PFK2 activity in mutants exposed to isotonic conditions?

SELECTED READINGS

Behrens, M., Briand, L., de March, C.A., Matsunami, H., Yamashita, A., Meyerhof, W., and Weyand, S., Structure–function relationships of olfactory and taste receptors, *Chemical Senses* 43, 81–87, doi: 10.1093/chemse/bjx083 (2018). [A brief review of current understanding of sensory receptors.]

Chen, Q., Iverson, T.M., and Gurevich, V.V., Structural basis of arrestin-dependent signal transduction, *Trends Biochem. Sci.* 43, 412–423, doi: 10.1016/j.tibs.2018.03.005 (2018). [Describes the conformational changes in arrestin on binding to a receptor and to other proteins.]

Hilger, D., Masureel, M., and Kobilka, B.K., Structure and dynamics of GPCR signaling complexes, *Nat. Struct. Mol. Biol.* 25, 4–12, doi: 10.1038/s41594-017-0011-7 (2018). [Summarizes key features of GPCR interactions with G proteins and arrestin.]

Lemmon, A. and Schlessinger, J., Cell signaling by receptor tyrosine kinases, *Cell* 141, 1117–1134 (2010). [Describes the many events that occur following ligand binding to the receptor tyrosine kinases.]

Taylor, S.S. and Kornev, A.P., Protein kinases: Evolution of dynamic regulatory proteins, *Trends Biochem. Sci.* 36, 65–77 (2011). [Reviews the general structure of protein kinases and their mechanism of activation by phosphorylation of an activation loop.]

Thai, D.M., Glukhova, A., Sexton, P.M., and Christopoulos, A., Structural insights into G–protein–coupled receptor allostery. *Nature* 559, 45–52, doi: 10.1038/s41586-018-0259-z (2018). [Argues that although GPCRs can be considered as allosteric proteins that switch between active and inactive states, the proteins behave more dynamically, with many conformational states.]

Valentino, R.J. and Volkow, N.D., Untangling the complexity of opioid receptor function, *Neuropsychopharmacology* 43, 2514–2520, doi: 10.1038/s41386-018-0225-3 (2018). [Summarizes the three types of opioid receptors and their modes of signal transduction.]

CHAPTER 10 CREDITS

Figure 10.3 Image based on 2RH1. Cherezov, V., Rosenbaum, D.M., Hanson, M.A., Rasmussen, S.G.F., Thian, F.S., Kobilka, T.S., Choi, H.J., Kuhn, P., Weis, W.I., Kobilka, B.K., Stevens, R.C., High resolution crystal structure of human B2-adrenergic G protein-coupled receptor, *Science* 318, 1258–1265 (2007).

Figure 10.4 Image from Koehl, A., Hu, H., Feng, D. et al., 2019. Structural insights into the activation of metabotropic glutamate receptors, *Nature* 566, 79–84. Reproduced with permission of Springer Nature.

Figure 10.5a Image based on 3SN6. Rasmussen, S.G., DeVree, B.T., Zou, Y., Kruse, A.C., Chung, K.Y., Kobilka, T.S., Thian, F.S., Chae, P.S., Pardon, E., Calinski, D., Mathiesen, J.M., Shah, S.T., Lyons, J.A., Caffrey, M., Gellman, S.H., Steyaert, J., Skiniotis, G., Weis, W.I., Sunahara, R.K., Kobilka, B.K., Crystal structure of the beta2 adrenergic receptor-Gs protein complex, *Nature* 477, 549–555 (2011).

Figure 10.5b Image based on 1GP2. Wall, M.A., Coleman, D.E., Lee, E., Iniguez-Lluhi, J.A., Posner, B.A., Gilman, A.G., Sprang, S.R.,

The structure of the G protein heterotrimer $G_{i\alpha1}\beta_{1\gamma2}$, Cell 83, 1047–1058 (1995).

Figure 10.7 Image based on 3TNP. Zhang, P., Smith-Nguyen, E.V., Keshwani, M.M., Deal, M.S., Kornev, A.P., Taylor, S.S., Structure and allostery of the PKA RIIbeta tetrameric holoenzyme, *Science* 335, 712–716 (2012).

Figure 10.8 Image based on 1ATP. Zheng, J., Trafny, E.A., Knighton, D.R., Xuong, N.H., Taylor, S.S., Ten Eyck, L.F., Sowadski, J.M., 2.2 A refined crystal structure of the catalytic subunit of cAMP-dependent protein kinase complexed with MnATP and a peptide inhibitor, *Acta Crystallogr. D* 49, 362–365 (1993).

Figure 10.9 Image based on 4ZWJ. Kang, Y., Zhou, X.E., Gao, X., He, Y., Liu, W., Ishchenko, A., Barty, A., White, T.A., Yefanov, O., Han, G.W., Xu, Q., de Waal, P.W., Ke, J., Tan, M.H., Zhang, C., Moeller, A., West, G.M., Pascal, B.D., Van Eps, N., Caro, L.N., Vishnivetskiy, S.A., Lee, R.J., Suino-Powell, K.M., Gu, X., Pal, K., Ma, J., Zhi, X., Boutet, S., Williams, G.J., Messerschmidt, M., Gati, C., Zatsepin, N.A., Wang, D., James, D., Basu, S., Roy-Chowdhury, S., Conrad, C.E., Coe, J., Liu, H., Lisova, S., Kupitz, C., Grotjohann, I., Fromme, R., Jiang, Y., Tan, M., Yang, H., Li, J., Wang, M., Zheng, Z., Li, D., Howe, N., Zhao, Y., Standfuss, J., Diederichs, K., Dong, Y., Potter, C.S., Carragher, B., Caffrey, M., Jiang, H., Chapman, H.N., Spence, J.C., Fromme, P., Weierstall, U., Ernst, O.P., Katritch, V., Gurevich, V.V., Griffin, P.R., Hubbell, W.L., Stevens, R.C., Cherezov, V., Melcher, K., Xu, H.E., Crystal structure of rhodopsin bound to arrestin by femtosecond X-ray laser, *Nature* 523, 561–567 (2015).

Figure 10.10 Image at top based on 3CLN. Babu, Y.S., Bugg, C.E., Cook, W.J., Structure of calmodulin refined at 2.2 A resolution, *J. Mol. Biol.* 204, 191–204 (1988). Image at bottom based on 2BBM. Ikura, M., Clore, G.M., Gronenborn, A.M., Zhu, G., Klee, C.B., Bax, A., Solution structure of a calmodulin-target peptide complex by multidimensional NMR, *Science* 256, 632–638 (1992).

Figure 10.11 Image from Kim, S.K., Chen, Y., Abrol, R., Goddard, W.A., Guthrie, B.. Activation mechanism of the G protein-coupled sweet receptor heterodimer with sweeteners and allosteric agonists, *Proc. Natl. Acad. Sci.* March 7, 2017, 114, 2568–2573. Reproduced with permission of National Academy of Sciences.

Figure 10.12b Image based on 4ZXB. Croll, T.I., Smith, B.J., Margetts, M.B., Whittaker, J., Weiss, M.A., Ward, C.W., Lawrence, M.C., Higher-resolution structure of the human insulin receptor ectodomain: Multi-modal inclusion of the insert domain, *Structure* 24, 469–476 (2016).

Figure 10.13 adapted from Gutmann, T., Schäfer, I.B., Poojari, C., Brankatschk, B., Vattulainen, I., Strauss, M., Coskun, Ü., Cryo-EM structure of the complete and ligand-saturated insulin receptor ectodomain, *J. Cell. Biol.* 219, e201907210 (2020) and Uchikawa, E., Choi, E., Shang, G., Yu, H., Bai, X-c., Activation mechanism of the insulin receptor revealed by cryo-EM structure of the fully liganded receptor–ligand complex, *eLife*, 8, e48630 (2019).

Figure 10.14 Image of inactive kinase domain based on 1IRK., Hubbard, S.R., Wei, L., Ellis, L., Hendrickson, W.A., Crystal structure of the tyrosine kinase domain of the human insulin receptor, *Nature* 372, 746–754 (1994). Image of active kinase domain based on 1IR3. Hubbard, S.R., Crystal structure of the activated insulin receptor tyrosine kinase in complex with peptide substrate and ATP analog, *EMBO J.* 16, 5572–5581 (1997).

Figure 10.17 Image based on 1GLU. Luisi, B.F., Xu, W.X., Otwinowski, Z., Freedman, L.P., Yamamoto, K.R., Sigler, P.B., Crystallographic analysis of the interaction of the glucocorticoid receptor with DNA, *Nature* 352, 497–505 (1991).

Carbohydrates

A corn kernel consists of a hard pericarp (hull) surrounding a solid core that is mostly starch with some water. In popcorn kernels, heating generates steam that ruptures the pericarp and forces the starch to puff out. Why does ordinary corn not pop? In popcorn kernels, the pericarp contains highly ordered cellulose fibers that efficiently transfer heat to the starch within and allow the temperature and pressure to build up. In ordinary corn kernels, a less organized pericarp begins to burn and leak before the internal temperature or pressure rises enough to puff up the starch.

Do You Remember?

- Cells contain four major types of biological molecules and three major types of polymers (Section 1.3).
- The polar water molecule forms hydrogen bonds with other molecules (Section 2.1).
- Membrane glycoproteins face the cell exterior (Section 8.4).

Of all the classes of biological molecules, carbohydrates are the simplest, in terms of their atomic composition and molecular structure. They are typically constructed of C, H, and O atoms and adhere to the molecular formula $(CH_2O)_n$, where $n \geq 3$ (hence the name carbohydrate). Even carbohydrate derivatives—many of which include groups containing nitrogen, phosphorus, and other elements—are easy to recognize by their large number of hydroxyl (—OH) groups. Despite this regularity, carbohydrates participate in activities ranging from energy metabolism to cellular structure. This chapter surveys simple and complex carbohydrates and examines some of their biological functions.

11.1 Monosaccharides

KEY CONCEPTS

Recognize monosaccharides and their derivatives.

- Distinguish aldoses and ketoses.
- Recognize enantiomers, epimers, and anomers.
- Identify the parent sugar in carbohydrate derivatives.

Carbohydrates, also known as sugars or saccharides, occur as **monosaccharides** (simple sugars), small polymers (**disaccharides, trisaccharides,** and so on), and larger **polysaccharides** (sometimes called complex carbohydrates). The simplest sugars are the three-carbon compounds glyceraldehyde and dihydroxyacetone:

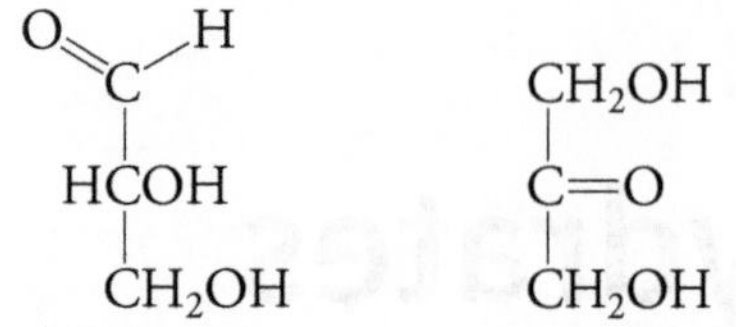

Glyceraldehyde   Dihydroxyacetone

A sugar such as glyceraldehyde, in which the carbonyl group is part of an aldehyde, is known as an **aldose,** and a carbohydrate such as dihydroxyacetone, in which the carbonyl group is part of a ketone, is known as a **ketose.** In longer ketoses, the carbonyl group is usually located at the second carbon (C2).

Monosaccharides can also be described according to the number of carbon atoms they contain; for example, the three-carbon compounds shown above are **trioses. Tetroses** contain four carbons, **pentoses** five, **hexoses** six, and so on. The aldopentose ribose is a component of ribonucleic acid (RNA; its derivative 2′-deoxyribose occurs in deoxyribonucleic acid, DNA). By far the most abundant monosaccharide is glucose, an aldohexose. It is the preferred metabolic fuel for most types of cells; it is stored in significant amounts in plants and animals, and it is a key component of plant cell walls. A common ketohexose is fructose, which is also a metabolic fuel.

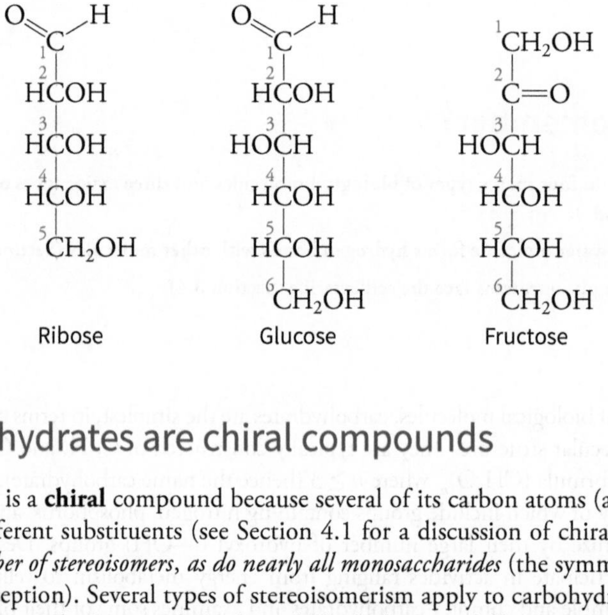

Ribose   Glucose   Fructose

Most carbohydrates are chiral compounds

Note that glucose is a **chiral** compound because several of its carbon atoms (all except C1 and C6) bear four different substituents (see Section 4.1 for a discussion of chirality). As a result, *glucose has a number of stereoisomers, as do nearly all monosaccharides* (the symmetric dihydroxyacetone is one exception). Several types of stereoisomerism apply to carbohydrates.

Like the amino acids (Section 4.1), glyceraldehyde has two different structures that exhibit mirror symmetry. *Such pairs of structures, known as **enantiomers,** cannot be superimposed by rotation.* By convention, these structures are given the designations L and D, derived from the Latin *laevus,* "left," and *dexter,* "right." The enantiomeric forms of larger monosaccharides are given the D or L designation by comparing their structures to D- and L-glyceraldehyde. In a D **sugar,** the asymmetric carbon farthest from the carbonyl group (that would be C5 in glucose) has the same spatial arrangement as the chiral carbon of D-glyceraldehyde. In an L **sugar,** that carbon has the same arrangement as in L-glyceraldehyde. Thus, every D sugar is the mirror image of an L sugar.

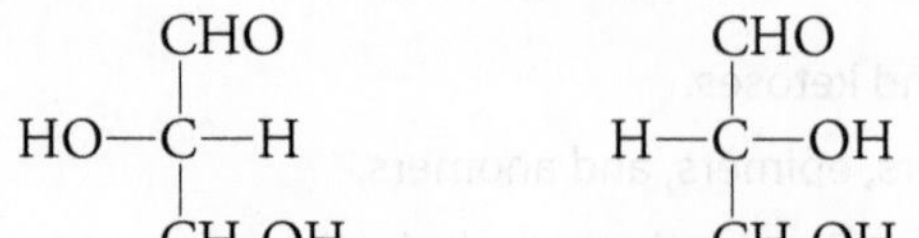

L-Glyceraldehyde   D-Glyceraldehyde

Although enantiomers behave identically in a strictly chemical sense, they are not biologically equivalent. This is because biological systems, which are built of other chiral compounds, such as L-amino acids, can distinguish D and L sugars. Most naturally occurring sugars have the D configuration, so the D and L prefixes are often omitted from their names.

Glucose, in addition to its enantiomeric carbon (C1), has four other asymmetric carbons, so there are stereoisomers based on the configuration at each of these positions. *Carbohydrates that differ in configuration at one of these carbons are known as* **epimers.** For example, the common monosaccharide galactose is an epimer of glucose, at position C4:

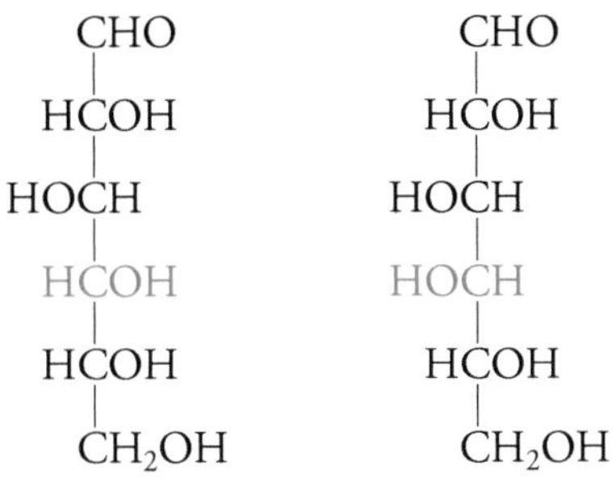

Both ketoses and aldoses have epimeric forms. Like enantiomers, epimers are not biologically interchangeable: An enzyme whose active site accommodates glucose may not recognize galactose at all.

Cyclization generates α and β anomers

The numerous hydroxyl groups that characterize carbohydrate structures also provide multiple points for chemical reactions to occur. One such reaction is an intramolecular rearrangement in which the sugar's carbonyl group reacts with one of its hydroxyl groups to form a cyclic structure (**Fig. 11.1**). The cyclic sugars are represented as **Haworth projections** in which the darker horizontal lines correspond to bonds above the plane of the paper, and the lighter lines correspond to bonds behind the plane of the paper. A simple rule makes it easy to convert a structure from its linear **Fischer projection** (in which horizontal bonds are above the plane of the paper and vertical bonds are behind it) to a Haworth projection: Groups projecting to the right in a Fischer projection will point down in a Haworth projection, and groups projecting to the left will point up.

As a result of the cyclization reaction, the hydroxyl group attached to what was the carbonyl carbon (C1 in the case of glucose) may point either up or down. In the **α anomer,** this hydroxyl group lies on the opposite side of the ring from the CH_2OH group of the chiral carbon that determines the D or L configuration (in the α anomer of glucose, the hydroxyl group points down; see Fig. 11.1). In the **β anomer,** the hydroxyl group lies on the same side of the ring as the CH_2OH

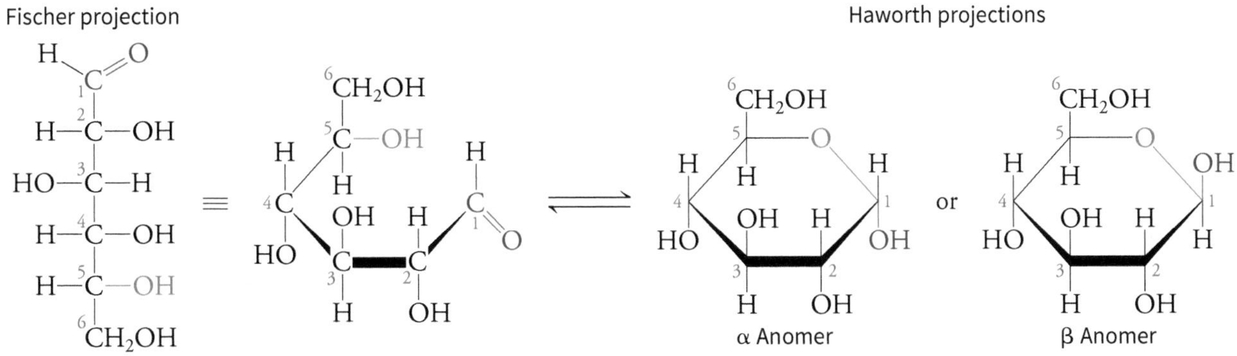

Figure 11.1 Representation of glucose.   In a linear Fischer projection, the horizontal bonds point out of the page and the vertical bonds point below the page. Glucose cyclizes to form a six-membered ring represented by a Haworth projection, in which the heaviest bonds point out from the plane of the page. The α and β anomers freely interconvert.

group of the chiral carbon that determines the D or L configuration (in the β anomer of glucose, the hydroxyl group points up; see Fig. 11.1).

Unlike enantiomers and epimers, which are not interchangeable, **anomers** *in an aqueous solution freely interconvert between the α and β forms, unless the hydroxyl group attached to the anomeric carbon is linked to another molecule.* In fact, a solution of glucose molecules consists of about 64% β anomer, about 36% α anomer, and only trace amounts of the linear or open-chain form.

Hexoses and pentoses, which also undergo cyclization, do not form planar structures, as a Haworth projection might suggest. Instead, the sugar ring puckers so that each C atom can retain its tetrahedral bonding geometry. The substituents of each carbon may point either above the ring (axial positions) or outward (equatorial positions). Glucose can adopt a chair conformation in which all its bulky ring substituents (the —OH and —CH₂OH groups) occupy equatorial positions.

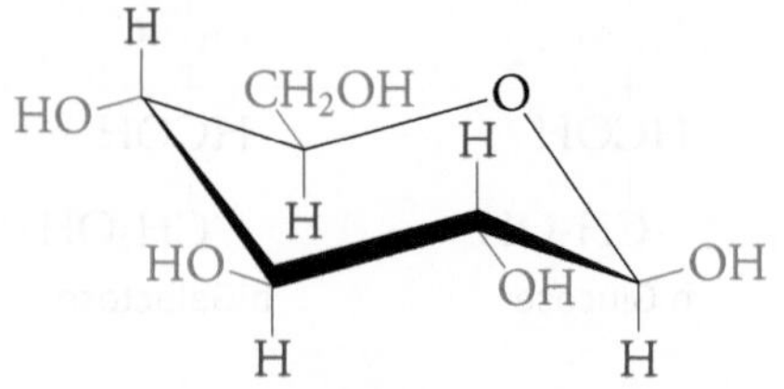

Glucose

In all other hexoses, some of these groups must occupy the more crowded—and therefore less stable—axial positions. The greater stability of glucose may be one reason for its abundance among monosaccharides.

Monosaccharides can be derivatized in many different ways

The anomeric carbon of a monosaccharide is easy to recognize: It is the carbonyl carbon in the straight-chain form of the sugar, and it is the carbon bonded to both the ring oxygen and a hydroxyl group in the cyclic form of the sugar. The anomeric carbon can undergo oxidation, so it can reduce substances such as Cu(II) to Cu(I). This chemical reactivity, often assayed using a copper-containing solution known as Benedict's reagent, can distinguish a free monosaccharide, called a **reducing sugar,** from a monosaccharide in which the anomeric carbon has already condensed with another molecule. For example, when a glucose molecule (a reducing sugar) reacts with methanol (CH_3OH), the result is a **nonreducing sugar** (**Fig. 11.2**). Because the anomeric carbon is involved in the reaction, the methyl group can end up in either the α or β position. The bond that links the anomeric carbon to the other group is called a **glycosidic bond** and a molecule consisting of a sugar linked to another molecule is called a **glycoside.** Glycosidic bonds link the monomers

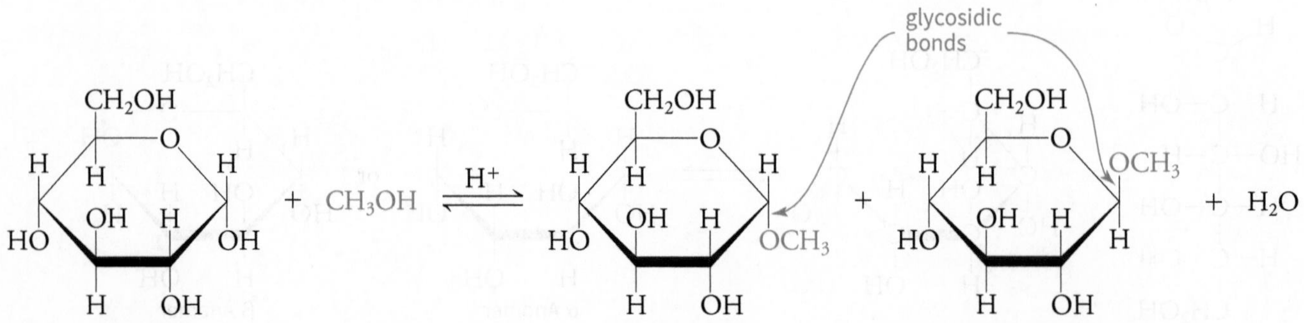

Figure 11.2 Reaction of glucose with methanol.   The addition of methanol to the anomeric carbon blocks the ability of glucose to function as a reducing sugar. The glycosidic bond that forms between the anomeric carbon and the oxygen of methanol may have the α or β configuration.

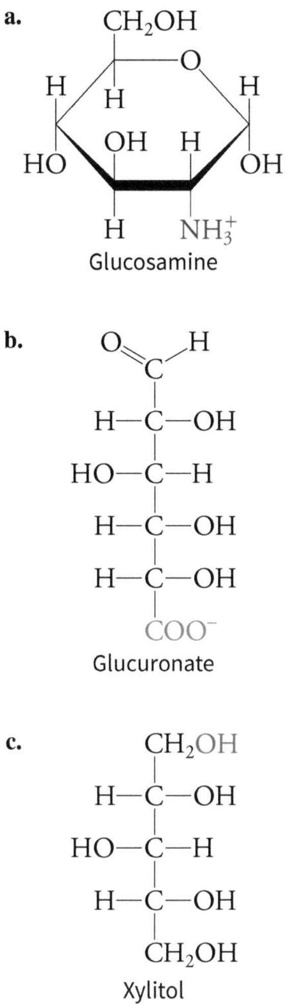

Figure 11.3 Some monosaccharide derivatives. **a.** In an amino sugar, —NH$_3^+$ replaces an —OH group. **b.** Oxidation reactions yield sugars with carboxylate groups. **c.** Reduction reactions yield sugars with additional hydroxyl groups.

Question **Identify the net charge of each sugar.**

a.

Glucosamine

in oligo- and polysaccharides (Section 11.2) and also link the ribose groups to the purine and pyrimidine bases of nucleotides (Section 3.1).

Phosphorylated sugars, including glyceraldehyde-3-phosphate and fructose-6-phosphate, appear as intermediates in the metabolic pathways for breaking down glucose (glycolysis; Section 13.1) and synthesizing it (photosynthesis; Section 16.3).

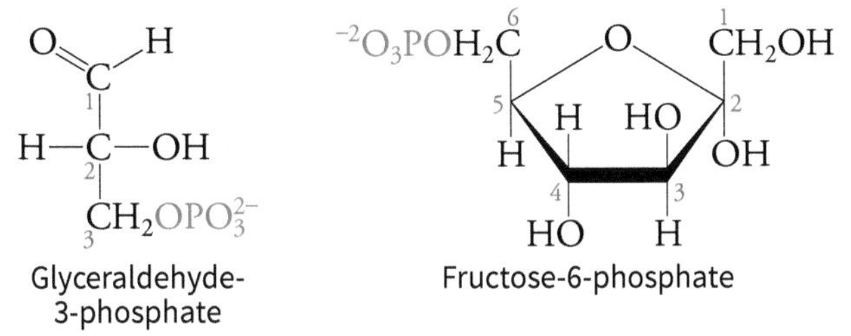

Glyceraldehyde-
3-phosphate

Fructose-6-phosphate

b.

Glucuronate

Other metabolic processes replace a hydroxyl group with an amino group to produce an amino sugar, such as glucosamine (**Fig. 11.3a**). Reactions between a sugar's carbonyl group and a protein's amino group generate compounds that give cooked foods color and flavor (**Box 11.A**).

Oxidation of a sugar's carbonyl and hydroxyl groups can yield uronic acids (sugars containing carboxylic acid groups; Fig. 11.3b), and reduction can yield molecules such as xylitol, a sweetener used in "sugarless" foods (Fig. 11.3c). One metabolically essential carbohydrate-modifying reaction is the one catalyzed by ribonucleotide reductase, which reduces the 2'-OH group of ribose to convert a ribonucleotide to a deoxyribonucleotide for DNA synthesis (Section 18.5):

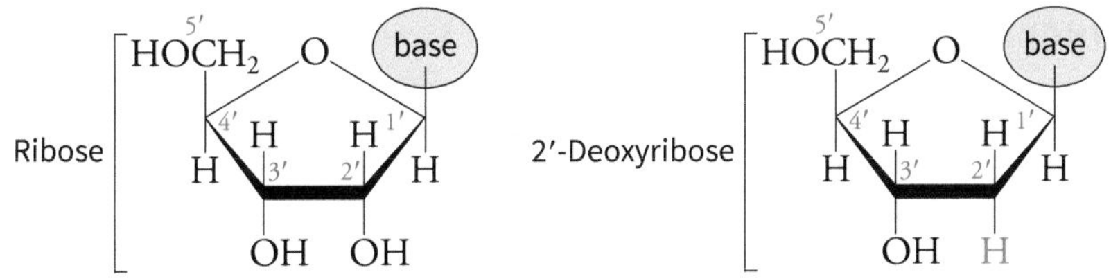

c.

Xylitol

Ribose   2'-Deoxyribose

Box 11.A   The Maillard Reaction

In cellular biochemistry, enzymes typically recognize the cyclic forms of carbohydrates as substrates. However, the linear forms, with their free carbonyl groups, are reactive in nonbiological situations. In the **Maillard reaction,** a monosaccharide's carbonyl group reacts with an unprotonated amino group. This condensation reaction (water is lost) yields an imine (also called a Schiff base), which is unstable and tends to rearrange:

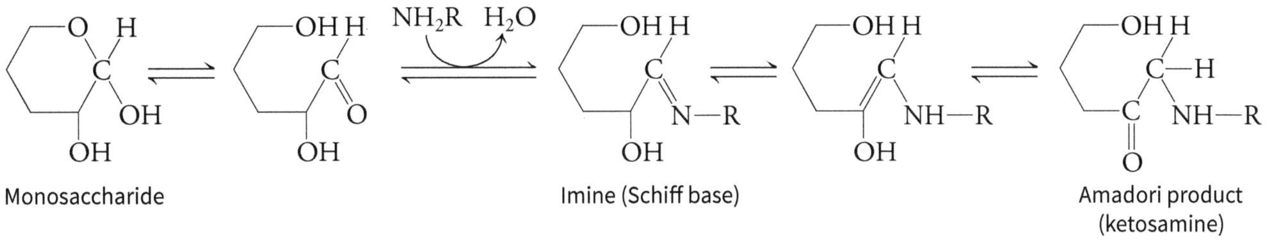

The resulting ketosamine is known as an Amadori product, whose carbonyl group can then react with other molecules.

Subsequent chemical reactions, including oxidation, cyclization, and degradation, generate even more products, including large cross-linked complexes. The Maillard reaction, and the additional reactions that are set in motion, occur slowly at room temperature but more rapidly with heat. Because nearly all foods contain carbohydrates as well as proteins (a source of amino groups), cooking generates hundreds of new compounds, which lend their colors, flavors, and aromas to food. This bit of chemistry is the reason why toast does not taste like bread. It also explains why a roasting step is

Monosaccharide   Imine (Schiff base)   Amadori product
(ketosamine)

needed to "bring out" the flavor of coffee beans and cocoa beans during processing. However, numerous studies also show that some Maillard products are toxic to cells.

Sucrose, which lacks a free carbonyl group, can participate in the Maillard reaction at high temperatures, when the disaccharide breaks down to monosaccharides. When the temperature rises even higher, the sugars can caramelize, which occurs when they undergo dehydration and cyclization and react with each other. Like the Maillard reaction, caramelization involves a variety of chemical reactions, so the ultimate result is a complex mix of products. Molasses, which gives brown sugar its color and flavor, is a sucrose solution that has been heated to allow the Maillard reaction, as well as caramelization, to occur.

In vivo, the Maillard reaction is less benign. In individuals with diabetes, which is characterized by high levels of circulating glucose, the sugar may react more extensively than normal with the body's proteins, generating adducts that are known as advanced glycation end products. Two of these are shown here:

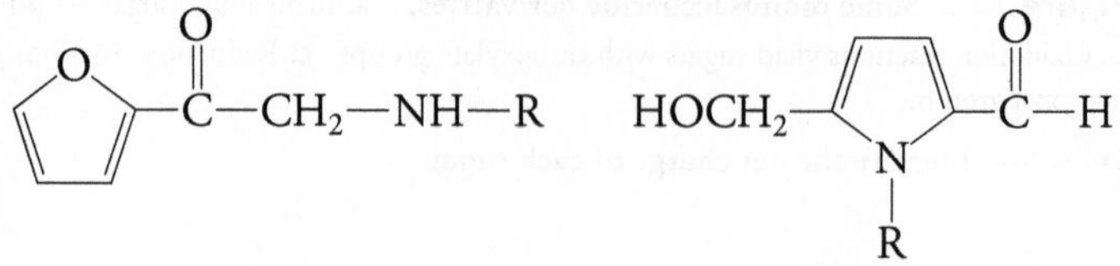

Furosine   Pyralline

When the affected proteins accumulate in cells with low turnover (for example, neurons), cellular functions may be compromised. Even without abnormally high levels of glucose, the Maillard reaction may slowly contribute to organ degeneration as bodies age. Some studies show a correlation between developing a degenerative disease and consuming foods with high levels of Maillard products, but there is little information about safe levels of consumption or ideal cooking practices that would yield desirable flavors without generating harmful products.

Concept Check

1. Draw the straight-chain form of D-glucose, a ketose isomer of glucose, the L enantiomer of glucose, and one of its epimers.
2. Explain why the α and β anomers of a monosaccharide can interconvert.
3. List some types of monosaccharide derivatives.

11.2 Disaccharides, Oligosaccharides, and Polysaccharides

KEY CONCEPTS

Relate the structures of polysaccharides to their biological functions.

- Explain why monosaccharides can be linked in multiple ways.
- Recognize lactose and sucrose.
- Describe the structures and functions of starch, glycogen, cellulose, chitin, and biofilms.

Monosaccharides are the building blocks of polysaccharides, in which glycosidic bonds link successive residues. Each glycosidic bond results from condensation between the hydroxyl group of the anomeric carbon and a second hydroxyl group. Unlike amino acids and nucleotides—the other polymer-forming biological molecules—which are linked in only one configuration, monosaccharides can be hooked together in a variety of ways to produce a dizzying array of chains. *Although each monosaccharide has only one reactive anomeric carbon, it contains several free —OH groups that can participate in a condensation reaction, which permits different bonding arrangements and allows for branching.* While this expands the structural repertoire of carbohydrates, it makes studying them difficult.

In the laboratory, carbohydrate chains, or **glycans,** can be sequenced using mass spectrometry (Section 4.6), although the results are sometimes ambiguous due to the inability to distinguish isomers, which have the same mass. Glycan three-dimensional structures are typically studied using NMR techniques (Section 4.6), since these yield an average conformation for the molecules, which tend to be highly flexible in solution. Due to the challenges of defining carbohydrate sequences and structures, **glycomics,** the systematic study of carbohydrates, is not as fully developed as genomics or proteomics.

The most complex glycans are the oligosaccharides that are commonly linked to other molecules, for example, in glycoproteins. Polysaccharides, some of which are truly enormous molecules, generally do not exhibit the heterogeneity and complexity of oligosaccharides. Instead, they tend to consist of one or a pair of monosaccharides that are linked over and over in the same fashion.

This sort of structural homogeneity is well suited to the function of polysaccharides as fuel-storage molecules and architectural elements. We begin our survey with the simplest of the polysaccharides, the disaccharides.

Lactose, maltose, and sucrose are the most common disaccharides

A glycosidic bond links two monosaccharides to generate a disaccharide. In nature, disaccharides occur as intermediates in the digestion of polysaccharides and as a source of metabolic fuel. For example, lactose, secreted into the milk of lactating mammals, consists of galactose and glucose residues:

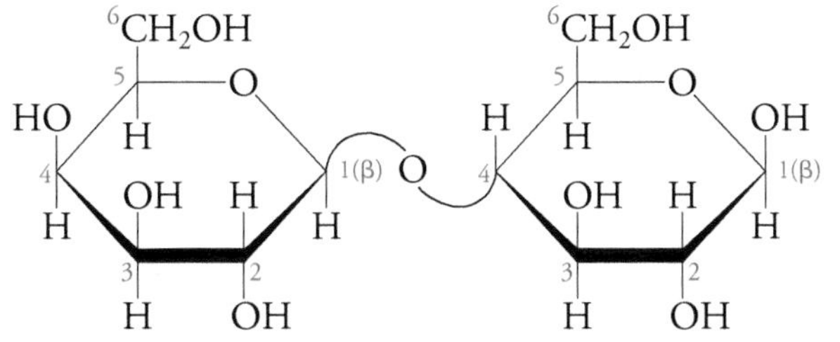

Galactose   Glucose

Lactose

The curved lines in the diagram are ordinary covalent bonds; drawing the linkages in this fashion makes it easier to inspect the monosaccharides side by side.

Note that the anomeric carbon (C1) of galactose is linked to C4 of glucose via a β-glycosidic bond. If the two sugars were linked by an α-glycosidic bond, or if the galactose anomeric carbon were linked to a different glucose carbon, the result would be an entirely different disaccharide. Lactose serves as a major food for newborn mammals. Most adult mammals, including humans, produce very little lactase (also called β-galactosidase), the enzyme that breaks the glycosidic bond of lactose, and therefore cannot efficiently digest this disaccharide.

In addition to lactose, milk contains hundreds of different oligosaccharides. Many of these cannot be digested by mammalian enzymes and therefore cannot be a food source for the newborn. Instead, these carbohydrates appear to sustain certain types of bacteria that are essential for establishing a healthy microbial community in the newborn's intestine.

Maltose or maltobiose, or malt sugar it is a disaccharide made up of 2 α-D-glucose molecules in which (C1) of the first glucose molecule is linked to the (C4) of the second molecule of glucose with an α (1→4) glycosidic linkage:

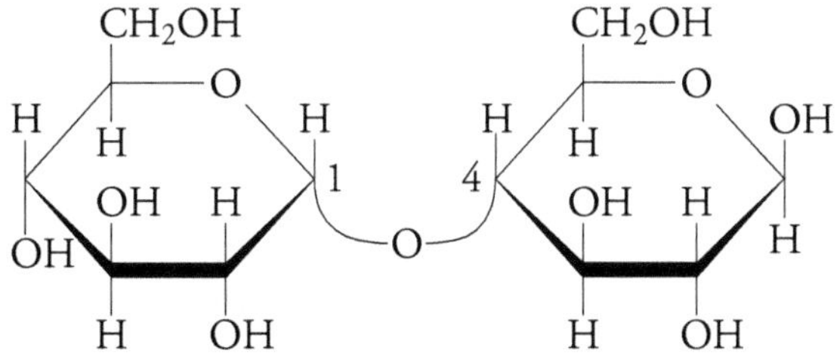

Maltose

In humans, maltose is hydrolyzed by maltase present in the lining of the small intestine to give two molecules of glucose. It is obtained from softened grains in water and germinating seeds.

Sucrose, or table sugar, is the most abundant disaccharide in nature:

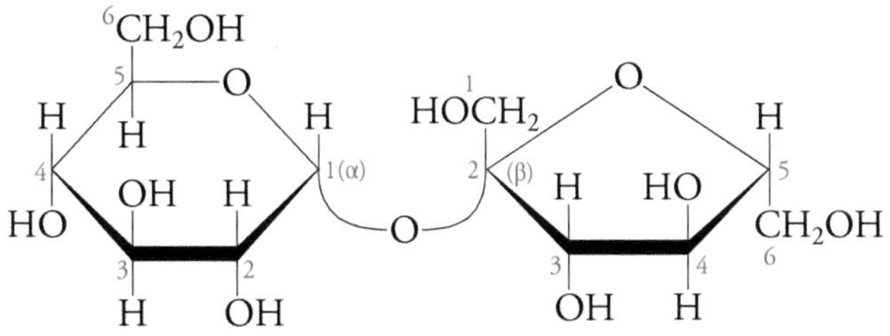

Glucose   Fructose

Sucrose

In this molecule, the anomeric carbon of glucose (in the α configuration) is linked to the anomeric carbon of fructose (in the β configuration). Sucrose is the major form in which newly synthesized carbohydrates are transported from a plant's leaves, where most photosynthesis occurs, to other plant tissues to be used as a fuel or stored as starch for later use.

Starch and glycogen are fuel-storage molecules

Starch and glycogen are polymers of glucose residues linked by glycosidic bonds designated α(1 → 4); in other words, the anomeric carbon (carbon 1) of one residue is linked by an α-glycosidic bond to carbon 4 of the next residue:

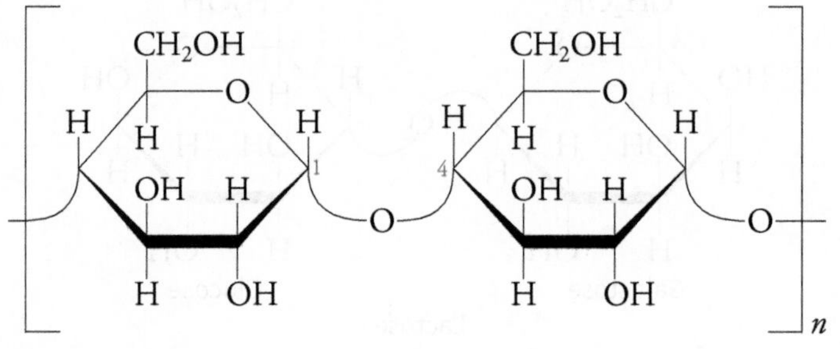

Plants manufacture a linear form of starch, called amylose, which can consist of several thousand glucose residues. Amylopectin, an even larger molecule, includes α(1 → 6) glycosidic linkages every 24 to 30 glucose residues to generate a branched polymer:

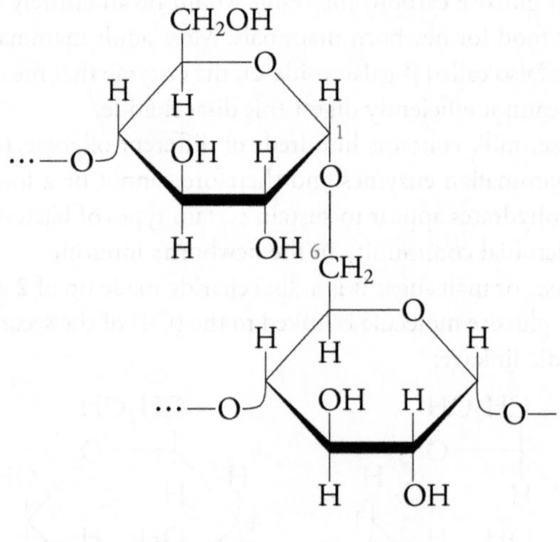

Compiling many monosaccharide residues in a single polysaccharide is an efficient way to store glucose, the plant's primary metabolic fuel. The α-linked chains curve into helices so that the entire molecule forms a relatively compact particle (**Fig. 11.4**).

Animals store glucose in the form of glycogen, a polymer that resembles amylopectin but with branches every 12 residues or so. Due to its highly branched structure, a glycogen molecule can be quickly assembled or disassembled according to the metabolic needs of the cell because the enzymes that add or remove glucose residues work from the ends of the branches. In the mammalian liver, which can store significant amounts of glycogen, individual glycogen molecules aggregate by an unknown mechanism to form larger granules that include some of the glycogen-metabolizing enzymes.

Cellulose and chitin provide structural support

Cellulose, like amylose, is a linear polymer containing thousands of glucose residues. However, *the residues are linked by β(1 → 4) rather than α(1 → 4) glycosidic bonds:*

Figure 11.4 Structure of amylose. This unbranched polysaccharide (six residues are shown) forms a large left-handed helix. C atoms are gray, O atoms red, and hydroxyl H atoms white (other H atoms are not shown).

Question In what part of a cell would amylose be located?

CH₂OH ... O ... H ... H ... OH ... H ... 1 ... O ... 4 ... H ... H ... OH ... n

This simple difference in bonding has profound structural consequences: Whereas starch molecules form compact granules inside the cell, cellulose forms extended fibers that lend rigidity and strength to plant cell walls. Individual cellulose polymers form bundles with extensive hydrogen bonding within and between adjacent chains (**Fig. 11.5**). Cellulose is a relatively long and stiff molecule, so it would be difficult for a cell to synthesize the polymer in the cytosol and then secrete it. In fact, cellulose synthase is an integral membrane protein. Its active site has easy access to the intracellular pool of glucose, it forms a channel for the polymer to pass to the extracellular side of the membrane, and it uses an α helix as a ratchet to push the growing chain through the channel after the addition of each glucose residue.

Plant cell walls include other polymers that, together with cellulose, yield a strong but resilient substance. Recovering the carbohydrates from materials such as wood remains a challenge for the biofuels industry (**Box 11.B**).

Only a few animals can synthesize cellulose, and most cannot digest it in order to use its glucose residues as an energy source. Organisms such as termites and ruminants (grazing mammals), who do derive energy from cellulose-rich foods, harbor microorganisms that produce cellulases capable of hydrolyzing the β(1 → 4) bonds between glucose residues. Humans lack these microorganisms, so although up to 80% of the dry weight of plants consists of glucose, much of this is not available for metabolism in humans. However, its bulk, called fiber, is required for the normal function of the digestive system.

The exoskeletons of insects and crustaceans and the cell walls of many fungi contain a cellulose-like polymer called chitin, in which the β(1 → 4)-linked residues are the glucose derivative *N*-acetylglucosamine (glucosamine with an acetyl group linked to its amino group):

CH₂OH ... O ... H ... H ... OH ... H ... 1 ... O ... 4 ... H ... H ... OH ... NHCCH₃ ... O ... n

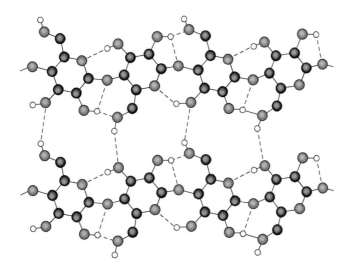

Figure 11.5 Cellulose structure. Glucose residues are represented as hexagons, with C atoms gray and O atoms red. Not all H atoms (small circles) are shown. Hydrogen bonds (dashed lines) link residues in the same and adjacent chains so that a bundle of cellulose polymers forms an extended rigid fiber.

Box 11.B   Cellulosic Biofuel

Cellulose is by far the most abundant polysaccharide in nature, with chains containing thousands of monosaccharide residues. For this reason, cellulose-rich materials, including wood and agricultural waste, are sources of sugars that can be converted to biofuels such as ethanol, potentially replacing petroleum-derived fuels. Unfortunately, cellulose does not exist in pure form in nature; plant cell walls typically contain other polymers, including hemicellulose, pectin, and lignin.

Hemicellulose is the name given to a class of polysaccharides whose chains are shorter than cellulose (500–3000 residues) and may be branched. Hemicellulose is a heteropolymer, indicating that it contains a variety of monomeric units, in this case 5- and 6-carbon sugars. Xylose is the most abundant of these:

$$
\begin{array}{c}
\text{CHO} \\
\text{HCOH} \\
\text{HOCH} \\
\text{HCOH} \\
\text{CH}_2\text{OH}
\end{array}
$$

Xylose

While cellulose forms rigid fibers and hemicellulose forms a network, the spaces in between these polymers are occupied by pectin, a heteropolymer containing galacturonate and rhamnose residues, among others:

Galacturonate   Rhamnose

The large number of hydroxyl groups makes pectin highly hydrophilic, so it "holds" a great deal of water and has the physical properties of a gel.

Lignin—in contrast to cellulose, hemicellulose, and pectin—is not a polysaccharide at all. It is a highly heterogeneous, difficult-to-characterize polymer built from aromatic (phenolic) compounds. With few hydroxyl groups, it is relatively hydrophobic. Lignin is covalently linked to hemicellulose chains, so it contributes to the mechanical strength of cell walls.

All of the components of wood, including lignin, represent a large amount of stored free energy, which can be released by combustion (for example, when wood burns). The industrial conversion of this stored energy into other types of fuels is known as bioconversion. The first step is the hardest: separating the polysaccharides from lignin. Physical methods such as grinding and pulverizing consume energy, but chemical methods—which may include strong acids or organic solvents—come with their own hazards. An additional drawback is the generation of reaction products that can inhibit subsequent steps in biofuel production, which depend on living organisms or enzymes derived from them.

Once freed from lignin, the carbohydrate polymers are accessible to hydrolytic enzymes, most of which originate in bacteria and fungi that are adept at degrading plant materials. The result is a mixture of monosaccharides. Fungi such as yeast efficiently ferment the glucose to ethanol (Section 13.1), which can be distilled and used as a fuel. Other organisms can convert monosaccharides such as xylose to ethanol, but these pathways are not always efficient and they may yield other substances, such as lactate and acetate, as end products. The most promising approach appears to be bioengineering microorganisms to carry out polysaccharide hydrolysis and then convert the resulting monosaccharides to ethanol, which is relatively stable and easy to transport and store. Alternative strategies use microorganisms to convert the sugar mixture into hydrocarbons that can be used in place of diesel fuel.

Question How do xylose, galacturonate, and rhamnose differ from glucose?

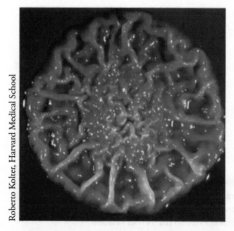

Figure 11.6 A *Pseudomonas aeruginosa* biofilm.   These pathogenic bacteria growing on the surface of an agar plate form a biofilm with a complex three-dimensional shape.

Bacterial polysaccharides form a biofilm

Prokaryotes do not synthesize cellulosic cell walls (see Section 11.3) or store fuel as starch or glycogen, but they do produce extracellular polysaccharides that provide a protective matrix for their growth. *A **biofilm** is attached to a surface and harbors a community of embedded bacteria that contribute to biofilm production and maintenance* (**Fig. 11.6**). The extracellular material of the biofilm includes an assortment of highly hydrated polysaccharides, such as those containing glucuronate and *N*-acetylglucosamine, along with DNA and lipids. A biofilm can be difficult to characterize because it typically houses a mixture of species and the proportions of its component polysaccharides depend on many environmental factors.

The gel-like consistency of a biofilm, such as the plaque that forms on teeth, prevents bacterial cells from being washed away and protects them from desiccation. Biofilms that develop on medical apparatus, such as catheters, are problematic because they offer a foothold for pathogenic organisms and create a barrier for antibiotics and cells of the immune system.

1. Explain why it is possible for two monosaccharides to form more than one type of disaccharide.
2. Summarize the physiological roles of lactose, sucrose, starch, glycogen, cellulose, and chitin.
3. Describe how the physical properties of polysaccharides—such as overall size, shape, branching, composition, and hydrophilicity—relate to their biological functions.

11.3 Glycoconjugates: Glycoproteins, Glycolipids, and Proteoglycans

KEY CONCEPTS

Describe the structures and functions of glycoproteins.

- Distinguish *N*- and *O*-linked oligosaccharides.
- Summarize the functions of oligosaccharide markers.
- Explain how proteoglycans function in shock absorption and microbial defense.
- Describe peptidoglycan as a strong, elastic, and porous casing for bacterial cells.

Because there are so many different monosaccharides and so many ways to link them, the number of possible structures for oligosaccharides with even just a few residues is enormous. Organisms take advantage of this complexity to mark various structures—mainly proteins and lipids—with unique oligosaccharides. Most of the proteins that are secreted from eukaryotic cells or remain on their surface are glycoproteins in which one or more **oligosaccharide** chains are covalently attached to the polypeptide chain shortly after its synthesis.

Oligosaccharides are *N*-linked or *O*-linked

In eukaryotes, the oligosaccharides attached to glycoproteins are usually linked either to an asparagine side chain (***N*-linked oligosaccharides**) or to a serine or threonine side chain (***O*-linked oligosaccharides**).

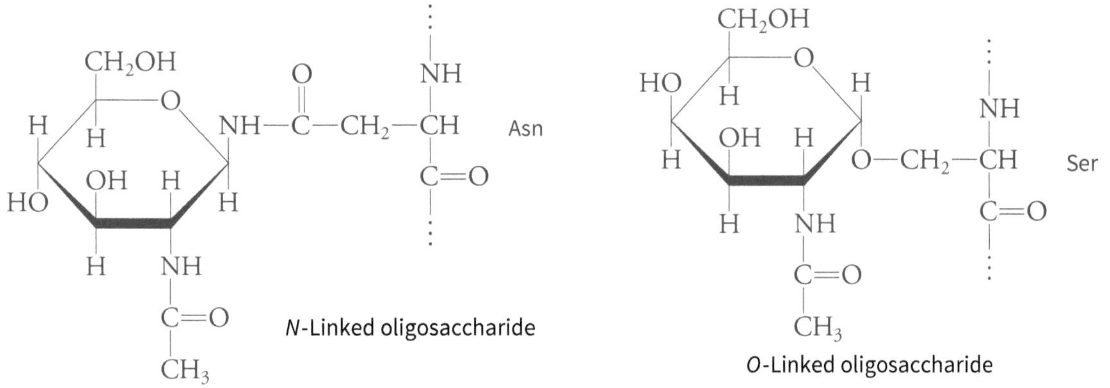

N-Linked oligosaccharide

O-Linked oligosaccharide

N-Glycosylation begins while a protein is being synthesized by a ribosome associated with the rough endoplasmic reticulum (ER). As the protein is translocated into the ER lumen (the internal space), an oligosaccharide chain of 14 residues is attached to an asparagine residue (**Fig. 11.7**). When the newly synthesized protein leaves the ER and traverses the Golgi apparatus (a series of membrane-bounded compartments), *enzymes known as* **glycosidases** *remove various monosaccharide residues and other enzymes, called* **glycosyltransferases,** *add new monosaccharides.*

Figure 11.7 Processing of an N-linked oligosaccharide. Enzymes in the Golgi apparatus process the 14-residue oligosaccharide that is attached to the newly synthesized protein in the ER. Only one of many possible mature oligosaccharides is shown.

Question How many different glycosyltransferases would be required to synthesize the oligosaccharide shown here?

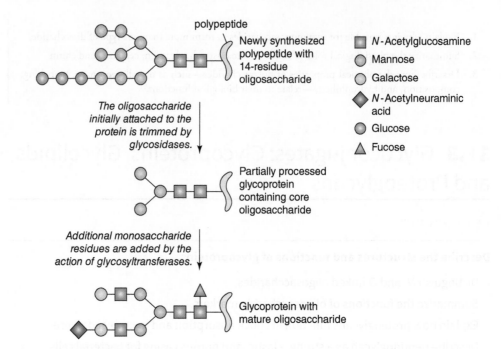

These processing enzymes are highly specific for the identities of the monosaccharides and the positions of the glycosidic bonds. N-linked oligosaccharides end up retaining a common five-residue core consisting of three mannose and two N-acetylglucosamine residues.

Apparently, the amino acid sequence or local structure of the protein, as well as the set of processing enzymes present in the cell, roughly determine which sugars are added and deleted. The net result is a great deal of heterogeneity in the oligosaccharide chains attached to different glycoproteins or even to different molecules of the same glycoprotein. An example of an N-linked oligosaccharide is shown in **Figure 11.8**. Despite the variability, attaching oligosaccharides to proteins is essential, an idea supported by the documentation of at least 100 congenital disorders linked to faulty glycosylation.

O-Linked oligosaccharides are built, one residue at a time, primarily in the Golgi apparatus, through the action of glycosyltransferases. Unlike N-linked oligosaccharides, the O-linked oligosaccharides do not undergo processing by glycosidases. Glycoproteins with O-linked saccharide chains tend to have many such groups, and the glycan chains tend to be longer than those of N-linked oligosaccharides.

Oligosaccharide groups are biological markers

Because oligosaccharides are highly hydrophilic and are conformationally flexible, they occupy a large effective volume above the protein's surface. An oligosaccharide group can therefore help protect the protein by preventing the approach of degradative molecules. Alternatively, the presence of an oligosaccharide may limit the possible conformations available to the polypeptide and help lock it into a functional shape. For some N-glycosylated proteins, oligosaccharide processing and chaperone-assisted protein folding (Section 4.3) are interlinked, so that an improperly folded protein cannot be glycosylated and *vice versa*. The cell pays for this protein quality control system by the metabolically expensive process of adding sugars and then removing them.

In some cases, oligosaccharide groups constitute a sort of intracellular addressing system so that newly synthesized proteins can be delivered to their proper cellular location, such as a lysosome. In other cases, the oligosaccharide groups act as recognition and attachment points for interactions between different types of cells. For example, the familiar A, B, and O blood types are determined by the presence of different oligosaccharides on the surface of red blood cells (**Box 11.C**). Circulating white blood cells latch on to glycoproteins on the cells lining the blood vessels in

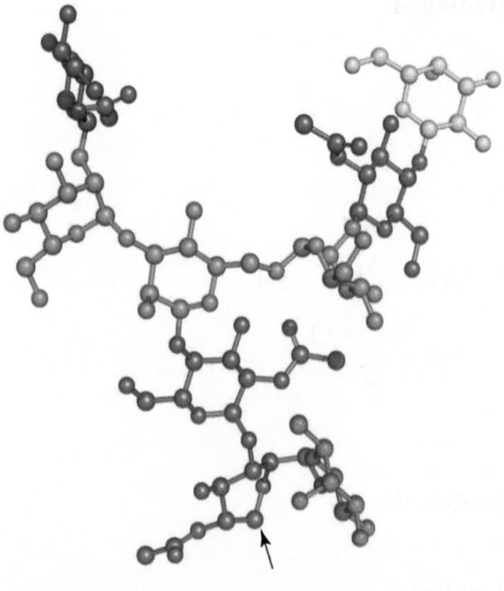

Figure 11.8 Structure of an N-linked oligosaccharide. The nine oligosaccharide residues (colored as in Fig. 11.7) experience considerable conformational flexibility, so the structure shown here is just one of many possible. The arrow indicates the point where the glycan is attached to the N atom of an Asn side chain.

Box 11.C   The ABO Blood Group System

The carbohydrates on the surface of red blood cells and other human cells form 15 different blood group systems. The best known and one of the clinically important carbohydrate-classification schemes is the ABO blood group system, which has been known for over a century. Biochemically, the ABO system involves the oligosaccharides attached to sphingolipids and proteins on red blood cells and other cells.

In individuals with type A blood, the oligosaccharide has a terminal *N*-acetylated galactose group. In type B individuals, the terminal sugar is galactose. Neither of these groups appears in the oligosaccharides of type O individuals.

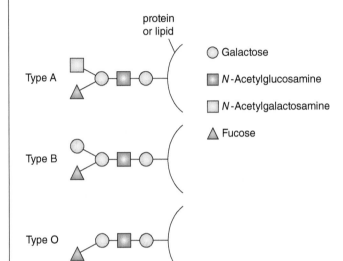

Blood groups are genetically determined: Type A and B individuals have slightly different versions of the gene for a glycosyltransferase that adds the final monosaccharide residue to the oligosaccharide. Type O individuals have a mutation such that they lack the enzyme entirely and therefore produce an oligosaccharide without the final residue.

Type A individuals develop antibodies that recognize and cross-link red blood cells bearing the type B oligosaccharide. Type B individuals develop antibodies to the type A oligosaccharide. Therefore, a transfusion of type B blood cannot be given to a type A individual, and vice versa. Individuals with type AB blood bear both types of oligosaccharides and therefore do not develop antibodies to either type. They can receive transfusions of either type A or type B blood. Type O individuals develop both anti-A and anti-B antibodies. If they receive a transfusion of type A, type B, or type AB blood, their antibodies will react with the transfused cells, which causes them to lyse or to clump together and block blood vessels. On the other hand, type O individuals are universal donors: Type A, B, or AB individuals can safely receive type O blood (these individuals do not develop antibodies to the O-type oligosaccharide because it occurs naturally in these individuals as a precursor of the A-type and B-type oligosaccharides).

Question In rare cases, individuals do not produce an A-, B-, or O-type oligosaccharide at all. Can these individuals receive type A, B, or O blood transfusions? Can they donate blood to others?

order to leave the bloodstream and migrate to sites of injury or infection. Unfortunately, many viruses and pathogenic bacteria also recognize specific carbohydrate groups on cell surfaces and attach themselves to these sites before invading the host cell.

Intracellular eukaryotic proteins are also glycosylated, most often by the addition of a single *N*-acetylglucosamine to serine or threonine side chains in cytosolic, mitochondrial, and nuclear proteins. This protein modification is much more common than previously thought and appears to influence a protein's localization, rate of degradation, or interactions with other molecules—all of which can in turn regulate metabolic activity and patterns of gene expression. In this respect, intracellular protein glycosylation is analogous to the protein phosphorylation and dephosphorylation that occur during signal transduction (Chapter 10). *The monosaccharide groups are derived directly from glucose (the cell's major fuel) and glutamine (a key amino acid and a cellular source of amino groups) and therefore serve as a sort of nutrient sensor to help coordinate the cell's activities with its food supply.* This regulatory scheme could become a target for drugs to treat conditions where fuel metabolism is abnormal, such as in diabetes and cancer.

Proteoglycans contain long glycosaminoglycan chains

Proteoglycans are glycoproteins in which the protein chain serves mainly as an attachment site for enormous linear *O*-linked polysaccharides called ***glycosaminoglycans.*** Most glycosaminoglycan chains consist of a repeating disaccharide of an amino sugar (often *N*-acetylated) and a uronic acid (a sugar with a carboxylate group). After synthesis, various hydroxyl groups may be enzymatically sulfated (an —OSO_3^- group added). The repeating disaccharide of the glycosaminoglycan known as chondroitin sulfate is shown in **Figure 11.9**. The polysaccharide

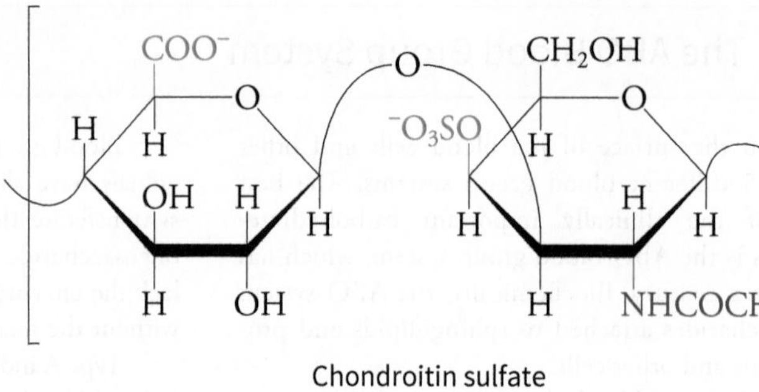

Figure 11.9 The repeating disaccharide of chondroitin sulfate. A chondroitin sulfate chain may include hundreds of these disaccharide units and the degree of sulfation may vary along its length.

Question What ions are likely to be associated with chondroitin sulfate?

Chondroitin sulfate

chain may contain hundreds of disaccharide units, and the degree of sulfation may vary along its length.

Proteoglycans may be transmembrane proteins or lipid-linked (Section 8.3), but *the glycosaminoglycan chains are invariably on the extracellular side of the plasma membrane.* Extracellular proteoglycans and glycosaminoglycan chains that are not attached to a protein scaffold play an important structural role in connective tissue.

The many hydrophilic groups on the glycosaminoglycans attract water molecules, so glycosaminoglycans are highly hydrated and occupy the spaces between cells and other components of the extracellular matrix, such as collagen fibrils (Section 5.3). Under mechanical pressure, some of the water can be squeezed out of the glycosaminoglycans, which allows connective tissue and other structures to accommodate the body's movements. Pressure also brings the negatively charged sulfate and carboxylate groups of the polysaccharides close together. When the pressure abates, the glycosaminoglycans quickly spring back to their original shape as the repulsion between anionic groups is relieved and water is drawn back into the molecule. This spongelike action of glycosaminoglycans in the spaces of the joints provides shock absorption.

The proteoglycans known as mucins form the protective mucus lining the respiratory, gastrointestinal, and reproductive tracts. These proteins, half of whose residues may be serine and threonine, serve as scaffolding for numerous glycosaminoglycan chains, so that the entire molecule is often enormous (up to 10,000,000 D, with carbohydrate accounting for as much as 80% of the total mass). The viscous, tangled mucin chains act as lubricants and also trap bacteria and other foreign particles to prevent their access to the body. In the respiratory tract, the cilia protruding from epithelial cells create a current that moves the overlying mucus layer and any captured particles out of the airways so that they can be swallowed (**Fig. 11.10**).

Bacterial cell walls are made of peptidoglycan

A network of cross-linked carbohydrate chains and peptides constitutes the cell walls of bacteria. This material, called **peptidoglycan,** *surrounds the plasma membrane of the cell and determines its overall shape.* The carbohydrate component in many species is a repeating $\beta(1 \rightarrow 4)$-linked disaccharide with a typical total length of about 20 to 40 disaccharides:

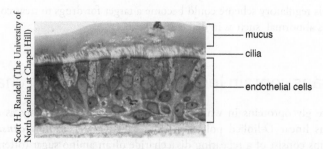

Scott H. Randell (The University of North Carolina at Chapel Hill)

Figure 11.10 The airway surface. The outward-facing cell surface is ciliated. A mucus layer lies above the cilia so that ciliary movement sweeps away the mucus and trapped particles.

Peptides of four or five amino acids covalently cross-link the saccharide chains in three dimensions to form a structure as thick as 250 Å in some species. Antibiotics of the penicillin family block the formation of the peptide cross-links, thereby killing bacteria without harming their eukaryotic hosts. Archaebacterial cell walls contain proteins and sometimes polysaccharides but not peptidoglycans.

The overall structure of the peptidoglycan cell wall is not known. However, a model of peptidoglycan from the bacterium *Escherichia coli* suggests that the glycan chains are parallel to the cell surface and are arranged with enough space in between to allow nutrients and wastes to easily diffuse to and from the cell surface (**Fig. 11.11**). Although the carbohydrate chains are relatively stiff, calculations indicate that the peptide cross-links are more elastic and could accommodate cell shape changes of 25 to 33%.

The peptidoglycan layer is relatively thick in **Gram-positive bacteria;** these species get their name because the cell wall retains crystal violet stain during the Gram-staining process. The cell wall of **Gram-negative bacteria** does not retain the purple dye because the peptidoglycan layer is relatively thin (**Fig. 11.12**). However, this weak cell wall is surrounded by a second, outer membrane that contains unusual lipids, many with long saccharide chains, and is much stiffer than a typical lipid bilayer. The extra coating provides structural support and some protection against antibiotics.

Some bacteria supplement their thick cells walls (Gram-positive species) or outer membranes (Gram-negative species) with another protective layer of polysaccharides that are anchored to the cell wall peptidoglycan. For example, the pathogenic Gram-positive *Streptococcus pneumoniae* has an outer capsule consisting of chains of repeating di- to octasaccharides. The residues and linkages vary among the 98 different pneumococcal strains. Because they trigger a strong antibody response in animals, portions of these capsule polysaccharides are used to make vaccines to protect against *S. pneumoniae* infections. The so-called 23-valent vaccine includes oligosaccharides from 23 strains. One challenge in formulating effective vaccines is that a given antigenic saccharide sequence needs to be at least four residues long, and to promote high-avidity binding (see Section 5.5), the vaccine must include multiple antibody-binding sites.

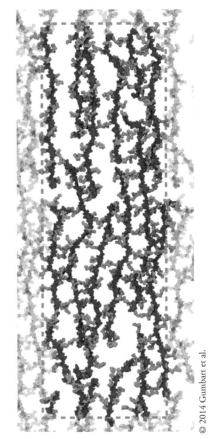

Figure 11.11 Model of *E. coli* peptidoglycan. In this model, looking down on the cell surface, a single layer of peptidoglycan is shown with carbohydrate strands in blue and cross-linking peptides in green. The scale bar is 10 nm.

© 2014 Gumbart et al.

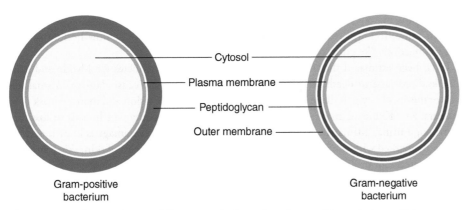

Figure 11.12 Gram-positive and Gram-negative bacteria. Gram-positive species have a thick peptidoglycan cell wall, whereas Gram-negative species have a thin cell wall and an outer membrane.

At the same time, polysaccharides that are too large can physically interfere with antibody binding. Consequently, vaccine developers continue to work on finding the optimal formula for pneumococcal vaccines.

Concept Check

1. Summarize the processing of an *N*-linked oligosaccharide.
2. List some functions of the carbohydrate groups of glycoproteins.
3. Compare *N*- and *O*-linked oligosaccharide structures and functions.
4. Distinguish glycosylation and glycation.
5. Relate the structures of proteoglycans to their functions.
6. Compare proteoglycans and peptidoglycans to polysaccharides such as starch and cellulose.

11.4 Clinical Connection: Glucose and Diabetes Mellitus

KEY CONCEPTS

Related to Glucose and its levels in Diabetes Mellitus correlation with Insulin.

- Explain the role of insulin absorption in diabetes mellitus,
- Diabetes Type I and Type II
- Role of lecithin in the absorption of carbohydrates, and maintenance of blood glucose levels.

Diabetes mellitus is an endocrine disorder characterized by high glucose levels in the blood called **hyperglycemia**. It mainly depends on whether cells of the pancreas are unable to produce insulin sufficiently, or body cells are not able to take up insulin properly produced in the body. It may lead to serious health problems if untreated for a long time. It is mainly of two types Type I and Type II diabetes.

Diabetes Type I is a condition developed due to the impairment of insulin secretion also called insulin-dependent diabetes. Because insulin hormone helps to allow blood glucose to enter into cells for energy production in the body.

Diabetes Type II develops mainly in two conditions, when cells in muscle, fat, and the liver become resistant to insulin causing impairment of glucose take-up into the cells. In the second one pancreas fails to make ample insulin to maintain blood glucose within normal levels.

Symptoms mainly include feeling more thirsty, frequent urination, loss of weight, ketone bodies in urine, feeling tired and weak, mood changes, blurry vision, slow-healing sores, and prone to infections like gum, skin, and vaginal infections.

It has been estimated that nearly 500 million people have diabetes worldwide and it is the eighth leading cause of death. The primary complications mostly cause damage in small blood vessels of the eyes known as diabetic retinopathy results in loss of vision and increased risk of glaucoma cataracts. Kidney damage also known as diabetic nephropathy which leads to loss of proteins through urine, ultimately causes acute kidney diseases. Nerve damage is known as diabetic neuropathy marked with tingling, numbness, unusual sweating, dryness of skin, skin damage, and diabetic-related foot problems – ulcers.

The following discussion focused on the most common conditions of diabetes mellitus:

Insulin hormone is involved in the regulation of blood glucose uptake into body cells and organs like the liver, muscle, and adipose tissue. Low levels of insulin hormone or the sensitivity loss of insulin receptors form the basis for all types of diabetes mellitus.

Three main sources of glucose to the body:

- Food absorption in the intestine
- Glycogen breakdown
- Storage of glucose in the liver.

Insulin hormone is crucial in controlling blood glucose levels by inhibiting the glycogen breakdown or gluconeogenesis in the body, it stimulates glucose transport into fat bodies and muscle cells, and stores glucose in the glycogen form.

After consuming food there is an increase in the glucose levels in blood, as a result, insulin is released from the beta cells (β-cells) of islets of Langerhans. About two-thirds of insulin is utilized by body cells for glucose absorption from the blood, used as a fuel for metabolic activities and storage. Low levels of glucose affect the release of insulin from the β-cells and glucose conversion from glycogen. Proper absorption of glucose into cells and storage of glucose in the liver and muscles requires insulin. If insulin is not available sufficiently it causes high levels of glucose in the blood, low protein synthesis, and defective metabolic processes.

If glucose levels remain high for a long time in blood results in the excretion of glucose in urine called as glycosuria. This leads to a rise of urinary osmotic pressure and inhibits water reabsorption in the kidney characterized by increased urination and loss of fluids known as polyuria. Signs of diabetes also include dehydration, heavy thirst—polydipsia, and glucose deficiency within the cells will increase the rate of food consumption—polyphagia.

Diabetes can be diagnosed by the following tests:

- Fasting Blood Sugar test (FBS) – in this test, blood sample is collected from the patient after prolonged fasting preferably before breakfast (<110 mg/dL).
- Random Blood Sugar test (RBS) – in this test, blood sample is collected from the patient after 2 hours from the time of food consumption (<140 mg/dL).
- Oral Glucose Tolerance Test (OGTT) – in this test, patient is advised to take oral consumption of 75 grams of glucose and the blood is collected after 2 hours to measure the response of cells to glucose (<140 mg/dL).
- Glycated Hemoglobin (HbA1c) – average glucose metabolism rate is measured for 90 days ≥ 48 mmol/mol (≥ 6.5 DCCT %).

There is no prevention for diabetes unless it is under control by employing various lifestyle changes including treatment, dietary changes, and exercise.

In addition, **Lectin** also plays an important role in type I diabetes because lectins mimic insulin; it binds to the insulin receptor sites on cells, thereby blocking the action of insulin. As a result, we can suffer from poor energy utilization. Glucose is shifted into fat bodies, which makes insulin resistant, as a diabetic precursor. Lectins are non-enzymatic binding proteins of carbohydrates present in all life domains. It brings about the identification, interaction, and reversible binding of specific sugar moieties. In turn helps to stabilize blood glucose levels by decreasing the rate of digestion, absorption, and assimilation of carbohydrates.

Concept Check

1. If a person is suffering from severe muscle cramps and tingling. Explain what type of diabetes?
2. If a person suffering from frequent urination and severe loss of proteins from the body. Explain.

SUMMARY

11.1 Monosaccharides

- Carbohydrates, which have the general formula $(CH_2O)_n$, exist as monosaccharides and polysaccharides of different sizes. The monosaccharides may be aldoses or ketoses and exist as enantiomers, which are mirror images, and as epimers, which differ in configuration at individual carbon atoms.

- Cyclization of a monosaccharide produces α and β anomers. Formation of a glycosidic bond prevents the interconversion of the α and β forms.

- Monosaccharide derivatives include phosphorylated sugars; sugars with amino, carboxylate, and extra hydroxyl groups; and deoxy sugars.

11.2 Disaccharides, Oligosaccharides, and Polysaccharides

- Lactose consists of galactose linked β(1 → 4) to glucose. Sucrose consists of glucose linked α(1 → 2) to fructose.
- Starch is a linear polymer of α(1 → 4)-linked glucose residues; glycogen also contains α(1 → 6) branch points. Cellulose consists of β(1 → 4)-linked glucose residues; in chitin, the residues are *N*-acetylglucosamine.
- A bacterial biofilm is a community of cells embedded in an extra-cellular polysaccharide matrix.

11.3 Glycoconjugates: Glycoproteins, Glycolipids, and Proteoglycans

- Oligosaccharides are attached to proteins as *N*- and *O*-linked oligosaccharides. *N*-Linked oligosaccharides undergo processing by glycosidases and glycosyltransferases. The carbohydrate chains of glycoproteins function as protection and as recognition markers.
- Proteoglycans consist mainly of long glycosaminoglycan chains that can function as shock absorbers, lubricants, and protective coatings.
- Peptidoglycan in bacterial cell walls is constructed of cross-linked oligosaccharides and peptides.

11.4 Clinical Connection: Glucose and Diabetes Mellitus/Clinical Connection: Lectin

- Diabetes mellitus is a disease that affects how the body uses blood sugar (glucose), which is the main source of energy for the cells and the brain.
- There are two main types of diabetes mellitus: Type I and Type II. Type I diabetes is also called insulin-dependent diabetes,
- Some of the common symptoms of diabetes mellitus are increased thirst, frequent urination, weight loss, ketones in urine, fatigue, mood changes, blurred vision, slow-healing wounds, and infections.

KEY TERMS

carbohydrate	tetrose	Haworth projection	glycoside	*O*-linked	Gram-negative
monosaccharide	pentose	Fischer projection	Maillard reaction	oligosaccharide	hyperglycemia
disaccharide	hexose	α anomer	glycan	glycosidase	diabetes type I
trisaccharide	chirality	β anomer	glycomics	glycosyltransferase	diabetes type II
polysaccharide	enantiomers	anomers	biofilm	proteoglycan	lectin
aldose	D sugar	reducing sugar	oligosaccharide	glycosaminoglycan	
ketose	L sugar	nonreducing sugar	*N*-linked	peptidoglycan	
triose	epimers	glycosidic bond	oligosaccharide	Gram-positive	

BIOINFORMATICS

Brief Bioinformatics Exercises

11.1 Drawing Monosaccharides

11.2 Introduction to Glycomics

11.3 Oligosaccharides, Polysaccharides, and Glycoproteins

Bioinformatics Projects

Glycomics and the H1N1 Flu

PROBLEMS

11.1 Monosaccharides

1. Classify the following sugars as adloses or ketoses:

a.
```
      CHO
      |
HO—C—H
      |
HO—C—H
      |
 H—C—OH
      |
    CH₂OH
```

b.
```
    CH₂OH
      |
     C=O
      |
HO—C—H
      |
    CH₂OH
```

c.
```
      CHO
      |
HO—C—H
      |
HO—C—H
      |
HO—C—H
      |
 H—C—OH
      |
    CH₂OH
```

2. Glucose can be described as an aldohexose. Use similar terminology to describe the following sugars:

a.
```
      CHO
      |
 H—C—OH
      |
 H—C—OH
      |
 H—C—OH
      |
    CH₂OH
```

b.
```
    CH₂OH
      |
     C=O
      |
 H—C—OH
      |
    CH₂OH
```

c.

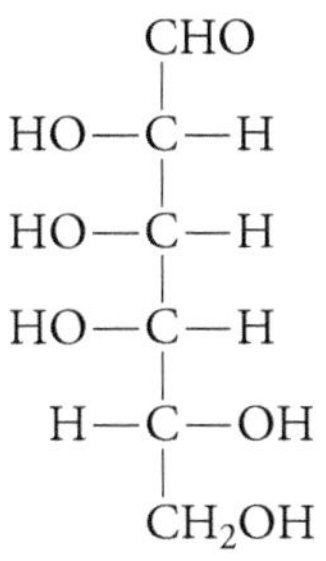

```
      CHO
HO—C—H
HO—C—H
HO—C—H
 H—C—OH
     CH₂OH
```

d.

```
     CH₂OH
      C=O
 H—C—OH
 H—C—OH
 H—C—OH
     CH₂OH
```

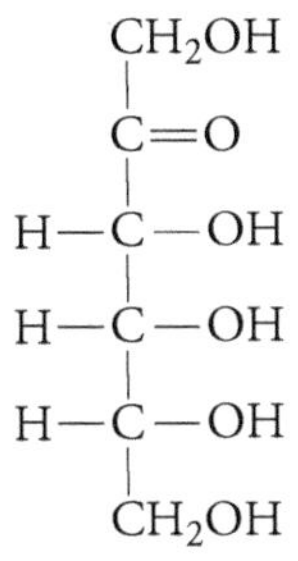

3. Which of the sugars shown in Problem 2 has the greatest number of stereoisomers? (*Hint*: The number of stereoisomers is equal to 2^n where n is the number of chiral carbons.)

4. Identify the monosaccharide(s) present in coenzyme A, NAD, and FAD (see Fig. 3.2).

5. Identify the following sugars as D or L:

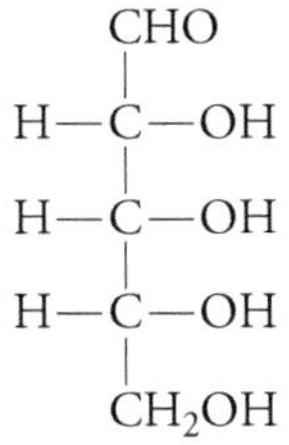

a.

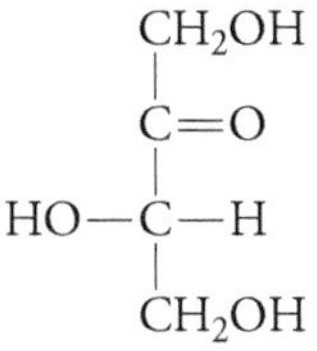

```
     CH₂OH
      C=O
HO—C—H
     CH₂OH
```

b.

```
      CHO
 H—C—OH
 H—C—OH
 H—C—OH
     CH₂OH
```

c.

```
     CH₂OH
 H—C—OH
 H—C—OH
 H—C—OH
 H—C—OH
     CH₂OH
```

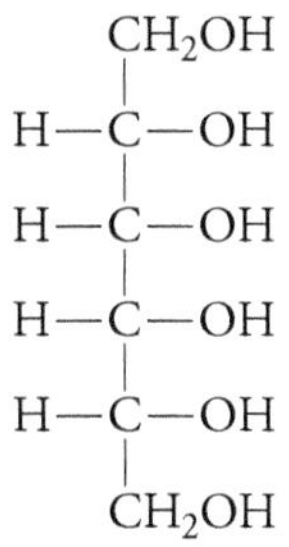

d.

```
      CHO
      C=O
 H—C—OH
HO—C—H
HO—C—H
     CH₂OH
```

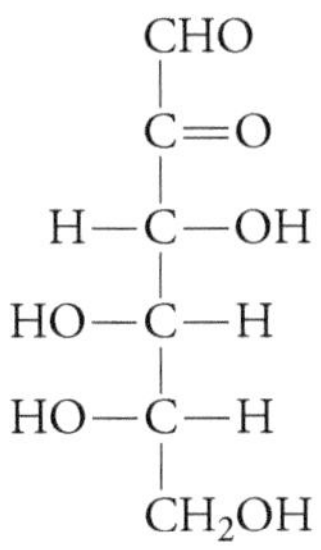

6. Which type of isomer is represented by each pair of sugars?

a. D-gulose and D-idose

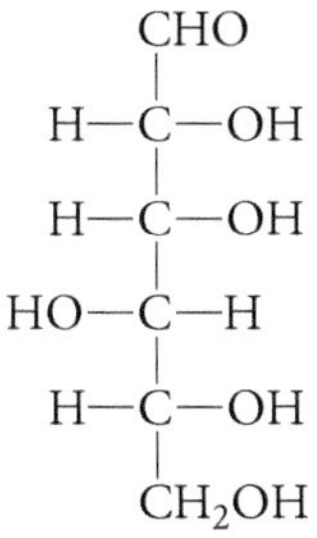

```
      CHO                CHO
 H—C—OH            HO—C—H
 H—C—OH             H—C—OH
HO—C—H             HO—C—H
 H—C—OH             H—C—OH
     CH₂OH              CH₂OH
   D-Gulose           D-Idose
```

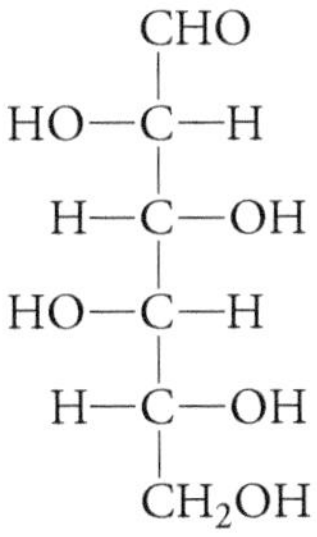

b. D-fructose and L-fructose

7. a. Mannose is the C2 epimer of glucose. Draw the Fischer structure of mannose. **b.** Some proteins have a mannose-6-phosphate "marker" that directs them to be transported to lysosomes. Draw the Fischer structure of mannose-6-phosphate. **c.** Draw the Haworth projection of α-D-mannose-6-phosphate.

8. Which of the following are structural isomers of glucose? **a.** glucose-6-phosphate, **b.** fructose, **c.** galactose, and **d.** ribose.

9. Tagatose is an "artificial" sweetener that is similar to sucrose in sweetness and replaces sucrose in Pepsi Slurpees®. Tagatose is a C4 epimer of fructose. **a.** Draw the Fischer structure of D-tagatose. **b.** L-Tagatose is also about as sweet as sugar, but it costs more to manufacture than its D isomer; thus an early decision to market L-tagatose was abandoned. Draw the Fischer structure of L-tagatose. **c.** Only about 30% of tagatose is absorbed in the small intestine; effectively, tagatose has 30% of the calories of sucrose. Why is tagatose absorbed less efficiently?

10. D-Psicose, the C3 epimer of fructose, has no calories and has also been found to have hypoglycemic effects, which might be useful in treating diabetes. Draw the structure of D-psicose.

11. The monosaccharide sorbose is a C3 epimer of D-tagatose (see Problem 9). Along with tagatose and psicose (see Problem 10), sorbose is a "rare sugar" that can be used as a sweetener by diabetics because it inhibits the enzyme that hydrolyzes sucrose (see Section 11.2) and thus suppresses the rise in blood glucose that follows a meal. Draw the structure of D-sorbose.

12. Which is more stable: the α or the β anomer of galactose?

13. The monosaccharide allose, like sorbose (see Problem 11), is a rare sugar that the FDA recently approved for consumption. Allose is not digested by humans and, like sorbose, has a favorable postprandial blood glucose profile. Allose is a C3 epimer of glucose. Draw its Fischer structure.

14. Carry out a cyclization reaction with galactose and draw the Haworth projections of the two possible reaction products. Name the products.

15. Like glucose, ribose can undergo a cyclization reaction with its aldehyde group and the C5 hydroxyl group to form a six-membered ring. Draw the Haworth projections of the two possible reaction products and name the products.

16. Ribose can undergo a cyclization reaction different from the one presented in Problem 15. In this case, the aldehyde group reacts with the C4 hydroxyl group. **a.** Draw the structures of the two possible reaction products. What is the size of the ring? **b.** Which anomer is found in RNA?

17. a. Like glucose, fructose can undergo a cyclization reaction. The most common reaction involves the ketone group and the C5 hydroxyl group. Draw the structures of the two possible reaction products. What is the size of the ring? **b.** Repeat the exercise described in part **a** but use the C6 hydroxyl group instead. What is the size of the resulting ring?

18. D-Xylose and L-arabinose are key five-carbon sugars in plant cell walls. Their Fischer structures are shown below. **a.** D-Xylose undergoes a cyclization reaction to form a six-membered ring. Draw the Haworth projection of β-D-xylose. **b.** L-Arabinose undergoes a cyclization reaction to form a five-membered ring. Draw the Haworth projection of α-L-arabinose.

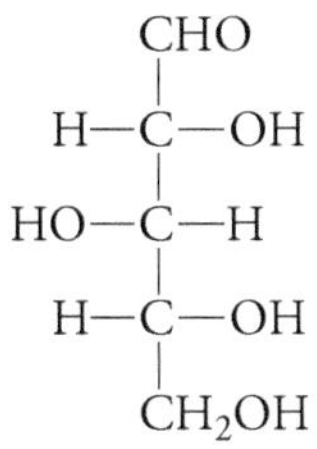

```
      CHO                CHO
 H—C—OH            H—C—OH
HO—C—H             HO—C—H
 H—C—OH            HO—C—H
     CH₂OH              CH₂OH
   D-Xylose          L-Arabinose
```

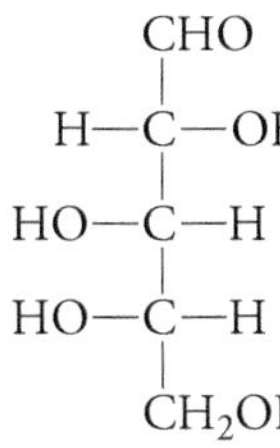

19. An enzyme recognizes only the α anomer of glucose as a substrate and converts it to product. If the enzyme is added to a mixture of the α and β anomers, explain why all the sugar molecules in the sample will eventually be converted to product.

20. As described in the chapter, a solution of glucose molecules consists of about 64% β anomer and about 36% α anomer. Why is the mixture not 50% β anomer and 50% α anomer? In other words, why is the β anomer favored?

21. a. Oxidation of the aldehyde group of glucose yields gluconate. Draw the Fischer structure of gluconate. **b.** Gluconate cyclizes to form a cyclic ester called a lactone. Draw the Haworth projection of the lactone. **c.** Reduction of the aldehyde group of glucose yields sorbitol. Draw the Fischer structure of sorbitol.

22. Benedict's solution is an alkaline copper sulfate solution, which is used to detect the presence of aldehyde groups. In the presence of Benedict's solution, the aldehyde group is oxidized and the aqueous blue Cu^{2+} ion is reduced to a red Cu_2O precipitate. Sugars such as glucose, which produce the red precipitate when Benedict's solution is added, are called reducing sugars because they can reduce Cu^{2+} to Cu^+. Which of the following carbohydrates would give a positive reaction with Benedict's reagent? **a.** galactose, **b.** sorbitol (see Solution 21c), **c.** β-ethylglucoside, **d.** gluconate (see Solution 21a).

23. In algae, the breakdown of certain monosaccharides yields 2-keto-3-deoxygluconate (KDG). Use what you know about glucose derivatives to deduce the structure of KDG.

24. Use the examples in the textbook to deduce the structures of the following sugars: **a.** fructose-1,6-bisphosphate; **b.** galactosamine; **c.** N-acetylglucosamine. **d.** the Fischer structure and the Haworth projection of the α-anomer of galacturonate.

25. When glucose from the blood enters a cell, intracellular enzymes convert it to glucose-6-phosphate. This strategy "traps" the glucose inside the cell. Explain.

26. Some pathogenic bacteria release monosaccharides that can be detected by white blood cells. These monosaccharides are not found in humans, so the enzymes that catalyze their synthesis in bacteria are attractive drug targets. Using the name of the monosaccharide and examples provided in the chapter, draw the Haworth projection of a bacterial heptose named 2-O-acetyl-6-deoxy-α-D-mannoheptopyranose. (*Hint:* See Solution 7a for the structure of mannose; the term "hepto" indicates seven carbons, and "pyran" indicates that a six-membered ring is formed.)

27. Glucuronate undergoes epimerization to iduronate during the synthesis of glycosaminoglycans (see Section 11.3). **a.** Draw the structure of iduronate, the C5 epimer of glucuronate. What kind of monosaccharide is formed in the epimerization reaction? **b.** The iduronate subsequently becomes sulfated on C2. Draw the structure of 2-O-sulfo-α-iduronate.

28. As described in Section 7.4, some drugs are hydroxylated by cytochrome P450. A drug molecule may also be modified by liver glucuronosyltransferases, which link C1 of glucuronate to the drug by a glycosidic bond. Draw the structure of glucuronidated acetaminophen.

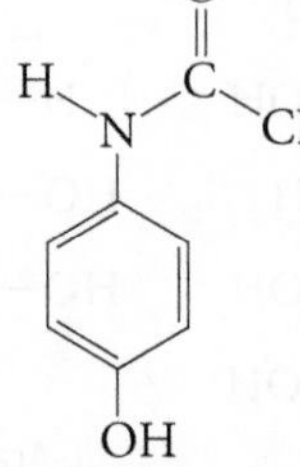

Acetaminophen

29. a. If methanol is added to a sugar in the presence of an acid catalyst, only the anomeric hydroxyl group becomes methylated, as shown in Figure 11.2. If a stronger methylating agent is used, such as methyl iodide, CH_3I, all hydroxyl groups become methylated. Draw the Haworth projection of the product that results when methyl iodide is added to a solution of α-D-glucose. **b.** If the product that formed in part **a** is treated with a strong aqueous acid solution, the glycosidic methyl group is readily hydrolyzed, but the methyl ethers are not. Draw the Fischer structure of the product that results when the compound formed in part **a** is treated with strong aqueous acid.

30. When an unknown monosaccharide is treated with methyl iodide followed by strong aqueous acid (see Problem 29), the product is 2,3,5,6-tetra-O-methyl-D-glucose. Draw a Haworth projection of the unmodified monosaccharide.

31. A glucose derivative named hexadecyl-β-D-glucopyranoside has been demonstrated to have liquid crystal properties. The compound forms when the hydroxyl group of hexadecyl alcohol reacts with the anomeric carbon of glucose to form a glycosidic bond. Draw the structure of the compound. (*Hint: Pyran* indicates that a six-membered ring forms.)

32. The Fischer structure of N-acetylneuraminic acid (NANA) is shown below. It is synthesized from N-acetylmannosamine and pyruvic acid. This monosaccharide can undergo a cyclization reaction to form a six-membered ring. Draw the structure of the α-anomer of NANA.

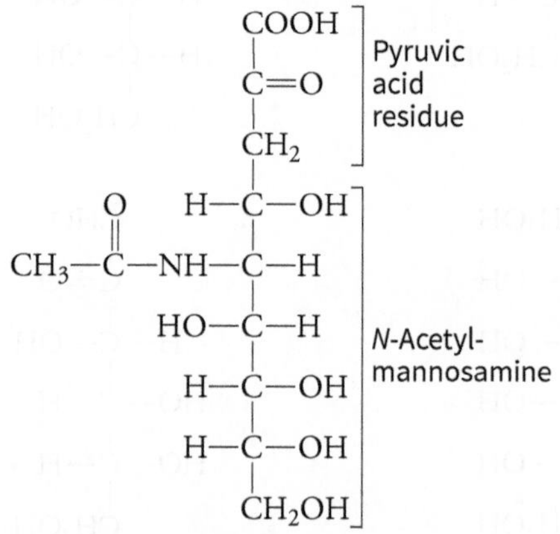

N-Acetylneuraminic acid
(Fischer structure)

33. Bakers dip pretzel dough into a solution of NaOH or KOH before baking. Explain why this step is necessary for achieving a rich brown crust on the pretzel.

34. Explain why boiled potato pieces (the basis of mashed potatoes) have a different taste than roasted potato pieces.

11.2 Disaccharides, Oligosaccharides, and Polysaccharides

35. Explain why lactose is a reducing sugar, whereas sucrose is not.

36. As described in the chapter, lactase is the enzyme that catalyzes the hydrolysis of lactose. The enzyme is expressed in newborns, but its activity typically decreases as the organism matures, and adults become *lactose-intolerant*. However, adult members of populations whose ancestors domesticated cows (Northern Europe) or goats (Africa) are lactose-tolerant. Explain why.

37. Cellobiose is a disaccharide composed of two glucose monomers linked by a β(1→4) glycosidic bond. Draw the structure of cellobiose. Is cellobiose a reducing sugar?

38. The structure of trehalose is shown. **a.** Identify its monosaccharide residues and the type of glycosidic bond that links them. **b.** Is trehalose a reducing sugar? **c.** Trehalose accumulates in some plants under dehydrating conditions. The trehalose stabilizes dehydrated enzymes, proteins, and lipid membranes when the plant is desiccated. These plants are often called "resurrection plants" because the plants resume their normal metabolism when water becomes available. How does trehalose stabilize cellular molecules?

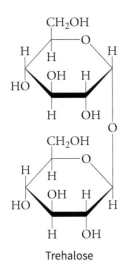

Trehalose

39. Trehalase is an enzyme that catalyzes hydrolysis of the bond that links the two monosaccharide residues of trehalose (see Problem 38). Draw the structures of the trehalase reaction products.

40. One of the major nonlactose oligosaccharides in human milk is 2′-fucosyllactose, in which fucose forms a glycosidic bond to C2 of the galactose residue. Fucose is a form of L-galactose that is missing the C6 hydroxyl group. Draw the structure of 2′-fucosyllactose.

41. The disaccharide gentobiose is a component of amygdalin, a poisonous cyanogenic glycoside found in the kernels of apricot seeds. Draw the structure of gentobiose, which consists of two glucose residues linked by a β(1→6) glycosidic bond.

42. The sugar alcohol sorbitol (see Solution 21c) is sometimes used instead of sucrose to sweeten processed foods. Unfortunately, the consumption of large amounts of sorbitol may lead to gastrointestinal distress in susceptible individuals, especially children. Sorbitol-sweetened foods have a lower calorie count than sucrose-sweetened foods. Explain why.

43. The disaccharide isomaltulose is used as a sugar substitute for diabetics because its glycosidic bond is hydrolyzed slowly, minimizing the rise in blood glucose concentration after eating. **a.** Draw the structure of isomaltulose, which consists of α(1→6)-linked glucose and fructose. **b.** Is isomaltulose a reducing sugar? **c.** Compare the structures of isomaltulose and sucrose.

44. Melibiose is a disaccharide found in artichokes. It consists of galactose and glucose linked by an α(1→6) glycosidic bond. **a.** Draw the structure of melibiose. **b.** Is melibiose a reducing sugar? **c.** Compare the structures of melibiose and lactose.

45. Propose a structure for a disaccharide that is consistent with the following information: Complete hydrolysis yields only D-glucose; it reduces Cu^{2+} to Cu_2O (see Problem 22), and it is hydrolyzed by α-glucosidase but not β-glucosidase.

46. An unknown disaccharide is subjected to methylation with CH_3I followed by acid hydrolysis (see Problem 27). The reaction yields 2,3,4,6-tetra-*O*-methyl-D-glucose and 2,3,4-tri-*O*-methyl-D-glucose. **a.** Draw the structure of the unknown disaccharide. **b.** Compare this structure to the structure of gentobiose that you drew in Solution 43.

47. The human genome codes for glycoside hydrolases that act on starch, glycogen, sucrose, and lactose. How many different types of glycosidic bonds must these enzymes break in order to generate monosaccharide products?

48. The presence of the oligosaccharide raffinose (below) in beans is blamed for causing flatulence when beans are consumed. Bacteria in the large intestine act on undigested raffinose and produce gas as a metabolic by-product. Which glycosidic bond(s) in raffinose are humans able to hydrolyze? Which bond(s) cannot be hydrolyzed?

Raffinose

49. Nigerotriose consists of three glucose residues linked by α(1→3) glycosidic bonds. **a.** Draw the structure of nigerotriose. **b.** Are humans able to hydrolyze the glycosidic bonds in nigerotriose?

50. Scientists in Singapore injected soybeans with an enzyme extract from the fungus *R. oligosporus* before allowing the soybeans to germinate. After three days, the concentration of raffinose (see Problem 48) decreased dramatically. (As an added bonus, the concentration of cancer-fighting isoflavones in the soybeans increased.) What enzymes must have been present in the fungus?

51. In a homopolymer, all the monomers are the same, and in a heteropolymer, the monomers are different. Which of the polysaccharides discussed in this chapter are homopolymers and which are heteropolymers?

52. How many reducing ends are in a molecule of potato amylopectin that contains 100,000 residues with a branch every 50 residues?

53. Instead of starch, some plants produce inulin, which is a polymer of β(2→1)-linked fructose residues. Chicory root is a particularly rich source of inulin. **a.** Draw the structure of an inulin disaccharide. **b.** Why do some food manufacturers add inulin to their products (yogurt, ice cream, and even drinks) to boost their "fiber" content?

54. When amylose is suspended in water in the presence of I_2, a blue color results due to the ability of I_3^-, an ion that forms in aqueous iodine solutions, to occupy the interior of the helix. A drop of yellow iodine solution on a potato slice turns blue, but when a drop of iodine is placed on an apple slice, the color of the solution remains yellow. Explain why.

55. Individuals with irritable bowel syndrome may be sensitive to FODMAPs (fermentable oligosaccharides, disaccharides, monosaccharides, and polyols). An example of a fermentable oligosaccharide is the β-glucan polysaccharide found in oats. How many different bacterial hydrolases are required to convert this polysaccharide to its constituent monosaccharides?

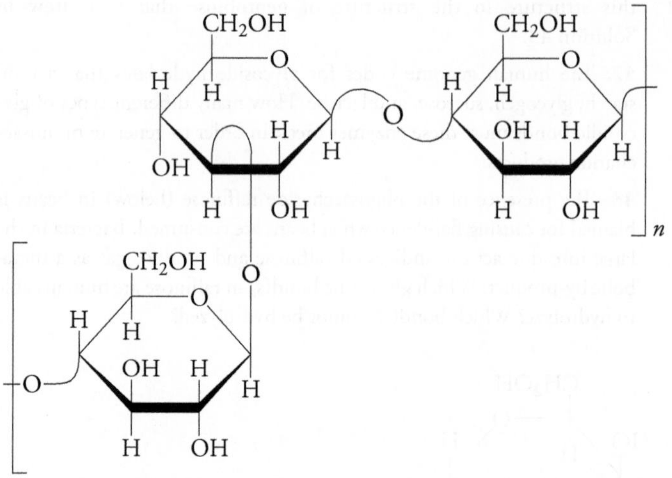

56. The repeating structure of the polysaccharide pectin is shown. Identify its constituent monosaccharides and the type of glycosidic bond that links them.

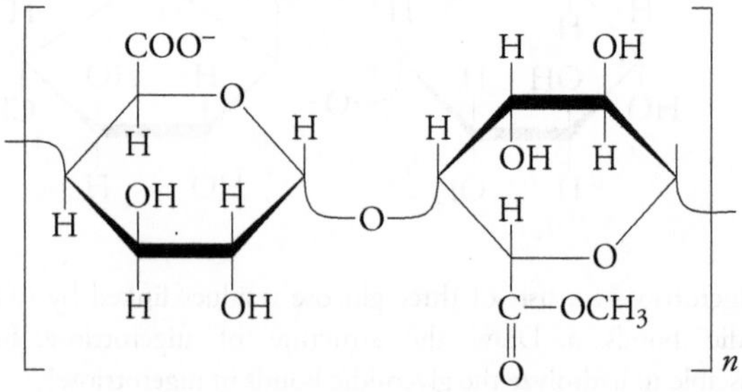

57. Explain why pectin (see Problem 56 and Box 11.B) is sometimes added to fruit extracts to make jams and jellies.

58. Hemicellulose, another polysaccharide produced in plants, is not a cellulose derivative but is a random heteropolymer of various monosaccharides. D-Xylose residues (see Box 11.B) are present in the largest amount and are linked with $\beta(1\rightarrow4)$ glycosidic bonds. Draw the structure of a hemicellulose disaccharide unit containing D-xylose.

59. Explain why a growing plant cell, which produces cellulose, must also produce cellulase.

60. A friend tells you that she plans to eat a lot of celery because she heard that the digestion of celery consumes more calories than the food contains, and this will help her to lose weight. How do you respond?

61. Stonewashed cotton clothing, which appears faded and weathered, can be prepared by tumbling a garment with water and stones. Alternatively, the garment can be briefly soaked in a solution containing cellulase. **a.** Explain what cellulase does. **b.** What would happen if the cellulase treatment were prolonged?

62. Some bacteria synthesize cellulose in which every other monosaccharide residue has a phosphoethanolamine group (see Section 8.1) attached to C6. Draw the structure of this modified residue.

63. Conventional cellulose and cellulose modified with phosphoethanolamine groups (see Problem 62) differ in their physical properties. Describe the difference and explain why the modified cellulose is superior to conventional cellulose for forming biofilms.

64. Brown algae are an attractive source of biofuels, since these organisms do not require land, fertilizer, or fresh water and lack lignin. One major polysaccharide in brown algae is laminarin, which consists of glucose residues linked by $\beta(1\rightarrow3)$ glycosidic bonds. Draw the structure of the disaccharide unit of laminarin.

65. Mycobacteria contain several unusual polymethylated polysaccharides. One consists of repeating units of 3-O-methylmannose linked with $\alpha(1\rightarrow4)$ glycosidic bonds. Draw the structure of the disaccharide unit of this polysaccharide. (*Hint:* See Solution 7a for the structure of mannose.)

66. The repeating disaccharide unit of the polysaccharide heparin is shown below. An affinity chromatography method (see Section 4.6) of purifying DNA-binding proteins was developed by immobilizing heparin on a solid support. DNA-binding proteins bind to heparin while proteins with no affinity for DNA elute from the column in the flow-through. Why do DNA-binding proteins have an affinity for heparin?

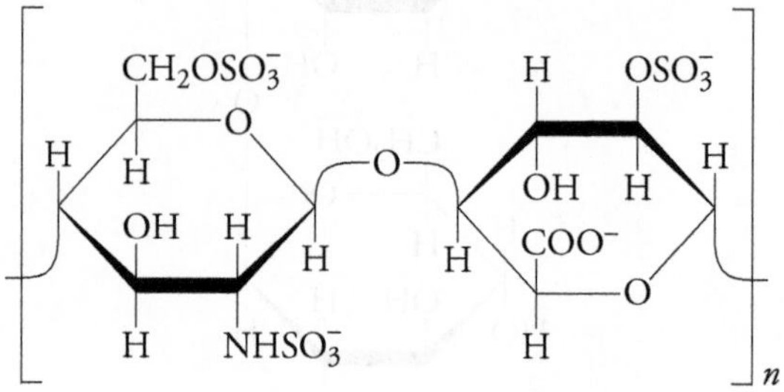

67. A newly discovered disaccharide named matriglycan links cell-surface receptors to the membrane in muscle cells and is essential for maintaining cellular integrity. The repeating disaccharide unit is expressed as GlcA$\beta(1\rightarrow3)$-Xyl$\alpha(1\rightarrow3)$. Draw the structure of the disaccharide unit. (Note that "GlcA" is glucuronate and "Xyl" is xylose.)

11.3 Glycoconjugates: Glycoproteins, Glycolipids, and Proteoglycans

68. Identify the *N*- and *O*-linked saccharides shown at the start of Section 11.3. Are the linkages α- or β-glycosidic bonds?

69. Carbohydrates present in many species with repeating $\beta(1\rightarrow4)$-glycosidic linked 20 to 40 disaccharides. Explain.

70. Collagen isolated from a deep-sea hydrothermal vent worm contains a glycosylated threonine residue in the "Y" position of the (Gly-X-Y)$_n$ repeating triplet (see Section 5.3). A galactose residue is covalently attached to the threonine via a β-glycosidic bond. Draw the structure of the galactosylated threonine residue.

71. In the past, it was believed that glycosylated proteins were found only in eukaryotic cells, but recently glycoproteins have been found in prokaryotic cells. The flagellin protein from the Gram-positive bacterium *L. monocytogenes* is glycosylated with a single *N*-acetylglucosamine at up to six different serine or threonine residues. Draw the structure of this linkage.

72. Give the structures of oligosaccharides that play a key role in the ABO blood group system.

73. Draw the structures of glycoproteins containing serine and threonine amino acids.

74. Glycoproteins that are *N*-linked to β-rhamnose (see Box 11.B) have been found in archaea and eubacteria. Draw the structure of this linkage.

75. A viral glycoprotein has $\alpha(1\rightarrow2)$-linked mannose oligomers, which are not found in mammalian cells. The enzyme responsible for synthesizing this linkage is a potential antiviral target. Draw the structure of this linkage. (*Hint:* See Solution 7a for the structure of mannose.)

76. A common *O*-glycosidic attachment of an oligosaccharide to a glycoprotein is β-galactosyl-$(1\rightarrow3)$-α-*N*-acetylgalactosyl-Ser. Draw the structure of the oligosaccharide and its linkage to the glycoprotein.

77. A transferase enzyme highly expressed in cancer cells catalyzes the synthesis of a glycan that is found in cancer cells but not in normal cells, which makes the enzyme a potential drug target for cancer therapy. Draw the structure of the trisaccharide unit of the glycan: GlcNAc($\beta1\rightarrow6$)-Man($\alpha1\rightarrow6$)-Man($\beta1\rightarrow4$). (*Hint:* See Solution 7a for the structure of mannose.)

78. Identify the parent monosaccharides of the chondroitin sulfate disaccharide (Fig. 11.9) and identify the linkages between them.

79. Calculate the net charge of a chondroitin sulfate molecule containing 150 disaccharide units.

80. a. Identify the monosaccharide precursor of the repeating disaccharide in peptidoglycan. **b.** In one type of peptide cross-link in peptidoglycan, an alanine residue forms an amide bond with the repeating disaccharide. Identify the location of this linkage.

81. Lysozyme, an enzyme present in tears and mucus secretions, is a $\beta(1\rightarrow4)$ glycosidase. Explain how lysozyme helps prevent bacterial infections.

82. Transmembrane proteins destined for the plasma membrane are post-translationally processed. Oligosaccharide chains are added in the ER to selected asparagine residues and then the oligosaccharide chains are processed in the Golgi apparatus (Fig. 11.7). One of the sugars added to the core oligosaccharide is *N*-acetylneuraminic acid, also called sialic acid (Problem 32). Tumor cells frequently overexpress sialic acid on their surfaces, which might contribute to a tumor cell's ability to detach from the tumor and travel through the bloodstream to form additional tumors. The sialic acid residues on tumor cells are typically not recognized by the immune system, which compounds the problem. **a.** How does the presence of sialic acid on the cell surface facilitate detachment? **b.** What strategies would you use to design therapeutic agents to kill these types of tumor cells?

83. Keratan sulfate is composed of repeating disaccharides of *N*-acetylglucosamine and galactose linked via a $\beta(1\rightarrow3)$ glycosidic bond. An enzyme catalyzes the addition of sulfate groups to both monosaccharides at the C6 positions in infants during the early postnatal period. Draw the structure of the repeating unit of keratan sulfate.

11.4 Clinical Connection: Glucose and Diabetes Mellitus/Clinical Connection: Lectin

84. Patients with hyperglycemic conditions should consume what type of diet to control their blood glucose levels immediately.

85. Fatty liver is more frequent in patients suffering from diabetes mellitus. Explain.

86. Scarring of tissues is most common in patients with abnormalities of carbohydrate metabolism. Explain.

87. Would pancreas transplantation be the solution to hyperglycemia?

SELECTED READINGS

Reily, C., Stewart, T.J., Renfrow, M.B., and Novak, J., Glycosylation in health and disease, *Nat. Rev. Nephrol.* 15, 346–366, doi: 10.1038/s41581-019-0129-4 (2019). [Includes a review of different types of glycosylation and the role of protein glycosylation in a variety of diseases.]

Seeberger, P.H., Monosaccharide diversity, Chapter 2 in Varki, A., Cummings, R.D., Esko, J.D., et al. (eds), *Essentials of Glycobiology*, 3rd edition, Cold Spring Harbor Laboratory Press, doi: 10.1101/glycobiology.3e.002 (2017). [A primer on basic carbohydrate chemistry.]

Springer, S.A. and Gagneux, P., Glycomics: Revealing the dynamic ecology and evolution of sugar molecules, *J. Proteomics* 135, 90–100, doi: 10.1016/j.jprot.2015.11.022 (2016). [Introduces the field of glycomics, with a focus on the diversity of glycans.]

Townsley, L. and Shank, E.A., Natural product antibiotics: cues for modulating bacterial biofilm formation, *Trends Microbiol.* 25, 1016–1026, doi: 10.1016/j.tim.2017.06.003. (2017). [Includes details about biofilm composition and the regulation of biofilm synthesis.]

Turner, R.D., Vollmer, W., and Foster, S.J., Different walls for rods and balls: The diversity of peptidoglycan, *Molec. Microbiol.* 91, 862–874, doi: 10.1111/mmi.12513 (2014). [Summarizes the formation and structure of bacterial cell walls.]

Varki, A., Biological roles of glycans, *Glycobiology* 27, 3–49, doi: 10.1093/glycob/cww086 (2016). [A review of the biological roles of carbohydrates within and between species, by an expert in the field.]

Woods, R.J., Predicting the structures of glycans, glycoproteins, and their complexes, *Chem. Rev.* 118, 8005–8024, doi: 10.1021/acs.chemrev.8b0003 (2018). [Describes approaches for understanding the structures and dynamics of oligosaccharides and their interactions with proteins.]

CHAPTER 11 CREDITS

Figure 11.8 Image based on 1IGY. Harris, L.J., Skaletsky, E., McPherson, A., Crystallographic structure of an intact IgG1 monoclonal antibody, *J. Mol. Biol.* 275, 861–872 (1998).

Figure 11.11 Image from Gumbart, J.C., Beeby, M., Jensen, G.J., and Roux, B., *PLoS Comput. Biol.* 10(2): e1003475. doi:10.1371/journal.pcbi.1003475 (2014) Licensed under CC-BY 4.0.

Metabolism and Bioenergetics

Karlos Lomsky/Adobe Stock

One of the champions of metabolic efficiency is the wildebeest, a large mammal that can walk 80 km in searing temperatures for over five days—without drinking. A combination of efficient muscles and a slow gait prevents overheating and minimizes the loss of water in evaporative cooling.

Do You Remember?

- Living organisms obey the laws of thermodynamics (Section 1.4).
- Nucleotides and nucleotide derivatives serve a variety of functions (Section 3.1).
- Amino acids are linked by peptide bonds to form a polypeptide (Section 4.1).
- Allosteric regulators can inhibit or activate enzymes (Section 7.3).
- Lipids are predominantly hydrophobic molecules that can be esterified but cannot form polymers (Section 8.1).
- Monosaccharides can be linked by glycosidic bonds in various arrangements (Section 11.2).

The diversity of life-forms on earth makes it impossible to examine all the molecular components and chemical reactions that occur in each organism. However, it turns out that organisms share some basic cellular structures, make the same types of molecules, and use similar enzymes to build and break down those molecules. In the coming chapters, we will focus on these common pathways that transform raw material and energy in cells. We begin by introducing the notions of metabolic fuels and metabolic pathways and then turn to explore different forms of energy in biological systems.

12.1 Food and Fuel

KEY CONCEPTS

Summarize the pathways for digesting and mobilizing metabolic fuels.

- Distinguish autotrophs and heterotrophs.
- List the monomer and polymer form for each major type of metabolic fuel.
- For each fuel, summarize the process of digestion, absorption, storage, and mobilization.

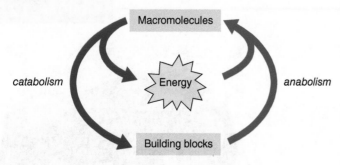

Figure 12.1 Catabolism and anabolism. Catabolic (degradative) reactions yield energy and small molecules that can be used for anabolic (synthetic) reactions. Metabolism is the sum of all catabolic and anabolic processes.

The biochemical activity of cells can be summarized as shown in **Figure 12.1.** Cells break down or **catabolize** large molecules to release energy and small molecules. The cells then use the energy and small molecules to rebuild larger molecules, a process called **anabolism.** The set of all catabolic and anabolic activities constitutes an organism's **metabolism.**

Some microorganisms, known as **chemoautotrophs** (from the Greek *trophe,* "nourishment"), obtain virtually all their metabolic building materials and free energy from the simple inorganic compounds CO_2, N_2, H_2, and S_2. **Photoautotrophs,** such as the familiar green plants, need little more than CO_2, H_2O, a source of nitrogen, and sunlight. In contrast, **heterotrophs,** a group that includes animals, directly or indirectly obtain all their building materials and energy from organic compounds produced by chemo- or photoautotrophs.

As heterotrophs, mammals rely on food produced by other organisms. After food is digested and absorbed, it becomes a source of metabolic energy and materials to support the animal's growth and other activities. The human diet includes the four types of biological molecules introduced in Section 1.3 and described in more detail in subsequent chapters. These molecules are often present as macromolecular polymers, namely proteins, nucleic acids, polysaccharides, and triacylglycerols (technically, fats are not polymers since the monomeric units are not linked to each other but to glycerol). *Digestion reduces the polymers to their monomeric components: amino acids, nucleotides, monosaccharides, and fatty acids.* The breakdown of nucleotides does not yield significant amounts of metabolic energy, so we will devote more attention to the catabolism of other types of biomolecules.

Cells take up the products of digestion

Mammalian digestion takes place extracellularly in the mouth, stomach, and small intestine and is catalyzed by hydrolytic enzymes (**Fig. 12.2**). For example, salivary amylase begins to break down starch, which consists of linear polymers of glucose residues (amylose) and branched polymers

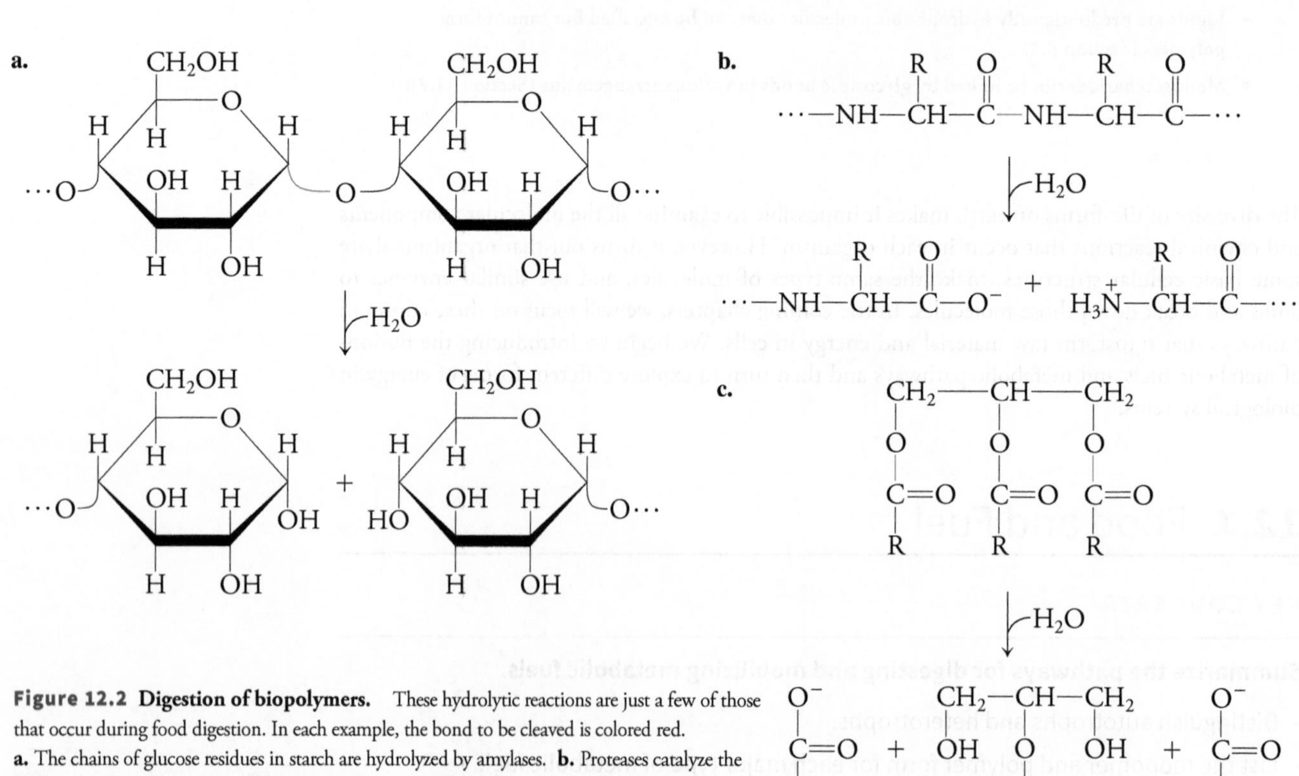

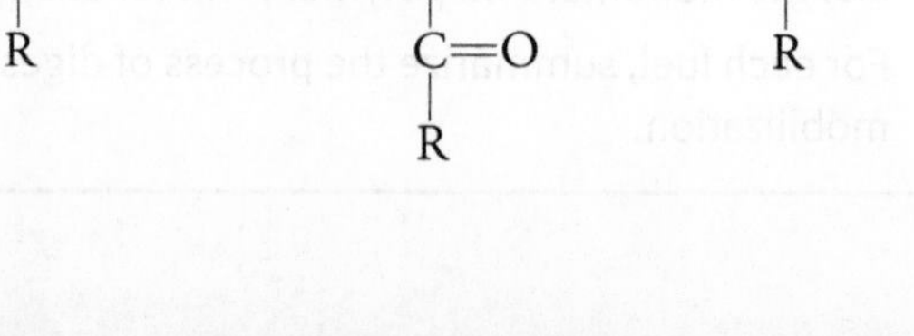

Figure 12.2 Digestion of biopolymers. These hydrolytic reactions are just a few of those that occur during food digestion. In each example, the bond to be cleaved is colored red. **a.** The chains of glucose residues in starch are hydrolyzed by amylases. **b.** Proteases catalyze the hydrolysis of peptide bonds in proteins. **c.** Lipases hydrolyze the ester bonds linking fatty acids to the glycerol backbone of triacylglycerols.

Question Explain why the reactions shown here are thermodynamically favorable.

(amylopectin; Section 11.2). Gastric and pancreatic proteases (including trypsin, chymotrypsin, and elastase) degrade proteins to small peptides and amino acids. Lipases synthesized by the pancreas and secreted into the small intestine catalyze the release of fatty acids from triacylglycerols. Water-insoluble lipids do not freely mix with the other digested molecules but instead form micelles (Fig. 2.9).

The products of digestion are absorbed by the cells lining the intestine. Monosaccharides enter the cells via active transporters such as the Na^+-glucose system diagrammed in Figure 9.19. Similar symport systems bring amino acids and di- and tripeptides into the cells. Some highly hydrophobic lipids diffuse through the cell membrane; others require transporters. Inside the cell, the triacylglycerol digestion products re-form triacylglycerols, and some fatty acids are linked to cholesterol to form cholesteryl esters, for example,

Cholesteryl stearate

Triacylglycerols and cholesteryl esters are packaged, together with certain proteins, to form **lipoproteins.** These particles, known specifically as chylomicrons, are released into the lymphatic circulation before entering the bloodstream for delivery to tissues.

Water-soluble substances such as amino acids and monosaccharides leave the intestinal cells and enter the portal vein, which drains the intestine and other visceral organs and leads directly to the liver. *The liver therefore receives the bulk of a meal's nutrients and catabolizes them, stores them, or releases them back into the bloodstream.* The liver also takes up chylomicrons and repackages the lipids with different proteins to form other lipoproteins, which circulate throughout the body, carrying cholesterol, triacylglycerols, and other lipids (lipoproteins are discussed in greater detail in Section 17.1).

The allocation of resources following a meal varies with the individual's needs at that time and with the type of nutrients consumed. Fortunately, the body does this efficiently, regardless of what food was eaten. The sheer abundance of dietary advice—often conflicting—indicates that scientists have yet to identify the ideal human diet. The best they can do is identify the body's overall needs and make broad recommendations. One challenge in monitoring nutrient intake is that very few foods are composed of pure substances, and "serving size" can be interpreted in many ways. In addition, processed foods may combine different amounts of raw ingredients, not to mention nonfood chemicals (preservatives, flavorings, and so on). In any case, consideration of how eating patterns have varied across centuries and across continents indicates that the human body must be remarkably versatile in converting a variety of raw materials into the molecular building blocks and metabolic energy required to sustain life.

Monomers are stored as polymers

Immediately following a meal, the circulating concentrations of monomeric compounds are relatively high. All cells can take up these materials to some extent to fulfill their immediate needs, but *some tissues are specialized for the long-term storage of nutrients.* For example, fatty acids are used to build triacylglycerols, many of which travel in the form of lipoproteins to **adipose tissue.** Here, adipocytes take up the triacylglycerols and store them as intracellular fat globules. Because the mass of lipid is hydrophobic and does not interfere with activities in the aqueous cytoplasm, the fat globule can be enormous, occupying most of the volume of the adipocyte (**Fig. 12.3**).

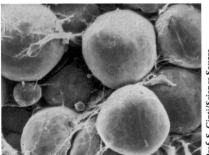

Figure 12.3 Adipocytes. These cells, which make up adipose tissue, contain a small amount of cytoplasm surrounding a large globule of triacylglycerols (fat).

Prof. S. Cinti/Science Source

Figure 12.4 Glycogen structure.
a. Schematic diagram of a glycogen molecule. Each hexagon represents a glucose monomer, and branches occur every 8 to 14 residues. **b.** Electron micrograph of a liver cell showing glycogen granules (colored pink). Mitochondria are green, and a fat globule is yellow.

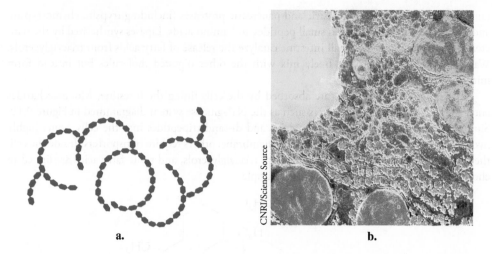

a.

b.

Virtually all cells can take up monosaccharides and immediately catabolize them to produce energy. Some tissues, primarily liver and muscle (which makes up a significant portion of the human body), polymerize glucose to form glycogen. Glycogen is a highly branched polymer with a compact shape. Several glycogen molecules may clump together to form granules that are visible by electron microscopy (**Fig. 12.4**). Glycogen's branched structure means that a single molecule can be expanded quickly, by adding glucose residues to its many branches, and degraded quickly, by simultaneously removing glucose from the ends of many branches. Glucose that does not become part of glycogen can be catabolized to two-carbon acetyl units and converted into fatty acids for storage as triacylglycerols.

Amino acids can be used to build polypeptides. A protein is not a dedicated storage molecule for amino acids, as glycogen is for glucose and triacylglycerols are for fatty acids, so excess amino acids cannot be saved for later. However, in certain cases, such as during starvation, proteins are catabolized to supply the body's energy needs. If the intake of amino acids exceeds the body's immediate protein-building needs, the excess amino acids can be broken down and converted to carbohydrate (which can be stored as glycogen) or converted to acetyl units (which can then be converted to fat).

Amino acids and glucose are both required to synthesize nucleotides. Aspartate, glutamine, and glycine supply some of the carbon and nitrogen atoms used to build the purine and pyrimidine bases (Section 18.5). The ribose-5-phosphate component of nucleotides is derived from glucose by a pathway that converts the six-carbon sugar to a five-carbon sugar (Section 13.4). To summarize, the allocation of resources within a cell depends on the type of tissue and its need to build cellular structures, provide energy, or stockpile resources in anticipation of future needs.

Fuels are mobilized as needed

Amino acids, monosaccharides, and fatty acids are known as **metabolic fuels** *because they can be broken down by processes that make energy available for the cell's activities.* After a meal, glucose and amino acids are catabolized to release their stored energy. When these fuel supplies are exhausted, the body **mobilizes** its storage resources; that is, it converts its polysaccharide and triacylglycerol storage molecules (and sometimes proteins) to their respective monomeric units. Most of the body's tissues prefer to use glucose as their primary metabolic fuel, and the central nervous system can run on almost nothing else. In response to this demand, the liver mobilizes glucose by breaking down glycogen.

In general, depolymerization reactions are hydrolytic, but in the case of glycogen, the molecule that breaks the bonds between glucose residues is not water but phosphate. Thus, the degradation of glycogen is called **phosphorolysis** (**Fig. 12.5**). This reaction is catalyzed by glycogen phosphorylase, which releases residues from the ends of branches in the glycogen polymer.

The phosphate group of glucose-1-phosphate is removed before glucose is released from the liver into the circulation. Other tissues absorb glucose from the blood. In the disease **diabetes mellitus,** this does not occur, and the concentration of circulating glucose may become elevated.

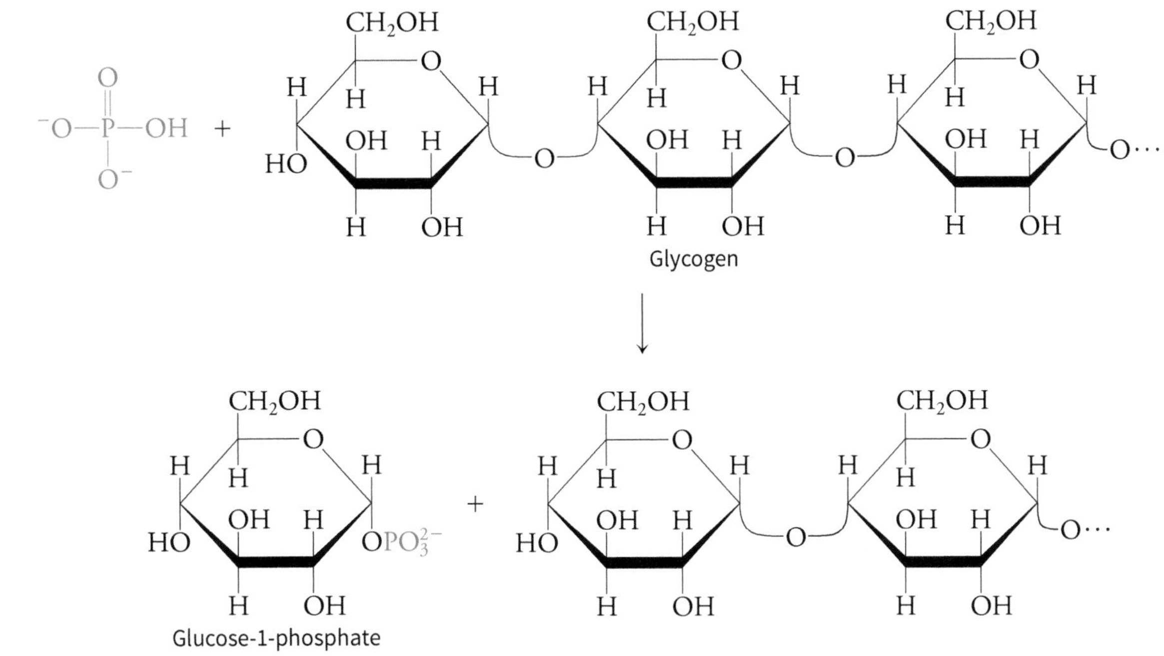

Figure 12.5 Glycogen phosphorolysis. This mechanism of glycogen degradation generates glucose-1-phosphate.

Only when the supply of glucose runs low does adipose tissue mobilize its fat stores. A lipase hydrolyzes triacylglycerols so that fatty acids can be released into the bloodstream. These free fatty acids are not water-soluble and therefore bind to circulating proteins. Except for the heart, which uses fatty acids as its primary fuel, the body does not have a budget for burning fatty acids. In general, as long as dietary carbohydrates and amino acids can meet the body's energy needs, stored fat will not be mobilized, even if the diet includes almost no fat. This feature of mammalian fuel metabolism is a source of misery for many dieters!

Amino acids are not mobilized to generate energy except during a fast, when glycogen stores are depleted (in this situation, the liver can also convert some amino acids into glucose). However, cellular proteins are continuously degraded and rebuilt with the changing demand for particular enzymes, transporters, cytoskeletal elements, and so on. There are two major mechanisms for degrading unneeded proteins. In the first, the **lysosome,** an organelle containing proteases and other hydrolytic enzymes, breaks down proteins that are enclosed in a membranous vesicle. Membrane proteins and extracellular proteins taken up by endocytosis are degraded by this pathway, but intracellular proteins that become enclosed in vesicles can also be broken down by lysosomal enzymes during autophagy (Section 9.4).

A second pathway for degrading intracellular proteins requires a protein complex known as the **proteasome**. A protein to be degraded includes a characteristic sequence known as a **degron,** often positioned at the N-terminus, that is recognized by one or more of a set of 800 enzymes. These enzymes attach a small 76-residue protein, called ubiquitin, that is ubiquitous (hence its name) and highly conserved in eukaryotes. First, the C-terminus of ubiquitin is covalently linked to a lysine side chain in the protein and then additional ubiquitin molecules are added to the first, each one linked via its C-terminus to a lysine side chain of the preceding ubiquitin. A chain of at least four ubiquitins is required to mark a protein for destruction by a proteasome.

The polyubiquitin tag is recognized by the proteasome's regulatory particle, which carries out a series of actions. It disrupts the protein's tertiary structure, it translocates the polypeptide into the proteasome core, and it detaches the ubiquitin chain (the ubiquitins are not degraded and can be reused). The unfolding and translocating activities are carried out by a ring of six ATP-hydrolyzing subunits that convert the chemical energy of the ATP → ADP reaction to mechanical energy in order to force the protein substrate into a more linear form that can be fed into the inner cavity of the proteasome (**Fig. 12.6**).

The core of the proteasome is barrel-shaped, forming a chamber with six separate active sites that carry out peptide bond hydrolysis (**Fig. 12.7**). In this enclosed space, the unfolded polypeptide substrate is broken down to peptides of about 8–10 residues that can diffuse out of

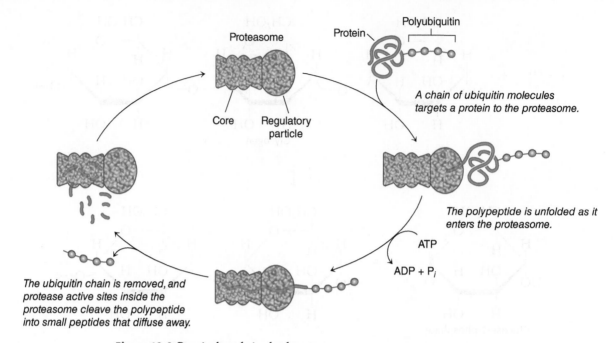

A chain of ubiquitin molecules targets a protein to the proteasome.

The polypeptide is unfolded as it enters the proteasome.

ATP

ADP + Pᵢ

The ubiquitin chain is removed, and protease active sites inside the proteasome cleave the polypeptide into small peptides that diffuse away.

Figure 12.6 Protein degradation by the proteasome.

Question **Draw the linkage between a protein's C-terminus and a ubiquitin Lys residue.**

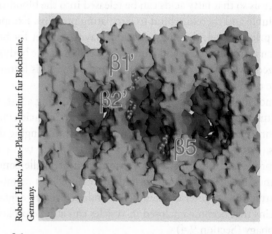

Robert Huber, Max-Planck-Institut fur Biochemie, Germany.

Figure 12.7 Structure of the yeast proteasome core.   This cutaway view shows the inner chamber, where proteolysis occurs. The red structures mark the locations of three of the six protease active sites.

the proteasome. The proteasome takes about 20 seconds to deconstruct an average 300-residue protein. The peptides are further degraded by cytosolic peptidases so that the amino acids can be catabolized or recycled. Although the degron-recognition process is not completely understood (and some degrons may trigger autophagy instead), the system is sophisticated enough to allow unneeded or defective proteins to be destroyed while sparing the cell's essential proteins. Although prokaryotes do not use ubiquitin, they contain proteasome-like structures to degrade proteins.

Concept Check

1. Explain the relationship between autotrophs and heterotrophs.
2. Review the steps by which nutrients from food molecules reach the body's tissues.
3. Name the major metabolic fuels and describe how they are stored.
4. Describe how metabolic fuels are mobilized.
5. Summarize the mechanisms for intracellular protein degradation.

12.2 Free Energy Changes in Metabolic Reactions

KEY CONCEPTS

Analyze the energy changes that occur during metabolic reactions.

- Distinguish the actual and standard free energy change for a reaction.
- Relate the free energy change to the concentrations of reactants.
- Explain what occurs when reactions are coupled.
- Identify molecules that function as energy currency.
- Explain how certain reactions can control flux through a pathway.

We have introduced the idea that catabolic reactions tend to release energy and anabolic reactions tend to consume it (see Fig. 12.1), but, in fact, *all reactions in vivo occur with a net decrease in free energy; that is,* ΔG *is always less than zero* (free energy is discussed in Section 1.4). In a cell, metabolic reactions are not isolated but are linked, so the free energy of a thermodynamically favorable reaction can be used to pull a second, unfavorable reaction forward. How can energy be transferred from one reaction to another? Free energy is not a substance or the property of a single molecule, so it is misleading to refer to a molecule or a bond within that molecule as having a large amount of free energy. Rather, *free energy is a property of a system, and it changes when the system undergoes a chemical reaction.*

The free energy change depends on reactant concentrations

The change in free energy of a system is related to the concentrations of the reacting substances. When a reaction such as $A + B \rightleftharpoons C + D$ is at equilibrium, the concentrations of the four reactants define the **equilibrium constant, K_{eq},** for the reaction:

$$K_{eq} = \frac{[C]_{eq}[D]_{eq}}{[A]_{eq}[B]_{eq}} \tag{12.1}$$

(the brackets indicate the molar concentration of each substance). Recall that at equilibrium, the rates of the forward and reverse reactions are balanced, so there is no net change in the concentration of any reactant. Equilibrium does *not* mean that the concentrations of the reactants and products are equal.

When the system is not at equilibrium, the reactants experience a driving force to reach their equilibrium values. This force is the **standard free energy change** for the reaction, $\Delta G^{\circ\prime}$, which is defined as

$$\Delta G^{\circ\prime} = -RT \ln K_{eq} \tag{12.2}$$

where R is the gas constant ($8.3145 \ \text{J} \cdot \text{K}^{-1} \cdot \text{mol}^{-1}$) and T is the temperature in Kelvin. The larger the value of the free energy change, the farther the reaction is from equilibrium and the stronger the tendency for the reaction to proceed. Recall from Section 1.4 that free energy has units of joules per mole. One joule is equivalent to 0.239 calories, and 1 cal = 4.1484 J (1000 cal = 1 kcal = 1 Calorie). Equation 12.2 can be used to calculate $\Delta G^{\circ\prime}$ from K_{eq} and vice versa (see Sample Calculation 12.1).

SAMPLE CALCULATION 12.1

Problem Calculate $\Delta G^{\circ\prime}$ for a reaction at 25°C when $K_{eq} = 5.0$.

Solution Use Equation 12.2:

$$\Delta G^{\circ\prime} = -RT \ln K_{eq}$$
$$= -(8.3145 \ \text{J} \cdot \text{K}^{-1} \cdot \text{mol}^{-1})(298 \ \text{K}) \ln 5.0$$
$$= -4000 \ \text{J} \cdot \text{mol}^{-1} = -4.0 \ \text{kJ} \cdot \text{mol}^{-1}$$

Table 12.1 Biochemical Standard State

Temperature	25°C (298 K)
Pressure	1 atm
Reactant concentration	1 M
pH	7.0 ($[H^+] = 10^{-7}$ M)
Water concentration	55.5 M

By convention, measurements of standard free energy are valid under **standard conditions,** where the temperature is 25°C (298 K) and the pressure is 1 atm (these conditions are indicated by the degree symbol after ΔG). For a chemist, standard conditions specify an initial activity of 1 for each reactant (activity is the reactant's concentration corrected for its nonideal behavior). However, these conditions are impractical for biochemists since most biochemical reactions occur near neutral pH (where $[H^+] = 10^{-7}$ M rather than 1 M) and in aqueous solution (where $[H_2O] = 55.5$ M). The biochemical standard conditions are summarized in **Table 12.1.** Biochemists use a prime symbol to indicate the standard free energy change for a reaction under biochemical standard conditions.

In most equilibrium expressions, $[H^+]$ and $[H_2O]$ are set to 1 so that these terms can be ignored. Because biochemical reactions typically involve dilute solutions of reactants, molar concentrations can be used instead of activities.

Like K_{eq}, $\Delta G^{\circ\prime}$ is a constant for a particular reaction. It may be a positive or negative value, and it indicates whether the reaction can proceed spontaneously ($\Delta G^{\circ\prime} < 0$) or not ($\Delta G^{\circ\prime} > 0$) under standard conditions. In a living cell, reactants and products are almost never present at standard-state concentrations and the temperature may not be 25°C, yet reactions do occur with some change in free energy. Thus, it is important to distinguish the standard free energy change of a reaction from its actual free energy change, ΔG. ΔG is a function of the actual concentrations of the reactants and the temperature (37°C or 310 K in humans). ΔG is related to the standard free energy change for the reaction:

$$\Delta G = \Delta G^{\circ\prime} + RT \ln \frac{[C][D]}{[A][B]} \qquad (12.3)$$

Here, the bracketed quantities represent the actual, nonequilibrium concentrations of the reactants. The concentration term in Equation 12.3 is sometimes called the **mass action ratio.**

When the reaction is at equilibrium, $\Delta G = 0$ and

$$\Delta G^{\circ\prime} = -RT \ln \frac{[C]_{eq}[D]_{eq}}{[A]_{eq}[B]_{eq}} \qquad (12.4)$$

which is equivalent to Equation 12.2. Note that Equation 12.3 shows that *the criterion for spontaneity for a reaction is ΔG, a property of the actual concentrations of the reactants, not the constant $\Delta G^{\circ\prime}$.* Thus, a reaction with a positive standard free energy change (a reaction that cannot occur when the reactants are present at standard concentrations) may proceed *in vivo*, depending on the concentrations of reactants in the cell (see Sample Calculation 12.2). Keep in mind that

SAMPLE CALCULATION 12.2

Problem The standard free energy change for the reaction catalyzed by phosphoglucomutase is –7.1 kJ · mol⁻¹. Calculate the equilibrium constant for the reaction. Calculate ΔG at 37°C when the concentration of glucose-1-phosphate is 1 mM and the concentration of glucose-6-phosphate is 25 mM. Is the reaction spontaneous under these conditions?

Glucose-1-phosphate ⇌ Glucose-6-phosphate

Solution The equilibrium constant K_{eq} can be derived by rearranging Equation 12.2.

$$K_{eq} = e^{-\Delta G^{\circ\prime}/RT}$$
$$= e^{-(-7100\ J\cdot mol^{-1})/(8.3145\ J\cdot K^{-1}\cdot mol^{-1})(298\ K)}$$
$$= e^{2.87} = 17.6$$

At 37°C, $T = 310$ K.

$$\Delta G = \Delta G^{\circ\prime} + RT \ln \frac{[\text{glucose-6-phosphate}]}{[\text{glucose-1-phosphate}]}$$
$$= -7100\ J\cdot mol^{-1} +$$
$$(8.3145\ J\cdot K^{-1}\cdot mol^{-1})(310\ K)\ln(0.025/0.001)$$
$$= -7100\ J\cdot mol^{-1} + 8300\ J\cdot mol^{-1}$$
$$= 1200\ J\cdot mol^{-1} = 1.2\ kJ\cdot mol^{-1}$$

The reaction is not spontaneous because ΔG is greater than zero.

thermodynamic spontaneity does not imply a rapid reaction. Even a substance with a strong tendency to undergo reaction ($\Delta G \ll 0$) will usually not react until acted upon by an enzyme that catalyzes the reaction.

Unfavorable reactions are coupled to favorable reactions

A biochemical reaction may at first seem to be thermodynamically forbidden because its free energy change is greater than zero. Yet the reaction can proceed *in vivo* when it is coupled to a second reaction whose value of ΔG is very large and negative so that the *net* change in free energy for the combined reactions is less than zero. *ATP is often involved in such coupled processes because its reactions occur with a relatively large negative change in free energy.*

Adenosine triphosphate (ATP) contains two phosphoanhydride bonds (**Fig. 12.8**). Cleavage of either of these bonds—that is, transfer of one or two of its phosphoryl groups to another molecule—is a reaction with a large negative standard free energy change (under physiological conditions, ΔG is even more negative). As a reference point, biochemists use the reaction in which the third phosphoryl group is transferred to water—in other words, hydrolysis of one phosphoanhydride bond:

$$ATP + H_2O \rightarrow ADP + P_i$$

This is a spontaneous reaction with a $\Delta G^{\circ\prime}$ value of -30 kJ $\cdot$ mol^{-1}. Hydrolysis of ATP's other phosphoanhydride bond is even more favorable ($\Delta G^{\circ\prime} = -45.6$ kJ $\cdot$ mol^{-1}).

The following example illustrates the role of ATP in a coupled reaction. Consider the phosphorylation of glucose by inorganic phosphate (HPO_4^{2-} or P_i), a thermodynamically unfavorable reaction ($\Delta G^{\circ\prime} = +13.8$ kJ $\cdot$ mol^{-1}):

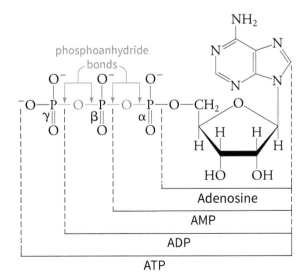

Glucose   Glucose-6-phosphate

Figure 12.8 Adenosine triphosphate.   The three phosphate groups are sometimes described by the Greek letters α, β, and γ. The linkage between the first (α) and second (β) phosphoryl groups, and between the second (β) and third (γ), is a phosphoanhydride bond.

Question **How does hydrolysis of a phosphoanhydride bond affect the net charge of a nucleotide?**

When this reaction is combined with the ATP hydrolysis reaction, the values of $\Delta G°'$ for each reaction are added:

$$
\begin{array}{lr}
 & \Delta G°' \\
\text{glucose} + P_i \rightarrow \text{glucose-6-phosphate} + H_2O & +13.8 \text{ kJ} \cdot \text{mol}^{-1} \\
\text{ATP} + H_2O \rightarrow \text{ADP} + P_i & -30.5 \text{ kJ} \cdot \text{mol}^{-1} \\
\hline
\text{glucose} + \text{ATP} \rightarrow \text{glucose-6-phosphate} + \text{ADP} & -16.7 \text{ kJ} \cdot \text{mol}^{-1}
\end{array}
$$

The net chemical reaction, the phosphorylation of glucose, is thermodynamically favorable ($\Delta G < 0$). *In vivo,* this reaction is catalyzed by hexokinase (introduced in Section 6.3), and a phosphoryl group is transferred from ATP directly to glucose. The ATP is not actually hydrolyzed and there is no free phosphoryl group floating around the enzyme. However, writing out the two coupled reactions, as shown above, makes it easier to see what is going on thermodynamically.

Some biochemical processes appear to occur with the concomitant hydrolysis of ATP to ADP + P_i, for example, the operation of myosin and kinesin (Section 5.4) or the Na,K-ATPase ion pump (Section 9.3). However, a closer look reveals that in all these processes, ATP actually transfers a phosphoryl group to a protein. Later, the phosphoryl group is transferred to water, so the net reaction takes the form of ATP hydrolysis. The same ATP "hydrolysis" effect applies to some reactions in which the AMP moiety of ATP (rather than a phosphoryl group) is transferred to a substance, leaving inorganic pyrophosphate (PP_i). Cleavage of the phosphoanhydride bond of PP_i also has a large negative value of $\Delta G°'$.

Because the ATP hydrolysis reaction appears to drive many thermodynamically unfavorable reactions, it is tempting to think of ATP as an agent that transfers packets of energy around the cell. This is one reason why ATP is commonly called the energy currency of the cell. The general role of ATP in linking exergonic ATP-producing processes to endergonic ATP-consuming processes can be diagrammed as

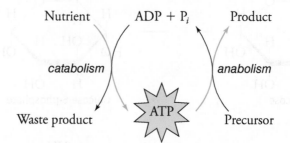

In this scheme, it appears that the "energy" of the catabolized nutrient is transferred to ATP and then the "energy" of ATP is transferred to another product in a biosynthetic reaction. However, energy is not a tangible item and there is nothing magic about ATP. The two phosphoanhydride bonds of ATP are sometimes called "high-energy" bonds, but they are no different from other covalent bonds. What matters is that *a reaction in which phosphoryl groups are transferred to another molecule—breaking these bonds—is a process with a large negative free energy change.* Using the simple example of ATP hydrolysis, we can state that a large amount of free energy is released when ATP is hydrolyzed because the products of the reaction have less free energy than the reactants. It is worth examining two reasons why this is so.

1. *The ATP hydrolysis products are more stable than the reactants.* At physiological pH, ATP has three to four negative charges (its pK is close to 7) and the anionic groups repel each other. In the products ADP and P_i, separation of the charges relieves some of this unfavorable electrostatic repulsion.

2. *A compound with a phosphoanhydride bond experiences less resonance stabilization than its hydrolysis products.* **Resonance stabilization** reflects the degree of electron delocalization in a molecule and can be roughly assessed by the number of different ways of depicting the molecule's structure. There are fewer equivalent ways of arranging the bonds of the terminal phosphoryl group of ATP than there are in free P_i:

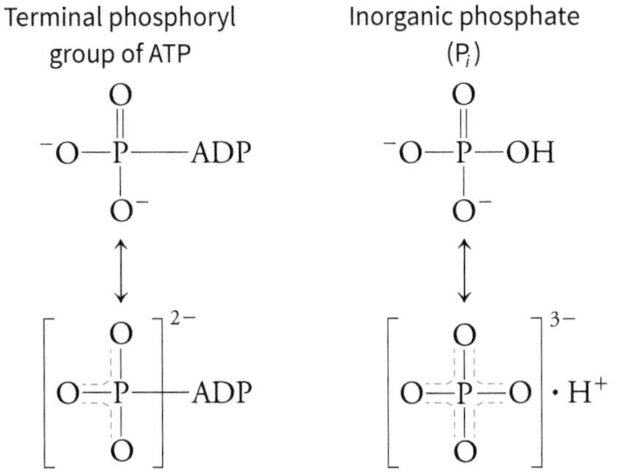

Terminal phosphoryl group of ATP

Inorganic phosphate (P_i)

To summarize, *ATP functions as an energy currency because its hydrolysis reaction is highly exergonic ($\Delta G \ll 0$).* The favorable ATP reaction (ATP → ADP) can therefore pull another, unfavorable reaction with it, provided that the sum of the free energy changes for both reactions is less than zero. In effect, the cell "spends" ATP to make another process happen.

Energy can take different forms

ATP is not the only substance that functions as energy currency in the cell. Other compounds that participate in reactions with large negative changes in free energy can serve the same purpose. For example, a number of phosphorylated compounds other than ATP can give up their phosphoryl group to another molecule. **Table 12.2** lists the standard free energy changes for some of these reactions in which the phosphoryl group is transferred to water.

Although hydrolysis of the bond linking the phosphate group to the rest of the molecule could be a wasteful process (the product would be free phosphate, P_i), the values listed in the table are a guide to how such compounds would behave in a coupled reaction, such as the hexokinase reaction described above. For example, phosphocreatine has a standard free energy of hydrolysis of -43.1 kJ·mol^{-1}:

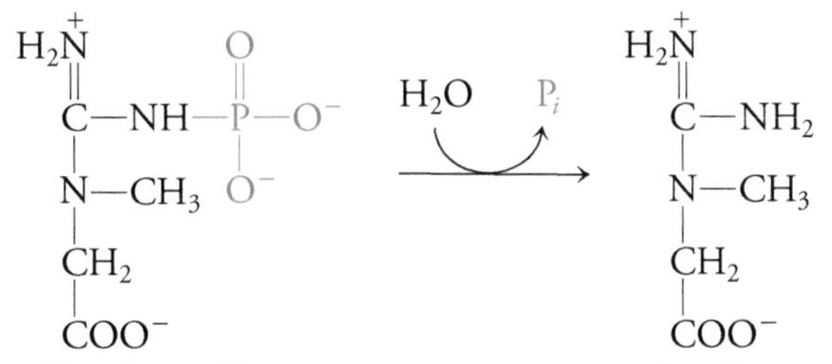

Phosphocreatine   Creatine

Creatine has lower energy than phosphocreatine since it has two, rather than one, resonance forms; this resonance stabilization contributes to the large negative free energy change when phosphocreatine transfers its phosphoryl group to another compound. In muscles, phosphocreatine transfers a phosphoryl group to ADP to boost the production of ATP during periods of high demand.

Like ATP, other nucleoside triphosphates have large negative standard free energies of hydrolysis. GTP rather than ATP serves as the energy currency for reactions that occur during cellular signaling (Section 10.2) and protein synthesis (Section 22.3). In the cell, nucleoside triphosphates are freely interconverted by reactions such as the one catalyzed by nucleoside diphosphate kinase, which transfers a phosphoryl group from ATP to a nucleoside diphosphate (NDP):

$$ATP + NDP \rightleftharpoons ADP + NTP$$

Table 12.2 Standard Free Energy Change for Phosphate Hydrolysis

Compound	$\Delta G^{\circ\prime}$ (kJ·mol^{-1})
Phosphoenolpyruvate	−61.9
1,3-Bisphosphoglycerate	−49.4
ATP → AMP + PP$_i$	−45.6
Phosphocreatine	−43.1
ATP → ADP + P$_i$	−30.5
Glucose-1-phosphate	−20.9
PP$_i$ → 2 P$_i$	−19.2
Glucose-6-phosphate	−13.8
Glycerol-3-phosphate	−9.2

Because the reactants and products are energetically equivalent, $\Delta G°'$ values for these reactions are near zero, and the reactions can proceed in either direction.

Another class of compounds that can release a large amount of energy upon hydrolysis are **thioesters,** such as acetyl-CoA. Coenzyme A is a nucleotide derivative with a side chain ending in a sulfhydryl (SH) group (see Fig. 3.2a). An acyl or acetyl group (the "A" for which coenzyme A was named) is linked to the sulfhydryl group by a thioester bond. Hydrolysis of this bond has a $\Delta G°'$ value of -31.5 kJ·mol^{-1}, comparable to that of ATP hydrolysis:

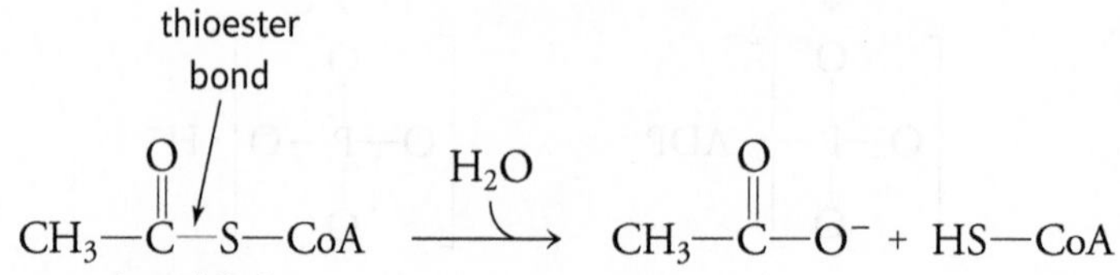

Hydrolysis of a thioester is more exergonic than the hydrolysis of an ordinary (oxygen) ester because thioesters have less resonance stability than oxygen esters, owing to the larger size of an S atom relative to an O atom. An acetyl group linked to coenzyme A can be readily transferred to another molecule because formation of the new linkage is powered by the favorable free energy change of breaking the thioester bond.

We have already seen that in oxidation–reduction reactions, cofactors such as NAD$^+$ and ubiquinone can collect electrons. The reduced cofactors are a form of energy currency because their subsequent reoxidation by another compound occurs with a negative change in free energy. Ultimately, the transfer of electrons from one reduced cofactor to another and finally to oxygen, the final electron acceptor in many organisms, releases enough energy to drive the synthesis of ATP.

Keep in mind that free energy changes occur not just as the result of chemical changes such as phosphoryl-group transfer or electron transfer. As decreed by the first law of thermodynamics (Section 1.4), *energy can take many forms.* We will see that ATP production in cells depends on the energy of an electrochemical gradient, that is, an imbalance in the concentration of a substance (in this case, protons) on the two sides of a membrane. The energy released in dissipating this gradient (allowing the system to move toward equilibrium) is converted to the mechanical energy of an enzyme that synthesizes ATP. In photosynthetic cells, the chemical reactions required to generate carbohydrates are ultimately driven by the energy changes of reactions in which light-excited molecules return to a lower-energy state.

Regulation occurs at the steps with the largest free energy changes

In a series of reactions that make up a metabolic pathway, some reactions have ΔG values near zero. These near-equilibrium reactions are not subject to a strong driving force to proceed in either direction. Rather, flux can go forward or backward, according to slight fluctuations in the concentrations of reactants and products. When the concentrations of metabolites change, the enzymes that catalyze these near-equilibrium reactions tend to act quickly to restore the near-equilibrium state.

Reactions with large negative changes in free energy have a longer way to go to reach equilibrium; these are the reactions that experience the greatest "urge" to proceed forward. However, the enzymes that catalyze these reactions do not allow the reaction to reach equilibrium because they work too slowly. Often the enzymes are already saturated with substrate, so the reactions cannot go any faster (when $[S] \gg K_M$, $v \approx V_{max}$; Section 7.2). The rates of these far-from-equilibrium reactions limit flux through the entire pathway because the reactions function like dams:

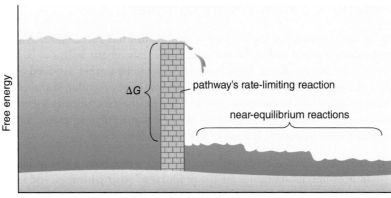

Cells can regulate flux through a pathway by adjusting the rate of a reaction with a large free energy change. This can be done by increasing the amount of enzyme that catalyzes that step or by altering the intrinsic activity of the enzyme through allosteric mechanisms (see Fig. 7.17). As soon as more metabolite has gone past the dam, the near-equilibrium reactions go with the flow, allowing the pathway intermediates to move toward the final product. Most metabolic pathways do not have a single flow-control point, as the dam analogy might suggest. Instead, flux is typically controlled at several points to ensure that the pathway can work efficiently as part of the cell's entire metabolic network.

Concept Check

1. Explain why free energy changes must be negative for reactions *in vivo*.
2. What is the relationship between kinases and phosphatases?
3. Relate the standard free energy change to a reaction's equilibrium constant.
4. Describe the difference between ΔG and $\Delta G^{\circ\prime}$.
5. Explain why it is misleading to refer to ATP as a high-energy molecule.
6. Explain why cleavage of one of ATP's phosphoanhydride bonds releases large amounts of free energy.
7. List the types of energy "currencies" used by cells.
8. Describe the reasons why cells control metabolic reactions that have large negative free energy changes.

12.3 Metabolic Pathways

KEY CONCEPTS

Recognize the common chemical features of metabolic pathways.

- Recognize common metabolites.
- Identify oxidized and reduced partners in chemical reactions.
- Explain why metabolic pathways are connected, regulated, and cell-specific.
- List some vitamins and their biochemical roles.

The interconversion of a biopolymer and its monomeric units is usually accomplished in just one or a few enzyme-catalyzed steps. In contrast, many steps are required to break down the monomeric compounds or build them up from smaller precursors. These series of reactions are known as **metabolic pathways.** A metabolic pathway can be considered from many viewpoints: as a series of **intermediates** or **metabolites,** as a set of enzymes that catalyze the reactions by which metabolites are interconverted, as an energy-producing or energy-requiring phenomenon, or as a dynamic process whose activity can be turned up or down. As we explore metabolic pathways in the coming chapters, we will use each of these perspectives.

Metabolic pathways are complex

So far we have sketched the outlines of mammalian fuel metabolism, in which macromolecules are stored and mobilized so that their monomeric units can be broken down into smaller intermediates. These intermediates can be further degraded (oxidized) and their electrons collected by cofactors. We have also briefly mentioned anabolic (synthetic) reactions in which the common two- and three-carbon intermediates give rise to larger compounds. At this point, we can present this information in schematic form in order to highlight some important features of metabolism (**Fig. 12.9**).

1. *Metabolic pathways are all connected.* In a cell, a metabolic pathway does not operate in isolation; its substrates are the products of other pathways, and vice versa. For example, the NADH and QH_2 generated by the citric acid cycle can be considered as the starting materials for oxidative phosphorylation.

2. *Pathway activity is regulated.* Cells do not synthesize polymers when monomers are in short supply. Conversely, they do not catabolize fuels when the need for ATP is low. The **flux,** or flow of intermediates, through a metabolic pathway is regulated in various ways according to substrate availability and the cell's need for the pathway's products. The activity of one or more enzymes in a pathway may be controlled by allosteric effectors (Sections 5.1 and 7.3). These changes in turn may reflect extracellular signals that activate intracellular kinases, phosphatases, and second messengers (Section 10.1). Regulation of pathways is especially critical when the simultaneous operation of two opposing processes, such as fatty acid synthesis and degradation, would be wasteful.

3. *Not every cell carries out every pathway.* Figure 12.9 is a composite of a number of metabolic processes, and a given cell or organism may undertake only a subset of these. Mammals do not perform photosynthesis, and only the liver and kidney can synthesize glucose from noncarbohydrate precursors.

4. *Each cell has a unique metabolic repertoire.* In addition to the pathways outlined in Figure 12.9, which are centered on fuel metabolism, cells carry out a plethora of biosynthetic reactions that are not explicitly shown. Such pathways contribute to the unique metabolic capabilities of different cells and organisms (**Box 12.A**).

5. *Organisms may be metabolically interdependent.* Photosynthetic plants and the heterotrophs that consume them are an obvious example of metabolic complementarity, but there are numerous other examples, especially in the microbial world. Certain organisms that release methane as a waste product live in close proximity to methanotrophic species (which consume CH_4 as a fuel); neither organism can survive without the other. Humans also exhibit interspecific cooperativity: Thousands of different microbial species, amounting to some 100 trillion cells, can live in or on the human body (the microbiota is introduced in Section 1.2). Collectively, these organisms express millions of different genes and carry out a correspondingly rich set of metabolic activities.

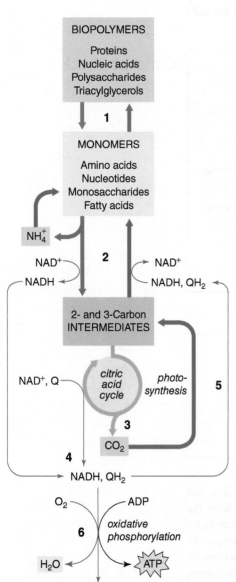

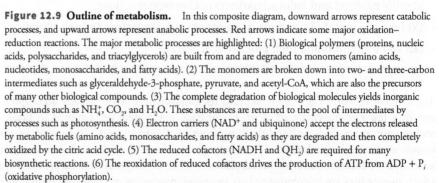

Figure 12.9 Outline of metabolism. In this composite diagram, downward arrows represent catabolic processes, and upward arrows represent anabolic processes. Red arrows indicate some major oxidation–reduction reactions. The major metabolic processes are highlighted: (1) Biological polymers (proteins, nucleic acids, polysaccharides, and triacylglycerols) are built from and are degraded to monomers (amino acids, nucleotides, monosaccharides, and fatty acids). (2) The monomers are broken down into two- and three-carbon intermediates such as glyceraldehyde-3-phosphate, pyruvate, and acetyl-CoA, which are also the precursors of many other biological compounds. (3) The complete degradation of biological molecules yields inorganic compounds such as NH_4^+, CO_2, and H_2O. These substances are returned to the pool of intermediates by processes such as photosynthesis. (4) Electron carriers (NAD^+ and ubiquinone) accept the electrons released by metabolic fuels (amino acids, monosaccharides, and fatty acids) as they are degraded and then completely oxidized by the citric acid cycle. (5) The reduced cofactors (NADH and QH_2) are required for many biosynthetic reactions. (6) The reoxidation of reduced cofactors drives the production of ATP from ADP + P_i (oxidative phosphorylation).

Question Is the citric acid cycle a process of carbon oxidation or reduction? Is photosynthesis a process of carbon oxidation or reduction?

Box 12.A   The Transcriptome, the Proteome, and the Metabolome

Modern biologists have developed research tools that use the power of computers to collect enormous data sets and analyze them. Such endeavors provide great insights but also have limitations. As we saw in Section 3.4, genomics, the study of an organism's complete set of genes, yields a glimpse of that organism's overall metabolic repertoire. However, what the organism, or a single cell, is actually doing at a particular moment depends in part on which genes are active.

A cell's population of mRNA molecules represents genes that are turned on, or transcribed. The study of these mRNAs is known as **transcriptomics.** Identifying and quantifying all the mRNA transcripts (the **transcriptome**) from a single cell type can be done by assembling short strands of DNA with known sequences on a solid support, then allowing them to hybridize, or form double-stranded structures, with fluorescent-labeled mRNAs from a cell preparation. The strength of fluorescence indicates how much mRNA binds to a particular complementary DNA sequence. The collection of DNA sequences is called a **microarray** or **DNA chip** because thousands of sequences fit in a few square centimeters. The microarray may represent an entire genome or just a few selected genes. Each bright spot in the DNA chip shown here represents a DNA sequence to which a fluorescent mRNA molecule has bound.

Volker Steger/Science Photo Library

Biologists use DNA chips to identify genes whose expression changes under certain conditions or at different developmental stages.

Unfortunately, the correlation between the amount of a particular mRNA and the amount of its protein product is not perfect; some mRNAs are rapidly degraded, whereas others are translated many times, yielding large quantities of the corresponding protein. Hence, a more reliable way to assess gene expression is through **proteomics**—by examining a cell's **proteome,** the complete set of proteins that are synthesized by the cell at a particular point in its life cycle. However, this approach is limited by the technical problems of detecting minute quantities of thousands of different proteins. Mass spectrometry, a popular technique (Section 4.6), necessarily breaks proteins into small fragments, which complicates analysis.

Where genomics, transcriptomics, and proteomics fall short, **metabolomics** steps in, attempting to pin down the actual metabolic activity in a cell or tissue by identifying and quantifying all its metabolites, that is, its **metabolome.** This is no trivial task, since a bacterial cell may contain thousands of different types of compounds, and an animal cell as many as one million. Moreover, metabolite concentrations range over many orders of magnitude, and some molecules have lifetimes of only a few seconds.

Researchers sometimes distinguish between primary and secondary metabolites. Primary metabolites belong to the central metabolic pathways that operate in most organisms, and secondary metabolites, which perform more specialized functions, are produced only in certain cells or certain types of organisms. In humans, endogenous metabolites (those produced by the body) are probably far outnumbered by exogenous metabolites from food and other sources, including toxins, preservatives, drugs, and their degradation products. Of course, many of the metabolites that can be detected in the human body originate from the microbiota. Odd compounds can also be generated by nonenzymatic processes such as the Maillard reaction (Box 11.A).

Metabolites are typically detected through nuclear magnetic resonance (NMR) spectroscopy or mass spectrometry. In the example shown, approximately 20 metabolites are visible in an ^1H NMR spectrum of a 10-μL sample of rat brain.

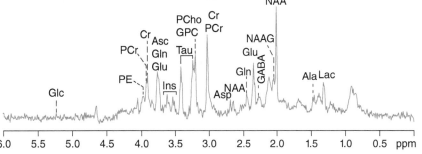

Metabolite profile of rat brain.

Source: Raghavendra Rao, University of Minnesota, Minneapolis.

In practice, metabolomics analysis often involves comparisons, such as before and after a drug treatment or normal versus diseased tissue. Consequently, only the substances whose levels fluctuate need to be identified.

As has been done for genomics and proteomics and other areas of bioinformatics, metabolomic data are deposited in publicly accessible databases for retrieval and analysis, for example, the Human Metabolome Database (hmdb.ca). One hope for metabolomics is that disease diagnosis could be streamlined by obtaining a complete metabolic profile of a patient's urine or blood. The unique "fingerprint" of a disease such as cancer could then be monitored to observe the patient's response to treatment. In one study of 226 circulating "biomarkers," high levels of 14 substances were correlated with a higher risk of death in the following few years. Industrial applications of metabolomics include monitoring biological processes such as winemaking and bioremediation (using microorganisms to detoxify contaminated environments).

The field of metabolomics also encompasses databases devoted to the relationships among metabolites, that is, the metabolic pathways in which the metabolites participate as intermediates. With hundreds of pathways, representing thousands of different organisms, such databases can be searched for specific compounds (including drugs), and links to enzyme and sequence databases can begin to paint a holistic view of the biochemical processes that occur in cells.

One ongoing challenge is that identifying the enzymes and measuring the concentrations of intermediates in a pathway does not indicate the flux through that pathway. Studying the enzymes' kinetics and regulation (as described in Chapter 7) can potentially reveal the pathway's importance to the organism. In a similar way, a map may indicate the locations of roads, but planning an efficient trip also requires information about traffic volumes and speed limits.

Question Compare the metabolomic complexity of a single-celled prokaryote and that of a multicellular eukaryote.

An overview such as Figure 12.9 does not convey the true complexity of cellular metabolism, which takes place in a milieu crowded with multiple substrates, competing enzymes, and layers of regulatory mechanisms. Moreover, Figure 12.9 does not include any of the reactions involved in transmitting and decoding genetic information (these topics are covered in the final section of this book). However, a diagram such as Figure 12.9 is a useful tool for mapping the relationships among metabolic processes, and we will refer back to it in the coming chapters. Online databases provide additional information about metabolic pathways, enzymes, intermediates, and metabolic diseases.

Many metabolic pathways include oxidation–reduction reactions

In general, *the catabolism of amino acids, monosaccharides, and fatty acids is a process of oxidizing carbon atoms, and the synthesis of these compounds involves carbon reduction.* Recall from Section 1.4 that **oxidation** is defined as the loss of electrons and **reduction** is the gain of electrons. Oxidation–reduction, or **redox,** reactions are like acid–base reactions: They involve pairs of reactants. As one compound becomes more oxidized (gives up electrons or loosens its hold on them), another compound becomes reduced (receives the electrons or tightens its grip on them).

In biological systems, unaccompanied electrons are rare. Keep in mind that electrons typically remain associated with molecules, usually as a pair of electrons that constitute a covalent bond. As in any chemical reaction—which is a matter of old bonds breaking and new bonds forming—we can see where electrons travel during a redox reaction by comparing the bonding arrangements in the reactant and product molecules.

For the metabolic reactions that we are concerned with, the oxidation of carbon atoms frequently appears as the replacement of C—H bonds (in which the C and H atoms share the bonding electrons equally) with C—O bonds (in which the more electronegative O atom "pulls" the electrons away from the carbon atom). Carbon has given up some of its electrons, even though the electrons are still participating in a covalent bond.

The transformation of methane to carbon dioxide represents the conversion of carbon from its most reduced state to its most oxidized state:

$$\text{H}-\underset{\underset{\text{H}}{|}}{\overset{\overset{\text{H}}{|}}{\text{C}}}-\text{H} \longrightarrow \text{O}=\text{C}=\text{O}$$

Similarly, oxidation occurs during the catabolism of a fatty acid, when saturated methylene ($-\text{CH}_2-$) groups are converted to CO_2 and when the carbons of a carbohydrate (represented as CH_2O) are converted to CO_2:

$$\text{H}-\overset{|}{\underset{|}{\text{C}}}-\text{H} \longrightarrow \text{O}=\text{C}=\text{O}$$

$$\text{H}-\overset{|}{\underset{|}{\text{C}}}-\text{OH} \longrightarrow \text{O}=\text{C}=\text{O}$$

The reverse of either of these processes—converting the carbons of CO_2 to the carbons of fatty acids or carbohydrates—is a reduction process (this is what occurs during photosynthesis, for example). In reduction processes, the carbon atoms regain electrons as C—O bonds are replaced by C—H bonds.

Turning CO_2 into carbohydrate (CH_2O) requires the input of energy (think: sunlight). Therefore, *the reduced carbons of the carbohydrate represent a form of stored energy.* This energy is recovered when cells break the carbohydrate back down to CO_2. Of course, such a metabolic conversion does not happen all at once but takes place through many enzyme-catalyzed steps.

In following metabolic pathways that include oxidation–reduction reactions, we can examine the redox state of the carbon atoms and we can also trace the path of the electrons that are transferred. In some cases, this is straightforward, as when an oxidized metal ion such as iron gains an electron (represented as e^-) to become reduced:

$$Fe^{3+} + e^- \rightarrow Fe^{2+}$$

However, in many cases, electron transfer means that one substance ends up with one less covalent bond and another substance ends up with one more bond (since each bond represents a pair of electrons). Sometimes the electrons travel along with protons as H atoms, or a pair of electrons travels with a proton as a hydride ion (H^-).

When a metabolic fuel molecule is oxidized, its electrons may be transferred to a compound such as nicotinamide adenine dinucleotide (NAD^+) or nicotinamide adenine dinucleotide phosphate ($NADP^+$). The structure of these nucleotides is shown in Figure 3.2b. NAD^+ and $NADP^+$ are called **cofactors** or **coenzymes,** organic compounds that allow an enzyme to carry out a particular chemical reaction (Section 6.2). The redox-active portion of NAD^+ and $NADP^+$ is the nicotinamide group, which accepts a hydride ion to form NADH or NADPH:

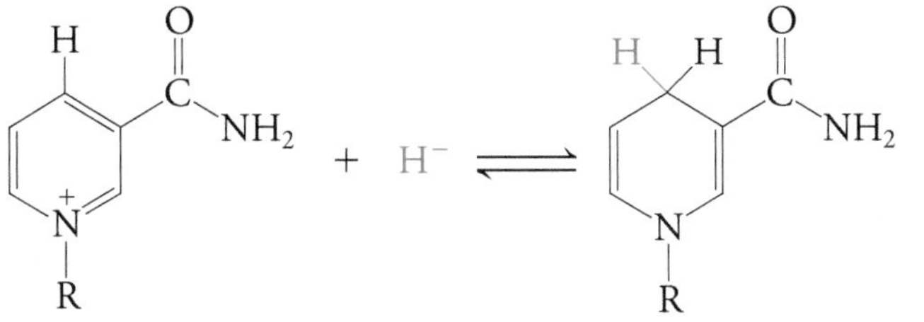

NAD(P)$^+$
(*oxidized*)

NAD(P)H
(*reduced*)

This reaction is reversible, so the reduced cofactors can become oxidized by giving up a hydride ion. In general, NAD^+ participates in catabolic reactions and $NADP^+$ in anabolic reactions. Because these electron carriers are soluble in aqueous solution, they can travel throughout the cell, shuttling electrons from reduced compounds to oxidized compounds.

Many cellular oxidation–reduction reactions take place at membrane surfaces, for example, in the inner membranes of mitochondria and chloroplasts in eukaryotes and in the plasma membrane of prokaryotes. In these cases, a membrane-associated enzyme may transfer electrons from a substrate to a lipid-soluble electron carrier such as ubiquinone (coenzyme Q, abbreviated Q; see Section 8.1). Ubiquinone's hydrophobic tail, containing 10 five-carbon isoprenoid units in mammals, allows it to diffuse within the membrane. *Ubiquinone can take up one or two electrons* (in contrast to NAD^+, which is strictly a two-electron carrier). A one-electron reduction of ubiquinone (addition of an H atom) produces a semiquinone, a stable free radical (shown as QH·). A two-electron reduction (two H atoms) yields ubiquinol (QH_2):

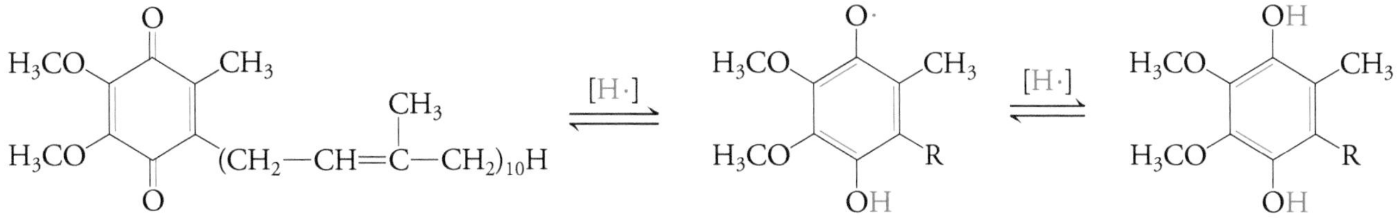

Ubiquinone (Q)

Ubisemiquinone (QH·)

Ubiquinol (QH_2)

The reduced ubiquinol can then diffuse within the membrane to donate its electrons in another oxidation–reduction reaction.

Catabolic pathways, such as the citric acid cycle, generate considerable amounts of reduced cofactors. Some of them are reoxidized in anabolic reactions. The rest are reoxidized by a process that is accompanied by the synthesis of ATP from ADP and P_i. In mammals, the reoxidation of NADH and QH_2 and the concomitant production of ATP require the reduction of O_2 to H_2O. This pathway is known as **oxidative phosphorylation.**

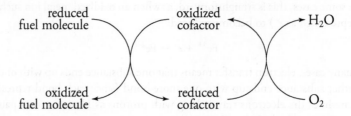

In effect, NAD⁺ and ubiquinone collect electrons (and hence energy) from reduced fuel molecules. When the electrons are ultimately transferred to O_2, this energy is harvested in the form of ATP.

Metabolic pathways may be anabolic, catabolic, anaplerotic, or cataplerotic

The living being undergone a complex web of metabolic events to sustain a life cycle known as metabolism. These responses can be divided into catabolic and anabolic pathways. Additionally, some pathways, such as anaplerotic pathways, play a significant role in replenishing intermediates, while other pathways, such as cataplerotic pathways, act as consumers of intermediates.

Anabolic pathways also known as biosynthetic pathways, are collections of chemical processes that create complex compounds out of simpler ones. Through this pathway, many large, complex molecules, such as proteins, nucleic acids, polysaccharides, and lipids, are produced, which require energy to form. An energy input is necessary for anabolic reactions to continue because they are endergonic processes. The classic examples of anabolic processes include protein synthesis (translation), DNA replication, fatty acid synthesis, and photosynthesis (in plants). Anabolic pathways are essential for the development and upkeep of living things. They make sure that cells are produced and that tissues are repaired. Without anabolic pathways, cells could not produce the intricate chemicals that are necessary for life activities. Catabolic pathways are metabolic processes that convert complex molecules into simpler ones. These reactions release energy and are exergonic processes. Catabolic pathways are essential to extract energy from nutrients and degrade cellular waste products. This energy is subsequently used by anabolic pathways to create new biomolecules. Glycolysis and the citric acid cycle are well-known examples.

Some metabolic pathways, like the citric acid cycle, have anaplerotic mechanisms that replenish their intermediates. The replenishment of these intermediates, frequently depleted during biosynthetic reactions, is made possible via anaplerotic pathways. The maintenance of the flow of carbon via the core metabolic pathways depends on anaplerotic pathways. For instance, the citric acid cycle's oxaloacetate is replenished through pyruvate carboxylation. Cataplerotic pathways, on the other hand, are metabolic procedures that eliminate intermediates from primary metabolic pathways. They serve as drains, removing intermediates for use in other processes, such as biosynthetic reactions. As an illustration, consider how phosphoenolpyruvate carboxykinase (see pages 366 and 367) changes oxaloacetate into phosphoenolpyruvate.

Common intermediates in major metabolic pathways

One of the challenges of studying metabolism is dealing with the large number of reactions that occur in a cell—involving thousands of different intermediates. However, *a handful of metabolites appear as precursors or products in the pathways that lead to or from virtually all other types of biomolecules.* These intermediates are worth examining at this point, since they will reappear several times in the coming chapters.

In **glycolysis,** the pathway that degrades the monosaccharide glucose, the six-carbon sugar is phosphorylated and split in half, yielding two molecules of glyceraldehyde-3-phosphate (**Fig. 12.10**). This compound is then converted in several more steps to another three-carbon molecule, pyruvate. The decarboxylation of pyruvate (removal of a carbon atom as CO_2) yields acetyl-CoA, in which a two-carbon acetyl group is linked to the carrier molecule coenzyme A (CoA; Fig. 3.2a).

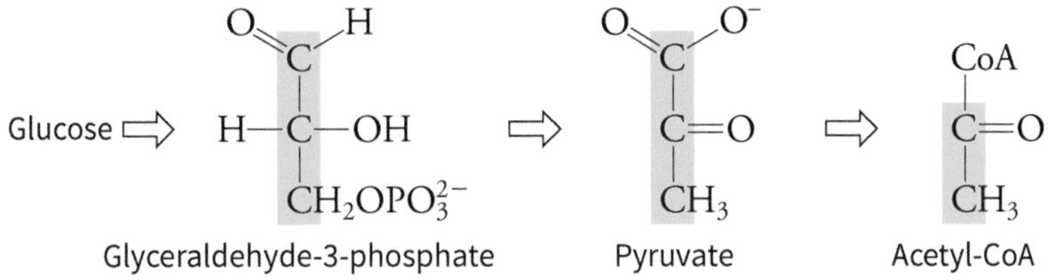

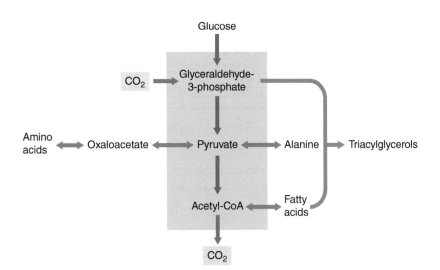

Figure 12.10 Some intermediates resulting from glucose catabolism.

Question **Compare the oxidation states of the carbons in glyceraldehyde-3-phosphate and in pyruvate.**

Glyceraldehyde-3-phosphate, pyruvate, and acetyl-CoA are key players in other metabolic pathways. For example, glyceraldehyde-3-phosphate is the metabolic precursor of the three-carbon glycerol backbone of triacylglycerols. In plants, it is a product of carbon "fixation" during photosynthesis; in this case, two molecules of glyceraldehyde-3-phosphate combine to form a six-carbon monosaccharide. Pyruvate can undergo a reversible amino-group transfer reaction to yield alanine:

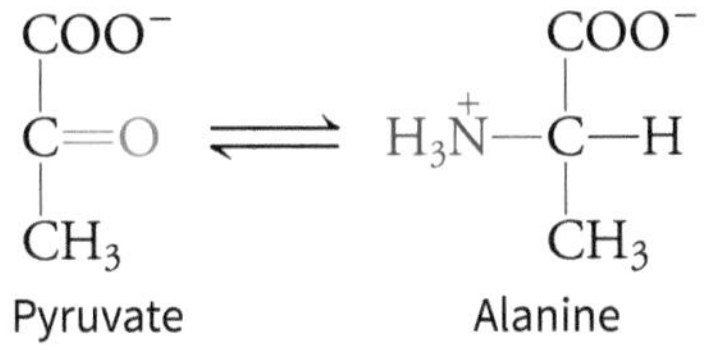

This makes pyruvate both a precursor of an amino acid and the degradation product of one. Pyruvate can also be carboxylated to yield oxaloacetate, a four-carbon precursor of several other amino acids:

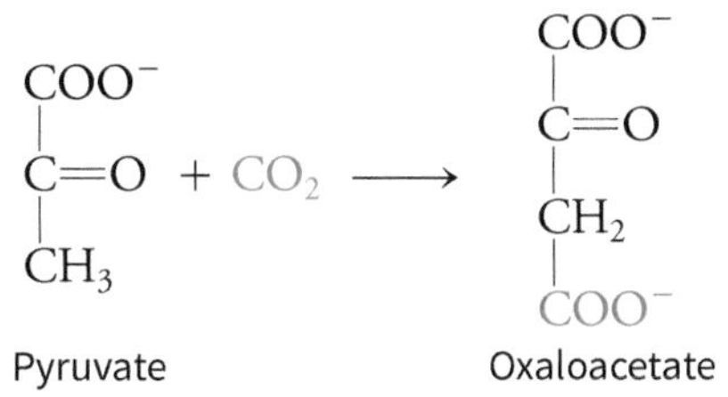

Fatty acids are built by the sequential addition of two-carbon units derived from acetyl-CoA; fatty acid breakdown yields acetyl-CoA. These relationships are summarized in **Figure 12.11**. If not used to synthesize other compounds, two-carbon intermediates can be broken down to CO_2 by the **citric acid cycle,** a metabolic pathway essential for the catabolism of all metabolic fuels.

Figure 12.11 Some of the metabolic roles of the common intermediates.

Question Without looking at the text, draw the structures of glyceraldehyde-3-phosphate, pyruvate, and oxaloacetate.

Human metabolism depends on vitamins

Humans and other mammals lack many of the biosynthetic pathways that occur in plants and microorganisms and so rely on other species to provide certain raw materials. Some amino acids and unsaturated fatty acids are considered **essential** because the human body cannot synthesize them and must obtain them from food (**Table 12.3**). **Vitamins** likewise are compounds that humans need but cannot make. Presumably, the pathways for synthesizing these substances, which require many specialized enzymes, are not necessary for heterotrophic organisms and have been lost through evolution.

The word *vitamin* comes from *vital amine,* a term coined by Casimir Funk in 1912 to describe organic compounds that are required in small amounts for normal health. It turns out that most vitamins are not amines, but the name has stuck. **Table 12.4** lists the vitamins and their metabolic roles. Vitamins A, D, E, and K are lipids; their functions were described in Box 8.B. Many of the water-soluble vitamins are the precursors of coenzymes, which we will describe as we encounter them in the context of their particular metabolic reactions. Vitamins are a diverse group of compounds, whose discoveries and functional characterization have provided some of the more colorful stories in the history of biochemistry.

Many vitamins were discovered through studies of nutritional deficiencies. One of the earliest links between nutrition and disease was observed centuries ago in sailors suffering from scurvy, an illness caused by vitamin C deficiency (Box 5.B). A study of the disease beriberi led to the discovery of the first B vitamin. Beriberi, characterized by leg weakness and swelling, is caused by a deficiency of thiamine (vitamin B_1).

Thiamine (vitamin B_1)

Thiamine acts as a prosthetic group in some essential enzymes, including the one that converts pyruvate to acetyl-CoA. Rice husks are rich in thiamine, and individuals whose diet consists largely of polished (huskless) rice can develop beriberi. The disease was originally thought to be infectious, until the same symptoms were observed in chickens and prisoners fed a diet of polished rice. Thiamine deficiency can occur in chronic alcoholics and others with a limited diet and impaired nutrient absorption.

Niacin, a component of NAD^+ and $NADP^+$, was first identified as the factor missing in the vitamin-deficiency disease pellagra.

Table 12.3 Some Essential Substances for Humans

Amino acids	Fatty acids		Other	
Histidine	Linoleate	$CH_3(CH_2)_4(CH=CHCH_2)_2(CH_2)_6COO^-$	Choline	$(CH_3)_3N^+CH_2CH_2OH$
Isoleucine	Linolenate	$CH_3CH_2(CH=CHCH_2)_3(CH_2)_6COO^-$		
Leucine				
Lysine				
Methionine				
Phenylalanine				
Threonine				
Tryptophan				
Valine				

Table 12.4 Vitamins and Their Roles

Vitamin	Coenzyme product	Biochemical function	Human deficiency disease	Text reference
Water-Soluble				
Ascorbic acid (C)	Ascorbate	Cofactor for hydroxylation of collagen	Scurvy	Box 5.B
Biotin (B_7)	Biocytin	Cofactor for carboxylation reactions	*	Section 13.1
Cobalamin (B_{12})	Cobalamin coenzymes	Cofactor for alkylation reactions	Anemia	Section 17.2
Folic acid	Tetrahydrofolate	Cofactor for one-carbon transfer reactions	Anemia	Section 18.2
Lipoic acid	Lipoamide	Cofactor for acyl transfer reactions	*	Section 14.1
Nicotinamide (niacin, B_3)	Nicotinamide coenzymes (NAD^+, $NADP^+$)	Cofactor for oxidation–reduction reactions	Pellagra	Fig. 3.2, Section 12.2
Pantothenic acid (B_5)	Coenzyme A	Cofactor for acyl transfer reactions	*	Fig. 3.2, Section 12.3
Pyridoxine (B_6)	Pyridoxal phosphate	Cofactor for amino-group transfer reactions	*	Section 18.1
Riboflavin (B_2)	Flavin coenzymes (FAD, FMN)	Cofactor for oxidation–reduction reactions	*	Fig. 3.2
Thiamine (B_1)	Thiamine pyrophosphate	Cofactor for aldehyde transfer reactions	Beriberi	Sections 12.2, 14.1
Fat-Soluble				
Vitamin A (retinol)		Light-absorbing pigment	Blindness	Box 8.B
Vitamin D		Hormone that promotes Ca^{2+} absorption	Rickets	Box 8.B
Vitamin E (tocopherol)		Antioxidant	*	Box 8.B
Vitamin K (phylloquinone)		Cofactor for carboxylation of blood coagulation proteins	Bleeding	Box 8.B

*Deficiency in humans is rare or unobserved.

Niacin

The symptoms of pellagra, including diarrhea and dermatitis, can be alleviated by boosting the intake of the essential amino acid tryptophan, which humans can convert to niacin. Niacin deficiency was once common in certain populations whose diet consisted largely of maize (corn). This grain is low in tryptophan and its niacin is covalently bound to other molecules; hence, it is not easily absorbed during digestion. In South America, where maize originated, the kernels are traditionally prepared by soaking or boiling them in an alkaline solution, a treatment that releases niacin and prevents pellagra. Unfortunately, this food-preparation custom did not spread to other parts of the world that adopted maize farming.

Most vitamins are readily obtained from a balanced diet, although poor nutrition, particularly in impoverished parts of the world, still causes vitamin-deficiency diseases. Intestinal bacteria, as well as plant- and animal-derived foods, are the natural sources of vitamins. However, plants do not contain cobalamin, so individuals who follow a strict vegetarian diet are at higher risk for developing a cobalamin deficiency. In addition to the cobalt that is part of cobalamin, animals require small amounts of several other metals, including copper, iron, manganese, molybdenum, and zinc (**Box 12.B**).

Box 12.B	Iron Metabolism

Like macronutrients (proteins, carbohydrates, and fats), the micro-nutrient metals can be studied in terms of their acquisition, storage, and transport. Here we focus on the biochemistry of iron. Although abundant on Earth, iron is precious in biological systems. It plays a part in almost every aspect of metabolism, but it is not always read-ily available. Humans obtain iron from heme groups and iron-con-taining proteins that are present in food. Iron from nonbiological sources is mostly ferric (Fe^{3+}) iron, which is insoluble in body fluids at neutral pH. Redox enzymes on the surface of intestinal cells must reduce Fe^{3+} to the more soluble ferrous (Fe^{2+}) iron for absorption. However, Fe^{2+} can react with O_2 to generate toxic free radicals, so it must be handled carefully.

Inside cells, free iron is stored by a protein called ferritin, whose 24 subunits form a hollow sphere. Ferritin oxidizes the Fe^{2+} to Fe^{3+} for storage, so the iron ends up more or less as a lump containing as many as 4500 iron ions as well as phosphate and hydroxide ions. The iron can be mobilized when the ferritin cage is degraded by autophagy. Spe-cialized chaperones may escort the released Fe ions to specific proteins that require iron for their function.

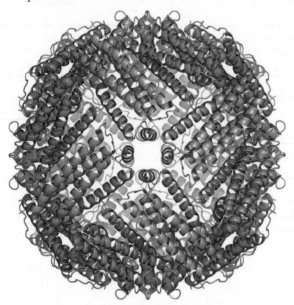

Ferritin with 24 bound Fe^{3+} ions (orange).

Iron can exit the cell through the membrane protein ferroportin, at which point it is loaded onto a soluble protein called transferrin in order to travel through the bloodstream. Macrophages in the spleen and liver (which perform all kinds of clean-up jobs) are particularly adept at extracting the iron from worn-out red blood cells and packaging it with transferrin for recycling. Trans-ferrin binds two iron ions (as Fe^{3+}), along with carbonate (CO_3^{2-}) ions, although it is normally only about 20–40% saturated with iron. This factor, plus the extraordinarily high binding affinity (the transferrin–Fe dissociation constant is about 10^{-20} M) means that transferrin can whisk away any iron ions that make their way into the bloodstream, leaving virtually no free iron ions floating around the body.

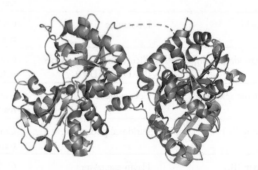

Transferrin with two bound Fe^{3+} ions.

Cells need iron, so they display transferrin receptors. Cells that are developing into new erythrocytes express high levels of the trans-ferrin receptor in order to acquire the iron needed for heme and hemoglobin synthesis. Following receptor–transferrin endocytosis (Section 9.4), the endosomal vesicle becomes acidic, which causes the carrier protein to release the iron. The Fe^{3+} must be reduced to Fe^{2+} before it can be used by the cell. All of the steps described above are regulated by feedback mechanisms that adjust the production of ferritin, ferroportin, transferrin, and other proteins in order to balance iron supply and demand.

Because microbes also need iron, sequestering iron (making it even less available) is part of the body's response to infections. Less iron is absorbed from food, transferrin levels drop, and more iron accumulates in ferritin inside cells. This may lead to anemia, the price of preventing the growth of pathogens. To evade the body's defensive moves, some bacteria are equipped with transferrin receptors. Other bacteria produce small molecules known as siderophores, which bind iron even more avidly than transferrin (with dissociation constants as low as 10^{-49} M!). In an evolutionary arms race, the body fights back by releasing siderocalins, small proteins that snatch up siderophores to starve the bacteria of essential iron.

Concept Check

1. Without looking at the text, draw a diagram showing the metabolic relationships of glyceraldehyde-3-phosphate, pyruvate, and acetyl-CoA.

2. Explain how cofactors such as NAD^+ and ubiquinone participate in metabolic reactions.

3. Explain the importance of reoxidizing NADH and QH_2 by molecular oxygen.

4. Summarize the main features of metabolic pathways.

5. Explain the relationship between vitamins and coenzymes.

6. Give examples of well-known catabolic pathways and explain their significance.

SUMMARY

12.1 Food and Fuel

- Polymeric food molecules such as starch, proteins, and triacylglycerols are broken down to their monomeric components (glucose, amino acids, and fatty acids), which are absorbed. These materials are stored as polymers in a tissue-specific manner.

- Metabolic fuels are mobilized from glycogen, fat, and proteins as needed.

12.2 Free Energy Changes in Metabolic Reactions

- The standard free energy change for a reaction is related to the equilibrium constant, but the actual free energy change also depends on the actual cellular concentrations of reactants and products.

- A thermodynamically unfavorable reaction may proceed when it is coupled to a favorable process involving ATP, whose phosphoanhydride bonds release a large amount of energy when cleaved.

- Other forms of cellular energy currency include phosphorylated compounds, thioesters, and reduced cofactors.

- Cells regulate metabolic pathways at the steps that are farthest from equilibrium.

12.3 Metabolic Pathways

- Series of reactions known as metabolic pathways break down and synthesize biological molecules. Several pathways make use of the same small molecule intermediates.

- During the oxidation of amino acids, monosaccharides, and fatty acids, electrons are transferred to carriers such as NAD^+ and ubiquinone. Reoxidation of the reduced cofactors drives the synthesis of ATP by oxidative phosphorylation.

- Metabolic pathways form a complex network, but not all cells or organisms carry out all possible metabolic processes. Humans rely on other organisms to supply vitamins and other essential materials.

KEY TERMS

catabolism	mobilization	standard free energy	intermediate	proteome	oxidative
anabolism	phosphorolysis	change ($\Delta G°'$)	metabolite	metabolomics	phosphorylation
metabolism	diabetes mellitus	standard conditions	flux	metabolome	glycolysis
chemoautotroph	lysosome	mass action ratio	transcriptomics	oxidation	citric acid cycle
photoautotroph	proteasome	resonance	transcriptome	reduction	essential compound
heterotroph	degron	stabilization	microarray (DNA	redox reaction	vitamin
lipoprotein	equilibrium constant	thioester	chip)	cofactor	
adipose tissue	(K_{eq})	metabolic pathway	proteomics	coenzyme	
metabolic fuel					

BIOINFORMATICS

Brief Bioinformatics Exercises

12.1 Metabolism and the BRENDA and KEGG Databases

12.2 Cofactor Chemistry

Bioinformatics Projects

Metabolic Enzymes, Microarrays, and Proteomics

Metabolomics Databases and Tools

PROBLEMS

12.1 Food and Fuel

1. Classify the following organisms as chemoautotrophs, photoautotrophs, or heterotrophs: **a.** *Hydrogenobacter*, which converts molecular hydrogen and oxygen to water, **b.** *Arabidopsis thaliana*, a green plant, **c.** the nitrosifying bacteria, which oxidize NH_3 to nitrite, **d.** *Saccharomyces cerevisiae*, yeast, **e.** *Caenorhabditis elegans*, a nematode worm, **f.** the *Thiothrix* bacteria, which oxidize hydrogen sulfide, **g.** Cyanobacteria (erroneously termed "blue-green algae" in the past).

2. The purple nonsulfur bacteria obtain their cellular energy from a photosynthetic process that does not produce oxygen. These bacteria

also require an organic carbon source. Using the terms in this chapter, coin a new term that describes the trophic strategy of this organism.

3. Digestion of carbohydrates begins in the mouth, where salivary amylases act on dietary starch. When the food is swallowed and enters the stomach, carbohydrate digestion ceases (it resumes in the small intestine). Why does carbohydrate digestion not occur in the stomach?

4. When complex carbohydrates like starch are digested, they yield monosaccharides. Explain how monosaccharides are transported into the cells lining the intestine for absorption and what provides the energy for this process.

5. Pancreatic amylase, which is similar to salivary amylase, is secreted by the pancreas into the small intestine. The active site of pancreatic amylase accommodates five glucosyl residues and hydrolyzes the glycosidic bond between the second and third residues. The enzyme cannot accommodate branched chains. **a.** What type of glycosidic bond is hydrolyzed by amylase? **b.** What are the main products of amylase digestion? **c.** What are the products of amylopectin digestion?

6. Starch digestion is completed by the enzymes isomaltase (or α-dextrinase), which catalyzes the hydrolysis of α(1→6) glycosidic bonds, and maltase, which hydrolyzes α(1→4) bonds. Why are these enzymes needed in addition to α-amylase?

7. White flour is highly processed to remove the germ and the bran from the grain, leaving the starch-rich endosperm. Whole-wheat flour is less processed and retains all of the elements of the whole grain. The starch in white bread is digested more quickly and efficiently than the starch in whole-wheat bread. Explain the reason for this observation.

8. The K_M and k_{cat} values for isomaltase and maltase (see Problem 6) are shown in the table. Calculate the catalytic efficiency (see Section 7.2) of each enzyme. Considering only these parameters, which enzyme would make the greater contribution to starch digestion in the small intestine?

	Maltase	Isomaltase
K_M (mM^{-1})	0.4	61
k_{cat} (s^{-1})	63	3.4

9. You carry out an experiment in which you determine that isomaltase makes a greater contribution to starch digestion than maltase. Assuming that your results are correct, how can you explain these results along with the data in Solution 8?

10. Some compounds, like sugar alcohols, are absorbed differently in the digestive system than monosaccharides. Discuss the absorption mechanism for sugar alcohols and compare it to the monosaccharide mechanism. Which process occurs more rapidly, and why?

11. Alcohol (ethanol) absorption varies depending on the presence of food. Explain how ethanol is absorbed in both the stomach and the small intestine and how the presence of food affects its absorption.

12. Monosaccharides, the products of polysaccharide and disaccharide digestion, enter the cells lining the intestine via a specialized transport system. What is the source of free energy for this transport process?

13. Unlike the monosaccharides described in Problem 12, sugar alcohols such as sorbitol (see Solution 11.21c) are absorbed via passive diffusion. Why? Which process occurs more rapidly, passive diffusion or passive transport?

14. Use what you know about the properties of alcohol (ethanol) to describe how it is absorbed in both the stomach and the small intestine. What effect does the presence of food have on the absorption of ethanol?

15. Nucleic acids that are present in food are hydrolyzed by digestive enzymes. What type of mechanism most likely mediates the entry of the reaction products into intestinal cells?

16. Hydrolysis of proteins begins in the stomach, catalyzed by the hydrochloric acid secreted into the stomach by parietal cells. **a.** How does the low pH of the stomach affect protein structure in such a way that the proteins are prepared for hydrolytic digestion? **b.** Draw the reaction that shows the hydrolysis of a peptide bond catalyzed by hydrochloric acid.

17. In addition to hydrochloric acid, the gastric protease pepsin catalyzes the hydrolysis of peptide bonds in dietary proteins. Like the serine proteases (see Section 6.4), pepsin is synthesized as a zymogen and is inactive at its site of synthesis, where the pH is 7. Pepsin becomes activated when secreted into the stomach, where it encounters a pH of ~2.

Pepsinogen contains a "basic peptide" that blocks its active site at pH 7. The basic peptide dissociates from the active site at pH 2 and is cleaved, resulting in the formation of the active form of the enzyme. What amino acid residues are found in the active site of pepsin? Why does the basic peptide bind tightly to the active site at pH 7 and why does it dissociate at the lower pH?

18. The hydrolysis of peptide bonds in the stomach is catalyzed by both hydrochloric acid (see Problem 16) and the gastric protease pepsin (see Problem 17). Peptide bond hydrolysis continues in the small intestine, catalyzed by the pancreatic enzymes trypsin and chymotrypsin. At what pH does pepsin function optimally; that is, at what pH is the V_{max} for pepsin greatest? Is the pH optimum for pepsin different from that for trypsin and chymotrypsin? Explain.

19. Scientists have recently discovered why the botulinum toxin survives the acidic environment of the stomach. The toxin forms a complex with a second nontoxic protein that acts as a shield to protect the botulinum toxin from being digested by stomach enzymes. Upon entry into the small intestine, the two proteins dissociate and the botulinum toxin is released. What is the likely interaction between the botulinum toxin and the nontoxic protein, and why does the complex form readily in the stomach but not in the small intestine?

20. Free amino acid transport from the intestinal lumen into intestinal cells requires Na$^+$ ions. Draw a diagram that illustrates amino acid transport into these cells.

21. In oral rehydration therapy (ORT), patients suffering from diarrhea are given a solution consisting of a mixture of glucose and electrolytes. Some formulations also contain amino acids. Why are electrolytes added to the mixture?

22. Triacylglycerol digestion begins in the stomach. Gastric lipase catalyzes hydrolysis of the fatty acid from the third glycerol carbon. **a.** Draw the reactants and products of this reaction. **b.** Conversion of the triacylglycerol to a diacylglycerol and a fatty acid promotes emulsification of fats in the stomach; that is, the products are more easily incorporated into micelles. Explain why.

23. Most of the fatty acids produced in the reaction described in Problem 22 form micelles and are absorbed as such, but a small percentage of fatty acids are free and are transported into the intestinal epithelial cells without the need for a transport protein. Explain why a transport protein is not required.

24. The cells lining the small intestine absorb cholesterol but not cholesteryl esters. Draw the reaction catalyzed by cholesteryl esterase that produces cholesterol from cholesteryl stearate.

25. Some cholesterol is converted back to cholesteryl esters in the epithelial cells lining the small intestine (the reverse of the reaction described in Problem 24). Both cholesterol and cholesteryl esters are packaged into particles called chylomicrons, which consist of lipid and protein. Use what you know about the physical properties of cholesterol and cholesteryl esters to describe their locations in the chylomicron particle.

26. **a.** Consider the physical properties of a polar glycogen molecule and an aggregation of hydrophobic triacylglycerols. On a per-weight basis, why is fat a more efficient form of energy storage than glycogen? **b.** Explain why there is an upper limit to the size of a glycogen molecule but there is no upper limit to the amount of triacylglycerols that an adipocyte can store.

27. Glycogen can be expanded quickly, by adding glucose residues to its many branches, and degraded quickly, by simultaneously removing glucose from the ends of these branches. Are the enzymes that catalyze these processes specific for the reducing or nonreducing ends of the glycogen polymer? Explain.

28. The phosphorolysis reaction that removes glucose residues from glycogen in the liver yields as its product glucose-1-phosphate. Glucose-1-phosphate is isomerized to glucose-6-phosphate; then the phosphate group is removed in a hydrolysis reaction. Why is it necessary to remove the phosphate group before the glucose exits the cell to enter the circulation?

29. Hydrolytic enzymes encased within the membrane-bound lysosomes all work optimally at pH ~5. This feature serves as a cellular "insurance policy" in the event of lysosomal enzyme leakage into the cytosol. Explain.

30. Proteins destined for degradation by the proteasome are "tagged" by covalently linking the C-terminus of ubiquitin to a Lys side chain on the doomed protein. **a.** Draw the structure of this linkage. What type of linkage is this? **b.** Linkages between ubiquitin and other amino acids, such as Ser residues, have been discovered. Draw the structure of this linkage. What type of linkage is this?

31. a. One protease active site in the proteasome has an Asp-His-Ser catalytic triad. What type of enzyme is this? **b.** Proteasomal proteases can be inhibited by compounds such as ixazomib (shown below). How does this compound inhibit protease activity? **c.** Ixazomib can be used as an anticancer drug. Why would inhibiting proteasomal proteases be detrimental to the cancer cell?

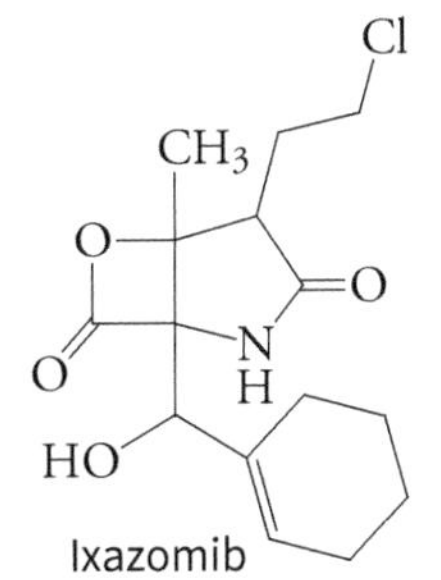

Ixazomib

12.2 Free Energy Changes in Metabolic Reactions

32. a. Calculate $\Delta G^{\circ\prime}$ for a reaction at 25°C when $K_{eq} = 4$. **b.** If the reaction were carried out at 37°C, how would $\Delta G^{\circ\prime}$ change? **c.** Is the reaction spontaneous?

33. Consider two reactions: $A \rightleftharpoons B$ and $C \rightleftharpoons D$. K_{eq} for the $A \rightleftharpoons B$ reaction is 10, and K_{eq} for the $C \rightleftharpoons D$ reaction is 0.1. You place 1 mM A in tube 1 and 1 mM C in tube 2 and allow the reactions to reach equilibrium. **a.** Without doing any calculations, determine whether the concentration of B in Tube 1 will be greater than or less than the concentration of D in tube 2. **b.** Calculate the $\Delta G^{\circ\prime}$ values for the reactions. Assume a temperature of 37°C.

34. Consider the reaction $E \rightleftharpoons F$, $K_{eq} = 1$. **a.** Without doing any calculations, what can you conclude about the $\Delta G^{\circ\prime}$ value for the reaction? **b.** You place 1 mM F in a tube and allow the reaction to reach equilibrium. Determine the final concentrations of E and F. **c.** Determine the direction the reaction will proceed if you place 5 mM E and 2 mM F in a test tube. What are the final concentrations of E and F?

35. For the reaction $R \rightleftharpoons P$, $\Delta G^{\circ\prime} = 4.1$ kJ · mol⁻¹. Calculate K_{eq} for the reaction.

36. a. Calculate K_{eq} for the reverse of the reaction described in Problem 35. **b.** Compare your answer with your answer to Problem 35. What is the relationship between the sign of $\Delta G^{\circ\prime}$ and the ratio of products to reactants at equilibrium?

37. Calculate ΔG for the $A \rightleftharpoons B$ reaction described in Problem 33 when the concentrations of A and B are 0.8 mM and 0.2 mM, respectively. In which direction will the reaction proceed? Assume a temperature of 37°C.

38. Calculate ΔG for the $C \rightleftharpoons D$ reaction described in Problem 33 when the concentrations of C and D are 0.8 mM and 0.2 mM, respectively. In which direction will the reaction proceed?

39. Calculate ΔG for the $E \rightleftharpoons F$ reaction described in Problem 34 under the conditions described in Problem 34. Assume a temperature of 37°C. Is the reaction spontaneous? Is this consistent with your answer to Problem 34?

40. Calculate ΔG for the $R \rightleftharpoons P$ reaction described in Problem 35 when the concentrations of R and P are 0.9 mM and 0.1 mM, respectively. Assume a temperature of 37°C. In which direction will the reaction proceed?

41. a. The $\Delta G^{\circ\prime}$ value for a hypothetical reaction is 15 kJ · mol⁻¹ at 25°C. Compare the K_{eq} for this reaction with the K_{eq} for a reaction whose $\Delta G^{\circ\prime}$ value is twice as large. **b.** Carry out the same exercise for a hypothetical reaction whose $\Delta G^{\circ\prime}$ value is –15 kJ · mol⁻¹.

42. Calculate ΔG for the phosphoglucomutase reaction shown in Sample Calculation 12.2 at 37°C when the initial concentration of glucose-1-phosphate (G1P) is 5 mM and the initial concentration of glucose-6-phosphate (G6P) is 20 mM. Is the reaction spontaneous under these conditions?

43. Calculate the ratio of the concentration of glyceraldehyde-3-phosphate to the concentration of dihydroxyacetone phosphate that gives a free energy change of 2 kJ · mol⁻¹.

44. a. Use the standard free energies provided in Table 12.2 to calculate $\Delta G^{\circ\prime}$ for the isomerization of glucose-1-phosphate to glucose-6-phosphate. Is this value consistent with the value of $\Delta G^{\circ\prime}$ shown in Sample Calculation 12.2? Is this reaction spontaneous under standard conditions? **b.** Is the reaction spontaneous at 37°C when the concentration of glucose-6-phosphate is 10 mM and the concentration of glucose-1-phosphate is 0.5 mM?

45. Use the standard free energies provided in Table 12.2 to calculate $\Delta G^{\circ\prime}$ for the synthesis of ATP from ADP when phosphoenolpyruvate (PEP) donates a phosphate group to ADP. **a.** Is this reaction favorable under standard conditions at 25°C? **b.** Is this reaction favorable when [PEP] = 10 mM, [ATP] = 5 mM, [ADP] = 5 μM, [pyruvate] = 20 mM, and the temperature is 37°C?

46. Use the standard free energies provided in Table 12.2 to calculate $\Delta G^{\circ\prime}$ for the synthesis of glycerol-3-phosphate (G3P) when ATP serves as the phosphoryl donor. **a.** Is this reaction favorable under standard conditions at 25°C? **b.** Is this reaction favorable when [glycerol] = 3 mM, [ATP] = 5 mM, [ADP] = 5 μM, [G3P] = 1 mM, and the temperature is 37°C?

47. Construct a graph plotting free energy versus the reaction coordinate for the following reactions: **a.** glucose + P_i → glucose-6-phosphate, **b.** $ATP + H_2O → ADP + P_i$, and **c.** the coupled reaction.

48. a. Which of the compounds listed in Table 12.2 could be involved in a reaction coupled to the synthesis of ATP from $ADP + P_i$? **b.** Which of the compounds listed in Table 12.2 could be involved in a reaction coupled to the hydrolysis of ATP to $ADP + P_i$?

49. $\Delta G^{\circ\prime}$ for the hydrolysis of ATP under standard conditions at pH 7 and in the presence of magnesium ions is −30.5 kJ · mol⁻¹. **a.** How would this value change if ATP hydrolysis were carried out at a pH of less than 7? Explain. **b.** How would this value change if magnesium ions were not present?

50. Creatine undergoes reversible phosphorylation in muscles: ATP + creatine $\rightleftharpoons$ phosphocreatine + ADP. Some studies (but not all) show that creatine supplementation increases performance in high-intensity exercises lasting less than 30 seconds. Would you expect creatine supplements to affect endurance exercise?

51. Calculate ΔG for the hydrolysis of ATP under cellular conditions, where [ATP] = 3 mM, [ADP] = 1 mM, and [P_i] = 5 mM. Assume a temperature of 37°C.

52. The standard free energy change for the reaction catalyzed by triose phosphate isomerase is 7.9 kJ · mol⁻¹.

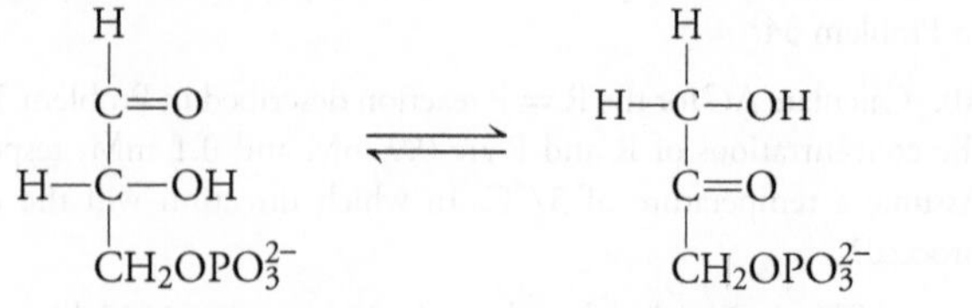

Glyceraldehyde-3-phosphate   Dihydroxyacetone phosphate

a. Calculate the equilibrium constant for the reaction at 25°C. **b.** Calculate ΔG at 37°C when the concentration of glyceraldehyde-3-phosphate (GAP) is 0.2 mM and the concentration of dihydroxyacetone phosphate (DHAP) is 0.6 mM. **c.** Is the reaction spontaneous under these conditions? Would the reverse reaction be spontaneous?

53. An apple contains about 72 Calories. Express this quantity in terms of ATP equivalents (that is, the number of ATP → ADP + P_i reactions). (*Note:* A nutritional "Calorie" is equivalent to 1 kilocalorie and 1 cal = 4.184 J.)

54. A large hot chocolate with whipped cream purchased at a national coffee chain contains 760 Calories. Express this quantity in terms of ATP equivalents (see Problem 53).

55. A moderately active adult female weighing 125 pounds must consume 2200 Calories of food daily **a.** If this energy is used to synthesize ATP, calculate the number of moles of ATP that would be synthesized each day under standard conditions (assuming 33% efficiency). **b.** Calculate the number of grams of ATP that would be synthesized each day. The molar mass of ATP is 505 g · mol⁻¹. What is the mass of ATP in pounds (2.2 kg = 1 lb)? **c.** There is approximately 40 g of ATP in the adult 125-lb female. Consider this, and your answer to part **b**, and suggest an explanation that is consistent with these findings.

56. a. How many apples (see Solution 53) would be required to provide the amount of ATP calculated in Problem 55? **b.** How many large hot chocolate drinks (see Solution 54) would be required?

57. The complete oxidation of glucose releases a considerable amount of energy. $\Delta G°'$ for the reaction shown is −2850 kJ · mol⁻¹.

$$C_6H_{12}O_6 + 6\ O_2 \rightarrow 6\ CO_2 + 6\ H_2O$$

How many moles of ATP could be produced under standard conditions from the oxidation of one mole of glucose, assuming about 33% efficiency?

58. The oxidation of palmitate, a 16-carbon saturated fatty acid, releases 9781 kJ · mol⁻¹.

$$C_{16}H_{32}O_2 + 23\ O_2 \rightarrow 16\ CO_2 + 16\ H_2O$$

a. How many moles of ATP could be produced under standard conditions from the oxidation of one mole of palmitate, assuming 33% efficiency? **b.** Calculate the number of ATP molecules produced per carbon for glucose (see Solution 57) and palmitate. Explain the reason for the difference.

59. Calculate ΔG for converting substance P to substance Q at 37°C when the initial concentration of P is 10 mM, and the initial concentration of Q is 2 mM. Is the reaction spontaneous under these conditions? The standard free energy change for the reaction is −7100 Jmol⁻¹.

60. Calculate the ratio of the Q to the P concentration (see Problem 59) that gives a free energy change of −4.0 kJmol⁻¹. Assume a temperature of 37°C.

61. Citrate is isomerized to isocitrate in the citric acid cycle (Chapter 14). The reaction is catalyzed by the enzyme aconitase. $\Delta G°'$ for the reaction is 5 kJ · mol⁻¹. The properties of the reaction are studied *in vitro*, where 1 M citrate and 1 M isocitrate are added to an aqueous solution of the enzyme at 25°C. **a.** What is the K_{eq} for the reaction at 25°C? **b.** What are the equilibrium concentrations of the reactant and product? **c.** What is the preferred direction of the reaction under standard conditions? **d.** The aconitase reaction is the second step of an eight-step pathway and occurs in the direction shown in the figure. How can you reconcile these facts with your answer to part **c**?

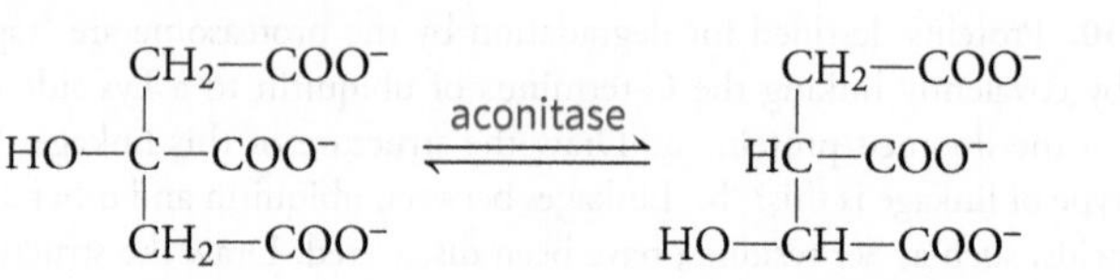

Citrate   Isocitrate

62. The equilibrium constant for the conversion of glucose-6-phosphate (G6P) to fructose-6-phosphate (F6P) is 0.41. The reaction is reversible and is catalyzed by the enzyme phosphoglucose isomerase.

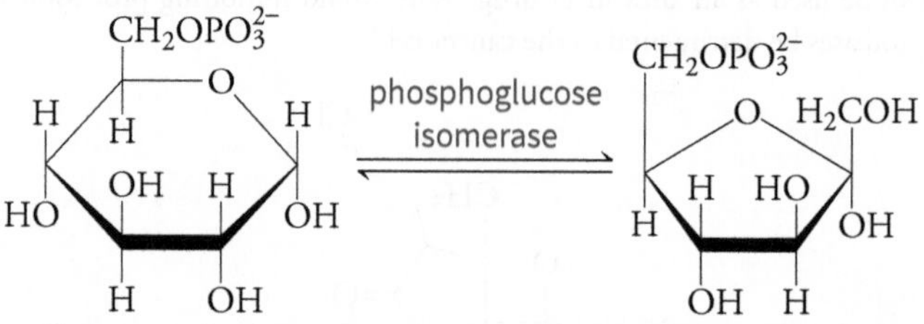

Glucose-6-phosphate   Fructose-6-phosphate

a. What is $\Delta G°'$ for this reaction? Would this reaction proceed in the direction written under standard conditions? **b.** What is ΔG for this reaction at 37°C when the concentration of glucose-6-phosphate is 2.0 mM and the concentration of fructose-6-phosphate is 0.5 mM? Would the reaction proceed in the direction written under these cellular conditions?

63. The phosphorylation of glucose to glucose-6-phosphate, the first step of glycolysis (Chapter 13), can be described by the equation glucose + P_i ⇌ glucose-6-phosphate + H_2O. **a.** Calculate the equilibrium constant for this reaction. **b.** What would the equilibrium concentration of glucose-6-phosphate (G6P) be under cellular conditions (both glucose and phosphate concentrations are 5 mM) if glucose was phosphorylated according to this reaction? Does this reaction provide a feasible route for producing glucose-6-phosphate for glycoly sis? **c.** One way to increase the amount of product is to increase the concentrations of the reactants. This would decrease the mass action ratio (see Equation 12.3) and would theoretically make the reaction as written more favorable. What concentration of glucose would be required to achieve a glucose-6-phosphate concentration of 250 μM? Is this strategy physiologically feasible, given that the solubility of glucose in aqueous medium is less than 1 M? **d.** Another way to promote the formation of glucose-6-phosphate from glucose is to couple the phosphorylation of glu cose to the hydrolysis of ATP as shown in Section 12.2. Calculate K_{eq} for the reaction in which glucose is converted to glucose-6-phos phate with concomitant ATP hydrolysis. **e.** When the ATP-dependent phosphorylation of glucose is carried out, what concentration of glucose is needed to achieve a 250 μM intracellular concentration of glucose-6-phosphate when the concentrations of ATP and ADP are 5.0 mM and 1.25 mM, respectively? **f.** Which route is more feasible to accomplish the phosphorylation of glucose to glucose-6-phosphate: the direct phosphorylation by P_i or the coupling of this phosphorylation to ATP hydrolysis? Explain.

64. Fructose-6-phosphate is phosphorylated to fructose-1,6-bis phosphate in the third step of the glycolytic pathway (Chapter 14). The phosphorylation of fructose-6-phosphate is described by the following equation:

fructose-6-phosphate + P_i ⇌ fructose-1,6-bisphosphate

$$\Delta G^{\circ\prime} = 47.7 \text{ kJ} \cdot \text{mol}^{-1}$$

a. What is the ratio of fructose-1,6-bisphosphate (F16BP) to fructose-6-phosphate (F6P) at equilibrium if the concentration of phosphate in the cell is 5 mM? **b.** Suppose that the phosphorylation of fructose-6-phosphate is coupled to the hydrolysis of ATP. Write the new equation that describes this process and calculate $\Delta G^{\circ\prime}$ for the reaction. **c.** What is the ratio of fructose-1,6-bisphosphate to fructose-6-phosphate at equilibrium for the reaction you wrote in part b if the equilibrium concentration of ATP = 3 mM and [ADP] = 1 mM? **d.** Compare your answers to parts **a** and **c**. What is the likely path for the cellular synthesis of fructose-1,6-bisphosphate? **e.** One can envisage two mechanisms for coupling ATP hydrolysis to the phosphorylation of fructose-6-phosphate (F6P) to fructose-1,6-bisphosphate (F16BP), yielding the same overall reaction:

Mechanism I: ATP is hydrolyzed as F6P is transformed to F16BP:

$$\text{F6P} + P_i \rightleftharpoons \text{F16BP}$$
$$\text{ATP} + H_2O \rightleftharpoons \text{ADP} + P_i$$

Mechanism II: ATP transfers its γ phosphate directly to F6P in one step, producing F16BP:

$$\text{F6P} + \text{ATP} + H_2O \rightleftharpoons \text{F16BP} + \text{ADP}$$

Choose one of the above mechanisms as the more biochemically feasible and provide a rationale for your choice.

65. Glyceraldehyde-3-phosphate (GAP) is converted to 3-phosphoglycerate (3PG) in one of the steps in the glycolytic pathway.

Consider these two scenarios:

I. GAP is oxidized to 1,3-BPG ($\Delta G^{\circ\prime} = 6.7$ kJ · mol^{-1}), which is subsequently hydrolyzed to yield 3PG ($\Delta G^{\circ\prime} = -49.3$ kJ · mol^{-1}).

II. GAP is oxidized to 1,3-BPG, which subsequently transfers its phosphate to ADP, yielding ATP ($\Delta G^{\circ\prime} = -18.8$ kJ · mol^{-1}).

Write the overall equations for the two scenarios. Which is more likely to occur in the cell, and why?

66. The conversion of glutamate to glutamine is unfavorable. In order for this transformation to occur in the cell, it must be coupled to the hydrolysis of ATP. Consider two possible mechanisms:

I: An ammonia group is transferred to glutamate and ATP is hydrolyzed:

$$\text{glutamate} + NH_3 \rightleftharpoons \text{glutamine}$$
$$\text{ATP} + H_2O \rightleftharpoons \text{ADP} + P_i$$

II: A phosphoryl group is transferred from ATP to glutamate, followed by the transfer of an ammonia group to glutamate, with release of free phosphate:

$$\text{glutamate} + \text{ATP} \rightleftharpoons \gamma\text{-glutamylphosphate} + \text{ADP}$$

$$\gamma\text{-glutamylphosphate} + H_2O + NH_3 \rightleftharpoons \text{glutamine} + P_i$$

Write the overall equation for the reaction for each mechanism. Is one mechanism more likely than the other? Or are both mechanisms equally feasible for the conversion of glutamate to glutamine? Explain.

67. $\Delta G^{\circ\prime}$ for the formation of UDP–glucose from glucose-1-phosphate and UTP is about zero. Yet the production of UDP–glucose is highly favorable. What is the driving force for this reaction?

$$\text{glucose-1-phosphate} + \text{UTP} \rightleftharpoons \text{UDP–glucose} + PP_i$$

68. Palmitate is activated in the cell by forming a thioester bond to coenzyme A. $\Delta G^{\circ\prime}$ for the synthesis of palmitoyl-CoA from palmitate and coenzyme A is 31.5 kJ · mol^{-1}.

a. What is the ratio of products to reactants at equilibrium? Is the reaction favorable? Explain. **b.** Suppose the synthesis of palmitoyl-CoA were coupled with ATP hydrolysis. Write the new equation for the activation of palmitate when coupled with ATP hydrolysis to ADP. Calculate $\Delta G^{\circ\prime}$ for the reaction. What is the ratio of products to reactants at equilibrium for the reaction? Is the reaction favorable? Compare your answer to the answer you obtained in part **a**. **c.** Suppose the reaction described in part **a** were coupled with ATP hydrolysis to AMP. Write the new equation for the activation of palmitate when coupled with ATP hydrolysis to AMP. Calculate $\Delta G^{\circ\prime}$ for the reaction. What is the ratio of products to reactants at equilibrium for the reaction? Is the reaction favorable? Compare your answer to the answer you obtained in part **b**.

d. Pyrophosphate, PP_i, is hydrolyzed to $2P_i$. The activation of palmitate, as described in part **c**, is coupled to the hydrolysis of pyrophosphate. Write the equation for this coupled reaction and calculate $\Delta G^{\circ\prime}$. What is the ratio of products to reactants at equilibrium for the reaction? Is the reaction favorable? Compare your answer to the answers you obtained in parts **b** and **c**.

69. DNA containing broken phosphodiester bonds ("nicks") can be repaired by the action of a ligase enzyme. Formation of a new phosphodiester bond in DNA requires the free energy of ATP, which is hydrolyzed to AMP:

$$\text{ATP} + \text{nicked bond} \xrightleftharpoons{\text{ligase}} \text{AMP} + PP_i + \text{phosphodiester bond}$$

The equilibrium constant expression for this reaction can be rearranged to define a constant, C, as follows:

$$K_{eq} = \frac{[\text{phosphodiester bond}][\text{AMP}][\text{PP}_i]}{[\text{nick}][\text{ATP}]}$$

$$\frac{[\text{nick}]}{[\text{phosphodiester bond}]} = \frac{[\text{AMP}][\text{PP}_i]}{K_{eq}[\text{ATP}]}$$

$$C = \frac{[\text{PP}_i]}{K_{eq}[\text{ATP}]}$$

$$\frac{[\text{nick}]}{[\text{phosphodiester bond}]} = C[\text{AMP}]$$

a. The ratio of nicked bonds to phosphodiester bonds at various concentrations of AMP was determined. Using the data provided, construct a plot of [nick]/[phosphodiester bond] versus [AMP] and determine the value of C from the plot.

[AMP] (mM)	[Nick]/[Phosphodiester bond]
10	4.01×10^{-5}
20	5.47×10^{-5}
30	8.67×10^{-5}
40	9.30×10^{-5}
50	1.13×10^{-4}

b. Determine the value of K_{eq} for the reaction in which the concentrations of PP_i and ATP are 1.0 mM and 14 µM, respectively. **c.** What is the value of $\Delta G°'$ for the reaction? **d.** What is the value of $\Delta G°'$ for nicked bond → phosphodiester bond reaction? (*Note:* The $\Delta G°'$ for the hydrolysis of ATP to AMP and PP_i is −48.5 kJ · mol^{-1} under the conditions used in the experiment.) **e.** The $\Delta G°'$ for the hydrolysis of a typical phosphomonoester bond is −13.8 kJ · mol^{-1}. Compare the stability of the phosphodiester bond in DNA to the stability of a typical phosphomonoester bond.

12.3 Metabolic Pathways

70. The common intermediates listed in the table—acetyl-CoA, glyceraldehyde-3-phosphate (GAP), and pyruvate—appear as reactants or products in several pathways. Place a check mark in the box that indicates the appropriate pathway—glycolysis, citric acid cycle, fatty acid metabolism, triacylglycerol (TAG) synthesis, photosynthesis, and transamination—for each reactant.

	Acetyl-CoA	GAP	Pyruvate
Glycolysis			
Citric acid cycle			
Fatty acid metabolism			
TAG synthesis			
Photosynthesis			
Transamination			

71. For each of the (unbalanced) reactions shown below, tell whether the reactant is being oxidized or reduced. **a.** A reaction from the alcoholic fermentation pathway, a catabolic process:

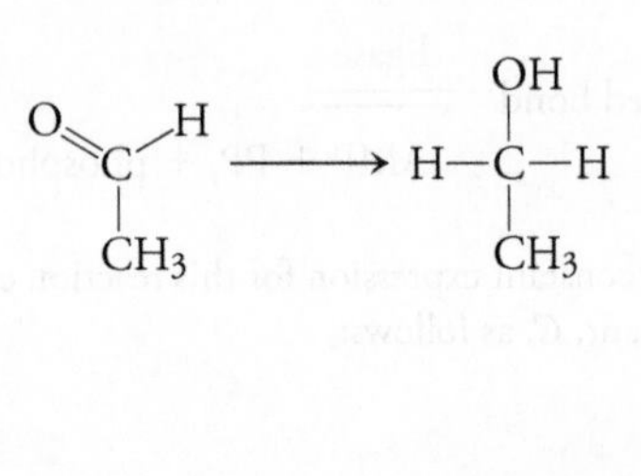

b. A reaction from the anabolic fatty acid synthesis pathway:

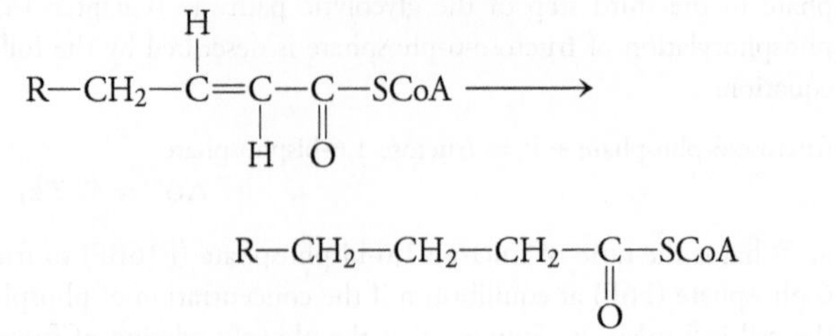

c. A reaction associated with the citric acid cycle, which plays an important role in catabolism:

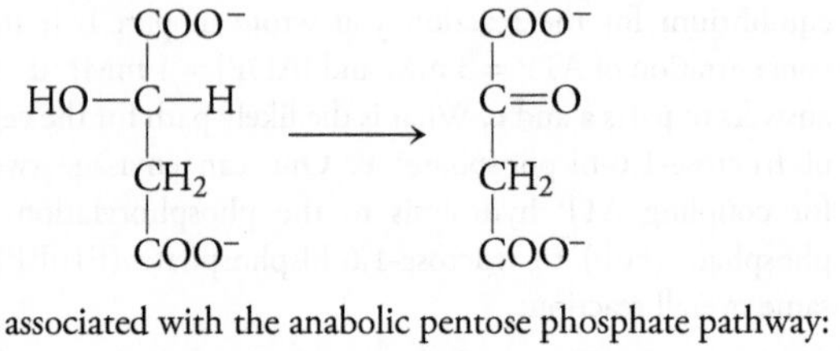

d. A reaction associated with the anabolic pentose phosphate pathway:

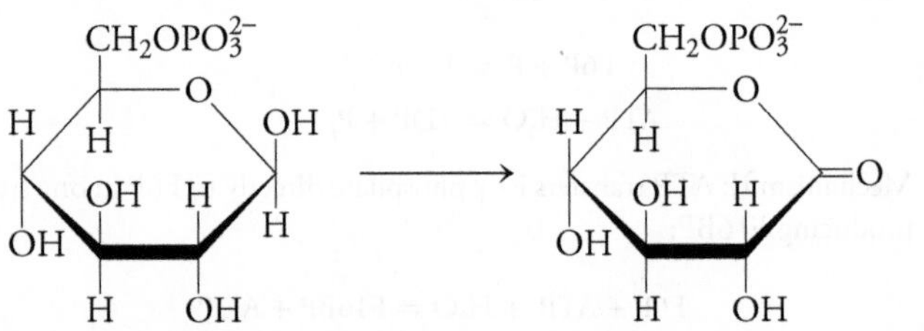

72. For each of the reactions shown in Problem 71, identify the cofactor as NAD$^+$, NADP$^+$, NADH, or NADPH.

73. For each of the (unbalanced) reactions shown below, tell whether the reactant is being oxidized or reduced. **a.** A reaction associated with the catabolic glycolytic pathway:

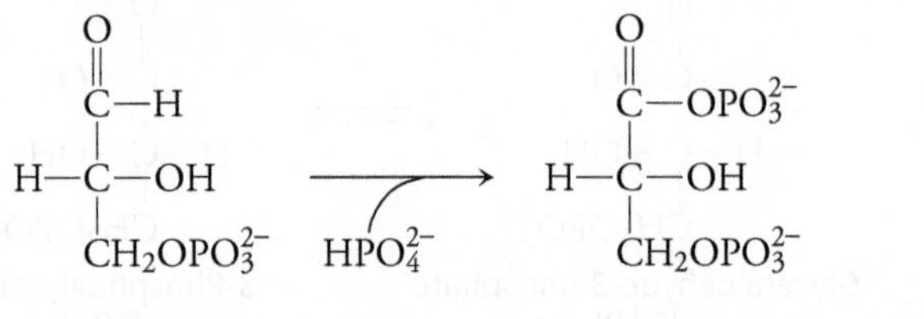

b. A reaction that follows the catabolic glycolytic pathway if oxygen is unavailable:

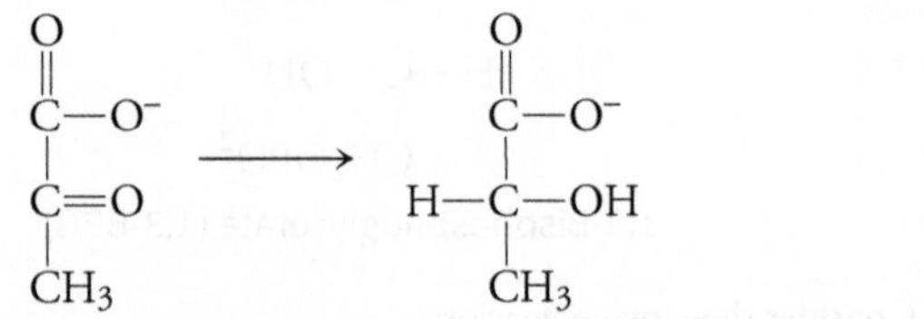

74. a. For each of the reactions shown in Problem 73, identify the cofactor as NAD$^+$, NADP$^+$, NADH, or NADPH. **b.** The reaction shown in Problem 73a is one of the 10 steps in glycolysis. The reaction shown in Problem 73b occurs at the end of glycolysis if oxygen is unavailable. Why might these two reactions be linked to one another?

75. A potential way to reduce the concentration of methane, a greenhouse gas, is to take advantage of sulfate-reducing bacteria. **a.** Complete the chemical equation for methane consumption by these organisms:

$$CH_4 + SO_4^{2-} \rightarrow \underline{\hspace{1cm}} + HS^- + H_2O$$

Identify the reaction component that undergoes **b.** oxidation and **c.** reduction.

76. Why do anabolic pathways require an input of energy, and what are some examples of anabolic processes? How do anabolic pathways contribute to the development and maintenance of living organisms?

77. Explain the function of anaplerotic pathways and their significance for metabolic processes.

78. What are cataplerotic pathways, and how do they differ from anaplerotic pathways?

79. Prokaryotic species A and B live close together because they are metabolically interdependent. **a.** If species A converts CH_4 to CO_2, what process would species B need to undertake: the conversion of S^{2-} to SO_4^{2-} or the conversion of SO_4^{2-} to S^{2-}? **b.** If species A converts CH_4 to CO_2, what process would species B need to undertake: the conversion of Fe^{3+} to Fe^{2+} or the conversion of Fe^{2+} to Fe^{3+}?

80. Vitamin B_{12} is synthesized by certain gastrointestinal bacteria and is also found in foods of animal origin such as meat, milk, eggs, and fish. When vitamin B_{12}–containing foods are consumed, the vitamin is released from the food and binds to a salivary vitamin B_{12}–binding protein called haptocorrin. The haptocorrin–vitamin B_{12} complex passes from the stomach to the small intestine, where the vitamin is released from the haptocorrin and then binds to intrinsic factor (IF). The IF–vitamin B_{12} complex then enters the cells lining the intestine by receptor-mediated endocytosis. Using this information, make a list of individuals most at risk for vitamin B_{12} deficiency.

81. A vitamin K–dependent carboxylase enzyme catalyzes the γ-carboxylation of specific glutamate residues in blood coagulation proteins. **a.** Draw the structure of a γ-carboxyglutamate residue. **b.** Why does this post-translational modification assist coagulation proteins in binding the Ca^{2+} ions required for blood clotting?

82. Food scientists at Ohio State University gave volunteers salsa or salad with or without lipid-rich avocados, then drew blood samples periodically and measured levels of β-carotene (see Box 8.B). They found that serum β-carotene levels were 2 to 15 times greater when the volunteers consumed the food with avocado. Explain these results.

83. Hartnup disease is a hereditary disorder caused by a defective transporter for nonpolar amino acids. **a.** The symptoms of the disease (photosensitivity and neurological abnormalities) can be prevented through dietary adjustments. What sort of diet would be effective? **b.** Patients with Hartnup disease often exhibit pellagra-like symptoms. Explain.

84. Refer to Table 12.4 and identify the vitamin required to accomplish each of the following reactions:

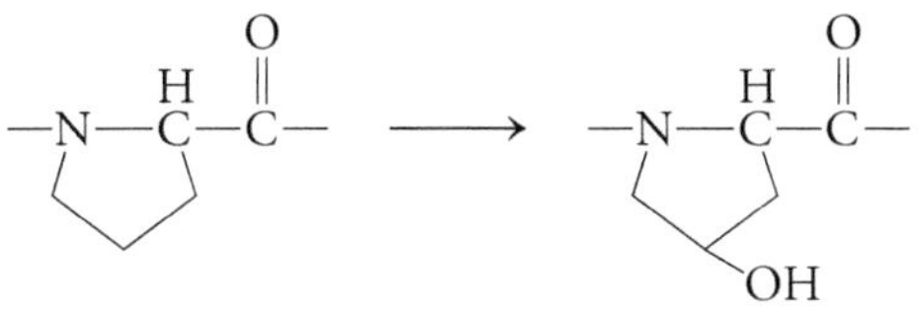

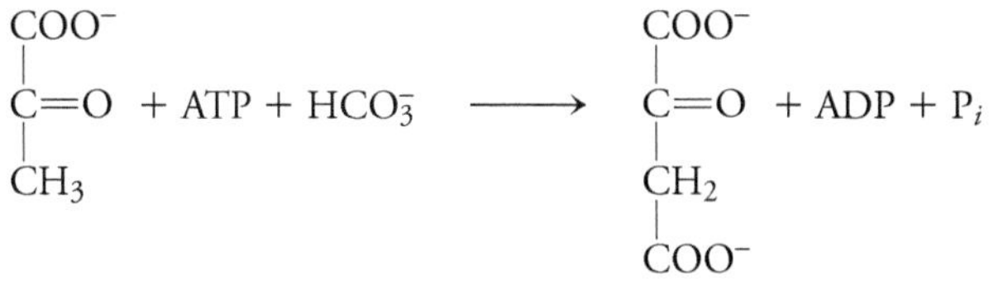

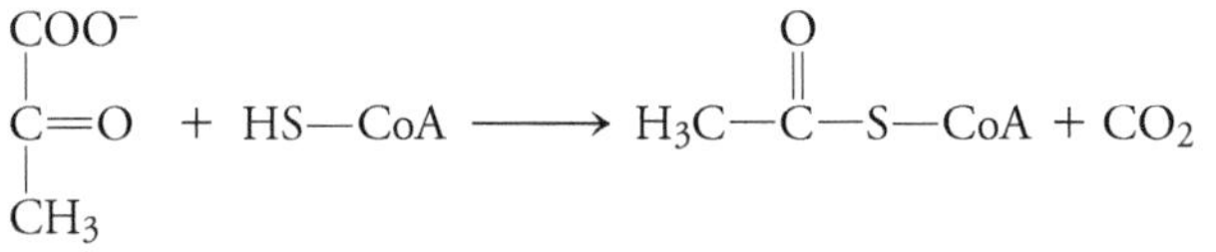

85. The microbial enzyme lactate racemase contains a nickel-based prosthetic group. **a.** To which common coenzyme is this prosthetic group related? **b.** Which two amino acid side chains hold the group in place?

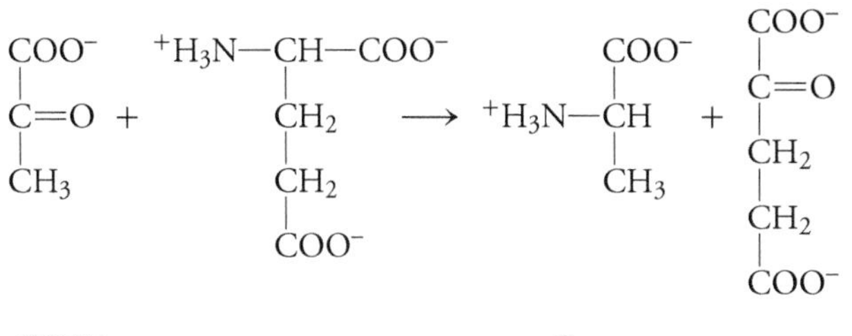

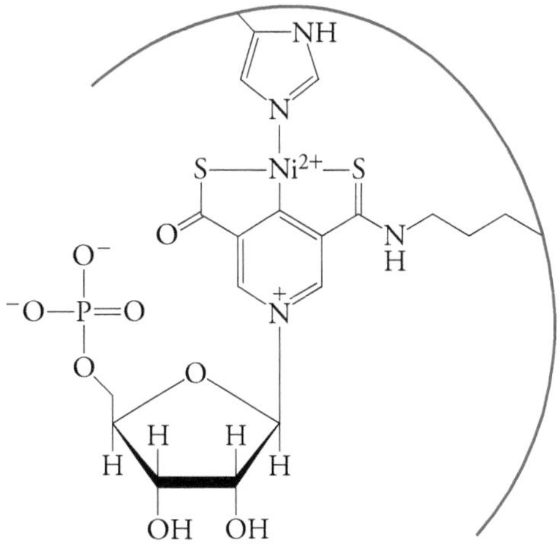

86. Why is niacin technically not a vitamin?

SELECTED READINGS

Coffey, R. and Ganz, T., Iron homeostasis: An anthropocentric perspective, *J. Biol. Chem.* 292, 12727–12734, doi: 10.1074/jbc.R117.781823 (2017). [Describes the storage and transport of iron and how these processes are regulated in humans.]

Coll-Martínez, B. and Crosas, B., How the 26S proteasome degrades ubiquitinated proteins in the cell, *Biomolecules* 9, 395, doi: 10.3390/biom9090395 (2019). [Provides structural and functional information about the proteasome, based on 30 years of research studies.]

Kim, M.-S. et al., A draft map of the human proteome, *Nature* 509, 575–581 (2014). [Describes a dataset of 17,000 proteins representing 30 different human tissues.]

Ludwig, D.S., Willett, W.C., Volek, J.S., and Neuhouser, M.L., Dietary fat: From foe to friend? *Science* 362, 764–770, doi: 10.1126/science.aau2096 (2018). [Summarizes some of the controversies around high-carbohydrate and high-fat diets and the links to disease.]

Martin, W.F. and Thauer, R.K., Energy in ancient metabolism, *Cell* 168, 953–955, doi: 10.1016/j.cell.2017.02.032 (2017). [Discusses ATP and other, more ancient energy currencies.]

Wishart, D., Metabolomics for investigating physiological and pathophysiological processes, *Physiol. Rev.* 99, 1819–1875, doi: 10.1152/physrev.00035.2018 (2019). [Reviews many practical aspects of metabolomics, including linking specific metabolites to diseases.]

CHAPTER 12 CREDITS

Box 12.B Image of ferritin based on 1FHA. Lawson, D.M., Artymiuk, P.J., Yewdall, S.J., Smith, J.M., Livingstone, J.C., Treffry, A., Luzzago, A., Levi, S., Arosio, P., Cesareni, G., Thomas, C.D., Shaw, W.V., Harrison, P.M., Solving the structure of human H ferritin by genetically engineering intermolecular crystal contacts, *Nature* 349, 541–544 (1991).

Image of transferrin based on 1H76. Hall, D.R., Hadden, J.M., Leonard, G.A., Bailey, S., Neu, M., Winn, M., Lindley, P.F., The crystal and molecular structures of diferric porcine and rabbit serum transferrins at resolutions of 2.15 and 2.60A, respectively, *Acta Crystallogr.* D 58, 70 (2002).

Carbohydrate Metabolism

photosomething/Adobe Stock

Kimchi and other fermented foods take advantage of microbial metabolism. Salting draws water and nutrients out of raw vegetables and then sugar is added to initiate anaerobic glycolysis carried out by bacteria. The resulting lactate, along with other metabolites, gives the food its characteristic flavor and prevents the growth of organisms that would otherwise completely degrade and spoil the food.

Do You Remember?

- Enzymes accelerate chemical reactions using acid–base catalysis, covalent catalysis, and metal ion catalysis (Section 6.2).
- Glucose polymers include the fuel-storage polysaccharides starch and glycogen and the structural polysaccharide cellulose (Section 11.2).
- Coenzymes such as NAD^+ and ubiquinone collect electrons from compounds that become oxidized (Section 12.3).
- A reaction with a large negative change in free energy can be coupled to another, unfavorable reaction (Section 12.2).
- A reaction that breaks a phosphoanhydride bond in ATP occurs with a large change in free energy (Section 12.2).
- Nonequilibrium reactions often serve as metabolic control points (Section 12.2).

Glucose occupies a central position in the metabolism of most cells. It is a major source of metabolic energy (in some cells, it is the only source), and it provides the precursors for the synthesis of other biomolecules. Recall that glucose is stored in polymeric form as starch in plants and as glycogen in animals (Section 11.2). The breakdown of these polymers provides glucose monomers that can be catabolized to release energy. In this chapter, we will examine the major metabolic pathways involving glucose, including the interconversion of the monosaccharide glucose with glycogen, the degradation of glucose to the three-carbon intermediate pyruvate, the synthesis of glucose from smaller compounds, and the conversion of glucose to the five-carbon monosaccharide ribose. For all the pathways, we will present the intermediates and some of the relevant enzymes. We will also examine the thermodynamics of reactions that release or consume large amounts of free energy and discuss how some of these reactions are regulated.

13.1 Glycolysis

KEY CONCEPTS

Describe the substrates, products, and chemical reaction for each step of glycolysis.

- Identify the energy-consuming and energy-generating steps of glycolysis.
- List the flux-control points for the pathway.
- Describe the metabolic uses of pyruvate.

Pathways dealing with carbohydrates, highlighted in **Figure 13.1**, are part of the larger metabolic scheme introduced in Figure 12.9. **Glycolysis,** the conversion of glucose to pyruvate, is a good place to begin a study of metabolic pathways. As a result of many years of research, we know a great deal about the pathway's intermediates and the enzymes that mediate their chemical transformations. We have also learned that glycolysis, along with other metabolic pathways, exhibits the following properties:

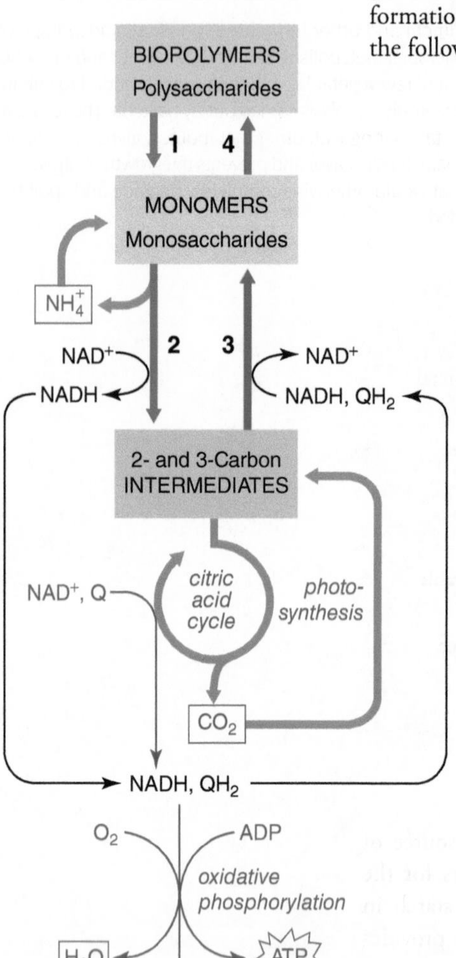

1. Each step of the pathway is catalyzed by a distinct enzyme.
2. The energy consumed or released in certain reactions is transferred by molecules such as ATP and NADH.
3. The rate of the pathway can be controlled by altering the activity of individual enzymes.

If metabolic processes did not occur via multiple enzyme-catalyzed steps, cells would have little control over the amount and type of reaction products and no way to manage energy. For example, the combustion of glucose and O_2 to CO_2 and H_2O—if allowed to occur in one grand explosion—would release about 2850 $kJ \cdot mol^{-1}$ of energy all at once. In the cell, *glucose combustion requires many steps so that the cell can recover its energy in smaller, more useful quantities.*

Glycolysis, representing the first ten steps of this process, appears to be an ancient metabolic pathway. The fact that it does not require molecular oxygen suggests that it evolved before photosynthesis increased the level of atmospheric O_2. Overall, glycolysis is a series of enzyme-catalyzed steps in which a six-carbon glucose molecule is broken down into two three-carbon pyruvate molecules. This catabolic pathway is accompanied by the phosphorylation of two molecules of ADP (to produce 2 ATP) and the reduction of two molecules of NAD^+. The net equation for the pathway (ignoring water and protons) is

$$\text{glucose} + 2\,NAD^+ + 2\,ADP + 2\,P_i \rightarrow 2\,\text{pyruvate} + 2\,NADH + 2\,ATP$$

It is convenient to divide the 10 reactions of glycolysis into two phases. In the first (Reactions 1–5), the hexose is phosphorylated and cleaved in half. In the second (Reactions 6–10), the three-carbon molecules are converted to pyruvate (**Fig. 13.2**).

As you examine each of the reactions of glycolysis described in the following pages, note how the reaction substrates are converted to products by the action of an enzyme (and note how the enzyme's name often reveals its purpose). Pay attention also to the free energy change of each reaction.

Figure 13.1 Glucose metabolism in context. (1) The polysaccharide glycogen is degraded to glucose, which is then catabolized by the glycolytic pathway (2) to the three-carbon intermediate pyruvate. Gluconeogenesis (3) is the pathway for the synthesis of glucose from smaller precursors. Glucose can then be reincorporated into glycogen (4). The conversion of glucose to ribose, a component of nucleotides, is not shown in this diagram.

Energy is invested at the start of glycolysis

The early reactions of glycolysis can be considered as preparation for the later, energy-producing reactions. In fact, two of the early reactions require the investment of energy in the form of ATP.

Glucose

ATP ⤵
1 hexokinase
ADP ↙

Glucose-6-phosphate

2 phosphoglucose
isomerase

Fructose-6-phosphate

ATP ⤵
3 phosphofructokinase
ADP ↙

Fructose-1,6-bisphosphate

4 aldolase

Glyceraldehyde-   Dihydroxyacetone
3-phosphate   +   phosphate

5 triose phosphate
isomerase

2 Pᵢ + 2 NAD⁺
glyceraldehyde-3-
6 phosphate
dehydrogenase
2 NADH + 2 H⁺

2 1,3-Bisphosphoglycerate

2 ADP ⤵
7 phosphoglycerate
kinase
2 ATP ↙

2 3-Phosphoglycerate

8 phosphoglycerate
mutase

2 2-Phosphoglycerate

2 H₂O ↙
9 enolase

2 Phosphoenolpyruvate

2 ADP ⤵
10 pyruvate kinase
2 ATP ↙

2 Pyruvate

Figure 13.2 The reactions of glycolysis. The substrates, products, and enzymes corresponding to the 10 steps of the pathway are shown. Shading indicates the substrates (blue) and products (green) of the pathway as a whole.

Question Next to each reaction, write the term that describes the type of chemical change that occurs.

1. Hexokinase

In the first step of glycolysis, the enzyme hexokinase transfers a phosphoryl group from ATP to the C6 OH group of glucose to form glucose-6-phosphate:

Glucose $\quad\quad\quad\quad\quad\quad$ Glucose-6-phosphate

A **kinase** is an enzyme that transfers a phosphoryl group to or from ATP (or another nucleoside triphosphate) to or from another substance.

Recall from Section 6.3 that the hexokinase active site closes around its substrates so that a phosphoryl group is efficiently transferred from ATP to glucose. The standard free energy change for this reaction, which cleaves one of ATP's phosphoanhydride bonds, is -16.7 kJ·mol^{-1} (ΔG, the actual free energy change for the reaction inside a cell, has a similar value). The magnitude of this free energy change means that the reaction proceeds in only one direction; the reverse reaction is extremely unlikely since its standard free energy change would be $+16.7$ kJ·mol^{-1}. Consequently, hexokinase is said to catalyze a **metabolically irreversible reaction** that helps direct glucose toward its fate. *Many metabolic pathways have a similar irreversible step near the start that commits a metabolite to proceed through the pathway.*

2. Phosphoglucose Isomerase

The second reaction of glycolysis is an isomerization reaction in which glucose-6-phosphate is converted to fructose-6-phosphate:

Glucose-6-phosphate $\quad\quad\quad\quad\quad\quad$ Fructose-6-phosphate

Because fructose is a six-carbon ketose (Section 11.1), it forms a five-membered ring. This isomerization reaction is necessary in order to place a carbonyl group next to the C3—C4 bond that will be cleaved in step 4.

The standard free energy change for the phosphoglucose isomerase reaction is $+2.2$ kJ · mol^{-1}, but the reactant concentrations *in vivo* yield a ΔG value of about -1.4 kJ · mol^{-1}. A value of ΔG near zero indicates that the reaction operates close to equilibrium (at equilibrium, $\Delta G = 0$). *Such **near-equilibrium reactions** are considered to be freely reversible, since a slight excess of products can easily drive the reaction in reverse by mass action effects.* In a metabolically irreversible reaction, such as the hexokinase reaction, the concentration of product could never increase enough to compensate for the reaction's large value of ΔG.

3. Phosphofructokinase

The third reaction of glycolysis consumes a second ATP molecule in the phosphorylation of fructose-6-phosphate to yield fructose-1,6-bisphosphate.

Fructose-6-phosphate $\quad\quad\quad\quad\quad\quad$ Fructose-1,6-bisphosphate

Phosphofructokinase operates in much the same way as hexokinase, and the reaction it catalyzes is irreversible, with a $\Delta G^{\circ\prime}$ value of -17.2 kJ·mol^{-1}.

In cells, the activity of phosphofructokinase is regulated. We have already seen how the activity of a bacterial phosphofructokinase responds to allosteric effectors (Section 7.3). ADP binds to the enzyme and causes a conformational change that promotes fructose-6-phosphate binding, which in turn promotes catalysis. This mechanism is useful because the concentration of ADP in the cell is a good indicator of the need for ATP, which is a product of glycolysis. Phosphoenolpyruvate, the product of step 9 of glycolysis, binds to bacterial phosphofructokinase and causes it to assume a conformation that destabilizes fructose-6-phosphate binding, thereby diminishing catalytic activity. Thus, when the glycolytic pathway is producing plenty of phosphoenolpyruvate and ATP, the phosphoenolpyruvate can act as a feedback inhibitor to slow the pathway by decreasing the rate of the reaction catalyzed by phosphofructokinase (**Fig. 13.3a**). Citrate, an intermediate of the citric acid cycle (which completes glucose catabolism), is also a feedback inhibitor of phosphofructokinase.

The most potent activator of phosphofructokinase in mammals is the compound fructose-2,6-bisphosphate, which is synthesized from fructose-6-phosphate by an enzyme known as phosphofructokinase-2. (The glycolytic enzyme is therefore sometimes called phosphofructokinase-1.)

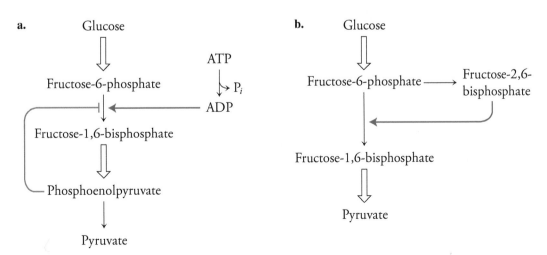

Fructose-6-phosphate   Fructose-2,6-bisphosphate

The activity of phosphofructokinase-2 is hormonally stimulated when the concentration of glucose in the blood is high. The resulting increase in fructose-2,6-bisphosphate concentration activates phosphofructokinase to increase the flux (rate of flow) of glucose through the glycolytic pathway (**Fig. 13.3b**).

The phosphofructokinase reaction is the primary control point for glycolysis. It is the slowest reaction of glycolysis, so the rate of this reaction largely determines the flux of glucose through the entire pathway. In general, a **rate-determining reaction**—such as the phosphofructokinase reaction—operates far from equilibrium; that is, it has a large negative free energy change and is irreversible under metabolic conditions. The rate of the reaction can be altered by allosteric effectors but not by fluctuations in the concentrations of its substrates or products. Thus, it acts as a one-way valve. In contrast, a near-equilibrium reaction—such as the phosphoglucose isomerase reaction—cannot

Figure 13.3 Regulation of phosphofructokinase. **a.** Regulation in bacteria. ADP, produced when ATP is consumed elsewhere in the cell, stimulates the activity of phosphofructokinase (green arrow). Phosphoenolpyruvate, a late intermediate of glycolysis, inhibits phosphofructokinase (red symbol), thereby decreasing the rate of the entire pathway. **b.** Regulation in mammals.

serve as a rate-determining step for a pathway because it can respond to small changes in reactant concentrations by operating in reverse.

4. Aldolase

Reaction 4 converts the hexose fructose-1,6-bisphosphate to two three-carbon molecules, each of which bears a phosphate group.

$$
\begin{array}{c}
\overset{1}{C}H_2OPO_3^{2-} \\
| \\
\overset{2}{C}=O \\
| \\
HO-\overset{3}{C}-H \\
| \\
H-\overset{4}{C}-OH \\
| \\
H-\overset{5}{C}-OH \\
| \\
\overset{6}{C}H_2OPO_3^{2-}
\end{array}
\quad \underset{aldolase}{\rightleftharpoons} \quad
\begin{array}{c}
\overset{3}{C}H_2OPO_3^{2-} \\
| \\
\overset{2}{C}=O \\
| \\
HO-\overset{1}{C}H_2
\end{array}
\quad + \quad
\begin{array}{c}
O\quad H \\
\diagdown\,/ \\
\overset{1}{C} \\
| \\
H-\overset{2}{C}-OH \\
| \\
\overset{3}{C}H_2OPO_3^{2-}
\end{array}
$$

Fructose-1,6-bisphosphate — Dihydroxyacetone phosphate — Glyceraldehyde-3-phosphate

This reaction is the reverse of an aldol (aldehyde–alcohol) condensation, so the enzyme that catalyzes the reaction is called aldolase. It is worth examining its mechanism. The active site of mammalian aldolase contains two catalytically important residues: a lysine residue that forms a Schiff base (imine) with the substrate and an ionized aspartate residue that acts as a base catalyst (**Fig. 13.4**). (Bacterial aldolase uses lysine and an ionized tyrosine residue.)

Early studies of aldolase implicated a cysteine residue in catalysis because iodoacetate, a reagent that reacts with the cysteine side chain, also inactivates the enzyme:

$$
\begin{array}{c}
C=O \\
| \\
HC-CH_2-SH \\
| \\
NH \\
Cys
\end{array}
\quad + \quad ICH_2COO^- \quad \xrightarrow{\ HI\ } \quad
\begin{array}{c}
C=O \\
| \\
HC-CH_2-SCH_2COO^- \\
| \\
NH
\end{array}
$$

Iodoacetate

Researchers used iodoacetate to help identify the intermediates of glycolysis: In the presence of iodoacetate, fructose-1,6-bisphosphate accumulates because the next step is blocked. Acetylation of the cysteine residue, which turned out not to be part of the active site, probably interferes with a conformational change that is necessary for aldolase activity.

The $\Delta G^{\circ\prime}$ value for the aldolase reaction is +22.8 kJ·mol^{-1}, indicating that the reaction is unfavorable under standard conditions. However, the reaction proceeds *in vivo* (ΔG is actually less than zero) because the products of the reaction are quickly whisked away by subsequent reactions. In essence, the rapid consumption of glyceraldehyde-3-phosphate and dihydroxyacetone phosphate "pulls" the aldolase reaction forward.

5. Triose Phosphate Isomerase

The products of the aldolase reaction are both phosphorylated three-carbon compounds, but only one of them—glyceraldehyde-3-phosphate—proceeds through the remainder of the pathway. Dihydroxyacetone phosphate is converted to glyceraldehyde-3-phosphate by triose phosphate isomerase:

$$
\begin{array}{c}
H \\
| \\
H-C-OH \\
| \\
C=O \\
| \\
CH_2OPO_3^{2-}
\end{array}
\quad \underset{\substack{triose\ phosphate \\ isomerase}}{\rightleftharpoons} \quad
\begin{array}{c}
O\quad H \\
\diagdown\,/ \\
C \\
| \\
H-C-OH \\
| \\
CH_2OPO_3^{2-}
\end{array}
$$

Dihydroxyacetone phosphate — Glyceraldehyde-3-phosphate

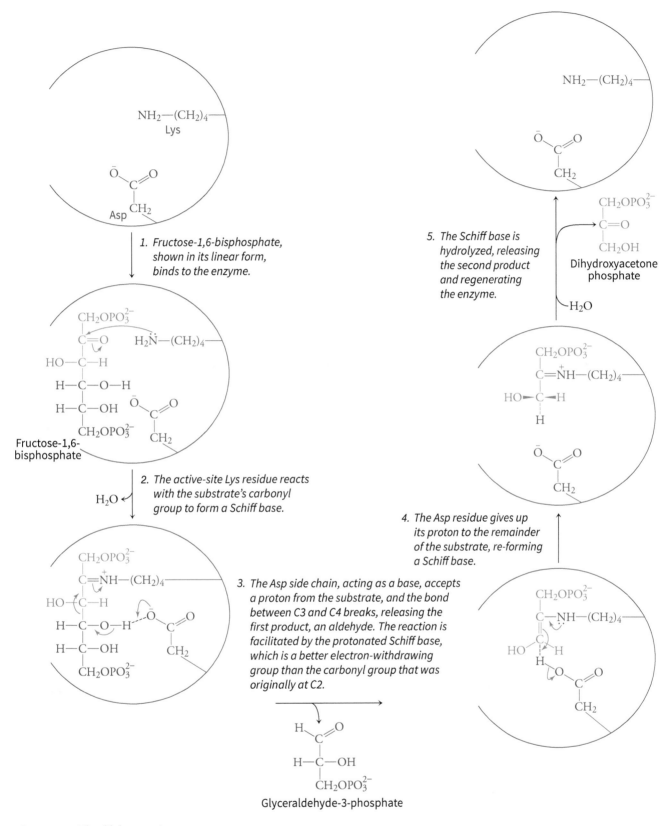

Figure 13.4 The aldolase reaction.

Question **Does this reaction follow an ordered or ping pong mechanism (see Section 7.2)?**

Triose phosphate isomerase was introduced in Section 7.2 as an example of a catalytically perfect enzyme, one whose rate is limited only by the rate at which its substrates can diffuse to its active site. The catalytic mechanism of triose phosphate isomerase may involve low-barrier hydrogen bonds (which also help stabilize the transition state in serine proteases; see Section 6.3).

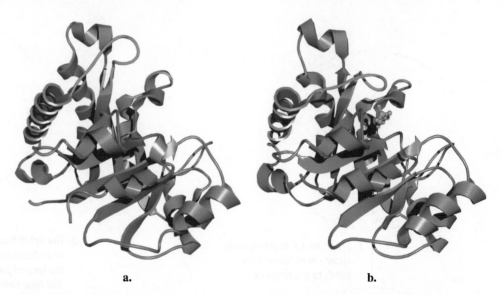

Figure 13.5 Conformational changes in yeast triose phosphate isomerase. **a.** The 11-residue loop is highlighted in green. **b.** The loop closes after substrate binds. In this model, the transition state analog 2-phosphoglycolate (orange) occupies the active site. Triose phosphate isomerase is actually a homodimer; only one subunit is pictured here.

In addition, the catalytic power of triose phosphate isomerase depends on the movement of a protein loop comprising residues 166–176. After a substrate binds, the loop closes over the active site and stabilizes the reaction's transition state (**Fig. 13.5**).

The standard free energy change for the triose phosphate isomerase reaction is slightly positive, even under physiological conditions ($\Delta G^{\circ\prime} = +7.9 \cdot \text{kJ} \cdot \text{mol}^{-1}$ and $\Delta G = +4.4 \text{ kJ} \cdot \text{mol}^{-1}$), but the reaction proceeds because glyceraldehyde-3-phosphate is quickly consumed in the next reaction, so more dihydroxyacetone phosphate is constantly being converted to glyceraldehyde-3-phosphate.

Despite the catalytic "perfection" of triose phosphate isomerase, mistakes do occur. The interconversion of glyceraldehyde-3-phosphate and dihydroxyacetone phosphate proceeds via an enediol intermediate, which can break down to form methylglyoxal.

$$
\begin{array}{ccc}
\underset{\text{Enediol intermediate}}{
\begin{array}{l}
\text{H}\diagdown\text{C}\diagup\text{OH} \\[2pt]
\text{C}-\text{OH} \\[2pt]
\text{CH}_2\text{OPO}_3^{2-}
\end{array}}
&
\xrightarrow{\ \text{HPO}_4^{2-}\ }
&
\underset{\text{Methylglyoxal}}{
\begin{array}{l}
\text{H}\diagdown\text{C}\diagup\text{O} \\[2pt]
\text{C}=\text{O} \\[2pt]
\text{CH}_3
\end{array}}
\end{array}
$$

Methylglyoxal, which is also a product of other enzymatic and nonenzymatic processes, can react with lipids, nucleotides, and the amino groups of proteins. This molecular damage seems to be an unavoidable consequence of carrying out glycolysis. Not surprisingly, most organisms have a set of highly conserved enzymes to detoxify methylglyoxal or eliminate its reaction products. In fact, many metabolic enzymes produce small amounts of toxic by-products that must be dealt with by other "repair" enzymes.

ATP is generated near the end of glycolysis

So far, the reactions of glycolysis have consumed 2 ATP, but this investment pays off later in glycolysis when 4 ATP are produced, for a net gain of 2 ATP. Reactions 6–10 all involve three-carbon intermediates, but keep in mind that *each glucose molecule that enters the pathway yields two of these three-carbon units.*

Some species convert glucose to glyceraldehyde-3-phosphate by different pathways than the one presented above. However, the last five reactions of glycolysis, which convert glyceraldehyde-3-phosphate to pyruvate, are the same in all organisms. This suggests that glycolysis may have evolved from the "bottom up"; that is, it first evolved as a pathway for extracting free energy from abiotically produced small molecules, before cells developed the ability to synthesize larger molecules such as hexoses.

6. Glyceraldehyde-3-Phosphate Dehydrogenase

In the sixth reaction of glycolysis, glyceraldehyde-3-phosphate is both oxidized and phosphorylated:

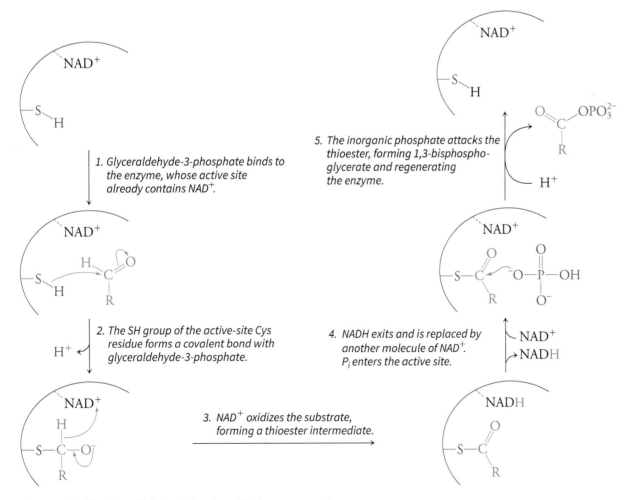

Unlike the kinases that catalyze Reactions 1 and 3, glyceraldehyde-3-phosphate dehydrogenase does not use ATP as a phosphoryl-group donor; it adds inorganic phosphate to the substrate. This reaction is also an oxidation–reduction reaction in which the aldehyde group of glyceraldehyde-3-phosphate is oxidized and the cofactor NAD^+ is reduced to NADH. In effect, glyceraldehyde-3-phosphate dehydrogenase catalyzes the removal of an H atom (actually, a hydride ion); hence the name "dehydrogenase." Note that the reaction product NADH must eventually be reoxidized to NAD^+ or else glycolysis will come to a halt. In fact, the reoxidation of NADH, which is a form of "energy currency," generates ATP in oxidative phosphorylation (Chapter 15).

An active-site cysteine residue participates in the glyceraldehyde-3-phosphate dehydrogenase reaction (**Fig. 13.6**). The enzyme is inhibited by arsenate (AsO_4^{3-}), which competes with $P_i (PO_4^{3-})$ for binding in the enzyme active site.

Figure 13.6 The glyceraldehyde-3-phosphate dehydrogenase reaction.

Question Identify the reactant that undergoes oxidation and the reactant that undergoes reduction.

7. Phosphoglycerate Kinase

The product of Reaction 6, 1,3-bisphosphoglycerate, is an acyl phosphate.

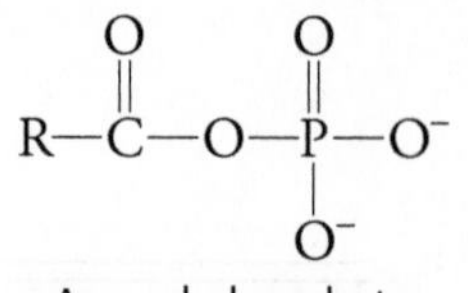

An acyl phosphate

The subsequent removal of its phosphoryl group releases a large amount of free energy in part because the reaction products are more stable (the same principle contributes to the large negative value of ΔG for reactions involving cleavage of ATP's phosphoanhydride bonds; see Section 12.2). *The free energy released in this reaction is used to drive the formation of ATP, as 1,3-bisphosphoglycerate donates its phosphoryl group to ADP:*

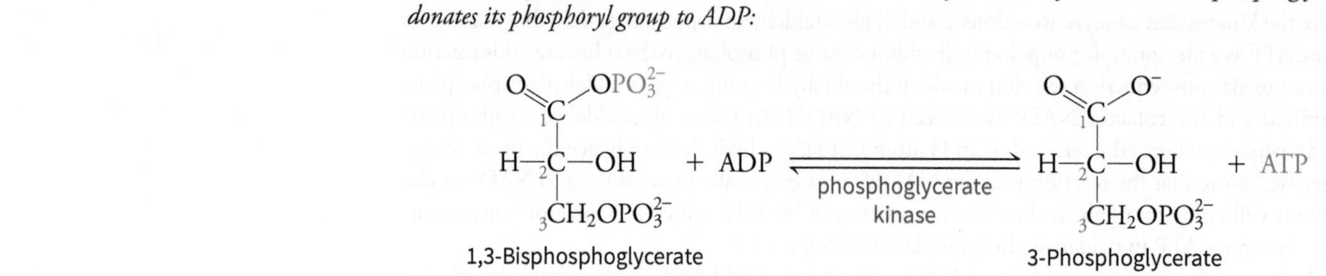

1,3-Bisphosphoglycerate   3-Phosphoglycerate

Note that the enzyme that catalyzes this reaction is called a kinase since it transfers a phosphoryl group between ATP and another molecule.

The standard free energy change for the phosphoglycerate kinase reaction is $-18.8 \text{ kJ} \cdot \text{mol}^{-1}$. This strongly exergonic reaction helps pull the glyceraldehyde-3-phosphate dehydrogenase reaction forward, since its standard free energy change is greater than zero ($\Delta G^{\circ\prime} = +6.7 \text{ kJ} \cdot \text{mol}^{-1}$). This pair of reactions provides a good example of the coupling of a thermodynamically favorable and unfavorable reaction so that both proceed with a net decrease in free energy: $-18.8 \text{ kJ} \cdot \text{mol}^{-1} + 6.7 \text{ kJ} \cdot \text{mol}^{-1} = -12.1 \text{ kJ} \cdot \text{mol}^{-1}$. Under physiological conditions, ΔG for the paired reactions is close to zero.

8. Phosphoglycerate Mutase

In the next reaction, 3-phosphoglycerate is converted to 2-phosphoglycerate:

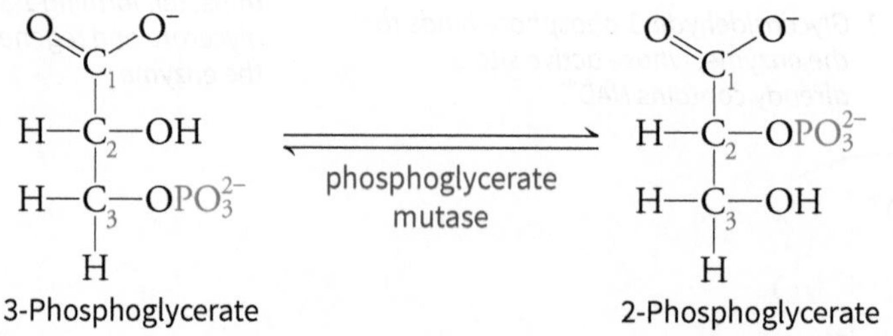

3-Phosphoglycerate   2-Phosphoglycerate

Although the reaction appears to involve the simple intramolecular transfer of a phosphoryl group, the reaction mechanism is a bit more complicated and requires an enzyme active site that contains a phosphorylated histidine residue. The phospho-His transfers its phosphoryl group to 3-phosphoglycerate to generate 2,3-bisphosphoglycerate, which then gives a phosphoryl group back to the enzyme, leaving 2-phosphoglycerate and phospho-His:

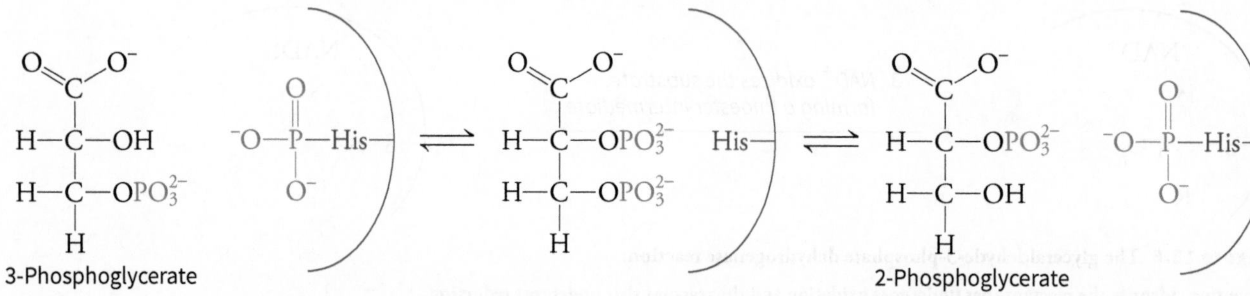

3-Phosphoglycerate   2-Phosphoglycerate

As can be guessed from its mechanism, the phosphoglycerate mutase reaction is freely reversible *in vivo*.

9. Enolase   Enolase catalyzes a dehydration reaction, in which water is eliminated:

2-Phosphoglycerate   Phosphoenolpyruvate

The enzyme active site includes an Mg^{2+} ion that apparently coordinates with the OH group at C3 and makes it a better leaving group. Fluoride ion and P_i can form a complex with the Mg^{2+} and thereby inhibit the enzyme. In early studies demonstrating the inhibition of glycolysis by F^-, 2-phosphoglycerate, the substrate of enolase, accumulated. The concentration of 3-phosphoglycerate also increased in the presence of F^- since phosphoglycerate mutase readily converted the excess 2-phosphoglycerate back to 3-phosphoglycerate.

10. Pyruvate Kinase   The tenth reaction of glycolysis is catalyzed by pyruvate kinase, which converts phosphoenolpyruvate to pyruvate and transfers a phosphoryl group to ADP to produce ATP:

Phosphoenolpyruvate   Pyruvate

The reaction actually occurs in two parts. First, ADP attacks the phosphoryl group of phosphoenolpyruvate to form ATP and enolpyruvate:

Phosphoenolpyruvate   ADP   Enolpyruvate

Removal of phosphoenolpyruvate's phosphoryl group is not a particularly exergonic reaction. When written as a hydrolytic reaction (transfer of the phosphoryl group to water), the $\Delta G^{\circ\prime}$ value is -16 kJ·mol^{-1}. This is not enough free energy to drive the synthesis of ATP from ADP + P_i (this reaction requires $+30.5$ kJ·mol^{-1}). However, the second half of the pyruvate kinase reaction is highly exergonic. This is the **tautomerization** (isomerization through the shift of an H atom) of enolpyruvate to pyruvate.

Enolpyruvate   Pyruvate

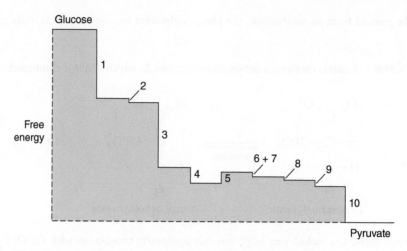

Figure 13.7 Graphical representation of the free energy changes of glycolysis. Three steps have large negative values of ΔG; the remaining steps are near equilibrium ($\Delta G \approx 0$). The height of each step corresponds to its ΔG value in heart muscle, and the numbers correspond to glycolytic enzymes. Keep in mind that ΔG values vary slightly among tissues.

$\Delta G^{\circ\prime}$ for this step is -46 kJ·mol^{-1}, so $\Delta G^{\circ\prime}$ for the net reaction (hydrolysis of phosphoenolpyruvate followed by tautomerization of enolpyruvate to pyruvate) is -61.9 kJ·mol^{-1}, more than enough free energy to drive the synthesis of ATP.

Three of the ten reactions of glycolysis (the reactions catalyzed by hexokinase, phosphofructokinase, and pyruvate kinase) have large negative values of ΔG. In theory, any of these far-from-equilibrium reactions could serve as a flux-control point for the pathway. The other seven reactions function near equilibrium ($\Delta G \approx 0$) and can therefore accommodate flux in either direction. The free energy changes for the ten reactions of glycolysis are shown graphically in **Figure 13.7**.

We have already discussed the mechanisms for regulating phosphofructokinase activity, the major control point for glycolysis. Hexokinase also catalyzes an irreversible reaction and is subject to inhibition by its product, glucose-6-phosphate. However, hexokinase cannot be the sole control point for glycolysis because glucose can also enter the pathway as glucose-6-phosphate, bypassing the hexokinase reaction. In addition, glycolysis is not the only fate for glucose-6-phosphate, as we will see. Finally, it would not be efficient for the pyruvate kinase reaction to be the primary regulatory step for glycolysis because it occurs at the very end of the 10-step pathway. Even so, pyruvate kinase activity can be adjusted. In some organisms, fructose-1,6-bisphosphate activates pyruvate kinase at an allosteric site. This is an example of **feed-forward activation**: Once a monosaccharide has entered glycolysis, fructose-1,6-bisphosphate helps ensure rapid flux through the pathway.

The glycolytic pathway can be considered to be a sort of pipe, with an entrance for glucose (or glucose-6-phosphate) and an exit for pyruvate. The control points for the pathway are like one-way valves that prevent backflow. Between those control points, intermediates can move in either direction. The pipe never runs dry because the pyruvate gets used up and more glucose is always available. In addition, intermediates can enter or leave the pathway at any point. Even the simplest cells contain many copies of each glycolytic enzyme, acting on a pool of millions of substrate molecules, so their collective behavior is what we refer to when we discuss flux through the pathway.

To sum up the last five reactions of glycolysis: Glyceraldehyde-3-phosphate is converted to pyruvate with the synthesis of 2 ATP (in Reactions 7 and 10). Since each molecule of glucose yields two molecules of glyceraldehyde-3-phosphate, Reactions 6–10 must be doubled, so the yield is 4 ATP. Two molecules of ATP are invested in Reactions 1 and 3, bringing the net yield to 2 ATP produced per glucose molecule. Two NADH are also generated for each glucose molecule. Monosaccharides other than glucose are metabolized in a similar fashion to yield ATP (**Box 13.A**).

Box 13.A Catabolism of Sugars Other than Hexoses

Though starch is the major source of carbohydrates, the human diet also contains many carbohydrates other than glucose and its polymers. For example, lactose, a disaccharide of glucose and galactose, is present in milk and milk products (Section 11.2). Following uptake, lactose is hydrolyzed into glucose and galactose in the small intestine by lactase, and the two monosaccharides are absorbed, transported to the liver, and metabolized. Galactose undergoes phosphorylation and is interconverted by a multi-step isomerization reaction to enter the glycolytic pathway as glucose-6-phosphate. Its energy yield is the same as that of glucose. For galactose catabolism, there is no dedicated pathway. Hence, galactose is converted to a metabolic derivative of glucose, glucose-6-phosphate by an *interconversion pathway* comprising four steps. Galactokinase converts galactose to galactose-1-phosphate. Galactose-1-phosphate is subsequently first linked to an uridyl group from uridine diphosphate glucose (UDP-glucose) by galactose-1–phosphate uridyl transferase, and glucose 1-phosphate is released. The configuration of the hydroxyl group at C4 is inverted by UDP-galactose-4-epimerase which alters the configuration of the hydroxyl group at C4 to form UDP-glucose. The epimerase reaction is reversible and is of immense physiological significance as UDP-galactose is required by galactosyl transferases while synthesizing galactose containing polysaccharides and glycoproteins. In the last step of galactose catabolism glucose 1-phosphate, formed in the galactose-1-phosphate uridyl transferase step, is isomerized by *phosphoglucomutase* to form glucose-6-phosphate.

Interference with galactose metabolism like deficiency of glactose-1-phosphate uridyl transferase, leads to galactosemia. The pathophysiology involves elevated blood galactose levels and its presence in urine. Etiologically, the affected infants fail to thrive with severe vomiting and diarrhea after milk consumption, along with hepatomegaly and progressive cirrhosis. Effective remedial includes removal of galactose from diet though central nervous system malfunction and ovarian failure (females) often persist. The incidence of cataracts is directly correlated with milk consumption. In the absence or decline of transferase activity in the lens of the eye, aldose reductase reduces galactose to galactitol, an osmotically active sugar alcohol that stimulates diffusion of water into the lens culminating in cataracts.

The milk sugar lactose is the source of galactose which is produced along with glucose by the action of lactase on lactose. A gastrointestinal ailment, hypolactasia, or lactose intolerance is caused by a deficiency of lactase. Though depletion of lactase action during development is common for mammals, in matured individuals the action becomes one-tenth of that of infant stages when milk usually is the primary diet. Such depletion is not prominent in populations like Northern Europeans who can consume milk in adulthood without gastrointestinal difficulty. Cattle domestication along with the advantage of acquiring energy from milk sugars might confer a selected advantage to genomic variations linked to lactase production in adulthood.

Sucrose, the other major dietary disaccharide, is composed of glucose and fructose (Section 11.2); it is present in a variety of foods of plant origin. Like lactose, sucrose is hydrolyzed in the small intestine as well, and its monosaccharide components (glucose and fructose) are absorbed. Fructose is enriched in many common foods, particularly fruits and honey. It is highly soluble and tastes sweeter than sucrose, and its industrial production is less expensive. Thus, fructose is a preferred sweetener for manufacturing soft drinks and sweet processed foods. Overconsumption of fructose has been proposed to be directly correlated to the obesity epidemic. Fructose is metabolized primarily by the liver, but liver hexokinase present (glucokinase) has a very low affinity for fructose. Fructose, therefore, enters glycolysis through a different route.

First, fructokinase phosphorylates fructose to yield fructose-1-phosphate. Fructose-1-phosphate aldolase then splits it into glyceraldehyde and dihydroxyacetone phosphate. Dihydroxyacetone phosphate is converted to glyceraldehyde-3-phosphate by triose phosphate isomerase and can proceed through the second phase of glycolysis. The glyceraldehyde can be phosphorylated to glyceraldehyde-3-phosphate. It is also the precursor for glycerol-3-phosphate which forms the backbone of triacylglycerols. Thus, fructose consumption is linked to fat deposition. A second potential hazard of fructose catabolism is that it bypasses the phosphofructokinase catalyzed step of glycolysis, a major regulatory point. Thus, the system loses regulation over fuel metabolism. Though fructose consumption has been decreasing in the last two decades, the prevalence of obesity remains high. Also, in a typical diet, individuals consume about five times more glucose than fructose. So effectively, total carbohydrate consumption might simply be accountable for fat accumulation rather than consumption of a particular sugar.

Question **When its concentration is extremely high, fructose is converted to fructose-1-phosphate much faster than it can be cleaved by the aldolase. How would this affect the cell's ATP supply?**

Some cells convert pyruvate to lactate or ethanol

What happens to the pyruvate generated by the catabolism of glucose? It can be further broken down or used to synthesize other compounds. The fate of pyruvate depends on the cell type and the need for metabolic free energy and molecular building blocks. For example, pyruvate may be converted to a two-carbon acetyl group linked to the carrier coenzyme A. The acetyl group may be further broken down by the citric acid cycle or used to synthesize fatty acids. In muscle, pyruvate is reduced to lactate. Yeast degrade pyruvate to ethanol and CO_2. Pyruvate can also be carboxylated to produce the four-carbon oxaloacetate. These options are summarized in **Figure 13.8**.

During exercise, pyruvate may be temporarily converted to lactate. In a highly active muscle cell, glycolysis rapidly provides ATP to power muscle contraction, but the pathway also consumes NAD^+ at the glyceraldehyde-3-phosphate dehydrogenase step. The two NADH molecules generated for each glucose molecule catabolized can be reoxidized in the presence of oxygen. However, this process is too slow to replenish the NAD^+ needed for the rapid production of ATP by glycolysis. *To regenerate NAD^+, the enzyme lactate dehydrogenase reduces pyruvate to lactate:*

Pyruvate + NADH + H⁺ ⇌ (lactate dehydrogenase) Lactate + NAD⁺

This reaction, sometimes called the eleventh step of glycolysis, allows the muscle to function anaerobically for a minute or two. The net reaction for anaerobic glucose catabolism is

$$\text{glucose} + 2\,\text{ADP} + 2\,\text{P}_i \rightarrow 2\,\text{lactate} + 2\,\text{ATP}$$

Lactate represents a sort of metabolic dead end: Its only options are to be eventually converted back to pyruvate (the lactate dehydrogenase reaction is reversible) or to be exported from the cell. The liver takes up lactate from the blood, oxidizes it back to pyruvate, and then converts the pyruvate to glucose. The fuel produced in this manner may eventually make its way back to the muscle to help power continued contraction. When the muscle is functioning aerobically, NADH produced by the glyceraldehyde-3-phosphate dehydrogenase reaction is reoxidized by oxygen and the lactate dehydrogenase reaction is not needed. Cancer cells sustain high rates of lactate production, even under aerobic conditions, as part of the metabolic alterations that allow rapid cell growth (Section 19.4).

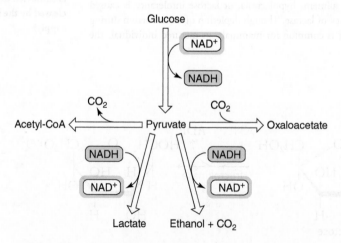

Figure 13.8 Fates of pyruvate. Pyruvate can be converted to other 2-, 3-, and 4-carbon molecules.

Question Beside each arrow, write the names of the enzymes that catalyze the process.

Organisms such as yeast growing under anaerobic conditions can regenerate NAD^+ by producing alcohol. In the mid-1800s, Louis Pasteur called this process **fermentation,** meaning life without air, although yeast also ferment sugars in the presence of O_2. Alcoholic fermentation is a two-step process. First, pyruvate decarboxylase (an enzyme not present in animals) catalyzes the removal of pyruvate's carboxylate group to produce acetaldehyde. Next, alcohol dehydrogenase reduces acetaldehyde to ethanol.

Pyruvate → (pyruvate decarboxylase, CO_2) → Acetaldehyde → (alcohol dehydrogenase, NADH NAD$^+$) → Ethanol

Ethanol is considered to be a waste product of sugar metabolism; its accumulation is toxic to other organisms (**Box 13.B**), including the yeast that produce it. This is one reason why the alcohol content of yeast-fermented beverages such as wine is limited to about 13%. "Hard" liquor must be distilled to increase its ethanol content.

Yeast and other fungi, as well as bacteria, do more than convert glucose to ethanol or lactate. Microbial fermentation transforms a variety of animal and plant products to foods such as kimchi,

---

Box 13.B   Alcohol Metabolism

Unlike yeast, mammals do not produce ethanol, although it is naturally present in many foods and is produced in small amounts by intestinal microorganisms. The liver is equipped to metabolize ethanol, a small, weakly polar substance that is readily absorbed from the gastrointestinal tract and transported by the bloodstream. First, alcohol dehydrogenase converts ethanol to acetaldehyde. This is the reverse of the reaction yeast use to produce ethanol. A second reaction converts acetaldehyde to acetate:

Ethanol → (alcohol dehydrogenase, NAD^+ → NADH, H^+) → Acetaldehyde

Acetaldehyde → (acetaldehyde dehydrogenase, OH^- NAD^+ → NADH, H^+) → Acetate

Note that both of these reactions require NAD^+, a cofactor used in many other oxidative cellular processes, including glycolysis. The liver uses the same two-enzyme pathway to metabolize the excess ethanol obtained from alcoholic beverages. Ethanol itself is mildly toxic, and the physiological effects of alcohol also reflect the toxicity of acetaldehyde and acetate in tissues such as the liver and brain.

Ethanol induces vasodilation, apparent as flushing (warming and reddening of the skin due to increased blood flow). At the same time, the heart rate and respiration rate become slightly lower. The kidneys increase the excretion of water as ethanol interferes with the ability of the hypothalamus (a region of the brain) to properly sense osmotic pressure.

Ethanol also stimulates signaling from certain neurotransmitter receptors that function as ligand-gated ion channels (Section 9.2) to inhibit neuronal signaling, producing a sedative effect. Sensory, motor, and cognitive functions are impaired, leading to delayed reaction time, loss of balance, and blurred vision. Some of the symptoms of ethanol intoxication can be experienced even at low doses, when the blood alcohol concentration is less than 0.05%. At high doses, usually at blood concentrations above 0.25%, ethanol can cause loss of consciousness, coma, and death.

The mostly pleasant responses to moderate ethanol consumption are followed by a period of recovery, when the concentrations of ethanol metabolites are relatively high. The unpleasant symptoms of a hangover in part reflect the chemistry of producing acetaldehyde and acetate. Their production in the liver consumes NAD^+, thereby lowering the cell's NAD^+:NADH ratio. Without sufficient NAD^+, the liver's ability to produce ATP by glycolysis is diminished (since NAD^+ is required for the glyceraldehyde-3-phosphate dehydrogenase reaction). Acetaldehyde itself can react with liver proteins, inactivating them. Acetate (acetic acid) production lowers blood pH.

Long-term, excessive alcohol consumption exacerbates the toxic effects of ethanol and its metabolites. For example, a shortage of liver NAD^+ slows fatty acid breakdown (like glycolysis, a process that requires NAD^+) and promotes fatty acid synthesis, leading to fat accumulation in the liver. Over time, cell death causes permanent loss of function in the central nervous system. The death of liver cells and their replacement by fibrous scar tissue causes liver cirrhosis.

Question **Normally, the liver converts lactate, produced mainly by muscles, back to pyruvate so that the pyruvate can be converted to glucose by gluconeogenesis (Section 13.2). How do the activities of alcohol dehydrogenase and acetaldehyde dehydrogenase contribute to hypoglycemia?**

cheese, salami, and soy sauce. Even *Cacao* seeds undergo fermentation during chocolate production. In many cases, fermentation involves a community of organisms, acting together or in succession, to break down food molecules. In traditional cheesemaking, lactate (lactic acid) produced by bacterial fermentation of milk lowers the pH and causes the denaturation and solidification of milk proteins to produce soft, spreadable cheeses. Hard cheeses develop more slowly, as the proteins are catabolized by bacterial enzymes over months to produce more flavorful compounds. The interior of an aging cheese is usually anaerobic, but in veined cheeses like blue cheese, aerobic fungi grow—and produce characteristic by-products—in air channels within the cheese.

Pyruvate is the precursor of other molecules

Although glycolysis is an oxidative pathway, its end product pyruvate is still a relatively reduced molecule. The further catabolism of pyruvate begins with its oxidative decarboxylation to form a two-carbon acetyl group linked to coenzyme A.

$$
\underset{\text{Pyruvate}}{\text{CH}_3-\overset{\overset{\displaystyle O}{\|}}{C}-COO^-} \quad \xrightarrow[\text{dehydrogenase}]{\overset{\displaystyle \text{HSCoA} \;\; \text{NAD}^+ \;\; \text{NADH} \;\; CO_2}{\text{pyruvate}}} \quad \underset{\text{Acetyl-CoA}}{\text{CH}_3-\overset{\overset{\displaystyle O}{\|}}{C}-SCoA}
$$

The resulting acetyl-CoA is a substrate for the citric acid cycle (Chapter 14), which converts the acetyl carbon atoms to CO_2. The complete oxidation of the six glucose carbons to CO_2 ($\Delta G^{\circ\prime} = -2850 \text{ kJ} \cdot \text{mol}^{-1}$) releases much more free energy than the conversion of glucose to lactate ($\Delta G^{\circ\prime} = -196 \text{ kJ} \cdot \text{mol}^{-1}$). Much of this energy is recovered in the synthesis of ATP by the reactions of the citric acid cycle and oxidative phosphorylation (Chapter 15), pathways that require the presence of molecular oxygen.

Pyruvate is not always destined for catabolism. *Its carbon atoms provide the raw material for synthesizing a variety of molecules*, including, in the liver, more glucose (discussed in the following section). Fatty acids, the precursors of triacylglycerols and many membrane lipids, can be synthesized from the two-carbon units of acetyl-CoA derived from pyruvate. This is how fat is produced from excess carbohydrate.

Pyruvate is also the precursor of oxaloacetate, a four-carbon molecule that is an intermediate in the synthesis of several amino acids. It is also one of the intermediates of the citric acid cycle. Oxaloacetate is synthesized by the action of pyruvate carboxylase:

$$
\underset{\text{Pyruvate}}{\overset{\displaystyle COO^-}{\underset{\displaystyle CH_3}{\overset{\displaystyle |}{\underset{|}{C=O}}}}} + CO_2 + ATP \xrightarrow[\text{carboxylase}]{\text{pyruvate}} \underset{\text{Oxaloacetate}}{\overset{\displaystyle COO^-}{\underset{\displaystyle COO^-}{\overset{\displaystyle |}{\underset{|}{\underset{CH_2}{\overset{C=O}{|}}}}}}} + ADP + P_i
$$

The pyruvate carboxylase reaction is interesting because of its unusual chemistry. The enzyme has a biotin prosthetic group that acts as a carrier of CO_2. Biotin is considered a vitamin, but a deficiency is rare because it is present in many foods and is synthesized by intestinal bacteria. The biotin group is covalently linked to an enzyme lysine residue to form the functional biocytin cofactor:

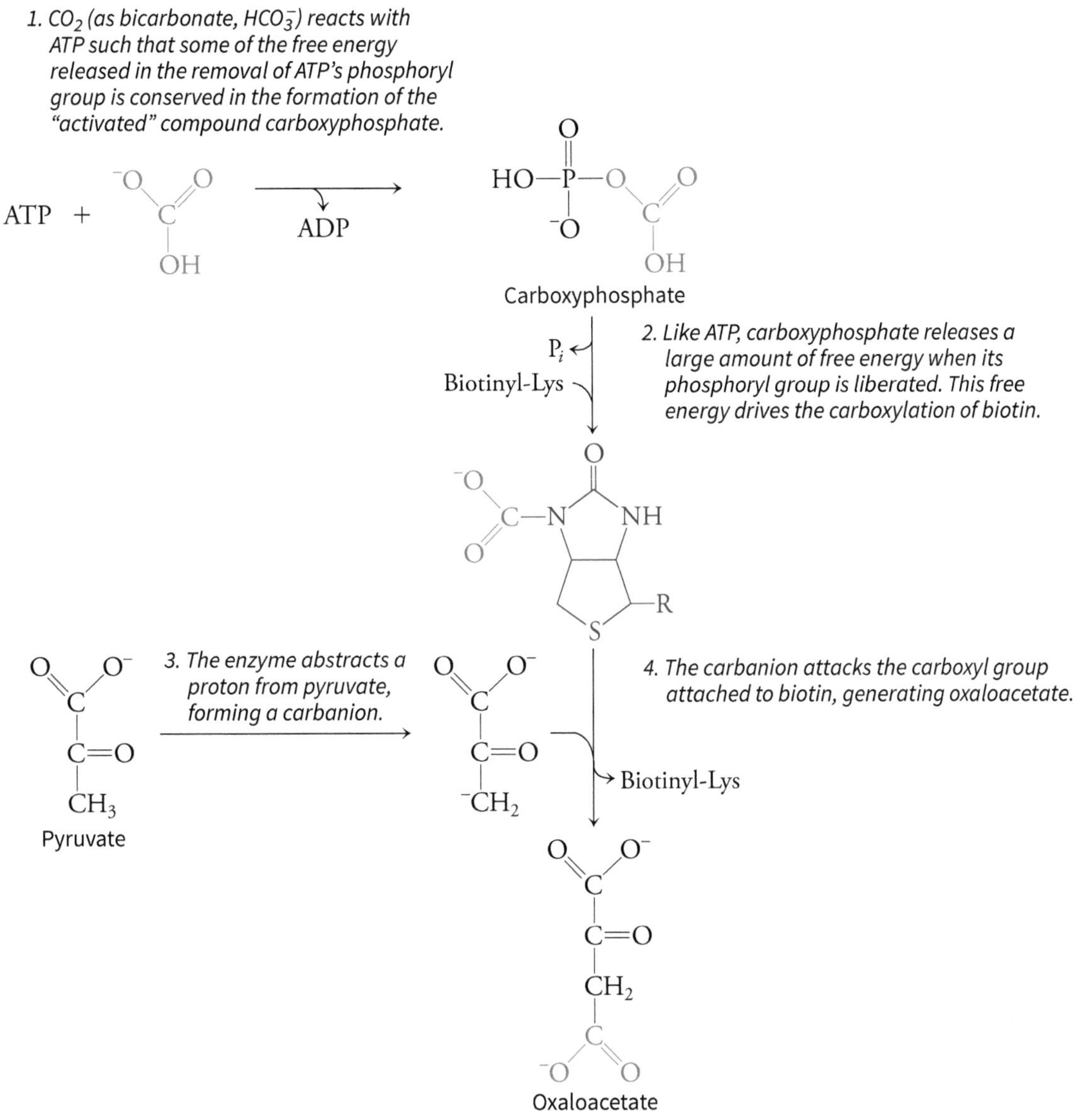

1. CO₂ (as bicarbonate, HCO₃⁻) reacts with ATP such that some of the free energy released in the removal of ATP's phosphoryl group is conserved in the formation of the "activated" compound carboxyphosphate.

ATP +

Carboxyphosphate

2. Like ATP, carboxyphosphate releases a large amount of free energy when its phosphoryl group is liberated. This free energy drives the carboxylation of biotin.

Biotinyl-Lys

3. The enzyme abstracts a proton from pyruvate, forming a carbanion.

Pyruvate

4. The carbanion attacks the carboxyl group attached to biotin, generating oxaloacetate.

Biotinyl-Lys

Oxaloacetate

Figure 13.9 The pyruvate carboxylase reaction.

The lysine side chain and its attached biotin group form a 14-Å-long flexible arm that swings between two active sites in the enzyme. In one active site, a CO_2 molecule is first "activated" by its reaction with ATP, and is then transferred to the biotin. The second active site transfers the carboxyl group to pyruvate to produce oxaloacetate (**Fig. 13.9**).

Concept Check

1. Write the net equation for glycolysis.
2. Draw the structures of the substrates and products of the 10 glycolytic reactions and name the enzyme that catalyzes each step.
3. List the glycolytic reactions that consume ATP or generate ATP.
4. Calculate the net yield of ATP and NADH per glucose molecule.
5. Identify the reactions that serve as flux-control points for glycolysis.
6. List the possible metabolic fates of pyruvate.
7. Metabolic function of lactate dehydrogenase.

13.2 Gluconeogenesis

KEY CONCEPTS

Describe the substrates, products, and reactions of gluconeogenesis.

* List the enzymes that are unique to gluconeogenesis or are shared with glycolysis.
* Explain how the rates of gluconeogenesis and glycolysis are related.

Conversion of pyruvate to glucose

We have already alluded to the ability of the liver to synthesize glucose from noncarbohydrate precursors via the pathway of gluconeogenesis, which operates when the liver's supply of glycogen is exhausted. This pathway also occurs to a limited extent in kidneys. Certain tissues, such as the central nervous system and red blood cells, which burn glucose as their primary metabolic fuel, rely on the liver to supply them with newly synthesized glucose. Gluconeogenesis is the reversal of glycolysis, where two pyruvate molecules are converted to one glucose molecule. Although some of the steps of gluconeogenesis are catalyzed by the reversal of the action of some glycolytic enzymes, ΔG for glycolysis is ~ -84 kJ mol^{-1}. Hexokinase (1), phosphofructokinase (3), and pyruvate kinase (10) catalyze the irreversible steps of glycolysis leading to such reduction of ΔG. For gluconeogenesis, these steps are strategically bypassed (**Fig. 13.10**).

1. Conversion of Pyruvate to phosphoenol by pyruvate carboxylase and phosphoenolpyruvate carboxykinase

Pyruvate carboxylase synthesizes oxaloacetate from pyruvate by carboxylation at C3. The enzyme utilizes one molecule of ATP which activates the enzyme by binding at the N-terminal ATP binding domain. At the C-terminal biotin prosthetic group is covalently linked with to e-amino group of a specific lysine side chain. The carboxylation of pyruvate takes place in three stages: The HCO_3^- is activated to carboxyphosphate to the N-1 atom of the biotin ring to form the carboxybiotin enzyme intermediate. From carboxybiotin, the activated carboxyl group is then transferred to pyruvate forming oxaloacetate. The carboxylation of biotin requires allosteric activation of the enzyme by acetyl CoA.

Pyruvate carboxylase is localized in the mitochondria and since the rest of the enzymes of gluconeogenesis are localized in the cytosol, oxaloacetate produced by pyruvate carboxylase is transported to the cytoplasm by exploiting the malate aspartate shuttle. Mitochondrial NADH-linked malate dehydrogenase reduces oxaloacetate to malate, that is transported across the mitochondrial membrane to cytosol and reoxidized to oxaloacetate by cytoplasmic malate dehydrogenase. Subsequently, oxaloacetate is decarboxylated and phosphorylated by *phosphoenolpyruvate carboxykinase* to yield phosphoenolpyruvate. The reaction requires hydrolysis of one GTP molecule to become energetically favorable.

Phosphoenolpyruvate is metabolized in the reverse direction of glycolysis by the enzymes that catalyze reactions 8 to 4 of glycolysis which occur near equilibrium under physiological conditions until it reaches the next irreversible step (phosphofructokinase).

Pyruvate   Oxaloacetate   Phosphoenolpyruvate

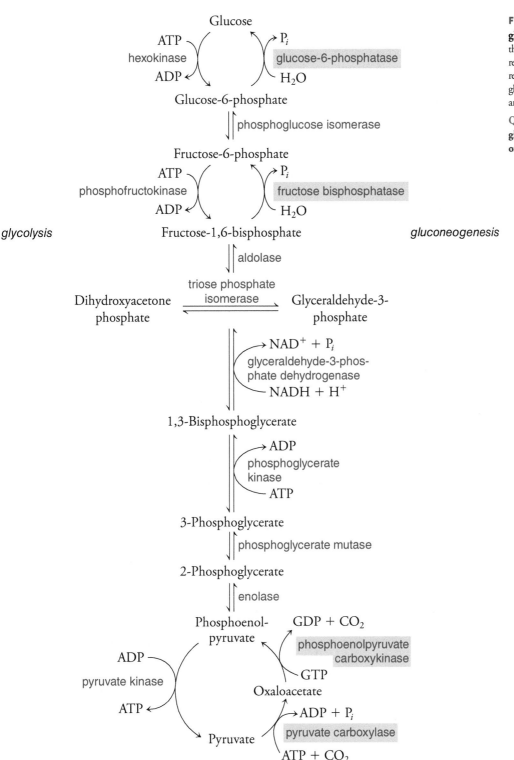

Figure 13.10 The reactions of gluconeogenesis. The pathway uses the seven glycolytic enzymes that catalyze reversible reactions. The three irreversible reactions of glycolysis are bypassed in gluconeogenesis by the four enzymes that are highlighted in blue.

Question Compare the ATP yield of glycolysis with the ATP consumption of gluconeogenesis.

2. Fructose-6-phosphate is formed from fructose-1,6-bisphosphate by hydrolysis of the phosphate ester at carbon

Fructose-1,6-bisphosphate produced by aldolase is hydrolyzed to fructose-6-phosphate and P_i. The hydrolase responsible for this step is fructose-1,6-bisphosphatase. The enzyme is allosterically regulated by fructose-2,6-bisphosphate and is a critical control point for glucose production by gluconeogenesis. The fructose-6-phosphate generated by fructose-1,6-bisphosphatase is readily converted into glucose-6-phosphate. In majority of the tissues, gluconeogenesis ends here.

3. Glucose is formed by hydrolysis of glucose-6-phosphate in a reaction catalyzed by glucose-6-phosphatase

Glucose-6-phosphate is hydrolyzed to glucose by glucose-6-phosphatase after being transported into the lumen of the endoplasmic reticulum. Glucose-6-phosphatase is a membrane-bound enzyme and requires a Ca^{2+} binding stabilizing for phosphatase activity. Glucose and P_i generated by the reaction are shuttled back to the cytosol by respective transporters.

Gluconeogenic enzymes and glycolytic enzymes

Glycolysis and gluconeogenesis are harmonized in a way within a cell so that either of the pathways is active at a time. Simultaneous activation of both would be futile with hydrolysis of four nucleotide triphosphates per reaction cycle. Overall, both pathways are highly exergonic, and hence there is no apparent thermodynamic barrier as such. In this context, several effective enzymes and the activity of each enzyme in each of the pathways are precisely regulated.

One of the critical nodes of such regulation is the interconversion of fructose-6-phosphate and fructose-1,6-bisphosphate. The AMP-ATP ratio acts as an energy charge indicator as a high level of AMP signals the need for ATP generation. Conversely, high levels of ATP and citrate indicate that biosynthetic and metabolic precursors are abundant. Phosphofructokinase activity is fostered by AMP and is dampened by citrate and ATP. On the contrary, citrate activates fructose-1,6-bisphosphatase and AMP inhibits the enzyme. Hence, glycolysis is triggered under energy depleted conditions whereas gluconeogenesis is triggered when energy levels are high barring such metabolic intermediates, in the liver fructose-2,6-bisphosphate acts as a signal molecule to modulate the activities of the two enzymes. Fructose-2,6-bisphosphate allosterically activates phosphofructokinase and inhibits fructose-1,6-bisphosphatase. The level of F-2,6-BP is low during starvation and high in the fed state, owing to the antagonistic effects of glucagon and insulin on the production and degradation of this signal molecule. Fructose-2,6-bisphosphate strongly stimulates phosphofructokinase and inhibits fructose-1,6-bisphosphatase. Hence, glycolysis is accelerated, and gluconeogenesis is diminished when glucose is abundant. During starvation, the level of F-2,6-BP declines and gluconeogenesis predominates. Glucose formed by the liver under those conditions is essential for the functioning of the brain and muscle.

The interconversion of phosphoenolpyruvate and pyruvate is also explicitly regulated. Pyruvate kinase is controlled by allosteric effectors like ATP, alanine, and by phosphorylation. Conversely, pyruvate carboxylase is activated by acetyl CoA and inhibited by ADP. Likewise, ADP inhibits phosphoenolpyruvate carboxykinase. Hence, elevated levels of ATP favor gluconeogenesis.

The level and the activities of the enzymes are regulated by glucoregulatory hormones as well, via transcriptional, post-transcriptional, translational, and post-translational regulation. Insulin enhances the expression of phosphofructokinase, pyruvate kinase, and PFK-2-FBPase-2, a bifunctional enzyme that makes and degrades fructose-2,6-bisphosphate. Glucagon, on the flip side, limits the expression of these enzymes and increases the production of phosphoenolpyruvate carboxykinase and fructose-1,6-bisphosphatase. The generation and degradation of fructose-2,6-bisphosphate are stimulated by insulin and glucagon respectively. Though controlling gene expression is slower than allosteric regulation, a combination of both precisely regulates glycolytic and glucogenic enzymes to respond according to the condition.

Gluconeogenesis is regulated at the fructose bisphosphatase step

Gluconeogenesis is energetically expensive. Producing 1 glucose from 2 pyruvate consumes 6 ATP, 2 each at the steps catalyzed by pyruvate carboxylase, phosphoenolpyruvate carboxykinase, and phosphoglycerate kinase. If glycolysis occurred simultaneously with gluconeogenesis, there would be a net consumption of ATP:

glycolysis	glucose + 2 ADP + 2 P_i → 2 pyruvate + 2 ATP
gluconeogenesis	2 pyruvate + 6 ATP → glucose + 6 ADP + 6 P_i
net	4 ATP → 4 ADP + 4 P_i

To avoid this waste of metabolic free energy, gluconeogenic cells (mainly liver cells) carefully regulate the opposing pathways of glycolysis and gluconeogenesis according to the cell's energy needs. *The major regulatory point is centered on the interconversion of fructose-6-phosphate and fructose-1,6-bisphosphate.* We have already seen that fructose-2,6-bisphosphate is a potent allosteric activator of phosphofructokinase, which catalyzes step 3 of glycolysis. Not surprisingly, fructose-2,6-bisphosphate is a potent *inhibitor* of fructose bisphosphatase, which catalyzes the opposing gluconeogenic reaction.

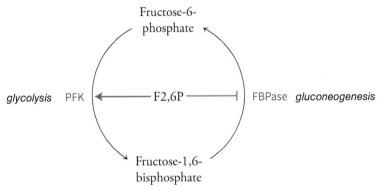

This mode of allosteric regulation is efficient because *a single compound can control flux through two opposing pathways in a reciprocal fashion.* Thus, when the concentration of fructose-2,6-bisphosphate is high, glycolysis is stimulated and gluconeogenesis is inhibited, and vice versa.

Many cells that do not carry out gluconeogenesis do contain the gluconeogenic enzyme fructose bisphosphatase. What is the reason for this? When both fructose bisphosphatase (FBPase) and phosphofructokinase (PFK) are active, the net result is the hydrolysis of ATP:

PFK	fructose-6-phosphate + ATP → fructose-1,6-bisphosphate + ADP
FBPase	fructose-1,6-bisphosphate + H_2O → fructose-6-phosphate + P_i
net	ATP + H_2O → ADP + P_i

This combination of metabolic reactions is called a **futile cycle** since it seems to have no useful result. However, Eric Newsholme realized that such futile cycles could actually provide a means for fine-tuning the output of a metabolic pathway. For example, flux through the phosphofructokinase step of glycolysis is diminished by the activity of fructose bisphosphatase. An allosteric compound such as fructose-2,6-bisphosphate modulates the activity of both enzymes so that as the activity of one enzyme increases, the activity of the other one decreases. This dual regulatory effect results in a greater possible range of net flux than if the regulator merely activated or inhibited a single enzyme. Similarly, a car's speed is easier to control if it has both an accelerator and a brake.

Concept Check

1. List the reactions of gluconeogenesis that are catalyzed by glycolytic enzymes.
2. List the enzymes that are unique to gluconeogenesis and explain why they are needed.
3. Describe the fructose-6-phosphate futile cycle and explain its purpose.

13.3 Glycogen Synthesis and Degradation

KEY CONCEPTS

Compare the processes of glycogen synthesis and degradation.

- Identify the substrates and products for each process.
- Compare the energy needs of each pathway.
- List the metabolic fates of glucose-6-phosphate.

In animals, dietary glucose and the glucose produced by gluconeogenesis are stored in the liver and other tissues as glycogen. Later, glucose units can be removed from the glycogen polymer by phosphorolysis (see Section 12.1). Because glycogen degradation is thermodynamically spontaneous, glycogen synthesis requires the input of free energy. The two opposing pathways use different sets of enzymes so that each process can be thermodynamically favorable under cellular conditions.

Glycogen synthesis consumes the energy of UTP

The monosaccharide unit that is incorporated into glycogen is glucose-1-phosphate, which is produced from glucose-6-phosphate (the penultimate product of gluconeogenesis) by the action of the enzyme phosphoglucomutase:

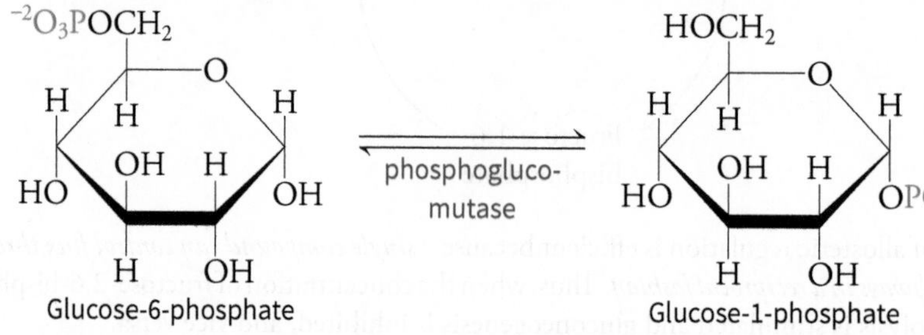

In mammalian cells, glucose-1-phosphate is then "activated" by reacting with UTP to form UDP–glucose (like GTP, UTP is energetically equivalent to ATP).

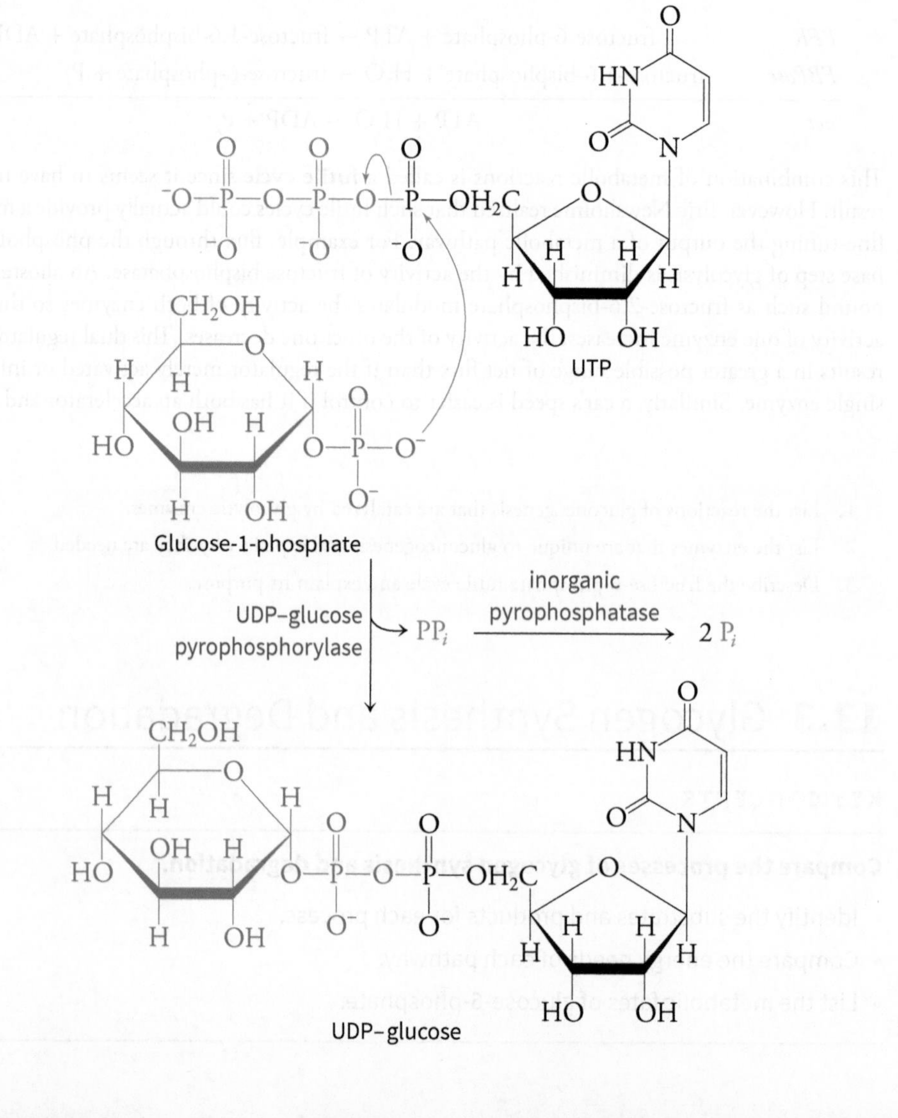

This process is a reversible phosphoanhydride exchange reaction ($\Delta G \approx 0$). Note that the two phosphoanhydride bonds of UTP are conserved, one in the product PP_i and one in UDP–glucose. However, the PP_i is rapidly hydrolyzed by inorganic pyrophosphatase to 2 P_i in a highly exergonic reaction ($\Delta G°' = -19.2$ kJ·mol^{-1}). Thus, cleavage of a phospho-anhydride bond makes the formation of UDP–glucose an exergonic, irreversible process—that is, *PP_i hydrolysis "drives" a reaction that would otherwise be near equilibrium.* The hydrolysis of PP_i by inorganic pyrophosphatase is a common strategy in biosynthetic reactions; we will see this reaction again in the synthesis of other polymers, namely DNA, RNA, and polypeptides.

Finally, glycogen synthase transfers the glucose unit to the C4 OH group at the end of one of glycogen's branches to extend the linear polymer of $\alpha(1 \rightarrow 4)$-linked residues.

A separate enzyme, called a transglycosylase or branching enzyme, cleaves off a seven-residue segment and reattaches it to a glucose C6 OH group to create an $\alpha(1 \rightarrow 6)$ branch point.

The steps of glycogen synthesis can be summarized as follows:

UDP–glucose pyrophosphorylase	glucose-1-phosphate + UTP $\rightleftharpoons$ UDP–glucose + PP_i
pyrophosphatase	$PP_i + H_2O \rightarrow 2\ P_i$
glycogen synthase	UDP–glucose + glycogen (*n* residues) $\rightarrow$ glycogen (*n* +1 residues) + UDP
net	glucose-1-phosphate + glycogen + UTP + $H_2O \rightarrow$ glycogen + UDP + 2 P_i

The energetic cost of adding one glucose-1-phosphate molecule to glycogen is the cleavage of one phosphoanhydride bond of UTP. Adding a glucose molecule would cost two phos-phoanhydride bonds, since ATP would be spent to convert glucose to glucose-1-phosphate. Nucleotides are also required for the synthesis of other saccharides. For example, lactose is synthesized from glucose and UDP–galactose. In plants, starch is synthesized using ADP–glucose, and cellulose is synthesized using CDP–glucose as starting materials.

Glycogen synthase cannot build a new glycogen molecule from scratch; it can only extend a pre-existing chain of glucose residues. The first residues of a new chain are actually assembled by a pair of small proteins called glycogenin. Each glycogenin attaches one or two glucose residues (donated by ADP–glucose) to a tyrosine side chain on the other glycogenin protein. Then each glycogenin extends its own glucose chain by adding another dozen or so glucose residues. At this point, glycogen synthase and the branching enzyme take over, expanding the glycogen molecule by tens of thousands of glucose residues to form a spherical glycogen granule (**Fig. 13.11**).

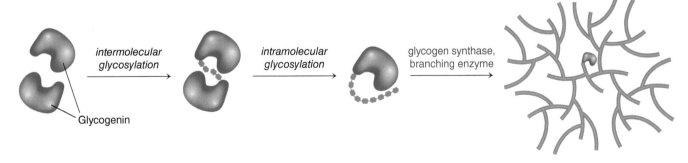

Figure 13.11 Growth of a glycogen particle. Glycogenin initiates glucose polymerization by intermolecular glycosylation followed by intramolecular glycosylation. Glycogen synthase then extends the chain, and branching enzyme creates branches.

Question Draw the 1-*O*-linked glucose–tyrosine complex.

Glycogen phosphorylase catalyzes glycogenolysis

Glycogen breakdown follows a different set of steps than glycogen synthesis. *In **glycogenolysis**, glycogen is phosphorolyzed, not hydrolyzed, to yield glucose-1-phosphate.* However, a debranching enzyme removes $\alpha(1 \rightarrow 6)$-linked residues by hydrolysis. In the liver, phosphoglucomutase converts glucose-1-phosphate to glucose-6-phosphate, which is transported into the endoplasmic reticulum and hydrolyzed by glucose-6-phosphatase to release free glucose.

$$\text{glycogen} \xrightarrow[\substack{\text{glycogen} \\ \text{phosphorylase}}]{P_i} \text{glucose-1-phosphate} \xrightleftharpoons[\substack{\text{phosphogluco-} \\ \text{mutase}}]{} \text{glucose-6-phosphate} \xrightarrow[\substack{\text{glucose-6-} \\ \text{phosphatase}}]{H_2O \quad P_i} \text{glucose}$$

This glucose leaves the cell and enters the bloodstream. Only gluconeogenic tissues such as the liver can make glucose available to the body at large. Other tissues that store glycogen, such as muscle, lack glucose-6-phosphatase and so break down glycogen only for their own needs. In these tissues, the glucose-1-phosphate liberated by phosphorolysis of glycogen is converted to glucose-6-phosphate, which then enters glycolysis at the phosphoglucose isomerase reaction (step 2). The hexokinase reaction (step 1) is skipped, thereby sparing the consumption of ATP. Consequently, *glycolysis using glycogen-derived glucose has a higher net yield of ATP than glycolysis using glucose supplied by the bloodstream.*

Because the mobilization of glucose must be tailored to meet the energy demands of a particular tissue or the entire body, the activity of glycogen phosphorylase is carefully regulated by a variety of mechanisms linked to hormonal signaling. Likewise, the activity of glycogen synthase is subject to hormonal control, so that the two opposing pathways of degradation and synthesis do not occur at the same time. In Chapter 19 we will examine some of the mechanisms for regulating different aspects of fuel metabolism, including glycogen synthesis and degradation. Some disorders of glycogen metabolism are discussed in Section 13.5.

Concept Check

1. Describe the role of UTP in glycogen synthesis.
2. Explain why only some tissues contain glucose-6-phosphatase.
3. Explain the advantage of breaking down glycogen by phosphorolysis rather than hydrolysis. Explain why only some tissues contain glucose-6-phosphatase.

13.4 The Pentose Phosphate Pathway

KEY CONCEPTS

Describe the substrates, products, and reactions of the pentose phosphate pathway.

- Identify the oxidation–reduction reactions of the pentose phosphate pathway.
- Explain how the pathway responds to the cell's need for ribose groups.

We have already seen that glucose catabolism can lead to pyruvate, which can be further oxidized to generate more ATP or used to synthesize amino acids and fatty acids. Glucose is also a precursor of the ribose groups used for nucleotide synthesis. The **pentose phosphate pathway,** which converts glucose-6-phosphate to ribose-5-phosphate, is an oxidative pathway that occurs in all cells. But unlike glycolysis, the pentose phosphate pathway generates NADPH rather than NADH. The two cofactors are not interchangeable and are easily distinguished by degradative enzymes (which generally use NAD^+ as an oxidizing agent) and biosynthetic enzymes (which generally use NADPH as a reducing agent). The pentose phosphate pathway is by no means a minor feature of glucose metabolism. As much as 30% of glucose in the liver may be catabolized by the pentose phosphate pathway. This pathway can be divided into two phases: a series of oxidative reactions followed by a series of reversible interconversion reactions.

The oxidative reactions of the pentose phosphate pathway produce NADPH

The starting point of the pentose phosphate pathway is glucose-6-phosphate, which can be derived from free glucose, from the glucose-1-phosphate produced by glycogen phosphorolysis, or from gluconeogenesis. In the first step of the pathway, glucose-6-phosphate dehydrogenase catalyzes the metabolically irreversible transfer of a hydride ion from glucose-6-phosphate to $NADP^+$, forming a lactone and NADPH:

Glucose-6-phosphate → 6-Phosphoglucono-δ-lactone

The lactone intermediate is hydrolyzed to 6-phosphogluconate by the action of 6-phosphoglu-conolactonase, although this reaction can also occur in the absence of the enzyme:

6-Phosphoglucono-δ-lactone → 6-Phosphogluconate

In the third step of the pentose phosphate pathway, 6-phosphogluconate is oxidatively decarbox-ylated in a reaction that converts the six-carbon sugar to a five-carbon sugar and reduces a second $NADP^+$ to NADPH:

6-Phosphogluconate → Ribulose-5-phosphate

The two molecules of NADPH produced for each glucose molecule that enters the pathway are used primarily for biosynthetic reactions, such as fatty acid synthesis and the synthesis of deoxynucleotides.

Isomerization and interconversion reactions generate a variety of monosaccharides

The ribulose-5-phosphate product of the oxidative phase of the pentose phosphate pathway can isomerize to ribose-5-phosphate:

Ribulose-5-phosphate   Ribose-5-phosphate

Ribose-5-phosphate is the precursor of the ribose unit of nucleotides. In many cells, this marks the end of the pentose phosphate pathway, which has the net equation

$$\text{glucose-5-phosphate} + 2\ NADP^+ + H_2O \rightarrow \text{ribose-5-phosphate} + 2\ NADPH + CO_2 + 2\ H^+$$

Not surprisingly, the activity of the pentose phosphate pathway is high in rapidly dividing cells that must synthesize large amounts of DNA. In fact, the pentose phosphate pathway not only produces ribose, it also provides a reducing agent (NADPH) required for the reduction of ribose to deoxyribose. Ribonucleotide reductase carries out the reduction of nucleoside diphosphates (NDPs):

NDP   dNDP

The enzyme, which is oxidized in the process, is restored to its original state by a series of reactions in which NADPH is reduced (Section 18.5).

In some cells, however, the need for NADPH for other biosynthetic reactions is greater than the need for ribose-5-phosphate. In this case, *the excess carbons of the pentose are recycled into intermediates of the glycolytic pathway so that they can be degraded to pyruvate or used in gluconeogenesis,* depending on the cell type and its metabolic needs.

Ribose-5-phosphate and xylulose-5-phosphate, formed from ribulose-5-phosphate by epimerization, participate in a set of reversible reactions that transform three five-carbon sugars into two six-carbon sugars (fructose-6-phosphate) and one three-carbon sugar (glyceraldehyde-3-phosphate, **Fig. 13.12**). These molecular rearrangements are catalyzed by the enzymes transaldolase, which transfers a three-carbon segment, and transketolase, which transfers a two-carbon segment (the reaction catalyzed by transketolase was introduced in Section 7.2). Because all these interconversions are reversible, *glycolytic intermediates can also be siphoned from glycolysis or gluconeogenesis to synthesize ribose-5-phosphate.* Thus, the cell can use some or all of the steps of the pentose phosphate pathway in order to generate NADPH, to produce ribose, and to interconvert other monosaccharides.

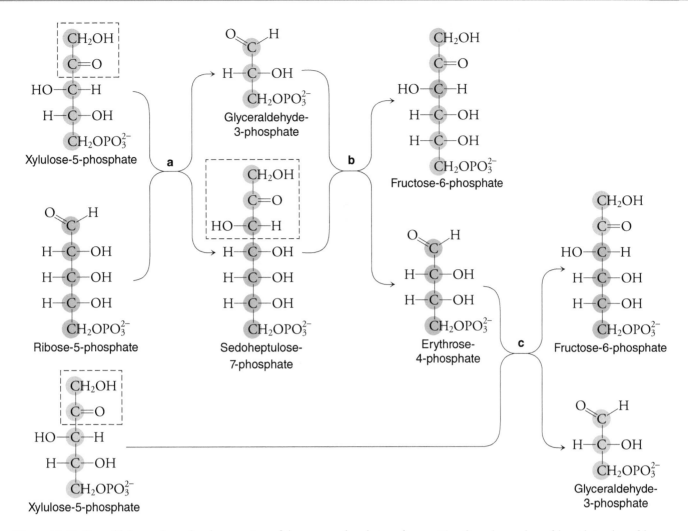

Figure 13.12 Transaldolase and transketolase reactions of the pentose phosphate pathway. Three five-carbon products of the oxidative phase of the pentose phosphate pathway are converted to other monosaccharides by reversible reactions. In Reactions a and c, transketolase transfers a two-carbon unit (boxed) and in Reaction b, transaldolase transfers a three-carbon unit (boxed).

Question **Make a similar color-coded diagram to show how glycolytic intermediates could be transformed to ribose-5-phosphate.**

A summary of glucose metabolism

Although our coverage of glucose metabolism is far from exhaustive, this chapter describes quite a few enzymes and reactions, which are compiled in **Figure 13.13**. As you examine this diagram, keep in mind the following points, which also apply to the metabolic pathways we will encounter in subsequent chapters:

1. A metabolic pathway is a series of enzyme-catalyzed reactions, so the pathway's substrate is converted to its product in discrete steps.

2. A monomeric compound such as glucose is interconverted with its polymeric form (glycogen), with other monosaccharides (fructose-6-phosphate and ribose-5-phosphate, for example), and with smaller metabolites such as the three-carbon pyruvate.

3. Although anabolic and catabolic pathways may share some steps, their irreversible steps are catalyzed by enzymes unique to each pathway.

4. Certain reactions consume or produce energy in the form of ATP. In most cases, these are phosphoryl-group transfer reactions.

5. Some steps are oxidation–reduction reactions that require or generate a reduced cofactor such as NADH or NADPH.

Figure 13.13 Summary of glucose metabolism. This diagram includes the pathways of glycogen synthesis and degradation, glycolysis, gluconeogenesis, and the pentose phosphate pathway. Dotted lines are used where the individual reactions are not shown. Filled gold symbols indicate ATP production; shadowed gold symbols indicate ATP consumption. Filled and shadowed red symbols represent the production and consumption of the reduced cofactors NADH and NADPH.

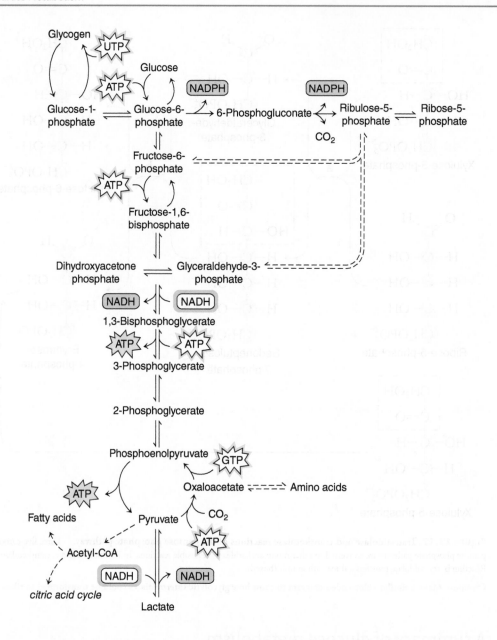

Concept Check

1. List the products of the pentose phosphate pathway and describe how the cell uses them.
2. Explain why cells maintain a high NAD$^+$/NADH ratio and a low NADP$^+$/NADPH ratio.
3. Explain how the cell catabolizes excess ribose groups.

13.5 Clinical Connection: Disorders of Carbohydrate Metabolism

KEY CONCEPTS

Relate enzyme deficiencies to defects in carbohydrate metabolism.

- Explain why red blood cells are so susceptible to defects in glucose metabolic pathways.
- Describe the symptoms of glycogen storage diseases affecting liver and muscle.

Although there is not enough space to describe all the known enzyme deficiencies that affect carbohydrate metabolic pathways in humans, a few disorders are worth highlighting. As with many metabolic diseases, the discovery and study of carbohydrate metabolic disorders has helped shed light on how the normal metabolic pathways function. Keep in mind that an enzyme deficiency may result from a genetic variation that limits production of the protein, directly impacts its catalytic power, or affects its regulation.

Deficiencies of glycolytic enzymes usually have severe consequences, particularly in tissues that rely heavily on glycolysis for ATP production. For example, red blood cells, which lack mitochondria to produce ATP by oxidative phosphorylation, use the ATP generated by glycolysis to power the Na,K-ATPase that maintains the cells' ion concentration gradients (Section 9.3). Low glycolytic activity reduces the supply of ATP, and the resulting ion imbalance leads to osmotic swelling and bursting of the red blood cells.

In addition to anemia (the loss of red blood cells), other abnormalities may develop when glycolytic enzymes are defective. In a pyruvate kinase deficiency, several of the intermediates upstream of phosphoenolpyruvate (the pyruvate kinase substrate) accumulate because they are in equilibrium. Red blood cells normally convert one of these—1,3-bisphosphoglycerate—to 2,3-bisphosphoglycerate (BPG), which binds to hemoglobin to decrease its oxygen affinity (Section 7.1). The elevated concentrations of glycolytic intermediates resulting from a pyruvate kinase deficiency actually boost BPG production, allowing red blood cells to deliver O_2 more efficiently, which helps offset the anemia caused by the enzyme deficiency. A shortage of hexokinase, however, slows the entire glycolytic pathway, which limits the production of BPG in red blood cells and thereby reduces the amount of O_2 delivered to tissues.

Defects in pathways that metabolize sugars other than glucose have variable effects. Individuals with fructose intolerance lack fructose-1-phosphate aldolase (see Box 13.A). The resulting accumulation of fructose-1-phosphate ties up the liver's phosphate supply, which hinders ATP production from ADP. As in the disorders already described, one of the first casualties is the Na,K-ATPase, whose inadequate activity leads to cell death.

An inability to convert galactose to glucose can be deadly, especially in infancy, when the major carbohydrate source is lactose, a disaccharide of glucose and galactose. High concentrations of galactose that cannot be metabolized contribute to side reactions such as the addition of sugar derivatives to proteins. Damage to proteins in nerve cells causes growth retardation and abnormal brain development. Fortunately, an early diagnosis and a galactose-free diet can avoid such damage.

A deficiency of glucose-6-phosphate dehydrogenase, the first enzyme of the pentose phosphate pathway, is the most common human enzyme deficiency. This defect decreases the cellular production of the reducing agent NADPH, which participates in certain oxidation–reduction processes and helps protect cells from oxidative damage. Red blood cells, where the O_2 concentration is high, are most at risk. Glucose-6-phosphate dehydrogenase deficiency affects about 500 million people, mostly in Africa, tropical South America, and southeast Asia—areas with historically high rates of malaria. The enzyme defect causes anemia, but the release of heme from damaged red blood cells triggers anti-inflammatory responses that increase an individual's chances of surviving malaria. The same effect is observed in individuals carrying the hemoglobin S variant (Section 5.2).

Glycogen storage diseases affect liver and muscle

The **glycogen storage diseases (GSDs)** are a set of inherited disorders of glucose and glycogen metabolism, not all of which result in glycogen accumulation, as the name might suggest. The symptoms of the glycogen storage diseases vary, depending on whether the affected tissue is liver or muscle or both. In general, the disorders that affect the liver cause hypoglycemia (too little glucose in the blood) and an enlarged liver. Glycogen storage diseases that affect primarily muscle are characterized by muscle weakness and cramps. The incidence of glycogen storage diseases is estimated to be as high as 1 in 20,000 births, although some disorders are not apparent until adulthood. **Table 13.1** lists the enzyme deficiency associated with each type of glycogen storage disease.

Table 13.1 Glycogen Storage Diseases

Type	Enzyme deficiency
GSD0	Glycogen synthase
GSD1	Glucose-6-phosphatase
GSD2	α-1,4-Glucosidase
GSD3	Amylo-1,6-glucosidase (debranching enzyme)
GSD4	Amylo-(1,4 → 1,6)-transglycosylase (branching enzyme)
GSD5	Muscle glycogen phosphorylase
GSD6	Liver glycogen phosphorylase
GSD7	Phosphofructokinase
GSD9	Phosphorylase kinase
GSD10	Phosphoglycerate mutase
GSD11	Lactate dehydrogenase
GSD12	Aldolase
GSD13	Enolase
GSD14	Phosphoglucomutase
GSD15	Glycogenin

A defect of glucose-6-phosphatase (GSD1) affects both gluconeogenesis and glycogenolysis, since the phosphatase catalyzes the final step of gluconeogenesis and makes free glucose available from glycogenolysis. The enlarged liver and hypoglycemia can lead to a host of other symptoms, including irritability, lethargy, and, in severe cases, death. A related defect is the deficiency of the transport protein that imports glucose-6-phosphate into the endoplasmic reticulum, where the phosphatase is located.

Glycogen storage disease 3 results from a deficiency of the glycogen debranching enzyme. This condition accounts for about one-quarter of all cases of glycogen storage disease and usually affects both liver and muscle. The symptoms include muscle weakness and liver enlargement due to the accumulation of glycogen that cannot be efficiently broken down. The symptoms of GSD3 often improve with age and disappear by early adulthood.

The most common type of glycogen storage disease is GSD9. In this disorder, the kinase that activates glycogen phosphorylase is defective. Symptoms range from severe to mild and may fade with time. The complexity of this disease reflects the fact that the phosphorylase kinase consists of four subunits, with isoforms that are differentially expressed in the liver and other tissues.

In the past, glycogen storage diseases were diagnosed on the basis of symptoms, blood tests, and painful biopsies of liver or muscle to assess its glycogen content. Current diagnostic methods are centered on analyzing the relevant genes for mutations, a noninvasive approach. Treatment of glycogen storage diseases typically includes a regimen of frequent, small, carbohydrate-rich meals to alleviate hypoglycemia. However, because dietary therapy does not completely eliminate the symptoms of some glycogen storage diseases, and because the metabolic abnormalities, such as chronic hypoglycemia and liver damage, can severely impair physical growth as well as cognitive development, liver transplant has proved to be an effective treatment. The glycogen storage diseases are single-gene defects, which makes them attractive targets for gene therapy (see Section 3.3).

Concept Check

1. Make a list of enzyme deficiencies and their physiological effects.
2. Compare the impact of various disorders on red blood cells, liver cells, and muscle cells.

SUMMARY

13.1 Glycolysis

- The pathway of glucose catabolism, or glycolysis, is a series of enzyme-catalyzed steps in which energy is conserved as ATP or NADH.

- The 10 reactions of glycolysis convert the six-carbon glucose to two molecules of pyruvate and produce two molecules of NADH and two molecules of ATP. The first and third reactions (catalyzed by hexokinase and phosphofructokinase) require the investment of two ATP. The irreversible reaction catalyzed by phosphofructokinase is the rate-determining step and the major control point for glycolysis. Later steps (catalyzed by phosphoglycerate kinase and pyruvate kinase) generate four ATP per glucose.

- Pyruvate may be reduced to lactate or ethanol, further oxidized by the citric acid cycle, or converted to other compounds.

13.2 Gluconeogenesis

- The pathway of gluconeogenesis converts two molecules of pyruvate to one molecule of glucose at a cost of six molecules of ATP. The pathway uses seven glycolytic enzymes, and the activities of pyruvate carboxylase, phosphoenolpyruvate carboxykinase, fructose bisphosphatase, and glucose-6-phosphatase bypass the three irreversible steps of glycolysis.

- A futile cycle involving phosphofructokinase and fructose bisphosphatase helps regulate the flux through glycolysis and gluconeogenesis.

13.3 Glycogen Synthesis and Degradation

- Glucose residues are incorporated into glycogen after first being activated by attachment to UDP.

- Phosphorolysis of glycogen produces phosphorylated glucose that can enter glycolysis. In the liver, this glucose is dephosphorylated and exported.

13.4 The Pentose Phosphate Pathway

- The pentose phosphate catabolic pathway for glucose yields NADPH and ribose groups. The five-carbon sugar intermediates can be converted to glycolytic intermediates.

13.5 Clinical Connection: Disorders of Carbohydrate Metabolism

- Disorders are associated with deficiencies of glycolytic enzymes and enzymes that metabolize fructose and galactose.

- Glycogen storage diseases cause hypoglycemia, muscle weakness, and liver damage.

KEY TERMS

glycolysis	near-equilibrium	tautomerization	fermentation	glycogenolysis	glycogen storage dis-
kinase	reaction	feed-forward	gluconeogenesis	pentose phosphate	ease (GSD)
metabolically irre-	rate-determining	activation	futile cycle	pathway	
versible reaction	reaction				

BIOINFORMATICS

Brief Bioinformatics Exercises

13.1 Hexokinase Structure and Ligand Binding

13.2 Glycolysis and the KEGG Database

PROBLEMS

13.1 Glycolysis

1. Cite examples of the following reactions from the steps of glycolysis. Mention cofactors involved in each of the reactions. **a.** carbon–carbon bond cleavages **b.** phosphorylations, **c.** oxidation–reductions, **d.** isomerizations, and **e.** dehydrations.

2. Elaborate metabolically irreversible reactions of glycolysis. What is the significance of such reactions?

3. Except during starvation, the brain oxidizes glucose as its sole metabolic fuel and consumes up to 40% of the body's circulating glucose. Brain hexokinase has a K_M for glucose that is 100 times lower than the concentration of circulating glucose (5 mM). What is the advantage of this low K_M?

4. There are four isozymes of hexokinase termed hexokinases I, II, and III (which have K_M values of ~0.02 mM) and hexokinase IV (which has a K_M value of ~5 mM). Normal hepatocytes (liver cells) express low amounts of hexokinases I and II, but upon transformation to cancer cells, the expression of hexokinase II (and to some extent hexokinase I) increases while hexokinase IV expression is silenced. How does this strategy promote the survival of the cancer cell?

5. Residue Asn 204 in the glucose binding site of hexokinase IV (see Problem 4) was mutated, in two separate experiments, to either Ala or Asp. The Asn → Ala mutant had a K_M nearly 50-fold greater than the wild-type enzyme, and the Asn → Asp mutant had a 140-fold greater K_M value than the wild-type enzyme. What do these experiments reveal about the intermolecular interactions between the enzyme and the glucose substrate?

6. The V_{max} and K_M values for an unusual hexokinase found in *Trypanosoma cruzi* (the causative agent of Chagas disease) are shown in the presence and absence of a bisphonate inhibitor (structure shown).

	Without inhibitor	With inhibitor
K_M (mM)	90	125
V_{max} (μmol $\cdot$ min^{-1} $\cdot$ mL^{-1})	0.30	0.12

A bisphonate compound

a. What type of inhibitor is bisphonate? **b.** The parasite hexokinase, unlike the mammalian enzyme, is not inhibited by glucose-6-phosphate but is inhibited by pyrophosphate (PP$_i$). Is this observation consistent with your answer to part **a**? **c.** Might bisphonate be a good candidate for a drug to treat the disease?

7. 1,5-Anhydroglucitol, a glucose derivative missing the hydroxyl group on the anomeric carbon, is present in small amounts in nearly all foods. Draw the structure of the product that results when hexokinase uses 1,5-anhydroglucitol as a substrate.

8. Mannose, the C2 epimer of glucose, enters mammalian cells via glucose transporters and becomes a substrate for hexokinase. Explain why treating cells with mannose can impair glycolysis, particularly in cells with low phosphomannose isomerase activity.

9. The standard free energy change for the phosphoglucose isomerase reaction is +2.2 kJ $\cdot$ mol^{-1}, but the reactant concentrations in vivo yield a ΔG value of about −1.4 kJ $\cdot$ mol^{-1}. Calculate the ratio of fructose-6-phosphate (F6P) to glucose-6-phosphate (G6P) under **a.** standard conditions at 25°C and **b.** in vivo conditions at 37°C. Explain in which direction the reaction should proceed *in vivo*?

10. The radioactively labeled compound [^{18}F]fluorodeoxyglucose (FDG) is used to measure glucose uptake in cells. FDG, a derivative of glucose in which the C2 hydroxyl is replaced with fluorine, is phosphorylated to FDG-6-phosphate upon entering the cell but cannot proceed any further through glycolysis. The phosphorylated FDG remains in the cell and its presence can be detected and quantitated, providing information on the rate of glucose uptake. (Cancer cells take up glucose particularly rapidly.) **a.** Why does phosphorylation trap the FDG in the cell? **b.** Why is FDG unable to proceed though glycolysis following its phosphorylation? (*Hint:* Refer to the mechanism of phosphoglucose isomerase reaction.)

11. ADP stimulates the activity of phosphofructokinase (PFK), yet it is a product of the reaction and not a reactant. Explain this apparent contradictory regulatory strategy.

12. Often the "T" and "R" nomenclature are used to describe the low- and high-affinity conformations of allosteric enzymes. An allosteric inhibitors stabilize the T form, which has a low affinity for its substrate, while activators stabilize the high-affinity R form. Comment on the impact of the following effectors on 'T-R' equilibrium of PFK **a.** fructose-2,6-bisphosphate (mammals), **b.** phosphoenolpyruvate (PEP; bacteria), **c.** ADP (bacteria).

13. Explain the impact of each of the effectors on K_M of PFK for fructose 6-phosphate.

14. Refer to Figure 7.16. Why does the conformational change that results when Arg 162 changes places with Glu 161 result in a form of phosphofructokinase that has a low affinity for its substrate?

15. *Saccharomyces cerevisiae* deficient in PFK cannot grow on medium containing glucose as the sole carbon source, though it can grow on media containing glycerol as the carbon source. Suggest a possible reason for such an observation.

16. Researchers isolated a yeast phosphofructokinase mutant in which a serine at the fructose-2,6-bisphosphate (F26BP) binding site was replaced with an aspartate residue. The amino acid substitution completely abolished the binding of F26BP to PFK. There was a dramatic decline in glucose consumption and ethanol production in the mutant compared to control yeast. **a.** Propose a hypothesis that explains why the mutant PFK cannot bind F26BP. **b.** What does the decline of glucose consumption and ethanol production in the yeast reveal about the role of F26BP in glycolysis?

17. Elaborate the reaction mechanism of aldolase. **a.** Explain the role of the Lys side chain in catalysis. **b.** Compare the pK of the Lys side chain compared to that of free Lys. **c.** Explain the role of the Asp side chain in catalysis. **d.** comment on the change of pK of the Asp side chain after the formation of the Schiff base. **e.** Predict the effect of Asp → Ala and Lys → Ala mutation on aldolase activity.

18. Does the aldolase enzyme mechanism (see Fig. 13.4) use acid catalysis, base catalysis, covalent catalysis, or some combination of these strategies (see Section 6.2)? Explain.

19. The aldolase reaction has a standard free energy change of 22.8 kJ · mol^{-1}. **a.** Is the reaction favorable under standard conditions? **b.** What is the actual free energy change, ΔG, at 37°C, if the concentration of fructose-1,6-bisphosphate (F16BP) is 21 μM, the concentration of glyceraldehyde-3-phosphate (GAP) is 4.5 μM, and the concentration of dihydroxyacetone phosphate (DHAP) is 24 μM? Is the forward or the reverse direction of this reaction favored under these conditions? Explain.

20. The $\Delta G°'$ value for the hexokinase reaction is −16.7 kJ · mol^{-1}, while the ΔG value under cellular conditions is similar.

21. Some bacteria degrade glucose via the Entner–Douderoff pathway, outlined below, rather than by the glycolytic pathway presented in the chapter. (The structure of 2-keto-3-deoxygluconate (KDG) is shown in Solution 11.23.) **a.** In which step does a bacterial aldolase catalyze a reaction similar to the aldolase reaction in glycolysis in humans? **b.** The compound shown below inhibits the bacterial aldolase. Propose a hypothesis to explain how the inhibitor interferes with the enzyme's activity. **c.** The inhibitor has a K_I of 200 μM. The K_M for the substrate is 50 μM. Is the compound an effective inhibitor of the bacterial aldolase?

Glucose
↓ NADP$^+$
↘ NADPH
Gluconate
↓
↘ H$_2$O
2-Keto-3-deoxygluconate (KDG)
↓ ATP
↘ ADP
2-Keto-3-deoxy-6-phosphogluconate (KDPG)
↙   ↘
Pyruvate   Glyceraldehyde-3-phosphate
↓
1,3-BPG
↓ *4 steps*
Pyruvate

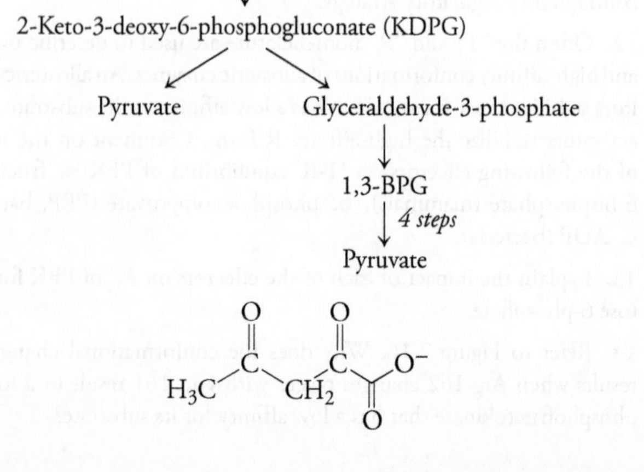

22. How the products of the aldolase reaction are interconverted? What is the ratio of the products of the aldolase reaction (in cells at 37°C under non-equilibrium conditions, $\Delta G°' = +7.9$ · kJ · mol^{-1} and $\Delta G = +4.4$ kJ · mol^{-1})? In the context of the ratio determined, how does the conversion of DHAP to GAP occur readily in cells?

23. The term "turbo design" has been used to describe pathways such as glycolysis that have one or more ATP-consuming steps followed by one or more ATP-producing steps with a net yield of ATP production for the pathway overall. Mathematical models have shown that "turbo" pathways have the risk of substrate-accelerated death unless there is a "guard at the gate," that is, a mechanism for inhibiting an early step of the pathway. In yeast, hexokinase is inhibited by a complex mechanism mediated by trehalose-6-phosphate synthase (TPSI). Mutant yeast in which TPSI is defective (there is no "guard at the gate") die if grown under conditions of high glucose concentration. Explain why.

24. Biochemists use transition state analogs to determine the structure of a short-lived intermediate in an enzyme-catalyzed reaction. Because an enzyme binds tightly to the transition state, a compound that resembles the transition state should be a potent competitive inhibitor. Phosphoglycohydroxamate binds 150 times more tightly than dihydroxyacetone phosphate to triose phosphate isomerase. Based on this information, propose a structure for the intermediate of the triose phosphate isomerase reaction.

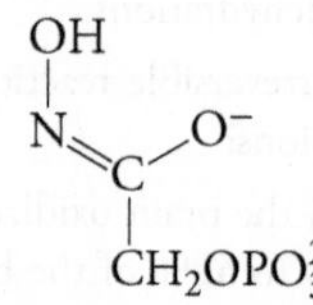

OH
|
N═C─O$^-$
|
CH$_2$OPO$_3^{2-}$

Phosphoglycohydroxamate

25. Organisms such as yeast growing under anaerobic conditions can convert pyruvate to alcohol in a process called fermentation, as described in the text. Instead of being converted to lactate, pyruvate is converted to ethanol in a two-step reaction. Why is the second step of this process essential to the yeast cell?

26. Cancer cells have elevated levels of glyceraldehyde-3-phosphate dehydrogenase (GAPDH), which may account for the high rate of glycolysis in these cells. The compound methylglyoxal has been shown to inhibit GAPDH in cancer cells but not in normal cells. This observation may lead to the development of drugs for treating cancer. **a.** Propose a hypothesis to explain why GAPDH levels in cancer cells are elevated. **b.** Why might methylglyoxal inhibit GAPDH in cancer cells but not in normal cells?

27. Glyceraldehyde-3-phosphate can react with arsenate, AsO$_4^{3-}$, a phosphate analog, in the GAPDH reaction to form 1-arseno-3-phosphoglycerate, which is unstable and spontaneously hydrolyzes to form 3-phosphoglycerate (3PG), as shown. What is the effect of arsenate on cells undergoing glycolysis?

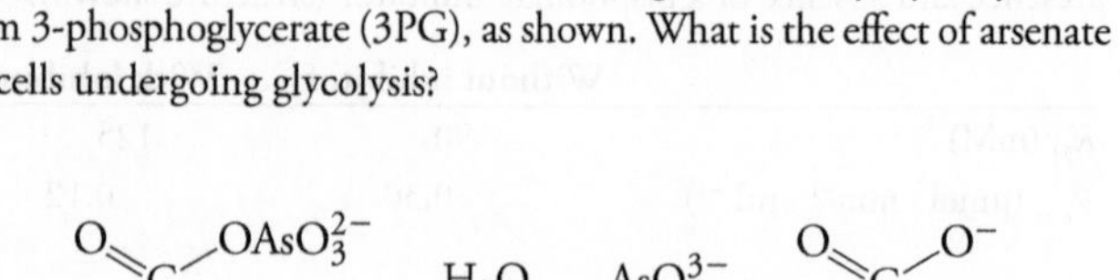

O   OAsO$_3^{2-}$ O   O$^-$
\\   /   \\   /
C   H$_2$O   AsO$_4^{3-}$ C
| ─────────────────────→ |
H─C─OH   H─C─OH
| |
CH$_2$OPO$_3^{2-}$ CH$_2$OPO$_3^{2-}$

1-arseno-3-phosphoglycerate   3PG

28. GAPDH activity to generate 1, 3 BPG is controlled by the NADH/NAD$^+$ ratio in bacteria. Hypothesize how GAPDH activity is affected by the ratio considering the production of 1, 3 BPG.

O   O
\\   \\
C   C
/ \ / \
H$_3$C   CH$_2$ O$^-$
|
O

29. The thermophilic archaebacterium *Thermoproteus tenax* expresses two GAPDH enzymes: one that carries out the reaction shown in the chapter (although NADP⁺ is the reactant rather than NAD⁺) and a second GAPDH enzyme that catalyzes the reaction shown here.

$$\text{Glyceraldehyde-3-phosphate + NAD}^+ \xrightarrow{\substack{\textit{T. tenax} \\ \text{GAPDH}}}$$
$$\text{3-phosphoglycerate + NADH + H}^+$$

a. How do the properties of the two enzymes differ? **b.** In an experiment, the activity of the second *T. tenax* GAPDH enzyme was measured in the presence of various metabolites. The results are shown in the figure. Estimate the K_M values for the enzyme in the presence and absence of effectors and classify the effectors as activators or inhibitors of the enzyme. **c.** Propose a hypothesis that explains why these effectors act as either activators or inhibitors.

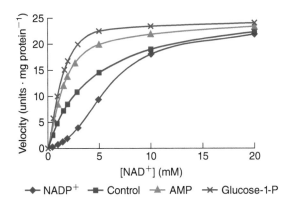

30. Phosphoglycerate kinase in red blood cells is bound to the plasma membrane. This allows the kinase reaction to be coupled to the Na,K-ATPase pump. How does the proximity of the enzyme to the membrane facilitate the action of the pump?

31. Vanadate, VO_4^{3-}, inhibits GAPDH, not by acting as a phosphate analog (see Problem 27), but by interacting with essential —SH groups on the enzyme. What happens to cellular levels of phosphate, ATP, and 2,3-bisphosphoglycerate (see pathway below) when red blood cells are incubated with vanadate?

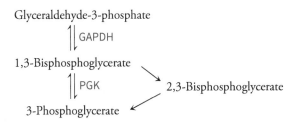

32. The mechanism of plant phosphoglycerate mutase is different from the mechanism of mammalian phosphoglycerate mutase presented in the chapter. 3-Phosphoglycerate (3PG) binds to the plant enzyme, transfers its phosphate to the enzyme, and then the enzyme transfers the phosphate group back to the substrate to form 2-phosphoglycerate (2PG). What is the fate of the [³²P] label when [³²P]-labeled 3PG is added to **a.** cultured hepatocytes or **b.** plant cells?

33. Explain why iodoacetate was useful for determining the order of intermediates in glycolysis but provided misleading information about an enzyme's active site.

34. Several studies have shown that aluminum inhibits phosphofructokinase in liver cells. **a.** Compare the production of pyruvate by perfused livers in control and aluminum-treated rats using fructose as an energy source (see Box 13.A). **b.** What would the experimental results be if glucose were used instead of fructose?

35. Name some glycolytic enzymes that can be inhibited by iodoacetate.

36. Enzymes occasionally display weak "side" activities. Draw the structure of the product of the reaction that results when pyruvate kinase, operating in reverse, uses lactate as a substrate. What are the other substrate and product of this reaction?

37. The pyruvate kinase in red blood cells has a $\Delta G^{\circ\prime}$ value of -31.4 kJ · mol⁻¹. **a.** What is the equilibrium constant, K_{eq}, for the reaction at 25°C? **b.** Calculate the ΔG value for the reaction under the following conditions: [phosphoenolpyruvate, PEP] = 12.5 μM, [pyruvate] = 25.5 μM, [ATP] = 0.5 mM, and [ADP] = 0.05 mM. Is the reaction favorable under these conditions?

38. In an experiment, the activity of a pyruvate kinase isoenzyme common in cancer cells was measured in the presence and in the absence of 1.2 mM oxalate. The kinetic data were graphed as a Lineweaver–Burk plot whose *x* and *y* intercepts are given in the table. **a.** Calculate K_M and V_{max} in the presence and in the absence of the inhibitor. **b.** Calculate K_I for the inhibitor.

	x intercept (mM⁻¹)	*y* intercept (units⁻¹ · mg)
Without inhibitor	−0.5	0.004
With inhibitor	−0.04	0.004

39. About 15% of ingested ethanol is metabolized by a cytochrome P450 (see Section 7.4) and chronic alcohol consumption induces the expression of this enzyme. Explain how this changes the effectiveness of therapeutic drugs.

40. How can you test whether any of the glycolytic enzymes in a pathogenic bacteria are druggable targets for developing novel antimicrobials?

41. Explain the biochemical basis for methanol toxicity (refer to Box 13.B).

42. Drinking methanol can cause blindness and death, depending on the dosage. The causative agent is formaldehyde derived from methanol. **a.** Draw the balanced chemical reaction for the conversion of methanol to formaldehyde. **b.** Why would administering whiskey (ethanol) to a person poisoned with methanol be a good antidote?

43. a. Explain why alcohol consumption is associated with increased risk of developing hypothermia. **b.** Drinking a glass of water for each alcoholic drink is a popular hangover-prevention strategy. Explain how increased water consumption might relieve some of the negative effects of alcohol consumption.

44. Assuming a standard free energy change of 30.5 kJ · mol⁻¹ for the synthesis of ATP from ADP and P_i, how many molecules of ATP could theoretically be produced by **a.** homolactic fermentation (the catabolism of glucose to lactate, $\Delta G^{\circ\prime} = -196$ kJ · mol⁻¹) and **b.** by catabolism of glucose to CO_2 ($\Delta G^{\circ\prime} = -2850$ kJ · mol⁻¹), assuming 33% efficiency?

45. a. How many molecules of ATP could theoretically be produced (see Problem 44) by the catabolism of glucose to CO_2 ($\Delta G^{\circ\prime} = -2850$ kJ · mol⁻¹), assuming 33% efficiency? **b.** Compare your answer to Solution 44. Does this explain the Pasteur effect (the observation, first made by Louis Pasteur, that glucose consumption in yeast dramatically decreases in the presence of oxygen)?

46. Studies have shown that the halophilic organism *Halococcus saccharolyticus* degrades glucose via the Entner–Doudoroff pathway (see Problem 21). **a.** What is the ATP yield per mole of glucose for this pathway? **b.** Describe (in general) what kinds of reactions would need to follow the Entner–Doudoroff pathway in this organism.

47. Trypanosomes living in the bloodstream obtain all their energy from glycolysis. They take up glucose from the host's blood and excrete pyruvate as a waste product. In this part of their life cycle, trypanosomes do not carry out any oxidative phosphorylation, but they do use another oxygen-dependent pathway, which is absent in mammals, to oxidize NADH. **a.** Why is the oxygen-dependent pathway necessary? **b.** Would this pathway be necessary if the trypanosome excreted lactate rather than pyruvate? **c.** Why would the oxygen-dependent pathway be a good target for antiparasitic drugs?

13.2 Gluconeogenesis

48. Which of the 11 reactions of gluconeogenesis are **a.** oxidation–reductions, **b.** carbon–carbon bond formations, **c.** phosphorylations, **d.** hydrations, **e.** isomerizations, **f.** hydrolysis reactions, **g.** decarboxylations, and **h.** carboxylations?

49. Which of the following substrates can be converted to glucose via the gluconeogenesis pathway? **a.** pyruvate, **b.** glycerol, **c.** oxaloacetate, **d.** acetyl-CoA, **e.** alanine, **f.** leucine (which is degraded to acetyl-CoA), **g.** lactate.

50. Biotin deficiencies are rare because biotin is present in many foods and is also synthesized by intestinal bacteria. Biotin deficiencies have been observed, however, in individuals consuming large quantities of raw eggs. Explain why. (*Hint:* See Problem 4.85.)

51. a. Use the equations provided to determine $\Delta G^{\circ\prime}$ for the favorable synthesis of oxaloacetate (OAA):

$$\text{pyruvate} + CO_2 \rightarrow H^+ + OAA \qquad \Delta G^{\circ\prime} = 31.8 \text{ kJ} \cdot \text{mol}^{-1}$$

$$ADP + P_i + H^+ \rightarrow ATP + H_2O \qquad \Delta G^{\circ\prime} = 30.5 \text{ kJ} \cdot \text{mol}^{-1}$$

b. How is the synthesis of oxaloacetate from pyruvate accomplished in cells? **c.** Calculate ΔG for this reaction under the following conditions: 37°C, pH 7, [Pyruvate] = [CO_2] = 2.4 mM, [OAA] = 1.2 mM, [ATP] = 1.5 mM, [P_i] = 2.5 mM, and [ADP] = 0.8 mM. Is the reaction favorable under these conditions?

52. Flux through the opposing pathways of glycolysis and gluconeogenesis is controlled in several ways. **a.** Explain how the activation of pyruvate carboxylase by acetyl-CoA affects glucose metabolism. **b.** Pyruvate can undergo a reversible amino-group transfer reaction to yield alanine (see Section 12.3). Alanine is an allosteric effector of pyruvate kinase. Would you expect alanine to stimulate or inhibit pyruvate kinase? Explain.

53. A physician diagnoses an infant patient with a pyruvate carboxylase deficiency in part by measuring the patient's blood levels of lactate and pyruvate. **a.** How would the patient's [lactate]/[pyruvate] ratio compare to the ratio in a normal infant? **b.** The diagnosis can be confirmed by giving the patient an alanine injection. What happens to the injected alanine in the patient and how does this differ in a normal patient?

54. A liver biopsy of a four-year-old boy indicated that fructose-1,6-bisphosphatase enzyme activity was 20% of normal. The patient's blood glucose levels were normal at the beginning of a fast but then decreased suddenly. Pyruvate and alanine concentrations were also elevated, as was the glyceraldehyde-3-phosphate/dihydroxyacetone phosphate ([GAP]/[DHAP]) ratio. Explain the reason for these symptoms.

55. Insulin is one of the major hormones that regulates gluconeogenesis. Insulin acts in part by decreasing the transcription of genes coding for certain gluconeogenic enzymes. For which genes would you expect insulin to suppress transcription?

56. Type 2 diabetes is characterized by insulin resistance, in which insulin is unable to perform its many functions. What symptoms would you expect in a Type 2 diabetic patient if insulin is unable to perform the function described in Problem 55?

57. The concentration of fructose-2,6-bisphosphate (F26BP) is regulated in the cell by a homodimeric enzyme with two catalytic activities: a kinase that phosphorylates fructose-6-phosphate to form F26BP and a phosphatase that catalyzes the hydrolysis of the C2 phosphate group. **a.** Which enzyme activity, the kinase or the phosphatase, would you expect to be active under fasting conditions? Explain. **b.** Which hormone is likely to be responsible for inducing this activity? **c.** Consult Section 10.2 and propose a mechanism for this induction.

58. Brazilin, a compound found in aqueous extracts of sappan wood, has been used to treat diabetics in Korea. Brazilin increases the activity of the kinase enzyme that produces F26BP (see Problem 57), and the compound also stimulates the activity of pyruvate kinase. **a.** What is the effect of adding brazilin to hepatocytes (liver cells) in culture? **b.** Why would brazilin be an effective treatment for diabetes?

59. Metformin is a drug that decreases the expression of phosphoenolpyruvate carboxykinase. Explain why metformin would be helpful in treating diabetes.

60. The "carbon skeletons" of most amino acids can be converted to glucose, a process that may require many enzymatic steps. Which amino acids can enter the gluconeogenic pathway directly after undergoing deamination (a reaction in which the carbon with the amino group becomes a ketone)?

61. Draw a diagram that illustrates how alanine (see Problem 60) released from the muscle is converted back to glucose in the liver. What is the physiological cost if this cycle runs for a prolonged period of time?

62. Draw a diagram that illustrates how lactate released from the muscle is converted back to glucose in the liver. What is the cost (in ATP) of running this cycle?

63. The *Leishmania* parasite carries out a modified form of gluconeogenesis in a specialized organelle called the glycosome. Gluconeogenesis is essential to parasite survival, so understanding the pathway could assist in the development of drugs to treat parasitic infections. Draw a diagram of the reactions that occur in the glycosome, given the following information: PEPCK and fructose bisphosphatase are present; glucose-6-phosphatase and pyruvate carboxylase are not present. Pyruvate phosphate dikinase (PPDK), which converts pyruvate to phosphoenolpyruvate (PEP), is located in the glycosome. A glycerol-3-kinase (GK) enzyme is also located in the glycosome. Malate dehydrogenase (MDH), which catalyzes the reversible interconversion of malate and oxaloacetate (OAA), is found in the cytosol as well as the glycosome.

64. The purpose of gluconeogenesis in the *Leishmania* parasite (see Problem 63) is to synthesize fructose-6-phosphate, a precursor for mannogen, a homopolysaccharide consisting of β(1 → 2) mannose residues. Using the diagram you constructed in Problem 57, explain how each of these compounds can be used to synthesize mannogen: **a.** glycerol, **b.** aspartate, and **c.** alanine. (*Note:* The glycosome includes transporters for glycerol, pyruvate, and malate.)

13.3 Glycogen Synthesis and Degradation

65. Explain the role of glycogenin in glycogen synthesis.

66. a. From the details provided for the reactions engaged in glycogen synthesis, a. outline the overall reaction and calculate a $\Delta G^{\circ\prime}$ value for it. **b.** estimate the K_{eq} at 25°C for the reaction.

67. Beer is produced from raw materials such as wheat and barley. Explain why the grains are allowed to sprout, a process in which their starch is broken down to glucose, before fermentation begins.

68. Some bread manufacturers add amylase to bread dough prior to the fermentation process. What role does this enzyme (see Section 12.1) play in the bread-making process?

69. Glycogen is degraded via a phosphorolysis process, which produces glucose-1-phosphate. What advantage does this process have over a simple hydrolysis, which would produce glucose instead of phosphorylated glucose?

70. The equation for the degradation of glycogen is shown below. **a.** What is the ratio of $[P_i]/[G1P]$ under standard conditions? **b.** What is the value of ΔG under cellular conditions when the $[P_i]/[G1P]$ ratio is 25/1? **c.** What advantage does degradation by phosphorolysis have over a simple hydrolysis, which would produce glucose instead of glucose-1-phosphate?

$$\text{glycogen } (n) + P_i \xrightarrow{\text{phosphorylase}} \text{glycogen } (n-1) + \text{G1P}$$
$$\Delta G^{\circ\prime} = +3.1 \text{ kJ} \cdot \text{mol}^{-1}$$

71. The mechanism of the phosphoglucomutase enzyme is similar to that of the plant mutase described in Problem 32 and is shown below. On occasion, the glucose-1,6-bisphosphate dissociates from the enzyme. Why does the dissociation of glucose-1,6-bisphosphate inhibit the enzyme?

Glucose-6-phosphate

Glucose-1,6-bisphosphate

Glucose-1-phosphate

72. Glucagon, a hormone secreted by the pancreas in the fasted state, binds to G–protein coupled receptors in the liver. As a result of glucagon action, glycogen phosphorylase is phosphorylated. Consult Section 10.2 and describe the steps leading up to the phosphorylation of glycogen phosphorylase. Is the phosphorylated form of the enzyme active or inactive?

73. In addition to phosphorylating glycogen phosphorylase, glucagon signaling (see Problem 72) results in the phosphorylation of

glycogen synthase. Is the phosphorylated form of the enzyme active or inactive?

74. Trehalose, a disaccharide consisting of two glucose residues (see Problem 11.38), is one of the major sugars in the insect hemolymph (the fluid that circulates through the insect's body). Trehalose serves as a storage form of glucose and also helps protect the insect from desiccation and freezing. Its concentration in the hemolymph must be closely regulated. Trehalose is synthesized in the insect fat body, which plays a role in metabolism analogous to the vertebrate liver. Studies of the insect *Manduca sexta* show that during starvation, hemolymph glucose concentration decreases, which results in an increase in fat body glycogen phosphorylase activity and a decrease in the concentration of fructose-2,6-bisphosphate. What effect do these changes have on hemolymph trehalose concentration in the fasted insect?

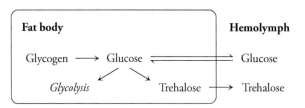

75. The glycolytic pathway in the thermophilic archaebacterium *Thermoproteus tenax* differs from the pathway presented in this chapter. The phosphofructokinase (PFK) reaction in *T. tenax* is reversible and depends on pyrophosphate rather than ATP. In addition, *T. tenax* has two glyceraldehyde-3-phosphate dehydrogenase (GADPH) isozymes. The "phosphorylating GAPDH" is similar to the enzyme described in this chapter. The second isozyme is the irreversible "nonphosphorylating GAPDH," which catalyzes the reaction described in Problem 29. *T. tenax* relies on glycogen stores as a source of energy. What is the ATP yield for one mole of glucose oxidized by the pathway that uses the nonphosphorylating GAPDH enzyme?

13.4 The Pentose Phosphate Pathway

76. Most metabolic pathways include an enzyme-catalyzed reaction that commits a metabolite to continue through the pathway. **a.** Identify the first committed step of the pentose phosphate pathway. Explain your reasoning. **b.** Hexokinase catalyzes an irreversible reaction at the start of glycolysis. Does this step commit glucose to continue through glycolysis?

77. A given metabolite may follow more than one metabolic pathway. List all the possible fates of glucose-6-phosphate in **a.** a liver cell and **b.** a muscle cell.

78. Experiments were carried out in cultured cells to determine the relationship between glucose-6-phosphate dehydrogenase (G6PD) activity and rates of cell growth. Cells were cultured in a medium supplemented with serum, which contains growth factors that stimulate G6PD activity. Predict how the cellular $NADPH/NADP^+$ ratio would change under the following circumstances: **a.** Serum is withdrawn from the medium. **b.** DHEA, an inhibitor of glucose-6-phosphate dehydrogenase, is added. **c.** The oxidant H_2O_2 is added. **d.** Serum is withdrawn and H_2O_2 is added.

79. Write a mechanism for the nonenzymatic hydrolysis of 6-phosphogluconolactone to 6-phosphogluconate.

80. Enzymes in the soil fungus *Aspergillus nidulans* use NADPH as a coenzyme when converting nitrate to ammonium ions. When the fungus was cultured in a growth medium containing nitrate, the activities of several enzymes involved in glucose metabolism increased. What enzymes are good candidates for regulation under these conditions? Explain.

81. Enzymes of the pentose phosphate pathway are attractive drug targets to treat cancer. Explain why decreasing the activity of the pentose phosphate pathway would be detrimental to a cancer cell.

82. Predict the structure of the product of the following reaction:

$$\text{Erythrose-4-phosphate} + P_i + \text{NAD+} \xrightarrow{\text{GAPDH}} ? + \text{NADH} + H^+$$

83. Name and draw the structure of the product of the transaldolase-catalyzed condensation of dihydroxyacetone phosphate (DHAP) and erythrose-4-phosphate (E4P).

84. Which pathways would "collaborate" to maximize a cell's production of NADPH from glucose-6-phosphate, assuming that the cell needed NADPH but not ribose?

85. Which pathways would "collaborate" to maximize a cell's production of ribose-5-phosphate from glucose-6-phosphate, assuming that the cell needed ribose-5-phosphate but not NADPH?

86. Several studies have shown that the metabolite glucose-1,6-bisphosphate (G16BP) regulates several pathways of carbohydrate metabolism by inhibiting or activating key enzymes. The effect of G16BP on several enzymes is summarized in the table below. What pathways are active when G16BP is present? What pathways are inactive? What is the overall effect? Explain.

Enzyme	Effect of G16BP
Hexokinase	Inhibits
Phosphofructokinase (PFK)	Activates
Pyruvate kinase (PK)	Activates
Phosphoglucomutase	Activates
6-Phosphogluconate dehydrogenase	Inhibits

87. Xylulose-5-phosphate acts as an intracellular signaling molecule that activates kinases and phosphatases in liver cells. As a result of this signaling, there is an increase in the activity of the enzyme that produces fructose-2,6-bisphosphate, and the expression of genes for lipid synthesis is increased. What is the net effect of these responses?

13.5 Clinical Connection: Disorders of Carbohydrate Metabolism

88. Red blood cells synthesize and degrade 2,3-bisphosphoglycerate (BPG) as a detour from the glycolytic pathway, as shown in Problem 31. BPG decreases the oxygen affinity of hemoglobin by binding in the central cavity of the deoxygenated form of hemoglobin (see Fig. 5.11). This encourages delivery of oxygen to tissues. A defect in one of the glycolytic enzymes may affect levels of BPG. The plot shows oxygen-binding curves for normal erythrocytes and for hexokinase-deficient and pyruvate kinase–deficient erythrocytes. Identify which curve corresponds to which enzyme deficiency.

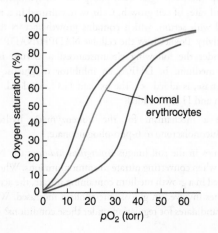

89. a. What happens to the [ADP]/[ATP] and [NAD⁺]/[NADH] ratios in red blood cells with a pyruvate kinase deficiency (see Problem 88)? **b.** One of the symptoms of a pyruvate kinase deficiency is hemolytic anemia, in which red blood cells swell and eventually lyse. Explain why the enzyme deficiency brings about this symptom.

90. Galactose is metabolized by three enzymes, as shown below. Ultimately, galactose is converted to glucose-6-phosphate, which is connected to pathways of carbohydrate metabolism, as discussed in Solution 77. An inability to digest galactose could arise from a deficiency in any of the three enzymes shown. **a.** A deficiency in galactokinase results in the accumulation of galactose, which can serve as a substrate for a reductase enzyme that converts galactose to galactitol, a toxic derivative that is excreted in the urine. Using what you learned about monosaccharide derivatives in Section 11.1, draw the structure of galactitol. **b.** Galactose can also serve as a substrate for dehydrogenase enzymes that oxidize the aldehyde group to form galactonate, which is less toxic than galactitol. Draw its structure. **c.** A deficiency of the uridyltransferase leads to the accumulation of galactose-1-phosphate, which has been shown to inhibit glucose-6-phosphatase, phosphoglucomutase, and glycogen phosphorylase. What symptoms arise when these three enzymes are inhibited as a result of a uridyltransferase deficiency?

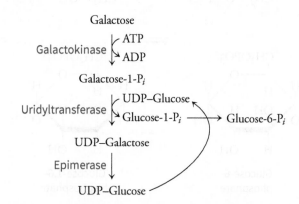

91. Individuals with fructose intolerance lack fructose-1-phosphate aldolase, a liver enzyme essential for catabolizing fructose. In the absence of fructose-1-phosphate aldolase, fructose-1-phosphate accumulates in the liver and inhibits glycogen phosphorylase and fructose-1,6-bisphosphatase. **a.** Explain why individuals with fructose intolerance exhibit hypoglycemia (low blood sugar). **b.** Administering glycerol and dihydroxyacetone phosphate does not alleviate the hypoglycemia, but administering galactose does relieve the hypoglycemia. Explain.

92. Glycogen storage disease 7 (GSD7) results from a deficiency of muscle phosphofructokinase. Patients with this genetic disease may have muscle PFK levels that are 1–5% of normal. Why do these patients suffer from myoglobinuria (myoglobin in the urine) and muscle cramping during exercise?

93. Patients with McArdle's disease have normal liver glycogen content and structure, but their muscles are easily fatigued. Identify the type of glycogen storage disease as listed in Table 13.1.

94. A patient with McArdle's disease (see Solution 93) performs ischemic (anaerobic) exercise for as long as he is able to do so. The patient's blood is withdrawn every few minutes during the exercise period and tested for lactate. The patient's samples are compared with control samples from a patient who does not suffer from a glycogen storage disease. The results are shown in the figure. Why does the lactate concentration increase in the normal patient? Why is there no corresponding increase in the patient's lactate concentration?

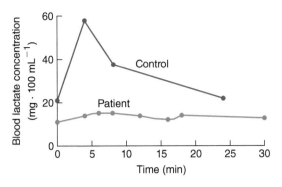

95. During even mild exertion, individuals with McArdle's disease (see Solution 93) experience painful muscle cramps. Yet the muscles in these individuals contain normal amounts of glycogen. What does this observation reveal about the pathways for glycogen degradation and glycogen synthesis?

96. Patients with von Gierke's disease (glycogen storage disease 1 or GSD1) have a deficiency of glucose-6-phosphatase. One of the most prominent symptoms of the disease is a protruding abdomen due to an enlarged liver. **a.** Explain why the liver is enlarged in patients with von Gierke's disease. **b.** Some patients with von Gierke's disease also have enlarged kidneys. Explain why.

97. a. Does a patient with McArdle's disease (see Problems 93–95) suffer from hypoglycemia, hyperglycemia, or neither? **b.** Does a patient with von Gierke's disease (see Problem 96) suffer from hypoglycemia, hyperglycemia, or neither?

98. Would a small feeding of cornstarch administered at bedtime help relieve the symptoms of a child with glycogen storage disease 0 (GSD0, see Table 13.1)? Explain why or why not.

99. Reduced glutathione (GSH), a tripeptide containing a Cys residue (see Solutions 4.17–4.18), is found in red blood cells, where it reduces organic hydroperoxides (ROOH) formed in cellular structures exposed to high concentrations of reactive oxygen. In converting the peroxides to less harmful alcohols (ROH), GSH becomes oxidized to GSSG.

$$2\,GSH + R{-}O{-}OH \xrightarrow{\text{Glutathione peroxidase}} GSSG + R{-}OH + H_2O$$

Organic hydroperoxide

Reduced glutathione also plays a role in maintaining normal red blood cell structure and keeping the iron ion of hemoglobin in the +2 oxidation state. Oxidized glutathione (GSSG) is regenerated as shown in the following reaction:

$$GSSG + NADPH + H^+ \xrightarrow{\text{Glutathione reductase}} 2\,GSH + NADP^+$$

Seemingly unrelated activities, such as taking the antimalarial drug primaquine or consuming fava beans, stimulate peroxide formation. Why do individuals with a glucose-6-phosphate dehydrogenase deficiency suffer from hemolytic anemia if they take antimalarial drugs or consume fava beans?

SELECTED READINGS

Hannou, S.A., Haslam, D.E., McKeown, N.M., and Herman, M.A., Fructose metabolism and metabolic disease, *J. Clin. Invest.* 128, 545–555, doi: 10.1172/JCI96702 (2018). [Reviews the links between fructose and dysregulation of carbohydrate and lipid metabolism.]

Kanungo, S., Wells, K., Tribett, T., and, El-Gharbawy, A., Glycogen metabolism and glycogen storage disorders, *Ann. Transl. Med.* 6, 474, doi: 10.21037/atm.2018.10.59 (2018). [Reviews the defects, symptoms, and treatments for genetic diseases of glucose and glycogen metabolism.]

Lenzen, S., A fresh view of glycolysis and glucokinase regulation: History and current status, *J. Biol. Chem.* 289, 12189–12194 (2014). [Includes some history of glycolysis research, with an emphasis on regulation of Steps 1 and 3 of glycolysis.]

Prats, C., Graham, T.E., and Shearer, J., The dynamic life of the glycogen granule, *J. Biol. Chem.* 293, 7089–7098, doi: 10.1074/jbc. R117.802843 (2018). [Describes the synthesis, degradation, and structure of glycogen granules.]

Rajas, F., Gautier-Stein, A., and Mithieux, G., Glucose-6-phosphate, a central hub for liver carbohydrate metabolism, *Metabolites* 9, 282, doi: 10.3390/metabo9120282 (2019). [Summarizes the metabolic fates of glucose-6-phosphate in health and disease.]

Stincone, A., Prigione, A., Cramer, T., Wamelink, M.M.C., Campbell, K., Cheung, E., Olin-Sandoval, V., Grüning, N.-M., Krüger, A., Tauqeer Alam, M., Keller, M.A., Breitenbach, M., Brindle, M., Rabinowitz, J.D., and Ralser, M., The return of metabolism: Biochemistry and physiology of the pentose phosphate pathway, *Biol. Rev.* 90, 927–963, doi: 10.1111/brv.12140 (2015). [A thorough review of the pathway, its evolutionary history, regulation, and role in disease.]

CHAPTER 13 CREDITS

Figure 13.5a Image based on 1YPI. Lolis, E., Alber, T., Davenport, R.C., Rose, D., Hartman, F.C., Petsko, G.A., Structure of yeast triosephosphate isomerase at 1.9-A resolution, *Biochemistry* 29, 6609–6618 (1990).

Figure 13.5b Image based on 2YPI, Lolis, E., Petsko, G.A., Crystallographic analysis of the complex between triosephosphate isomerase and 2-phosphoglycolate at 2.5-A resolution, implications for catalysis, *Biochemistry* 29, 6619–6625 (1990).

Figure 13.7 Data from Newsholme, E.A. and Start, C., *Regulation in Metabolism,* p. 97, Wiley (1973).

The Citric Acid Cycle

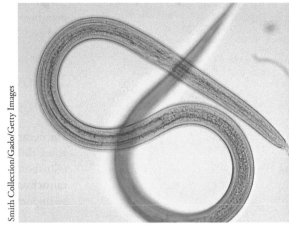

Smith Collection/Gado/Getty Images

Parasites, such as the roundworm shown here, produce and release a variety of metabolites, including the citric acid cycle intermediate succinate. Specialized epithelial cells in the host animal's intestine, lungs, nose, and other locations detect the extracellular succinate and respond by increasing their antiparasite defenses.

Do You Remember?

- Enzymes accelerate chemical reactions using acid–base catalysis, covalent catalysis, and metal ion catalysis (Section 6.2).
- Coenzymes such as NAD^+ and ubiquinone collect electrons from compounds that become oxidized (Section 12.3).
- Metabolic pathways in cells are connected and are regulated (Section 12.3).
- Many vitamins, substances that humans cannot synthesize, are components of coenzymes (Section 12.3).
- Pyruvate can be converted to lactate, acetyl-CoA, or oxaloacetate (Section 13.1).

The citric acid cycle logically follows glycolysis in an overview of cellular energy metabolism, but the cycle does much more than just continue the breakdown of glucose. Occupying a central place in the metabolism of most cells, the citric acid cycle processes the remnants of all types of metabolic fuels, including fatty acids and amino acids, so that their energy can be used to synthesize ATP. The citric acid cycle also operates anabolically, supplying the precursors for biosynthetic pathways. We will use pyruvate, the end product of glycolysis, as the starting point for our study of the citric acid cycle. We will then examine the eight steps of the citric acid cycle and discuss how this sequence of reactions might have evolved. Finally, we will consider the citric acid cycle as a multifunctional pathway with links to other metabolic processes.

14.1 The Pyruvate Dehydrogenase Reaction

KEY CONCEPTS

Summarize the reactions carried out by the pyruvate dehydrogenase complex.

- List the substrates, products, and cofactors of the pyruvate dehydrogenase reaction.
- Explain the advantages of a multienzyme complex.

The end product of glycolysis is the three-carbon compound pyruvate. In aerobic organisms, these carbons are ultimately oxidized to 3 CO_2 (although the oxygen atoms come not from molecular oxygen but from water and phosphate). The first molecule of CO_2 is released when pyruvate is decarboxylated to an acetyl unit. The second and third CO_2 molecules are products of the citric acid cycle.

The pyruvate dehydrogenase complex contains multiple copies of three different enzymes

The decarboxylation of pyruvate is catalyzed by the pyruvate dehydrogenase complex. In eukaryotes, this enzyme complex, and the enzymes of the citric acid cycle itself, are located inside the mitochondrion (an organelle surrounded by a double membrane and whose interior is called the **mitochondrial matrix**). Accordingly, pyruvate produced by glycolysis in the cytosol must first be imported into the mitochondria by a specific transport protein.

For convenience, the three kinds of enzymes that make up the pyruvate dehydrogenase complex are called E1, E2, and E3. Together *they catalyze the oxidative decarboxylation of pyruvate and the transfer of the acetyl unit to coenzyme A*:

$$\text{pyruvate} + \text{CoA} + \text{NAD}^+ \rightarrow \text{acetyl-CoA} + CO_2 + \text{NADH}$$

The structure of coenzyme A, a nucleotide derivative containing the vitamin pantothenate, is shown in Figure 3.2a.

In some bacteria, the 4600-kD pyruvate dehydrogenase complex consists of a cubic core of 24 E2 subunits (**Fig. 14.1a**), which are surrounded by an outer shell of 24 E1 and 12 E3 subunits. In mammals and some other bacteria, the enzyme complex is even larger, with 42–48 E1, 60 E2, and 6–12 E3 plus additional proteins that hold the complex together and regulate its enzymatic activity. The 60-subunit E2 core of the pyruvate dehydrogenase complex from *Bacillus stearothermophilus* is shown in Figure 14.1b.

a.   **b.**

Figure 14.1 Models of the E2 core of the pyruvate dehydrogenase complex.   a. In the *Azotobacter vinelandii* complex, 24 E2 polypeptides are arranged in a cube. **b.** The 60 subunits of the E2 core from *B. stearothermophilus* form a dodecahedron, a shape with 12 pentagonal faces.

Pyruvate dehydrogenase converts pyruvate to acetyl-CoA

The operation of the pyruvate dehydrogenase complex requires several coenzymes, whose functional roles in the five-step reaction are described below.

1. In the first step, which is catalyzed by E1 (also called pyruvate dehydrogenase), pyruvate is decarboxylated. This reaction requires the cofactor thiamine pyrophosphate (TPP; **Fig. 14.2**). TPP attacks the carbonyl carbon of pyruvate, and the departure of CO_2 leaves a hydroxyethyl group attached to TPP. This carbanion is stabilized by the positively charged thiazolium ring group of TPP:

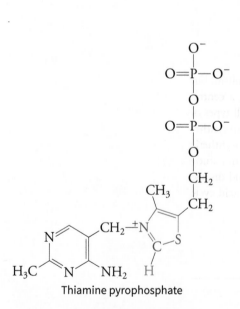

Pyruvate

Figure 14.2 Thiamine pyrophosphate (TPP).   This cofactor is the phosphorylated form of thiamine, also known as vitamin B_1 (see Section 12.3). The central thiazolium ring (blue) is the active portion. An acidic proton (red) dissociates, and the resulting carbanion is stabilized by the nearby positively charged nitrogen. TPP is a cofactor for several different decarboxylases.

Thiamine pyrophosphate

Hydroxyethyl–TPP

2. The hydroxyethyl group is then transferred to E2 of the pyruvate dehydrogenase complex. The hydroxyethyl acceptor is a lipoamide prosthetic group (**Fig. 14.3**). The transfer reaction regenerates the TPP cofactor of E1 and oxidizes the hydroxyethyl group to an acetyl group:

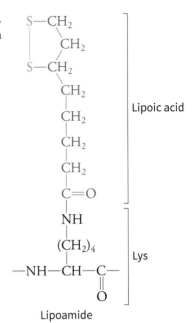

Figure 14.3 Lipoamide. This prosthetic group consists of lipoic acid (a vitamin) linked via an amide bond to the ε-amino group of a protein lysine residue. The active portion of the 14-Å-long lipoamide is the disulfide bond (red), which can be reversibly reduced.

3. Next, E2 transfers the acetyl group to coenzyme A, producing acetyl-CoA and leaving a reduced lipoamide group.

Recall that acetyl-CoA is a thioester, a form of energy currency (see Section 12.2). Some of the energy released in the oxidation of the hydroxyethyl group to an acetyl group is conserved in the formation of acetyl-CoA.

4. The final two steps of the reaction restore the pyruvate dehydrogenase complex to its original state. E3 reoxidizes the lipoamide group of E2 by transferring electrons to a Cys–Cys disulfide group in the enzyme.

5. Finally, NAD^+ reoxidizes the reduced cysteine sulfhydryl groups. This electron-transfer reaction is facilitated by an FAD prosthetic group (the structure of FAD, a nucleotide derivative, is shown in Fig. 3.2c).

During the five-step reaction (summarized in **Fig. 14.4**), the long lipoamide group of E2 acts as a swinging arm that visits the active sites of E1, E2, and E3 within the multienzyme complex. The arm picks up an acetyl group from an E1 subunit and transfers it to coenzyme A in an E2 active site. The arm then swings to an E3 active site, where it is reoxidized. Some other multienzyme complexes also include swinging arms, often attached to hinged protein domains to maximize their mobility.

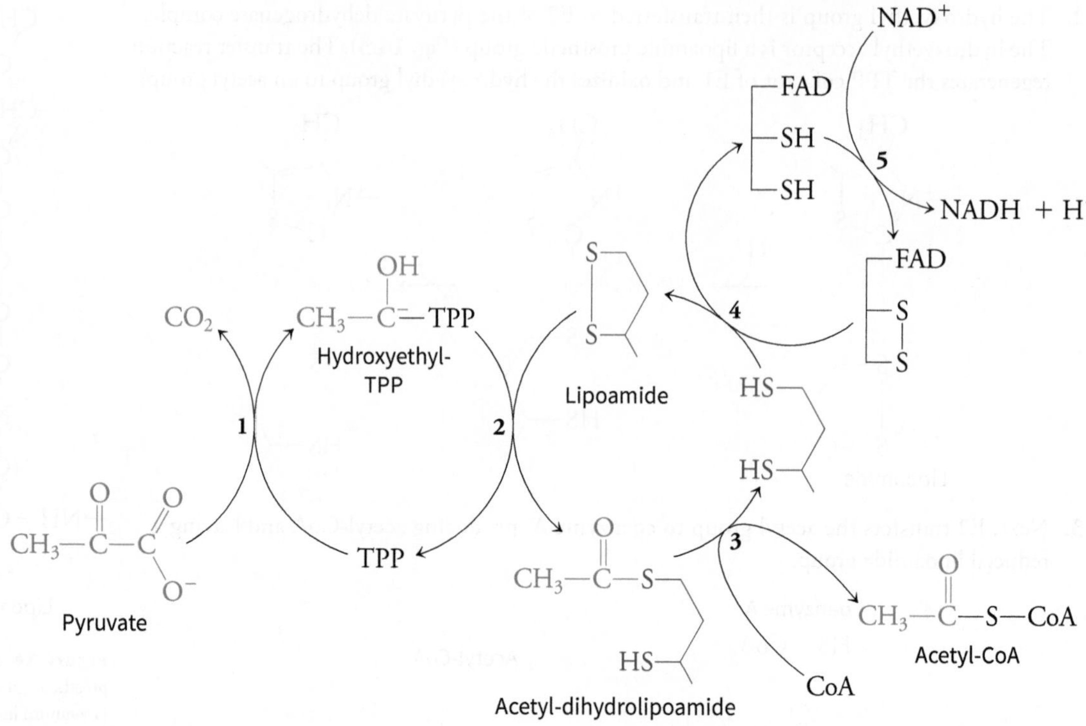

Figure 14.4 Reactions of the pyruvate dehydrogenase complex. In these five reactions, an acetyl group from pyruvate is transferred to CoA, CO_2 is released, and NAD^+ is reduced to NADH.

Question Without looking at the text, write the net equation for the reactions shown here. How many vitamin-derived cofactors are involved?

A **multienzyme complex** such as the pyruvate dehydrogenase complex can carry out a multi-step reaction sequence efficiently because *the product of one reaction can quickly become the substrate for the next reaction without diffusing away or reacting with another substance.* There is also evidence that the individual enzymes of glycolysis and the citric acid cycle associate loosely with each other so that the close proximity of their active sites can increase flux through their respective pathways.

Flux through the pyruvate dehydrogenase complex is regulated by product inhibition: Both NADH and acetyl-CoA act as inhibitors. The activity of the complex is also regulated by hormone-controlled phosphorylation and dephosphorylation, which suits its function as the gatekeeper for the entry of a metabolic fuel into the citric acid cycle.

Concept Check

1. Describe the functional importance of the coenzymes that participate in the reactions carried out by the pyruvate dehydrogenase complex.
2. Discuss the advantages of a multienzyme complex.

14.2 The Eight Reactions of the Citric Acid Cycle

KEY CONCEPTS

Describe the substrate, product, and type of chemical reaction for each step of the citric acid cycle.

The starting material for the **citric acid cycle** is an acetyl-CoA molecule that may be derived from a carbohydrate via pyruvate, as just described, or from another metabolic fuel. The carbon skeletons of amino acids are broken down to either pyruvate or acetyl-CoA, and fatty acids are broken down to acetyl-CoA. In some tissues, the bulk of acetyl-CoA entering the citric acid cycle comes from fatty acids rather than carbohydrates or amino acids. Whatever their source, the citric acid cycle converts all these two-carbon acetyl groups into CO_2 and therefore represents the final stage in fuel oxidation (**Fig. 14.5**). As the carbons become fully oxidized to CO_2, their energy is conserved and subsequently used to produce ATP.

The eight reactions of the citric acid cycle take place in the cytosol of prokaryotes and in the mitochondria of eukaryotes. Unlike a linear pathway such as glycolysis (see Fig. 13.2) or gluconeogenesis (see Fig. 13.10), the citric acid cycle always returns to its starting position, essentially behaving as a multistep catalyst.

The cycle as a whole is highly exergonic, and free energy is conserved at several steps in the form of a nucleotide triphosphate (GTP) and reduced cofactors. *For each acetyl group that enters the citric acid cycle, two molecules of fully oxidized CO_2 are produced, representing a loss of four pairs of electrons.* These electrons are transferred to 3 NAD^+ and 1 ubiquinone (Q) to produce 3 NADH and 1 QH_2. The net equation for the citric acid cycle is therefore

$$acetyl\text{-}CoA + GDP + P_i + 3\ NAD^+ + Q \rightarrow 2\ CO_2 + CoA + GTP + 3\ NADH + QH_2$$

In this section we examine the sequence of eight enzyme-catalyzed reactions of the citric acid cycle, focusing on a few interesting reactions. The entire pathway, summarized in **Figure 14.6**, is also known as the Krebs cycle (after Hans Krebs, who worked out the sequence of reactions in the 1930s) and the tricarboxylic acid cycle (because citrate has three carboxyl groups).

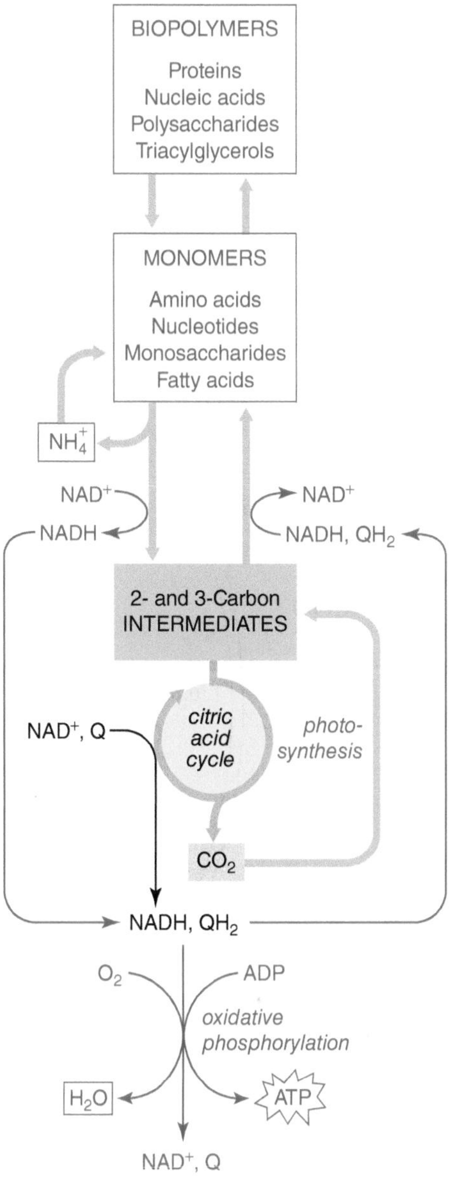

1. Citrate synthase adds an acetyl group to oxaloacetate

In the first reaction of the citric acid cycle, the acetyl group of acetyl-CoA condenses with the four-carbon compound oxaloacetate to produce the six-carbon compound citrate:

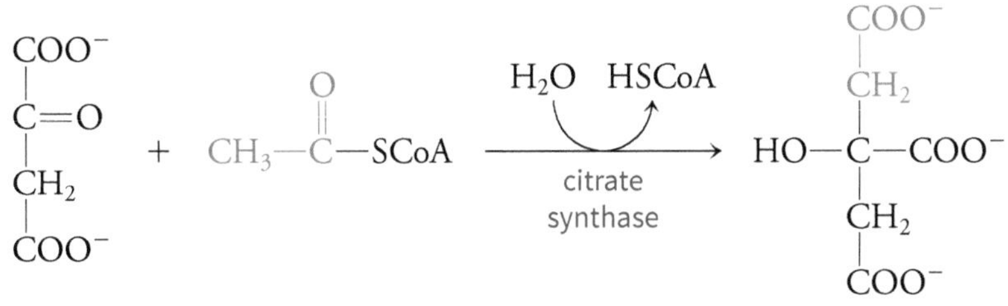

Figure 14.5 The citric acid cycle in context. The citric acid cycle is a central metabolic pathway whose starting material is two-carbon acetyl units derived from amino acids, monosaccharides, and fatty acids. These are oxidized to the waste product CO_2, with the reduction of the cofactors NAD^+ and ubiquinone (Q).

Citrate synthase, a dimer, undergoes a large conformational change when oxaloacetate binds. (**Fig. 14.7**). This shift creates a binding site for acetyl-CoA and explains why oxaloacetate must bind to the enzyme before acetyl-CoA can bind.

Citrate synthase is one of the few enzymes that can synthesize a carbon–carbon bond without using a metal ion cofactor. Its mechanism is shown in **Figure 14.8**. The first reaction intermediate may be stabilized by the formation of low-barrier hydrogen bonds, which are stronger than ordinary hydrogen bonds (see Section 6.3). The coenzyme A released during the final step can be reused by the pyruvate dehydrogenase complex or used later in the citric acid cycle to synthesize the intermediate succinyl-CoA.

The reaction catalyzed by citrate synthase is highly exergonic ($\Delta G^{\circ\prime} = -31.5\ kJ \cdot mol^{-1}$), equivalent to the free energy of hydrolyzing the thioester bond of acetyl-CoA). We will see later why the efficient operation of the citric acid cycle requires that this step have a large negative free energy change.

Figure 14.6 Reactions of the citric acid cycle.

Question Identify the number of carbon atoms in each intermediate.

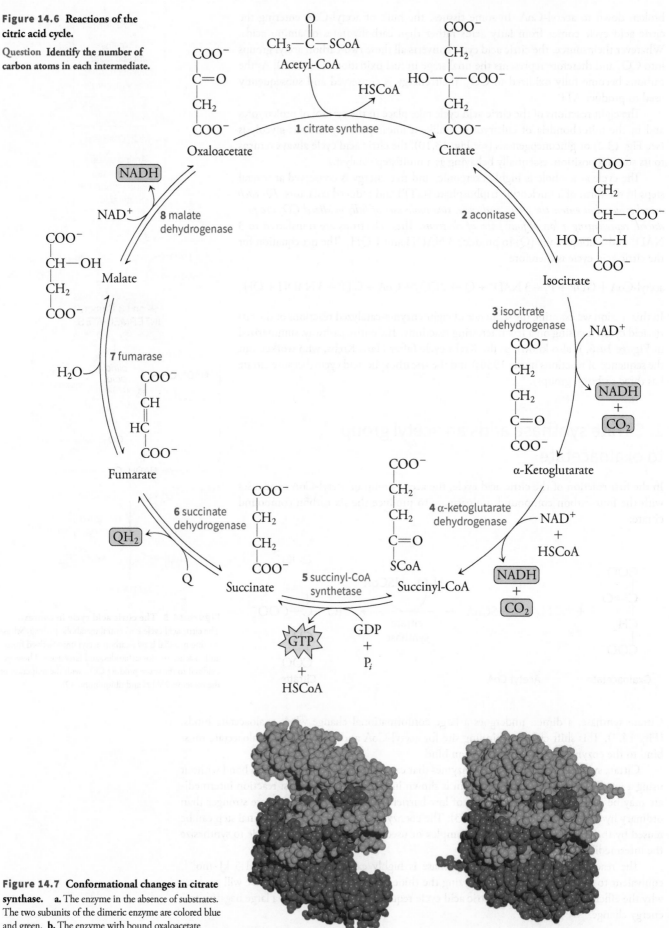

Figure 14.7 Conformational changes in citrate synthase. a. The enzyme in the absence of substrates. The two subunits of the dimeric enzyme are colored blue and green. **b.** The enzyme with bound oxaloacetate (red, mostly buried) and an acetyl-CoA analog (orange).

a.

b.

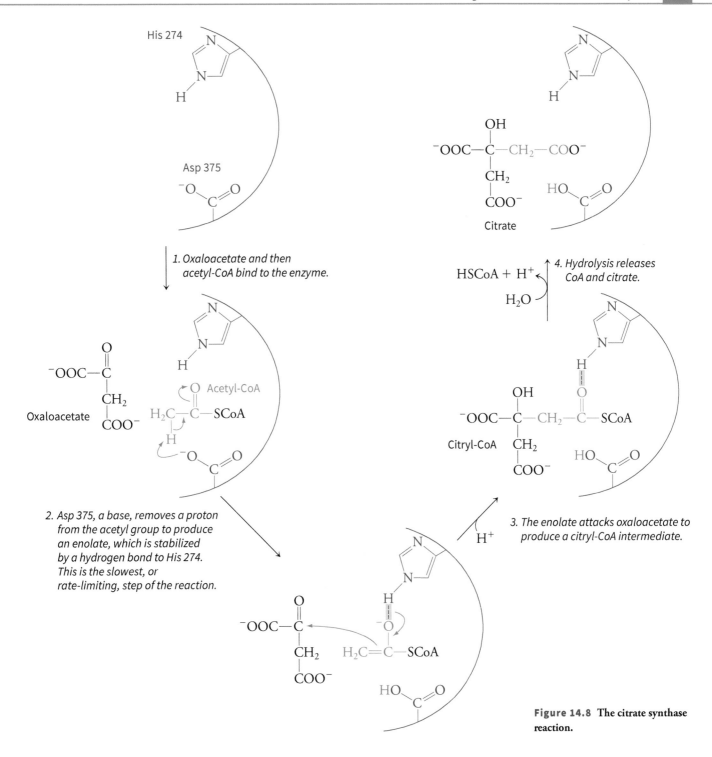

1. Oxaloacetate and then acetyl-CoA bind to the enzyme.

2. Asp 375, a base, removes a proton from the acetyl group to produce an enolate, which is stabilized by a hydrogen bond to His 274. This is the slowest, or rate-limiting, step of the reaction.

3. The enolate attacks oxaloacetate to produce a citryl-CoA intermediate.

4. Hydrolysis releases CoA and citrate.

Figure 14.8 **The citrate synthase reaction.**

2. Aconitase isomerizes citrate to isocitrate

The second enzyme of the citric acid cycle catalyzes the reversible isomerization of citrate to isocitrate:

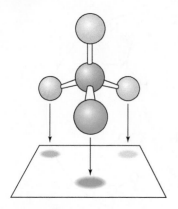

Figure 14.9 Stereochemistry of aconitase. The three-point attachment of citrate to the enzyme allows only one carboxymethyl group (shown in green) to react.

The enzyme is named after the reaction intermediate.

Citrate is a symmetrical molecule, yet only one of its two carboxymethyl arms ($-CH_2-COO^-$) undergoes dehydration and rehydration during the aconitase reaction. This stereochemical specificity long puzzled biochemists, including Hans Krebs, who first described the citric acid cycle. Eventually, Alexander Ogston pointed out that although citrate is symmetrical, its two carboxymethyl groups are no longer identical when it is bound to an asymmetrical enzyme (**Fig. 14.9**). In fact, a three-point attachment is not even necessary for an enzyme to distinguish two groups in a molecule such as citrate, which are related by mirror symmetry. You can prove this yourself with a simple organic chemistry model kit. By now you should appreciate that biological systems, including enzyme, are inherently chiral (also see Section 4.1).

3. Isocitrate dehydrogenase releases the first CO_2

The third reaction of the citric acid cycle is the oxidative decarboxylation of isocitrate to α-ketoglutarate. The substrate is first oxidized in a reaction accompanied by the reduction of NAD^+ to NADH. Then the carboxylate group β to the ketone function (that is, two carbon atoms away from the ketone) is eliminated as CO_2. An Mn^{2+} ion in the active site helps stabilize the negative charges of the reaction intermediate.

The CO_2 molecules generated by isocitrate dehydrogenase—along with the CO_2 generated in the following reaction and the CO_2 produced by the decarboxylation of pyruvate—diffuse out of the cell and are carried in the bloodstream to the lungs, where they are breathed out. Note that these CO_2 molecules are produced through oxidation–reduction reactions. The carbons are oxidized, while NAD^+ is reduced. O_2 is not directly involved in this process.

4. α-Ketoglutarate dehydrogenase releases the second CO_2

α-Ketoglutarate dehydrogenase, like isocitrate dehydrogenase, catalyzes an oxidative decarboxylation reaction. It also transfers the remaining four-carbon fragment to CoA:

The free energy of oxidizing α-ketoglutarate is conserved in the formation of the thioester succinyl-CoA. α-Ketoglutarate dehydrogenase is a multienzyme complex that resembles the pyruvate dehydrogenase complex in both structure and enzymatic mechanism. In fact, the same E3 enzyme is a member of both complexes.

The isocitrate dehydrogenase and α-ketoglutarate dehydrogenase reactions both release CO_2. These two carbons are not the ones that entered the citric acid cycle as acetyl-CoA; those acetyl carbons become part of oxaloacetate and are lost in subsequent rounds of the cycle (**Fig. 14.10**). *However, the net result of each round of the citric acid cycle is the loss of two carbons as CO_2 for each acetyl-CoA that enters the cycle.*

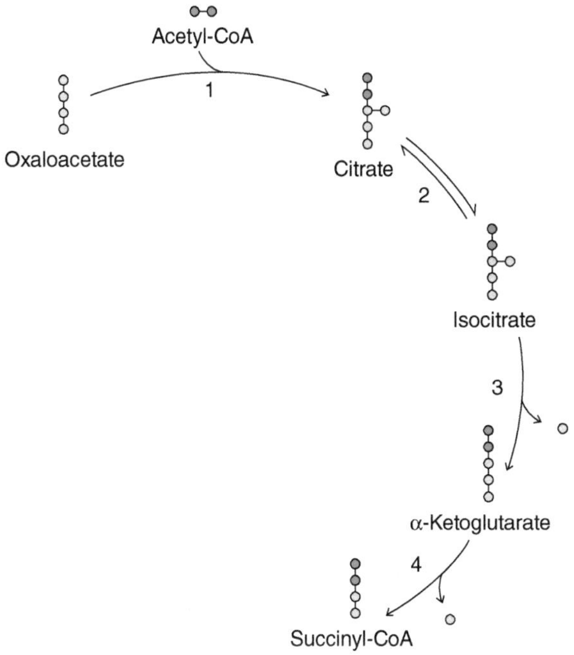

5. Succinyl-CoA synthetase catalyzes substrate-level phosphorylation

The thioester succinyl-CoA releases a large amount of free energy when it is hydrolyzed ($\Delta G^{\circ\prime} = -32.6$ kJ·mol⁻¹). This is enough free energy to drive the synthesis of a nucleoside triphosphate from a nucleo-side diphosphate and P_i ($\Delta G^{\circ\prime} = 30.5$ kJ·mol⁻¹). The change in free energy for the net reaction is near zero, so the reaction is reversible. In fact, the enzyme is named for the reverse reaction. Succinyl-CoA synthetase in the mammalian citric acid cycle generates GTP, whereas the plant and bacterial enzymes generate ATP (recall that GTP is energetically equivalent to ATP). An exergonic reaction coupled to the transfer of a phosphoryl group to a nucleoside diphosphate is termed **substrate-level phosphorylation** to distinguish it from oxidative phosphorylation (Section 15.3) and photophosphorylation (Section 16.2), which are more indirect ways of synthesizing ATP.

How does succinyl-CoA synthetase couple thioester cleavage to the synthesis of a nucleoside triphosphate? The reaction is a series of phosphoryl-group transfers that involve an active-site histidine residue (**Fig. 14.11**). After the succinyl group is phosphorylated, the phosphoryl group is transferred to the side chain of His 246 close by. A protein loop containing the phospho-His reaction intermediate must then move a large distance—about 35 Å—to deliver the phosphoryl group to the waiting nucleoside diphosphate (**Fig. 14.12**).

Figure 14.10 Fates of carbon atoms in the citric acid cycle. The two carbon atoms that are lost as CO_2 in the reactions catalyzed by isocitrate dehydrogenase (step 3) and α-ketoglutarate dehydrogenase (step 4) are not the same carbons that entered the cycle as acetyl-CoA (red).

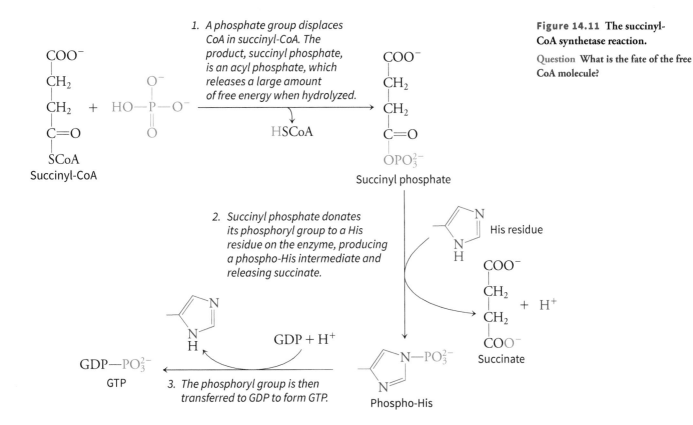

1. *A phosphate group displaces CoA in succinyl-CoA. The product, succinyl phosphate, is an acyl phosphate, which releases a large amount of free energy when hydrolyzed.*

2. *Succinyl phosphate donates its phosphoryl group to a His residue on the enzyme, producing a phospho-His intermediate and releasing succinate.*

3. *The phosphoryl group is then transferred to GDP to form GTP.*

Figure 14.11 The succinyl-CoA synthetase reaction.

Question What is the fate of the free CoA molecule?

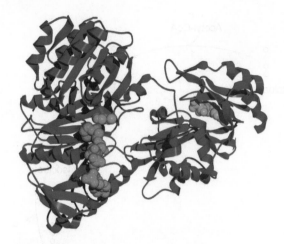

Figure 14.12 Substrate binding in succinyl-CoA synthetase. The substrate succinyl-CoA is represented by coenzyme A (red), the side chain of His 246 is green, and the nucleoside diphosphate awaiting phosphorylation (ADP) is orange.

6. Succinate dehydrogenase generates ubiquinol

The final three reactions of the citric acid cycle convert succinate back to the cycle's starting substrate, oxaloacetate. Succinate dehydrogenase catalyzes the reversible dehydrogenation of succinate to fumarate. This oxidation–reduction reaction requires an FAD prosthetic group, which is reduced to $FADH_2$ during the reaction:

$$Succinate \xrightarrow[\text{succinate dehydrogenase}]{Enz\text{-}FAD \quad Enz\text{-}FADH_2} Fumarate$$

To regenerate the enzyme, the $FADH_2$ group must be reoxidized. Since succinate dehydrogenase is embedded in the inner mitochondrial membrane (it is the only one of the eight citric acid cycle enzymes that is not soluble in the mitochondrial matrix), it can be reoxidized by the lipid-soluble electron carrier ubiquinone (see Section 12.3) rather than by the soluble cofactor NAD^+. Ubiquinone (abbreviated Q) acquires two electrons to become ubiquinol (QH_2).

$$Enzyme\text{-}FADH_2 \xrightleftharpoons{Q \quad QH_2} Enzyme\text{-}FAD$$

7. Fumarase catalyzes a hydration reaction

In the seventh reaction, fumarase (also known as fumarate hydratase) catalyzes the reversible hydration of a double bond to convert fumarate to malate:

$$Fumarate \xrightleftharpoons[\text{fumarase}]{H_2O} Malate$$

8. Malate dehydrogenase regenerates oxaloacetate

The citric acid cycle concludes with the regeneration of oxaloacetate from malate in an NAD^+-dependent oxidation reaction:

$$Malate \xrightarrow[\text{malate dehydrogenase}]{NAD^+ \quad NADH + H^+} Oxaloacetate$$

The standard free energy change for this reaction is $+29.7 \ kJ \cdot mol^{-1}$, indicating that the reaction has a low probability of occurring as written. However, the product oxaloacetate is a substrate for the next reaction (Reaction 1 of the citric acid cycle). The highly exergonic—and therefore highly favorable—citrate synthase reaction helps pull the malate dehydrogenase reaction forward. This is the reason for the apparent waste of free energy released by cleaving the thioester bond of acetyl-CoA in the first reaction of the citric acid cycle.

Concept Check

1. List the sources of the acetyl groups that enter the citric acid cycle.
2. Write the net equation for the citric acid cycle.
3. Draw the structures of the substrates and products of the eight reactions and name the enzyme that catalyzes each step.
4. Identify the steps that generate ATP, CO_2, and reduced cofactors.

14.3 Thermodynamics of the Citric Acid Cycle

KEY CONCEPTS

Explain how the citric acid cycle recovers energy for the cell.

- Calculate the ATP yield for one round of the cycle.
- Identify the irreversible steps that regulate flux through the cycle.
- Describe how the cycle reactions can operate in reverse.

Because the eighth reaction of the citric acid cycle returns the system to its original state, *the entire pathway acts in a catalytic fashion to dispose of carbon atoms derived from amino acids, carbohydrates, and fatty acids.* Albert Szent-Györgyi discovered the catalytic nature of the pathway by observing that small additions of organic compounds such as succinate, fumarate, and malate stimulated O_2 uptake in a tissue preparation. Because the O_2 consumption was much greater than would be required for the direct oxidation of the added substances, he inferred that the compounds acted catalytically.

The citric acid cycle is an energy-generating catalytic cycle

We now know that oxygen is consumed during oxidative phosphorylation, the process that reoxidizes the reduced cofactors (NADH and QH_2) that are produced by the citric acid cycle. Although the citric acid cycle generates one molecule of GTP (or ATP), considerably more ATP is generated when the reduced cofactors are reoxidized by O_2. Each NADH yields approximately 2.5 ATP and each QH_2 yields approximately 1.5 ATP (we will see in Section 15.3 why these values are not whole numbers). *Every acetyl unit that enters the citric acid cycle can therefore generate a total of 10 ATP equivalents.* The energy yield of a molecule of glucose, which generates two acetyl units, can be calculated (**Fig. 14.13**).

A muscle operating anaerobically produces only 2 ATP per glucose, but under aerobic conditions when the citric acid cycle is fully functional, each glucose molecule generates about 32 ATP equivalents. This general phenomenon is called the **Pasteur effect,** after Louis Pasteur, who first observed that the rate of glucose consumption by yeast cells decreased dramatically when the cells were shifted from anaerobic to aerobic growth conditions.

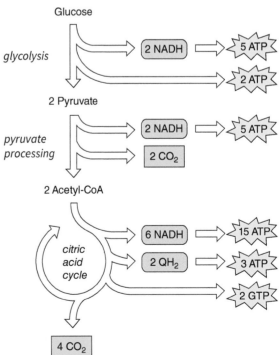

Figure 14.13 ATP yield from glucose. Two rounds of the citric acid cycle are required to fully oxidize one molecule of glucose.

Question Trace the fate of the six glucose carbon atoms.

The citric acid cycle is regulated at three steps

Flux through the citric acid cycle is regulated primarily at the cycle's three metabolically irreversible steps: those catalyzed by citrate synthase (Reaction 1), isocitrate dehydrogenase (Reaction 3), and α-ketoglutarate dehydrogenase (Reaction 4). The major regulators are shown in **Figure 14.14**.

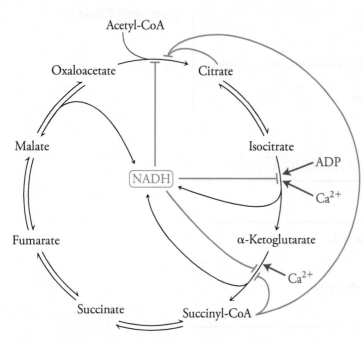

Figure 14.14 Regulation of the citric acid cycle.   Inhibition is represented by red symbols, activation by green symbols.

Neither acetyl-CoA nor oxaloacetate is present at concentrations high enough to saturate citrate synthase, so flux through the first step of the citric acid cycle depends largely on the substrate concentrations. The product of the reaction, citrate, inhibits citrate synthase (citrate also inhibits phosphofructokinase, thereby decreasing the supply of acetyl-CoA produced by glycolysis). Succinyl-CoA, the product of Reaction 4, inhibits the enzyme that produces it. It also acts as a feedback inhibitor by competing with acetyl-CoA in Reaction 1.

The activity of isocitrate dehydrogenase is inhibited by its reaction product, NADH. NADH also inhibits α-ketoglutarate dehydrogenase and citrate synthase. Both dehydrogenases are activated by Ca^{2+} ions, which generally signify the need to generate cellular energy. ADP, also representing the need for more ATP, activates isocitrate dehydrogenase.

Changes in enzyme reaction rates also regulate the flow of acetyl carbons through the cycle by altering the concentrations of cycle intermediates. Because the entire cycle acts as a catalyst, more intermediates means that more acetyl groups can be processed, just as a city can move more commuters by adding buses during rush hour. Not surprisingly, citric acid cycle defects have serious consequences (**Box 14.A**).

Box 14.A   Mutations in Citric Acid Cycle Enzymes

Possibly because the citric acid cycle is a central metabolic pathway, severe defects in any of its components are expected to be incompatible with life. However, researchers have documented mutations in the genes for several of the cycle's enzymes, including α-ketoglutarate dehydrogenase, succinyl-CoA synthetase, and succinate dehydrogenase. These defects, which are all rare, typically affect the central nervous system, causing symptoms such as movement disorders and neurodegeneration. A rare form of fumarase deficiency results in brain malformation and developmental disabilities.

Some citric acid cycle enzyme mutations are linked to cancer. One possible explanation is that a defective enzyme contributes to **carcinogenesis** (the development of cancer, or uncontrolled cell growth) by causing the accumulation of particular metabolites, which are responsible for altering the cell's activities. For example, normal cells respond to a drop in oxygen availability (hypoxia) by activating transcription factors known as hypoxia-inducible factors (HIFs; Section 19.2). These proteins interact with DNA to turn on the expression of genes for glycolytic enzymes and a growth factor that promotes the development of new blood vessels. When the fumarase gene is defective, fumarate accumulates and inhibits a

protein that destabilizes HIFs. As a result, the fumarase deficiency promotes glycolysis (an anaerobic pathway) and the growth of blood vessels. These two adaptations favor tumors, whose growth depends on high glycolytic flux and a constant supply of nutrients delivered by the bloodstream.

Defects in isocitrate dehydrogenase also promote cancer in an indirect fashion. Many cancerous cells exhibit a mutation in one of the two genes for the enzyme, suggesting that the unaltered copy is necessary for maintaining the normal activity of the citric acid cycle, while the mutated copy plays a role in carcinogenesis. The mutated isocitrate dehydrogenase no longer carries out the usual reaction (converting isocitrate to α-ketoglutarate) but instead converts α-ketoglutarate to 2-hydroxyglutarate in an NADPH-dependent manner. The mechanism whereby 2-hydroxyglutarate contributes to carcinogenesis is not clear, but its involvement is bolstered by the observation that individuals who harbor other mutations that lead to 2-hydroxyglutarate accumulation have an increased risk of developing brain tumors.

Question   How would a fumarase deficiency affect the levels of pyruvate, fumarate, and malate?

The citric acid cycle probably evolved as a synthetic pathway

A circular pathway such as the citric acid cycle must have evolved from a linear set of pre-existing biochemical reactions. Clues to its origins can be found by examining the metabolism of organisms that resemble earlier life-forms. Such organisms emerged before atmospheric oxygen was available and may have used sulfur as their ultimate oxidizing agent, reducing it to H_2S. Their modern-day counterparts are anaerobic autotrophs that harvest free energy by pathways that are independent of the pathways of carbon metabolism. These organisms therefore do not use the citric acid cycle to generate reduced cofactors that are subsequently oxidized by molecular oxygen. However, all organisms must synthesize small molecules that can be used to build proteins, nucleic acids, carbohydrates, and so on.

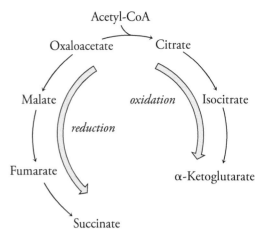

Figure 14.15 Pathways that might have given rise to the citric acid cycle. The pathway starting from oxaloacetate and proceeding to the right is an oxidative biosynthetic pathway, whereas the pathway that proceeds to the left is a reductive pathway.

Even organisms that do not use the citric acid cycle contain genes for some citric acid cycle enzymes. For example, the cells may condense acetyl-CoA with oxaloacetate, leading to α-ketoglutarate, which is a precursor of several amino acids. They may also convert oxaloacetate to malate, proceeding to fumarate and then to succinate. Together, these two pathways resemble the citric acid cycle, with the right arm following the usual oxidative sequence of the cycle and the left arm following a reversed, reductive sequence (**Fig. 14.15**). The reductive sequence of reactions might have evolved as a way to regenerate the cofactors reduced during other catabolic reactions (for example, the NADH produced by the glyceraldehyde-3-phosphate dehydrogenase reaction of glycolysis; see Section 13.1).

It is easy to theorize that the evolution of an enzyme to interconvert α-ketoglutarate and succinate could have connected the two "half pathways" (Fig. 14.15) to create a cyclic pathway similar to the modern citric acid cycle. Interestingly, *E. coli*, which uses the citric acid cycle under aerobic growth conditions, uses an interrupted citric acid cycle like the one diagrammed in Figure 14.15 when it is growing anaerobically.

Since the final four reactions of the modern citric acid cycle are metabolically reversible, the primitive citric acid cycle might easily have accommodated one-way flux in the clockwise direction, forming an oxidative cycle. If the complete cycle proceeded in the counterclockwise direction, the result would have been a reductive biosynthetic pathway (**Fig. 14.16**). This pathway, which would incorporate, or "fix," atmospheric CO_2 into biological molecules, may have preceded the modern CO_2-fixing pathway found in green plants and some photosynthetic bacteria (described in Section 16.3).

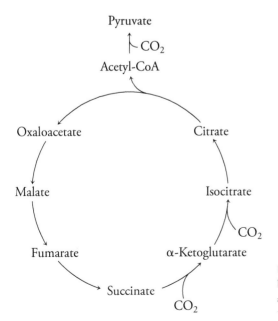

Figure 14.16 A proposed reductive biosynthetic pathway based on the citric acid cycle. This pathway might have operated to incorporate CO_2 into biological molecules.

Concept Check

1. Identify the products of the citric acid cycle that represent forms of energy currency for the cell.
2. Compare the ATP yield for glucose degradation via glycolysis and via glycolysis plus the citric acid cycle.
3. Describe how substrates and products of the citric acid cycle regulate flux through the pathway.
4. Explain how primitive oxidative and reductive biosynthetic pathways might have combined to generate a circular metabolic pathway.

14.4 Anabolic and Catabolic Functions of the Citric Acid Cycle

KEY CONCEPTS

Explain how the citric acid cycle connects with other metabolic processes.

- Identify the cycle intermediates that are precursors for the synthesis of other compounds.
- Describe how citric acid cycle intermediates are replenished.

The citric acid cycle does not operate like a simple pipeline, where one substance enters at one end and another emerges from the opposite end. In mammals, six of the eight citric acid cycle intermediates (all except isocitrate and succinate) are the precursors or products of other pathways. For this reason, it is impossible to designate the citric acid cycle as a purely catabolic or anabolic pathway.

Citric acid cycle intermediates are precursors of other molecules

Intermediates of the citric acid cycle can be siphoned off to form other compounds (**Fig. 14.17**). For example, succinyl-CoA is used for the synthesis of heme. The five-carbon α-ketoglutarate

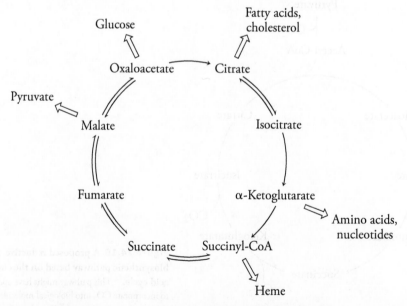

Figure 14.17 Citric acid cycle intermediates as biosynthetic precursors.

(sometimes called 2-oxoglutarate) can undergo reductive amination by glutamate dehydrogenase to produce the amino acid glutamate:

$$
\underset{\alpha\text{-Ketoglutarate}}{\begin{array}{l} COO^- \\ | \\ CH_2 \\ | \\ CH_2 \\ | \\ C{=}O \\ | \\ COO^- \end{array}} \quad + \quad NH_4^+ \quad \underset{\substack{\text{glutamate} \\ \text{dehydrogenase}}}{\overset{NADH + H^+ \quad NAD^+}{\rightleftharpoons}} \quad \underset{\text{Glutamate}}{\begin{array}{l} COO^- \\ | \\ CH_2 \\ | \\ CH_2 \\ | \\ H{-}C{-}NH_3^+ \\ | \\ COO^- \end{array}} \quad + \quad H_2O
$$

Glutamate is a precursor of the amino acids glutamine, arginine, and proline. Glutamine in turn is a precursor for the synthesis of purine and pyrimidine nucleotides. We have already seen that oxaloacetate is a precursor of monosaccharides (Section 13.2). Consequently, *any of the citric acid cycle intermediates, which can be converted by the cycle to oxaloacetate, can ultimately serve as gluconeogenic precursors.*

Citrate produced by the condensation of acetyl-CoA with oxaloacetate can be transported out of the mitochondria to the cytosol by a specific transport protein. ATP-citrate lyase then catalyzes the reaction

$$ATP + citrate + CoA \rightarrow ADP + P_i + oxaloacetate + acetyl\text{-}CoA$$

The resulting acetyl-CoA is used for fatty acid and cholesterol synthesis, which take place in the cytosol. Because the rate of lipid synthesis is high in rapidly growing cancer cells, inhibitors of ATP-citrate lyase are being developed as anti-cancer drugs. Note that the ATP-citrate lyase reaction undoes the work of the exergonic citrate synthase reaction. This seems wasteful, but cytosolic ATP-citrate lyase is essential because acetyl-CoA, which is produced in the mitochondria, cannot cross the mitochondrial membrane to reach the cytosol, whereas citrate can. The oxaloacetate product of the ATP-citrate lyase reaction can be converted to malate by a cytosolic malate dehydrogenase operating in reverse. Malate is then decarboxylated by the action of malic enzyme to produce pyruvate:

$$
\underset{\text{Malate}}{\begin{array}{l} COO^- \\ | \\ CH{-}OH \\ | \\ CH_2 \\ | \\ COO^- \end{array}} \quad \underset{\text{malic enzyme}}{\overset{NADP^+ \quad NADPH}{\rightleftharpoons}} \quad \underset{\text{Pyruvate}}{\begin{array}{l} COO^- \\ | \\ C{=}O \\ | \\ CH_3 \end{array}} \quad + \quad CO_2
$$

Pyruvate can reenter the mitochondria, with the help of the pyruvate transporter, and be converted back to oxaloacetate to complete the cycle shown in **Figure 14.18**. In plants, isocitrate is diverted from the citric acid cycle in a biosynthetic pathway known as the **glyoxylate pathway** (**Box 14.B**).

Anaplerotic reactions replenish citric acid cycle intermediates

Intermediates that are diverted from the citric acid cycle for other purposes can be replenished through **anaplerotic reactions** (from the Greek *ana*, "up," and *plerotikos*, "to fill"; **Fig. 14.19**). One of the most important of these reactions is catalyzed by pyruvate carboxylase (this is also the first step of gluconeogenesis; Section 13.2):

$$pyruvate + CO_2 + ATP + H_2O \rightarrow oxaloacetate + ADP + P_i$$

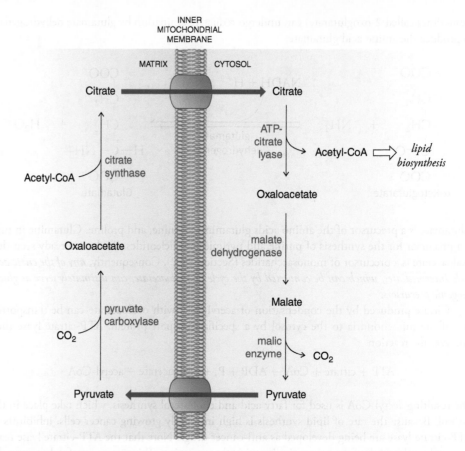

Figure 14.18 The citrate transport system. This set of transport proteins and enzymes allows carbon atoms from mitochondrial acetyl-CoA to be transferred to the cytosol for the synthesis of fatty acids and cholesterol.

Question Can the same CO_2 molecule released in the malic enzyme reaction be used in the pyruvate carboxylase reaction?

Box 14.B   The Glyoxylate Pathway

Plants and some bacterial cells contain certain enzymes that act together with some citric acid cycle enzymes to convert acetyl-CoA to oxaloacetate, a gluconeogenic precursor. Animals lack the enzymes to do this and therefore cannot undertake the net synthesis of carbohydrates from two-carbon precursors. In plants, the glyoxylate pathway includes reactions that take place in the mitochondria and the **glyoxysome,** an organelle that, like the peroxisome, contains enzymes that carry out some essential metabolic processes.

In the glyoxysome, acetyl-CoA condenses with oxaloacetate to form citrate, which is then isomerized to isocitrate, as in the citric acid cycle. However, the next step is not the isocitrate dehydrogenase reaction but a reaction catalyzed by the glyoxysome enzyme isocitrate lyase, which converts isocitrate to succinate and the two-carbon compound glyoxylate. Succinate continues as usual through the mitochondrial citric acid cycle to regenerate oxaloacetate.

In the glyoxysome, the glyoxylate condenses with a second molecule of acetyl-CoA in a reaction catalyzed by the glyoxysome enzyme malate synthase to form the four-carbon compound malate.

Malate can then be converted to oxaloacetate for gluconeogenesis. The two reactions that are unique to the glyoxylate pathway are shown in green in the figure; reactions that are identical to those of the citric acid cycle are shown in blue.

In essence, the glyoxylate pathway bypasses the two CO_2-generating steps of the citric acid cycle (catalyzed by isocitrate dehydrogenase and α-ketoglutarate dehydrogenase) and incorporates a second acetyl unit (at the malate synthase step). The net result of the glyoxylate pathway is the production of a four-carbon compound that can be used to synthesize glucose. This pathway is highly active in germinating seeds, where stored oils (triacylglycerols) are broken down to acetyl-CoA. The glyoxylate pathway thus provides a route for synthesizing glucose from fatty acids. Because animals lack isocitrate lyase and malate synthase, they cannot convert excess fat to carbohydrate.

Question Write the net equation for the glyoxylate cycle as shown here.

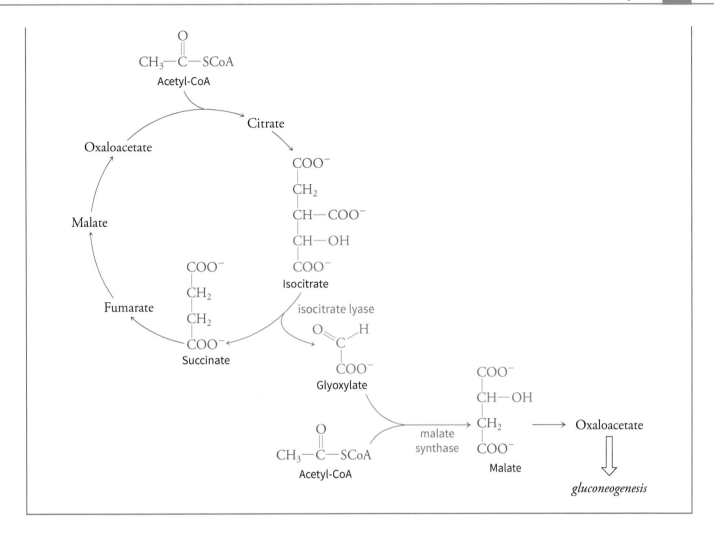

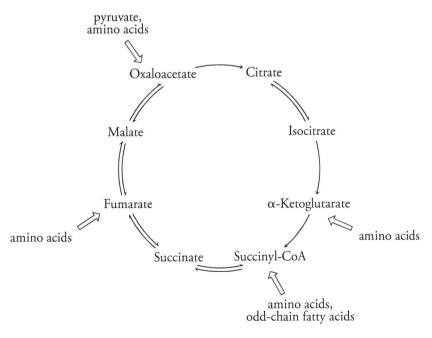

Figure 14.19 Anaplerotic reactions of the citric acid cycle.

Acetyl-CoA activates pyruvate carboxylase, so when the activity of the citric acid cycle is low and acetyl-CoA accumulates, more oxaloacetate is produced. The concentration of oxaloacetate is normally low since the malate dehydrogenase reaction is thermodynamically unfavorable and the citrate synthase reaction is highly favorable. The replenished oxaloacetate is converted to citrate, isocitrate, α-ketoglutarate, and so on, so the concentrations of all the citric acid cycle intermediates increase and the cycle can proceed more quickly. *Since the citric acid cycle acts as a catalyst, increasing the concentrations of its components increases flux through the pathway.*

The degradation of fatty acids with an odd number of carbon atoms yields the citric acid cycle intermediate succinyl-CoA. Other anaplerotic reactions include the pathways for the degradation of some amino acids, which produce α-ketoglutarate, succinyl-CoA, fumarate, and oxaloacetate. Some of these reactions are transaminations, such as

Aspartate + Pyruvate ⇌ Oxaloacetate + Alanine

Because transamination reactions have ΔG values near zero, the direction of flux into or out of the pool of citric acid cycle intermediates depends on the relative concentrations of the reactants.

In vigorously exercising muscle, the concentrations of citric acid cycle intermediates increase about three- to four-fold within a few minutes. This may help boost the energy-generating activity of the citric acid cycle, but it cannot be the sole mechanism, since flux through the citric acid cycle actually increases as much as 100-fold due to the increased activity of the three enzymes at the control points: citrate synthase, isocitrate dehydrogenase, and α-ketoglutarate dehydrogenase. The increase in citric acid cycle intermediates may actually be a mechanism for accommodating the large increase in pyruvate that results from rapid glycolysis at the start of exercise. Not all of this pyruvate is converted to lactate (Section 13.1); some is shunted into the pool of citric acid cycle intermediates via the pyruvate carboxylase reaction. Some pyruvate also undergoes a reversible reaction catalyzed by alanine aminotransferase:

Pyruvate + Glutamate ⇌ Alanine + α-Ketoglutarate

The resulting α-ketoglutarate then augments the pool of citric acid cycle intermediates, thereby increasing the ability of the cycle to oxidize the extra pyruvate.

Note that any compound that enters the citric acid cycle as an intermediate is not itself oxidized; it merely boosts the catalytic activity of the cycle, whose net reaction is still the oxidation of the two carbons of acetyl-CoA.

Concept Check

1. List the citric acid cycle intermediates that are used for the synthesis of amino acids, glucose, and fatty acids.
2. Explain what the ATP-citrate lyase reaction accomplishes.
3. Explain why the concentration of oxaloacetate remains low.
4. Explain why synthesizing more oxaloacetate increases flux through the cycle.
5. Write equations for all the anaplerotic reactions.

SUMMARY

14.1 The Pyruvate Dehydrogenase Reaction

- In order for pyruvate, the product of glycolysis, to enter the citric acid cycle, it must undergo oxidative decarboxylation catalyzed by the multienzyme pyruvate dehydrogenase complex, which yields acetyl-CoA, CO_2, and NADH.

14.2 The Eight Reactions of the Citric Acid Cycle

- The eight reactions of the citric acid cycle function as a multistep catalyst to convert the two carbons of acetyl-CoA to 2 CO_2.

14.3 Thermodynamics of the Citric Acid Cycle

- The electrons released in the oxidative reactions of the citric acid cycle are transferred to 3 NAD^+ and to ubiquinone. The reoxidation of the reduced cofactors generates ATP by oxidative phosphorylation. In addition, succinyl-CoA synthetase yields one molecule of GTP or ATP.

- The regulated reactions of the citric acid cycle are its irreversible steps, catalyzed by citrate synthase, isocitrate dehydrogenase, and α-ketoglutarate dehydrogenase.

- The citric acid cycle most likely evolved from biosynthetic pathways leading to α-ketoglutarate or succinate.

14.4 Anabolic and Catabolic Functions of the Citric Acid Cycle

- Six of the eight citric acid cycle intermediates serve as precursors of other compounds, including amino acids, monosaccharides, and lipids. Anaplerotic reactions convert other compounds into citric acid cycle intermediates, thereby allowing increased flux of acetyl carbons through the pathway.

KEY TERMS

mitochondrial matrix	citric acid cycle	Pasteur effect	glyoxylate pathway	glyoxysome
multienzyme complex	substrate-level phosphorylation	carcinogenesis	anaplerotic reaction	

BIOINFORMATICS

Brief Bioinformatics Exercises

14.1 Viewing and Analyzing the Pyruvate Dehydrogenase Complex

14.2 The Citric Acid Cycle and the KEGG Database

PROBLEMS

14.1 The Pyruvate Dehydrogenase Reaction

1. Outline possible transformations of pyruvate in **a.** yeast cells **b.** mammalian cells.

2. Determine which one of the five steps of the pyruvate dehydrogenase complex reaction is metabolically irreversible and explain why.

3. The product of the pyruvate dehydrogenase complex, acetyl-CoA, is released in step 3 of the overall reaction. What is the purpose of steps 4 and 5?

4. Beriberi is a disease that results from a dietary lack of thiamine, the vitamin that serves as the precursor for thiamine pyrophosphate (TPP). There are two metabolites that accumulate in individuals with beriberi, especially after ingestion of glucose. Which metabolites accumulate and why?

5. Most cases of pyruvate dehydrogenase deficiency disease that have been studied to date involve a mutation in the E1 subunit of the enzyme. The disease is extremely difficult to treat successfully, but physicians who identify patients with a pyruvate dehydrogenase deficiency administer thiamine as a first course of treatment. Explain why.

6. Arsenite is toxic in part because it binds to sulfhydryl compounds such as dihydrolipoamide, as shown in the figure. How would the presence of arsenite affect the citric acid cycle?

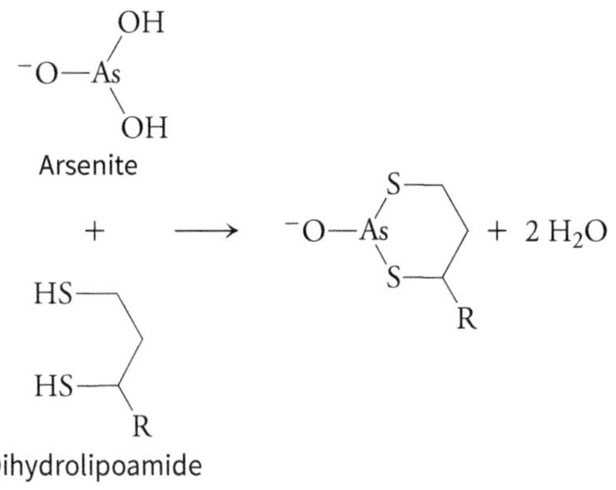

Arsenite

Dihydrolipoamide

7. Using the pyruvate dehydrogenase complex reaction as a model, reconstruct the TPP-dependent yeast pyruvate decarboxylase reaction in alcoholic fermentation (see Box 13.B).

8. The compound α-ketoisovalerate was shown to inhibit pyruvate dehydrogenase activity. Further investigation showed that when cultured cells were exposed to α-ketoisovalerate that was radioactively labeled at C2 (marked with * in the figure), the label became incorporated into the lipoamide prosthetic group of E2 of the pyruvate dehydrogenase complex. Draw the structure of the modified enzyme. What type of inhibitor is α-ketoisovalerate (see Section 7.3)?

$$CH_3-CH-\overset{\overset{\displaystyle O}{\|}}{\underset{*}{C}}-COO^-$$
$$\underset{\displaystyle CH_3}{|}$$

α-Ketoisovalerate

9. How is the activity of the pyruvate dehydrogenase complex affected by **a.** a high [NADH]/[NAD⁺] ratio or **b.** a high [acetyl-CoA]/[HSCoA] ratio?

10. In addition to the allosteric regulation described in Problem 9, the activity of the pyruvate dehydrogenase complex is controlled by phosphorylation. Pyruvate dehydrogenase kinase (PDH kinase) catalyzes the phosphorylation of a specific Ser residue on the E1 subunit of the enzyme, rendering it inactive. Pyruvate dehydrogenase phosphatase (PDH phosphatase) reverses the inhibition by catalyzing the removal of this phosphate group. The PDH kinase is highly regulated and its activity is influenced by various cellular metabolites. Indicate whether the following would activate or inhibit PDH kinase: **a.** NAD⁺, **b.** NADH, **c.** coenzyme A, **d.** acetyl-CoA, **e.** ADP.

11. The PDH kinase and PDH phosphatase enzymes (see Problem 10) are controlled by cytosolic Ca²⁺ levels. In the muscle, Ca²⁺ levels rise when the muscle contracts. Which of these two enzymes is inhibited by Ca²⁺ and which is activated by Ca²⁺?

12. How would the activities of PDH kinase and PDH phosphatase enzymes (see Problem 10) be affected by the presence of a calmodulin antagonist (see Fig. 10.10)?

13. Explain why PDH kinase–deficient mice (see Problem 10) have blood glucose levels that are lower than normal.

14. Explain why the activity of the pyruvate dehydrogenase complex is suppressed in individuals consuming a low carbohydrate–high fat diet.

15. 'The assembly of E1, E2 and E3 in pyruvate dehydrogenase complex does not always represent a 1:1:1 stoichiometry' –elaborate with proper example.

16. A second strategy to treat a pyruvate dehydrogenase deficiency disease (see Problem 5) involves administering dichloroacetate, a compound that inhibits PDH kinase (see Problem 10). How might this strategy be effective?

17. Hypoxia inducible factors (HIF) represents a family of transcription factors that modulate the expression of genes during hypoxic and normoxic conditions. Expressions of gene X and gene Y were analyzed by western blotting performed with lysates from hepatocytes cultured in normoxic and hypoxic conditions. Differences in expression for the proteins were observed. However, in cells with a mutated HIF, the difference was abolished for gene X (as depicted in the figure below), while it remained unaltered for gene Y similar to non-mutated cells (not shown). Present a possible explanation for the observation.

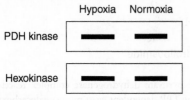

18. Enlist the sources of acetyl-CoA. Elaborate on how the 2-carbon intermediate is fluxed into the TCA cycle.

14.2 The Eight Reactions of the Citric Acid Cycle

19. Identify the "2- and 3-carbon intermediates" in Figure 14.5. What biomolecules are the source of these intermediates and what is their importance to the citric acid cycle?

20. Define substrate-level phosphorylation and outline the steps at which substrate-level phosphorylation occurs while completely oxidizing glucose through glycolysis and subsequent TCA cycle.

21. Twenty-four electrons are transmitted to ETC from the complete oxidation of glucose through glycolysis and the TCA cycle through the generation of reducing equivalent. Mention the steps at which reducing equivalents are generated.

22. Name an enzyme that can catalyze a carbon–carbon bond formation without using a metal ion cofactor. Briefly delineate the reaction mechanism.

23. Does the citrate synthase enzyme mechanism (Fig. 14.8) use an acid catalysis strategy, a base catalysis strategy, a covalent catalysis strategy, or some combination of these strategies (see Section 6.2)? Explain.

24. Predict the impact of the following site-directed mutagenesis on the activity of citrate synthase compared to the wild-type (nonmutated) version of the enzyme: **a.** Asp375Glu **b.** His274Ala **c.** His274Arg

25. Acetonyl-CoA is a structural analogue of acetyl-CoA and can inhibit citrate synthase. From the Lineweaver–Burk plot provided alongside, determine what kind of inhibitor action is demonstrated by acetonyl-CoA on citrate synthase.

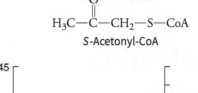

S-Acetonyl-CoA

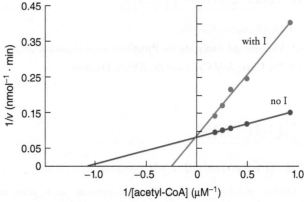

26. Animals that have ingested the leaves of the poisonous South African plant *Dichapetalum cymosum* exhibit a 10-fold increase in levels of cellular citrate. The plant contains fluoroacetate, which is converted to fluoroacetyl-CoA. Describe the mechanism that leads to increased levels of citrate in animals that have ingested this poisonous plant. (*Note:* Fluoroacetyl-CoA is not an inhibitor of citrate synthase.)

27. The compound carboxymethyl-CoA (shown below) is a competitive inhibitor of citrate synthase and is a proposed transition state analog. Propose a structure for the reaction intermediate derived from acetyl-CoA in the rate-limiting step of the reaction, just prior to its reaction with oxaloacetate.

$$CoA-S-CH_2-\overset{\overset{\displaystyle OH}{|}}{\underset{\underset{\displaystyle O}{\|}}{C}}$$

Carboxymethyl-CoA
(transition state analog)

28. Outline the first isomerization reaction in the TCA cycle and mention its significance in the context of subsequent steps.

29. The $\Delta G^{\circ\prime}$ value for the aconitase reaction is 5.8 kJ · mol^{-1}. What is the ratio of [isocitrate]/[citrate] under standard conditions? Given your answer, how can you account for the fact that this reaction occurs readily in cells?

30. Kinetic studies with aconitase revealed that *trans*-aconitate is a competitive inhibitor of the enzyme if *cis*-aconitate is used as the substrate. However, if citrate is used as the substrate, *trans*-aconitate is a noncompetitive inhibitor. Propose a hypothesis that explains this observation.

31. Can a loss of function mutation for aconitase, in *Saccharomyces cerevisiae*, result in deviation from the Pasteur effect? Comment with proper reasoning.

32. During bacterial infections, macrophages (a type of white blood cell) produce large amounts of an immunoregulatory molecule called itaconate, which is generated by the decarboxylation of aconitate. Draw the structure of itaconate.

$$
\begin{array}{c}
\text{COO}^- \\
| \\
\text{CH}_2 \\
| \\
\text{C}-\text{COO}^- \\
\| \\
\text{CH} \quad \text{H}^+ \\
| \\
\text{COO}^- \\
\text{Aconitate}
\end{array}
$$

33. In addition to regulating immune responses, itaconate (see Solution 32) can condense with coenzyme A to yield itaconyl-CoA, which has antibacterial activity. Draw the structure of itaconyl-CoA.

34. The $\Delta G^{\circ\prime}$ value for the isocitrate dehydrogenase reaction is -21 kJ · mol^{-1}. What is K_{eq} for this reaction at 25°C?

35. The crystal structure of isocitrate dehydrogenase shows a cluster of highly conserved amino acids in the substrate binding pocket—three arginines, a tyrosine, and a lysine. Why are these residues conserved and what is a possible role for their side chains in substrate binding?

36. Three isozymes for isocitrate dehydrogenase has been identified. The localization and co-factor preference are enlisted in the following table:

Isozyme of IDH	Localization	Co-enzyme
IDH1	Cytosol	NADP$^+$
IDH2	Mitochondria	NADP$^+$
IDH3	Mitochondria	NAD$^+$

In cancer cells, an arginine to histidine substitution in IDH1 active site and another arginine to lysine substitution in IDH2 are quite common. Explain whether such substitution can affect substrate binding in the mutated versions compared to the nonmutated enzyme.

37. The isocitrate dehydrogenase enzymes described in Problem 35 are "gain-of-function" mutants in which the normal catalytic properties of the enzyme are lost and a new biochemical activity is gained as a result of the mutation. The IDH1 mutant uses NADPH as a cofactor to catalyze the conversion of α-ketoglutarate to 2-hydroxyglutarate. **a.** Draw the structure of the product. **b.** What is the source of the NADPH and what are the implications for the cell if this reaction occurs at a high rate?

38. Using the pyruvate dehydrogenase complex reaction as a model, draw the intermediates of the α-ketoglutarate dehydrogenase reaction. Describe what happens in each of the five reaction steps.

39. Using the mechanism you drew for Problem 38, explain how succinyl phosphonate (below) inhibits α-ketoglutarate dehydrogenase.

$$
\begin{array}{c}
\text{O}^- \\
| \\
^-\text{O}-\text{P}=\text{O} \\
| \\
\text{C}=\text{O} \\
| \\
\text{CH}_2 \\
| \\
\text{CH}_2 \\
| \\
\text{COO}^-
\end{array}
$$
Succinyl phosphonate

40. Succinyl-CoA synthetase is also called succinate thiokinase. Why is the enzyme considered to be a kinase?

41. Malonate is a competitive inhibitor of succinate dehydrogenase. What citric acid cycle intermediates accumulate if malonate is present in a preparation of isolated mitochondria?

42. Malate dehydrogenase and lactate dehydrogenase can both use α-ketoglutarate as a substrate. Draw the product of those reactions.

43. The $\Delta G^{\circ\prime}$ value for the fumarase reaction is -3.4 kJ · mol^{-1}, but the ΔG value is close to zero. What is the ratio of fumarate to malate under cellular conditions at 37°C? Is this reaction likely to be a control point for the citric acid cycle?

44. A mutant bacterial fumarase was constructed by replacing the Glu (E) at position 315 with Gln (Q). The kinetic parameters of the mutant and wild-type enzymes are shown in the table. Explain the significance of the changes.

	Wild-type enzyme	E315Q mutant enzyme
V_{max} (μmol · min^{-1} · mg^{-1})	173	16
K_M (mM)	0.50	0.40
k_{cat}/K_M (M^{-1} · s^{-1})	2.8×10^6	2.1×10^5

45. Determine the fates of the labelled carbon (C^{14}) when a suspension of actively respiring mitochondria was incubated with: **a.** pyruvate labelled with C^{14} at C1 **b.** acetyl-CoA labelled with C^{14} at C1 **c.** oxaloacetate labelled with C^{14} at C1.

46. The complex metabolic pathways in the parasite *Trypanosoma brucei* (the causative agent of sleeping sickness) were elucidated in part by adding radiolabeled metabolites to cultured parasites. In the parasite, glucose is converted to phosphoenolpyruvate (PEP) in the cytosol. PEP then enters an organelle called the glycosome and is converted to oxaloacetate (OAA); OAA is then converted to malate, and malate to fumarate. Fumarate reductase catalyzes the conversion of fumarate to succinate; the succinate is then secreted from the glycosome. **a.** If C1 of glucose is labeled, what carbons in succinate are labeled? **b.** If citrate becomes radioactively labeled, what can you conclude about the connection between glycosomal and mitochondrial pathways in the parasite?

14.3 Thermodynamics of the Citric Acid Cycle

47. Which of the eight reactions of the citric acid cycle are **a.** phosphorylations, **b.** isomerizations, **c.** oxidation–reductions, **d.** hydrations, **e.** carbon–carbon bond formations, and **f.** decarboxylations?

48. Studies indicate that many animal tissues use circulating lactate, rather than pyruvate produced by intracellular glycolysis, as a substrate for the citric acid cycle. Compare the ATP yield for the complete oxidation of lactate and the complete oxidation of pyruvate by the citric acid cycle.

49. Flux through the citric acid cycle is regulated by the simple mechanisms of **a.** substrate availability, **b.** product inhibition, and **c.** feedback inhibition. Give examples of each.

50. Predict the effect of the following metabolites on the activity of citrate synthase: **a.** NADH, **b.** citrate, **c.** succinyl-CoA.

51. Citrate competes with oxaloacetate for binding to citrate synthase. Isocitrate dehydrogenase is activated by Ca^{2+} ions, which are released when a muscle contracts. How do these two regulatory strategies assist the cell in making the transition from the rested state (low citric acid cycle activity) to the exercise state (high citric acid cycle activity)?

52. Reactions 8 and 1 of the citric acid cycle can be considered to be coupled because the exergonic hydrolysis of the thioester bond of acetyl-CoA in Reaction 1 drives the regeneration of oxaloacetate in Reaction 8. **a.** Write the equation for the overall coupled reaction and calculate its $\Delta G°'$. **b.** What is the equilibrium constant for the coupled reaction? Compare this equilibrium constant with the equilibrium constant of Reaction 8 alone.

53. Administering high concentrations of oxygen (hyperoxia) is effective in treating lung injuries but at the same time can also be quite damaging. **a.** Lung aconitase activity is dramatically decreased during hyperoxia. How would this affect the concentration of citric acid cycle intermediates? **b.** The decreased aconitase activity and decreased mitochondrial respiration in hyperoxia are accompanied by elevated rates of glycolysis and the pentose phosphate pathway. Explain why.

54. The scientists who carried out the hyperoxia experiments described in Problem 53 noted that they could mimic this effect by administering either fluoroacetate or fluorocitrate to cells in culture. Explain. (*Hint:* Fluoroacetate can react with coenzyme A to form fluoro-acetyl-CoA.)

55. In bacteria, isocitrate dehydrogenase is regulated by phosphorylation of a specific Ser residue in the enzyme active site. X-ray structures of the phosphorylated and the nonphosphorylated enzyme show no significant conformational differences. **a.** How does phosphorylation regulate isocitrate dehydrogenase activity? **b.** To confirm their hypothesis, the investigators constructed a mutant enzyme in which the Ser residue was replaced with Asp. The mutant was unable to bind isocitrate. Are these results consistent with the hypothesis you proposed in part **a**?

56. The expression of several enzymes changes when yeast grown on glucose are abruptly shifted to a 2-carbon food source such as acetate. **a.** Why does the level of expression of isocitrate dehydrogenase increase when the yeast are shifted from glucose to acetate? **b.** The metabolism of a yeast mutant with a nonfunctional isocitrate dehydrogenase enzyme was compared to that of a wild-type yeast. The yeast were grown on glucose and then abruptly shifted to acetate as the sole carbon source. The [NAD⁺]/[NADH] ratio was measured over a period of 48 hours. The results are shown below. Why does the ratio increase slightly at 36 hours for the wild-type yeast? Why is there a more dramatic increase in the ratio for the mutant?

57. Malate dehydrogenase is more active in cells oxidizing glucose aerobically than in cells oxidizing glucose anaerobically. Explain why.

58. Succinyl-CoA inhibits both citrate synthase and α-ketoglutarate dehydrogenase. How is succinyl-CoA able to inhibit both enzymes?

59. Why is it advantageous for citrate, the product of Reaction 1 of the citric acid cycle, to inhibit phosphofructokinase, which catalyzes the third reaction of glycolysis?

60. Citrate can inhibit phosphofructokinase (PFK1), a major rate limiting step for glucose utilization. Comment on how the regulation is crucial for coordination of the pathways.

61. Succinyl-CoA synthetase is a dimer composed of an α subunit and a β subunit. A single gene codes for the α subunit protein. Two genes code for two different β subunit proteins—one subunit is specific for GDP, and one is specific for ADP. **a.** The β subunit specific for ADP is expressed in "catabolic tissues" such as brain and muscles, whereas the β subunit specific for GDP is expressed in "anabolic tissues" such as liver and kidneys. Propose a hypothesis to explain this observation. **b.** Individuals who are born with a mutation in the gene coding for the α subunit of the enzyme experience severe lactic acidosis and usually die within a few days of birth. Why is this mutation so deleterious? **c.** Individuals who are born with a mutation in the gene coding for the ADP-specific β subunit of the enzyme experience normal to moderately elevated concentrations of lactate and usually survive to their early 20s. Why is the prognosis for these patients better than for patients with a mutation in the gene for the α subunit?

62. Why would a deficiency of succinate dehydrogenase lead to a shortage of coenzyme A?

63. Fumaric aciduria, is an autosomal recessive disorder that develops due to deficiency of fumarase. The manifestation includes the build-up of fumaric acid in the urine, lactic acid in the blood, and depletion of malate. Propose a possible reason for the manifestation.

64. A patient with an α-ketoglutarate dehydrogenase deficiency exhibits a small increase in blood pyruvate level and a large increase in blood lactate level, resulting in a [lactate]/[pyruvate] ratio that is many times greater than normal. Explain the reason for these symptoms.

65. Certain microorganisms with an incomplete citric acid cycle decarboxylate α-ketoglutarate to produce succinate semialdehyde (shown below). A dehydrogenase then converts succinate semialdehyde to succinate. These reactions can be combined with other standard citric acid cycle reactions to create a pathway from citrate to oxaloacetate. How does this alternative pathway compare to the standard citric acid cycle in its ability to make free energy available to the cell?

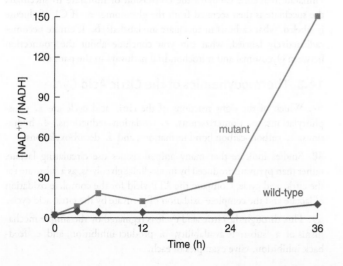

14.4 Anabolic and Catabolic Functions of the Citric Acid Cycle

66. a. Why is the reaction catalyzed by pyruvate carboxylase the most important anaplerotic reaction of the citric acid cycle? **b.** Why is the activation of pyruvate carboxylase by acetyl-CoA a good regulatory strategy?

67. Many amino acids are broken down to intermediates of the citric acid cycle. **a.** Why can't these amino acid "remnants" be completely oxidized to CO_2 by the citric acid cycle? **b.** Explain why amino acids that are broken down to pyruvate can be completely oxidized by the citric acid cycle.

68. Describe how the aspartate + pyruvate transamination reaction could function as an anaplerotic reaction for the citric acid cycle.

69. A bacterial mutant with low levels of isocitrate dehydrogenase is able to grow normally when the culture medium is supplemented with glutamate. Explain why.

70. A list of precursors and their breakdown products is provided below. Categorize the precursors as 'gluconeogenic' or 'non-gluconeogenic'.

Precursor	Product
Leucine	Acetyl-CoA and acetoacetate (metabolic equivalent to two acetyl-CoA)
Tryptophan	Alanine and acetoacetate
Phenylalanine	Acetoacetate and fumarate
Palmitate (fatty acid, 16:0)	Acetyl-CoA
Pentadecanoate (fatty acid, 15:0)	Acetyl-CoA and propionyl CoA (subsequently converted to succinyl-CoA)

71. Pancreatic islet cells cultured in the presence of 1–20 mM glucose showed increased activities of pyruvate carboxylase and the E1 subunit of the pyruvate dehydrogenase complex proportional to the increase in glucose concentration. Explain why.

72. A physician is attempting to diagnose a neonate with a pyruvate carboxylase deficiency. An injection of alanine normally leads to a gluconeogenic response, but in the patient no such response occurs. Explain.

73. The physician treats the patient described in Problem 72 by administering glutamine. Explain why glutamine supplements are effective in treating the disease.

74. Physicians often attempt to treat a pyruvate carboxylase deficiency by administering biotin. Explain why this strategy might be effective.

75. Patients with a pyruvate dehydrogenase deficiency and patients with a pyruvate carboxylase deficiency (see Problems 72 to 74) both have high blood levels of pyruvate and lactate. Explain why.

76. The udder of a ewe uses the majority of the glucose synthesized by the ewe, both for production of the milk sugar lactose and for production of triacylglycerols. During the winter, milk production decreases due to a poor food supply. Administration of propionate (which is converted to succinyl-CoA in ruminants) is an appropriate remedy. How does this treatment work?

77. Oxygen does not appear as a reactant in any of the citric acid cycle reactions, yet it is essential for the proper functioning of the cycle. Explain why.

78. The activity of isocitrate dehydrogenase in *E. coli* is regulated by the covalent attachment of a phosphate group, which inactivates the enzyme. When acetate is the food source for a culture of *E. coli*, isocitrate dehydrogenase is phosphorylated. **a.** Draw a diagram showing how acetate is metabolized in *E. coli*. **b.** When glucose is added to the culture, the phosphate group is removed from isocitrate dehydrogenase.

79. Yeast are unusual in their ability to use ethanol as a gluconeogenic substrate. Ethanol is converted to glucose with the assistance of the glyoxylate pathway. Describe how the ethanol → glucose conversion takes place.

80. Animals lack a glyoxylate pathway and cannot convert fats to carbohydrates. If an animal is fed a fatty acid with all of its carbons replaced by the isotope ^{14}C, some of the labeled carbons later appear in glucose. How is this possible?

81. Succinate dehydrogenase is not considered to be part of the glyoxylate pathway (see Box 14.B), yet it is vital to the proper functioning of the pathway. Why?

82. Although mammals do not possess the glyoxylate pathway (see Box 14.B), glyoxylate is generated as a metabolite in glycine and hydroxyproline metabolism. In the liver, glyoxylate is oxidized by lactate dehydrogenase. Draw the structure of the product.

83. The activity of the purine nucleotide cycle (shown below) in muscles increases during periods of high activity. Explain how the cycle contributes to the ability of the muscle cell to generate energy during intense exercise. IMP is inosine monophosphate.

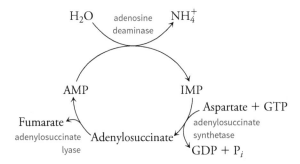

84. Metabolites in rat muscle were measured before and after exercising. After exercise, the rat muscle showed an increase in oxaloacetate concentration, a decrease in phosphoenolpyruvate concentration, and no change in pyruvate concentration. Explain.

85. The plant metabolite hydroxycitrate is advertised as an agent that prevents fat buildup. **a.** How does this compound differ from citrate? **b.** Hydroxycitrate inhibits the activity of ATP-citrate lyase. What kind of inhibition is likely to occur? **c.** Why might inhibition of ATP-citrate lyase block the conversion of carbohydrates to fats? **d.** The synthesis of what other compounds would be inhibited by hydroxycitrate?

$$CH_2-COO^-$$
$$|$$
$$HO-C-COO^-$$
$$|$$
$$HO-CH-COO^-$$

Hydroxycitrate

86. Yeast cells that are grown on nonfermentable substrates and then abruptly switched to glucose exhibit substrate-induced inactivation of several enzymes. Which enzymes would glucose cause to be inactivated and why?

87. Phagocytes protect the host against damage caused by invading microorganisms by engulfing a foreign microbe and forming a membrane-bound structure around it called a phagosome. The phagosome then fuses with a lysosome, a cellular organelle that contains proteolytic enzymes that destroy the pathogen. However, some microbes, such as *Mycobacterium tuberculosis*, survive dormant inside the phagosome for a prolonged period of time. In this scenario, levels of bacterial isocitrate lyase, malate synthase, citrate synthase, and malate dehydrogenase increase inside the phagosome to levels as much as 20 times above normal. **a.** What pathway(s) does *M. tuberculosis* employ while in the phagosome and why are these pathways essential to its survival? **b.** What might be good drug targets for treating a patient infected with *M. tuberculosis*?

88. Bacteria and plants (but not animals) possess the enzyme phosphoenolpyruvate carboxylase (PPC), which catalyzes the reaction shown. **a.** What is the importance of this reaction to the organism? **b.** PPC is allosterically activated by both acetyl-CoA and fructose-1,6-bisphosphate. Explain these regulatory strategies.

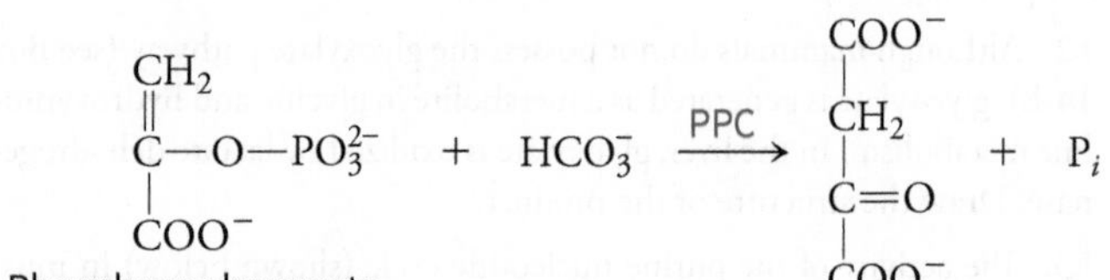

$$
\begin{array}{c}
CH_2 \\
\| \\
C-O-PO_3^{2-} \\
| \\
COO^-
\end{array}
+ \ HCO_3^- \ \xrightarrow{PPC} \
\begin{array}{c}
COO^- \\
| \\
CH_2 \\
| \\
C=O \\
| \\
COO^-
\end{array}
+ \ P_i
$$

Phosphoenolpyruvate   Oxaloacetate

89. The pharmaceutical, cosmetics, and food industries synthesize succinic acid by "green" or environmentally responsible methods that involve bacteria instead of petrochemicals. Industrial succinate production by bacteria occurs under anaerobic conditions, in which malate dehydrogenase activity increases. **a.** Draw a reaction scheme outlining how phosphoenolpyruvate is converted to succinate. Include the names of all reactants, products, and enzymes. **b.** Why is it essential that the production of succinate take place under anaerobic conditions?

90. Experiments with cancer cells grown in culture show that glutamine is consumed at a high rate and used for biosynthetic reactions, aside from protein synthesis. One possible pathway involves the conversion of glutamine to glutamate and then to α-ketoglutarate. The α-ketoglutarate can then be used to produce pyruvate for gluconeogenesis. **a.** Describe the types of reactions that convert glutamine to α-ketoglutarate. **b.** Give the sequence of enzymes that can convert α-ketoglutarate to pyruvate.

91. Many cancer cells carry out glycolysis at a high rate but convert most of the resulting pyruvate to lactate rather than to acetyl-CoA. Acetyl-CoA, however, is required for the synthesis of fatty acids, which are needed in large amounts by rapidly growing cancer cells. In these cells, the isocitrate dehydrogenase reaction apparently operates in reverse. Explain why this reaction could facilitate the conversion of amino acids such as glutamate into fatty acids.

SELECTED READINGS

Cavalcanti, J.H.F., Esteves-Ferreira, A.A., Quinhones, C.G.S., Pereira-Lima, I.A., Nunes-Nesi, A., Fernie, A.R., and Araújo, W.L., Evolution and functional implications of the tricarboxylic acid cycle as revealed by phylogenetic analysis, *Genome Biol. Evol.* 6, 2830–2848, doi: 10.1093/gbe/evu221 (2014). [Includes descriptions of each of the cycle's enzymes, from an evolutionary perspective.]

Gray, L.R., Tompkins, S.C., and Taylor, E.B., Regulation of pyruvate metabolism and human disease. *Cell Mol. Life Sci.* 71, 2577–2604 (2014). [Includes useful diagrams of pathways involving pyruvate and discusses diseases related to them.]

Martínez-Reyes, I. and Chandel, N.S., Mitochondrial TCA cycle metabolites control physiology and disease, *Nature Communications* 11, 102, doi: 10.1038/s41467-019-13668-3 (2020). [Summarizes the additional roles of citric acid cycle intermediates in signaling and regulating gene expression.]

Muchowska, K.B., Varma, S.J., and Moran, J., Synthesis and breakdown of universal metabolic precursors promoted by iron, *Nature* 569, 104–107, doi: 10.1038/s41586-019-1151-1 (2019). [Describes a scenario for the prebiotic evolution of the citric acid cycle and glyoxylate cycle.]

Patel, M.S., Nemeria, N.S., Furey, W., and Jordan, F., The pyruvate dehydrogenase complexes: Structure-based function and regulation, *J. Biol. Chem.* 289, 16615–16623 (2014). [Compares the structure and chemical reactions of the human and *E. coli* enzyme complexes.]

CHAPTER 14 CREDITS

Figure 14.1a Image based on 1EAA. Mattevi, A., Hol, W.G.J., Atomic structure of the cubic core of the pyruvate dehydrogenase multienzyme complex, *Biochemistry* 32, 3887–3901 (1993).

Figure 14.1b Image based on 1B5S. Izard, T., Aevarsson, A., Allen, M.D., Westphal, A.H., Perham, R.N., de Kok, A., Hol, W.G., Principles of quasi-equivalence and Euclidean geometry govern the assembly of cubic and dodecahedral cores of pyruvate dehydrogenase complexes, *Proc. Natl. Acad. Sci. USA* 96, 1240–1245 (1999).

Figure 14.7a Image based on 5CSC. Liao, D.-I., Karpusas, M., Remington, S.J., Crystal structure of an open conformation of citrate synthase from chicken heart at 2.8-A resolution, *Biochemistry* 30, 6031–6036 (1991).

Figure 14.7b Image based on 5CTS. Karpusas, M., Branchaud, B., Remington, S.J., Proposed mechanism for the condensation reaction of citrate synthase: 1.9-A structure of the ternary complex with oxaloacetate and carboxymethyl coenzyme A, *Biochemistry* 29, 2213–2219 (1990).

Figure 14.12 Image based on 1CQI. Joyce, M.A., Fraser, M.E., James, M.N., Bridger, W.A., Wolodko, W.T., ADP-binding site of Escherichia coli succinyl-CoA synthetase revealed by X-ray crystallography, *Biochemistry* 39, 17–25 (2000).

Oxidative Phosphorylation

elleon/Shutterstock.com

To survive winter's cold, honeybees form a cluster whose center is about 20°C higher than the outside. A high rate of oxidative metabolism in the innermost bees supports the shivering movements of their flight muscles, which generates heat to keep the colony warm.

Do You Remember?

- Living organisms obey the laws of thermodynamics (Section 1.4).
- Transporters obey the laws of thermodynamics, providing a way for solutes to move down their concentration gradients or using ATP to move substances against their gradients (Section 9.1).
- Coenzymes such as NAD⁺ and ubiquinone collect electrons from compounds that become oxidized (Section 12.3).
- A reaction that hydrolyzes a phosphoanhydride bond in ATP occurs with a large negative change in free energy (Section 12.2).

The early stages of oxidation of metabolic fuels such as glucose, fatty acids, and amino acids, as well as the oxidation of acetyl carbons to CO_2 via the citric acid cycle, yield the reduced cofactors NADH and ubiquinol (QH_2). These compounds are forms of energy currency (see Section 12.2), not because they are chemically special but because their reoxidation—ultimately by molecular oxygen in aerobic organisms—is an exergonic process. That free energy is harvested to synthesize ATP, a phenomenon called oxidative phosphorylation. To understand oxidative phosphorylation, we need to first examine why and how electrons flow from reduced cofactors to O_2. Then we can explore how the chemical energy of the redox reactions is conserved in the formation of a transmembrane gradient of protons, another type of energy that drives the rotation of ATP synthase so that it can build ATP from ADP and P_i.

15.1 The Thermodynamics of Oxidation–Reduction Reactions

KEY CONCEPTS

Summarize the thermodynamics of oxidation–reduction reactions.

- Use standard reduction potential and concentration to calculate a substance's tendency to become reduced.

- Predict the direction of electron transfer in a mixture of two substances.
- Convert the change in reduction potential to the change in free energy for a reaction.

In the scheme introduced in Figure 12.9, **oxidative phosphorylation** represents the final phase of the catabolism of metabolic fuels and the major source of the cell's ATP (**Fig. 15.1**). Oxidative phosphorylation differs from the conventional biochemical reactions we have focused on in the last two chapters. In particular, ATP synthesis is not directly coupled to a single discrete chemical reaction, such as a kinase-catalyzed reaction. Rather, oxidative phosphorylation is a more indirect process of energy transformation.

The flow of electrons from reduced compounds such as NADH and QH_2 to an oxidized compound such as O_2 is a thermodynamically favorable process. The free energy changes for the movements of electrons through a series of electron carriers can be quantified by considering the reduction potentials of the chemical species involved in each transfer.

Oxidation–reduction reactions (or redox reactions, introduced in Section 12.3) are similar to other chemical reactions in which a portion of a molecule—electrons in this case—is transferred. In any oxidation–reduction reaction, one reactant (called the **oxidizing agent** or **oxidant**) is reduced as it gains electrons. The other reactant (called the **reducing agent** or **reductant**) is oxidized as it gives up electrons:

$$A_{oxidized} + B_{reduced} \rightleftharpoons A_{reduced} + B_{oxidized}$$

For example, in the succinate dehydrogenase reaction (step 6 of the citric acid cycle; see Section 14.2), the two electrons of the reduced $FADH_2$ prosthetic group of the enzyme are transferred to ubiquinone (Q) so that $FADH_2$ is oxidized and ubiquinone is reduced:

$$\underset{(reduced)}{FADH_2} + \underset{(oxidized)}{Q} \rightleftharpoons \underset{(oxidized)}{FAD} + \underset{(reduced)}{QH_2}$$

In this reaction, the two electrons are transferred as H atoms (an H atom consists of a proton and an electron, or H^+ and e^-). In oxidation–reduction reactions involving the cofactor NAD^+, the electron pair takes the form of a hydride ion (H^-, a proton with two electrons). In biological systems, electrons usually travel in pairs, although, as we will see, they may also be transferred one at a time. Note that the change in oxidation state of a reactant may be obvious, such as when Fe^{3+} is reduced to Fe^{2+}, or it may require closer inspection of the molecule's structure, such as when succinate is oxidized to fumarate (Section 14.2). Often, oxidation can be recognized as the replacement of C—H bonds with C—O bonds, for example,

$$H-\overset{|}{\underset{|}{C}}-OH \quad \underset{\xleftarrow{reduction}}{\xrightarrow{oxidation}} \quad \overset{|}{\underset{|}{C}}=O$$

Here, the carbon atom "loses" electrons to the more electronegative oxygen atom.

Reduction potential indicates a substance's tendency to accept electrons

The tendency of a substance to accept electrons (to become reduced) or to donate electrons (become oxidized) can be quantified. Although an oxidation–reduction reaction necessarily requires both an oxidant and a reductant, it is helpful to consider just one

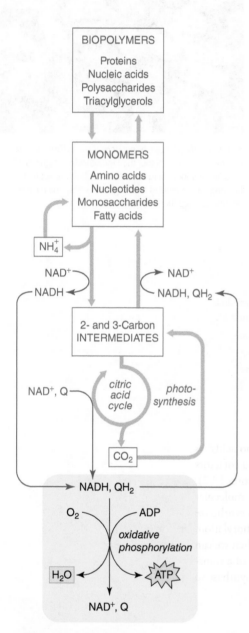

Figure 15.1 Oxidative phosphorylation in context. The reduced cofactors NADH and QH_2, which are generated in the oxidative catabolism of amino acids, monosaccharides, and fatty acids, are reoxidized by a process that requires molecular oxygen. The energy of this process is conserved in a manner that powers the synthesis of ATP from $ADP + P_i$.

substance at a time, that is, a **half-reaction.** For example, the half-reaction for ubiquinone (by convention, written as a reduction reaction) is

$$Q + 2H^+ + 2e^- \rightleftharpoons QH_2$$

(the reverse equation would describe an oxidation half-reaction).

The affinity of a substance such as ubiquinone for electrons is its **standard reduction potential ($\mathcal{E}°'$),** which has units of volts (note that the degree and prime symbols indicate a value under standard biochemical conditions where the pressure is 1 atm, the temperature is 25°C, the pH is 7.0, and all species are present at concentrations of 1 M). *The greater the value of $\mathcal{E}°'$, the greater the tendency of the oxidized form of the substance to accept electrons and become reduced.* The standard reduction potentials of some biological substances are given in **Table 15.1.**

Like a ΔG value, *the actual reduction potential depends on the actual concentrations of the oxidized and reduced species.* The actual **reduction potential ($\mathcal{E}$)** is related to the standard reduction potential ($\mathcal{E}°'$) by the **Nernst equation:**

$$\mathcal{E} = \mathcal{E}°' - \frac{RT}{n\mathcal{F}} \ln \frac{[A_{reduced}]}{[A_{oxidized}]} \tag{15.1}$$

R (the gas constant) has a value of $8.3145 \text{ J} \cdot \text{K}^{-1} \cdot \text{mol}^{-1}$, T is the temperature in Kelvin, n is the number of electrons transferred (one or two in most of the reactions we will encounter), and $\mathcal{F}$ is the **Faraday constant** ($96,485 \text{ J} \cdot \text{V}^{-1} \cdot \text{mol}^{-1}$; it is equivalent to the electrical charge of one mole of electrons). At 25°C (298 K), the Nernst equation reduces to

$$\mathcal{E} = \mathcal{E}°' - \frac{0.026 \text{ V}}{n} \ln \frac{[A_{reduced}]}{[A_{oxidized}]} \tag{15.2}$$

In fact, for many substances in biological systems, the concentrations of the oxidized and reduced species are similar, so the logarithmic term is small (recall that $\ln 1 = 0$) and $\mathcal{E}$ is close to $\mathcal{E}°'$ (Sample Calculation 15.1).

Table 15.1 Standard Reduction Potentials of Some Biological Substances

Half-reaction	$\mathcal{E}°'$ (V)
$\frac{1}{2}O_2 + 2H^+ + 2e^- \rightleftharpoons H_2O$	0.815
$SO_4^{2-} + 2H^+ + 2e^- \rightleftharpoons SO_3^{2-} + H_2O$	0.48
$NO_3^- + 2H^+ + 2e^- \rightleftharpoons NO_2^- + H_2O$	0.42
Cytochrome a_3 $(Fe^{3+}) + e^- \rightleftharpoons$ cytochrome a_3 (Fe^{2+})	0.385
Cytochrome a $(Fe^{3+}) + e^- \rightleftharpoons$ cytochrome a (Fe^{2+})	0.29
Cytochrome c $(Fe^{3+}) + e^- \rightleftharpoons$ cytochrome c (Fe^{2+})	0.235
Cytochrome c_1 $(Fe^{3+}) + e^- \rightleftharpoons$ cytochrome c_1 (Fe^{2+})	0.22
Cytochrome b $(Fe^{3+}) + e^- \rightleftharpoons$ cytochrome b (Fe^{2+}) (*mitochondrial*)	0.077
Ubiquinone $+ 2H^+ + 2e^- \rightleftharpoons$ ubiquinol	0.045
Fumarate$^- + 2H^+ + 2e^- \rightleftharpoons$ succinate$^-$	0.031
$FAD + 2H^+ + 2e^- \rightleftharpoons FADH_2$ (*in flavoproteins*)	~ 0.
Oxaloacetate$^- + 2H^+ + 2e^- \rightleftharpoons$ malate$^-$	-0.166
Pyruvate$^- + 2H^+ + 2e^- \rightleftharpoons$ lactate$^-$	-0.185
Acetaldehyde $+ 2H^+ + 2e^- \rightleftharpoons$ ethanol	-0.197
$S + 2H^+ + 2e^- \rightleftharpoons H_2S$	-0.23
Lipoic acid $+ 2H^+ + 2e^- \rightleftharpoons$ dihydrolipoic acid	-0.29
$NAD^+ + H^+ + 2e^- \rightleftharpoons NADH$	-0.315
$NADP^+ + H^+ + 2e^- \rightleftharpoons NADPH$	-0.320
Acetoacetate$^- + 2H^+ + 2e^- \rightleftharpoons$ 3-hydroxybutyrate$^-$	-0.346
Acetate$^- + 3H^+ + 2e^- \rightleftharpoons$ acetaldehyde $+ H_2O$	-0.581

SAMPLE CALCULATION 15.1

Problem   Calculate the reduction potential of fumarate ($\mathcal{E}^{\circ\prime} = 0.031$ V) at 25°C when [fumarate] = 40 μM and [succinate] = 200 μM.

Solution   Use Equation 15.2. Fumarate is the oxidized compound and succinate is the reduced compound.

$$\mathcal{E} = \mathcal{E}^{\circ\prime} - \frac{0.026\text{ V}}{n} \ln \frac{[A_{reduced}]}{[A_{oxidized}]}$$

$$= 0.031\text{ V} - \frac{0.026\text{ V}}{2} \ln \frac{(2 \times 10^{-4})}{(4 \times 10^{-5})}$$

$$= 0.031\text{ V} - 0.021\text{ V} = 0.010\text{ V}$$

The free energy change can be calculated from the change in reduction potential

Knowing the reduction potentials of different substances is useful for predicting the movement of electrons between the two substances. When the substances are together in solution or connected by wire in an electrical circuit, *electrons flow spontaneously from the substance with the lower reduction potential to the substance with the higher reduction potential.* For example, in a system containing Q/QH$_2$ and NAD$^+$/NADH, we can predict whether electrons will flow from QH$_2$ to NAD$^+$ or from NADH to Q. Using the standard reduction potentials given in Table 15.1, we note that $\mathcal{E}^{\circ\prime}$ for NAD$^+$ (–0.315 V) is lower than $\mathcal{E}^{\circ\prime}$ for ubiquinone (0.045 V). Therefore, NADH will tend to transfer its electrons to ubiquinone; that is, NADH will be oxidized and Q will be reduced.

A complete oxidation–reduction reaction is just a combination of two half-reactions. For the NADH–ubiquinone reaction, the net reaction is the ubiquinone reduction half-reaction (the half-reaction as listed in Table 15.1) combined with the NADH oxidation half-reaction (the reverse of the half-reaction listed in Table 15.1). Note that because the NAD$^+$ half-reaction has been reversed to indicate oxidation, we have also reversed the sign of its $\mathcal{E}^{\circ\prime}$ value:

NADH $\rightleftharpoons$ NAD$^+$ + H$^+$ + 2e^-	$\mathcal{E}^{\circ\prime} = +0.315$ V
Q + 2 H$^+$ + 2e^- $\rightleftharpoons$ QH$_2$	$\mathcal{E}^{\circ\prime} = 0.045$ V
net: NADH + Q + H$^+$ $\rightleftharpoons$ NAD$^+$ + QH$_2$	$\Delta\mathcal{E}^{\circ\prime} = +0.360$ V

When the two half-reactions are added, their reduction potentials are also added, yielding a $\Delta\mathcal{E}^{\circ\prime}$ value. Keep in mind that the reduction potential is a property of the half-reaction and is independent of the direction in which the reaction occurs. Reversing the sign of $\mathcal{E}^{\circ\prime}$, as shown above, is just a shortcut to simplify the task of calculating $\Delta\mathcal{E}^{\circ\prime}$. Another method for calculating $\Delta\mathcal{E}^{\circ\prime}$ uses the following equation:

$$\Delta\mathcal{E}^{\circ\prime} = \mathcal{E}^{\circ\prime}_{(e^- \text{ acceptor})} - \mathcal{E}^{\circ\prime}_{(e^- \text{ donor})} \qquad (15.3)$$

Not surprisingly, *the larger the difference in $\mathcal{E}$ values (the greater the $\Delta\mathcal{E}$ value), the greater the tendency of electrons to flow from one substance to the other, and the greater the change in free energy of the system.* ΔG is related to $\Delta\mathcal{E}$ as follows:

$$\Delta G^{\circ\prime} = -n\mathcal{F}\Delta\mathcal{E}^{\circ\prime} \quad \text{or} \quad \Delta G = -n\mathcal{F}\Delta\mathcal{E} \qquad (15.4)$$

Accordingly, an oxidation–reduction reaction with a large positive $\Delta\mathcal{E}$ value has a large negative value of ΔG (see Sample Calculation 15.2). Depending on the relevant reduction potentials, an oxidation–reduction reaction can release considerable amounts of free energy. This is what happens in the mitochondria, where the reduced cofactors generated by the oxidation of metabolic fuels are reoxidized. The free energy released in this process powers ATP synthesis by oxidative phosphorylation. **Figure 15.2**

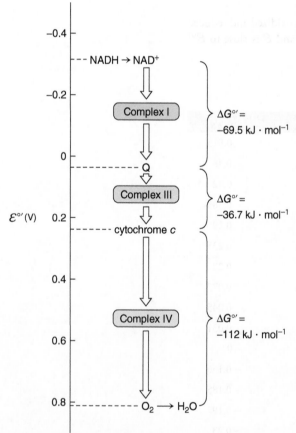

Figure 15.2 Overview of mitochondrial electron transport. The reduction potentials of the key electron carriers are indicated. The oxidation–reduction reactions mediated by Complexes I, III, and IV release free energy.

Question Determine the total free energy change for the oxidation of NADH by O$_2$.

SAMPLE CALCULATION 15.2

Problem Calculate the standard free energy change for the oxidation of malate by NAD^+. Is the reaction spontaneous under standard conditions?

Solution

Method 1
Write the relevant half-reactions, reversing the malate half-reaction (so that it becomes an oxidation reaction) and reversing the sign of its $\mathcal{E}^{\circ\prime}$:

malate $\rightarrow$ oxaloacetate $+ 2H^+ + e^-$	$\mathcal{E}^{\circ\prime} = +0.166$ V
$NAD^+ + H^+ + 2e^- \rightarrow NADH$	$\mathcal{E}^{\circ\prime} = -0.315$ V

net: malate $+ NAD^+ \rightarrow$ oxaloacetate $+ NADH + H^+$ $\Delta\mathcal{E}^{\circ\prime} = -0.149$ V

Method 2
Identify the electron acceptor (NAD^+) and electron donor (malate). Substitute their standard reduction potentials into Equation 15.3:

$$\Delta\mathcal{E}^{\circ\prime} = \mathcal{E}^{\circ\prime}_{(e^- \text{ acceptor})} - \mathcal{E}^{\circ\prime}_{(e^- \text{ donor})}$$

$$= -0.315 \text{ V} - (-0.166 \text{ V})$$

$$= -0.149 \text{ V}$$

Both Methods
The $\Delta\mathcal{E}^{\circ\prime}$ for the net reaction is -0.149 V. Use Equation 15.4 to calculate $\Delta G^{\circ\prime}$:

$$\Delta G^{\circ\prime} = -n\mathcal{F}\Delta\mathcal{E}^{\circ\prime}$$

$$= -(2)(96{,}485 \text{ J} \cdot \text{V}^{-1} \cdot \text{mol}^{-1})(-0.149 \text{ V})$$

$$= +28{,}750 \text{ J} \cdot \text{mol}^{-1} = +28.8 \text{ kJ} \cdot \text{mol}^{-1}$$

The reaction has a positive value of $\Delta G^{\circ\prime}$ and so is not spontaneous. (*In vivo*, this endergonic reaction occurs as step 8 of the citric acid cycle and is coupled to step 1, which is exergonic.)

---

shows the major mitochondrial electron transport components arranged by their reduction potentials. Each stage of electron transfer, from NADH to O_2, the final electron acceptor, occurs with a negative change in free energy.

Concept Check

1. Explain why an oxidation–reduction reaction must include both an oxidant and a reductant.
2. When two reactants are mixed together, explain how you can predict which one will become reduced and which one will become oxidized.
3. Explain how adding the $\mathcal{E}^{\circ\prime}$ values for two half-reactions yields a value of $\Delta\mathcal{E}^{\circ\prime}$ and $\Delta G^{\circ\prime}$ for an oxidation–reduction reaction.
4. Select the two half-reactions from Table 15.1 that would be most likely to form a freely reversible (near-equilibrium) redox reaction.

15.2 Mitochondrial Electron Transport

KEY CONCEPTS

Map the path of electrons through the redox groups of the electron transport pathway.

- Explain why the mitochondrion includes a variety of transport systems.
- Identify the sources of electrons for Complexes I, III, and IV.
- Describe the mechanisms for transporting protons across the mitochondrial membrane.
- Explain why respiratory complexes may not actually form a chain.

In aerobic organisms, the NADH and ubiquinol produced by glycolysis, the citric acid cycle, fatty acid oxidation, and other metabolic pathways are ultimately reoxidized by molecular oxygen, a process called **cellular respiration.** The standard reduction potential of $+0.815$ V for the reduction of O_2 to H_2O indicates that O_2 is a more effective oxidizing agent than any other biological compound (see Table 15.1). The oxidation of NADH by O_2, that is, the transfer of electrons from NADH directly to O_2, would release a large amount of free energy, but this reaction does not occur in a single step. Instead, *electrons are shuttled from NADH to O_2 in a multistep process that offers several opportunities to conserve the free energy of oxidation.* In eukaryotes, all the steps of oxidative phosphorylation are carried out by a series of electron carriers that include small molecules as well as the prosthetic groups of large integral membrane proteins in mitochondria (in prokaryotes, similar electron carriers are located in the plasma membrane). The following sections describe how electrons flow through this respiratory **electron transport chain** from reduced cofactors to oxygen.

Mitochondrial membranes define two compartments

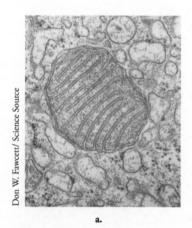

- Outer membrane
- Inner membrane
- Intermembrane space
- Matrix

Figure 15.3 Model of mitochondrial structure. The relatively impermeable inner mitochondrial membrane encloses the protein-rich matrix. The intermembrane space has an ionic composition similar to that of the cytosol.

In accordance with its origin as a bacterial symbiont, the **mitochondrion** (plural, *mitochondria*) has two membranes. The outer membrane, analogous to the outer membrane of some bacteria, is relatively porous due to the presence of porin-like proteins that permit the transmembrane diffusion of substances with masses up to about 10 kD (see Section 9.2 for an example of porin structure and function). The inner membrane has a convoluted architecture that encloses a space called the **mitochondrial matrix.** Because the inner mitochondrial membrane prevents the transmembrane movements of ions and small molecules (except via specific transport proteins), the composition of the matrix differs from that of the space between the inner and outer membranes. In fact, the ionic composition of the **intermembrane space** is considered to be equivalent to that of the cytosol (**Fig. 15.3**).

Mitochondria are customarily shown as bean-shaped organelles with the inner mitochondrial membrane forming a system of baffles called **cristae** (**Fig. 15.4a**). However, **electron tomography,** a technique for visualizing cellular structures in three dimensions by analyzing micrographs of sequential cell slices, reveals that mitochondria are highly variable structures. For example, the cristae may be irregular and bulbous rather than planar and may make several tubular connections with the rest of the inner mitochondrial membrane (Fig. 15.4b). Moreover, a cell may contain hundreds to thousands of discrete bacteria-shaped mitochondria, or a single tubular organelle may take the form of an extended network with many branches and interconnections (Fig. 15.4c). Individual mitochondria can move around the cell and undergo fusion (joining) and fission (separating).

Reflecting its ancient origin as a free-living organism, the mitochondrion has its own genome and protein-synthesizing machinery consisting of mitochondrially encoded rRNA and tRNA. The mitochondrial genome encodes 13 proteins, all of which are components of the respiratory chain complexes. This is only a small subset of the approximately 1500 proteins required for mitochondrial function; the other respiratory chain proteins, matrix enzymes, transporters, and so on are encoded by the cell's nuclear genome, synthesized in the cytosol, and imported into the mitochondria via translocases in the outer membrane (the TOM complex) and inner membrane (the TIM complex). Lipids synthesized by the endoplasmic reticulum appear to be transferred to mitochondria at points where "tether" proteins temporarily hold the two organelles together.

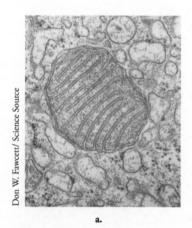

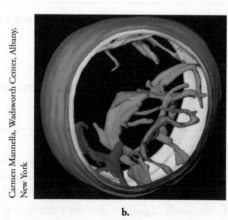

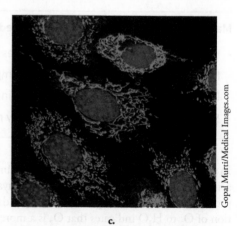

a.   b.   c.

Figure 15.4 Images of mitochondria. **a.** Conventional electron micrograph showing cristae as a system of planar baffles. **b.** Three-dimensional reconstruction of a mitochondrion by electron tomography, showing irregular tubular cristae. **c.** Fluorescence micrograph of cultured human cells. The cytosol contains a network of mitochondria (red). The nucleus is blue.

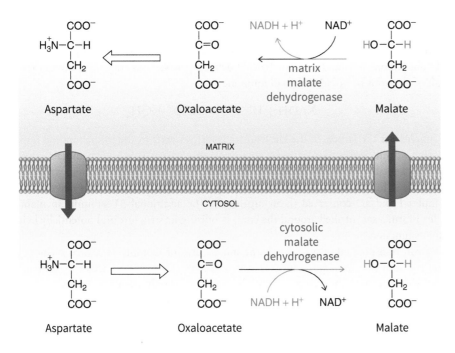

Figure 15.5 The malate–aspartate shuttle system.   Cytosolic oxaloacetate is reduced to malate for transport into mitochondria. Malate is then reoxidized in the matrix. The net result is the transfer of "reducing equivalents" from the cytosol to the matrix. Mitochondrial oxaloacetate can be exported back to the cytosol after being converted to aspartate by an aminotransferase.

Much of the cell's NADH and QH_2 is generated by the citric acid cycle inside mitochondria. Fatty acid oxidation also takes place largely inside mitochondria and yields NADH and QH_2. These reduced cofactors transfer their electrons to the protein complexes of the respiratory electron transport chain, which are tightly associated with the inner mitochondrial membrane. However, NADH produced by glycolysis and other oxidative processes in the cytosol cannot directly reach the respiratory chain. There is no transport protein that can ferry NADH across the inner mitochondrial membrane. Instead, "reducing equivalents" are imported into the matrix by the chemical reactions of systems such as the malate–aspartate shuttle system (**Fig. 15.5**).

Mitochondria also need a mechanism to export ATP and to import ADP and P_i, since most of the cell's ATP is generated in the matrix by oxidative phosphorylation and is consumed in the cytosol. A transport protein called the ADP/ATP carrier, or adenine nucleotide translocase, exports ATP and imports ADP, binding one or the other and changing its conformation to release the bound nucleotide on the other side of the membrane (**Fig. 15.6a**). Inorganic phosphate, a substrate for oxidative phosphorylation, is imported from the cytosol in symport with H^+ (Fig. 15.6b).

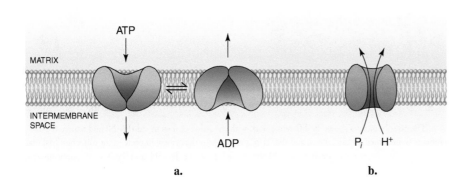

a.   b.

Figure 15.6 Mitochondrial transport systems.   a. The ADP/ATP carrier binds either ATP or ADP and changes its conformation to release the nucleotide on the opposite side of the inner mitochondrial membrane. This translocase can therefore export ATP and import ADP. **b.** A P_i–H^+ symport protein permits the simultaneous movement of inorganic phosphate and a proton into the mitochondrial matrix.

Question Does the activity of either of these transporters contribute to the mitochondrial membrane potential?

Complex I transfers electrons from NADH to ubiquinone

The path electrons travel through the respiratory chain begins with Complex I, also called NADH:ubiquinone oxidoreductase or NADH dehydrogenase. This enzyme catalyzes the transfer of a pair of electrons from NADH to ubiquinone:

$$NADH + H^+ + Q \rightleftharpoons NAD^+ + QH_2$$

Complex I is the largest of the electron transport proteins in the mitochondrial respiratory chain, with 45 different subunits and a total mass of about 970 kD in mammals. The crystal structure of a smaller (536-kD) bacterial Complex I reveals an L-shaped protein with numerous transmembrane helices and a peripheral arm (**Fig. 15.7**). Fourteen subunits make up the core of Complex I and are conserved in all organisms. The additional 31 subunits in mammalian Complex I form a sort of shell around the core, stabilizing its structure and potentially helping to regulate its function.

Electron transport takes place in the peripheral arm of Complex I, which includes several prosthetic groups that undergo reduction as they receive electrons and become oxidized as they give up their electrons to the next group. All these groups, or **redox centers,** appear to have reduction potentials approximately between the reduction potentials of NAD$^+$ ($\mathcal{E}°' = -0.315$ V) and ubiquinone ($\mathcal{E}°' = +0.045$ V). This allows them to form a chain where the electrons travel a path of increasing reduction potential. The redox centers do not need to be in intimate contact with each other, as they would be if the transferred group were a larger chemical entity. An electron can move between redox centers up to 14 Å apart by "tunneling" through the covalent bonds of the protein.

The two electrons donated by NADH are first picked up by flavin mononucleotide (FMN; **Fig. 15.8**) near the far end of the Complex I arm. This noncovalently bound prosthetic group, which is similar to FAD, then transfers the electrons, one at a time, to a second type of redox center, an iron–sulfur (Fe–S) cluster. Depending on the species, Complex I bears 8 to 10 of these prosthetic groups, which contain equal numbers of iron and sulfide ions (**Fig. 15.9**). Unlike the electron carriers we have introduced so far, Fe–S clusters are one-electron carriers. They have an oxidation state of either +3 (oxidized) or +2 (reduced), regardless of the number of Fe atoms

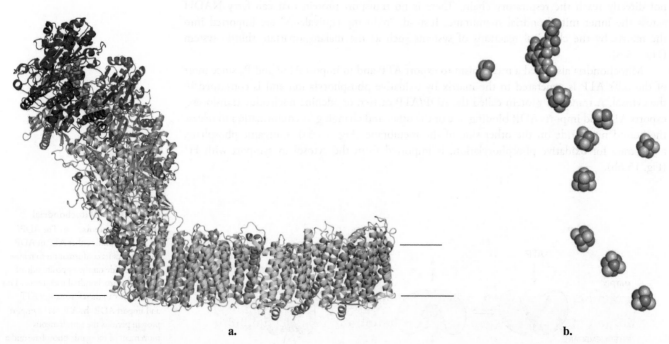

a.   **b.**

Figure 15.7 Structure of bacterial Complex I. **a.** The 16 subunits are shown in different colors, with the redox centers in red (FMN) and orange (Fe–S clusters). The horizontal portion is embedded in the membrane (represented by black lines) and the "arm" projects into the cytosol in bacteria (or mitochondrial matrix in eukaryotes). **b.** Arrangement of the redox centers in Complex I. Atoms are color coded: C gray, N blue, O red, P orange, Fe gold, and S yellow. Electrons flow through the groups from upper left to lower right.

Question **Identify the FMN group, the 4Fe–4S clusters, and the 2Fe–2S clusters in part b.**

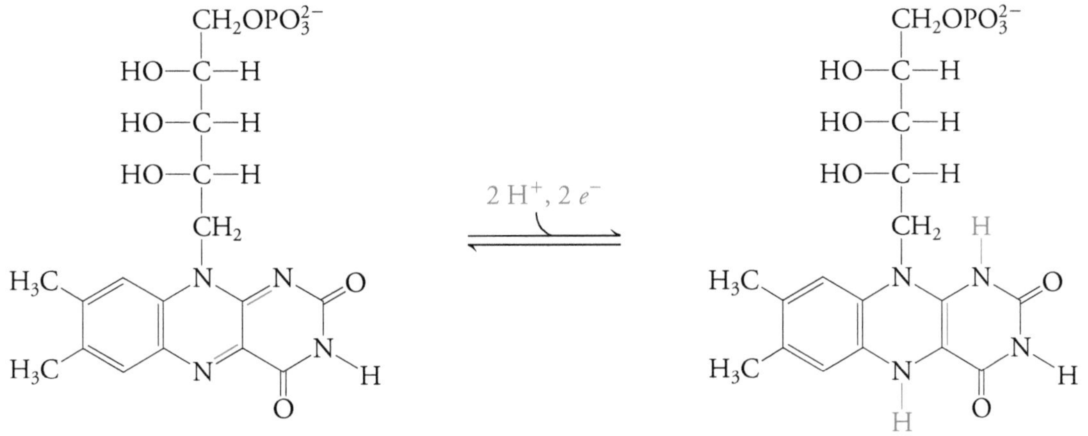

Flavin mononucleotide (FMN)   FMNH₂

$2\,H^+, 2\,e^-$

Figure 15.8 Flavin mononucleotide (FMN).   This prosthetic group resembles flavin adenine dinucleotide (FAD; see Fig. 3.2c) but lacks the AMP group of FAD. The transfer of two electrons and two protons to FMN yields FMNH₂.

in the cluster (each cluster is a conjugated structure that functions as a single unit). Electrons travel between several Fe–S clusters before reaching ubiquinone. Like FMN, ubiquinone is a two-electron carrier (see Section 12.3), but it accepts one electron at a time from an Fe–S donor. Iron–sulfur clusters may be among the most ancient of electron carriers, dating from a time when the earth's abundant iron and sulfur were major players in prebiotic chemical reactions. The ubiquinone binding site is located in the Complex I arm not far from the membrane surface.

As electrons are transferred from NADH to ubiquinone, Complex I transfers four protons from the matrix to the intermembrane space. Comparisons with other transport proteins and detailed analysis of cryo-EM and crystal structures indicate the presence of four proton-translocating "channels" in the membrane-embedded arm of Complex I. When redox groups in the peripheral arm are transiently reduced and reoxidized, the protein undergoes conformational changes that are transmitted from the peripheral arm to the membrane arm in part by a horizontally oriented helix that lies within the membrane portion of the complex. These conformational changes do not open passageways, as occurs in Na⁺ and K⁺ channel proteins (Section 9.2). Instead, each proton passes from one side of the membrane to the other via a **proton wire,** a series of hydrogen-bonded protein groups plus water molecules that form a chain through which a proton can be rapidly relayed. (Recall from Fig. 2.13 that protons readily jump between water molecules.) Note that in this relay mechanism, the protons taken up from the matrix are not the same ones that are released into the intermembrane space. The reactions of Complex I are summarized in **Figure 15.10.**

Other oxidation reactions contribute to the ubiquinol pool

The reduced quinone product of the Complex I reaction joins a pool of quinones that are soluble in the inner mitochondrial membrane by virtue of their long hydrophobic isoprenoid tails (see Section 12.3). *The pool of reduced quinones is augmented by the activity of other oxidation–reduction reactions.* One of these is catalyzed by succinate dehydrogenase, which carries out step 6 of the citric acid cycle (see Section 14.2).

$$\text{succinate} + Q \rightleftharpoons \text{fumarate} + QH_2$$

Succinate dehydrogenase is the only one of the citric acid cycle enzymes that is not soluble in the mitochondrial matrix; it is embedded in the inner membrane. Like the other respiratory complexes, it contains several redox centers, including an FAD group. Succinate dehydrogenase is also called Complex II of the mitochondrial respiratory chain. However, it is more like a tributary because it does not undertake proton translocation and therefore does

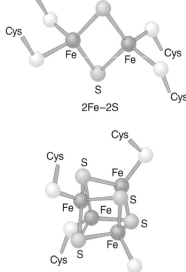

Figure 15.9 Iron–sulfur clusters. Although some Fe–S clusters contain up to eight Fe atoms, the most common are the 2Fe–2S and 4Fe–4S clusters. In all cases, the iron–sulfur clusters are coordinated by the S atoms of cysteine side chains. These prosthetic groups undergo one-electron redox reactions.

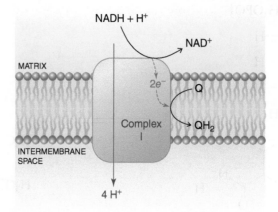

Figure 15.10 Complex I function. As two electrons from the water-soluble NADH are transferred to the lipid-soluble ubiquinone, four protons are translocated from the matrix into the intermembrane space.

not directly contribute the free energy of its oxidation–reduction reaction toward ATP synthesis. Nevertheless, it does feed reducing equivalents as ubiquinol into the electron transport chain (**Fig. 15.11a**).

A major source of ubiquinol is fatty acid oxidation, another energy-generating catabolic pathway that takes place in the mitochondrial matrix. A membrane-bound fatty acyl-CoA dehydrogenase catalyzes the oxidation of a C—C bond in a fatty acid attached to coenzyme A. The electrons removed in this dehydrogenation reaction are transferred to ubiquinone (Fig. 15.11b). As we will see in Section 17.2, the complete oxidation of a fatty acid also produces NADH that is reoxidized by the mitochondrial electron transport chain, starting with Complex I.

Electrons from cytosolic NADH can also enter the mitochondrial ubiquinol pool through the actions of a cytosolic and a mitochondrial glycerol-3-phosphate dehydrogenase (Fig. 15.11c). First, the cytosolic enzyme uses NADH to reduce dihydroxyacetone phosphate to glycerol-3-phosphate.

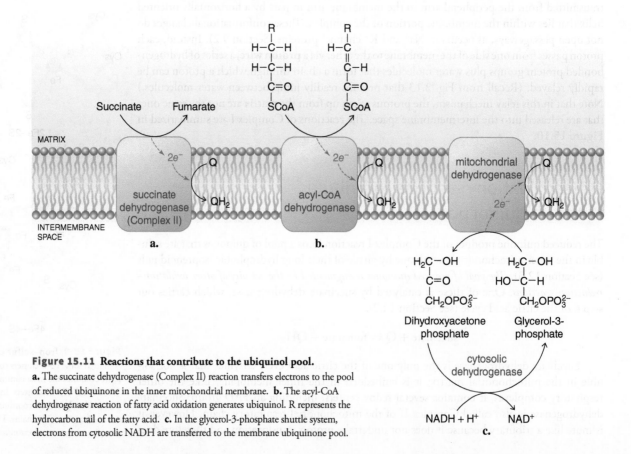

Figure 15.11 Reactions that contribute to the ubiquinol pool.
a. The succinate dehydrogenase (Complex II) reaction transfers electrons to the pool of reduced ubiquinone in the inner mitochondrial membrane. **b.** The acyl-CoA dehydrogenase reaction of fatty acid oxidation generates ubiquinol. R represents the hydrocarbon tail of the fatty acid. **c.** In the glycerol-3-phosphate shuttle system, electrons from cytosolic NADH are transferred to the membrane ubiquinone pool.

Then the mitochondrial enzyme, embedded in the inner membrane, reoxidizes the glycerol-3-phosphate, transferring two electrons to ubiquinone. This system, which shuttles electrons from NADH to ubiquinol, bypasses Complex I.

Complex III transfers electrons from ubiquinol to cytochrome c

Ubiquinol is reoxidized by Complex III, an integral membrane protein with 11 subunits in each of its two monomeric units. Complex III, also called ubiquinol:cytochrome c oxidoreductase or cytochrome bc_1, transfers electrons to the peripheral membrane protein cytochrome c. **Cytochromes** are proteins with heme prosthetic groups. The name *cytochrome* literally means "cell color"; cytochromes are largely responsible for the purplish-brown color of mitochondria. Cytochromes are commonly named with a letter (*a*, *b*, or *c*) indicating the exact structure of the porphyrin ring of their heme group (**Fig. 15.12**). The structure of the heme group and the surrounding protein microenvironment influence the protein's absorption spectrum. They also determine the reduction potentials of cytochromes, which range from about −0.080 V to about +0.385 V.

Unlike the heme prosthetic groups of hemoglobin and myoglobin, the heme groups of cytochromes undergo reversible one-electron reduction, with the central Fe atom cycling between the Fe^{3+} (oxidized) and Fe^{2+} (reduced) states. Consequently, the net reaction for Complex III, in which two electrons are transferred, includes two cytochrome c proteins:

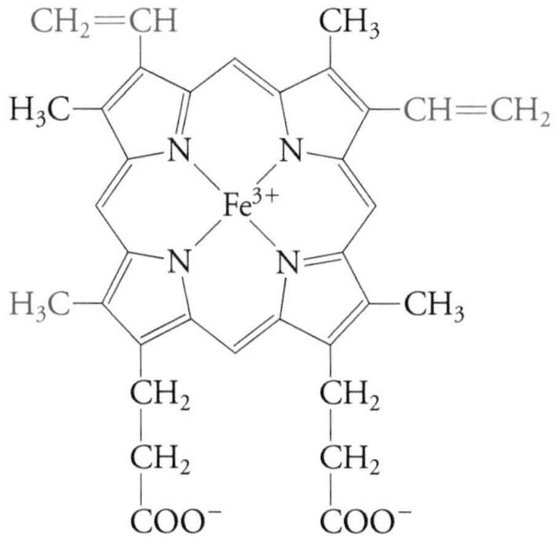

$$QH_2 + 2 \text{ cytochrome } c \text{ (Fe}^{3+}) \rightleftharpoons Q + 2 \text{ cytochrome } c \text{ (Fe}^{2+}) + 2H^+$$

Complex III itself contains two cytochromes (cytochrome *b* and cytochrome c_1) that are integral membrane proteins. These two proteins, along with an iron–sulfur protein (also called the Rieske protein), form the functional core of Complex III (these same three subunits are the only ones that have homologs in the corresponding bacterial respiratory complex). Altogether, each monomer of Complex III is anchored in the membrane by 14 transmembrane α helices (**Fig. 15.13**).

The flow of electrons through Complex III is complicated, in part because the two electrons donated by ubiquinol must split up in order to travel through a series of one-electron carriers

Figure 15.12 The heme group of a *b* cytochrome.   The planar porphyrin ring surrounds a central Fe atom, shown here in its oxidized (Fe^{3+}) state. The heme substituent groups that are colored blue differ in the *a* and *c* cytochromes (the heme group of hemoglobin and myoglobin has the *b* structure; see Section 5.1).

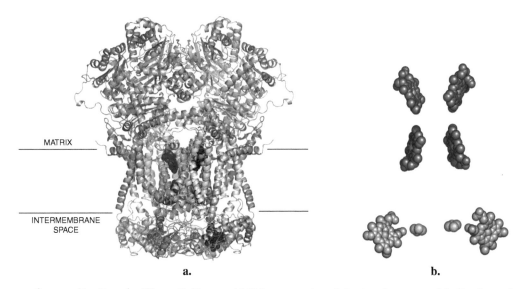

a.   **b.**

Figure 15.13 Structure of mammalian Complex III.   a. Backbone model. Eight transmembrane helices in each monomer of the dimeric complex are contributed by cytochrome *b* (light blue with heme groups dark blue). The iron–sulfur protein (green with Fe–S clusters orange) and cytochrome c_1 (pink with heme groups purple) project into the intermembrane space. **b.** Arrangement of prosthetic groups. The two heme groups of each cytochrome *b* (blue) and the heme group of cytochrome c_1 (purple), along with the iron–sulfur clusters (Fe atoms orange), provide a pathway for electrons between ubiquinol (in the membrane) and cytochrome c (in the intermembrane space).

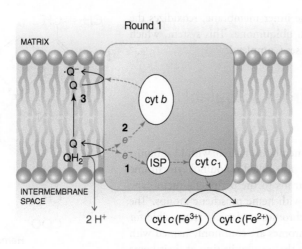

Round 1

1. In the first round, QH₂ donates one electron to the iron–sulfur protein (ISP). The electron then travels to cytochrome c_1 and then to cytochrome c.

2. QH₂ donates its other electron to cytochrome b. The two protons from QH₂ are released into the intermembrane space.

3. The oxidized ubiquinone diffuses to another quinone-binding site, where it accepts the electron from cytochrome b, becoming a half-reduced semiquinone ($\cdot Q^-$).

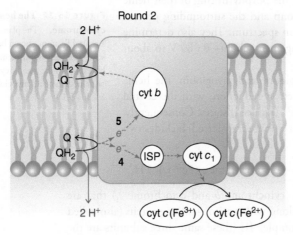

Round 2

4. In the second round, a second QH₂ surrenders its two electrons to Complex III and its two protons to the intermembrane space. One electron goes to reduce cytochrome c.

5. The other electron goes to cytochrome b and then to the waiting semiquinone produced in the first part of the cycle. This step regenerates QH₂, using protons from the matrix.

Figure 15.14 The Q cycle.

Question Write an equation for round 1 and for round 2.

that includes the 2Fe–2S cluster of the iron–sulfur protein, cytochrome c_1, and cytochrome b (which actually contains two heme groups with slightly different reduction potentials). Except for the 2Fe–2S cluster, all the redox centers are arranged in such a way that electrons can tunnel from one to another. The iron–sulfur protein must change its conformation by rotating and moving about 22 Å in order to pick up and deliver an electron. Further complicating the picture is the fact that each monomeric unit of Complex III has two active sites where quinone cofactors undergo reduction and oxidation.

The circuitous route of electrons from ubiquinol to cytochrome c is described by the two-round **Q cycle,** diagrammed in **Figure 15.14.** *The net result of the Q cycle is that two electrons from QH₂ reduce two molecules of cytochrome c. In addition, four protons are translocated to the intermembrane space, two from QH₂ in the first round of the Q cycle and two from QH₂ in the second round.* This proton movement contributes to the transmembrane proton gradient. The reactions of Complex III are summarized in **Figure 15.15.**

Many prokaryotes lack a homolog of mitochondrial Complex III and instead use other machinery to oxidize quinol. Although the conventional and alternative complexes are not structurally similar, both include heme groups and iron–sulfur clusters. Some alternative Complex III enzymes are able to transfer electrons directly to Complex IV without a mobile electron carrier such as cytochrome c.

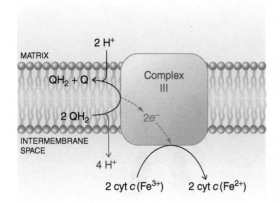

Figure 15.15 Complex III function. For every two electrons that pass from ubiquinol to cytochrome c, four protons are translocated to the intermembrane space.

Question How does the proton-translocating mechanism of Complex III differ from the one in Complex I?

Complex IV oxidizes cytochrome *c* and reduces O_2

Just as ubiquinone ferries electrons from Complex I and other enzymes to Complex III in mitochondria, cytochrome *c* ferries electrons between Complexes III and IV, carrying just one at a time. Unlike ubiquinone and the other proteins of the respiratory chain, cytochrome *c* is soluble in the intermembrane space (**Fig. 15.16**). Because this small peripheral membrane protein is central to the metabolism of many organisms, analysis of its sequence has played a large role in elucidating evolutionary relationships.

Complex IV, also called cytochrome *c* oxidase, is the last enzyme to deal with the electrons derived from the oxidation of metabolic fuels. Four electrons delivered by cytochrome *c* are consumed in the reduction of molecular oxygen to water:

$$4 \text{ cytochrome } c \text{ (Fe}^{2+}) + O_2 + 4 \text{ H}^+ \rightarrow 4 \text{ cytochrome } c \text{ (Fe}^{3+}) + 2 \text{ H}_2\text{O}$$

The redox centers of mammalian Complex IV include heme groups and copper ions situated among the 13 subunits in each half of the dimeric complex (**Fig. 15.17**).

Within each monomer, an electron travels from cytochrome *c* to the Cu_A redox center, which has two copper ions, and then to a heme *a* group. From there it travels to a binuclear center consisting of the iron atom of heme a_3 and a copper ion (Cu_B). The four-electron reduction of O_2 occurs at the Fe–Cu binuclear center, with help from a nearby tyrosine side chain. Note that the chemical reduction of O_2 to H_2O consumes four protons from the mitochondrial matrix. One possible mechanism, based on spectroscopic and other evidence, is shown in **Figure 15.18**.

Cytochrome *c* oxidase also relays four additional protons from the matrix to the intermembrane space, most likely during steps 3, 4, 5, and 6 of the mechanism shown in Figure 15.18. In other words, two protons are translocated for every pair of electrons traveling through the electron transport chain. Complex IV appears to harbor two proton wires. One delivers H^+ ions from the matrix to the oxygen-reducing active site. The other one spans the 50-Å distance between the matrix and intermembrane faces of the protein. Protons are relayed through the proton wires when the protein changes its conformation in response to changes in its oxidation state. *The production of water and the proton relays both deplete the matrix proton concentration and thereby contribute to the formation of a proton gradient across the inner mitochondrial membrane* (**Fig. 15.19**).

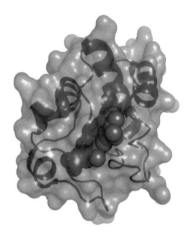

Figure 15.16 Cytochrome *c*. The protein is shown as a gray transparent surface over its ribbon backbone. The heme group (pink) lies in a deep pocket.

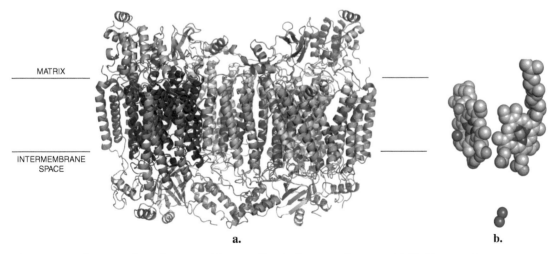

a.   b.

Figure 15.17 Structure of cytochrome *c* oxidase. a. The 13 subunits in each monomeric half of the mammalian complex comprise 28 transmembrane α helices. **b.** The heme groups (C atoms gray, N blue, O red, and Fe gold) and copper ions (brown) from one half of the complex are shown in space-filling form.

Question **Locate the copper ion and heme iron of the binuclear center where O_2 is reduced.**

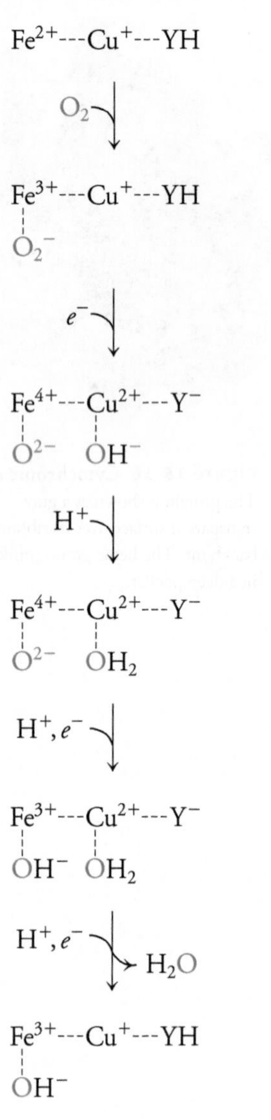

Fe2+---Cu+---YH
O2
Fe3+---Cu+---YH
O2−
e−
Fe4+---Cu2+---Y−
O2− OH−
H+
Fe4+---Cu2+---Y−
O2− OH2
H+, e−
Fe3+---Cu2+---Y−
OH− OH2
H+, e−
H2O
Fe3+---Cu+---YH
OH−
H+, e−
H2O
Fe2+---Cu+---YH

Figure 15.18 A proposed model for the cytochrome *c* oxidase reaction. The Fe–Cu binuclear center and tyrosine side chain (Y) are indicated.

½O2 + 2 H+
H2O
MATRIX
Complex IV
2e−
INTERMEMBRANE SPACE
2 H+
2 cyt c (Fe3+)
2 cyt c (Fe2+)

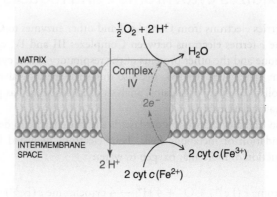

Figure 15.19 Complex IV function. For every two electrons donated by cytochrome *c*, two protons are translocated to the intermembrane space. Two protons from the matrix are also consumed in the reaction $\frac{1}{2}O_2 \rightarrow H_2O$ (the full reduction of O_2 requires four electrons).

Respiratory complexes associate with each other

Complexes I–IV are located mostly in the planar portions of cristae, where they have some lateral mobility. However, the three complexes that contribute to the transmembrane proton gradient—Complexes I, III, and IV—are able to associate with each other in **supercomplexes** whose composition seems to vary with the type of cell and the physiological conditions. Supercomplexes containing a Complex III dimer along with Complex I or Complex IV have been described in detail. An assembly of all three complexes is known as a **respirasome**, since it is capable of catalyzing the complete process of electron transfer from NADH to O_2 (**Fig. 15.20**).

Supercomplexes have been observed in bacteria, yeast, and mammals, which argues that bringing respiratory enzymes close together is functionally significant. Although it is tempting to speculate that proximity increases the efficiency of electron transfer, the complexes are still somewhat separated, with membrane phospholipids between them. Moreover, the complexes appear to be dynamic. For example, the large Complex I is the least mobile and is found mostly in supercomplexes, whereas Complex IV exists largely as monomers but can form isolated dimers

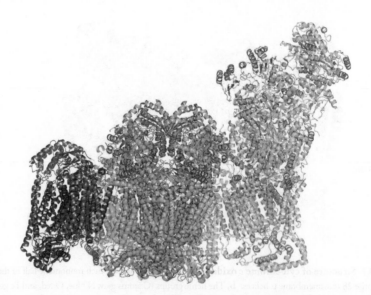

Figure 15.20 A respirasome. This supercomplex of Complex I (green), a Complex II dimer (cyan), and monomeric Complex IV (blue) is based on the cyro-EM study of a respirasome from porcine heart.

Question Add lines to show the location of the mitochondrial membrane.

(as shown in Fig. 15.17) as well as supercomplexes. Supercomplex formation may be a way to stabilize the structures of the large proteins or to maximize the density of the electron-transporting machinery in the membrane.

Even if the respiratory complexes do not actually form a chain, they provide a pathway for the reduced cofactors NADH and ubiquinol to surrender their electrons to O_2. The oxidized cofactors rejoin the pool of cofactors in the cell, ready to accept additional electrons by participating in redox reactions such as those of the citric acid cycle. In this way, NAD^+ and ubiquinone function as shuttles, accepting electrons (becoming reduced) then giving them up (becoming oxidized) over and over again. Because the electron carriers are regenerated with each reaction cycle, electron flow continues, provided there is plenty of metabolic fuel (the source of electrons) and O_2 (the final electron acceptor). However, the electron transport system can also produce damaging by-products (**Box 15.A**).

Box 15.A   Reactive Oxygen Species

Although electron transport and oxidative phosphorylation generate considerable amounts of ATP, processes that depend on O_2 have some disadvantages. Aerobic organisms—in addition to the challenges of acquiring O_2 from the air, water, or soil they live in—must also deal with the possibility of oxidative damage. The very machinery that makes aerobic metabolism possible also produces toxic molecules known as **reactive oxygen species,** including superoxide ($\cdot O_2^-$), hydrogen peroxide (H_2O_2), and the hydroxyl radical ($\cdot OH$).

Electrons occasionally leak from the electron transport chain, primarily in Complexes I and III, and react with O_2 to generate reactive oxygen species, including **free radicals**. For example,

$$O_2 + e^- \rightarrow \cdot O_2^-$$

A free radical is an atom or molecule with a single unpaired electron and is highly reactive as it seeks another electron to form a pair. Although a free radical is short-lived (the half-life of $\cdot O_2^-$ is 10^{-6} seconds), it can damage nearby proteins, lipids, and nucleic acids by stealing their electrons. Presumably, the most damage is inflicted on mitochondrial components. The accumulation of cellular damage from free radicals may be responsible for the degeneration of tissues that occurs with aging.

The primary cellular defense against $\cdot O_2^-$ is the enzyme superoxide dismutase, which rapidly converts $\cdot O_2^-$ to the less-toxic H_2O_2:

$$2 \cdot O_2^- + 2H^+ \rightarrow H_2O_2 + O_2$$

Other cellular components, such as ascorbate (see Box 5.B) and α-tocopherol (see Box 8.B) may protect cells from oxidative damage by scavenging free radicals. Findings such as these have led to the popular belief that dietary supplements with antioxidant activity can prevent aging. Following a similar line of reasoning, caloric restriction may lower free radical production by decreasing the availability of fuel molecules that undergo oxidative metabolism. Caloric restriction extends the lifespans of some experimental animals, but a similar effect has not been confirmed in humans. Furthermore, a high rate of fuel oxidation in active animal muscles is not always accompanied by high production of reactive oxygen species, which leaves the link between oxidative metabolism and free radicals in question.

To make matters more complicated, many cells deliberately synthesize reactive oxygen and nitrogen species as signaling molecules. For example, an H_2O_2 molecule can diffuse through membranes and selectively oxidize Cys sulfhydryl groups, which often exist as thiolate groups ($-S^-$), in cellular proteins. This modification may alter protein activity.

Restoring the modified Cys residues and other intentionally or accidentally oxidized cellular components often involves the tripeptide glutathione (GSH, γ-Glu-Cys-Gly), in which the side chain of glutamate is linked to the amino group of cysteine:

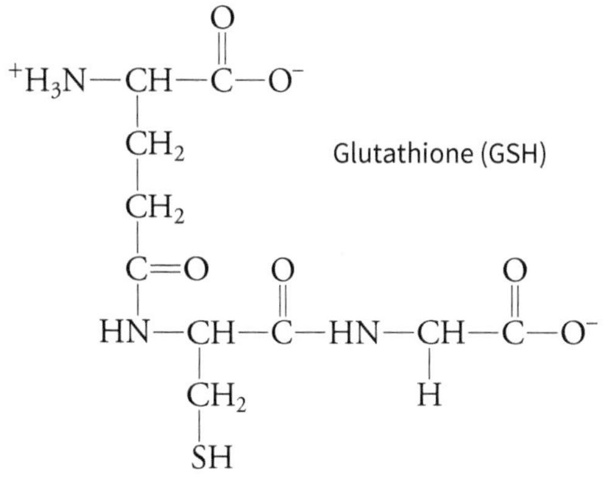

Glutathione (GSH)

Enzymes called glutathione peroxidases use glutathione to convert hydrogen peroxide to water and to convert other organic peroxides ($R-O-O-R$) and sulfenic acids ($R-S-O-H$) to less toxic products. In the process, two glutathione tripeptides become linked through a disulfide bond between their Cys residues, to form GSSG (see Problem 4.18). This oxidized glutathione must then be reduced by glutathione reductase in a reaction that requires NADPH as an electron donor:

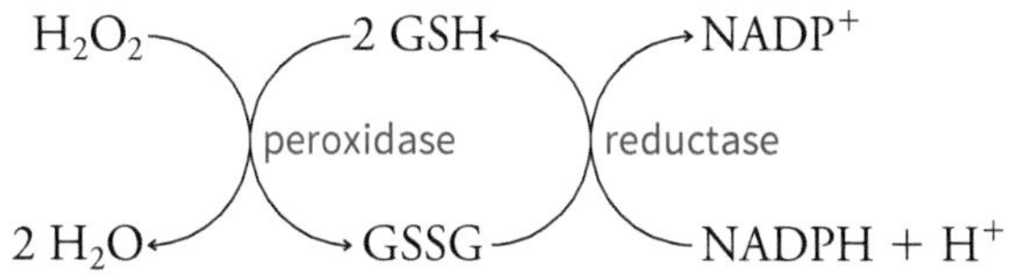

The high cellular concentration of glutathione—typically in the millimolar range—seems excessive for the job of scavenging free radicals and repairing other oxidized molecules, and it is possible that glutathione plays a more general role in regulating the overall redox balance in cells.

Concept Check

1. Describe the compartments of a mitochondrion.
2. List the transport proteins that occur in the inner mitochondrial membrane.
3. Draw a simple diagram showing the electron-transport complexes and the mobile carriers that link them.
4. List the different types of redox groups in the respiratory electron transport chain and identify them as one- or two-electron carriers.
5. How are electrons delivered to complexes I, III, and IV?
6. Explain why O_2 is the final electron acceptor in the chain.
7. Describe the operation of a proton wire.
8. Write an equation to describe the overall redox reaction carried out by each mitochondrial complex.
9. Compare the arrangement of an electron transport chain and a supercomplex.

15.3 ATP Synthase

KEY CONCEPTS

Describe the structure and operation of ATP synthase.

- Recognize the structural components of ATP synthase.
- Identify the energy transformations that occur in ATP synthase.
- Describe the binding change mechanism.
- Explain why P:O ratios are nonintegral.
- Explain why oxidative phosphorylation is coupled to electron transport.

The protein that taps the electrochemical proton gradient to phosphorylate ADP is known as the F-ATP synthase (or Complex V). One part of the protein, called F_0, functions as a transmembrane channel that permits H^+ to flow back into the matrix, following its gradient. The F_1 component catalyzes the reaction $ADP + P_i \rightarrow ATP + H_2O$ (**Fig. 15.21**). This section describes the structures of the two components of ATP synthase and shows how their activities are linked so that exergonic H^+ transport can be coupled to endergonic ATP synthesis.

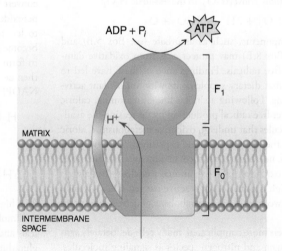

Figure 15.21 ATP synthase function. As protons flow through the F_0 component from the intermembrane space to the matrix, the F_1 component catalyzes the synthesis of ATP from $ADP + P_i$.

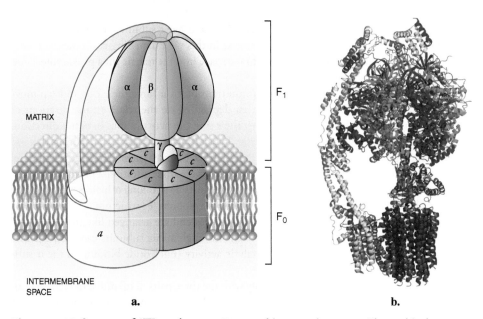

Figure 15.22 Structure of ATP synthase. a. Diagram of the mammalian enzyme. The α and β subunits are connected via a central shaft to the membrane-embedded ring of 8 *c* subunits. A peripheral stalk links the *a* subunit to the catalytic domain. **b.** Ribbon model of the yeast enzyme, with subunits colored as in a. The yeast *c* ring contains 10 subunits.

Proton translocation rotates the c ring of ATP synthase

Not surprisingly, the overall structure of ATP synthase is conserved among different species, although the number of individual protein subunits varies. The F_1 component contains three α and three β subunits surrounding a central shaft. The membrane-embedded portion of ATP synthase includes a variety of subunits. Some of these make up a stalk that extends from the cytoplasmic side of F_0 to the top of the F_1 component in the matrix (**Fig. 15.22**). A number of membrane-embedded *c* subunits—ranging from 8 in mammalian mitochondria to as many as 17 in some bacteria—form a ring that rotates within the lipid bilayer.

In all species, proton transport through ATP synthase requires rotation of the *c* ring past the stationary *a* subunit. The carboxylate side chain of a highly conserved aspartate or glutamate residue on each *c* subunit serves as a proton binding site (**Fig. 15.23**). When properly positioned at

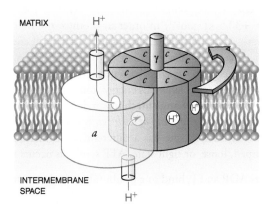

Figure 15.23 Mechanism of proton transport by ATP synthase. When a *c* subunit (pink) binds a proton from one side of the membrane, it moves away from the *a* subunit (blue). Because the *c* subunits form a ring, rotation brings another *c* subunit toward the *a* subunit, where it releases its bound proton to the opposite side of the membrane.

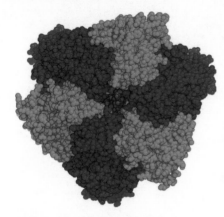

Figure 15.24 Structure of the F₁ component of ATP synthase. The alternating α (blue) and β (green) subunits form a hexamer around the end of the γ shaft (purple). This view is looking from the matrix down onto the top of the ATP synthase.

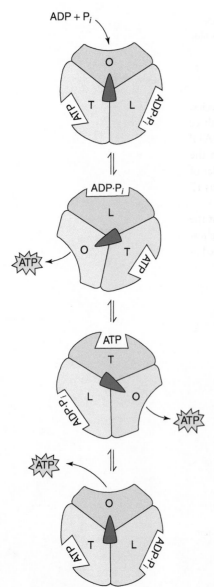

the *a* subunit, a *c* subunit takes up a proton from the intermembrane space. Proton binding neutralizes the carboxylate group, freeing it from the electrostatic attraction of a positively charged arginine residue on the *a* subunit. As a result, the protonated *c* subunit moves away, and this slight rotation of the *c* ring brings another *c* subunit into position so that it can release its bound proton into the matrix. In this way, the *c* ring continues to rotate, translocating protons from the intermembrane space into the matrix (in a prokaryote, from outside the cell into the cytosol).

The favorable thermodynamics of proton translocation force the *c* ring to keep moving in one direction. Experiments show that depending on the relative concentrations of protons on the two sides of the membrane, the *c* ring can actually spin in either direction. Related proteins, known as P- and V-ATPases, in fact function as active transporters that use the free energy of the ATP hydrolysis reaction to drive ion movement across the membrane.

Attached to the *c* ring and rotating along with it are three additional protein subunits. Two of these are small, but the one known as γ consists of two long α helices arranged as a bent coiled coil that protrudes into the center of the globular F₁ structure. The three α and three β subunits of F₁ have similar tertiary structures and are arranged like the sections of an orange around the γ subunit (**Fig. 15.24**). Although all six subunits can bind adenine nucleotides, only the β subunits have catalytic activity (nucleotide binding to the α subunits may play a regulatory role).

The γ subunit interacts asymmetrically with the three pairs of αβ units. In fact, each αβ unit has a slightly different conformation, and the three units cannot simultaneously adopt the same conformation. *The three αβ pairs change their conformations as the γ subunit rotates* (it is like a shaft driven by the *c* ring "rotor"). The αβ hexamer itself does not rotate, since it is held in place by the peripheral stalk that is anchored to the *a* subunit (see Fig. 15.22a).

As each proton moves across the membrane, the *c* ring and γ subunit rotate. Although the *c* ring may spin smoothly, the γ subunit lurches in steps of 120°, interacting successively with each of the three αβ pairs in one full rotation of 360°. Electrostatic interactions between the γ and β subunits apparently act as a catch that holds the γ subunit in place while translocation of protons builds up strain. The γ subunit is probably too stiff to absorb much rotational strain. Instead, the strain seems to be transmitted to the α and β subunits, which wobble somewhat before the γ subunit suddenly snaps into position at the next β subunit.

ATP synthase has 8 *c* subunits in mammals, 10 in yeast, and 15 in many bacteria. Thus, 8, 10, and 15 proton translocations are required for one complete rotation, and the *c* ring rotates in increments of 24° to 45° per proton translocated, depending on the number of *c* subunits.

The binding change mechanism explains how ATP is made

At the start of the chapter, we pointed out that ATP synthase catalyzes a highly endergonic reaction ($\Delta G°' = +30.5 \text{ kJ} \cdot \text{mol}^{-1}$) in order to produce the bulk of a cell's ATP supply. This enzyme operates in an unusual fashion, using mechanical energy (rotation) to form a chemical bond (the attachment of a phosphoryl group to ADP). In other words, the enzyme converts mechanical energy to the chemical energy of ATP. The interaction between the γ subunit and the αβ hexamer explains this energy transduction.

According to the **binding change mechanism** described by Paul Boyer, *rotation-driven conformational changes alter the affinity of each catalytic β subunit for an adenine nucleotide.* At any moment, each catalytic site has a different conformation (and binding affinity), referred to as the open, loose, or tight state. ATP synthesis occurs as follows (**Fig. 15.25**):

1. The substrates ADP and P₍ bind to a β subunit in the loose state.

Figure 15.25 The binding change mechanism. The diagram shows the catalytic (β) subunits of the F₁ component of ATP synthase from the same perspective as in Fig. 15.24. Each of the three β subunits adopts a different conformation: open (O), loose (L), or tight (T). The substrates ADP and P₍ bind to a loose site, ATP is synthesized when the site becomes tight, and ATP is released when the subunit becomes open. The conformational shifts are triggered by the 120° rotation of the γ subunit, arbitrarily represented by the purple shape.

2. The substrates are converted to ATP as rotation of the γ subunit causes the β subunit to shift to the tight conformation.

3. The product ATP is released after the next rotation, when the β subunit shifts to the open conformation.

Because the three β subunits of ATP synthase act cooperatively, they all change their conformations simultaneously as the γ subunit turns. A full rotation of 360° is required to restore the enzyme to its initial state, but each rotation of 120° results in the release of ATP from one of the three active sites.

Theoretical calculations show that the rotation-based mechanism of ATP synthase is more thermodynamically efficient than a back-and-forth mechanism like those used by other enzymes and transport proteins. This is because a spinning protein can move in increments, each step requiring a small free energy payment (the binding of one proton), whereas a conventional enzyme with an alternating-conformation mechanism would require the simultaneous binding of ~3 protons, which has a higher free-energy cost and would therefore be slower.

Experiments with the isolated F_1 component of ATP synthase show that in the absence of F_0, F_1 functions as an ATPase, hydrolyzing ATP to ADP + P_i (a thermodynamically favorable reaction). In the intact ATP synthase, dissipation of the proton gradient is tightly coupled to ATP synthesis with near 100% efficiency. Consequently, *in the absence of a proton gradient, no ATP is synthesized because there is no free energy to drive the rotation of the γ subunit.* Agents that dissipate the proton gradient can therefore "uncouple" ATP synthesis from electron transport, the source of the proton gradient (**Box 15.B**).

The P:O ratio describes the stoichiometry of oxidative phosphorylation

Since the γ shaft of ATP synthase is attached to the *c*-subunit rotor, 3 ATP molecules are synthesized for every complete *c*-ring rotation. However, the number of protons translocated per ATP depends on the number of *c* subunits. For mammalian ATP synthase, which has 8 *c* subunits, the stoichiometry is 8 H^+ per 3 ATP, or 2.7 H^+ per ATP. Such nonintegral values would be difficult to reconcile with most biochemical reactions, but they are consistent with the chemiosmotic theory: *Chemical energy (from the respiratory oxidation–reduction reactions) is transduced to a protonmotive force, then to the mechanical movement of a rotary engine (the c ring and its attached γ shaft), and finally back to chemical energy in the form of ATP.* Note that ATP synthase does not build an ATP molecule from scratch; it simply forms a bond between the second and third phosphate groups. When free energy–requiring cellular reactions consume ATP, this same bond is most often the one that is broken.

Box 15.B   Uncoupling Agents Prevent ATP Synthesis

When the metabolic need for ATP is low, the oxidation of reduced cofactors proceeds until the transmembrane proton gradient builds up enough to halt further electron transport. When the protons reenter the matrix via the F_0 component of ATP synthase, electron transport resumes. However, if the protons leak back into the matrix by a route other than ATP synthase, then electron transport will continue without any ATP being synthesized. ATP synthesis is said to be "uncoupled" from electron transport, and the agent that allows the proton gradient to dissipate in this way is called an **uncoupler.** Some small molecules that act as uncouplers are poisons, but physiological uncoupling does occur. Dissipating a proton gradient prevents ATP synthesis, but it allows oxidative metabolism to continue at a high rate. The by-product of this metabolic activity is heat.

Uncoupling for **thermogenesis** (heat production) occurs in specialized adipose tissue known as brown fat (its dark color is due to the relatively high concentration of cytochrome-containing mitochondria; ordinary adipose tissue is lighter). The inner membrane of the mitochondria in brown fat contains a transmembrane proton channel called a UCP (uncoupling protein). Protons translocated to the intermembrane space during respiration can reenter the mitochondrial matrix via the uncoupling protein, bypassing ATP synthase. The free energy of respiration is therefore given off as heat rather than used to synthesize ATP. Brown fat is abundant in hibernating mammals and newborn humans, and the activity of the UCP is under the control of hormones that also mobilize the stored fatty acids to be oxidized in the brown fat mitochondria.

Question   Why would increasing the activity of UCP promote weight loss?

The relationship between respiration (the activity of the electron transport complexes) and ATP synthesis is traditionally expressed as a **P:O ratio,** that is, the number of phosphorylations of ADP relative to the number of oxygen atoms reduced. For example, the oxidation of NADH by O_2 (carried out by the sequential activities of Complexes I, III, and IV) translocates 10 protons into the intermembrane space. The movement of these 10 protons back into the matrix via the F_0 component would theoretically drive the synthesis of about 3.7 ATP since 1 ATP can be made for every 2.7 protons translocated, at least in mammalian mitochondria:

$$\frac{1 \text{ ATP}}{2.7\,H^+} \times 10\,H^+ = 3.7$$

Thus, the P:O ratio would be about 3.7 (3.7 ATP per $\frac{1}{2}O_2$ reduced). For an electron pair originating as QH_2, only 6 protons would be translocated (by the activities of Complexes III and IV), and the P:O ratio would be approximately 2.2:

$$\frac{1 \text{ ATP}}{2.7\,H^+} \times 6\,H^+ = 2.2$$

In vivo, the P:O ratios are actually a bit lower than the theoretical values, because some of the protons translocated during electron transport do leak across the membrane or are consumed in other processes, such as the transport of P_i into the mitochondrial matrix (see Fig. 15.6). Consequently, experimentally determined P:O ratios are closer to 2.5 when NADH is the source of electrons and 1.5 for ubiquinol. These values are the basis for our tally of the ATP yield for the complete oxidation of glucose by glycolysis and the citric acid cycle (see Fig. 14.13).

The rate of oxidative phosphorylation reflects the need for ATP

The processes of electron transport and oxidative phosphorylation, like other metabolic pathways, maintain a steady state. In animal cells, the proton gradient does not build up beyond the typical 0.75-pH-unit difference across the membrane, because as soon as protons are pumped into the intermembrane space, they return to the matrix via ATP synthase. Similarly, the cell's concentration of ATP remains more or less constant because ATP is regenerated as quickly as it is converted to ADP. However, *the rates of ATP consumption and synthesis may fluctuate dramatically*, depending on the cell's activity level (**Box 15.C**).

Box 15.C   Powering Human Muscles

Cells cannot stockpile ATP; its concentration remains remarkably stable (between 2 and 5 mM in most cells) under widely varying levels of demand. Human muscle cells illustrate the multiple ways that animal cells can generate ATP for different needs. For example, a single burst of activity by the actin–myosin system (Section 5.4) can be powered by the ATP already available in the cell. This ATP supply is boosted by phosphocreatine (introduced in Section 12.2). In resting muscles, when the demand for ATP is low, creatine kinase catalyzes the transfer of a phosphoryl group from ATP to creatine to produce phosphocreatine:

ATP + creatine ⇌ phosphocreatine + ADP

This reaction runs in reverse when ADP concentrations rise, as they do when muscle contraction converts ATP to ADP + P_i. Phosphocreatine therefore acts as a sort of phosphoryl-group reservoir to maintain the supply of ATP. Without phosphocreatine, a muscle would exhaust its stock of ATP before it could be replenished by other, slower processes.

Muscle activity lasting up to a few seconds requires phosphocreatine, but the amount of phosphocreatine itself is limited,

so continued muscle contraction must rely on ATP produced by glycolysis, using glucose obtained from the muscle's store of glycogen. The end product of this pathway is lactate, the conjugate base of a weak acid, and muscle pain sets in as the acid accumulates and the pH begins to drop. Up to this point, the muscle functions mostly anaerobically (without the participation of O_2).

To continue its activity, the muscle must boost aerobic (O_2-dependent) metabolism and further oxidize fuels via the citric acid cycle. Sources of acetyl-CoA for the citric acid cycle include additional glucose, delivered by the bloodstream, and fatty acids. Recall that the citric acid cycle generates reduced cofactors that must be reoxidized by molecular oxygen. Aerobic metabolism of glucose and fatty acids is slower than anaerobic glycolysis, but it generates considerably more ATP. Some forms of physical activity and the systems that power them are diagrammed below.

A casual athlete can detect the shift from mostly anaerobic to mostly aerobic metabolism after about a minute and a half. In world-class athletes, the breakpoint occurs at about 150 seconds, which corresponds roughly to the finish line in a 1000-meter race.

The muscles of sprinters have a high capacity for anaerobic ATP generation, whereas the muscles of distance runners are better adapted to produce ATP aerobically. Such differences in energy metabolism are visibly manifest in the flight muscles of birds. Migratory birds such as geese, which power their long flights primarily with fatty acids, have large numbers of mitochondria to carry out oxidative phosphorylation. The reddish-brown mitochondria give the flight muscles a dark color. Birds that rarely fly, such as chickens, have fewer mitochondria and lighter-colored muscles. When these birds do fly, it is usually only a short burst of activity that is powered by anaerobic mechanisms.

Question **Why do some athletes believe that creatine supplements boost their performance?**

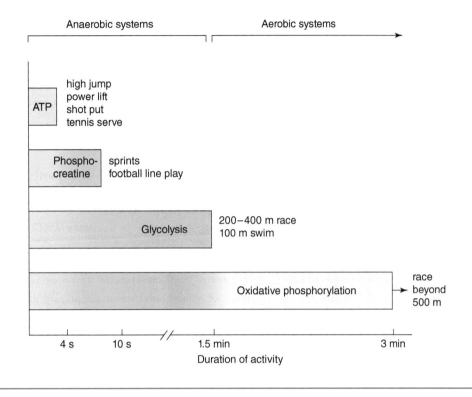

The close coupling between generation of the proton gradient and ATP synthesis allows oxidative phosphorylation to be regulated by the availability of reduced cofactors (NADH and QH_2) produced from metabolic fuels. This supply-and-demand system differs from other metabolic pathways, where control is exerted at one or a few irreversible, highly exergonic steps. The rate-limiting step of oxidative phosphorylation, the reaction catalyzed by Complex IV, is not a major control point, although there is evidence that ATP, the product of oxidative phosphorylation, allosterically inhibits Complex IV. Complexes I–IV and ATP synthase undergo phosphorylation catalyzed by protein kinases (Section 10.3), but the functional significance of these modifications is not yet understood.

In eukaryotes, ATP synthase itself is regulated by a small protein called inhibitory factor 1 (IF1). Different forms of IF1—all less than 100 residues—are intrinsically disordered (see Section 4.3) at relatively high matrix pH values, when the electron transport chain is pumping protons into the intermembrane space and the proton gradient is steep. When the pH drops, a sign that the protonmotive force is waning, IF1 dimerizes and forms extended α helices that insert in between the α and β subunits of F_1 and contact the γ shaft (**Fig. 15.26**). IF1 binding prevents ATP synthase from carrying out the binding change mechanism.

IF1 binding to ATP synthase may also stabilize ATP synthase supercomplexes containing two or four copies of the enzyme. The V-shaped synthase dimer appears to cause the inner mitochondrial membrane to bend by about 70–90° (**Fig. 15.27**). Rows of ATP synthase dimers may be responsible for the sharply curved edges and tubular shapes of cristae (see Fig. 15.4). Alternatively, these arrays might be just an efficient way to pack ATP synthase—which accounts for about 20% of mitochondrial membrane protein—into a tight space.

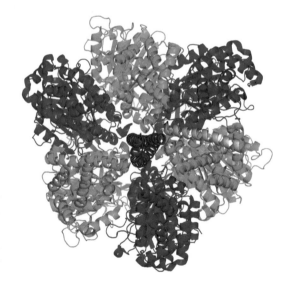

Figure 15.26 IF1 bound to the F_1 portion of ATP synthase. A 41-residue helical segment of IF1 (gold) inserts between an α (blue) and β (green) subunit. The view is from the bottom of F_1 (as shown in Fig. 15.24), with the γ subunit in purple.

Figure 15.27 ATP synthase dimer.   The orientation of the two enzymes (orange and yellow) induces a sharp bend in the membrane (black lines).

Concept Check

1. Draw a simple diagram of ATP synthase and indicate which parts are stationary and which rotate.
2. Explain how ATP synthase dissipates the proton gradient.
3. Recount how the three conformational states of the β subunits of ATP synthase are involved in ATP synthesis.
4. Explain how ATP synthase could operate in reverse to hydrolyze ATP.
5. Explain why the number of protons translocated per ATP synthesized varies among species.
6. Explain why the availability of reduced substrates is the primary mechanism for regulating oxidative phosphorylation.
7. What is a P:O ratio and why is it non-integral?

15.4 Chemiosmosis

KEY CONCEPTS

Explain how the protonmotive force links electron transport and ATP synthesis.

- Describe the formation of the proton gradient.
- Relate the pH difference of the proton gradient to the free energy change.

The electrons collected from metabolic fuels during their oxidation are consumed in the reduction of O_2 to H_2O. However, their free energy has been conserved. How much free energy is potentially available? Using the ΔG values calculated from the standard reduction potentials of the substrates and products of Complexes I, III, and IV (presented graphically in Fig. 15.2), we can see that each of the three respiratory complexes theoretically releases enough free energy to drive the endergonic phosphorylation of ADP to form ATP ($\Delta G^{\circ\prime} = +30.5$ kJ·mol^{-1}).

Complex I: NADH → QH$_2$	$\Delta G^{\circ\prime} = -69.5$ kJ·mol^{-1}
Complex III: QH$_2$ → cytochrome c	$\Delta G^{\circ\prime} = -36.7$ kJ·mol^{-1}
Complex IV: cytochrome c → O$_2$	$\Delta G^{\circ\prime} = -112.0$ kJ·mol^{-1}
NADH → O$_2$	$\Delta G^{\circ\prime} = -218.2$ kJ·mol^{-1}

Recall that energy cannot be created or destroyed, but it can be transformed. Understanding oxidative phosphorylation requires recognizing energy in several different forms along the way from metabolic fuels to ATP.

Chemiosmosis links electron transport and oxidative phosphorylation

Until the 1960s, the connection between respiratory electron transport (measured as O_2 consumption) and ATP synthesis was a mystery. Many biochemists believed that fuel oxidation generated a "high-energy" intermediate capable of phosphorylating ADP, but they were unable to identify any such compound. The actual connection between oxidation–reduction chemistry and ATP synthesis was discovered by Peter Mitchell, a somewhat unconventional scientist who was inspired by his work on mitochondrial phosphate transport and recognized the importance of compartmentation in biological systems. Mitchell's **chemiosmotic theory** proposed that the proton-translocating activity of the electron transport complexes in the inner mitochondrial membrane generates a proton gradient across the membrane. The protons cannot diffuse back into the matrix because the membrane is impermeable to ions. *The imbalance of protons represents a source of energy, also called a **protonmotive force**, that can drive the activity of an ATP synthase.*

We now know that for each pair of electrons that flow through Complexes I, III, and IV, 10 protons are translocated from the matrix to the intermembrane space (which is ionically equivalent to the cytosol). In bacteria, electron transport complexes in the plasma membrane translocate protons from the cytosol to the cell exterior. Mitchell's theory of chemiosmosis actually explains more than just aerobic respiration. It also applies to systems where the energy from sunlight is used to generate a transmembrane proton gradient (this aspect of photosynthesis is described in Section 16.2).

The proton gradient is an electrochemical gradient

When the mitochondrial complexes translocate protons across the inner mitochondrial membrane, the concentration of H^+ outside increases and the concentration of H^+ inside decreases (**Fig. 15.28**). *This imbalance of protons, a nonequilibrium state, has an associated energy (the force that would restore the system to equilibrium).* The energy of the proton gradient has two components, reflecting the difference in the concentration of the chemical species and the difference in electrical charge of the positively charged protons (for this reason, the mitochondrial proton gradient is referred to as an electrochemical gradient rather than a simple concentration gradient). The free energy change for generating the chemical imbalance of protons is

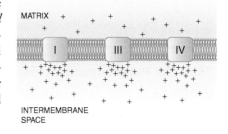

$$\Delta G = RT \ln \frac{[H^+]_{out}}{[H^+]_{in}} \tag{15.5}$$

The pH ($-\log [H^+]$) of the intermembrane space (*out*) is typically about 0.75 units less than the pH of the matrix (*in*).

The free energy change for generating the electrical imbalance of protons is

$$\Delta G = ZF\Delta\psi \tag{15.6}$$

Figure 15.28 Generation of a proton gradient. During the oxidation–reduction reactions catalyzed by mitochondrial Complexes I, III, and IV, protons (represented by positive charges) are translocated out of the matrix into the intermembrane space. This creates an imbalance in both proton concentration and electrical charge.

where Z is the ion's charge (+1 in this case) and $\Delta\psi$ is the membrane potential caused by the imbalance in positive charges (see Section 9.1). For mitochondria, $\Delta\psi$ is positive, usually 150 to 200 mV. This value indicates that the intermembrane space or cytosol is more positive than the matrix (recall from Section 9.1 that for a whole cell, the cytosol is more negative than the extracellular space and $\Delta\psi$ is negative).

Combining the chemical and electrical effects gives an overall free energy change for transporting protons from the matrix (*in*) to the intermembrane space (*out*):

$$\Delta G = RT \ln \frac{[H^+]_{out}}{[H^+]_{in}} + ZF\Delta\psi \tag{15.7}$$

Typically, the free energy change for translocating one proton out of the matrix is about $+20 \text{ kJ} \cdot \text{mol}^{-1}$ (see Sample Calculation 15.3 for a detailed application of Equation 15.7). This is a thermodynamically costly event. Passage of the proton back *into* the matrix, following its electrochemical gradient, would have a free energy change of about $-20 \text{ kJ} \cdot \text{mol}^{-1}$. This event is thermodynamically favorable, but it does not provide enough free energy to drive the synthesis of ATP. However, the 10 protons translocated for each pair of electrons transferred from NADH to O_2 have an associated protonmotive force of over 200 kJ · mol^{-1}, enough to drive the phosphorylation of several molecules of ADP.

SAMPLE CALCULATION 15.3

Problem Calculate the free energy change for translocating a proton out of the mitochondrial matrix, where $pH_{matrix} = 7.8$, $pH_{cytosol} = 7.15$, $\Delta\psi = 170$ mV, and $T = 25°C$.

Solution Since $pH = -\log[H^+]$ (Equation 2.4), the logarithmic term of Equation 15.7 can be rewritten. Equation 15.7 then becomes

$$\Delta G = 2.303 \, RT(pH_{in} - pH_{out}) + Z\mathcal{F}\Delta\psi$$

Substituting known values gives

$$\Delta G = 2.303(8.3145 \, J \cdot K^{-1} \cdot mol^{-1})(298 \, K)(7.8 - 7.15)$$
$$+ (1)(96,485 \, J \cdot V^{-1} \cdot mol^{-1})(0.170 \, V)$$
$$= 3700 \, J \cdot mol^{-1} + 16,400 \, J \cdot mol^{-1}$$
$$= +20.1 \, kJ \cdot mol^{-1}$$

Concept Check

1. Describe the importance of mitochondrial structure for generating the protonmotive force.
2. Identify the source of the protons for the transmembrane gradient.
3. Explain why the proton gradient has a chemical and an electrical component.

SUMMARY

15.1 The Thermodynamics of Oxidation–Reduction Reactions

- The electron affinity of a substance participating in an oxidation–reduction reaction, which involves the transfer of electrons, is indicated by its reduction potential, $\mathcal{E}°'$.

- The difference in reduction potential between substances undergoing oxidation and reduction is related to the free energy change for the reaction.

15.2 Mitochondrial Electron Transport

- Oxidation of reduced cofactors generated by metabolic reactions takes place in the mitochondrion. Shuttle systems and transport proteins allow the transmembrane movement of reducing equivalents, ATP, ADP, and P_i.

- The electron transport chain consists of a series of integral membrane protein complexes that contain multiple redox groups, including iron–sulfur clusters, flavins, cytochromes, and copper ions, and that are linked by mobile electron carriers. Starting from NADH, electrons travel a path of increasing reduction potential through Complex I, ubiquinone, Complex III, cytochrome c, and then to Complex IV, where O_2 is reduced to H_2O.

- As electrons are transferred, protons are translocated to the intermembrane space via proton wires in Complexes I and IV and by the action of the Q cycle associated with Complex III.

15.3 ATP Synthase

- The energy of the proton gradient is tapped as protons spontaneously flow through ATP synthase. Proton transport allows rotation of a ring of integral membrane c subunits. The linked γ subunit thereby rotates, triggering conformational changes in the F_1 portion of ATP synthase.

- According to the binding change mechanism, the three functional units of the F_1 portion cycle through three conformational states to sequentially bind ADP and P_i, convert the substrates to ATP, and release ATP.

- The P:O ratio quantifies the link between electron transport and oxidative phosphorylation in terms of the ATP synthesized and the O_2 reduced. Because these processes are coupled, the rate of oxidative phosphorylation is controlled primarily by the availability of reduced cofactors from fuel metabolism.

15.4 Chemiosmosis

- The chemiosmotic theory describes how proton translocation during mitochondrial electron transport generates an electrochemical gradient whose free energy drives ATP synthesis.

KEY TERMS

oxidative phosphorylation

oxidizing agent (oxidant)

reducing agent (reductant)

half-reaction

standard reduction potential ($\mathcal{E}^{\circ\prime}$)

reduction potential ($\mathcal{E}$)

Nernst equation

Faraday constant ($\mathcal{F}$)

cellular respiration

electron transport chain

mitochondrion

mitochondrial matrix

intermembrane space

cristae

electron tomography

redox center

proton wire

cytochrome

Q cycle

supercomplex

respirasome

reactive oxygen species

free radical

binding change mechanism

uncoupler

thermogenesis

P:O ratio

chemiosmotic theory

protonmotive force

Z

BIOINFORMATICS

Brief Bioinformatics Exercises

15.1 Viewing and Analyzing Complexes I–IV

15.2 Diversity of the Electron-Transport Chain and the KEGG Database

PROBLEMS

15.1 The Thermodynamics of Oxidation–Reduction Reactions

1. Calculate the standard free energy change for the reduction of a. Pyru vate by NADH. b. Oxygen by cytochrome a_3. Consult Table 15.1 for the relevant half-reactions. Are these reactions spontaneous under standard conditions?

2. Identify the oxidized and reduced forms from the following pairs: a. malate/oxaloacetate; b. pyruvate/lactate; c. NADH/NAD$^+$; d. fumarate/succinate; e. Fe(CN)$_6^{3-}$/Fe(CN)$_6^{2-}$; f. H$_2$O$_2$/O$_2$; g. FMN/FMNH$_2$ (see Fig 15.8).

3. Identify the oxidized and reduced forms from the following pairs: a. O$_2$/H$_2$O, b. NO$_2$/NO$_3^-$, c. dihydrolipoic acid/lipoic acid, d. NADP$^+$/NADPH.

4. Identify the oxidized and reduced forms from the following pairs: a. ethanol/acetate, b. glucose-6-phosphate/6-phosphogluconate, c. 1,3-bisphosphoglycerate/glyceraldehyde-3-phosphate, d. ubiquinone/ubiquinol.

5. Calculate the reduction potential of NADP$^+$ at 25°C when [NADP$^+$] = 50 µM and [NADPH] = 500 µM.

6. Calculate the reduction potential of FAD at 25°C when the concentration of FAD is 1 µM and the concentration of FADH$_2$ is 14 µM.

7. Calculate the reduction potential of fumarate at 37°C when [fumarate] = 80 µM and [succinate] = 100 µM.

8. Calculate the reduction potential of ubiquinone (Q) at 37°C when the concentration of Q is 40 µM and the concentration of ubiquinol (QH$_2$) is 10 µM.

9. a. Calculate the standard reduction potential of substance A when $\mathcal{E}$ = 0.49 V at 25°C, [A$_{reduced}$] = 10 µM, and [A$_{oxidized}$] = 250 µM. Assume that n = 2. b. Calculate the standard reduction potential of substance B when $\mathcal{E}$ = −0.22 V at 25°C; [B$_{reduced}$] = 60 µM, and [B$_{oxidized}$] = 3 µM. Assume that n = 2.

10. a. Calculate the standard reduction potential of substance C when $\mathcal{E}$ = 0.29 V at 25°C, [C$_{reduced}$] = 5 µM, and [C$_{oxidized}$] = 0.1 µM. Assume that n = 1. b. Consult Table 15.1. What is a possible identity of substance C?

11. a. Calculate the standard reduction potential of substance D when $\mathcal{E}$ = −0.334 V at 25°C, [D$_{reduced}$] = 0.1 mM, and [D$_{oxidized}$] = 0.3 mM. Assume that n = 2. b. Consult Table 15.1. What is a possible identity of substance D?

12. Calculate the standard free energy change for the reduction of pyruvate by NADH. Consult Table 15.1 for the relevant half-reactions. Is this reaction spontaneous under standard conditions?

13. Calculate the standard free energy change for the oxidation of FADH$_2$ by oxygen. Consult Table 15.1 for the relevant half-reactions. Is this reaction spontaneous under standard conditions?

14. Calculate the standard free energy change for the reduction of acetoacetate by NADH. Consult Table 15.1 for the relevant half-reactions. Is this reaction spontaneous under standard conditions?

15. Calculate the standard free energy change for the reduction of cytochrome c by cytochrome c_1. Consult Table 15.1 for the relevant half-reactions. Is this reaction spontaneous under standard conditions?

16. Calculate the standard free energy change for the reduction of cytochrome a_3 by cytochrome a. Consult Table 15.1 for the relevant half-reactions. Is this reaction spontaneous under standard conditions?

17. Calculate the standard free energy change for the reduction of oxygen by cytochrome a_3. Consult Table 15.1 for the relevant half-reactions. Is this reaction spontaneous under standard conditions?

18. Calculate the standard free energy change for the oxidation of malate by ubiquinone. Is the reaction spontaneous under standard conditions?

19. In one of the final steps of the pyruvate dehydrogenase reaction (see Section 14.1), E3 reoxidizes the lipoamide group of E2, then NAD$^+$ reoxidizes E3. Calculate $\Delta G^{\circ\prime}$ for the electron transfer from dihydrolipoic acid to NAD$^+$.

20. Each electron from cytochrome c is donated to a Cu$_A$ redox center in Complex IV. The $\mathcal{E}^{\circ\prime}$ value for the Cu$_A$ redox center is 0.245 V. Calculate $\Delta G^{\circ\prime}$ for this electron transfer.

21. Acetaldehyde may be oxidized to acetate. Would NAD$^+$ be an effective oxidizing agent? Explain, supporting your answer with appropriate calculations.

22. Acetoacetate may be reduced to 3-hydroxybutyrate. What serves as a better reducing agent, **a.** NADH or **b.** $FADH_2$? Explain, supporting your answer with appropriate calculations.

23. For every two QH_2 that enter the Q cycle, one is regenerated and the other passes its two electrons to two cytochrome c_1 centers. The overall equation is

$$QH_2 + 2 \text{ cyt } c_1 \text{ (Fe}^{3+}) \rightarrow 2H^+ \rightarrow Q + 2 \text{ cyt } c_1 \text{ (Fe}^{2+}) + 4H^+$$

Calculate the free energy change associated with the Q cycle.

24. Why is succinate oxidized by FAD instead of by NAD^+? Explain, supporting your answer with the appropriate calculations.

25. **a.** What is the $\Delta \mathcal{E}$ value for the oxidation of ubiquinol (QH_2) by cytochrome c at 25°C when the ratio of QH_2/Q is 10 and the ratio of cyt c (Fe^{3+})/cyt c (Fe^{2+}) is 5? **b.** Calculate ΔG for the reaction described in part **a.** Is the reaction spontaneous under these conditions?

26. An iron–sulfur protein in Complex III donates an electron to cytochrome c_1. The reduction half-reactions and $\mathcal{E}^{\circ\prime}$ values are shown below. Write the balanced equation for the reaction and calculate the standard free energy change. How can you account for the fact that this reaction occurs spontaneously in the cell?

$$\text{FeS } (ox) + e^- \rightarrow \text{FeS } (red) \qquad \mathcal{E}^{\circ\prime} = 0.280 \text{ V}$$
$$\text{cyt } c_1 \text{ (Fe}^{3+}) + e^- \rightarrow \text{cyt } c_1 \text{ (Fe}^{2+}) \quad \mathcal{E}^{\circ\prime} = 0.215 \text{ V}$$

27. If the ΔG value for the reaction described in Problem 26 is -6.0 kJ · mol^{-1}, what is the value of $\Delta \mathcal{E}$?

15.2 Mitochondrial Electron Transport

28. Calculate the overall efficiency of oxidative phosphorylation, assuming standard conditions, by comparing the **a.** free energy potentially available from the oxidation of NADH by O_2 and **b.** the free energy required to synthesize 2.5 ATP from 2.5 ADP.

29. **a.** Write the equation for the overall reaction that occurs in Complex I. **b.** Calculate $\Delta G^{\circ\prime}$ for the reaction. **c.** Using the percent efficiency calculated in Problem 28, calculate the number of ATP generated by Complex I.

30. **a.** Write the equation for the overall reaction that occurs in Complex II. **b.** Calculate $\Delta G^{\circ\prime}$ for the reaction. **c.** Using the percent efficiency calculated in Problem 28, calculate the number of ATP generated by Complex II.

31. **a.** Write the equation for the overall reaction that occurs in Complex III. **b.** Calculate $\Delta G^{\circ\prime}$ for the reaction. **c.** Using the percent efficiency calculated in Problem 28, calculate the number of ATP generated by Complex III.

32. **a.** Write the equation for the overall reaction that occurs in Complex IV. **b.** Calculate $\Delta G^{\circ\prime}$ for the reaction. **c.** Using the percent efficiency calculated in Problem 28, calculate the number of ATP generated by Complex IV.

33. Use your answers to Problems 29–32 to determine the amount of ATP generated when **a.** NADH and **b.** $FADH_2$ are oxidized in the electron transport chain. **c.** Do these values agree with those presented in Figure 14.13?

34. The sequence of events in electron transport was elucidated in part by the use of inhibitors that block electron transfer at specific points along the chain. For example, adding rotenone (a plant toxin) or amytal (a barbiturate) blocks electron transport in Complex I; antimycin A (an antibiotic) blocks electron transport in Complex III; and cyanide (CN$^-$) blocks electron transport in Complex IV by binding to the Fe^{2+} in the Fe–Cu binuclear center of cytochrome a_3. What happens to oxygen consumption when these inhibitors are added to a suspension of respiring mitochondria?

35. What is the redox state of the electron carriers in the electron transport chain when each of the inhibitors described in Problem 34 is added separately to a mitochondrial suspension?

36. **a.** The investigators who first studied the effect of rotenone on electron transport (see Problem 34) noted that the inhibitor blocked the oxidation of malate but not succinate. How did these results assist in the identification of rotenone as a Complex I inhibitor? **b.** If the same experiments described in part **a** are carried out in the presence of antimycin A, what are the expected results?

37. When the antifungal agent myxothiazol is added to a suspension of respiring mitochondria, the QH_2/Q ratio increases. Where in the electron transport chain does myxothiazol inhibit electron transfer?

38. What is the effect of added succinate on rotenone-blocked, antimycin A–blocked, or cyanide-blocked mitochondria (see Problem 34)? In other words, can succinate help "bypass" any of these blocks? Explain.

39. **a.** The compound tetramethyl-p-phenylenediamine (TMPD) donates a pair of electrons directly to Complex IV. Can TMPD act as a bypass for the rotenone-blocked, antimycin-blocked, or cyanide-blocked mitochondria described in Problem 34? **b.** Ascorbate (vitamin C) can donate a pair of electrons to cytochrome c. Can ascorbate act as a bypass for the rotenone-blocked, antimycin-blocked, or cyanide-blocked mitochondria described in Problem 34?

40. If cyanide poisoning (see Problem 34) is diagnosed immediately, it can be treated by administering nitrites that can oxidize the Fe^{2+} in hemoglobin to Fe^{3+}. Why is this treatment effective?

41. The effect of the drug fluoxetine (Prozac®; Box 9.B) on isolated rat brain mitochondria was examined by measuring the rate of electron transport (units not given) in the presence of various combinations of substrates and inhibitors (see Problems 34–37). **a.** How do pyruvate, malate, and succinate serve as substrates for electron transport? **b.** What is the effect of fluoxetine on electron transport?

	Rate of electron transport		
[Fluoxetine] (mM)	**pyruvate + malate**	**succinate + rotenone**	**ascorbate + TMPD**
0	200	150	240
0.25	60	145	150

42. Complex I, succinate dehydrogenase, acyl-CoA dehydrogenase, and glycerol-3-phosphate dehydrogenase (see Fig. 15.11) are all flavoproteins; that is, they contain an FMN or FAD prosthetic group. Explain the function of the flavin group in these enzymes. Why are the flavoproteins ideally suited to transfer electrons to ubiquinone?

43. What side chains would you expect to find as part of a proton wire in a proton-translocating membrane protein?

44. Ubiquinone is not anchored in the mitochondrial membrane but is free to diffuse laterally throughout the membrane among the electron transport chain components. What aspects of its structure account for this behavior?

45. Explain why the ubiquinone-binding site of Complex I (Fig. 15.7) is located at the end of the peripheral arm closest to the membrane.

46. Cytochrome c is easily dissociated from isolated mitochondrial membrane preparations, but the isolation of cytochrome c_1 requires the use of strong detergents. Explain why.

47. Release of cytochrome c from the mitochondrion to the cytosol is one of the signals that induces apoptosis, a form of programmed cell death. What structural features of cytochrome c allow it to play this role?

48. In coastal marine environments, high concentrations of nutrients from terrestrial runoff may lead to algal blooms. When the nutrients are depleted, the algae die and sink and are degraded by other microorganisms. The algal die-off may be followed by a sharp drop in oxygen in the depths, which can kill fish and bottom-dwelling invertebrates. Why do these "dead zones" form?

49. Chromium is most toxic and highly soluble in its oxidized Cr(VI) state but is less toxic and less soluble in its more reduced Cr(III) state. Efforts to detoxify Cr-contaminated groundwater involve injecting chemical reducing agents underground. Another approach is bioremediation, which involves injecting molasses or cooking oil into the contaminated groundwater. Explain how these substances would promote the reduction of Cr(VI) to Cr(III).

50. At one time, it was believed that myoglobin functioned simply as an oxygen-storage protein. New evidence suggests that myoglobin plays a much more active role in the muscle cell. The phrase *myoglobin-facilitated oxygen diffusion* describes myoglobin's role in transporting oxygen from the muscle cell sarcolemma to the mitochondrial membrane surface. Mice in which the myoglobin gene was knocked out had higher tissue capillary density, elevated red blood cell counts, and increased coronary blood flow. Explain the reasons for these compensatory mechanisms in the knockout mice.

51. The myoglobin and cytochrome *c* oxidase content were determined in several animals, as shown in the table. What is the relationship between the two proteins? Explain.

	Myoglobin content, mmol · kg^{-1}	Cytochrome *c* oxidase activity
Hare	0.1	900
Sheep	0.19	950
Ox	0.31	1200
Horse	0.38	1800

52. Myoglobin has an affinity for membrane phospholipids. **a.** Which amino acid side chains are expected to interact favorably with phospholipid head groups? **b.** An experiment showed that the p_{50} value of myoglobin increased when myoglobin was incubated with a suspension of mitochondria. What is the physiological benefit of the p_{50} increase that occurs when myoglobin binds to phospholipids?

53. Myoglobin is not confined to muscle cells. Tumor cells, which generally exist in hypoxic (low oxygen) conditions because of limited blood flow, express myoglobin. How does this adaptation increase the chances of tumor cell survival?

54. Cancer cells, even when sufficient oxygen is available, produce large amounts of lactate. It has been observed that the concentration of fructose-2,6-bisphosphate is much higher in cancer cells than in normal cells. Why would this result in anaerobic metabolism being favored, even when oxygen is available?

55. Hawkmoths are nectar-feeding insects with a high rate of aerobic metabolism. Some of the sugar they ingest is not catabolized by glycolysis to make ATP but instead is directed into the pentose phosphate pathway. Explain why this metabolic diversion helps the insect avoid damage from reactive oxygen species generated during electron transport.

56. A group of elderly patients who did not exercise regularly were asked to participate in a 12-week exercise program. Data collected from the patients are shown in the table. What was the result of the exercise intervention and why did this occur?

	Prior to exercise intervention	Post 12-week exercise intervention
Total mitochondrial DNA (copies per diploid genome)	1300	1900
Complex II activity	0.13	0.20
Complex I–IV activity	0.51	1.00

57. Amyotrophic lateral sclerosis (ALS) is a neurodegenerative disease that causes muscle paralysis and eventually death. Researchers measured the activity of the electron transport chain complexes in various regions of the nervous system in patients with ALS. In a certain region of the spinal cord, Complex I showed decreased activity but not decreased concentration. How does this contribute to progression of the disease?

15.3 ATP Synthase

58. In experimental systems, the F_0 component of ATP synthase can be reconstituted into a membrane. F_0 can then act as a proton channel that is blocked when the F_1 component is added to the system. What molecule must be added to the system in order to restore the proton-translocating activity of F_0? Explain.

59. Oligomycin is an antibiotic that blocks proton transfer through the F_0 proton channel of ATP synthase. What is the effect on **a.** ATP synthesis, **b.** electron transport, and **c.** oxygen consumption when oligomycin is added to a suspension of respiring mitochondria?

60. Dicyclohexylcarbodiimide (DCCD) is a reagent that reacts with Asp or Glu residues. Explain why the reaction of DCCD with just one *c* subunit completely blocks both the ATP-synthesizing and ATP-hydrolyzing activity of ATP synthase.

61. In the 1950s, experiments with isolated mitochondria showed that organic compounds are oxidized and O_2 is consumed only when ADP is included in the preparation. When the ADP supply runs out, oxygen consumption halts. Explain these results.

62. a. Calculate the ratio of protons translocated to ATP synthesized for yeast ATP synthase, which has 10 *c* subunits, and for spinach chloroplast ATP synthase, which has 14 *c* subunits. **b.** Would you expect the yeast or the chloroplast to have a higher P:O ratio?

63. The ATP synthase from bovine heart mitochondria has 8 *c* subunits. What is the P:O ratio for NADH?

64. Experiments indicate that the *c* ring of ATP synthase spins at a rate of 6000 rpm. How many ATP molecules are generated each second?

65. The side chain of a glutamate residue normally has a pK value of about 5.0 in solution. The pH of the cytosol is about 7.0. **a.** Does this information explain how a Glu side chain on a *c* subunit of ATP synthase could acquire a cytosolic proton and then release it into the matrix (where the pH is about 8.0)? **b.** Within a few angstroms of the Glu side chain are two additional Glu side chains. Explain how these two residues could promote protonation of the first Glu residue.

66. How much ATP can be obtained by the cell from the complete oxidation of one mole of glucose? Compare this value with the amount of ATP obtained when glucose is anaerobically converted to lactate or ethanol. Do organisms that can completely oxidize glucose have an advantage over organisms that cannot?

67. During anaerobic fermentation in yeast, the majority of the available glucose is oxidized via the glycolytic pathway and the rest enters the pentose phosphate pathway to generate NADPH and ribose.

This occurs during aerobic respiration as well, except that the percentage of glucose entering the pentose phosphate pathway is much greater in aerobic respiration than during anaerobic fermentation. Explain why.

68. When cells cannot carry out oxidative phosphorylation, they synthesize ATP through substrate-level phosphorylation. **a.** Which enzymes of glycolysis and the citric acid cycle catalyze substrate-level phosphorylation? **b.** Compare the amount of ATP generated per mole of glucose via substrate-level phosphorylation and oxidative phosphorylation.

69. The glycerol-3-phosphate shuttle can transport cytosolic NADH equivalents into the mitochondrial matrix (see Fig. 15.11c). In this shuttle, the protons and electrons are donated to FAD, which is reduced to $FADH_2$. These protons and electrons are subsequently donated to coenzyme Q in the electron transport chain. **a.** How much ATP is generated per mole of glucose when the glycerol-3-phosphate shuttle is used? **b.** What is the P:O ratio for cytosolic NADH?

70. The complete oxidation of glucose yields $-2850 \text{ kJ} \cdot \text{mol}^{-1}$ of free energy. Incomplete oxidation by conversion to lactate yields $-196 \text{ kJ} \cdot \text{mol}^{-1}$ and by alcoholic fermentation yields $-235 \text{ kJ} \cdot \text{mol}^{-1}$. **a.** Calculate the overall efficiencies of glucose oxidation by these three processes. **b.** Do organisms that can completely oxidize glucose have an advantage over organisms that cannot?

71. A culture of yeast grown under anaerobic conditions is exposed to oxygen, resulting in a dramatic decrease in glucose consumption by the cells. This phenomenon is referred to as the Pasteur effect. **a.** Explain the Pasteur effect. **b.** The [NADH]/[NAD+] and [ATP]/[ADP] ratios also change when an anaerobic culture is exposed to oxygen. Explain how these ratios change and what effect this has on glycolysis and the citric acid cycle in the yeast.

72. Experiments in the late 1970s attributed the Pasteur effect (see Problem 71) to the stimulation of hexokinase and phosphofructokinase under anaerobic conditions. Upon exposure to oxygen, the stimulation of these enzymes ceases. Why are these enzymes more active in the absence of oxygen?

73. Consider the adenine nucleotide translocase and the $P_i - H^+$ symport protein that import ADP and P_i, the substrates for oxidative phosphorylation, into the mitochondrion (see Fig. 15.6). **a.** How does the activity of the adenine nucleotide translocase affect the electrochemical gradient across the mitochondrial membrane? **b.** How does the activity of the $P_i - H^+$ symport protein affect the gradient? **c.** What can you conclude about the thermodynamic force that drives the two transport systems?

74. A hormone signal leads to the activation of protein kinase A (see Fig. 10.7), which phosphorylates a Ser residue in IF1. Phosphorylated IF1 cannot bind to ATP synthase. Does the hormone signal instruct the cell to increase or decrease the rate of metabolic fuel oxidation? Explain.

75. Some evidence suggests that about 2% of the time, ATP synthase translocates a K^+ ion rather than a proton into the mitochondrial matrix. Do the cytosolic concentrations of K^+ (140 mM) and H^+ (see Sample Calculation 15.3) support or contradict this claim?

76. In addition to its effects on electron transport, fluoxetine (see Problem 41) can also inhibit ATP synthase. Why might long-term use of fluoxetine be a concern?

77. Mutations that impair ATP synthase function are rare. Laboratory studies indicate that adding α-ketoglutarate boosts ATP production in ATP synthase–deficient cells, but only when aspartate is also added to the cells. Explain.

78. In yeast, pyruvate may be converted to ethanol in a two-step pathway catalyzed by pyruvate decarboxylase and alcohol dehydrogenase (see Section 13.1). Pyruvate may also be converted to acetyl-CoA by pyruvate dehydrogenase. Yeast mutants in which the pyruvate decarboxylase gene is missing (*pdc*–) are useful for studying the regulation of the pyruvate dehydrogenase enzyme. When wildtype yeast were pulsed with glucose, glycolytic flux increased dramatically and the rate of respiration increased. But when the same experiments were performed with *pdc*– mutants, only a small increase in glycolysis was observed, and pyruvate was the main product excreted by the yeast cells. Explain these results.

79. Ascorbate (vitamin C) can donate a pair of electrons to cytochrome *c*. What is the P:O ratio for ascorbate?

80. Consider the ADP/ATP carrier and the P_i–H^+ symport protein that import ADP and P_i, the substrates for oxidative phosphorylation, into the mitochondrion (see Fig. 15.6). **a.** How does the activity of the ADP/ATP carrier affect the electrochemical gradient across the mitochondrial membrane? **b.** How does the activity of the P_i–H^+ symport protein affect the gradient? **c.** What can you conclude about the thermodynamic force that drives the two transport systems?

81. The ADP/ATP carrier, which exchanges cytoplasmic ADP and mitochondrial ATP, can also function as a passive proton transporter. **a.** Would the carrier protein augment or diminish the proton-motive force? **b.** Researchers found that nucleotide transport inhibits proton transport by the carrier protein. Could this competitive effect help link the rate of oxidative phosphorylation to the cell's need for ATP?

82. The compounds atractyloside and bongkrekic acid both bind tightly to and inhibit the ADP/ATP carrier. How do these compounds affect ATP synthesis? Electron transport?

83. Dinitrophenol was introduced as a "diet pill" in the 1920s. Its use was discontinued because the side effects were fatal in some cases. What was the rationale for believing that DNP would be an effective diet aid?

84. The compound carbonylcyanide-*p*-trifluoromethoxy phenyl-hydrazone (FCCP) is an uncoupler similar to DNP. Describe how FCCP acts as an uncoupler.

FCCP

85. Dinitrophenol was introduced as a "diet pill" in the 1920s. Its use was discontinued because the side effects were fatal in some cases. What was the rationale for believing that DNP would be an effective diet aid?

86. A patient seeks treatment because her metabolic rate is twice normal and her temperature is elevated. A biopsy reveals that her muscle mitochondria are structurally unusual and not subject to normal respiratory controls. Electron transport takes place regardless of the concentration of ADP. **a.** What is the P:O ratio (compared to normal) for NADH that enters the electron transport chain in the mitochondria of this patient? **b.** Why are the patient's metabolic rate and temperature elevated? **c.** Will this patient be able to carry out strenuous exercise?

87. UCP1 is an uncoupling protein in brown fat (Box 15.B). Experiments were carried out using UCP1-knockout mice (animals missing the gene for UCP1). **a.** Oxygen consumption increased over twofold when a β_3 adrenergic agonist that stimulates UCP1 was injected into normal mice. This was not observed when the agonist was injected into the knockout mice. Explain these results. **b.** In one experiment, normal mice and UCP1-knockout mice were placed in a cold (5°C) room overnight. The normal mice were able to maintain their body temperature at 37°C even after 24 hours in the cold. However, the body temperatures of the cold-exposed knockout mice decreased by 10°C or more. Explain.

88. The Eastern skunk cabbage can maintain its temperature 15–35°C higher than ambient temperature during the months of February and March, when ambient temperatures range from −15 to +15°C. Thermogenesis in the skunk cabbage is critical to the survival of the plant since the spadix (a flower component) is not frost-resistant. An uncoupling protein is responsible for the observed thermogenesis. **a.** The spadix relies on the skunk cabbage's massive root system, which stores appreciable quantities of starch. Why is a large quantity of starch required for the skunk cabbage to carry out sustained thermogenesis for weeks rather than hours? **b.** Oxygen consumption by the skunk cabbage increases as the temperature decreases, nearly doubling with every 10°C drop in ambient temperature. Oxygen consumption was observed to decrease during the day, when temperatures were close to 30°C, and increase at night. What is the biochemical explanation for these observations?

89. The gene that codes for an uncoupling protein (see Box 15.B) in potatoes was isolated. The results of a Northern blot analysis (which detects mRNA) are shown below. What is your interpretation of these results? How does the mRNA level affect thermogenesis in the potato?

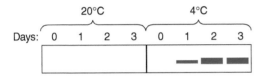

90. Glutamate can be used as an artificial substrate for mitochondrial respiration, as shown in the diagram. When ceramide is added to a mitochondrial suspension respiring in the presence of glutamate, respiration decreases, leading scientists to hypothesize that ceramide might regulate mitochondrial function *in vivo*.

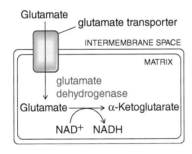

a. How does glutamate act as a substrate for mitochondrial respiration? **b.** Ceramide-induced inhibition of respiration could be due to several different factors. List several possibilities. **c.** Mitochondria treated with ceramide were exposed to an uncoupler, but the respiration rate did not increase. What site(s) of inhibition can be ruled out? **d.** In another experiment, mitochondria were subjected to a freeze–thaw cycle that rendered the inner mitochondrial membrane permeable to NADH. NADH could then be added to a mitochondrial suspension as a substrate for electron transport. When NADH was added, ceramide decreased the respiration rate to the same extent as when glutamate was the substrate. What site(s) of inhibition can be ruled out?

91. Rapidly growing bacterial cells tend to rely on glycolysis and fermentation rather than oxidative phosphorylation to generate ATP, even when O_2 is abundant. One group of researchers noted that glycolysis/fermentation requires fewer proteins than oxidative phosphorylation. Could this observation explain why rapidly growing cells prefer glycolysis over more efficient oxidative phosphorylation?

15.4 Chemiosmosis

92. a. How many protons are transferred from the matrix to the intermembrane space during the transport of two electrons through each of the complexes? **b.** What is the free energy change associated with the passage of a pair of electrons from NADH to O_2 if the passage of one proton from the matrix to the intermembrane space costs $20.1 \text{ kJ} \cdot \text{mol}^{-1}$, as shown in Sample Calculation 15.3? **c.** Compare your answer in part **b** to your answer to Problem 28a and comment on the significance of the comparison.

93. a. What is the free energy change associated with the passage of a pair of electrons from $FADH_2$ to O_2 if the passage of one proton from the matrix to the intermembrane space costs $20.1 \text{ kJ} \cdot \text{mol}^{-1}$ as shown in Sample Calculation 15.3? **b.** Compare your answer in part **a** to your answer to Problem 14 and comment on the significance of the comparison.

94. What is the free energy change for generating the electrical imbalance of protons: **a.** In neuroblastoma cells, where $\Delta\psi$ is 81 mV? **b.** In respiring mitochondria in culture, where $\Delta\psi$ is 150 mV?

95. Calculate the free energy change for translocating a proton out of the mitochondrial matrix, where $pH_{matrix} = 7.6$, $pH_{cytosol} = 7.2$, $\Delta\psi = 200$ mV, and $T = 37°C$.

96. What size of pH gradient (the difference between pH_{matrix} and $pH_{cytosol}$) would correspond to a free energy change of $30.5 \text{ kJ} \cdot \text{mol}^{-1}$? Assume that $\Delta\psi = 170$ mV and $T = 25°C$.

97. What size of pH gradient (the difference between pH_{matrix} and $pH_{cytosol}$) would correspond to a free energy change of $29.2 \text{ kJ} \cdot \text{mol}^{-1}$? Assume that $\Delta\psi = 170$ mV and $T = 37°C$.

98. Several key experimental observations were important in the development of the chemiosmotic theory. Explain how each of these observations is consistent with the chemiosmotic theory as described by Peter Mitchell. **a.** The pH of the intermembrane space is lower than the pH of the mitochondrial matrix. **b.** Oxidative phosphorylation does not occur in mitochondrial preparations to which detergents have been added.

99. Mitchell's original chemiosmotic hypothesis relies on the impermeability of the inner mitochondrial membrane to ions other than H^+, such as Na^+ and Cl^-. **a.** Why was this thought to be important? **b.** Could ATP still be synthesized if the membrane were permeable to other ions?

100. Nigericin is an antibiotic that integrates into membranes and functions as a K^+/H^+ antiporter. Another antibiotic, valinomycin, is similar, but it allows the passage of K^+ ions. When both antibiotics are added simultaneously to suspensions of respiring mitochondria, the electrochemical gradient completely collapses. **a.** Draw a diagram of a mitochondrion in which nigericin and valinomycin have integrated into the inner mitochondrial membrane, in a manner that is consistent with the experimental results. **b.** Explain why the electrochemical gradient dissipates. What happens to ATP synthesis?

101. How does transport of inorganic phosphate from the intermembrane space to the mitochondrial matrix affect the pH difference across the inner mitochondrial membrane?

102. Pioglitazone, a drug used to treat diabetes, causes some membrane-embedded portions of mitochondrial Complex I to separate from the rest of the protein that includes the matrix "arm." Predict the effect of pioglitazone on electron transport and ATP production.

103. Metformin, another diabetes drug, suppresses the activity of mitochondrial glycerol-3-phosphate dehydrogenase. Predict the effect of metformin on electron transport and ATP production.

SELECTED READINGS

Ishigami, I., Lewis-Ballester, A., Echelmeier, A., Brehm, G., Zatsepin, N.A., Grant, T.D., Coe, J.D., Lisova, S., Nelson, G., Zhang, S., Dobson, Z.F., Boutet, S., Sierra, R.G., Batyuk, A., Fromme, P., Fromme, R., Spence, J.C. H., Ros, A., Yeh, S-R., and Rousseau, D.L., Snapshot of an oxygen intermediate in the catalytic reaction of cytochrome *c* oxidase, *Proc. Natl. Acad. Sci. USA* **116**, 3572–3577, doi: 10.1073/pnas.1814526116 (2019). [Provides structural data in support of a mechanism in which O_2 reduction is coupled to proton transport in Complex IV.]

Milenkovic, D., Blaza, J.N., Larsson, N.-G., and Hirst, J., The enigma of the respiratory chain supercomplex, *Cell Metabolism* **25**, 765–776, doi: 10.1016/j.cmet.2017.03.009 (2017). [Describes the supercomplexes and discusses possible reasons for their existence.]

Parey, K., Brandt, U., Xie, H., Mills, D.J., Siegmund, K., Vonck, J., Kühlbrandt, W., and Zickermann, V., Cryo-EM structure of respiratory complex I at work, *eLife* **7**, doi: 10.7554/eLife.39213 (2018). [Presents detailed views of Complex I.]

Shinzawa-Itoh, K., Sugimura, T., Misaki, T., Tadehara, Y., Yamamoto, S., Hanada, M., Yano, N., Nakagawa, T., Uene, S., Yamada, T., Aoyama, H., Yamashita, E., Tsukihara, T., Yoshikawa, S., and Muramoto, K., Monomeric structure of an active form of bovine cytochrome c oxidase, *Proc. Natl. Acad. Sci. USA* **116**, 19945–19951, doi: 10.1073/pnas.1907183116 (2019). [Describes the structure and functional significance of the cytochrome c oxidase (Complex IV) monomer.]

Srivastava, A.P., Luo, M., Zhou, W., Symersky, J., Bai, D., Chambers, M.G., Faraldo-Gómez, J.D., Liao, M., and Mueller, D.M., High-resolution cryo-EM analysis of the yeast ATP synthase in a lipid membrane, *Science* **360**, doi: 10.1126/science.aas9699 (2018). [Presents the structure of ATP synthase and explains possible mechanisms for proton translocation.]

Youle, R.J., Mitochondria—Striking a balance between host and endosymbiont, *Science* **365**, 655, doi: 10.1126/science.aaw9855 (2019). [Outlines the challenges for eukaryotic cells to mange mitochondrial gene expression and avoid mitochondrial dysfunction.]

CHAPTER 15 CREDITS

Table 15.1 Data mostly from Loach, P.A., in Fasman, G.D. (ed.), *Handbook of Biochemistry and Molecular Biology* (3rd ed.), Physical and Chemical Data, Vol. I, pp. 123–130, CRC Press (1976).

Figure 15.7 Image based on 4HEA. Baradaran, R., Berrisford, J.M., Minhas, G.S., Sazanov, L.A., Crystal structure of the entire respiratory complex I, *Nature* **494**, 443–448 (2013).

Figure 15.13 Image based on 1BE3. Iwata, S., Lee, J.W., Okada, K., Lee, J.K., Iwata, M., Rasmussen, B., Link, T.A., Ramaswamy, S., Jap, B.K., Complete structure of the 11-subunit bovine mitochondrial cytochrome bc1 complex, *Science* **281**, 64–71 (1998).

Figure 15.16 Image based on 5CYT. Takano, T., Refinement of myoglobin and cytochrome c, in *Methods and Applications in Crystallographic Computing*, S.R. Hall and T. Shida, ed., 262 (1984).

Figure 15.17 Image based on 2OCC. Yoshikawa, S., Shinzawa-Itoh, K., Nakashima, R., Yaono, R., Yamashita, E., Inoue, N., Yao, M., Fei, M.J., Libeu, C.P., Mizushima, T., Yamaguchi, H., Tomizaki, T., Tsukihara, T., Redox-coupled crystal structural changes in bovine heart cytochrome c oxidase, *Science* **280**, 1723–1729 (1998).

Figure 15.20 Image based on 5GPN. Gu, J., Wu, M., Guo, R., Yan, K., Lei, J., Gao, N., Yang, M., The architecture of the mammalian respirasome, *Nature* **537**, 639–643 (2016).

Figure 15.22b Image based on 6CP6. Srivastava, A.P., Luo, M., Zhou, W., Symersky, J., Bai, D., Chambers, M.G., Faraldo-Gomez, J.D., Liao, M., Mueller, D.M., High-resolution cryo-EM analysis of the yeast ATP synthase in a lipid membrane, *Science* **360**, 619 (2018).

Figures 15.24 and 15.26 Images based on 4TSF. Bason, J.V., Montgomery, M.G., Leslie, A.G., Walker, J.E., Pathway of binding of the intrinsically disordered mitochondrial inhibitor protein to F1-ATPase, *Proc. Natl. Acad. Sci. USA* **111**, 11305 (2014).

Box 15.C Figure adapted from McArdle, W.D., Katch, F.I., and Katch, V.L., *Exercise Physiology* (2nd ed.), p. 348, Lea & Febiger (1986).

Figure 15.27 Image based on 6J5K. Gu, J., Zhang, L., Zong, S., Guo, R., Liu, T., Yi, J., Wang, P., Zhuo, W., Yang, M., Cryo-EM structure of the mammalian ATP synthase tetramer bound with inhibitory protein IF1, *Science* **364**, 1068–1075 (2019).

Photosynthesis

Cells on the surface of the petals of the California poppy synthesize extra cellulose so that their outer cell walls form thick, triangular ridges. These ridges act like prisms to focus light onto pigments at the base of the cell, producing intense color with a silk-like reflectiveness that attracts pollinators.

Do You Remember?

- Glucose polymers include the fuel-storage polysaccharides starch and glycogen and the structural polysaccharide cellulose (Section 11.2).
- Coenzymes such as NAD^+ and ubiquinone collect electrons from compounds that become oxidized (Section 12.3).
- Electrons are transferred from a substance with a lower reduction potential to a substance with a higher reduction potential (Section 15.1).
- The formation of a transmembrane proton gradient during electron transport provides the free energy to synthesize ATP (Section 15.4).

Every year, plants and bacteria convert an estimated 6×10^{16} grams of carbon to organic compounds by photosynthesis. About half of this activity occurs in forests and savannas, and the rest occurs in the ocean and under ice—wherever water, carbon dioxide, and light are available. The organic materials produced by photosynthetic organisms sustain them as well as the organisms that feed on them. We will begin by examining the absorption of light energy and then look at the electron transport complexes that convert solar energy to biologically useful forms of energy such as ATP and the reduced cofactor NADPH. Finally, we will see how plants use these energy currencies to synthesize carbohydrates.

16.1 Photosynthesis: An Overview

KEY CONCEPTS

Describe the structure and purpose of pigment molecules.

- Relate a pigment's color to the energy of the light it absorbs.
- List the ways that absorbed energy can be dissipated.
- Explain how absorbed light energy is transferred to the reaction center.

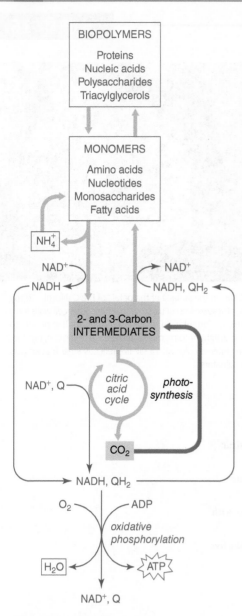

Figure 16.1 Photosynthesis in context. Photosynthetic organisms incorporate atmospheric CO_2 into three-carbon compounds that are the precursors of biological molecules such as carbohydrates and amino acids. Photosynthesis requires light energy to drive the production of the ATP and NADPH consumed in the biosynthetic reactions.

Chloroplasts and solar energy

The ability to use sunlight as an energy source evolved about 3.5 billion years ago. Before that, cellular metabolism probably centered around the inorganic reductive reactions associated with hydrothermal vents. The first **photosynthetic** organisms produced various pigments (light-absorbing molecules) to capture solar energy and thereby drive the reduction of metabolites. The descendants of some of these organisms are known today as purple bacteria and green sulfur bacteria. By about 2.5 billion years ago, the cyanobacteria had evolved. These organisms absorb enough solar energy to undertake the energetically costly oxidation of water to molecular oxygen. In fact, the dramatic increase in the level of atmospheric oxygen (from an estimated 1% to the current level of about 20%) around 2.1 to 2.4 billion years ago is attributed to the rise of cyanobacteria. Modern plants are the result of the symbiosis of early eukaryotic cells with cyanobacteria.

Although the apparatus and reactions of photosynthesis are not found in all organisms, they can be placed in the context of the metabolic scheme outlined in Chapter 12 (**Fig. 16.1**). As you examine the harvest of solar energy and its use in incorporating CO_2 into three-carbon compounds, you will see that significant portions of these processes resemble metabolic pathways that you have already encountered.

Photosynthesis in green plants takes place in **chloroplasts,** discrete organelles that are descended from cyanobacteria. Like mitochondria, chloroplasts contain their own DNA, in this case coding for 100 to 200 chloroplast proteins. DNA in the cell's nucleus contains close to a thousand more genes whose products are essential for photosynthesis.

The chloroplast is enclosed by a porous outer membrane and an ion-impermeable inner membrane (**Fig. 16.2**). The inner compartment, called the **stroma,** is analogous to the mitochondrial matrix and is rich in enzymes, including those required for carbohydrate synthesis. Within the stroma is a membranous structure called the **thylakoid.** Unlike the planar or tubular mitochondrial cristae (see Fig. 15.4), the thylakoid membrane folds into stacks of flattened vesicles and encloses a compartment called the thylakoid lumen. The energy-transducing reactions of photosynthesis take place in the thylakoid membrane. The analogous reactions in photosynthetic bacteria typically take place in folded regions of the plasma membrane.

Pigments absorb light of different wavelengths

Light can be considered as both a wave and a particle, the **photon.** The energy (E) of a photon depends on its wavelength, as expressed by **Planck's law:**

$$E = \frac{hc}{\lambda} \tag{16.1}$$

where h is Planck's constant (6.626×10^{-34} J·s), c is the speed of light (2.998×10^8 m·s^{-1}), and λ is the wavelength (about 400 to 700 nm for visible light; see Sample Calculation 16.1). *This energy is absorbed by the photosynthetic apparatus of the chloroplast and transduced to chemical energy.*

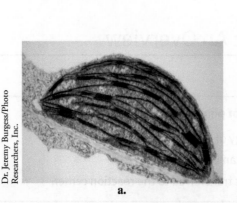

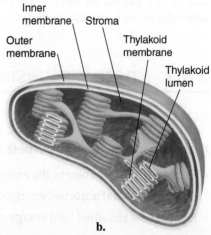

Figure 16.2 The chloroplast. **a.** Electron micrograph of a chloroplast from tobacco. **b.** Drawing of a chloroplast. The stacked thylakoid membranes are known as grana (singular, *granum*).

Question Compare these images to the images of a mitochondrion in Figure 15.4.

Dr. Jeremy Burgess/Photo Researchers, Inc.

SAMPLE CALCULATION 16.1

Problem   Calculate the energy of a photon with a wavelength of 550 nm.

Solution

$$E = \frac{hc}{\lambda}$$

$$E = \frac{(6.626 \times 10^{-34} \text{ J} \cdot \text{s}) (2.998 \times 10^{8} \text{ m} \cdot \text{s}^{-1})}{550 \times 10^{-9} \text{ m}}$$

$$E = 3.6 \times 10^{-19} \text{ J}$$

Chloroplasts contain a variety of light-absorbing groups called pigments or **photorecep-tors** (**Fig. 16.3**). Chlorophyll is the principal photoreceptor. It appears green because it absorbs both blue and red light. Chlorophyll resembles the heme groups of hemoglobin and cytochromes

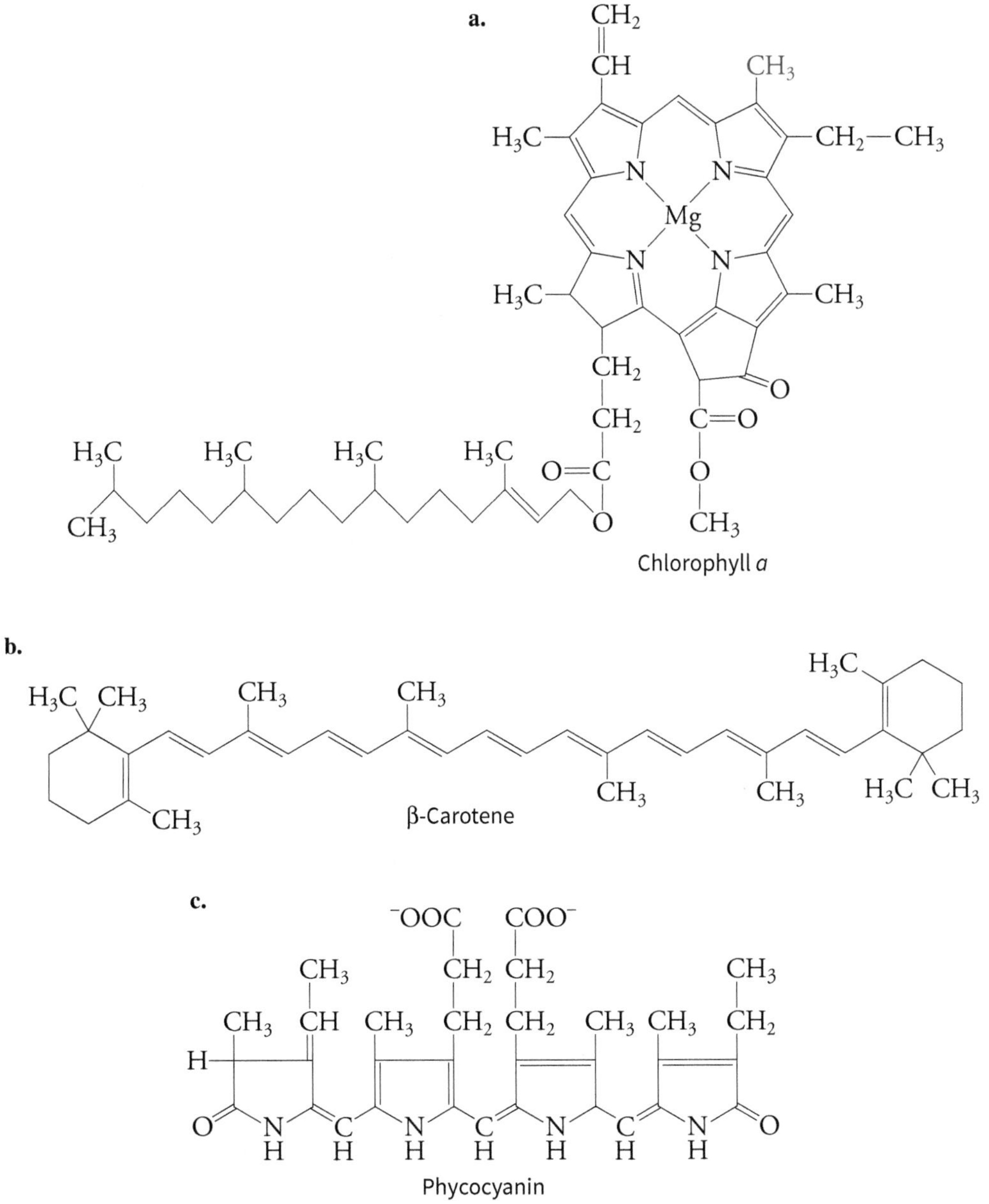

Chlorophyll *a*

β-Carotene

Phycocyanin

Figure 16.3 Some common chloroplast photoreceptors.   a. Chlorophyll *a*. In chlorophyll *b*, a methyl group (blue) is replaced by an aldehyde group. **b.** The carotenoid β-carotene, a precursor of vitamin A (see Box 8.B). **c.** Phycocyanin, a linear tetrapyrrole. It resembles an unfolded chlorophyll molecule.

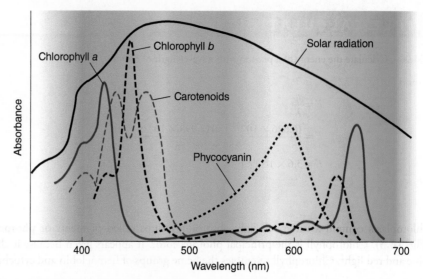

Figure 16.4 Visible light absorption by some photosynthetic pigments. The wavelengths of absorbed light correspond to the peak of the solar energy that reaches the earth.

Question **Use this diagram to explain the color of each type of pigment molecule.**

(see Fig. 15.12), but it has a central Mg^{2+} rather than an Fe^{2+} ion; it includes a fused cyclopentane ring, and it has a long lipid side chain. The second most common pigments are the red carotenoids, which absorb blue light. Pigments such as phycocyanin, which absorb longer-wavelength red light, are common in aquatic systems because water absorbs blue light. Together, these and other types of pigments absorb all the wavelengths of visible light (**Fig. 16.4**).

Each photosynthetic pigment is a highly conjugated molecule. When it absorbs a photon of the appropriate wavelength, one of its delocalized electrons is promoted to a higher-energy orbital and the molecule is said to be excited. The excited molecule can return to its low-energy, or ground, state by several mechanisms (**Fig. 16.5**):

1. The absorbed energy can be lost as heat.

2. The energy can be given off as light, or **fluorescence.** For thermodynamic reasons, the emitted photon has a lower energy (longer wavelength) than the absorbed photon.

3. The energy can be transferred to another molecule. This process is called **exciton transfer** (an exciton is the packet of transferred energy) or resonance energy transfer, since the molecular orbitals of the donor and recipient groups must be oscillating in a coordinated manner in order to transfer energy.

4. An electron from the excited molecule can be transferred to another molecule. In this process, called **photooxidation,** the excited molecule becomes oxidized and the recipient molecule becomes reduced. A different electron-transfer reaction is required to restore the photooxidized molecule to its original reduced state.

All of these energy-transferring processes occur in chloroplasts to some extent, but *exciton transfer and photooxidation are the most important for photosynthesis.*

Light-harvesting complexes transfer energy to the reaction center

The primary reactions of photosynthesis occur at specific chlorophyll molecules called **reaction centers.** However, chloroplasts contain many more chlorophyll molecules and other pigments than reaction centers. *Many of these extra, or **antenna,** pigments are located in membrane proteins*

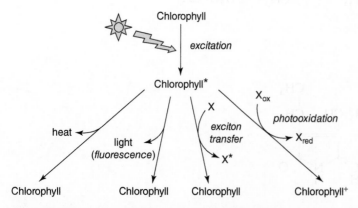

Figure 16.5 Dissipation of energy in a photoexcited molecule.
A pigment molecule such as chlorophyll is excited by absorbing a photon. The excited molecule (chlorophyll*) can return to its ground state by one of several mechanisms.

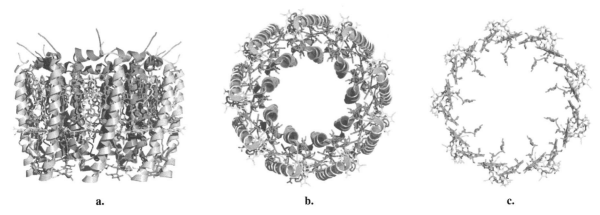

Figure 16.6 A light-harvesting complex from *Rhodopseudomonas acidophila*. The nine pairs of subunits (light and dark gray) are mostly buried in the membrane and form a scaffold for two rings of chlorophyll molecules (yellow and green) and carotenoids (red). **a.** Side view. The extracellular side is at the top. **b.** Top view. **c.** Top view showing only the chlorophyll molecules.

called **light-harvesting complexes.** Over 30 different kinds of light-harvesting complexes have been characterized, and they are remarkable for their regular geometry. For example, one light-harvesting complex in purple photosynthetic bacteria consists of 18 polypeptide chains holding two concentric rings of chlorophyll molecules, plus carotenoids (**Fig. 16.6**). This artful arrangement of light-absorbing groups is essential for the function of the light-harvesting complex. The pigments are all within a few angstroms of each other and overlap, so that excitation energy can be delocalized over the entire ring.

The protein microenvironment of each photoreceptor influences the wavelength (and therefore the energy) of the photon it can absorb (just as the cytochrome protein structure influences the reduction potential of its heme group; see Section 15.2). Consequently, *the various light-harvesting complexes with their multiple pigments can absorb light of many different wavelengths.* Within a light-harvesting complex, the precisely aligned pigment molecules can quickly transfer their energy to other pigments. *Exciton transfer eventually brings the energy to the chlorophyll at the reaction center* (**Fig. 16.7**). Without light-harvesting complexes to collect and concentrate light, the reaction center chlorophyll could collect only a small fraction of the incoming solar radiation. Even so, a leaf captures only about 1% of the available solar energy.

During periods of high light intensity, some accessory pigments may function to dissipate excess solar energy as heat so that it does not damage the photosynthetic apparatus by inappropriate photooxidation. Various pigment molecules may also act as photosensors to regulate the plant's growth rate and shape and to coordinate the plant's activities—such as germination, flowering, and dormancy—according to daily or seasonal light levels.

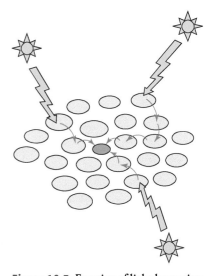

Figure 16.7 Function of light-harvesting complexes. A typical photosynthetic system consists of a reaction center (dark green) surrounded by light-harvesting complexes (light green), whose multiple pigments absorb light of different wavelengths. Exciton transfer funnels this captured solar energy to the chlorophyll at the reaction center.

Concept Check

1. Correlate a photon's energy to its wavelength.
2. Explain why it is advantageous for photosynthetic pigments to absorb different colors of light.
3. Use nonscientific terms to explain why leaves are green.
4. Describe the four mechanisms by which a photoexcited molecule can return to its ground state.
5. Explain the function of a light-harvesting complex.

16.2 The Light Reactions

KEY CONCEPTS

Trace the energy transformations that take place during the light reactions of photosynthesis.

- Recount the changes in reduction potential and free energy that occur during photooxidation.

- Describe the substrates, products, and driving force for the water-splitting reaction.
- List the order of electron carriers from H_2O to $NADP^+$.
- Describe the events of photophosphorylation.
- Compare linear and cyclic electron flow.

In plants and cyanobacteria, the energy captured by the antenna pigments of the light-harvesting complexes is funneled to two photosynthetic reaction centers. *Excitation of the reaction centers drives a series of oxidation–reduction reactions whose net results are the oxidation of water, the reduction of $NADP^+$, and the generation of a transmembrane proton gradient that powers ATP synthesis.* These events are known as the **light reactions** of photosynthesis. (Most photosynthetic bacteria undertake similar reactions but have a single reaction center and do not produce oxygen.) The two photosynthetic reaction centers that mediate light energy transduction are part of protein complexes called Photosystem I and Photosystem II. These, along with other integral and peripheral proteins of the thylakoid membrane, operate in a series, much like the mitochondrial electron transport chain.

Photosystem II is a light-activated oxidation–reduction enzyme

In plants and cyanobacteria, the light reactions begin with Photosystem II (the number indicates that it was the second to be discovered). This integral membrane protein complex is dimeric, with more bulk on the lumenal side of the thylakoid membrane than on the stromal side. The cyanobacterial Photosystem II monomer contains at least 19 subunits (14 of them integral membrane proteins). Its numerous prosthetic groups include light-absorbing pigments and redox-active cofactors (**Fig. 16.8**).

Plant Photosystem II contains additional proteins but has the same core subunits and the same overall structure as the cyanobacterial system. In the thylakoid membrane, Photosystem II is surrounded by light-harvesting complexes to form a supercomplex (**Fig. 16.9**). The fully assembled supercomplex houses about 400 pigment molecules, mostly chlorophylls and carotenoids. Some light-harvesting proteins are mobile within the membrane and may dissociate from Photosystem II as light conditions change.

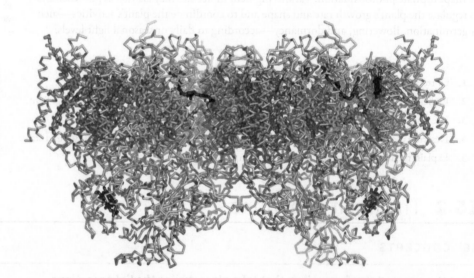

Figure 16.8 Structure of cyanobacterial Photosystem II. The proteins are shown as gray ribbons, and the various prosthetic groups are shown as stick models and color-coded: chlorophyll, green; pheophytin, orange; β-carotene, red; heme, purple; and quinone, blue. The stroma is at the top and the thylakoid lumen at the bottom.

Question Add lines to show the location of the lipid bilayer.

Several dozen chlorophyll molecules in the core of Photosystem II function as internal antennas, funneling energy to the two reaction centers, each of which includes a pair of chlorophyll molecules known as P680 (680 nm is the wavelength of one of their absorption peaks). The reaction center chlorophylls overlap so that they are electronically coupled and function as a single unit. When P680 is excited, as indicated by the notation P680*, it quickly gives up an electron, dropping to a lower-energy state, P680+. In other words, *light has oxidized P680*. The photooxidized chlorophyll molecule must be reduced in order to return to its original state.

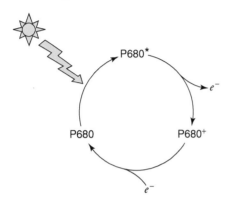

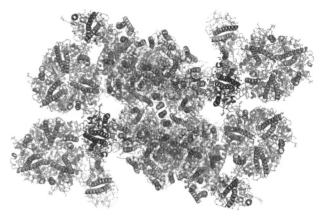

Figure 16.9 Plant Photosystem II supercomplex. The core dimeric photosystem, viewed from the stromal side, is gray, and the various types of light-harvesting complexes are shown in color.

The two P680 groups are located near the lumenal side of Photosystem II. The electron given up by each photooxidized P680 travels through several redox groups (**Fig. 16.10**). Although the prosthetic groups in Photosystem II are arranged more or less symmetrically, they do not all directly participate in electron transfer. First, the electron travels to one of two pheophytin groups, which are essentially chlorophyll molecules without the central Mg^{2+} ion. Next, the electron is transferred to a tightly bound plastoquinone molecule and then to a loosely bound plastoquinone on the stromal side of Photosystem II. An iron atom may assist the final electron transfer. Plastoquinone (PQ) is similar to mammalian mitochondrial ubiquinone (see Section 12.3).

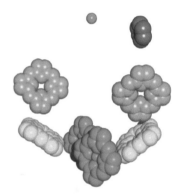

Figure 16.10 Arrangement of prosthetic groups in Photosystem II. The green chlorophyll groups constitute the photooxidizable P680. The two "accessory" chlorophyll groups (yellow) do not undergo oxidation or reduction. Pheophytin is orange, plastoquinone is blue, and an iron atom is red. The lipid tails of the prosthetic groups are not shown.

Plastoquinone

It functions in the same way as a two-electron carrier. The fully reduced plastoquinol (PQH_2) joins a pool of plastoquinones that are soluble in the thylakoid membrane. Two electrons (two photooxidations of P680) are required to fully reduce plastoquinone to PQH_2. This reaction also consumes two protons, which are taken from the stroma.

The oxygen-evolving complex of Photosystem II oxidizes water

O_2, a waste product of photosynthesis, is generated from H_2O by a lumenal portion of Photosystem II called the oxygen-evolving center. This reaction can be written as

$$2\,H_2O \rightarrow O_2 + 4\,H^+ + 4\,e^-$$

The electrons derived from H_2O are used to restore photooxidized P680 to its reduced state.
    The catalyst for the water-splitting reaction is a cofactor with the composition Mn_4CaO_5 that is held in place by Asp, Glu, and His side chains (**Fig. 16.11**). This unusual inorganic

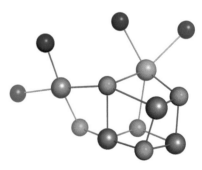

Figure 16.11 Structure of the Mn_4CaO_5 cluster. Atoms are color-coded: Mn purple, Ca green, and O red. One or more of the four H_2O molecules associated with the cluster (oxygen atoms in blue) may be substrates for the water-splitting reaction.

cofactor occurs in all Photosystem II complexes, which suggests a unique chemistry that has remained unaltered for about 2.5 billion years. No synthetic catalyst can match the manganese cluster in its ability to extract electrons from water to form O_2. The water-splitting reaction is rapid, with about 50 O_2 produced per second per Photosystem II, and generates most of the earth's atmospheric O_2.

During water oxidation, the manganese cluster undergoes multiple changes in its oxidation state, somewhat reminiscent of the changes in the Fe–Cu binuclear center of cytochrome c oxidase (mitochondrial Complex IV; see Fig. 15.18), which carries out the reverse reaction. The four water-derived protons are released into the thylakoid lumen, contributing to a drop in pH relative to the stroma. A tyrosine radical (Y·) in Photosystem II transfers each of the four water-derived electrons to P680$^+$ (a tyrosine radical also plays a role in electron transfer in cytochrome c oxidase; see Section 15.2).

Tyrosine radical

The oxidation of water is a thermodynamically demanding reaction because O_2 has an extremely high reduction potential (+0.815 V) and electrons spontaneously flow from a group with a lower reduction potential to a group with a higher reduction potential (see Section 15.1). In fact, photooxidized P680 is the most powerful biological oxidant, with a reduction potential of about +1.15 V.

Upon photoexcitation, the reduction potential of P680 (now P680*) is dramatically diminished, to about −0.8 V. *This low reduction potential allows P680* to surrender an electron to a series of groups with increasingly positive reduction potentials* (**Fig. 16.12**). Recall that the lower the reduction potential, the higher the energy. The overall result is that the input of solar energy

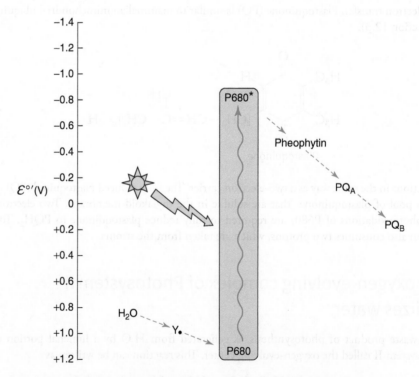

Figure 16.12 Reduction potential and electron flow in Photosystem II. Electrons flow spontaneously from a group with a lower reduction potential to a group with a higher reduction potential. The transfer of electrons from H_2O to plastoquinone is made possible by the excitation of P680 (wavy arrow), which dramatically lowers its reduction potential.

Question Will pheophytin be in the oxidized or reduced state when it is dark?

allows an electron to travel a thermodynamically favorable path from water to plastoquinone. *Four photooxidation events in Photosystem II are required to oxidize two H_2O molecules and produce one O_2 molecule.* **Figure 16.13** summarizes the functions of Photosystem II.

Cytochrome b_6f links Photosystems I and II

After they leave Photosystem II as plastoquinol, electrons reach a second membrane-bound protein complex known as cytochrome b_6f. This complex resembles mitochondrial Complex III (also called cytochrome bc_1)—from the entry of electrons in the form of a reduced quinone, through the circular flow of electrons among its redox groups, to the final transfer of electrons to a mobile electron carrier.

The cytochrome b_6f complex contains eight subunits in each of its monomeric halves (**Fig. 16.14**). Three subunits bear electron-transporting prosthetic groups. One of these subunits is cytochrome b_6, which is homologous to mitochondrial cytochrome b. The second is cytochrome f, whose heme group is actually of the c type. Although it shares no sequence homology with mitochondrial cytochrome c_1, it functions similarly. The photosynthetic complex also contains a Rieske iron–sulfur protein with a 2Fe–2S group that behaves like its mitochondrial counterpart. However, the cytochrome b_6f complex also contains subunits with prosthetic groups that are absent in the mitochondrial complex: a chlorophyll molecule and a β-carotene. These light-absorbing molecules do not appear to participate in electron transfer and may instead help regulate the activity of cytochrome b_6f by registering the amount of available light.

Electron flow in the cytochrome b_6f complex follows a cyclic pattern that is probably identical to the Q cycle in mitochondrial Complex III (see Fig. 15.14). However, in chloroplasts, the final electron acceptor is not cytochrome c but plastocyanin, a small protein with an active-site copper ion (**Fig. 16.15**). Plastocyanin functions as a one-electron carrier by cycling between the Cu^+ and Cu^{2+} oxidation states. Like cytochrome c, plastocyanin is a peripheral membrane protein; it picks up electrons at the lumenal surface of cytochrome b_6f and delivers them to another integral membrane protein complex, in this case Photosystem I.

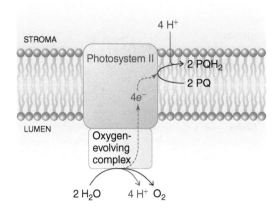

Figure 16.13 Photosystem II function. For every oxygen molecule produced, two plastoquinone molecules are reduced.

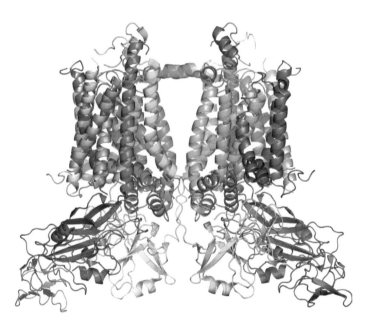

Figure 16.14 Structure of spinach cytochrome b_6f. Each subunit of the dimeric complex is a different color. The prosthetic groups are not shown.

Question **Compare this structure to the functionally similar mitochondrial cytochrome bc_1 (Complex III) in Figure 15.13.**

Figure 16.15 Plastocyanin. The redox-active copper ion (green) is coordinated by a Cys, a Met, and two His residues (yellow).

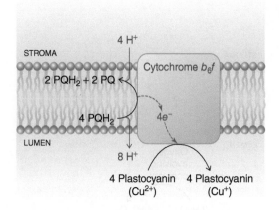

Figure 16.16 Cytochrome b_6f function.
The stoichiometry shown for the cytochrome b_6f Q cycle reflects the four electrons released by the oxygen-evolving complex of Photosystem II.

The net result of the cytochrome b_6f Q cycle is that for every two electrons emanating from Photosystem II, four protons are released into the thylakoid lumen. Since the oxidation of 2 H₂O is a four-electron reaction, *the production of one molecule of O_2 causes the cytochrome b_6f complex to produce eight lumenal H^+* (**Fig. 16.16**). The resulting pH gradient between the stroma and the lumen is a source of energy that drives ATP synthesis, as described below.

A second photooxidation occurs at Photosystem I

Photosystem I, like Photosystem II, is a large protein complex containing multiple pigment molecules. The Photosystem I in the cyanobacterium *Synechococcus* is a symmetric trimer with 12 proteins in each monomer (**Fig. 16.17**). Ninety-six chlorophyll molecules and 22 carotenoids operate as a built-in light-harvesting complex.

In plants, Photosystem I is a monomer that shares a core structure with the cyanobacterial complex. Four Photosystem I–specific light-harvesting complexes associate with one face of the plant photosystem to form a supercomplex (**Fig. 16.18**). A light-harvesting complex normally associated with Photosystem II can dock to the opposite face of Photosystem I under some conditions. Photosystem I requires slightly longer-wavelength light than Photosystem II, which can lead to unequal excitation, yet the two complexes must work in concert. When Photosystem II is overexcited, the level of oxidized plastoquinone decreases. This activates a kinase to phosphorylate the Photosystem II light-harvesting complexes, some of which then move through the membrane to function as antenna complexes for Photosystem I.

In the core of Photosystem I, a pair of chlorophyll molecules constitute the photoactive group known as P700 (it has a slightly longer-wavelength absorbance maximum than P680). Like P680, P700 undergoes exciton transfer from an antenna pigment. P700* gives up an electron to achieve a low-energy oxidized state, P700⁺. The group is then reduced by accepting an electron donated by plastocyanin.

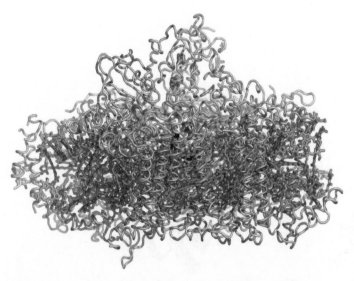

Figure 16.17 Structure of cyanobacterial Photosystem I monomer.
The protein is shown as a gray ribbon, and the various prosthetic groups are color-coded: chlorophyll, green; β-carotene, red; phylloquinone, blue; and Fe–S clusters, orange. In this side view, the stroma is at the top.

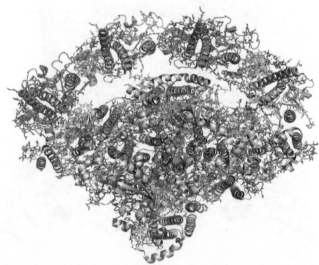

Figure 16.18 Plant Photosystem I supercomplex. The photosystem, viewed from the stroma, is gray and the four copies of the light-harvesting complex are in color.

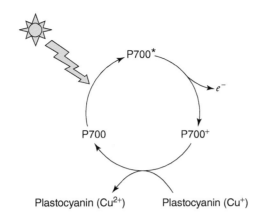

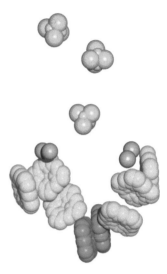

Figure 16.19 Prosthetic groups in Photosystem I. The groups include P700 (the green chlorophyll molecules), "accessory" chlorophylls (yellow), quinones (marked by blue spheres), and 4Fe–4S clusters (orange). The lipid tails of the prosthetic groups are not shown.

P700 is not a particularly good reducing agent (its reduction potential is relatively high, about +0.45 V). However, excited P700 (P700*) has an extremely low $\mathcal{E}^{\circ\prime}$ value (about −1.3 V), so electrons can spontaneously flow from P700* to the other redox groups of Photosystem I. These groups include four additional chlorophyll molecules, quinones, and iron–sulfur clusters of the 4Fe–4S type (**Fig. 16.19**). As in Photosystem II, these prosthetic groups are arranged with approximate symmetry. However, in Photosystem I, all the redox groups appear to undergo oxidation and reduction.

Each electron given up by photooxidized P700 eventually reaches ferredoxin, a small peripheral protein on the stromal side of the thylakoid membrane. Ferredoxin undergoes a one-electron reduction at a 2Fe–2S cluster (**Fig. 16.20**). Reduced ferredoxin participates in two different electron transport pathways in the chloroplast, which are known as linear and cyclic electron flow.

In **linear electron flow,** ferredoxin serves as a substrate for ferredoxin–NADP$^+$ reductase. This stromal enzyme uses two electrons (from two separate ferredoxin molecules) to reduce NADP$^+$ to NADPH (**Fig. 16.21**). *The net result of linear electron flow is therefore the transfer of electrons from water, through Photosystem II, cytochrome b$_6$f, Photosystem I, and then on to NADP$^+$.* For every two water molecules converted to O_2 by Photosystem II, two NADPH are produced. Photosystem I does not contribute to the transmembrane proton gradient except by consuming stromal protons in the reduction of NADP$^+$ to NADPH.

When plotted according to reduction potential, the electron-carrying groups of the pathway from water to NADP$^+$ form a diagram called the **Z-scheme** of photosynthesis (**Fig. 16.22**). The zigzag pattern is due to the two photooxidation events, which markedly decrease the reduction potentials of P680 and P700. The input of light energy ensures that electrons

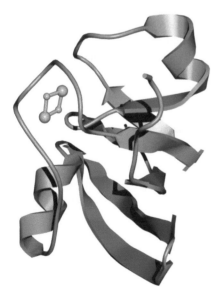

Figure 16.20 Ferredoxin. The 2Fe–2S cluster is shown in orange.

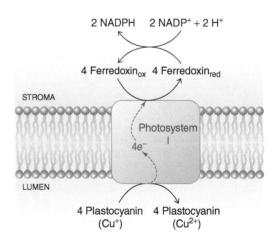

Figure 16.21 Linear electron flow through Photosystem I. Electrons donated by plastocyanin are transferred to ferredoxin and used to reduce NADP$^+$. The stoichiometry reflects the four electrons released by the oxidation of 2 H$_2$O in Photosystem II.

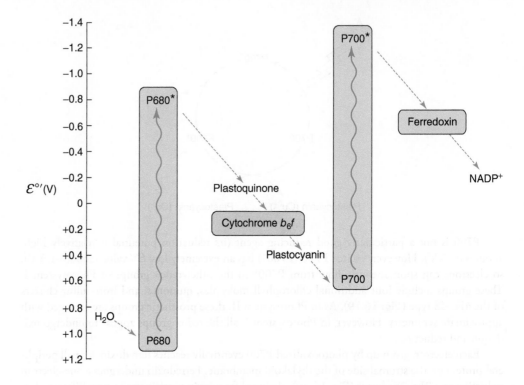

follow a thermodynamically favorable pathway to groups with increasing reduction potential. Note that the four-electron process of producing one O_2 and two NADPH is accompanied by the absorption of eight photons (four each at Photosystem II and Photosystem I).

In **cyclic electron flow,** electrons from Photosystem I do not reduce $NADP^+$ but instead return to the cytochrome b_6f complex. Formation of a supercomplex—containing Photosystem I, its light-harvesting complexes, and cytochrome b_6f—may facilitate this detour. From cytochrome b_6f, the electrons are transferred to plastocyanin and flow back to Photosystem I to reduce photooxidized $P700^+$. Meanwhile, plastoquinol molecules circulate between the two quinone-binding sites of cytochrome b_6f so that protons are translocated from the stroma to the lumen, in the Q cycle (**Fig. 16.23**).

Cyclic electron flow requires the input of light energy at Photosystem I but not Photosystem II. During cyclic flow, no free energy is recovered in the form of the reduced cofactor NADPH, but free energy is conserved in the formation of a transmembrane proton gradient by the activity of the cytochrome b_6f complex. Consequently, *cyclic electron flow augments ATP generation by chemiosmosis* (in some bacteria with just a single reaction center, electrons flow through a similar pathway that does not produce O_2 or NADPH). By varying the proportion of electrons that

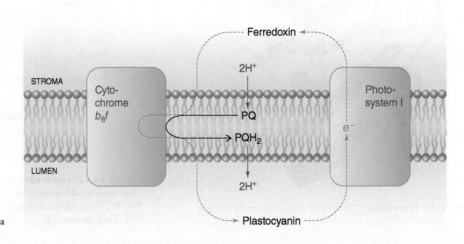

Figure 16.23 Cyclic electron flow. Electrons circulate between Photosystem I and the cytochrome b_6f complex. No NADPH or O_2 is produced, but the activity of cytochrome b_6f builds up a proton gradient that drives ATP synthesis.

follow the linear and cyclic pathways through Photosystem I, a photosynthetic cell can vary the proportions of ATP and NADPH produced by the light reactions.

Chemiosmosis provides the free energy for ATP synthesis

Chloroplasts and mitochondria use the same mechanism to synthesize ATP: *They couple the dissipation of a transmembrane proton gradient to the phosphorylation of ADP.* In photosynthetic organisms, this process is called **photophosphorylation.** Chloroplast ATP synthase is highly homologous to mitochondrial and bacterial ATP synthases. The CF_1CF_0 complex ("C" indicates chloroplast) consists of a proton-translocating integral membrane component (CF_0) mechanically linked to a soluble CF_1 component where ATP synthesis occurs by a binding change mechanism (as described in Fig. 15.25). The movement of protons from the thylakoid lumen to the stroma provides the free energy to drive ATP synthesis (**Fig. 16.24**).

As in mitochondria, the proton gradient has both chemical and electrical components. In chloroplasts, the pH gradient (about 3.5 pH units) is much larger than in mitochondria (about 0.75 units). However, in chloroplasts, the electrical component is less than in mitochondria because of the permeability of the thylakoid membrane to ions such as Mg^{2+} and Cl^-. Diffusion of these ions tends to minimize the difference in charge due to protons.

Assuming linear electron flow, 8 photons are absorbed (4 by Photosystem II and 4 by Photosystem I) to generate 4 lumenal protons from the oxygen-evolving complex and 8 protons from the cytochrome $b_6 f$ complex. Theoretically, these 12 protons can drive the synthesis of about 3 ATP, which is consistent with experimental results showing approximately 3 ATP generated for each molecule of O_2. The c ring of chloroplast ATP synthase contains 15 subunits, almost twice the number in mammalian mitochondrial ATP synthase (8). ATP synthesis in chloroplasts therefore requires more protons, but this inefficiency may also allow ATP to be produced more quickly than in mitochondria.

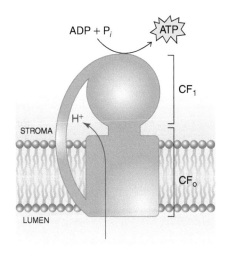

Figure 16.24 Photophosphorylation. As protons traverse the CF_0 component of chloroplast ATP synthase (following their concentration gradient from the lumen to the stroma), the CF_1 component carries out ATP synthesis.

Concept Check

1. Summarize the functions of Photosystem II, the oxygen-evolving complex, plastoquinone, the cytochrome $b_6 f$ complex, plastocyanin, Photosystem I, and ferredoxin.

2. Explain how photon absorption drives electron transfer from water to plastoquinone.

3. Describe the Z-scheme of photosynthesis and explain its zigzag shape.

4. Add an arrow to Fig. 16.22 to represent cyclic electron flow.

5. Discuss the yields of O_2, NADPH, and ATP in cyclic and linear electron flow.

6. Compare and contrast the chloroplast light reactions and mitochondrial electron transport.

7. Compare photophosphorylation and oxidative phosphorylation.

8. Draw a diagram to explain the interdependence of photosynthesis and cellular respiration.

9. Which reactions contribute to the pH gradient that drives the activity of chloroplast ATP synthase?

16.3 Carbon Fixation

KEY CONCEPTS

Describe the steps of carbon fixation by the Calvin cycle.

- Distinguish the two activities of rubisco.
- Identify the functions of the other Calvin cycle enzymes.
- Explain how the "dark" reactions are linked to the light reactions.
- List the metabolic fates of newly synthesized glyceraldehyde-3-phosphate.

The production of ATP and NADPH by the photoactive complexes of the thylakoid membrane (or bacterial plasma membrane) is only part of the story of photosynthesis. The rest of this chapter focuses on the use of the products of the light reactions in the so-called **dark reactions.** These reactions, which occur in the chloroplast stroma, incorporate atmospheric carbon dioxide in biologically useful organic molecules, a process called **carbon fixation.** Photosynthetic organisms have evolved six different pathways for fixing CO_2, but the focus here is on the most widely used pathway.

Rubisco catalyzes CO_2 fixation

Carbon dioxide is fixed by the action of ribulose bisphosphate carboxylase/oxygenase, or rubisco. *This enzyme adds CO_2 to a five-carbon sugar and then cleaves the product to two three-carbon units* (**Fig. 16.25**). This reaction itself does not require ATP or NADPH, but the reactions that transform the rubisco reaction product, 3-phosphoglycerate, to the three-carbon sugar glyceraldehyde-3-phosphate require both ATP and NADPH, as we will see.

Three-carbon compounds are the biosynthetic precursors of monosaccharides, amino acids, and—indirectly—nucleotides. They also give rise to the two-carbon acetyl units used to build fatty acids. The metabolic importance of these small molecular building blocks is one reason why the scheme shown in Figure 16.1 presents photosynthesis as a process in which CO_2 is converted to two- and three-carbon intermediates.

Rubisco is a notable enzyme, in part because its activity directly or indirectly sustains most of the earth's biomass. Plant chloroplasts are packed with the enzyme, which accounts for about half of the chloroplast's protein content. One reason for its high concentration is that it is not a particularly efficient enzyme. Its catalytic output is only about three CO_2 fixed per second.

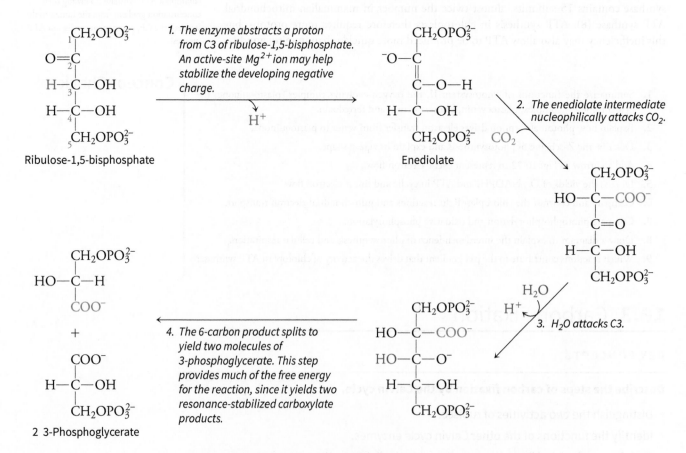

Figure 16.25 The rubisco carboxylation reaction.

Question Before this reaction was understood, scientists believed that carbon fixation involved the reaction of CO_2 with a two-carbon molecule. Explain.

Bacterial rubisco is usually a small dimeric enzyme, whereas the plant enzyme is a large multimer of eight large and eight small subunits (**Fig. 16.26**). In some archaebacteria, rubisco has 10 identical subunits. Enzymes with multiple catalytic sites typically exhibit cooperative behavior and are regulated allosterically, but this does not seem to be true for plant rubisco, whose eight active sites operate independently. Multimerization may simply be an efficient way to pack more active sites into the limited space of the chloroplast.

Despite its metabolic importance, rubisco is not a highly specific enzyme. It also acts as an oxygenase (as reflected in its name) by reacting with O_2, which chemically resembles CO_2. The products of the oxygenase reaction are a three-carbon and a two-carbon compound:

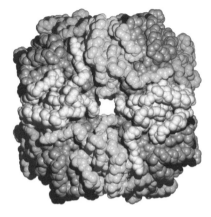

Figure 16.26 Spinach rubisco. The complex has a mass of approximately 550 kD. The eight catalytic sites are located in the large subunits (dark colors). Only four of eight small subunits (light colors) are visible in this image.

$$CH_2OPO_3^{2-}$$
$$O=C$$
$$H-C-OH \quad + \quad O_2 \quad \xrightarrow[\text{rubisco}]{H_2O}$$
$$H-C-OH$$
$$CH_2OPO_3^{2-}$$
Ribulose bisphosphate

$$CH_2OPO_3^{2-}$$
$$^-O-C=O$$
2-Phosphoglycolate

$+$

$$COO^-$$
$$H-C-OH$$
$$CH_2OPO_3^{2-}$$
3-Phosphoglycerate

The 2-phosphoglycolate product of the rubisco oxygenation reaction is subsequently metabolized by a pathway that consumes ATP and NADPH and produces CO_2. This process, called **photorespiration,** uses the products of the light reactions and therefore wastes some of the free energy of captured photons.

Oxygenase activity is a feature of all known rubisco enzymes and must play an essential role that has been conserved throughout plant evolution. *Photorespiration apparently provides a mechanism for plants to dissipate excess free energy under conditions where the CO_2 supply is insufficient for carbon fixation.* Photorespiration may consume significant amounts of ATP and NADPH at high temperatures, which favor oxygenase activity. Some plants have evolved a mechanism, called the **C_4 pathway,** to minimize photorespiration (**Box 16.A**).

Box 16.A   The C_4 Pathway

On hot, bright days, the light reactions produce O_2, the substrate for photorespiration, and CO_2 supplies are low as plants close their stomata (pores in the leaf surface) to avoid evaporative water loss. This combination of events can bring photosynthesis to a halt. Some plants avoid this possibility by stockpiling CO_2 in four-carbon molecules so that photosynthesis can proceed even while stomata are closed.

The mechanism for storing carbon begins with the condensation of bicarbonate (HCO_3^-) with phosphoenolpyruvate to yield oxaloacetate, which is then reduced to malate (see diagram). These four-carbon acids give the C_4 pathway its name. The subsequent oxidative decarboxylation of malate regenerates CO_2 and NADPH to be used in the Calvin cycle. The three-carbon remnant, pyruvate, is recycled back to phosphoenolpyruvate.

Because the C_4 pathway and the rubisco reaction compete for CO_2, they take place in different types of cells or at different times of day. For example, in some plants, carbon accumulates in mesophyll cells, which are near the leaf surface and lack rubisco. The C_4 compounds then enter bundle sheath cells in the leaf interior, which contain abundant rubisco. In other plants, the C_4 pathway occurs at night, when the stomata are open and water loss is minimal, and carbon is fixed by rubisco during the day.

The C_4 pathway is energetically expensive, so it requires lots of sunlight. Consequently, C_4 plants grow more slowly than conventional, or C_3, plants when light is limited, but they have the advantage in hot, dry climates. About 5% of the earth's plants, including the economically important maize (corn), sugarcane, and sorghum, use the C_4 pathway.

The recognition of climate change has led to predictions that C_4 "weeds" may overtake economically important C_3 plants as temperatures increase. In fact, the increase in atmospheric CO_2 that is driving global warming appears to promote the growth of C_3 plants, which can obtain CO_2 more easily without losing too much water via their stomata. However, if water is limited, C_4 plants may still have a competitive edge, as they are adapted not just for hot environments but for arid ones.

Question Operation of the C_4 pathway consumes ATP. Using the pathway shown here, indicate where this occurs.

$$\begin{array}{c} COO^- \\ | \\ C{-}OPO_3^{2-} \\ || \\ CH_2 \end{array}$$

Phosphoenolpyruvate

$$\begin{array}{c} COO^- \\ | \\ C{=}O \\ | \\ CH_3 \end{array}$$

Pyruvate

$CO_2 \xrightarrow[\text{anhydrase}]{\text{carbonic}} HCO_3^-$ $\xrightarrow[\text{carboxylase}]{\text{phosphoenolpyruvate}}$ P_i

malic enzyme $\rightarrow$ NADPH + CO_2 $\Longrightarrow$ Calvin cycle / NADP$^+$

$$\begin{array}{c} COO^- \\ | \\ C{=}O \\ | \\ CH_2 \\ | \\ COO^- \end{array}$$

Oxaloacetate

$\xrightarrow[\text{NADPH \quad NADP}^+]{\text{malate dehydrogenase}}$

$$\begin{array}{c} COO^- \\ | \\ CHOH \\ | \\ CH_2 \\ | \\ COO^- \end{array}$$

Malate

Photorespiration and its significance

Apart from oxidative phosphorylation, illuminated plants can consume O_2 and evolve CO_2 through a distinct pathway known as photorespiration that commences between three organelles chloroplasts, leaf peroxisomes, and mitochondria. In the presence of elevated levels of O_2 and reduced levels of CO_2, photorespiration can outpace photosynthetic CO_2 fixation. Hence, photorespiration is seemingly a futile process against the biosynthetic gain attained by photosynthesis. The basis of photorespiration is the bifunctionality of Rubisco. For the oxygenase reaction catalyzed by the enzyme, O_2 competes with CO_2, and reacts with RuBP to yield 3-phosphoglycerate (3PG) and 2-phosphoglycolate. 2-Phosphoglycolate is converted to glycolate by 2-phosphoglycolate phosphatase. From chloroplast, glycolate is exported to glyoxisome (peroxisomes), and in the glyoxisomal matrix, it is oxidized to glyoxylate and H_2O_2 by glycolate oxidase. Glyoxisomal catalase catalyzes the disproportionation reaction and converts H_2O_2 to H_2O and O_2. Glycolate oxidase converts a portion of glyoxylate to oxalate and the remainder portion is utilized by alanine glyoxylate transaminase to produce glycine. The glycine is transported to mitochondria, where two molecules of glycine are converted to one molecule of serine and one molecule of CO_2. The serine is transported to glyoxisome, where serine glutamate transaminase converts serine to hydroxypyruvate. Hydroxypyruvate reductase 1 reduces hydroxypyruvate to glycerate using NADH. Glycerate is phosphorylated in the cytosol to 3PG by glycerate kinase, which re-enters the chloroplast and is reconverted to RuBP through the Calvin-Benson cycle. Effectively the photorespiration cycle results in wasteful utilization of part of ATP and NADPH generated by photophosphorylation.

For healthy plants, the rate of photosynthesis predominates over CO_2 release by photorespiration. However, the oxygenase activity of Rubisco is enhanced with elevated temperature. On a hot bright day when the level of CO_2 has been depleted at the chloroplast and the level of O_2 has been raised by photosynthesis, the CO_2 compensation point rises and the rate of photorespiration may approaches that of photosynthesis. Thus plants possessing a Rubisco variant with depleted oxygenase activity would have higher photosynthetic efficiency with a reduced need for water. Such plants would also require less fertilizer as they would require less Rubisco.

Despite being of no significant metabolic function, the oxygenase activity of Rubisco is retained in all the photosynthetic organisms where it is present. It is an evolutionary conundrum in terms of functional optimization of the enzyme toward CO_2 fixation. It is hypothesized that photosynthesis evolved when there was high CO_2 content in the earth's atmosphere compared to O_2 resulting in little or no oxygenase action. Another possibility is that photorespiration might protect the chloroplasts from photo-oxidative damage when sufficient CO_2 is unavailable. Experimental support for such a hypothesis has been availed with the observation that when chloroplast or leaves are illuminated in the absence of CO_2, photosynthetic capacity is promptly and permanently lost.

The Calvin cycle rearranges sugar molecules

If rubisco is responsible for fixing CO_2, what is the origin of its other substrate, ribulose bisphosphate? The answer—elucidated over many years by Melvin Calvin, James Bassham, and Andrew Benson—is a metabolic pathway known as the **Calvin cycle.** Early experiments to study the fate of ^{14}C-labeled CO_2 in algae showed that, within a few minutes, the cells had synthesized a complex mixture of sugars, all containing the radioactive label. Rearrangements among these sugar molecules generate the five-carbon substrate for rubisco.

The Calvin cycle actually begins with a sugar monophosphate, ribulose-5-phosphate, which is phosphorylated in an ATP-dependent reaction (**Fig. 16.27**). The resulting bisphosphate is a substrate for rubisco, as we have already seen. Each 3-phosphoglycerate product of the rubisco reaction is then phosphorylated, again at the expense of ATP. This phosphorylation reaction (step 3 of Fig. 16.27) is identical to the phosphoglycerate kinase reaction of glycolysis (see Section 13.1). Next, bisphosphoglycerate is reduced by the chloroplast enzyme glyceraldehyde-3-phosphate dehydrogenase, which resembles the glycolytic enzyme but uses NADPH rather than NADH. The NADPH is produced during the light reactions of photosynthesis.

Some glyceraldehyde-3-phosphate is siphoned from the Calvin cycle for metabolic fates such as glucose or amino acid synthesis. Recall from Section 13.2 that the pathway from glyceraldehyde-3-phosphate to glucose consists of reactions that require no further input of ATP. Glyceraldehyde-3-phosphate can also be converted to pyruvate and then to oxaloacetate, both of which can undergo transamination to generate amino acids. Additional reactions lead to other metabolites.

The glyceraldehyde-3-phosphate that is not used for biosynthetic pathways enters a series of isomerization and group-transfer reactions that regenerate ribulose-5-phosphate to complete the Calvin cycle (**Fig. 16.28**). These interconversion reactions are similar to those of the pentose phosphate pathway (Section 13.4). Consequently, if the Calvin cycle starts with three five-carbon

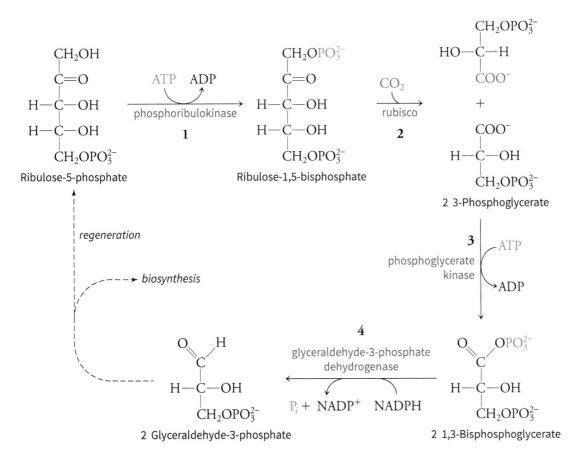

Figure 16.27 Initial reactions of the Calvin cycle.   Note that ATP and NADPH, products of the light-dependent reactions, are consumed in the process of converting CO_2 to glyceraldehyde-3-phosphate.

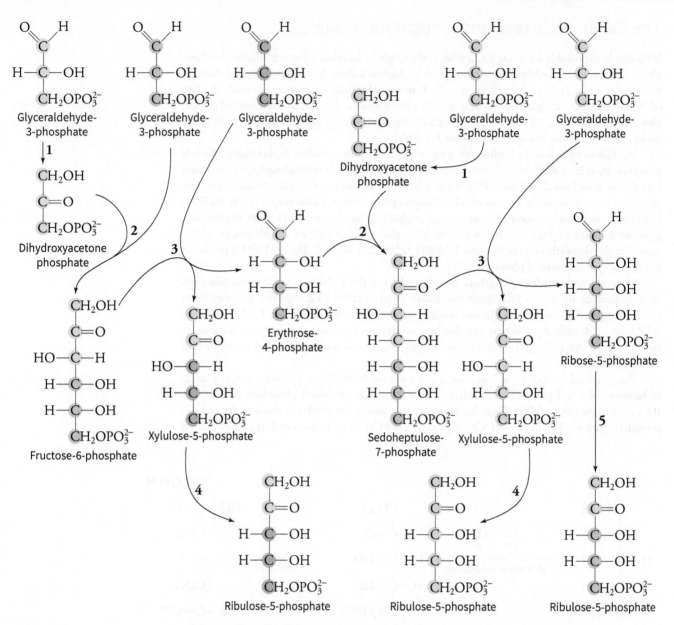

Figure 16.28 Regeneration reactions of the Calvin cycle. Reaction 1 is catalyzed by triose phosphate isomerase, 2 by aldolase and an isomerase, 3 by transketolase (see Section 7.2), 4 by an epimerase, and 5 by an isomerase.

Question Add boxes around the 2-carbon fragments transferred by transketolase.

ribulose molecules, so that three CO_2 molecules are fixed (Fig. 16.27), the products are six three-carbon glyceraldehyde-3-phosphate molecules, five of which are recycled to form three ribulose molecules (Fig. 16.28), leaving the sixth (representing the three fixed CO_2) as its net product. The net equation for the Calvin cycle, including the ATP and NADPH cofactors, is

$$3\ CO_2 + 9\ ATP + 6\ NADPH \rightarrow glyceraldehyde\text{-}3\text{-}phosphate + 9\ ADP + 8\ P_i + 6\ NADP^+$$

Fixing a single CO_2 therefore requires 3 ATP and 2 NADPH—approximately the same quantity of ATP and NADPH produced by the absorption of eight photons. The relationship between the number of photons absorbed and the amount of carbon fixed or oxygen released is known as the **quantum yield** of photosynthesis. Keep in mind that the exact number of carbons fixed per photon absorbed depends on factors such as the number of protons translocated per ATP synthesized by the chloroplast ATP synthase and the ratio of cyclic to linear electron flow in Photosystem I.

The availability of light regulates carbon fixation

Plants must coordinate the light reactions with carbon fixation. During the day, both processes occur. At night, when the photosystems are inactive, the plant turns off the "dark" reactions to conserve ATP and NADPH while it turns on pathways to regenerate these cofactors by metabolic pathways such as glycolysis and the pentose phosphate pathway. It would be wasteful for these catabolic processes to proceed simultaneously with the Calvin cycle. Thus, the "dark" reactions do not actually occur in the dark!

All the mechanisms for regulating the Calvin cycle are directly or indirectly linked to the availability of light energy. Some of the regulatory mechanisms are highlighted here. For example, a catalytically essential Mg^{2+} ion in the rubisco active site is coordinated in part by a carboxylated lysine side chain that is produced by the reaction of CO_2 with the ε-amino group:

$$-(CH_2)_4-NH_2 + CO_2 \rightleftharpoons -(CH_2)_4-NH-COO^- + H^+$$
$$\text{Lys}$$

By forming the $Mg^{2}+$-binding site, this "activating" CO_2 molecule promotes the ability of rubisco to fix additional substrate CO_2 molecules. The carboxylation reaction is favored at high pH, a signal that the light reactions are working (depleting the stroma of protons) and that ATP and NADPH are available for the Calvin cycle.

Magnesium ions also directly activate rubisco and several of the Calvin cycle enzymes. During the light reactions, the rise in stromal pH triggers the flux of Mg^{2+} ions from the lumen to the stroma (this ion movement helps balance the charge of the protons that are translocated in the opposite direction). Some of the Calvin cycle enzymes are also activated when the ratio of reduced ferredoxin to oxidized ferredoxin is high, another signal that the photosystems are active.

Calvin cycle products are used to synthesize sucrose and starch

Many of the three-carbon sugars produced by the Calvin cycle are converted to sucrose or starch. The polysaccharide starch is synthesized in the chloroplast stroma as a temporary storage depot for glucose. It is also synthesized as a long-term storage molecule elsewhere in the plant, including leaves, seeds, and roots. In the first stage of starch synthesis, two molecules of glyceraldehyde-3-phosphate are converted to glucose-6-phosphate by reactions analogous to those of mammalian gluconeogenesis (see Fig. 13.10). Phosphoglucomutase then carries out an isomerization reaction to produce glucose-1-phosphate. Next, this sugar is "activated" by its reaction with ATP to form ADP–glucose:

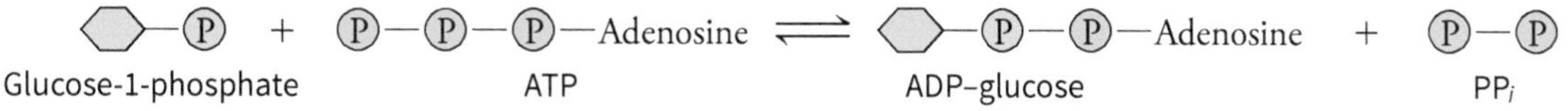

(Recall from Section 13.3 that glycogen synthesis uses the chemically related nucleotide sugar UDP–glucose.) Starch synthase then transfers the glucose residue to the end of a starch polymer, forming a new glycosidic linkage.

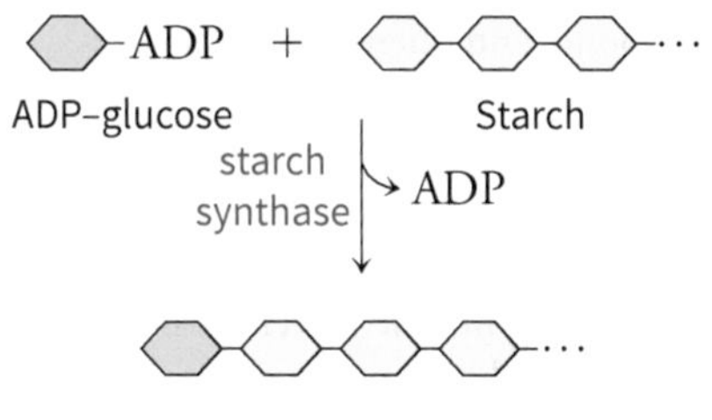

The overall reaction is driven by the exergonic hydrolysis of the PP_i released in the formation of ADP–glucose. Thus, one phosphoanhydride bond is consumed in lengthening a starch molecule by one glucose residue.

Sucrose, a disaccharide of glucose and fructose, is synthesized in the cytosol. Glyceraldehyde-3-phosphate or its isomer dihydroxyacetone phosphate is transported out of the chloroplast by an antiport protein that exchanges phosphate for a phosphorylated three-carbon sugar. Two of these sugars combine to form fructose-6-phosphate, and two others combine to form glucose-1-phosphate, which is subsequently activated by UTP. Next, fructose-6-phosphate reacts with UDP–glucose to produce sucrose-6-phosphate. Finally, a phosphatase converts the phosphorylated sugar to sucrose:

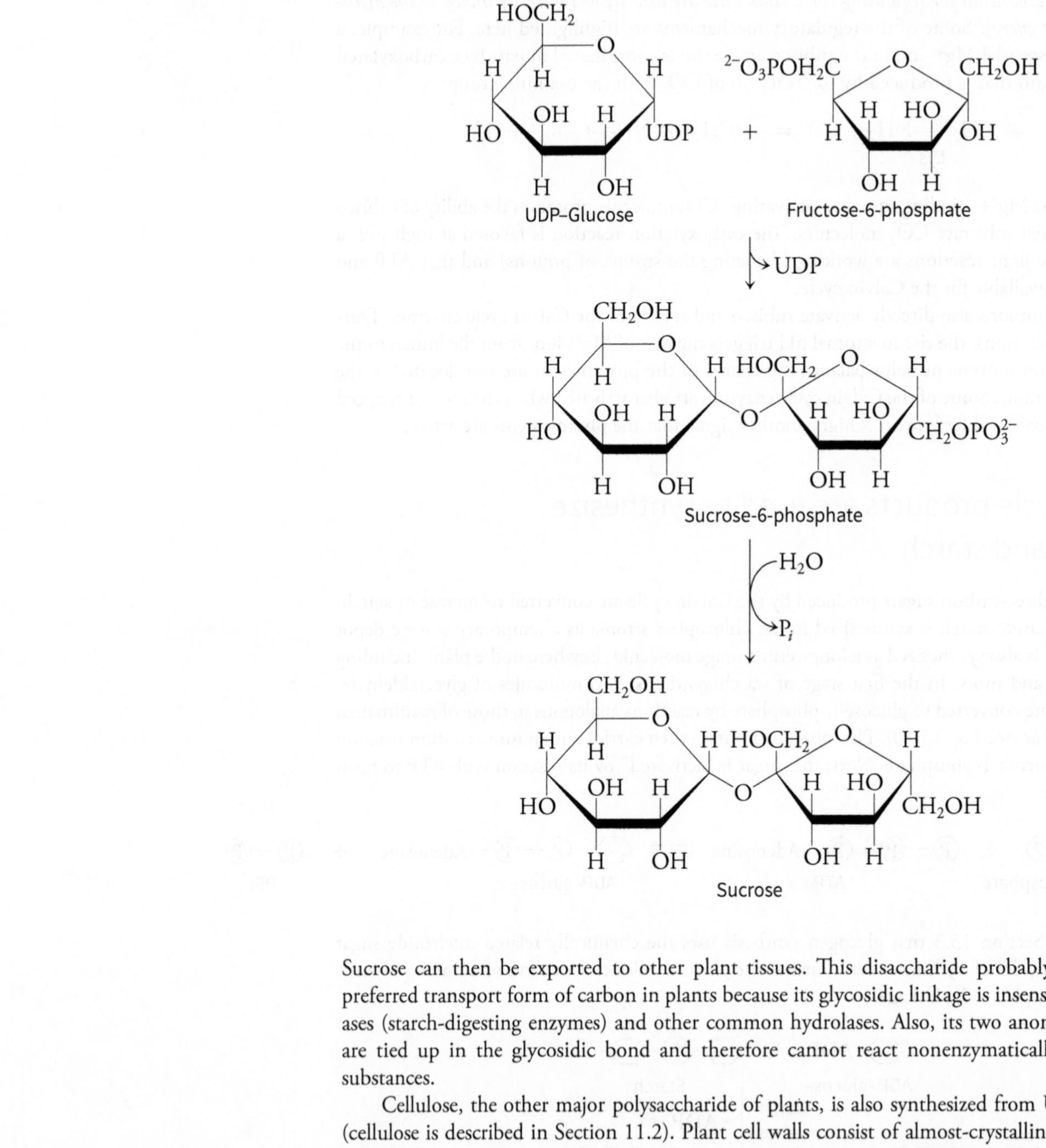

UDP–Glucose   Fructose-6-phosphate

Sucrose-6-phosphate

Sucrose

Sucrose can then be exported to other plant tissues. This disaccharide probably became the preferred transport form of carbon in plants because its glycosidic linkage is insensitive to amylases (starch-digesting enzymes) and other common hydrolases. Also, its two anomeric carbons are tied up in the glycosidic bond and therefore cannot react nonenzymatically with other substances.

Cellulose, the other major polysaccharide of plants, is also synthesized from UDP–glucose (cellulose is described in Section 11.2). Plant cell walls consist of almost-crystalline cables, each containing approximately 36 cellulose polymers, which are embedded in an amorphous matrix of other polysaccharides (see Box 11.B; synthetic materials such as fiberglass are built on the same principle). Unlike starch in plants or glycogen in mammals, cellulose is synthesized by enzyme complexes in the plant plasma membrane and is extruded into the extracellular space.

Concept Check

1. List the reactants and products of the two reactions catalyzed by rubisco.
2. Compare the physiological implications of carbon fixation and photorespiration.
3. Explain how the carbon from a molecule of fixed CO_2 becomes incorporated into other compounds such as monosaccharides.
4. Identify the source of the ribulose-1,5-bisphosphate used for carbon fixation by rubisco.
5. Compare the Calvin cycle to the pentose phosphate pathway.
6. Describe some of the mechanisms for regulating the activity of the "dark" reactions.
7. Summarize the role of nucleotides in the synthesis of starch and sucrose.

SUMMARY

16.1 Photosynthesis: An Overview

- Plant chloroplasts contain pigments that absorb photons and release the energy, primarily by transferring it to another molecule (exciton transfer) or giving up an electron (photooxidation). Light-harvesting complexes act to capture and funnel light energy to the photosynthetic reaction centers.

16.2 The Light Reactions

- In the so-called light reactions of photosynthesis, electrons from the photooxidized P680 reaction center of Photosystem II pass through several prosthetic groups and then to plastoquinone. The P680 electrons are replaced when the oxygen-evolving complex of Photosystem II converts water to O_2, a four-electron oxidation reaction.
- Electrons flow next to a cytochrome b_6f complex that carries out a proton-translocating Q cycle, and then to the protein plastocyanin.
- A second photooxidation at P700 of Photosystem I allows electrons to flow to the protein ferredoxin and finally to $NADP^+$ to produce NADPH.

- The free energy of light-driven electron flow, particularly cyclic flow, is also conserved in the formation of a transmembrane proton gradient that drives ATP synthesis in the process called photophosphorylation.

16.3 Carbon Fixation

- The enzyme rubisco "fixes" CO_2 by catalyzing the carboxylation of a five-carbon sugar. Rubisco also acts as an oxygenase in the process of photorespiration.
- The reactions of the Calvin cycle use the products of the light reactions (ATP and NADPH) to convert the product of the rubisco reaction to glyceraldehyde-3-phosphate and to regenerate the five-carbon carboxylate receptor. These "dark" reactions are regulated according to the availability of light energy.
- Chloroplasts convert the glyceraldehyde-3-phosphate product of photosynthesis into glucose residues for incorporation into starch, sucrose, and cellulose.

KEY TERMS

photosynthesis, chloroplast, stroma, thylakoid, photon, Planck's law, photoreceptor, fluorescence, exciton transfer, photooxidation, reaction center, antenna pigment, light-harvesting complex, light reactions, linear electron flow, Z-scheme, cyclic electron flow, photophosphorylation, dark reactions, carbon fixation, photorespiration, C_4 pathway, Calvin cycle, quantum yield

BIOINFORMATICS

Brief Bioinformatics Exercises

16.1 Viewing and Analyzing Photosystems I and II
16.2 Photosynthesis and the KEGG Database

PROBLEMS

16.1 Photosynthesis: An Overview

1. Indicate with a C or an M whether the following occur in chloroplasts, mitochondria, or both: **a.** proton translocation, **b.** photophosphorylation, **c.** photooxidation, **d.** quinones, **e.** oxygen reduction, **f.** water oxidation, **g.** electron transport, **h.** oxidative phosphorylation, **i.** carbon fixation, **j.** NADH oxidation, **k.** Mn cofactor, **l.** heme groups, **m.** binding change mechanism, **n.** iron–sulfur clusters, **o.** NADP+ reduction, and **p.** gluconeogensis.

2. Compare and contrast the structures of chloroplasts and mitochondria.

3. The thylakoid membrane contains some unusual lipids. One of these is galactosyl diacylglycerol; a β-galactose residue is attached to the first glycerol carbon. Draw the structure of β-galactosyl diacylglycerol.

4. Thylakoid membranes contain lipids with a high degree of unsaturation. What does this tell you about the character of the thylakoid membrane?

5. Calculate the energy per photon and per mole of photons with a wavelength of 680 nm.

6. Calculate the energy per photon and per mole of photons with wavelengths of **a.** 400 nm and **b.** 600 nm. What is the relationship between wavelength and energy?

7. Phycocyanins absorb light at 620 nm. **a.** What color is this light? What color are phycocyanins? **b.** Calculate the energy per photon of light at this wavelength.

8. Chlorophyll *a* absorbs light at 430 nm and 662 nm. **a.** What colors are associated with each of these wavelengths? **b.** What color is chlorophyll *a*? **c.** Calculate the energy per photon of light at these wavelengths. Which wavelength of light has greater energy?

9. At what wavelength would a mole of photons have an energy of 500 kJ?

10. At what wavelength would a mole of photons have an energy of **a.** 350 kJ and **b.** 500 kJ?

11. Assuming 100% efficiency, calculate the number of moles of ATP that could be generated per mole of photons with the energy calculated in Problem 5.

12. Assuming 100% efficiency, calculate the number of moles of ATP that could be generated per mole of photons with the energies calculated in Problem 6.

13. Assuming 100% efficiency, calculate the number of moles of ATP that could be generated per mole of photons with the energy calculated in Problem 7.

14. Assuming 100% efficiency, calculate the number of moles of ATP that could be generated per mole of photons with the energies calculated for the **a.** 430 nm light and **b.** 662 nm light in Problem 8. **c.** What is the relationship between wavelength and amount of ATP synthesized?

15. Red tides result from algal blooms that cause seawater to become visibly red. In the photosynthetic process, red algae take advantage of wavelengths not absorbed by other organisms. Describe the photosynthetic pigments of the red algae.

16. Some photosynthetic bacteria live in murky ponds where visible light does not penetrate easily. What wavelengths might the photosynthetic pigments in these organisms absorb?

17. Compare the structures of chlorophyll *a* (Fig. 16.3) and the reduced form of heme *b* (Fig. 15.12).

18. You are investigating the functional similarities of chloroplast cytochrome *f* and mitochondrial cytochrome *c₁*. **a.** Which would provide more useful information: the amino acid sequences of the proteins or models of their three-dimensional shapes? Explain. **b.** Would it be better to examine high-resolution models of the two apoproteins (the polypeptides without their heme groups) or low-resolution models of the holoproteins (polypeptides plus heme groups)?

19. Under conditions of very high light intensity, excess absorbed solar energy is dissipated by the action of "photoprotective" proteins in the thylakoid membrane. Explain why it is advantageous for these proteins to be activated by a buildup of a proton gradient across the membrane.

20. Of the four mechanisms for dissipating light energy shown in Figure 16.5, which would be best for "protecting" the photosystems from excess light energy as described in Problem 19?

16.2 The Light Reactions

21. The three electron-transporting complexes of the thylakoid membrane can be called plastocyanin–ferredoxin oxidoreductase, plastoquinone–plastocyanin oxidoreductase, and water–plastoquinone oxidoreductase. What are the common names of these enzymes and in what order do they act?

22. Photosystem II is located mostly in the tightly stacked regions of the thylakoid membrane, whereas Photosystem I is located mostly in the unstacked regions (see Fig. 16.2). Why might it be important for the two photosystems to be separated?

23. The herbicide 3-(3,4-dichlorophenyl)-1,1-dimethylurea (DCMU) blocks electron flow from Photosystem II to Photosystem I. What is the effect on oxygen production and photophosphorylation when DCMU is added to plants?

24. a. When the antifungal agent myxothiazol is added to a suspension of chloroplasts, the QH_2/Q ratio increases. Where does myxothiazol inhibit electron transfer? **b.** Compare the effect of myxothiazol on chloroplasts (part **a**) and mitochondria (Problem 15.37).

25. a. Plastoquinone is not firmly anchored to any thylakoid membrane component but is free to diffuse laterally throughout the membrane among the photosynthetic components. What aspects of its structure account for this behavior? **b.** What component of the mitochondrial electron transport chain exhibits similar behavior?

26. The Photosystem II reaction center contains a protein called D1 that contains the PQ_B binding site. The D1 protein in the singlecelled alga *Chlamydomonas reinhardtii* is predicted to have five hydrophobic membrane-spanning helical regions. A loop between the fourth and fifth segments is located in the stroma. This loop is believed to lie along the membrane surface, and it contains several highly conserved amino acid residues. D1 proteins with mutations at the Ala 251 position were evaluated for photosynthetic activity and herbicide susceptibility. The results are shown in the table. What are the essential properties of the amino acid at position 251 in the D1 protein?

Amino acid at position 251	Characteristics
Ala	Wild-type
Cys	Similar to wild-type
Gly, Pro, Ser	Impaired in photosynthesis
Ile, Lau, Val	Impaired in photoautotrophic growth Impaired in photosynthesis Resistant to herbicides
Arg, Glan, Glu, His, and Asp	Not photosynthetically competent

27. Use Equation 15.4 to calculate the free energy change for transforming one mole of P680 to P680*.

28. Determine the wavelength of the photons whose absorption would supply the free energy to transform one mole of P680 to P680* (see Problem 27).

29. a. Calculate the free energy of translocating a proton out of the stroma when the lumenal pH is 3.5 units lower than the stromal pH, the temperature is 25°C, and $\Delta\psi$ is −50 mV. b. Compare the free energy translocating a proton out of a mitochondrion where the pH difference is 0.75 units and $\Delta\psi$ is 200 mV. Compare both types of translocation. Are the processes exergonic or endergonic? c. Which contributes a larger component of the free energy for each process, the pH difference or the membrane potential?

30. Calculate the standard free energy change for the oxidation of one molecule of water by $NADP^+$.

31. Calculate the energy available in two photons of wavelength 700 nm. Compare this value with the standard free energy change you calculated in Problem 30. Do two photons supply enough energy to drive the oxidation of one molecule of water by $NADP^+$?

32. Photophosphorylation in chloroplasts is similar to oxidative phosphorylation in mitochondria. What is the final electron acceptor in photosynthesis? What is the final electron acceptor in mitochondrial electron transport?

33. If radioactively labeled water ($H_2^{18}O$) is provided to a plant, where does the label appear?

34. In the 1940s, Robert Hill conducted an experiment in which he illuminated chloroplasts in the absence of CO_2 but in the presence of the electron acceptor $Fe(CN)_6^{3-}$. Oxygen was evolved. What was the other product of the reaction? What is the significance of this experiment?

35. Cyclic electron flow in some organisms can occur by more than one mechanism. For example, electrons may travel from ferredoxin to plastoquinone via a dehydrogenase complex that resembles mitochondrial Complex I in overall structure and function. a. Sketch a diagram, similar to Figure 16.12, that shows the relative reduction potentials of the components in this pathway. b. Explain the advantage of using the dehydrogenase complex rather than a single oxidoreductase enzyme to transfer electrons from ferredoxin to plastoquinone.

36. Predict the effect of an uncoupler such as dinitrophenol (see Box 15.B) on production of a. ATP and b. NADPH by a chloroplast.

37. Antimycin A (an antibiotic) blocks electron transport in Complex III of the electron transport chain in mitochondria. How would the addition of antimycin A to chloroplasts affect chloroplast ATP synthesis and NADPH production?

38. a. Explain the term quantum yield. b. Does the quantum yield of photosynthesis increase or decrease for systems where c. the CF_o component of ATP synthase contains more c subunits and d. the proportion of cyclic electron flow through Photosystem I increases?

39. Oligomycin inhibits the proton channel (F_o) of the ATP synthase enzyme in mitochondria but does not inhibit CF_o. When oligomycin is added to plant cells undergoing photosynthesis, the cytosolic ATP/ADP ratio decreases whereas the chloroplastic ATP/ADP ratio is unchanged or even increases. Explain these results.

40. The compound paraquat accepts electrons from ferredoxin and then reduces oxygen to superoxide ($\cdot O_2^-$). What are the consequences for a plant treated with paraquat?

41. If Lys and Arg residues are modified to Ala in CF_1, predict the possible outcome of the addition of ADP and P_i in a reaction mix containing non-mutated and mutated versions of CF_1.

16.3 Carbon Fixation

42. Defend or refute this statement: The "dark" reactions are so named because these reactions occur only at night.

43. Examine the net equation for the light and "dark" reactions of photosynthesis—that is, the incorporation of one molecule of CO_2 into carbohydrate, which has the chemical formula $(CH_2O)_n$.

$$CO_2 + 2\,H_2O \rightarrow (CH_2O) + O_2 + H_2O$$

How would this equation differ for a bacterial photosynthetic system in which H_2S rather than H_2O serves as a source of electrons?

44. Melvin Calvin and his colleagues noted that when $^{14}CO_2$ was added to algal cells, a single compound was radiolabeled within 5 seconds of exposure. What is the compound and where does the radioactive label appear?

45. As described in the chapter, rubisco is not a particularly specific enzyme. Scientists have wondered why millions of years of evolution failed to produce a more specific enzyme. What advantage would be conferred upon a plant that evolved a rubisco enzyme that was able to react with CO_2 but not oxygen?

46. A tiny acorn grows into a massive oak tree. Using what you know about photosynthesis, what accounts for the increase in mass?

47. The $\Delta G^{\circ\prime}$ value for the rubisco reaction is −35.1 kJ·mol^{-1} and the ΔG value is −45.1 kJ·mol^{-1}. What is the ratio of products to reactants under normal cellular conditions? Assume a temperature of 25°C.

48. Efforts to engineer a more efficient rubisco, one that could fix CO_2 more quickly than three per second, could improve farming by allowing plants to grow larger and/or faster. Explain why the engineered rubisco might also decrease the need for nitrogen-containing fertilizers.

49. Some plants synthesize the sugar 2-carboxyarabinitol-1-phosphate (CA1P, shown below). This compound binds very tightly to rubisco, with a K_d of 32 nM, which inhibits enzyme activity. a. What is the probable mechanism of action of the inhibitor? b. Why do plants synthesize CA1P at night and break it down during the day? c. The enzyme that breaks down CA1P is a phosphatase that catalyzes hydrolysis of the phosphate at C1 of CA1P. How would NADPH, ribulose-5-phosphate, and P_i affect the activity of the phosphatase?

$$
\begin{array}{c}
CH_2OPO_3^{2-} \\
|\\
HO-C-COO^- \\
|\\
H-C-OH \\
|\\
H-C-OH \\
|\\
CH_2OH
\end{array}
$$

2-Carboxyarabinitol-
1-phosphate

50. An "activating" CO_2 reacts with a lysine side chain on rubisco to carboxylate it. This modification is essential for rubisco activity. a. Explain why the carboxylation reaction is favored at high pH. b. How does this activation step help link carbon fixation to the light reactions? c. How does the activation step "cooperate" with CA1P (see Problem 49) to regulate the activity of rubisco?

51. Chloroplast phosphofructokinase (PFK) is inhibited by ATP and NADPH. How does this observation support the hypothesis that the light reactions are linked to the regulation of glycolysis?

52. Crabgrass, a C_4 plant, remains green during a long spell of hot dry weather when C_3 grasses turn brown. Explain this observation.

53. Chloroplasts contain thioredoxin, a small protein with two cysteine residues that can form an intramolecular disulfide bond. The sulfhydryl/disulfide interconversion in thioredoxin is catalyzed by an enzyme known as ferredoxin–thioredoxin reductase. This enzyme, along with some of the Calvin cycle enzymes, also includes two Cys residues that undergo sulfhydryl/disulfide transitions. Show how disulfide interchange reactions involving thioredoxin could coordinate the activity of Photosystem I with the activity of the Calvin cycle.

54. The inner chloroplast membrane is impermeable to large polar and ionic compounds such as NADH and ATP. However, the membrane has an antiport protein that facilitates the passage of dihydroxyacetone phosphate or 3-phosphoglycerate in exchange for P_i. This system permits the entry of P_i for photophosphorylation and the exit of the products of carbon fixation. Draw a diagram that shows how the same antiport could "transport" ATP and reduced cofactors from the chloroplast to the cytosol.

55. The sedoheptulose bisphosphatase (SBPase) enzyme in the Calvin cycle catalyzes the removal of a phosphate group from C1 of sedoheptulose-1,7-bisphosphate (SBP) to produce sedoheptulose-7-phosphate (S7P). The $\Delta G^{\circ\prime}$ value for this reaction is -14.2 $kJ \cdot mol^{-1}$ and the ΔG value is -24.2 $kJ \cdot mol^{-1}$. What is the ratio of products to reactants under normal cellular conditions? Is this enzyme likely to be regulated in the Calvin cycle? Assume a temperature of 25°C.

56. Phosphoenolpyruvate carboxylase (PEPC) catalyzes the carboxylation of phosphoenolpyruvate (PEP) to oxaloacetate (OAA). The enzyme is commonly found in plants but is absent in animals. The reaction is shown. **a.** Why is PEPC referred to as an anaplerotic enzyme? **b.** Acetyl-CoA is an allosteric regulator of PEPC. Does acetyl-CoA activate or inhibit PEPC? Explain.

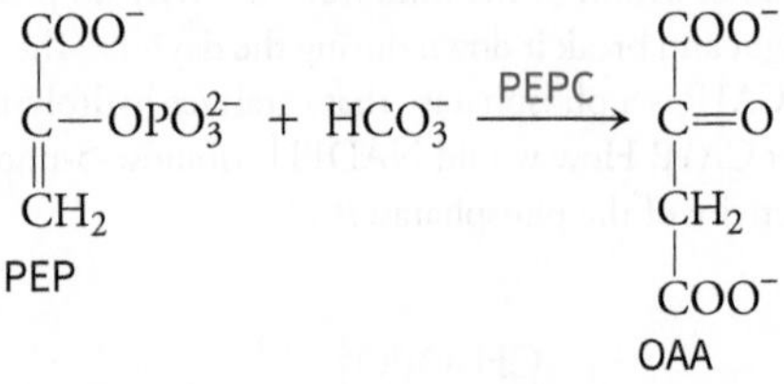

57. In germinating oil seeds, triacylglycerols are rapidly converted to sucrose and protein. What is the role of PEPC (see Problem 56) in this process?

58. The enzyme that "activates" glucose for starch synthesis is called ADP-glucose pyrophosphorylase (AGPase). Use the following observations to explain the role of inorganic phosphate (P_i) in regulating the activity of AGPase: **a.** Plants grown in a phosphate-deficient medium showed a ten-fold increase in starch accumulation. **b.** Leaf discs incubated with mannose (a C2 epimer of glucose) showed a fifteen-fold increase in starch accumulation.

59. AGPase (see Problem 58) is allosterically regulated by 3-phosphoglycerate (3PG). Is 3PG an allosteric activator or an inhibitor? Explain.

60. The Monsanto Company constructed a transgenic "Roundup Ready" soybean cultivar in which the bacterial gene for the enzyme EPSPS is inserted into the plant genome. EPSPS catalyzes an important step in the synthesis of aromatic amino acids, as shown in the diagram. The herbicide Roundup® contains glyphosate, a compound that competitively inhibits plant EPSPS but not the bacterial form of the enzyme. Explain the strategy for using glyphosate-containing herbicide on a soybean crop to kill weeds.

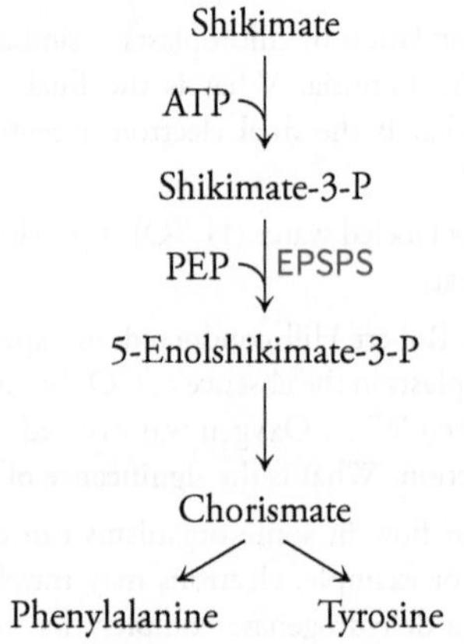

61. Is Roundup® toxic to humans? Explain why or why not.

62. Novartis constructed a corn cultivar that contains a gene from the bacterium *Bacillus thuringiensis* (Bt) that codes for an endotoxin protein. When insects of the order Lepidoptera (which includes the European corn borer but also unfortunately includes the Monarch butterfly) eat the corn, the ingested endotoxin enters the high pH environment of the insect's midgut. Under these conditions, the endotoxin forms a pore in the membrane of the cells lining the midgut, causing ions to flow into the cells and ultimately resulting in the death of the organism. Is the Bt endotoxin toxic to humans? Explain why or why not.

SELECTED READINGS

Bathellier, C., Tcherkez, G., Lorimer, G.H., and Farquhar, G.D., Rubisco is not really so bad, *Plant Cell Environ.* 41, 705–716, doi: 10.1111/pce.13149 (2018). [Argues that rubisco's speed, rate enhancement, and substrate specificity are in line with other enzymes.]

Cardona, T., Shao, S., and Nixon, P.J., Enhancing photosynthesis in plants: The light reactions, *Essays Biochem.* 62, 85–94, doi: 10.1042/EBC20170015 (2018). [Discusses ways to increase the efficiency of photosynthesis through genetic engineering and also includes an overview of the light-dependent reaction pathway.]

Johnson, M.P., Photosynthesis, *Essays Biochem.* 60, 255–273, doi: 10.1042/EBC20160016 (2016). [A very readable summary, with helpful diagrams.]

Martin, W.F., Bryant, D.A., and Beatty, J.T., A physiological perspective on the origin and evolution of photosynthesis, *FEMS Microbiol. Rev.* 42, 205–231, doi: 10.1093/femsre/fux056 (2018). [Outlines the features of early photosynthetic electron transport systems.]

CHAPTER 16 CREDITS

Figure 16.6 Image based on 1KZU. Prince, S.M., Papiz, M.Z., Freer, A.A., McDermott, G., Hawthornthwaite-Lawless, A.M., Cogdell, R.J., Isaacs, N.W., Apoprotein structure in the LH2 complex from Rhodopseudomonas acidophila strain 10050: Modular assembly and protein pigment interactions, *J. Mol. Biol.* 268, 412–423 (1997).

Figures 16.8 and 16.10 Images based on 1S5L. Ferreira, K.N., Iverson, T.M., Maghlaoui, K., Barber, J., Iwata, S., Architecture of the photosynthetic oxygen-evolving center, *Science* 303, 1831–1838 (2004).

Figure 16.9 Image based on 5XNL. Su, X., Ma, J., Wei, X., Cao, P., Zhu, D., Chang, W., Liu, Z., Zhang, X., Li, M., Structure and assembly mechanism of plant C2S2M2-type PSII-LHCII supercomplex, *Science* 357, 815–820 (2017).

Figure 16.11 Image based on 3WU2. Umena, Y., Kawakami, K., Shen, J.-R., Kamiya, N., Crystal structure of oxygen-evolving photosystem II at a resolution of 1.9 A, *Nature* 473, 55–60 (2011).

Figure 16.14 Image based on 6RQF. Malone, L.A., Qian, P., Mayneord, G.E., Hitchcock, A., Farmer, D.A., Thompson, R.F., Swainsbury, D.J.K., Ranson, N.A., Hunter, C.N., Johnson, M.P., Cryo-EM structure of the spinach cytochrome b6f complex at 3.6 Angstrom resolution, *Nature* 575, 535–539 (2019).

Figure 16.15 Image based on 1PLC. Guss, J.M., Bartunik, H.D., Freeman, H.C., Accuracy and precision in protein structure analysis: Restrained least-squares refinement of the structure of poplar plastocyanin at 1.33 A resolution, *Acta Crystallogr. B* 48, 790–811 (1992).

Figures 16.17 and 16.19 Images based on 1JB0. Jordan, P., Fromme, P., Witt, H.T., Klukas, O., Saenger, W., Krauss, N., Three-dimensional structure of cyanobacterial Photosystem I at 2.5 A resolution, *Nature* 411, 909–917 (2001).

Figure 16.18 Image based on 4XK8. Qin, X., Suga, M., Kuang, T., Shen, J.R., Structural basis for energy transfer pathways in the plant PSI-LHCI supercomplex, *Science* 348, 989–995 (2015).

Figure 16.20 Image based on 1CZP. Morales, R., Charon, M.H., Hudry-Clergeon, G., Petillot, Y., Norager, S., Medina, M., Frey, M., Refined X-ray structures of the oxidized, at 1.3 A, and reduced, at 1.17 A, [2Fe-2S] ferredoxin from the cyanobacterium Anabaena PCC7119 show redox-linked conformational changes, *Biochemistry* 38, 15764–15773 (1999).

Figure 16.26 Image based on 1RCX. Taylor, T.C., Andersson, I., The structure of the complex between rubisco and its natural substrate ribulose 1,5-bisphosphate, *J. Mol. Biol.* 265, 432–444 (1997).

Lipid Metabolism

About half the mass of an almond seed is lipid, 80% of which contains monounsatuated fatty acids such as oleate. Monounsaturated fatty acids are associated with a healthy balance among circulating lipoproteins, and inside cells, these fatty acids interact with regulatory molecules that promote mitochondiral biogenesis and oxidative metabolism.

Do You Remember?

- Lipids are predominantly hydrophobic molecules that can be esterified but cannot form polymers (Section 8.1).
- Metabolic fuels can be mobilized by breaking down glycogen, triacylglycerols, and proteins (Section 12.1).
- A few metabolites appear in several metabolic pathways (Section 12.3).
- Cells also use the energy of other phosphorylated compounds, thioesters, reduced cofactors, and electrochemical gradients (Section 12.2).

Like carbohydrates, lipids are metabolic fuels and therefore can be examined in terms of their synthesis, storage, mobilization, and catabolism—pathways that intersect with the processes we have already studied. In this chapter, we will investigate the breakdown and synthesis of fatty acids and related molecules. Unlike other molecules, lipids are insoluble in water, so we will begin by looking at how they travel between organs.

17.1 Lipid Digestion and Transport

KEY CONCEPTS

Summarize the roles of lipoproteins in lipid metabolism.

- Explain why lipoproteins are needed to transport lipids.
- Relate a lipoprotein's density to its lipid content.
- Describe the functions of LDL and HDL in cholesterol transport.

Figure 17.1 An atherosclerotic plaque in an artery. Note the thickening of the vessel wall.

Approximately half of all deaths in the United States are linked to the vascular disease **atherosclerosis** (a term derived from the Greek *athero,* "paste," and *sclerosis,* "hardness"). Atherosclerosis is a slow progressive disease in which white blood cells known as macrophages take up circulating lipids—particularly oxidized lipids—and lodge themselves in the walls of large blood vessels. The engorged macrophages participate in an inflammatory response and produce signals that recruit additional white blood cells, thereby perpetuating the inflammation.

The damaged vessel wall forms a plaque with a core of cholesterol, cholesteryl esters, and remnants of dead macrophages, surrounded by proliferating smooth muscle cells that may undergo calcification, as occurs in bone formation. This accounts for the "hardening" of the arteries. Although a very large plaque can occlude the lumen of the artery (**Fig. 17.1**), blood flow is usually not completely blocked unless the plaque ruptures, triggering formation of a blood clot that can prevent circulation to the heart (a heart attack) or brain (a stroke).

Lipoproteins in lipid transport

The lipids that lead to atherosclerosis are packaged in lipoproteins known as LDL (for low-density lipoproteins). **Lipoproteins** (particles consisting of lipids and specialized proteins) are the primary form of circulating lipid in animals (**Fig. 17.2**). The proteins wrap around a core of lipids in a flexible arrangement that accommodates changes in the size of the particle. The proteins help target the particle to cell surfaces and modulate the activities of enzymes that act on the component lipids. The various types of lipoproteins differ in size, lipid composition, protein composition, and density (a function of the relative proportions of lipid and protein).

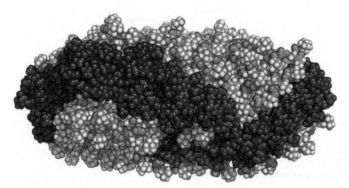

Figure 17.2 Model of a lipoprotein. In this computer simulation of an early HDL, two copies of the APOA1 protein (purple and blue) encircle a 100-Å disk containing 20 cholesterol molecules and 200 phospholipids (with C green, H white, O red, and P orange).

Question Explain why the proteins are amphipathic.

Recall from Section 12.1 that dietary lipids travel from the intestine to other tissues as chylomicrons. These lipoproteins are relatively large (1000 to 5000 Å in diameter) with a protein content of only 1% to 2%. Their primary function is to transport dietary triacylglycerols to adipose tissue and cholesterol to the liver. The liver repackages the cholesterol and other lipids—including triacylglycerols, phospholipids, and cholesteryl esters—into other lipoproteins known as VLDL (very-low-density lipoproteins). VLDL have a triacylglycerol content of about 50% and a diameter of about 500 Å. As they circulate in the bloodstream, VLDL give up triacylglycerols to the tissues, becoming smaller, denser, and richer in cholesterol and cholesteryl esters. After passing through an intermediate state (IDL, or intermediate-density lipoproteins), they become LDL, about 200 Å in diameter and about 45% cholesteryl ester (**Table 17.1**).

High concentrations of circulating LDL, measured as serum cholesterol (popularly called "bad cholesterol"), are a major factor in atherosclerosis. Some high-fat diets (especially those rich in saturated fats) may contribute to atherosclerosis by boosting LDL levels, but genetic factors, smoking, and infection also increase the risk of atherosclerosis. The disease is less likely to occur in individuals who consume low-cholesterol diets and who have high levels of HDL (high-density lipoproteins, sometimes called "good" cholesterol). HDL particles are even smaller and denser than LDL (see Table 17.1), and their primary function is to transport the body's excess cholesterol back to the liver. HDL therefore counter the atherogenic tendencies of LDL.

Table 17.1 Characteristics of Lipoproteins

Lipoprotein	Diameter (Å)	Density (g·cm⁻³)	% Protein	% Triacylglycerol	% Cholesterol and cholesteryl ester
Chylomicrons	1000–5000	<0.95	1–2	85–90	4–8
VLDL	300–800	0.95–1.006	5–10	50–65	15–25
IDL	250–350	1.006–1.019	10–20	20–30	40–45
LDL	180–250	1.019–1.063	20–25	7–15	45–50
HDL	50–120	1.063–1.210	40–55	3–10	15–20

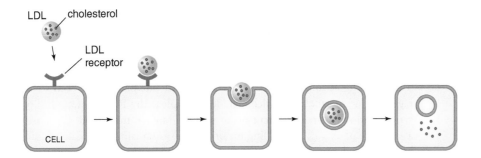

Figure 17.3 Summary of lipoprotein function.

The roles of the various lipoproteins are summarized in **Figure 17.3**. Briefly, the large chylomicrons, which are mostly lipid, transport dietary lipids to the liver and other tissues. The liver produces triacylglycerol-rich VLDL and releases them into the circulation. As the VLDL give up their triacylglycerols, they become cholesterol-rich LDL, which are taken up by cells. High-density lipoproteins (HDL), the smallest and densest of the lipoproteins, transport cholesterol from the tissues back to the liver.

All cells in the body can synthesize cholesterol (Section 17.4), which is an essential membrane component, but LDL are also a major source of this lipid. When LDL proteins dock with the LDL receptor on the cell surface, the lipoprotein–receptor complex undergoes endocytosis (Section 9.4). Inside the cell, the lipoprotein is degraded and cholesterol enters the cytosol. The LDL receptor itself is not degraded but cycles back to the cell surface, ready for another round of receptor-mediated endocytosis of an LDL particle.

The role of LDL in delivering cholesterol to cells is dramatically illustrated by the disease familial hypercholesterolemia, which is due to a genetic defect in the LDL receptor. The cells of homozygotes are unable to take up LDL, so the concentration of serum cholesterol is about three times higher than normal. This contributes to atherosclerosis, and many individuals die of the disease before age 30. Some other cases of **hypercholesterolemia** (high concentrations of cholesterol in the blood) can be treated with drugs that block the activity of a protein called PCSK9. This protein normally binds to the LDL receptor, making it more likely to be degraded than recycled to the cell surface. Inhibiting PCSK9 leads to more LDL receptors on the cell surface and more uptake of circulating cholesterol-rich LDL.

High-density lipoproteins (HDL) are essential for removing excess cholesterol from cells. The efflux of cholesterol requires the close juxtaposition of the cell membrane and an HDL particle as well as specific cell-surface proteins. One of these is an ABC transporter (Section 9.3) that acts as a floppase (see Fig. 8.15) to move cholesterol from the cytosolic leaflet to the extracellular leaflet, from which it can diffuse into the HDL particle.

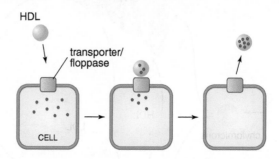

Defects in the gene for the transporter cause Tangier disease, which is characterized by accumulations of cholesterol in tissues and a high risk of heart attack.

Concept Check

1. Draw a diagram showing how cholesterol and other lipids are transported in the body.
2. Describe the relationship between a lipoprotein's protein and lipid content and its density.
3. Explain why both LDL and HDL are needed for cholesterol homeostasis.

17.2 Fatty Acid Oxidation

KEY CONCEPTS

Describe the chemical reactions required to oxidize fatty acids.

- Explain how fatty acids are activated by ATP.
- Describe how fatty acyl groups are imported into mitochondria.
- List the substrates and products of each cycle of β oxidation.
- Compare the pathways for oxidizing saturated, unsaturated, and odd-chain fatty acids.
- Summarize the role of peroxisomes in fatty acid oxidation.

The opposing actions of LDL and HDL are just one part of the body's efforts to regulate lipid metabolism, which includes multiple pathways. For example, lipids are synthesized from smaller precursors as well as obtained from food. Cells use lipids as a source of energy, as building materials, and as signaling molecules. Lipids may be catabolized quickly or stored long-term in adipose tissue. For the most part, the pathways responsible for synthesizing and degrading lipids fall within the shaded portions of the metabolic map shown in **Figure 17.4.**

The degradation (oxidation) of fatty acids releases free energy. In this section we describe how cells obtain, activate, and oxidize fatty acids. In humans, *dietary triacylglycerols are the primary source of fatty acids used as metabolic fuel.* The triacylglycerols are carried by lipoproteins to tissues, where hydrolysis separates their fatty acids from the glycerol backbone.

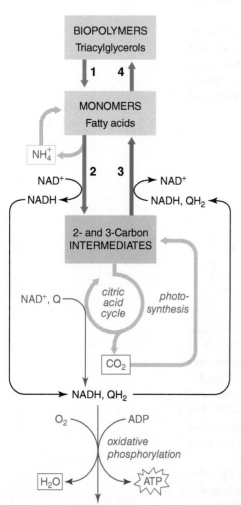

Figure 17.4 Lipid metabolism in context. Triacylglycerols, the "polymeric" form of fatty acids, are hydrolyzed to release fatty acids (1) that are oxidatively degraded to the two-carbon intermediate acetyl-CoA (2). Acetyl-CoA is also the starting material for the reductive biosynthesis of fatty acids (3), which can then be stored as triacylglycerols (4) or used in the synthesis of other lipids. Acetyl-CoA is also the precursor of lipids that are not built from fatty acids.

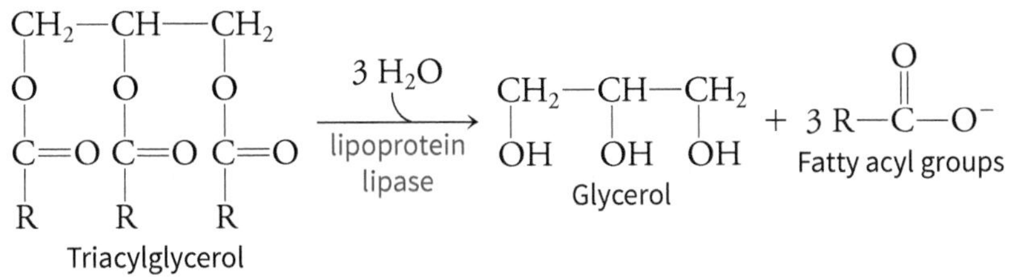

Hydrolysis occurs extracellularly, catalyzed by lipoprotein lipase. This enzyme is synthesized primarily by muscle and adipose cells, then secreted and taken up by the endothelial cells that line capillaries. There, a lipid-linked binding protein holds the lipase onto the outer cell surface as it acts on the triacylglycerol components of passing lipoproteins. Drugs that increase the production of lipoprotein lipase can help lower dangerously high levels of circulating lipids.

Triacylglycerols that are stored in adipose tissue are mobilized (their fatty acids are released to be used as fuel) by an intracellular hormone-sensitive lipase. The mobilized fatty acids travel through the bloodstream, not as part of lipoproteins, but bound to albumin, a 66-kD protein that accounts for about half of the serum protein (it also binds metal ions and hormones, serving as an all-purpose transport protein).

The concentration of free fatty acids in the body is very low because these molecules are detergents (which form micelles; see Section 2.2) and can disrupt cell membranes. After they enter cells, probably with the assistance of transport proteins, the fatty acids are either broken down for energy or re-esterified to form triacylglycerols or other complex lipids (as described in Section 17.4).

Fatty acids are activated before they are degraded

To be oxidatively degraded, a fatty acid must first be activated. Activation is a two-step reaction catalyzed by acyl-CoA synthetase. First, the fatty acid displaces the diphosphate group of ATP, then coenzyme A (HSCoA) displaces the AMP group to form an acyl-CoA:

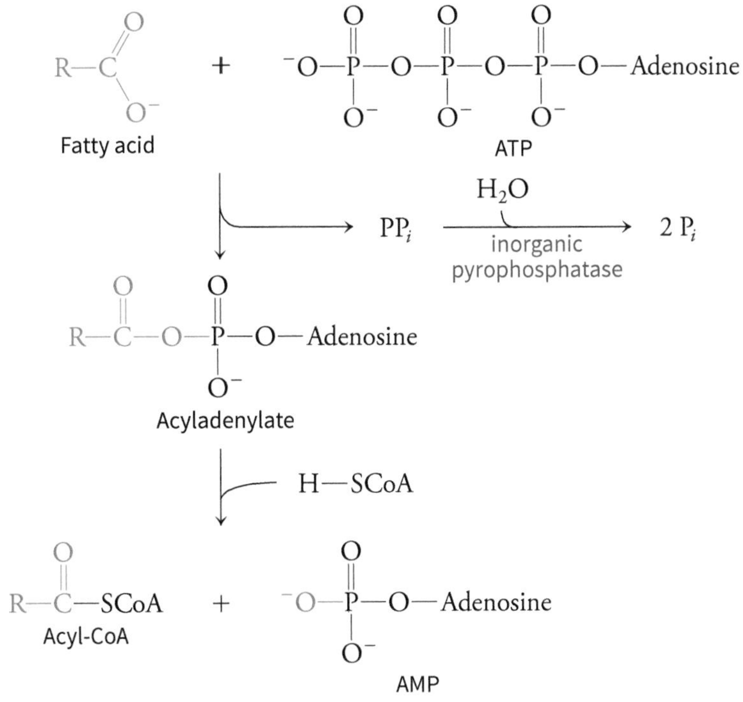

The acyladenylate product of the first step has a large free energy of hydrolysis so its formation conserves the free energy of the cleaved phosphoanhydride bond in ATP. The second step, transfer of the acyl group to CoA (the same molecule that carries acetyl groups as acetyl-CoA), likewise conserves free energy in the formation of a thioester bond (see Section 12.2). Consequently, the overall reaction

$$\text{fatty acid} + \text{CoA} + \text{ATP} \rightleftharpoons \text{acyl-CoA} + \text{AMP} + \text{PP}_i$$

has a free energy change near zero. However, subsequent hydrolysis of the product PP_i (by the ubiquitous enzyme inorganic pyrophosphatase) is highly exergonic, and this reaction makes the formation of acyl-CoA spontaneous and irreversible. This sequence of steps is another example of cells using the free energy of ATP hydrolysis to drive a reaction (the attachment of the fatty acid to coenzyme A) that would not occur on its own. Note, again, that ATP is not simply hydrolyzed in isolation; it participates in the reaction via formation of the acyladenylate intermediate. A similar ATP-dependent process occurs in activating ribose groups for nucleotide synthesis (Section 18.5) and in activating amino acids to be attached to transfer RNA molecules for protein synthesis (Section 22.1).

Most cells contain a set of acyl-CoA synthetases specific for fatty acids that are short (C_2–C_3), medium (C_4–C_{12}), long ($\geq C_{12}$), or very long ($\geq C_{22}$). The enzymes that are specific for the longest acyl chains may function in cooperation with a membrane transport protein so that the fatty acid is activated as it enters the cell. Once the large and polar coenzyme A is attached, the fatty acid is unable to diffuse back across the membrane and remains inside the cell to be metabolized.

In animals, fatty acids are activated in the cytosol, but the rest of the oxidation pathway occurs in the mitochondria. Because there is no transport protein for CoA adducts, acyl groups must cross the inner mitochondrial membrane via a shuttle system. First, a cytosolic carnitine acyltransferase transfers an acyl group from CoA to carnitine, a small quaternary amine (**Fig. 17.5**). The carnitine transporter then allows the acyl-carnitine to enter the mitochondrial matrix, where a mitochondrial carnitine acyltransferase transfers the acyl group from carnitine to a mitochondrial CoA molecule. The acyl group is now ready to be oxidized. Free carnitine returns to the cytosol via the transport protein to complete the cycle. Individuals whose diets include animal products obtain most of the carnitine they need from food, but human cells—mainly in the liver—can also synthesize carnitine from amino acids. Severe deficiencies of any of the components of the carnitine shuttle system can drastically reduce the ability to oxidize fatty acids.

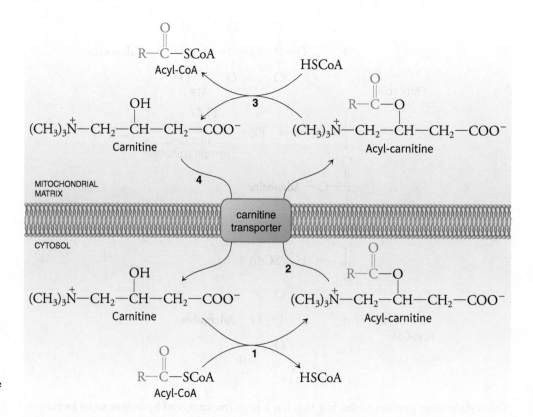

Figure 17.5 The carnitine shuttle system. (1) The acyl group is transferred from CoA to carnitine, (2) enters the mitochondrion as acyl-carnitine, and (3) is transferred back to CoA, leaving carnitine (4) to return to the cytosol.

Question Why do acyl groups move into the mitochondrion, not out of it?

Each round of β oxidation has four reactions

The pathway known as **β oxidation** degrades an acyl-CoA in a way that produces acetyl-CoA molecules for further oxidation and energy production by the citric acid cycle. In fact, in some tissues or under certain conditions, β oxidation supplies far more acetyl groups to the citric acid cycle than does glycolysis. β Oxidation also feeds electrons directly into the mitochondrial electron transport chain (see Fig. 15.11), which generates ATP by oxidative phosphorylation.

β Oxidation is a spiral pathway. *Each round consists of four enzyme-catalyzed steps that yield one molecule of acetyl-CoA and an acyl-CoA shortened by two carbons, which becomes the starting substrate for the next round.* Seven rounds of β oxidation degrade the C_{16} fatty acid palmitate to eight molecules of acetyl-CoA:

Figure 17.6 shows the reactions of β oxidation. The β indicates that oxidation occurs at the β position, which is the carbon atom that is two away from the carbonyl carbon (C3 is the β carbon). Note that acetyl units are not lost from the methyl end of the fatty acid but from the activated, CoA end.

The oxidation of fatty acids by the successive removal of two-carbon units was discovered in 1904, and the enzymatic steps were elucidated in the 1950s. However, β oxidation still offers surprises in the details. For example, many enzymes are required to fully degrade an acyl-CoA to acetyl-CoA. Each of the four steps shown in Figure 17.6 appears to be catalyzed by two to five different enzymes with different chain-length specificities, as in the acyl-CoA synthetase reaction. The existence of some of these isozymes was inferred from studies of patients with disorders of fatty acid oxidation. One of these often-fatal diseases is due to a deficiency of

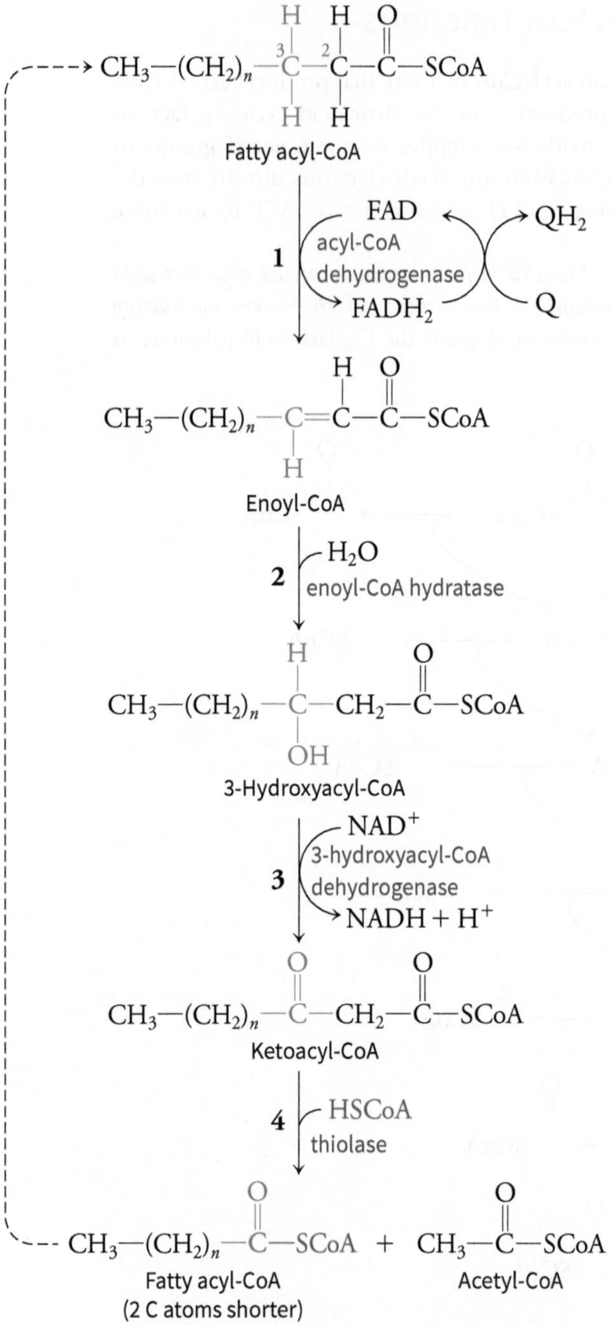

CH₃—(CH₂)ₙ—C—C—C—SCoA

Fatty acyl-CoA

1 FAD ← QH₂
acyl-CoA
dehydrogenase
FADH₂ ← Q

CH₃—(CH₂)ₙ—C=C—C—SCoA

Enoyl-CoA

2 H₂O
enoyl-CoA hydratase

CH₃—(CH₂)ₙ—C—CH₂—C—SCoA
OH

3-Hydroxyacyl-CoA

3 NAD⁺
3-hydroxyacyl-CoA
dehydrogenase
NADH + H⁺

CH₃—(CH₂)ₙ—C—CH₂—C—SCoA

Ketoacyl-CoA

4 HSCoA
thiolase

CH₃—(CH₂)ₙ—C—SCoA + CH₃—C—SCoA

Fatty acyl-CoA   Acetyl-CoA
(2 C atoms shorter)

1. Oxidation of acyl-CoA at the 2,3 position is catalyzed by an acyl-CoA dehydrogenase to yield a 2,3-enoyl-CoA. The two electrons removed from the acyl group are transferred to an FAD prosthetic group. A series of electron-transfer reactions eventually transfers the electrons to ubiquinone (Q).

2. The second step is catalyzed by a hydratase, which adds the elements of water across the double bond produced in the first step.

3. The hydroxyacyl-CoA is oxidized by another dehydrogenase. In this case, NAD⁺ is the cofactor.

4. The final step, thiolysis, is catalyzed by a thiolase and releases acetyl-CoA. The remaining acyl-CoA, two carbons shorter than the starting substrate, undergoes another round of the four reactions (dotted line).

Figure 17.6 The reactions of β oxidation.

Question Which steps of the pathway are redox reactions?

medium-chain acyl-CoA dehydrogenase; affected individuals cannot degrade acyl-CoAs having 4 to 12 carbons, and derivatives of these molecules accumulate in the liver and are excreted in the urine.

β Oxidation is a major source of cellular energy, especially during a fast, when carbohydrates are not available. Each round of β oxidation produces one QH₂, one NADH, and one acetyl-CoA. The citric acid cycle oxidizes the acetyl-CoA to produce an additional three NADH, one QH₂, and one GTP. Oxidation of all the reduced cofactors yields approximately 13 ATP: 3 from the two QH₂ and 10 from the four NADH (recall from the discussion of P:O ratios that oxidative phosphorylation is an indirect process, so the amount of ATP produced per pair of electrons entering the electron transport chain is not a whole number; see Section 15.3). A total of 14 ATP are generated from each round of β oxidation:

One round of β oxidation	Citric acid cycle	Oxidative phosphorylation
1 QH$_2$		→ 1.5 ATP
1 NADH		→ 2.5 ATP
1 Acetyl-CoA →	3 NADH	→ 7.5 ATP
	1 QH$_2$	→ 1.5 ATP
	1 GTP	→ 1 ATP

Total   14 ATP

β Oxidation is regulated primarily by the availability of free CoA (to make acyl-CoA) and by the ratios of NAD$^+$/NADH and Q/QH$_2$ (these reflect the state of the oxidative phosphorylation system). In addition, most of the enzymes of β oxidation are subject to product inhibition. Ketoacyl-CoA, the product of the pathways's third reaction, also inhibits the enzymes that catalyze the first two reactions.

Degradation of unsaturated fatty acids requires isomerization and reduction

Common fatty acids such as oleate and linoleate contain *cis* double bonds that present obstacles to the enzymes that catalyze β oxidation.

For linoleate, the first three rounds of β oxidation proceed as usual, but the acyl-CoA that begins the fourth round has a 3,4 double bond (originally the 9,10 double bond). Furthermore, this molecule is a *cis* enoyl-CoA, but enoyl-CoA hydratase (the enzyme that catalyzes step 2 of β oxidation) recognizes only the *trans* configuration. This metabolic obstacle is removed by the enzyme enoyl-CoA isomerase, which converts the *cis* 3,4 double bond to a *trans* 2,3 double bond so that β oxidation can continue.

A second obstacle for linoleate oxidation arises after the first reaction of the fifth round of β oxidation. Acyl-CoA dehydrogenase introduces a 2,3 double bond as usual, but the original 12,13 double bond of linoleate is now at the 4,5 position. The resulting dienoyl-CoA is not a good substrate for the next enzyme, enoyl-CoA hydratase. The dienoyl-CoA must undergo an NADPH-dependent reduction to convert its two double bonds to a single *trans* 3,4 double bond. This product must then be isomerized to produce the *trans* 2,3 double bond that is recognized by enoyl-CoA hydratase.

Carbon compounds with double bonds are slightly more oxidized than saturated compounds (see Table 1.3), so less energy is released in converting them to CO_2. Accordingly, a diet rich in unsaturated fatty acids contains fewer calories than a diet rich in saturated fatty acids. The bypass reactions described above provide the molecular explanation of why *unsaturated fatty acids yield less free energy than saturated fatty acids*. First, the enoyl-CoA isomerase reaction bypasses the QH_2-producing acyl-CoA dehydrogenase step, so 1.5 fewer ATP are produced. Second, the NADPH-dependent reductase consumes 2.5 ATP equivalents because NADPH is energetically equivalent to NADH.

Oxidation of odd-chain fatty acids yields propionyl-CoA

Most fatty acids have an even number of carbon atoms (this is because they are synthesized by the addition of two-carbon acetyl units, as we will see later in this chapter). However, some plant and bacterial fatty acids that make their way into the human system have an odd number of carbon atoms. *The final round of β oxidation of these molecules leaves a three-carbon fragment, propionyl-CoA, rather than the usual acetyl-CoA.*

Propionyl-CoA

This intermediate can be further metabolized by the sequence of steps outlined in **Figure 17.7**. At first, this pathway seems longer than necessary. For example, adding a carbon to C3 of the propionyl group would immediately generate succinyl-CoA. However, such a reaction is not chemically favored, because C3 is too far from the electron-delocalizing effects of the CoA thioester. Consequently, propionyl-CoA carboxylase must add a carbon to C2, and then methylmalonyl-CoA mutase must rearrange the carbon skeleton to produce succinyl-CoA. Note that succinyl-CoA is not the end point of the pathway. Because it is a citric acid cycle intermediate, it acts catalytically and is not consumed by the cycle (see Section 14.2). *The complete catabolism of the carbons derived from propionyl-CoA requires that the succinyl-CoA be converted to pyruvate and then to acetyl-CoA, which enters the citric acid cycle as a substrate.*

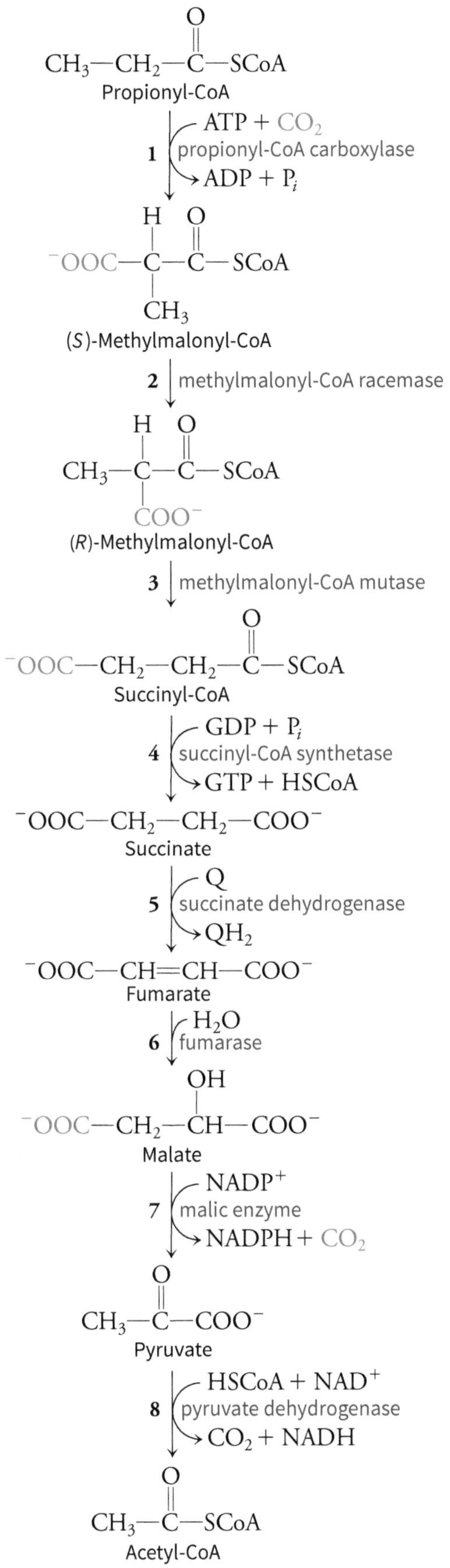

Figure 17.7 Catabolism of propionyl-CoA.

Question How many ATP equivalents can be produced from propionyl-CoA?

1. Propionyl-CoA carboxylase adds a carboxyl group at C2 of the propionyl group to form a four-carbon methylmalonyl group.

2. A racemase interconverts the two different methylmalonyl-CoA stereoisomers (the two configurations are indicated by the R and S symbols).

3. Methylmalonyl-CoA mutase rearranges the carbon skeleton to generate succinyl-CoA.

4–6. Succinyl-CoA, a citric acid cycle intermediate, is converted to malate by reactions 5–7 of the citric acid cycle (see Fig. 14.6).

7. After being exported from the mitochondria, malate is decarboxylated by malic enzyme to produce pyruvate in the cytosol.

8. Pyruvate, imported back into the mitochondria, can then be converted to acetyl-CoA by the pyruvate dehydrogenase complex.

Methylmalonyl-CoA mutase, which catalyzes step 3 of Figure 17.7, is an unusual enzyme because it mediates a rearrangement of carbon atoms and requires a prosthetic group derived from the vitamin cobalamin (vitamin B_{12}). The cofactor includes a hemelike ring structure with

Figure 17.8 The cobalamin-derived cofactor. The unusual C—Co bond is purple.

Question Locate the nucleotides in this structure.

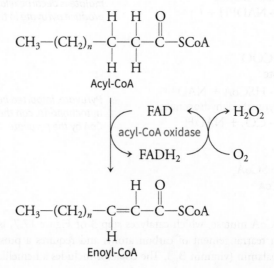

a central cobalt ion (**Fig. 17.8**). One of the Co ligands is a carbon atom, an extremely rare instance of a carbon–metal bond in a biological system. Only about a dozen enzymes are known to use cobalamin cofactors. The small amounts of cobalamin required for human health are usually easily obtained from a diet that contains animal products. Vegans, however, are advised to consume B_{12} supplements. A disorder of vitamin B_{12} absorption causes the disease pernicious anemia.

Some fatty acid oxidation occurs in peroxisomes

The majority of a mammalian cell's fatty acid oxidation occurs in mitochondria, but a small percentage is carried out in organelles known as **peroxisomes** (**Fig. 17.9**). In plants, all fatty acid oxidation occurs in peroxisomes and glyoxysomes. Peroxisomes are enclosed by a single membrane and contain a variety of degradative and biosynthetic enzymes. The peroxisomal β oxidation pathway differs from the mitochondrial pathway in the first step. An acyl-CoA oxidase catalyzes this reaction:

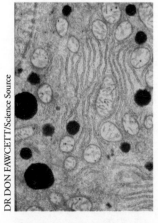

Figure 17.9 Peroxisomes. Nearly all eukaryotic cells contain these single-membrane-bound organelles (dark structures), which are similar to plant glyoxysomes (see Box 14.B).

DR DON FAWCETT/Science Source

The enoyl-CoA product of the reaction is identical to the product of the mitochondrial acyl-CoA dehydrogenase reaction (see Fig. 17.6), but the electrons removed from the acyl-CoA are transferred not to ubiquinone but directly to molecular oxygen to produce hydrogen peroxide, H_2O_2. This reaction product, which gives the peroxisome its name, is subsequently broken down by the peroxisomal enzyme catalase:

$$2 H_2O_2 \rightarrow 2 H_2O + O_2$$

The second, third, and fourth reactions of fatty acid oxidation are the same as in mitochondria.

Because the peroxisomal oxidation enzymes are specific for very-long-chain fatty acids (such as those containing over 20 carbons) and bind short-chain fatty acids with low affinity, *the peroxisome serves as a chain-shortening system.* The partially degraded fatty acyl-CoAs then make their way to the mitochondria for complete oxidation.

The peroxisome is also responsible for degrading some branched-chain fatty acids, which are not recognized by the mitochondrial enzymes. One such nonstandard fatty acid is phytanate,

Phytanate

which is derived from the side chain of chlorophyll molecules (see Fig. 16.3) and is present in all plant-containing diets. Phytanate must be degraded by peroxisomal enzymes because the methyl group at C3 prevents dehydrogenation by 3-hydroxyacyl-CoA dehydrogenase (step 3 of the standard β oxidation pathway). A deficiency of any of the phytanate-degrading enzymes results in Refsum's disease, a degenerative neuronal disorder characterized by an accumulation of phytanate in the tissues. The importance of peroxisomal enzymes in lipid metabolism (both catabolic and anabolic) is confirmed by the fatal outcome of most diseases stemming from deficient peroxisomal enzymes or improper synthesis of the peroxisomes themselves (**Box 17.A**).

| **Box 17.A** | α and ω oxidation of fatty acids |

α-oxidation of fatty acids

Branched-chain fatty acids with a methyl group on β-carbon like phytanic acid undergo an α-oxidation pathway in peroxisomes. In this reaction, phytanic acid is hydroxylated at α-carbon by phytanoyl CoA α-hydroxylase converting it into pristanic acid. Pristanic acid is activated to its CoA derivative and undergoes β-oxidation. Individuals with genetic defects in phytanoyl CoA α-hydroxylase develop Refsum's disease with neurological symptoms.

ω-oxidation of fatty acids

If there is a defect in one of the enzymes of β-oxidation, then ω-oxidation contributes towards oxidation of fatty acids. The site of ω-oxidation is the smooth ER of the liver and kidney. It normally occurs for medium-chain fatty acids (C_6-C_{12}). The ω carbon (most distal from carboxylic acid) gets hydroxylated by cytochrome P450 followed by oxidation of the alcohol group to aldehyde by the enzyme alcohol dehydrogenase. This is followed by the oxidation of the aldehyde group to a carboxylic acid by the enzyme aldehyde dehydrogenase resulting in a dicarboxylic acid. The dicarboxylic acid can attach with coenzyme A from either end or undergo a normal β-oxidation pathway in mitochondria.

Question **What are some of the key enzymes involved in the ω-oxidation and α-oxidation of fatty acids?**

Concept Check

1. Write a net equation for the activation of a fatty acid.
2. Make a list of all the enzymes required to fully oxidize linoleate.
3. Identify the energy-generating steps of fatty acid oxidation and explain how they contribute to ATP synthesis.
4. Compare the contributions of the cytosol, mitochondria, and peroxisomes to fatty acid catabolism.
5. What is the role of peroxisomes in fatty acid catabolism?

17.3 Fatty Acid Synthesis

KEY CONCEPTS

Describe the chemical reactions required to synthesize fatty acids and ketone bodies.

- Explain the purpose of the acetyl-CoA carboxylase reaction.
- List the substrates and products for the seven steps of fatty acid synthesis.
- Compare fatty acid synthesis and fatty acid oxidation.
- Summarize the ways that palmitate can be modified.
- Explain how fatty acid synthesis is regulated.
- Describe the steps of ketone body synthesis and degradation.

At first glance, fatty acid synthesis appears to be the exact reverse of fatty acid oxidation. For example, fatty acyl groups are built and degraded two carbons at a time, and several of the reaction intermediates in the two pathways are similar or identical. However, *the pathways for fatty acid synthesis and degradation must differ for thermodynamic reasons,* as we saw for glycolysis and gluconeogenesis. Since fatty acid oxidation is a thermodynamically favorable process, simply reversing the steps of this pathway would be energetically unfavorable.

In mammalian cells, the opposing metabolic pathways of fatty acid synthesis and degradation are entirely separate. β Oxidation takes place in the mitochondrial matrix, and synthesis occurs in the cytosol. Furthermore, the two pathways use different cofactors. β Oxidation funnels electrons to ubiquinone and NAD⁺, but in fatty acid synthesis, NADPH is the source of reducing power. β Oxidation requires two ATP equivalents (two phosphoanhydride bonds) to "activate" the acyl group, but the biosynthetic pathway consumes one ATP for every two carbons incorporated into a fatty acid. Finally, in β oxidation, the acyl group is attached to coenzyme A, but a growing fatty acyl chain is bound by an acyl-carrier protein (ACP; **Fig. 17.10**). Both ACP and coenzyme A include a pantothenate (vitamin B$_5$) derivative ending with a sulfhydryl group that forms a thioester with an acyl or acetyl group. In CoA, the pantothenate derivative is esterified to an adenine nucleotide; in ACP, the group is esterified to a serine OH group of a polypeptide (in mammals, ACP is part of a larger multifunctional protein, fatty acid synthase). In this section, we focus on the reactions of fatty acid synthesis, comparing and contrasting them to β oxidation.

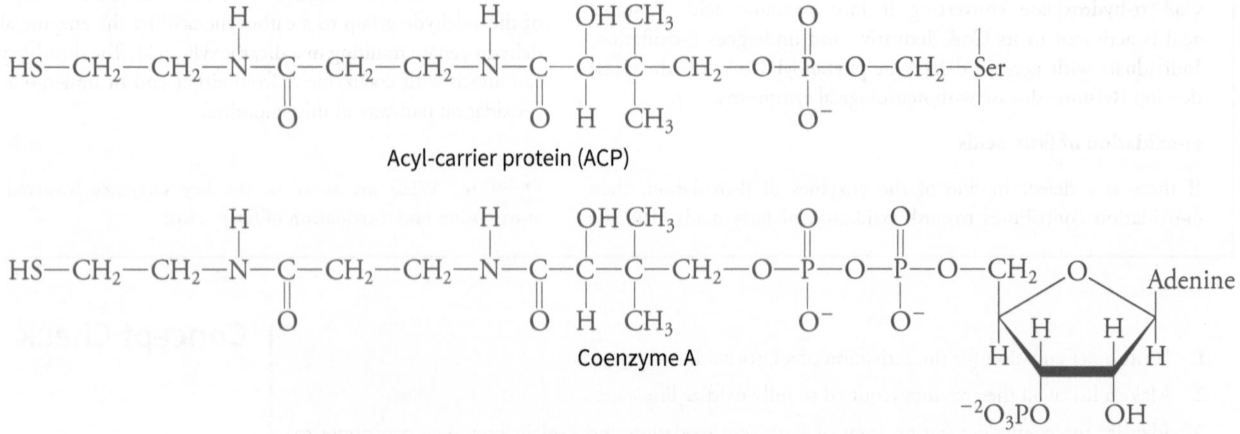

Figure 17.10 Acyl-carrier protein and coenzyme A.

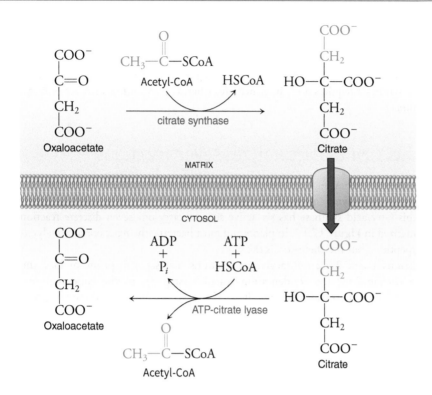

Figure 17.11 The citrate transport system. The citrate transport protein, along with mitochondrial citrate synthase and cytoplasmic ATP-citrate lyase, provides a route for transporting acetyl units from the mitochondrial matrix to the cytosol.

Question Add to the diagram the steps by which oxaloacetate carbons are returned to the matrix via the pyruvate transporter (see Fig. 14.18).

Acetyl-CoA carboxylase catalyzes the first step of fatty acid synthesis

The starting material for fatty acid synthesis is acetyl-CoA, which may be generated in the mitochondria by the action of the pyruvate dehydrogenase complex (Section 14.1). However, just as cytosolic acyl-CoA cannot directly enter the mitochondria to be oxidized, mitochondrial acetyl-CoA cannot exit to the cytosol for biosynthetic reactions. *The transport of acetyl groups to the cytosol involves citrate, which has a transport protein.* Citrate synthase (the enzyme that catalyzes the first step of the citric acid cycle; see Fig. 14.6) combines acetyl-CoA with oxaloacetate to produce citrate, which then leaves the mitochondria. ATP-citrate lyase "undoes" the citrate synthase reaction to produce acetyl-CoA and oxaloacetate in the cytosol (**Fig. 17.11**). Note that ATP is consumed in the ATP-citrate lyase reaction to drive the formation of a thioester bond.

The first step of fatty acid synthesis is the carboxylation of acetyl-CoA, an ATP-dependent reaction carried out by acetyl-CoA carboxylase. This enzyme catalyzes the rate-controlling step for the fatty acid synthesis pathway. The acetyl-CoA carboxylase mechanism is similar to that of propionyl-CoA carboxylase (step 1 in Fig. 17.7) and pyruvate carboxylase (see Fig. 13.9). First, CO_2 (as bicarbonate, HCO_3^-) is "activated" by its attachment to a biotin prosthetic group in a reaction that converts ATP to $ADP + P_i$:

$$\text{biotin} + HCO_3^- + ATP \rightarrow \text{biotin}\!-\!COO^- + ADP + P_i$$

Next, the carboxybiotin prosthetic group transfers the carboxylate group to acetyl-CoA to form the three-carbon malonyl-CoA and regenerate the enzyme:

Malonyl-CoA is the donor of the two-carbon acetyl units that are used to build a fatty acid. The carboxylate group added by the carboxylation reaction is lost in a subsequent decarboxylation reaction. This sequence of carboxylation followed by decarboxylation also occurs in the conversion of pyruvate to phosphoenolpyruvate in gluconeogenesis (see Section 13.2). Note that fatty acid synthesis requires a C_3 intermediate, whereas β oxidation involves only two-carbon acetyl units.

Fatty acid synthase catalyzes seven reactions

The protein that carries out the main reactions of fatty acid synthesis in animals is a 540-kD **multifunctional enzyme** made of two identical polypeptides (**Fig. 17.12**). Each polypeptide of this fatty acid synthase has six active sites to carry out seven discrete reactions, which are summarized in **Figure 17.13**. In plants and most bacteria, the reactions are catalyzed by separate polypeptides, but the chemistry is the same.

Reactions 1 and 2 are transacylation reactions that serve to prime or load the enzyme with the reactants for the condensation reaction (step 3). In the condensation reaction, decarboxylation of the malonyl group allows C2 to attack the acetyl thioester group to form acetoacetyl-ACP:

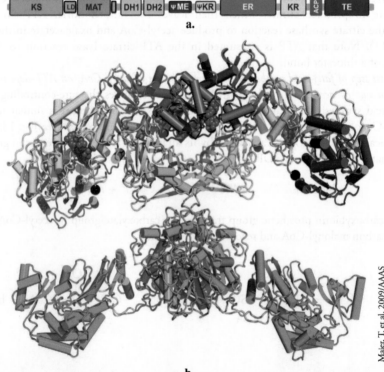

This chemistry explains the necessity for carboxylating an acetyl group to a malonyl group: C2 of an acetyl group would not be sufficiently reactive.

Figure 17.12 Mammalian fatty acid synthase. a. Domain organization of the fatty acid synthase polypeptide. The domains labeled KS, MAT, DH1/DH2, ER, KR, and TE are the six enzymes. ACP is the acyl-carrier protein. The domains labeled ψME and ψKR have no enzymatic activity. **b.** Three-dimensional structure of the fatty acid synthase dimer, with the domains of each monomer colored as in part **a**. The ACP and TE (thioesterase) modules are missing in this model. The NADP⁺ molecules are shown in blue, and black spheres represent the anchor points for the acyl-carrier protein domains.

Question Provide an example of another multifunctional enzyme.

Maier, T. et al. 2009/AAAS

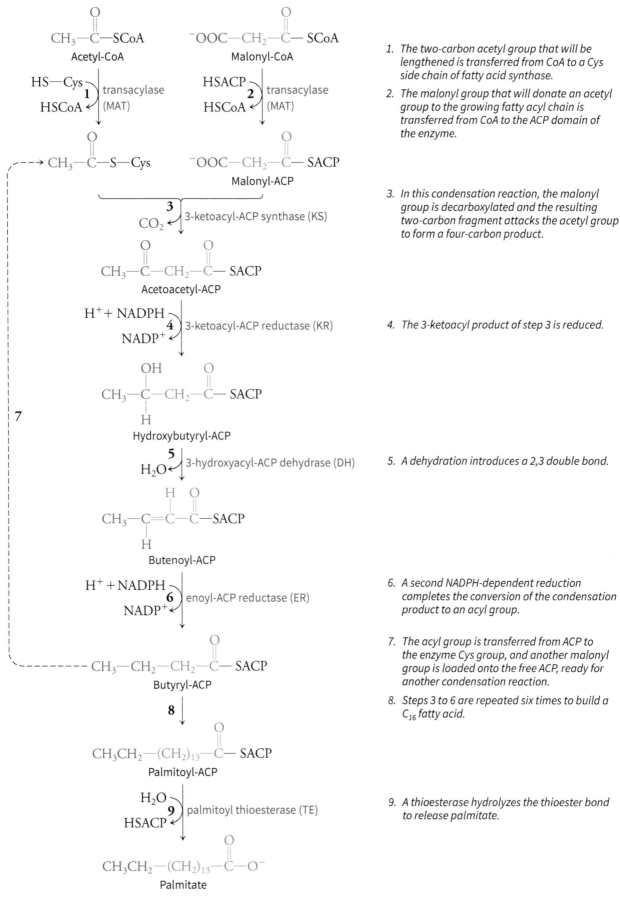

1. The two-carbon acetyl group that will be lengthened is transferred from CoA to a Cys side chain of fatty acid synthase.

2. The malonyl group that will donate an acetyl group to the growing fatty acyl chain is transferred from CoA to the ACP domain of the enzyme.

3. In this condensation reaction, the malonyl group is decarboxylated and the resulting two-carbon fragment attacks the acetyl group to form a four-carbon product.

4. The 3-ketoacyl product of step 3 is reduced.

5. A dehydration introduces a 2,3 double bond.

6. A second NADPH-dependent reduction completes the conversion of the condensation product to an acyl group.

7. The acyl group is transferred from ACP to the enzyme Cys group, and another malonyl group is loaded onto the free ACP, ready for another condensation reaction.

8. Steps 3 to 6 are repeated six times to build a C_{16} fatty acid.

9. A thioesterase hydrolyzes the thioester bond to release palmitate.

Figure 17.13 Fatty acid synthesis. The steps show how fatty acid synthase carries out the synthesis of the C_{16} fatty acid palmitate, starting from acetyl-CoA. The abbreviation next to each enzyme corresponds to a structural domain as shown in Figure 17.12.

The hydroxyacyl product of Reaction 4 is chemically similar to the hydroxyacyl product of step 2 of β oxidation, but the intermediates of the two pathways have opposite configurations (see Section 4.1):

$$CH_3-(CH_2)_n-\underset{\underset{H}{|}}{\overset{\overset{OH}{|}}{C}}-CH_2-\overset{\overset{O}{\|}}{C}-SACP \qquad CH_3-(CH_2)_n-\underset{\underset{OH}{|}}{\overset{\overset{H}{|}}{C}}-CH_2-\overset{\overset{O}{\|}}{C}-SCoA$$

3-Hydroxyacyl-ACP intermediate of
fatty acid synthesis
(D configuration)

3-Hydroxyacyl-ACP intermediate of
β oxidation
(L configuration)

Note also that growth of the acyl chain—like chain shortening in β oxidation—occurs at the thioester end of the molecule, not at the methyl end.

The NADPH required for the two reduction steps of fatty acid synthesis (steps 4 and 6) is supplied mostly by the pentose phosphate pathway (see Section 13.4). The synthesis of one molecule of palmitate (the usual product of fatty acid synthase) requires the production of 7 malonyl-CoA, at a cost of 7 ATP. The seven rounds of fatty acid synthesis consume 14 NADPH, which is equivalent to 14×2.5, or 35, ATP, bringing the total cost to 42 ATP. Still, this energy investment is much less than the free energy yield of oxidizing palmitate.

During fatty acid synthesis, the long flexible arm of the pantothenate derivative in ACP (see Fig. 17.10) shuttles intermediates between the active sites of fatty acid synthase (the lipoamide group in the pyruvate dehydrogenase complex functions similarly; see Section 14.1). In the fatty acid synthase dimer, two fatty acids can be built simultaneously.

Packaging several enzyme activities into one multifunctional protein like mammalian fatty acid synthase allows the enzymes to be synthesized and controlled in a coordinated fashion. Also, the product of one reaction can quickly diffuse to the next active site. Bacterial and plant fatty acid synthase systems may lack the efficiency of a multifunctional protein, but because the enzymes are not locked together, a wider variety of fatty acid products can be more easily made. In mammals, fatty acid synthase produces mostly the 16-carbon saturated fatty acid palmitate.

Other enzymes elongate and desaturate newly synthesized fatty acids

Some sphingolipids contain C_{22} and C_{24} fatty acyl groups. *These and other long-chain fatty acids are generated by enzymes known as elongases,* which extend the C_{16} fatty acid produced by fatty acid synthase. Elongation can occur in either the endoplasmic reticulum or mitochondria. The endoplasmic reticulum reactions use malonyl-CoA as the acetyl-group donor and are chemically similar to those of fatty acid synthase. In the mitochondria, fatty acids are elongated by reactions that more closely resemble the reversal of β oxidation but use NADPH.

Desaturases introduce double bonds into saturated fatty acids. These reactions take place in the endoplasmic reticulum, catalyzed by membrane-bound enzymes. The electrons removed in the dehydrogenation of the fatty acid are eventually transferred to molecular oxygen to produce H_2O. The most common unsaturated fatty acids in animals are palmitoleate (a C_{16} molecule) and oleate (a C_{18} fatty acid; see Section 8.1), both with one *cis* double bond at the 9,10 position. *Trans* fatty acids are relatively rare in plants and animals, but they are abundant in some prepared foods, which has produced confusion among individuals concerned with eating the "right" kinds of fats (**Box 17.B**).

Elongation can follow desaturation (and vice versa), so animals can synthesize a variety of fatty acids with different chain lengths and degrees of unsaturation. However, mammals cannot introduce double bonds at positions beyond C9 and therefore cannot synthesize fatty acids such as linoleate and linolenate. These molecules are precursors of the C_{20} fatty acid arachidonate and other lipids with specialized biological activities (**Fig. 17.14**). *Mammals must therefore obtain linoleate and linolenate from their diet.* These **essential fatty acids** are abundant in fish and plant oils. Unsaturated fatty acids with a double bond three carbons from the end, omega-3 fatty acids, may have health benefits (Box 8.A). A deficiency of essential fatty acids resulting from a very-low-fat diet may elicit symptoms such as slow growth and poor wound healing.

Box 17.B   Fats, Diet, and Heart Disease

Years of study have established a link between elevated LDL levels and atherosclerosis and indicate that certain diets contribute to the formation of fatty deposits that clog arteries and cause cardiovascular disease. Considerable research has been devoted to showing how dietary lipids influence serum lipid levels. For example, early studies showed that diets rich in saturated fats increased blood cholesterol (that is, LDL), whereas diets in which unsaturated vegetable oils replaced the saturated fats had the opposite effect. These and other findings led to recommendations that individuals at risk for atherosclerosis avoid butter, which is rich in saturated fat as well as cholesterol, and instead use substitutes prepared from cholesterol-free vegetable oils.

Processing liquid plant oils (triacylglycerols containing unsaturated fatty acids) often includes a hydrogenation step to chemically saturate the carbons of the fatty acyl chains. In this procedure, some of the original *cis* double bonds are converted to *trans* double bonds. In clinical studies, *trans* fatty acids are comparable to saturated fatty acids in their tendency to increase LDL levels and decrease HDL levels. Dietary guidelines now warn against the excessive

intake of *trans* fatty acids in the form of hydrogenated vegetable oils (small amounts of *trans* fatty acids also occur naturally in some animal fats).

Linking specific types of dietary fats to human health and disease has always been a risky venture because quantitative information comes mainly from epidemiological and clinical studies, which are typically time-consuming and often inconclusive or downright contradictory. Scientists still do not fully understand *how* the consumption of specific fatty acids—saturated or unsaturated, *cis* or *trans*—influences lipoprotein metabolism. Other dietary factors also play a role. For example, one consequence of low-fat diets is that individuals consume relatively more carbohydrates. Furthermore, when they reduce their meat intake (an obvious source of fat), people eat more fruits and vegetables, which may have health-enhancing effects of their own.

Question   Rank the "healthiness" of the following sources of fatty acids: animal fat, olive oil, and hydrogenated soybean oil.

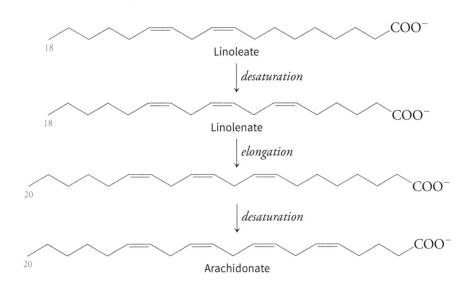

Figure 17.14 Synthesis of arachidonate. Linoleate (or linolenate) is elongated and desaturated to produce arachidonate, a C_{20} fatty acid with four double bonds.

Fatty acid synthesis can be activated and inhibited

Under conditions of abundant metabolic fuel, the products of carbohydrate and amino acid catabolism are directed toward fatty acid synthesis, and the resulting fatty acids are stored inside cells as triacylglycerols. These form a lipid droplet that is coated with a phospholipid monolayer.

The rate of fatty acid synthesis is controlled by acetyl-CoA carboxylase, which catalyzes the first step of the pathway. This enzyme is inhibited by its reaction product (malonyl-CoA) and the ultimate product of fatty acid synthesis (palmitoyl-CoA), and it is allosterically activated by citrate (which signals abundant acetyl-CoA). The enzyme is also subject to allosteric regulation by hormone-stimulated phosphorylation and dephosphorylation. Curiously, human acetyl-CoA carboxylase polymerizes into long helical filaments that are either more active or less active, depending on the presence of allosteric regulators and the degree of phosphorylation.

The concentration of malonyl-CoA is also critical for preventing the wasteful simultaneous activity of fatty acid synthesis and fatty acid oxidation. Malonyl-CoA is the source of acetyl groups that are incorporated into fatty acids, and it also blocks β oxidation by inhibiting carnitine acyltransferase, the enzyme involved in shuttling acyl groups into the mitochondria

Figure 17.15 Some control mechanisms in fatty acid metabolism. Red symbols indicate inhibition and the green symbol indicates activation.

Question Name all the enzymes required to convert excess glucose into fatty acids.

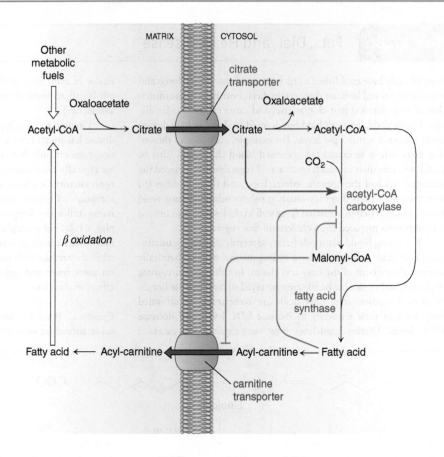

(see Fig. 17.5). Consequently, *when fatty acid synthesis is under way, no acyl groups are transported into the mitochondria for oxidation.* Some of the mechanisms that regulate fatty acid metabolism are summarized in **Figure 17.15**.

There are both natural and synthetic inhibitors of fatty acid synthase, such as the widely used antibacterial agent triclosan and drugs that target pathogens more specifically (**Box 17.C**). Fatty acid synthase inhibitors are of great scientific and popular interest, given that excess body weight (due to fat) is a major health problem, affecting about two-thirds of the population of the United States. Because many tumors sustain high levels of fatty acid synthesis, fatty acid synthase inhibitors may also be useful for treating cancer.

Acetyl-CoA can be converted to ketone bodies

During a prolonged fast, when glucose is unavailable from the diet and liver glycogen has been depleted, many tissues depend on fatty acids released from stored triacylglycerols to meet their energy needs. However, the brain does not burn fatty acids because they pass poorly through the blood–brain barrier. Gluconeogenesis helps supply the brain's energy needs, but the liver also produces **ketone bodies** to supplement gluconeogenesis. The ketone bodies—acetoacetate and 3-hydroxybutyrate (also called β-hydroxybutyrate)—are synthesized from acetyl-CoA in liver mitochondria by a process called **ketogenesis.** Because ketogenesis uses fatty acid–derived acetyl groups, it helps spare amino acids that would otherwise be diverted to gluconeogenesis.

The assembly of ketone bodies is somewhat reminiscent of the synthesis of fatty acids or the oxidation of fatty acids in two-carbon steps (**Fig. 17.16**). In fact, the hydroxymethylglutaryl- CoA intermediate is chemically similar to the 3-hydroxyacyl intermediates of β oxidation and fatty acid synthesis.

Box 17.C   Inhibitors of Fatty Acid Synthesis

Because fatty acid synthesis is an essential metabolic activity, inhibiting the process in pathogenic organisms—but not in their mammalian hosts—is a useful strategy for preventing or curing certain infectious diseases. For example, many cosmetics, toothpastes, antiseptic soaps, and even plastic toys and kitchenware contain the compound 5-chloro-2-(2,4-dichlorophenoxy)-phenol, better known as triclosan.

Triclosan was long believed to act as a general microbicide, an agent that kills nonspecifically, much like household bleach or ultraviolet light. Such microbicides are effective because it is difficult for bacteria to evolve specific resistance mechanisms. However, triclosan actually operates more like an antibiotic with a specific biochemical target, in this case, enoyl-ACP reductase, which catalyzes step 6 of fatty acid synthesis (see Fig. 17.13).

The enzyme's natural substrate has a K_M of about 22 μM, but the dissociation constant for the inhibitor is 20 to 40 pM, indicating extremely tight binding. In the active site, one of the phenyl rings of triclosan, whose structure mimics the structure of the reaction intermediate, stacks on top of the nicotinamide ring of the NADH cofactor. Triclosan also binds through van der Waals interactions and hydrogen bonds with amino acid residues in the active site.

The antibiotic isoniazid has been used to treat *Mycobacterium tuberculosis* infections. Inside the bacterial cell, isoniazid is oxidized and the reaction product combines with NAD+ to generate a compound that inhibits one of the cell's enoyl-ACP reductases. The target enzyme is specific for extremely long-chain fatty acids, which are incorporated into mycolic acids, the waxlike components of the mycobacterial cell wall.

Drugs such as isoniazid must be taken for many months, since mycobacteria replicate very slowly and can remain dormant inside host cells, protected from drugs as well as the host's immune system.

Some fungal species are susceptible to cerulenin, which inhibits 3-ketoacyl-ACP synthase (step 3 of fatty acid synthesis; see Fig. 17.13) by blocking the reaction of malonyl-ACP, that is, the condensation step. Cerulenin is also effective against *M. tuberculosis*, inhibiting the production of long-chain fatty acids required for cell-wall synthesis. The drug, which contains a reactive epoxide group, reacts irreversibly with the enzyme's active-site cysteine residue, forming a C2—S covalent bond. Cerulenin's hydrocarbon tail occupies the site that would normally accommodate the growing fatty acyl chain.

Question   **What types of mutations might allow bacteria to become resistant to triclosan?**

Triclosan   Isoniazid   Cerulenin

As we will see, the first steps of ketogenesis are identical to the first steps of cholesterol synthesis (Section 17.4).

Because they are small and water-soluble, ketone bodies are transported in the bloodstream without specialized lipoproteins, and they can easily pass into the central nervous system. During periods of high ketogenic activity, such as in diabetes, ketone bodies may be produced faster than they are consumed. Some of the excess acetoacetate breaks down to acetone, which gives the breath a characteristic sweet smell. Ketone bodies are also acids, with a pK of about 3.5. Their overproduction can lead to a drop in the pH of the blood, a condition called ketoacidosis. Mild symptoms may also develop in some individuals following a high-protein, low-carbohydrate diet, when ketogenesis increases to offset the shortage of dietary carbohydrates.

Ketone bodies produced by the liver enter other cells by passive transport through monocarboxylate transport proteins. Once inside mitochondria, the ketone bodies are converted back to acetyl-CoA to be oxidized by the citric acid cycle (**Fig. 17.17**). The liver itself cannot catabolize ketone bodies because it lacks one of the required enzymes, succinyl-CoA transferase (also called 3-ketoacyl-CoA transferase).

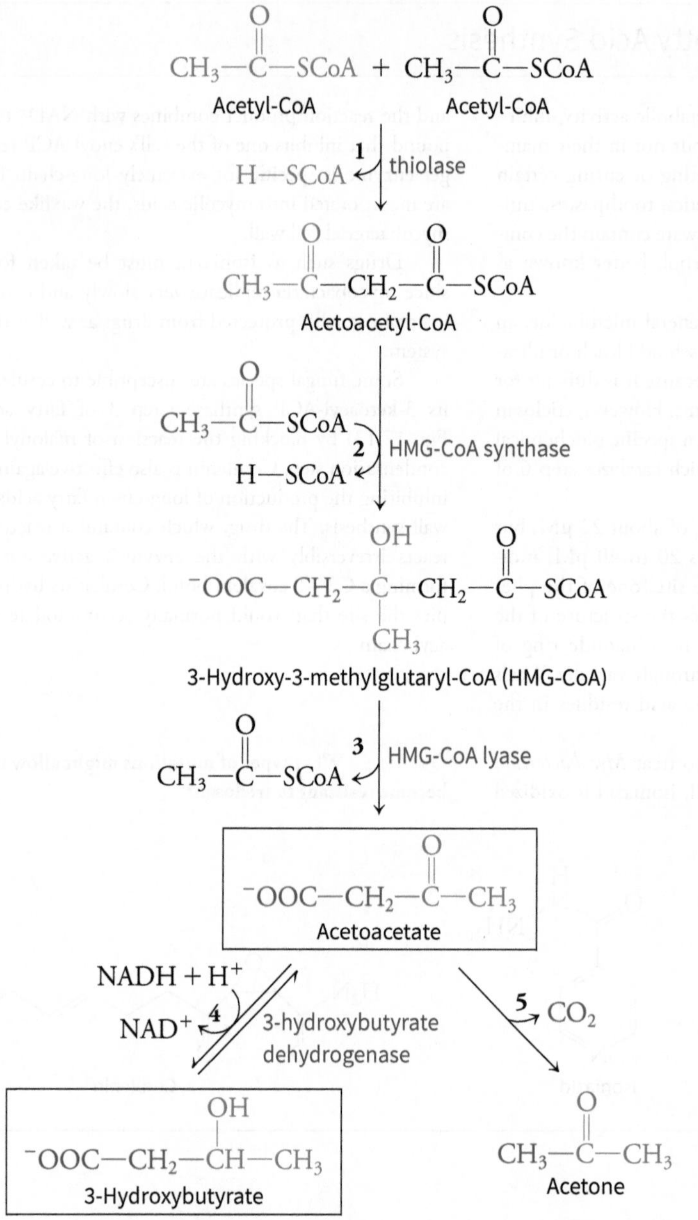

1. Two molecules of acetyl-CoA condense to form acetoacetyl-CoA. The reaction is catalyzed by a thiolase, which breaks a thioester bond.

2. The four-carbon acetoacetyl group condenses with a third molecule of acetyl-CoA to form the six-carbon 3-hydroxymethylglutaryl-CoA (HMG-CoA).

3. HMG-CoA is then degraded to the ketone body acetoacetate and acetyl-CoA.

4. Acetoacetate undergoes reduction to produce another ketone body, 3-hydroxybutyrate.

5. Some acetoacetate may also undergo nonenzymatic decarboxylation to acetone and CO_2.

Figure 17.16 Ketogenesis. The ketone bodies are boxed.

Concept Check

1. Explain why the pathways of fatty acid synthesis and degradation must differ.
2. Describe the role of malonyl-CoA in fatty acid synthesis.
3. Construct a table to compare fatty acid synthesis and oxidation with respect to cellular location, cofactors, use of ATP, acyl group carrier, and role of acetyl-CoA.
4. Explain the purpose of elongases and desaturases.
5. List the metabolites that regulate fatty acid synthesis.
6. Describe the conditions under which ketone bodies are made and used.

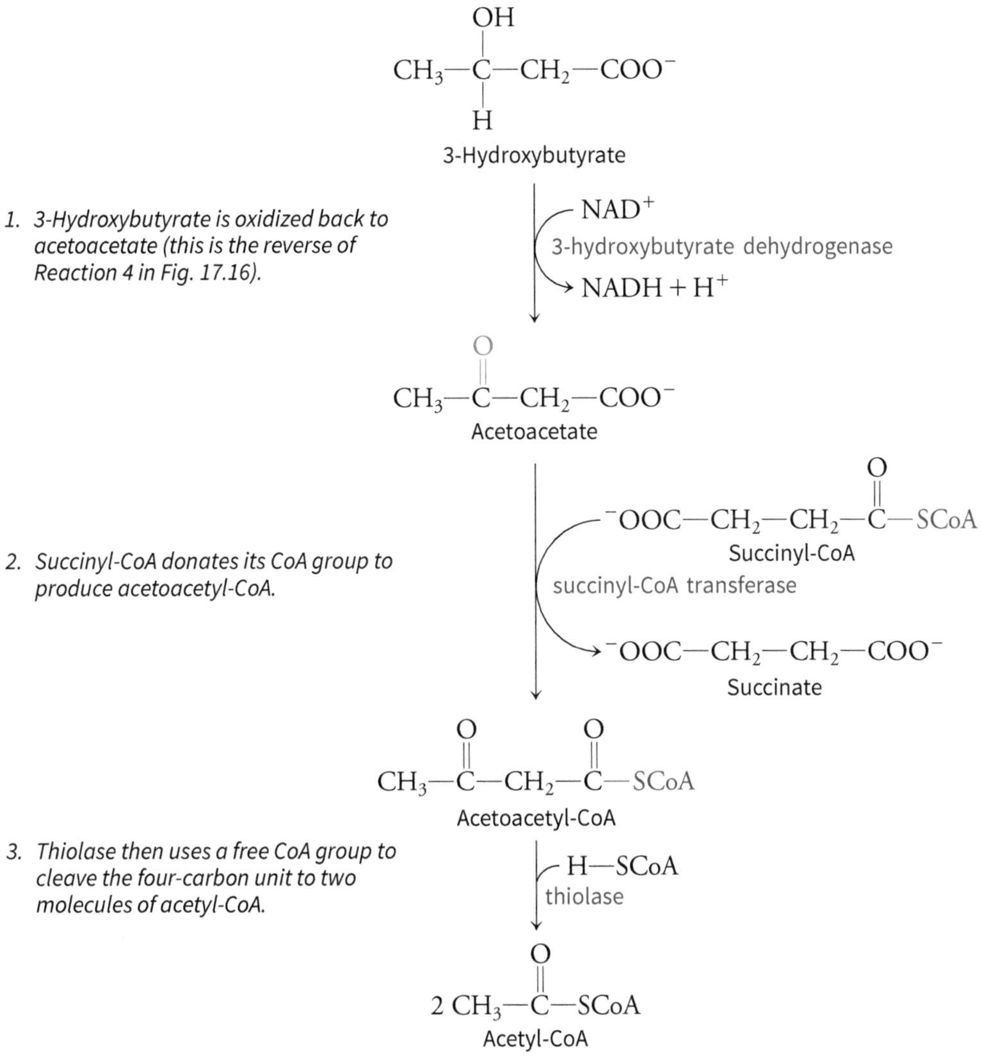

$$CH_3-\underset{\underset{H}{|}}{\overset{\overset{OH}{|}}{C}}-CH_2-COO^-$$

3-Hydroxybutyrate

1. 3-Hydroxybutyrate is oxidized back to acetoacetate (this is the reverse of Reaction 4 in Fig. 17.16).

NAD$^+$
3-hydroxybutyrate dehydrogenase
NADH + H$^+$

$$CH_3-\overset{\overset{O}{\|}}{C}-CH_2-COO^-$$

Acetoacetate

2. Succinyl-CoA donates its CoA group to produce acetoacetyl-CoA.

$$^-OOC-CH_2-CH_2-\overset{\overset{O}{\|}}{C}-SCoA$$

Succinyl-CoA
succinyl-CoA transferase

$$^-OOC-CH_2-CH_2-COO^-$$

Succinate

$$CH_3-\overset{\overset{O}{\|}}{C}-CH_2-\overset{\overset{O}{\|}}{C}-SCoA$$

Acetoacetyl-CoA

3. Thiolase then uses a free CoA group to cleave the four-carbon unit to two molecules of acetyl-CoA.

H—SCoA
thiolase

$$2\ CH_3-\overset{\overset{O}{\|}}{C}-SCoA$$

Acetyl-CoA

Figure 17.17 Catabolism of ketone bodies.

Question **Compare this pathway to ketogenesis. Which steps are similar?**

17.4 Synthesis of Other Lipids

KEY CONCEPTS

Summarize the synthesis of triacylglycerols, phospholipids, and cholesterol.

- Explain how acyl groups are activated for transfer.
- Describe the roles of CTP in glycerophospholipid synthesis.
- Identify the regulated step of cholesterol synthesis.
- Describe the metabolic fates of cholesterol.

Lipid metabolism encompasses many chemical reactions involving fatty acids, which are structural components of other lipids such as triacylglycerols, glycerophospholipids, and sphingolipids. Fatty acids such as arachidonate are also the precursors of eicosanoids that function as signaling molecules (Section 10.4). This section covers the biosynthesis of some of the major types of lipids, including the synthesis of cholesterol from acetyl-CoA.

Triacylglycerols and phospholipids are built from acyl-CoA groups

Cells have a virtually unlimited capacity for storing fatty acids in the form of triacylglycerols, which aggregate in the cytoplasm to form droplets surrounded by a layer of amphipathic phospholipids. *Triacylglycerols are synthesized by attaching fatty acyl groups to a glycerol backbone derived from phosphorylated glycerol or from glycolytic intermediates,* for example, dihydroxyacetone phosphate:

$$
\begin{array}{ccc}
\text{CH}_2\text{—OH} & & \text{CH}_2\text{—OH} \\
| & \xrightarrow[\text{dehydrogenase}]{\text{glycerol-3-phosphate}} & | \\
\text{C}=\text{O} & & \text{CH—OH} \\
| & & | \\
\text{CH}_2\text{—O—PO}_3^{2-} & & \text{CH}_2\text{—O—PO}_3^{2-}
\end{array}
$$

NADH + H⁺ → NAD⁺

Dihydroxyacetone phosphate   Glycerol-3-phosphate

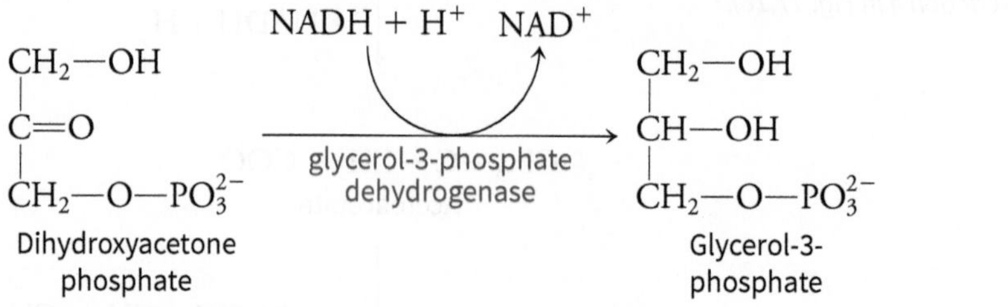

The fatty acyl groups are first activated to CoA thioesters in an ATP-dependent manner:

$$\text{fatty acid} + \text{CoA} + \text{ATP} \rightleftharpoons \text{acyl-CoA} + \text{AMP} + \text{PP}_i$$

This reaction is catalyzed by acyl-CoA synthetase, the same enzyme that activates fatty acids for oxidation. Triacylglycerols are assembled as shown in **Figure 17.18**. An acyltransferase appends a fatty acyl group to C1 of glycerol-3-phosphate. A second acyltransferase reaction adds an acyl group to C2, yielding phosphatidate. A phosphatase removes P_i to produce diacylglycerol. The addition of a third acyl group yields a triacylglycerol. The acyltransferases that add fatty acids to the glycerol backbone are not highly specific with respect to chain length or degree of unsaturation of the fatty acyl group, but human triacylglycerols usually contain palmitate at C1 and unsaturated oleate at C2.

The triacylglycerol biosynthetic pathway also provides the precursors for glycerophospholipids. *These amphipathic phospholipids are synthesized from phosphatidate or diacylglycerol by pathways that include an activating step in which the nucleotide cytidine triphosphate (CTP) is cleaved.* In some cases, the phospholipid head group is activated; in other cases, the lipid tail portion is activated.

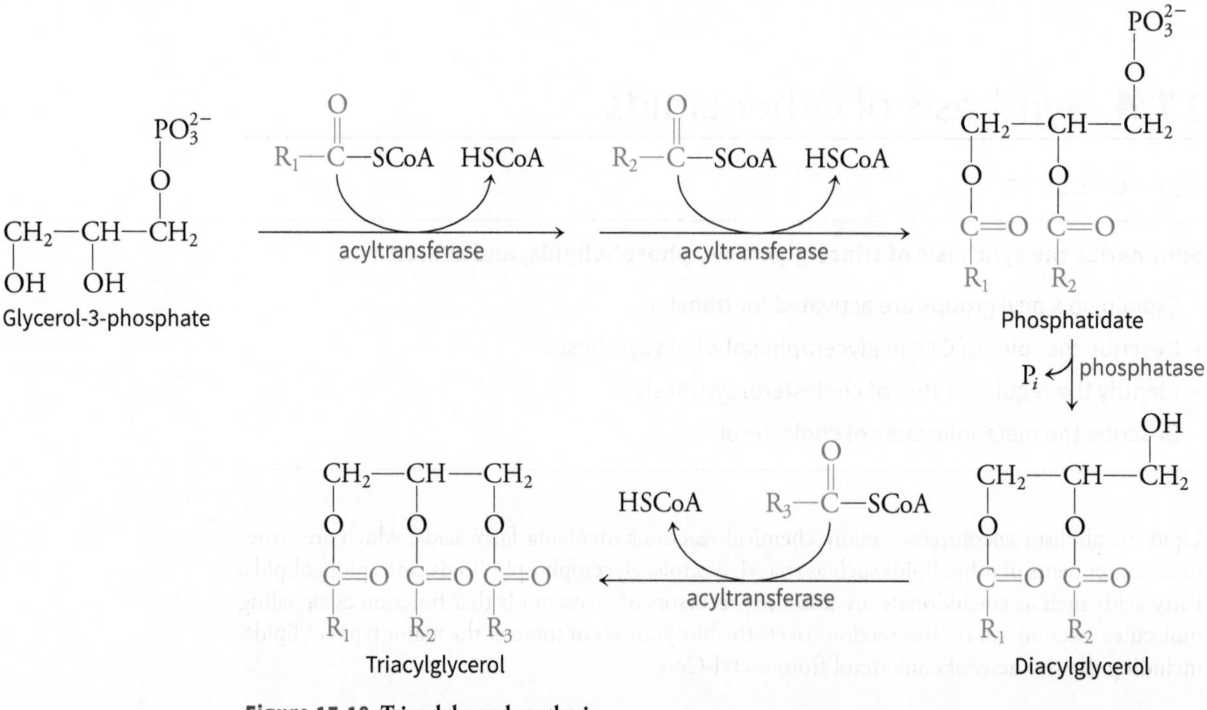

Figure 17.18 Triacylglycerol synthesis.

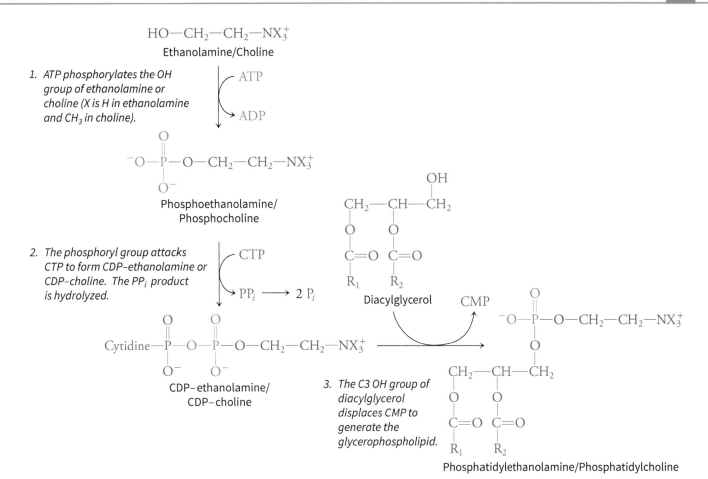

Figure 17.19 Synthesis of phosphatidylethanolamine and phosphatidylcholine.

Figure 17.19 shows how the head groups ethanolamine and choline are activated before being added to diacylglycerol to produce phosphatidylethanolamine and phosphatidylcholine. Similar chemistry involving nucleotide sugars is used in the synthesis of glycogen from UDP–glucose (see Section 13.3) and starch from ADP–glucose (see Section 16.3).

Phosphatidylserine is synthesized from phosphatidylethanolamine by a head-group exchange reaction in which serine displaces the ethanolamine head group:

In the synthesis of phosphatidylinositol, the diacylglycerol component is activated, rather than the head group, so that the inositol head group adds to CDP–diacylglycerol (**Fig. 17.20**).

Glycerophospholipids (and sphingolipids) are components of cellular membranes. New membranes are formed by inserting proteins and lipids into pre-existing membranes, mainly in the endoplasmic reticulum. The newly synthesized membrane components reach their final cellular destinations primarily via vesicles that bud off the endoplasmic reticulum and, in some

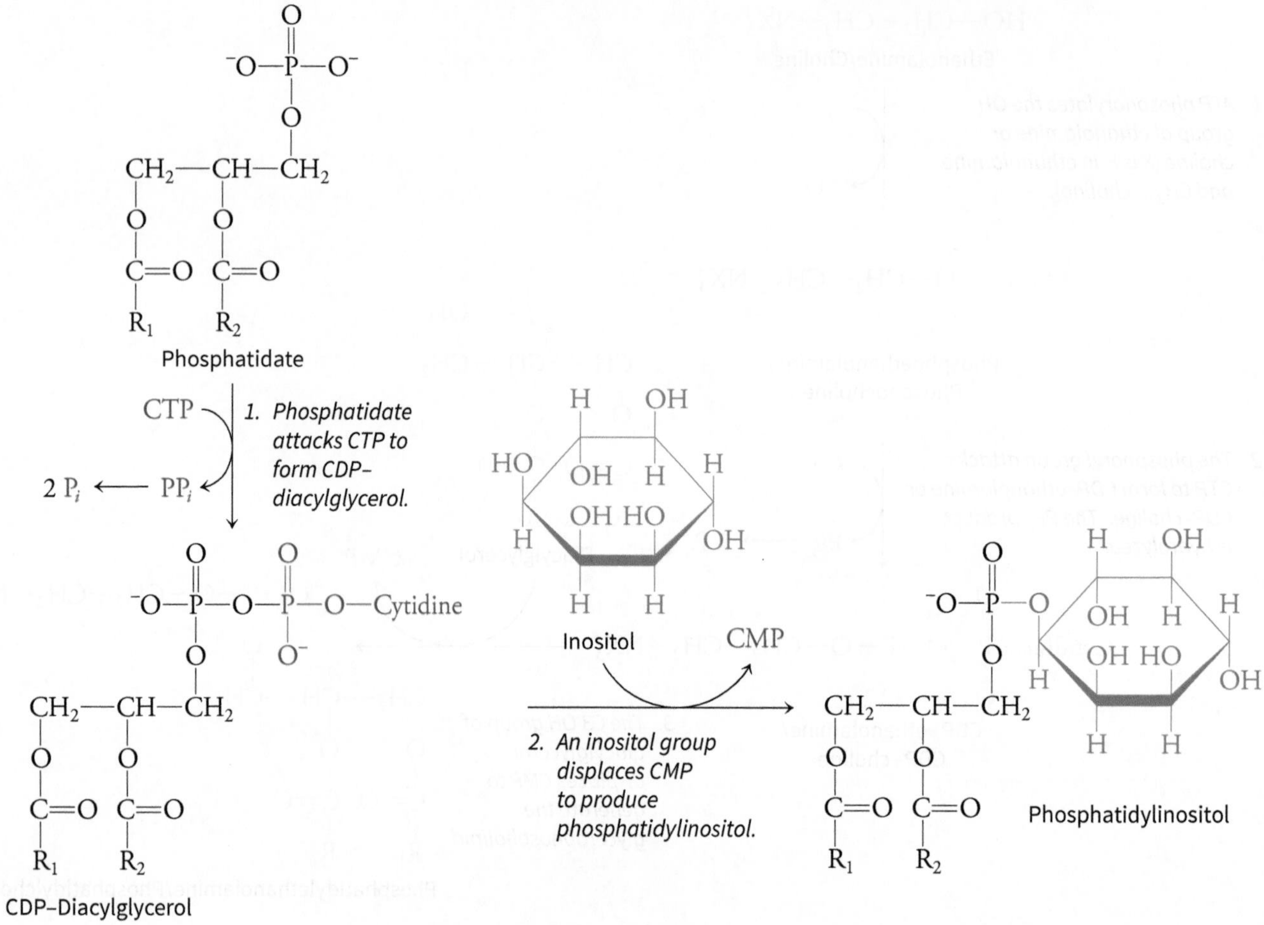

Figure 17.20 Phosphatidylinositol synthesis.

cases, by diffusing at points where two membranes make physical contact. Glycerophospholipids may undergo remodeling through the action of phospholipases and acyltransferases that remove and reattach different fatty acyl groups.

Cholesterol synthesis begins with acetyl-CoA

Cholesterol molecules, like fatty acids, are built from two-carbon acetyl units. In fact, *the first two steps of cholesterol synthesis are the same as the first two steps of ketogenesis.* However, ketone bodies are synthesized in the mitochondria (and only in the liver), and cholesterol is synthesized in the cytosol. The reactions of cholesterol biosynthesis and ketogenesis diverge after the production of HMG-CoA. In ketogenesis, this compound is cleaved to produce acetoacetate (see Fig. 17.16). In cholesterol synthesis, the thioester group of HMG-CoA is reduced to an alcohol. HMG-CoA reductase catalyzes a four-electron reductive deacylation to yield the six-carbon compound mevalonate (**Fig. 17.21**).

In the next four steps of cholesterol synthesis, mevalonate acquires two phosphoryl groups and is decarboxylated to produce the five-carbon compound isopentenyl pyrophosphate:

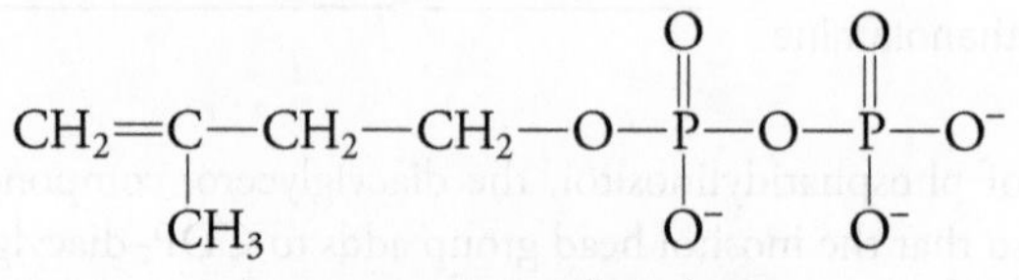

Isopentenyl pyrophosphate

This isoprene derivative is the precursor of cholesterol as well as other isoprenoids, such as ubiquinone, the C_{15} farnesyl group that is attached to some lipid-linked membrane proteins, and pigments

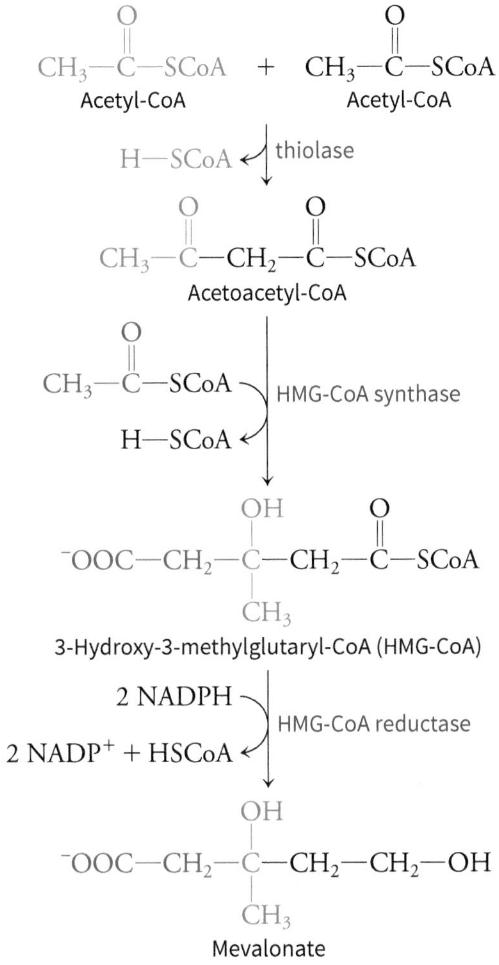

Figure 17.21 The first steps of cholesterol biosynthesis.

Question **Which reactions are shared with ketogenesis (Fig. 17.16)?**

such as β-carotene. Isoprenoids are an extremely diverse group of compounds, particularly in plants, with about 25,000 characterized to date.

In cholesterol synthesis, six isoprene units condense to form the C_{30} compound squalene. Cyclization of this linear molecule leads to a structure with four rings, resembling cholesterol (**Fig. 17.22**). A total of 21 reactions are required to convert squalene to cholesterol. NADH or NADPH is required for several steps.

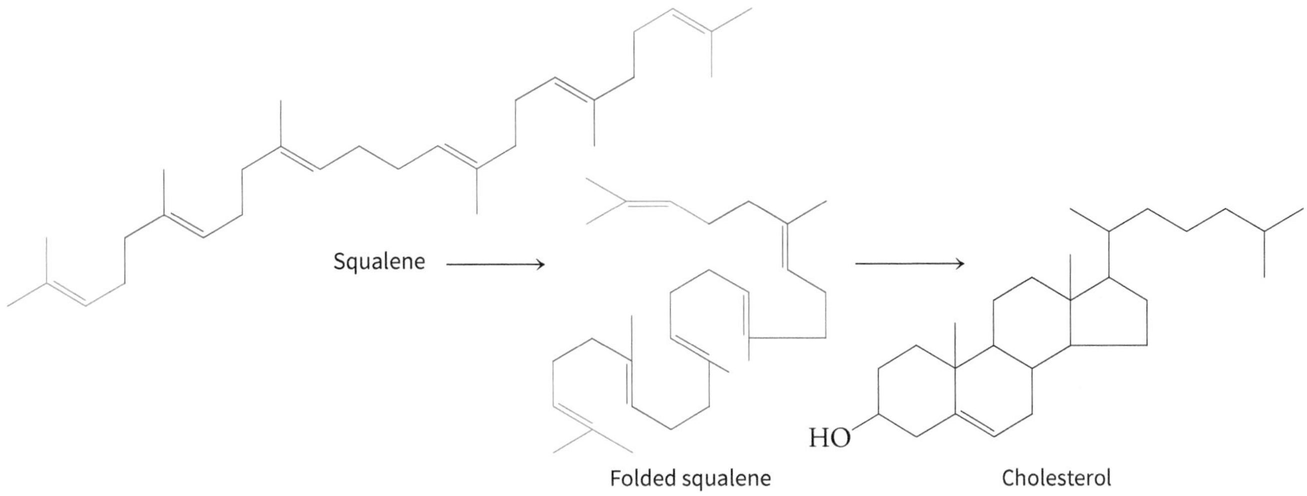

Figure 17.22 Conversion of squalene to cholesterol.   The six isoprene units of squalene are shown in different colors. The molecule folds and undergoes cyclization. Additional reactions convert the C_{30} squalene to cholesterol, a C_{27} molecule.

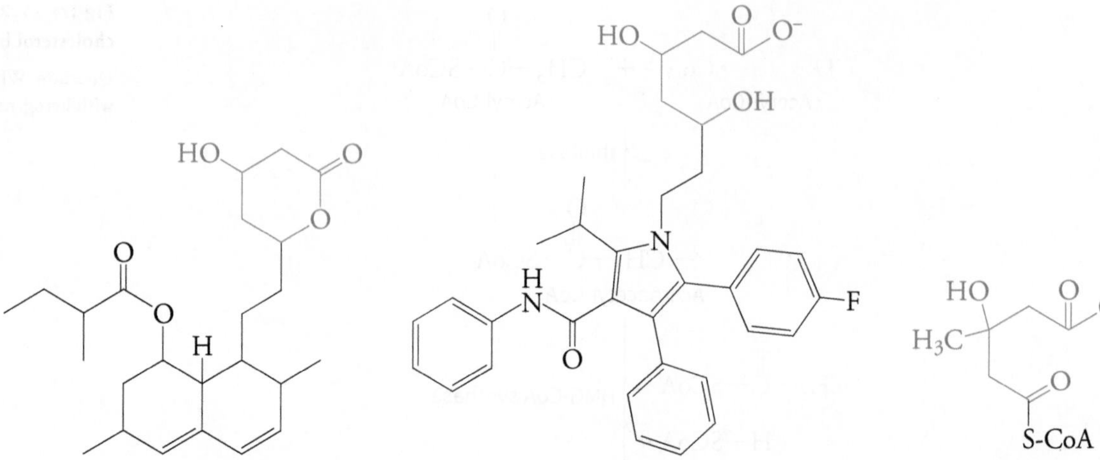

Lovastatin Atorvastatin HMG-CoA

Figure 17.23 Some statins. These inhibitors of HMG-CoA reductase have a bulky hydrophobic group plus an HMG-like group (colored red).

The rate-determining step of cholesterol synthesis (a pathway with over 30 steps) and the major control point is the conversion of HMG-CoA to mevalonate by HMG-CoA reductase. This enzyme is one of the most highly regulated enzymes known. For example, the rates of its synthesis and degradation are tightly controlled, and the enzyme is subject to inhibition by phosphorylation of a serine residue.

Synthetic inhibitors known as statins bind extremely tightly to HMG-CoA reductase, with K_I values in the nanomolar range (Section 7.3). The substrate HMG-CoA has a K_M of about 4 μM. All the statins have an HMG-like group that acts as a competitive inhibitor of HMG-CoA binding to the enzyme (**Fig. 17.23**). Their rigid hydrophobic groups also prevent the enzyme from forming a structure that would accommodate the pantothenate moiety of CoA. The physiological effect of the statins is to lower serum cholesterol levels by blocking mevalonate synthesis. Cells must then obtain cholesterol from circulating lipoproteins. Unfortunately, since mevalonate is also the precursor of other isoprenoids such as ubiquinone, the long-term use of statins can have negative side effects.

Newly synthesized cholesterol has several fates:

1. It can be incorporated into a cell membrane.
2. It may be acylated to form a cholesteryl ester for storage or, in the liver, for packaging in VLDL.

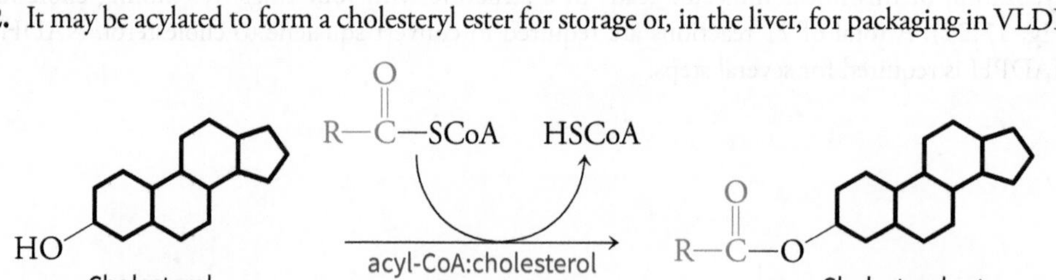

Cholesterol Cholesteryl ester

3. It is a precursor of steroid hormones such as testosterone and estrogen in the appropriate tissues.
4. It is a precursor of bile acids such as cholate:

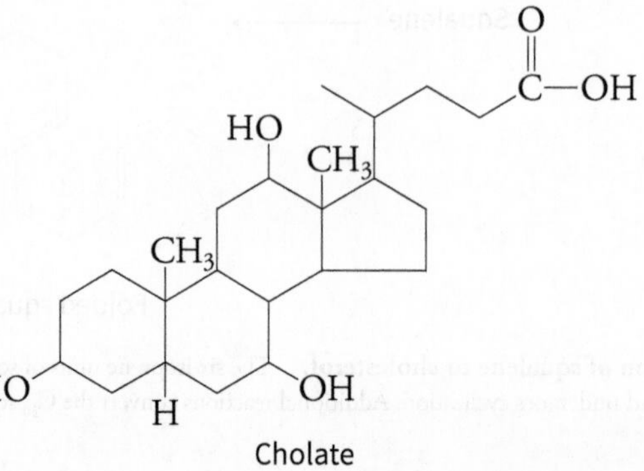

Cholate

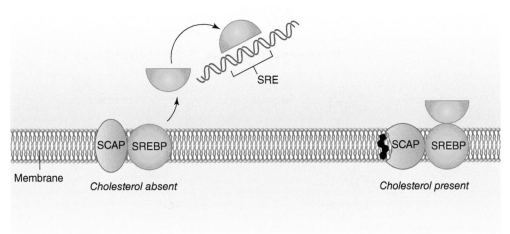

Figure 17.24 The SREBP regulatory system. When cholesterol levels are low, SCAP (the SREBP cleavage-activation protein) promotes the release of the DNA-binding regulatory portion of SREBP, which binds to the sterol regulatory element (SRE) to turn on gene expression.

Bile acids are synthesized in the liver, stored in the gallbladder, and secreted into the small intestine. There, they aid digestion by acting as detergents to solubilize dietary fats and make them more susceptible to lipases. Although bile acids are mostly reabsorbed and recycled through the liver for reuse, some are excreted from the body. *This is virtually the only route for cholesterol disposal.*

Because cells do not break down cholesterol and because the accumulation of cholesterol is potentially toxic (it could disrupt membrane structure), the body must coordinate cholesterol production and transport among tissues. The levels of HMG-CoA reductase (involved in cholesterol synthesis), the LDL receptor (involved in cholesterol uptake), and at least 20 other proteins related to cholesterol metabolism are regulated by a feedback system (**Fig. 17.24**). When cellular cholesterol levels are low, a membrane-embedded sterol-sensing protein known as SCAP allows another protein, the sterol regulatory element binding protein (SREBP), to be proteolyzed. This step releases a soluble DNA-binding protein that travels to the nucleus and binds to the sterol regulatory element, a DNA sequence that is part of the cholesterol-sensitive genes, to promote gene expression. When cellular cholesterol levels return to normal, SCAP binds cholesterol and is no longer able to facilitate the release of the DNA-binding fragment of SREBP. As a result, expression of the cholesterol-sensitive genes drops.

A summary of lipid metabolism

The processes of breaking down and synthesizing lipids illustrate some general principles related to how the cell carries out opposing metabolic pathways. The diagram in **Figure 17.25** includes the major lipid metabolic pathways covered in this chapter. Several features are worth noting:

1. The pathways for fatty acid catabolism and synthesis as well as the synthesis of some other compounds all converge at the common intermediate acetyl-CoA, which is also a product of carbohydrate metabolism (see Section 14.1) and a key player in amino acid metabolism (which will be covered in Chapter 18).

2. The pathways for fatty acid degradation and fatty acid synthesis have a certain degree of symmetry, with similar intermediates and a role for thioesters, but the pathways have very different free energy considerations. β Oxidation produces reduced cofactors and requires only two ATP equivalents; fatty acid synthesis consumes NADPH and requires the input of ATP in each round. Other metabolic pathways, including cholesterol synthesis, consume reduced cofactors generated by catabolic reactions.

3. The catabolic pathway of β oxidation, the conversion of acetyl-CoA to ketone bodies, the oxidation of acetyl-CoA via the citric acid cycle, and the reoxidation of reduced cofactors occur in mitochondria (although some lipid metabolic reactions also occur in peroxisomes). In contrast, many lipid biosynthetic reactions take place in the cytosol or in association with the endoplasmic reticulum. Various pathways therefore require transmembrane transport systems and/or separate pools of substrates and cofactors.

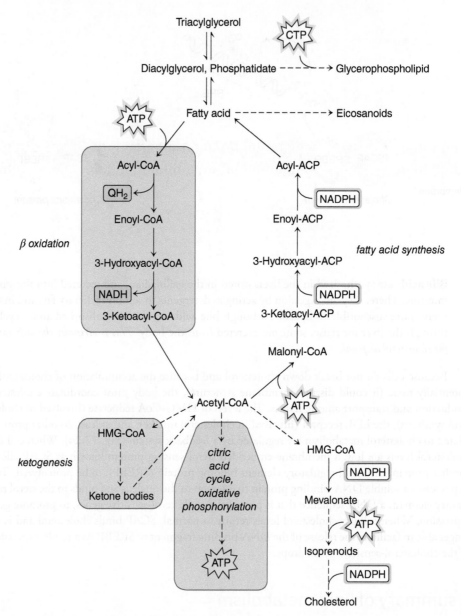

Figure 17.25 Summary of lipid metabolism. Only the major pathways covered in this chapter are included. Open gold symbols indicate ATP consumption; filled gold symbols indicate ATP production. Open and filled red symbols represent the consumption and production of reduced cofactors (NADH, NADPH, and QH₂). The shaded portions of the diagram indicate reactions that occur in mitochondria.

4. Although the central pathways of lipid metabolism, as outlined in Figure 17.25, comprise just a few different reactions, complexity is introduced in the form of isozymes with different acyl-chain-length specificities; in additional enzymes to deal with odd-chain, branched, and unsaturated fatty acids; and in tissue-specific reactions leading to particular products such as eicosanoids or isoprenoids.

Concept Check

1. Summarize the role of coenzyme A in triacylglycerol biosynthesis.
2. Describe how CTP can activate either diacylglycerol or a lipid head group.
3. Compare cholesterol synthesis and ketogenesis.
4. List the metabolic fates of newly synthesized cholesterol.
5. Without looking at the text, draw a diagram showing the major pathways of lipid metabolism.
6. Summarize the roles of LDL and HDL in cholesterol metabolism.

SUMMARY

17.1 Lipid Digestion and Transport

- Lipoproteins transport lipids, including cholesterol, in the bloodstream. High levels of LDL are associated with the development of atherosclerosis.

17.2 Fatty Acid Oxidation

- Fatty acids released from triacylglycerols by the action of lipases are activated by their attachment to CoA in an ATP-dependent reaction.

- In the process of β oxidation, a series of four enzymatic reactions degrades a fatty acyl-CoA two carbons at a time, producing one QH_2, one NADH, and one acetyl-CoA, which can be further oxidized by the citric acid cycle. Reoxidation of the reduced cofactors generates considerable ATP.

- Oxidation of unsaturated and odd-chain fatty acids requires additional enzymes. Very-long-chain and branched fatty acids are oxidized in peroxisomes.

17.3 Fatty Acid Synthesis

- Fatty acids are synthesized by a pathway that resembles the reverse of β oxidation. In the first step of fatty acid synthesis, acetyl-CoA carboxylase catalyzes an ATP-dependent reaction that converts acetyl-CoA to malonyl-CoA, which becomes the donor of two-carbon groups.

- Mammalian fatty acid synthase is a multifunctional enzyme in which the growing fatty acyl chain is attached to acyl-carrier protein rather than CoA. Elongases and desaturases may modify newly synthesized fatty acids.

- The liver can convert acetyl-CoA to ketone bodies to be used as metabolic fuels in other tissues.

17.4 Synthesis of Other Lipids

- Triacylglycerols are synthesized by attaching three fatty acyl groups to a glycerol backbone. Intermediates of the triacylglycerol pathway are the starting materials for the synthesis of phospholipids.

- Cholesterol is synthesized from acetyl-CoA. The rate-determining step of this pathway is the target of drugs known as statins.

KEY TERMS

atherosclerosis	hypercholesterolemia	peroxisome	multifunctional	essential fatty acid	ketogenesis
lipoprotein	β oxidation		enzyme	ketone bodies	bile acid

BIOINFORMATICS

Brief Bioinformatics Exercises

17.1 Viewing and Analyzing Fatty Acid Synthase

17.2 Lipid Metabolism and the KEGG Database

Bioinformatics Project

Drug Design and Cholesterol Medications

PROBLEMS

17.1 Lipid Digestion and Transport

1. **a.** Use the information in Table 17.1 to explain the density rankings of the lipoproteins. **b.** Draw a diagram of lipoprotein transport, using the information presented in Section 17.1.

2. What amino acid side chains of the APOA1 protein would be expected to be in contact with the lipid core of the lipoprotein shown in Figure 17.2? Which of the lipid components of lipoproteins are amphipathic and which are completely nonpolar? **a.** cholesterol, **b.** cholesteryl esters, **c.** triacylglycerols, and **d.** phospholipids. Based on the structures of the lipid components of lipoproteins, describe how the lipids are arranged in the lipoprotein particle and how is this arrangement consistent with the hydrophobic effect (see Section 2.2)?

3. Premenopausal women typically have higher HDL levels than men. **a.** Physicians inform their patients of the total cholesterol concentration in the blood, but the HDL/LDL cholesterol ratio is also reported to the patient. Why is the HDL/LDL cholesterol ratio more informative? **b.** Which is more desirable, a high HDL/LDL cholesterol ratio or a low ratio? Explain why. **c.** Compare the risk of heart disease for men and women. **d.** Why is HDL level alone not a good indicator of the risk of developing heart disease?

4. What types of mutations would result in the inability of an individual's cells to take up LDL from the bloodstream? How is the concentration of LDL in the bloodstream affected in individuals with **a.** loss-of-function or **b.** gain-of-function mutations in the PCSK9 gene?

5. Explain why the phrase *reverse cholesterol transport* is often used to describe the physiological function of HDL.

6. HDL has been observed to increase the concentration of nitric oxide in vascular endothelial cells (see Table 10.1). Is this consistent with the physiological benefits of HDL as described in Solution 3?

7. Individuals who are unable to produce chylomicrons display symptoms consistent with vitamin A deficiency. Explain.

17.2 Fatty Acid Oxidation

8. **a.** Biodiesel, a fuel typically derived from plant oils, can be manufactured by treating the oil with a methanol/KOH mixture to produce fatty acid methyl esters. Indicate the structures of the products generated by treating the triacylglycerol 1-palmitoyl-2,3-dioleoylglycerol with methanol/KOH. **b.** The methanol/KOH reaction described must be carried out in "dry" methanol. If any H_2O were present, what would happen?

9. Triacylglycerols are mobilized from adipocytes during the fasting state. Epinephrine binds to G protein–coupled cell-surface receptors on the adipocytes and initiates a series of events that leads to the activation of hormone-sensitive lipase. Use what you know about the signaling properties of epinephrine (Section 10.2) to explain how hormone-sensitive lipase is activated.

10. One of the products of adipocyte triacylglycerol degradation (see Problem 9), in addition to the fatty acids, is glycerol. Glycerol is released from the adipocyte and travels to the liver. What is the fate of the glycerol? Why is this an advantage in the fasted state?

11. The overall reaction for the activation of a fatty acid to fatty acyl-CoA, with concomitant hydrolysis of ATP to AMP, has a free energy change of about zero. The reaction is favorable because of subsequent hydrolysis of pyrophosphate to orthophosphate (the reaction has a $\Delta G^{\circ\prime}$ value of -19.2 kJ $\cdot$ mol^{-1}). Write the equation for the coupled reaction and calculate **a.** $\Delta G^{\circ\prime}$ and **b.** the equilibrium constant for the reaction. Assume a temperature of 25°C. **c.** Is the reaction spontaneous under standard conditions?

12. The reactions in Figure 17.5 are all reversible. Why do the reactions tend to run in one direction (i.e., favoring the delivery of acyl-CoA to the mitochondrial matrix)?

13. A deficiency of carnitine results in muscle cramps, which are exacerbated by fasting or exercise. Give a biochemical explanation for the muscle cramping and explain why cramping increases during fasting and exercise.

14. Muscle biopsy and enzyme assays of a carnitine-deficient individual show that medium-chain (C_8–C_{10}) fatty acids can be metabolized normally, despite the carnitine deficiency. What does this tell you about the role of carnitine in fatty acid transport across the inner mitochondrial membrane?

15. **a.** Which intermediates accumulate in individuals with a deficiency of medium-chain acyl-CoA dehydrogenase (MCAD)? **b.** Why are hypoglycemia and lethargy two symptoms experienced by individuals with an MCAD deficiency? **c.** How should a patient with a medium-chain acyl-CoA dehydrogenase deficiency be treated?

16. The first three reactions of β oxidation are similar to three reactions of the citric acid cycle. Which reactions are these and why are they similar?

17. During β oxidation, methylene (—CH_2—) groups in a fatty acid are oxidized to carbonyl (C=O) groups, yet no oxygen is consumed by the reactions of β oxidation. How is this possible?

18. The β oxidation pathway was elucidated in part by Franz Knoop in 1904. He fed dogs various fatty acid phenyl derivatives and then analyzed their urine for the resulting metabolites. What metabolites were produced when the dogs were fed **a.** phenylpropionate and **b.** phenylbutyrate?

$$\text{⬡—CH}_2\text{—CH}_2\text{—COO}^-$$

Phenylpropionate

$$\text{⬡—CH}_2\text{—CH}_2\text{—CH}_2\text{—COO}^-$$

Phenylbutyrate

19. A deficiency of phytanate-degrading enzymes in peroxisomes results in Refsum's disease, a neuronal disorder caused by phytanate accumulation. Patients with Refsum's disease cannot convert phytanate to pristanate because they lack the enzymes involved in the α oxidation reaction in which the α-carboxylate group is lost as CO_2 to form pristanate. Show how pristanate is oxidized via β oxidation and list the products of pristanate oxidation.

Pristanate

20. How many molecules of ATP are generated when **a.** palmitate (16:0) **b.** stearate (18:0, see Table 8.1) are completely oxidized via β oxidation in mitochondria **c.** arachidate (20:0) **d.** behenate (22:0, see Table 8.1) are completely oxidized via β oxidation in mitochondria **e.** palmitoleate (16:1, see Table 8.1) is completely oxidized via β oxidation **f.** oleate (18:1, see Table 8.1) is completely oxidized via β oxidation **g.** linoleate (18:2, see Table 8.1) is completely oxidized via β oxidation and **h.** α-linolenate (18:3, see Table 8.1) is completely oxidized via β oxidation?

21. **a.** How many molecules of ATP are generated when a fully saturated 17-carbon fatty acid is oxidized via β oxidation? **b.** Compare this value to the ATP generated by oxidation of palmitate (Problem 20a) and oleate (Problem 20f). **c.** Is the C_{17} fatty acid considered to be glucogenic?

22. How many molecules of ATP are generated when oxidation of a fully saturated C_{24} fatty acid begins in the peroxisome and is completed by the mitochondrion when 12 carbons remain?

23. A vitamin B_{12} deficiency leads to pernicious anemia. Often the disease is caused not by the lack of the vitamin itself but by the lack of a protein called intrinsic factor, which is secreted by gastric parietal cells. Intrinsic factor binds to vitamin B_{12} and facilitates its absorption by the small intestine. Use this information to devise a treatment for a patient diagnosed with pernicious anemia.

24. If you were a physician and wanted to test a patient for pernicious anemia (see Problem 23), what metabolite would you measure in the patient's blood or urine, and why?

25. Both fatty acid oxidation and glucose oxidation by glycolysis generate large amounts of ATP. Explain why a cell preparation containing all the enzymes required for either pathway cannot generate ATP when a fatty acid or glucose is added unless a small amount of ATP is also added.

26. The complete oxidation to CO_2 of glucose and palmitate releases considerable free energy: $\Delta G^{\circ\prime} = -2850$ kJ $\cdot$ mol^{-1} for glucose oxidation and $\Delta G^{\circ\prime} = -9781$ kJ $\cdot$ mol^{-1} for palmitate. For each fuel molecule, compare the ATP yield per carbon atom **a.** in theory

and **b.** *in vivo.* **c.** What do these results tell you about the relative efficiency of oxidizing carbohydrates and fatty acids? **d.** Is it possible to oxidize fatty acids both anaerobically and aerobically, as it is for glucose?

17.3 Fatty Acid Synthesis

27. Compare fatty acid degradation and fatty acid synthesis with respect to the following: **a.** cellular location, **b.** acyl group carrier, **c.** electron carrier, **d.** ATP requirement, **e.** unit product/unit donor, **f.** configuration of hydroxyacyl intermediate, and **g.** end of the fatty acyl chain where shortening/growth occurs.

28. Draw the mechanism for the carboxylation of acetyl-CoA to malonyl-CoA catalyzed by acetyl-CoA carboxylase.

29. What do acetyl-CoA carboxylase, pyruvate carboxylase, and propionyl-CoA carboxylase have in common?

30. The activity of acetyl-CoA carboxylase is regulated by hormone-controlled phosphorylation and dephosphorylation. Based on what you know about signaling via epinephrine (Section 10.2), describe the effect of epinephrine on acetyl-CoA carboxylase and fatty acid metabolism. Is this consistent with epinephrine's effect on glycogen metabolism?

31. Acetyl-CoA carboxylase is regulated by phosphorylation by several cellular kinases, including protein kinase A (see Section 10.2 and Problem 30) and a kinase called AMP-dependent protein kinase (AMPK). As its name suggests, AMPK is activated by AMP. How does AMPK allow the cell to regulate the fatty acid biosynthetic pathway in response to cellular ATP levels?

32. Palmitoyl-CoA is an allosteric effector of acetyl-CoA carboxylase. **a.** Does palmitoyl-CoA act as an activator or an inhibitor? **b.** Is the K_I (see Equation 7.30) for the regulation of acetyl-CoA carboxylase by palmitoyl-CoA higher or lower when the enzyme is phosphorylated (see Problem 30)?

33. There are two isozymes (see Section 6.1) of acetyl-CoA carboxylase, ACC1 and ACC2. Both catalyze the formation of malonyl-CoA from acetyl-CoA and bicarbonate. ACC1 (the isoform discussed in the chapter) is located in the cytosol whereas ACC2 is located in the mitochondrial matrix. What are the physiological roles of these two isozymes?

34. Mice that are deficient in acetyl-CoA carboxylase are thinner than normal and exhibit continuous fatty acid oxidation. Explain these observations.

35. During fatty acid synthesis, why is the condensation of an acetyl group and a malonyl group energetically favorable, whereas the condensation of two acetyl groups would be unfavorable?

36. What is the cost of synthesizing palmitate from acetyl-CoA?

37. On what carbon atoms does the $^{14}CO_2$ used to synthesize malonyl-CoA from acetyl-CoA appear in palmitate?

38. Why does triclosan inhibit bacterial fatty acid synthase but not mammalian fatty acid synthase?

39. The greater-than-normal fatty acid synthesis activity observed in cancer cells has led some researchers to investigate fatty acid synthase as an anti-tumor drug target. **a.** A series of potential inhibitors of fatty acid synthase were synthesized with a common structure as shown below; only the alkyl chain (R) varied in length. Why were these compounds effective inhibitors of the synthase?

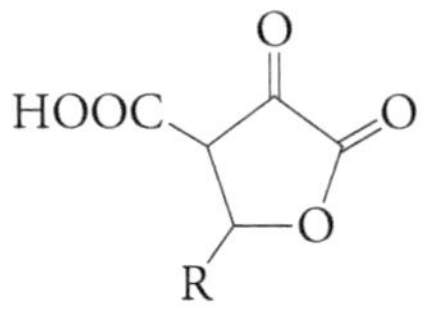

b. Each compound was tested for its ability to inhibit fatty acid synthase activity in both normal cells and breast cancer cells. ID_{50} values (the inhibitor concentration required to inhibit cell growth by 50%) are shown in the table. Which inhibitors are most effective? What are the characteristics of the effective inhibitors? (*Hint:* Calculate the ratio of the ID_{50} values for normal and breast cancer cells.) An effective inhibitor must also be soluble in aqueous solution, so consider solubility as an additional factor in your answer.

Compound	Alkyl side chain (R)	Breast cancer cells ID_{50} ($\mu g \cdot mL^{-1}$)	Normal cells ID_{50} ($\mu g \cdot mL^{-1}$)
A	$-C_{13}H_{27}$	3.9	10.6
B	$-C_{11}H_{23}$	4.8	29.0
C	$-C_8H_{17}$	5.0	21.3
D	$-C_6H_{13}$	8.4	12.4

40. Does fatty acid synthase activity increase or decrease under the following conditions? **a.** high-carbohydrate diet (liver fatty acid synthase), **b.** high-fat diet (liver fatty acid synthase), **c.** mid to late pregnancy (mammary gland fatty acid synthase). Explain.

41. A tuberculosis patient typically harbors 10^{12} *M. tuberculosis* cells. **a.** If mutations that confer resistance to isoniazid (see Box 17.C) occur with a frequency of 1 in 10^8, approximately how many bacterial cells will be resistant to isoniazid? **b.** Assuming a similar rate for mutations that confer resistance to other antibiotics, approximately how many bacterial cells will be resistant to a combination of three drugs? **c.** Do the results of your calculations help explain why tuberculosis is treated with a "cocktail" of drugs rather than a single drug?

42. Many bacteria can convert saturated fatty acids to unsaturated fatty acids by a dehydration reaction that does not require O_2. Explain why this reaction might be a potential target for antibiotics.

43. Draw the structures of the following fatty acids, given the shorthand form described in Problem 8.1: **a.** oleate (18:1*n*-9), **b.** linoleate (18:2*n*-6), **c.** α-linolenate (18:3*n*-3), and **d.** palmitoleate (16:1*n*-7). Which are essential for humans?

44. Why is docosahexaenoic acid (DHA, 22:6*n*-3), a fatty acid commonly found in fish, added to baby formula?

45. Isolated heart cells in culture undergo contraction even in the absence of glucose and fatty acids if they are supplied with acetoacetate. **a.** How does this compound act as a metabolic fuel? **b.** Even with plentiful acetoacetate, the rate of flux through the citric acid cycle gradually drops off unless pyruvate is added to the cultured cells. Explain.

46. When glucose is unavailable, the liver begins to break down fatty acids to supply the rest of the body with metabolic fuel. Explain why fatty acid–derived acetyl-CoA is not catabolized by the citric acid cycle but is instead diverted to ketogenesis when no glucose is available.

47. Discuss the energetic costs of converting two acetyl-CoA to the ketone body 3-hydroxybutyrate in the liver and then converting the 3-hydroxybutyrate back to two acetyl-CoA in the muscle.

48. Compare the ATP yield from the complete oxidation of the two acetyl-CoA generated by **a.** the first two cycles of β oxidation of a fatty acid and by **b.** the breakdown of 3-hydroxybutyrate. Which process is more efficient?

49. The "keto diet" is a dietary regimen in which carbohydrates are severely restricted, protein intake is low, and fat comprises 70–90% of daily calories. The diet causes "nutritional ketosis," which is a different metabolic state than the ketosis that occurs in an untreated diabetic. **a.** The normal ketone body concentration is less than 0.6 mM. Explain why an individual on the keto diet may have a blood ketone body concentration of up to 3 mM. **b.** What biochemical pathways are active in an individual adhering to the keto diet? **c.** What pathways are largely inactive? **d.** Protein intake is kept low because a high-protein diet would inhibit ketosis. Explain why. **e.** Why might the keto diet lead to weight loss?

50. When beginning the keto diet (see Problem 49), individuals often complain of fatigue, dizziness, constipation, bad breath, and difficulty tolerating exercise. (Proponents of the diet refer to this suite of symptoms as "keto flu.") Provide biochemical explanations for these symptoms.

51. Two siblings born two years apart were separately diagnosed with a pyruvate carboxylase deficiency. Both neonates died less than a month after they were born. Blood samples taken before the death of the infants showed a high concentration of ketone bodies in the blood. Explain why the ketone body concentration was elevated.

52. A symptom of a pyruvate carboxylase deficiency (see Problem 51) is a decreased β-hydroxybutyrate/acetoacetate ratio. Explain.

53. The activity of phosphofructokinase (PFK; Section 13.1) is inhibited by long-chain fatty acids. Inhibition involves the acylation of several amino acid side chains in PFK and is reversed if a thioesterase is added. **a.** What amino acid is acylated by the long-chain fatty acids? Draw the structure of the linkage. **b.** Propose a hypothesis to explain why acylation inhibits PFK. **c.** What happens to the concentration of citrate under these conditions?

54. The Randle hypothesis proposed in the 1960s envisions glucose catabolism and fatty acid catabolism competing for mitochondrial oxidation. Many experiments have been carried out *in vivo* to test the hypothesis. Does the regulatory strategy described in Problem 53 support or refute the Randle hypothesis?

17.4 Synthesis of Other Lipids

55. The glycerol-3-phosphate required for the first step of triacylglycerol synthesis can be obtained from either glucose or pyruvate. Explain how this occurs. Can all cells obtain glycerol-3-phosphate from either precursor?

56. A fatty acid inhibitor called Compound D (see Problem 39) was tested for its ability to inhibit triacylglycerol synthesis in leukemia cells. The cells were incubated with ^{14}C-labeled acetate, and the amount of radioactivity in various types of cellular lipids was measured. What is your interpretation of the results shown in the graph?

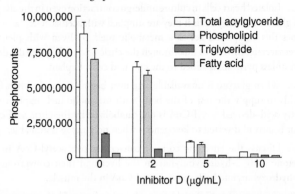

57. Manufacturers of cooking oil can chemically convert triacylglycerols to diacylglycerols and they claim that the oil containing diacylglycerols is less "fattening" than oil containing triacylglycerols. **a.** What kind of chemical reaction occurs and what is its purpose? **b.** Is this claim accurate?

58. Refer to Figure 17.19 to answer the following questions: **a.** What type of enzyme catalyzes the reaction shown in Step 1? **b.** What drives the reaction shown in Step 2 to completion? **c.** How many phosphoanhydride bonds must be hydrolyzed in order to provide the free energy required to synthesize phosphatidylcholine from choline and diacylglycerol?

59. The malaria parasite can synthesize large amounts of phosphatidylcholine even in the absence of choline. To do this, it relies on an enzyme called phosphoethanolamine methyltransferase. **a.** Describe the

reaction catalyzed by this enzyme. **b.** What information would you need in order to assess whether this methyltransferase would be a suitable antimalarial drug target?

60. Cancer cells appear to increase the expression of the enzyme that catalyzes the first step in phospholipid synthesis (see Fig. 17.19). What is the advantage of this increased expression to the cancer cell?

61. A recently developed magnetic resonance spectroscopy technique to quantitate choline and choline derivatives allowed cancer researchers to observe that tumor cells accumulate phosphocholine derivatives. In addition to greater expression of choline kinase (see Solution 60), researchers hypothesized that increased expression of phospholipases C and D could contribute to the higher concentration of phosphocholine. Explain.

62. The first four steps of sphingolipid synthesis are shown below. Identify the step catalyzed by each of the following enzymes: **a.** acyl-CoA transferase, **b.** 3-ketosphinganine synthase, **c.** dihydroceramide dehydrogenase, **d.** 3-ketosphinganine reductase.

63. Genetically engineered bacteria can convert acetyl-CoA to butanol, a biofuel that has about 84% of the energy of gasoline. Identify the step catalyzed by each of the following enzymes: **a.** butyryl-CoA dehydrogenase, **b.** thiolase, **c.** butanol dehydrogenase, **d.** crotonase (crotonic acid is a butenoic acid), **e.** 3-hydroxybutyryl-CoA dehydrogenase, **f.** butyraldehyde dehydrogenase.

64. In a site-directed mutagenesis experiment, an essential serine of HMG-CoA reductase was replaced with alanine. In normal cells, HMG-CoA reductase levels decrease when the cells are incubated with media containing LDL particles, but in cells expressing the mutant enzyme, enzyme activity did not change. What do these results indicate regarding the regulation of HMG-CoA reductase?

65. HMG-CoA reductase is phosphorylated by the same kinase that phosphorylates acetyl-CoA carboxylase (see Problem 30). How does this strategy assist the cell in regulating the two biosynthetic pathways?

66. Fungi produce ergosterol rather than cholesterol. List the ways that ergosterol differs from cholesterol.

Ergosterol

67. Proteins called SREBPs (sterol regulatory element binding proteins; see Fig. 17.24) are cellular cholesterol sensors that regulate cholesterol uptake and biosynthesis according to the needs of the cell. In the absence of cholesterol, an SREBP residing in the endoplasmic reticulum is proteolytically cleaved to release a large soluble N-terminal domain that includes a structural motif found in many DNA-binding proteins. **a.** Why is it important that the SREBP be an integral membrane protein? **b.** Why is proteolysis of the SREBP required? **c.** How might the SREBP regulate the transcription of enzymes related to cholesterol metabolism?

68. Individuals with a mutation in the gene for apolipoprotein B-100 produce very low levels of this protein, which is a component of LDL. **a.** Explain why these individuals exhibit accumulation of fat in the liver. **b.** Would such individuals exhibit hypercholesterolemia or hypocholesterolemia?

SELECTED READINGS

Evans, R.D. and Hauton, D., The role of triacylglycerol in cardiac energy provision, *Biochim. Biophys. Acta* 1861, 1481–1491, doi: 10.1016/j.bbalip.2016.03.010 (2016). [Discusses the uptake, storage, breakdown, and oxidation of triacylglycerols in heart tissue.]

Merritt, J.L., 2nd, Norris, M., and Kanungo, S., Fatty acid oxidation disorders, *Ann. Transl. Med.* 6, 473, doi: 10.21037/atm.2018.10.57 (2018). [Summarizes the symptoms and treatments of deficiencies of a variety of proteins involved in fatty acid catabolism.]

Pownall, H.J. and Gotto, A.M., Cholesterol: Can't live with it, can't live without it, *Methodist Debakey Cardiovasc. J.* 15, 9–15 (2019). [Summarizes the historical and current understanding of the relationship between cholesterol and cardiovascular disease.]

Puchalska, P. and Crawford, P.A., Multi-dimensional roles of ketone bodies in fuel metabolism, signaling, and therapeutics, *Cell Metab.* 25, 262–284, doi: 10.1016/j.cmet.2016.12.022 (2017). [A thorough review of ketone body metabolism and possible links to cancer and heart and liver disease.]

Rhoads, J.P. and Major, A.S., How oxidized low-density lipoprotein activates inflammatory responses, *Crit. Rev. Immunol.* 38, 333–342, doi: 10.1615/CritRevImmunol.2018026483 (2018). [Summarizes some key experimental studies of atherosclerosis, with an emphasis on immune system responses.]

CHAPTER 17 CREDITS

Figure 17.2 Based on data from Pourmousa, M., Song, H.D., He, Y., Heinecke, J.W., Segrest, J.P., and Pastor, R.W., *Proc. Natl. Acad. Sci. USA* 115, 5163–5168 (2018).

Figure 17.12 Image from Maier, T., Leibundgut, M., Ban, N., The crystal structure of a mammalian fatty acid synthase, Science 321: 5894, 2008 © AAAS. Reproduced with permission of American Association for the Advancement of Science - AAAS.

Nitrogen Metabolism

The amino acid theanine, a minor component of tea leaves (*Camellia sinensis*), acts on receptors in the brain. It is at least partly responsible for the calming effect of drinking tea. Studies with theanine alone, apart from other tea ingredients such as caffeine, suggest that it may also boost memory and cognition.

Do You Remember?

- Nucleotides consist of a purine or pyrimidine base, deoxyribose or ribose, and phosphate (Section 3.1).
- The 20 amino acids differ in the chemical characteristics of their R groups (Section 4.1).
- A few metabolites appear in several metabolic pathways (Section 12.3).
- Many vitamins, substances that humans cannot synthesize, are components of coenzymes (Section 12.3).
- The citric acid cycle supplies precursors for the synthesis of other compounds (Section 14.4).

The metabolic pathways we have examined so far are centered on carbon. In this chapter, we include another essential element, the nitrogen that is a component of amino acids and nucleotides. In addition to examining how amino groups, the most common biological form of nitrogen, are acquired and disposed of, we will survey the pathways for synthesizing and breaking down amino acids and nucleotides.

18.1 Nitrogen Fixation and Assimilation

KEY CONCEPTS

Describe the chemical reactions of nitrogen fixation and assimilation.

- Explain the function of nitrogenase in the nitrogen cycle.
- Distinguish the activities of glutamine synthetase and glutamate synthase.
- Recount the steps of the transamination reaction.

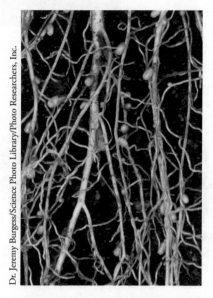

Figure 18.1 Root nodules from clover. Legumes (such as beans, clover, and alfalfa) and some other plants harbor nitrogen-fixing bacteria in root nodules. The symbiotic relationship revolves around the ability of the bacteria to fix nitrogen and the ability of the plant to make other nutrients available to the bacteria.

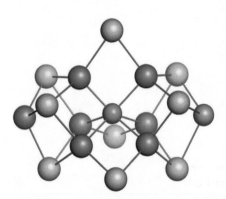

Figure 18.2 Model of the FeMo cofactor of nitrogenase. This prosthetic group in the enzyme nitrogenase consists of iron atoms (orange), sulfur atoms (yellow), and a molybdenum atom (cyan). A carbon atom (green) is liganded to six iron atoms.

Approximately 78% of the air we breathe is nitrogen (N_2), but we cannot use this form of nitrogen for the synthesis of amino acids, nucleotides, and other nitrogen-containing biomolecules. Instead, we—along with most macroscopic and many microscopic life-forms—depend on the activity of a few types of microorganisms that can "fix" gaseous N_2 by transforming it into biologically useful forms. The low availability of fixed nitrogen—mainly as nitrate and ammonia—is believed to limit the biological productivity in much of the world's oceans. It also limits the growth of terrestrial organisms, which is why farmers use fertilizer (a source of fixed nitrogen, among other things) to promote crop growth.

Nitrogenase converts N_2 to NH_3

The known **nitrogen-fixing** organisms, or **diazotrophs,** include certain marine cyanobacteria and bacteria that colonize the root nodules of leguminous plants (**Fig. 18.1**). These bacteria make the enzyme nitrogenase, which carries out the energetically expensive reduction of N_2 to NH_3. Nitrogenase is a metalloprotein containing iron–sulfur centers and a cofactor with both iron and molybdenum, which resembles an elaborate Fe–S cluster (**Fig. 18.2**). Some diazotrophs use a cofactor with vanadium or iron rather than molybdenum. The industrial fixation of nitrogen also involves metal catalysts, but this nonbiological process requires temperatures of 300 to 500°C and pressures of over 300 atm in order to break the triple bond between the two nitrogen atoms.

Biological N_2 reduction consumes large amounts of ATP and requires a strong reducing agent such as ferredoxin (see Section 16.2) to donate electrons. The net reaction is

$$N_2 + 8\,H^+ + 8\,e^- + 16\,ATP + 16\,H_2O \longrightarrow 2\,NH_3 + H_2 + 16\,ADP + 16\,P_i$$

Note that eight electrons are required for the nitrogenase reaction, although N_2 reduction formally requires only six electrons; the two extra electrons are used to produce H_2. *In vivo,* the inefficiency of the reaction boosts the ATP toll to about 20 or 30 per N_2 reduced. The exact steps of the nitrogenase reaction are not known, but hydrogen atoms and N_2 appear to interact with two iron atoms near the center of the FeMo cofactor (the uppermost two Fe atoms in Fig. 18.2).

Oxygen inactivates nitrogenase, so many nitrogen-fixing bacteria are confined to anaerobic habitats or carry out nitrogen fixation when O_2 is scarce. Some plant leaves contain nitrogen-fixing bacteria that limit their oxygen exposure by forming a biofilm (Section 11.2); these organisms may be an unrecognized source of fixed nitrogen for the plants.

Biologically useful nitrogen also originates from nitrate (NO_3^-), which is naturally present in water and soils. *Nitrate is reduced to NH_3 by plants, fungi, and many bacteria.* First, nitrate reductase catalyzes the two-electron reduction of nitrate to nitrite (NO_2^-):

$$NO_3^- + 2\,H^+ + 2\,e^- \longrightarrow NO_2^- + H_2O$$

Next, nitrite reductase converts nitrite to ammonia:

$$NO_2^- + 8\,H^+ + 6\,e^- \longrightarrow NH_4^+ + 2\,H_2O$$

Under physiological conditions, ammonia exists primarily in the protonated form, NH_4^+ (the ammonium ion), which has a pK of 9.25.

Nitrate is also produced by certain bacteria that oxidize NH_4^+ to NO_2^- and then NO_3^-, a process called **nitrification.** Still other organisms convert nitrate back to N_2, which is called **denitrification.** All the reactions we have discussed so far constitute the earth's **nitrogen cycle** (**Fig. 18.3**).

Ammonia is assimilated by glutamine synthetase and glutamate synthase

The enzyme glutamine synthetase is found in all organisms. In microorganisms, it is a metabolic entry point for fixed nitrogen. In animals, it helps mop up excess ammonia, which is toxic. In the

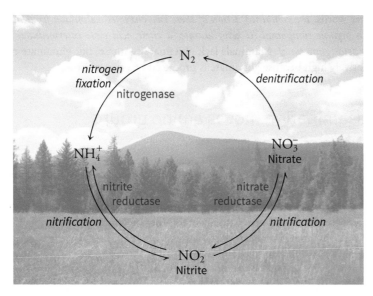

Figure 18.3 The nitrogen cycle.   Nitrogen fixation converts N_2 to the biologically useful NH_4^+. Nitrate can also be converted to NH_4^+. Ammonia is transformed back to N_2 by nitrification followed by denitrification.

Question Indicate which processes are oxidations and which are reductions.

first step of the reaction, ATP donates a phosphoryl group to glutamate. Then ammonia reacts with the reaction intermediate, displacing P_i to produce glutamine:

$$\text{Glutamate} \xrightarrow[\text{ATP ADP}]{} [\text{intermediate}] \xrightarrow[\text{NH}_4^+ \; P_i]{} \text{Glutamine}$$

The name *synthetase* indicates that ATP is consumed in the reaction.

Glutamine, along with glutamate, is usually present in organisms at much higher concentrations than the other amino acids, which is consistent with its role as a carrier of amino groups. Not surprisingly, *the activity of glutamine synthetase is tightly regulated to maintain a supply of accessible amino groups*. For example, the dodecameric glutamine synthetase from *E. coli* is regulated allosterically and by covalent modification (**Fig. 18.4**). The enzyme's symmetrical arrangement of subunits is a general feature of enzymes that are regulated by allosteric effectors: Changes in activity at one of the active sites can be efficiently communicated to the other active sites.

The glutamine synthetase reaction that introduces fixed nitrogen (ammonia) into biological compounds requires a nitrogen-containing compound (glutamate) as a substrate. So what is the source of the nitrogen in glutamate? In bacteria and plants, the enzyme glutamate synthase catalyzes the reaction

$$\alpha\text{-Ketoglutarate} + \text{Glutamine} \xrightarrow[\text{glutamate synthase}]{\text{NADPH} + \text{H}^+ \quad \text{NADP}^+} 2 \text{ Glutamate}$$

(a reaction catalyzed by a synthase does not require ATP). The net result of the glutamine synthetase and glutamate synthase reactions is

$$\alpha\text{-ketoglutarate} + NH_4^+ + NADPH + ATP \longrightarrow \text{glutamate} + NADP^+ + ADP + P_i$$

In other words, *the combined action of these two enzymes assimilates fixed nitrogen (NH_4^+) into an organic compound (α-ketoglutarate, a citric acid cycle intermediate) to produce an amino acid (glutamate).* Mammals lack glutamate synthase, but glutamate concentrations are relatively high because glutamate is produced by other reactions.

Transamination moves amino groups between compounds

Because reduced nitrogen is so precious but free ammonia is toxic, amino groups are transferred from molecule to molecule, with glutamate often serving as an amino-group donor. We saw some of these **transamination** reactions in Section 14.4 when we examined how citric acid cycle intermediates participate in other metabolic pathways.

A transaminase (also called an aminotransferase) catalyzes the transfer of an amino group to an α-keto acid. For example,

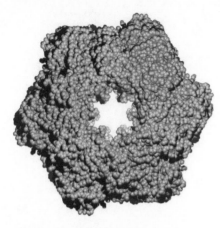

Figure 18.4 *E. coli* **glutamine synthetase.** The 12 identical subunits of this enzyme are arranged in two stacked rings of 6 subunits (only the upper ring is visible here).

$$\underset{\substack{\text{Glutamate}\\\text{(amino acid)}}}{\begin{array}{c}\text{COO}^-\\|\\H_3\overset{+}{N}-C-H\\|\\CH_2\\|\\CH_2\\|\\COO^-\end{array}} + \underset{\substack{\text{Pyruvate}\\(\alpha\text{-keto acid})}}{\begin{array}{c}\text{COO}^-\\|\\O=C\\|\\CH_3\end{array}} \underset{\text{transaminase}}{\rightleftharpoons} \underset{\substack{\alpha\text{-Ketoglutarate}\\(\alpha\text{-keto acid})}}{\begin{array}{c}\text{COO}^-\\|\\O=C\\|\\CH_2\\|\\CH_2\\|\\COO^-\end{array}} + \underset{\substack{\text{Alanine}\\\text{(amino acid)}}}{\begin{array}{c}\text{COO}^-\\|\\H_3\overset{+}{N}-C-H\\|\\CH_3\end{array}}$$

During such an amino-group transfer reaction, the amino group is transiently attached to a prosthetic group of the enzyme. This group is pyridoxal-5′-phosphate (PLP), a derivative of pyridoxine (an essential nutrient also known as vitamin B_6):

Pyridoxal-5′-phosphate (PLP)

Pyridoxine (vitamin B_6)

Enzyme

Enzyme–PLP Schiff base

PLP is covalently attached to the enzyme via a protonated Schiff base (imine) linkage to the ε-amino group of a lysine residue (at right). The amino acid substrate of the transaminase displaces this Lys amino group, which then acts as an acid–base catalyst. The steps of the reaction are diagrammed in **Figure 18.5**.

The transamination reaction is freely reversible, so transaminases participate in pathways for amino acid synthesis as well as degradation. Note that if the α-keto acid produced in step 4 reenters the active site, then the amino group that was removed from the starting amino acid is restored. However, most transaminases accept only α-ketoglutarate or oxaloacetate as the α-keto acid substrate for the second part of the reaction (steps 5 to 7). This means that most transaminases generate glutamate or aspartate. Lysine is the only amino acid that cannot be transaminated. The presence of transaminases in muscle and liver cells makes them useful markers of tissue damage (**Box 18.A**).

PLP is a cofactor for about 4% of all enzymes. Some of these enzymes catalyze the decarboxylation of amino acids, racemization (conversion from the ʟ to the ᴅ configuration), or the elimination of a side chain (as in the serine → glycine reaction). In all cases, the PLP group acts to delocalize the electrons of the reaction intermediate generated by removing one of the groups attached to the α carbon of the amino acid.

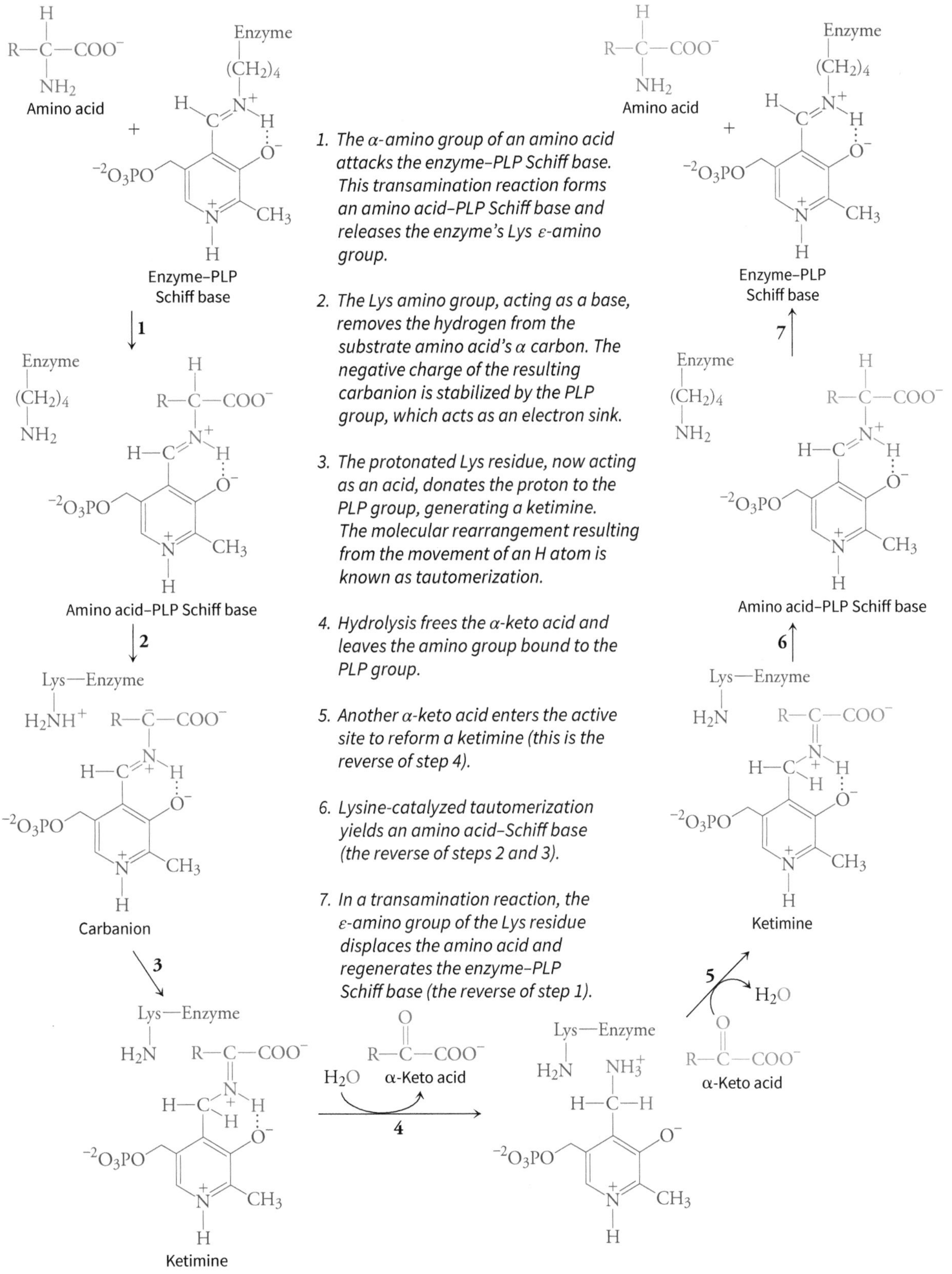
Amino acid
Enzyme–PLP Schiff base
1. The α-amino group of an amino acid attacks the enzyme–PLP Schiff base. This transamination reaction forms an amino acid–PLP Schiff base and releases the enzyme's Lys ε-amino group.
2. The Lys amino group, acting as a base, removes the hydrogen from the substrate amino acid's α carbon. The negative charge of the resulting carbanion is stabilized by the PLP group, which acts as an electron sink.
3. The protonated Lys residue, now acting as an acid, donates the proton to the PLP group, generating a ketimine. The molecular rearrangement resulting from the movement of an H atom is known as tautomerization.
4. Hydrolysis frees the α-keto acid and leaves the amino group bound to the PLP group.
5. Another α-keto acid enters the active site to reform a ketimine (this is the reverse of step 4).
6. Lysine-catalyzed tautomerization yields an amino acid–Schiff base (the reverse of steps 2 and 3).
7. In a transamination reaction, the ε-amino group of the Lys residue displaces the amino acid and regenerates the enzyme–PLP Schiff base (the reverse of step 1).
Amino acid–PLP Schiff base
Carbanion
Ketimine
α-Keto acid

Figure 18.5 PLP-catalyzed transamination.

| Box 18.A | Transaminases in the Clinic |

Assays of transaminase activity in the blood are the basis of the widely used clinical measurements known as AST (aspartate aminotransferase; also known as serum glutamate–oxaloacetate transaminase, or SGOT) and ALT (alanine transaminase; also known as serum glutamate–pyruvate transaminase, or SGPT). In clinical lab tests, blood samples are added to a mixture of the enzymes' substrates. The reaction products, whose concentrations are proportional to the amount of enzyme present, are then detected by secondary reactions that generate colored products easily quantified by spectrophotometry. Prepackaged kits give reliable results in a matter of minutes.

The concentration of AST in the blood increases after a heart attack, when damaged heart muscle leaks its intracellular contents.

Typically, AST concentrations rise in the first hours after a heart attack, peak in 24 to 36 hours, and return to normal within a few days. However, since many tissues contain AST, monitoring cardiac muscle damage more commonly relies on measurements of cardiac troponins (proteins specific to heart muscle). ALT is primarily a liver enzyme, so it is useful as a marker of liver damage resulting from infection, trauma, or chronic alcohol abuse. Certain drugs, including the cholesterol-lowering statins (Section 17.4), sometimes increase AST and ALT levels to such an extent that the drugs must be discontinued.

Question **Identify the substrates and products for the AST and ALT reactions.**

Concept Check

1. Sketch a diagram of the nitrogen cycle and indicate where nitrogenase acts.
2. Make a list of the processes that can generate ammonia.
3. Write equations for the reactions catalyzed by glutamine synthetase and glutamate synthase.
4. Describe the function of the PLP cofactor.
5. Explain why transaminases catalyze reversible reactions.
6. Draw a diagram to trace a nitrogen atom from N_2 to the amino group of threonine.

18.2 Amino Acid Biosynthesis

KEY CONCEPTS

Summarize the pathways for synthesizing the essential and nonessential amino acids.

- Distinguish essential and nonessential amino acids.
- Explain the importance of transamination in amino acid synthesis.
- Identify common metabolites that are used to synthesize amino acids.
- Describe the types of reactions that convert amino acids to neurotransmitters.

Amino acids are synthesized from intermediates of glycolysis, the citric acid cycle, and the pentose phosphate pathway. Their amino groups are derived from the nitrogen carrier molecules glutamate and glutamine. Using the metabolic scheme introduced in Chapter 12, we can show how amino acid biosynthesis and other reactions of nitrogen metabolism are related to the other pathways we have examined (**Fig. 18.6**).

Humans can synthesize only some of the 20 amino acids that are commonly found in proteins. These are known as **nonessential** amino acids. The other amino acids are said to be **essential** because humans cannot synthesize them and must obtain them from their food. The ultimate sources of the essential amino acids are plants and microorganisms, which produce all the enzymes necessary to undertake the synthesis of these compounds. The essential and nonessential amino acids for humans are listed in **Table 18.1**. This classification scheme can be

Table 18.1 Essential and Nonessential Amino Acids

Essential	Nonessential
Histidine	Alanine
Isoleucine	Arginine
Leucine	Asparagine
Lysine	Aspartate
Methionine	Cysteine
Phenylalanine	Glutamate
Threonine	Glutamine
Tryptophan	Glycine
Valine	Proline
	Serine
	Tyrosine

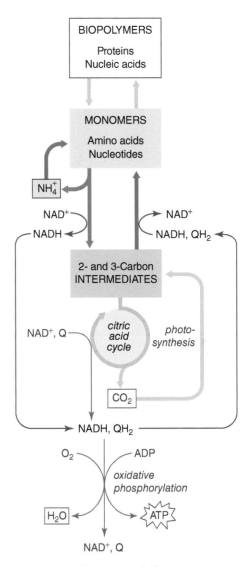

Figure 18.6 Nitrogen metabolism in context. Amino acids are synthesized mostly from three-carbon intermediates of glycolysis and from intermediates of the citric acid cycle. Amino acid catabolism yields some of the same intermediates, as well as the two-carbon acetyl-CoA. Amino acids are also the precursors of nucleotides.

somewhat confusing. For example, some nonessential amino acids, such as arginine, may be essential for young children; that is, dietary sources must supplement what the body can produce on its own. Human cells cannot synthesize histidine, so it is classified as an essential amino acid, even though a dietary requirement has never been defined (probably because sufficient quantities are naturally supplied by intestinal microorganisms). Tyrosine can be considered essential in that it is synthesized directly from the essential amino acid phenylalanine. Likewise, cysteine synthesis depends on the availability of sulfur provided by the essential amino acid methionine.

Several amino acids are easily synthesized from common metabolites

We have already seen that *some amino acids can be produced by transamination reactions.* In this way, alanine is produced from pyruvate, aspartate from oxaloacetate, and glutamate from α-ketoglutarate. Glutamine synthetase catalyzes the amidation of glutamate to produce glutamine. Asparagine synthetase, which uses glutamine as an amino-group donor rather than ammonia, converts aspartate to asparagine:

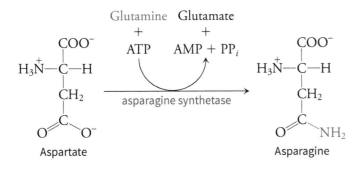

To summarize, three common metabolic intermediates (pyruvate, oxaloacetate, and α-ketoglutarate) give rise to five nonessential amino acids by simple transamination and amidation reactions.

Slightly longer pathways convert glutamate to proline and arginine, which each have the same five-carbon core:

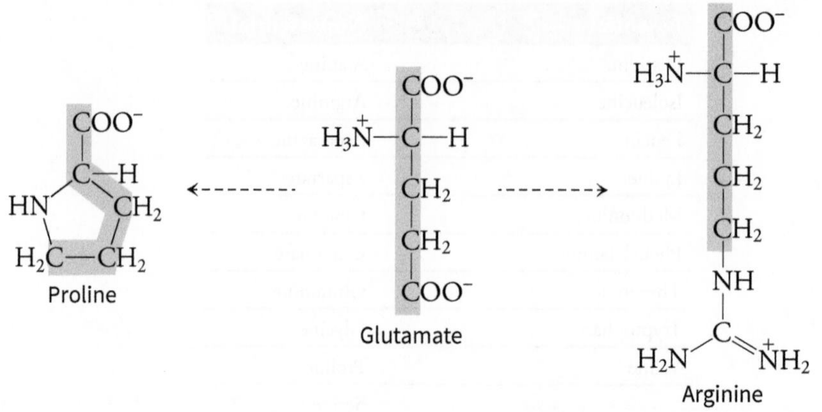

Serine is derived from the glycolytic intermediate 3-phosphoglycerate in three steps:

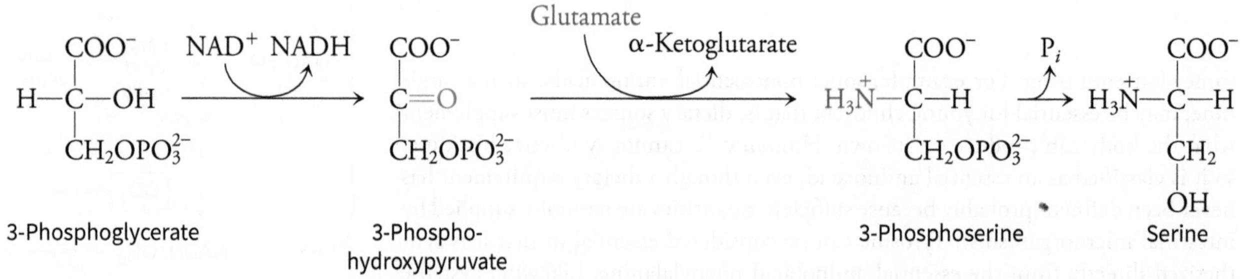

Serine, a three-carbon amino acid, gives rise to the two-carbon glycine in a reaction catalyzed by serine hydroxymethyltransferase (the reverse reaction converts glycine to serine). This enzyme uses a PLP-dependent mechanism to remove the hydroxymethyl ($-CH_2OH$) group attached to the α carbon of serine; this one-carbon fragment is then transferred to the cofactor tetrahydrofolate:

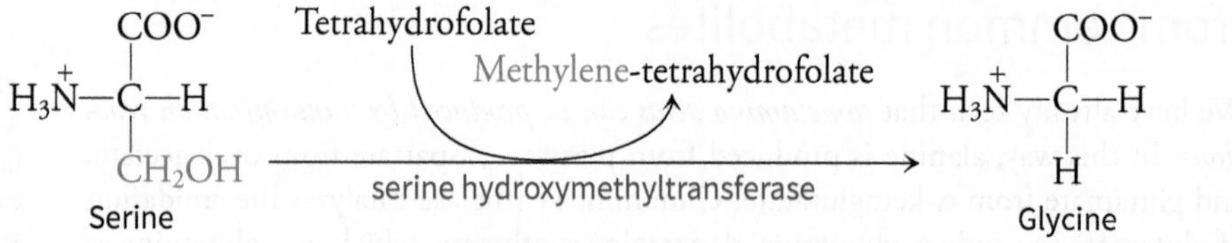

Tetrahydrofolate functions as a carrier of one-carbon units in several reactions of amino acid and nucleotide metabolism (**Fig. 18.7**). Mammals cannot synthesize folate (the oxidized form of tetrahydrofolate) and must therefore obtain it as a vitamin from their diet. Folate is abundant in

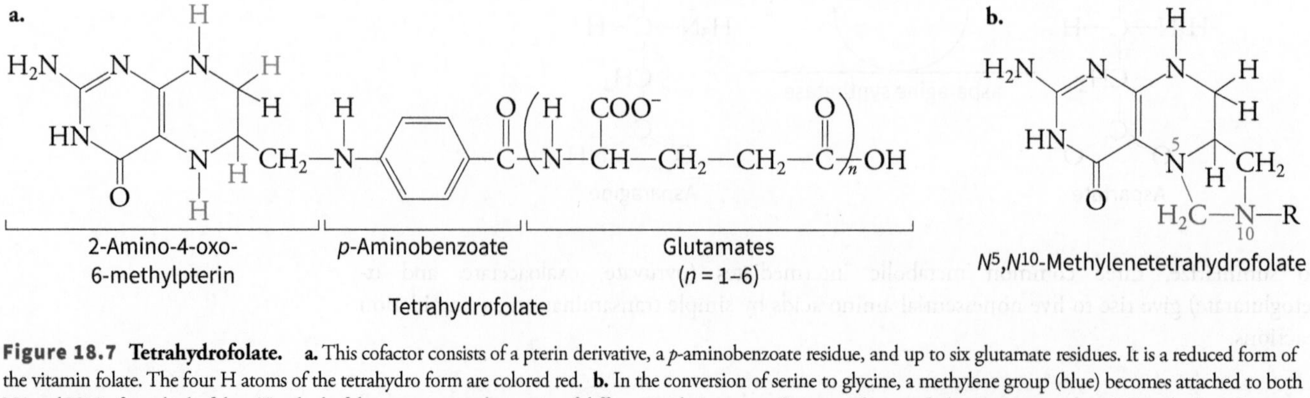

Figure 18.7 Tetrahydrofolate. **a.** This cofactor consists of a pterin derivative, a *p*-aminobenzoate residue, and up to six glutamate residues. It is a reduced form of the vitamin folate. The four H atoms of the tetrahydro form are colored red. **b.** In the conversion of serine to glycine, a methylene group (blue) becomes attached to both N5 and N10 of tetrahydrofolate. Tetrahydrofolate can carry carbon units of different oxidation states. For example, a methyl group can attach to N5, and a formyl group ($-HCO$) can attach at N5 or N10.

foods such as fortified cereal, fruits, and vegetables. The requirement for folate increases during the first few weeks of pregnancy, when the fetal nervous system begins to develop. Supplemental folate appears to prevent certain neural tube defects such as spina bifida, in which the spinal cord remains exposed.

Amino acids with sulfur, branched chains, or aromatic groups are more difficult to synthesize

We have just described how a few metabolites—pyruvate, 3-phosphoglycerate, oxaloacetate, and α-ketoglutarate—are converted in a few enzyme-catalyzed steps to nine different amino acids. Synthesis of the other amino acids (the essential amino acids and those derived directly from them) also begins with common metabolites. However, these biosynthetic pathways tend to be more complicated. At some point in their evolution, animals lost the ability to synthesize these amino acids, probably because the pathways were energetically expensive and the compounds were already available in food. In general, humans cannot synthesize branched-chain amino acids or aromatic amino acids and cannot incorporate sulfur into compounds such as methionine. In this section, we will focus on a few interesting points related to the synthesis of essential amino acids.

The bacterial pathway for producing sulfur-containing amino acids begins with serine and uses sulfur that comes from inorganic sulfide:

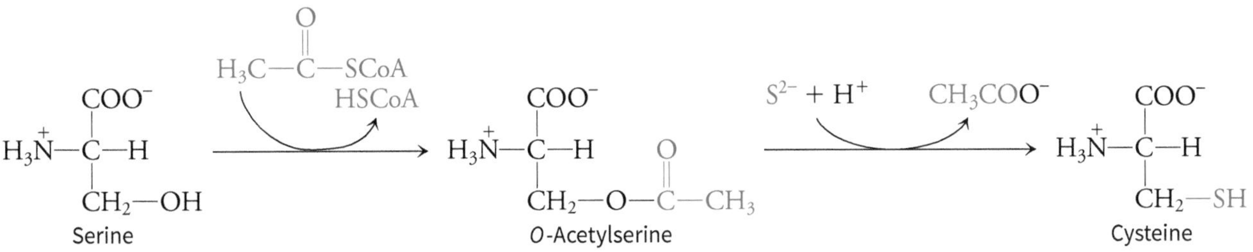

Cysteine can then donate its sulfur atom to a four-carbon compound derived from aspartate, forming the nonstandard amino acid homocysteine. The final step of methionine synthesis is catalyzed by methionine synthase, which adds to homocysteine a methyl group carried by tetrahydrofolate:

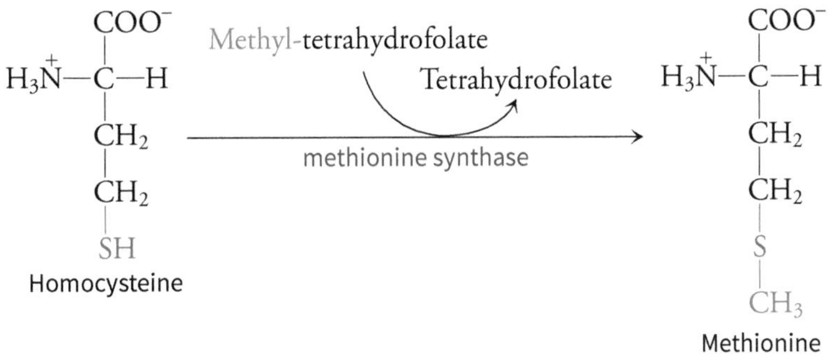

In humans, serine reacts with homocysteine (derived from the metabolism of methionine) to yield cysteine:

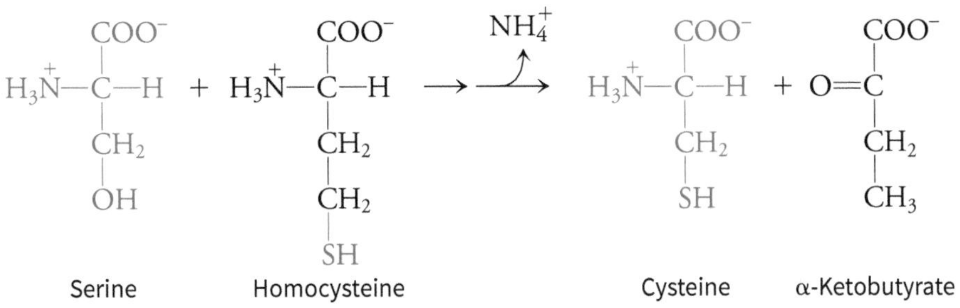

Box 18.B   Homocysteine, Methionine, and One-Carbon Chemistry

Tetrahydrofolate is not the only cofactor that shuttles one-carbon groups among metabolites in human cells. In fact, after tetrahydrofolate (THF) hands off a methyl group to homocysteine to form methionine, the methionine can react with ATP to form S-adenosylmethionine, a donor of methyl groups throughout the cell. A hydrolysis reaction then regenerates homocysteine, and the cycle continues (see diagram).

The sulfonium ion makes S-adenosylmethionine an efficient methylating agent. One of its main tasks is to convert phosphatidylethanoloamine (a glycerophospholipid with ethanolamine in its head group; see Fig. 17.19) to phosphatidylcholine (mammals cannot synthesize choline itself). In addition, S-adenosylmethionine is the source of methyl groups that are used to covalently modify cysteine residues in order to "mark" DNA without changing its nucleotide sequence (Section 21.1).

Ultimately, the one-carbon groups are derived mostly from serine, during the serine hydroxymethyltransferase reaction, and from glycine breakdown (Section 18.3). Not surprisingly, a cell's consumption of these two nonessential amino acids can vary considerably, depending on the cell's overall level of activity. Homocysteine, however, appears to be more difficult to regulate.

High levels of homocysteine in the blood are associated with neurological and cardiovascular diseases. The link was first discovered in individuals with homocystinuria, a disorder in which excess homocysteine is excreted in the urine. These individuals develop atherosclerosis as children, apparently because the homocysteine directly damages the walls of blood vessels, even in the absence of elevated LDL levels (see Section 17.1). A deficiency of folate, the vitamin precursor of tetrahydrofolate, can increase the level of circulating homocysteine by preventing its conversion to methionine. Methionine synthase, the enzyme responsible for this reaction, also requires cobalamin (methylmalonyl-CoA mutase is the only other mammalian enzyme that requires a vitamin B_{12} cofactor; Section 17.2). The disruption of one-carbon metabolism can therefore reflect inadequate intake of vitamins as well as enzyme deficiencies.

This pathway is the reason why cysteine is considered a nonessential amino acid, although its sulfur atom must ultimately come from methionine, an essential amino acid. The sulfur-containing amino acids also play an important role in one-carbon chemistry (**Box 18.B**).

Aspartate, the precursor of methionine, is also the precursor of the essential amino acids threonine and lysine. Since these amino acids are derived from another amino acid, they already have an amino group. The branched-chain amino acids (valine, leucine, and isoleucine) are synthesized by pathways that use pyruvate as the starting substrate. These amino acids require a step catalyzed by a transaminase (with glutamate as a substrate) to introduce an amino group.

In plants and bacteria, the pathway for synthesizing the aromatic amino acids (phenylalanine, tyrosine, and tryptophan) begins with the condensation of the C_3 compound phosphoenolpyruvate (a glycolytic intermediate) and erythrose-4-phosphate (a four-carbon intermediate of the Calvin cycle and the pentose phosphate pathway). The seven-carbon reaction product then cyclizes and undergoes additional modifications, including the addition of three more carbons from phosphoenolpyruvate, before becoming chorismate, the last common intermediate in the

Box 18.C Glyphosate, the Most Popular Herbicide

Glycine phosphonate, also known as glyphosate or Roundup (its trade name), competes with the second phosphoenolpyruvate in the pathway leading to chorismate:

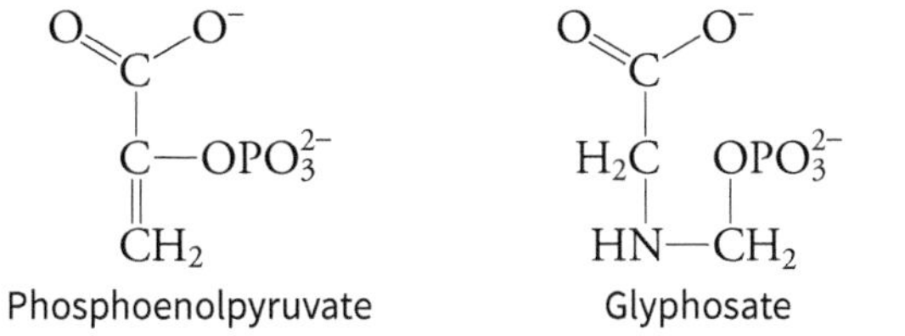

Phosphoenolpyruvate   Glyphosate

Because plants cannot manufacture aromatic amino acids without chorismate, glyphosate acts as an herbicide. Used widely in agriculture as well as home gardens, it has become the most popular herbicide in the United States, replacing other, more toxic compounds. Glyphosate that is not directly absorbed by the plant appears to bind tightly to soil particles and then is rapidly broken down by bacteria.

Consequently, glyphosate has less potential to contaminate water supplies than do more stable herbicides. Some studies suggest that glyphosate is not entirely harmless in animals, but there is no clear evidence that it causes cancer.

Farmers can take advantage of glyphosate's weed-killing properties by planting glyphosate-resistant crops and then spraying the field with glyphosate when weeds emerge and begin to compete with the crop plants. Such "Roundup-Ready" species include soybeans, corn (maize), and cotton. These plants have been genetically engineered to express a bacterial version of the enzyme that uses phosphoenolpyruvate but is not inhibited by glyphosate. Predictably, the use of glyphosate selects for herbicide resistance, so many types of weeds have already evolved resistance to glyphosate.

Question In addition to plants, what other types of organisms synthesize chorismate as a precursor of aromatic amino acids? How would glyphosate affect them?

---

synthesis of the three aromatic amino acids. *Because animals do not synthesize chorismate, this pathway is an obvious target for agents that can inhibit plant metabolism without affecting animals* (**Box 18.C**). A summary of the chorismate pathway is shown:

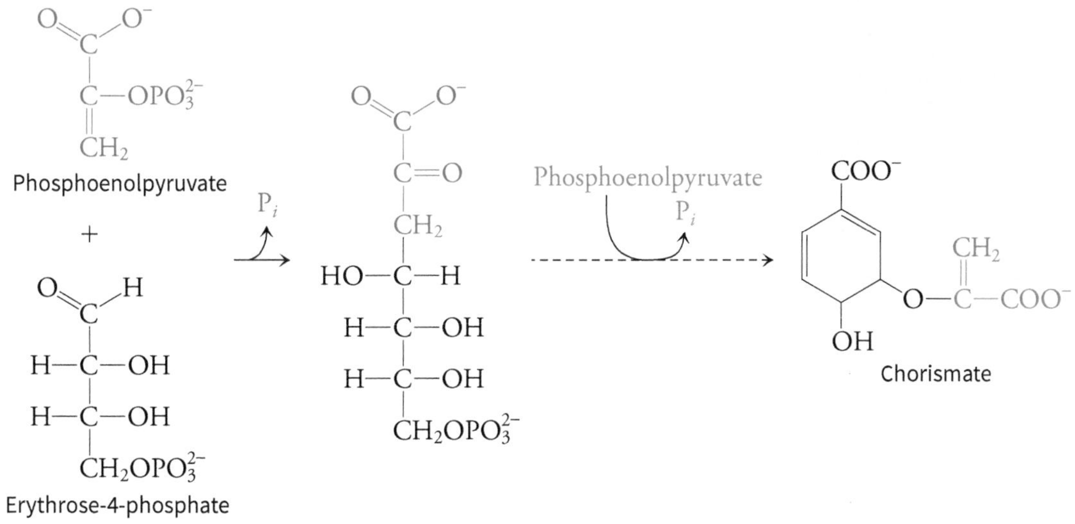

Phenylalanine and tyrosine are derived from chorismate by diverging pathways. In humans, tyrosine is generated by hydroxylating phenylalanine, which is why tyrosine is not considered an essential amino acid.

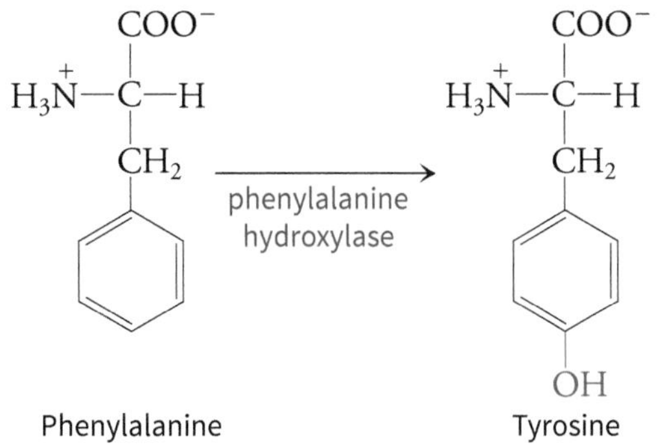

The final two reactions of the tryptophan biosynthetic pathway (which has 13 steps altogether) are catalyzed by tryptophan synthase, a bifunctional enzyme with an $\alpha_2\beta_2$ quaternary structure. The α subunit cleaves indole-3-glycerol phosphate to indole and glyceraldehyde-3-phosphate; then the β subunit adds serine to indole to produce tryptophan:

Indole-3-glycerol phosphate → (Glyceraldehyde 3-phosphate) → Indole → (Serine, H_2O) → Tryptophan

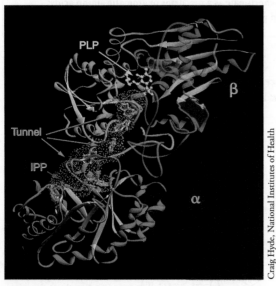

Figure 18.8 Tryptophan synthase. Only one α subunit (blue and tan) and one β subunit (yellow, orange, and tan) are shown. Indolepropanol phosphate (IPP; red) marks the active site of the α subunit. The β active site is marked by its PLP cofactor (yellow). The surface of the tunnel between the two active sites is outlined with yellow dots. Several indole molecules (green) are included in the model to show how this intermediate can pass between the active sites.

Craig Hyde, National Institutes of Health

Indole, the product of the α-subunit reaction and the substrate for the β-subunit reaction, never leaves the enzyme. Instead, it diffuses directly from one active site to the other without entering the surrounding solvent. The X-ray structure of the enzyme reveals that the active sites in adjacent α and β subunits are 25 Å apart but are connected by a tunnel through the protein that is large enough to accommodate indole (**Fig. 18.8**). *The movement of a reactant between two active sites is called* **channeling,** *and it increases the rate of a metabolic process by preventing the loss of intermediates.* Channeling is known to occur in a few other multifunctional enzymes.

All but one of the 20 standard amino acids are synthesized entirely from precursors produced by the main carbohydrate-metabolizing pathways. The exception is histidine, to which ATP provides one nitrogen and one carbon atom. Glutamate and glutamine donate the other two nitrogen atoms, and the remaining five carbons are derived from a phosphorylated monosaccharide, 5-phosphoribosyl pyrophosphate (PRPP):

ATP (Ribose triphosphate) + 5-Phosphoribosyl pyrophosphate → (Glutamine, Glutamate) → Histidine

5-Phosphoribosyl pyrophosphate is also the source of the ribose group of nucleotides. This suggests that histidine might have been one of the first amino acids synthesized by an early life-form making the transition from an all-RNA metabolism to an RNA-and-protein-based metabolism.

Amino acids are the precursors of some signaling molecules

Many amino acids that are ingested or built from scratch are used to manufacture a cell's proteins, but some also have essential functions as precursors of other compounds, including **neurotransmitters.** Communication in the complex neuronal circuitry of the nervous system relies on small chemical signals that are released by one neuron and taken up by another (see Section 9.4). Common neurotransmitters include the amino acids glycine and glutamate and a glutamate derivative (its carboxylate group has been removed) known as γ-aminobutyric acid (GABA) or γ-aminobutyrate.

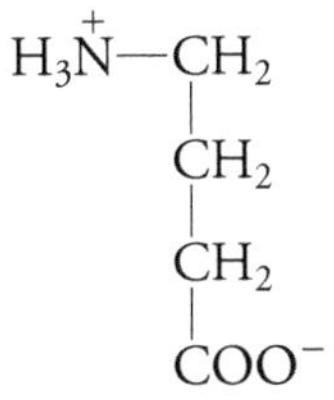

$$H_3\overset{+}{N}-CH_2$$
$$CH_2$$
$$CH_2$$
$$COO^-$$

γ-Aminobutyrate

Several other amino acid derivatives also function as neurotransmitters. For example, tyrosine gives rise to dopamine, norepinephrine, and epinephrine. These compounds are called catecholamines, reflecting their resemblance to catechol.

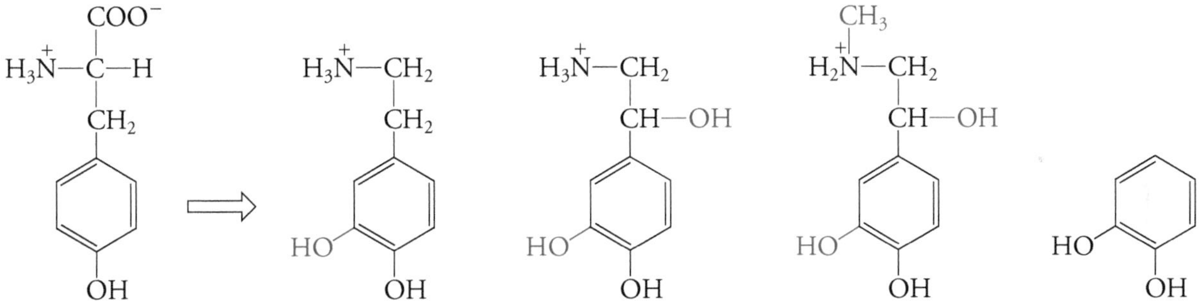

Tyrosine Dopamine Norepinephrine Epinephrine Catechol

A deficiency of dopamine produces the symptoms of Parkinson's disease: tremor, rigidity, and slow movements. As we saw in Section 10.2, catecholamines are also produced by other tissues and function as hormones.

Tryptophan is the precursor of the neurotransmitter serotonin:

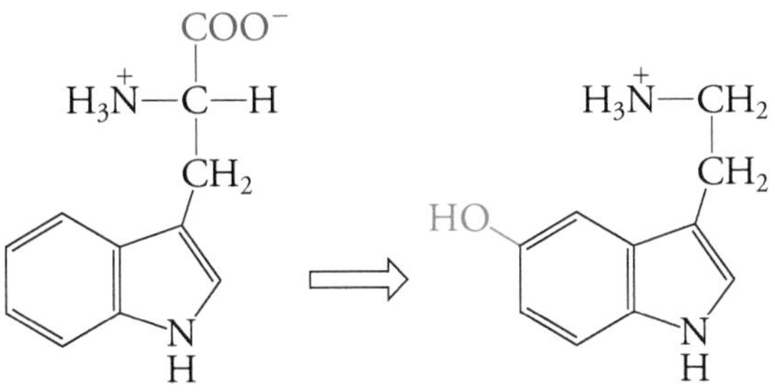

Tryptophan Serotonin

Low levels of serotonin in the brain have been linked to conditions such as depression, aggression, and hyperactivity. The antidepressive effect of drugs such as Prozac® results from their ability to

Box 18.D   Nitric Oxide

In the 1980s, vascular biologists were investigating the nature of an endothelial cell–derived "relaxation factor" that caused blood vessels to dilate. This substance diffused quickly, acted locally, and disappeared within seconds. To the surprise of many, the mysterious factor turned out to be the free radical nitric oxide (·NO). Although NO was known to elicit vasodilation, it had not been considered a good candidate for a biological signaling molecule because its unpaired electron makes it extremely reactive and it breaks down to yield the corrosive nitric acid.

NO is a signaling molecule in a wide array of tissues. At low concentrations it induces blood vessel dilation; at high concentrations (along with oxygen radicals) it kills pathogens. NO is synthesized from arginine by nitric oxide synthase, an enzyme whose cofactors include FMN, FAD, tetrahydrobiopterin (discussed in Section 18.3), and a heme group. The first step of NO production is a hydroxylation reaction. In the second step, one electron oxidizes N-hydroxyarginine (see diagram).

NO is unusual among signaling molecules for several reasons: It cannot be stockpiled for later release; it diffuses into cells, so it does not need a cell-surface receptor; and it needs no degradative enzyme

because it breaks down on its own. NO is produced only when and where it is needed. A free radical gas such as NO cannot be directly introduced into the body, but an indirect source of NO has been clinically used for over a century. Individuals who suffer from angina pectoris, a painful condition caused by obstruction of the coronary blood vessels, can relieve their symptoms by taking nitroglycerin:

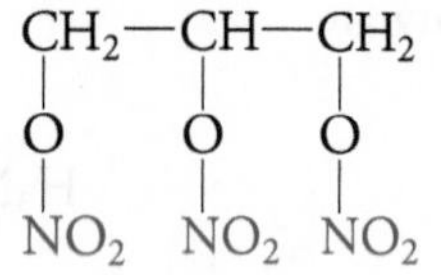

$$
\begin{array}{ccc}
CH_2 & CH & CH_2 \\
| & | & | \\
O & O & O \\
| & | & | \\
NO_2 & NO_2 & NO_2
\end{array}
$$

Nitroglycerin

In vivo, nitroglycerin yields NO, which rapidly stimulates vasodilation, temporarily relieving the symptoms of angina.

Question Explain why blood vessels express constant amounts of nitric oxide synthase whereas white blood cells must be induced to produce the enzyme.

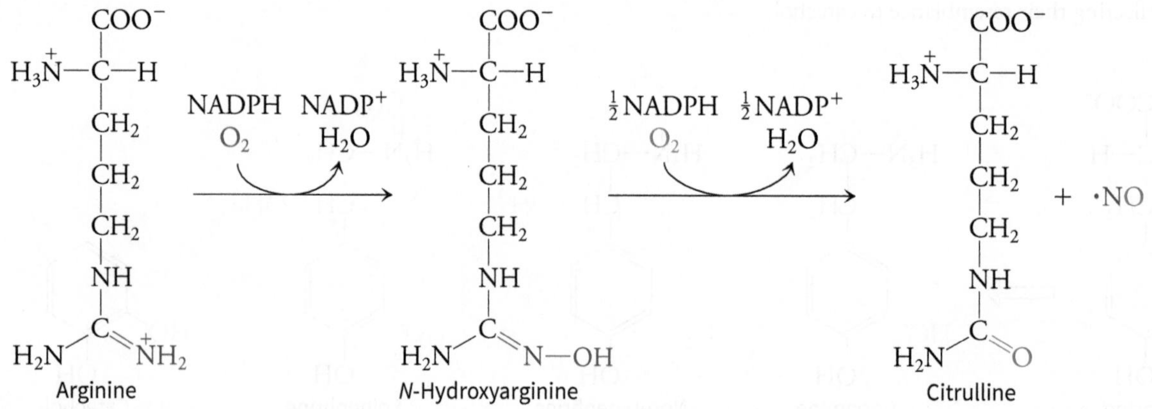

increase serotonin levels by blocking the reabsorption of the released neurotransmitter (see Box 9.B). Serotonin is the precursor of melatonin (at right). This tryptophan derivative is synthesized in the pineal gland and retina. Its concentration is low during the day, rising during darkness. Because melatonin appears to govern the synthesis of some other neurotransmitters that control circadian (daily) rhythms, it has been touted as a cure for sleep disorders and jet lag.

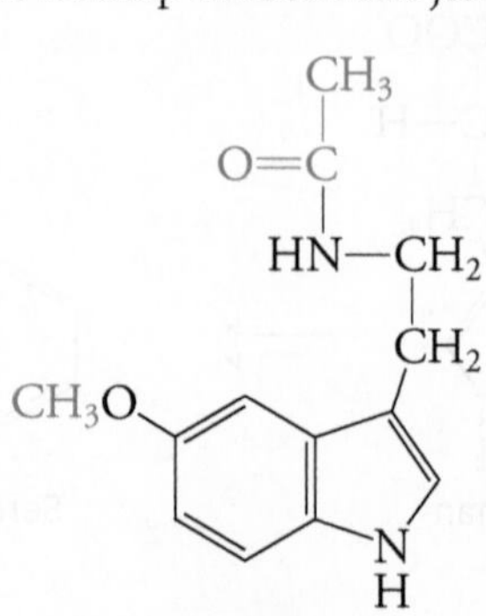

Melatonin

Arginine is the precursor of a signaling molecule that was discovered to be the free radical gas nitric oxide (NO; **Box 18.D**).

1. List the metabolites that are used as precursors for the nonessential amino acids.
2. Make a list of amino acids that cannot be synthesized by humans.
3. Identify the amino acids that are synthesized by simple transamination reactions.
4. Describe the role of tetrahydrofolate in amino acid biosynthesis.
5. Explain how histidine synthesis differs from the synthesis of other amino acids.
6. List some amino acids that give rise to neurotransmitters and signaling molecules.
7. Why do herbicides target the pathway for synthesizing aromatic amino acids?

18.3 Amino Acid Catabolism

KEY CONCEPTS

Summarize the pathways for degrading amino acids.

- Distinguish glucogenic and ketogenic amino acids.
- Identify the end products of amino acid catabolism.
- Summarize the role of coenzyme A in amino acid degradation.

Like monosaccharides and fatty acids, *amino acids are metabolic fuels that can be broken down to release free energy.* In fact, amino acids, not glucose, are the major fuel for the cells lining the small intestine. These cells absorb dietary amino acids and break down almost all of the available glutamate and aspartate and a good portion of the glutamine supply (note that these are all non-essential amino acids).

Other tissues, mainly the liver, also catabolize amino acids originating from the diet and from the normal turnover of intracellular proteins. As described in Section 12.1, protein degradation is accomplished by the activity of proteasomes as well as lysosomal and cytosolic enzymes. When necessary, such as during a prolonged fast, amino acids can be mobilized through the breakdown of muscle tissue, which accounts for about 40% of the total protein in the body.

In the liver, amino acids undergo transamination reactions to remove their α-amino groups, and their carbon skeletons then enter the central pathways of energy metabolism (principally the citric acid cycle). However, *the catabolism of amino acids in the liver is not complete.* There is simply not enough oxygen available for the liver to completely oxidize all the carbon to CO_2; even if there were, the liver would not need all the ATP that would be produced as a result. Instead, the amino acids are partially oxidized to substrates for gluconeogenesis (or ketogenesis). Glucose can then be exported to other tissues or stored as glycogen.

The reactions of amino acid catabolism, like those of amino acid synthesis, are too numerous to describe in full here, and the catabolic pathways do not necessarily mirror the anabolic pathways, as they do in carbohydrate and fatty acid metabolism. In this section, we will focus on some general principles and a few interesting chemical aspects of amino acid catabolism. In the following section we will see how organisms dispose of the nitrogen component of catabolized amino acids.

Amino acids are glucogenic, ketogenic, or both

It is useful to classify amino acids in humans as **glucogenic** (giving rise to gluconeogenic precursors such as citric acid cycle intermediates) or **ketogenic** (giving rise to acetyl-CoA, which can be used for ketogenesis or fatty acid synthesis, but not gluconeogenesis). As shown in **Table 18.2**, *all but leucine and lysine are at least partly glucogenic, most of the nonessential amino acids are glucogenic, and the large skeletons of the aromatic amino acids are both glucogenic and ketogenic.*

Three amino acids are converted to gluconeogenic substrates by simple transamination (the reverse of their biosynthetic reactions): alanine to pyruvate, aspartate to oxaloacetate, and glutamate to α-ketoglutarate. Glutamate can also be **deaminated** in an oxidation reaction catalyzed by

Table 18.2 Catabolic Fates of Amino Acids

Glucogenic	Both glucogenic and ketogenic	Ketogenic
Alanine	Isoleucine	Leucine
Arginine	Phenylalanine	Lysine
Asparagine	Threonine	
Aspartate	Tryptophan	
Cysteine	Tyrosine	
Glutamate		
Glutamine		
Glycine		
Histidine		
Methionine		
Proline		
Serine		
Valine		

glutamate dehydrogenase (discussed in the next section). Asparagine undergoes a simple hydrolytic deamidation to aspartate, which is then transaminated to oxaloacetate:

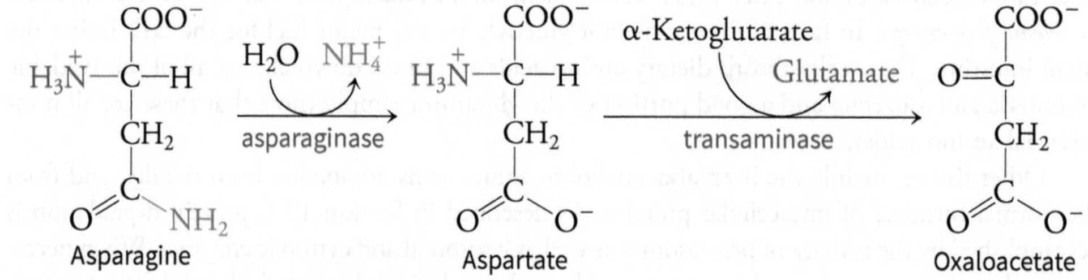

Similarly, glutamine is deamidated by a glutaminase to glutamate, and the glutamate dehydrogenase reaction yields α-ketoglutarate. Serine is converted to pyruvate:

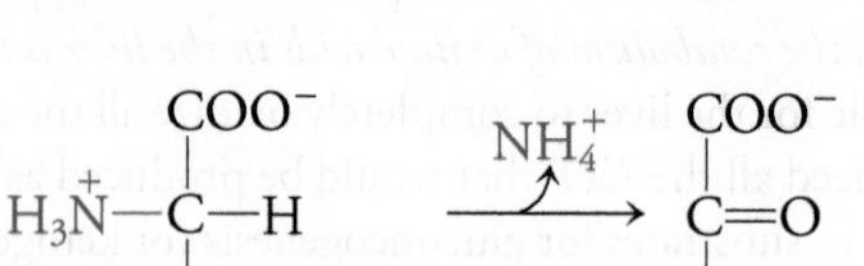

Note that in this reaction and in the conversion of asparagine and glutamine to their acid counterparts, the amino group is released as NH_4^+ rather than being transferred to another compound.

Arginine and proline (which are synthesized from glutamate) as well as histidine are catabolized to glutamate, which is then converted to α-ketoglutarate. Amino acids of the glutamate "family"—arginine, glutamate, glutamine, histidine, and proline—constitute about 25% of dietary amino acids, so their potential contribution to energy metabolism is significant.

Cysteine is converted to pyruvate by a process that releases ammonia as well as sulfur:

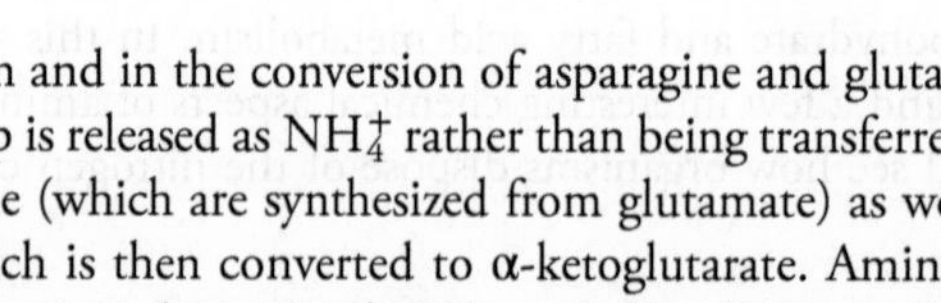

The products of the reactions listed so far—pyruvate, oxaloacetate, and α-ketoglutarate—are all gluconeogenic precursors. Threonine is both glucogenic and ketogenic because it is broken down to acetyl-CoA and glycine:

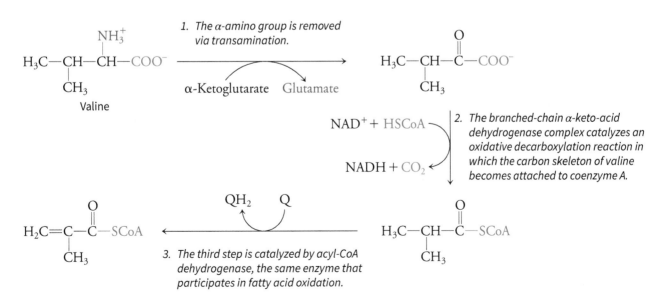

The acetyl-CoA is a precursor of ketone bodies (see Section 17.3), and the glycine is potentially glucogenic—if it is first converted to serine by the action of serine hydroxymethyltransferase. The major route for glycine disposal, however, is catalyzed by a multiprotein complex known as the glycine cleavage system:

The degradation pathways for the remaining amino acids are more complicated. For example, the branched-chain amino acids—valine, leucine, and isoleucine—undergo transamination to their α-keto-acid forms and are then linked to coenzyme A in an oxidative decarboxylation reaction. This step is catalyzed by the branched-chain α-keto-acid dehydrogenase complex, a multienzyme complex that resembles the pyruvate dehydrogenase complex (see Section 14.1) and even shares some of the same subunits.

The initial reactions of valine catabolism are shown in **Fig. 18.9**. Subsequent steps yield the citric acid cycle intermediate succinyl-CoA. Isoleucine is degraded by a similar pathway that yields succinyl-CoA and acetyl-CoA. Leucine degradation yields acetyl-CoA and the ketone body

Figure 18.9 The initial steps of valine degradation.

Question List all the cofactors involved in this process. (*Hint:* Step 2 is carried out by a multienzyme complex.)

acetoacetate. Lysine degradation, which follows a different pathway from the branched-chain amino acids, also yields acetyl-CoA and acetoacetate. The degradation of methionine produces succinyl-CoA.

Finally, the cleavage of the aromatic amino acids—phenylalanine, tyrosine, and tryptophan—yields the ketone body acetoacetate as well as a glucogenic compound (alanine or fumarate). The first step of phenylalanine degradation is a hydroxylation reaction that produces tyrosine, as we have already seen. This reaction is worth noting because it uses the cofactor tetrahydrobiopterin (which, like folate, contains a pterin group):

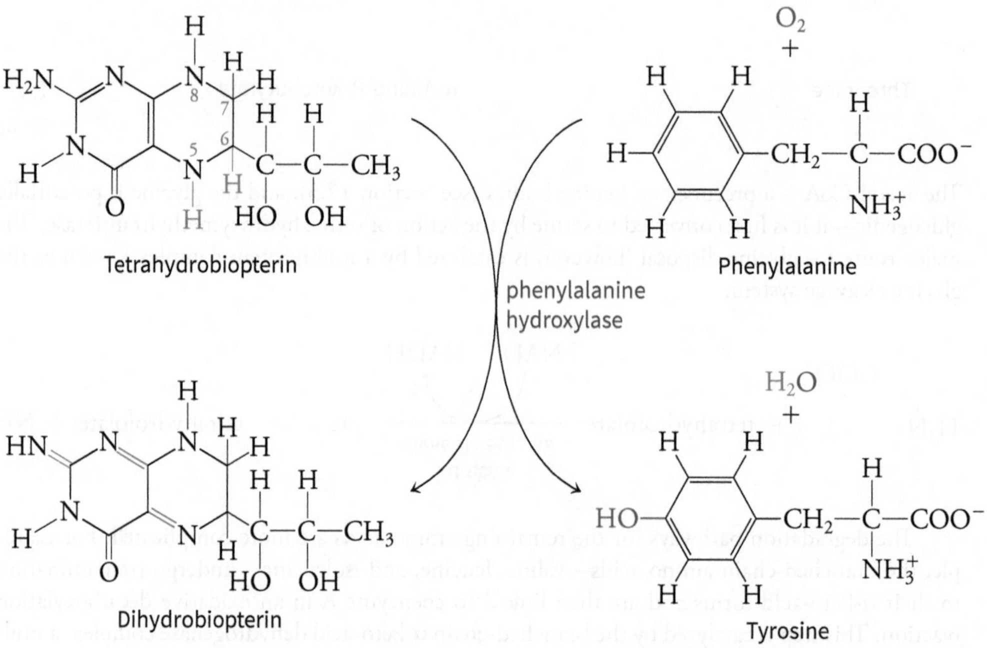

The tetrahydrobiopterin is oxidized to dihydrobiopterin in the phenylalanine hydroxylase reaction. This cofactor must be subsequently reduced to the tetrahydro form by a separate NADH-dependent enzyme. Another step of the phenylalanine (and tyrosine) degradation pathway is also notable because a deficiency of the enzyme was one of the first-characterized metabolic diseases (Box 18.E).

Box 18.E   Diseases of Amino Acid Metabolism

Inherited diseases result from defective genes, and malfunctioning genes underlie many noninherited diseases as well. The link between genes and disease was first recognized by the physician Archibald Garrod, who coined the term *inborn error of metabolism* in 1902. Garrod's insights came from his studies of individuals with alcaptonuria. Their urine turned black upon exposure to air because it contained homogentisate, a product of tyrosine catabolism. Garrod concluded that this inherited condition resulted from the lack of a specific enzyme. We now know that homogentisate is excreted because the enzyme that breaks it down, homogentisate dioxygenase, is missing or defective.

Garrod also described a number of other inborn errors of metabolism, including albinism, cystinuria (excretion of cysteine in the urine), and several other disorders that were not life-threatening and left easily detected clues in the patient's urine. Of course, many "inborn errors" are catastrophic. For example, phenylketonuria (PKU) results from a deficiency of phenylalanine hydroxylase (see the diagram). Without

this enzyme, phenylalanine cannot be broken down, although it can undergo transamination. The resulting α-keto-acid derivative phenylpyruvate accumulates and is excreted in the urine, giving it a mousy odor. If not treated, PKU causes neurological impairment. Fortunately, the disease can be detected in newborns. Afflicted individuals develop normally if they consume a diet that is low in phenylalanine.

Maple syrup urine disease results from a deficiency of the branched-chain α-keto-acid dehydrogenase complex. The deaminated forms of the branched-chain amino acids accumulate and are excreted in the urine, giving it a smell like maple syrup. Like PKU, this disorder can be treated by a controlled diet.

Defects in the glycine cleavage system cause hyperglycinemia and lead to severe neurological abnormalities, perhaps due to glycine's role as a neurotransmitter. It cannot be treated by dietary adjustments.

Not all disorders of amino acid metabolism have obvious symptoms and easily measured excesses of certain metabolites.

Many more disorders have come to light through DNA sequencing techniques, which can detect enzyme abnormalities even when the levels of circulating amino acids or their metabolites are near normal. Furthermore, metabolic pathways interconnect in unexpected ways. For example, high levels of circulating branched-chain amino acids (leucine, isoleucine, and valine) are strongly associated with cardiovascular disease and type 2 diabetes, but researchers are not sure whether disruption of amino acid metabolism is a cause or an effect of these other diseases.

Question Individuals with PKU must consume a certain amount of phenylalanine. Describe the three metabolic fates of this amino acid.

| Phenylalanine | Tyrosine | p-Hydroxyphenylpyruvate | Homogentisate | 4-Maleyl-acetoacetate |

Concept Check

1. List the end-products of catabolism of the 20 amino acids.
2. Summarize the role of the liver in catabolizing amino acids.
3. List the cofactors involved in amino acid breakdown.
4. Distinguish the glucogenic and ketogenic amino acids.

18.4 Excretion of Excess Nitrogen: The Urea Cycle

KEY CONCEPTS

Describe the chemical reactions of the urea cycle.

- Explain the purpose of the urea cycle.
- Identify the sources of the urea amino groups.

When the supply of amino acids exceeds the cell's immediate needs for protein synthesis or other amino acid–consuming pathways, the carbon skeletons are broken down and the nitrogen disposed of. All amino acids except lysine can be deaminated by the action of transaminases, but this merely transfers the amino group to another molecule; it does not eliminate it from the body.

Some catabolic reactions do release free ammonia, which can be excreted as a waste product in the urine. In fact, the kidney is a major site of glutamine catabolism, and the resulting NH_4^+ facilitates the excretion of metabolic acids such as H_2SO_4 that arise from the catabolism of methionine and cysteine. However, ammonia production is not feasible for disposing of large amounts of excess nitrogen. First, high concentrations of NH_4^+ in the blood cause alkalosis. Second, ammonia is highly toxic. It easily enters the brain, where it activates the NMDA receptor, whose normal agonist is the neurotransmitter glutamate. The activated receptor is an ion channel that normally opens to allow Ca^{2+} and Na^+ ions to enter the cell and K^+ ions to exit the cell. However, the large Ca^{2+} influx triggered by ammonia binding to the receptor results in neuronal cell death, a phenomenon called excitotoxicity. Humans and many other organisms have therefore evolved safer ways to deal with excess amino groups.

Approximately 80% of the body's excess nitrogen is excreted in the form of urea,

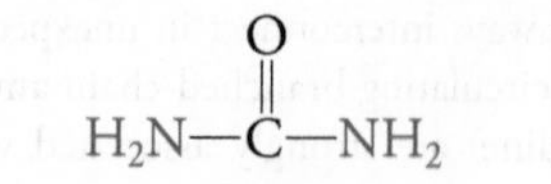

$$H_2N-\overset{\overset{\displaystyle O}{\|}}{C}-NH_2$$
Urea

which is produced in the liver by the reactions of the **urea cycle.** This catabolic cycle was elucidated in 1932 by Hans Krebs and Kurt Henseleit; Krebs went on to outline another circular pathway—the citric acid cycle—in 1937.

Glutamate supplies nitrogen to the urea cycle

Because many transaminases use α-ketoglutarate as the amino-group acceptor, glutamate is one of the most abundant amino acids inside cells. Glutamate can be deaminated to regenerate α-ketoglutarate and release NH_4^+ in an oxidation–reduction reaction catalyzed by glutamate dehydrogenase:

$$^-OOC-CH_2-CH_2-\overset{\overset{\displaystyle NH_3^+}{|}}{\underset{\underset{\displaystyle H}{|}}{C}}-COO^- \quad \underset{NAD(P)^+}{\overset{NAD(P)H + H^+}{\rightleftharpoons}}$$
Glutamate

$$\left[^-OOC-CH_2-CH_2-\overset{\overset{\displaystyle NH_2^+}{\|}}{C}-COO^- \right] \quad \overset{H_2O \quad NH_4^+}{\rightleftharpoons}$$

$$^-OOC-CH_2-CH_2-\overset{\overset{\displaystyle O}{\|}}{C}-COO^-$$
α-Ketoglutarate

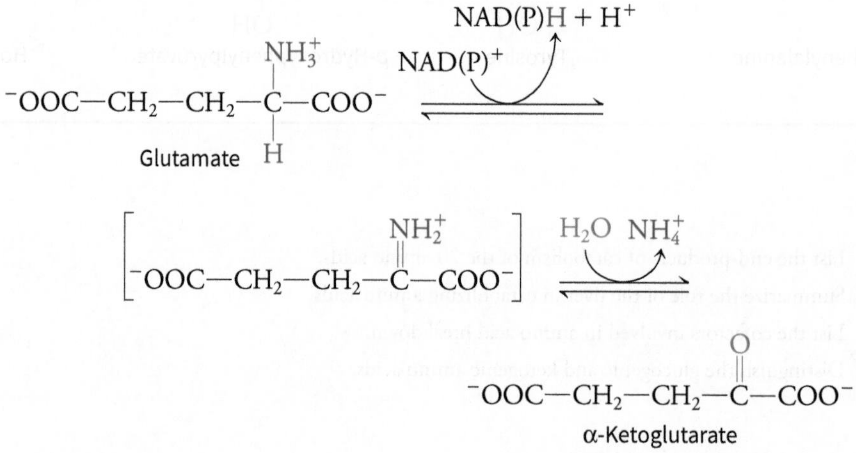

This mitochondrial enzyme is unusual: It is one of only a few enzymes that can use either NAD^+ or $NADP^+$ as a cofactor. The glutamate dehydrogenase reaction is a major route for feeding amino acid–derived amino groups into the urea cycle and, not surprisingly, is subject to allosteric activation and inhibition.

The starting substrate for the urea cycle is an "activated" molecule produced by the condensation of bicarbonate and ammonia, as catalyzed by carbamoyl phosphate synthetase (**Fig. 18.10**).

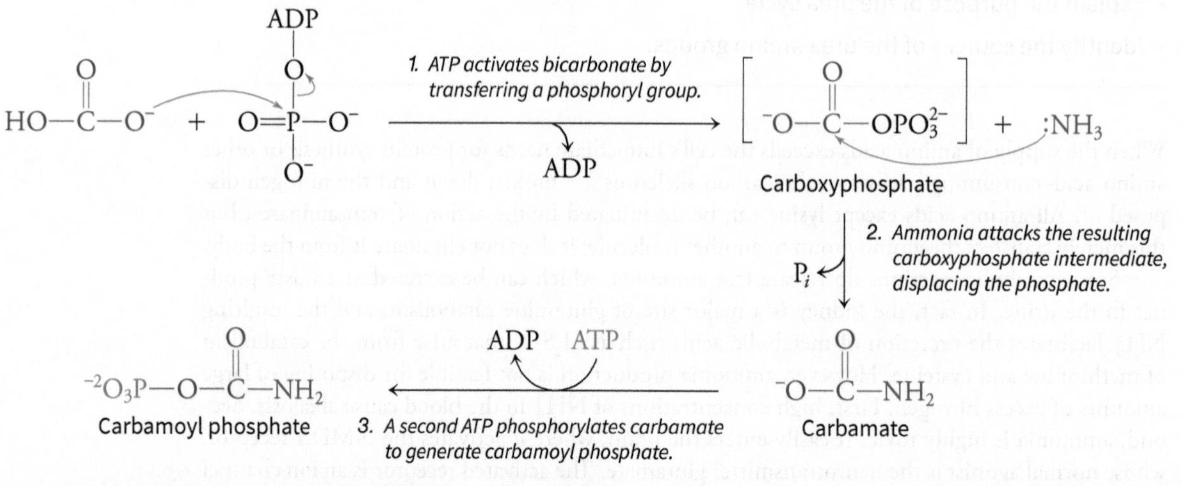

Figure 18.10 The carbamoyl phosphate synthetase reaction.

Question Bicarbonate (as CO_2) and ammonia can easily diffuse into mitochondria. Why doesn't the reaction product diffuse out?

The NH_4^+ may be contributed by the glutamate dehydrogenase reaction or another process that releases ammonia. The bicarbonate is the source of the urea carbon. Note that the phosphoanhydride bonds of two ATP molecules are consumed in the energetically costly production of carbamoyl phosphate.

The urea cycle consists of four reactions

The four enzyme-catalyzed reactions of the urea cycle proper are shown in **Figure 18.11**. The cycle also provides a means for synthesizing arginine: The five-carbon ornithine is derived from glutamate, and the urea cycle converts it to arginine. However, the arginine needs of children exceed the biosynthetic capacity of the urea cycle, so arginine is classified as an essential amino acid.

The fumarate generated in step 3 of the urea cycle is converted to malate and then oxaloacetate, which is used for gluconeogenesis. The aspartate substrate for Reaction 2 is equivalent to oxaloacetate that has undergone transamination. Combining these ancillary reactions with those

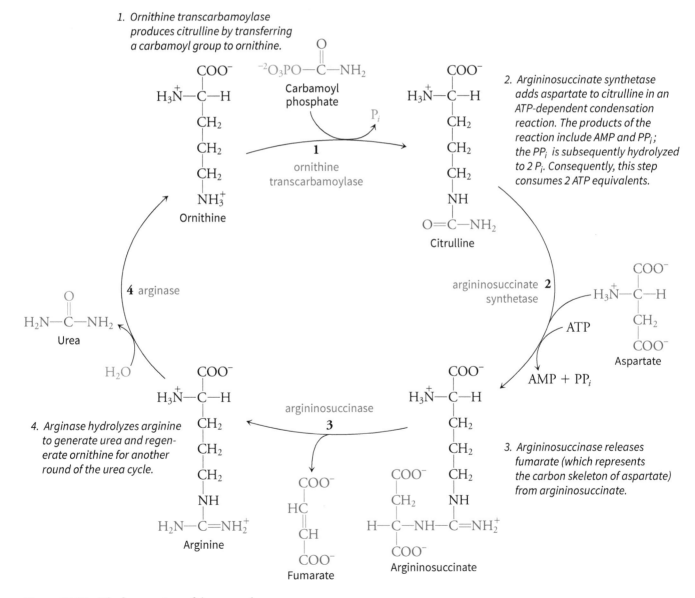

Figure 18.11   The four reactions of the urea cycle.

Question **Which steps are likely to be irreversible *in vivo*?**

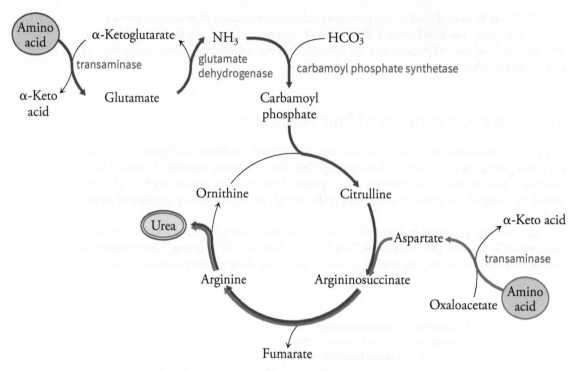

Figure 18.12 The urea cycle and related reactions. Two routes for the disposal of amino groups are highlighted. The blue pathway shows how an amino group from an amino acid enters the urea cycle via glutamate and carbamoyl phosphate. The red pathway shows how an amino group from an amino acid enters via aspartate.

Question Identify all the products of this metabolic scheme. Which ones are recycled?

of the urea cycle, the carbamoyl phosphate synthetase reaction, and the glutamate dehydrogenase reaction yields the pathway outlined in **Figure 18.12.** *The overall effect is that transaminated amino acids donate amino groups, via glutamate and aspartate, to urea synthesis.* Because the liver is the only tissue that can carry out urea synthesis, amino groups to be eliminated travel through the blood to the liver mainly as glutamine, which accounts for up to one-quarter of circulating amino acids.

Like many other metabolic loops, the urea cycle involves enzymes located in both the mitochondria and cytosol. Glutamate dehydrogenase, carbamoyl phosphate synthetase, and ornithine transcarbamoylase are mitochondrial, whereas argininosuccinate synthetase, argininosuccinase, and arginase are cytosolic. Consequently, citrulline is produced in the mitochondria but must be transported to the cytosol for the next step, and ornithine produced in the cytosol must be imported into the mitochondria to begin a new round of the cycle.

The carbamoyl phosphate synthetase reaction and the argininosuccinate synthetase reactions each consume 2 ATP equivalents, so the cost of the urea cycle is 4 ATP per urea. However, when considered in context, operation of the urea cycle is often accompanied by ATP synthesis. The glutamate dehydrogenase reaction produces NADH (or NADPH), whose energy is conserved in the synthesis of 2.5 ATP by oxidative phosphorylation. Catabolism of the carbon skeletons of the amino acids that donated their amino groups via transamination also yields ATP.

The rate of urea production is controlled largely by the activity of carbamoyl phosphate synthetase. This enzyme is allosterically activated by *N*-acetylglutamate, which is synthesized from glutamate and acetyl-CoA:

When amino acids are undergoing transamination and being catabolized, the resulting increases in the cellular glutamate and acetyl-CoA concentrations boost production of N-acetylglutamate. This stimulates carbamoyl phosphate synthetase activity, and flux through the urea cycle increases. Such a regulatory system allows the cell to efficiently dispose of the nitrogen released from amino acid degradation.

Urea is relatively nontoxic and easily transported through the bloodstream to the kidneys for excretion in the urine. However, the polar urea molecule requires large amounts of water for its efficient excretion. This presents a problem for flying vertebrates such as birds and for reptiles that are adapted to arid habitats. These organisms deal with waste nitrogen by converting it to urate via purine synthesis (Section 18.5). The relatively insoluble urate is excreted as a semisolid paste, which conserves water.

Bacteria, fungi, and some other organisms use an enzyme called urease to break down urea:

$$H_2N - \overset{\overset{\displaystyle O}{\|}}{C} - NH_2 \;+\; H_2O \;\xrightarrow{urease}\; 2\,NH_3 \;+\; CO_2$$

Urease has the distinction of being the first enzyme to be crystallized (in 1926). It helped promote the theory that catalytic activity was a property of proteins. This premise is only partly true, as we have seen, since many enzymes contain metal ions or organic cofactors (urease itself contains two catalytic nickel atoms).

Concept Check

1. Explain the importance of glutamate dehydrogenase and carbamoyl phosphate synthetase for nitrogen disposal.
2. Summarize the reactions that occur in the four steps of the urea cycle.
3. Compare the solubility and toxicity of ammonia, urate, and urea.
4. Describe some situations when flux through the urea cycle would be very high or very low.
5. Why do mammals eliminate waste nitrogen as urea rather than as ammonia? Why do some organisms excrete uric acid?
6. How are amino groups from amino acids incorporated into urea?

18.5 Nucleotide Metabolism

KEY CONCEPTS

Describe the key reactions in the synthesis and degradation of nucleotides and deoxynucleotides.

- Compare the role of 5-phosphoribosyl pyrophosphate in purine and pyrimidine nucleotide synthesis.
- Summarize the ribonucleotide reductase reaction.
- Explain the importance of the thymidylate synthase reaction.
- List the products of nucleotide degradation.

A discussion of nitrogen metabolism would not be complete without considering nucleotides, whose *nitrogenous bases are synthesized mainly from amino acids*. The human body can also recycle nucleotides from nucleic acids and nucleotide cofactors that are broken down. Although food supplies nucleotides, the biosynthetic and recycling pathways are so efficient that there is no true dietary requirement for purines and pyrimidines. In this section we will take a brief look at the biosynthesis and degradation of purine and pyrimidine nucleotides in mammals.

Purine nucleotide synthesis yields IMP and then AMP and GMP

Purine nucleotides (AMP and GMP) are synthesized by building the purine base onto a ribose-5-phosphate molecule. In fact, the first step of the pathway is the production of 5-phosphoribosyl pyrophosphate (which is also a precursor of histidine):

Ribose-5-phosphate   5-Phosphoribosyl pyrophosphate

A similar ATP-dependent step "activates" fatty acids for degradation (Section 17.2) and prepares amino acids to be attached to tRNAs for protein synthesis (Section 22.1).

The next 10 steps of the purine synthesis pathway require as substrates glutamine, glycine, aspartate, and bicarbonate, plus one-carbon formyl (—HC=O) groups donated by tetrahydrofolate. The pathway enzymes appear to associate as a complex, called the **purinosome,** located near mitochondria. The purinosome generates inosine monophosphate (IMP), a nucleotide whose base is the purine hypoxanthine:

Inosine monophosphate (IMP)

IMP is the substrate for two short pathways that yield AMP and GMP. In AMP synthesis, an amino group from aspartate is transferred to the purine; in GMP synthesis, glutamate is the source of the amino group (**Fig. 18.13**). Kinases then catalyze phosphoryl-group transfer reactions to convert the nucleoside monophosphates to diphosphates and then triphosphates (ATP and GTP).

Figure 18.13 indicates that GTP participates in AMP synthesis and ATP participates in GMP synthesis. High concentrations of ATP therefore promote GMP production, and high concentrations of GTP promote AMP production. *This reciprocal relationship is one mechanism for balancing the production of adenine and guanine nucleotides.* (Because most nucleotides are destined for DNA or RNA synthesis, they are required in roughly equal amounts.) The pathway leading to AMP and GMP is also regulated by feedback inhibition at several points, including the first step, the production of 5-phosphoribosyl pyrophosphate from ribose-5-phosphate, which is inhibited by both ADP and GDP.

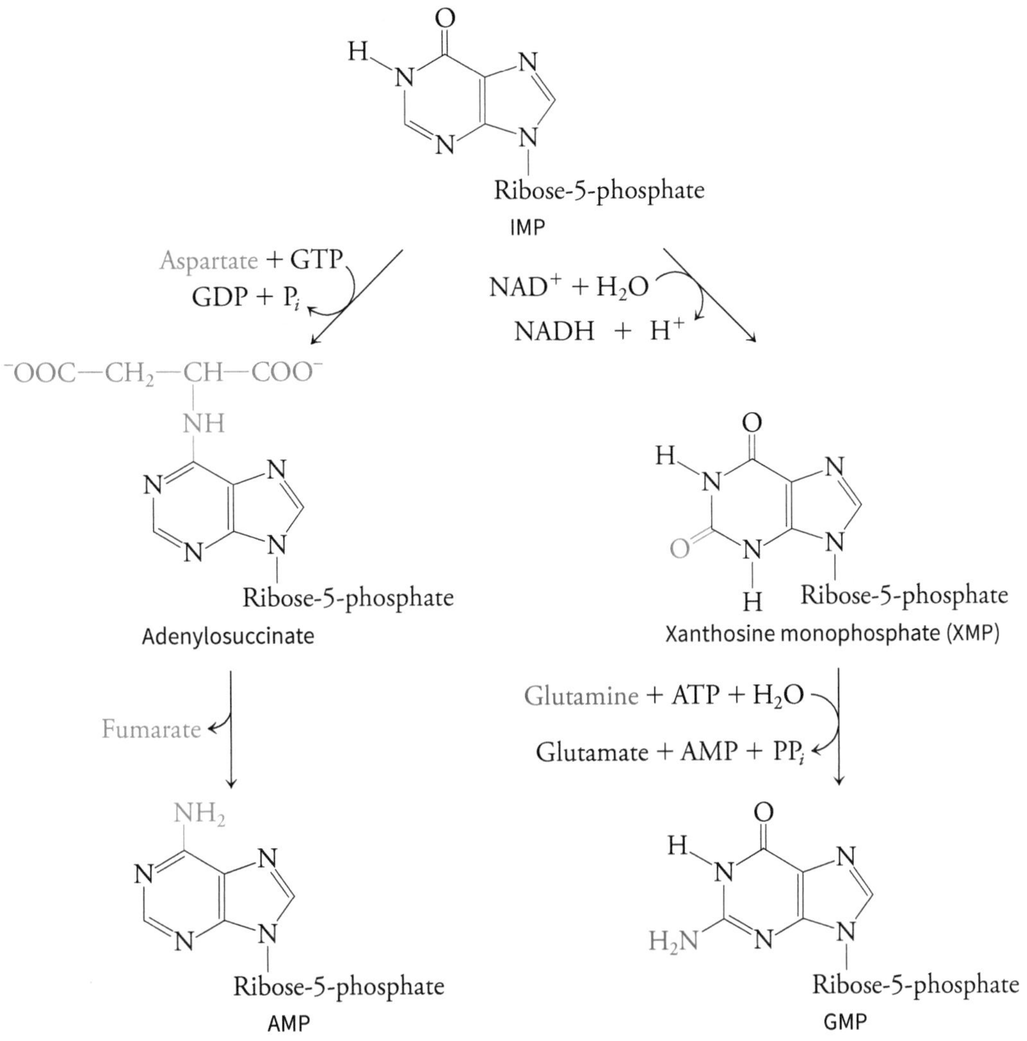

Figure 18.13 AMP and GMP synthesis from IMP.

Pyrimidine nucleotide synthesis yields UTP and CTP

In contrast to purine nucleotides, pyrimidine nucleotides are synthesized as a base that is subsequently attached to 5-phosphoribosyl pyrophosphate to form a nucleotide. The six-step pathway that yields uridine monophosphate (UMP) requires glutamine, aspartate, and bicarbonate.

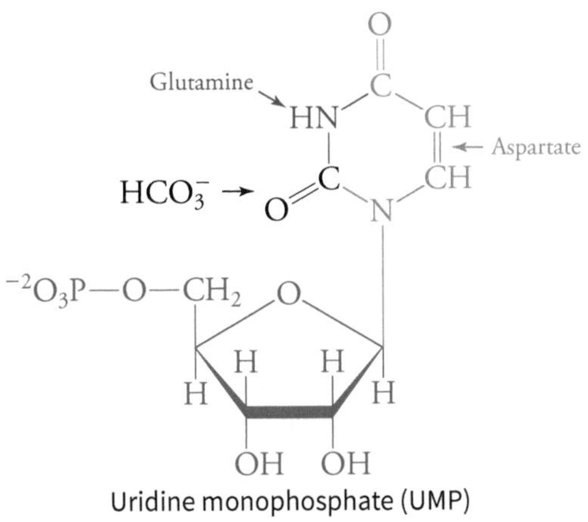

Uridine monophosphate (UMP)

UMP is phosphorylated to yield UDP and then UTP. CTP synthetase catalyzes the amination of UTP to CTP, using glutamine as the donor:

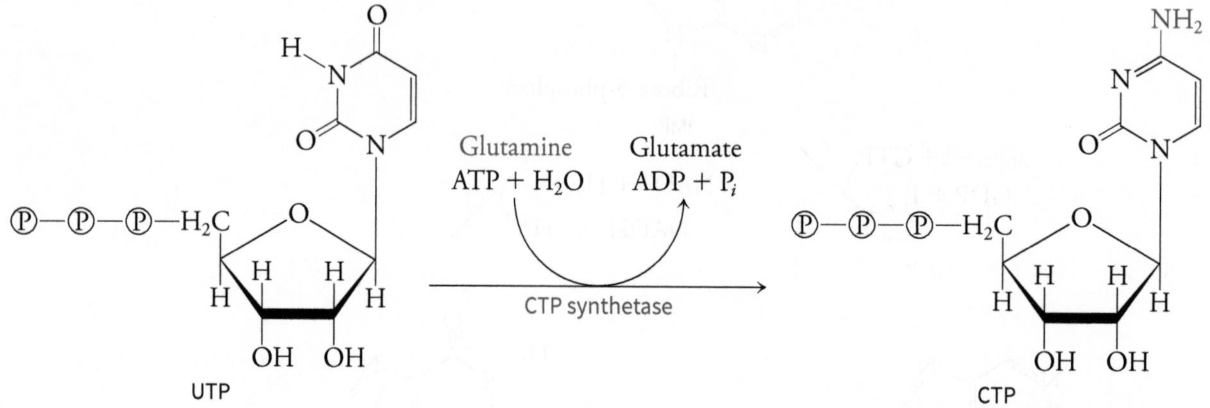

The UMP synthetic pathway in mammals is regulated primarily through feedback inhibition by UMP, UDP, and UTP. ATP activates the enzyme that catalyzes the first step; this helps balance the production of purine and pyrimidine nucleotides.

Ribonucleotide reductase converts ribonucleotides to deoxyribonucleotides

So far, we have accounted for the synthesis of ATP, GTP, CTP, and UTP, which are substrates for the synthesis of RNA. DNA, of course, is built from deoxynucleotides. In deoxynucleotide synthesis, each of the four nucleoside triphosphates (NTPs) is converted to its diphosphate (NDP) form, ribonucleotide reductase replaces the 2′ OH group with H, and then the resulting deoxynucleoside diphosphate (dNDP) is phosphorylated to produce the corresponding triphosphate (dNTP).

Ribonucleotide reductase is an essential enzyme that carries out a chemically difficult reaction using a mechanism that involves free radicals. Three types of ribonucleotide reductases, which differ in their catalytic groups, have been described. Class I enzymes (the type that occurs in mammals and most bacteria) have two Fe^{3+} ions and an unusually stable tyrosine radical (most free radicals, which have one unpaired electron, are highly reactive and short-lived).

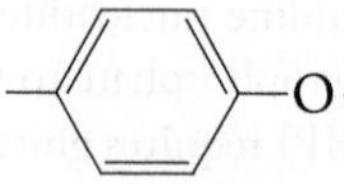

Tyrosine radicals are also features of the active sites of cytochrome c oxidase (mitochondrial Complex IV) and Photosystem II in plants. A possible reaction mechanism for ribonucleotide reductase is shown in **Figure 18.14**.

The final step of the reaction, which regenerates the enzyme, requires the small protein thioredoxin. The oxidized thioredoxin must then undergo reduction to return to its original state. *This reaction uses NADPH, which is therefore the ultimate source of reducing power for the synthesis of deoxyribonucleotides.* Recall that the pentose phosphate pathway, which provides the ribose-5-phosphate for nucleotide synthesis, also generates NADPH (Section 13.4).

Not surprisingly, the activity of ribonucleotide reductase is tightly regulated so that the cell can balance the levels of ribo- and deoxyribonucleotides as well as the proportions of each of the four deoxyribonucleotides. Control of the enzyme involves two regulatory sites that are distinct from the substrate-binding site. For example, ATP binding to the so-called activity site activates the enzyme. Binding of the deoxyribonucleotide dATP decreases enzyme activity. Several nucleotides bind to the so-called substrate specificity site. Here, ATP binding induces the enzyme to act on pyrimidine nucleotides, and dTTP binding causes the enzyme to prefer GDP as a substrate.

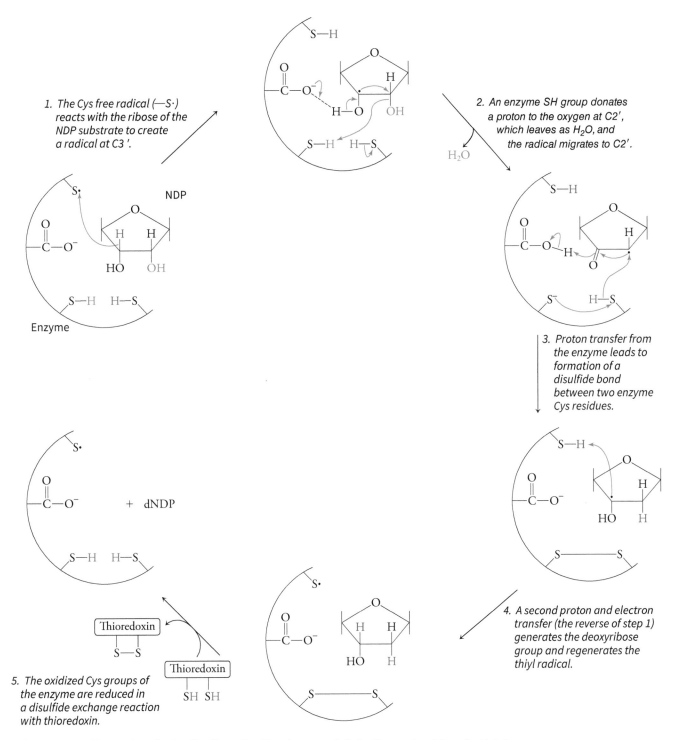

1. The Cys free radical (—S·) reacts with the ribose of the NDP substrate to create a radical at C3′.

2. An enzyme SH group donates a proton to the oxygen at C2′, which leaves as H_2O, and the radical migrates to C2′.

3. Proton transfer from the enzyme leads to formation of a disulfide bond between two enzyme Cys residues.

4. A second proton and electron transfer (the reverse of step 1) generates the deoxyribose group and regenerates the thiyl radical.

5. The oxidized Cys groups of the enzyme are reduced in a disulfide exchange reaction with thioredoxin.

Figure 18.14 Proposed mechanism for ribonucleotide reductase. Only the ribose portion of the nucleotide is shown.

These mechanisms, in concert with other mechanisms for balancing the amounts of the various nucleotides, help make all four deoxynucleotides available for DNA synthesis.

Thymidine nucleotides are produced by methylation

The ribonucleotide reductase reaction, followed by kinase-catalyzed phosphorylation, generates dATP, dCTP, dGTP, and dUTP. However, dUTP is not used for DNA synthesis. Instead, *it is rapidly converted to thymine nucleotides* (which helps prevent the accidental incorporation of uracil into DNA). First, dUTP is hydrolyzed to dUMP. Next, thymidylate synthase adds a methyl group to dUMP to produce dTMP, using methylene-tetrahydrofolate as a one-carbon donor.

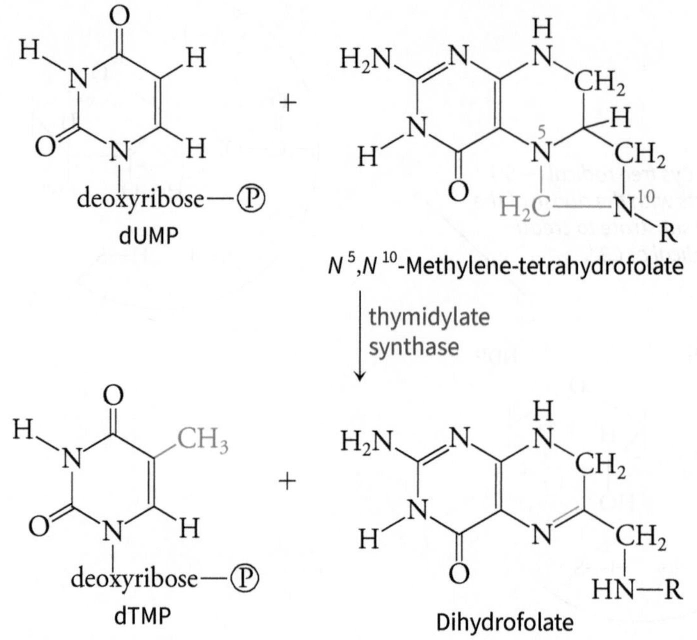

dUMP   + N^5,N^{10}-Methylene-tetrahydrofolate

thymidylate synthase

dTMP   +   Dihydrofolate

The serine hydroxymethyltransferase reaction, which converts serine to glycine (Section 18.2), is the main source of methylene-tetrahydrofolate.

In converting the methylene group (—CH_2—) of the cofactor to the methyl group (—CH_3) attached to thymine, thymidylate synthase oxidizes the tetrahydrofolate cofactor to dihydrofolate. An NADPH-dependent enzyme called dihydrofolate reductase must then regenerate the reduced tetrahydrofolate cofactor. Finally, dTMP is phosphorylated to produce dTTP, the substrate for DNA polymerase.

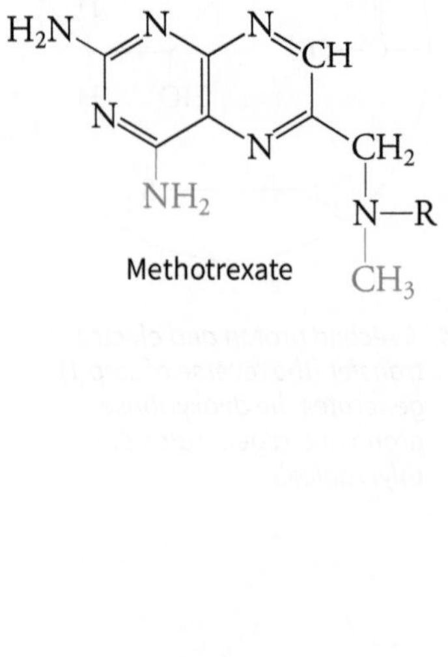

Because cancer cells undergo rapid cell division, the enzymes of nucleotide synthesis, including thymidylate synthase and dihydrofolate reductase, are highly active. Compounds that inhibit either of these reactions can therefore act as anticancer agents. For example, the dUMP analog 5-fluorodeoxyuridylate, introduced in Section 7.3, inactivates thymidylate synthase. "Antifolates" such as methotrexate (at right) are competitive inhibitors of dihydrofolate reductase because they compete with dihydrofolate for binding to the enzyme. In the presence of methotrexate, a cancer cell cannot regenerate the tetrahydrofolate required for dTMP production, and the cell dies. Most noncancerous cells, which grow much more slowly, are not as sensitive to the effect of the drug.

Methotrexate

Nucleotide degradation produces urate or amino acids

Nucleotides that are obtained from food or synthesized by cells can be broken down, releasing ribose groups and a purine or pyrimidine that can be further catabolized and excreted (purines) or used as a metabolic fuel (pyrimidines). At several points in the degradation pathways, intermediates may be redirected toward the synthesis of new nucleotides by so-called **salvage pathways.** For example, a free adenine base can be reattached to ribose by the reaction

$$\text{adenine} + \text{5-phosphoribosyl pyrophosphate} \rightleftharpoons \text{AMP} + \text{PP}_i$$

Degradation of a nucleoside monophosphate begins with dephosphorylation to produce a nucleoside. In a subsequent step, a phosphorylase breaks the glycosidic bond between the base and the ribose by adding phosphate (a similar phosphorolysis reaction occurs during glycogenolysis; Section 13.3):

The phosphorylated ribose can be catabolized or salvaged and converted to 5-phosphoribosyl pyrophosphate for synthesis of another nucleotide. The fate of the base depends on whether it is a purine or a pyrimidine.

The purine bases are eventually converted to urate in a process that may require deamination and oxidation, depending on whether the original base was adenine, guanine, or hypoxanthine. Uric acid has a pK of 5.4, so it exists mainly as an anion at physiological pH.

However, urate is poorly soluble and, in humans, is excreted in the urine. Excess urate may precipitate as crystals of sodium urate in the kidneys (kidney "stones"). Deposits of urate in the joints, primarily the knees and toes, cause a painful condition called gout. Other organisms can further catabolize urate to generate more soluble waste products such as urea and ammonia.

The pyrimidines cytosine, thymine, and uracil undergo deamination and reduction, after which the pyrimidine ring is opened. Further catabolism produces the nonstandard amino acid β-alanine (from cytosine and uracil) or β-aminoisobutyrate (from thymine), both of which feed into other metabolic pathways (**Box 18.F**).

Consequently, pyrimidine catabolism contributes to the pool of cellular metabolites for both anabolic and catabolic processes. In contrast, purine catabolism generates a waste product that is excreted from the body.

Box 18.F   Inborn Error of Nucleotide Metabolism

Lesch-Nyhan syndrome is an X-linked rare genetic disorder affecting males and females are generally carriers. It is characterized by hyperuricemia and elevated uric acid levels and clinically manifests as gout, renal disease, spasticity, speech deficit, and mental retardation. The disease is caused by the absence of the Hypoxanthine-guanine phosphoribosyltransferase (HGPRT) enzyme which participates in the salvage pathway of purine metabolism and redirects hypoxanthine and guanine to form inosinic and guanylic acid. The absence of this pathway results in elevated levels of uric acid and many of the symptoms associated with Lesch-Nyhan syndrome.

Question **What is the role of Hypoxanthine-guanine phosphoribosyltransferase (HGPRT) enzyme in hybridoma technology?**

Concept Check

1. List the molecules that are used to build purine and pyrimidine nucleotides.
2. Explain why humans don't require purines and pyrimidines in their diet.
3. Compare the metabolic fates of IMP and UTP.
4. What does ribonucleotide reductase do?
5. Explain why dihydrofolate reductase is as important as thymidylate synthase.
6. Describe the metabolic fates of ribose, purines, and pyrimidines.

SUMMARY

18.1 Nitrogen Fixation and Assimilation

- Nitrogen-fixing organisms convert N_2 to NH_3 in the ATP-consuming nitrogenase reaction. Nitrate and nitrite can also be reduced to NH_3.
- Ammonia is incorporated into glutamine by the action of glutamine synthetase.
- Transaminases use a PLP prosthetic group to catalyze the reversible interconversion of α-amino acids and α-keto acids.

18.2 Amino Acid Biosynthesis

- In general, the nonessential amino acids are synthesized from common metabolic intermediates such as pyruvate, oxaloacetate, and α-ketoglutarate.
- The essential amino acids, which include the sulfur-containing, branched-chain, and aromatic amino acids, are synthesized by more elaborate pathways in bacteria and plants.
- Amino acids are the precursors of some neurotransmitters and hormones.

18.3 Amino Acid Catabolism

- Following removal of their amino groups by transamination, amino acids are broken down to intermediates that can be converted to glucose or acetyl-CoA for use in the citric acid cycle, fatty acid synthesis, or ketogenesis.

18.4 Excretion of Excess Nitrogen: The Urea Cycle

- In mammals, excess amino groups are converted to urea for disposal. The urea cycle is regulated at the carbamoyl phosphate synthetase step, an entry point for ammonia. Other organisms convert excess nitrogen to compounds such as urate.

18.5 Nucleotide Metabolism

- The synthesis of nucleotides requires glutamate, glycine, and aspartate as well as ribose-5-phosphate. The pathways for purine and pyrimidine biosynthesis are regulated to balance the production of the various nucleotides.
- A ribonucleotide reductase uses a free radical mechanism to convert nucleotides to deoxynucleotides.
- Thymidine production requires a methyl group donated by the cofactor tetrahydrofolate.
- In humans, purines are degraded to urate for excretion, and pyrimidines are converted to β-amino acids.

KEY TERMS

nitrogen fixation	nitrogen cycle	essential amino acid	glucogenic amino	deamination	purinosome
diazotroph	transamination	channeling	acid	urea cycle	salvage pathway
nitrification	nonessential amino	neurotransmitter	ketogenic amino acid		
denitrification	acid				

BIOINFORMATICS

Brief Bioinformatics Exercises

18.1 Viewing and Analyzing Pyridoxal Phosphate Enzymes

18.2 Amino Acid Metabolism and the KEGG Database

PROBLEMS

18.1 Nitrogen Fixation and Assimilation

1. Determine the oxidation state of nitrogen in the following compounds or ions: **a.** NO_3^-, **b.** NO_2^-, **c.** N_2, **d.** NH_3, **e.** NH_4^+, **f.** NO, and **g.** N_2O.

2. Consult your answer to Problem 1 and Figure 18.3 to determine how the oxidation state of nitrogen changes during **a.** nitrification and **b.** denitrification.

3. $\Delta G^{\circ\prime}$ for the half-reaction

$$N_2 + 6\,H^+ + 6\,e^- \longrightarrow 2\,NH_3$$

is -0.34 V. The reduction potential of the nitrogenase component that donates electrons for nitrogen reduction is about -0.29 V. ATP hydrolysis apparently induces a conformational change in the protein that alters its reduction potential by about 0.11 V. Does this change increase or decrease $\mathcal{E}^{\circ\prime}$ of the electron donor and why is this change necessary?

4. Over the past 20 years, scientists have discovered prokaryotic species that can oxidize ammonia to N_2 and convert ammonia directly to nitrate. **a.** Draw a diagram that includes these processes along with those illustrated in Figure 18.3. **b.** If each process in your diagram corresponds to one organism, which organism(s) could be added to farm plots to avoid the need for chemical fertilizers?

5. How would you expect NH_4^+ to regulate the activity of nitrogenase?

6. Plants whose root nodules contain nitrogen-fixing bacterial symbionts synthesize a heme-containing protein, called leghemoglobin, which structurally resembles myoglobin. What is the function of this protein in the root nodules?

7. Photosynthetic cyanobacteria carry out nitrogen fixation in specialized cells that have Photosystem I but lack Photosystem II. Explain the reason for this strategy.

8. *E. coli* glutamine synthetase is regulated by adenylylation; that is, an AMP group is covalently attached to a tyrosine side chain. The enzyme is less active in the adenylylated form. The reaction is catalyzed by an adenylyltransferase (ATase) enzyme. **a.** Write a balanced equation for this reaction and show the structure of the adenylylated tyrosine side chain. **b.** Would you expect α-ketoglutarate to stimulate or inhibit the adenylylation of the enzyme?

9. Investigators who first studied the de-adenylylation of glutamine synthetase (the reverse of the reaction described in Problem 8) noted that the reaction is catalyzed by the same ATase enzyme that catalyzes the forward reaction; however, the reverse reaction requires inorganic phosphate (P_i). **a.** What does this information reveal about the way in which the de-adenylylation reaction is carried out? **b.** What reaction covered in a previous chapter is similar? **c.** The ATase reaction is allosterically regulated by glutamine. How would you expect glutamine to influence the activity of the enzyme, i.e., does it act as an activator or an inhibitor of the enzyme, and in which direction?

10. Draw the structures of the α-keto acid products of the following transamination reactions in which the amino acid reacts with α-ketoglutarate to form glutamate and an α-keto acid: **a.** glycine, **b.** arginine, **c.** serine, **d.** phenylalanine **e.** aspartate, and **f.** alanine.

11. Draw the structures of the α-keto acid products of the following transamination reactions: **a.** glutamate + oxaloacetate. **b.** What do all of the products have in common?

12. Which amino acid generates each of the following products in a transamination reaction with α-ketoglutarate?

13. In the *de novo* synthesis of an amino acid in bacteria, a transaminase enzyme catalyzes the last step in a multistep pathway in which *p*-hydroxyphenylpyruvate reacts with glutamate. Draw the structures and name the products of this reaction.

p-Hydroxyphenylpyruvate

14. The transamination reaction is an example of an enzyme that uses a ping pong mechanism (see Section 7.2). Describe the steps involved.

15. Cancer cells show a much higher rate of glutamine utilization than normal cells. **a.** What are two strategies that cancer cells might employ to increase the cellular glutamine content? **b.** How could you use this information to devise a therapeutic agent for treating cancer?

16. Serine hydroxymethyltransferase catalyzes the conversion of threonine to glycine in a PLP-dependent reaction. The mechanism is slightly different from that shown in the transamination reaction in Figure 18.5. The degradation of threonine to glycine begins with a threonine C_α—C_β bond cleavage. Draw the structure of the threonine–Schiff base intermediate that forms in this reaction and show how C_α—C_β bond cleavage occurs.

17. a. Write equations for the reactions catalyzed by AST and ALT (see Box 18.A). **b.** Explain how AST contributes to the transport of NADH from the cytosol to the mitochondrial matrix (see Fig. 15.5). **c.** Explain how ALT contributes to the glucose–alanine cycle (see Solution 13.55 and Fig. 19.4).

18. Alcoholics may suffer from malnutrition and vitamin deficiencies. How might this affect the results of the ALT test (see Box 18.A)? If you were the clinician running the test, how could you alter the assay procedure to account for this?

19. The highly versatile prokaryotic cell can incorporate ammonia into amino acids using two different mechanisms, depending on the concentration of available ammonia. **a.** One method involves coupling glutamine synthetase, glutamate synthase, and transamination reactions. Write the overall balanced equation for this process. **b.** A second method involves coupling the freely reversible glutamate dehydrogenase reaction and a transamination reaction. Write the overall balanced equation for this process.

20. Refer to your answer to Problem 23. **a.** Which process is used when the concentration of available ammonium ions is low? When the concentration of ammonium ions is high? (*Hint:* The K_M of glutamine synthetase for ammonia is lower than the K_M of glutamate dehydrogenase for the ammonium ion.) **b.** Explain why the prokaryotic cell is at a disadvantage when the concentration of ammonium ions is low.

18.2 Amino Acid Biosynthesis

21. Glutamine synthetase and asparagine synthetase catalyze reactions to produce glutamine and asparagine, respectively. It is reasonable to expect that these enzymes are similar, but they catalyze the formation of their amino acid products very differently. **a.** Compare and contrast the two enzymes in terms of energy sources and nitrogen donors. **b.** Draw the structure of the reaction intermediate in the asparagine synthetase reaction and compare its structure to the reaction intermediate in the glutamine synthetase reaction presented in the chapter.

22. Asparagine synthetase can catalyze the synthesis of asparagine from aspartate using ammonium ions rather than glutamate as the nitrogen donor (see Solution 25). **a.** Write a balanced equation for this reaction. **b.** Compare this reaction to the glutamine synthetase reaction.

23. a. From what pathway is 3-phosphoglycerate derived in order to synthesize serine? **b.** What types of enzymes catalyze each of the three steps in the serine biosynthetic pathway?

24. The expression of phosphoglycerate dehydrogenase, which catalyzes the first step of serine biosynthesis, is dramatically elevated in some cancers, which are highly metabolically active. Experimental results suggest that in addition to supplying extra serine, the elevated dehydrogenase activity increases flux through the citric acid cycle. Explain. (*Hint:* 3-phosphohydroxypyruvate is an α-keto acid.)

25. Serine hydroxymethyltransferase catalyzes the reaction shown in the chapter, in which serine and tetrahydrofolate are converted to glycine and methylene-tetrahydrofolate. The methylene-tetrahydrofolate is then used as a reactant in nucleotide synthesis. What types of cells have a high expression of serine hydroxymethyltransferase?

26. The concentrations of several metabolites are elevated in the urine of individuals suffering from prostate cancer. One of these metabolites is sarcosine (*N*-methylglycine). **a.** Draw its structure. **b.** Folate deficiency causes sarcosine levels in the blood to increase. Propose an explanation for this observation.

27. In many bacteria, the pathway for synthesizing cysteine and methionine consists of enzymes that contain relatively few Cys and Met residues. Explain why this provides an advantage to the bacteria.

28. "Branched-chain" amino acids are synthesized by multistep pathways in which the last reaction is catalyzed by a transaminase, using the α-keto acids shown below and glutamate as a nitrogen donor. Which amino acids are produced from the following α-keto acids? Name the α-keto acid product.

a.

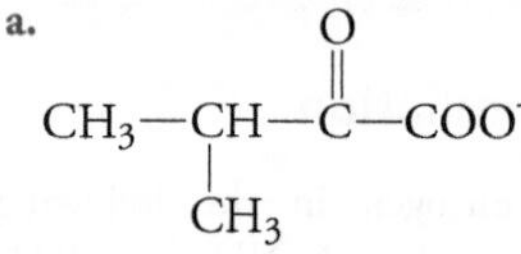

$$CH_3—\underset{\underset{\textstyle CH_3}{|}}{CH}—\overset{\overset{\textstyle O}{||}}{C}—COO^-$$

α-Ketoisovalerate

b.

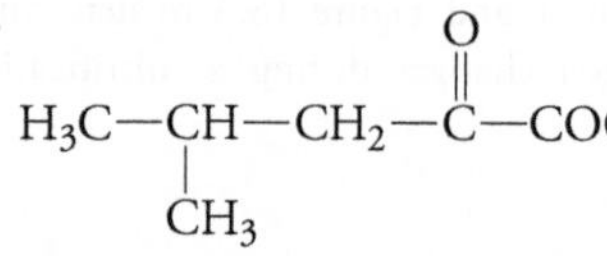

$$H_3C—\underset{\underset{\textstyle CH_3}{|}}{CH}—CH_2—\overset{\overset{\textstyle O}{||}}{C}—COO^-$$

α-Ketoisocaproate

c.

$$H_3C—CH_2—\underset{\underset{\textstyle CH_3}{|}}{CH}—\overset{\overset{\textstyle O}{||}}{C}—COO^-$$

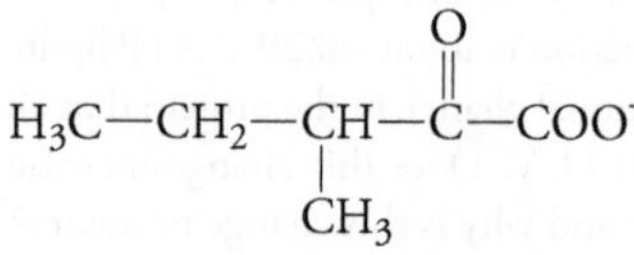

α-Keto-β-methylvalerate

29. Yeast express a transaminase that uses alanine and glyoxylate as substrates. **a.** Draw the structures of the reactants and products of the reaction catalyzed by this enzyme. Name the products. **b.** In certain yeast mutants, the phosphatase that converts phosphoserine to serine is nonfunctional. The mutants are able to synthesize serine if acetate is provided as the carbon source, but not if glucose is the source. Wild-type yeast are able to synthesize serine from either acetate or glucose. Explain these observations.

30. Taurine, a naturally occurring compound that is added to some energy drinks, is used in the synthesis of bile acids (Section 17.4); it may also help regulate cardiovascular function and lipoprotein metabolism. From what amino acid is taurine derived, and what types of reactions are required to convert this amino acid to taurine?

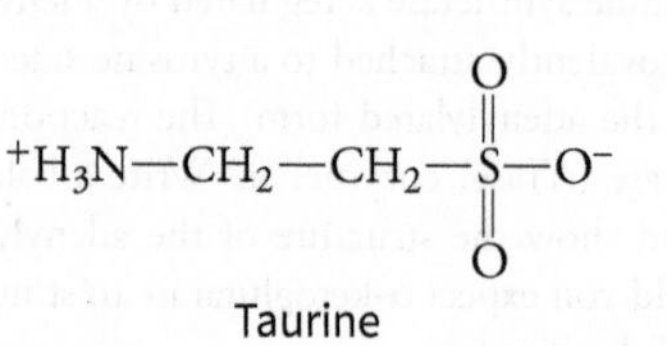

$$^+H_3N—CH_2—CH_2—\overset{\overset{\textstyle O}{||}}{\underset{\underset{\textstyle O}{||}}{S}}—O^-$$

Taurine

31. Sulfonamides (sulfa drugs) act as antibiotics by inhibiting the synthesis of folate in bacteria. **a.** Which portion of the folate molecule does the sulfonamide resemble? **b.** Why do sulfonamides kill bacteria without harming their mammalian hosts?

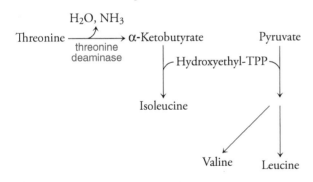

A sulfonamide

32. Theanine, an amino acid found in tea leaves, is synthesized from ethylamine (derived from alanine) and glutamate. It is also known as γ-glutamylethylamide and N^5-ethyl-glutamine. Draw its structure.

33. Animal cells can use dietary amino acids that circulate in the blood. However, many mammalian cells lack an aspartate transporter and must synthesize their own aspartate from oxaloacetate. Under most conditions, α-ketoglutarate can be converted to oxaloacetate by the enzymes of the citric acid cycle. However, if the level of NAD^+ is low, the α-ketoglutarate dehydrogenase reaction is slow and the cell uses an NADPH-dependent isocitrate dehydrogenase instead. Explain how this step could make oxaloacetate available for aspartate synthesis, especially when oxidized cofactors are scarce.

34. A dilute solution of gelatin, which is derived from the protein collagen (see Section 5.3), is sometimes given to ill children who have not been able to consume solid food for several days. **a.** Explain why gelatin is not a good source of essential amino acids. **b.** What is the advantage of giving gelatin, rather than a sugar solution, to someone who has not eaten for several days?

35. A person whose diet is poor in just one of the essential amino acids may enter a state of negative nitrogen balance, in which nitrogen excretion is greater than nitrogen intake. Explain why this occurs, even when the supply of other amino acids is high.

36. a. Is tyrosine an essential amino acid? **b.** Would tyrosine be an essential amino acid in a patient with a phenylalanine hydroxylase deficiency?

37. Threonine deaminase catalyzes the committed step of the biosynthetic pathway leading to the branched-chain amino acid isoleucine. The enzyme catalyzes the dehydration and deamination of threonine to α-ketobutyrate. This pathway is linked to pathways that produce valine and leucine, as shown in the figure.

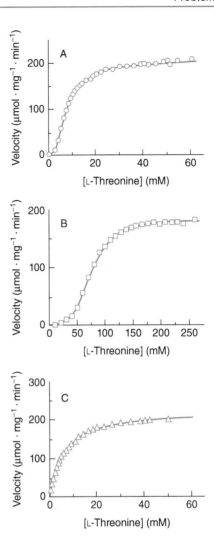

Threonine deaminase is a tetramer, has a low affinity, or "T," form and a high affinity, or "R," form, and is allosterically regulated. The activity of threonine deaminase was measured at increasing concentrations of its substrate, threonine (curve A). Measurements were also made in the presence of isoleucine (curve B) and valine (curve C). **a.** Using the plots provided, determine the values of V_{max} and K_M for threonine deaminase for each reaction condition. **b.** How does isoleucine affect threonine deaminase activity? To which form of the enzyme does isoleucine bind? **c.** How does valine (a product of a parallel pathway) affect threonine deaminase activity? To which form of the enzyme does valine bind?

38. Bacteria synthesize the essential amino acids lysine, methionine, and threonine using aspartate as a substrate, as summarized below. Certain enzymes in this pathway serve as regulatory points so that the cell can maintain appropriate concentrations of each amino acid. The amino acids themselves serve as allosteric enzyme regulators. Which enzyme(s) is(are) good candidates for regulation of the synthesis of **a.** lysine, **b.** methionine, and **c.** threonine?

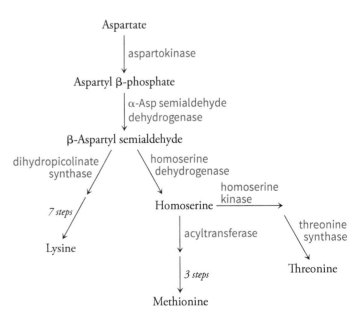

39. Based on the structures of the neurotransmitters shown in the chapter, name the types of enzymes required to accomplish the following transformations: **a.** tyrosine → dopamine → norepinephrine → epinephrine and **b.** tryptophan → serotonin → melatonin.

40. Some invertebrates synthesize octopamine, a neurotransmitter that is derived from tyrosine by decarboxylation followed by β-hydroxylation. Draw the structure of octopamine and describe how it differs from norepinephrine.

18.3 Amino Acid Catabolism

41. The catabolic pathways for the 20 amino acids vary considerably, but all amino acids are degraded to one of seven metabolites: pyruvate, α-ketoglutarate, succinyl-CoA, fumarate, oxaloacetate, acetyl-CoA, or acetoacetate. What is the fate of each of these metabolites?

42. Although amino acids are classified as glucogenic, ketogenic, or both, it is possible for all their carbon skeletons to be broken down to acetyl-CoA: i.e., all amino acids are ketogenic. Explain.

43. Peptides containing citrulline residues are often detected in inflammation, such as occurs in rheumatoid arthritis. **a.** From which amino acid are citrulline residues derived and what kind of reaction generates them? **b.** Why are citrullinated peptides likely to trigger an autoimmune response?

44. Urinary 3-methylhistidine (a modified amino acid found primarily in actin) is used as an indicator of the rate of muscle degradation. When actin is degraded, 3-methylhistidine is excreted because it cannot be re-used for protein synthesis. Why? Explain why monitoring 3-methylhistidine levels offers only an approximation of the rate of muscle degradation.

45. Mouse embryonic stem cells are small but divide extremely rapidly. To maintain high metabolic flux, these cells require a high concentration of threonine and express high levels of threonine dehydrogenase, which catalyzes the first step of threonine breakdown. Explain how threonine catabolism contributes to citric acid cycle activity and nucleotide biosynthesis.

46. Isoleucine is degraded to acetyl-CoA and propionyl-CoA by a pathway in which the first steps are identical to those of valine degradation (Fig. 18.9) and the last steps are identical to those of fatty acid oxidation. **a.** Draw the intermediates of isoleucine degradation and indicate the enzyme that catalyzes each step. **b.** Which reaction in the degradation scheme is analogous to the reaction catalyzed by pyruvate dehydrogenase? **c.** What reaction is analogous to the reaction catalyzed by acyl-CoA dehydrogenase in fatty acid oxidation? **d.** What is the fate of the propionyl-CoA that is produced by isoleucine degradation?

47. Maple syrup urine disease (MSUD) is an inborn error of metabolism that results in the excretion of branched chain α-keto acids into the urine, imparting a maple syrup–like smell to the urine. **a.** What enzyme is nonfunctional in these patients? **b.** Severe neurological symptoms develop if the disease is not treated but, if treated, patients can live a fairly normal life. How would you treat the disease?

48. The effects of the hormones glucagon and insulin on the activity of phenylalanine hydroxylase were investigated, with and without preincubation with phenylalanine (phenylalanine converts the enzyme from the dimeric to the tetrameric form). The results are shown below. **a.** What hormone stimulates the enzyme? **b.** What is the role of phenylalanine in regulating enzyme activity? **c.** A separate experiment showed that the amount of radioactively labeled phosphate incorporated into the enzyme with glucagon treatment was nearly sevenfold greater than in controls. Explain this observation. **d.** Is phenylalanine hydroxylase more active in the fed or fasting state?

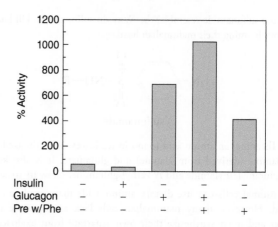

Insulin	−	+	−	−	−
Glucagon	−	−	+	+	−
Pre w/Phe	−	−	−	+	+

49. Phenylketonuria (PKU) is an inherited disease that results from the lack of phenylalanine hydroxylase (see Problem 48), which catalyzes the first step in phenylalanine degradation. In PKU patients, phenylalanine accumulates in the blood and is transaminated to phenylpyruvate, a phenylketone. **a.** Draw the structure of phenylpyruvate. **b.** Why do children with a deficiency of tetrahydrobiopterin excrete large quantities of the phenylketone? **c.** Why are patients with PKU given a low phenylalanine diet and not a phenylalanine-free diet? **d.** Why should patients with PKU avoid the artificial sweetener aspartame (see Problem 4.15)? **e.** Why should PKU patients increase their dietary tyrosine?

50. Nonketotic hyperglycinemia (NKH) is an inborn error of metabolism characterized by high levels of glycine in the blood, urine, and cerebrospinal fluid. Babies with this disease suffer from hypotonia, seizures, and intellectual disability. What enzyme is most likely to be nonfunctional in patients with NKH?

51. In mammals, metabolic fuels can be stored: glucose as glycogen and fatty acids as triacylglycerols. What type of molecule could be considered as a sort of storage depot for amino acids? How does it differ from other fuel-storage molecules?

18.4 Excretion of Excess Nitrogen: The Urea Cycle

52. List all the reactions shown in this chapter that generate free ammonia.

53. Which three mammalian enzymes can potentially "mop up" excess NH_4^+?

54. a. As described in the text, ammonia activation of the NMDA receptor in the brain leads to neuronal cell death. One cause of death is depletion of ATP in the post-synaptic cell. Explain why over-stimulation of the NMDA receptor leads to loss of ATP. (*Hint:* The Ca^{2+} ions that entered the cytosol upon stimulation must be returned to the mitochondrial matrix and Na^+ ions must be ejected from the cell.) **b.** Ammonia-induced activation of NMDA receptors in neurons results in the formation of NO (see Box 18.C), which inhibits glutamine synthetase. What effect does this have on the neurons?

55. Glutamate dehydrogenase is allosterically regulated by a variety of metabolites. Predict the effect of each of the following on the deamination activity of glutamate dehydrogenase: **a.** GTP, **b.** ADP, **c.** NADH.

56. N-acetylglutamate synthase, which catalyzes the synthesis of the allosteric regulator N-acetylglutamate, is itself allosterically regulated by arginine. Is arginine an allosteric activator or inhibitor of N-acetylglutamate synthase? Explain.

57. Identify the sources of the two nitrogen atoms in urea.

58. Which of the metabolites listed are consumed in the urea cycle? Which are not consumed? **a.** NH_4^+, **b.** CO_2, **c.** ATP, **d.** ornithine, **e.** citrulline, **f.** argininosuccinate, **g.** arginine, and **h.** aspartate.

59. Urea cycle dysfunction can result from deficiencies in carbamoyl phosphate synthetase, *N*-acetylglutamate synthase (see Problem 56), the urea cycle enzymes themselves, and the transport protein that transports ornithine from the cytosol to the mitochondrial matrix. A complete deficiency of any of these enzymes usually causes death soon after birth, but a partial deficiency may be tolerated. **a.** Explain why hyperammonemia (high levels of ammonia in the blood) accompanies a urea cycle enzyme deficiency. **b.** What dietary adjustments might minimize ammonia toxicity?

60. Arginine supplements are used to treat all urea cycle deficiencies (see Problem 59) except one. Which enzyme deficiency does not respond to arginine supplementation? Explain the reasoning behind this therapeutic strategy for the other deficiencies.

61. A deficiency of carbamoyl phosphate synthetase is the most serious of the urea cycle deficiency disorders (see Problem 59). Explain why.

62. A deficiency of *N*-acetylglutamate synthase (see Problem 56) has the same symptoms as a carbamoyl phosphate synthetase deficiency (see Problem 61). Explain why.

63. The most common urea cycle defect is a deficiency of ornithine transcarbamoylase (OTC). Physicians treat OTC-deficient patients with a dietary regimen in which protein and glutamine are restricted. Supplements in the form of citrulline and arginine are provided. Explain the rationale behind this therapy.

64. An inborn error of metabolism results in a deficiency of argininosuccinase. **a.** What metabolites accumulate? **b.** What metabolites decrease in concentration? **c.** What could be added to the diet to boost urea production?

65. The drug Ucephan, a mixture of sodium salts of phenylacetate and benzoate (shown below), is used to treat the hyperammonemia associated with a urea cycle enzyme deficiency (see Problem 59). Phenylacetate reacts with glutamine, and benzoate reacts with glycine; the reaction products are excreted in the urine. Draw the structures of the products. What is the biochemical rationale for this treatment?

66. Production of the enzymes that catalyze the reactions of the urea cycle can increase or decrease according to the metabolic needs of the organism. High levels of these enzymes are associated with high-protein diets as well as starvation. Explain this paradox.

67. Vigorous exercise is known to break down muscle proteins. What is the probable metabolic fate of the resulting free amino acids?

68. a. The kidneys help regulate acid–base balance in humans by releasing amino groups from glutamine so that the resulting ammonium ions can neutralize metabolic acids (Section 2.5). Which two kidney enzymes are responsible for removing glutamine's amino groups? Write the reaction catalyzed by each enzyme and write the net reaction. **b.** Recall that ketone bodies are produced during starvation (Section 17.3). Under these conditions, the kidneys increase their uptake of glutamine. Explain how this helps counteract the effects of ketosis.

69. Gastric ulcers result from infection by *Helicobacter pylori*. To survive in the extreme acidity of the stomach, the bacteria express high levels of the enzyme urease. **a.** Why is urease activity essential for *H. pylori* survival? **b.** Why is it important for at least some urease to be associated with the bacterial cell surface?

18.5 Nucleotide Metabolism

70. Purine nucleotide synthesis, outlined below, is a highly regulated process. The main objective is to provide the cell with approximately equal concentrations of ATP and GTP for DNA synthesis. **a.** How do ADP and GDP regulate ribose phosphate pyrophosphokinase? **b.** Amidophosphoribosyl transferase catalyzes the committed step of the IMP synthetic pathway. How might 5-phosphoribosyl pyrophosphate (PRPP), AMP, ADP, ATP, GMP, GDP, and GTP affect the activity of this enzyme? Explain your reasoning.

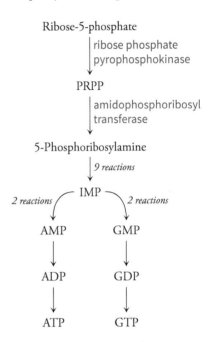

71. The purine nucleotide synthesis pathway shown in Problem 70 is the *de novo* synthetic pathway and is virtually identical in all organisms. Most organisms have additional salvage pathways in which purines released from degradative processes are recycled to form their respective nucleotides. (Some organisms that do not have the *de novo* pathway rely exclusively on a salvage pathway to synthesize purine nucleotides.) One salvage pathway involves the conversion of adenine and 5-phosphoribosyl pyrophosphate (PRPP) to AMP, catalyzed by the enzyme adenine phosphoribosyltransferase (APRT). If adenine is present in large amounts, it is converted to dihydroxyadenine, which can form kidney stones. **a.** A mutation in the APRT gene results in a 10-fold increase in the K_M value for one of the APRT substrates. What is the consequence of this mutation? **b.** How would you treat the condition resulting from this mutation?

$$\text{Adenine} + \text{PRPP} \underset{}{\overset{\text{APRT}}{\rightleftharpoons}} \text{AMP} + \text{PP}_i$$

$$\downarrow \begin{array}{c}\text{xanthine}\\\text{dehydrogenase}\end{array}$$

Dihydroxyadenine

72. A second purine salvage pathway (see Problem 71) is catalyzed by hypoxanthine-guanine phosphoribosyl transferase (HGPRT), which catalyzes the following reactions with 5-phosphoribosyl pyrophosphate (PRPP):

$$\text{Hypoxanthine} + \text{PRPP} \xrightleftharpoons{\text{HGPRT}} \text{IMP} + \text{PP}_i$$

$$\text{Guanine} + \text{PRPP} \xrightleftharpoons{\text{HGPRT}} \text{GMP} + \text{PP}_i$$

Some intracellular protozoan parasites have high levels of HGPRT. Inhibitors of this enzyme could potentially be used to block parasite growth. What is the metabolic effect of inhibiting the parasite's HGPRT and what does this tell you about the parasite's metabolic capabilities?

73. Lesch–Nyhan syndrome is a disease caused by a severe deficiency in HGPRT activity (see Problem 72). The disease is characterized by the accumulation of excessive amounts of urate, a product of nucleotide degradation, which causes neurological abnormalities and destructive behavior, including self-mutilation. Explain why the absence of HGPRT causes urate to accumulate.

74. Antibody-producing B lymphocytes use both the *de novo* (see Problem 70) and salvage pathways (see Problem 71) to synthesize nucleotides, but their survival rate in culture is only 7–10 days. Myeloma cells lack the HGPRT enzyme (see Problem 72) and survive indefinitely in culture. The preparation of long-lived antibody-producing hybridoma cells involves fusing a lymphocyte and a myeloma cell to produce a hybridoma, and then selecting for hybridomas that can grow in HAT medium. The medium contains hypoxanthine, aminopterin (an antibiotic that inhibits enzymes of the *de novo* nucleotide synthetic pathway), and thymidine. How does the HAT medium select for hybridoma cells?

75. Multifunctional enzymes are common in eukaryotic metabolism. Explain why it would be an advantage for dihydrofolate reductase (DHFR) and thymidylate synthase (TS) activities to be part of the same protein.

76. Although methotrexate is used for cancer chemotherapy as described in the chapter, it is also prescribed to treat some autoimmune diseases such as rheumatoid arthritis. Explain why this strategy is effective.

77. The compound 5-fluorouracil (shown below) is often used topically to treat minor skin cancers. When 5-fluorouracil is added to cells in culture, the concentration of dUTP increases, while dTTP is depleted. How do you account for these observations and how does 5-fluorouracil kill cancer cells?

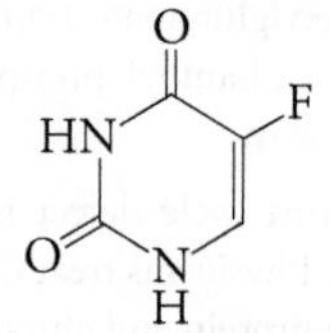

5-Fluorouracil

SELECTED READINGS

Aliu, E., Kanungo, S., and Arnold, G.L., Amino acid disorders, *Ann. Transl. Med.* 6, 471, doi: 10.21037/atm.2018.12.12 (2018). [Reviews the biochemistry, symptoms, and treatment of the most common disorders of amino acid metabolism.]

Brosnan, J.T., Glutamate, at the interface between amino acid and carbohydrate metabolism, *J. Nutr.* 130, 988S–990S (2000). [Summarizes the metabolic roles of glutamate as a fuel and as a participant in deamination and transamination reactions.]

Lane, A.N. and Fan, T.W.-M., Regulation of mammalian nucleotide metabolism and biosynthesis, *Nuc. Acids Res.* 43, 2466–2485 (2015). [Includes overviews of the major metabolic pathways for synthesizing nucleotides.]

Liang, J., Han, Q., Tan, Y., Ding, H., and Li, J., Current advances on structure-function relationships of pyridoxal 5′-phosphate-dependent enzymes, *Front. Mol. Biosci.* 6, 4, doi: 10.3389/fmolb.2019.00004 (2019). [Explains how different enzymes use the same PLP prosthetic group to carry out a variety of reactions involving amino acids.]

Seefeldt, L.C., Peters, J.W., Beratan, D.N., Bothner, B., Minteer, S.D., Raugei, S., and Hoffman, B.M., Control of electron transfer in nitrogenase, *Curr. Opin. Chem. Biol.* 47, 54–59, doi: 10.1016/j.cbpa.2018.08.011 (2018). [Discusses some features of the enzyme mechanism, including the timing of ATP hydrolysis and electron transfer.]

CHAPTER 18 CREDITS

Figure 18.2 Image based on 1M1N. Einsle, O., Tezcan, F.A., Andrade, S.L., Schmid, B., Yoshida, M., Howard, J.B., Rees, D.C., Nitrogenase MoFe-protein at 1.16 A resolution: A central ligand in the FeMo-cofactor, *Science* 297, 1696–1700 (2002).

Figure 18.4 Image based on 2GLS. Yamashita, M.M., Almassy, R.J., Janson, C.A., Cascio, D., Eisenberg, D., Refined atomic model of glutamine synthetase at 3.5 A resolution, *J. Biol. Chem.* 264, 17681–17690 (1989).

Regulation of Mammalian Fuel Metabolism

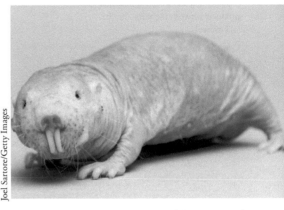

The naked mole rat, *Heterocephalus glaber*, is unusual among mammals in its ability to survive short periods of anoxia. Other adaptations to subterranean life include a low metabolic rate, tolerance to high levels of CO_2, and lack of thermoregulation. For a rodent, it has an extraordinarily long lifespan—up to 32 years—although it shows few signs of aging and is resistant to cancer.

Do You Remember?

- Allosteric regulators can inhibit or activate enzymes (Section 7.3).
- G protein–coupled receptors and receptor tyrosine kinases are the two major types of receptors that transduce extracellular signals to the cell interior (Section 10.1).
- Metabolic fuels can be mobilized by breaking down glycogen, triacylglycerols, and proteins (Section 12.1).
- Metabolic pathways in cells are connected and are regulated (Section 12.3).

As a machine obeying the laws of thermodynamics, the human body is remarkably flexible in managing its resources. Humans and other mammals rely on different organs that are specialized for using, storing, and interconverting metabolic fuels. The exchange of materials between organs and communication between organs allows the body to operate as a unified whole. However, for the same reasons, disorders in one aspect of fuel metabolism can have body-wide consequences. We begin this chapter by reviewing the roles of different organs. Then we explore how metabolic processes are coordinated and how they are disrupted in disease.

19.1 Integration of Fuel Metabolism

KEY CONCEPTS

Summarize the metabolic functions of liver, kidney, muscle, and adipose tissue.

- Identify the major sources of fuel in each organ.
- Describe the mobilization of stored fuel in each organ.
- Trace the movement of metabolites between organs.

Figure 19.1 Cellular locations of major metabolic pathways.
The diagram includes some transport proteins that transfer substrates and products between the mitochondria and cytosol.

Question For each transporter, indicate the primary direction of solute movement and identify its metabolic purpose.

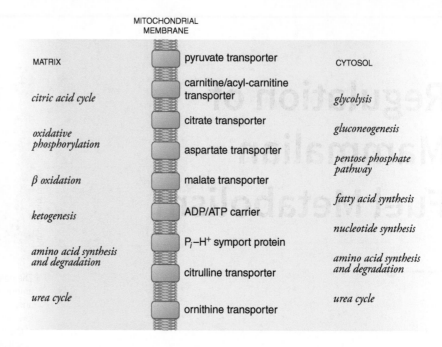

In examining the various pathways for the catabolism and anabolism of the major metabolic fuels and building blocks in mammals—carbohydrates, fatty acids, amino acids, and nucleotides—we have seen that biosynthetic and degradative pathways differ for thermodynamic reasons. These pathways are also regulated so that their simultaneous operation does not waste resources. One form of regulation is the compartmentation of opposing processes. For example, fatty acid oxidation takes place in mitochondria, whereas fatty acid synthesis takes place in the cytosol. In fact, in mammalian cells, most metabolic pathways occur in only one compartment. Exceptions include amino acid degradation and the urea cycle, which require enzymes located in the mitochondrial matrix and cytosol. The locations of the major metabolic pathways are shown in **Figure 19.1**. Other pathways occur in the peroxisome, endoplasmic reticulum, Golgi apparatus, and lysosome. The movement of materials between cellular compartments requires an extensive set of membrane transporters, some of which are included in Figure 19.1.

Organs are specialized for different functions

Compartmentation also takes the form of organ specialization: *Different tissues have different roles in energy storage and use.* For example, the liver carries out most metabolic processes as well as liver-specific functions such as gluconeogenesis, ketogenesis, and urea production. Adipose tissue is specialized to store about 95% of the body's triacylglycerols. Some tissues, such as red blood cells, do not store glycogen or fat and rely primarily on glucose supplied by the liver.

The functions of some organs as fuel depositories or fuel sources depend on whether the body is experiencing abundance (for example, immediately following a meal) or deprivation (after many hours of fasting). The major metabolic functions of some organs, including their reciprocating roles in storing and mobilizing fuels, are diagrammed in **Figure 19.2**.

Following a meal, the liver takes up glucose and converts it to glycogen for storage. Excess glucose and amino acids are catabolized to acetyl-CoA, which is used to synthesize fatty acids. The fatty acids are esterified to glycerol, and the resulting triacylglycerols, along with dietary triacylglycerols, are exported to other tissues. During a fast, the liver mobilizes glucose from glycogen stores and releases it into the circulation for other tissues to use. Triacylglycerols are broken down to acetyl-CoA, which can be converted to ketone bodies to power the brain and heart when glucose is in short supply. Amino acids derived from proteins can be converted to glucose by gluconeogenesis (the nonglucogenic amino acids can be converted to ketone bodies). The liver also deals with lactate and alanine produced by muscle activity, converting these molecules to glucose and disposing of amino groups through urea synthesis.

Muscle cells take up glucose when it is available and store it as glycogen, although there is a limit to how much glycogen can be stockpiled. During exercise, the glycogen is quickly broken

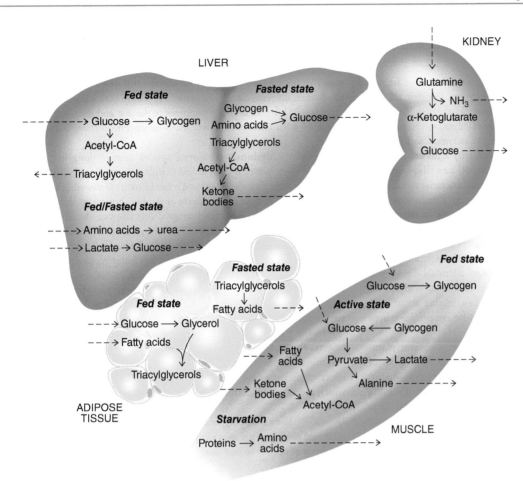

Figure 19.2 The major metabolic roles of the liver, kidney, muscle, and adipose tissue. The liver is the most metabolically active organ in the body, followed by adipose tissue, and then muscle.

Question Identify the processes that are mainly catabolic and mainly anabolic.

down for glycolysis, a pathway for the rapid—if inefficient—production of ATP. Muscle cells can also burn fatty acids and ketone bodies. During intense prolonged activity, muscle cells export lactate and alanine (see below). Muscle protein can be tapped as a source of metabolic fuel during starvation, when amino acids are needed to generate glucose.

Unlike skeletal muscle cells, heart muscle cells must maintain a near-constant level of activity. The heart is specialized to burn fatty acids as its primary fuel and is rich in mitochondria to carry out this aerobic activity. Rather than rely on the unpredictable delivery of fatty acids from lipoproteins, heart muscle cells stockpile fuel in the form of triacylglycerols.

Adipocytes also store triacylglycerols after taking up circulating fatty acids plus glucose, which is converted to glycerol. The adipocyte's fat globule takes up most of the cell's volume (see Fig. 12.3) but shrinks as fatty acids are mobilized in times of need.

In addition to eliminating wastes and maintaining acid–base balance (see Section 2.5), the kidneys play a minor role in fuel metabolism. The removal of amino groups from glutamine leaves α-ketoglutarate, which can be converted to glucose (the liver and kidney are the only organs that carry out gluconeogenesis).

Metabolites travel between organs

The body's organs are connected to each other by the circulatory system so that metabolites synthesized by one organ, such as glucose produced by the liver, can easily reach other tissues. Amino acids released from various tissues travel to the liver or kidney for disposal of their amino groups. Materials are also exchanged between the body's organs and the microorganisms that inhabit the intestines (**Box 19.A**).

Some metabolic pathways are circuits that include interorgan transport. For example, the **Cori cycle** (named after Carl and Gerty Cori, who first described it) is a metabolic pathway involving the muscles and liver. During periods of high activity, muscle glycogen is broken

Box 19.A   The Intestinal Microbiota Contribute to Metabolism

The trillions of microbial organisms—the microbiota—that are part of the human body shape its overall metabolism (Section 1.2 and Box 12.A). For example, bacteria and fungi in the small intestine ferment undigested carbohydrates, mainly plant polysaccharides that cannot be broken down by human digestive enzymes, and produce propionate and butyrate. These short-chain fatty acids are absorbed by the host and are used as fuel or transformed into triacylglycerols for long-term storage. Some of the same small bacterial metabolites also appear to promote the activity of regulatory T cells, which are white blood cells that prevent an immune response to commensal bacteria (the organisms that live in a human body but do not cause disease). Intestinal bacteria also produce vitamin K (Box 8.B), biotin (Section 13.1), and folate (Section 18.2), some of which can be taken up and used by the host. The importance of microbial digestion is illustrated by mice grown in a germ-free environment. Without the normal bacterial partners, the mice need to consume about 30% more food than normal animals whose digestive systems have been colonized by microorganisms.

Studies show that switching between meat-rich and plant-based diets rapidly changes the proportions of different types of intestinal microbes in humans, suggesting that the human–microbial ecosystem has evolved to let humans take advantage of varying food availability. However, there is also evidence that thin and obese individuals harbor different proportions of two major types of bacteria, the Bacteroidetes and the Firmicutes. However, the microbes from obese individuals are not necessarily more efficient at extracting nutrients from food so that the host absorbs and stores the excess as fat. Rather, different classes of microbes appear to play a role in promoting or inhibiting immune responses in the gut, and the resulting degree of inflammation influences fuel-use patterns throughout the host.

Similarly, **dysbiosis** (abnormal species composition) in the intestinal microbiota has been linked to inflammatory bowel disease. In addition to genetic variations and other factors, the disease is associated with the growth of facultative anaerobes (organisms that can function either with or without oxygen) that displace the obligate anaerobes (organisms that cannot live in the presence of O_2) that are the main source of short-chain fatty acids. Permanently correcting the ecological imbalance could be helpful but is not yet practical.

The species composition of the intestinal microbial community differs between individuals, oscillates over a 24-hour period, and may change even without significant shifts in diet. All this makes it difficult to accurately assess the effects of particular nutrients, including "probiotic" foods that contain live organisms. Intentional manipulations of microbiota tend to produce mixed results, although transplanting a sample of the fecal microbiota from a healthy donor is an effective treatment for recurrent infections of *Clostridium difficile*, which causes severe diarrhea.

Question Formulate a hypothesis to explain the link between **increased antibiotic use over the past 60 years and the increasing incidence of obesity in humans.**

down to glucose, which undergoes glycolysis to produce the ATP required for muscle contraction. The rapid catabolism of muscle glucose exceeds the capacity of the mitochondria to reoxidize the resulting NADH and so generates lactate as an end product. This three-carbon molecule is excreted from the muscle cells and travels via the bloodstream to the liver. There, lactate dehydrogenase converts lactate back to pyruvate, which can then be used to synthesize glucose by gluconeogenesis. The newly synthesized glucose can then return to the muscle cells to sustain ATP production even after the muscle glycogen has been depleted (**Fig. 19.3**).

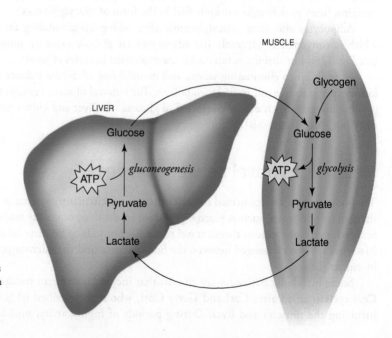

Figure 19.3 The Cori cycle. The product of muscle glycolysis is lactate, which travels to the liver. The input of energy in the liver (in the form of ATP) is recovered when glucose returns to the muscles to be catabolized.

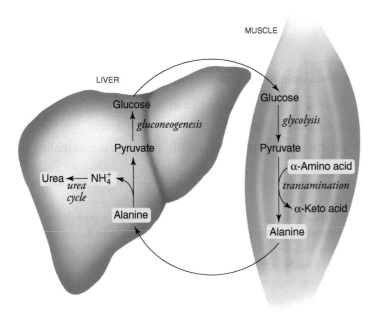

Figure 19.4 The glucose–alanine cycle. The pyruvate produced by muscle glycolysis undergoes transamination to alanine, which delivers amino groups to the liver. The carbon skeleton of alanine is converted back to glucose to be used by the muscles, and the nitrogen is converted to urea for disposal.

Question What is the fate of the α-keto acids in muscle?

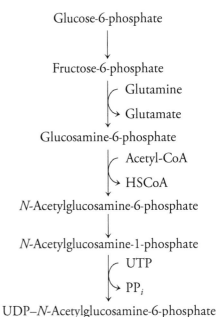

Figure 19.5 The hexosamine pathway.

Question Identify the substrates that correspond to each class of biological molecule.

The energy to drive gluconeogenesis in the liver is derived from ATP produced by the oxidation of fatty acids. In effect, *the Cori cycle transfers free energy from the liver to the muscles.*

A second interorgan pathway, the **glucose–alanine cycle,** also links the muscles and liver. During vigorous exercise, muscle proteins break down, and the resulting amino acids undergo transamination to generate intermediates to boost the activity of the citric acid cycle (Section 14.4). Transamination reactions convert pyruvate, a product of glycolysis, to alanine, which travels by the blood to the liver. The amino group is used for urea synthesis in the liver (Section 18.4) and the resulting pyruvate is converted back to glucose by the reactions of gluconeogenesis. As in the Cori cycle, the glucose returns to the muscle cells to complete the metabolic loop (**Fig. 19.4**). *The net effect of the glucose–alanine cycle is to transport nitrogen from muscles to the liver.*

Even within a single cell, separate metabolic pathways are integrated, both in the way that they are regulated (described in the next section) and in the way that they share intermediates. For example, glucose-6-phosphate is a substrate for glycogen synthesis, glycolysis, and the pentose phosphate pathway (see Figure 13.13). Another option for glucose-6-phosphate is the hexosamine pathway, which generates the monosaccharide units needed to synthesize glycoproteins (Section 11.3). The synthesis of UDP–*N*-acetylglucosamine, the pathway product, illustrates the interdependence of metabolic pathways for all four classes of biological molecules: carbohydrates, amino acids, lipids, and nucleotides (**Fig. 19.5**).

Concept Check

1. Explain the importance of intracellular transport systems in coordinating metabolic activities.
2. Make a list of the major metabolic process in liver, kidney, muscle, and adipose tissue.
3. Identify the unique metabolic functions of the liver.
4. Draw a diagram showing fuel distribution to each organ in the fed state.
5. Draw a diagram showing fuel mobilization in each organ in the fasted state.
6. Describe the individual steps and the net effect of the Cori cycle and the glucose–alanine cycle.

19.2 Hormonal Control of Fuel Metabolism

KEY CONCEPTS

Describe the effects of hormone signaling on fuel metabolism.

- Summarize the effects of insulin on muscle, adipose tissue, and liver.
- Recount how epinephrine and glucagon signaling lead to fuel mobilization.
- Describe the roles of AMP-dependent protein kinase, mTOR, and HIF-α in metabolic regulation.

Individual cells or organs must regulate the activities of their respective pathways according to their metabolic needs and the availability of fuel and building materials, which are supplied intermittently. Metabolic processes are not at equilibrium but can be continually adjusted to maintain a steady state. The human body buffers itself against fluctuations in the fuel supply by storing metabolic fuels, mobilizing them as needed, and replenishing them after the next meal. Maintaining a steady supply of glucose is especially critical for the brain, which exerts a large and relatively constant demand for glucose, regardless of how the dietary intake of carbohydrates varies or how much carbohydrate is oxidized to support other activities.

How does the body control the level of glucose and other fuels from hour to hour or day to day? The activities of organs that store and release fuels are coordinated by **hormones,** which are substances produced by one tissue that affect the functions of other tissues throughout the body. The most important hormones involved in fuel metabolism are insulin, glucagon, and the catecholamines epinephrine and norepinephrine, but a host of substances produced by multiple organs participate in a network that regulates appetite, fuel allocation, and body weight.

The ability of a cell to respond to an extracellular signal depends on cell-surface receptors that recognize the hormone and transmit a signal to the cell interior. Intracellular responses to the hormone include changes in enzyme activity and gene expression. The major types of signal transduction pathways are described in Chapter 10.

Insulin is released in response to glucose

Insulin plays a large role in regulating fuel metabolism by stimulating activities such as glucose uptake and inhibiting processes such as glycogen breakdown. A lack of insulin or an inability to respond to it results in the disease **diabetes mellitus** (Section 19.3). Immediately following a meal, blood glucose concentrations may rise to about 8 mM, from a normal concentration of about 3.6 to 5.8 mM. The increase in circulating glucose triggers the release of the hormone insulin, a 51–amino acid polypeptide (**Fig. 19.6**). Insulin is synthesized in the β cells of pancreatic islets, which are small clumps of cells that produce hormones rather than digestive enzymes (**Fig. 19.7**). The hormone is named after the Latin *insula,* "island."

The mechanism that triggers the release of insulin from the β cells is not well understood. The pancreatic cells do not express a glucose receptor on their surface, as might be expected. Instead, *the cellular metabolism of glucose itself seems to generate the signal to release insulin.* In liver and pancreatic β cells, the glycolytic degradation of glucose begins with a reaction catalyzed by glucokinase (an isozyme of hexokinase; see Section 13.1):

$$\text{glucose} + \text{ATP} \rightarrow \text{glucose-6-phosphate} + \text{ADP}$$

The hexokinases in other cell types have a relatively low K_M for glucose (less than 0.1 mM), which means that the enzymes are saturated with substrate at physiological glucose concentrations. Glucokinase, in contrast, has a high K_M of 5–10 mM, *so it is never saturated and its activity is maximally sensitive to the concentration of available glucose* (**Fig. 19.8**).

Interestingly, the velocity versus substrate curve for glucokinase is not hyperbolic, as might be expected for a monomeric enzyme such as glucokinase. Instead, the curve is sigmoidal, which is typical of allosteric enzymes with multiple active sites operating cooperatively (see Section 7.2).

Figure 19.6 Structure of human insulin. This two-chain hormone is colored by atom type: C gray, O red, N blue, H white, and S yellow.

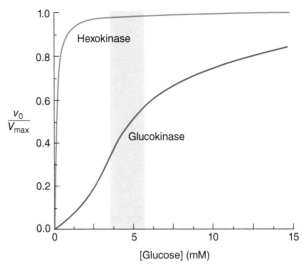

Figure 19.8 Activities of glucokinase and hexokinase. Both enzymes catalyze the ATP-dependent phosphorylation of glucose as the first step of glycolysis. Glucokinase has a high K_M, so its reaction velocity changes in response to changes in glucose concentrations. In contrast, hexokinase is saturated with glucose at physiological concentrations (shaded region).

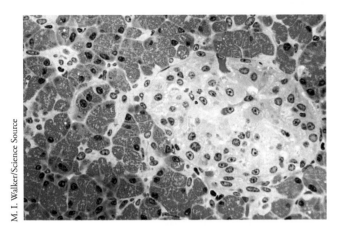

Figure 19.7 Pancreatic islet cells. The pancreatic islets of Langerhans (named for their discoverer) consist of two types of cells. The β cells produce insulin and the α cells produce glucagon. Most other pancreatic cells produce digestive enzymes.

The sigmoidal kinetics of glucokinase, which has only one active site, may be due to a substrate-induced conformational change such that at the end of the catalytic cycle, the enzyme briefly maintains a high affinity for the next glucose molecule. At high glucose concentrations, this would mean a high reaction velocity; at low glucose concentrations, the enzyme would operate more slowly because it reverts to a low-affinity conformation before binding another glucose substrate.

The role of glucokinase as a pancreatic glucose sensor is supported by the fact that mutations in the glucokinase gene cause a rare form of diabetes. However, other cellular factors may be involved, particularly in the mitochondria of the β cells. The glucose sensor responsible for triggering insulin release may also depend on the mitochondrial $NAD^+/NADH$ or ADP/ATP ratios. For this reason, age-related declines in mitochondrial function may be a factor in the development of diabetes in the elderly.

Once released into the bloodstream, insulin can bind to receptors on cells in muscle and other tissues. Insulin binding to its receptor stimulates the tyrosine kinase activity of the receptor's intracellular domains (Section 10.3). These kinases phosphorylate each other as well as tyrosine residues in other proteins, including IRS-1 and IRS-2 (insulin receptor substrates 1 and 2). The IRS proteins then trigger additional events in the cell, not all of which have been fully characterized.

Insulin promotes fuel use and storage

Only cells that bear insulin receptors can respond to the hormone, and the cells' response is tissue specific. In general, *insulin signals fuel abundance: It decreases the metabolism of stored fuel while promoting fuel storage.* The effects of insulin on various tissues are summarized in **Table 19.1**.

In tissues such as muscle and adipose tissue, insulin stimulates glucose transport into cells by several fold. The V_{max} for glucose transport increases, not because insulin alters the intrinsic catalytic activity of the transporter but because insulin increases the number of transporters at

Table 19.1 Summary of Insulin Action

Target tissue	Metabolic effect
Muscle and most other tissues	Promotes glucose transport into cells
	Stimulates glycogen synthesis
	Suppresses glycogen breakdown
Adipose tissue	Activates extracellular lipoprotein lipase
	Increases level of acetyl-CoA carboxylase
	Stimulates triacylglycerol synthesis
	Suppresses lipolysis
Liver	Promotes glycogen synthesis
	Promotes triacylglycerol synthesis
	Suppresses gluconeogenesis

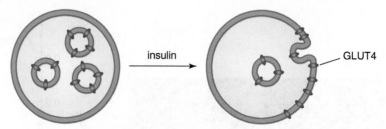

Figure 19.9 Effect of insulin on GLUT4.   Insulin triggers vesicle fusion so that the glucose transport protein GLUT4 is translocated from intracellular vesicles to the plasma membrane. This increases the rate at which the cells take up glucose.

the cell surface. These transporters, named GLUT4, are situated in the membranes of intracellular vesicles. When insulin binds to the cell, the vesicles fuse with the plasma membrane. *This translocation of transporters to the cell surface increases the rate at which glucose enters the cell* (**Fig. 19.9**). GLUT4 is a passive transporter, operating as shown in Figure 9.12. When the insulin stimulus is removed, endocytosis returns the transporters to intracellular vesicles.

Insulin stimulates fatty acid uptake as well as glucose uptake. When the hormone binds to its receptors in adipose tissue, it activates the extracellular protein lipoprotein lipase, which hydrolyzes triacylglycerols in circulating lipoproteins so that the released fatty acids can be taken up for storage by adipocytes.

The insulin signaling pathway also alters the activity of glycogen-metabolizing enzymes. Glycogen metabolism is characterized by a balance between glycogen synthesis and glycogen degradation. Synthesis is carried out by the enzyme glycogen synthase, which adds glucose units donated by UDP–glucose to the ends of the branches of a glycogen polymer (see Section 13.3):

$$\text{UDP–glucose} + \text{glycogen}_{(n\text{ residues})} \rightarrow \text{UDP} + \text{glycogen}_{(n+1\text{ residues})}$$

Glycogen phosphorylase mobilizes glucose residues from glycogen by phosphorolysis (cleavage through addition of a phosphoryl group rather than water):

$$\text{glycogen}_{(n\text{ residues})} + \text{P}_i \rightarrow \text{glycogen}_{(n-1\text{ residues})} + \text{glucose-1-phosphate}$$

This reaction, followed by an isomerization reaction, yields glucose-6-phosphate, the first intermediate of glycolysis and the substrate for additional pathways.

Glycogen synthase is a homodimer and glycogen phosphorylase is a heterodimer. Both enzymes are regulated by allosteric effectors. For example, glycogen synthase is activated by glucose-6-phosphate. AMP activates glycogen phosphorylase and ATP inhibits it. These effects are consistent with the role of glycogen phosphorylase in making glucose available to boost cellular ATP production. However, *the primary mechanism for regulating glycogen synthase and glycogen phosphorylase is through covalent modification (phosphorylation and dephosphorylation) that is under hormonal control.* Both enzymes undergo reversible phosphorylation at specific serine residues. Phosphorylation (transfer of a phosphoryl group from ATP to the enzyme) deactivates glycogen synthase and activates glycogen phosphorylase. Removal of the phosphoryl groups has the opposite effect: Dephosphorylation activates glycogen synthase and deactivates glycogen phosphorylase (**Fig. 19.10**).

Covalent modification is a type of allosteric regulation (Section 7.3). The attachment or removal of the highly anionic phosphoryl group triggers a conformational shift between a more active (*a* or R) state and a less active (*b* or T) state. *The reciprocal regulation of glycogen synthase and glycogen phosphorylase promotes metabolic efficiency, since the two enzymes catalyze key reactions in opposing metabolic pathways.* The advantage of this regulatory system is that a single kinase can tip the balance between glycogen synthesis and degradation. Similarly, a single phosphatase can tip the balance in the opposite direction. Covalent modifications, such as phosphorylation and dephosphorylation, permit a much wider range of enzyme activities than could be accomplished solely through the allosteric effects of metabolites whose cellular concentrations do not vary much. Insulin signaling activates phosphatases that dephosphorylate

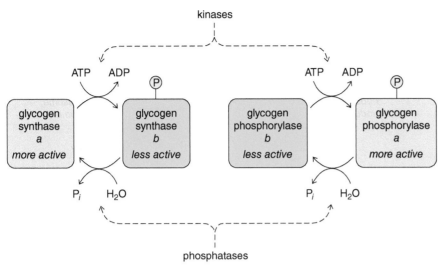

Figure 19.10 The reciprocal regulation of glycogen synthase and glycogen phosphorylase. The more active form of each enzyme is known as the *a* form (indicated in green), and the less active form is known as the *b* form (in red).

Question Would insulin signaling lead to activation of the kinases or of the phosphatases in this diagram?

(activate) glycogen synthase and dephosphorylate (deactivate) glycogen phosphorylase. As a result, glycogen synthesis accelerates and the rate of glycogenolysis decreases when glucose is abundant.

mTOR responds to insulin signaling

Among the cellular components activated by insulin signaling is a Ser/Thr kinase called mechanistic (or mammalian) target of rapamycin (mTOR), which is a component of two macromolecular complexes known as mTORC1 and mTORC2. mTORC2 responds mostly to extracellular growth factors (such as insulin), but mTORC1 is sensitive to these external signals as well as a host of intracellular factors such as the presence of various types of nutrients. As a result, mTORC1 integrates different types of information about the cell's overall energy status and the need for cell growth and proliferation. For example, insulin signaling leads to the phosphorylation and activation of mTORC1, and free amino acids are involved in mechanisms that position the complex at certain locations within the cell, such as the lysosomal membrane.

The downstream targets of mTORC1 include a ribosomal protein and a translation initiation factor. Phosphorylation of these substrates by the mTOR kinase promotes ribosome production and increases the rate of protein synthesis. mTOR also helps activate the SREBP transcription factor that switches on genes involved in cholesterol synthesis (Section 17.4). In addition, mTOR promotes the synthesis of other lipids and nucleotides and stimulates glycolysis. Other targets of mTOR phosphorylation suppress protein turnover by autophagy (Section 9.4) and proteasome activity (Section 12.1). Collectively, all these mTOR effects direct cellular resources into pathways that support growth (**Fig. 19.11**).

mTOR is inhibited by rapamycin, a bacterial compound with antifungal, immunosuppressant, and anticancer properties. Rapamycin binds to a single subunit of the mTORC1 complex and does not directly affect mTORC2. Its wide range of clinical applications reflects the central role of mTOR in linking a cell's overall activity to environmental conditions. Rapamycin blocks the growth and proliferation of T cells that are essential for an immune response, and it helps tame the over-exuberant expansion of cancer cells. Interestingly, suppressing mTOR activity extends the lifespan of model organisms, although this effect has not been confirmed for humans.

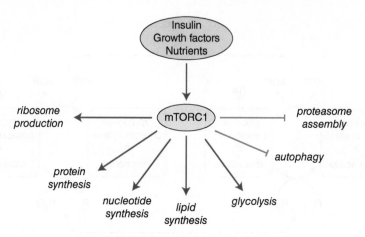

Figure 19.11 mTOR activity. As part of the mTORC1 complex, mTOR phosphorylates and activates proteins related to cell growth and inhibits protein degradation.

Glucagon and epinephrine trigger fuel mobilization

Of course, most mammalian cells do not receive unlimited nutrients and growth-promoting signals. For example, within hours of a meal, the insulin signal is gone and dietary glucose has already been taken up by cells and consumed as fuel, stored as glycogen, or converted to fatty acids for long-term storage. At this point, the liver must begin mobilizing glucose in order to keep the blood glucose concentration constant. This phase of fuel metabolism is governed not by insulin but by other hormones, mainly glucagon and the catecholamines epinephrine and norepinephrine.

Glucagon, a 29-residue peptide hormone, is synthesized and released by the α cells of pancreatic islets when the blood glucose concentration begins to drop below about 5 mM (**Fig. 19.12**). Catecholamines are tyrosine derivatives (see Section 18.2) that are synthesized by the central nervous system as neurotransmitters and by the adrenal glands as hormones. Glucagon, epinephrine, and norepinephrine bind to receptors with seven membrane-spanning segments. Hormone binding triggers a conformational change that activates an associated G protein, which goes on to activate other cellular components such as an adenylate cyclase, which produces the second messenger cyclic AMP (cAMP; Section 10.2). cAMP activates protein kinase A.

In contrast to insulin, *glucagon stimulates the liver to generate glucose by glycogenolysis and gluconeogenesis, and it stimulates* **lipolysis** *in adipose tissue.* Muscle cells do not express a glucagon receptor but do respond to catecholamines, which elicit the same overall effects as glucagon. Thus, epinephrine stimulation of muscle cells activates glycogenolysis, which makes more glucose available to power muscle contraction.

One of the intracellular targets of protein kinase A is phosphorylase kinase, the enzyme that phosphorylates (deactivates) glycogen synthase and phosphorylates (activates) glycogen phosphorylase. Consequently, hormones such as glucagon and epinephrine, which lead to cAMP production, promote glycogenolysis and inhibit glycogen synthesis. Although phosphorylase kinase is activated by protein kinase A, it is maximally active when Ca^{2+} ions are also present. Ca^{2+} concentrations increase during signaling via the phosphoinositide pathway, which responds to the catecholamine hormones (Section 10.2).

In adipocytes, protein kinase A phosphorylates an enzyme known as hormone-sensitive lipase, thereby activating it. This lipase catalyzes the rate-limiting step of lipolysis, the conversion of stored triacylglycerols to diacylglycerols and then to monoacylglycerols, which releases fatty acids. *Hormone stimulation not only increases the lipase catalytic activity, it also relocates the lipase from the cytosol to the fat droplet of the adipocyte.* Co-localization with its substrate, possibly by interacting with a lipid-binding protein, boosts the rate at which fatty acids are

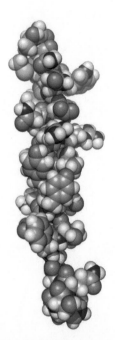

Figure 19.12 Structure of glucagon. The atoms of the 29-residue peptide are colored by type: C gray, O red, N blue, H white, and S yellow.

mobilized. Thus, glucagon and epinephrine promote the breakdown of both glycogen and fat. These responses are summarized in **Figure 19.13**.

Additional hormones influence fuel metabolism

In addition to well-known endocrine organs such as the pancreas (the source of insulin and glucagon) and adrenal glands (the source of epinephrine and norepinephrine), many other tissues produce hormones that help regulate all aspects of food acquisition and use (**Table 19.2**). In fact, adipose tissue, once thought to be a relatively inert fat-storage site, actively communicates with the rest of the body.

Adipose tissue produces the hormone leptin, a 146-residue polypeptide that functions as a signal of satiety (satisfaction or fullness). It acts on the hypothalamus, a part of the brain, to suppress appetite. The level of leptin is proportional to the amount of adipose tissue: The more fat that accumulates in the body, the stronger the appetite-suppressing signal.

Like leptin, adiponectin is released by adipose tissue, but this 247-residue polypeptide exists as an assortment of multimers with different receptor-binding properties. Adiponectin exerts its effects on a variety of tissues by activating an AMP-dependent protein kinase (see below). The effects of adiponectin include increased combustion of glucose and fatty acids. It also increases the sensitivity of tissues to insulin.

Adipocytes also release a 108-residue hormone called resistin, which blocks the activity of insulin. Levels of resistin increase during obesity, which would help explain the link between weight gain and decreased responsiveness to insulin (Section 19.3).

The digestive system produces at least 20 different peptide hormones with varying functions. Several of these signal that a meal has been consumed. For example, incretins are released by the intestine and enhance insulin secretion by the pancreas. The oligopeptide known as PYY_{3-36}, whose release is triggered especially by a high-protein meal, acts on the hypothalamus to suppress appetite. The level of ghrelin, a 28-residue peptide produced by the stomach, increases during fasting and decreases immediately following a meal; this is the only gastrointestinal hormone that stimulates appetite.

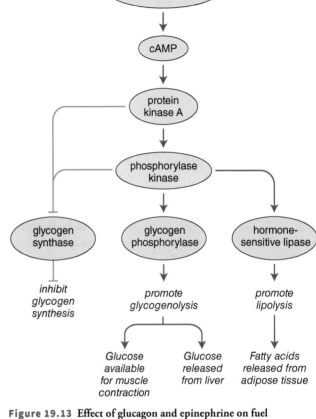

Figure 19.13 Effect of glucagon and epinephrine on fuel metabolism. Green arrows represent activation events and red symbols represent inhibition.

Question **Explain how these events account for the fight-or-flight response triggered by epinephrine (adrenaline).**

Table 19.2 Some Hormones that Regulate Fuel Metabolism

Hormone	Source	Action
Adiponectin	Adipose tissue	Activates AMPK (promotes fuel catabolism)
Leptin	Adipose tissue	Signals satiety
Resistin	Adipose tissue	Blocks insulin activity
Neuropeptide Y	Hypothalamus	Stimulates appetite
Cholecystokinin	Intestine	Suppresses appetite
Incretin	Intestine	Promotes insulin release; inhibits glucagon release
PYY_{3-36}	Intestine	Suppresses appetite
Amylin	Pancreas	Signals satiety
Ghrelin	Stomach	Stimulates appetite

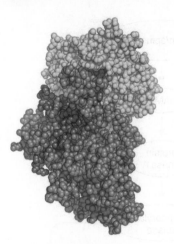

Figure 19.14 Structure of AMPK. The active conformation of the catalytic subunit (green) is maintained by AMP (red) binding to one of the regulatory subunits (yellow).

AMP-dependent protein kinase acts as a fuel sensor

So far we have looked at a variety of signals that regulate fuel intake, storage, and mobilization to help the body maintain homeostasis. Individual cells also have a fuel gauge to adjust their activity on a finer scale. The AMP-dependent protein kinase (AMPK) responds to the cell's balance of ATP, ADP, and AMP to activate and inhibit a number of enzymes involved in different metabolic pathways. *AMP and ADP, representing the cell's need for energy, activate AMPK, and ATP, representing a state of energy sufficiency, inhibits the kinase.*

AMPK is a highly conserved Ser/Thr kinase consisting of a catalytic subunit and two regulatory subunits (**Fig. 19.14**). Like many other kinases, AMPK is activated by phosphorylation of a specific threonine residue in the catalytic subunit. AMP or ADP binding to the regulatory portion of AMPK prevents dephosphorylation of this residue, thereby maintaining the kinase in an active state. AMP also acts as an allosteric activator of the kinase so that overall its activity increases about 2000-fold. ATP, which competes with AMP and ADP for binding to the regulatory sites, inhibits AMPK. This multipart regulatory scheme allows AMPK to respond to a wide range of cellular energy states.

In addition to responding to intracellular energy deficits, AMPK responds to hormones such as leptin and adiponectin. *As a result of AMPK activity, the cell switches off ATP-consuming anabolic pathways and switches on ATP-generating catabolic pathways.* For example, in exercising muscle, AMPK phosphorylates and activates the enzyme that produces fructose-2,6-bisphosphate, an allosteric activator of phosphofructokinase, so that glycolytic flux increases (Section 13.1). In adipose tissue, AMPK phosphorylates and inactivates acetyl-CoA carboxylase, the enzyme that generates malonyl-CoA, to suppress fatty acid synthesis (Section 17.3). Since malonyl-CoA inhibits fatty acid transport into mitochondria, AMPK increases the rate of mitochondrial β oxidation in tissues such as muscle. AMPK activation also promotes the production of new mitochondria. Some metabolic effects of AMPK are listed in **Table 19.3**.

Fuel metabolism is also controlled by redox balance and oxygen

Fluctuations in a cell's fuel supply are not the only factors that govern a cell's metabolic activity. Two other stressors—the cell's redox balance and the availability of oxygen—must also be considered. As we have seen, almost every metabolic pathway includes oxidation–reduction reactions. In general, biosynthetic processes, including fatty acid synthesis and deoxyribonucleotide synthesis, tend to require NADPH as a source of reducing power. *Cells therefore maintain a high NADPH:NADP$^+$ ratio*, with the pentose phosphate pathway supplying most of the NADPH. Conversely, catabolic pathways such as glycolysis, the citric acid cycle, and fatty acid oxidation use NAD$^+$ as an electron acceptor. Accordingly, *cells maintain a high NAD$^+$:NADH ratio*. NAD$^+$ is regenerated mainly by the mitochondrial electron transport chain but also by lactate dehydrogenase in the cytosol.

A cell's catabolic and anabolic processes must occur in a way that avoids perturbing the optimal ratios of oxidized and reduced cofactors. For example, a cell can metabolize glucose to pyruvate (which consumes NAD$^+$) or to lactate (which regenerates NAD$^+$) in order to fine-tune

Table 19.3 Effects of AMP-Dependent Protein Kinase

Tissue	Response
Hypothalamus	Increases food intake
Liver	Increases glycolysis Increases fatty acid oxidation Decreases glycogen synthesis Decreases gluconeogenesis
Muscle	Increases fatty acid oxidation Increases mitochondrial biogenesis
Adipose tissue	Decreases fatty acid synthesis Increases lipolysis

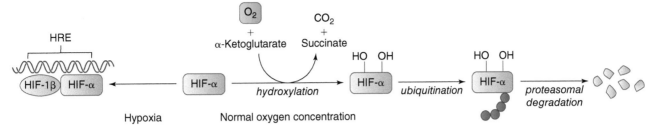

Figure 19.15 Regulation by hypoxia-inducible factor. Under low-oxygen conditions, HIF-α is able to bind to hypoxia response elements (HREs) in genes.

the NAD^+:NADH ratio. The lactate can then be released from the cell and used elsewhere in the body as a metabolic fuel. This feature of metabolism helps explain why the circulating lactate concentration is comparable to the glucose concentration in many mammals.

Because most of the body's ATP is made by oxidative phosphorylation, cells need to sense the level of O_2 and take steps to conserve ATP if the oxygen concentration dips. This could occur as the result of inflammation or a cardiovascular event that limits blood flow. One of the cell's oxygen sensors is a protein called hypoxia-inducible factor alpha (HIF-α). In the presence of O_2, enzymes catalyze the hydroxylation of two proline side chains in HIF-α in reactions that are accompanied by the oxidative decarboxylation of α-ketoglutarate (**Fig. 19.15**). The modified HIF-α then becomes ubiquitinated; the addition of a polyubiquitin chain marks the protein for destruction by a proteasome (Section 12.1).

Under hypoxic (low-oxygen) conditions, HIF-α is not hydroxylated or degraded and instead forms a complex with the HIF-1β protein, moves to the nucleus, and interacts with DNA sequences known as hypoxia response elements. HIF protein binding to these DNA sequences turns on the expression of the associated genes, which code for a variety of proteins, including glucose transporters, glycolytic enzymes, and signaling molecules that promote the growth of new blood vessels and the production of red blood cells.

Concept Check

1. Summarize the metabolic effects of insulin signaling on muscle cells and adipocytes.
2. Compare glucokinase and hexokinase.
3. Describe how insulin increases the rate of glucose entry into cells.
4. Explain why mTOR is involved in so many different cellular activities.
5. Explain how phosphorylation and dephosphorylation reciprocally regulate glycogen synthase and glycogen phosphorylase.
6. Summarize the metabolic effects of glucagon and epinephrine on liver cells and adipocytes.
7. Summarize the metabolic effects of hormones produced by adipose tissue and the digestive system.
8. Explain how AMPK functions as a cell's energy sensor.
9. List the metabolic activities that are stimulated or suppressed by AMPK.
10. List different situations that affect a cell's metabolism.

19.3 Disorders of Fuel Metabolism

KEY CONCEPTS

Compare the metabolic changes that occur in starvation, obesity, and diabetes.

- Describe how fuel use changes during starvation.
- Compare brown and white fat and their contributions to obesity.
- Describe the causes and symptoms of type 1 and type 2 diabetes.

The multifaceted regulation of mammalian fuel metabolism offers many opportunities for things to go wrong. Excessive intake and storage of fuel can cause obesity. Starvation results from insufficient food. The faulty regulation of carbohydrate and lipid metabolism can lead to diabetes. In this section we examine some of the biochemistry behind these conditions.

The body generates glucose and ketone bodies during starvation

Most tissues in the body use glucose as their preferred fuel and turn to fatty acids only when the glucose supply diminishes. Except in the intestine, amino acids are not a primary fuel. However, when no food is available for an extended period, the body must make adjustments to mobilize different types of fuels. An average adult can survive a famine lasting up to a few months, an adaptation likely shaped by seasonal food shortages during human evolution. Starvation in children, of course, may severely impact development (**Box 19.B**).

The liver and muscles store less than a day's supply of glucose in the form of glycogen. As glycogen stores are depleted, muscle switches from burning glucose to burning fatty acids. Insulin secretion ceases with the drop in circulating glucose, so insulin-responsive tissues are not stimulated to take up glucose. This means that more glucose is available for tissues such as the brain, which stores very little glycogen and cannot use fatty acids as fuel.

The liver and kidneys respond to the continued demand for glucose by increasing the rate of gluconeogenesis, using noncarbohydrate precursors such as amino acids (derived from protein degradation) and glycerol (from fatty acid breakdown). After several days, the liver begins to convert mobilized fatty acids to acetyl-CoA and then to ketone bodies. These small water-soluble fuels are used by a variety of tissues, including the heart and brain. The gradual switch from glucose to ketone bodies prevents the body from using up its proteins to supply gluconeogenic precursors. During a 40-day fast, the concentration of circulating fatty acids varies about 15-fold, and the concentration of ketone bodies increases about 100-fold. In contrast, the concentration of glucose in the blood varies by no more than threefold. These patterns of fuel use are summed up in **Table 19.4**.

Table 19.4 Source of Metabolic Fuels Under Different Conditions

	Carbohydrates (%)	Fatty acids (%)	Amino acids (%)
Immediately after a meal	50	33	17
After an overnight fast	12	70	18
After a 40-day fast	0	95[a]	5

[a]This value reflects a high concentration of fatty acid–derived ketone bodies.

Box 19.B   Marasmus and Kwashiorkor

Chronic malnutrition takes a toll on human life in many ways. For example, it exacerbates infectious diseases that would not necessarily be fatal in well-nourished individuals. Severely malnourished children also fail to reach their full potential in terms of body size and cognitive development, even if their food intake later increases to normal levels. Plentiful glucose for the brain is especially critical in infancy, when a child's brain is relatively large and its liver (where glycogen is stored) is relatively small.

There are two major forms of severe malnutrition, **marasmus** and **kwashiorkor,** which may also occur in combination. In marasmus, inadequate intake of metabolic fuels of all types causes wasting. Individuals with this condition are emaciated, with very little muscle mass and essentially no subcutaneous fat. Similar symptoms also develop in some chronic diseases such as cancer, tuberculosis, and AIDS.

Kwashiorkor results from inadequate protein intake, which may or may not be accompanied by inadequate energy intake. A child with kwashiorkor typically has thin limbs, reddish hair, and a swollen belly. Without an adequate supply of amino acids, the liver makes too little albumin, a protein that helps retain fluid inside blood vessels. When the concentration of albumin drops, fluid enters tissues by osmosis. This swelling (edema) also occurs in other diseases that impair liver function. The liver is enlarged in kwashiorkor due to the deposition of fat. Depigmentation of the hair and skin occurs because tyrosine, which is derived from the essential amino acid phenylalanine, is also the precursor of melanin, a brown pigment molecule.

Question   **Explain why kwashiorkor sometimes develops in infants who are fed rice milk rather than breast milk.**

Obesity has multiple causes

Obesity has become an enormous public health problem. In addition to its impact on the quality of life, it is physiologically costly: Masses of fat prevent the lungs from fully expanding; the heart must work harder to circulate blood through an abnormally large body; and the additional weight stresses hip, knee, and ankle joints. Obesity also increases the risk of cardiovascular disease, diabetes, and cancer. It is not an exaggeration to describe obesity as an epidemic, since it affects an estimated one-third of the adult population of the United States.

Like many conditions, obesity has no single cause. It is a complex disorder involving appetite and metabolism and reflecting environmental as well as genetic factors. Despite a high degree of heritability for obesity, only a few rare cases can be explained by a single defective gene. Although obesity is often considered to be a "lifestyle" disease, simple overeating or lack of exercise cannot account for more than about a 15-pound weight gain. Instead, obesity seems to result from the readjustment of multiple regulatory mechanisms and can be considered to be a chronic illness rather than a lifestyle choice.

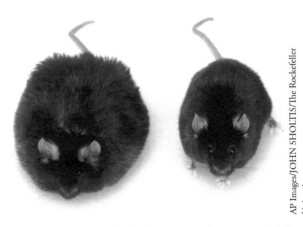

Figure 19.16 Normal and obese mice. The mouse on the left lacks a functional gene for leptin and is several times the size of a normal mouse (right).

The human body appears to have a **set-point** for body weight that remains constant and relatively independent of energy intake and expenditure, even over many decades. The hormone leptin may help establish the set-point, since the absence of leptin causes severe obesity in rodents and humans (**Fig. 19.16**). However, the majority of obese humans do not appear to lack leptin, so they may instead be suffering from leptin resistance due to a defect somewhere in the leptin signaling pathway. When leptin is less effective at suppressing the appetite, the individual gains weight. Eventually, the increase in leptin concentration resulting from the increase in adipose tissue mass succeeds in signaling satiety, but the result is a high set-point (a higher body weight that must be maintained). This may be one reason why overweight people who manage to shed a few pounds often regain the lost weight and return to the original set-point.

It turns out that humans have several types of fat, including subcutaneous fat (beneath the skin), visceral fat (surrounding the abdominal organs), and brown fat. **Brown adipose tissue,** named for its high mitochondrial content, is specialized for generating heat to maintain body temperature. Brown fat is prominent in newborns and hibernating mammals (see Box 15.B), but it also occurs at least in small amounts in adult humans, mainly in the neck and in the body cavity. Developmentally and metabolically, brown adipose tissue more closely resembles muscle than ordinary white adipose tissue (**Fig. 19.17**). Instead of one large fat globule, that occupies most of the cell's volume, brown adipose tissue contains many small fat droplets, which are a source of fatty acids that are oxidized to generate heat. Brown adipocytes also contain relatively more mitochondria than white adipocytes.

The hormone norepinephrine binds to receptors on brown adipocytes, and signal transduction via protein kinase A activates a lipase that liberates fatty acids from triacylglycerols. The uncoupling protein (UCP) is expressed in the mitochondria of brown adipose tissue, so fuel oxidation occurs without ATP synthesis. A compelling hypothesis is that lean individuals have a higher capacity to burn off excess fuel in this manner instead of storing it in white adipose tissue. In fact, the amount of brown fat appears to be inversely proportional to the degree of obesity. Intriguingly, hormonal stimulation of white fat causes some cells (called beige fat) to develop the

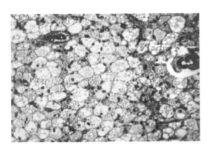

a.

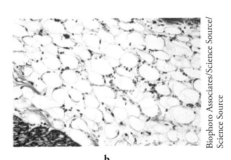

b.

Figure 19.17 Brown and white adipose tissue. a. In brown adipose tissue, cells contain more mitochondria and numerous small fat globules. **b.** In white adipose tissue, each cell contains a large fat globule, and there is little cytoplasm.

characteristics of brown fat, which suggests that there may be a way to manipulate the body's fat deposits to treat severe obesity.

Diabetes is characterized by hyperglycemia

Another well-characterized disorder of fuel metabolism is diabetes mellitus, which affects about 10% of the population of the United States. Worldwide, the disease affects about 500 million people, killing about 4 million each year. Only about half of those affected by diabetes know that they have the disease.

The words *diabetes* (meaning "to run through") and *mellitus* ("honey") describe an obvious symptom of the disease. Diabetics excrete large amounts of urine containing high concentrations of glucose (the kidneys work to eliminate excess circulating glucose by excreting it in urine, a process that requires large amounts of water).

Type 1 diabetes (juvenile-onset or insulin-dependent diabetes) is an autoimmune disease in which the immune system destroys pancreatic β cells. Symptoms first appear in childhood as insulin production begins to drop off. At one time, the disease was invariably fatal. This changed dramatically in 1922, when Frederick Banting and Charles Best administered a pancreatic extract to save the life of a severely ill diabetic boy. Banting and Best already knew that a preparation of pancreatic tissue could improve the symptoms of dogs whose pancreas had been surgically removed to induce diabetes (**Fig. 19.18**).

Modern treatments for type 1 diabetes rely on highly purified insulin produced through genetic engineering (Section 20.6) rather than pancreatic extracts, and the drug can be administered via a syringe or a small pump. An ongoing challenge lies in tailoring the delivery of insulin to the body's needs over the course of a typical 24-hour cycle of eating and fasting. Diabetics must check their circulating glucose levels several times a day, either by testing a tiny drop of blood or by using a continuous monitoring system with a thin probe inserted into the skin. Truly noninvasive devices for measuring glucose are being tested.

Freeing patients from frequent needle sticks, whether for monitoring blood glucose or injecting insulin, is the goal of pancreatic islet cell transplants, which have seen some success. Gene therapy (Section 3.3) to treat diabetes is an elusive goal because the insulin gene must be introduced into the body in such a way that the gene's expression is glucose-sensitive.

By far the most common form of diabetes, accounting for up to 95% of all cases, is type 2 diabetes (also known as adult-onset or non-insulin-dependent diabetes). These cases are characterized by **insulin resistance,** which is the failure of the body to respond to normal or even elevated concentrations of the hormone. Only a small fraction of patients with type 2 diabetes bear genetic defects in the insulin receptor, as might be expected; in the majority of cases, the underlying cause is not known.

The primary feature of untreated diabetes is chronic **hyperglycemia** (high levels of glucose in the blood). *The loss of responsiveness of tissues to insulin means that cells fail to take up glucose.* The body's metabolism responds as if no glucose were available, so liver gluconeogenesis increases, further promoting hyperglycemia. Glucose circulating at high concentrations can participate in nonenzymatic glycosylation of proteins (see Box 11.A). This process is slow, but the modified proteins may gradually accumulate and damage tissues with low turnover rates, such as neurons.

Tissue damage also results from the metabolic effects of hyperglycemia. Since muscle and adipose tissue are unable to increase their uptake of glucose in response to insulin, glucose tends to enter other tissues. Inside these cells, aldose reductase catalyzes the conversion of glucose to sorbitol:

Figure 19.18 Banting and Best. Frederick Banting (right) and Charles Best (left) with a research subject.

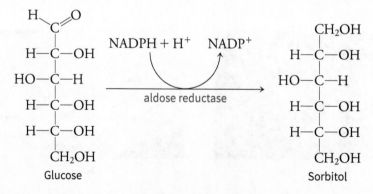

Because aldose reductase has a relatively high K_M for glucose (about 100 mM), flux through this reaction is normally very low. However, under hyperglycemic conditions, sorbitol accumulates and may alter the cell's osmotic balance. This may alter kidney function and may trigger protein precipitation in other tissues. The accumulation of sorbitol in the lens of the eye leads to swelling and precipitation of lens proteins. The resulting opacification, a cataract, can cause blurred vision or complete loss of sight (Fig. 19.19). Neurons and cells lining blood vessels may be similarly damaged, increasing the likelihood of neuropathies and circulatory problems that in severe cases result in kidney failure, heart attack, stroke, or the amputation of extremities.

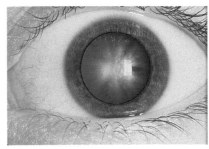

Figure 19.19 Photo of a diabetic cataract.

Although commonly considered a disorder of glucose metabolism, *diabetes is also a disorder of fat metabolism,* since insulin normally stimulates triacylglycerol synthesis and suppresses lipolysis in adipocytes. Uncontrolled diabetics tend to metabolize fatty acids rather than carbohydrates, and the resulting production of ketone bodies may give the breath a sweet odor. Overproduction of ketone bodies leads to diabetic ketoacidosis (Section 2.5).

Not all diabetics experience the same symptoms, which has led some researchers to subdivide patients into clusters. For example, some diabetics fail to produce insulin but do not have an autoimmune disease as in classical type 1 diabetes. Other patients have mild forms of diabetes as a result of being overweight or simply old. However, patients who produce insulin but exhibit severe insulin resistance are the most likely to go on to develop life-threatening complications.

A variety of drugs have been developed to help compensate for the physiological effects of insulin resistance; within each class of drugs there are multiple options with slightly different pharmacokinetics (Table 19.5). For example, metformin improves diabetic symptoms by activating AMPK in liver and other tissues. Liver glucose production is suppressed by the decreased expression of the gluconeogenic enzymes phosphoenolpyruvate carboxykinase and glucose-6-phosphatase (Section 13.2). Metformin also increases glucose uptake and fatty acid oxidation in muscle.

Drugs of the thiazolidinedione class, such as rosiglitazone, act via intracellular receptors known as peroxisome proliferator–activated receptors. These receptors, which normally respond to lipid signals, are transcription factors that alter gene expression (Section 10.4). Thiazolidinediones increase adiponectin levels and decrease resistin levels (in fact, research on the pharmacology of these drugs led to the discovery of resistin). The net result is an increase in insulin sensitivity.

Many diabetic patients use a combination of drugs to help lower blood glucose levels. The most widely prescribed drugs can be taken orally, although they all have side effects. For example, the increased risk of heart attacks has severely restricted the use of rosiglitazone.

Exercise—in addition to insulin signaling—stimulates the repositioning of GLUT4 transporters in the plasma membrane of muscle cells, but by a different mechanism, so that the effects of muscle activity and insulin are somewhat additive. This explains why an exercise program may help lower blood glucose levels in insulin-resistant diabetics.

Obesity, diabetes, and cardiovascular disease are linked

In diabetes, the body behaves as if it were starving. Paradoxically, about 85% of patients with type 2 diabetes are obese, and obesity—particularly when abdominal fat deposits are large—is

Table 19.5 Some Antidiabetic Drugs

Class	Example	Mechanism of action
Biguanides	Metformin	Stimulates AMPK; reduces glucose release from liver; increases glucose uptake by muscle
Sulfonylureas	Glipizide	Blocks a K^+ channel in β cells, leading to increased production and secretion of insulin
Thiazolidinediones	Rosiglitazone	Binds to peroxisome proliferator–activated receptors to activate gene transcription; increases insulin sensitivity

Dr. Manuel Datiles III, Cataract and Cornea Section, OGCSB, National Eye Institute, National Institutes of Health

strongly correlated with the development of the disease. Some researchers use the term **metabolic syndrome** to refer to a set of symptoms that appear to be linked: high concentrations of circulating glucose and triacylglycerols, low levels of HDL cholesterol, abdominal fat deposits, and hypertension (high blood pressure). Metabolic syndrome is not a discrete disease—some individuals show only some of the symptoms—but it is strongly associated with other well-defined diseases. Individuals with metabolic syndrome are more likely to later develop type 2 diabetes. Their atherosclerosis and hypertension increase their risk of a heart attack or stroke, and they experience higher rates of kidney disease and cancer.

What links the various disorders? Causes and effects are not always clear-cut. For example, sensitivity to insulin may decline for years before type 2 diabetes is diagnosed, and liver damage may appear and contribute to additional disruptions. Furthermore, most individuals over age 65 suffer from multiple cardiometabolic disorders. However, one common culprit appears to be a high proportion of visceral fat (assessed as a high waist-to-hip ratio). This type of fat exhibits a different hormone profile than subcutaneous fat. Specifically, visceral fat produces less leptin and adiponectin (hormones that increase insulin sensitivity) and more resistin (which promotes insulin resistance). Visceral fat—or the white blood cells associated with it—also produce inflammatory signals, which are a normal part of the body's defense system. Chronic inflammation, though, may damage nearby tissues, as in atherosclerosis. Inflammatory signals from fat tissue or produced in response to an unbalanced intestinal microbial community (see Box 19.A) may also lead to the phosphorylation of IRS-1, a modification that prevents its activation by the insulin receptor kinase and contributes to insulin resistance.

Even in nonobese individuals, a rich diet may overstimulate mTOR, which promotes lipogenesis. The newly synthesized lipids or a high-fat diet can trigger changes in lipoprotein homeostasis. High levels of circulating fatty acids lead to fat accumulation in muscle tissue as well as adipose tissue, impairing GLUT4 translocation and impeding glucose uptake. As lipids infiltrate the liver, nonalcoholic fatty liver disease develops, and the accompanying inflammation can damage the liver. Dead cells are replaced by fibrous tissue, and the liver may eventually fail. In the meantime, circulating fatty acids trigger gluconeogenesis in the liver, further contributing to hyperglycemia. Pancreatic β cells respond to the high levels of glucose by secreting more insulin, but higher concentrations of the hormone seem only to desensitize the insulin-signaling apparatus—a vicious cycle that is the essence of insulin resistance.

Despite the multiorgan pathology of metabolic syndrome, symptoms usually improve when the individual loses weight, even if the loss is modest. If lifestyle changes related to diet and exercise are not effective, drugs that increase insulin sensitivity can prevent progression to both diabetes and serious cardiovascular disorders.

Concept Check

1. Summarize the metabolic changes that occur during starvation.
2. Explain how a signaling molecule such as leptin could help determine the set-point for body weight.
3. Compare type 1 and type 2 diabetes.
4. Explain why hyperglycemia is a symptom of diabetes.
5. Describe the action of some antidiabetic drugs.
6. Explain how obesity is related to type 2 diabetes.

19.4 Clinical Connection: Cancer Metabolism

KEY CONCEPTS

Relate metabolic changes to the rapid growth of cancer cells.

The typical patterns of fuel use described in Fig. 19.2 are altered in cancer cells, whose metabolism must support rapid growth and cell division. Normal differentiated cells typically grow slowly, if at all, and rely on oxidative phosphorylation to meet their energy needs. In contrast, most cancer cells are characterized by rapid, uncontrolled proliferation and carry out glycolysis at a high rate. One explanation is that cancer cells, particularly in the center of a solid tumor that is not well served by blood vessels, are starved for oxygen and the hypoxia-inducible factor promotes the use of glucose. However, cancer cells actually consume large quantities of glucose even when oxygen is plentiful.

Heightened glucose uptake by cancer cells is the basis for the PET (positron emission tomography) scan used to locate tumors and monitor their growth. About an hour before entering the scanner, a patient is given an injection of 2-deoxy-2-[^{18}F]fluoroglucose (fluorodeoxyglucose, or FDG), which is taken up by all glucose-metabolizing cells. Decay of the ^{18}F isotope emits a positron (which is like an electron with a positive charge) that is ultimately detected as a flash of light. The PET scanner generates a two- or three-dimensional map showing the location of the tracer, that is, tissues with a high rate of glucose uptake (**Fig. 19.20**).

Aerobic glycolysis supports biosynthesis

Aerobic glycolysis, known as the **Warburg effect,** has puzzled biochemists since Otto Warburg described it in the 1920s. One would expect that cancer cells would outcompete normal cells by burning fuel and generating ATP using more efficient pathways. In fact, cancer cells do perform oxidative phosphorylation, yet the cells still consume large amounts of glucose and eliminate waste carbons in the form of lactate rather than CO_2. Why do cancer cells use glucose in an apparently inefficient manner? Glycolysis is not just an ATP-generating pathway; *it also converts glucose carbons to the precursors for anabolic processes such as the synthesis of fatty acids, amino acids, and nucleotides, all of which are needed in large amounts as cells divide.* Excreting lactate also allows the cells to use the lactate dehydrogenase reaction to control their redox balance.

The need for biosynthetic precursors also seems to explain why—despite the high rate of glucose catabolism—many cancer cells express a variant pyruvate kinase that has lower enzymatic activity. The resulting bottleneck in glycolytic flux may help divert some glucose carbons into other pathways. For example, the glycolytic intermediate 3-phosphoglycerate can be diverted to generate serine, which, when converted to glycine, supplies one-carbon groups for other metabolic processes, including thymidine synthesis, that support cell growth and division. Glycine itself is a precursor of adenine and guanine nucleotides. Serine allosterically activates the pyruvate kinase variant, so that glucose is converted to pyruvate when serine is plentiful.

Glycolytically generated pyruvate is a precursor of alanine, which is, after leucine, the most abundant amino acid in proteins. Pyruvate can also be processed to acetyl-CoA, which combines with oxaloacetate to form citrate. However, rather than continuing through the citric acid cycle, the citrate exits the mitochondria, and ATP-citrate lyase converts it back to acetyl groups destined for fatty acid synthesis. Both citrate and fatty acyl-CoA molecules inhibit the activity of phosphofructokinase, thereby directing relatively more glucose through the pentose phosphate pathway, which yields the ribose and NADPH required for nucleotide synthesis.

Cancer cells consume large amounts of glutamine

Glutamine, the most abundant amino acid in the body, supports the growth of cancer cells by providing a source of nitrogen for purine and pyrimidine synthesis. In addition, glutamate derived from glutamine can be deaminated by glutamate dehydrogenase to produce α-ketoglutarate. Although the increased α-ketoglutarate can increase flux through the citric acid cycle for the purpose of oxidative phosphorylation, some citric acid cycle intermediates have other fates. For example, malate can be converted by malic enzyme to pyruvate; this reaction also generates NADPH for biosynthetic pathways. Oxaloacetate can be transaminated to aspartate, which is a precursor for purine nucleotides. Oxaloacetate can also condense with acetyl-CoA to generate citrate to supply cytosolic acetyl groups for fatty acid synthesis.

Glutamate dehydrogenase is a major control point for cancer cell metabolism. The enzyme is activated by ADP and inhibited by GTP, which helps tie amino acid metabolism to the cell's energy budget. Leucine stimulates glutamate dehydrogenase activity, which could help balance the

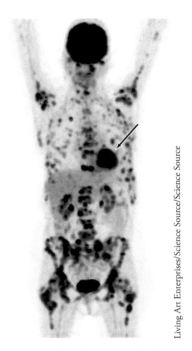

Figure 19.20 PET scan. The dark areas of the PET scan indicate tissues that rely heavily on glucose metabolism (brain, heart) or accumulate the radioactive tracer (bladder). The small dark areas are tumors.

Living Art Enterprises/Science Source/Science Source

Figure 19.21 Cancer metabolism. Some of the biosynthetic pathways that require glucose and glutamine are highlighted.

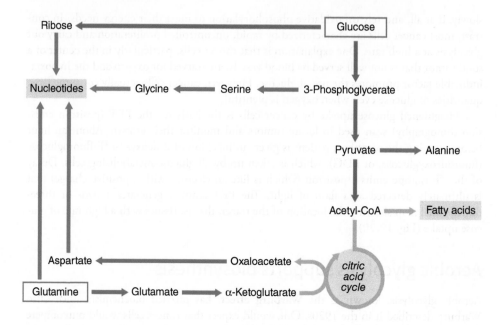

amino acid supply in cancer cells. Palmitoyl-CoA inhibits glutamate dehydrogenase activity, so that when fatty acid synthesis rates are high and acyl-CoA groups accumulate, less α-ketoglutarate is added to the citric acid cycle.

When fatty acid oxidation rates are high, the glutamate dehydrogenase inhibition is relieved and glutamate can replenish α-ketoglutarate to increase the flux of fat-derived acetyl-CoA through the citric acid cycle. Glutamate dehydrogenase operating in reverse (α-ketoglutarate → glutamate) also allows cancer cells to scavenge free ammonia, giving them a survival advantage.

Some of the metabolic patterns in cancer cells are summarized in **Figure 19.21.** Cancer researchers are investigating the possibility of using dietary modifications (such as restricting the intake of glucose, fructose, and amino acids) or specific inhibitors to cripple cancer cell metabolism. However, most cancers comprise a heterogeneous population of cells that change over time. Studies with other anticancer therapies have already demonstrated that although a treatment may eliminate susceptible cancer cells, it selects for resistant cells that continue to proliferate.

The same types of metabolic adjustments that occur in cancer cells also occur in some other types of rapidly dividing cells, including white blood cells responding to an infection. In this situation, a handful of cells can expand to 10,000 in a few days. Drugs that interfere with glycolysis have shown some promise in treating autoimmune disorders, which are characterized by inappropriately responding white blood cells. Interestingly, virus-infected cells may exhibit higher metabolic activity than normal. Some viruses carry genes encoding proteins that reprogram their host cell's metabolism, often to support the enhanced nucleotide synthesis required for efficient viral replication (Box 3.C).

Eighty percent of individuals with advanced cancer develop **cachexia,** the extreme loss of weight and muscle mass that is also common in the late stages of some other chronic illnesses such as tuberculosis and AIDS. Cachexia cannot be reversed by consuming more food, suggesting that the high rate of protein catabolism that results in muscle wasting is the result of reprogramming the entire body's metabolism, which includes adipose tissue and the brain's appetite-control center. During cachexia, white fat undergoes "browning" so that the body burns up its stored fuel. Halting this process would be one way to improve the quality of life in patients with end-stage illness, even if the underlying disease cannot be cured.

Concept Check

1. Explain how aerobic glycolysis differs from anaerobic glycolysis.
2. Summarize how the metabolism of glucose and glutamine supports the synthesis of nucleotides, lipids, and amino acids.

SUMMARY

19.1 Integration of Fuel Metabolism

- The liver is specialized to store glucose as glycogen, to synthesize triacylglycerols, to carry out gluconeogenesis, and to synthesize ketone bodies and urea. The muscles synthesize glycogen and can use glucose, fatty acids, and ketone bodies as fuel. Adipose tissue stores fatty acids as triacylglycerols.

- Pathways such as the Cori cycle and the glucose–alanine cycle link different organs.

19.2 Hormonal Control of Fuel Metabolism

- Insulin, which is synthesized by the pancreas in response to glucose, binds to a receptor tyrosine kinase. Cellular responses to insulin include increased uptake of glucose and fatty acids and the activation of the mTOR kinase.

- The balance between glycogen synthesis and degradation depends on the relative activities of glycogen synthase and glycogen phosphorylase, which are controlled by hormone-triggered phosphorylation and dephosphorylation.

- Glucagon and catecholamines lead to the activation of cAMP-dependent protein kinase, which promotes glycogenolysis in liver and muscle, and lipolysis in adipose tissue.

- Adipose tissue is the source of the hormones leptin, adiponectin, and resistin, which help regulate appetite, fuel combustion, and insulin resistance. The stomach, intestines, and other organs also produce hormones that regulate appetite.

- AMP is an allosteric activator of AMPK, whose activity switches on pathways such as glycolysis and fatty acid oxidation.

- A cell's metabolic activities are sensitive to redox balance and oxygen.

19.3 Disorders of Fuel Metabolism

- In starvation, glycogen stores are depleted, but the liver makes glucose from amino acids and converts fatty acids into ketone bodies.

- The cause of obesity is not clear but may involve a failure in leptin signaling that raises the body weight set-point.

- The most common form of diabetes is characterized by insulin resistance, the inability to respond to insulin. The resulting hyperglycemia can lead to tissue damage.

- Metabolic disturbances resulting from obesity may lead to insulin resistance, a condition termed metabolic syndrome.

19.4 Clinical Connection: Cancer Metabolism

- Cancer cell metabolism is characterized by increased consumption of glucose for aerobic glycolysis and increased consumption of glutamine. In cancer cells, metabolic flux is directed toward the biosynthesis of amino acids, nucleotides, and fatty acids to support rapid cell growth and division.

KEY TERMS

Cori cycle	glucose–alanine cycle	lipolysis	set-point	hyperglycemia	Warburg effect
microbiome	hormone	marasmus	brown adipose tissue	metabolic syndrome	cachexia
dysbiosis	diabetes mellitus	kwashiorkor	insulin resistance		

BIOINFORMATICS

Brief Bioinformatics Exercises

19.1 Insulin Secretion and the KEGG Database

19.2 Insulin Control of Glucose Metabolism and the KEGG Database

19.3 Glucagon Control of Glycogen and Glucose Metabolism

PROBLEMS

19.1 Integration of Fuel Metabolism

1. Name the two small metabolites at the "crossroads" of metabolism. How are these metabolites connected to the metabolic pathways we have studied?

2. After a meal, dietary glucose from the portal vein enters hepatocytes via GLUT2 transporters and is immediately phosphorylated. Explain the importance of this process.

3. The metabolite glucose-6-phosphate (G6P) is linked to several pathways in carbohydrate metabolism. Describe how glucose-6-phosphate is linked to these pathways.

4. What fuels are mobilized, and by which tissues, in the fasted state?

5. Incubating brain slices in a medium containing ouabain (a Na, K-ATPase inhibitor) decreases respiration by 50%. What does this tell you about ATP use in the brain? What pathways are involved in producing ATP in the brain?

6. Red blood cells lack mitochondria. Describe the metabolic pathways involved in ATP production in the red blood cell. What is the ATP yield per glucose molecule?

7. During exercise, the concentration of AMP in muscle cells increases (see Solution 5). AMP is a substrate for the adenosine deaminase reaction:

$$AMP + H_2O \rightarrow IMP + NH_4^+$$

AMP is subsequently regenerated by a process in which the amino group of aspartate becomes attached to the purine ring of IMP and fumarate is released in a series of reactions known as the purine nucleotide cycle. **a.** What is the likely fate of the fumarate product? **b.** Why doesn't the muscle cell increase the concentration of citric acid cycle intermediates by converting aspartate to oxaloacetate by a simple transamination reaction? **c.** Ammonium ions stimulate the activity of phosphofructokinase and pyruvate kinase. Explain how adenosine deaminase activity could promote ATP production in active muscle.

8. How can glucose provide all of the substrates needed for fatty acid biosynthesis?

9. What is the "energy cost" in ATP of running the Cori cycle? How is the ATP obtained?

10. How does the liver deal with lactate production by muscle cells during intense physical activity?

11. The lactate dehydrogenase (LDH) that converts pyruvate to lactate (Section 13.1) is a cytosolic enzyme. Another LDH isozyme is located in the mitochondrial matrix, where it catalyzes the conversion of lactate back to pyruvate. **a.** Explain why the activity of mitochondrial LDH depends on the presence of a transport protein such as the lactate–oxaloacetate antiporter. **b.** When lactate is added to cultured cells under conditions where cytosolic LDH is inactive, oxaloacetate appears in the cytosol. List the enzymes and additional substrates involved in converting lactate to oxaloacetate. **c.** Draw a simple diagram of the metabolic sequence described in parts **a** and **b,** which makes up part of a "revised" Cori cycle. **d.** Is there an advantage for liver cells to use mitochondrial rather than cytosolic LDH during the operation of the Cori cycle?

12. An infant was diagnosed at the age of 3 months with a pyruvate carboxylase deficiency. **a.** Which metabolites are elevated in this patient? Which metabolites are deficient? **b.** Explain why the patient suffers from lactic acidosis and ketosis. **c.** Acetyl-CoA was added to the patient's cultured fibroblasts to see whether pyruvate carboxylase activity could be detected. What was the rationale behind this experiment?

13. An infant with a pyruvate carboxylase deficiency (see Problem 12) suffers from poor muscle tone that results from a lack of the neurotransmitters glutamate, aspartate, and γ-aminobutyric acid (GABA). Why would a pyruvate carboxylase deficiency result in the decreased synthesis of these neurotransmitters?

14. Treating farm animals such as cows and chickens with low doses of antibiotics promotes weight gain. **a.** Propose an explanation for this observation. **b.** The widespread use of antibiotics in animals has been implicated in the rise of antibiotic resistance in human pathogenic bacteria. How might this happen?

15. The caterpillar (larva) of the blue *Morpho* butterfly contains about 20 mg of fatty acids, compared to about 7 mg in the adult butterfly. What does this tell you about the energy source for metamorphosis (when the insect does not eat)?

16. During early lactation, cows cannot eat enough to supply the nutrients required for milk production. A recent study showed that enzymes involved in glycogen synthesis and the citric acid cycle decrease during the early lactation period, whereas glycolytic enzyme activity, lactate production, and fatty acid degradation activity increase. What metabolic strategies are used by the cows to obtain the nutrients required for milk production in the absence of sufficient food intake? Be sure to specify which tissues are involved.

17. The final product of the hexosamine pathway, UDP–*N*-acetylglucosamine-6-phosphate, is considered to be a "nutrient sensor." Explain why.

19.2 Hormonal Control of Fuel Metabolism

18. Estimate the K_M values for hexokinase and glucokinase from Figure 19.8. Compare these values to the normal blood glucose concentration, which ranges from 3.6 to 5.8 mM but can reach over 8 mM immediately after a meal.

19. Why does the sigmoidal behavior of glucokinase, the liver isozyme of hexokinase (see Fig. 19.8), help the liver to adjust its metabolic activities to the amount of available glucose?

20. Explain why a tyrosine phosphatase might be involved in limiting the signaling effect of insulin.

21. Why is insulin required for triacylglycerol synthesis in adipocytes?

22. Insulin activates cAMP phosphodiesterase. Explain how this augments insulin's metabolic effect.

23. Glycogen phosphorylase cleaves glucose residues from glycogen via a phosphorolytic rather than a hydrolytic cleavage. **a.** Write an equation for each process. **b.** What is the metabolic advantage of phosphorolytic cleavage?

24. a. Why is the liver sometimes referred to as the body's "glucose buffer"? **b.** Why is the liver able to supply body tissues with glucose but muscle is unable to do so?

25. Glucose-6-phosphate is an allosteric regulator of both glycogen phosphorylase and glycogen synthase. How would you expect glucose-6-phosphate to regulate the activities of these two enzymes? Explain your rationale.

26. In addition to being regulated by phosphorylation (see Fig. 19.10), glycogen phosphorylase is regulated by acetylation on a specific Lys side chain. **a.** Draw the structure of an acetylated lysine residue. **b.** Acetylation of lysine promotes the association of glycogen phosphorylase with a specific phosphatase. Does acetylation promote the activation or the inhibition of the enzyme? **c.** Would you expect high concentrations of glucose to stimulate or inhibit acetylation?

27. Intestinal microbes produce imidazole propionate (below), which activates mTOR and may contribute to the development of diabetes. Identify the amino acid precursor of this metabolite and describe how it is derived.

Imidazole propionate

28. The protein known as ChREBP binds to DNA sequences called carbohydrate response elements to turn on the expression of the genes for pyruvate kinase, acetyl-CoA carboxylase, and fatty acid synthase. ChREBP is activated by glucose-6-phosphate, but its activity is also governed by phosphorylation. Is the activity of ChREBP increased or decreased by the actions of protein kinase A (PKA) and AMP-dependent protein kinase (AMPK)?

29. Glycogen synthase kinase 3 (GSK3) can phosphorylate glycogen synthase in muscle cells. Activation of the insulin receptor leads to the activation of protein kinase B (Akt; see Section 10.2), which phosphorylates GSK3. How does insulin affect glycogen metabolism through GSK3?

30. In addition to its role in activating AMPK, adiponectin blocks the phosphorylation of glycogen synthase kinase 3 (GSK3; see Problem 29). What is the effect of adiponectin on glycogen metabolism?

31. AMPK affects gene transcription. Predict the effect of AMPK activation on the expression of **a.** muscle GLUT4 and **b.** liver glucose-6-phosphatase.

32. The compound 5-aminoimidazole-4-carboxamide ribonucleoside (AICAR) is taken up into cells and phosphorylated, forming a ribonucleoside monophosphate that binds to AMPK as AMP does. A Northern blot (which detects mRNA) analysis of PEPCK is carried out on extracts of AICAR-treated cells. What are the expected results?

33. Hyperthyroidism is a condition that occurs when the thyroid secretes an excess of hormones, leading to an increase in the metabolic rate of all the cells in the body. Hyperthyroidism increases the demand for glucose and for substrates for oxidative phosphorylation. Which metabolic pathways are active in the fasted state in the liver, muscle, and adipose tissue in order to meet this demand?

34. Phosphorylase kinase is the one of the most complex enzymes known. It consists of four copies each of four different subunits, denoted as $\alpha_4\beta_4\gamma_4\delta_4$. The γ subunit contains the catalytic site. The α and β subunits can be phosphorylated. The δ subunit is calmodulin (see Section 10.2). What does this information tell you about the regulation of this enzyme's activity?

35. How would phosphorylation by AMPK affect the activity of **a.** glycogen synthase, **b.** phosphorylase kinase, **c.** HMG-CoA reductase, and **d.** hormone-sensitive lipase?

36. Inexperienced athletes might consume a meal high in glucose just before a race, but veteran marathon runners know that doing so would impair their performance. Explain.

37. The results of a study showed that glucagon leads to an increase in the rate of glucose-6-phosphate hydrolysis. **a.** How could this explain the following results: Phosphoenolpyruvate concentration increases twofold, glucose-6-phosphate concentration decreases by 60%, and hepatic glucose concentration increases twofold in the presence of glucagon and exogenously administered dihydroxyacetone phosphate. **b.** Inhibition of glucose-6-phosphate hydrolysis results in both activation of gluconeogenesis and inhibition of glycolysis. Explain.

38. Write the balanced chemical equations for the complete combustion (to CO_2 and H_2O) of **a.** glucose and **b.** palmitate ($C_{16}H_{32}O_2$). **c.** The respiratory exchange ratio is used to compare the amount of exhaled CO_2 to the amount of inhaled O_2. Use the equations you wrote in parts **a** and **b** to calculate the respiratory exchange ratio for each substrate.

39. The respiratory exchange ratio (see Problem 38) is approximately 0.8 in a human at rest. **a.** Does the ratio increase or decrease when an individual uses relatively more fatty acids than glucose as fuel? Would you expect the respiratory exchange ratio to increase or decrease during **b.** low-intensity exercise or **c.** short-term, high-intensity exercise?

40. A 15-year-old male patient sees a physician because his parents are concerned about his inability to perform any kind of strenuous exercise without suffering painful muscle cramps. **a.** The patient's response to glucagon is tested by injecting a high dose of glucagon intravenously and then drawing samples of blood periodically and measuring the glucose content. After the glucagon injection, the patient's blood sugar rises dramatically. Is this the response you would expect in a normal person? Explain. **b.** The patient's liver is normal in size, but his muscles are flabby and poorly developed. Liver and muscle biopsies reveal that glycogen content in the liver is normal, but muscle glycogen content is

elevated. The biochemical structure of glycogen in both tissues appears to be normal. Consult Table 13.1. What type of glycogen storage disease does this patient have?

41. How do leptin and ghrelin participate in energy metabolism?

42. A patient with a glycogen storage disease (see Problem 40) performs 30 minutes of ischemic (anaerobic) exercise and blood is withdrawn every few minutes and analyzed for alanine. In a normal person, the concentration of alanine in blood increases during ischemic exercise, but in the patient, alanine decreases during exercise, leading you to believe that his muscle cells are taking up alanine rather than releasing it. **a.** Why would blood alanine concentrations increase in a normal person? Why do blood alanine concentrations decrease in the patient? **b.** The patient is advised to avoid strenuous exercise. If he does wish to perform light or moderate exercise, he is advised to consume sports drinks containing glucose or fructose frequently while exercising. Why would this help alleviate the muscle cramps suffered during exercise?

43. Glucose-6-phosphatase is associated with the endoplasmic reticulum (ER), with the active site of the enzyme facing the lumen (inside) of the ER. Yet many of the pathways associated with the enzyme occur in the cytosol. Draw a diagram to illustrate how transport proteins mediate the movement of substrates and products across the endoplasmic reticulum membrane.

44. A physician examines a patient and suspects that she might have von Gierke's disease, which is usually characterized by a lack of glucose-6-phosphatase activity (GSD1). To confirm her hypothesis, the physician provides the patient with a carbohydrate meal, and then gives the patient a glucagon injection 2 hours later. A blood test following the glucagon injection confirms the disease. **a.** How does the patient respond to the glucagon test and how does her response differ from that of a normal patient? **b.** What proteins are possibly defective in the patient (*Hint:* see Problem 43)?

19.3 Disorders of Fuel Metabolism

45. What metabolites can serve as substrates for gluconeogenesis?

46. Explain why fasting increases the liver concentrations of phosphoenolpyruvate carboxykinase and glucose-6-phosphatase.

47. After several days of starvation, the ability of the liver to metabolize acetyl-CoA via the citric acid cycle is severely compromised. Explain why.

48. During early starvation, gluconeogenic activity in the liver is high, but if starvation is prolonged, the major site of gluconeogenesis switches from the liver to the kidney (see Fig. 19.2). What metabolite increases as a result of this switch? What metabolite decreases?

49. Explain how the reactions of the glucose–alanine cycle would operate during starvation.

50. a. During a 24-hour fast, a person utilizes protein at a rate of 75 g · day^{-1}. If a non-obese person has 6000 g of protein reserves and if death occurs when 50% of the protein reserves have been utilized, how prolonged can the fast be before death occurs? **b.** In fact, during starvation, protein utilization does not progress at a rate of 75 g · day^{-1} but dramatically slows down to 20 g · day^{-1} as the fast increases in duration. What body fuels are utilized during a prolonged fast to conserve body protein?

51. In the first few days of starvation, urea production doubles, but if the fast is prolonged, urea cycle activity diminishes dramatically. Explain why.

52. Hyperglycemic conditions affect the physiology of eye cells. Explain.

53. Individuals who are trying to lose weight are advised to consume fewer calories as well as to exercise. Why would exercising (keeping muscles active) help promote the loss of stored fat from adipose tissue?

54. Adipocytes secrete leptin, a hormone that suppresses appetite. Leptin exerts its effects through the central nervous system and also directly on target tissues by binding to specific receptors. Leptin can inhibit insulin secretion but can also act as an insulin mimic by activating some of the same intracellular signaling components as insulin. For example, leptin can induce tyrosine phosphorylation of insulin receptor substrate-1 (IRS-1). Using this information, predict leptin's effect on the following: **a.** Glucose uptake by skeletal muscle. **b.** Hepatic glycogenolysis and liver glycogen phosphorylase activity. **c.** cAMP phosphodiesterase activity.

55. Insulin resistance is characterized by the failure of insulin-sensitive tissues to respond to the hormone, which normally results in the uptake of glucose and its subsequent conversion to a storage form, such as glycogen or triacylglycerols. Why might inhibiting glycogen synthase kinase 3 (GSK3, see Problem 29) be useful for treating diabetes?

56. a. Describe in detail the signaling events that occur when norepinephrine, a ligand that binds to G protein–coupled receptors (see Section 10.2), elicits the release of fatty acids from triacylglycerols in brown fat. **b.** Provide the biochemical details to describe how fatty acids can generate heat in brown fat in the presence of an uncoupling protein.

57. Adults have deposits of brown adipose tissue located mainly in the muscles of the lower neck and collarbone. **a.** Brown adipose tissue expresses more cytochrome c than white adipose tissue. What is the purpose of the elevated cytochrome c? **b.** Investigators measured the uptake of labeled glucose into brown fat in volunteers under room-temperature conditions and while the volunteers placed one foot in 7–9°C water. There was a 15-fold increase in uptake of labeled glucose by brown fat when the subjects were exposed to the colder temperature. Explain.

58. Decreased carnitine acyltransferase activity along with reduced activity of the mitochondrial electron transport chain have been observed in obese individuals. Explain the significance of these observations.

59. Some obese patients with type 2 diabetes undergo gastric bypass surgery, in which the upper part of the stomach is reconnected to the lower part of the small intestine. In some patients, the surgery appears to cure the symptoms of diabetes even before the patient has lost any weight. Propose an explanation for this observation.

60. The activity of acetyl-CoA carboxylase is stimulated by a fat-free diet and inhibited in starvation and diabetes. Explain.

61. How the fat-rich diet, non-alcoholic fatty liver (NAFLD), and insulin resistance are interconnected?

62. The properties of acetyl-CoA carboxylase (ACC) were studied to see whether the enzyme might be a possible drug target to treat obesity. Mammals have two forms of acetyl-CoA carboxylase, one in the liver and adipose tissue cytosol (ACC1) and one in the mitochondrial matrix of heart and muscle (ACC2). Both are sensitive to regulation by malonyl-CoA. **a.** How does malonyl-CoA regulate the activity of acetyl-CoA carboxylase? **b.** Experiments with ACC2-knockout mice (in which the ACC2 gene is not expressed but the ACC1 gene is still functional) showed a 20% reduction in liver glycogen compared to control mice. Explain. **c.** In the knockout mice, the concentration of fatty acids in the blood was lower but the concentration of triacylglycerols was higher than in the wild-type mice. Explain.

63. A fatty acid synthase (FAS) inhibitor ("compound C" in Problem 17.39) was investigated as a possible candidate for a weight-loss drug. **a.** Mice were injected intraperitoneally with compound C and radioactively labeled acetate. What is the fate of the label? **b.** Mice

receiving intraperitoneal injections of compound C reduced their food intake by more than 90% and lost nearly one-third of their body weight, although they gained back the weight when the drug was withdrawn. The investigators measured brain concentrations of neuropeptide Y (NPY), a compound known to act on the hypothalamus to increase appetite during starvation. Predict the effect of compound C on brain levels of NPY. **c.** Because hepatic malonyl-CoA levels were high in the inhibitor-treated mice but not in control mice, the investigators hypothesized that malonyl-CoA inhibits feeding. If their hypothesis is correct, predict what would happen if mice were pretreated with an acetyl-CoA carboxylase inhibitor prior to injection with compound C. **d.** What other cellular metabolites accumulate when concentrations of malonyl-CoA rise? (These molecules are candidates for signaling molecules, which might stimulate a biochemical pathway that decreases appetite.)

64. PTP-1B is a phosphatase that dephosphorylates the insulin receptor and might also dephosphorylate IRS-1. **a.** After feeding, mice deficient in PTP-1B reduce circulating blood glucose levels, using half as much insulin as normal mice. Explain this observation. **b.** What intracellular changes are observed in muscle cells when insulin is injected into the PTP-1B–deficient mice? **c.** How would you use this information to design a drug to treat diabetes? Are there any concerns involved in using the drug you have designed?

65. There is convincing evidence that AMPK can phosphorylate acetyl-CoA carboxylase. It is also possible that AMPK phosphorylates protein kinase B (see Section 10.2), which increases the rate of translocation of GLUT4 vesicles to the plasma membrane. Given this information, how does metformin (Table 19.5) treat the symptoms of metabolic syndrome?

19.4 Clinical Connection: Cancer Metabolism

66. The compound 2-deoxy-D-glucose is structurally similar to glucose and can be taken up by cells via glucose transporters. **a.** Once inside the cells, the compound is converted to 2-deoxy-D-glucose-6-phosphate by hexokinase. Draw the structures of the substrate and product of the reaction. **b.** When 2-deoxy-D-glucose is added to cultured cancer cells, the intracellular concentration of ATP rapidly decreases. In a separate experiment, antimycin A (which prevents the transfer of electrons to cytochrome c in the electron transport chain) is added to the cells, but in this case there is no effect on ATP production. Explain these results.

67. Positron emission tomography (PET) using glucose as a tracer is not useful for visualizing brain tumors because the background level of glucose uptake by the brain is already high. **a.** Based on what you know about cancer biology, propose an alternative substance that could be used to visualize brain tumors. **b.** Explain your choice of tracer.

68. Some cancer cells produce large amounts of the enzyme that converts fructose-6-phosphate to the allosteric regulator fructose-2, 6-bisphosphate. How does this benefit cancer growth?

69. In cancer cells with low pyruvate kinase activity, phosphoenolpyruvate accumulates and donates its phosphoryl group to the active-site histidine in phosphoglycerate mutase. The phospho-His then spontaneously undergoes hydrolysis. What is accomplished by this alternative pathway for phosphoenolpyruvate?

70. Lactate stimulates growth in the cells that form new blood vessels. Does this observation help explain tumor growth?

71. Changes in biochemical pathways in cancer cells are mediated in part by hypoxia-inducible factors (HIFs), which respond to the low-oxygen environment of the cancer cell. Under hypoxic conditions, hypoxia-inducible factors translocate to the nucleus, where they bind to specific regions of the DNA to regulate gene expression. What happens to the expression of pyruvate dehydrogenase kinase (PDK; see Problem 14.10) under hypoxic conditions, according to the data shown in the figure? How does this contribute to the Warburg effect?

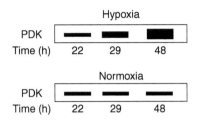

72. A type of renal cancer is characterized by a deficiency in fumarase. **a.** How would a fumarase deficiency affect the levels of pyruvate, fumarate, and malate? **b.** Why does a patient with a fumarase deficiency have the same symptoms as a patient with a succinate dehydrogenase deficiency?

73. Cancer researchers examined the relationship between fumarase and lactate dehydrogenase activities and cell survival. They isolated cells from fumarase-deficient renal cancer patients (FH63 cells) and then constructed a cell line in which they "knocked down" the expression of LDH in these cells (indicated as ΔLDH). The survival of the FH63 cells—with and without LDH—was compared to control cells and cells missing the LDH gene. The results are shown below. **a.** How does the percentage of apoptosis (programmed cell death) vary among the different types of cells? **b.** What strategies do the different types of cells use to survive?

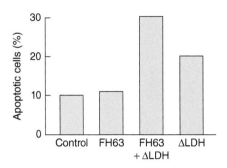

74. A study of glioma tumor cells revealed an Arg → His mutation in an isozyme of isocitrate dehydrogenase. **a.** What single nucleotide change could result in this mutation? **b.** The mutant isocitrate dehydrogenase produces 2-hydroxyglutarate, which accumulates in the cancer cell. Draw the structure of this metabolite. **c.** How does the reaction catalyzed by the mutant enzyme differ from the reaction catalyzed by the normal enzyme? **d.** How does operation of the mutant enzyme affect the cell's supply of reduced cofactors?

75. A strain of herpes virus is associated with a type of cancer called Kaposi's sarcoma. Some herpes virus infections increase the activity of pyruvate carboxylase and glutamate-oxaloacetate transaminase in the host cell. How do these changes promote viral replication?

SELECTED READINGS

Cantley, J. and Ashcroft, F.M., Q&A: Insulin secretion and type 2 diabetes: Why do β-cells fail? *BMC Biology* 13, 33, doi: 10.1186/s12915-015-0140-6 (2015). [A highly readable review of insulin production and the possible causes and treatments of diabetes.]

Haeusler, R.A., McGraw, T.E., and Accili, D., Biochemical and cellular properties of insulin receptor signaling, *Nat. Rev. Mol. Cell. Biol.* 19, 31–44, doi: 10.1038/nrm.2017.89 (2018). [Offers a thorough description of the direct and indirect effects of insulin on cells.]

Hosios, A.M. and Vander Heiden, M.G., The redox requirements of proliferating mammalian cells, *J. Biol. Chem.* 293, 7490–7498, doi: 10.1074/jbc.TM117.000239 (2018). [Provides an overview of the major metabolic pathways, with a focus on the participation of NAD⁺ and NADPH.]

Rajas, F., Gautier-Stein, A., and Mithieux, G., Glucose-6 phosphate, a central hub for liver carbohydrate metabolism, *Metabolites* 9, 282,

doi: 10.3390/metabo9120282 (2019). [Summarizes the metabolic fates of glucose and their regulation in health and disease.]

Saxton, R.A. and Sabatini, D.M., mTOR signaling in growth, metabolism, and disease, *Cell* 168, 960–976, doi: 10.1016/j.cell.2017.02.004 (2018). [A thorough review of the structure and many activities of mTOR.]

Vettore, L., Westbrook, R.L., and Tennant, D.A., New aspects of amino acid metabolism in cancer, *Br. J. Cancer* 122, 150–156, doi: 10.1038/s41416-019-0620-5 (2020). [A brief review of ways that cancer cells adjust their metabolism to take advantage of nutrients provided by other cells and to maintain redox homeostasis.]

Zmora, N., Suez, J., and Elinav, E., You are what you eat: Diet, health and the gut microbiota, *Nat. Rev. Gastroenterol. Hepatol.* 16, 35–56, doi: 10.1038/s41575-018-0061-2 (2019). [Outlines some of the challenges in defining the "normal" microbiota and identifying the beneficial and harmful effects of microbial metabolism.]

CHAPTER 19 CREDITS

Figure 19.6 Image based on 1AI0. Chang, X., Jorgensen, A.M., Bardrum, P., Led, J.J., Solution structures of the R6 human insulin hexamer, *Biochemistry* 36, 9409–9422 (1997).

Figure 19.12 Image based on 1GCN. Sasaki, K., Dockerill, S., Adamiak, D.A., Tickle, I.J., Blundell, T., X-ray analysis of glucagon and its relationship to receptor binding, *Nature* 257, 751–757 (1975).

Figure 19.14 Image based on 4CFH. Xiao, B., Sanders, M.J., Underwood, E., Heath, R., Mayer, F., Carmena, D., Jing, C., Walker, P.A., Eccleston, J.F., Haire, L.F., Saiu, P., Howell, S.A., Aasland, R., Martin, S.R., Carling, D., Gamblin, S.J., Structure of mammalian AMPK and its regulation by ADP, *Nature* 472, 230 (2011).

DNA Replication and Repair

Genetic variations in crop species, such as lentils (*Lens culinaris*), result from DNA mutations that occur naturally or are induced by radiation or chemical treatment. High-throughput DNA sequencing techniques can identify the genetic changes associated with favorable traits, accelerating the development of higher-yielding crops.

Do You Remember?

- A DNA molecule contains two antiparallel strands that wind around each other to form a double helix in which A and T bases in opposite strands, and C and G bases in opposite strands, pair through hydrogen bonding (Section 3.1).
- Double-stranded nucleic acids are denatured at high temperatures; at lower temperatures, complementary polynucleotides anneal (Section 3.2).
- Changes in a DNA sequence can cause disease (Section 3.3).
- A reaction that hydrolyzes a phosphoanhydride bond in ATP occurs with a large change in free energy (Section 12.2).

A human cell contains 46 separate DNA molecules—chromosomes—comprising over 6 billion base pairs. Lined up end-to-end, these molecules would be slightly longer than 2 m, but the average diameter of a mammalian nucleus is only 6 μm (0.000006 m). Fitting all the DNA into the nucleus would be like stuffing a 50-mile-long strand of hair into a backpack. Not surprisingly, cells have elaborate mechanisms for keeping DNA neatly packaged as well as unpacking it so that it can be copied. DNA replication can be considered to be part of the central dogma of molecular biology: the information in a parent DNA molecule must be copied to produce two identical DNA molecules that are passed on to daughter cells when the parent cell divides. In this chapter, we will look at the process of DNA replication in detail, as well as some of the challenges cells face in accurately replicating DNA, repairing damaged DNA, and storing it safely. The Tools and Techniques section at the end of the chapter describes some of the methods used to manipulate and sequence DNA.

20.1 DNA Packaging: Nucleosome and Chromatin

KEY CONCEPTS

Describe how DNA is packaged in cells.

- Visualize supercoiling as a result of over- or underwinding DNA.
- Describe how topoisomerases alter DNA supercoiling.
- Explain how nucleosomes and higher-order structures condense DNA.

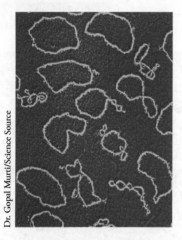

Figure 20.1 Supercoiled DNA molecules. The circular DNA molecules are slightly underwound, so they coil up on themselves, forming supercoils.

During replication and repair, DNA molecules in bacteria and eukaryotes are relatively extended and accessible to the proteins that carry out these processes. However, for much of the time, chromosomes are kept in a more compact arrangement that allows large amounts of genetic information to be packed into a small space and kept safe.

DNA is negatively supercoiled

In order to facilitate processes such as replication and transcription, which require opening up the double helix, the DNA molecules in cells are slightly underwound. In other words, the two strands of the helix make fewer than the expected number of helical turns around each other. In order to maintain a conformation close to the stable B form, the DNA molecule twists up on itself, much like the cord of an old-fashioned telephone. This phenomenon, termed **supercoiling,** is readily apparent in small circular DNA molecules (**Fig. 20.1**).

The geometry of a DNA molecule can be described by the branch of mathematics known as topology. For example, consider a strip of paper. When the strip is looped once, it is said to have one writhe. If the ends of the paper are gently pulled in opposite directions, the strip is deformed into a shape called a twist:

A writhe ⇌ A twist

Twisting the strip further introduces more writhes; twisting it in the opposite direction removes writhes, or introduces negative writhes. The same topological terms apply to DNA: *Each writhe, or supercoil, in DNA is the result of overtwisting or undertwisting the DNA helix.* The twisted molecule, like the piece of paper, prefers to writhe, since this is more energetically favorable.

To demonstrate supercoiling for yourself, cut a flat rubber band to obtain a linear piece a few inches long. Hold the ends apart, twist one of them, and then bring the ends closer together. You will see the twists collapse into writhes (supercoils). The same thing happens if you let the rubber band relax and then twist it in the opposite direction. Note that the twisted rubber band (representing a double-stranded DNA molecule) adopts a more compact shape. Naturally occurring DNA molecules are negatively supercoiled, which allows them to take up less space. It also means that if the DNA were stretched out (that is, its writhes converted to twists), the two strands would unwind slightly.

Topoisomerases alter DNA supercoiling

Normal replication and transcription require that cells actively maintain the supercoiling state of DNA by adding or removing supercoils. Of course, cells cannot grab the ends of a long DNA molecule in order to twist or untwist it. Instead, topological changes occur when a topoisomerase cuts one or both strands of the supercoiled DNA, alters the structure, and then reconnects the broken strands. Type I topoisomerases cut one DNA strand, and type II enzymes cut both strands of DNA and require ATP.

Type I topoisomerases occur in all cells and alter supercoiling by altering the DNA's helical twisting. Type IA enzymes nick the DNA (cut the backbone of one strand), pass the intact strand through the break, then seal the broken strand in order to unwind the DNA by one turn (**Fig. 20.2**). Type IB enzymes also cut one strand but hold the DNA on one side of the nick while allowing the DNA on the other side of the nick to rotate one or more turns before the broken strand is sealed. In both cases, unwinding is driven by the strain in the supercoiled DNA, so a type I topoisomerase can "relax" both negatively and positively supercoiled DNA.

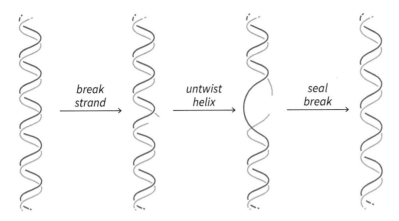

Figure 20.2 Action of a type I topoisomerase.   The supercoiling of a DNA segment is decreased when a DNA strand is broken, the helix untwisted by passing the other strand through the break, and then sealed.

Question **Count the number of helical turns in the structures shown here.**

When a type I topoisomerase cleaves one strand of the DNA, an active-site tyrosine residue forms a covalent bond with the backbone phosphate at one side of the nick:

Formation of this diester linkage conserves the free energy of the broken phosphodiester bond of the DNA strand, so strand cleavage and subsequent sealing do not require any other source of free energy.

Type II topoisomerases directly alter the number of writhes in a supercoiled DNA molecule by passing one DNA segment through another, a process that requires a double-strand break (**Fig. 20.3**). The type II enzymes are dimeric, with two active-site Tyr residues that form covalent bonds to the 5′ phosphate groups of the broken DNA strands. The enzyme must undergo conformational changes in order to cleave a DNA molecule and hold the ends apart while another DNA segment passes through the break. ATP hydrolysis appears to produce the free energy to restore the enzyme to its starting conformation. Type II topoisomerases, which can relieve both negative and positive supercoiling, participate in DNA replication by relaxing positive supercoils ahead of the replication fork. They also untangle the products of replication, which remain linked after replication forks meet up:

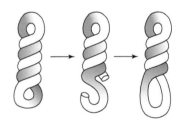

Figure 20.3 Action of a type II topoisomerase.   The twisted worm shape represents a supercoiled double-stranded DNA molecule. A type II topoisomerase breaks both strands of the DNA, passes another segment of DNA through the break, then seals the break.

Question **Does supercoiling increase or decrease in this diagram?**

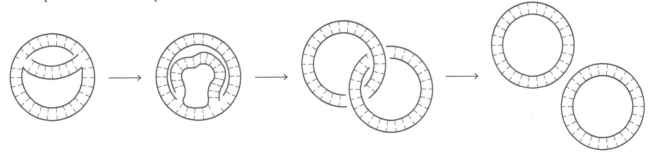

Bacteria contain a type II topoisomerase called DNA gyrase, which can introduce additional negative supercoils into DNA, so its net effect is to further underwind the DNA helix. A number of antibiotics inhibit DNA gyrase without affecting eukaryotic type II topoisomerases. For example, ciprofloxacin acts on DNA gyrase to enhance the rate of DNA cleavage or reduce the rate of sealing broken DNA. The result is a large number of DNA breaks that interfere with the transcription and replication required for normal cell growth and division, and the bacteria die.

Ciprofloxacin

Eukaryotic DNA is packaged in nucleosomes

Eukaryotic cells lack DNA gyrase but maintain negative supercoiling by wrapping DNA in **nucleosomes.** These complexes of DNA and protein are the basic unit of eukaryotic DNA packaging. *The core of a nucleosome consists of eight **histone** proteins: two each of the histones known as H2A, H2B, H3, and H4. Approximately 146 base pairs of DNA make about 1.65 turns around the histone octamer* (**Fig. 20.4**). A complete nucleosome contains the core structure plus histone H1, a small protein that binds outside the core. Neighboring nucleosomes are separated by short stretches of DNA of 20 to 40 bp. Winding DNA in nucleosomes helps protect the DNA from chemical damage and generates the negative supercoils needed to unwind DNA to initiate replication.

The histone proteins interact with DNA in a sequence-independent manner, primarily via hydrogen bonding and ionic interactions with the sugar–phosphate backbone. Although prokaryotes lack histones, other DNA-binding proteins may help package DNA in bacterial cells. The histones are among the most highly conserved proteins known, in keeping with their essential function in packaging the genetic material in all eukaryotic cells. Each histone pairs with another, and the set of eight forms a compact structure. However, the tails of the histones, which are flexible and charged, extend outward from the core of the nucleosome (see Fig. 20.4). These histone tails are subject to covalent modification, which helps control gene expression (Section 21.1).

During DNA replication, nucleosomes are disassembled as the DNA spools through the replication machinery. Histone chaperones, along with some components of the replisome, help reassemble nucleosomes on newly replicated DNA, using the displaced histones as well as newly synthesized histones imported from the cytoplasm to the nucleus.

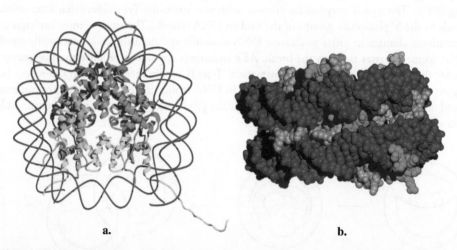

a.   b.

Figure 20.4 Structure of the nucleosome core.   a. Top view.   **b.** Side view (space-filling model). The DNA (dark blue) winds around the outside of the histone octamer.

The nucleosomes themselves form higher-order structures that are not entirely understood, although chromosomes seem to be organized in long loops. Some of the cell's DNA appears to be highly condensed at the edges of the nucleus. This DNA is known as **heterochromatin,** which is transcriptionally silent. **Euchromatin** is less condensed and appears to be transcribed at a higher rate. The two forms of **chromatin** are distinguishable by electron microscopy (**Fig. 20.5**).

When the cell is ready to divide, the DNA is further condensed. Nucleosomes compact the DNA only by a factor of about 30 to 40, but the chain of nucleosomes itself can coil into a solenoid (coil) with a diameter of about 30 nm (**Fig. 20.6**). The DNA in the 30-nm fiber is presumably well protected from nuclease attack and is inaccessible to the proteins that carry out replication and transcription. During cell division, chromosomes condense even further so that each one has an average length of about 10 μm and a diameter of 1 μm. This makes sense for a dividing cell, since fully extended DNA molecules would become hopelessly tangled rather than segregating neatly to form two equivalent sets of chromosomes (see Box 3.A).

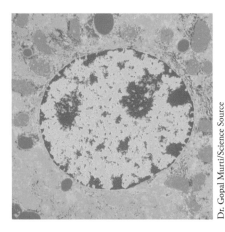

Figure 20.5 A eukaryotic nucleus.
Heterochromatin is the darkly staining material (red in this colorized electron micrograph); euchromatin stains more lightly (yellow).

Dr. Gopal Murti/Science Source

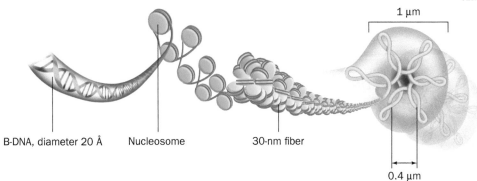

1 μm

B-DNA, diameter 20 Å   Nucleosome   30-nm fiber

0.4 μm

Figure 20.6 Levels of chromatin structure.   The DNA helix (blue) is wrapped around a histone octamer (orange) to form a nucleosome; nucleosomes aggregate to form the 30-nm fiber; this packs into loops in the fully condensed chromosome. The approximate diameter of each structure is indicated.

Concept Check

1. Describe the relationship between writhing (supercoiling) and helical twisting.
2. Relate nucleosome structure to DNA supercoiling.
3. Explain how topoisomerases alter DNA supercoiling by transiently cutting one or both strands of DNA.
4. Explain why negative supercoiling of DNA assists processes such as replication.
5. Describe the action of type I and type II topoisomerases and their source of free energy.
6. Explain the role of histones in packaging DNA.
7. Compare the structure and activity of heterochromatin and euchromatin.
8. How could covalent modification of histone or DNA alter gene expression?

20.2 The DNA Replication Machinery

KEY CONCEPTS

Summarize the actions of the enzymes and other proteins involved in synthesizing the leading and lagging strands.

- Explain why DNA replication is semiconservative.
- Relate the structure of helicase to its function.
- Explain why DNA replication requires an RNA primer.

- Compare synthesis of the leading and lagging strands of DNA.
- Describe the factors responsible for the processivity and accuracy of replication.
- Describe how an endonuclease and ligase yield continuous strands of DNA.

When Watson and Crick described the complementary, double-stranded nature of DNA in 1953, they recognized that DNA could be duplicated by a process involving separation of the strands followed by the assembly of two new complementary strands. This mechanism of copying, or **replication,** was elegantly demonstrated by Matthew Meselson and Franklin Stahl in 1958. They grew bacteria in a medium containing the heavy isotope ^{15}N in order to label the cells' DNA. The bacteria were then transferred to fresh medium containing only ^{14}N, and the replicated DNA was isolated and sedimented according to its density in an ultracentrifuge. Meselson and Stahl found that the first generation of replicated DNA had a lower density than the parental DNA but a higher density than DNA containing only ^{14}N. From this, they concluded that DNA is a hybrid containing one parental (heavy) strand and one new (light) strand. In other words, DNA is replicated **semiconservatively.** Because Meselson and Stahl did not observe any all-heavy DNA in the first generation, they were able to discount the possibility that DNA was copied in a way that left intact—or conserved—the original double-stranded molecule (**Fig. 20.7**).

DNA polymerase, the enzyme that catalyzes the polymerization of deoxynucleotides, is just one of the proteins involved in replicating double-stranded DNA. The entire process—separating the two template strands, initiating new complementary polynucleotide chains, and extending them—is carried out by a complex of enzymes and other proteins known as the **replisome.** In this section we examine the structures and functions of the major players in DNA replication in bacteria and in eukaryotes.

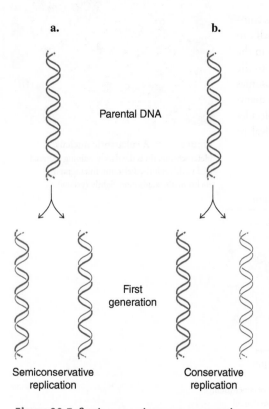

Figure 20.7 Semiconservative versus conservative DNA replication.   a. In semiconservative replication, each DNA molecule contains one parental (heavy) and one new (light) polynucleotide strand. **b.** If DNA replication were conservative, the parental DNA (both strands heavy) would persist, while new DNA would consist of two light strands.

Question What would the second generation of DNA look like for each mode of replication?

DNA replication requires many enzymes and proteins

In the circular chromosomes of bacteria, DNA replication begins at a particular site called the origin. Here, proteins bind to the DNA and melt it open in an ATP-dependent manner. Polymerization then proceeds in both directions from this point until the entire chromosome (4.6×10^6 bp in *E. coli*) has been replicated. The point where the parental strands separate and the new strands are synthesized is known as the **replication fork.** One origin gives rise to two replication forks.

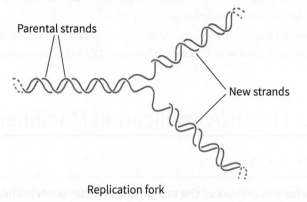

Replication fork

Archaebacterial chromosomes have up to four origins of replication, and the much larger chromosomes of eukaryotes have multiple origins. In yeast, these sites are about 40 kb apart and have a characteristic sequence. In mammals, the origins are farther apart and their locations appear to be determined by features of chromosome organization rather than sequence.

At one time, complexes of DNA polymerase and other replication proteins were thought to move along DNA like a train on tracks. This "locomotive" model of DNA replication requires that the large replisome move along the relatively thin template strand, rotating around it while generating a double-stranded helical product. In fact, cytological studies indicate that DNA replication (as well as transcription) occurs in "factories" at discrete sites. For example, in bacteria, DNA polymerase and associated factors appear to be immobilized in one or two complexes near the plasma membrane. In the eukaryotic nucleus, newly synthesized DNA appears at 100 to 150 spots, each one representing several hundred replication forks (**Fig. 20.8**). *According to the **factory model of replication,** the protein machinery is stationary and the DNA is reeled through it.* In eukaryotes, this organization, which depends on interactions with the nucleoskeleton, presumably facilitates the synchronous elongation of many DNA segments, permitting the efficient replication of enormous eukaryotic genomes.

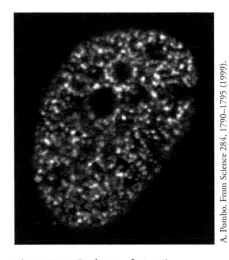

Figure 20.8 Replication foci. The fluorescent patches (foci) in the nucleus of a eukaryotic cell mark the presence of newly synthesized DNA.

Helicases convert double-stranded DNA to single-stranded DNA

In order to replicate a DNA molecule, its two strands, which are coiled around each other, must be separated so that each one can serve as a template for DNA synthesis. One of the first proteins to bind to the origin of replication is an enzyme called a **helicase,** which catalyzes unwinding of the strands of the double helix. In eukaryotes, an inactive form of the helicase binds to the DNA at each of many potential origins well before replication begins. When the appropriate signals are received from internal as well as external signaling pathways, the cell commits to replicating its DNA in preparation for cell division. In response to kinase-catalyzed phosphorylation events and the addition of more proteins, some fraction of the helicases become activated. After this point, no additional helicases can bind to the DNA. It is believed that by separating the helicase loading and activation steps, the cell can ensure that all its DNA is replicated more or less simultaneously (since all the helicases are activated at the same time) but only once per cell cycle (since only the helicases already poised at the origins are activated).

Most helicases are hexameric proteins shaped roughly like a donut that circles one DNA strand (**Fig. 20.9**). Helicase is a motor protein (see Section 5.4) that uses the free energy of ATP hydrolysis to move along the DNA strand, pushing away the complementary DNA strand to open up the helix. For each ATP hydrolyzed, up to five base pairs of DNA are separated. The helicase appears to operate in a rotary fashion, with conformational changes driven by ATP hydrolysis—somewhat reminiscent of the binding change mechanism of the F_1 component of ATP synthase (Section 15.3). Some nonreplicative helicases are monomers that use the free energy of the ATP reaction to pry apart the two DNA strands.

As replication proceeds, the parental DNA strands are continuously separated by the action of a helicase at each replication fork. However, strand separation forces the helix to wind more tightly ahead of the replication fork:

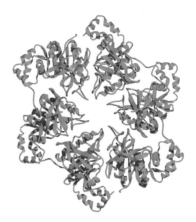

Figure 20.9 Structure of a hexameric helicase. This helicase, from the bacteriophage T7, forms a hexameric ring around a single strand of DNA and pushes double-stranded DNA apart.

Question What would happen if an activated helicase could dissociate from DNA and re-bind to another origin?

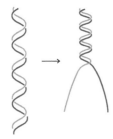

The strain generated by the extra twisting is relieved by a **topoisomerase,** an enzyme that cuts the DNA and allows it to "relax" before the cuts are sealed. These enzymes are described in more detail in Section 20.5.

As single-stranded DNA is exposed at the replication fork, it associates with a protein known as single-strand binding protein (SSB). SSB coats the DNA strands to protect them from nucleases and to prevent them from reannealing or forming secondary structures that might impede

A. Pombo. From Science 284, 1790–1795 (1999).

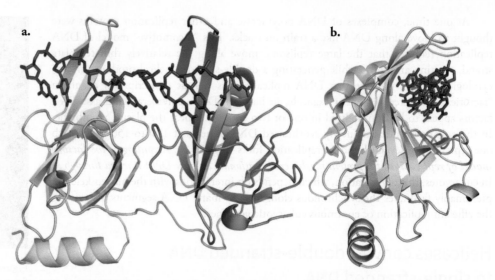

Figure 20.10 DNA-binding domains of replication protein A. Two of the four DNA-binding domains are shown in yellow and green in this model, with a bound octanucleotide (polydeoxycytidine, shown in purple). **a.** Front view. **b.** Side view.

Question **Which types of amino acids would you expect to find in the DNA-binding site of this protein?**

replication. *E. coli* SSB is a tetramer with a positively charged cleft that accommodates loops of single-stranded—but not double-stranded—DNA. The eukaryotic SSB, called replication protein A, is a larger protein that includes four DNA-binding domains separated by flexible regions (**Fig. 20.10**). Both prokaryotic and eukaryotic SSBs can adopt different conformations and bind DNA in several different ways. Studies of replication protein A suggest that DNA binding occurs cooperatively such that the first protein–DNA interaction is relatively weak (with a K_d in the μM range) but allows the other domains to "unroll" along the DNA to form a stable complex that protects about 30 nucleotides and has an overall dissociation constant of about 10^{-9} M. As the template DNA is reeled through the polymerase, SSB is displaced and is presumably redeployed as helicase exposes more single-stranded DNA.

DNA polymerase faces two problems

The mechanism of DNA polymerase and the double-stranded structure of DNA present two potential obstacles to the efficient replication of DNA. First, *DNA polymerase can only extend a preexisting chain; it cannot initiate polynucleotide synthesis.* However, an RNA polymerase can do this, so DNA chains *in vivo* begin with a short stretch of RNA that is later removed and replaced with DNA:

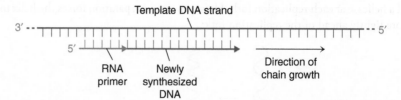

This stretch of RNA, about 25–29 nucleotides in *E. coli,* is known as a **primer,** and the enzyme that produces it during DNA replication is known as a **primase.** As we will see, primase is required throughout DNA replication, not just at the start, and associates with helicase (forming a complex that is sometimes called a primosome). The active site of the primase is narrow (about 9 Å in diameter) at one end, where the single-stranded DNA template is threaded through. The other end of the active site is wide enough to accommodate a DNA–RNA hybrid helix (which has an A-DNA–like conformation; see Fig. 3.5).

The second problem facing DNA polymerase is that the antiparallel template DNA strands are replicated simultaneously by a pair of polymerase enzymes. However, each DNA polymerase catalyzes a reaction in which the 3′ OH group at the end of the growing DNA chain attacks the phosphate group

Figure 20.11 Mechanism of DNA polymerase. The 3′ OH group (at the 3′ end of a growing polynucleotide chain) is a nucleophile that attacks the phosphate group of an incoming deoxynucleoside triphosphate (dNTP) that base pairs with the template DNA strand. Formation of a new phosphodiester bond eliminates PP_i.

Question **Why is water also a reaction product?**

of a free nucleotide that base pairs with the template DNA strand (**Fig. 20.11**). For this reason, *a polynucleotide chain is said to be synthesized in the 5′ → 3′ direction.* The reactants and products of the polymerization reaction have similar free energies, so the reaction is reversible. However, the subsequent hydrolysis of PP_i makes the reaction irreversible *in vivo.* RNA polymerase follows the same mechanism.

Because the two template DNA strands are antiparallel, the synthesis of two new DNA strands requires that the template strands be pulled in opposite directions through the replication machinery so that the DNA polymerases can continually add nucleotides to the 3′ end of each new strand.

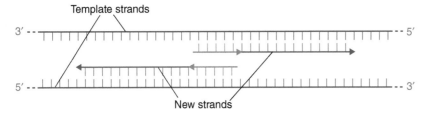

This awkward situation does not occur in cells. Instead, *the two polymerases are stationary, working side-by-side, and one template DNA strand periodically loops out.* In this scenario, one strand of DNA, called the **leading strand,** can be synthesized in one continuous piece. It is initiated by the action of a primase, then extended in the 5′ → 3′ direction by the action of one DNA polymerase enzyme. The other strand, called the **lagging strand,** is synthesized in pieces, or **discontinuously.** Its template is repeatedly looped out so that its polymerase can also operate in the 5′ → 3′ direction. Thus, *the lagging strand consists of a series of polynucleotide segments, which are called **Okazaki fragments** after their discoverer* (**Fig. 20.12**).

Bacterial Okazaki fragments are about 500 to 2000 nucleotides long; in eukaryotes, they are about 100 to 200 nucleotides long. Each Okazaki fragment has a short stretch of RNA at its 5′ end, since each segment is initiated by a separate priming event. This explains why primase is required throughout replication: Although the leading strand theoretically requires only one priming event, the discontinuously synthesized lagging strand requires multiple primers.

The mechanism for continuous synthesis of the leading strand and discontinuous synthesis of the lagging strand means that the lagging-strand template must periodically be repositioned. Each time an Okazaki fragment is completed, a polymerase begins extending the RNA primer of the next Okazaki fragment. Other protein components of the replication complex assist in this repositioning in order to coordinate the activities of the two DNA polymerases at the replication

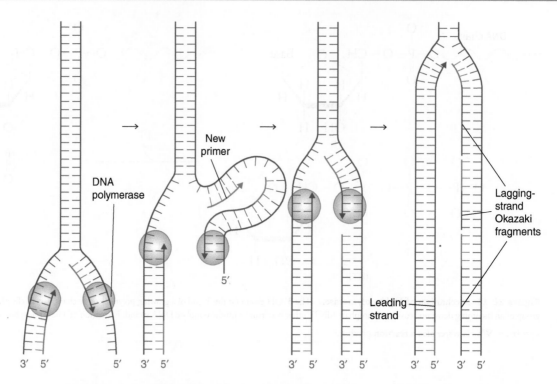

Figure 20.12 A model for DNA replication.   Two DNA polymerase enzymes (green) are positioned at the replication fork to make two complementary strands of DNA. The leading and lagging strands both start with RNA primers (red) and are extended by DNA polymerase in the $5' \rightarrow 3'$ direction. Because the two template strands are antiparallel, the lagging-strand template loops out. The leading strand is therefore synthesized continuously, while the lagging strand is synthesized as a series of Okazaki fragments.

fork. In *E. coli* and possibly other bacteria, three polymerases localize to the replication fork. One synthesizes the leading strand, and the other two apparently take turns synthesizing Okazaki fragments; sharing the work appears to increase replication efficiency.

In a cell, the activity of the replisome is almost certainly more complicated than is described here. For example, replication may slow down momentarily as replisome components are swapped out. Encounters with a damaged DNA template and collisions with the repair machinery or a transcribing RNA polymerase can also cause the replication fork to "stall" or "collapse." In such situations, the helicase may need to be reloaded and new primers made before the polymerases can resume their work.

DNA polymerases share a common structure and mechanism

Most DNA polymerases are shaped somewhat like a hand, with domains corresponding to palm, fingers, and thumb. These structures are likely the result of convergent evolution, as only the palm domains exhibit strong homology. In fact, comparative studies of bacterial, archaebacterial, and eukaryotic DNA polymerases suggest that DNA polymerases evolved from RNA polymerases—another piece of evidence that the first cells used RNA rather than DNA.

One of the best known polymerases is *E. coli* DNA polymerase I, the first such enzyme to be characterized (**Fig. 20.13**). The template strand and the newly synthesized DNA strand, which form a double helix, lie across the palm, in a cleft lined with basic residues.

The polymerase active site, at the bottom of the cleft, contains two metal ions (usually Mg^{2+} but in some cases Mn^{2+}) about 3.6 Å apart. These metal ions are coordinated by aspartate side chains of the enzyme and by the phosphate groups of the substrate nucleoside triphosphate. One of the metal ions interacts with the $3'$ O atom at the end of the primer or growing DNA strand to enhance its nucleophilicity as it attacks the incoming nucleotide:

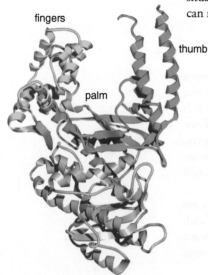

Figure 20.13 *E. coli* DNA polymerase I.   This model shows the so-called Klenow fragment of DNA polymerase I (residues 324 to 928). A loop at the end of the thumb is missing in this model.

The second metal ion helps stabilize the nucleotide substrate. A third metal ion with lower affinity may interact with the PP_i reaction product.

The deoxy carbon at the 2′ position of the incoming nucleotide lies in a hydrophobic pocket. This binding site allows the polymerase to discriminate against ribonucleotides, which bear a 2′ OH group.

After each polymerization event, the enzyme must advance the template strand by one nucleotide. Most DNA polymerases are **processive** enzymes, which means that they undergo several catalytic cycles (about 10 to 15 for *E. coli* DNA polymerase *in vitro*) before dissociating from their substrates. *E. coli* DNA polymerase is even more processive *in vivo*, polymerizing as many as 5000 nucleotides before releasing the DNA. *This enhanced processivity is due to an accessory protein that forms a sliding clamp around the DNA and helps hold DNA polymerase in place.* In *E. coli* the clamp is a dimeric protein, and in eukaryotes the clamp is a trimer (**Fig. 20.14**). Both types of clamps have a hexagonal ring structure and similar dimensions. The inner surface of the ring is positively charged and encloses a space with a diameter of about 35 Å, more than large enough to accommodate double-stranded DNA or a DNA–RNA hybrid helix with a diameter of 26 Å. Without the sliding clamp to enhance the processivity of DNA polymerase, replication would be much less efficient.

The sliding clamp is positioned on the DNA with the help of a clamp-loading complex that uses the free energy of ATP hydrolysis to twist open the ring and allow it to encircle the DNA (the hexameric helicase is loaded onto the DNA in a similar fashion). For the lagging-strand DNA polymerase, the clamp must be reloaded at the start of each Okazaki fragment. This occurs about once every second in *E. coli*.

E. coli has five different DNA polymerase enzymes (I–V), and there are at least 14 eukaryotic DNA polymerases (designated by Greek letters), not including those found in mitochondria and chloroplasts. Why so many? First, DNA replication requires polymerases with different activities to completely synthesize the leading and lagging strands. Other polymerases perform specialized roles in DNA repair pathways, most of which involve the excision and replacement of damaged DNA (Section 20.4).

For example, in *E. coli,* DNA polymerase III is the main replication polymerase for the leading and lagging strands, and DNA polymerase I participates in replacing RNA primers (described below). Polymerases II, IV, and V are used for DNA repair. DNA polymerase III is fast and highly processive, polymerizing about 1 kb of DNA per second at each replication fork. At that rate, the *E. coli* chromosome can be completely copied in 30 minutes.

In humans, DNA synthesis is about 20 times slower than in *E. coli,* and even with multiple origins of replication, it takes an estimated 8 hours to copy the largest chromosome of 250 Mb (Mb = millions of base pairs). Synthesis of both the leading and

a.

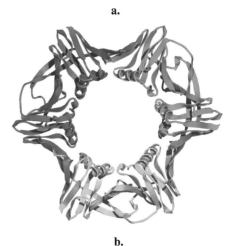

b.

Figure 20.14 DNA polymerase-associated clamps. **a.** In *E. coli*, the β subunit of DNA polymerase III forms a dimeric clamp. **b.** In humans, the clamp is a trimer, called the proliferating cell nuclear antigen (PCNA).

Question **What structural changes must occur each time the clamp is loaded onto DNA?**

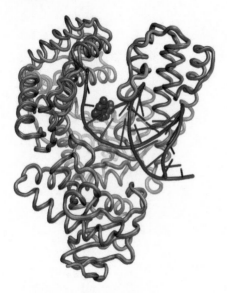

Figure 20.15 Open and closed conformations of a DNA polymerase. The structure of the polymerase from *Thermus aquaticus* was determined in the absence (magenta trace) and presence (green trace) of a nucleotide substrate analog (shown in space-filling form, purple). The models include a segment of DNA representing the template and primer strands (purple).

lagging strands begins with DNA polymerase α, which consists of two DNA polymerase subunits and two primase subunits. After the complex builds primers of about 7 to 12 ribonucleotides, the polymerases extend the primers with another 10 to 25 deoxyribonucleotides. At this point, DNA polymerase δ takes over synthesis of the lagging strand, building each Okazaki fragment in turn. After the initial work of polymerase α, DNA polymerase ε completes the leading strand. Unlike other polymerases, DNA polymerase ε does not strictly require an external sliding clamp to be processive, because its structure includes an extra loop of protein that circles the DNA as a built-in clamp.

DNA polymerase proofreads newly synthesized DNA

During polymerization, the incoming nucleotide base pairs with the template DNA so that the new strand will be complementary to the original. The polymerase accommodates base pairs snugly—recall that all possible pairings (A:T, T:A, C:G, and G:C) have the same overall geometry (see Section 3.2). The tight fit minimizes the chance of mispairings. In fact, structural studies indicate that DNA polymerase shifts between an "open" conformation and a "closed" conformation, analogous to the fingers moving closer to the thumb, when a nucleotide substrate is bound (Fig. 20.15). This conformational change may facilitate the polymerization of a tightly bound (correctly paired) nucleotide, or it could reflect a mechanism for quickly releasing a mismatched nucleotide before catalysis occurs.

If the wrong nucleotide does become covalently linked to the growing chain, the polymerase can detect the distortion it creates in the newly generated double helix. Many DNA polymerases contain a second active site that catalyzes hydrolysis of the nucleotide at the 3′ end of the growing DNA strand. *This 3′ → 5′ exonuclease excises misincorporated nucleotides, thereby acting as a **proofreader** for DNA polymerase* (Fig. 20.16). Recall that an exonuclease removes residues from the end of a polymer; an endonuclease cleaves within the polymer. In *E. coli* DNA polymerase I, the 3′ → 5′ exonuclease active site is located about 25 Å from the polymerase active site, indicating that the enzyme–DNA complex must undergo a large conformational change in order to shift from polymerization to nucleotide hydrolysis. After the mispaired nucleotide is removed, the DNA moves back to the polymerization active site, and the polymerase can continue generating an accurately base-paired DNA product.

Mismatched nucleotides are the most common polymerization error, but insertions and deletions of nucleotides, known as **indels,** can also occur. DNA polymerase is most likely to slip while copying repetitive template sequences, which occur in all types of organisms. The proofreading exonuclease can correct these mistakes—all of which distort the newly made double helix—to limit the overall error rate of DNA polymerase to about one in 10^6 bases.

Following replication, misincorporated bases can be removed through various DNA repair mechanisms, which further reduces the error rate of replication to about one in 10^{10} nucleotides in all types of organisms. This high degree of fidelity is absolutely essential for the accurate transmission of biological information from one cell to its two daughter cells. In multicellular organisms, many cell divisions occur during the transformation of a fertilized egg into a mature organism that can produce gametes (see Box 3.A). Because errors can occur with each round of DNA replication, the overall rate of nucleotide changes increases; in humans, it is about one in 10^8 nucleotides per generation.

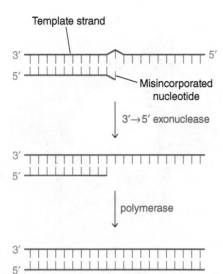

Figure 20.16 Proofreading during polymerization. DNA polymerase detects a distortion that results from the incorporation of a mismatched nucleotide. The 3′ → 5′ exonuclease activity hydrolyzes the nucleotide at the 3′ end of the new strand, and the polymerase resumes its activity.

An RNase and a ligase are required to complete the lagging strand

In bacteria as well as eukaryotes, replication forks advance until they essentially run into each other. In a circular bacterial chromosome, a set of *Ter* sequences (*Ter* for termination) are located roughly opposite the replication origin. The first replication fork that arrives at a *Ter* sequence pauses until the other replication fork meets it.

In linear eukaryotic chromosomes, two approaching replication forks merge so that each leading-strand polymerase continues adding nucleotides until it runs into the lagging strand of the other replication fork:

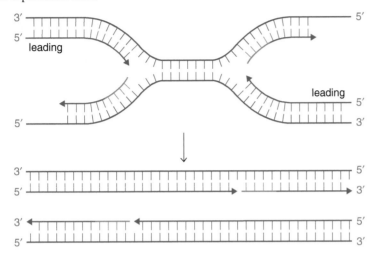

However, replication is not complete until these segments are connected.

In addition, the lagging strand, which is synthesized one Okazaki fragment at a time, must be made whole. *As the polymerase finishes each Okazaki fragment, its RNA primer, along with some of the adjoining DNA, is hydrolytically removed and replaced with DNA; then the backbone is sealed to generate a continuous DNA strand.* This process also increases the accuracy of DNA replication: Primase has low fidelity, so the RNA primers tend to contain errors, as do the first few deoxynucleotides added to the primer by the action of the DNA polymerase. For example, in humans, DNA polymerase α lacks exonuclease activity and therefore cannot proofread its work.

In many cells, an exonuclease known as RNase H (H stands for hybrid) operates in the $5' \rightarrow 3'$ direction to excise nucleotides at the primer end of an Okazaki fragment. Nucleotide hydrolysis may continue until DNA polymerase, now in the process of completing another Okazaki fragment, "catches up" with RNase H (the polymerase is faster than the exonuclease). In extending the newer Okazaki fragment, the polymerase replaces the excised ribonucleotides of the older Okazaki fragment with deoxyribonucleotides, leaving a single-strand **nick** between the two lagging-strand segments (**Fig. 20.17**).

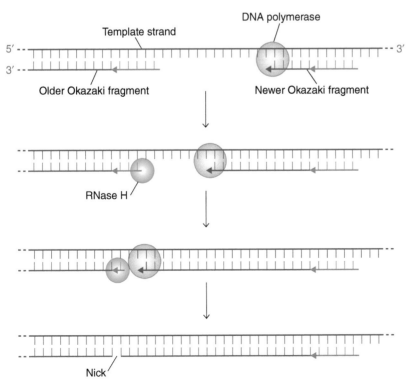

Figure 20.17 Primer excision. RNase H removes the RNA primer and some of the adjoining DNA in the older Okazaki fragment, allowing DNA polymerase to accurately replace these nucleotides. The nick can then be sealed.

The *E. coli* DNA polymerase I polypeptide actually includes a $5' \to 3'$ exonuclease activity (this is in addition to the $3' \to 5'$ proofreading exonuclease), so that, at least *in vitro*, a single protein can remove ribonucleotides from the previous Okazaki fragment as it extends the next Okazaki fragment. The combined activities of removing and replacing nucleotides have the net effect of moving the nick in the $5' \to 3'$ direction. This phenomenon is known as **nick translation.** DNA polymerase cannot seal the nick; this is the function of yet another enzyme.

The discontinuous segments of the lagging strand are joined by the action of **DNA ligase.** The reaction, which results in the formation of a phosphodiester bond, consumes the free energy of a similar bond in a nucleotide cofactor. Prokaryotes use NAD^+ for the reaction, yielding as products AMP and nicotinamide mononucleotide; eukaryotes use ATP and produce AMP and PP_i.

Eukaryotes   Adenine—Ribose—Ⓟ—Ⓟ—Ⓟ

ATP

↓

Adenine—Ribose—Ⓟ   +   Ⓟ—Ⓟ

AMP PP_i

Prokaryotes   Adenine—Ribose—Ⓟ—Ⓟ—Ribose—Nicotinamide

Nicotinamide adenine dinucleotide (NAD^+)

↓

Adenine—Ribose—Ⓟ   +   Ⓟ—Ribose—Nicotinamide

AMP   Nicotinamide mononucleotide

Ligation of Okazaki fragments yields a continuous lagging strand, completing the process of DNA replication.

Figure 20.18 is a composite, showing the major proteins that participate in DNA replication in *E. coli*. For clarity, the components of the replisome are separated. In a cell, they

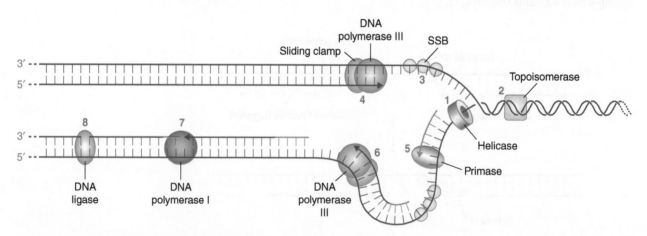

Figure 20.18 Overview of replication in *E. coli*.

1. Helicase unwinds the two strands of parental DNA.
2. Topoisomerase relieves the overwinding ahead of the replication fork.
3. SSB coats the exposed single strands.
4. One DNA polymerase III extends a primer continuously to build the leading strand. The sliding clamp increases the polymerase's processivity.
5. The lagging-strand template loops out to allow primase to synthesize a new RNA primer.
6. A second DNA polymerase III extends each RNA primer to complete an Okazaki fragment.
7. When DNA polymerase III reaches the previous Okazaki fragment, it is replaced by DNA polymerase I, which removes the RNA primer of the older Okazaki fragment and extends the newer Okazaki fragment with DNA.
8. DNA ligase seals the remaining nick between Okazaki fragments to generate a continuous lagging strand.

are close together; in fact, one subunit of DNA polymerase III associates with helicase. In addition, some proteins are not shown, such as the machinery that reloads the sliding clamp onto the lagging-strand template each time a new Okazaki fragment begins. All of the activities of building primers, extending them, replacing RNA, and sealing nicks take place in a small area. The functions of each enzyme are coordinated so that some of the steps that are separated in Figure 20.18 actually occur simultaneously. For example, Okazaki fragments are joined while the DNA is being replicated, not after termination occurs.

Concept Check

1. Make a list of all the proteins described for DNA replication in *E. coli* and in a human. Describe the function of each protein.

2. Use different colors to draw a diagram of replicating DNA, including the template strands, primers, and newly synthesized strands.

3. Describe how two DNA polymerases, operating in the 5′ → 3′ direction, replicate the two antiparallel template strands.

4. Explain why the factory model for DNA replication is superior to the locomotive model.

5. Explain why DNA polymerization must be primed by RNA.

6. How does the polymerase proofread its mistakes?

7. Explain why an RNase, a polymerase, and a ligase are required to complete lagging-strand synthesis.

8. Compare the reactions catalyzed by DNA polymerase and DNA ligase with respect to substrates, energy source, and type of bond generated.

20.3 Telomeres

KEY CONCEPTS

Explain the synthesis and purpose of telomeres.

- Explain why DNA polymerase cannot replicate the 3′ ends of chromosomes.
- Describe the function of telomerase RNA.
- Relate telomerase function to cell immortality.

Bacterial DNA replication produces two identical circular DNA molecules. Eukaryotic DNA replication yields two identical linear DNA molecules that remain attached at their **centromeres,** giving rise to the familiar X-shaped chromosome that becomes visible during cell division (see Box 3.A).

It seems reasonable to assume that eukaryotic DNA polymerases proceed to the very ends of the linear chromosomes. In fact, the leading-strand DNA polymerase can copy the 5′ end of the parental DNA template, since it extends a new complementary strand in the 5′ → 3′ direction.

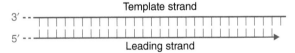

Template strand
3′ --
5′ --
Leading strand

However, replication of the extreme 3′ ends of the parental DNA strands presents a problem, for the same reason. Even if an RNA primer were paired with the 3′ end of a template strand, DNA polymerase would not be able to replace the ribonucleotides with deoxynucleotides.

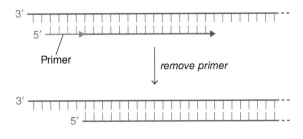

3′
5′
Primer
remove primer

3′
5′

The 3′ end of each parental (template) DNA strand would then extend past the end of each new strand and would be susceptible to nucleases. Consequently, *each round of DNA replication would lead to chromosome shortening:*

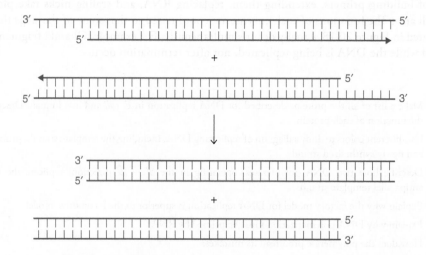

Eukaryotic cells counteract the potential loss of genetic information by actively extending the ends of chromosomes. The resulting structure, which consists of short tandem (side-by-side) repeating DNA sequences, is called a **telomere.** The proteins associated with the telomeric DNA protect the end of the chromosome from nucleolytic degradation and prevent the joining of two chromosomes through end-to-end ligation (a reaction that normally occurs to repair broken DNA molecules).

Telomerase extends chromosomes

The telomere proteins include an enzyme complex known as **telomerase,** which was first described by Elizabeth Blackburn. *Telomerase repeatedly adds a sequence of six nucleotides to the 3′ end of a DNA strand, using an enzyme-associated RNA molecule as a template* (**Fig. 20.19**). The catalytic subunit of telomerase is a **reverse transcriptase,** a homolog of a viral enzyme that copies the viral RNA genome into DNA (**Box 20.A**).

Although the telomerase catalytic core is highly conserved among eukaryotes, its RNA subunit varies from ~150 to ~2000 nucleotides in different species. In humans, the RNA is a sequence of 451 bases, including the six that serve as a template for the addition of DNA repeats with the sequence TTAGGG. Similar G-rich sequences extend the 3′ ends of chromosomes in all eukaryotes. In order to synthesize telomeric DNA, the template RNA must be repeatedly realigned with

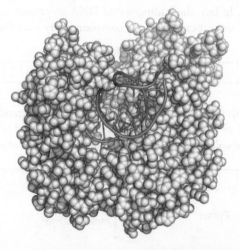

Figure 20.19 Telomerase. This model of an insect telomerase includes a segment of the RNA template (green) and telomeric DNA (gold).

Box 20.A   HIV Reverse Transcriptase

The human immunodeficiency virus (HIV, introduced in Box 7.A) is a retrovirus, a virus whose RNA genome must be copied to DNA inside the host cell. After entering a cell, the HIV particle disassembles. The 9-kb viral RNA is then transcribed into DNA by the action of the viral enzyme reverse transcriptase. Another viral enzyme, an integrase, incorporates the resulting DNA into the host genome. Expression of the viral genes produces 15 different proteins, some of which must be processed by HIV protease to achieve their mature forms. Eventually, new viral particles are assembled and bud off from the host cell, which dies. Because HIV preferentially infects cells of the immune system, cell death leads to an almost invariably fatal immunodeficiency.

HIV reverse transcriptase, which synthesizes DNA from an RNA template (a contradiction of the central dogma outlined in Section 3.3), resembles other polymerases in having fingers, thumb, and palm domains. These parts of the protein (colored red in the model shown here) comprise a polymerase active site that can use either DNA or RNA as a template (no other enzyme has this dual specificity). A separate domain (green) contains an RNase active site that degrades the RNA template.

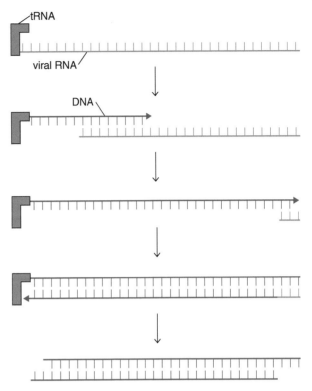

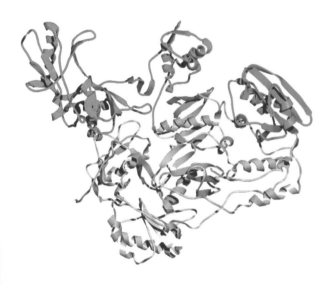

Reverse transcription occurs as follows: The enzyme binds to the RNA template and generates a complementary DNA strand. In a host cell, DNA synthesis is primed by a transfer RNA molecule. As polymerization proceeds, the RNase active site degrades the RNA strand of the RNA–DNA hybrid molecule, leaving a single DNA strand that then serves as a template for the reverse transcriptase polymerase active site to use in synthesizing a second strand of DNA. The result is a double-stranded DNA molecule (see diagram).

Viral reverse transcriptase has proved to be more than a biological curiosity. It has become a valuable laboratory tool, allowing researchers to purify messenger RNA transcripts from cells, transform them to DNA (called **cDNA** for complementary DNA), and then quantify them, sequence them, or use them to direct protein synthesis.

Reverse transcriptase activity can be blocked by two different types of drugs. Nucleoside analogs such as Zidovudine and Zalcitabine (below) readily enter cells and are phosphorylated. The resulting nucleotides bind in the reverse transcriptase active site and are linked, via their 5′ phosphate group, to the growing DNA chain. However, because they lack a 3′ OH group, further addition of nucleotides is impossible.

Reverse transcriptase can also be inhibited by non-nucleoside analogs such as nevirapine, a noncompetitive inhibitor that binds to a hydrophobic patch on the surface of reverse transcriptase near the base of the thumb domain. This does not interfere with RNA or nucleotide binding, but it does inhibit polymerase activity, probably by restricting thumb movement. HIV infections are typically treated with a "cocktail" of drugs that often includes a reverse transcriptase inhibitor along with a protease inhibitor (see Box 7.A).

Question Explain why the drugs described here interfere only minimally with nucleic acid metabolism in the human host.

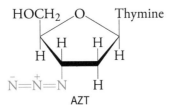

AZT
(3′-Azido-2′,3′-dideoxythymidine, Zidovudine)

ddC
(2′,3′-Dideoxycytidine, Zalcitabine)

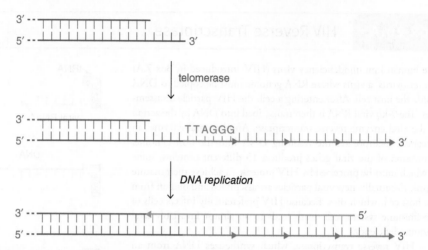

Figure 20.20 Synthesis of telomeric DNA.   Telomerase extends the 3′ end of a DNA strand by adding TTAGGG repeats (brown segments). DNA polymerase can then extend the complementary strand by the normal mechanism for lagging-strand synthesis (orange segment).

Question Explain why primase, RNase H, and DNA ligase are required to complete the lagging strand.

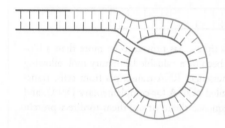

Figure 20.21 A T-loop. Telomeric DNA folds back on itself, and the G-rich single strand invades the double helix.

the end of the preceding hexanucleotide extension. The precise alignment depends on base pairing between telomeric DNA and a nontemplate portion of the telomerase RNA. When the 3′ end of a DNA strand has been extended by telomerase, it can then serve as a template for the conventional synthesis of a C-rich extension of the complementary DNA strand (**Fig. 20.20**).

In humans, telomeric DNA is 4 to 15 kb long, depending on the tissue and the age of the individual, and includes a 3′ single-strand overhang of 100–300 nucleotides. This single strand of DNA folds back on itself to form a structure called a T-loop (**Fig. 20.21**). Multiple copies of a protein complex known as shelterin bind to the telomere with a stoichiometry of one shelterin for every 100 bp. The six proteins of the shelterin complex stabilize the T-loop, prevent detection by DNA repair enzymes, and interact with telomerase. When the telomere is relatively short, shelterin promotes telomerase activity, but when the telomere is relatively long, telomerase activity is inhibited. In this way, shelterin helps regulate telomere length, which is an indicator of the cell's longevity.

Is telomerase activity linked to cell immortality?

Cells that normally undergo a limited number of cell divisions appear to contain no active telomerase. Consequently, the size of the telomeres decreases with each replication cycle, until the cells reach a senescent stage and no longer divide. For most human cells, this point is reached after about 35–50 cell divisions. Not surprisingly, short telomeres are associated with age-related diseases such as atherosclerosis and Alzheimer's disease, and telomere shortening appears to be accelerated by environmental factors and cellular stress. Telomere shortening seems to operate as a timing mechanism: When a cell perceives that its telomeres are too short, it stops dividing.

Cells that are "immortal"—such as unicellular organisms, many stem cells, and the reproductive cells of multicellular organisms—appear to have active telomerase. These findings are consistent with the role of telomerase in maintaining the ends of chromosomes over many rounds of replication. Telomerase also appears to be active in some cancer cells derived from tissues that are normally senescent (no longer dividing). This is the basis for the use of telomerase inhibitors as anti-cancer drugs (**Box 20.B**).

Paradoxically, many cancer cells seem to have relatively short telomeres. Tumors are heterogeneous, so "older" tumor cells might exhibit short telomeres, while more "stemlike" tumor cells retain long telomeres. Another possibility is that shorter telomeres are more accessible to DNA-repair enzymes, so that the chromosome ends may be trimmed or joined together, leading to further genetic changes that are a feature of cancer cells (Section 20.5).

| Box 20.B | Telomeres in Cancer |

Organisms cannot escape from cellular senescence and their intrinsic mortality because every cell has a limited capacity to divide. Leonard Hayflick studied cell division in culture and noticed that the number of divisions that a cell can undergo is a feature of the origin of the cell. This limited capacity of cell division is known as the "Hayflick limit." It means every cell has an intrinsic countdown clock that limits its cell division and protects multicellular organisms from cancer. Important features of cancerous cells are that they are immortal and undergo uncontrolled **proliferation**. The discovery of telomeres and their role in replication helps us understand the involvement of active telomerase in malignant transformation. The telomere length in young people is longer than that of older people. These observations support the idea that telomeres are associated with the process of aging. Also, cells having a high proliferative capacity such as stem cells or cancer cells have high telomerase activity as compared to normal cells that have limited telomerase activity. Many experiments support the idea of the involvement of high telomerase activity in the malignant transformation of cells, where expression of telomerase in normal cells results in their conversion to immortal cells. These observations suggest that telomerase activity may represent a novel approach to control the growth capacity of uncontrolled proliferative cells or have lost normal growth control. The involvement of telomerase in cellular senescence, and cell division has led to the development of many chemotherapeutic agents that can act as a telomerase inhibitor.

Question **Why is telomerase an attractive target in cancer treatment?**

Concept Check

1. Explain why DNA replication leads to chromosome shortening.
2. Describe the structure of telomeric DNA.
3. Summarize the function of the RNA component of telomerase.
4. Explain the relationship between cell immortality and telomerase activity.

20.4 DNA Damage and Repair

KEY CONCEPTS

Describe the enzymes and other proteins that repair damaged DNA.

- List the causes of DNA damage.
- Summarize the steps involved in direct repair, base excision repair, and nucleotide excision repair.
- Distinguish end-joining and recombination repair.
- Explain why DNA repair systems may introduce rather than prevent mutations.

An alteration in a cell's DNA, whether from a polymerization error or another cause, becomes permanent—that is, a **mutation**—unless it is repaired. In a unicellular organism, altered DNA is replicated and passed on to the daughter cells when the cell divides. In a multicellular organism, a mutation is passed to offspring only if the altered DNA is present in the reproductive cells. Mutations that arise in other types of cells affect only the progeny of those cells within the parent organism.

A genetic change can affect the expression of genes positively or negatively (or it may have no measurable effect on the organism). The accumulation of genetic changes over an individual's lifetime may contribute to the gradual loss of functionality associated with aging. In this section, we examine different types of DNA damage and the mechanisms cells use to restore DNA integrity.

DNA damage is unavoidable

DNA damage is a fact of life. Even with its proofreading activity, DNA polymerase makes mistakes by introducing mismatched nucleotides, skipping nucleotides, adding extra nucleotides, or incorporating uridine rather than thymidine. Nonreplicating DNA can also be altered.

Cellular metabolism itself exposes DNA to the damaging effects of reactive oxygen species (for example, the superoxide anion $\cdot O_2^-$, the hydroxyl radical $\cdot OH$, or H_2O_2) that are normal by-products of oxidative metabolism. Over 100 different oxidative modifications of DNA have been catalogued. For example, guanine is commonly oxidized to 8-oxoguanine (oxoG):

Guanine   8-Oxoguanine
                              (oxoG)

When the modified DNA strand is replicated, the oxoG can base pair with either an incoming C or A. Ultimately, the original G:C base pair can become a T:A base pair. A nucleotide substitution such as this is called a **point mutation.** Changing a purine (or pyrimidine) to another purine (or pyrimidine) is known as a **transition mutation;** a **transversion** occurs when a purine replaces a pyrimidine or vice versa.

Other nonenzymatic reactions disintegrate DNA under physiological conditions. For example, hydrolysis of the *N*-glycosidic bond linking a base to a deoxyribose group yields an **abasic site** (also called an apurinic or apyrimidinic or AP site).

Deamination reactions can alter the identities of bases. This is particularly dangerous in the case of cytosine, since deamination (actually an oxidative deamination) yields uracil:

Cytosine   Uracil

Recall that uracil has the same base-pairing propensities as thymine, so the original C:G base pair could give rise to a T:A base pair after DNA replication. Since DNA has evolved to contain thymine rather than uracil, a uracil base resulting from cytosine deamination can be recognized and corrected before the change becomes permanent.

In addition to the cellular factors described above, *environmental agents such as ultraviolet light, ionizing radiation, and certain chemicals can physically damage DNA.* For example, ultraviolet (UV) light induces covalent linkages between adjacent thymine bases:

DNA backbone with   Thymine dimer
thymine residues

This brings the bases closer together, which distorts the helical structure of DNA. Thymine dimers can thereby interfere with normal replication and transcription.

Ionizing radiation also damages DNA either through its direct action on the DNA molecule or indirectly by inducing the formation of free radicals, particularly the hydroxyl radical, in the surrounding medium. This can lead to strand breakage. Many thousands of chemicals, both natural and manufactured, can potentially react with the DNA molecule and cause mutations. Such compounds are known as **mutagens,** or as **carcinogens** if the mutations lead to cancer.

The unavoidable nature of many DNA lesions has driven the evolution of mechanisms to detect and remedy errors. A cell containing a point mutation or a small insertion or deletion may suffer no ill effects, particularly if the mutation is in a part of the genome that does not contain an essential gene. However, more serious lesions, such as single- or double-strand breaks, usually bring replication or translation to a halt, and the cell must deal with these show-stoppers immediately.

Repair enzymes restore some types of damaged DNA

In a few cases, repair of damaged DNA is a simple process involving one enzyme. For example, in bacteria and some other organisms (but not mammals), UV-induced thymine dimers can be restored to their monomeric form by the action of a light-activated enzyme called DNA photolyase.

Mammals can reverse other simple forms of DNA damage, such as the methylation of a guanine residue, which yields O^6-methylguanine (this modified base can pair with either cytosine or thymine):

$$O^6\text{-Methylguanine}$$

A methyltransferase removes the offending methyl group, transferring it to one of its cysteine residues. This permanently inactivates the protein. Apparently, the expense of sacrificing the methyltransferase is justified by the highly mutagenic nature of O^6-methylguanine.

In bacteria as well as eukaryotes, *nucleotide mispairings are corrected shortly after DNA replication by a **mismatch repair** system.* A protein, called MutS in bacteria, monitors newly synthesized DNA and binds to the mispair. Although it binds only 20 times more tightly to the mispair than to a normal base pair, MutS undergoes a conformational change and causes the DNA to bend (**Fig. 20.22**). These changes apparently induce an endonuclease to cleave the strand with the incorrect base at a site as far as 1000 bases away. A third protein then unwinds the helix so that the defective segment of DNA can be destroyed and replaced with accurately paired nucleotides by DNA polymerase. How does the endonuclease know which strand contains the incorrect base? Cellular DNA is normally methylated (Section 20.1), and the endonuclease can select the newly synthesized strand because it has not yet undergone methylation. The mismatch repair system lowers the error rate of DNA replication about 1000-fold.

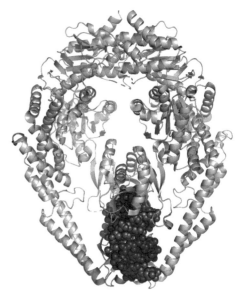

Figure 20.22 The mismatch repair protein MutS bound to DNA.   Two subunits of MutS protein (gray) encircle the DNA (blue and purple) at the site of a mispaired nucleotide.

Base excision repair corrects the most frequent DNA lesions

*Modified bases that cannot be directly repaired can be removed and replaced in a process known as **base excision repair.*** This pathway begins with a glycosylase, which removes the damaged base. An endonuclease then cleaves the backbone, and the gap is filled in by DNA polymerase (**Fig. 20.23**).

The structures and mechanisms of several DNA glycosylases have been described in detail. For example, there is a glycosylase that recognizes oxoG. Uracil-DNA glycosylase recognizes and removes the uracil bases that are mistakenly incorporated into DNA during replication or that result from cytosine deamination. When a glycosylase binds DNA, the damaged base typically flips out from the helix so that it can bind in a cavity on the protein surface, and various protein side chains take the place of the damaged base and form hydrogen bonds with its complement on the other DNA strand (**Fig. 20.24**).

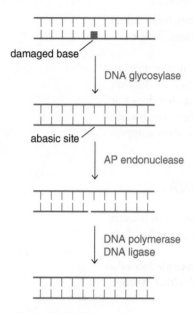

Figure 20.23 Base excision repair. The phosphodiesterase activity of DNA polymerase removes the ribose–phosphate group before replacing it with a nucleotide.

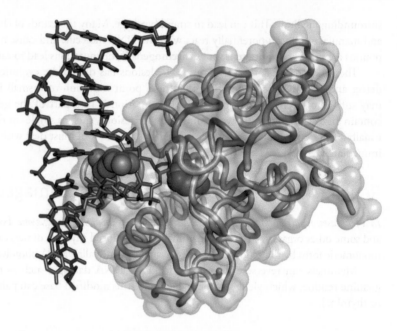

Figure 20.24 Uracil-DNA glycosylase bound to DNA. The enzyme is shown in gray, and the DNA substrate is shown with blue and purple strands. The flipped-out uracil (whose glycosidic bond has already been hydrolyzed) is shown in red. An Arg side chain that takes the place of the uracil is shown in orange.

After excising the offending base, DNA glycosylases appear to remain bound to the DNA at the abasic site, possibly to help recruit the next enzyme of the repair pathway. This enzyme, usually called the AP endonuclease, nicks the DNA backbone at the 5′ side of the abasic ribose. The nuclease inserts two protein loops into the major and minor grooves of the DNA and bends the DNA by about 35° to expose the abasic site. A backbone structure with a base attached cannot enter the active-site pocket. During the hydrolysis reaction, an Mg^{2+} ion in the active site stabilizes the anionic leaving group (**Fig. 20.25**).

In the next step of base excision repair, a DNA polymerase (such as DNA polymerase β in eukaryotes) binds to the nicked DNA, removes the deoxyribose-phosphate group, and then fills in the one-nucleotide gap. In some cases, the DNA polymerase may push aside and replace as many as 10 nucleotides on the damaged strand. The displaced single strand can then be cleaved off by an endonuclease. Finally, the remaining nick is sealed by DNA ligase.

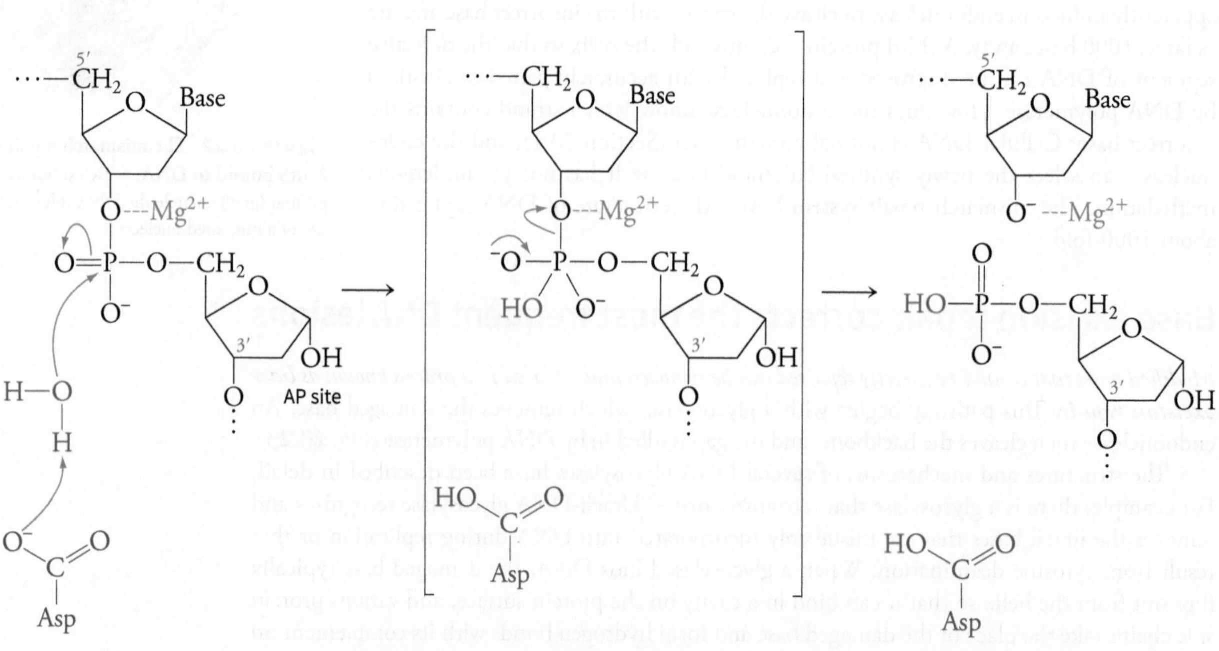

Figure 20.25 The AP endonuclease reaction.

Nucleotide excision repair targets the second most common form of DNA damage

Nucleotide excision repair, as its name suggests, is similar to base excision repair but mainly targets DNA damage resulting from insults such as ultraviolet light or oxidation. In nucleotide excision repair, a segment containing the damaged nucleotide and about 30 of its neighbors is removed, and the resulting gap is filled in by a DNA polymerase that uses the intact complementary strand as a template (**Fig. 20.26**). Many of the 30 or so proteins that are involved in this pathway in humans have been identified through mutations that are manifest as two genetic diseases.

The rare hereditary disease Cockayne syndrome is characterized by neural underdevelopment, failure to grow, and sensitivity to sunlight. It results from a mutation in any of several genes that encode proteins participating in a pathway for recognizing an RNA polymerase that has stalled in the act of transcribing a gene into messenger RNA. Stalling occurs when the DNA template is damaged and distorted so that it blocks the progress of the RNA polymerase. The polymerase must be removed so that the damage can be addressed by the nucleotide excision repair system. Defects in the Cockayne syndrome genes prevent the cell from recognizing and removing the stalled RNA polymerase. Consequently, *the DNA never has a chance to be repaired and the cell dies.* The death of transcriptionally active cells may account for the developmental symptoms of Cockayne syndrome.

Like Cockayne syndrome, the disease xeroderma pigmentosum is characterized by high sensitivity to sunlight, but individuals with xeroderma pigmentosum are about 1000 times more likely to develop skin cancer and do not suffer from developmental problems. Xeroderma pigmentosum is caused by a mutation in one of the genes that participate directly in nucleotide excision repair. Whereas Cockayne syndrome gene products appear to detect DNA damage that prevents transcription, the xeroderma pigmentosum proteins are responsible for repairing the damage. The failure to repair UV-induced lesions explains the high incidence of skin cancer.

When the cell attempts to replicate damaged DNA that has not been repaired, it may rely on one of its nonstandard DNA polymerases. For example, eukaryotic DNA polymerase η can bypass DNA lesions such as UV-induced thymine dimers by incorporating two adenine bases in the new strand. Although it is useful as a translesion polymerase, DNA polymerase η is relatively inaccurate and has no proofreading exonuclease activity. It inserts an incorrect base on average every 30 nucleotides. This may not be problematic, as the errors can be detected and corrected by the mismatch repair system described above.

The existence of error-prone polymerases provides a fail-safe mechanism for replicating stretches of DNA that cannot be navigated by the standard replication machinery. In fact, synthesis of these alternative polymerases increases when bacterial cells experience DNA damage. Evidently, the possibility of introducing small errors during replication is acceptable when the only other option is cell death.

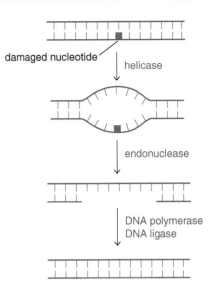

Figure 20.26 Nucleotide excision repair.

Question **Compare this process to base excision repair (Fig. 20.23).**

Double-strand breaks can be repaired by joining the ends

Segments of a DNA double helix that have been completely severed by the effects of radiation or free radicals can be rejoined by **nonhomologous end-joining,** a process that does not require the presence of a homologous DNA molecule. In mammals, nearly all double-strand breaks are repaired by nonhomologous end-joining.

The first step in this repair pathway is the recognition of the broken DNA ends by a dimeric protein called Ku (**Fig. 20.27**). When Ku binds the cut DNA, it undergoes a conformational change so that it can recruit a nuclease, which trims up to 10 residues from the ends of the DNA molecule. The protein–DNA complex may then be joined by a DNA polymerase such as polymerase μ, which can extend the ends of the DNA either with or without a template. Template-independent polymerization, along with a tendency for polymerase μ to slip, means that the break site may end up with additional nucleotides that were not

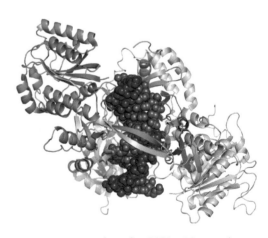

Figure 20.27 Ku bound to DNA. The two subunits of the Ku heterodimer are shown in light and dark green, and the DNA strands are shown in blue and purple.

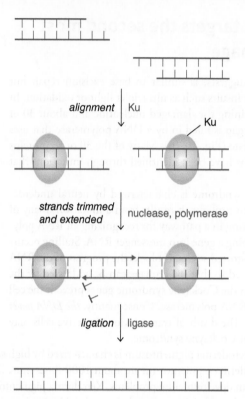

Figure 20.28 Nonhomologous end-joining. Ku recognizes the ends of broken DNA and aligns them. The activities of a nuclease, polymerase, and ligase generate an unbroken DNA molecule that may differ in sequence from the original.

present in the DNA before it broke. A DNA ligase finishes the repair job by joining the two backbones of the DNA segment (**Fig. 20.28**).

Two Ku–DNA complexes associate with each other so that, ideally, the proper halves of a broken DNA molecule can be stitched back together. However, nonhomologous end-joining is also responsible for combining DNA segments that do not belong together, leading to chromosomal rearrangements. In addition, the Ku–DNA complex can interact with the nuclease, polymerase, and ligase in any order, so a DNA break could potentially be repaired in many different ways, with or without the addition and removal of nucleotides. Consequently, *nonhomologous end-joining is inherently mutagenic,* but, as in other forms of DNA repair, this may be a small price to pay for restoring a continuous double-stranded DNA. The imperfect nature of repair by end-joining is the basis for editing genes using the CRISPR system (**Box 20.C**). A sloppy DNA repair process is also a key part of the mechanism for generating additional sequence diversity when developing B lymphocytes assemble DNA segments to build an immunoglobulin gene (Section 5.5).

Recombination also restores broken DNA molecules

In some organisms, double-strand breaks can be repaired through **recombination,** a process that also occurs in the absence of DNA damage as a mechanism for shuffling genes between homologous chromosomes during meiosis. Recombination repair of a broken chromosome can occur at any time in a diploid organism (which has two sets of homologous chromosomes) but can occur only after DNA replication in an organism with only one chromosome. Recombination requires another intact homologous double-stranded molecule as well as nucleases, polymerases, ligases, and other proteins (**Fig. 20.29**).

In recombination repair, a single strand from the damaged DNA molecule changes places with a homologous strand in another DNA molecule. In order for a single strand of DNA to "invade" a double-stranded DNA molecule (step 2 in Fig. 20.29), the single strand must first be coated with an ATP-binding protein, called RecA in *E. coli* and Rad51 in humans. RecA binds to a DNA strand in a cooperative fashion, beginning at the break point. This binding unwinds and stretches the DNA by about 50%, but not uniformly: Sets of three nucleotides retain a near-standard conformation (with about 3.4 Å between bases) but are separated from the next

Box 20.C   Gene Editing with CRISPR

DNA is constantly and randomly changing as a result of natural processes, but alterations can be introduced in a targeted manner in the laboratory. One of the most efficient techniques for modifying genes takes advantage of a bacterial system that evolved to destroy bacteriophages. A bacterial cell carries traces of its ancestors' encounters with various bacteriophages in the form of multiple short segments of phage DNA that have been inserted into the bacterial chromosome among a set of **clustered regularly interspersed short palindromic repeats**, or **CRISPRs.**

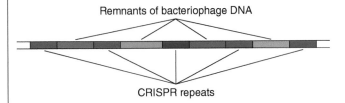

Remnants of bacteriophage DNA

CRISPR repeats

When this DNA is transcribed, the resulting RNA is cleaved into short segments, each corresponding to about 30 nucleotides of phage DNA. These RNAs then bind to a nuclease called Cas9 (for *CRISPR-associated*) and guide it to complementary DNA sequences, which presumably represent bacteriophages trying to infect the cell. The Cas9 nuclease cleaves both strands of this DNA, destroying the incoming bacteriophage.

Without the CRISPR RNA guide, Cas9 cannot cleave DNA, but *it will cleave whatever DNA is complementary to the guide RNA*. This creates the ability to use the **CRISPR-Cas9 system** as a gene-editing tool. Using recombinant DNA technology, an engineered guide RNA "gene" and the Cas9 gene are introduced into a host cell. The cell synthesizes the encoded guide RNA and the Cas9 protein. The RNA, containing a 22–23-bp segment that is complementary to a target gene, positions Cas9 to cleave both strands of the DNA at the points marked by arrowheads.

Cells have various mechanisms for repairing broken DNA. For example, nonhomologous end-joining (left side of diagram) knits the severed pieces back together—but imperfectly, so the repair does not restore the original DNA sequence. When this occurs in the middle of a gene, the result is almost always an unreadable gene. The CRISPR-Cas9 system can therefore "knock out" a specific gene in an organism.

To "knock in" or correct a defective gene, researchers deploy the CRISPR-Cas9 system to cleave the target gene near the defect, but they also introduce the correct segment of that same gene. The cell then undertakes a more elaborate recombination repair process to rebuild the cleaved DNA using the intact—but altered—sequence as a blueprint (right side of diagram). The result is a functional gene containing the correct DNA sequence.

Because the CRISPR-Cas9 gene-editing system can permanently disable a defective gene or replace a defective gene with a normal copy at the editing site, it has become the preferred approach for genetically modifying organisms. For example, CRISPR-Cas9 technology can generate animal models for human genetic diseases. It is also an option for gene therapy, which introduces a functional gene into an individual in order to compensate for or correct a malfunctioning gene (Section 3.3).

The CRISPR-Cas9 method shares some of the limitations of traditional gene therapy, such as the need for a suitable vector that can deliver its payload to the intended tissues. In addition, Cas9 sometimes cleaves DNA sequences that are not perfectly complementary to its guide RNA. Such "off target" gene inactivation could have disastrous unintended consequences. A number of variations that potentially make CRISPR technology safer have been developed. For example, engineered variants of the Cas9 enzyme can promote the deamination of C to U, which, when repaired, gives rise to a T:A base pair where the DNA originally had a C:G base pair. This base-editing process avoids the potentially dangerous cleavage of both DNA strands. Another technique combines Cas9 with reverse transcriptase and an RNA template. The reverse transcriptase uses the RNA template to synthesize a correct DNA strand, which displaces the defective DNA strand. The cell's DNA-repair enzymes finish the repair, and the result is a modified gene—accomplished without the risky two-strand cuts. Some CRISPR technologies affect the target gene's transcription without any cutting at all.

As with other developments in genetic engineering, researchers and clinicians must carefully weigh the ethical implications of editing genes. Because CRISPR techniques make permanent genetic changes, the research community recognizes the need to proceed cautiously. A cell can transmit its modified DNA to its daughter cells when the cell divides, so it is prudent to limit gene alterations to somatic (body) cells and avoid altering the DNA in reproductive cells that pass DNA to the next generation.

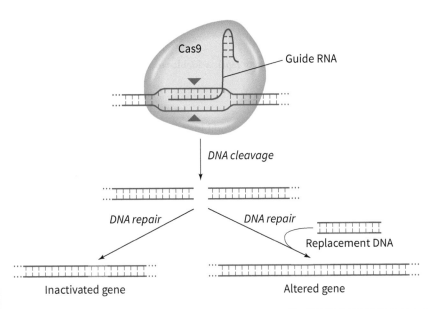

Cas9

Guide RNA

DNA cleavage

DNA repair   *DNA repair*

Replacement DNA

Inactivated gene   Altered gene

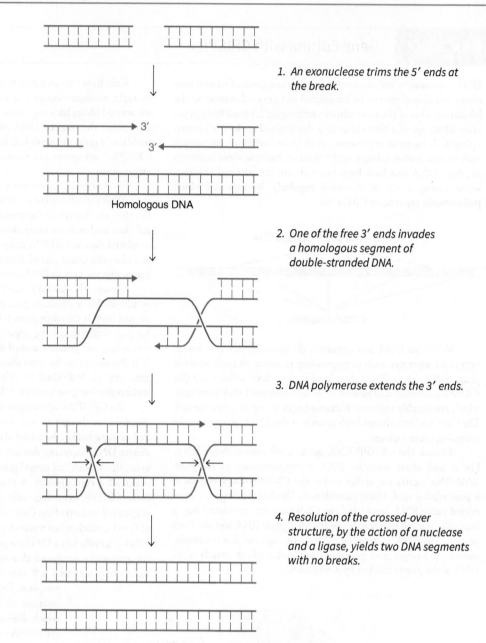

1. An exonuclease trims the 5' ends at the break.

Homologous DNA

2. One of the free 3' ends invades a homologous segment of double-stranded DNA.

3. DNA polymerase extends the 3' ends.

4. Resolution of the crossed-over structure, by the action of a nuclease and a ligase, yields two DNA segments with no breaks.

Figure 20.29 Homologous recombination to repair a double-strand break.

Figure 20.30 Conformation of DNA bound to RecA. The single strand of DNA is shown in space-filling form with atoms color coded: C gray, N blue, O red, and P orange. Brackets indicate sets of three nucleotides that retain a near–B-DNA conformation. The DNA backbone stretches in between these triplets so that the strand extends to about 1.5 times its original length. The RecA protein subunits are not shown.

triplet by about 7.8 Å (**Fig. 20.30**). During recombination, the RecA–DNA filament aligns with double-stranded DNA containing a complementary strand. The elongated structure of the RecA filament may induce a similar shape change in the double-stranded DNA so that base pairs in the double-stranded DNA are disrupted and bases become unstacked. These changes would facilitate the strand swapping that occurs during recombination. At this point, hydrolysis of the ATP bound to RecA allows the protein to release the displaced single strand and a new double-stranded DNA. When the CRISPR-Cas9 system (Box 20.C) is used to edit rather than inactivate a gene, a DNA segment with the desired gene sequence is included so that it can serve as a template for recombination repair.

The mechanism of recombination is highly conserved among prokaryotes and eukaryotes, which argues for its essential role in maintaining the integrity of DNA as a vehicle of genetic information. The proteins that carry out recombination appear to function **constitutively** (that is, the genes are always expressed), which is consistent with proposals that DNA strand breaks are unavoidable. However, *many DNA repair mechanisms are induced only when the relevant form* *of DNA damage is detected.* This makes sense, since the repair enzymes might otherwise interfere with normal replication. In fact, activation of the repair pathways usually halts DNA synthesis, an advantage when error-prone DNA polymerases—which would be a liability in normal replication—are active.

20.5 Clinical Connection: Cancer as a Genetic Disease

KEY CONCEPTS

Describe the molecular mechanisms that can lead to cancer.

- Explain why several events are necessary for carcinogenesis.
- Relate faulty DNA repair to cancer.

In eukaryotes, cells with badly damaged DNA tend to be so impaired that they undergo **apoptosis,** or programmed cell death, making room for their replacement by healthy cells. But in some cases, cells with damaged DNA do not die but instead escape the normal growth-control mechanisms and proliferate excessively, resulting in cancer. Some of the metabolic capabilities of cancer cells are outlined in Section 19.4.

Cancer is one of the most common diseases, affecting one in three people and killing one in four. The clinical picture of cancer varies tremendously, depending on the tissue affected, but all cancer cells share an ability to proliferate uncontrollably and ignore the usual signals to differentiate or undergo apoptosis. Eventually, cancer may kill by invading and eroding the surrounding normal tissue (**Fig. 20.31**).

Three factors contribute to the development of cancer: inherited genetic variations, environmental agents, and random mutations. Only a few percent of cancers are classified as hereditary, but there are over 20 different kinds of these diseases, and they shed considerable light on specific molecular mechanisms of **carcinogenesis** (cancer development). The linkage between environmental factors and cancer is supported by epidemiological studies, which have shown, for example, that sunlight increases the risk of cutaneous melanoma (skin cancer) and smoking and asbestos exposure promote the development of lung cancer. Viral infections contribute to certain types of cancer, for example, liver cancer from hepatitis B virus and cervical cancer from human papillomaviruses. Chronic bacterial infections may also lead to cancer.

The analysis of tumor genomes has confirmed the importance of genetic alterations. For example, melanomas and lung tumors contain an average of about 200 mutations, consistent with the known role of DNA-damaging events in these cancers (ultraviolet light and smoking). The link between cancer and DNA changes means that cancer can be considered to be a disease of the genes. Furthermore, replication errors occur every time a cell divides, so mutations—and tumors—tend to occur more frequently in tissues where stem cells (the cells that give rise to other cells) divide more frequently.

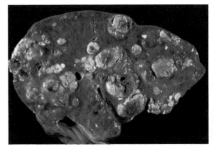

Figure 20.31 Tumors in human liver.
The white bulges are masses of cancerous cells.

Tumor growth depends on multiple events

Because cell growth and division are tightly regulated, a single genetic change is unlikely to send a cell into a pattern of uncontrolled proliferation that leads to cancer. The ability of a cell to grow and divide depends on numerous factors, including the balance between growth-promoting signals and apoptotic signals, the state of the telomeres, contacts with neighboring cells, and the delivery of oxygen and nutrients to support expansion. In general, *several regulatory pathways must be over-ridden by separate genetic events.* This is known as the multiple-hit hypothesis for carcinogenesis.

In this scenario, a tumor gradually evolves as a harmless cell becomes a small clump of cells that continue to multiply until they become a mass of cells (a tumor) that may be able to spread to other tissues as a malignant tumor. About 90% of cancers form solid tumors; the others (for example, leukemia) are dispersed. Hundreds of genes are known to be associated with cancer, yet a typical cancer cell contains mutations ("hits") in only four or five of these genes. Each of these genetic changes contributes only a small growth advantage, but over time, following the principle of natural selection, the mass of altered cells can become large. This explains why most cancers develop later in life: it usually takes years for the mutations to accumulate and for the cancerous cells to overcome growth-limiting checkpoints.

Aside from the critical mutations that drive cancer development, additional mutations occur that may have no bearing on a tumor's growth. A current challenge in cancer research is to distinguish the handful of "driver" mutations from the hundreds or even thousands of "passenger" mutations. Databases, such as the one sponsored by the National Cancer Institute (portal.gdc.cancer.gov), allow researchers to collect and share genetic information for different types of cancers, which could shed more light on the early development of cancer. Such information could also lead to more effective interventions. Most cancer treatments address the late stages of the disease, when tumors are already too large or too dispersed to be efficiently eliminated.

We have already seen that oncogenic mutations in growth-signaling pathways can produce constitutively active receptors and kinases so that cells grow and divide even in the absence of a growth signal (see Box 10.C). Inactivating genetic events also contribute to cancer. For example, the childhood cancer known as retinoblastoma, which is characterized by retinal tumors, is linked to mutations in a gene known as a **tumor suppressor gene.** Children who inherit a defective copy of the tumor suppressor gene are at higher risk of developing cancer. In theory, any combination of changes in oncogenes or tumor suppressor genes could lead to a tumor.

In most cases, there is no defined sequence of events for transforming a normal cell to a cancer cell. In addition, cancer is not a single disease: No two individuals harbor the same mutations. Even within a single tumor, there may be considerable genetic variation among its billions of cells. Most surprisingly, normal cells show similar behavior. They appear to accumulate genetic changes—passenger mutations as well as cancer-driving mutations—but do not become cancerous. This observation suggests that cells may actually tolerate a certain level of mutation in order to avoid the high cost of repairing damaged DNA.

DNA repair pathways are closely linked to cancer

Genetic changes can take the form of point mutations, large or small deletions, chromosomal rearrangements, or inappropriate methylation that leads to gene silencing (Section 21.1). In many tumors, the state of the DNA goes from bad to worse, reflecting failure in the mechanisms that detect damaged DNA and promote its restoration. These steps require a host of proteins, including kinases and other intracellular signaling components that respond rapidly (usually within minutes) to DNA damage.

One key player in the damage response pathway is the protein kinase known as ATM, which is defective in ataxia telangiectasia. This disease is characterized by neurodegeneration, premature aging, and a propensity to develop cancer. ATM is a large protein (350,000 D) that includes a DNA-binding domain. It is activated in response to DNA damage, such as double-strand breaks. Among the substrates for ATM's kinase activity are proteins involved in initiating cell division, the protein BRCA1 (whose gene is commonly mutated in breast cancer), and the tumor suppressor known as p53.

Breast cancer strikes about one in nine women in developed countries. About 10% of breast cancers have a familial form, and about half of these exhibit mutations in the *BRCA1* or *BRCA2* genes. A woman bearing one of these mutated genes has a 70% chance of developing breast cancer. BRCA1 and BRCA2 are large multidomain proteins that function in part as scaffolding proteins to link proteins that detect DNA damage to proteins that can repair the damage or halt the cell cycle. For example, BRCA2 binds to Rad51, a protein necessary for the recombination repair pathway. The loss of BRCA1 or BRCA2 increases the likelihood that a cell would attempt to divide without first repairing damaged DNA.

The tumor suppressor gene p53 is found to be mutated in at least half of all human tumors. The level of p53 in the cell is controlled by its rate of degradation (it is ubiquitinated and targeted to a proteasome for destruction; Section 12.1). The concentration of p53 increases when its degradation is slowed. This can occur by a decrease in the rate at which ubiquitin is attached to it or when it is phosphorylated by the action of a kinase such as ATM. Thus, *DNA damage, which activates ATM, leads to an increase in the cellular concentration of p53* (**Fig. 20.32**).

Other modifications, such as acetylation and glycosylation, also increase the activity of p53, which responds not just to DNA damage but also to other forms of cellular stress such as low oxygen and elevated temperatures. Covalent modification of p53 triggers a conformational change that allows the protein to bind to specific DNA sequences in order to promote the transcription of several dozen different genes. p53 binds as a tetramer (a dimer of dimers) to DNA, encircling it (**Fig. 20.33**). Portions of p53 are intrinsically disordered (Section 4.3), which presumably facilitates its abilty to interact with multiple partners.

p53 stimulates production of a protein that blocks the cell's progress toward cell division. This regulatory mechanism buys time for the cell to repair DNA using enzymes whose synthesis is also stimulated by p53. Moreover, activated p53 turns on the gene for a subunit of ribonucleotide reductase (see Section 18.5), which promotes synthesis of the deoxynucleotides required for DNA repair.

Some of p53's other target genes encode proteins that carry out apoptosis, a multistep process in which the cell's contents are apportioned into membrane-bounded vesicles that are subsequently engulfed by macrophages. For multicellular organisms, death by apoptosis is often a better option than allowing a malfunctioning cell to persist.

It is possible that p53's ultimate effects are dose-dependent, with lower doses (reflecting mild DNA damage) pausing the cell cycle and higher doses (reflecting severe, irreparable DNA damage) leading to cell death. *The position of p53 at the interface of pathways related to DNA repair, cell cycle control, and apoptosis indicates why the loss of the p53 gene is so strongly associated with the development of cancer.*

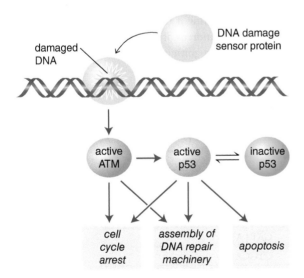

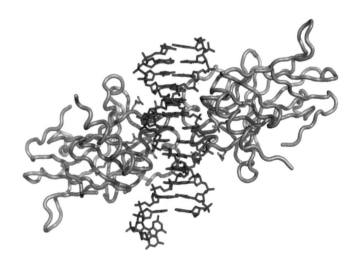

Figure 20.32 Overview of p53 function. DNA damage (via ATM) and other factors regulate the level of p53 activity, which in turn affects the cell's ability to complete the cell cycle, repair damaged DNA, or undergo apoptosis.

Figure 20.33 p53–DNA complex. This model shows a p53 dimer (gray) bound to DNA (blue). The six residues that are most commonly mutated in p53 from cancer cells are shown in red. These residues either interact directly with the DNA or are involved in stabilizing p53 structure.

Concept Check

1. Explain how mutations, environmental factors, and random events can contribute to the development of cancer.
2. Summarize the multiple hit hypothesis for carcinogenesis.
3. Describe how defects in the genes for BRCA1 or p53 can lead to cancer.

20.6 Tools and Techniques: Manipulating DNA

KEY CONCEPTS

Describe how researchers manipulate and sequence DNA.

- Summarize the roles of restriction enzymes, DNA ligase, and vectors in generating recombinant DNA molecules.
- Explain how researchers clone a gene using a plasmid.
- Describe how DNA polymerase makes PCR and DNA sequencing possible.
- Summarize the steps involved in sequencing DNA.

Molecular biologists have devised clever procedures for manipulating DNA in the laboratory and in living organisms. *Many of the techniques take advantage of naturally occurring enzymes that cut, copy, and link nucleic acids.* These techniques also exploit the ability of nucleic acids to interact with complementary molecules. In this section we focus on current methods for generating recombinant DNA, amplifying specific DNA segments, and determining the sequence of nucleotides in DNA.

Cutting and pasting generates recombinant DNA

Long DNA molecules tend to break from mechanical stress during handling, but researchers can use bacterial **restriction endonucleases** (or **restriction enzymes**) to cut DNA in well-defined ways. Restriction enzymes catalyze the hydrolysis of phosphodiester bonds at specific nucleotide sequences, thereby destroying foreign DNA that enters the cell, such as bacteriophage (viral) DNA. In this way, the bacterial cell "restricts" the growth of the phage. The bacterial cell protects its own DNA from endonucleolytic digestion by methylating it (adding a —CH_3 group) at the same sites recognized by its restriction endonucleases. Hundreds of these enzymes have been discovered; some are listed in **Table 20.1** along with their recognition sequences and cleavage sites.

Table 20.1 Recognition and Cleavage Sites of Some Restriction Endonucleases

Enzyme	Recognition/cleavage site [a]
AluI	AG | CT
MspI	C | CGG
AsuI	G | GNCC [b]
EcoRI	G | AATTC
EcoRV	GAT | ATC
PstI	CTGCA | G
SauI	CC | TNAGG
NotI	GC | GGCCGC

[a] The sequence of one of the two DNA strands is shown. The vertical bar indicates the cleavage site.

[b] N represents any nucleotide.

Restriction enzymes typically recognize a four- to eight-base sequence that is identical, when read in the same $5' \rightarrow 3'$ direction, on both strands. DNA with this form of symmetry is said to be **palindromic** (words such as *madam* and *noon* are palindromes). One restriction enzyme isolated from *E. coli* is known as EcoRI (the first three letters are derived from the genus and species names). It makes two cuts (shown by the red arrows) in its 6-bp recognition sequence:

$$5' - GAATTC - 3'$$
$$3' - CTTAAG - 5'$$
$$\longrightarrow$$
$$-G \qquad AATTC-$$
$$-CTTAA \qquad G-$$

Because the EcoRI cleavage sites are symmetrical but staggered, the enzyme generates DNA fragments with single-stranded extensions known as **sticky ends.** In contrast, the *E. coli* restriction enzyme EcoRV cleaves both strands of DNA at the center of its recognition sequence so that the resulting DNA fragments have **blunt ends:**

$$5' - GATATC - 3'$$
$$3' - CTATAG - 5'$$
$$\longrightarrow$$
$$-GAT \qquad ATC-$$
$$-CTA \qquad TAG-$$

Restriction enzymes have many uses in the laboratory. For example, they are indispensable for reproducibly breaking large pieces of DNA into smaller pieces of manageable size. **Restriction digests** of well-characterized DNA molecules, such as the 48,502-bp *E. coli* bacteriophage λ, yield fragments of predictable size that can be separated by electrophoresis, a procedure in which molecules move through a gel-like matrix under the influence of an electric field (Section 4.6). Because all the DNA segments have a uniform charge density, they are separated on the basis of their size (the smallest molecules move fastest; **Fig. 20.34**).

A segment of DNA cut by a restriction enzyme or obtained by a method such as chemical synthesis or PCR (see below) can be joined to another DNA molecule. When different samples of DNA are digested with the same sticky end–generating restriction endonuclease, all the fragments have identical sticky ends. If the fragments are mixed together, the sticky ends can find their complements and re-form base pairs. The discontinuities in the sugar–phosphate backbone can then be mended by DNA ligase. Ligases that act on blunt-ended DNA segments are also available. *The cutting-and-pasting process allows a desired DNA segment to be incorporated into a carrier DNA molecule called a* **vector,** *leaving an unbroken, fully base-paired recombinant DNA molecule* (**Fig. 20.35**).

Some commonly used vectors are derived from **plasmids,** which are small, circular DNA molecules present in many bacterial cells. A single cell may contain multiple copies of a plasmid, which replicates independently of the bacterial chromosome and usually does not contain genes that are essential for the host's normal activities. However, plasmids often do carry genes for specialized functions, such as resistance to certain antibiotics (these genes often encode proteins that inactivate the antibiotics). An antibiotic resistance gene allows the **selection** of cells that harbor the plasmid: Only cells that contain the plasmid can survive in the presence of the antibiotic.

A recombinant plasmid—one that contains a foreign DNA sequence—can be introduced into a bacterial host, where it multiplies as the bacteria multiply. This is one way to produce large

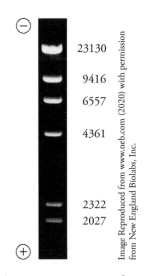

23130
9416
6557
4361
2322
2027

Figure 20.34 Digestion of bacteriophage λ DNA by the restriction enzyme HindIII. The digested DNA was applied to the top of the gel, and the negatively charged DNA fragments moved downward through the gel toward the positive electrode. The fragments were visualized with a fluorescent DNA-binding dye. The numbers indicate the number of base pairs in each fragment.

Question Explain why the bands at the top appear brighter than the bands at the bottom.

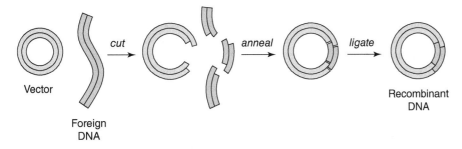

Vector

Foreign DNA

cut

anneal

ligate

Recombinant DNA

Figure 20.35 Production of a recombinant DNA molecule. A circular vector and a sample of DNA are cut with the same restriction enzyme, generating complementary sticky ends, so that the fragment of foreign DNA can be ligated into the vector.

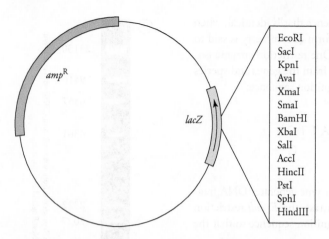

Figure 20.36 Map of a cloning vector. This 2743-bp circular DNA molecule, called pGEM-3Z, has a gene for resistance to ampicillin and a site comprising recognition sequences for different restriction enzymes. Inserting a foreign DNA segment at this site interrupts the *lacZ* gene, which encodes the enzyme β-galactosidase.

amounts of a desired DNA segment. (The segment can be harvested by recovering the plasmids and treating them with the same restriction enzyme used to insert the foreign DNA.) *DNA that is copied in this manner is said to be cloned.* Note that a **clone** is simply an identical copy of an original. The term is used to refer either to a gene that has been amplified, as described here, or to a cell or organism that is genetically identical to its parent.

An example of a plasmid used as a cloning vector is shown in **Figure 20.36**. This plasmid contains a gene (called *amp*^R) for resistance to the antibiotic ampicillin. Adding ampicillin to the culture medium eliminates cells that lack the plasmid. The vector also includes a gene (called *lacZ*), which encodes the enzyme β-galactosidase (it catalyzes the hydrolysis of certain galactose derivatives). The *lacZ* gene has been engineered to contain several restriction sites, any one of which can be used as an insertion point for a piece of foreign DNA with compatible sticky ends. Interrupting the *lacZ* gene with foreign DNA prevents the synthesis of the β-galactosidase protein. By adding a β-galactosidase substrate that yields a colored or fluorescent product, researchers can distinguish bacterial colonies that contain the foreign DNA (they lack β-galactosidase activity) from colonies that simply carry the empty vector (they retain β-galactosidase activity). Using this selection method, researchers can grow large quantities of the bacteria in order to harvest the cloned DNA.

A wide variety of vectors have been developed to accommodate different sizes of DNA inserts and to deliver that DNA to different types of hosts, including human cells. If a gene that has been isolated and cloned in a host cell is also expressed (transcribed and translated), the gene product can be isolated from the cultured cells or from the medium in which they grow. Similar steps are used to generate transgenic organisms (Box 3.B). In addition, genetic engineers can intentionally modify a segment of DNA using CRISPR-Cas technology (Box 20.C) or a variation of PCR (described below) in order to study how genes work, to make new gene products, and even to change the genetic makeup of an entire organism.

The polymerase chain reaction amplifies DNA

Cloning DNA in cultured cells is a laborious process. A much more efficient technique for amplifying a particular DNA sequence was developed by Kary Mullis in 1985: the **polymerase chain reaction (PCR).** *One of the advantages of PCR over traditional cloning techniques is that the starting material need not be pure* (this makes the technique ideal for analyzing complex mixtures such as tissues or biological fluids).

PCR takes advantage of the ability of DNA polymerase to synthesize a new strand of DNA by following an existing strand as a template. Because DNA polymerase cannot begin a new nucleotide strand (it can only extend a pre-existing chain), a short single-stranded DNA primer that base pairs with the template strand is added to the reaction mixture. This means that the polymerase copies only what it is "told" to copy.

The PCR reaction mixture contains a DNA sample, DNA polymerase, all four deoxy- nucleotide substrates for the polymerase, and two synthetic oligonucleotide primers that are complementary to the 3′ ends of the two strands of the target DNA sequence. Keep in mind that the reaction mixture actually contains millions of molecules of each of these substances. In the first step of PCR, the sample is heated to ~95°C to separate the DNA strands. Next, the temperature is lowered to about 55°C, cool enough for the primers to hybridize with the DNA strands. The temperature is then increased to about 75°C, and the DNA polymerase synthesizes new DNA strands by extending the primers (**Fig. 20.37**). The three steps—strand separation, primer binding, and primer extension—are repeated as many as 40 times. Because the primers represent the two ends of the target DNA, this sequence is preferentially amplified so that *it doubles in concentration with each reaction cycle.* For example, 20 cycles of PCR can theoretically yield $2^{20} = 1,048,576$ copies of the target sequence in a matter of hours.

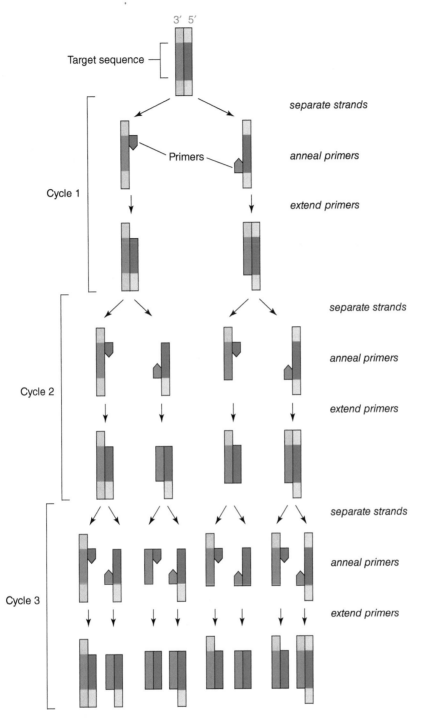

Figure 20.37 The polymerase chain reaction. Each cycle consists of separation of DNA strands, binding of primers to the 3′ ends of the target sequence, and extension of the primers by DNA polymerase.

Question **Indicate the temperature at which each step takes place.**

PCR relies on bacterial DNA polymerases that can withstand the high temperatures required for strand separation (these temperatures inactivate most enzymes). Commercial PCR kits usually contain DNA polymerase from a species such as *Thermus aquaticus* ("Taq," which lives in hot springs), since its enzymes perform optimally at high temperatures.

One of the limitations of PCR is that choosing appropriate primers requires some knowledge of the DNA sequence to be amplified; otherwise the primers will not anneal with complementary sequences in the DNA and the polymerase will not be able to make any new DNA strands. However, since no new DNA is synthesized *unless* the primers can bind to the DNA sequence,

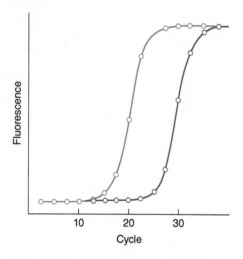

Figure 20.38 Quantitative reverse transcriptase PCR. The graph shows the amount of fluorescent DNA segments arising from high (red) and 1000-fold lower (blue) concentrations after multiple cycles of amplification.

PCR can be used to verify the presence of that sequence. Practical applications of PCR include identifying human genetic defects and diagnosing infections. PCR is the most efficient way to detect the presence of bacteria and viruses that are difficult or dangerous to work with. Blood banks use PCR to screen for certain pathogens in donated blood.

In the forensics laboratory, **DNA fingerprinting** methods use PCR to analyze segments of repetitive DNA, most often short tandem repeats of four nucleotides, which vary in size among individuals. Because the first step of fingerprinting is PCR, only a tiny amount of DNA is needed—about 1 μg, or the amount present in a single hair. The amplified PCR products are separated according to size by electrophoresis. If multiple sequences are simultaneously examined, the probability of two unrelated individuals yielding the same results is less than one in a million or more.

Researchers who want to do more than just detect a specific DNA sequence can perform **quantitative PCR (qPCR,** sometimes called real-time PCR). In this technique, the polymerase chain reaction continually generates new DNA sequences that bind to fluorescent probes, so that the amount of the DNA sequence can be monitored over time (rather than at the end of many reaction cycles, as in standard PCR). This approach is useful for quantifying infectious agents such as bacteria and viruses. Real-time PCR methods are also used to assess the level of gene expression in cells: Cellular mRNA is first reverse-transcribed to DNA and then a specific DNA sequence is amplified by PCR, a technique known as reverse transcriptase quantitative PCR or **RT-qPCR** (**Fig. 20.38**). A gene's level of expression is sometimes reported relative to that of a gene that encodes a protein such as actin (Section 5.3), which is typically produced at constant levels in cells and can therefore serve as a sort of benchmark.

DNA sequencing uses DNA polymerase to make a complementary strand

Determining the sequence of nucleotides in a DNA molecule involves using DNA polymerase to make copies. Various methods have been devised to detect the newly made DNA, such as labeling it with radioactive or fluorescent tags. The key, of course, *is to identify the added nucleotides one at a time in order to reconstruct the sequence of the original (template) DNA.* The newest, most efficient DNA sequencing methods, sometimes called next-generation or high-throughput systems, require specialized instruments, but they operate extremely quickly, sequence thousands of DNA segments simultaneously, and can handle huge tasks, such as sequencing an organism's entire genome within hours or days.

In the **Illumina sequencing** method, the DNA to be sequenced is broken into fragments of a few hundred bases. Specific oligonucleotide sequences, called adapters, are then attached to each end of the DNA strands. The adapters allow the DNA fragments to be immobilized in an array on a solid support. The adapters also provide a uniform binding site for primers so that the anchored DNA strands, whatever their sequence, can be amplified by PCR. To determine the sequences of the resulting DNA strands, a new primer and DNA polymerase are added, along with a solution containing all four deoxynucleotides, each with a different fluorescent group at its 3′ position. A nucleotide that can pair with the base in the immobilized template strand is incorporated into a new DNA chain. Unreacted nucleotides are washed away, and the nucleotides that remain can be identified by their fluorescence. Before the next batch of nucleotides is delivered, the fluorescent groups on the growing DNA strands, which block the 3′ OH group needed for polymerization, are detached so that DNA polymerase can attach new fluorescent nucleotides to be detected in the next reaction cycle (**Fig. 20.39**). The order of appearance of fluorescent signals corresponds to the nucleotide sequence of the growing DNA chain in each cluster of DNA molecules (**Fig. 20.40**).

Other DNA-sequencing technologies involve similar amplification and polymerase-catalyzed synthesis of a new strand. For example, the pyrophosphate (PP$_i$) released after the incorporation of a given nucleotide into a new strand can be detected by additional reactions that produce a flash of light. Alternatively, the H$^+$ released during polymerization can be measured as a decrease in pH. Most of these sequencing technologies can analyze multiple DNA sequences simultaneously but produce "reads" of a few hundred nucleotides at most. Deducing the sequence of a long segment of DNA requires that the original sample be randomly broken into multiple short segments that are individually sequenced

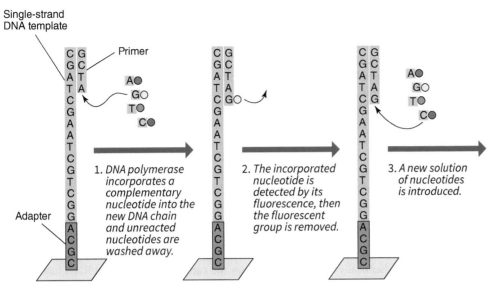

Figure 20.39 Illumina sequencing reaction. DNA polymerase uses immobilized single strands of DNA as templates to add a fluorescent nucleotide to a new strand. A detector records the fluorescence to identify the nucleotide added in that round of polymerization.

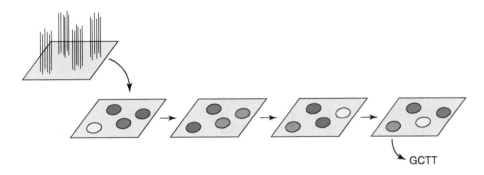

Figure 20.40 Illumina sequencing results. Each spot in the array represents a cluster of amplified DNA segments, which are the template strands. A fluorescent signal identifies the newest nucleotide added to the growing strands.

Question **What are the sequences of the new DNA strands in the other spots?**

(this is known as a "shotgun" approach). The original sequence can then be reconstructed by computer analysis of the overlapping sequences.

Original sequence

T G C C G A T A A G C T G C G G A T T C C G T A A G T A C A G

Overlapping fragments

T G C C G A T A A G C T G C G G A T T C C G T A A G T A C A G

T G C C G A T A A G C T G C G G A T T C C G T A A G T A C A G

T G C C G A T A A G C T G C G G A T T C C G T A A G T A C A G

The high redundancy of the raw data helps minimize sequencing errors. Finished sequence data are customarily deposited in a database such as GenBank (ncbi.nlm.nih.gov/genbank/) and are available for further *in silico* analysis.

The high-throughput sequencing technologies described above are efficient but are not always suitable for analyzing lengthy repetitive regions of chromosomes, including centromeres and telomeres, which must be sequenced by other methods. Still, automated sequencing combined with sophisticated cell-sorting techniques makes it possible to isolate the mRNA from a single cell, reverse-transcribe it to DNA, amplify it by PCR, and then sequence it. Such information can reveal subtle differences between cells, even from the same tissue.

Some sequencing tasks, such as the rapid identification of organisms and whole-genome analysis, can be carried out using **nanopore sequencing.** In this technique, the DNA is not amplified by PCR. Instead, long DNA molecules are modified by attaching an adaptor segment, which helps deliver the DNA to a protein channel (that is, a "nanopore") embedded in an artificial membrane. A voltage difference across the membrane generates a current so that a single, negatively charged DNA strand moves through the pore. As the nucleotide residues pass through the channel, the current fluctuates slightly, depending on the size and shape of each base. Monitoring the current gives a readout of the nucleotide sequence (**Fig. 20.41**). To keep the DNA strand moving uniformly through the pore, a helicase acts as a ratcheting device. Nanopore sequencing has a high error rate (about 10%), but its advantages over other sequencing technologies include speed, smaller and less expensive instruments (there is no need to detect light output), and long reads of tens to thousands of kilobases. Furthermore, because the sample is not copied for analysis, the nanopore technique can directly detect chemically modified nucleotides and can be adapted to read RNA sequences.

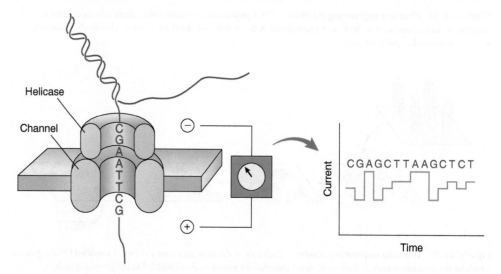

Figure 20.41 Nanopore sequencing. Unamplified DNA moves through a protein channel under the influence of an electric field. A helicase ensures that a single strand passes through at a regular rate, and small changes in the current indicate the identities of the nucleotide bases.

Concept Check

1. Make a list of the reagents and equipment needed to carry out each of the procedures outlined in this section.

2. Summarize the steps involved in each process.

3. Explain why PCR is part of many other methods for working with DNA.

4. Using Fig. 20.37, draw the products of the fourth PCR cycle.

5. Make a list of the practical applications of the techniques described in this section.

SUMMARY

20.1 DNA Packaging: Nucleosome and Chromatin

- DNA unwinding is facilitated by the negative supercoiling (underwinding) of DNA molecules. Enzymes called topoisomerases can add or remove supercoils by temporarily introducing breaks in one or both DNA strands.

- Eukaryotic DNA is wound around a core of eight histone proteins to form a nucleosome, which represents the first level of DNA compaction in the nucleus.

20.2 The DNA Replication Machinery

- DNA replication requires a host of enzymes and other proteins located in a stationary factory. Helicases separate the two DNA strands at a replication fork, and SSB binds to the exposed single strands.

- DNA polymerase can only extend a preexisting chain and therefore requires an RNA primer synthesized by a primase. Two polymerases operate side-by-side to replicate DNA, so the leading strand of DNA is synthesized continuously while the lagging strand is synthesized discontinuously as a series of Okazaki fragments. The RNA primers of the Okazaki fragments are removed, the gaps are filled in by DNA polymerase, and the nicks are sealed by DNA ligase.

- DNA polymerase and other polymerases catalyze a reaction in which the 3′ OH group of the growing chain nucleophilically attacks the phosphate group of an incoming nucleotide that base pairs with the template strand. A sliding clamp promotes the processive activity of DNA polymerase.

- Many DNA polymerases contain a second active site that hydrolytically excises a mismatched nucleotide.

20.3 Telomeres

- In eukaryotes, the extreme 3′ end of a DNA strand cannot be replicated, so the enzyme telomerase adds repeating sequences to the 3′ end to create a structure known as a telomere.

20.4 DNA Damage and Repair

- Normal replication errors, spontaneous deamination, radiation, and chemical damage can cause mutations in DNA.

- Mechanisms for repairing damaged DNA include direct repair, mismatch repair, base excision repair, nucleotide excision repair, end-joining, and recombination.

20.5 Clinical Connection: Cancer as a Genetic Disease

- Cancer is uncontrolled cell division triggered by several genetic changes. These mutations occur as a result of inheritance, environmental factors, or random chance.

- Cellular pathways that detect and repair damaged DNA or prevent the cell from dividing are commonly disrupted in cancer.

20.6 Tools and Techniques: Manipulating DNA

- DNA molecules can be reproducibly fragmented by the action of restriction enzymes, which cleave DNA at specific sequences.

- DNA fragments can be joined to each other to generate recombinant DNA molecules that are then introduced into host cells.

- A DNA segment can be amplified many times by the polymerase chain reaction, in which a DNA polymerase makes complementary copies of the target DNA.

- Modern techniques for determining the sequence of nucleotides in a segment of DNA immobilize DNA segments, amplify them, use DNA polymerase to add nucleotides to a new complementary chain, and detect each added nucleotide.

KEY TERMS

replication	lagging strand	reverse transcriptase	nucleotide excision	supercoiling	plasmid
semiconservative	discontinuous	cDNA	repair	nucleosome	selection
replication	synthesis	proliferation	nonhomologous	histone	clone
DNA polymerase	Okazaki fragment	mutation	end-joining	heterochromatin	polymerase chain
replisome	processivity	point mutation	recombination	euchromatin	reaction (PCR)
replication fork	proofreading	transition mutation	CRISPR	chromatin	DNA fingerprinting
factory model of	indel	transversion	CRISPR-Cas9	restriction	quantitative PCR
replication	nick	mutation	system	endonuclease	(qPCR)
helicase	nick translation	abasic site	constitutive	palindrome	RT-qPCR
topoisomerase	DNA ligase	mutagen	apoptosis	sticky ends	Illumina sequencing
primer	centromere	carcinogen	carcinogenesis	blunt ends	nanopore sequencing
primase	telomere	mismatch repair	tumor suppres-	restriction digest	
leading strand	telomerase	base excision repair	sor gene	vector	

BIOINFORMATICS

Brief Bioinformatics Exercises

20.2 DNA Replication and the KEGG Database

20.3 Viewing and Analyzing Uracil–DNA Glycosylase in DNA Repair

20.4 Restriction Enzyme Mapping

PROBLEMS

20.1 DNA Packaging: Nucleosome and Chromatin

1. a. Why do type II topoisomerase enzymes require ATP whereas type I topoisomerases do not? **b.** Explain why a type I topoisomerase sometimes participates in nucleotide excision repair.

2. A variety of compounds inhibit topoisomerases. Novobiocin is an antibiotic and, like ciprofloxacin, it inhibits DNA gyrase. Doxorubicin and etoposide are anticancer drugs that inhibit eukaryotic topoisomerases. What properties distinguish the antibiotics from the anticancer drugs?

3. The anticancer drug doxorubicin (see Problem 2) is more correctly termed a topoisomerase II poison because it blocks catalysis after the DNA is cleaved and before it is re-ligated. In contrast, the anticancer drug aclarubicin is termed a "catalytic inhibitor" because it prevents the binding of topoisomerase II to the DNA. What distinguishes an inhibitor from a poison?

4. What is the role of DNA gyrase during bacterial DNA replication?

5. DNA replication of the circular chromosome in *E. coli* begins when a protein called DnaA binds to the replication origin on the DNA. **a.** DnaA initiates replication only when the DNA that constitutes the DNA replication origin is negatively supercoiled. Why? **b.** Is the replication origin likely be richer in G:C or A:T base pairs?

6. The percentages of arginine and lysine residues in the histones of calf thymus DNA are shown in the table below. **a.** Why do histones have a large number of Lys and Arg residues? **b.** How does this explain why a 0.5-M NaCl solution is effective at dissociating histones from DNA in a sample of chromatin?

Histone	% Arg	% Lys
H1	1	29
H2A	9	11
H2B	6	16
H3	13	10
H4	14	11

7. Cows and peas diverged from a common ancestor over a billion years ago, yet they have histone H4 proteins that differ by only two amino acids residues out of 102 total residues. The changes are conservative—Val 60 in cow H4 is replaced by Ile in peas; Lys 77 in cow H4 is replaced by Arg in peas. Propose a hypothesis that explains why H4 is so highly evolutionarily conserved.

8. Lysine residues in histone tails may be acetylated on the ε-amino group. Would you expect to find more acetylated lysine residues in euchromatin or in heterochromatin? Explain.

9. Chemical methylating agents can modify guanosine to form N7-methyl-2′-deoxyguanosine (MdG). Unlike O^6-methylguanine (see Problem 58), MdG was initially considered to be a benign lesion because it does not perturb the three-dimensional structure of DNA very much, and hydrolysis of the base (which would result in an abasic site) occurs very slowly (on the order of days). However, studies using DNA bound to histones (rather than pure DNA) showed that the MdG base, rather than undergoing hydrolysis at the *N*-glycosidic bond, instead forms cross-links with lysine residues in histones. Why did the researchers conclude that this chemical modification is more toxic than originally believed?

10. During sperm development, about 95% of the cell's DNA is associated with small proteins known as protamines rather than histones. Protamine–DNA complexes pack more compactly than histone–DNA complexes. **a.** Explain the advantage of replacing histones with protamines during sperm development. **b.** What type of genes would you expect to find in the remaining nucleosomes? (*Hint:* These genes are not transcribed until after fertilization.)

11. A sample of chromatin is briefly treated with micrococcal nuclease, which catalyzes the hydrolysis of double-stranded DNA. **a.** Analysis by agarose gel electrophoresis indicates that the products are double-stranded DNA segments about 200-bp long. Explain these results. **b.** If the digestion is allowed to proceed for an extended period, 146-bp DNA bands are observed on the agarose gel. Explain.

12. Why are Okazaki fragments 500–2000 bp long in bacteria but only 100–200 bp long in eukaryotic cells?

13. Why is 0.5 M NaCl effective at dissociating histones from DNA in a sample of chromatin?

14. *In vivo*, DNA undergoes transient local melting, producing small "bubbles" of single-stranded DNA. Does this DNA "breathing" activity explain why the rate of C → T mutations is lower when the DNA is part of a nucleosome?

20.2 The DNA Replication Machinery

15. Semiconservative replication is shown in Figure 20.1. Draw a diagram that illustrates the composition of DNA daughter molecules for the second, third, and fourth generations.

16. What is the composition of DNA daughter molecules for the second generation if a hypothetical DNA replicates conservatively (see Fig. 20.7)? Compare the composition to that of a second generation of actual DNA that replicates semiconservatively (see Solution 15).

17. DNA replication of the circular chromosome in *E. coli* begins when a protein called DnaA binds to the replication origin on the DNA. Is the replication origin likely to be richer in G:C or A:T base pairs?

18. The DnaA protein (see Problem 17) binds to both ADP and ATP. Both complexes can bind to the *E. coli* replication origin, but only the DnaA–ATP complex melts the DNA duplex. Explain why.

19. DNA helicases can be visualized as molecular motors that convert the chemical energy of NTP hydrolysis into mechanical energy for separating DNA strands. The bacteriophage T7 genome encodes a protein that assembles into a hexameric ring with helicase activity. **a.** In order to unwind DNA, the helicase requires two single-stranded DNA tails at one end of a double-stranded DNA segment. However, the helicase appears to bind to only one of the single strands, apparently by encircling it. Is this consistent with its ability to unwind a double-stranded DNA helix? **b.** Kinetic measurements indicate that the T7 helicase moves along the DNA at a rate of 132 bases per second. The protein hydrolyzes 49 dTTP per second. What is the relationship between dTTP hydrolysis and DNA unwinding? **c.** What does the structure of T7 helicase suggest about its processivity?

20. At eukaryotic origins of replication, helicase cannot be activated until the polymerase is also positioned on the DNA. Explain what would happen if the helicase became active in the absence of DNA polymerase.

21. Bacterial helicases move along a DNA strand in the 5′ → 3′ direction, whereas eukaryotic helicases operate in the opposite direction. Which type of helicase binds to the leading-strand template? Which binds to the lagging strand template?

22. Temperature-sensitive mutations allow a protein to function at a low temperature (the permissive temperature) but not a high (nonpermissive) temperature. Temperature-sensitive mutations in some replication proteins result in the immediate cessation of bacterial growth; for other mutant proteins, growth comes to a halt more gradually when the bacteria are exposed to a nonpermissive temperature. What happens to DNA replication and bacterial growth when the temperature is suddenly increased and the temperature-sensitive mutation is in **a.** helicase or **b.** DnaA (see Problem 17)? **c.** In an experiment, cell extracts from the helicase mutants and DnaA mutants are combined. What happens to DNA replication at the nonpermissive temperature?

23. Based on your knowledge of replication proteins, compare DNA polymerase and single-strand binding protein (SSB) with respect to **a.** affinity for DNA and **b.** cellular concentration.

24. Explain why treatment of a preparation of *E. coli* single strand binding protein (SSB) with acetic anhydride produces a chemically modified protein that is unable to bind to DNA. (*Hint:* Acetic anhydride reacts with primary amino groups.)

25. *E. coli* primase begins polymerizing nucleotides at a 5′-TTC-3′ sequence in DNA. What are the first three bases in the resulting primer?

26. In the chain-termination method of DNA sequencing (described in Section 3.4), the Klenow fragment of DNA polymerase I (see Fig. 20.13) is used to synthesize a complementary strand using single-stranded DNA as a template. Along with a primer and the four dNTP substrates, a small amount of 2′, 3′-dideoxynucleoside triphosphate (ddNTP) is added to the reaction mixture. When the ddNTP is incorporated into the growing nucleotide chain, polymerization stops. Explain.

2′,3′-Dideoxynucleoside
triphosphate (ddNTP)

27. You have discovered a drug that inhibits the activity of the enzyme inorganic pyrophosphatase. What effect would this drug have on DNA synthesis? (*Hint:* See Fig. 20.11.)

28. A reaction mixture contains purified DNA polymerase, the four dNTPs, and one of the DNA molecules whose structures are represented below. Which reaction mixture produces PP$_i$?

29. DNA polymerases include two Mg^{2+} (or in some cases Mn^{2+}) ions in the active site. **a.** Describe how the metal ions could enhance the nucleophilicity of the 3′ OH group that attacks the α-phosphate of an incoming nucleotide. **b.** The polymerization mechanism includes the formation of a pentacovalent phosphorus. Sketch this structure. How could a metal ion contribute to transition state stabilization during DNA polymerization?

30. Early experiments showed that the *E. coli* sliding clamp protein associates with a circular DNA molecule with a half-life of 72 minutes. However, when the circular DNA was treated with an endonuclease to linearize it, the half-life for the DNA–sliding clamp association was too short to measure. How did these experiments demonstrate that the sliding clamp actually slides along DNA?

31. Which eukaryotic DNA polymerase would you expect to have greater processivity: polymerase α, δ, and ε? Explain.

32. The "flap" endonuclease recognizes the junction between RNA and DNA near the 5′ end of an Okazaki fragment. What feature(s) of this structure might the enzyme recognize?

33. In eukaryotes, DNA replication proceeds at a rate of about 1000 bp per minute. Approximately how many origins of replication are theoretically needed in order to replicate all 46 human chromosomes within 10 hours?

34. An *in vitro* system was developed in which simian virus 40 (SV40) can be replicated using purified mammalian proteins. Make a list of the proteins required to replicate SV40 DNA in a cell-free system.

35. Make a list of strategies that a cell uses to minimize the incorporation of mispaired nucleotides during DNA replication.

36. Why would it not make sense for the cell to wait to combine Okazaki fragments into one continuous lagging strand until the entire DNA molecule had been replicated?

37. Explain how DNase (an endonuclease that cleaves the backbone of one strand of a DNA molecule), *E. coli* DNA polymerase I (which includes 5′ → 3′ exonuclease activity), and DNA ligase could be used in the laboratory to incorporate radioactive nucleotides into a DNA molecule.

38. Biotech companies sell the Klenow fragment of DNA polymerase I, which is missing 5′ → 3′ exonuclease activity but retains 5′ → 3′ polymerase activity and 3′ → 5′ exonuclease activity. Why is the Klenow fragment a better choice than wild-type DNA polymerase I for a laboratory task that involves the synthesis of double-stranded DNA from single-stranded DNA?

39. The mechanism of *E. coli* DNA ligase involves the transfer of the adenylyl (AMP) group of NAD$^+$ to the ε-amino group of the side chain of an essential lysine residue on the enzyme. The adenylyl group is subsequently transferred to the 5′ phosphate group of the nick. The first step is shown. Draw the complete mechanism of *E. coli* DNA ligase.

40. Why would ligase enzymes that use NAD$^+$ as a cofactor be attractive drug targets for treating diseases caused by bacteria?

20.3 Telomeres

41. Give the name of the enzyme that catalyzes each of the following reactions: **a.** makes a DNA strand from a DNA template, **b.** makes a DNA strand from an RNA template, and **c.** makes an RNA strand from a DNA template.

42. Human cells infected by HIV increase the expression of SAMDH1, a phosphohydrolase that removes triphosphate groups from dNTP substrates. Explain how SAMDH1 inhibits HIV replication.

43. In some species, G-rich telomeric DNA folds up on itself to form a four-stranded structure. In this structure, four guanine residues assume a hydrogen-bonded planar arrangement with an overall geometry that can be represented as

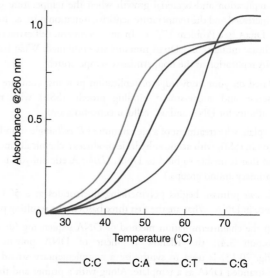

This "G quartet" may play a role as a negative regulator of telomerase activity. **a.** Draw the complete structure of the G quartet, including the hydrogen bonds between the purine bases. **b.** Show schematically how a single strand of four repeating TTAGGG sequences can fold to generate a structure with three stacked G quartets linked by TTA loops.

44. Why might a drug that induces the formation of a G quartet (see Problem 43) be effective as an antitumor agent?

45. Telomerase is a ribonucleoprotein. Explain.

46. In an experiment, the AAUCCC segment of the RNA template of telomerase is mutated. What cellular changes will be occur?

47. Use what you have learned in the first two sections of this chapter to compare and contrast DNA replication in prokaryotes and eukaryotes.

20.4 DNA Damage and Repair

48. Classify the following base changes as transitions or transversions: **a.** C → A, **b.** G → T, **c.** C → T, **d.** A → G, **e.** A → T, **f.** C → G.

49. Oligonucleotides with mismatched base pairs have different melting temperatures (T_m) than properly matched base pairs (see Fig. 3.8). **a.** Use the melting point curves shown in the figure below to estimate the melting points for a C:G base pair as well as mismatched C:T, C:A, and C:C base pairs. **b.** Which base pair is the least stable? **c.** Propose a hypothesis to explain why improperly paired bases are less thermally stable than their properly paired counterparts.

50. 8-Oxoguanine (oxoG) is produced when reactive oxygen species (ROS) react with guanine, so the oxidized base is a biomarker of oxidative stress. 8-Oxoguanine base-pairs with adenine. **a.** Draw the structure of an 8-oxoguanine:adenine base pair. (*Hint:* The 8-oxoguanine base pivots around the glycosidic bond in order to form two hydrogen bonds with adenine.) **b.** What type of mutation (transition or transversion) results if the damage is not repaired? **c.** If a specialized DNA glycosylase recognizes the oxoG and excises it, leaving an abasic site, how is the repair process completed? **d.** The specialized glycosylase is mutated in some colon cancers, which prevents the damage from being repaired. How does the mutation affect the DNA in the cancer cells?

51. The free deoxyribonucleotides dATP and dGTP are about 13,000 times more susceptible to oxidative damage (see Problem 39) than when they are incorporated into DNA. **a.** Explain why this minimizes the incidence of mutations. **b.** In eukaryotic cells, a specific triphosphatase hydrolyzes oxo-dGTP to its monophosphate form. Does this enzyme activity inhibit or promote DNA mutations? **c.** Explain how small molecules that inhibit the activity of the triphosphatase are able to prevent the growth of highly metabolically active cancer cells.

52. Oxidative deamination of nitrogenous bases can occur upon exposure to nitrous acid (HNO_2). Deamination of cytosine produces uracil, as described in the chapter, and deamination of adenine produces hypoxanthine. **a.** Draw the structure of hypoxanthine. **b.** Hypoxanthine can base-pair with cytosine. Draw the structure of this base pair. **c.** What is the consequence to the DNA if this deamination is not repaired?

53. Oxidative deamination (see Problem 52) of guanine produces xanthine. **a.** Draw the structure of xanthine. **b.** Xanthine base-pairs with cytosine. Would this cause a mutation? Explain.

54. Compounds such as ethidium bromide and acridine orange can be used to stain DNA bands in a gel. These compounds are known as intercalating agents, and they interact with the DNA by slipping in between stacked base pairs. This interaction increases the fl uorescence of the intercalating agent and allows for visualization of DNA bands following electrophoresis. Care must be taken when working with intercalating agents because they are powerful mutagens. Explain why.

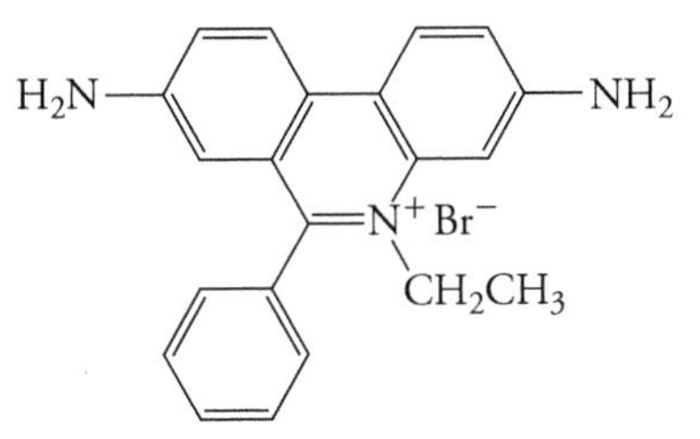

Ethidium bromide

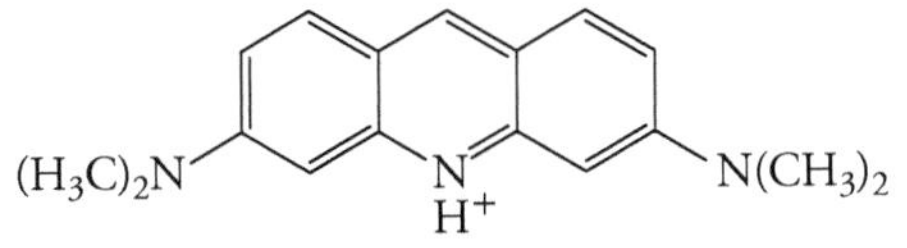

Acridine orange

55. DNA damage can be classified into the following types: **a.** mismatch (such as a mistake made during replication), **b.** base lesion (such as 8-oxoguanine), **c.** bulky adduct (such as thymine–thymine dimers), and **d.** single-strand breaks and double-strand breaks. What types of damage are likely to lead to failures in DNA replication and/or transcription? What types of damage are likely to lead to a mutation?

56. The compound 5-bromouracil, a thymine analog, can be incorporated into DNA in the place of thymine. 5-Bromouracil readily converts to an enol tautomer, which can base-pair with guanine. (The keto and enol tautomers freely interconvert through the movement of a hydrogen between an adjacent nitrogen and oxygen.) Draw the structure of the base pair formed by the enol form of 5-bromouracil and guanine. What kind of mutation can 5-bromouracil induce?

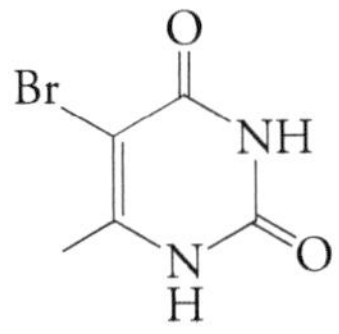

5-Bromouracil

57. The adenine analog 2-aminopurine (shown below) is a potent mutagen in bacteria. The 2-aminopurine substitutes for adenine during DNA replication and gives rise to mutations because it pairs with cytosine instead of thymine. Structural studies show that the "neutral wobble" base pair (so called because it is the dominant structure at neutral pH) is in equilibrium with a "protonated Watson–Crick" structure, which forms at lower pH. **a.** Two hydrogen bonds form between cytosine and 2-aminopurine in the "neutral wobble" pair. Draw the structure of this base pair. **b.** At lower pH, either the cytosine or the 2-aminopurine can become protonated. The two protonated forms are in equilibrium in the "protonated Watson–Crick" base pair structure; the proton essentially "shuttles" from one base to the other in the pair, and the hydrogen bond is maintained. Draw the two possible structures for the base pair, one with a protonated 2-aminopurine, and one with a protonated cytosine.

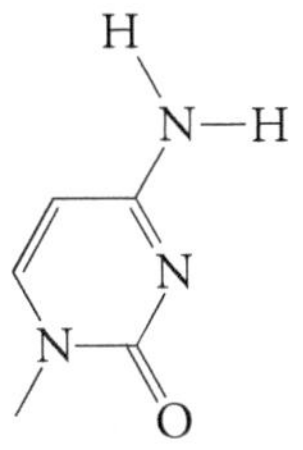

Cytosine 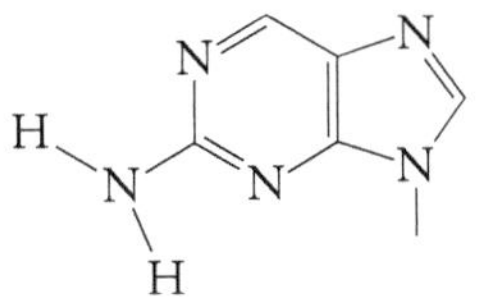   2-Aminopurine

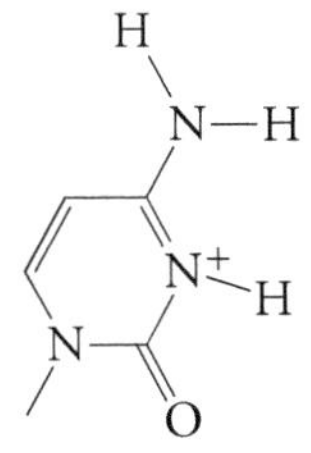

Protonated   Protonated
cytosine   2-aminopurine

58. Chemical methylating agents can react with guanine residues in DNA to produce O^6-methylguanine, which causes mutations because it can pair with thymine. The structure of O^6-methylguanine is shown in the text. This type of chemical damage is difficult to repair because the O^6-methylguanine:T base pair is structurally similar to a normal G:C base pair and the mismatch repair system has difficulty recognizing it. The resulting mutations can generate oncogenes. The O^6-methylguanine can form a "wobble" pair with thymine (similar to the 2-aminopurine:C base pair in Problem 57). The methylated guanine can also form a base pair with protonated cytosine at lower pH values. **a.** At physiological pH, the O^6-methylguanine:T wobble base pair predominates. Two hydrogen bonds form between the two bases. Draw the structure of this base pair. Does a mutation result? **b.** At lower pH values, cytosine becomes protonated (see Problem 38b). Draw the structure of the O^6-methyguanine:protonated cytosine base pair (three hydrogen bonds form between the two bases). Does a mutation result? **c.** Why might O^6-methylguanine be referred to as a "suicide substrate" for the methyltransferase that removes the methyl group?

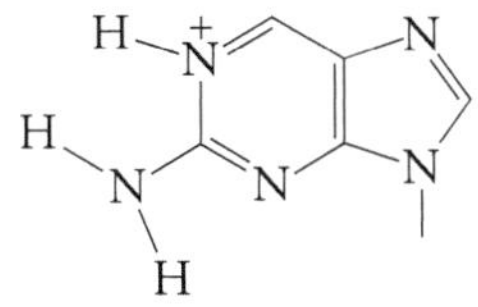

59. What is the most common DNA lesion in individuals with the disease xeroderma pigmentosum?

60. In bacteria, thymine dimers can be restored to their original form by DNA photolyases that cleave the dimer. The photolyases are so named because they are activated by absorption of light. Why is this a good biochemical strategy?

61. The evolution of DNA, which contains the bases A, C, G, and T, produced a molecule that is easy to repair. For example, deamination of adenine produces hypoxanthine (see Problem 52), deamination of guanine produces xanthine (see Problem 53), and deamination of cytosine produces uracil. Why are these deaminations quickly repaired?

62. Studies of bacteria and other organisms indicate that mutations occur more frequently at certain positions. These "hotspots" are due to the presence of 5-methylcytosine, which may undergo oxidative deamination. **a.** Draw the structure of deaminated 5-methylcytosine. **b.** By what other name is the base known? **c.** What kind of mutation results from 5-methylcytosine deamination? **d.** Why is the altered base difficult for the cell to repair? **e.** Describe how the cell might go about repairing the damage.

63. A strain of mutant bacterial cells lacks the enzyme uracil-DNA glycosylase. What is the consequence for the organism?

64. The incomplete removal of an RNA primer leaves ribonucleotides incorporated into DNA. List the enzyme activities that would be required to repair DNA containing a stretch of several ribonucleotides.

65. During replication in *E. coli*, certain types of DNA damage, such as thymine dimers, can be bypassed by DNA polymerase V. This polymerase tends to incorporate guanine residues opposite the damaged thymine residues and has a higher overall error rate than other polymerases. Thymine dimers can be bypassed by other polymerases, such as DNA polymerase III (which carries out most DNA replication in *E. coli*). DNA polymerase III incorporates adenine residues opposite the damaged thymine residues, but much more slowly than DNA polymerase V. Polymerase III is a highly processive enzyme, whereas polymerase V adds only 6–8 nucleotides before dissociating from the DNA. Explain how DNA polymerases III and V together carry out the efficient replication of UV-damaged DNA with minimal errors.

66. Eukaryotic cells contain a number of DNA polymerases. Several of these enzymes were tested for their ability to cleave nucleotides from the 3′ end of a DNA chain (3′ → 5′ exonuclease activity). The enzymes were also tested for the accuracy of DNA polymerization, expressed as the rate of base substitution. The results are summarized in the table.

Polymerase	3′ → 5′ exonuclease activity	Base substitution rate (× 10⁻⁵)
α	No	16
β	No	67
δ	Yes	1
ε	Yes	1
η	No	3500

a. Is there a correlation between the presence of 3′ → 5′ exonuclease activity and the error rate (base substitutions) during DNA polymerization? **b.** Express the error rate of each polymerase in terms of how often a wrong base is incorporated. **c.** Polymerization errors can result from the ability to insert an incorrect (mispaired) base or from the inability to efficiently insert the correct (template-matched) base. To distinguish which mechanism accounts for the high error rate of polymerase η, the catalytic efficiency (k_{cat}/K_M) of the polymerization reaction was determined for matched and unmatched bases. The results were compared to the catalytic efficiency of the HIV reverse transcriptase (HIV RT). The data are summarized below. Compare the efficiency of DNA polymerase η and HIV RT in incorporating correct bases and incorrect bases. What do these results reveal about the cause of errors during polymerization by polymerase η? **d.** The overexpression of genes that code for DNA polymerases similar to polymerase η has been observed in bacteria. What effect would this have on the mutation rate in these bacteria?

Polymerase	Template base	Incoming base	k_{cat}/K_M (μM⁻¹ · min⁻¹ × 10³)
polymerase η	T	A	420
	T	G	22
	T	C	1.6
	G	C	760
	G	G	8.7
HIV RT	T	A	800
	T	G	0.07

20.5 Clinical Connection: Cancer as a Genetic Disease

67. Some types of cancer are clearly linked to specific DNA repair pathways. For example, individuals with mutations in the human homolog of the bacterial MutS gene are predisposed to develop colon cancer. Explain why.

68. The product of the retinoblastoma gene, the Rb protein, binds to and inhibits the activity of a transcription factor that induces the expression of genes required for the cell to synthesize DNA. **a.** Explain how a mutation in the retinoblastoma gene promotes cancer. **b.** Explain why inheriting a defective gene for the retinoblastoma protein, BRCA1, or p53 does not guarantee that an individual will develop cancer.

69. Explain why mutations that affect the 3′ → 5′ exonuclease domain of DNA polymerase have been associated with some hereditary human cancers.

70. One of the proteins that senses damaged DNA may structurally resemble PCNA, the eukaryotic sliding-clamp protein that enhances the processivity of DNA polymerase (see Fig. 20.14). Explain why such a protein is suited for monitoring the chemical integrity of DNA.

71. Although DNA damage is a cause of cancer, some cancer chemotherapies are based on drugs that damage DNA. For example, the drug carboplatin introduces a platinum ion that cross-links adjacent guanine residues in a DNA strand. Explain why this leads to death of the cancer cell.

72. The Ras signaling pathway (Section 10.3) activates a transcription factor called Myc, which turns on a gene for a protein that blocks the activity of the protein that adds ubiquitin to p53. Is this mechanism consistent with the ability of excessive Ras signaling to promote cell growth?

73. The human immune system can recognize markers on the surface of cancerous cells so that these cells can be selectively killed. What feature of a transformed cell would change its appearance?

74. In the 1920s, Otto Warburg noted that cancer cells primarily use glycolysis rather than aerobic respiration to produce ATP (Section 19.4). In 2006, researchers discovered that p53 stimulates the expression of cytochrome *c* oxidase. How does the connection between p53 and cytochrome *c* oxidase explain Warburg's observation?

75. Why do unicellular organisms have no need for apoptosis?

20.6 Tools and Techniques: Manipulating DNA

76. Which is more likely to be called a "rare cutter": a restriction enzyme with a four-base recognition sequence or a restriction enzyme with an eight-base recognition sequence?

77. The sequence of a segment of the pET28 plasmid is shown below. Which of the restriction enzymes shown in Table 20.1 could cleave this DNA segment to insert a foreign gene into the plasmid?

TCCGAATTCGAGCTCCGTCGACAAGCTTGCGGCCGCACT

78. A sample of the DNA segment shown below is treated with the restriction enzyme EcoRI. Next, the sample is incubated with an exonuclease that acts only on single-stranded polynucleotides. What mononucleotides will be present in the reaction mixture?

TGCTTAGAATTCGGAACGA
ACGAATCTTAAGCCTTGCT

79. The plasmid pGEM-3Z shown in Figure 20.36 has three restriction sites for the enzyme HaeII, which cleaves at locations 323, 693, and 2564. Draw the result of agarose gel electrophoresis (see Figure 20.34) if the sample contains **a.** linear or **b.** circular plasmid digested with HaeII.

80. Restriction enzymes are used to construct restriction maps of DNA. These are diagrams of specific DNA molecules that show the sites where the restriction enzymes cleave the DNA. To construct a restriction map, purified samples of the DNA are treated with restriction enzymes, either alone or in combination, and then the reaction products are separated by agarose gel electrophoresis. Use the results of the agarose gel electrophoresis shown here to construct a restriction map for the sample of DNA.

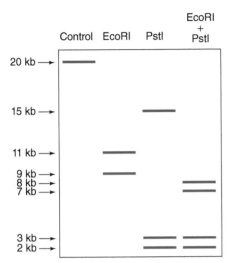

81. Design a 10-bp primer that could be used to amplify the following sequence of DNA:

5′-AGTCGATCCCTGATCGTACGCTAAGTTAACGT-3′

82. The primer used in sequencing a cloned DNA segment often includes the recognition sequence for a restriction endonuclease. Explain.

83. A portion of a gene is shown below. Design two 12-bp primers that could be used to amplify this gene segment.

5′-TTCGCCTCGGGGCTCCCCAGTCGCTGGTGCTGCTGAC-
GCTGCTAGCAG-3′

3′-AAGCGGAGCCCCGAGGGGTCAGCGACCACGACGACTGC-
GACGATCGTC-5′

84. a. Design two 18-bp PCR primers that could be used to amplify the gene shown below. **b.** Suppose you wanted to add EcoRI restriction sites to each end of the gene. How would you modify the sequences of the PCR primers you designed in part **a** to amplify the gene?

ATGGGCTCCATCGGTGCAGCAAGCATGGAA-(1104 bp)-
TTCTTTGGCAGATGTGTTTCCCCTTAAAAAGAA

85. Transgenic plants are constructed by inserting a new gene into the plant genome to give the plant a desirable characteristic, such as resistance to pesticides or frost, or to confer a nutritional benefit. Transformed plants contain the gene of interest, a promoter (to induce expression of the gene), and a terminator. The cauliflower mosaic virus (CaMV) is often used as a promoter in transgenic plants. A partial DNA sequence is shown below. Design a set of 18-bp primers that you could use to detect the presence of this promoter in a transgenic plant.

GTAGTGGGATTGTGCGTCATCCCTTACGTCAGT ··· (112 bases) ···
TCAACGATGGCCTTTCCTTTATCGCAATGATGGCATTTGTAGGAGC

86. *Taq* polymerase has no exonuclease activity. How would this affect laboratory procedures that use *Taq* polymerase?

87. Could you perform PCR with an ordinary DNA polymerase, that is, one that is destroyed by high temperatures? What modifications would you make in the PCR protocol?

88. Examine the sequence of the protein produced by the DNA sequence referred to in Problem 84. Assume that the corresponding DNA sequence is *not* known. Using the amino acid sequence as a guide, design a pair of nine-base deoxynucleotide primers that could be used for PCR amplification of the protein-coding portion of the gene. (*Hint:* DNA polymerase can extend a primer only from its 3′ end.) How many different pairs of primers could you choose from?

89. A researcher trying to identify the gene for a known protein might begin by looking closely at the protein's sequence in order to design a single-stranded oligonucleotide probe that will hybridize with the DNA of the gene. Why would the researcher focus on a segment of the protein containing a methionine (Met) or tryptophan (Trp) residue (see Table 3.3)?

90. a. The following DNA fragments were produced by using the "shotgun" method. **a.** Show how the fragments should be aligned to determine the sequence of the original DNA. **b.** Which portions of the reconstructed sequence are the most accurate (in other words, where are you confident that the sequence is correct)?

ACCGTGTTTCCGACCG
ATTGTTCCCACAGACCG
CGGCGAAGCATTGTTCC
TTGTTCCCACAGACCGTG

91. A group of investigators is interested in studying the gp41 protein from the human immunodeficiency virus (HIV). In order to do this, they use site-directed mutagenesis to synthesize a series of truncated proteins. A partial sequence of the gp41 protein is shown. Design an 18-bp "mismatched primer" that could be used to synthesize a truncated protein that would terminate after the Leu residue at position 700.

700
↓

··· WYIKLFIMIVGGLVGLRIVFAVLSIVNRVRGGYSP ···

92. In a CRISPR-Cas9 gene knock-in experiment, the replacement DNA sequence must differ from the original DNA sequence. Explain.

93. Researchers have devised a way to "turn on" specific genes by linking a transcription-activating protein to a version of Cas9 that cannot cleave DNA (it is called dCas9, where *d* stands for *dead*). What else is needed to turn a target gene on?

SELECTED READINGS

Doudna, J.A., The promise and challenge of therapeutic genome editing, *Nature* 578, 229–236, doi: 10.1038/s41586-020-1978-5 (2020).

[A pioneer of CRISPR technology discusses its applications as well as technical issues around its use in humans.]

Giani, A.M., Gallo, G.R., Gianfranceschi, L., and Formenti, G., Long walk to genomics: History and current approaches to genome sequencing and assembly, *Comp. Struct. Biotech. J.* 18, 9–19, doi: 10.1016/j.csbj.2019.11.002 (2019). [Reviews the overall approaches for determining DNA sequences.]

Greaves, M. Nothing in cancer makes sense except…, *BMC Biology* 16, 22, doi: 10.1186/s12915-018-0493-8 (2018). [A highly readable discussion of how the principles of evolutionary biology apply to the development and spread of cancerous cells.]

Her, J. and Bunting, S.F., How cells ensure correct repair of DNA double-strand breaks, *J. Biol. Chem.* 293, 10502–10511, doi: 10.1074/jbc.TM118.000371 (2018). [Presents the details of non-homologous end-joining and homologous (recombination) repair.]

Mueller, S.H., Spenkelink, L.M., and van Oijen, A.M., When proteins play tag: the dynamic nature of the replisome, *Biophys. Rev.* 11, 641–651, doi: 10.1—7/s12551-019-00569-4 (2019). [Reviews viral, bacterial, and eukaryotic replisome components and describes potential obstacles for the replisome.]

Oakley, A.J., A structural view of bacterial DNA replication, *Prot. Sci.* 28, 990–1004, doi: 10.1002/pro.3615 (2019). [Includes models and detailed descriptions of the various proteins of the bacterial replisome.]

Okamoto, K. and Seimiya, H., Revisiting telomere shortening in cancer, *Cells* 8, 107, doi: 10.3390/cells8020107 (2019). [A thorough review of telomeres and the possible roles of telomerase in cancer.]

CHAPTER 20 CREDITS

Figure 20.9 Image based on 1E0J. Singleton, M.R., Sawaya, M.R., Ellenberger, T., Wigley, D.B., Crystal structure of T7 gene 4 ring helicase indicates a mechanism for sequential hydrolysis of nucleotides, *Cell* 101, 589–600 (2000).

Figure 20.10 Image based on 1JMC. Bochkarev, A., Pfuetzner, R.A., Edwards, A.M., Frappier, L., Structure of the single-stranded-DNA-binding domain of replication protein A bound to DNA, *Nature* 385, 176–181 (1997).

Figure 20.13 Image based on 1KFD. Beese, L.S., Friedman, J.M., Steitz, T.A., Crystal structures of the Klenow fragment of DNA polymerase I complexed with deoxynucleoside triphosphate and pyrophosphate, *Biochemistry* 32, 14095–14101 (1993).

Figure 20.14a Image based on 2POL. Kong, X.P., Onrust, R., O'Donnell, M., Kuriyan, J., Three-dimensional structure of the beta subunit of E. coli DNA polymerase III holoenzyme: a sliding DNA clamp, *Cell* 69, 425–437 (1992).

Figure 20.14b Image based on 1AXC. Gulbis, J.M., Kelman, Z., Hurwitz, J., O'Donnell, M., Kuriyan, J., Structure of the C-terminal region of p21(WAF1/CIP1) complexed with human PCNA, *Cell* 87, 297–306 (1996).

Figure 20.15 Image of the open form based on 2KTQ and image of the closed form based on 3KTQ. Li, Y., Korolev, S., Waksman, G., Crystal structures of open and closed forms of binary and ternary complexes of the large fragment of Thermus aquaticus DNA polymerase I: structural basis for nucleotide incorporation, *EMBO J.* 17, 7514–7525 (1998).

Figure 20.19 Image based on 3KYL. Mitchell, M., Gillis, A., Futahashi, M., Fujiwara, H., Skordalakes, E., Structural basis for telomerase catalytic subunit TERT binding to RNA template and telomeric DNA, *Nat. Struct. Mol. Biol.* 17, 513–518 (2010).

Box 20.A Image based on 1BQN. Hsiou, Y., Das, K., Ding, J., Clark Jr., A.D., Kleim, J.P., Rosner, M., Winkler, I., Riess, G., Hughes, S.H., Arnold, E., Structures of Tyr188Leu mutant and wild-type HIV-1 reverse transcriptase complexed with the non-nucleoside inhibitor HBY 097: inhibitor flexibility is a useful design feature for reducing drug resistance, *J. Mol. Biol.* 284, 313–323 (1998).

Figure 20.22 Image based on 1EWQ. Obmolova, G., Ban, C., Hsieh, P., Yang, W., Crystal structures of mismatch repair protein MutS and its complex with a substrate DNA, *Nature* 407, 703–710 (2000).

Figure 20.24 Image based on 4SKN. Slupphaug, G., Mol, C.D., Kavli, B., Arvai, A.S., Krokan, H.E., Tainer, J.A., A nucleotide-flipping mechanism from the structure of human uracil-DNA glycosylase bound to DNA, *Nature* 384, 87–92 (1996).

Figure 20.27 Image based on 1JEY. Walker, J.R., Corpina, R.A., Goldberg, J., Structure of the Ku heterodimer bound to DNA and its implications for double-strand break repair, *Nature* 412, 607–614 (2001).

Figure 20.30 Image based on 3CMW. Chen, Z., Yang, H., Pavletich, N.P., Mechanism of homologous recombination from the RecA-ssDNA/dsDNA structures, *Nature* 453, 489–494 (2008).

Figure 20.33 Image based on 2GEQ. Ho, W.C., Fitzgerald, M.X., Marmorstein, R., Crystal structure of a p53 core dimer bound to DNA, *J. Biol. Chem.* 281, 20494–20502 (2006).

Figure 20.4 Image based on 1AOI. Luger, K., Mader, A.W., Richmond, R.K., Sargent, D.F., Richmond, T.J., Crystal structure of the nucleosome core particle at 2.8 A resolution, *Nature* 389, 251–260 (1997).

Table 20.1 Data from the Restriction Enzyme database: rebase.neb.com/rebase/rebase.html.

Transcription and RNA Processing

Arco Images GmbH/Hinze, K./Alamy Stock Photo

The sex of this red-eared turtle hatchling (*Trachemys scripta*) was determined by the temperature at which the egg developed. High temperatures suppress the expression of a transcription factor necessary to turn on genes for testis development, so the embryo develops as a female.

Do You Remember?

- DNA and RNA are polymers of nucleotides, each of which consists of a purine or pyrimidine base, deoxyribose or ribose, and phosphate (Section 3.2).
- The biological information encoded by a sequence of DNA is transcribed to RNA and then translated into the amino acid sequence of a protein (Section 3.3).
- Genes can be identified by their nucleotide sequences (Section 3.4).
- DNA replication is carried out by stationary protein complexes (Section 20.2).

Transcription is the fundamental mechanism by which a gene is expressed. It is the conversion of stored genetic information (DNA) to a more active form (RNA). The information contained in the sequence of deoxynucleotides in DNA is preserved in the sequence of ribonucleotides in the RNA transcript. Like DNA replication, transcription is characterized by template-directed nucleotide polymerization that requires a certain degree of accuracy. However, unlike DNA synthesis, RNA synthesis takes place selectively, on a gene-by-gene basis. Numerous protein cofactors interact with each other, with RNA polymerase, and with the DNA template to regulate where and when transcription takes place, to accurately transcribe the DNA, and to convert the initial RNA product to its mature, functional form.

21.1 Initiating Transcription

KEY CONCEPTS

Describe the roles of proteins and DNA sequences in initiating transcription.

- Define a gene.
- Explain how histone and DNA modifications affect gene expression.
- Describe the function of a promoter.

- Summarize the roles of eukaryotic transcription factors.
- Explain how distant sites can affect transcription initiation.
- Describe how the *lac* operon is switched on and off.

Transcription, unlike replication, must be highly selective, since only a small portion of the genome is involved. Most bacteria contain a few thousand genes, eukaryotic cells up to around 30,000, and these genes are usually separated by stretches of DNA that are not transcribed. Not surprisingly, the process of identifying a gene and transcribing it into RNA is complex.

What is a gene?

We can consider a **gene** as a segment of DNA that is transcribed for the purpose of expressing the encoded genetic information or transforming it to a form that is more useful to the cell. This definition of a gene requires some qualification:

1. *For protein-coding genes, the RNA transcript (called **messenger RNA** or **mRNA**) includes all the information specifying the sequence of amino acids in a polypeptide.* Keep in mind that not all RNA molecules are translated into protein. **Ribosomal RNA (rRNA), transfer RNA (tRNA),** and other types of RNA molecules carry out their functions without undergoing translation.

2. *Most RNA transcripts correspond to a single functional unit*, for example, one polypeptide. Some RNAs, particularly in prokaryotes, code for multiple proteins and result from the transcription of an **operon,** a set of contiguous genes whose products have related metabolic functions. In a few rare cases, a single mRNA carries information for two proteins in overlapping sequences of nucleotides.

3. *RNA transcripts typically undergo **processing**—the addition, removal, and modification of nucleotides—before becoming fully functional.* Because of variations in mRNA processing and post-translational modification, *several different forms of a protein may be derived from a single gene.*

4. Finally, *the proper transcription of a gene may depend on DNA sequences that are not transcribed but help position the **RNA polymerase** at the transcription start site or are involved in the regulation of gene expression.*

Most of this chapter focuses on the transcription of eukaryotic protein-coding genes. As introduced in Section 3.4, the human genome includes an estimated 21,000 such genes, which have an average length of about 27,000 bp (a typical bacterial gene is about 1000 bp). The actual fraction of protein-coding sequences is quite small, however (about 1.4% of the human genome), because a typical gene includes sequences that are removed from the RNA before it is translated into protein. Nevertheless, an estimated 80% of the human genome may undergo transcription to produce **noncoding RNAs (ncRNAs)** of various sizes. Some of these are listed in **Table 21.1**.

Table 21.1 Some Noncoding RNAs

Type	Typical size (Nucleotides)	Function
Ribosomal RNA (rRNA)	120–4718	Translation (ribosome structure and catalytic activity)
Transfer RNA (tRNA)	54–100	Delivery of amino acids to ribosome during translation
Small interfering RNA (siRNA)	20–25	Sequence-specific inactivation of mRNA
Micro RNA (miRNA)	20–25	Sequence-specific inactivation of mRNA
Large intervening noncoding RNA (lincRNA)	Up to 17,200	Transcriptional control
Small nuclear RNA (snRNA)	60–300	RNA splicing
Small nucleolar RNA (snoRNA)	70–100	Sequence-specific methylation of rRNA

The functions of some types of ncRNA molecules have been elucidated and are described later in this chapter, but others remain mysterious. It is possible that they represent transcriptional "noise," or random synthesis, which is consistent with the observation that they are degraded soon after their synthesis. However, evidence that some of these sequences are conserved among species indicates that they may play a role in large-scale regulation of transcriptional activity.

DNA packaging affects transcription

Recall from Section 20.1 that in eukaryotes, DNA is packaged in nucleosomes, which may form structures that further compact the DNA. Highly condensed chromatin tends to be transcriptionally inactive. The degree of DNA packaging is controlled in part by the covalent modification of histone proteins that form the core of each nucleosome (Fig. 20.4). Various groups, including methyl, acetyl, and phosphoryl groups, may be added to histone residues, particularly in the N-terminal regions that interact closely with the DNA in each nucleosome (**Fig. 21.1**). Some residues can be modified in multiple ways, and serine and threonine residues may also be modified by the addition of *N*-acetylglucosamine groups.

The addition or removal of the various histone-modifying groups can potentially introduce considerable variation in the fine structure of chromatin, offering a mechanism for promoting or preventing gene expression. For example, a lysine residue is positively charged and interacts strongly with the negatively charged DNA backbone, which helps stabilize nucleosome structure. Acetylation of the lysine side chain neutralizes it and weakens its interaction with the DNA, destabilizing the nucleosome and making it more accessible to the transcription machinery.

Acetyl-Lys

In general, acetylation of Lys 9 in histone 3 or acetylation of Lys 16 in histone 4 is associated with transcriptionally active chromatin. However, dimethylation of Lys 9 or trimethylation of Lys 27 in histone 3 is associated with transcriptionally silent chromatin.

Many histone modifications are interdependent. For example, in histone H3, methylation of Lys 9 inhibits phosphorylation of Ser 10, but phosphorylation of Ser 10 promotes acetylation of Lys 14. These relationships, along with the huge number of possible combinations of covalently

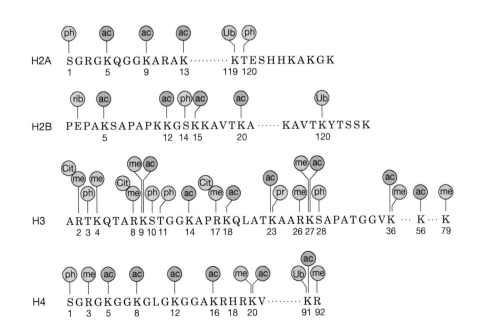

Figure 21.1 Histone modifications. Partial sequences of the four histones are shown using one-letter amino acid codes. The added groups are represented by colored symbols (ac = acetyl, me = methyl, ph = phosphoryl, pr = propionyl, rib = ADP-ribose, and ub = ubiquitin). Cit represents citrulline (deiminated arginine). This diagram is a composite; not all modifications occur in all organisms.

Question Which histone residues undergo acetylation? Methylation? Phosphorylation?

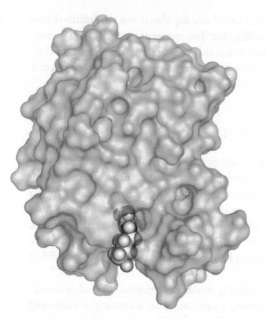

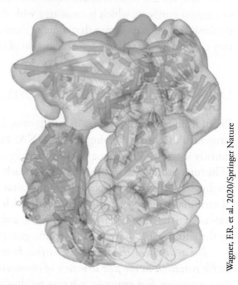

Wagner, F.R. et al. 2020/Springer Nature

Figure 21.2 A bromodomain.   The bromodomain (shown as a semitransparent surface model) includes a cavity for an acetyl-Lys side chain (space-filling model).

Figure 21.3 Chromatin-remodeling complex. This cryo-EM structure shows a nucleosome (yellow) surrounded by the five modules of the yeast RSC complex.

attached groups (as high as 10^{11} different arrangements, in theory) suggest the existence of a **histone code** that marks the transcription-readiness of different regions of a chromosome. Unfortunately, the effects of many of the histone modifications are not yet understood.

Some of the acetyltransferases, methyltransferases, and kinases that alter histones may act on many residues in different histone proteins; other enzymes that act as "writers" are highly specific for one residue in one histone. Because the modifying groups are often derived from common metabolites, such as acetyl-CoA, a cell's metabolic activities may influence the histone marks and thereby help control gene transcription.

In addition to altering nucleosome structure, the histone code can be interpreted by proteins that fine-tune the rate or timing of transcription. The proteins that function as "readers" of the histone code typically have domains that recognize specific chemical groups. For example, a tudor domain binds to dimethylarginine residues; a chromodomain binds to methylated lysine, and a bromodomain binds to acetylated lysine residues (**Fig. 21.2**).

Enzymes such as deacetylases, demethylases, and phosphatases—known as "erasers"—can remove the groups that modify histones. The time course for altering the histone code is highly variable, with some modifications occurring in a highly dynamic fashion and others remaining unchanged for the lifetime of the cell.

It is unlikely that histone modifications completely obliterate nucleosome structure (unwrapping all the DNA would be energetically costly). Instead, the chromatin is rearranged so that critical DNA sequences are exposed rather than wrapped tightly around histones in the nucleosome core. The work of rearranging the DNA is carried out by a protein complex, called RSC (remodels the structure of chromatin) in yeast (**Fig. 21.3**). Its structure includes flexible modules, suggesting that it interacts with DNA in a highly dynamic fashion. The complex uses the free energy of ATP hydrolysis to loosen a segment of DNA from the histone core. The loop of unattached DNA travels around the nucleosome as a wave so that a segment of DNA that was part of the nucleosome core is now available to interact with the transcription machinery (**Fig. 21.4**). In eukaryotes, transcription starts in nucleosome-free stretches of DNA.

Figure 21.4 Nucleosome sliding.   Chromatin-remodeling complexes may act by loosening a portion of DNA (blue) wound around a histone octamer (orange). The nucleosome appears to slide along the DNA, exposing different segments of DNA.

DNA also undergoes covalent modification

In many organisms, including plants and animals, DNA methyltransferases add methyl groups to cytosine residues, using *S*-adenosylmethionine as a donor (see Box 18.B). The methyl group projects out into the major groove of DNA and can potentially alter interactions with DNA-binding proteins.

5-Methylcytosine residue

In mammals, the methyltransferases target C residues next to G residues, so that about 80% of CG sequences (formally represented as CpG) are methylated. CpG sequences occur much less frequently than statistically predicted, but clusters of CpG (called **CpG islands**) are often located near the starting points of genes. Interestingly, these CpG islands are usually unmethylated. This suggests that *methylation may be part of the mechanism for marking or "silencing" DNA that contains no genes.*

After DNA is replicated, methyltransferases modify the new strand, using the methylation pattern of the parent DNA strand as a guide. In this way, the daughter cells produced by cell division continue the gene-expression program of the parent cell.

Variations in DNA methylation may be responsible for **imprinting,** in which the level of expression of a gene depends on its parental origin. Recall that an individual receives one copy of a gene from each parent (see Box 3.A). Normally, both copies (alleles) are expressed, but a gene that has been methylated may not be expressed. The gene behaves as if it has been "imprinted" and knows its parentage. Imprinting explains why some traits are transmitted in a maternal or paternal fashion, contrary to the Mendelian laws of inheritance.

Methylation and other DNA modifications are said to be **epigenetic** (from the Greek *epi*, "above") because they represent heritable information that goes beyond the sequence of nucleotides in the DNA. Stable histone modifications can also carry epigenetic information. Environmental factors, such as diet and exposure to smoke, leave epigenetic marks on DNA. In addition, whole-genome sequencing studies indicate that methylation patterns are altered in cancer and some other diseases, including type 2 diabetes (Section 19.3). In some cases, inhibiting DNA methyltransferase activity is a useful therapy. However, the connection between the methylation status of individual genes and complex metabolic disorders is difficult to unravel.

Transcription begins at promoters

Prokaryotes typically have compact genomes without much nontranscribed DNA, whereas in eukaryotes, protein-coding genes may be separated by large tracts of DNA. However, in both types of organisms, the efficient expression of genetic information usually involves the initiation of RNA synthesis at a particular site, known as a **promoter,** near the protein-coding sequence. The DNA sequence at the promoter is recognized by specific proteins, which either are part of the RNA polymerase protein or subsequently recruit the appropriate RNA polymerase to the DNA to begin RNA synthesis.

In bacteria such as *E. coli,* the most common promoter comprises a sequence of about 40 bases on the 5′ side of the transcription start site. By convention, such sequences are written for the coding, or nontemplate, DNA strand (see Section 3.3) so that the DNA sequence has the same sense and 5′ → 3′ directionality as the transcribed RNA. This *E. coli* promoter includes two conserved segments, whose **consensus sequences** (sequences indicating the nucleotides found most frequently at each position) are centered at positions −35 and −10 relative to the transcription start site (position +1; **Fig. 21.5**).

-35 region   -10 region   Transcription
                                                                                start site
                                                                                +1
5′···NNNTTGACANNNNNNNNNNNNNNNNNNNNNNNNTATAATNNNNNNNNNNN···3′

Figure 21.5 The *E. coli* promoter.   The first nucleotide to be transcribed is position +1. The two consensus promoter sequences, centered around positions −10 and −35, are shaded. N represents any nucleotide.

Question   Approximately how many turns of the DNA helix separate the −10 region from the −35 region and the transcription start site?

E. coli RNA polymerase is a five-subunit enzyme with a mass of about 450 kD and a subunit composition of $\alpha_2\beta\beta'\omega\sigma$. The σ subunit, or σ factor, recognizes the promoter, thereby precisely positioning the RNA polymerase enzyme to begin transcribing. Although bacterial cells contain only one type of core RNA polymerase ($\alpha_2\beta\beta'\omega$), they contain multiple σ factors (seven types in *E. coli*), each specific for a different promoter sequence. Since different genes may have similar promoters, *bacterial cells can regulate patterns of gene expression through the use of different σ factors.* Once transcription is under way, the σ factor is jettisoned, and the remaining subunits of RNA polymerase extend the transcript.

In eukaryotes, some promoters for protein-coding genes include a **TATA box,** an element that is also present in many bacterial promoters. Most eukaryotic genes lack this sequence but may instead exhibit a variety of other conserved promoter elements located both upstream (on the 5′ side) and downstream (on the 3′ side) of the transcription initiation site (**Fig. 21.6**). A given gene may contain several—although not all—of these elements, which may act synergistically. For example, the DPE (downstream promoter element) is always accompanied by an Inr (initiator) element.

The majority of mammalian genes lack well-defined promoters with a precise transcription start site, so transcription initiates over a stretch of 100–150 nucleotides. This DNA typically contains unmethylated CpG islands or other epigenetic marks, but little is known about how transcription factors and RNA polymerase interact with this DNA. Eukaryotic RNA polymerase does not include a subunit corresponding to the prokaryotic σ factor. Instead, RNA polymerase locates a transcription start site through a series of complex protein–protein and protein–DNA interactions. Some evidence suggests that the human genome contains as many as 500,000 potential transcription start sites (many more than the number of genes). Even more surprising is the observation that transcription at many of these sites can proceed in both directions, that is, using each DNA strand as a template. The purpose of all this transcriptional activity is not understood.

Transcription factors recognize eukaryotic promoters

*In eukaryotes, the initiation of transcription typically requires a set of six highly conserved proteins known as **general transcription factors.*** They are abbreviated as TFIIA, TFIIB, TFIID, TFIIE, TFIIF, and TFIIH (the II indicates that these transcription factors are specific for RNA polymerase II,

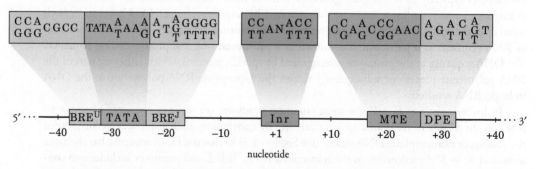

Figure 21.6 Sequences of some eukaryotic promoter elements.   The consensus sequence for each element in humans is given. Some positions can accommodate two or three different nucleotides. N represents any nucleotide.

Question   Compare these sequences to the *E. coli* promoter in Figure 21.5.

the enzyme that transcribes protein-coding genes). Some of these general transcription factors interact specifically with some of the promoter elements shown in Figure 21.6. For example, TFIIB binds to the BRE sequences, and the TATA-binding protein (TBP), a subunit of TFIID, binds to the TATA box. In fact, TBP also binds to promoter regions that lack an obvious TATA box.

The various transcription factors play numerous roles in preparing the DNA for transcription and recruiting RNA polymerase. TBP, a saddle-shaped protein about $32 \times 45 \times 60$ Å, consists of two structurally similar domains that sit astride the DNA at an angle (**Fig. 21.7**). This protein–DNA interaction introduces two sharp kinks into the DNA. The kinks are caused by the insertion of two phenylalanine side chains like a wedge between pairs of DNA nucleotides. This observation, along with other data, suggests that transcription initiation depends on DNA structural features, not just specific nucleotide sequences.

Once TBP is in place at the promoter, the conformationally altered DNA serves as a stage for the assembly of a preinitiation complex containing additional transcription factors and RNA polymerase (**Fig. 21.8**). TFIIB, for example, helps position the DNA near the polymerase active site. TFIIE joins the complex and recruits TFIIH, one subunit of which is a helicase that unwinds the DNA in an ATP-dependent manner. The result is an open structure called a transcription bubble that is stabilized in part by the binding of TFIIF, which interacts with the nontemplate DNA strand.

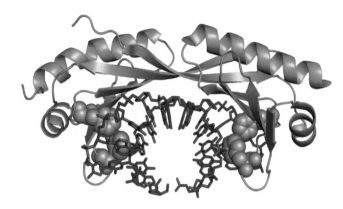

Figure 21.7 Structure of TBP bound to DNA. The TBP polypeptide (green) forms a pseudosymmetrical structure that straddles a segment of DNA (shown in blue and viewed end-on). The insertion of TBP Phe residues (orange) bends the DNA in two places.

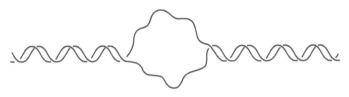

Transcription bubble

The structural changes in the DNA may extend beyond the immediate site of the transcription bubble. For instance, the TFIID component known as TAF1 has histone acetyltransferase activity, which may allow it to alter nucleosome packing by neutralizing lysine side chains. TAF1 may also diminish histone H1 cross-linking of neighboring nucleosomes by helping to link the small protein ubiquitin to H1, thereby marking it for proteolytic destruction by a proteasome (see Section 12.1).

In TAF1, two bromodomains are arranged side by side, which might allow the protein to bind cooperatively to a multiply acetylated histone protein. The histone acetyltransferase activity of TAF1 itself may lead to localized hyperacetylation, a positive feedback mechanism for promoting gene transcription.

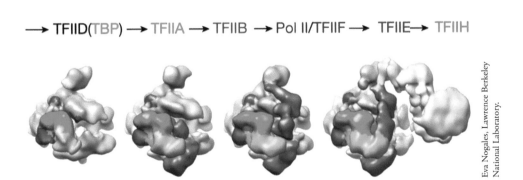

Figure 21.8 Assembly of the human preinitiation complex. This model is based on cryo-electron microscopy images, with each component in a different color. Pol II (the large gray shape) is RNA polymerase II.

Mediator integrates multiple regulatory signals

Additional sets of protein–protein and protein–DNA interactions may participate in the highly regulated expression of many eukaryotic genes. Whereas the rate of prokaryotic gene transcription varies about 1000-fold between the most and least frequently expressed genes, the gene transcription rate in eukaryotes may vary by as much as 10^9-fold. Some of this fine control is due to **enhancers,** DNA sequences that range from 50 to 1500 bp and are located up to 120 kb upstream or downstream of the transcription start site. A single gene may have more than one enhancer functionally associated with it, and hundreds of thousands of these elements appear to be scattered around the human genome.

The proteins that bind to enhancers are commonly called **activators,** although they are also known simply as transcription factors (hence the term *general transcription factor* for the ones that position RNA polymerase at the transcription start site). By facilitating (or inhibiting) transcription, these DNA-binding proteins can shape an organism's gene expression patterns in response to internal or external signals. We have already seen examples of transcription factors that are sensitive to the presence of cholesterol (Section 17.4) or oxygen (Section 19.2). Many transcription factors are linked to signal transduction pathways. For example, steroid hormones directly activate DNA-binding proteins (Section 10.4), and Ras signaling indirectly activates several transcription factors (Section 10.3). These proteins interact with DNA in a variety of ways (**Box 21.A**). The human genome includes about 1600 different transcription factors.

When an activator binds to the enhancer, a protein complex known as **Mediator** *links the enhancer-bound activator to the transcription machinery poised at the transcription start site.* Note that this interaction requires that the DNA form a loop in order to connect the enhancer and promoter region (**Fig. 21.9**). The packaging of DNA in nucleosomes may facilitate this long-range interaction by minimizing the length of the intervening DNA loop. In addition to enhancers, *a gene may have associated* **silencer** *sequences that bind proteins known as* **repressors.** Mediator may also relay silencer–repressor signals to the transcription machinery in order to repress gene transcription. The simple scheme shown in Figure 21.9 does not convey the complexity of many activator and repressor pathways, which may involve competition for binding sites among the many protein factors.

The Mediator complex, which contains 25 polypeptides in yeast and about 35 in mammals, interacts with RNA polymerase as well as with the general transcription factors (**Fig. 21.10**). As many as 60 proteins may ultimately congregate at a transcription initiation site. The three-lobed Mediator complex has some conformational flexibility, and many of its subunits include intrinsically disordered segments, suggesting that Mediator participates in dynamic interactions with its binding partners during the course of transcription initiation.

The multiplicity of activators and repressors, along with variability in Mediator sub-units, could constitute a sophisticated system for fine-tuning transcription, which could

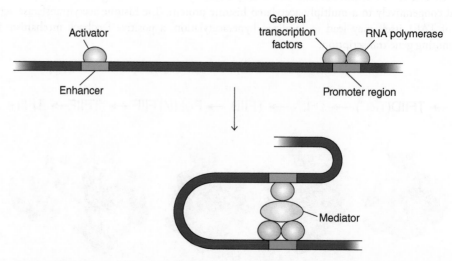

Figure 21.9 Overview of Mediator function. The Mediator complex interacts with an activator protein bound to a gene's enhancer sequence and with the general transcription factors and RNA polymerase at the gene's transcription start site to promote gene expression.

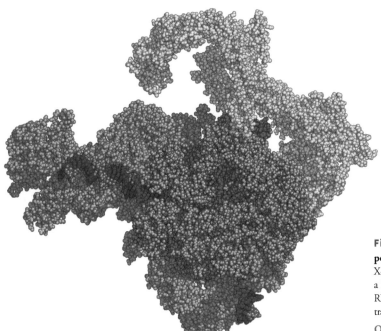

Figure 21.10 Mediator bound to RNA polymerase. This model, which is based on data from X-ray crystallography and cryo-electron microscopy, shows a 15-subunit yeast Mediator complex (yellow) surrounding RNA polymerase (green). Proteins representing the six general transcription factors are shown in orange, and the DNA is blue.

Question **Where would activators and repressors fit?**

help account for the different patterns of gene expression that distinguish the 200 or so cell types found in the human body. Although gene expression can be regulated at multiple points—transcription, RNA processing, protein synthesis, or protein processing—control of transcription is the largest source of variation in the amounts of proteins a cell produces. Furthermore, many genetic features that distinguish individuals result from variations in gene regulatory sequences rather than differences in the protein-coding segments of genes.

Box 21.A   DNA-Binding Proteins

The proteins that directly participate in or regulate processes such as DNA replication, repair, and transcription must interact intimately with the DNA double helix. In fact, many of the proteins that promote or suppress transcription recognize and bind to particular sequences in the DNA. However, there do not seem to be any strict rules by which certain amino acid side chains pair with certain nucleotide bases. In general, interactions are based on van der Waals interactions and hydrogen bonds, often with intervening water molecules.

An examination of the structures of a large variety of protein–DNA complexes in prokaryotic and eukaryotic cells reveals that the DNA-binding proteins fall into a limited number of classes, depending on the structural motif that contacts the DNA. Many of these motifs are likely the result of convergent evolution and may therefore represent the most stable and evolutionarily versatile ways for proteins to interact with DNA.

By far the most prevalent mode of protein–DNA interaction involves a protein α helix that binds in the major groove of DNA. This DNA-binding motif may take the form of a helix–turn–helix (HTH) structure in which two perpendicular α helices are connected by a small loop of at least four residues. The HTH motifs are colored red in the structure below, and the DNA is blue. In most cases, the side chains of one helix insert into the major groove and directly contact the exposed edges of bases in the DNA. Residues in the other helix and the turn may interact with the DNA backbone.

In prokaryotic and eukaryotic proteins, the HTH helices are usually part of a larger bundle of several α helices, which form a stable domain with a hydrophobic core. Prokaryotic transcription factors tend to be homodimeric proteins (as in the bacteriophage λ repressor) that recognize palindromic DNA sequences. In contrast, eukaryotic transcription factors more commonly are heterodimeric

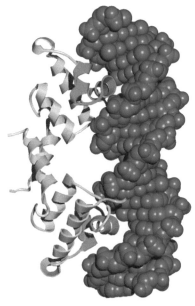

A portion of the bacteriophage λ repressor bound to DNA.

or contain multiple domains that recognize a nonsymmetrical series of binding sites. For this reason, eukaryotic DNA-binding proteins are able to interact with a wider variety of target DNA sequences.

Many eukaryotic transcription factors include a DNA-binding motif in which one zinc ion (sometimes two) is tetrahedrally coordinated by cysteine or histidine side chains. The metal ion stabilizes a small protein domain (which is sometimes involved in protein–protein rather than protein–DNA interactions). In most cases, the DNA-binding motif, known as a zinc finger (see Fig. 4.15), consists of two antiparallel β strands followed by an α helix. In the structure shown below, one of the three zinc fingers is colored red. The Zn^{2+} ions are represented by purple spheres. As in the HTH proteins, the helix of each zinc finger motif inserts into the major groove of DNA, where it interacts with a three-base-pair sequence.

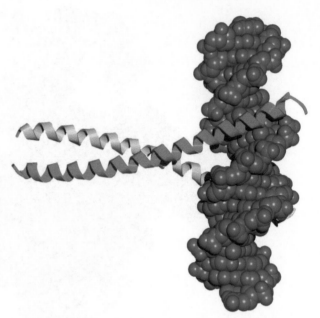

A portion of the yeast transcription factor GCN4.

Zinc finger domains from the mouse transcription factor Zif268.

Some homodimeric DNA-binding proteins in eukaryotes include a leucine zipper motif that mediates protein dimerization. Each subunit has an α helix about 60 residues long, which forms a coiled coil with its counterpart in the other subunit (see Section 5.3). Leucine residues, appearing at every seventh position, or about every two turns of the α helix, mediate hydrophobic contacts between the two helices (they do not actually interdigitate, as the term *zipper* might suggest). The DNA-binding portions of a leucine zipper protein are extensions of the dimerization helices that bind in the major groove, as shown here.

In a few proteins, a β sheet interacts with the DNA (TBP is one such protein; see Fig. 21.7). In a few cases, two antiparallel β strands constitute the DNA-binding segment, fitting into the major groove so that protein side chains can form hydrogen bonds with the functional groups on the edges of the DNA bases.

The DNA-binding proteins described here interact with a limited portion of the DNA (typically just a few base pairs), marking the positions of regulatory DNA sequences and making additional protein–protein contacts that control processes such as gene transcription. Proteins that carry out catalytic functions (polymerases, for example) interact with the DNA much more extensively—but in a sequence-independent manner—and they tend to envelop the entire DNA helix.

Question Explain why sequence-specific DNA-binding proteins usually interact with the major groove rather than with the minor groove.

Prokaryotic operons allow coordinated gene expression

Although prokaryotic cells regulate transcription initiation using less complicated mechanisms than eukaryotes use, functionally related genes may be organized in operons to ensure their coordinated expression in response to some metabolic signal. About 13% of prokaryotic genes are found in operons, including the three genes of the well-studied *E. coli lac* operon, whose protein products are involved in lactose metabolism.

When bacterial cells grown in the absence of the disaccharide lactose are transferred to a medium containing lactose, the synthesis of two proteins required for lactose catabolism is quickly enhanced. These proteins are lactose permease, a transporter that allows lactose to enter the cells, and β-galactosidase, an enzyme that catalyzes the hydrolysis of lactose to its component monosaccharides, which can then be oxidized to produce ATP:

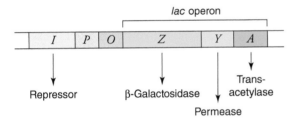

Lactose

H_2O — β-galactosidase

Galactose + Glucose

β-Galactosidase and lactose permease, along with a third enzyme (thiogalactoside transacetylase, whose physiological function is not clear), are encoded by the three-gene *lac* operon. The three genes (designated *Z*, *Y*, and *A*) are transcribed as a single unit from a promoter, *P*. Near the start of the β-galactosidase gene is a regulatory site called the operator, *O*, which binds a protein called the *lac* repressor. The repressor itself is a product of the *I* gene, which lies just upstream of the *lac* operon:

lac operon

| I | P | O | Z | Y | A |

Repressor   β-Galactosidase   Trans-acetylase

Permease

In the absence of lactose, the *Z*, *Y*, and *A* genes are not expressed because the *lac* repressor binds to the operator. The repressor is a tetrameric protein that functions as a dimer of dimers so that it can bind simultaneously to two segments of operator DNA (**Fig. 21.11**). The operator segments are separated by a 93-bp stretch of DNA, which forms a loop. The repressor does not interfere with RNA polymerase binding to the promoter DNA, but it prevents the polymerase from initiating transcription.

When the cell is exposed to lactose, a lactose isomer called allolactose acts as an inducer of the *lac* operon (allolactose is generated from lactose by trace amounts of β-galactosidase present in the bacterial cell):

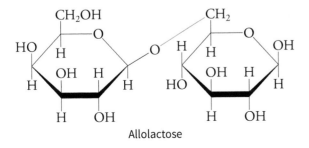

Allolactose

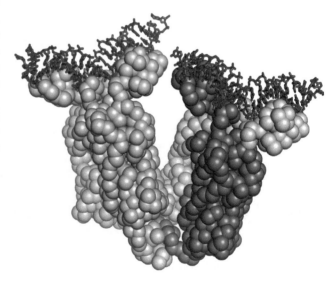

Figure 21.11 *lac* **repressor bound to DNA.** In this model, DNA segments are blue, the monomeric units of the repressor are light green, light blue, light pink, and dark pink (each amino acid is represented by a sphere).

The inducer binds to the *lac* repressor, triggering a conformational change that causes it to release its grip on the operator sequences. As a result, the promoter is freed for transcription, and the production of β-galactosidase and lactose permease can increase 1000-fold within a few minutes. This simple regulatory system ensures that the proteins required for metabolizing lactose are synthesized only when lactose is available as a metabolic fuel.

Genetic engineers have used the *lac* repressor gene (whose protein product can be activated by adding a lactose derivative) to "turn on" foreign genes that have been inserted near *lac* operator and promoter sequences (see Section 20.6).

In *E. coli* cells, *lac* operon transcription is also controlled by the catabolite activator protein (CAP), which is active in the presence of cyclic AMP, a signaling molecule that increases when the bacteria's main fuel—glucose—is scarce. The CAP–cAMP complex binds to the *lac* promoter to stimulate RNA polymerase. This system helps ensure that when glucose is absent and lactose is present, the cells can synthesize the proteins required for metabolizing lactose. Note that CAP is a positive regulator of gene expression, whereas the *lac* repressor is a negative regulator. Many operons and individual genes are similarly regulated by multiple factors that increase or decrease gene transcription.

Concept Check

1. Discuss reasons why it is difficult to define a gene.

2. For each of the histone modifications in Fig. 21.1, predict whether DNA binding would become looser, tighter, or remain unchanged.

3. Compare transcription initiation in prokaryotes and eukaryotes. Which prokaryotic elements also appear in eukaryotes?

4. Make a list of all the activities carried out by proteins during eukaryotic transcription initiation.

5. Summarize the roles of enhancers, silencers, activators, repressors, and mediators in regulating gene transcription.

6. Explain how hormone signaling could affect the activity of activators and repressors.

7. Explain how a mutation in a DNA segment that does not code for a protein could affect production of that protein.

8. Describe how genetic engineers could use elements of the *lac* operon to design a system for turning the expression of particular genes on or off.

21.2 RNA Polymerase

KEY CONCEPTS

Summarize the activity of RNA polymerase in elongation and termination.

- Describe the reaction catalyzed by RNA polymerase.
- Describe the changes that occur in the transition from initiation to elongation.
- Compare transcription termination in prokaryotes and eukaryotes.

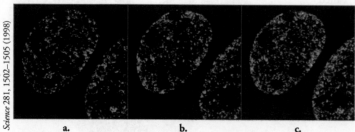

Ronald Berezney. From Wei, X. et al., *Science 281*, 1502–1505 (1998)

Figure 21.12 Spatial separation of transcription and replication.   a. In this fluorescence microscopy image of mouse nuclei, sites of DNA replication are green. **b.** Sites of RNA transcription are red. A single nucleus may contain 2000 to 3000 transcription sites or "factories." **c.** The merged images show little overlap of replication and transcription sites.

Like DNA replication, *RNA transcription is carried out by immobile protein complexes that reel in the DNA.* These transcription factories in eukaryotic nuclei can be visualized by immunofluorescence microscopy and are distinct from the replication factories where DNA is synthesized (**Fig. 21.12**). If the RNA polymerase were free to track along the length of a DNA molecule, rotating around the helical template, the newly synthesized RNA strand would become tangled with the DNA. In fact, except for a short 8- to 9-bp hybrid DNA–RNA helix at the polymerase active site, the newly synthesized RNA is released as a single-stranded molecule:

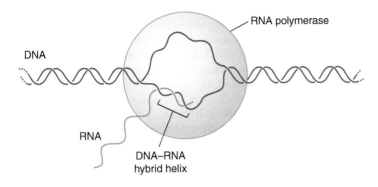

Bacterial cells contain just one type of RNA polymerase, but eukaryotic cells contain three (plus additional polymerases for chloroplasts and mitochondria). Eukaryotic RNA polymerase I is responsible for transcribing rRNA genes, which are present in multiple copies. RNA polymerase III mainly synthesizes tRNA molecules and other small RNAs. Protein-coding genes are transcribed by RNA polymerase II, which is the main focus of this section.

Once the transcription start site has been selected, with the help of transcription factors, RNA polymerase can begin transcribing the DNA. RNA polymerases I and III are able to melt apart about 15 bp of DNA on their own, but RNA polymerase II relies on the helicase activity of TFIIH. In this case, the initial transcription bubble is larger but shrinks to the standard 15-bp size after RNA polymerase II begins its work.

RNA polymerases have a common structure and mechanism

Mammalian RNA polymerase has a mass of over 500 kD and at first glance has the same fingers, thumb, and palm domains as DNA polymerase (**Fig. 21.13**). The highly conserved sequences of eukaryotic RNA polymerases indicate that they have virtually identical structures. The core structure of RNA polymerase and its catalytic mechanism are also very similar between eukaryotes and prokaryotes. The differences are mainly at the enzyme surface, where the proteins interact with transcription factors and other regulatory proteins. RNA viruses use a similar enzyme to copy their RNA genome (**Box 21.B**).

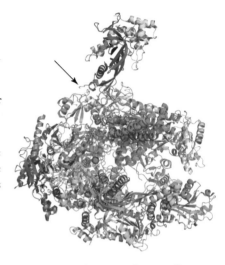

Figure 21.13 Structure of mammalian RNA polymerase II.   The model shows the 12 protein subunits in different colors. A magenta sphere represents an Mg^{2+} ion in the active site. An arrow indicates where the highly flexible C-terminal domain of the largest subunit would be located.

Question Compare this structure to that of DNA polymerase (**Fig. 20.13**). Locate the long green "bridge helix" to the right of the active site.

| **Box 21.B** | **RNA-Dependent RNA Polymerase** |

With a few exceptions, such as poly(A) polymerase (Section 21.3), nucleotide polymerases follow a template strand to synthesize a new polynucleotide chain. We have examined DNA-dependent DNA polymerase (Section 20.2) and DNA-dependent RNA polymerase, as well as RNA-dependent DNA polymerases (reverse transcriptase in Box 20.A and telomerase in Section 20.3). The remaining member of the polymerase club is RNA-dependent RNA polymerase, which uses an RNA strand as a template to synthesize a new complementary RNA strand.

Many RNA viruses (see Box 3.C) require an RNA-dependent RNA polymerase in order to replicate their genome. These virus-encoded enzymes have the fingers-palm-thumb structure common to all polymerases. Some of the viral polymerases have additional enzyme activities, but none of them has a built-in proofreader. As a result, RNA viruses have mutation rates as high as 10^{-4}, that is, one error for every 10,000 nucleotides polymerized. Presumably, this error rate is low enough for the virus to complete its infection cycle but high enough to permit mutations that allow the virus to escape its host organism's defenses.

RNA-dependent RNA polymerases are useful antiviral drug targets, because mammalian cells lack a similar enzyme. Many drugs that prevent the replication of RNA viruses are nucleoside analogs that are missing a 3′ OH group, so they can be incorporated into a polynucleotide chain by a polymerase but cannot participate in the next reaction cycle. Such chain-terminator drugs are also used against the human immunodeficiency virus (Box 20.A).

The coronavirus known as SARS-CoV-2 is not sensitive to most chain terminators, because in addition to coding for an RNA-dependent RNA polymerase, it has a gene for an exonuclease that can remove nucleotides from the 3′ end of a growing chain. The exonuclease is essentially an external polymerization proofreader that can excise a mispaired nucleotide residue as well as a chain-terminator nucleotide. The presence of this proofreading activity also explains why the mutation rate of coronaviruses is relatively low.

The antiviral drug remdesivir, originally developed for hepatitis virus, has been tested against SARS-CoV-2. Remdesivir is a phosphoramidate nucleotide analog:

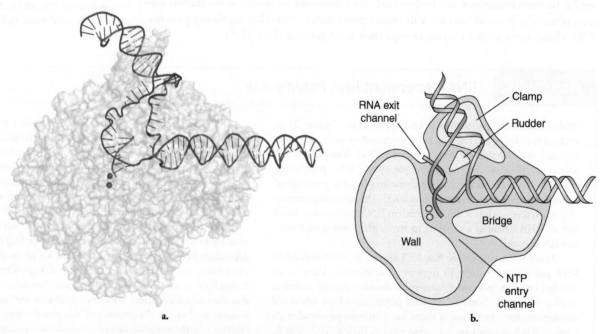

Remdesivir

The bulky group on the left masks the phosphate group, which would otherwise bear a negative charge, so that the drug can easily diffuse into cells, where it is converted to a triphosphate. The group on the right resembles adenine, so the coronavirus' RNA polymerase can use the compound as a substrate, pairing it with uracil in the template RNA strand. In fact, the viral polymerase incorporates the remdesivir nucleotide much more efficiently than a standard adenine nucleotide. The cyano ($C \equiv N$) group was added to the drug molecule to prevent it from interacting with the host cell's mitochondrial RNA polymerase.

Incorporating the remdesivir nucleotide, which has a 3′ OH group, into the new RNA chain does not appear to stop polymerization immediately. Instead, the polymerase proceeds to add about three more nucleotides before the odd shape of the remdesivir residue causes the enzyme to stop working. By that point, it is too late for the proofreading enzyme—which is an exonuclease, not an endonuclease—to remove the mismatched nucleotide. Although remdesivir may slow viral replication, it does not seem to be effective in preventing severe illness in humans.

The active site of RNA polymerase is located at the bottom of a positively charged cleft between the two largest subunits. The DNA to be transcribed enters the active-site cleft of RNA polymerase, and the two DNA strands separate to form the transcription bubble. The nontemplate (coding) strand lies outside of the active-site cavity, but the template strand threads through the polymerase, making an abrupt right-angle turn where it encounters a wall of protein. Here, the template base points away from the standard B-DNA conformation and toward the floor of the active site. This geometry allows the deoxynucleotide residue to base pair with an incoming ribonucleoside triphosphate, which enters the active site through a channel in the floor (**Fig. 21.14**).

Note that the incoming nucleotide has only a 25% chance of pairing correctly (since there are four possible ribonucleotide substrates: ATP, CTP, GTP, and UTP). When the correct nucleotide forms hydrogen bonds with the template base, a loop of protein called the trigger closes over it.

Figure 21.14 RNA polymerase with DNA and RNA. **a.** Surface view of human RNA polymerase II, looking from the bottom of the protein as depicted in Figure 21.13. The template DNA strand is blue, the nontemplate (coding) strand is purple, and a six-base segment of RNA is red. The magenta Mg^{2+} ions mark the location of the active site. **b.** Schematic diagram of RNA polymerase showing the approximate positions of internal structures that facilitate the polymerization reaction. DNA enters from the right, and the strands separate before reannealing on the left. Nucleotide substrates reach the active site through an opening at the bottom of the enzyme.

This movement brings a histidine side chain close enough to donate a proton to the second phosphate of the NTP substrate, a step necessary to incorporate a monophosphate residue into the growing RNA chain. A mispaired nucleotide or a dNTP does not allow the trigger loop to move close enough for catalysis to occur; in this way, selection of the correct nucleotide is linked to the polymerization reaction, with an error rate of only about 1 in 10,000.

Catalysis requires two metal ions (Mg^{2+}) coordinated by negatively charged side chains. Like DNA polymerase, *RNA polymerase catalyzes nucleophilic attack of the 3′ OH group of the growing polynucleotide chain on the 5′ phosphate of an incoming nucleotide* (see Fig. 20.11). The RNA molecule is therefore extended in the $5′ \rightarrow 3′$ direction. No primer is needed, so the RNA chain begins with the joining of two ribonucleotides. As the RNA strand is synthesized, it forms a hybrid double helix with the DNA template strand for eight or nine base pairs. The conformation of the hybrid is intermediate to the A form (as in double-stranded RNA) and the B form (as in double-stranded DNA):

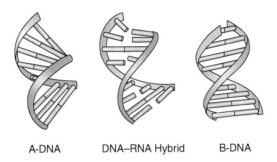

A-DNA   DNA–RNA Hybrid   B-DNA

RNA polymerase is a processive enzyme

During transcription, a portion of RNA polymerase known as the clamp (see Fig. 21.14) rotates by about 30° to close down snugly over the DNA template. *Clamp closure appears to promote the high processivity of RNA polymerase.* In experiments where RNA polymerase was immobilized and a magnetic bead was attached to the DNA, up to 180 rotations (representing thousands of base pairs at 10.4 bp per turn) were observed before the RNA polymerase slipped. This processivity is essential, since genes are usually thousands—sometimes millions—of nucleotides long, and the largest ones may require many hours to transcribe.

With each reaction cycle, the bridge helix located near the active site (see Figs. 21.13 and 21.14) appears to oscillate between a straight and bent conformation. This alternating movement seems to act as a ratchet to aid translocation of the template so that the next nucleotide can be added to the growing RNA chain. Throughout transcription, the sizes of the transcription bubble and the DNA–RNA hybrid helix remain constant. A protein loop known as the rudder (see Fig. 21.14b) may help separate the RNA and DNA strands so that a single RNA strand is extruded from the enzyme as the template and nontemplate DNA strands reanneal to restore the double-stranded DNA.

Like DNA polymerase, *RNA polymerase carries out proofreading.* If a deoxynucleotide or a mispaired ribonucleotide is mistakenly incorporated into RNA, the DNA–RNA hybrid helix becomes distorted. This causes polymerization to cease, and the newly synthesized RNA "backs out" of the active site through the channel by which ribonucleotides enter (**Fig. 21.15**). Eukaryotic transcription factor TFIIS binds to the RNA polymerase and stimulates it to act as an endonuclease to trim away the RNA containing the error. Transcription may resume if the 3′ end of the truncated transcript is then repositioned at the active site. Proofreading activity lowers the error rate to about 1 in 1,000,000.

Transcription elongation requires changes in RNA polymerase

In order for RNA polymerase II to efficiently transcribe a gene, it must move beyond the promoter and become part of an elongation complex. An RNA polymerase positioned at a promoter appears to initiate RNA synthesis repeatedly, producing and

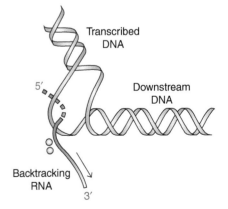

Transcribed DNA

Downstream DNA

5′

Backtracking RNA

3′

Figure 21.15 Schematic view of backtracking RNA in RNA polymerase. If polymerization stops due to a polymerization error, the 3′ end of the RNA transcript may back up into the channel for incoming nucleotides. The enzyme can then cleave off the 3′ end of the RNA and resume transcription.

Question Identify the oldest and newest segments of RNA.

releasing many short transcripts (up to about 12 nucleotides). This action most likely reflects the fact that the enzyme remains associated with Mediator and other factors that keep it anchored to the promoter. As a result, the template DNA enters the RNA polymerase active site but has nowhere to go after transcription starts. In prokaryotes, the buildup of strain—termed DNA scrunching—may provide the driving force for the polymerase to eventually escape the promoter and discard its σ factor. In eukaryotes, TFIIB occupies part of the RNA polymerase II active site and must be displaced to accommodate an RNA longer than a few residues. Moreover, the exit channel for the RNA is partially blocked when the polymerase is in initiation mode. RNA polymerase must shift to an elongation conformation in order to relieve these constraints and advance past the promoter.

In eukaryotes, the shift in RNA polymerase involves the C-terminal domain of the largest subunit of RNA polymerase (a structure that is intrinsically disordered and therefore not visible in the models shown in Figs. 21.13 and 21.14). The C-terminal domain of mammalian RNA polymerase contains 52 seven–amino acid pseudorepeats with the consensus sequence

Tyr–Ser–Pro–Thr–Ser–Pro–Ser

1   2   3   4   5   6   7

Serine residues 2 and 5 (and possibly 7) of each heptad can potentially be phosphorylated. During the initiation phase of transcription, Mediator triggers the phosphorylation process, which is initially carried out by a kinase subunit of TFIIH that mainly targets Ser 5. Later rounds of phosphorylation are catalyzed by other enzymes and mainly target Ser 2.

When RNA polymerase II becomes phosphorylated at its C-terminal domain, it can no longer bind a Mediator complex. At this point, the polymerase can abandon the transcription- initiating proteins and advance along the template DNA. In fact, when RNA polymerase "clears" the promoter, it leaves behind some general transcription factors, including TFIID (**Fig. 21.16**). These proteins, along with the Mediator complex, can reinitiate transcription by recruiting another RNA polymerase to the promoter.

Even after escaping the promoter, mammalian RNA polymerase II may pause about 30 to 50 nucleotides downstream of the transcription start site. Pausing, which can last anywhere from a few minutes to more than an hour, appears to depend on the binding of proteins such as NELF (negative elongation factor), which blocks the nucleotide entrance tunnel and causes the RNA–DNA hybrid helix to tilt so that a new ribonucleotide cannot be added. Shedding NELF and proceeding with transcription requires additional phosphorylation of the polymerase C-terminal domain and association with dedicated elongation factors. TFIIS may help get the paused polymerase back on track.

What is the purpose of pausing? The binding sites for initiation factors and elongation factors are mutually exclusive, so swapping them out may ensure that RNA polymerase II can finish transcribing a gene as efficiently as possible. Pausing may also help coordinate the RNA-modification reactions that occur co-transcriptionally (described in the next section). Several lines of evidence indicate that less-phosphorylated RNA polymerases associate with Mediator and other proteins in a liquid phase that is separate from the liquid phase containing more-phosphorylated RNA polymerases. These discrete phases, called condensates or membraneless organelles (see Section 4.3), correspond to some of the spots visible in Figure 21.12. Apparently, RNA polymerase II

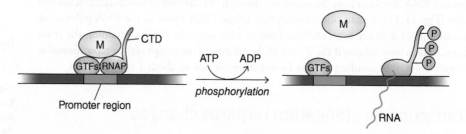

Figure 21.16 Promoter clearance.   During initiation, the Mediator complex (M) interacts with RNA polymerase (RNAP, with an unphosphorylated C-terminal domain, CTD). When the C-terminal domain undergoes phosphorylation, RNA polymerase dissociates from the Mediator complex and the general transcription factors (GTFs), which remain at the promoter.

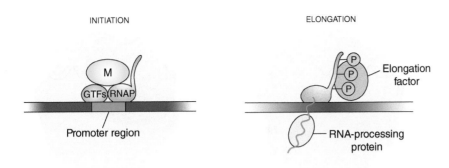

Figure 21.17 Liquid phases for transcription initiation and elongation. DNA, RNA polymerase, and the associated protein cofactors are components of separate liquid phases for the initiation and elongation stages of transcription. **Question** Why would phosphorylation of an intrinsically disordered segment help a protein move to a different liquid phase?

transitions between the phases according to its phosphorylation state. *The separate phases may prevent the initiation and elongation proteins from competing with each other for binding to RNA polymerase* (**Fig. 21.17**). In eukaryotes, RNA polymerase I is known to transcribe ribosomal RNA genes in another membraneless phase—the **nucleolus,** which is visible as a large dark spot in the nucleus in Fig. 1.8.

The rate of RNA synthesis seems to vary erratically from about 500 to 5000 nucleotides per minute, depending on the DNA sequence being transcribed. In prokaryotes, RNA polymerase may periodically slow down to permit protein synthesis to keep pace with transcription (both processes occur in the cytosol). The pace of a eukaryotic RNA polymerase may be modulated by the necessity of dealing with DNA that is wrapped around histones (only promoter DNA is entirely free of nucleosomes). Histone acetyltransferases associated with the RNA polymerase may alter the nucleosomes of a gene undergoing transcription. Consequently, *the first RNA polymerase to transcribe a gene acts as a "pioneer" polymerase that helps pave the way for additional rounds of transcription.*

Transcription is terminated in several ways

In both prokaryotes and eukaryotes, transcription termination involves cessation of RNA polymerization, release of the complete RNA transcript, and dissociation of the polymerase from the DNA template. In prokaryotes such as *E. coli,* termination occurs mainly by one of two mechanisms. In about half of *E. coli* genes, the 3′ end includes palindromic sequences followed by a stretch of T residues. The RNA corresponding to the gene can form a stem–loop or hairpin structure followed by a stretch of U residues. The other half of *E. coli* genes lack a hairpin sequence, and their termination depends on the action of a protein such as Rho, a hexameric helicase that may act by prying the nascent RNA away from the DNA and pushing the polymerase off the template. Both types of termination mechanisms can be explained in terms of destabilizing the DNA–RNA hybrid helix that forms in the transcription bubble during elongation. Hairpin formation or the ATP-dependent action of Rho exerts a force that causes the RNA polymerase to advance, extending the leading end of the transcription bubble without extending the RNA (**Fig. 21.18**). The hybrid helix, now shortened and consisting of relatively weak U:A base pairs, is easily disrupted, freeing the RNA transcript.

In eukaryotic protein-coding genes, transcription termination is somewhat imprecise. When RNA polymerase II has transcribed a DNA sequence that marks the end of a gene, it may pause, and a protein complex binds to the RNA and cuts it away from the transcription machinery. The RNA polymerase continues to transcribe the DNA for a short time, but an exonuclease quickly hydrolyzes the newly made RNA until it catches up with the polymerase and, like Rho protein (Fig. 21.18b), nudges the polymerase off the DNA template. Although the transcription termination point is not strictly defined, it does not really matter because the protein-coding portion of the mRNA has already been synthesized.

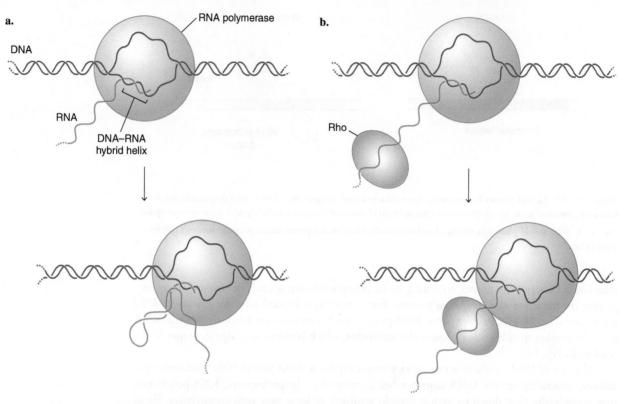

Figure 21.18 Mechanisms for transcription termination in prokaryotes. **a.** Formation of an RNA hairpin shortens the length of the DNA–RNA hybrid helix, which is rich in easily separated A:U base pairs. **b.** The movement of Rho along the RNA transcript pushes the polymerase forward, leaving a short hybrid helix from which the RNA more easily dissociates.

The production of an RNA molecule from a DNA template, a major step in the central dogma of molecular biology, makes a lot of sense. The RNA molecule may undergo extensive modification before it is used, and it is relatively quickly degraded, but the original information in the DNA remains intact and can be transcribed over and over again.

Concept Check

1. Describe the overall structure of RNA polymerase and the arrangement of DNA and RNA within the protein.
2. Compare RNA polymerase and DNA polymerase in terms of mechanism, accuracy, and processivity.
3. Explain what would happen if the DNA–RNA hybrid helix were too long or too short.
4. Describe the role of the C-terminal domain during transcription initiation, elongation, and termination.
5. Describe the events that accompany the switch from transcription initiation to elongation.
6. Explain what the "pioneer" polymerase accomplishes.

21.3 RNA Processing

KEY CONCEPTS

Describe the ways that RNA is processed.

- Recognize covalent modifications of RNA and explain their purpose.
- Describe the advantages of splicing protein-coding genes.
- Explain how RNA interference regulates gene expression.
- Review the features that make RNA structurally versatile.

In a prokaryotic cell, translation of an mRNA transcript begins even before it is fully synthesized. In a eukaryotic cell, however, transcription occurs in the nucleus (where the DNA is located), but translation takes place in the cytosol (where ribosomes are located). The separation of these processes gives eukaryotic cells two advantages: (1) They can modify the mRNA to produce a greater variety of gene products, and (2) the extra steps of RNA processing and transport provide additional opportunities for regulating gene expression.

In this section we examine some major types of RNA processing. Keep in mind that RNA is probably never found alone in the cell but rather interacts with a variety of proteins that covalently modify the transcript, splice out unneeded sequences, export the RNA from the nucleus, deliver it to a ribosome (if it is an mRNA), and eventually degrade it when it is no longer useful to the cell.

Eukaryotic mRNAs receive a 5′ cap and a 3′ poly(A) tail

mRNA processing begins well before transcription is complete, as soon as the transcript begins to emerge from RNA polymerase. Many of the various enzymes required for capping the 5′ end of the mRNA, for extending the 3′ end, and for splicing are recruited to the phosphorylated domain of RNA polymerase, *so processing is closely linked to transcription.* In fact, the presence of processing enzymes may actually promote transcriptional elongation.

At least three enzyme activities modify the 5′ end of the emerging mRNA to produce a structure called a **cap** *that protects the polynucleotide from 5′ exonucleases.* First, a triphosphatase removes the terminal phosphate from the 5′ triphosphate end of the mRNA. Next, a guanylyltransferase transfers a GMP unit from GTP to the remaining 5′ diphosphate. These two reactions, which are carried out by a bifunctional enzyme in mammals, create a 5′–5′ triphosphate linkage between two nucleotides. Finally, methyltransferases add a methyl group to the guanine and to the 2′ OH group of ribose residues (**Fig. 21.19**).

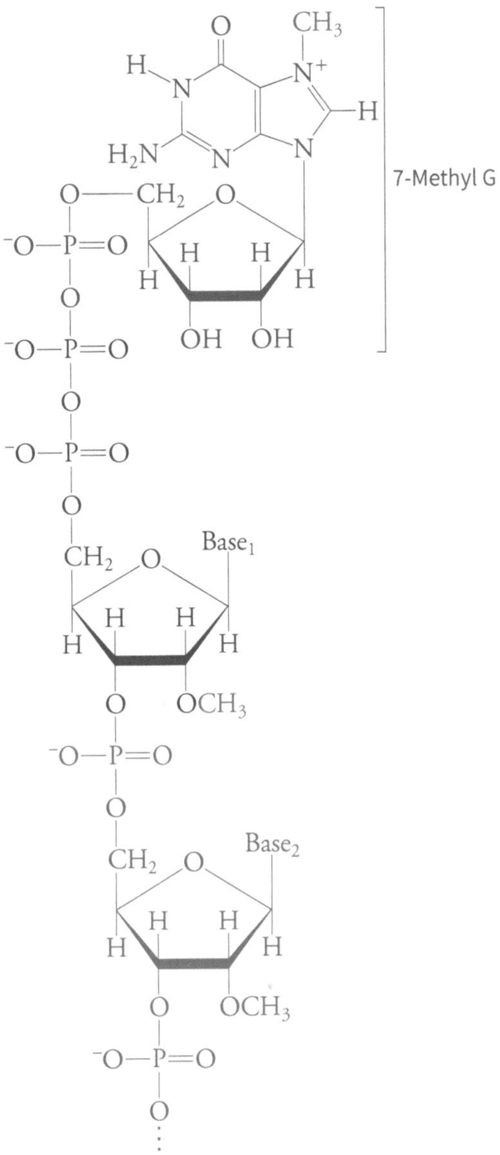

The 3′ end of an mRNA is also modified. Processing begins following the synthesis of the polyadenylation sequence AAUAAA, which is a signal for a protein complex to cleave the transcript and extend it by adding adenosine residues. In fact, the RNA cleavage reaction, which occurs while the RNA polymerase is still operating, triggers transcription termination.

The enzyme poly(A) polymerase generates a 3′ **poly(A) tail** (also called a polyadenylate tail) ranging from about 60 A residues in yeast to more than 200 in most animals. The enzyme resembles other polymerases in structure and catalytic mechanism, but it does not need a template to direct the addition of nucleotides.

Multiple copies of a binding protein associate with the mRNA tail. The poly(A)-binding protein consists of four copies of an RNA-binding domain (called an RBD, or RRM for RNA recognition motif) plus a C-terminal domain that mediates protein–protein contacts. A portion of the poly(A)-binding protein bound to RNA is shown in **Figure 21.20**. Each domain of about 80 amino acids can interact with two to six RNA nucleotides. Hence an mRNA's poly(A) tail can carry a large contingent of binding proteins, which help protect the 3′ end of the transcript from exonucleases. They may also provide a "handle" for proteins that deliver mRNA to ribosomes. Other RNA-binding proteins with different types of nucleotide-binding domains participate in RNA processing and other events.

Figure 21.19 An mRNA 5′ cap.

Question How many phosphoanhydride bonds are cleaved during cap construction?

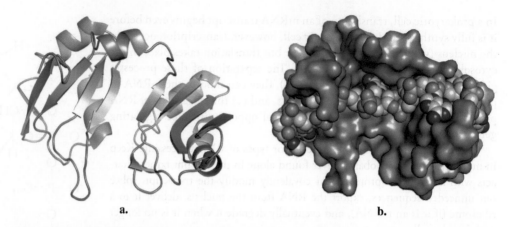

a.   b.

Figure 21.20 Poly(A)-binding protein bound to poly(A). **a.** Two of the RNA-binding domains from human poly(A)-binding protein are shown in ribbon form. **b.** Surface view of the protein with a 9-residue poly(A) nucleotide with atoms color-coded: C gray, O red, N blue, and P gold.

Splicing removes introns from eukaryotic RNA

Genes were once believed to be continuous stretches of DNA, but experimental work in the 1970s showed that hybridized DNA and mRNA molecules included large loops of unpaired DNA (**Fig. 21.21**). As the gene is being transcribed, portions of the sequence called **introns** (intervening sequences) are cut out of the RNA, and the remaining portions (expressed sequences, or **exons**) are joined together.

These mRNA **splicing** reactions do not occur in prokaryotes, whose protein-coding genes are continuous, but are the rule in complex eukaryotes. In simple organisms such as yeast, only about 5% of genes contain introns, but in humans, almost all genes contain at least one intron. A typical gene consists of eight exons with an average length of 145 bp that are separated by introns with an average length of 3365 bp.

Like capping, splicing commences before RNA polymerase has finished transcribing a gene, and some components of the splicing machinery assemble on the phosphorylated C-terminal domain of RNA polymerase. Most mRNA splicing is carried out by a **spliceosome,** a complex of five small RNA molecules (called **snRNAs,** for **small nuclear RNAs**) and over a hundred proteins (**Fig. 21.22**). Unlike the ribosome, another RNA–protein complex, the spliceosome does not exist as a preassembled complex. Instead, its separate components form a new complex at each intron to be excised.

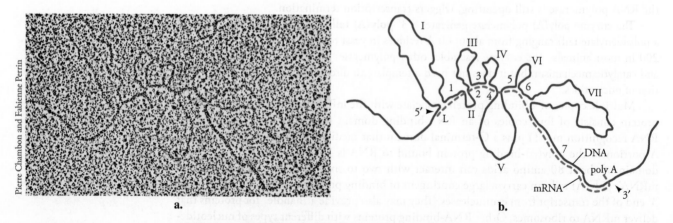

Figure 21.21 Hybridization of ovalbumin DNA and mRNA. The template DNA strand for the chicken ovalbumin gene was allowed to hybridize with the corresponding mRNA. Complementary sequences, representing exons, have annealed, while single-stranded DNA, coding for introns that have been spliced out of the mRNA, forms loops. **a.** Electron micrograph. **b.** Interpretive drawing. The mRNA is shown as a dashed line, the introns of the single-stranded DNA (blue line) are labeled I–VII, and the exons are labeled 1–7.

Figure 21.22 The yeast spliceosome. This model contains 37 of the >100 proteins (gray), three of the five snRNAs (green), and a portion of an intron (orange).

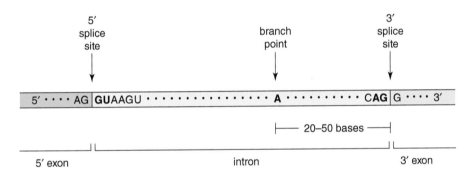

Figure 21.23 Consensus sequence at eukaryotic mRNA splice sites. Nucleotides shown in bold are invariant.

The spliceosome recognizes conserved sequences at the 5′ intron/exon junction and at a conserved A residue within the intron, called the branch point (**Fig. 21.23**). Recognition depends on base pairing between the conserved mRNA sequences and snRNA sequences. However, the sequence conservation is relatively weak, and introns can be enormous, ranging from less than a hundred nucleotides to a record of 2.4 million nucleotides. These factors make it difficult to identify introns and exons in genomic DNA sequences.

Splicing is a two-step transesterification process (**Fig. 21.24**). Each step requires an attacking nucleophile (a ribose OH group) and a leaving group (a phosphoryl group). Two catalytically essential Mg^{2+} ions enhance the nucleophilicity of the attacking hydroxyl group and stabilize the phosphate leaving group. Because there is no net change in the number of phosphodiester bonds, the cutting-and-pasting process, which is catalyzed by the snRNAs, needs no external source of free energy. However, proteins are essential for the overall reaction, which includes ATP hydrolysis–driven conformational changes to position the segments of RNA being processed.

Some types of introns, particularly in prokaryotic and protozoan rRNA genes, undergo self-splicing; that is, they catalyze their own transesterification reactions without the aid of proteins. These rRNA molecules were the first RNA enzymes (**ribozymes**) to be described, in 1982. One hypothesis for the evolutionary origin of splicing suggests that introns, and the splicing machinery itself, are the result of RNA molecules that spliced themselves *into* mRNA molecules, which were converted to DNA by the action of a reverse transcriptase (see Box 20.A) and then incorporated into the genome through recombination.

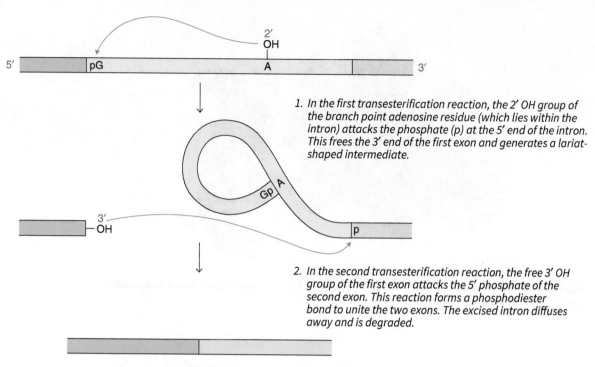

1. *In the first transesterification reaction, the 2' OH group of the branch point adenosine residue (which lies within the intron) attacks the phosphate (p) at the 5' end of the intron. This frees the 3' end of the first exon and generates a lariat-shaped intermediate.*

2. *In the second transesterification reaction, the free 3' OH group of the first exon attacks the 5' phosphate of the second exon. This reaction forms a phosphodiester bond to unite the two exons. The excised intron diffuses away and is degraded.*

Figure 21.24 mRNA splicing.

Introns typically comprise over 90% of a gene's total length, which means that a lot of RNA must be transcribed and then discarded. Moreover, a cell must spend energy to synthesize the RNA and proteins that make up the spliceosome and that destroy intronic RNA and incorrectly spliced transcripts. Finally, the complexity of the splicing process creates many opportunities for things to go wrong: A majority of mutations linked to inherited diseases involve defective splicing. So just what is the advantage of arranging a gene as a set of exons separated by introns?

One answer to this question is that splicing allows cells to increase variation in gene expression through alternative splicing. At least 95% of human protein-coding genes exhibit splice variants. Variation may result from selecting alternative sequences to serve as 5' or 3' splice sites, from skipping an exon, or from retaining an intron. Thus, certain exons present in the gene may or may not be included in the mature RNA transcript (**Fig. 21.25**). The signals that govern exon selection and splice sites probably involve RNA-binding proteins that recognize sequences or secondary structures within introns as well as exons. As a result of alternative splicing, *a given gene can generate more than one protein product*, and gene expression can be finely tailored to suit the needs of different types of cells. The evolutionary advantage of this regulatory flexibility clearly outweighs the cost of making the machinery that cuts and pastes

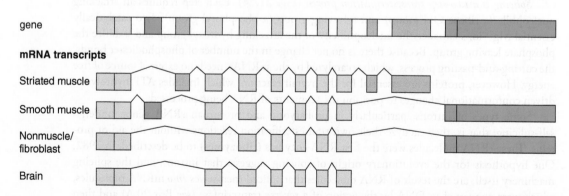

Figure 21.25 Alternative splicing.   The rat gene for the muscle protein α-tropomyosin (top) encodes 12 exons. The mature mRNA transcripts in different tissues consist of different combinations of exons (some exons are found in all transcripts), reflecting alternative splicing pathways.

Table 21.2 Protein-Coding Genes and Protein Count

Organism	Protein-coding genes	Estimated proteins
Saccharomyces cerevisiae (yeast)	6000	6000
Arabidopsis thaliana (plant)	25,300	48,300
Drosophila melanogaster (fruit fly)	13,000	30,600
Homo sapiens	21,000	118,400

Source: Data from NCBI https://www.ncbi.nlm.nih.gov/genome/.

RNA sequences. Alternative splicing also explains why humans are vastly more complex than organisms such as roundworms, which contain a comparable number of genes (**Table 21.2**).

mRNA turnover and RNA interference limit gene expression

Although mRNA accounts for only about 5% of cellular RNA (rRNA accounts for about 80% and tRNA for about 15%), it continuously undergoes synthesis and degradation. The lifespan of a given mRNA molecule is another regulated aspect of gene expression: mRNA molecules decay at different rates. In mammalian cells, mRNA lifespans range from less than an hour to about 24 hours. In eukaryotes, mRNA molecules must also be transported out of the nucleus (**Box 21.C**).

Box 21.C The Nuclear Pore Complex

Messenger RNA is synthesized and processed in the nucleus of eukaryotes, but ribosomes, where the mRNA is translated into protein, are in the cytosol. mRNA exits the nucleus via the **nuclear pore complex,** an assembly of proteins that spans the double membrane surrounding the nucleus. The yeast complex is built from about 550 copies of 30 different proteins collectively known as nucleoporins. The nuclear pore complexes in vertebrates have a similar structure but are larger. Some structural details suggest that the nuclear pore components share an evolutionary origin with clathrin and other membrane coat proteins (Section 9.4).

The nuclear pore complex consists largely of three symmetric protein rings: one facing the cytoplasm, one facing the nucleoplasm, and an inner ring positioned where the inner and outer nuclear membranes fuse together. The inner ring is reinforced by diagonal columns connected by cables, which apparently give the overall structure some flexibility.

Not shown in the model here are fibrous extensions that protrude into the cytoplasm and nucleus to form open cagelike structures. These structures may help organize traffic into and out of the nucleus.

The center of the nuclear pore complex is occupied by intrinsically disordered protein segments contributed by an assortment of nucleoporins. The highly concentrated polypeptides, which are rich in Phe-Gly repeats, form a sort of gel that prevents the contents of the nucleus from spilling out into the cytoplasm. Molecules with a mass less than about 40 kD can freely diffuse through the the gel-filled pore. Larger molecules must be escorted by carrier proteins, called transport factors. The carrier proteins may find their way by tracking along more rigid elements of the pore structure; they may also interact with the Phe-Gly repeats to "melt" into the gel. Without a transport factor to accompany them, large molecules are excluded from the pore, which is the only conduit between the nucleus and cytosol. A single nucleus may harbor a few thousand nuclear pore complexes.

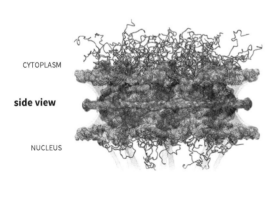

CYTOPLASM

side view

NUCLEUS

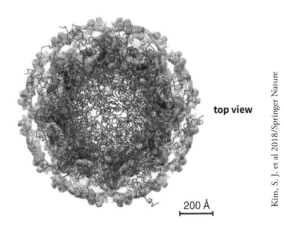

top view

200 Å

Kim, S. J. et al 2018/Springer Nature

The 40-nm inner diameter of the pore is large enough for several macromolecular "cargoes" to move simultaneously in the same or opposite directions. Diffusion is inherently random, so directionality is imposed on the transport process by additional proteins that use the free energy of ATP or GTP. By incorporating a nucleotide-hydrolysis step into the transport cycle, the movement becomes one-way. For example, proteins moving into the nucleus rely on the activity of a small G protein called Ran (other types of G proteins are described in Section 10.2). With the appropriate transport factors, proteins can also move from the nucleus to the cytosol. Messenger RNA moving out of the nucleus uses protein cofactors that hydrolyze ATP. Presumably, the transport factors that facilitate mRNA export are able to recognize fully processed mRNA, possibly when poly(A) polymerase finishes its work, so that only mature messages are sent to the cytosol to be translated into protein.

The rate of mRNA turnover depends in part on how rapidly its poly(A) tail is shortened by the activity of deadenylating exonucleases. Poly(A) tail shortening is followed by decapping, which allows exonucleases access to the 5′ end of the transcript and eventually leads to destruction of the entire message (**Fig. 21.26**). *In vivo*, the RNA cap and tail are close together because a protein involved in translation binds to both ends of the mRNA, effectively circularizing it. RNA-binding regulatory proteins are almost certainly involved in monitoring RNA integrity. For example, transcripts with a premature stop codon are preferentially degraded, thereby avoiding the waste of synthesizing a nonfunctional truncated polypeptide.

Sequence-specific degradation of certain RNAs, a phenomenon called **RNA interference (RNAi),** provides another mechanism for regulating gene expression after transcription has occurred. RNA interference was discovered by researchers who were attempting to boost gene expression in various types of cells by introducing extra copies of genetic information in the form of RNA. They observed that instead of increasing gene expression, the RNA—particularly if it was double-stranded—actually blocked production of the gene's product. This interference or gene-silencing effect results from the ability of the introduced RNA to target a complementary cellular mRNA for destruction. Endogenously produced RNAs, known as **small interfering RNAs (siRNAs)** and **micro RNAs (miRNAs),** appear to mediate RNA interference in virtually all types of eukaryotic cells, including human cells.

For siRNA, the RNA interference pathway begins with the production of double-stranded RNA, which may result when a single polynucleotide strand folds back on itself in a hairpin. A ribonuclease called Dicer cleaves the double-stranded RNA to generate segments of 20 to 25 nucleotides with a two-nucleotide overhang at each 3′ end (**Fig. 21.27**). These siRNAs bind to a multiprotein complex called the RNA-induced silencing complex (RISC), where one strand of the RNA (the "passenger" strand) is separated from the other by a helicase and/or degraded by a nuclease. The remaining strand serves as a guide for the RISC to identify and bind to a

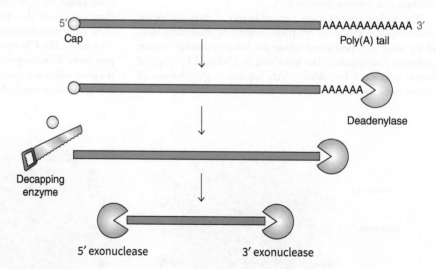

Figure 21.26 mRNA decay. A mature mRNA bears a 5′ cap and a 3′ poly(A) tail. After a deadenylase has shortened the tail, a decapping enzyme removes the methylguanosine cap at the 5′ end. The mRNA can then be degraded by exonucleases from both ends.

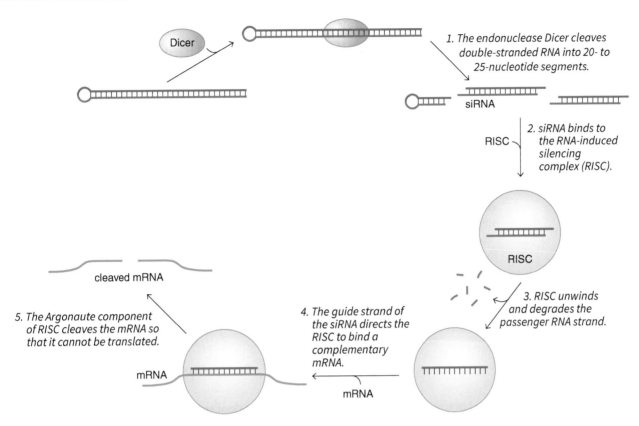

Figure 21.27 RNA interference.   The steps involving siRNA are shown. The miRNA pathway for inactivating mRNA is similar.

complementary mRNA molecule. The "Slicer" activity of the RISC, a protein known as Argonaute, then cleaves the mRNA, rendering it unfit for translation.

Like RNA splicing, RNA interference at first appears wasteful, but it provides cells with a mechanism for eliminating mRNAs—which could otherwise be translated into protein many times over—in a highly specific manner. It is believed that RNA interference originally evolved as an antiviral defense, since many viral life cycles include the formation of double-stranded RNAs.

In the miRNA pathway, RNA hairpins containing imperfectly paired nucleotides are processed by Dicer and other enzymes to double-stranded miRNAs that bind to the RISC. The passenger RNA strand is ejected and the remaining strand helps the RISC locate complementary target mRNAs. Whereas an siRNA specifically seeks and destroys an mRNA that is perfectly complementary, an miRNA can bind to a large number of target mRNAs—possibly hundreds—because it forms base pairs with a stretch of only six or seven nucleotides. The captured mRNAs are unavailable for translation and are susceptible to the standard mechanisms for RNA degradation diagrammed in Figure 21.26.

In addition to serving as a powerful laboratory technique for silencing genes in order to explore their functions, the RNA interference system is being exploited for practical purposes. RNA-based drugs have been developed to block the expression of genes responsible for some inherited diseases. siRNAs could also be used to turn off viral genes that are necessary for viral replication. Other diseases in which gene silencing would be desirable, such as cancer, are amenable to RNAi therapy, provided that the siRNA can be delivered selectively to cancerous cells. In general, introducing exogenous RNAs into cells is challenging, since nucleic acids don't easily cross cell membranes, and the presence of extracellular RNA may trigger the body's innate RNA-degrading antiviral defenses. Apples and potatoes have been engineered to use RNAi to prevent the synthesis of the oxidative enzymes that cause browning. It is hoped that these crops will be more acceptable to consumers who avoid traditional genetically modified foods that contain foreign genes (see Box 3.B).

rRNA and tRNA processing includes the addition, deletion, and modification of nucleotides

rRNA transcripts, which are generated mainly by RNA polymerase I in eukaryotes, must be processed to produce mature rRNA molecules. rRNA processing and all but the final stages of ribosome assembly take place in the nucleolus, a discrete liquid phase in the nucleus. The initial eukaryotic rRNA transcript is cleaved and trimmed by endo- and exonucleases to yield three rRNA molecules (**Fig. 21.28**). The rRNAs are known as 18S, 5.8S, and 28S rRNAs for their sedimentation coefficients (large molecules have larger sedimentation coefficients, a measure of how quickly they settle in an ultra-high-speed centrifuge).

rRNA transcripts may be covalently modified (in both prokaryotes and eukaryotes) by the conversion of some uridine residues to pseudouridine and by the methylation of certain bases and ribose 2′ OH groups.

Uridine   Pseudouridine (ψ)

This last type of modification is guided by a multitude of **small nucleolar RNA molecules** (called **snoRNAs**) that recognize and pair with specific 15-base segments in the rRNA sequences, thereby directing an associated protein methylase to each site. *Without the snoRNAs to mediate sequence-specific ribose methylation, the cell would require many different methylases in order to recognize all the different nucleotide sequences to be modified.*

A rapidly growing mammalian cell may synthesize as many as 7500 rRNA transcripts each minute, each of which associates with about 150 different snoRNAs. The processed rRNAs eventually combine with some 80 different ribosomal proteins to generate fully functional ribosomes, a task that requires careful coordination between RNA synthesis and ribosomal protein synthesis.

tRNA molecules, produced by the action of RNA polymerase III in eukaryotes, undergo nucleolytic processing and covalent modification. The initial tRNA transcripts are trimmed by ribonuclease P (see below).

Some tRNA transcripts undergo splicing to remove introns. In some bacteria, newly made tRNAs end with a 3′ CCA sequence, which serves as the attachment point for an amino acid that will be used for protein synthesis. In most organisms, however, the three nucleotides are added to the 3′ end of the immature molecule by the action of a nucleotidyl transferase.

Up to 25% of the nucleotides in tRNA molecules are covalently modified. The alterations range from simple additions of methyl groups to complex restructuring of the base. Some of the 200 or so known nucleotide modifications are shown in **Figure 21.29**. These are yet more examples of how cells alter genetic information as it is transcribed from relatively inert DNA to highly variable and much more dynamic RNA molecules.

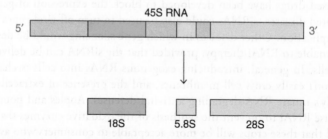

Figure 21.28 Eukaryotic rRNA processing. The initial transcript of about 13.7 kb has a sedimentation coefficient of 45S. Three smaller rRNA molecules (18S, 5.8S, and 28S) are derived from it by the action of nucleases.

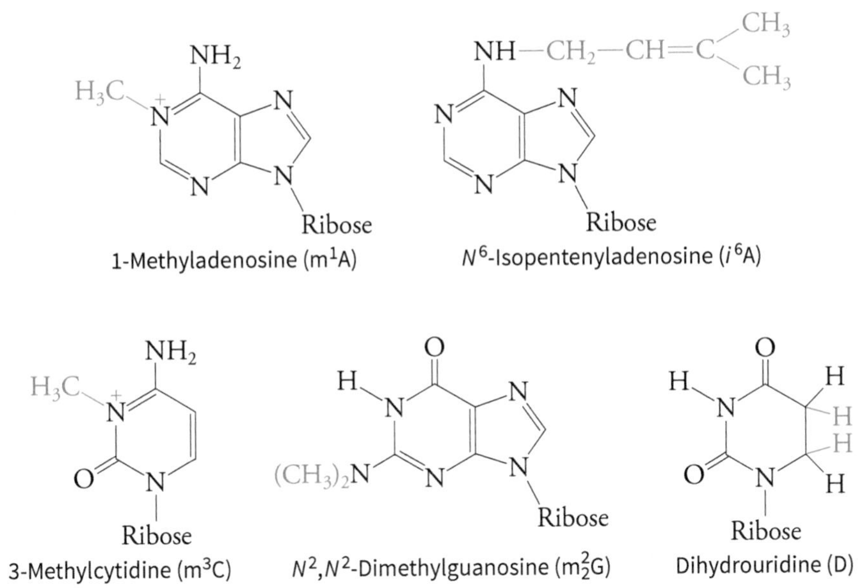

Figure 21.29 **Some modified nucleotides in tRNA molecules.**   The parent nucleotide is in black; the modification is shown in red.

RNAs have extensive secondary structure

Some of the modified nucleotides common in tRNAs also occur in other types of RNA, including mRNA. For example, N^6-methyladenosine occurs at thousands of highly conserved sites in mammalian mRNAs, which suggests that it has some functional significance. This modified nucleotide seems to function like the epigenetic marks in DNA, as it has reader, writer, and eraser enzymes associated with it. RNA methylation and other modifications could affect protein binding, RNA secondary structure, or liquid–liquid phase separation in the nucleus or cytosol.

Unlike DNA, whose conformational flexibility is considerably constrained by its double-stranded nature, single-stranded RNA molecules can adopt highly convoluted shapes through base pairing between different segments. In addition to the standard (Watson–Crick) types of base pairs, RNA accommodates nonstandard base pairs as well as hydrogen-bonding interactions among three bases (**Fig. 21.30**).

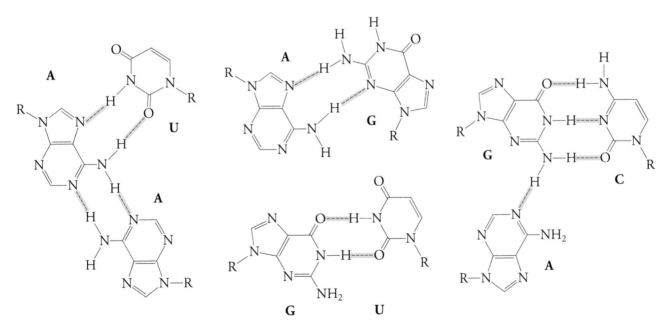

Figure 21.30 **Some nonstandard base pairs.**   R represents the ribose–phosphate backbone.

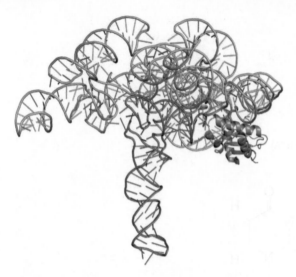

Figure 21.31 RNase P. In this model of the *Thermotoga maritima* enzyme, the 347-nucleotide RNA is gold and the product tRNA is red. The small protein component (117 amino acids) is green.

Base stacking stabilizes the RNA tertiary structure, achieving the same sort of balance between rigidity and flexibility exhibited by protein enzymes. *A folded RNA molecule can then bind substrates, orient them, and stabilize the transition state of a chemical reaction.*

All cells contain two essential ribozymes: the tRNA-processing RNase P and the ribosomal RNA that catalyzes peptide bond formation during protein synthesis. There are at least six other naturally occurring types of catalytic RNAs (such as those involved in splicing) and many more synthetic ribozymes. Self-splicing introns are not true catalysts, since they cannot participate in more than one reaction cycle, but the RNA component of RNase P, with its numerous base-paired stems and compact protein-like structure, is a true catalyst (**Fig. 21.31**). At one time it was believed that the enzyme's RNA molecule merely helped align the tRNA substrate for the protein to cleave, but the bacterial RNase P RNA is able to cleave its substrate in the absence of the RNase P protein.

The existence of RNA enzymes such as RNase P lends support to the theory of an early **RNA world** when RNA functioned as a repository of biological information (like modern DNA) as well as a catalyst (like modern proteins). Experiments with synthetic RNAs *in vitro* have demonstrated that RNA can catalyze a wide variety of chemical reactions, including the biologically relevant synthesis of glycosidic bonds (the type of bond that links the base and ribose in a nucleoside) and RNA-template–directed RNA synthesis. Apparently, most ribozymes that originated in an early RNA world were later supplanted by protein catalysts, leaving only a few examples of RNA's catalytic abilities (**Box 21.D**).

Box 21.D   Reverse transcription and cDNA

The retroviruses having RNA as genomic material are characterized by the presence of an enzyme known as reverse transcriptase. The existence of reverse transcriptase was independently identified by Howard Temin in 1962, and by David Baltimore in 1970 and they mentioned that retroviral virions have enzymatic properties that can convert RNA to DNA. Their discovery aroused much attention as genetic information can also occur in the reverse direction by the presence of reverse transcriptase converting RNA into DNA. Reverse transcription and integration are the defining features of retroviral replication during which viral encoded enzyme, reverse transcriptase, RNA-dependent DNA polymerase convert their RNA into DNA and integrate it into the host DNA. Retroviruses typically contain three genes namely *gag*, *pol*, and *env*. A transcript containing *gag* and *pol* undergoes the translation to form a single large polypeptide that converts into six different proteins with diversified functions. The core of the virus particle results from the product of the *gag* gene, whereas the *env* gene is associated with the formation of viral envelop protein. The *pol* gene encodes the protease that cuts the peptide chain into the integrase responsible for inserting the viral RNA into the host genome, and the reverse transcriptase. The sequence of long terminal repeat (LTR) containing a few hundred nucleotides has been reported at each end of the linear RNA genome that participates in facilitating the integration of the viral genome into the host DNA. These sequences also contain promoters for gene expression of virions. The reverse transcriptase contains Zn^{2+} required for its activity and shows similarity to other polymerases.

The process of reverse transcription begins when viral particles enter the cytoplasm of the target cell and generate a linear DNA duplex through an intricate series of steps. First, the reverse transcriptase catalyzes the formation of a DNA strand complementary to the viral RNA, then the degradation of the viral genomic RNA strand that is a part of the RNA-DNA hybrid, and finally the formation of a DNA strand against DNA complementary to the viral genome. The resulting dsDNA integrates into the genome of the host cell (eukaryotic) and undergoes transcription to form viral proteins and viral RNA that are assembled to form new RNA viruses. Reverse transcriptase has separate enzymatic active sites for DNA and RNA synthesis and for RNA degradation. The process of reverse transcription requires a primer and cellular tRNA which are obtained during earlier infection and carried in virus particles. The new DNA strand synthesis begins in a 5′ to 3′ direction and the tRNA base paired at its 3′ end is complementary to the viral genome. The error rate of reverse transcriptase is comparatively high about 1 per 20,000 nucleotides as compared to other polymerases. A high rate of mutation or error is one of the important factors for the frequent emergence of new strains of diseases resulting from retroviruses. Reverse transcriptase has an application in DNA cloning where the complementary DNA (cDNA) strands are synthesized into an mRNA template and used to clone cellular genes.

Question How does a single-stranded RNA of certain viruses give rise to double-stranded DNA?

Concept Check

1. Describe the structure and function of the mRNA 5′ cap and 3′ tail.

2. Draw diagrams to recount the events of RNA splicing.

3. Explain why splicing does not require free energy input.

4. What are the advantages and disadvantages of splicing protein-coding genes?

5. List the advantages of arranging genes as sets of exons and introns.

6. Summarize the steps of RNA interference.

7. Compare RNA polymerase, poly(A) polymerase, and the tRNA CCA-adding enzyme with respect to substrates, products, and requirement for a template.

8. Explain why the products of transcription exhibit much more variability than the genes that encode them.

SUMMARY

21.1 Initiating Transcription

- Transcription is the process of converting a segment of DNA into RNA. An RNA transcript may represent a protein-coding gene or it may participate in protein synthesis or other activities, including RNA processing.

- Gene expression may be regulated by altering histones through acetylation, phosphorylation, and methylation, by methylating DNA, and by rearranging nucleosomes.

- Transcription begins at a DNA sequence known as a promoter. A gene to be transcribed must be recognized by a regulatory factor such as the σ factor in prokaryotes.

- In eukaryotes, a set of general transcription factors interact with DNA at the promoter to form a complex that recruits RNA polymerase and may further alter chromatin structure.

- Regulatory DNA sequences may affect transcription through binding proteins that interact with RNA polymerase via the Mediator complex.

- The bacterial *lac* operon illustrates the regulation of transcription by a repressor protein.

21.2 RNA Polymerase

- Eukaryotic RNA polymerase II transcribes protein-coding genes. It requires no primer and polymerizes ribonucleotides to generate an RNA chain that forms a short double helix with the template DNA.

- The polymerase acts processively along the DNA template but reverses to allow the excision of a mispaired nucleotide.

- The elongation phase of transcription in eukaryotes is triggered by phosphorylation of the C-terminal domain of RNA polymerase II.

- Transcription termination in prokaryotes involves destabilization of the DNA–RNA hybrid helix. In eukaryotes, transcription termination is linked to RNA cleavage.

21.3 RNA Processing

- mRNA transcripts undergo processing that includes the addition of a 5′ cap structure and a 3′ poly(A) tail. mRNA splicing, carried out by RNA–protein complexes called spliceosomes, joins exons and eliminates introns.

- RNA interference is a pathway for inactivating mRNAs according to their ability to pair with a complementary siRNA or miRNA.

- rRNA and tRNA transcripts are processed by nucleases and enzymes that modify particular bases.

- The chemical and structural variability of RNA molecules makes it possible for some to function as enzymes.

KEY TERMS

transcription	operon	epigenetics	Mediator	splicing	small interfering
gene	RNA processing	promoter	silencer	spliceosome	RNA (siRNA)
messenger RNA	RNA polymerase	consensus sequence	repressor	small nuclear RNA	micro RNA
(mRNA)	noncoding RNA	TATA box	nucleolus	(snRNA)	(miRNA)
ribosomal RNA	(ncRNA)	general transcription	cap	ribozyme	small nucleolar RNA
(rRNA)	histone code	factor	poly(A) tail	nuclear pore complex	(snoRNA)
transfer RNA	CpG island	enhancer	intron	RNA interference	RNA world
(tRNA)	imprinting	activator	exon	(RNAi)	

BIOINFORMATICS

Brief Bioinformatics Exercises

21.1 Viewing and Analyzing RNA Polymerase

21.2 RNA Polymerase, Transcription, and the KEGG Database

21.3 Viewing and Analyzing the *lac* Repressor

PROBLEMS

21.1 Initiating Transcription

1. Why does the genome contain so many more genes for rRNA than mRNA?

2. Why is it effective for a bacterial cell to organize genes for related functions as an operon? How do eukaryotes achieve the same benefits?

3. Proteins can interact with DNA through relatively weak forces, such as hydrogen bonds and van der Waals interactions, as well as through stronger electrostatic interactions such as ion pairs. Which types of interactions predominate for sequence-specific DNA-binding proteins and for sequence-independent binding proteins?

4. Certain proteins that stimulate expression of a gene bind to DNA in a sequence-specific manner and induce conformational changes in the DNA. Describe the purpose of these two modes of interaction with the DNA.

5. Draw the structures of the amino acid side chains that correspond to the following histone modifications: **a.** acetylation of lysine, **b.** phosphorylation of serine, **c.** phosphorylation of histidine. **d.** How do these modifications change the character of their respective side chains? How is the binding affinity between the histone and DNA affected as a result?

6. A protein involved in reading the "histone code" binds to a trimethylated lysine in histone H3. **a.** Draw the structure of a trimethylated lysine residue. **b.** Describe the protein's putative binding site, given the observation that the protein binds to histone H3 only when the lysine is trimethylated.

7. Enzymes that catalyze histone acetylation (histone acetyltransferases, or HAT) are closely associated with transcription factors, which are proteins that promote transcription. Why is this a good biochemical strategy?

8. DNA methylation requires the methyl group donor S-adenosylmethionine, which is produced by the condensation of methionine with ATP (see Box 18.B). The sulfonium ion's methyl group is used in methyl-group transfer reactions. The demethylated S-adenosylmethionine is then hydrolyzed to produce adenosine and a nonstandard amino acid. Draw the structure of this amino acid. **a.** How does the cell convert this compound back to methionine to regenerate S-adenosylmethionine? **b.** The proper regulation of gene expression requires methylation as well as demethylation of cytosine residues in DNA. If a demethylase carries out a hydrolytic reaction to restore cytosine residues, what is the other reaction product?

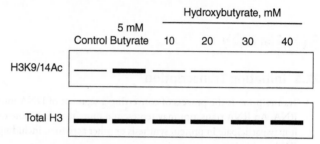

S-Adenosylmethionine

9. Explain why epigenetic marks such as methylation of cytosine residues are almost completely erased during the early stages of embryogenesis.

10. Mice that are homozygous for a mutation that renders the DNA methyltransferase enzyme nonfunctional usually die *in utero*. Why does this mutation have such serious consequences?

11. Despite their structural similarity, butyrate and β-hydroxybutyrate have different physiological roles. Butyrate is a short-chain fatty acid produced by colonic bacteria dining on dietary fermentable fiber, whereas β-hydroxybutyrate is a ketone body. **a.** Draw the structures of these two compounds. **b.** An embryonic kidney cell line was used as a model to investigate the hypothesis that these compounds act as histone deacetylase (HDAC) inhibitors. Cells were treated with 5 mM butyrate and increasing concentrations of β-hydroxybutyrate. Cell lysates were analyzed by immunoblotting in which histone H3 acetylated at Lys 9 and Lys 14 (H3K9/14Ac) was detected by the binding of specific anti-acetyl-Lys antibodies. The results are shown in the figure. Is the hypothesis correct?

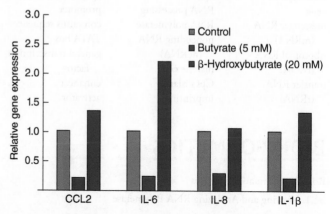

c. In a second experiment, butyrate and β-hydroxybutyrate were investigated for their ability to influence the transcription of the genes PGC1α and CPT1b. (PGC1α is a transcription factor involved in mitochondrial biogenesis, and CPT1b is a carnitine acyltransferase; see Fig. 17.5.) The results are shown in the table. How do butyrate and β-hydroxybutyrate influence transcription of these two genes and what are the physiological implications? How is an HDAC inhibitor able to influence gene transcription?

| | Relative gene expression | |
	PGC1α	CPT1b
Control	1.5	1.2
Butyrate	7.9	1.9
β-Hydroxybutyrate	1.0	1.3

12. The structurally similar compounds butyrate and β-hydroxybutyrate (see Problem 11) were investigated for their ability to influence transcriptional changes resulting in the secretion of cytokines CCL2, IL-6, IL-8, and IL-1β. These pro-inflammatory proteins are implicated in diseases such as atherosclerosis and autoimmunity. **a.** Use the results below to evaluate the ability of butyrate and β-hydroxybutyrate to elicit a pro-inflammatory response. **b.** How might these results be related to the ability of butyrate or β-hydroxybutyrate to act as an HDAC inhibitor (see Problem 11)? **c.** How might you respond to the claim that the currently popular "keto" diet (see Problems 17.49–17.50) is anti-inflammatory?

13. a. The antisense (noncoding) strand for the *E. coli rrnA1* gene is shown below. Write down the sequence of the mRNA transcript of it. **b.** Which region is AT-rich and why is this region composed of A:T and not G:C base pairs?

3′ AATGCTTGACTCTGTAGCGGGAAGGC 5′

14. Predict the effect of a mutation in one of the bases in either the –35 or the –10 region of the promoter.

15. Sp1 is a sequence-specific human DNA-binding protein that binds to a region on the DNA called the GC box, a promoter element with the sequence GGGCGG. Binding of Sp1 to the GC box enhances RNA polymerase II activity 50- to 100-fold. How would you use affinity chromatography (see Section 4.6) to purify Sp1?

16. a. The sense (coding strand) of the *E. coli* promoter for the *rrnA1* gene is shown below. The transcription initiation site is shown by +1. Identify the –35 and –10 regions for this gene. **b.** Predict the effect of a mutation in one of the bases in either the –35 or the –10 region of the promoter. **c.** Which region contains an AT-rich region and why is this region composed of A:T and not G:C base pairs?

AAAATAAATGCTTGACTCTGTAGCG-

+1
GGAAGGCGTATTATCCAACACCC

17. Discuss the significance of basic or general transcription factors, and DNA modifiers in eukaryotic transcription.

18. A rational drug design project was undertaken to identify small molecules that interfere with the binding of σ factor to RNA polymerase. Why would these small molecules serve as effective drugs to treat diseases caused by bacterial pathogens?

19. A histone lysine methyltransferase is upregulated in various types of cancers. How does this upregulation affect Lys 27 of histone 3? How does this affect the transcription of genes associated with this histone?

20. Three human TBP-associated factors (TAFs) contain protein domains that are homologous to those found in histones H2B, H3, and H4. Why is this finding not surprising?

21. The T7 bacteriophage RNA polymerase recognizes specific promoter sequences and melts open the DNA to form a transcription bubble without the need for transcription factors. **a.** Dissociation constants (K_d) were measured for the interaction between the polymerase and DNA segments containing the promoter sequences. In some cases, the DNA contained a bulge, caused by a mismatch of one, four, or eight bases, to mimic the intermediates in the formation of a transcription bubble. To which DNA segment does the polymerase bind most tightly? Explain in terms of the DNA structure.

DNA promoter segment	K_d (nM)
Fully base-paired	315
One-base bulge	0.52
Four-base bulge	0.0025
Eight-base bulge	0.0013

b. Use the data in the table to calculate the $\Delta G^{\circ\prime}$ for the binding of T7 RNA polymerase to fully base-paired DNA and to DNA with an eight-base bulge. (*Note:* The K_d is the inverse of K_{eq}.) Assume a temperature of 25°C. **c.** What do these results reveal about the thermodynamics of melting open a DNA helix for transcription? What is the approximate free energy cost of forming a transcription bubble equivalent to eight base pairs?

22. In bacteria, the core RNA polymerase binds to DNA with a dissociation constant of 5×10^{-12} M. The polymerase in complex with its σ factor has a dissociation constant of 10^{-7} M. Explain the reason for the difference in binding affinities.

23. One of the genes expressed by the *lac* operon is *lacY*, which encodes a lactose permease transporter that allows lactose to enter the cell. Why does the expression of this gene assist in the expression of the operon?

24. The genes of the *lac* operon are not expressed when the *lac* repressor binds to the operator. However, removal of the *lac* repressor is not sufficient to allow gene expression—a protein called catabolite activator protein (CAP) is also required to assist RNA polymerase and facilitate transcription. CAP can interact with the *lac* promoter only when it binds to its ligand cAMP. In *E. coli*, the intracellular concentration of cAMP falls when glucose is present. Describe the activity of the *lac* operon in each of the following scenarios: **a.** Both lactose and glucose are present. **b.** Glucose is present but lactose is absent. **c.** Both glucose and lactose are absent. **d.** Lactose is present and glucose is absent.

25. Researchers have isolated bacterial cells with mutations in various segments of the *lac* operon. What is the effect on gene expression if a mutation in the operator occurs so that the repressor cannot bind? What happens when lactose is added to the growth medium of these mutants?

26. The compound phenyl-β-D-galactose (phenyl-Gal) is not an inducer of the *lac* operon because it is unable to bind to the repressor. However, it can serve as a substrate for β-galactosidase, which cleaves phenyl-Gal to phenol and galactose. How can the addition of phenyl-Gal to growth medium distinguish between wild-type bacterial cells and cells that have a mutation in the *lacI* gene?

27. In bacterial cells, the genes that code for the enzymes of the tryptophan biosynthetic pathway are organized in an operon, as shown below. Another gene encodes a repressor protein that binds tryptophan. How does the repressor protein control the expression of the genes in the *trp* operon?

P	trpL	trpE	trpD	trpC	trpB	trpA

21.2 RNA Polymerase

28. RNA polymerase makes an error once in 10^6 bases while DNA polymerase makes an error once in 10^9 bases. **a.** Explain why the error rate for DNA polymerase is much lower than that of RNA polymerase. **b.** Why is the cell not harmed by the lower fidelity of RNA polymerase?

29. The promoters for genes transcribed by eukaryotic RNA polymerase I exhibit little sequence variation, yet the promoters for genes transcribed by eukaryotic RNA polymerase II are highly variable. Explain.

30. Explain why the adenosine derivative cordycepin inhibits RNA synthesis.

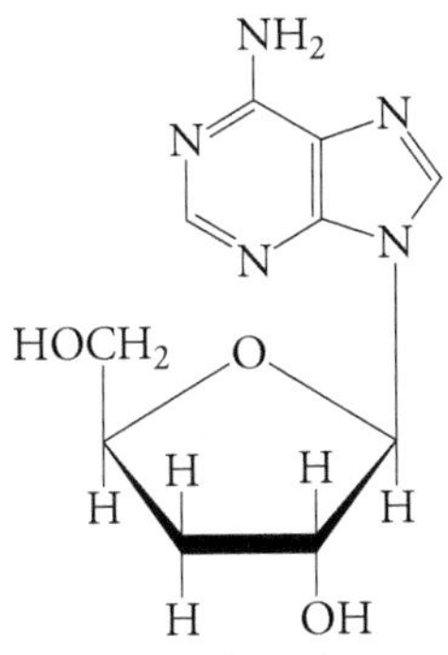

Cordycepin

31. The activity of RNA polymerase II is inhibited by the mushroom toxin α-amanitin ($K_d = 10^{-8}$ M). By contrast, RNA polymerase III is only moderately inhibited by the toxin ($K_d = 10^{-6}$ M), and RNA polymerase I is not affected at all. What would be the effect of adding 10 nM α-amanitin to cells in culture?

32. The three different eukaryotic RNA polymerases were discovered in the 1970s by researchers who loaded cell extracts onto a DEAE ion-exchange column (see Section 4.6) and then eluted the proteins with a salt gradient. Collected fractions were assayed for RNA polymerase activity in the presence and in the absence of Mg^{2+} ions and in the presence of the mushroom toxin α-amanitin. In addition to the difference in α-amanitin sensitivity described in Problem 31, the investigators noted that RNA polymerase I was fully active in the presence of 5 mM Mg^{2+} ions whereas polymerases II and III were only 50% active. How did these results support the conclusion that the three peaks constituted three different forms of RNA polymerase?

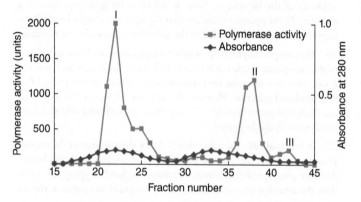

33. Actinomycin D, an antibiotic isolated from *Streptomyces* bacteria, intercalates between DNA bases and prevents the progression of RNA polymerases. Eukaryotic RNA polymerases (RNAP) differ in their sensitivity to actinomycin D, as shown below. Which enzyme is most sensitive? Which is least sensitive? Compare these results to the sensitivities of the eukaryotic RNA polymerases to α-amanitin (see Problem 31).

	IC_{50} (μg · mL^{-1})
RNAP I	0.05
RNAP II	0.5
RNAP III	5.0

34. A newly developed cancer drug selectively inhibits RNA polymerase I without affecting RNA polymerases II and III. Why might this drug be successful at treating cancer?

35. Triptolide is a cancer drug that inhibits the ATPase activity of TFIIH. Describe the consequences of treating cancer cells with triptolide. Does the drug affect transcription by all RNA polymerases equally?

36. Radioactively labeled γ-[^{32}P]GTP is added to a bacterial culture undergoing transcription. Is the resulting RNA labeled? If so, where?

37. The antibiotic rifampicin binds to the β subunit of bacterial RNA polymerase. In the presence of rifampicin, cultured bacterial cells are capable of synthesizing only short RNA oligomers. **a.** At what point in the transcription process does rifampicin exert its inhibitory effect? **b.** Why is rifampicin used to treat bacterial infections?

38. What is an abortive transcription and why it occurs?

39. What is the role of phosphorylation of serine residues (Ser 2 and Ser 5) at heptapeptide repeats of RNA polymerase CTD?

40. a. The C-terminal domain of RNA polymerase II projects away from the globular portion of the protein. Why? **b.** In an experiment, RNA polymerase II is truncated to produce a protein with a missing C-terminal domain (CTD). How would this affect cells?

41. The DNA sequence of a hypothetical *E. coli* terminator is shown below. N is an abbreviation for any of the four nucleotides. **a.** Write the sequence of the mRNA transcript that is made using the top strand as the coding strand. **b.** Draw the hairpin structure that would form in this RNA transcript.

5′ ··· NNAAGCGCCGNNNNCCGGCGCTTTTTTNNN ··· 3′
3′ ··· NNTTCGCGGCNNNNGGCCGCGAAAAAAANNN ··· 5′

42. Inosine triphosphate (ITP; Section 18.5) is added to a culture of bacteria, which use it in place of GTP. Inosine (I) base-pairs with cytidine (C), forming two hydrogen bonds. **a.** Write the sequence of the mRNA transcript that is made using the top strand of the gene shown in Problem 41 as the coding strand. **b.** Draw the hairpin structure that would form in this RNA transcript. Compare the stabilities of this RNA hairpin and the RNA hairpin you drew in Solution 41b. How is termination of transcription affected by the ITP substitution?

43. Mention the two modes of transcription termination in prokaryotes.

44. The addition of β,γ-imido nucleoside triphosphates to cells in culture has been shown to inhibit Rho-dependent termination. Explain why.

β,γ-Imido nucleoside triphosphate

45. In bacteria, the organization of functionally related genes in an operon allows the simultaneous regulation of expression of those genes. If the operon consists of genes encoding the enzymes for a biosynthetic pathway, then the pathway activity as a whole can be feedback-inhibited when the concentration of the pathway's final product accumulates. **a.** In one mode of feedback regulation, a repressor protein binds to a site in the operon (called the operator) to decrease the rate of transcription only when the repressor has bound a molecule representing the operon's ultimate metabolic product. Draw a diagram showing how such a regulatory system would work. **b.** Feedback regulation of gene expression can also occur after RNA synthesis has begun. In this case, the presence of the operon's ultimate product causes transcription to terminate prematurely or leads to an mRNA that cannot be translated. Draw a diagram illustrating this control mechanism. Assume that the feedback mechanism includes a protein to which the product binds. **c.** How would the feedback inhibition system in part **b** differ if no protein were involved?

46. In some bacteria, several genes required for the biosynthesis of the redox cofactor flavin adenine dinucleotide (FAD; Fig. 3.2c) are arranged in an operon. Comparisons of the sequences of this operon in different species reveal a conserved sequence in the untranslated region at the 5′ end of the operon's mRNA. The tertiary structure of an RNA molecule typically includes regions of base pairing and unpaired loops (stem–loop structures). By examining an RNA sequence and noting which positions are most conserved, it is possible to predict the stem–loop structure of the RNA. A portion of a conserved mRNA sequence called RFN, which regulates the expression of the FAD-synthesizing operon, is shown here.

··· G A U U C A G U U U A A G C U G A A G C ···

a. Draw the stem–loop structure for this RNA segment. **b.** In order to function as an FAD sensor, the RFN element (which consists of about 165 nucleotides) must alter its conformation when FAD binds. How could researchers assess RNA conformational changes? **c.** FAD can be considered as a derivative of flavin mononucleotide (FMN; a coenzyme that resembles FAD but lacks its AMP moiety), which in turn is derived from riboflavin (Fig. 3.2c). The ability of FAD, FMN, and riboflavin to bind to the RFN element was measured as a dissociation constant, K_d; results are shown in the table. Which compound is the most effective regulator of FAD biosynthesis in the cell? What portion of the FAD molecule is likely to be important for interacting with the mRNA?

Compound	K_d (nM)
FAD	300
FMN	5
Riboflavin	3000

47. A number of human neurological diseases result from the presence of trinucleotide repeats in certain protein-coding genes. The severity of each disease is correlated with the number of repeats, which may increase due to the slippage of DNA polymerase during replication. **a.** The most common repeated triplet is CAG, which is almost always located within an open reading frame. What amino acid is encoded by this triplet (see Table 3.3) and how would the repeats affect the protein? **b.** To test the effect of CAG repeats on transcription, researchers used a yeast expression system with genes engineered to contain CAG repeats. In addition to the expected transcripts corresponding to the known lengths of the genes, RNA molecules up to three times longer were obtained. Based on your knowledge of RNA synthesis and processing, what factors could account for longer-than-expected transcripts of a given gene? **c.** Unexpectedly long transcripts could result from slippage of RNA polymerase II during transcription of the CAG repeats. In this scenario, the polymerase temporarily ceases polymerization, slides backward along the DNA template, and then resumes transcription, in effect re-transcribing the same sequence. Slippage may be triggered by the formation of secondary structure in the DNA template strand. Draw a diagram showing how a DNA strand containing CAG repeats could form a secondary structure that might prevent the advance of RNA polymerase.

48. In *E. coli*, replication is several times faster than transcription. Occasionally, the replication fork catches up to an RNA polymerase that is moving in the same direction as replication fork movement. When this occurs, transcription stops, and the RNA polymerase is displaced from the template DNA. DNA polymerase can use the existing RNA transcript as a primer to continue replication. **a.** Draw a diagram of this process, showing how such collisions would produce a discontinuous leading strand. **b.** Explain why most *E. coli* genes are oriented such that replication and transcription proceed in the same direction.

49. A poly(A)-binding protein (PABP) has an affinity for RNA molecules with poly(A) tails. What is the effect of adding PABP to a cell-free system containing mRNA and RNases?

21.3 RNA Processing

50. In *E. coli*, mRNA degradation is carried out by an endonuclease, but the mRNA must first be modified by a 5′ pyrophosphohydrolase. What reaction does this enzyme catalyze?

51. The bacterial enzyme polynucleotide phosphorylase (PNPase) is a 3′→5′ exoribonuclease that degrades mRNA. The enzyme catalyzes a phosphorolysis reaction, as does glycogen phosphorylase

(see Section 13.3), rather than hydrolysis. **a.** Write an equation for the mRNA phosphorolysis reaction. **b.** *In vitro*, PNPase also catalyzes the reverse of the phosphorolysis reaction. What does this reaction accomplish and how does it differ from the reaction carried out by RNA polymerase? **c.** PNPase includes a binding site for long polyribonucleotides, which may promote the enzyme's processivity. Why would this be an advantage for the primary activity of PNPase *in vivo*?

52. Construct a table showing the template, substrates, and product for the following polymerases: **a.** DNA polymerase, **b.** human telomerase, **c.** RNApolymerase, **d.** poly(A)polymerase, and **e.** tRNA CCA-adding enzyme.

53. Tell whether the following elements are found on DNA or RNA: **a.** cap, **b.** CpG islands, **c.** –35 region, **d.** –10 region, **e.** poly(A) tail, **f.** TATA box, **g.** enhancer, **h.** 3′ CCA sequence, **i.** 5′ and 3′ splice sites, **j.** promoter, **k.** branch point.

54. How does RNA polymerase transcribe DNA that is present in the form of nucleosome/chromatin during the elongation phase?

55. Why are only mRNAs capped and polyadenylated? Why do these post-transcriptional modifications not take place on rRNA or tRNA?

56. Explain why capping the 5′ end of an mRNA molecule makes it resistant to 5′→3′ exonucleases. Why is it necessary for capping to occur before the mRNA has been completely synthesized?

57. Some short regulatory bacterial RNAs are capped at the 5′ end by the structure shown below. **a.** What metabolite is the source of the capping group? **b.** How might the cap structure link the activity of the attached RNA to the cell's overall metabolic state? **c.** Would the cap help protect the RNA from degradation by 5′ endonucleases?

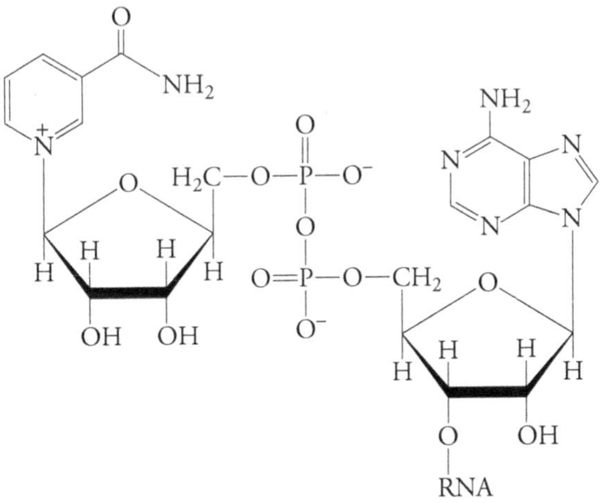

58. The complex of enzymes that cleaves newly synthesized RNA and attaches a poly(A) tail includes phosphatases. What do the phosphatases do and why is this important for transcription?

59. The poly(A) polymerase that modifies the 3′ end of mRNA molecules differs from other polymerases. The active sites of DNA and RNA polymerases are large enough to accommodate a double-stranded polynucleotide, but the active site of poly(A) polymerase is much narrower. **a.** Explain why. **b.** Explain how the substrate specificity of poly(A) polymerase differs from that of a conventional RNA polymerase.

60. What are the product of the following polymerases: **a.** RNA polymerase I, **b.** RNA polymerase II, and **c.** RNA polymerase III?

61. The only mRNA transcripts that lack poly(A) tails are those encoding histones. Why do these mRNA transcripts not require poly(A) tails?

62. ATP can be labeled with ^{32}P at any one of its three phosphate groups, designated α, β, and γ (see Fig. 12.8). A eukaryotic cell carrying out transcription and RNA processing is incubated with labeled ATP. Where will the radioactive isotope appear in RNA if the ATP is labeled with ^{32}P at the **a.** α position, **b.** β position, and **c.** γ position?

63. Genetic engineers must modify eukaryotic genes so that they can be expressed in bacterial host cells. Explain why the DNA from a eukaryotic gene cannot be placed directly into the bacteria but is first transcribed to mRNA and then reverse-transcribed back to cDNA.

64. Introns in eukaryotic protein-coding genes may be quite large, but almost none are smaller than about 65 bp. What are some reasons for this minimum intron size?

65. Introns are removed co-transcriptionally rather than post-transcriptionally. Why is this a good cellular strategy?

66. A portion of the β globin gene is shown below. The sequence on the left includes the 5′ splice site and the sequence on the right includes the 3′ splice site for the intron between exons 1 and 2. Identify the 5′ splice site and the 3′ splice site.

··· GGCAGGTTGGTA ···· ACCCTTAGGCTGCT ···

67. The gene for the ovalbumin Y protein contains seven exons. The sequence on the left includes the 5′ splice site and the sequence on the right includes the 3′ splice site for the first exon–intron junction. **a.** Identify the 5′ splice site and the 3′ splice site.

··· GAAGAAGGTAAGTT ···· CTTGCAGGTTCTT ···

b. Why does intron removal from ovalbumin Y mRNA not require the energy input of ATP hydrolysis?

68. The choice of exons in alternative mRNA splicing may reflect the rate of transcription. Propose an explanation that links the pace of RNA polymerase to the use of "strong" (common) or "weak" (rare) splice sites by the spliceosomes.

69. The ribozyme known as RNase P processes certain immature tRNA and rRNA molecules. Comparisons of RNase P RNAs from different species reveal conserved features that appear to be involved in endonuclease activity. For example, an unpaired uridine at position 69 is universally conserved. This residue does not pair with another nucleotide but forms a bulge in the RNA secondary structure. To test whether the identity or the geometry of U69 is critical for RNase P activity, several mutants were constructed, and their endonuclease activities measured. The results for each mutant are given as a rate constant (k) for catalysis and a dissociation constant (K_d) for substrate binding.

RNase P RNA	k (min^{-1})	K_d (nM)
Wild-type U69	0.26	1.7
U69 → G69	0.0034	73
U69 → C69	0.0056	3
U69 deletion	0.0056	7
U69 + U70	0.0054	181

a. What is the effect of mutating U69 to a G or a C residue? Do these data reveal whether the U69 bulge is more important for substrate binding or catalysis? **b.** What is the effect of increasing the size of the bulge by adding a second U residue (the U69 + U70 mutant) or removing the bulge by deleting the U69 residue? **c.** Like proteins, RNAs have primary, secondary, and tertiary structures. Use what you have learned about the primary, secondary, and tertiary structures of proteins (see Chapter 4) to describe the effect of base substitution on the structure and activity of the ribozyme RNase P.

70. Explain why the vast majority of nucleic acids with catalytic activity are RNA rather than DNA.

71. What are the consequences to the cell if mRNA leaves the nucleus before it is "export ready"?

72. Compare and contrast transcription in prokaryotic and eukaryotic cells.

73. RNA interference was investigated as a method to silence the gene for vascular endothelial growth factor (VEGF), a protein required for angiogenesis (development of blood vessels) in most cancers. The addition of siRNA specific for the VEGF gene almost completely eliminates the secretion of VEGF from prostate cancer cells in culture. A portion of the gene sequence (bases 189–207) is shown here. Design an siRNA targeted to this region of the gene.

5′ ··· GGAGTACCCTGATGAGATC ··· 3′

74. Some tRNA molecules include sulfur-containing nucleotides. Draw the structures of 4-thiouridine and 2-thiocytidine.

75. tRNA modifications are important for temperature adaptations in thermophiles (organisms that live in extreme heat) and psychrophiles (organisms that live in extreme cold). Provide a hypothesis that explains this observation.

76. Biochemistry textbooks published a few decades ago often used the phrase "one gene, one protein." **a.** Why is this phrase no longer accurate? **b.** Is your answer to part **a** consistent with the data shown in Table 21.2 in which the number of proteins expressed in humans is far greater than the number of protein-coding genes?

77. Based on the information in this chapter, give at least three reasons why a silent mutation in a gene (that is, a mutation that does not alter the amino acid sequence of the encoded protein) could decrease the amount of protein expressed.

SELECTED READINGS

Corbett, A.H., Post-transcriptional regulation of gene expression and human disease, *Curr. Opin. Cell Biol.* 52, 96–104, doi: 10.1016/j.ceb.2018.02.011 (2018). [Reviews the steps of post-transcriptional mRNA processing and disorders related to defects in specific proteins.]

Cramer, P., Organization and regulation of gene transcription, *Nature* 573, 48–54, doi: 10.1038/s41586-019-1517-4 (2019). [A succinct review of transcription that includes information about RNA polymerases I and III and emphasizes the importance of separate liquid phases for initiation and elongation.]

Gayon, J., From Mendel to epigenetics: History of genetics, *C. R. Biol.* 339, 225–230, doi: 10.1016/j.crvi.2016.05.009 (2016). [Provides some historical context around the difficulty of defining *gene*.]

Haberle, V. and Stark, A., Eukaryotic core promoters and the functional basis of transcription initiation, *Nat. Rev. Mol. Cell Biol.* 19, 621–637, doi: 10.1038/s41580-018-0028-8 (2018). [Discusses promoters, RNA polymerase pausing, and other events of transcription initiation.]

Murakami, K.S., Structural biology of bacterial RNA polymerase, *Biomolecules* 5, 848–864, doi: 10.3390/biom5020848 (2015). [Summarizes key features of the structure and function of the bacterial enzyme.]

Stewart, M., Polyadenylation and nuclear export of mRNAs, *J. Biol. Chem.* 294, 2977–2987, doi: 10.1074/jbc.REV118.005594 (2019).

[Reviews the events of transcription termination and movement of processed mRNA through the nuclear pore.]

Wilkinson, A.C., Nakauchi, H., and Göttgens, B., Mammalian transcription factor networks: Recent advances in interrogating biological complexity, *Cell Systems* 5, 319–331, doi: 10.1016/j.cels.2017.07.004 (2017). [Describes some practical and theoretical approaches to understanding the role of transcription factors in regulating gene expression.]

CHAPTER 21 CREDITS

Figure 21.1 Based on a diagram by Ali Shilatifard, St. Louis University School of Medicine.

Figure 21.2 Image based on 3UVW. Filippakopoulos, P., Picaud, S., Mangos, M., Keates, T., Lambert, J.P., Barsyte-Lovejoy, D., Felletar, I., Volkmer, R., Muller, S., Pawson, T., Gingras, A.C., Arrowsmith, C.H., Knapp, S., Histone recognition and large-scale structural analysis of the human bromodomain family, *Cell* 149, 214–231 (2012).

Figure 21.3 Image from Wagner, F.R., Dienemann, C., Wang, H., Stützer, A., Tegunov, D., Urlaub, H., & Cramer, P., Structure of SWI/SNF chromatin remodeller RSC bound to a nucleosome. *Nature,* 579(7799), 448–451. © 2020 Springer Nature.

Figure 21.7 Image based on 1YTB. Kim, Y., Geiger, J.H., Hahn, S., Sigler, P.B., Crystal structure of a yeast TBP/TATA-box complex, *Nature* 365, 512–520 (1993).

Box 21.A Image of λ repressor based on 1LMB. Beamer, L.J., Pabo, C.O., Refined 1.8 A crystal structure of the lambda repressor-operator complex, *J. Mol. Biol.* 227, 177–196 (1992).

Box 21.A Image of Zif268 based on 1AAY. Elrod-Erickson, M., Rould, M.A., Nekludova, L., Pabo, C.O., Zif268 protein-DNA complex refined at 1.6 A: a model system for understanding zinc finger-DNA interactions, *Structure* 4, 1171–1180 (1996).

Box 21.A Image of GCN4 based on 1DGC. Konig, P., Richmond, T.J., The X-ray structure of the GCN4-bZIP bound to ATF/CREB site DNA shows the complex depends on DNA flexibility, *J. Mol. Biol.* 233, 139–154 (1993).

Box 21.C Image from Kim, S.J., Fernandez-Martinez, J., Nudelman, I., Shi, Y., Zhang, W., Raveh, B., Herricks, T., Slaughter, B.D., Hogan, J.A., Upla, P., Chemmama, I.E., Pellarin, R., Echeverria, I., Shivaraju, M., Chaudhury, A.S., Wang, J., Williams, R., Unruh, J.R., Greenberg, C.H., Rout, M.P., Integrative structure and functional anatomy of a nuclear pore complex. *Nature,* 555(7697), 475–482. © 2018 Springer Nature.

Figure 21.10 Image based on 5OQM. Schilbach, S., Hantsche, M., Tegunov, D., Dienemann, C., Wigge, C., Urlaub, H., Cramer, P., Structures of transcription pre-initiation complex with TFIIH and Mediator, *Nature* 551, 204–209 (2017).

Figure 21.11 Image based on 1LBG. Lewis, M., Chang, G., Horton, N.C., Kercher, M.A., Pace, H.C., Schumacher, M.A., Brennan, R.G., Lu, P., Crystal structure of the lactose operon repressor and its complexes with DNA and inducer, *Science* 271, 1247–1254 (1996).

Figure 21.13 Image based on 5FLM. Bernecky, C., Herzog, F., Baumeister, W., Plitzko, J.M., Cramer, P., Structure of transcribing mammalian RNA polymerase II, *Nature* 529, 551 (2016).

Figure 21.14a Image based on 5IYD. He, Y., Yan, C., Fang, J., Inouye, C., Tjian, R., Ivanov, I., Nogales, E., Near-atomic resolution visualization of human transcription promoter opening, *Nature* 533, 359–365 (2016).

Figure 21.14b Based on a drawing by Roger Kornberg.

Figure 21.20 Image based on 1CVJ. Deo, R.C., Bonanno, J.B., Sonenberg, N., Burley, S.K., X-Ray crystal structure of the poly(A)-binding protein in complex with polyadenylate RNA, *Cell* 98, 835–845 (1999).

Figure 21.22 Image based on 3JB9. Yan, C., Hang, J., Wan, R., Huang, M., Wong, C., Shi, Y.. Structure of a yeast spliceosome at 3.6-Angstrom resolution, *Science* 349, 1182–1191 (2015).

Figure 21.25 Based on Breitbart, R.E., Andreadis, A., and Nadal-Ginard, B., *Annu. Rev. Biochem.* 56, 481 (1987).

Table 21.2 Data from NCBI, https://www.ncbi.nlm.nih.gov/genome/.

Figure 21.31 Image based on 3Q1Q. Reiter, N.J., Osterman, A., Torres-Larios, A., Swinger, K.K., Pan, T., Mondragon, A., Structure of a bacterial ribonuclease P holoenzyme in complex with tRNA, *Nature* 468, 784–789 (2010).

Protein Synthesis

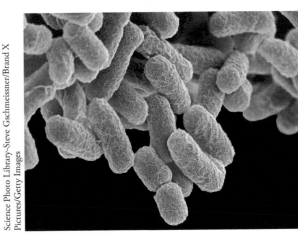

Science Photo Library-Steve Gschmeissner/Brand X Pictures/Getty Images

Each of these rapidly growing *E. coli* cells contains about 50,000 ribosomes, which account for about one-third of the cell's dry mass. These numbers reflect the fact that protein manufacturing is typically a cell's main job; they also mean that cells spend a significant part of their energy budget not only on synthesizing proteins but also on building the infrastructure to make those proteins.

Do You Remember?

- DNA and RNA are polymers of nucleotides, each of which consists of a purine or pyrimidine base, deoxyribose or ribose, and phosphate (Section 3.2).
- The biological information encoded by a sequence of DNA is transcribed to RNA and then translated into the amino acid sequence of a protein (Section 3.3).
- Amino acids are linked by peptide bonds to form a polypeptide (Section 4.1).
- Protein folding and protein stabilization depend on noncovalent forces (Section 4.3).
- rRNA and tRNA transcripts are modified to produce functional molecules (Section 21.3).

In the decade that followed the 1953 elucidation of DNA structure, nearly all the components required for expressing genetic information—that is, making a protein—were identified, including mRNA, tRNA, and ribosomes. We can examine how these molecules participate in protein synthesis by considering one step at a time, beginning with the attachment of a specific amino acid to the appropriate tRNA molecule. We can then look at how the tRNAs align with an mRNA sequences in a ribosome so that peptide bonds can link the amino acids in the order specified by the mRNA. We will also look at some of the steps required to convert a newly made polypeptide to a fully functional protein.

22.1 tRNA and the Genetic Code

KEY CONCEPTS

Describe the role of tRNA in reading the genetic code.

- Explain why the genetic code is redundant, unambiguous, and nonrandom.
- Identify the structural features of tRNAs.
- Describe the substrates, products, and catalytic activities of aminoacyl–tRNA synthetases.
- Explain how one tRNA anticodon can pair with more than one mRNA codon.

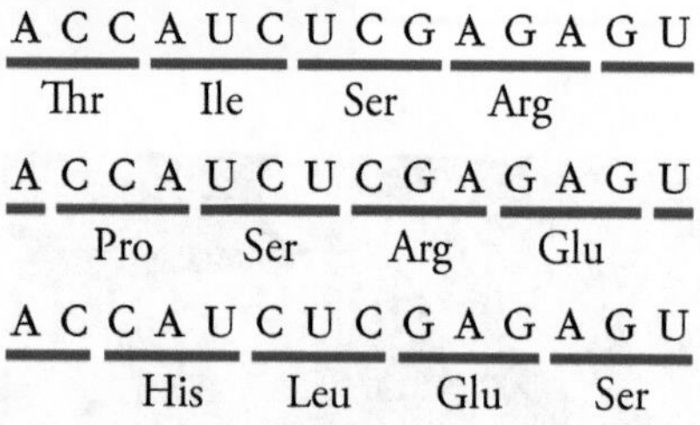

A C C A U C U C G A G A G U

Thr   Ile   Ser   Arg

A C C A U C U C G A G A G U

Pro   Ser   Arg   Glu

A C C A U C U C G A G A G U

His   Leu   Glu   Ser

Figure 22.1 Reading frames. Even with a nonoverlapping genetic code based on nucleotide triplets, a given nucleotide sequence has three possible reading frames.

In protein synthesis, the final step of the central dogma of molecular biology, a sequence of nucleotides (the first language) is translated to a sequence of amino acids (the second language). Soon after Francis Crick helped work out the structure of DNA, he hypothesized that **translation** required "adaptor" molecules (subsequently identified as tRNA) that carried amino acids and recognized genetic information in the form of nucleotides. The correspondence between DNA sequences and protein sequences was indisputable, but it required some biochemical detective work to discover the nature of the genetic code. Ultimately, *the genetic code was shown to be based on three-nucleotide **codons** that are read in a sequential and nonoverlapping manner.*

The genetic code is redundant

A triplet code is a mathematical necessity, since the number of possible combinations of three nucleotides of four different kinds (4^3, or 64) is more than enough to specify the 20 amino acids found in polypeptides (a doublet code, with 4^2, or 16, possibilities, would be inadequate). Genetic experiments with mutant bacteriophages demonstrated that triplet codons are read sequentially. For example, a mutation resulting from the deletion of a nucleotide within a gene can be corrected by a second mutation that inserts another nucleotide into the gene. The second mutation can restore gene function because it maintains the proper **reading frame** for translation. Since a given nucleotide sequence in an mRNA molecule can potentially have three different reading frames (**Fig. 22.1**), the selection of the proper one depends on the precise identification of a translation start site. In rare cases, a single RNA can be translated in different ways.

The genetic code, shown in **Table 22.1**, is said to be redundant because *several mRNA codons may correspond to the same amino acid.* In fact, most amino acids are specified by two or more codons (arginine, leucine, and serine each have six codons). Only methionine and tryptophan have only one codon each (they are also among the amino acids that occur least frequently in polypeptides; see Fig. 4.3). The methionine codon also functions as a translation initiation point. Three codons, known as stop or nonsense codons, signal translation termination. In Table 22.1, codons are shaded according to the overall hydrophobic, polar, or ionic character of the corresponding amino acid (using the scheme introduced in Fig. 4.2). Codons for chemically similar amino acids exhibit the same residue at the second position;

Table 22.1 The Standard Genetic Code

First position (5′ end)	Second position				Third position (3′ end)
	U	C	A	G	
U	UUU Phe	UCU Ser	UAU Tyr	UGU Cys	**U**
	UUC Phe	UCC Ser	UAC Tyr	UGC Cys	**C**
	UUA Leu	UCA Ser	UAA Stop	UGA Stop	**A**
	UUG Leu	UCG Ser	UAG Stop	UGG Trp	**G**
C	CUU Leu	CCU Pro	CAU His	CGU Arg	**U**
	CUC Leu	CCC Pro	CAC His	CGC Arg	**C**
	CUA Leu	CCA Pro	CAA Gln	CGA Arg	**A**
	CUG Leu	CCG Pro	CAG Gln	CGG Arg	**G**
A	AUU Ile	ACU Thr	AAU Asn	AGU Ser	**U**
	AUC Ile	ACC Thr	AAC Asn	AGC Ser	**C**
	AUA Ile	ACA Thr	AAA Lys	AGA Arg	**A**
	AUG Met	ACG Thr	AAG Lys	AGG Arg	**G**
G	GUU Val	GCU Ala	GAU Asp	GGU Gly	**U**
	GUC Val	GCC Ala	GAC Asp	GGC Gly	**C**
	GUA Val	GCA Ala	GAA Glu	GGA Gly	**A**
	GUG Val	GCG Ala	GAG Glu	GGG Gly	**G**

for example, U at the second position invariably specifies a hydrophobic amino acid. This apparently nonrandom pattern of codon–amino acid correspondence suggests that the genetic code might have evolved from a simpler system involving only two nucleotides and a handful of amino acids.

The genetic code is essentially universal (there are only a few minor variations in mitochondria and some unicellular eukaryotes). The common genetic code makes genetic engineering possible: a bacterium decodes a human gene in the same way a human cell does. The universal code also allows scientists to deduce evolutionary relationships based on DNA sequence differences (Section 1.2). This would not be possible if each organism had its own way of interpreting genetic information.

tRNAs have a common structure

Each tRNA interacts specifically with one codon via its **anticodon** sequence. A bacterial cell typically contains 30 to 40 different tRNAs, and humans produce over 300 different tRNAs in a tissue-specific manner. These numbers illustrate redundancy in biological systems, since only 20 different amino acids are routinely incorporated into polypeptides. tRNAs that bear the same amino acid but have different anticodons are called **isoacceptor tRNAs.** The hundreds of tRNAs in mammals that have the same anticodons but differ in other segments are known as **isodecoder tRNAs.** Nevertheless, the structures of all tRNA molecules are similar—even those that carry different amino acids.

Each tRNA molecule contains about 76 nucleotides (the range is 54 to 100), of which an average of 13 are post-transcriptionally modified (the structures of some of these modified nucleotides are shown in Fig. 21.29). Many of the tRNA bases pair intramolecularly, generating the short stems and loops of what is commonly called a cloverleaf secondary structure (**Fig. 22.2a**). A segment at the 5′ end of the tRNA pairs with bases near the 3′ end to form the acceptor stem (an amino acid attaches to the 3′ end). Several other base-paired stems end in small loops. The D loop often contains the modified base dihydrouridine (abbreviated D),

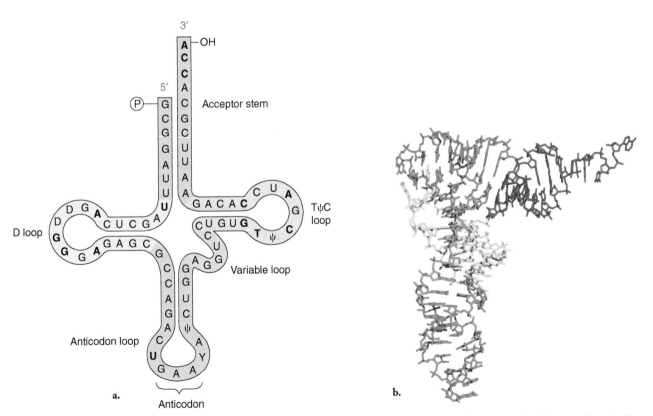

Figure 22.2 Structure of yeast tRNA^Phe. **a.** Secondary structure. The 76 nucleotides of this tRNA molecule, which can carry a phenylalanine residue at its 3′ end, form four base-paired stems arranged in a cloverleaf pattern. Invariant bases are shown in boldface. ψ is pseudouridine and Y is a guanosine derivative. Some C and G residues in this structure are methylated. **b.** Tertiary structure, with the various structures colored as in part **a.** The long arm of the L consists primarily of the anticodon loop and D loop, and the short arm is primarily made up of the TψC loop and acceptor stem.

Question **Why is it important that the bases in the anticodon loop point outward?**

and the TψC loop usually contains the indicated sequence (ψ is the symbol for the nucleotide pseudouridine; see Section 21.3). The variable loop, as its name implies, ranges from 3 to 21 nucleotides in different tRNAs. The anticodon loop includes the three nucleotides that pair with an mRNA codon.

The various elements of tRNA secondary structure fold into a compact L shape that is stabilized by extensive stacking interactions and nonstandard base pairs (Fig. 22.2b). Virtually all the bases are buried in the interior of the tRNA molecule, except for the anticodon triplet and the CCA sequence at the 3′ end. The narrow elongated structure of tRNA molecules allows them to align side-by-side so that they can interact with adjacent mRNA codons during translation. However, the tRNA anticodon is located a considerable distance (about 75 Å) from the 3′ aminoacyl group, whose identity is specified by that anticodon.

tRNA aminoacylation consumes ATP

Aminoacylation, the attachment of an amino acid to a tRNA, is catalyzed by an aminoacyl–tRNA synthetase (AARS). *To ensure accurate translation, the synthetase must attach the appropriate amino acid to the tRNA bearing the corresponding anticodon.* Many AARSs interact with the tRNA anticodon as well as the aminoacylation site at the other end of the tRNA molecule.

An AARS catalyzes the formation of an ester bond between an amino acid and an OH group of the ribose at the 3′ end of a tRNA to yield an aminoacyl–tRNA:

Aminoacyl–tRNA

The tRNA molecule is then said to be "charged" with an amino acid. The aminoacylation reaction has two steps and requires the free energy of ATP (**Fig. 22.3**). The overall reaction is

$$\text{amino acid} + \text{tRNA} + \text{ATP} \rightarrow \text{aminoacyl–tRNA} + \text{AMP} + \text{PP}_i$$

Similar enzyme-catalyzed reactions use ATP to "activate" fatty acids for oxidation (Section 17.2) and ribose groups for nucleotide synthesis (Section 18.5).

Most cells contain 20 different AARS enzymes, corresponding to the 20 standard amino acids (isoacceptor tRNAs are recognized by the same AARS). Although all AARSs catalyze the same reaction, they do not exhibit a conserved size or quaternary structure. Nevertheless, the enzymes fall into two groups based on several shared structural and functional features (**Table 22.2**). For example, the class I enzymes attach an amino acid to the 2′ OH group of the tRNA ribose, whereas the class II enzymes attach an amino acid to the 3′ OH group (this distinction is ultimately of no consequence, as the 2′-aminoacyl group shifts to the 3′ position before it takes part in protein synthesis).

Some bacteria appear to lack the full complement of 20 AARSs. The enzymes most commonly missing are GlnRS and AsnRS (which aminoacylate tRNA^Gln and tRNA^Asn). In these organisms, Gln–tRNA^Gln and Asn–tRNA^Asn are synthesized indirectly. First, GluRS and AspRS with relatively low tRNA specificity charge tRNA^Gln and tRNA^Asn with their corresponding acids (glutamate and aspartate). Next, an amidotransferase converts Glu–tRNA^Gln and Asp–tRNA^Asn to Gln–tRNA^Gln and Asn–tRNA^Asn using glutamine as an amino-group donor. In some microorganisms, this is the only pathway for producing asparagine.

Table 22.2 Classes of Aminoacyl–tRNA Synthetases

Amino acids		
Class I	Arg	Leu
	Cys	Met
	Gln	Trp
	Glu	Tyr
	Ile	Val
Class II	Ala	Lys
	Asn	Pro
	Asp	Phe
	Gly	Ser
	His	Thr

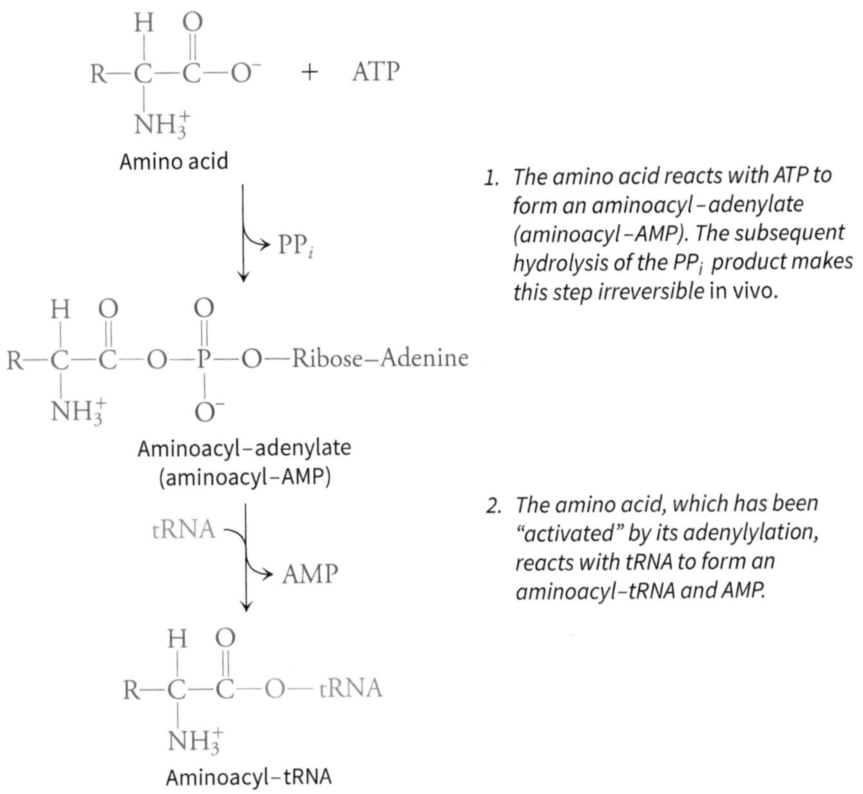

R—C—C—O⁻ + ATP

(with structure showing:)
H O
| ‖
R—C—C—O⁻ + ATP
|
NH₃⁺

Amino acid

↓ PP_i

H O   O
| ‖   ‖
R—C—C—O—P—O—Ribose—Adenine
| |
NH₃⁺   O⁻

Aminoacyl–adenylate
(aminoacyl–AMP)

tRNA ↘
      ↓ AMP

H O
| ‖
R—C—C—O—tRNA
|
NH₃⁺

Aminoacyl–tRNA

1. *The amino acid reacts with ATP to form an aminoacyl–adenylate (aminoacyl–AMP). The subsequent hydrolysis of the PP_i product makes this step irreversible in vivo.*

2. *The amino acid, which has been "activated" by its adenylylation, reacts with tRNA to form an aminoacyl-tRNA and AMP.*

Figure 22.3 The aminoacyl–tRNA synthetase reaction.

Question How many "high-energy" phosphoanhydride bonds break in this process? How many "high-energy" acyl-phosphate bonds form?

An AARS probably binds a tRNA relatively quickly and nonspecifically at first, via electrostatic interactions between positively charged side chains and tRNA phosphate groups. Specific enzyme–tRNA contacts form more slowly and involve discrete nucleotides (often chemically modified) or other structural features. Not all AARSs interact with their tRNA anticodons. For example, SerRS must be able to attach serine to tRNAs bearing six different anticodons; for SerRS, the main recognition feature lies in the variable arm of the tRNA^{Ser} molecules.

The structure of a complex of *E. coli* GlnRS and its cognate (matching) tRNA (tRNA^{Gln}) shows the extensive interaction between the protein and the concave face of the tRNA molecule (the inside of the L; **Fig. 22.4**). Most AARSs can activate an amino acid in the absence of a tRNA molecule, but GlnRS, GluRS, and ArgRS require a cognate tRNA molecule for aminoacyl–AMP formation. This suggests that the anticodon-recognition site and the aminoacylation active site somehow communicate with each other, which might help guarantee the attachment of the correct amino acid to the tRNA.

Editing increases the accuracy of aminoacylation

The accuracy of the AARS reaction depends on several factors. First, the amino acid binding sites of individual AARS enzymes may be tailored precisely to the geometry and electrostatic properties of a particular amino acid, making it less likely that one of the other 19 amino acids would be activated or transferred to a tRNA molecule. For example, TyrRS (the enzyme responsible for synthesizing Tyr–tRNA^{Tyr}) can distinguish between tyrosine and phenylalanine, which have similar shapes, because only tyrosine can form hydrogen bonds with the protein. PheRS uses the opposite strategy: An active-site Ala residue favors interaction with phenylalanine over tyrosine. The active site of ThrRS includes a zinc atom, which can interact with threonine but not with the similarly sized valine.

Kinetic studies indicate that even when different amino acids exhibit similar K_M values for a given AARS, the correct amino acid has a higher k_{cat} value; in other words,

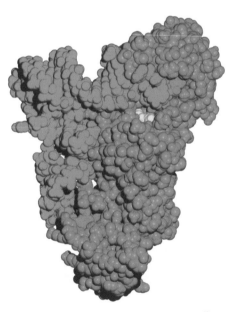

Figure 22.4 Structure of GlnRS with tRNA^{Gln}.
In this complex, the synthetase is green and the cognate tRNA is red. Both the 3′ (acceptor) end of the tRNA (top right) and the anticodon loop (lower left) are buried in the protein. ATP at the active site is shown in yellow.

the wrong amino acid might enter the enzyme's active site but reacts too slowly to generate a significant amount of misacylated tRNA molecules.

In some cases, *the accuracy of tRNA aminoacylation may be enhanced through proofreading by the AARS.* For example, both valine and isoleucine—which differ only by a single methylene group—can easily fit into the active site of IleRS. To prevent the synthesis of Val–tRNA^{Ile}, the enzyme relies on two active sites that operate by a "double-sieve" mechanism. The first active site activates isoleucine and presumably other amino acids that are chemically similar to and smaller than isoleucine (such as valine, alanine, and glycine) but excludes larger amino acids (such as phenylalanine and tyrosine). The second active site, which hydrolyzes aminoacylated tRNA^{Ile}, admits only aminoacyl groups that are smaller than isoleucine. Thus, the activating and editing active sites together ensure that IleRS produces only Ile–tRNA^{Ile} (Val–tRNA^{Ile} is made very rarely, about once every 50,000 reactions). The two active sites are on separate domains of the synthetase, so a newly aminoacylated tRNA must visit the proofreading hydrolytic active site before dissociating from the enzyme.

About half of the AARS enzymes use some sort of editing mechanism to ensure the high fidelity of aminoacylation. The editing activity may be part of the AARS protein, as with IleRS, or it may be a separate enzyme. For example, D aminoacyl–tRNA deacylases target tRNAs that have been charged with D amino acids, so that they can be destroyed before they reach a ribosome.

Bacterial AARSs are solitary proteins, but in eukaryotes, these enzymes tend to form complexes. In humans, the "multisynthetase" complex includes nine synthetases (ProRS and GluRS are actually on a single polypeptide chain) and three accessory proteins. The advantages of this arrangement might include an enhanced ability to match aminoacylation—and therefore protein synthesis—to the cell's amino acid supply or to the cell's overall metabolic state. In vertebrates, some AARSs exhibit a variety of functions apart from charging tRNAs; these "moonlighting" enzymes might provide additional ways to coordinate the cell's activities.

tRNA anticodons pair with mRNA codons

During translation, tRNA molecules align with mRNA codons, base pairing in an antiparallel fashion, for example

$$
\begin{array}{lcccc}
\text{tRNA anticodon} & 3' & -A-A-G- & 5' \\
 & & \vdots \ \ \vdots \ \ \vdots & \\
\text{mRNA codon} & 5' & -U-U-C- & 3'
\end{array}
$$

At first glance, this sort of specific pairing would require the presence of 61 different tRNA molecules, one to recognize each of the "sense" codons listed in Table 22.1. In fact, *many isoacceptor tRNAs can bind to more than one of the codons that specify their amino acid.* For example, yeast tRNA^{Ala} has the anticodon sequence 3'–CGI–5' (I represents the purine nucleotide inosine, a deaminated form of adenosine) and can pair with the alanine codons GCU, GCC, and GCA.

$$
\begin{array}{lccc}
\text{tRNA anticodon} & -C-G-I- & -C-G-I- & -C-G-I- \\
 & \vdots \ \ \vdots \ \ \vdots & \vdots \ \ \vdots \ \ \vdots & \vdots \ \ \vdots \ \ \vdots \\
\text{mRNA codon} & -G-C-U- & -G-C-C- & -G-C-A-
\end{array}
$$

As Crick originally proposed in the **wobble hypothesis,** the third codon position and the 5' anticodon position experience some flexibility, or wobble, in the geometry of their hydrogen bonding. The base pairs permitted by wobbling are given in **Table 22.3**. The wobble hypothesis explains why many bacterial cells can bind all 61 codons with a set of fewer than 40 tRNAs. The wobble position is a particularly popular location for modified bases in mammalian tRNAs. A few variant tRNAs allow nonstandard amino acids to be occasionally incorporated into polypeptides at positions corresponding to stop codons (**Box 22.A**).

Table 22.3 Allowed Wobble Pairs at the Third Codon–Anticodon Position

5' Anticodon base	3' Codon base
C	G
A	U
U	A, G
G	U, C
I	U, C, A

| Box 22.A | The Genetic Code Expanded |

In addition to the 20 standard amino acids listed in Figure 4.2, some amino acid variants can be incorporated into proteins during translation (keep in mind that a mature protein may contain a number of modified amino acids, but these changes almost always take place *after* the protein has been synthesized). Addition of a non-standard amino acid during protein synthesis requires a dedicated tRNA and a stop codon that can be reinterpreted. The expanded genetic code includes two naturally occurring amino acids, seleno-cysteine and pyrrolysine, plus a number of amino acids produced in the laboratory.

Selenocysteine occurs in a few proteins in both prokaryotes and eukaryotes, which explains why selenium is an essential trace element. Humans may produce as many as two dozen selenoproteins.

$$\begin{array}{c} | \\ NH \\ | \\ CH-CH_2-Se-H \\ | \\ C=O \\ | \end{array}$$

Selenocysteine (Sec) residue

Selenocysteine (Sec), which resembles cysteine, is generated from serine that has been attached to tRNA^{Sec} by the action of SerRS. A separate enzyme then converts Ser–tRNA^{Sec} to Sec–tRNA^{Sec}. This charged tRNA has an ACU anticodon (reading in the $3' \rightarrow 5'$ direction), which recognizes a UGA codon. Normally, UGA functions as a stop codon, but a hairpin secondary structure in the sele-noprotein's mRNA provides the contextual signal for selenocysteine to be accepted by the ribosome at that point.

A few prokaryotic species incorporate pyrrolysine (Pyl) into certain proteins. Synthesis of these proteins requires a 21st type of AARS that directly charges the tRNA^{Pyl} with pyrrolysine.

$$\begin{array}{c} | \\ NH \\ | \\ CH-CH_2-CH_2-CH_2-CH_2-NH-C(=O)-\text{(ring)}-CH_3 \\ | \\ C=O \\ | \end{array}$$

Pyrrolysine (Pyl) residue

The Pyl–tRNA^{Pyl} recognizes the stop codon UAG, which is reinter-preted as a Pyl codon with the help of a protein that recognizes sec-ondary structure in the mRNA bound to the ribosome.

In the laboratory, proteins containing unnatural amino acids can be synthesized in bacterial, yeast, and mammalian cells. These experimental systems rely on a pair of genetically engineered com-ponents: a tRNA that can "read" a stop codon and an AARS that can attach the unnatural amino acid to the tRNA. When the cell translates an mRNA containing the stop codon, the novel amino acid is incorporated at that codon. Dozens of amino acid derivatives with fluoride, reactive acetyl and amino groups, fluorescent tags, and other modifications have been introduced into specific proteins using this technology. Because the novel amino acids are genetically encoded, they appear only at the expected positions in the translated protein—a more reliable outcome than chemically modifying a pro-tein in a test tube.

Question **What is the disadvantage for a cell to have a variant tRNA that can insert an amino acid at a stop codon?**

Concept Check

1. Summarize the features of the genetic code.
2. Explain why a universal genetic code is useful for genetic engineers.
3. Draw a simple diagram of a tRNA molecule and label its parts.
4. Write an equation for each step of the aminoacyl–tRNA synthetase reaction and identify the energy-requiring step.
5. Explain why accurate aminoacylation is essential for accurate translation.
6. Explain why cells do not require 61 different codons.
7. Identify some codons whose meanings would change following a single-nucleotide substitution. Identify some codons whose meanings would not change.

22.2 Ribosome Structure

KEY CONCEPTS

Recognize the major features of the ribosome.

- Explain the importance of ribosomal RNA.
- Identify the three tRNA binding sites in the ribosome.

Table 22.4 Ribosome Components

	RNA	Polypeptides
E. coli ribosome (70S)		
Small subunit (30S)	16S	21
Large subunit (50S)	23S, 5S	31
Mammalian ribosome (80S)		
Small subunit (40S)	18S	33
Large subunit (60S)	28S, 5.8S, 5S	47

In order to synthesize a protein, genetic information (in the form of mRNA) and amino acids (attached to tRNA) must get together so that the amino acids can be covalently linked in the specified order. This is the job of the **ribosome,** and it is a huge job: A large mammalian cell may contain a few billion protein molecules.

The ribosome is mostly RNA

The ribosome is a large complex containing both RNA and protein. At one time, ribosomal RNA (rRNA) was believed to serve as a structural scaffolding for ribosomal proteins, which presumably carried out protein synthesis, but it is now clear that rRNA itself is central to ribosomal function.

A typical bacterial cell contains tens of thousands of ribosomes, a yeast cell around 200,000, and a large mammalian cell up to 10 million. These numbers account for the observation that at least 80% of a cell's RNA is located in ribosomes (tRNA comprises about 15% of cellular RNA; mRNA accounts for only a few percent of the total). A ribosome consists of a large and a small subunit containing rRNA molecules, all of which are described in terms of their sedimentation coefficients, S. Thus, the 70S bacterial ribosome has a large (50S) and a small (30S) subunit (the sedimentation coefficient indicates how quickly a particle settles during ultracentrifugation; it is related to the particle's mass). The 80S eukaryotic ribosome is made up of a 60S large subunit and a 40S small subunit. The compositions of prokaryotic and eukaryotic ribosomes are listed in **Table 22.4**. Regardless of its source, about two-thirds of the mass of a ribosome is due to the rRNA; the remainder is due to dozens of different proteins (80 unique proteins in eukaryotes). The core of the ribosome is probably the most highly conserved structure across all forms of life.

The structures of intact ribosomes from both prokaryotes and eukaryotes have been elucidated by X-ray crystallography—a monumental undertaking, given the ribosome's large size (about 2500 kD in bacteria and about 4300 kD in eukaryotes). The small ribosomal subunit from the heat-tolerant bacterium _Thermus thermophilus_ is shown in **Figure 22.5**. The overall shape of the subunit is defined by the 16S rRNA (1542 nucleotides in _E. coli_), which has numerous base-paired stems and loops that fold into several domains. This multidomain structure appears to confer some conformational flexibility on the 30S subunit—a requirement for protein synthesis. Twenty-one small polypeptides dot the surface of the structure.

Compared to the 30S subunit, the prokaryotic 50S subunit is solid and immobile. Its 23S rRNA (2904 nucleotides in _E. coli_) and 5S rRNA (120 nucleotides) fold into a single mass (**Fig. 22.6**). As in the small subunit, the ribosomal proteins associate with the surface of the rRNA, but the surfaces of the large and small subunits that make contact in the intact 70S ribosome are largely devoid of protein. _This highly conserved rRNA-rich subunit interface is the site where mRNA and tRNA bind during protein synthesis._

Eukaryotic ribosomes are about 40–50% larger than those from bacteria and contain many additional proteins and more extensive rRNA. The RNA sequences that have no counterparts in bacterial ribosomes are known as expansion segments; these structures, along with the unique eukaryotic protein components, surround a core structure that is shared with the simpler bacterial ribosome (**Fig. 22.7**). Eukaryotic rRNA includes a higher percentage of modified nucleotides than prokaryotic rRNA.

a.

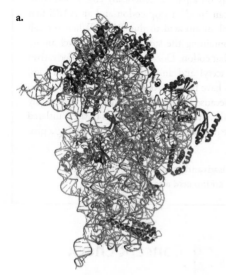

b.

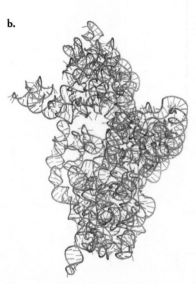

Figure 22.5 Structure of the 30S ribosomal subunit from _Thermus thermophilus._ a. The 30S subunit with the rRNA in gray and the proteins in purple. **b.** Structure of the 16S rRNA alone.

In eukaryotes, the large and small ribosomal subunits are assembled mostly in the **nucleolus,** a dense membraneless organelle inside the nucleus, where clustered repeats of several hundred rRNA genes are localized. In mammals, over 250 proteins participate in ribosome assembly by cleaving newly made rRNA (see Fig. 21.28), modifying the rRNA with the help of small nucleolar RNAs (snoRNAs), and arranging the ribosomal proteins (which must be imported into the nucleus via the nuclear pore complex; Box 21.C).

Theoretical studies suggest that a structure based on a few relatively large rRNA molecules plus numerous smaller proteins of similar size maximizes the efficiency of ribosome assembly. The cell can manufacture many small proteins in less time than it would take to make one large protein. The rRNAs are synthesized more quickly than the proteins, so there is no constraint on their size.

The large and small ribosomal subunits exit the nucleus and finish maturing in the cytosol. As described below, the two subunits must be separate before translation can begin. Excess ribosomes are destroyed by autophagy (Section 9.4), particularly during periods of low nutrient intake when protein synthesis is curtailed and the ribosomal proteins are needed as a source of amino acids for fuel.

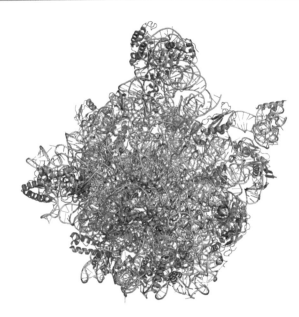

Figure 22.6 Structure of the 50S ribosomal subunit from *Haloarcula marismortui.* The 50S subunit is shown with rRNA in gray and proteins in green. Most of the ribosomal proteins are not visible in this view. The protein-free central area forms the interface with the 30S subunit.

Three tRNAs and one mRNA bind to the ribosome

Up to three tRNA molecules may bind to the ribosome at a given time (**Fig. 22.8**). The binding sites are known as the **A site** (for *aminoacyl*), which accommodates an incoming aminoacyl–tRNA; the **P site** (for *peptidyl*), which binds the tRNA with the growing polypeptide chain; and the **E site** (for *exit*), which temporarily holds a deacylated tRNA after peptide bond formation. The anticodon ends of the tRNAs extend into the 30S subunit to pair with mRNA codons, while their aminoacyl ends extend into the 50S subunit, which catalyzes peptide bond formation.

In bacteria, the two ribosomal subunits and the various tRNAs are held in place mainly by RNA–RNA contacts, with a number of stabilizing Mg^{2+} ions. In eukaryotic ribosomes, numerous proteins form intersubunit bridges. In both cases, the mRNA, which threads through the 30S subunit, makes a sharp bend between the codons in the A site and P site, where an Mg^{2+} ion interacts with mRNA backbone phosphate groups (see Fig. 22.8). The kink allows two tRNAs to fit side-by-side while interacting with consecutive mRNA codons. It may also help the ribosome maintain the reading frame by preventing it from slipping along the mRNA.

In a prokaryotic cell, DNA, mRNA, and ribosomes are in the same cellular compartment. But in eukaryotic cells, mRNA is synthesized and processed inside the nucleus and then exported

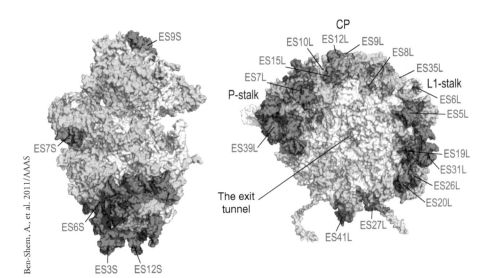

Figure 22.7 Eukaryotic ribosomal subunits. The solvent-exposed surfaces of the 40S subunit (left) and 60S subunit (right) are shown with the conserved ribosomal core structures in gray. Proteins that are unique to eukaryotes are shown in transparent yellow, and rRNA expansion segments are red. Newly synthesized proteins emerge from the ribosome through the exit tunnel.

Question Which areas of the ribosome appear to be most conserved between bacteria and eukaryotes? Why?

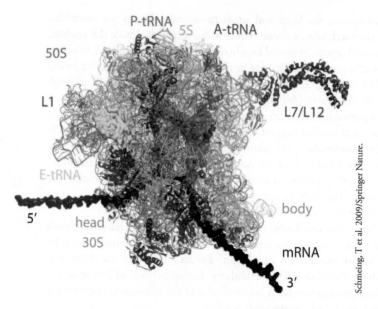

Schmeing, T et al. 2009/Springer Nature.

Figure 22.8 Model of the complete bacterial ribosome.   The large subunit is shown in shades of gold (rRNA) and brown (proteins), and the small subunit in shades of blue (rRNA) and purple (proteins). The three tRNAs are colored magenta (A site), green (P site), and yellow (E site). An mRNA molecule is shown in dark gray. Note that the anticodon ends of the tRNAs contact the mRNA in the small subunit, while their aminoacyl ends are buried in the large subunit, where peptide bond formation occurs.

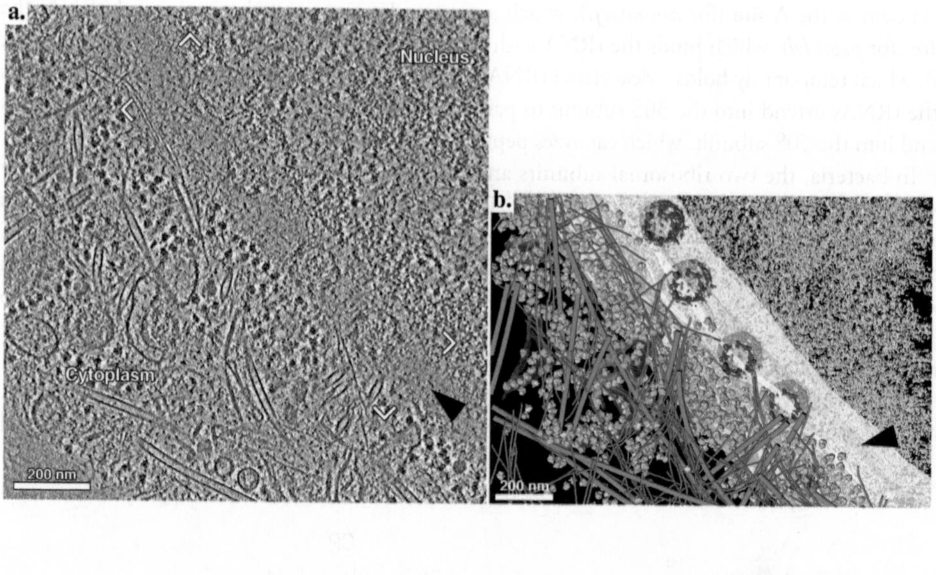

Figure 22.9 Cryo-electron tomography of a human cell.

The image in part **a,** from a cancerous HeLa cell, has been color-coded in part **b.** Gold = chromatin, light purple = endoplasmic reticulum, red = actin filaments, green = microtubules, blue = large ribosomal subunits, yellow = small ribosomal subunits, and dark purple = nuclear pore complexes. The arrowhead marks the nuclear envelope.

Wolfgang Baumeister, Max Planck Institute of Biochemistry

across the double nuclear membrane via the nuclear pore. Ribosomes in the cytoplasm then translate the mRNA into protein. Cryo-electron tomography, a method for reconstructing three-dimensional cellular structures by analyzing frozen slices by electron microscopy, has been used to visualize ribosomes and other structures (**Fig. 22.9**). The ribosomes appear to cluster near the nuclear envelope in a network of actin filaments and microtubules (see Section 5.3).

Ribosomes are not all identical and interchangeable; some show a preference for producing certain types of proteins, which are often related to a particular cellular function. Ribosome specialization might depend on the location of the ribosome or on the presence or absence of particular ribosomal proteins, which also exhibit variable chemical modifications. In mammals, defects in ribosomal proteins cause tissue-specific abnormalities, which would not be the case if all ribosomes in all cells were equally affected. Ribosomes that translate a preferred set of mRNAs, perhaps by recognizing specific mRNA sequences or structural features, could produce proteins at different rates or in different parts of the cell and would provide a way to coordinate the production of subunits in proteins with quaternary structure.

1. Describe the overall structure of a ribosome and its three tRNA binding sites.
2. Summarize the structural importance of ribosomal RNA and ribosomal proteins.
3. Compare bacterial and eukaryotic ribosomes.

22.3 Translation

KEY CONCEPTS

Summarize the events of translation initiation, elongation, and termination.

- Identify steps carried out by proteins and steps carried out by RNA.
- Explain why transpeptidation does not require free energy input.
- Explain how the ribosome maximizes the accuracy of translation.
- Describe the role of GTP in translation.
- List some factors that affect the translation rate.

Like DNA replication and RNA transcription, protein synthesis can be divided into separate phases for initiation, elongation, and termination. These stages require an assortment of accessory proteins that bind to tRNA and to the ribosome in order to enhance the speed and accuracy of translation.

Initiation requires an initiator tRNA

In both prokaryotes and eukaryotes, protein synthesis begins at an mRNA codon that specifies methionine (AUG). In bacterial mRNAs, this initiation codon lies about 10 bases downstream of a conserved sequence at the 5′ end of the mRNA, called a Shine–Dalgarno sequence (**Fig. 22.10**). This sequence base pairs with a complementary sequence at the 3′ end of the 16S rRNA, thereby positioning the initiation codon in the ribosome. Eukaryotic mRNAs lack a Shine–Dalgarno sequence that can pair with the 18S rRNA. Instead, translation usually begins at the first AUG codon of an mRNA molecule. In all types of organisms, a start codon may also have a CUG or GUG sequence.

The initiation codon is recognized by an initiator tRNA that has been charged with methionine. This tRNA does not recognize other Met codons that occur elsewhere in the coding sequence of the mRNA. In bacteria, chloroplasts, and mitochondria, the methionine attached to the initiator tRNA is modified by the transfer of a formyl group from tetrahydrofolate (see Section 18.2). The resulting aminoacyl group is designated fMet, and the initiator tRNA is known as $tRNA_f^{Met}$:

$$\begin{array}{c} CH_3 \\ | \\ S \\ | \\ CH_2 \\ | \\ CH_2 \quad O \\ | \quad \| \\ HC-NH-CH-C-O-tRNA_f^{Met} \\ \| \\ O \end{array}$$

N-Formylmethionine–$tRNA_f^{Met}$
($fMet$–$tRNA_f^{Met}$)

Because the amino group of fMet is derivatized, it cannot form a peptide bond. Consequently, fMet can be incorporated only at the N-terminus of a polypeptide. Later, the formyl group or the entire fMet residue may be removed. In eukaryotic and archaebacterial cells, the initiator tRNA, designated $tRNA_i^{Met}$, is charged with methionine but is not formylated.

mRNA

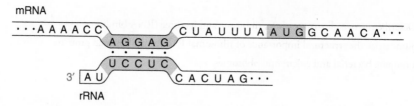

Figure 22.10 Alignment of a Shine–Dalgarno sequence with 16S rRNA. A segment of the rRNA (purple) pairs with the Shine–Dalgarno sequence (blue) in the mRNA, just upstream of the start codon (green). Note that the Shine–Dalgarno sequence shown is a consensus sequence that varies slightly from gene to gene.

Initiation in *E. coli* requires three **initiation factors (IFs)** called IF-1, IF-2, and IF-3. IF-3 binds to the small ribosomal subunit to promote the dissociation of the large and small subunits. fMet–tRNA$_f^{Met}$ binds to the 30S subunit with the assistance of IF-2, a GTP-binding protein. IF-1 sterically blocks the A site of the small subunit, thereby forcing the initiator tRNA into the P site. An mRNA molecule may bind to the 30S subunit either before or after the initiator tRNA has bound, indicating that a codon–anticodon interaction is not essential for initiating protein synthesis.

After the 30S–mRNA–fMet–tRNA$_f^{Met}$ complex has assembled, the 50S subunit associates with it to form the 70S ribosome. This change causes IF-2 to hydrolyze its bound GTP to GDP + P$_i$ and dissociate from the ribosome. The ribosome is now poised—with fMet–tRNA$_f^{Met}$ at the P site—to bind a second aminoacyl–tRNA in order to form the first peptide bond (**Fig. 22.11**).

Similar events occur during translation initiation in eukaryotes, when a 40S and a 60S subunit associate following the binding of Met–tRNA$_i^{Met}$ to an initiation codon. However, eukaryotes require at least 12 distinct initiation factors. Among these are proteins that recognize the 5′ cap and poly(A) tail of the mRNA (see Section 21.3) and interact so that the mRNA actually forms a circle. Initiation may also require the RNA helicase activity of an initiation factor called eIF-4A

Figure 22.11 Summary of translation initiation in *E. coli*. For simplicity, the ribosomal E site is not shown.

Question How must this diagram be modified in order to illustrate initiation in a eukaryote?

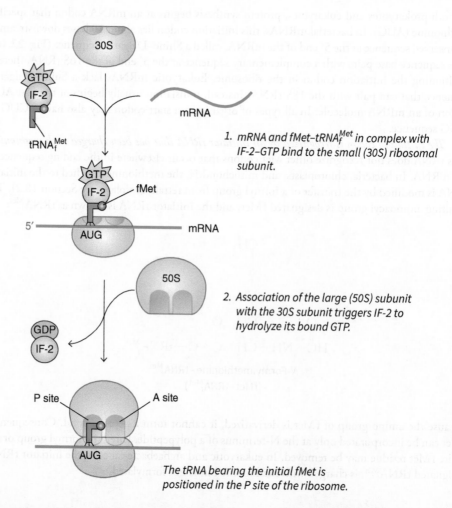

1. mRNA and fMet–tRNA$_f^{Met}$ in complex with IF-2–GTP bind to the small (30S) ribosomal subunit.

2. Association of the large (50S) subunit with the 30S subunit triggers IF-2 to hydrolyze its bound GTP.

The tRNA bearing the initial fMet is positioned in the P site of the ribosome.

to remove secondary structure in the mRNA that would impede translation. The 40S subunit scans the mRNA in an ATP-dependent manner until it encounters the first AUG codon, which is typically 50 to 70 nucleotides downstream of the 5′ cap (**Fig. 22.12**). The initiation factor eIF2 (the *e* signifies *eukaryotic*) hydrolyzes its bound GTP and dissociates, and the 60S subunit then joins the 40S subunit to form the intact 80S ribosome. IF-2 and eIF2 operate much like the heterotrimeric **G proteins** that participate in intracellular signal transduction pathways (see Section 10.2). In each case, *GTP hydrolysis induces conformational changes that trigger additional steps of the reaction sequence.*

The appropriate tRNAs are delivered to the ribosome during elongation

All tRNAs have the same size and shape so that they can fit into small slots in the ribosome. In each reaction cycle of the elongation phase of protein synthesis, an aminoacyl–tRNA enters the A site of the ribosome (the initiator tRNA is the only one that enters the P site without first binding to the A site). After peptide bond formation, the tRNA moves to the P site, and then to the E site. As Figure 22.8 shows, there is not much room to spare. In addition, all tRNAs must be able to bind interchangeably with protein cofactors.

Aminoacyl–tRNAs are delivered to the ribosome in a complex with a GTP-binding **elongation factor (EF)** known as EF-Tu in *E. coli*. EF-Tu is one of the most abundant *E. coli* proteins (about 100,000 copies per cell, enough to bind all the aminoacyl–tRNA molecules). An aminoacyl–tRNA can bind on its own to a ribosome *in vitro*, but EF-Tu increases the rate *in vivo*.

Because EF-Tu interacts with all 20 types of aminoacyl–tRNAs (representing more than 20 different tRNA molecules), it must recognize common elements of tRNA structure, primarily the acceptor stem and one side of the TψC loop (**Fig. 22.13**). A highly conserved protein pocket accommodates the aminoacyl group. Despite the differing chemical properties of their amino acids, *all aminoacyl–tRNAs bind to EF-Tu with approximately the same affinity* (uncharged tRNAs bind only weakly to EF-Tu). Apparently, the protein interacts with aminoacyl–tRNAs in a combinatorial fashion, offsetting less-than-optimal binding of an aminoacyl group with tighter binding of the acceptor stem and vice versa. This allows EF-Tu to deliver and surrender all 20 aminoacyl–tRNAs to a ribosome with the same efficiency.

EF-Tu does not know in advance which amino acid will be needed by the ribosome. In a cell, the incoming aminoacyl–tRNA is selected on the basis of its ability to recognize a complementary mRNA codon in the A site. Due to competition among all the aminoacyl–tRNA molecules in the cell, this is the rate-limiting step of protein synthesis. Before the 50S subunit catalyzes peptide bond formation, the ribosome must verify that the correct aminoacyl–tRNA is in place. This is accomplished by interactions between the rRNA, mRNA, and tRNA. For example, when tRNA binds to the A site of the 30S subunit, two highly conserved residues (A1492 and A1493) of the 16S rRNA "flip out" of an rRNA loop in order to form hydrogen bonds with various parts of the mRNA codon as it pairs with the tRNA anticodon. *These interactions physically link the two rRNA bases with the first two base pairs of the codon and anticodon so that they can sense a correct match between the mRNA and tRNA* (**Fig. 22.14**). Incorrect base pairing at the first or second codon position would prevent this three-way mRNA–tRNA–rRNA interaction. As expected from the wobble hypothesis (Section 22.1), the A1492/A1493 sensor can accommodate nonstandard base pairing at the third codon position.

As the rRNA nucleotides shift to confirm a correct codon–anticodon match, the conformation of the ribosome changes in such a way that the G protein EF-Tu is induced to hydrolyze its bound GTP. As a result of this reaction, EF-Tu dissociates from the ribosome, leaving behind the tRNA with its aminoacyl group to be incorporated into the growing polypeptide chain.

However, if the tRNA anticodon is not properly paired with the A-site codon, the 30S conformational change and GTP hydrolysis by EF-Tu do not occur. Instead, the aminoacyl–tRNA, along with EF-Tu–GTP, dissociates from the ribosome. Because a peptide bond cannot form until

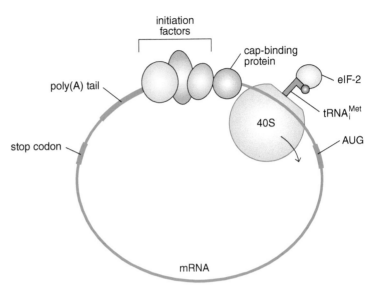

Figure 22.12 Circularization of eukaryotic mRNA at translation initiation. A number of initiation factors form a complex that links the 5′ cap and 3′ poly(A) tail of the mRNA. The small (40S) ribosomal subunit binds to the mRNA and locates the AUG start codon.

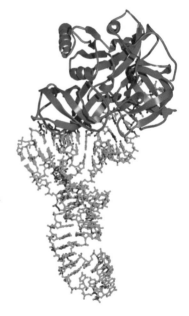

Figure 22.13 Structure of an EF-Tu–tRNA complex. The protein (blue) interacts with the acceptor end and TψC loop of an aminoacyl–tRNA (red).

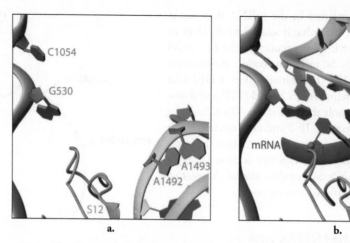

Figure 22.14 The ribosomal sensor for proper codon–anticodon pairing. These images show the A site of the 30S subunit in the **a.** absence and **b.** presence of mRNA and tRNA analogs. The rRNA is gray, with the "sensor" bases in red. The mRNA analog, representing the A-site codon, is purple, and the tRNA analog (labeled ASL) is gold. A ribosomal protein (S12) and two Mg²⁺ ions (magenta) are also visible. Note how rRNA bases A1492 and A1493 flip out to sense the codon–anticodon interaction.

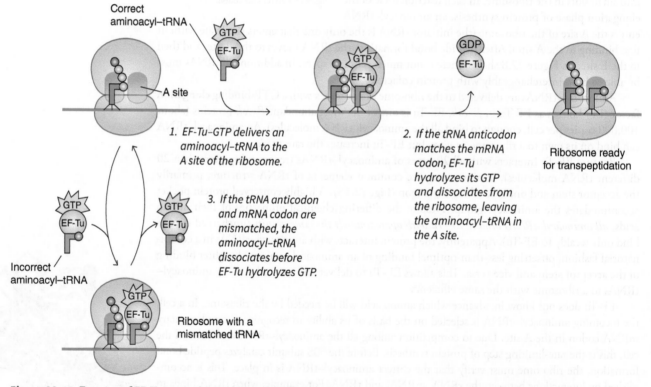

Figure 22.15 Function of EF-Tu in translation elongation in *E. coli*.

after EF-Tu hydrolyzes GTP, *EF-Tu ensures that polymerization does not occur unless the correct aminoacyl–tRNA is positioned in the A site.* The energetic cost of proofreading at the decoding stage of translation is the free energy of GTP hydrolysis (catalyzed by EF-Tu). The function of EF-Tu is summarized in **Figure 22.15**. In eukaryotes, elongation factor eEF1α performs the same service as the prokaryotic EF-Tu. The functional correspondence between some prokaryotic and eukaryotic translation cofactors is given in **Table 22.5**.

The ribosome itself performs a bit of proofreading. The departure of EF-Tu–GDP leaves behind the aminoacyl–tRNA, whose acceptor end can now slip all the way into the A site of the 50S ribosomal subunit. The 30S subunit closes in around the tRNA, but at this point, the only interactions that hold the aminoacyl–tRNA in place are codon–anticodon contacts. If there is still a slight mismatch, such as a G:U base pair at the first or second position, which is not detectable

Table 22.5 Prokaryotic and Eukaryotic Translation Factors

Prokaryotic protein	Eukaryotic protein	Function
IF-2	eIF2	Delivers initiator tRNA to P site of ribosome
EF-Tu	eEF1α	Delivers aminoacyl–tRNA to A site of ribosome during elongation
EF-G	eEF2	Binds to A site to promote translocation following peptide bond formation
RF-1, RF-2	eRF1	Binds to A site at a stop codon and induces peptide transfer to water

by the A1492/A1493 sensor, the strain of not being able to form a perfect Watson–Crick pair will be felt by the ribosome and possibly the tRNA itself, and the tRNA will slip out of the A site. In this way, the ribosome verifies correct codon–anticodon pairing twice for each aminoacyl–tRNA: when EF-Tu first delivers it to the ribosome and after EF-Tu departs. Ribosomal proofreading helps limit the error rate of translation to about 10^{-4} (one mistake for every 10^4 codons).

The peptidyl transferase active site catalyzes peptide bond formation

When the ribosomal A site contains an aminoacyl–tRNA and the P site contains a peptidyl–tRNA (or, prior to formation of the first peptide bond, an initiator tRNA), the peptidyl transferase activity of the large subunit catalyzes a **transpeptidation** reaction in which the free amino group of the aminoacyl–tRNA in the A site attacks the ester bond that links the peptidyl group to the tRNA in the P site (**Fig. 22.16**). This reaction lengthens the peptidyl

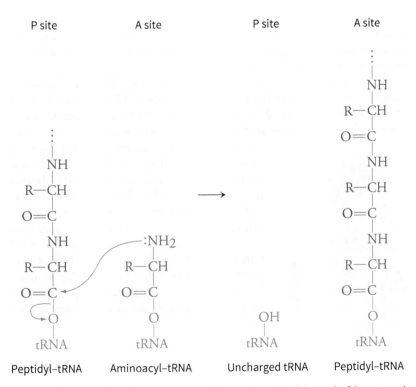

Figure 22.16 The peptidyl transferase reaction. Note that the nucleophilic attack of the aminoacyl group on the peptidyl group produces a free tRNA in the P site and a peptidyl–tRNA in the A site.

Question **Compare this reaction to the condensation reaction of two amino acids shown in Section 4.1.**

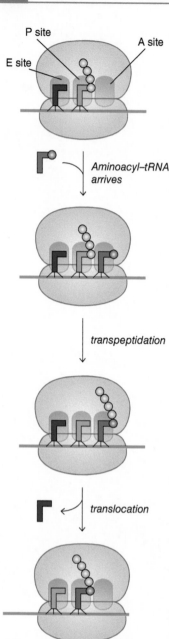

P site

A site

E site

Aminoacyl–tRNA
arrives

transpeptidation

translocation

Figure 22.17 Movement of tRNAs through the ribosome. The aminoacyl–tRNA enters the A site and, after transpeptidation, moves to the P site. The tRNA that previously carried the peptidyl group moves from the P site to the E site, displacing the tRNA from the previous cycle.

Question Sketch a diagram to show the next three steps of the process.

group by one amino acid at its C-terminal end. Thus, *a polypeptide grows in the N → C direction*. No external source of free energy is required for transpeptidation because the free energy of the broken ester bond of the peptidyl–tRNA is comparable to the free energy of the newly formed peptide bond. (Recall, though, that ATP was consumed in charging the tRNA with an amino acid.)

The peptidyl transferase active site lies in a highly conserved region of the bacterial 50S subunit, and the newly formed peptide bond is about 18 Å away from the nearest protein. Thus, the ribosome is a ribozyme (an RNA catalyst). How does rRNA catalyze peptide bond formation? Two highly conserved rRNA nucleotides, G2447 and A2451 in *E. coli*, do not function as acid–base catalysts, as was initially proposed. Rather, these residues help position the substrates for reaction, an example of induced fit (Section 6.3). Binding of a tRNA at the A site triggers a conformational change that exposes the ester bond of the peptidyl–tRNA in the P site. At other times, the ester bond must be protected so that it does not react with water, a reaction that would prematurely terminate protein synthesis. Proximity and orientation effects in the ribosome increase the rate of peptide bond formation about 10^7-fold above the uncatalyzed rate. Some antibiotics exert their effects by binding to the peptidyl transferase active site to directly block protein synthesis (**Box 22.B**).

During transpeptidation, the peptidyl group is transferred to the tRNA in the A site, and the P-site tRNA becomes deacylated. The new peptidyl–tRNA then moves into the P site, and the deacylated tRNA moves into the E site. The mRNA, which is still base paired with the peptidyl–tRNA anticodon, advances through the ribosome by one codon (**Fig. 22.17**). Experiments designed to assess the force exerted by the ribosome demonstrate that formation of a peptide bond causes the ribosome to loosen its grip on the mRNA. Cryo-EM studies of ribosomes frozen at different stages are revealing how events at the peptidyl transferase site in the large subunit are communicated to the mRNA decoding site in the small subunit. *The movement of tRNA and mRNA, which allows the next codon to be translated, is known as **translocation**.* This dynamic process requires the G protein called elongation factor G (EF-G) in *E. coli*.

EF-G bears a striking resemblance to the EF-Tu–tRNA complex (**Fig. 22.18**), and the ribosomal binding sites for the two proteins overlap. Structural studies show that the EF-G–GTP complex physically displaces the peptidyl–tRNA in the A site, causing it to translocate to the P site. This movement causes the deacylated tRNA in the P site to shift to the E site. EF-G binding to the ribosome stimulates its GTPase activity. After EF-G hydrolyzes its bound GTP, it dissociates from the ribosome, leaving a vacant A site available for the arrival of another aminoacyl–tRNA and another round of transpeptidation.

The GTPase activity of G proteins such as EF-Tu and EF-G allows the ribosome to cycle efficiently through all the steps of translation elongation. Because GTP hydrolysis is irreversible, the elongation reactions—aminoacyl–tRNA binding, transpeptidation, and translocation—proceed unidirectionally. The *E. coli* ribosomal elongation cycle is shown in **Figure 22.19**. Experimental evidence shows that bacterial RNA polymerase can sometimes bind to the ribosome, thereby feeding mRNA from the polymerase active site directly into the ribosome's decoding center. Not only does this help synchronize transcription and translation, but the one-way movement of the ribosome helps prevent backtracking by the RNA polymerase.

Eukaryotic cells contain elongation factors that function similarly to EF-Tu and EF-G (Table 22.5). These G proteins are continually recycled during protein synthesis. In some cases, accessory proteins help replace bound GDP with GTP to prepare the G protein for another reaction cycle.

Figure 22.18 Structure of EF-G from *T. thermophilus*.

Question Compare the size and shape of this complex to the size and shape of the complex containing EF-Tu and an aminoacyl–tRNA (see Fig. 22.13).

Box 22.B Antibiotic Inhibitors of Protein Synthesis

Antibiotics interfere with a variety of cellular processes, including cell-wall synthesis, DNA replication, and RNA transcription. Some of the most effective antibiotics, including many in clinical use, target protein synthesis. Because bacterial and eukaryotic ribosomes and translation factors differ, these antibiotics can kill bacteria without harming their mammalian hosts.

Puromycin, for example, resembles the 3′ end of Tyr–tRNA and competes with aminoacyl–tRNAs for binding to the ribosomal A site. Transpeptidation generates a puromycin–peptidyl group that cannot be further elongated because the puromycin "amino acid" group is linked by an amide bond rather than an ester bond to its "tRNA" group. As a result, peptide synthesis comes to a halt.

Puromycin

Tyrosyl-tRNA

The antibiotic chloramphenicol interacts with the active-site nucleotides, including the catalytically essential A2451, to prevent transpeptidation.

Chloramphenicol

Other antibiotics with more complicated structures interfere with protein synthesis through different mechanisms. For example, erythromycin physically blocks the tunnel that conveys the nascent polypeptide away from the active site. Six to eight peptide bonds form before the constriction of the exit tunnel blocks further chain elongation.

Streptomycin kills cells by binding tightly to the backbone of the 16S rRNA and stabilizing an error-prone conformation of the ribosome. In the presence of the antibiotic, the ribosome's affinity for aminoacyl–tRNAs increases, which increases the likelihood of codon–anticodon mispairing and therefore increases the error rate of translation. Presumably, the resulting burden of inaccurately synthesized proteins kills the cell.

The drugs described here, like all antibiotics, lose their effectiveness when their target organisms become resistant to them. For example, mutations in ribosomal components can prevent antibiotic binding. Alternatively, an antibiotic-susceptible organism may acquire a gene, often present on an extrachromosomal plasmid, whose product inactivates the antibiotic. Acquisition of an acetyltransferse gene leads to the addition of an acetyl group to chloramphenicol, which prevents its binding to the ribosome. Acquisition of a gene for an ABC transporter (Section 9.3) can hasten a drug's export from the cell, rendering it useless.

Question **Explain why some of the side effects of antibiotic use in humans can be traced to impairment of mitochondrial function.**

Release factors mediate translation termination

As the peptidyl group is lengthened by the transpeptidation reaction, it exits the ribosome through a tunnel in the center of the large subunit. The tunnel, about 100 Å long and 15 Å in diameter in the bacterial ribosome, shelters a polypeptide chain of up to 30 residues. The tunnel is defined by ribosomal proteins as well as the 23S rRNA. A variety of groups—including rRNA bases, backbone phosphate groups, and protein side chains—form a mostly hydrophilic surface for the tunnel. There are no large hydrophobic patches that could potentially impede the exit of a newly synthesized peptide chain.

Translation ceases when the ribosome encounters a stop codon (see Table 22.1). With a stop codon in the A site, the ribosome cannot bind an aminoacyl–tRNA but instead binds a protein known as a **release factor (RF).** In bacteria such as *E. coli*, RF-1 recognizes stop codons UAA and UAG, and RF-2 recognizes UAA and UGA. In eukaryotes, one protein—called eRF1—recognizes all three stop codons.

The release factor must specifically recognize the mRNA stop codon; it does this via an "anticodon" sequence of three amino acids, such as Pro–Val–Thr in RF-1 and Ser–Pro–Phe in RF-2, that interact with the first and second bases of the stop codon. At the same time, a

Figure 22.19 The *E. coli* ribosomal elongation cycle.

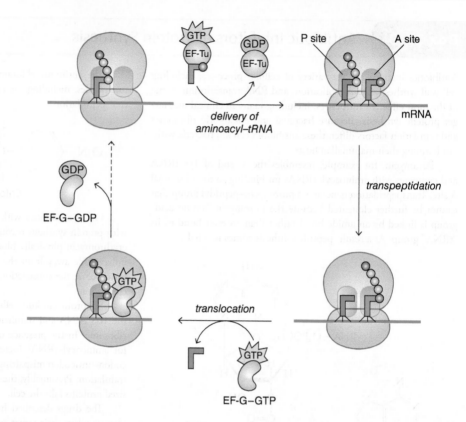

loop of the release factor with the conserved sequence Gly–Gly–Gln projects into the peptidyl transferase site of the 50S subunit (**Fig. 22.20**). The amide group of the Gln residue promotes transfer of the peptidyl group from the P-site tRNA to water, apparently by stabilizing the transition state of this hydrolysis reaction. The product of the reaction is an untethered polypeptide that can exit the ribosome. At one time, release factors were believed to act by mimicking tRNA molecules, as EF-G does. However, it now appears that release factors undergo conformational changes upon binding to the ribosome, so they don't operate straightforwardly as tRNA surrogates.

In *E. coli*, an additional RF (RF-3), which binds GTP, promotes the binding of RF-1 or RF-2 to the ribosome. In eukaryotes, eRF3 performs this role. Hydrolysis of the GTP bound to RF-3 (or eRF3) allows the release factors to dissociate (**Fig. 22.21**). This leaves the ribosome with a bound mRNA, an empty A site, and deacylated tRNAs in both the P and E sites.

In bacteria, preparing the ribosome for another round of translation is the responsibility of a **ribosome recycling factor (RRF)** that works in concert with EF-G. RRF apparently slips into the ribosomal A site. EF-G binding then translocates RRF to the P site, thereby displacing the deacylated tRNAs. Following GTP hydrolysis, EF-G and RRF dissociate from the ribosome, leaving it ready for a new round of translation initiation. In eukaryotes, a protein with ATPase activity binds to the ribosome containing eRF3 and helps dissociate the large and small subunits. Initiation factors appear to bind during this process, so translation termination and reinitiation are tightly linked.

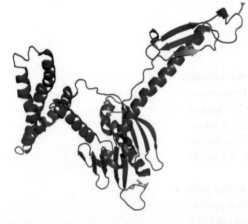

Figure 22.20 Structure of RF-1. The RF-1 protein is shown as a purple ribbon with its anticodon-binding Pro–Val–Thr (PVT) sequence in red and its peptidyl transferase–binding Gly–Gly–Gln (GGQ) sequence in orange. This image shows the structure of RF-1 as it exists when bound to a ribosome.

Translation is efficient and dynamic

A ribosome can extend a polypeptide chain by approximately 20 amino acids every second in bacteria and by an average of about 6 amino acids every second in eukaryotes. At these rates, most protein chains can be synthesized in under a minute. As we have seen, various G proteins trigger conformational changes that keep the ribosome operating efficiently through many elongation cycles. Cells also maximize the

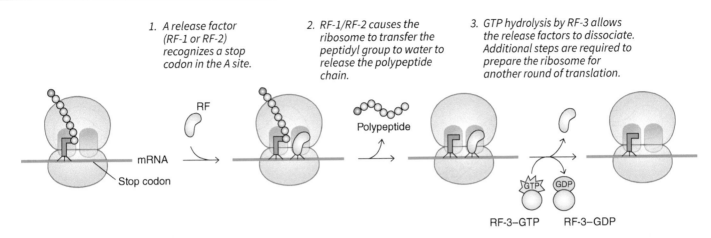

1. A release factor (RF-1 or RF-2) recognizes a stop codon in the A site.

2. RF-1/RF-2 causes the ribosome to transfer the peptidyl group to water to release the polypeptide chain.

3. GTP hydrolysis by RF-3 allows the release factors to dissociate. Additional steps are required to prepare the ribosome for another round of translation.

Figure 22.21 Translation termination in *E. coli*.

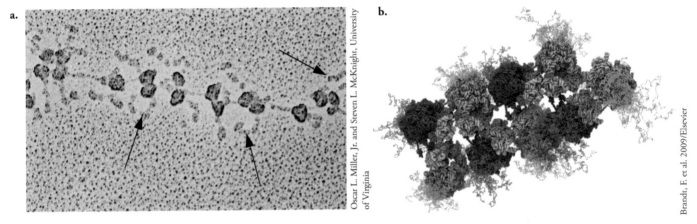

Figure 22.22 A polysome. **a.** In this electron micrograph, a single mRNA strand encoding silkworm fibroin is studded with ribosomes. Arrows indicate growing fibroin polypeptide chains. **b.** In this cryo-electron microscopy–based reconstruction of a polysome, ribosomes are blue and gold, and the emerging polypeptides are red and green.

Question **Identify the polysomes in Fig. 22.9.**

rate of protein synthesis by forming **polysomes.** These structures include a single mRNA molecule simultaneously being translated by multiple ribosomes (**Fig. 22.22**). As soon as the first ribosome has cleared the initiation codon, a second ribosome can assemble and begin translating the mRNA. The circularization of eukaryotic mRNAs (see Fig. 22.12) may promote repeated rounds of translation. Because the stop codon at the 3′ end of the coding sequence may be relatively close to the start codon at the 5′ end, the ribosomal components released at termination can be easily recycled for reinitiation.

The preceding sections have already mentioned some variations in the process of protein synthesis, such as ribosome heterogeneity and alternative start codons in mRNA. In fact, a large number of factors can impact the outcome of translation. Some of the dynamics of ribosome activity have been revealed through various structural and kinetic methods and by **ribosome profiling,** a technique based on sequencing and identifying mRNA segments that are being translated and are therefore protected from RNase digestion by the presence of an active ribosome.

In addition to chemical modifications to mRNA nucleotides, which is sometimes referred to as "RNA epigenetics," RNA secondary structure can affect the progress of a ribosome. Riboswitches illustrate how mRNA structure responds to the presence of a ligand to determine whether translation can occur. A **riboswitch** is a segment of mRNA, usually located near the 5′ end, that can change its conformation, for example, to expose a ribosome binding site (**Fig. 22.23**). Riboswitches occur in all types of organisms and may also affect transcription termination (in prokaryotes) or mRNA splicing (in eukaryotes). The ligands for the most common riboswitches are coenzymes,

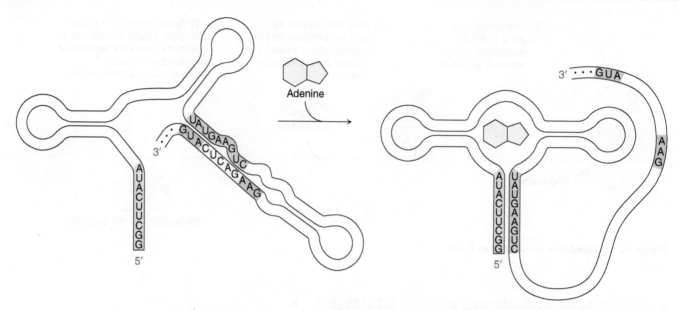

Figure 22.23 A bacterial adenine riboswitch.   In the absence of adenine, the Shine–Dalgarno sequence (blue) and start codon (green) are inaccessible; adenine binding stabilizes an alternate structure that allows translation to proceed.

whose presence prevents translation of a protein involved in the synthesis of that coenzyme (the example in Fig. 22.23 shows a ligand *promoting* translation).

Once it begins translating an mRNA, the progress of the ribosome is not necessarily uniform. The ribosome may slow or pause for a number of reasons, such as when it encounters a rarely used codon. Recall that each amino acid is specified by one to six different codons, which appear in genes at different frequencies, depending on the type of organism. Ribosome pausing may simply slow the rate of translation or it may lead to degradation of the mRNA. This outcome explains why many "silent" mutations that do not change the meaning of a codon can result in significant changes in the amount of a protein in an organism. Sequencing studies show that the proteins with the highest expression levels tend to use the most common codons, which favors fast translation, whereas less-abundant proteins tend to include more rare codons.

The sequence of the polypeptide product can also affect the rate of translation. For example, the transfer of a peptidyl group to proline is relatively slow, so the rate of transpeptidation drops when the ribosome reaches a stretch of proline codons. The nascent polypeptide may also interact with groups in the ribosome's exit channel, slowing its emergence into the cytosol. In rare cases, a ribosome may slip, particularly at a run of repeating nucleotides, resuming in a different reading frame and producing a polypeptide with a different sequence past that point.

The examples above are ample evidence that cells use a variety of ways to regulate gene expression, in addition to transcriptional control. As a result, some genes are rarely expressed while some gene products are made in large amounts. For example, in eukaryotes, an average of 1500 different proteins account for 90% of the total protein mass; in prokaryotes, the average is about 300. Many of these proteins are involved in energy-harvesting pathways and protein production. Surprisingly, about a quarter of the most abundant proteins have no assigned function. The activities of many low-abundance proteins are likewise mysterious.

Curiously, ribosomes often translate mRNA sequences outside of the canonical coding sequence defined by the start and stop codons. As many as half of mammalian mRNAs include an additional open reading frame—a sequence that may start with an alternative start codon and that potentially codes for a polypeptide—on the 5′ side of the main start codon. Translation starting from the upstream start codon can yield so-called microproteins that may have discrete functions. An alternative translation start site may also tie up ribosomes in a way that prevents them from translating the rest of the mRNA molecule.

1. Draw a diagram of a gene with and without introns and label the promoter, transcription termination site, start codon, and stop codon.
2. Make a list of all the substances that must interact with a ribosome in order to produce a polypeptide.
3. Compare the way a ribosome recognizes a start codon and a stop codon.
4. Describe the functions of bacterial IF-2, EF-Tu, EF-G, RF-1, and RRF.
5. Explain how the ribosome ensures that the correct amino acid is positioned in the ribosomal A site.
6. Compare translation initiation, elongation, and termination in bacteria and eukaryotes.
7. Summarize the different types of RNA–RNA interactions that occur during translation.
8. Without looking at the text, draw a diagram of a polysome, add labels for mRNA, polypeptides, and ribosomes, and indicate which ribosome was the first to arrive.
9. Describe the factors that affect the rate of translation or the amount or type of protein produced.
10. Explain how polysomes increase the rate of protein synthesis.

22.4 Post-Translational Modifications

KEY CONCEPTS

List the events that occur during post-translational processing.

- Describe what chaperones do and how they work.
- Recount the steps of producing a membrane or secreted protein.
- Recognize different types of post-translational modifications.

The polypeptide released from a ribosome is not yet fully functional. For example, it must fold to its native conformation; it may need to be transported to another location inside or outside the cell; and it may undergo **post-translational modification,** or **processing.** As we will see, many of these events are actually co-translational, not strictly post-translational, and the ribosome plays a part.

Chaperones promote protein folding

Studies of protein folding *in vitro* have revealed numerous insights into the pathways by which proteins (usually relatively small ones that have been chemically denatured) assume a compact globular shape with a hydrophobic core and a hydrophilic surface (see Section 4.3). Protein folding *in vivo* is not fully understood. For one thing, a protein can begin to fold as soon as its N-terminus emerges from the ribosome, even before it has been fully synthesized. The rate of translation can affect the packing options for newly formed polypeptide segments by making certain segments available at the proper intervals. In addition, a polypeptide must fold in an environment crowded with other proteins with which it might interact unfavorably. Finally, for proteins with quaternary structure, individual polypeptide chains must assemble with the proper stoichiometry and orientation. All of these processes may be facilitated in a cell by proteins known as **molecular chaperones.**

To prevent improper associations within or between polypeptide chains, chaperones bind to exposed hydrophobic patches on the protein surface (recall that hydrophobic groups tend to aggregate, which could lead to nonnative protein structure or protein aggregation and precipitation). *Many chaperones are ATPases that use the free energy of ATP hydrolysis to drive conformational changes that allow them to bind and release a polypeptide substrate while it assumes its native shape.* Some chaperones were originally identified as heat-shock proteins (Hsp) because their synthesis is induced by high temperatures—conditions under which proteins tend to denature (unfold) and aggregate.

The first chaperone a bacterial protein meets, called trigger factor, is poised just outside the ribosome's polypeptide exit tunnel, bound to a ribosomal protein (**Fig. 22.24**). When trigger factor binds to the ribosome, it opens up to expose a hydrophobic patch facing the

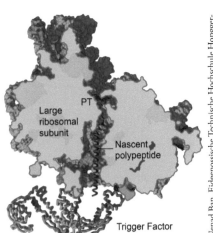

Figure 22.24 Trigger factor bound to a ribosome. Trigger factor, with its domains shown in different colors, binds to the 50S ribosomal subunit where a polypeptide (magenta helix) emerges. PT represents the peptidyl transferase active site.

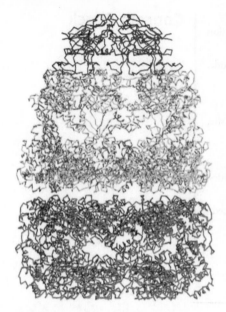

Figure 22.25 The GroEL/GroES chaperonin complex. The two seven-subunit GroEL rings, viewed from the side, are colored red and yellow. A seven-subunit GroES complex (blue) caps the so-called *cis* GroEL ring.

exit tunnel. Hydrophobic segments of the emerging polypeptide bind to this patch and are thereby prevented from sticking to each other or to other cellular components. Trigger factor may dissociate from the ribosome but remain associated with the nascent (newly formed) polypeptide until another chaperone takes over. Eukaryotes lack trigger factor, although they have other small heat-shock proteins that function in the same manner to protect newly made proteins. These chaperones are extremely abundant in cells, so there is at least one per ribosome.

Trigger factor may hand off a new polypeptide to another chaperone, such as DnaK in *E. coli* (it was named when it was believed to participate in DNA synthesis). DnaK and other heat-shock proteins in prokaryotes and eukaryotes interact with newly synthesized polypeptides and with existing cellular proteins and therefore can prevent as well as reverse improper folding. These chaperones, in a complex with ATP, bind to a short extended polypeptide segment with exposed hydrophobic groups (they do not recognize folded proteins, whose hydrophobic groups are sequestered in the interior). ATP hydrolysis causes the chaperone to release the polypeptide. As the polypeptide folds, the heat-shock protein may repeatedly bind and release it.

Ultimately, protein folding may be completed by multisubunit chaperones, called chaperonins, which form cagelike structures that physically sequester a folding polypeptide. The best-known chaperonin complex is the GroEL/GroES complex of *E. coli*. Fourteen GroEL subunits form two rings of seven subunits, with each ring enclosing a 45-Å-diameter chamber that is large enough to accommodate a folding polypeptide. Seven GroES subunits form a domelike cap for one GroEL chamber (**Fig. 22.25**). The GroEL ring nearest the cap is called the *cis* ring, and the other is the *trans* ring.

Each GroEL subunit has an ATPase active site. All seven subunits of the ring act in concert, hydrolyzing their bound ATP and undergoing conformational changes. *The two GroEL rings of the chaperonin complex act in a reciprocating fashion to promote the folding of two polypeptide chains in a safe environment* (**Fig. 22.26**). Note that seven ATP are consumed for each 10-second protein-folding opportunity. If the released substrate has not yet achieved its native conformation, it may rebind to the chaperonin complex. Only about 10% of bacterial proteins seem to require the GroEL/GroES chaperonin complex,

1. A GroEL ring binds 7 ATP and an unfolded polypeptide, which associates with hydrophobic patches on the GroEL subunits.

2. A GroES cap binds, triggering a conformational change that retracts the hydrophobic patches, thereby releasing the polypeptide into the GroEL chamber, where it can fold.

3. Within about 10 seconds, the cis GroEL ring hydrolyzes its 7 ATP.

6. The trans GroEL ring can now bind a GroES cap, and steps 3–5 repeat.

4. A second polypeptide substrate and 7 ATP bind to the trans GroEL ring.

5. The cis ring releases its GroES cap, the 7 ADP, and the folded substrate polypeptide.

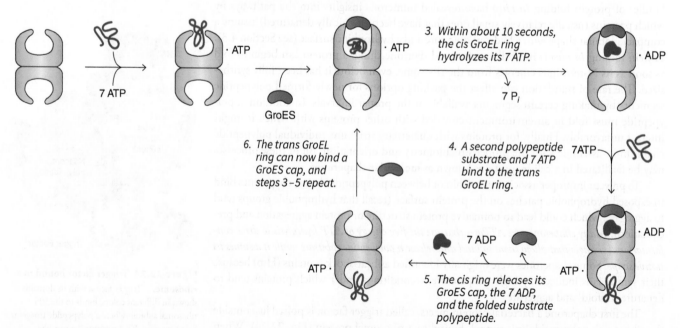

Figure 22.26 The chaperonin reaction cycle.

and most of these range in size from 10 to 55 kD (a protein larger than about 70 kD probably could not fit inside the protein-folding chamber). Immunocytological studies indicate that some proteins never stray far from a chaperonin complex, perhaps because they tend to unfold and must periodically restore their native structures.

In eukaryotes, a chaperonin complex known as CCT or TRiC functions analogously to bacterial GroEL, but it has eight subunits in each of its two rings, and its fingerlike projections take the place of the GroES cap. One of its primary functions is to fold the abundant cytoskeletal proteins actin and tubulin (Section 5.3).

The signal recognition particle targets some proteins for membrane translocation

For a cytosolic protein, the journey from a ribosome to the protein's final cellular destination is straightforward (in fact, the journey may be short, since some mRNAs are directed to specific cytosolic locations before translation commences). In contrast, an integral membrane protein or a protein that is to be secreted from the cell follows a different route since it must pass partly or completely through a membrane. These proteins account for about one-third of the proteins made by a cell.

In both prokaryotic and eukaryotic cells, *most membrane and secretory proteins are synthesized by ribosomes that dock with the plasma membrane (in prokaryotes) or endoplasmic reticulum (in eukaryotes).* The protein that emerges from the ribosome is inserted into or through the membrane co-translationally. The membrane translocation system is fundamentally similar in all cells and requires a ribonucleoprotein known as the **signal recognition particle (SRP).**

As in other ribonucleoproteins, including the ribosome and the spliceosome (see Section 21.3), the RNA component of the SRP is highly conserved and is essential for SRP function. The *E. coli* SRP consists of a single multidomain protein and a 4.5S RNA. The mammalian SRP contains a larger RNA and six different proteins, but its core is virtually identical to the bacterial SRP. The RNA component of the SRP includes several nonstandard base pairs, including G:A, G:G, and A:C, and interacts with several protein backbone carbonyl groups (most RNA–protein interactions involve protein side chains rather than the backbone).

How does the SRP recognize membrane and secretory proteins? Such proteins typically have an N-terminal **signal peptide** consisting of an α-helical stretch of 6 to 15 hydrophobic amino acids preceded by at least one positively charged residue. For example, human proinsulin (the polypeptide precursor of the hormone insulin) has the following signal sequence:

$$\text{M A L W M R L L P L L A L L A L W G P D P A A A F V N} \dots$$

The hydrophobic segment and a flanking arginine residue are shaded in green and pink.

The signal peptide binds to the SRP in a pocket formed mainly by a methionine-rich protein domain. The flexible side chains of the hydrophobic Met residues allow the pocket to accommodate helical signal peptides of variable sizes and shapes. In addition to the Met residues, the SRP binding pocket contains a segment of RNA, whose negatively charged backbone interacts electrostatically with the positively charged N-terminus and basic residue of the signal peptide (**Fig. 22.27**). Thus, the signal peptide makes hydrophobic as well as electrostatic contacts with the SRP protein and RNA.

Electron microscopic studies indicate that the SRP binds to the ribosome at the polypeptide exit tunnel. The SRP recognizes the signal peptide as it emerges, most likely competing with trigger factor or its eukaryotic counterpart. SRP binding to the signal peptide halts translation elongation, possibly via conformational changes in the ribosome resulting from SRP RNA–rRNA interactions. The stalled ribosome–SRP complex then docks at a receptor on the membrane in a process that requires GTP hydrolysis by the SRP. When translation resumes, the growing polypeptide is translocated through the membrane via a structure known as a **translocon.**

The translocon proteins, called SecY in prokaryotes and Sec61 in eukaryotes, form a transmembrane channel with a constricted pore that limits the diffusion of other substances across the membrane (**Fig. 22.28**). Insertion of the signal peptide itself nudges aside two Sec helices in order to open up the pore wide enough to allow a polypeptide segment

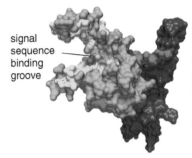

signal sequence binding groove

Figure 22.27 The SRP signal peptide binding domain. This model shows the molecular surface of a portion of the *E. coli* signal recognition particle. The protein is magenta with hydrophobic residues in yellow. Adjacent RNA phosphate groups are red, and the rest of the RNA is dark blue. A signal peptide binds in the SRP groove.

Batey, R. T. et al. 2000/AAAS.

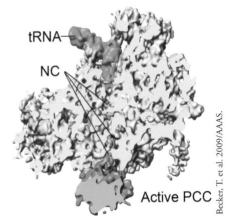

tRNA

NC

Active PCC

Figure 22.28 Ribosome bound to Sec61. In this cryo-EM-based image, a yeast ribosome (partially cut away) is bound to Sec 61 (pink), which is positioned at the end of the polypeptide exit tunnel. Portions of the nascent polypeptide (NC, green) are visible in the exit tunnel.

Becker, T. et al. 2009/AAAS.

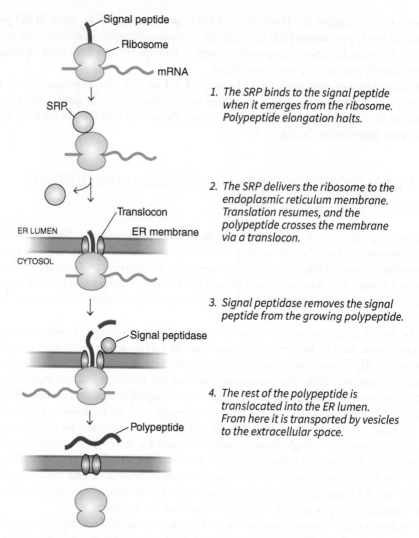

1. The SRP binds to the signal peptide when it emerges from the ribosome. Polypeptide elongation halts.

2. The SRP delivers the ribosome to the endoplasmic reticulum membrane. Translation resumes, and the polypeptide crosses the membrane via a translocon.

3. Signal peptidase removes the signal peptide from the growing polypeptide.

4. The rest of the polypeptide is translocated into the ER lumen. From here it is transported by vesicles to the extracellular space.

Figure 22.29 Membrane translocation of a eukaryotic secretory protein.

to pass through. In addition to allowing a secreted protein to pass entirely across a membrane in this manner, the translocon has a lateral opening that allows one or more hydrophobic protein segments to move sideways into the lipid bilayer, which explains how transmembrane proteins (Section 8.3) become incorporated into the membrane.

The driving force for co-translational protein translocation is provided by the ribosome: *the process of polypeptide synthesis simply pushes the chain through the channel.* In situations where a completed polypeptide is inserted across or into a membrane post-translationally (after it has left the ribosome), the transport system—which may involve the Sec translocon or other machinery— relies on some sort of motor protein that uses the free energy of the ATP hydrolysis reaction.

When the signal peptide emerges on the far side of the membrane, it may be cleaved off by an integral membrane protein known as a signal peptidase. This enzyme recognizes extended polypeptide segments such as those flanking the hydrophobic segment of a signal peptide, but it does not recognize α-helical structures, which are common in mature membrane proteins. The steps of translocating a eukaryotic secretory protein are summarized in **Figure 22.29**.

After translocation in eukaryotic cells, *chaperones and other proteins in the endoplasmic reticulum may help the polypeptide fold into its native conformation, form disulfide bonds, and assemble with other protein subunits.* Extracellular proteins are then transported from the endoplasmic reticulum through the Golgi apparatus and to the plasma membrane via vesicles. The proteins may undergo processing (described next) en route to their final destination.

Not all membrane and secreted proteins follow the pathway outlined above. For example, proteins whose C-terminal tail is anchored in a membrane (about 5% of eukaryotic membrane proteins) are synthesized by ribosomes in the cytosol, and chaperones direct the proteins to the appropriate membrane, where a translocon inserts the protein into the lipid bilayer. Some proteins destined for the extracellular environment are packaged in exosomes (see Box 9.C), whose membranes eventually disintegrate and release the protein. Approximately 900 of the 1000 mitochondrial proteins move from the cytosol across the outer membrane via a β-barrel transporter known as TOM (translocase of the outer membrane) and from there can then pass through TIM (translocase of the inner membrane).

Many proteins undergo covalent modification

Proteolysis is part of the maturation pathway of many proteins. For example, after it has entered the endoplasmic reticulum lumen and had its signal peptide removed and its cysteine side chains cross-linked as disulfides, the insulin precursor undergoes proteolytic processing. The prohormone is cleaved at two sites to generate the mature hormone (**Fig. 22.30**). Over 200 other types of post-translational modification have been documented.

Many extracellular eukaryotic proteins are **glycosylated** at asparagine, serine, or threonine side chains to generate glycoproteins (see Section 11.3). The short sugar chains (oligosaccharides) attached to glycoproteins may protect the proteins from degradation or mediate molecular recognition events. Methyl, acetyl, and propionyl groups may be added to various side chains. N-terminal groups are frequently modified by acetylation (up to 80% of human proteins), and C-terminal groups by amidation. Fatty acyl chains and other lipid groups are added to proteins to anchor them to membranes (Section 8.3). The addition and removal of phosphoryl groups is a powerful mechanism for allosterically regulating cellular signaling components (Section 10.2) and metabolic pathways (Section 19.2).

Modification of a protein by covalently attaching another protein occurs during protein degradation, when ubiquitin is covalently linked to a target protein (Section 12.1). A related protein, called SUMO (for *small ubiquitin-like modifier*), is also covalently attached to a target protein's lysine side chains, but rather than mark the protein for degradation, as ubiquitin does, SUMO is involved in various other processes, including protein transport into the nucleus.

All the post-translational modifications mentioned above are catalyzed by specific enzymes, which act more or less reliably depending on the nature of the modification and the cellular context. One consequence of post-translational processing therefore is that proteins may exhibit a great deal of variation beyond the sequence of amino acids that is specified by the genetic code (**Box 22.C**).

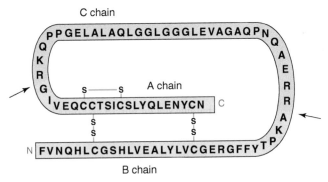

Figure 22.30 Conversion of proinsulin to insulin. The prohormone, with three disulfide bonds, is proteolyzed at two bonds (indicated by arrows) to eliminate the C chain. The mature insulin hormone consists of the disulfide-linked A and B chains.

Concept Check

1. Compare protein folding as catalyzed by heat-shock proteins and by the chaperonin complex.
2. Explain how the SRP recognizes a membrane or secretory protein.
3. Describe the functions of the translocon and the signal peptidase.
4. List the types of reactions that a polypeptide may undergo following its synthesis.

Box 22.C	Defects in the Ubiquitin-Proteasome Protein Degradation System May Lead to Diseases

The ubiquitin-proteasome system (UPS) is a complex and tightly regulated system that operates at the cellular level to remove unwanted or potentially harmful proteins rapidly. The central player in this system is ubiquitin, a small protein covalently attached to target proteins to mark them for degradation (see Section 12.1). This ubiquitination process involves a series of enzymatic reactions involving three groups of enzymes: E1, E2, and E3. Ubiquitin-activating enzyme (E1) activates ubiquitin molecules in an ATP-dependent manner. Ubiquitin-conjugating enzyme (E2) accepts the activated ubiquitin and transfers it to the target protein in the presence of the ubiquitin ligase (E3) enzyme. Once the target protein is polyubiquitinated, it is recognized and degraded by the proteasome, a large multisubunit complex responsible for protein degradation in the cytoplasm and nucleus.

$$\text{Ubiquitin} + \text{ATP} + \text{E1} \rightarrow \text{E1} - \text{Ubiquitin} + \text{AMP} + \text{PP}_i$$

$$\text{E1} - \text{Ubiquitin} + \text{E2} \rightarrow \text{E2} - \text{Ubiquitin} + \text{E1}$$

$$\text{E2} - \text{Ubiquitin} + \text{E3} + \text{Target protein} \rightarrow \text{Target protein} - \text{Ubiquitin} + \text{E2} + \text{E3}$$

$$\text{Polyubiquitinated target protein} + \text{Proteasome} \rightarrow \text{Ubiquitin} + \text{peptides}$$

Dysregulation or defects in the UPS have been implicated in the development and progression of many diseases, including neurodegenerative disorders, cancer, autoimmune diseases, and cardiovascular diseases. Neurodegenerative diseases such as Alzheimer's are characterized by the accumulation of misfolded proteins, including tau and beta-amyloid, in the brain. Proteins such as tau, when hyperphosphorylated, become resistant to ubiquitination and proteasomal degradation. The accumulation of these toxic protein aggregates can cause nerve cell damage and cognitive impairment. Similarly, Parkinson's disease is associated with the accumulation of alpha-synuclein aggregates in neurons. Defects in the UPS may contribute to impaired alpha-synuclein clearance. Mutations in Parkin and other UPS-related genes are associated with familial forms of Parkinson's disease, highlighting the direct involvement of UPS dysfunction in the pathogenesis of the disease. Malfunctioning UPS can also lead to oncoprotein accumulation, leading to uncontrolled cell proliferation and promoting tumorigenesis. For example, degradation of the tumor suppressor protein p53 is regulated by MDM2, an E3 ubiquitin ligase. High levels of MDM2 or mutations in p53 can lead to its degradation, allowing cells to escape apoptosis and lead to various types of cancer. On the other hand, UPS has also become a target for cancer treatment. Proteasome inhibitors are being developed to disrupt UPS function in cancer cells, accumulating toxic proteins and causing cell death. These drugs have shown promise in treating multiple myeloma and some lymphomas. UPS is also important in immune cell function and antigen presentation. UPS dysfunction can lead to abnormal immune responses, such as self-antigens presentation to T cells, which can cause autoimmune diseases such as rheumatoid arthritis and systemic lupus erythematosus (SLE). In these diseases, the immune system mistakenly attacks healthy tissues. Some pathogens have developed mechanisms to manipulate the UPS in their favor. They can hijack the UPS to deplete host immune factors or improve their survival. This interaction between UPS and pathogens can influence the severity and outcome of infectious diseases. Understanding the specific molecular mechanisms underlying UPS dysfunction in each disease context is crucial for developing targeted therapies to restore UPS function or exploit UPS vulnerabilities for therapeutic benefit. As research in this field continues to advance, it holds the potential to offer innovative treatments for a wide range of diseases.

Question: How does the ubiquitin-proteasome system lead to the development and progression of various diseases, including neurodegenerative disorders, cancer, autoimmune diseases, and infectious diseases, and how can this knowledge be leveraged for innovative therapeutic strategies across this spectrum of health conditions?

SUMMARY

22.1 tRNA and the Genetic Code

- The sequence of nucleotides in DNA is related to the sequence of amino acids in a protein by a triplet-based genetic code that must be translated by tRNA adaptors.

- tRNA molecules have similar L-shaped structures with a three-base anticodon at one end and an attachment site for a specific amino acid at the other end.

- Attachment of an amino acid to a tRNA is catalyzed by an aminoacyl–tRNA synthetase in a reaction that requires ATP. Various proofreading mechanisms ensure that the correct amino acid becomes linked to the tRNA.

22.2 Ribosome Structure

- The ribosome, the site of protein synthesis, consists of two subunits containing both rRNA and protein. The ribosome includes a binding site for mRNA and three binding sites (called the A, P, and E sites) for tRNA.

22.3 Translation

- Translation of mRNA requires an initiator tRNA bearing methionine (formyl-methionine in bacteria). Proteins known as initiation factors facilitate the separation of the ribosomal subunits and their reassembly with the initiator tRNA and an mRNA to be translated.

- During the elongation phase of protein synthesis, an elongation factor (EF-Tu in *E. coli*) interacts with aminoacyl–tRNAs and delivers them to the A site of the ribosome. Correct pairing between the mRNA codon and the tRNA anticodon allows the EF-Tu to hydrolyze its bound GTP and dissociate from the ribosome.

- Transpeptidation, or formation of a peptide bond, is catalyzed by rRNA in the large ribosomal subunit. The growing polypeptide chain becomes attached to the tRNA in the A site, which then moves to the P site. This movement is assisted by a GTP-binding protein elongation factor (EF-G in *E. coli*).

- Translation terminates when a release factor recognizes an mRNA stop codon in the A site of the ribosome. Additional factors prepare the ribosome for another round of translation.

- The efficiency of translation is affected by factors such as polysome formation, the presence of riboswitches, and the use of rare codons and alternative start codons.

22.4 Post-Translational Modifications

- Chaperone proteins bind newly synthesized polypeptides to facilitate their folding. Large chaperonin complexes form a barrel-shaped structure that encloses a folding protein.

- Proteins to be secreted must pass through a membrane. An RNA–protein complex called a signal recognition particle directs polypeptides bearing an N-terminal signal sequence to a membrane for translocation.

- Additional modifications to newly synthesized proteins include proteolytic processing and the attachment of carbohydrate, lipid, or other groups.

KEY TERMS

translation	wobble hypothesis	initiation factor (IF)	release factor (RF)	post-translational	translocon
codon	ribosome	G protein	ribosome recycling	processing	glycosylation
reading frame	nucleolus	elongation	factor (RRF)	molecular chaperone	
anticodon	A site	factor (EF)	polysome	signal recognition	
isoacceptor tRNA	P site	transpeptidation	ribosome profiling	particle (SRP)	
isodecoder tRNA	E site	translocation	riboswitch	signal peptide	

BIOINFORMATICS

Brief Bioinformatics Exercises

22.1 Viewing and Analyzing Transfer RNA

22.2 Ribosomes, Protein Processing, and the KEGG Database

PROBLEMS

22.1 tRNA and the Genetic Code

1. How many combinations of four nucleotides are possible with a hypothetical quadruplet code?

2. Synthetic biologists at the Scripps Institute in La Jolla, CA expanded the genetic repertoire by adding two new bases into living bacterial cells. The two bases are named d5SICS and dNaM, and they pair with one another. How many different amino acids could theoretically be encoded by a synthetic nucleic acid containing six different nucleotides?

3. Cells have a mechanism for tagging and destroying proteins containing a C-terminal poly(Lys) sequence. How are these proteins synthesized and why is destroying them helpful for the cell?

4. Cystic fibrosis is caused by a mutation in the 250,000-bp *CFTR* gene. The mature CFTR mRNA is only 6129 nucleotides. **a.** Why is the mature mRNA so much shorter than the gene from which it was transcribed? **b.** What is the minimum number of nucleotides required to encode the 1480-residue CFTR protein?

5. One portion of the normal *CFTR* gene has the sequence ···AATAT-AGATACAG···. In some individuals with cystic fibrosis, this portion of the gene has the sequence ···AATAGATACAG···. How has the DNA sequence changed and how does this affect the encoded protein?

6. A portion of DNA from a phage genome is shown. **a.** How many reading frames are possible? Specify the amino acid sequences for all the possible reading frames. **b.** Which frames represent open reading frames?

5′-CGATGAGCCTTTCAGCACCGC-
3′-GCTACTCGGAAAGTCGTGGCG-
                    TTAGTGAGGTTGCGCGCCACG
                    AATCACTCCAACGCGCGGTGC

7. Explain why a degenerate genetic code helps protect an organism from the effects of mutations.

8. What genetic experiments with mutant bacteriophages support the concept of sequential reading of triplet codons within genes?

9. Marshall Nirenberg and his colleagues deciphered the genetic code in the early 1960s. Their experimental strategy involved constructing RNA templates by using various ribonucleotides and polynucleotide phosphorylase, an enzyme that links available nucleotides together in random order. What protein sequence was obtained when the following templates were added to a cell-free translation system? **a.** poly(U), **b.** poly(C), **c.** poly(A).

10. The experimental strategy described in Problem 9 was used to synthesize poly(UA). Because the polymerization of nucleotides by

polynucleotide phosphorylase is random, all possible codons containing U and A could occur in the RNA template. **a.** What amino acids would be incorporated into a polypeptide synthesized by a cell-free translation system using this template? **b.** What amino acids would be incorporated into a protein when poly(UC) is used as a template? **c.** Repeat the exercise for poly (UG). **d.** How do these results show that the genetic code is redundant?

11. The elucidation of the genetic code was completed using H. Gobind Khorana's method of synthesizing polynucleotides with precise rather than random sequences, as described in Problem 10. **a.** What polypeptides are synthesized from a nonrandom template with the sequence poly(UAUC)? **b.** Explain why a single RNA template can yield more than one polypeptide.

12. What polypeptide is synthesized from a nonrandom poly(AUAG) template? Compare your results with your solution to Problem 11.

13. Organisms differ in their codon usage. For example, in yeast, only 25 of the 61 amino acid codons are used with high frequency. Cells contain high concentrations of the tRNAs that pair best with those codons, that is, form standard Watson–Crick base pairs. Explain how a point mutation in a gene, which does not change the identity of the encoded amino acid, could decrease the rate of synthesis of the protein corresponding to that gene.

14. a. Would you expect that the highly expressed genes in a yeast cell would have sequences corresponding to the cell's set of 25 preferred codons (see Problem 13)? Would you expect this to be the case for genes that are expressed only occasionally? **b.** The genomes of many bacterial species appear to contain genes acquired from other species, including mammals. Even when a gene's function cannot be identified, the gene's nonbacterial origin can be recognized. Explain.

15. Like the yeast described in Problem 13, *E. coli* also has codon preferences and does not use all 64 possible codons with the same frequency. Investigators interested in over-expressing human interleukin-2 (IL-2) in *E. coli* designed a synthetic DNA sequence that produces the correct amino acid sequence of the human protein but contains the codons preferred by *E. coli* rather than human cells. Shown below is a partial *E. coli* codon usage chart, with the fraction of codon use per amino acid. The human IL-2 gene sequence corresponding to the first 13 amino acids is also shown. **a.** Design a synthetic DNA that uses codons preferred by *E. coli* in order to maximize the protein yield in the expression system. **b.** What is the sequence of the protein produced by your synthetic DNA and is it the same as coded by the original gene sequence? **c.** If you wanted to maximize protein yield with the original DNA sequence, what alternative strategy could you use?

···GCACCTACTTCAAGTTCTACAAAGAAAACACAGCTACAA···

	UCU 0.11		
	UCA 0.15		
CUA 0.05	CCU 0.17	CAA 0.30	
CUG 0.46	CCG 0.55	CAG 0.70	
	ACU 0.16	AAA 0.73	AGU 0.14
	ACC 0.47	AAG 0.27	AGC 0.33
	ACA 0.13		
	GCC 0.31		
	GCA 0.21		

16. The genetic code is nearly universal, but there are some exceptions in the genetic code used in mitochondria, where UGA codes for Trp and CUN (where N is any nucleotide) codes for Thr. Shown below is a partial DNA sequence for subunit 1 of cytochrome *c* oxidase (Cox1),

a protein synthesized in yeast mitochondria. **a.** What is the sequence of the protein produced by this DNA sequence? **b.** The human mitochondrial genome produces 13 proteins, as well as the required cognate tRNAs. The remaining proteins required by the mitochondrion must be imported. Why is it advantageous for the mitochondrion to synthesize Cox1? **c.** Why does the mitochondrion need to express tRNAs in addition to the 13 proteins it expresses?

···TGATCACATCATATGTATATTG-

TAGGATTAGATGCAGATCTTAGA···

17. IleRS uses a double-sieve mechanism to accurately produce Ile–tRNA^{Ile} and prevent the synthesis of Val–tRNA^{Ile}. Which other pairs of amino acids differ in structure by a single carbon and might have AARSs that use a similar double-sieve proofreading mechanism?

18. An examination of AlaRS (the enzyme that attaches alanine to tRNA^{Ala}) suggests that the enzyme's aminoacylation active site cannot discriminate between Ala, Gly, and Ser. **a.** List all the products of the AlaRS reaction. **b.** A double-sieve mechanism like the one in IleRS cannot entirely solve the problem of misacylated tRNA^{Ala}. Explain. **c.** Many organisms express a protein called AlaXp, which is a soluble analog of the AlaRS editing domain that is able to hydrolyze Ser–tRNA^{Ala} but not Gly–tRNA^{Ala}. What is the purpose of AlaXp?

19. In some types of cells, an aminoacyl–tRNA synthetase appears to play a secondary role in promoting certain events in RNA splicing (see Section 21.3). **a.** What structural feature of the synthetase is most likely involved in both the aminoacylation and splicing reactions? **b.** The aminoacyl–tRNA synthetases are among the oldest proteins. Why is this relevant to the presence of multiple activities in a single synthetase molecule?

20. Why doesn't GlyRS need a proofreading domain?

21. A tRNA molecule cannot be aminoacylated unless it bears a 3′ CCA sequence. Many tRNA precursors are synthesized without this sequence, so a CCA-adding enzyme must append the three nucleotides to the 3′ end of the immature tRNA molecule. **a.** The CCA-adding enzyme does not require a polynucleotide template. What does this imply about the mechanism for adding the CCA sequence? **b.** What can you conclude about the substrate specificity of the CCA-adding enzyme? **c.** Most CCA-adding enzymes consist of a single polymerase domain, but in one species of bacteria the enzyme has two polymerase domains. Explain how this CCA-adding enzyme operates.

22. Name the amino acid that is attached to the 3′ end of the *E. coli* tRNA molecule with the anticodon sequence **a.** GUG, **b.** GUU, and **c.** CGU.

23. The 5′ nucleotide of a tRNA anticodon is often a nonstandard nucleotide such as a methylated guanosine. Why doesn't this interfere with the ribosome's ability to read the genetic code?

24. Draw the "wobble" base pair that forms between inosine (I) and adenosine.

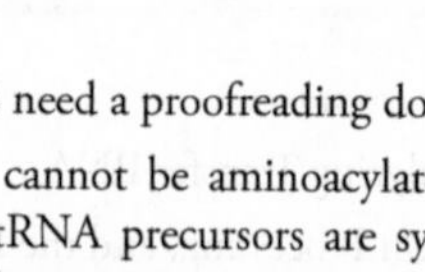

Inosine

25. *E. coli* produces a tRNA$^{Arg}_{ACG}$ (ACG is its 5′→3′ anticodon sequence). **a.** What codon does it recognize? **b.** What happens during protein synthesis if cellular enzymes convert the adenosine to inosine (see Problem 24)?

26. What anticodon would allow a tRNA molecule to recognize both lysine codons in *E. coli*?

27. A new tRNA discovered in *E. coli* contains a uridine modified to form uridine-5′-oxyacetic acid (cmo⁵U). The modified uridine can base pair with G, A, and U. What mRNA codons are recognized by tRNA$^{Leu}_{cmo^5UAG}$?

28. Some RNA transcripts are substrates for an adenosine deaminase. This "editing enzyme" converts adenosine residues to inosine residues, which can base pair with guanosine residues. Explain how the action of the deaminase could potentially increase the number of gene products obtained from a given gene.

29. Predict the effect on a protein's structure and function for all possible nucleotide substitutions at the first position of a Lys codon in the gene encoding the protein.

30. Protein engineers who study the effects of nonstandard amino acids on protein structure and function can use a cell-based system for synthesizing proteins containing nonstandard amino acids. In theory, a polypeptide containing the amino acid norleucine could be produced by cells growing in media containing high concentrations of norleucine and lacking norleucine's standard counterpart, leucine. Experimental results have shown that peptides containing norleucine in place of leucine are not produced unless the cells contain a mutant LeuRS. Explain these results.

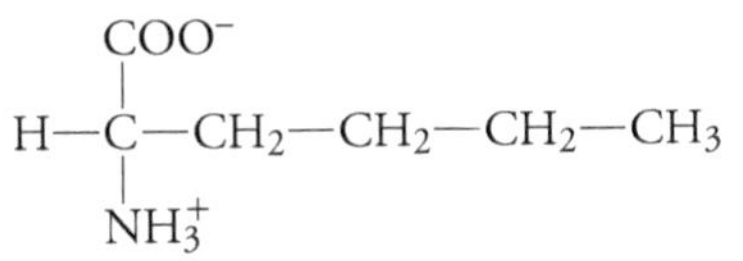

$$\underset{\overset{|}{NH_3^+}}{\overset{\overset{COO^-}{|}}{H-C}}-CH_2-CH_2-CH_2-CH_3$$

Norleucine

22.2 Ribosome Structure

31. The sequences for all the ribosomal proteins in *E. coli* have been elucidated and have been found to contain large amounts of lysine and arginine residues. Why is this finding not surprising? What kinds of interactions are likely to form between the ribosomal proteins and the ribosomal RNA?

32. In eukaryotes, the primary rRNA transcript is a 45S rRNA that includes the sequences of the 18S, 5.8S, and 28S rRNAs separated by short spacers (see Fig. 21.28). What is the advantage of this operon-like arrangement of rRNA genes?

33. The sequence of ribosomal RNA is highly conserved, even though there are many rRNA genes in the genome. How does this observation argue for a functional (not just structural) role for rRNA?

34. Ribosomal inactivating proteins (RIPs) are RNA *N*-glycosidases found in plants. They catalyze the hydrolysis of specific adenine residues in RNA. RIPs are highly toxic but might be useful as antitumor drugs because ribosome synthesis is upregulated in transformed cells. Give a general explanation that describes how RIPs inactivate ribosomes.

35. What is the effect of adding EDTA, a chelating agent specific for divalent cations, to a bacterial cell extract carrying out protein synthesis?

36. In an experiment, >95% of the proteins are extracted from the 50S ribosomal subunit. Only the 23S rRNA and some protein fragments remain. Peptidyl transferase activity is unaffected. What can you conclude from these results? Propose a role for the protein fragments left behind. Why might it have been difficult to remove them?

37. Ribosomal proteins can be separated by two-dimensional electrophoresis (a technique that separates proteins based on both charge and size). One of the proteins in the large ribosomal subunit is sometimes acetylated at its N-terminus. Explain why two-dimensional electrophoresis of ribosomal proteins yields two spots corresponding to this protein.

38. Like their protein counterparts, RNA molecules fold into a variety of structural motifs. Ribosomal proteins contain a so-called RNA-recognition motif. Rho factor and the poly(A) binding protein contain this same motif. Why is this observation not surprising?

39. Propose at least one hypothesis to explain why eukaryotic ribosomes are more complex than prokaryotic ribosomes.

22.3 Translation

40. In this chapter, we have frequently compared eukaryotic translation to *bacterial* translation rather than to *prokaryotic* translation in general. This distinction is intentional, because the protein-synthesizing machinery in Archaea is more similar to the eukaryotic system than to the system in Bacteria. Is this consistent with the evolutionary scheme outlined in Figure 1.6?

41. Complete the following table about eukaryotic replication, transcription, and translation.

Process	Replication	Transcription	Translation
Substrates			
Product			
Template or guide			
Primer			
Enzyme			
Cellular location			

42. The direction of protein synthesis was determined by carrying out an experiment in a cell-free system in which the mRNA consisted of a polymer of A residues with C at the 3′ end, as shown. **a.** What polypeptide was synthesized? **b.** What would the result be if the mRNA were read in the 3′ → 5′ direction? **c.** How does this directionality allow prokaryotes to begin translation before transcription is complete?

5′–A A A A ⋯ A A A C–3′

43. a. Mycobacteriophage genes occasionally begin with GUG or, even more rarely, UUG (rather than AUG). Which amino acids correspond to these codons? **b.** Yeast genes occasionally begin with CUG, ACG, or AUC. Which amino acids correspond to these codons?

44. The translation initiation sequence for the bacterial ribosomal protein L10 is shown below. Draw a diagram that shows how the Shine–Dalgarno sequence aligns with the appropriate sequence on the 16S rRNA. Identify the initiation codon.

5′–CUACCAGGAGCAAAGCUAAUGGCUUUA–3′

45. An *E. coli* phage replicase gene has the mRNA initiation sequence shown below. Indicate the correct start codon and the Shine–Dalgarno sequence upstream of it.

5′–UAACUAGGAUGAAAUGCAUGUCUAAG ⋯

46. Why does eukaryotic translation initiation require proteins that recognize the 5′ cap and the poly(A) tail on the RNA?

47. S1, a protein in the small ribosomal subunit, has a high affinity for single-stranded RNA and has been shown to be important in initiation. What role might S1 play during initiation?

48. What is the significance of the formylation of methionine in the initiator tRNA (tRNAfMet) in bacterial translation initiation? How does this modification affect the incorporation of methionine into the polypeptide chain?

49. What happens when colicin E3, which cleaves on the 5′ side of A1493 in the 16S rRNA, is added to a bacterial culture?

50. Explain why modification or mutagenesis of prokaryotic 16S rRNA at position 1492 or 1493 increases the error rate of translation.

51. What is the significance of limiting the translation error rate to 10^{-4} (one mistake for every 10^4 codons)? How does the combined action of EF-Tu and ribosomal proofreading contribute to maintaining the fidelity of protein synthesis?

52. A mutation in yeast eIF2 that stimulates premature GTPase activity results in initiation at non-AUG start codons. Explain how this observation clarifies the role of eIF2 in translation initiation.

53. In eukaryotes, cellular stressors such as misfolded proteins, oxidative stress, and nutrient shortages may lead to phosphorylation of a serine residue in the eIF2 protein, which blocks its ability to exchange GDP for GTP. How would this affect protein synthesis? How does this strategy ensure the survival of a cell under stress?

54. The pairing of an aminoacyl–tRNA with EF-Tu offers an opportunity for proofreading during translation. EF-Tu binds all 20 aminoacyl–tRNAs with approximately equal affinity so that it can deliver them and surrender them to the ribosome with the same efficiency. Based on the experimentally determined binding constants for EF-Tu and correctly charged and mischarged aminoacyl–tRNAs (shown here), explain how the tRNA–EF-Tu recognition system prevents the incorporation of the wrong amino acid in a protein.

Aminoacyl–tRNA	Dissociation constant (nM)
Ala-tRNA^{Ala}	6.2
Gln-tRNA^{Ala}	0.05
Gln-tRNA^{Gln}	4.4
Ala-tRNA^{Gln}	260

55. The affinity of a ribosome for a tRNA in the P site is about 50 times higher than for a tRNA in the A site. Explain why this promotes translational accuracy.

56. The bacterial elongation factors EF-Tu and EF-G are essential for translation *in vivo*, but bacterial ribosomes can translate mRNA into protein *in vitro* in the absence of EF-Tu and EF-G. Why are these factors not required *in vitro*? How does their absence affect the accuracy of translation?

57. Predict the effect on protein synthesis if EF-Tu were able to recognize and form a complex with fMet-tRNA$_f^{Met}$.

58. Identify the peptide encoded by the DNA sequence shown below (the top strand is the coding strand):

GCGATATGTGTGAACATGTGAACTCTTAGAC
CGCTATACACACTTGTACACTTGAGAATCTG

59. The sequence of a portion of the coding strand of a gene is shown, along with the sequence of a mutant form of the gene.

　　wild-type　　AC ATG CGA TCG CAA ATC CTA
　　mutant　　　AC ATG CAG ATC GCA AAT CCT

a. Give the polypeptide sequence that corresponds to the wild-type gene. **b.** How does the mutant gene differ from the wild-type gene, and how does the mutation affect the encoded polypeptide?

60. A DNA sequence from the beginning of a gene is shown. **a.** If this sequence is the coding strand, choose the appropriate reading frame and provide the sequence of the encoded peptide. **b.** What is the correct reading frame and the peptide sequence if the sequence is the noncoding strand?

ACAGACACCATGGTGCACCTGACTCCTG

61. Complete the same exercise described in Problem 60 when the DNA sequence is mutated as shown below. How does the mutation affect translation of the gene if the DNA sequence is **a.** the coding strand or **b.** the noncoding strand?

ACAGACACCATTGTGCACCTGACTCCTG

62. The ribosome incorporates the wrong amino acid into a polypeptide chain at a rate of about 10^{-4}. For an average-sized protein containing 400 amino acids, what percentage of protein molecules have the wrong sequence?

63. The gene for CFTR, the protein that is mutated in cystic fibrosis, contains 250,000 nucleotides. After splicing, the mature mRNA contains 6129 nucleotides. How many nucleotides are required to encode the 1480 residues of the protein? What is the function of the "extra" mRNA nucleotides?

64. Calculate the approximate energetic cost (in kJ) for the ribosomal synthesis of one mole of a 25-residue polypeptide. Why is the actual energetic cost *in vivo* probably higher than this value?

65. All cells contain an enzyme that hydrolyzes peptidyl–tRNA molecules that are not bound to a ribosome. Cells that are deficient in peptidyl–tRNA hydrolase grow very slowly. What is the function of this enzyme, which hydrolyzes the peptidyl group from the tRNA? What does this tell you about the ability of ribosomes to carry out protein synthesis?

66. In an experimental system, the rate of the peptidyl transferase reaction varies with the identity of the amino acid residue attached to the tRNA in the P site, as shown in the table. **a.** Explain why transpeptidation involving Pro is much slower than for other amino acids. **b.** What can you conclude about the electrostatic environment of the peptidyl transferase active site? **c.** For the nonpolar amino acids, what factor seems to facilitate transpeptidation?

Peptidyl group	Rate of peptidyl transfer (s^{-1})
Ala	57
Asp	8
Lys	100
Phe	16
Pro	0.14

67. The peptidyl transferase center of the ribosome actually catalyzes two reactions involving the ester bond linking a polypeptide to the tRNA: aminolysis and hydrolysis. Explain.

68. The rate of the peptidyl transferase reaction increases as the pH increases from 6 to 8. Use your knowledge of the peptidyl transferase reaction mechanism to explain this observation.

69. The events of transpeptidation resemble peptide bond hydrolysis in reverse (see Fig. 6.10). **a.** Draw the "tetrahedral intermediate" of the transpeptidation reaction. **b.** Researchers hypothesized that residue A2451 protonated at position N1 stabilizes the reaction intermediate. Draw this protonated adenine and explain how it might stabilize the tetrahedral intermediate. Would this catalytic mechanism be enhanced as the pH increases? (*Note:* This hypothesis turned out to be incorrect, as described in the chapter.)

70. Using what you have learned in the first three sections of this chapter, compare and contrast bacterial and eukaryotic translation.

71. In 1961, Howard Dintzis carried out a series of experiments demonstrating that translation occurs in the N → C terminal direction. He added [^{3}H]Leu to immature reticulocytes (cells that actively synthesize hemoglobin), then determined the extent of labeling in each isolated β-globin chain. Results of one of his experiments are shown in the graph. The data points represent protein fragments, with their distance from the N-terminus (expressed as a residue number) indicated on the x axis. **a.** How do the results confirm that translation occurs in the N → C direction? **b.** Why was it important to carry out the experiment in 4 minutes, which is less time than it takes the ribosome to synthesize the entire polypeptide chain? **c.** What would the curve look like if the reticulocytes were incubated with [³H]Leu for longer than it takes to synthesize the entire polypeptide chain?

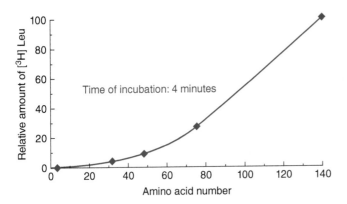

72. A group of investigators working around the same time as Dintzis (see Problem 71) carried out translation in reticulocytes and obtained results that supported his conclusion that translation proceeds in the N → C direction. The investigators added [¹⁴C]Val to reticulocytes in order to label globin chains that were being synthesized. They chose valine because it is the N-terminal residue of both α and β globins (see Fig. 5.5). After a short labeling period, the cells were lysed, and the ribosomes were isolated and then incubated with non-labeled valine in a cell-free protein translation system. The nascent hemoglobin chains were then isolated and analyzed to see what parts of the globin chains were radioactive. What were the results? What results would have been obtained had protein synthesis occurred in the C → N terminal direction?

73. A "nonsense suppressor" mutation results from a mutation in a tRNA anticodon sequence so that the tRNA can pair with a stop codon (also called a nonsense codon). **a.** What would be the effect of a nonsense suppressor mutation on protein synthesis in the cell? **b.** Would all of the cell's proteins be affected by the mutation? **c.** Could the aminoacyl–tRNA synthetase that aminoacylates the nonmutated tRNA play a role in minimizing the effects of a nonsense suppressor mutation?

74. Some bacterial genes that code for enzymes involved in the pathway for synthesizing an amino acid generate mRNAs whose 5′ untranslated segment includes a binding site for a tRNA corresponding to that amino acid. When the tRNA that interacts with the binding site is charged with the corresponding amino acid, translation is inhibited. Describe how this regulatory system works and why it makes sense for the bacterial cell.

75. How might a small molecule that inhibited the formation of a riboswitch serve as a drug to treat a human disease caused by a bacterial pathogen?

76. Many naturally occurring antibiotics target bacterial ribosomes in order to inhibit protein synthesis. In theory, any of the events of translation—including ribosome assembly, tRNA binding, codon–anticodon pairing, and peptide bond formation—are potential targets for antibiotics. The synthetic antibiotics known as oxazolidinones

inhibit protein synthesis. Experiments show that these compounds do not directly inhibit peptide bond formation. They do inhibit translation initiation, but only at very high concentrations. This suggests that the oxazolidinones act by interfering with another aspect of translation. To test different possibilities, researchers used an *E. coli lacZ* expression system (*lacZ* codes for the enzyme β-galactosidase; see Section 21.1) and measured β-galactosidase enzymatic activity in the presence and absence of an oxazolidinone. **a.** First, the researchers engineered a stop codon into the N-terminal region of the polypeptide sequence and then measured β-galactosidase activity. What is the expected effect of the stop codon? Why was the stop codon introduced near the N-terminus of the protein rather than the C-terminus? **b.** The researchers observed that the level of β-galactosidase expressed from *lacZ* containing a stop codon was extremely low but increased about eightfold in the presence of the oxazolidinone. What does this suggest about the action of the oxazolidinone? **c.** Next, the researchers used a *lacZ* gene containing either a one-nucleotide insertion or a one-nucleotide deletion. How would these mutations affect the measured β-galactosidase activity? **d.** When the insertion and deletion mutations were tested in the presence of the antibiotic, β-galactosidase activity levels increased 15–25 times relative to levels in the absence of the antibiotic. What does this suggest about the action of the oxazolidinone? **e.** Next, the researchers measured β-galactosidase activity using *lacZ* genes in which the codon for an active-site Glu residue was mutated. Which codons specify Glu (see Table 22.1)? Describe how a single-base substitution in a Glu codon can give rise to codons specific for Ala, Gln, Gly, and Val. **f.** All of the genes with mutations at the Glu codon yielded very low levels of β-galactosidase activity, and the oxazolidinone had no effect. What does this tell about the action of the oxazolidinone? **g.** Based on your understanding of oxazolidinone action, explain why the compound inhibits bacterial growth. **h.** Oxazolidinones bind to a single site on the ribosome that is in the 50S subunit near the peptidyl transferase active site. Is this information consistent with the experimental results described above?

22.4 Post-Translational Modifications

77. In prokaryotes, translation can begin even before an mRNA transcript has been completely synthesized. Why is co-transcriptional translation not possible in eukaryotes?

78. In 1957, Christian Anfinsen carried out a denaturation experiment *in vitro* with ribonuclease, a pancreatic enzyme consisting of a single 124-residue chain cross-linked by four disulfide bonds (see Problem 4.53). Urea (a denaturing agent) and 2-mercaptoethanol (a reducing agent) were added to a solution of purified ribonuclease, resulting in protein unfolding with a concomitant loss of biological activity. When urea and 2-mercaptoethanol were removed, the ribonuclease spontaneously folded back to its native conformation and regained full enzymatic activity. Why could proper protein folding occur in this experiment in the absence of molecular chaperones?

79. In another set of experiments, the lysine side chains on the surface of ribonuclease (see Problem 78) were covalently attached to an eight-residue chain of polyalanine. The presence of these polyalanine chains did not affect the ability of the ribonuclease to fold properly. What do these experiments reveal about the driving force for protein folding?

80. Chaperones are located not just in the cytosol but in the mitochondria as well. What is the role of mitochondrial chaperones?

81. The chaperone Hsp90 interacts with the tyrosine kinase domain of a growth factor receptor that has lost its ligand-binding domain but retains its tyrosine kinase domain (see Box 10.B). The antibiotic geldanamycin inhibits Hsp90 function. What is the effect of adding geldanamycin to cells expressing the abnormal growth factor receptor?

82. Multidomain proteins tend to fold better inside cagelike chaperonin structures (such as GroEL/GroES in *E. coli*) than with cytosolic chaperones. Explain why.

83. In immature red blood cells, globin synthesis is carefully regulated. The genes for α and β globin (see Section 5.1) are on separate chromosomes, and there are two α globin genes for every β globin gene. If too many β chains are produced, they form a functionally useless tetrameric hemoglobin. Excess α chains tend to precipitate and damage red blood cells. **a.** Explain why it is advantageous for the cell to synthesize a slight excess of α chains. **b.** Red blood cells express a protein that appears to stabilize the α chains and prevents their precipitation. Why is the stabilizing protein necessary? **c.** The disease β thalassemia results from a defect in a β globin gene. Heterozygotes (who have one normal and one abnormal gene) may develop a mild anemia, but homozygotes (who have two defective β globin genes) exhibit severe anemia. Explain why an extra copy of an α globin gene in an individual lacking one β globin gene would result in more severe anemia. **d.** Explain why a mutation in an α globin gene would reduce the severity of anemia in β thalassemia. **e.** Immature red blood cells produce a kinase that phosphorylates the ribosomal initiation factor eIF2. The phosphorylated eIF2 is unable to exchange bound GDP for GTP. How does this affect the rate of protein synthesis in the cell? **f.** In the presence of heme, the kinase is inactive. How does this mechanism regulate hemoglobin synthesis?

84. The N-terminal sequence of the secreted protein bovine proalbumin is shown below. Identify the essential features of the signal peptide in this protein.

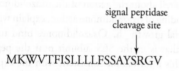

signal peptidase
cleavage site
↓

MKWVTFISLLLLLFSSAYSRGV

85. The mammalian signal recognition particle (SRP) consists of one molecule of RNA and six proteins. In addition to interacting with the signal peptide, what might be the role of the RNA in the SRP?

86. One of the six proteins in the mammalian signal recognition particle (SRP) contains a cleft lined with hydrophobic amino acids. Propose a role for this protein in the SRP.

87. Why does the mammalian signal recognition particle (SRP) bind to the nascent polypeptide as it emerges from the ribosome? Why doesn't the SRP wait until translation is complete before escorting the new polypeptide to the ER membrane?

88. A eukaryotic "cell-free" translation system contains all the components required for protein synthesis—ribosomes; tRNAs; aminoacyl–tRNA synthetases; initiation, elongation, and termination factors; amino acids; GTP; and Mg^{2+} ions. An exogenous mRNA added to this mixture can direct protein synthesis *in vitro*. When an mRNA encoding a secretory protein is added to this cell-free system, along with the SRP, the entire protein is synthesized. When microsomes (sealed vesicles derived from ER membranes) are subsequently added, the protein is not translocated into the microsomal lumen and the signal sequence is not removed. What does this observation reveal about the role of SRP in the synthesis of secretory proteins?

89. The elongation factor EF-Tu performs proofreading by monitoring codon–anticodon base pairing during translation. When a match is confirmed, a conformational change occurs, and EF-Tu hydrolyzes its bound GTP (see Fig. 22.15). Could the SRP use a similar mechanism to perform a proofreading function?

90. Antibiotics can lose effectiveness due to resistance mechanisms, including mutations in ribosomal components and the acquisition of resistance genes. What mechanisms can bacteria become resistant to antibiotics, and how do these mechanisms render antibiotics ineffective?

91. Polyglutamine diseases are neurodegenerative disorders caused by mutations in the DNA that produce CAG repeats (See Problem 21.47). Proteins translated from the mutated genes contain long stretches of glutamine residues that interfere with protein folding. Polyglutamine proteins tend to aggregate and form cytosolic and nuclear inclusion bodies. The result is a loss of neuron function, although the mechanism is unknown. Recent studies show that post-translational modification of the polyglutamine proteins may play a role in the progression of the disease. Interestingly, some post-translational modifications are neurotoxic while others are protective. **a.** Protein kinase B (see Section 10.2) phosphorylates a polyglutamine protein on an essential serine residue, resulting in decreased toxicity. Draw the structure of a phosphorylated serine residue. **b.** Proteins are marked for degradation by the attachment of the protein ubiquitin to a lysine side chain on the condemned protein (see Section 12.1). An isopeptide bond forms between the lysine side chain and the carboxyl terminal group of the ubiquitin. Ubiquitination enhances the degradation of the inclusion body protein. Draw the structure of the linkage between ubiquitin and a polyglutamine protein. **c.** Polyglutamine proteins interact with a histone acetyltransferase (see Section 21.1), sequestering the enzyme in inclusion bodies and hastening its degradation. What are the cellular consequences of this interaction?

92. The protein c-Myc is a leucine zipper protein (see Box 21.A) that regulates gene expression in cell proliferation and differentiation. Its activity was known to be regulated by phosphorylation of a specific threonine residue, but later studies showed that this same threonine could be modified by an *N*-acetylglucosamine residue and that phosphorylation and glycosylation were competitive processes. The specific threonine is mutated in some human lymphomas. Draw the structure of the *O*-glycosylated threonine residue.

93. In the *N*-myristoylation process, the fatty acid myristate (14:0) is attached to an N-terminal glycine residue of a protein during translation. Draw the structure of a myristoylated N-terminal glycine residue.

94. In the palmitoylation process, the fatty acid palmitate (16:0) is attached to the side chain of an internal cysteine residue of a protein. Draw the structure of a palmitoylated cysteine residue. What proteins involved in cell signaling pathways include this modification, and what is the role of the palmitate?

SELECTED READINGS

Çetin G, Klafack S, Studencka-Turski M, Krüger E, Ebstein F. The Ubiquitin–Proteasome System in Immune Cells. Biomolecules. 2021; 11(1):60. (https://doi.org/10.3390/biom11010060)

Choi, J., Grosely, R., Prabhakar, A., Lapointe, C.P., Wang, J., and Puglisi, J.D., How mRNA and nascent chain sequences regulate translation elongation, *Annu. Rev. Biochem.* 87, 421–449, doi: 10.1146/annurev-biochem-060815-014818 (2018). [Describes the approaches for studying translation dynamics and reviews the steps of elongation.]

Emmott, E., Jovanovic, M., and Slavov, N., Ribosome stoichiometry: From form to function, *Trends Biochem. Sci.* 44, 95–109, doi: 10.1016/j.tibs.2018.10.009 (2019). [Discusses approaches for exploring ribosome specialization.]

Hayer-Hartl, M., Bracher, A., and Hartl, F.U., The GroEL-GroES chaperonin machine: A nano-cage for protein folding, *Trends Biochem. Sci.* 41, 62–76, doi: 10.1016/j.tibs.2015.07.009 (2016). [Summarizes the key features of chaperonin action.]

Ross, Christopher A., and Cecile M. Pickart. "The ubiquitin–proteasome pathway in Parkinson's disease and other neurodegenerative diseases." Trends in cell biology 14.12 (2004): 703-711. (https://doi.org/10.1016/j.tcb.2004.10.006)

Rubio Gomez, M.A. and Ibba, M., Aminoacyl-tRNA synthetases, *RNA* 26, 910–936, doi: 10.1261/rna.071720.119 (2020). [A thorough review of the enzymes that link amino acids to tRNAs, including editing steps.]

Park, J., Cho, J. & Song, E.J. Ubiquitin–proteasome system (UPS) as a target for anticancer treatment. Arch. Pharm. Res. 43, 1144–1161 (2020). (https://doi.org/10.1007/s12272-020-01281-8)

Saier, M.H, Jr., Understanding the genetic code, *J. Bacteriol.* 201, e00091-19, doi: 10.1128/JB.00091-19 (2019). [A highly readable discussion of the evolution and conservation of the genetic code, with a focus on the importance of the second position.]

Stein, K.C. and Frydman, J., The stop-and-go traffic regulating protein biogenesis: How translation kinetics controls proteostasis, *J. Biol. Chem.* 294, 2076–2084, doi: 10.1074/jbc.REV118.002814 (2019). [Addresses the determinants of efficient translation and co-translational protein folding.]

Yusupova, G. and Yusupov, M., Crystal structure of eukaryotic ribosome and its complexes with inhibitors. *Phil. Trans. R. Soc. B* 372, 20160184, doi: 10.1098/rstb.2016.0184 (2017). [A brief and readable summary of ribosome structure and the effects of some inhibitors on ribosome function.]

CHAPTER 22 CREDITS

Figure 22.2b Image based on 4TRA. Westhof, E., Dumas, P., Moras, D., Restrained refinement of two crystalline forms of yeast aspartic acid and phenylalanine transfer RNA crystals *Acta Crystallogr.* A 44, 112–123 (1988).

Figure 22.4 Image based on 1QRT. Arnez, J.G., Steitz, T.A., Crystal structures of three misacylating mutants of Escherichia coli glutaminyl-tRNA synthetase complexed with tRNA(Gln) and ATP, *Biochemistry* 35, 14725–14733 (1996).

Figure 22.5 Image based on 1J5E. Wimberly, B.T., Brodersen, D.E., Clemons Jr., W.M., Morgan-Warren, R.J., Carter, A.P., Vonrhein, C., Hartsch, T., Ramakrishnan, V., Structure of the 30S ribosomal subunit, *Nature* 407, 327–339 (2000).

Figure 22.6 Image based on 1JJ2. Klein, D.J., Schmeing, T.M., Moore, P.B., Steitz, T.A., The kink-turn, a new RNA secondary structure motif, *EMBO J.* 20, 4214–4221 (2001).

Figure 22.7 Image from Ben-Shem, A., Garreau de Loubresse, N., Melnikov, S., Jenner, L., Yusupova, G., & Yusupov, M., The Structure of the Eukaryotic Ribosome at 3.0 A Resolution. *Science* 334(6062), 1524–1529. © 2011 American Association for the Advancement of Science.

Figure 22.8 Image from Schmeing, T., Ramakrishnan, V. What recent ribosome structures have revealed about the mechanism of translation. *Nature* 461, 1234–1242. © 2009 Springer Nature.

Figure 22.13 Image based on 1TTT. Nissen, P., Kjeldgaard, M., Thirup, S., Polekhina, G., Reshetnikova, L., Clark, B.F., Nyborg, J., Crystal structure of the ternary complex of Phe-tRNAPhe, EF-Tu, and a GTP analog, *Science* 270, 1464–1472 (1995).

Figure 22.14 Image from Ogle, J.M., Recognition of Cognate Transfer RNA by the 30S Ribosomal Subunit. *Science* 292(5518), 897–902. © 2001 American Association for the Advancement of Science.

Figure 22.18 Image based on 2BV3. Hansson, S., Singh, R., Gudkov, A.T., Liljas, A., Logan, D.T., Crystal structure of a mutant elongation factor G trapped with a GTP analogue, *FEBS Lett.* 579, 4492 (2005).

Figure 22.20 Image based on 3D5A Laurberg, M., Asahara, H., Korostelev, A., Zhu, J., Trakhanov, S., Noller, H.F., Crystal structure of a translation termination complex formed with release factor RF2, *Proc. Natl. Acad. Sci. USA* 105, 19684–19689 (2008).

Figure 22.22b Image from Brandt, F., Etchells, S.A., Ortiz, J.O., Elcock, A.H., Hartl, F.U., & Baumeister, W., The Native 3D Organization of Bacterial Polysomes. *Cell* 136(2), 261–271. © 2009 Elsevier.

Figure 22.25 Image based on 1AON. Xu, Z., Horwich, A.L., Sigler, P.B., The crystal structure of the asymmetric GroEL-GroES-(ADP)7 chaperonin complex, *Nature* 388, 741–750 (1997).

Figure 22.27 Image from Batey, R.T., Crystal Structure of the Ribonucleoprotein Core of the Signal Recognition Particle. *Science* 287(5456), 1232–1239. © 2000 American Association for the Advancement of Science.

Figure 22.28 Image from Becker, T., Bhushan, S., Jarasch, A., Armache, J.-P., Funes, S., Jossinet, F., Gumbart, J., Mielke, T., Berninghausen, O., Schulten, K., Westhof, E., Gilmore, R., Mandon, E. C., & Beckmann, R., Structure of Monomeric Yeast and Mammalian Sec61 Complexes Interacting with the Translating Ribosome. *Science* 326(5958), 1369–1373. © 2009 American Association for the Advancement of Science.

Numbers and Greek letters are alphabetized as if they were spelled out.

A

A-DNA. A conformation of DNA in which the double helix is wider than the standard B-DNA helix and in which base pairs are inclined to the helix axis.

A site. The ribosomal binding site that accommodates an aminoacyl–tRNA.

Abasic site. The deoxyribose residue remaining after the removal of a base from a DNA strand.

ABC transporter. A member of a family of structurally similar transmembrane proteins that use the free energy of ATP to drive conformational changes that move substances across the membrane.

Acid. A substance that can donate a proton.

Acid–base catalysis. A catalytic mechanism in which partial proton transfer from an acid or partial proton abstraction by a base lowers the free energy of a reaction's transition state.

Acid catalysis. A catalytic mechanism in which partial proton transfer from an acid lowers the free energy of a reaction's transition state.

Acid dissociation constant (K_a). The dissociation constant for an acid in water.

Acidic solution. A solution whose pH is less than 7.0 ($[H^+] > 10^{-7}$ M).

Acidosis. A pathological condition in which the pH of the blood drops below its normal value of 7.4.

Actin filament. A 70-Å-diameter cytoskeletal element composed of polymerized actin subunits. Also called a microfilament.

Action potential. The momentary reversal of membrane potential that occurs during transmission of a nerve impulse.

Activation energy (free energy of activation, $\Delta G^{\ddagger}$). The free energy of the transition state minus the free energies of the reactants in a chemical reaction.

Activator. A protein that binds at or near a gene so as to promote its transcription.

Active site. The region of an enzyme in which catalysis takes place.

Active transport. The transmembrane movement of a substance from low to high concentrations by a protein that couples this endergonic transport to an exergonic process such as ATP hydrolysis. See also secondary active transport.

Acyl group. A portion of a molecule with the formula —COR, where R is an alkyl group.

Adipose tissue. Tissue consisting of cells that are specialized for the storage of triacylglycerols. See also brown adipose tissue.

Aerobic. Occurring in or requiring oxygen.

Affinity. A measure of the strength of binding between two molecules, often expressed as a dissociation constant.

Affinity chromatography. A procedure in which a molecule is isolated by its ability to bind specifically to a second immobilized molecule.

Agonist. A substance that binds to a receptor so as to evoke a cellular response.

Aldose. A sugar whose carbonyl group is part of an aldehyde.

Alkalosis. A pathological condition in which the pH of the blood rises above its normal value of 7.4.

Allele. An alternate form of a gene; a diploid organism may contain two alleles for each gene.

Allosteric protein. A protein in which the binding of ligand at one site affects the binding of other ligands at other sites. Some enzymes are allosteric proteins. See also cooperative binding.

Allosteric regulation. Binding of an activator or inhibitor to one subunit of a multisubunit enzyme, which increases or decreases the catalytic activity of all the subunits. See also positive effector and negative effector.

α-amino acid. See amino acid.

α anomer. A sugar in which the OH substituent of the anomeric carbon is on the opposite side of the ring from the CH$_2$OH group of the chiral center that designates the D or L configuration.

α carbon. See Cα.

α helix. A regular secondary structure of polypeptides, with 3.6 residues per right-handed turn and hydrogen bonds between each backbone C=O group and the backbone N—H group that is four residues further.

Amino acid (α-amino acid). A compound consisting of a carbon atom to which are attached a primary amino group, a carboxylate group, a side chain (R group), and an H atom.

Amphiphilic (amphipathic). Having both polar and nonpolar regions and therefore being both hydrophilic and hydrophobic.

Amyloid deposit. An accumulation of certain types of insoluble protein aggregates in tissues (e.g., in the brain in Alzheimer's disease).

Anabolism. The reactions by which biomolecules are synthesized from simpler components.

Anaerobic. Occurring independently of oxygen.

Anaplerotic reaction. A reaction that replenishes the intermediates of a metabolic pathway.

Anemia. A condition caused by the insufficient production of or loss of red blood cells.

Anneal. To allow base pairing between complementary single polynucleotide strands so that double-stranded segments form.

Anomers. Sugars that differ only in the configuration around the carbonyl carbon that becomes chiral when the sugar cyclizes.

Antagonist. A substance that binds to a receptor but does not elicit a cellular response.

Antenna pigment. A molecule that transfers its absorbed energy to other pigment molecules and eventually to a photosynthetic reaction center.

Antibody. See immunoglobulin.

Anticodon. The sequence of three nucleotides in a tRNA that recognizes an mRNA codon through complementary base pairing.

Antigen. A substance, usually foreign to the body, that binds to a receptor on a B or T lymphocyte to trigger an immune response.

Antiparallel. Running in opposite directions.

Antiparallel β sheet. See β sheet.

Antiport. Transport that involves the simultaneous transmembrane movement of two molecules in opposite directions.

Antisense strand. See noncoding strand.

Apoptosis. Programmed cell death that results from extracellular or intracellular signals and involves the activation of enzymes that selectively degrade cellular structures.

Aquaporin. A membrane protein that facilitates the transmembrane movement of water molecules.

Archaea. One of the two major groups of prokaryotes.

Atherosclerosis. A disease characterized by the formation of cholesterol-containing fibrous plaques in the walls of blood vessels.

ATPase. An enzyme that catalyzes the hydrolysis of ATP to ADP + P$_i$.

Autoactivation. A process by which the product of an activation reaction also acts as a catalyst for the same reaction, so that it appears that the compound catalyzes its own activation.

Autolysosome. The organelle resulting from the fusion of an autophagosome and a lysosome during autophagy, inside of which unneeded cellular components are hydrolyzed.

Autophagosome. The double-membraned compartment that fully encloses cellular materials during autophagy.

Autophagy. The process of engulfing and degrading unneeded materials within a eukaryotic cell.

Autophosphorylation. The phosphorylation of a kinase by another molecule of the same kinase.

Avidity. A measure of the strength of an intermolecular interaction, such as antigen binding to an immunoglobin, that reflects the presence of multiple binding sites.

Axon. The extended portion of a neuron that conducts an action potential from the cell body to a synapse with a target cell.

B

B-DNA. The standard conformation of double-helical DNA.

B lymphocyte. A type of white blood that responds to the presence of an antigen by releasing immunoglobulins, which are soluble versions of the cell's antigen receptor.

Backbone. The atoms that form the repeating linkages between successive residues of a polymeric molecule, exclusive of the side chains.

Bacteria. One of the two major groups of prokaryotes.

Bacteriophage. A virus specific for bacteria. Also known as a phage.

Base. (1) A substance that can accept a proton. (2) A purine or pyrimidine component of a nucleoside, nucleotide, or nucleic acid.

Base catalysis. A catalytic mechanism in which partial proton abstraction by a base lowers the free energy of a reaction's transition state.

Base excision repair. A DNA repair pathway in which a damaged base is removed by a glycosylase so that the resulting abasic site can be repaired.

Base pair. The specific hydrogen-bonded association between nucleic acid bases. The standard base pairs are A:T and G:C. See also bp.

Basic solution. A solution whose pH is greater than 7.0 ($[H^+] < 10^{-7}$ M).

β anomer. A sugar in which the OH substituent of the anomeric carbon is on the same side of the ring as the CH$_2$OH of the chiral center that designates the D or L configuration.

β barrel. A protein structure consisting of a β sheet rolled into a cylinder.

β oxidation. A series of enzyme-catalyzed reactions in which fatty acids are progressively degraded by the removal of two-carbon units as acetyl-CoA.

β sheet. A regular secondary structure in which extended polypeptide chains form interstrand hydrogen bonds. In parallel β sheets, the polypeptide chains all run in the same direction; in antiparallel β sheets, neighboring chains run in opposite directions.

Bilayer. An ordered, two-layered arrangement of amphiphilic molecules in which polar segments are oriented toward the two solvent-exposed surfaces and the nonpolar segments associate in the center.

Bile acid. A cholesterol derivative that acts as a detergent to solubilize lipids for digestion and absorption.

Bimolecular reaction. A reaction involving two molecules, which may be identical or different.

Binding change mechanism. The mechanism whereby the subunits of ATP synthase adopt three successive conformations to convert $ADP + P_i$ to ATP as driven by the dissipation of the transmembrane proton gradient.

Biofilm. A complex of bacterial cells and a protective extracellular matrix containing polysaccharides.

Bioinformatics. The use of computers in collecting, storing, accessing, and analyzing biological data, such as molecular sequences and structures.

Biomineralization. The association of organic molecules such as proteins with inorganic crystals to form materials such as bones and shells.

Bisubstrate reaction. An enzyme-catalyzed reaction involving two substrates.

Blunt ends. The fully base-paired ends of a DNA fragment that are generated by a restriction endonuclease that cuts both strands at the same point.

Bohr effect. The decrease in O_2 binding affinity of hemoglobin in response to a decrease in pH.

bp. Base pair, the unit of length used for DNA molecules.

Brown adipose tissue. A type of adipose tissue in which fatty acid oxidation is uncoupled from ATP production so that the free energy of the fatty acids is released as heat.

Buffer. A solution of a weak acid and its conjugate base, which resists changes in pH upon the addition of acid or base.

C

C-terminus. The end of a polypeptide that has a free carboxylate group.

Cachexia. The severe loss of body mass that occurs in cancer and some other chronic diseases.

Cα. The alpha carbon, the carbon of an amino acid whose substituents are an amino group, a carboxylate group, an H atom, and a variable R group.

Calvin cycle. The sequence of photosynthetic reactions in which ribulose-5-phosphate is carboxylated, converted to three-carbon carbohydrate precursors, and regenerated.

cAMP. Cyclic AMP, an intracellular second messenger.

Cap. A 7-methylguanosine residue that is post-transcriptionally added to the 5′ end of a eukaryotic mRNA.

Capsid. The protein shell surrounding the nucleic acid of a virus.

Carbanion. A compound that bears a negative charge on a carbon atom.

Carbohydrate. A compound with the formula $(CH_2O)_n$, where $n \geq 3$. Also called a saccharide.

Carbon fixation. The incorporation of CO_2 into biologically useful organic molecules.

Carcinogen. An agent that causes a mutation in DNA that leads to cancer.

Carcinogenesis. The process of developing cancer.

Catabolism. The degradative metabolic reactions in which nutrients and cell constituents are broken down for energy and raw materials.

Catalyst. A substance that promotes a chemical reaction without undergoing permanent change. A catalyst increases the rate at which a reaction approaches equilibrium but does not affect the free energy change of the reaction.

Catalytic constant (k_{cat}). The ratio of the maximal velocity (V_{max}) of an enzyme-catalyzed reaction to the enzyme concentration. Also called a turnover number.

Catalytic perfection. A state achieved by an enzyme that operates at the diffusion-controlled limit.

Catalytic triad. The hydrogen-bonded Ser, His, and Asp residues that participate in catalysis in serine proteases.

cDNA. See complementary DNA.

Cellular respiration. The metabolic phenomenon whereby organic molecules are oxidized, with the electrons eventually transferred to molecular oxygen.

Central dogma of molecular biology. The idea that genetic information in the form of DNA is rewritten, or transcribed, to RNA, which then is translated to direct protein synthesis. Information passes from DNA to RNA to protein.

Centromere. The region of a replicated eukaryotic chromosome where the two identical DNA molecules are attached and where the spindle fibers attach during cell division.

C_4 pathway. A photosynthetic process used in some plants to concentrate CO_2 by incorporating it into oxaloacetate (a C_4 compound).

Channeling. The transfer of an intermediate product from one enzyme active site to another in such a way that the intermediate remains in contact with the protein.

Chaperone. See molecular chaperone.

Chemical labeling. A technique for identifying functional groups in a macromolecule by treating the molecule with a reagent that reacts with those groups.

Chemiosmotic theory. The postulate that the free energy of electron transport is conserved in the formation of a transmembrane proton gradient that can be subsequently used to drive ATP synthesis.

Chemoautotroph. An organism that obtains its building materials and free energy from inorganic compounds.

Chirality. The asymmetry or "handedness" of a molecule such that it cannot be superimposed on its mirror image.

Chloroplast. The plant organelle in which photosynthesis takes place.

Chromatin. The complex of DNA and protein that comprises the eukaryotic chromosomes.

Chromatography. A technique for separating the components of a mixture of molecules based on their partition between a mobile solvent phase and a porous matrix (stationary phase), often performed in a column.

Chromosome. The complex of protein and a single DNA molecule that comprises some or all of an organism's genome.

Citric acid cycle. A set of eight enzymatic reactions, arranged in a cycle, in which energy in the form of ATP, NADH, and QH_2 is recovered from the oxidation of the acetyl group of acetyl-CoA to CO_2.

Clinical trial. A three-phase series of tests of a drug's safety and effectiveness in human subjects.

Clone. An organism or collection of identical cells derived from a single parental cell.

Coagulation. The process of forming a blood clot.

Coding strand. The DNA strand that has the same sequence (except for the replacement of U with T) as the transcribed RNA; it is the nontemplate strand. Also called the sense strand.

Codon. The sequence of three nucleotides in DNA or RNA that specifies a single amino acid.

Coenzyme. A small organic molecule that is required for the catalytic activity of an enzyme. A coenzyme may be tightly associated with the enzyme as a prosthetic group.

Cofactor. A small organic molecule (coenzyme) or metal ion that is required for the catalytic activity of an enzyme.

Coiled coil. An arrangement of polypeptide chains in which two α helices wind around each other.

Competitive inhibition. A form of enzyme inhibition in which a substance competes with the substrate for binding to the enzyme active site and thereby appears to increase K_M.

Complement. (1) A molecule that pairs in a reciprocal fashion with another. (2) A set of circulating proteins that sequentially activate each other and lead to the formation of a pore in a microbial cell membrane.

Complementary DNA (cDNA). A segment of DNA synthesized from an RNA template.

Condensation reaction. The formation of a covalent bond between two molecules, during which the elements of water are lost.

Conformation. The three-dimensional shape of a molecule that it attained through rotation of its bonds.

Conjugate base. The compound that forms when an acid donates a proton.

Consensus sequence. A DNA or RNA sequence showing the nucleotides most commonly found at each position.

Conservative substitution. A change of an amino acid residue in a protein to one with similar properties (e.g., Leu to Ile or Asp to Glu).

Constant domain. A structural domain of an immunoglobulin heavy or light chain, whose sequence is the same in all immunoglobulins of the same class.

Constitutive. Being expressed at a continuous, steady rate rather than induced.

Convergent evolution. The independent development of similar characteristics in unrelated species.

Cooperative binding. A situation in which the binding of a ligand at one site on a macromolecule affects the affinity of other sites for the same ligand. See also allosteric protein.

Cori cycle. A metabolic pathway in which lactate produced by glycolysis in the muscles is transported via the bloodstream to the liver, where it is used for gluconeogenesis. The resulting glucose returns to the muscles.

Covalent catalysis. A catalytic mechanism in which the transient formation of a covalent bond between the catalyst and a reactant lowers the free energy of a reaction's transition state.

CpG island. A cluster of CG sequences that often marks the beginning of a gene in a mammalian genome.

CRISPR. Clustered regularly interspersed short palindromic repeats, short DNA segments involved in bacterial defense against bacteriophages.

CRISPR-Cas9 system. A gene-editing tool that uses the bacterial endonuclease Cas9, which is directed to cut a specific DNA segment that is complementary to a CRISPR-like guide RNA. The CRISPR-Cas9 system can be designed to inactivate a target gene or to replace it with an altered version of the gene.

Cristae. The infoldings of the inner mitochondrial membrane.

Cross-talk. The interactions of different signal transduction pathways through activation of the same signaling components.

Cryo-electron microscopy (cryo-EM). An application of electron microscopy in which high-resolution images of molecular structures are collected at very low temperatures.

Cyclic electron flow. The light-driven circulation of electrons between Photosystem I and cytochrome $b_6 f$, which leads to the production of ATP but not NADPH.

Cytochrome. A protein that carries electrons via a prosthetic Fe-containing heme group.

Cytokinesis. The splitting of the cell into two following mitosis.

Cytoskeleton. The network of intracellular fibers that gives a cell its shape and structural rigidity.

D

D sugar. A monosaccharide isomer in which the asymmetric carbon farthest from the carbonyl group has the same spatial arrangement as the chiral carbon of D-glyceraldehyde.

Dark reactions. The photosynthetic reactions in which NADPH and ATP produced by the light reactions are used to incorporate CO_2 into carbohydrates.

Deamination. The hydrolytic or oxidative removal of an amino group.

Degron. An amino acid sequence that functions as a signal for the polypeptide to be degraded by a proteasome or by autophagy.

ΔG. See free energy.

$\Delta G^{\ddagger}$. See activation energy.

$\Delta G^{\circ\prime}$. See standard free energy change.

$\Delta G_{reaction}$. The difference in free energy between the reactants and products of a chemical reaction; $\Delta G_{reaction} = \Delta G_{products} - \Delta G_{reactants}$.

$\Delta \psi$. See membrane potential.

Denaturation. The loss of ordered structure in a polymer, such as the disruption of native conformation in an unfolded polypeptide or the unstacking of bases and separation of strands in a nucleic acid.

Denitrification. The conversion of nitrate (NO_3^-) to nitrogen (N_2).

Deoxyhemoglobin. Hemoglobin that does not contain bound oxygen or is not in the oxygen-binding conformation.

Deoxynucleotide. A nucleotide in which the pentose is 2′-deoxyribose.

Deoxyribonucleic acid. See DNA.

Desensitization. A cell's adaptation to long-term stimulation through a reduced response to the stimulus.

Diabetes mellitus. A disease caused by a deficiency of insulin or the inability to respond to insulin, and characterized by elevated levels of glucose in the blood.

Diabetes type I. This is a condition developed due to the impairment of insulin secretion also called insulin-dependent diabetes.

Diabetes type II. It develops mainly in two conditions, when cells in muscle, fat and the liver become resistant to insulin causing impairment of glucose take-up into the cells.

Diazotroph. A bacterium that carries out nitrogen fixation, the conversion of N_2 to NH_3.

Dielectric constant. A measure of the ability of a substance to interfere with electrostatic interactions; a solvent with a high dielectric constant is able to dissolve salts by shielding the attractive electrostatic forces that would otherwise bring the ions together.

Diffraction pattern. The record of the radiation scattered from an object, for example, in X-ray crystallography.

Diffusion-controlled limit. The theoretical maximum rate of an enzymatic reaction in solution, about 10^8 to 10^9 $M^{-1} \cdot s^{-1}$.

Dimer. An assembly consisting of two monomeric units.

Diploid. Having two equivalent sets of chromosomes.

Dipole–dipole interaction. A type of van der Waals interaction between two strongly polar groups.

Disaccharide. A carbohydrate consisting of two monosaccharides.

Discontinuous synthesis. A mechanism whereby the lagging strand of DNA is synthesized as a series of fragments that are later joined.

Dissociation constant (K_d). The ratio of the products of the concentrations of the dissociated species to those of their parent compounds at equilibrium.

Disulfide bond. A covalent —S—S— linkage, often between two Cys residues in a protein.

Divergent evolution. The accumulation of changes in species that share an ancestor.

DNA (Deoxyribonucleic acid). A polymer of deoxynucleotides whose sequence of bases encodes genetic information in all living cells.

DNA barcoding. A technique for using species-specific gene sequences to identify different organisms represented in a sample of DNA.

DNA chip. See microarray.

DNA fingerprinting. A technique for distinguishing individuals on the basis of DNA polymorphisms, such as the number of short tandem repeats.

DNA ligase. An enzyme that catalyzes the formation of a phosphodiester bond to join two DNA segments.

DNA polymerase. See polymerase.

Domain. A stretch of polypeptide residues that fold into a globular unit with a hydrophobic core.

Dysbiosis. A change in the number of species or their relative abundance that affects the functionality of the microbiota.

E

ε. See reduction potential.

$\varepsilon^{\circ\prime}$. See standard reduction potential.

E site. The ribosomal binding site that accommodates a deacylated tRNA before it dissociates from the ribosome.

Eclipsed conformation. Where the torsion angle is 0°.

Edman degradation. A procedure for the stepwise removal and identification of the N-terminal residues of a polypeptide.

EF. See elongation factor.

Effector function. The activity of an immunoglobulin, following antigen binding to the Fab region(s), that depends on the Fc region of the protein.

Ehlers–Danlos syndrome. A genetic disease characterized by elastic skin and joint hyperextensibility, caused by mutations in genes for collagen or collagen-processing proteins.

EI complex. The noncovalent complex that forms between an enzyme and a reversible inhibitor.

Eicosanoids. Compounds derived from the C_{20} fatty acid arachidonic acid, which act in or near the cells that produce them and mediate pain, fever, and other physiological responses.

Electron tomography. A technique for reconstructing three-dimensional structures by analyzing electron micrographs of consecutive tissue slices.

Electron transport chain. The series of small molecules and protein prosthetic groups that transfer electrons from reduced cofactors such as NADH to O_2 during cellular respiration.

Electronegativity. A measure of an atom's affinity for electrons.

Electrophile. A compound containing an electron-poor center. An electrophile (electron-lover) reacts readily with a nucleophile (nucleus-lover).

Electrophoresis. A procedure in which macromolecules are separated on the basis of charge or size by their differential migration through a gel-like matrix under the influence of an applied electric field. In polyacrylamide gel electrophoresis (PAGE), the matrix is cross-linked polyacrylamide. In SDS-PAGE, the detergent sodium dodecyl sulfate is used to denature proteins.

Electrostatic catalysis. A catalytic mechanism in which sequestering the reacting groups away from the aqueous solvent lowers the free energy of a reaction's transition state.

Elongation factor (EF). A protein that interacts with tRNA and/or the ribosome during polypeptide synthesis.

Enantiomers. Stereoisomers that are nonsuperimposable mirror images of one another.

Endergonic reaction. A reaction that has an overall positive free energy change (a nonspontaneous process).

Endocytosis. The inward folding and budding of the plasma membrane to form a new intracellular vesicle. See also receptor-mediated endocytosis and pinocytosis.

Endonuclease. An enzyme that catalyzes the hydrolysis of the phosphodiester bonds between two nucleotide residues within a polynucleotide strand.

Endopeptidase. An enzyme that catalyzes the hydrolysis of a peptide bond within a polypeptide chain.

Endosome. The membrane-enclosed vesicle that results from the infolding of the plasma membrane during endocytosis.

Endothermic reaction. A reaction that absorbs heat from the surroundings so that its change in enthalpy (H) is greater than zero.

Enhancer. A eukaryotic DNA sequence located some distance from the transcription start site, where an activator of transcription may bind.

Enthalpy (*H*). A thermodynamic quantity that is taken to be equivalent to the heat content of a biochemical system.

Entropy (*S*). A measure of the degree of randomness or disorder of a system.

Enzyme. A biological catalyst. Most enzymes are proteins; a few are RNA.

Enzyme Immunoassays (EIA). Enzyme immunoassays utilize enzyme-labeled antibodies or antigens to detect the presence or quantity of specific substances in body fluids.

Epigenetics. The inheritance of patterns of gene expression mediated by chromosomal modifications that do not alter the DNA sequence.

Epimers. Sugars that differ only by the configuration at one C atom (excluding the anomeric carbon).

Equilibrium constant (K_{eq}). The ratio, at equilibrium, of the product of the concentrations of reaction products to that of the reactants.

ES complex. The noncovalent complex that forms between an enzyme and its substrate in the first step of an enzyme-catalyzed reaction.

Essential compound. An amino acid, fatty acid, or other compound that an animal cannot synthesize and must therefore obtain in its diet.

Euchromatin. The transcriptionally active, relatively uncondensed chromatin in a eukaryotic cell.

Eukarya. See eukaryote.

Eukaryote. An organism consisting of a cell (or cells) whose genetic material is contained in a membrane-bounded nucleus.

Evolution. Change over time, often driven by the process of natural selection.

Exciton transfer. A mode of decay of an energetically excited molecule, in which electronic energy is transferred to a nearby unexcited molecule.

Exergonic reaction. A reaction that has an overall negative free energy change (a spontaneous process).

Exocytosis. The fusion of an intracellular vesicle with the plasma membrane in order to release the contents of the vesicle outside the cell.

Exome. The set of exons, or expressed protein-coding gene sequences, in an organism's genome.

Exon. A portion of a gene that appears in both the primary and mature mRNA transcripts.

Exonuclease. An enzyme that catalyzes the hydrolytic excision of a nucleotide residue from the end of a polynucleotide strand.

Exopeptidase. An enzyme that catalyzes the hydrolytic excision of an amino acid residue from one end of a polypeptide chain.

Exosome. A vesicle released from a cell, possibly as a form of intercellular communication. Also known as a microvesicle.

Exothermic reaction. A reaction that releases heat to the surrounding so that its change in enthalpy (*H*) is less than zero.

Extracellular matrix. The extracellular proteins and polysaccharides that fill the space between cells and form connective tissue in animals.

Extrinsic protein. See peripheral membrane protein.

F

$\mathcal{F}$. See Faraday constant.

F-actin. The polymerized form of the protein actin. See also G-actin.

Fab fragment. One of the two antigen-binding portions of an immunoglobulin.

Factory model of replication. A model for DNA replication in which DNA polymerase and associated proteins remain stationary while the DNA template is spooled through them.

Faraday constant ($\mathcal{F}$). The charge of one mole of electrons, equal to 96,485 coulombs·mol⁻¹ or 96,485 J·V⁻¹·mol⁻¹.

Fatty acid. A carboxylic acid with a long-chain hydrocarbon side group.

Fc fragment. The portion of an immunoglobulin that is responsible for the protein's effector functions but does not bind antigen.

Feed-forward activation. The activation of a later step in a reaction sequence by the product of an earlier step.

Feedback inhibitor. A substance that inhibits the activity of an enzyme that catalyzes an early step of the substance's synthesis.

Fermentation. An anaerobic catabolic process.

Fibrous protein. A protein characterized by a stiff, elongated conformation, that tends to form fibers.

First-order reaction. A reaction whose rate is proportional to the concentration of a single reactant.

Fischer projection. A graphical convention for specifying molecular configuration in which horizontal lines represent bonds that extend above the plane of the paper and vertical bonds extend below the plane of the paper.

5′ end. The terminus of a polynucleotide whose C5′ is not esterified to another nucleotide residue.

Flip-flop. See transverse diffusion.

Flippase. A translocase that catalyzes the movement of membrane lipids from the non-cytoplasmic leaflet to the cytoplasmic leaflet.

Floppase. A translocase that catalyzes the movement of membrane lipids from the cytoplasmic leaflet to the non-cytoplasmic leaflet.

Fluid mosaic model. A model of biological membranes in which integral membrane proteins float and diffuse laterally in a fluid lipid layer.

Fluorescence. A mode of decay of an excited molecule, in which electronic energy is emitted in the form of a photon.

Flux. The rate of flow of metabolites through a metabolic pathway.

Fractional saturation (*Y*). The fraction of a protein's ligand-binding sites that are occupied by ligand.

Free energy (*G*). A thermodynamic quantity whose change indicates the spontaneity of a process. For spontaneous processes, $\Delta G < 0$, whereas for a process at equilibrium, $\Delta G = 0$.

Free energy of activation. See activation energy ($\Delta G^{\ddagger}$).

Free radical. A molecule with an unpaired electron.

Futile cycle. Two opposing metabolic reactions that function together to provide a control point for regulating metabolic flux.

G

G. See free energy.

G-actin. The monomeric form of the protein actin. See also F-actin.

G protein. A guanine nucleotide–binding and –hydrolyzing protein, involved in a process such as signal transduction or protein synthesis, that is inactive when it binds GDP and active when it binds GTP.

G protein–coupled receptor. A transmembrane protein that binds an extracellular ligand and transmits the signal to the cell interior by interacting with an intracellular G protein.

Gametes. The haploid cells produced by meiosis.

Gas constant (*R*). A thermodynamic constant equivalent to 8.3145 J·K⁻¹·mol⁻¹.

Gated channel. A transmembrane channel that opens and closes in response to a signal such as changing voltage, ligand binding, or mechanical stress.

Gel filtration chromatography. See sizeexclusion chromatography.

Gene. A unique sequence of nucleotides that encodes a polypeptide or RNA; it may include nontranscribed and nontranslated sequences that have regulatory functions.

Gene expression. The transformation by transcription and translation of the information contained in a gene to a functional RNA or protein product.

Gene therapy. The manipulation of genetic information in the cells of an individual in order to produce a therapeutic effect.

General transcription factor. One of a set of eukaryotic proteins that are typically required for the synthesis of mRNAs. See also transcription factor.

Genetic code. The correspondence between the sequence of nucleotides in a nucleic acid and the sequence of amino acids in a polypeptide; a series of three nucleotides (a codon) specifies an amino acid.

Genome. The complete set of genetic instructions in an organism.

Genome-wide association study (GWAS). An attempt to correlate genetic variations with a trait such as a particular disease.

Genomics. The study of the size, organization, and gene content of organisms' genomes.

Globin. The polypeptide component of myoglobin and hemoglobin.

Globular protein. A water-soluble protein characterized by a compact, highly folded structure.

Glucogenic amino acid. An amino acid whose degradation yields a gluconeogenic precursor. See also ketogenic amino acid.

Gluconeogenesis. The synthesis of glucose from noncarbohydrate precursors.

Glucose–alanine cycle. A metabolic pathway in which pyruvate produced by glycolysis in the muscles is converted to alanine and transported to the liver, where it is converted back to pyruvate for gluconeogenesis. The resulting glucose returns to the muscles.

Glycan. See polysaccharide.

Glycerophospholipid. An amphipathic lipid in which two hydrocarbon chains and a polar phosphate derivative are attached to a glycerol backbone.

Glycogen storage disease (GSD). An inherited defect in an enzyme or transporter that affects the formation, structure, or degradation of glycogen.

Glycogenolysis. The enzymatic degradation of glycogen to glucose-1-phosphate.

Glycolipid. A lipid to which carbohydrate is covalently attached.

Glycolysis. The 10-reaction pathway by which glucose is broken down to 2 pyruvate with the concomitant production of 2 ATP and the reduction of 2 NAD⁺ to 2 NADH.

Glycomics. The systematic study of the structures and functions of carbohydrates, including large glycans and the small oligosaccharides of glycoproteins.

Glycoprotein. A protein to which carbohydrate is covalently attached.

Glycosaminoglycan. An unbranched polysaccharide consisting of alternating residues of an amino sugar and a sugar acid.

Glycosidase. An enzyme that catalyzes the hydrolysis of glycosidic bonds.

Glycoside. A molecule containing a saccharide linked to another molecule by a glycosidic bond to the anomeric carbon.

Glycosidic bond. The covalent linkage between two monosaccharide units in a polysaccharide, or the linkage between the anomeric carbon of a saccharide and an alcohol or amine.

Glycosylation. The attachment of carbohydrate chains to a protein through *N*- or *O*-glycosidic linkages.

Glycosyltransferase. An enzyme that catalyzes the addition of a monosaccharide residue to a polysaccharide.

Glyoxylate pathway. A variation of the citric acid cycle in plants that allows acetyl-CoA to be converted quantitatively to gluconeogenic precursors.

Glyoxysome. A membrane-bounded plant organelle in which the reactions of the glyoxylate pathway take place.

Gout. An inflammatory disease, usually caused by impaired uric acid excretion and characterized by painful deposition of uric acid in the joints.

GPCR. See G protein–coupled receptor.

Gram-negative bacteria. Bacterial cells that have an outer membrane and are unable to retain the crystal violet stain.

Gram-positive bacteria. Bacterial cells that have a thick peptidoglycan cell wall and retain the crystal violet stain.

GSD. See glycogen storage disease.

Gustation. The sense of taste, involving the perception of acidity and saltiness by ion channels, or the perception of sweet, bitter, and umami substances by receptors in specialized cells.

GWAS. See genome-wide association study.

H

H. See enthalpy.

Half-reaction. The single oxidation or reduction process, involving the reduced and oxidized forms of a substance, that must be combined with another half-reaction to form a complete oxidation–reduction reaction.

Haploid. Having one set of chromosomes.

Haworth projection. A drawing of a sugar ring in which ring bonds that project in front of the plane of the paper are represented by heavy lines and ring bonds that project behind the plane of the paper are represented by light lines.

Helicase. An enzyme that unwinds DNA.

Heme. A protein prosthetic group that binds O_2 (in myoglobin and hemoglobin) or undergoes redox reactions (in cytochromes).

Henderson–Hasselbalch equation. The mathematical expression of the relationship between the pH of a solution of a weak acid and its pK: pH = pK + log ([A$^-$]/[HA]).

Hetero-. Different. In a heteropolymer, the subunits are not all identical.

Heterochromatin. Highly condensed, nonexpressed eukaryotic DNA.

Heterotroph. An organism that obtains its building materials and free energy from organic compounds produced by other organisms.

Hexose. A six-carbon sugar.

High-performance liquid chromatography (HPLC). An automated chromatographic procedure for fractionating molecules using high pressure and computer-controlled solvent delivery.

Highly repetitive DNA. See tandemly repeated DNA.

Histone code. The correlation between patterns of covalent modification of histone proteins and the transcriptional activity of the associated DNA.

Histones. Highly conserved basic proteins that form a core to which DNA is bound in a nucleosome.

Homeostasis. The maintenance of constant internal conditions.

Homo-. The same. In a homopolymer, all the subunits are identical.

Homologous chromosomes. Chromosomes with similar sequences in diploid cells; the result of combination of two haploid sets of chromosomes at fertilization.

Homologous genes. Genes that are related by evolution from a common ancestor.

Homologous proteins. Proteins that are related by evolution from a common ancestor.

Horizontal gene transfer. The transfer of genetic material between species.

Hormone. A substance that is secreted by one tissue and induces a physiological response in other tissues.

Hormone response element. A DNA sequence to which an intracellular hormone–receptor complex binds so as to regulate gene expression.

HPLC. See high-performance liquid chromatography.

Hydration. The molecular state of being surrounded by and interacting with solvent water molecules; that is, solvated by water.

Hydrogen bond. A partly electrostatic, partly covalent interaction between a donor group such as O—H or N—H and an electronegative acceptor atom such as O or N.

Hydrolysis. The cleavage of a covalent bond accomplished by adding the elements of water; the reverse of a condensation.

Hydronium ion. A proton associated with a water molecule, H_3O^+.

Hydrophilic. Having high enough polarity to readily interact with water molecules. Hydrophilic substances tend to dissolve in water.

Hydrophobic. Having insufficient polarity to readily interact with water molecules. Hydrophobic substances tend to be insoluble in water.

Hydrophobic effect. The tendency of water to minimize its contacts with nonpolar substances, thereby inducing the substances to aggregate.

Hypercholesterolemia. Elevated levels of cholesterol in the blood.

Hyperglycemia. Elevated levels of glucose in the blood.

Hypervariable loop. One of six surface loops with highly variable amino acid sequences that make up the unique antigen-binding site of an immunoglobulin.

I

IC$_{50}$. The concentration of an inhibitory substance necessary to achieve a 50% reduction in activity.

IF. See initiation factor.

Illumina sequencing. A procedure for determining the sequence of nucleotides in DNA by detecting the presence of a fluorescent marker attached to one of the four nucleotides added to a growing DNA strand.

Imine. A molecule with the formula $>$C $=$ NH.

Imino group. A portion of a molecule with the formula $>$C $=$ N—.

Immunization. The intentional exposure of the immune system to an antigen representing a pathogen in order to establish memory cells that can respond and prevent illness if the pathogen is later encountered. Also known as vaccination.

Immunoglobulin. A soluble version of the antigen receptor of a B lymphocyte, consisting of two heavy and two light chains, which can bind two antigen molecules as well as trigger additional responses. Also known as an antibody.

Imprinting. A heritable variation in the level of expression of a gene according to its parental origin.

In vitro. In the laboratory (literally, in glass).

In vivo. In a living organism.

Indel. An insertion or deletion of nucleotides in DNA, often as the result of an unrepaired DNA polymerase error during replication.

Induced fit. An interaction between a protein and its ligand that induces a conformational change in the protein that enhances the protein's interaction with the ligand.

Inhibition constant (K_I). The dissociation constant for the complex between an enzyme and a reversible inhibitor.

Initiation factor (IF). A protein that interacts with mRNA and/or the ribosome and that is required to initiate translation.

Insulin resistance. The inability of cells to respond to insulin.

Integral membrane protein. A membrane protein that is embedded in the lipid bilayer. Also called an intrinsic protein.

Intermediate. See metabolite.

Intermediate filament. A 100-Å-diameter cytoskeletal element consisting of coiled-coil polypeptide chains.

Intermembrane space. The compartment between the inner and outer mitochondrial membranes, which is equivalent to the cytosol in ionic composition.

Interspersed repetitive DNA. Sequences of DNA that are present at hundreds of thousands of copies in the human genome. Formerly known as moderately repetitive DNA.

Intrinsic protein. See integral membrane protein.

Intrinsically disordered protein. A protein whose tertiary structure includes highly flexible extended segments that can adopt various conformations.

Intron. A portion of a gene that is transcribed but excised by splicing prior to translation.

Invariant residue. A residue in a protein that is the same in all evolutionarily related proteins.

Ion exchange chromatography. A fractionation procedure in which charged molecules are selectively retained by a matrix bearing oppositely charged groups.

Ion pair. An electrostatic interaction between two ionic groups of opposite charge.

Ionic interaction. An electrostatic interaction between two groups that is stronger than a hydrogen bond but weaker than a covalent bond.

Ionization constant of water (K_w). A quantity that relates the concentrations of H^+ and OH^- in pure water: $K_w = [H^+][OH^-] = 10^{-14}$.

Ionophore It comprises diversified molecules that are able to transport specific ions across the membrane by forming a lipid ion-soluble complex.

Irregular secondary structure. A segment of a polymer in which each residue has a different backbone conformation; the opposite of regular secondary structure.

Irreversible inhibitor. A molecule that binds to and permanently inactivates an enzyme.

Isoacceptor tRNA. A tRNA that carries the same amino acid as another tRNA but has a different codon.

Isodecoder tRNA. A set of tRNAs that differ in sequence and structure but share the same anticodon and carry the same aminoacyl group.

Isoelectric point (pI). The pH at which a molecule has no net charge.

Isopeptide bond. An amide linkage between two amino acids involving an amino or carboxyl group in a side chain rather than at the α position.

Isoprenoid. A lipid constructed from fivecarbon units with an isoprene skeleton. Also called a terpenoid.

Isozymes. Different proteins that catalyze the same reaction.

K

K. See dissociation constant.

k. See rate constant.

K_a. See acid dissociation constant.

kb. Kilobase pairs; 1000 base pairs.

k_{cat}. See catalytic constant.

k_{cat}/K_M. The apparent second-order rate constant for an enzyme-catalyzed reaction; it indicates the enzyme's overall catalytic efficiency.

K_d. See dissociation constant.

K_{eq}. See equilibrium constant.

Ketogenesis. The synthesis of ketone bodies from acetyl-CoA.

Ketogenic amino acid. An amino acid whose degradation yields compounds that can be converted to fatty acids or ketone bodies but not to glucose. See also glucogenic amino acid.

Ketone bodies. Compounds (acetoacetate and 3-hydroxybutyrate) that are produced from acetyl-CoA by the liver and used as metabolic fuels in other tissues when glucose is unavailable.

Ketose. A sugar whose carbonyl group is part of a ketone.

K_I. See inhibition constant.

Kinase. An enzyme that transfers a phosphoryl group between ATP and another molecule.

Kinetics. The study of chemical reaction rates.

Kinetochore. A complex of proteins that is anchored to a chromosome and slides along a fraying microtubule, thereby pulling the chromosome toward one pole of a dividing cell.

K_M. See Michaelis constant.

K_w. See ionization constant of water.

Kwashiorkor. A form of severe malnutrition resulting from inadequate protein intake; marked by abdominal swelling and reddish hair. See also marasmus.

L

L **sugar.** A monosaccharide isomer in which the asymmetric carbon farthest from the carbonyl group has the same spatial arrangement as the chiral carbon of L-glyceraldehyde.

Lagging strand. The DNA strand that is synthesized as a series of discontinuous fragments that are later joined.

Lateral diffusion. The movement of a membrane component within one leaflet of a bilayer.

Le Châtelier's principle. The observation that a change in concentration, temperature, volume, or pressure in a system at equilibrium causes the equilibrium to shift in order to counteract the change.

Leading strand. The DNA strand that is synthesized continuously during DNA replication.

Lectin. It plays an important role in type I diabetes because lectins mimic insulin; it binds to the insulin receptor sites on cells, thereby blocking the action of insulin.

Ligand. (1) A small molecule that binds to a larger molecule. (2) A molecule or ion bound to a metal ion.

Ligase. See DNA ligase.

Light-harvesting complex. A pigment-containing protein that collects light energy in order to transfer it to a photosynthetic reaction center.

Light reactions. The photosynthetic reactions in which light energy is absorbed and used to generate NADPH and ATP.

Linear electron flow. The light-driven linear path of electrons from water through Photosystems II and I, which leads to the production of O_2, NADPH, and ATP. Also known as noncyclic electron flow.

Lineweaver–Burk plot. A rearrangement of the Michaelis–Menten equation that permits the determination of K_M and V_{max} from a linear plot.

Lipid. Any member of a broad class of macromolecules that are largely or wholly hydrophobic and therefore tend to be insoluble in water but soluble in organic solvents.

Lipid bilayer. See bilayer.

Lipid-linked protein. A protein that is anchored to a biological membrane via a covalently attached lipid.

Lipolysis. The degradation of a triacylglycerol so as to release fatty acids.

Lipoprotein. A globular particle, containing lipids and proteins, that transports lipids between tissues via the bloodstream.

Lock-and-key model. An early model of enzyme action, in which the substrate fit the enzyme like a key in a lock.

London dispersion forces. The weak van der Waals interactions between nonpolar groups as a result of fluctuations in their electron distributions that create a temporary separation of charge (polarity).

Low-barrier hydrogen bond. A short, strong hydrogen bond in which the proton is equally shared by the donor and acceptor atoms.

Lysosome. A membrane-bounded organelle in a eukaryotic cell that contains a battery of hydrolytic enzymes and that functions to digest ingested material and to recycle cell components.

Lysozyme. It primarily hydrolyzes the β-1, 4-glycosidic bonds in peptidoglycan, a major component of bacterial cell walls, breaking down the peptidoglycan network and causing the bacterial cell to lose its structural integrity.

M

Maillard reaction. The reaction of a monosaccharide carbonyl group with an unprotonated amino group to yield a reactive product that can generate additional reaction products, including substances that give flavor and color to cooked foods.

Major groove. The wider of the two grooves on a DNA double helix.

Marasmus. Body wasting due to inadequate intake of all types of foods. See also kwashiorkor.

Mass action ratio. The ratio of the product of the concentrations of reaction products to that of the reactants.

Mass spectrometry. A technique for identifying molecules by measuring the mass-to-charge ratios of gas-phase ions, such as peptide fragments.

Matrix. See mitochondrial matrix.

Mediator. A protein complex that interacts with transcription factors and RNA polymerase to regulate gene expression in eukaryotes.

Meiosis. A variation of mitosis to generate gametes with a haploid set of chromosomes.

Melting temperature (T_m). The midpoint temperature of the melting curve for the thermal denaturation of a macromolecule. For a lipid, the temperature of transition from an ordered crystalline state to a more fluid state.

Membrane potential ($\Delta\psi$). The difference in electrical charge across a membrane.

Membraneless organelle. An intracellular aggregation of molecules, including proteins or RNA, with a consistency that differs from the surrounding solution and that results from liquid–liquid phase separation.

Memory cell. A B or T lymphocyte that originates during an immune response to an antigen and that can mount a more effective response during a subsequent encounter with the same antigen.

Messenger RNA (mRNA). A ribonucleic acid whose sequence is complementary to that of a protein-coding gene in DNA.

Metabolic acidosis. A low blood pH caused by the overproduction or retention of hydrogen ions.

Metabolic alkalosis. A high blood pH caused by the excessive loss of hydrogen ions.

Metabolic fuel. A molecule that can be oxidized to provide free energy for an organism.

Metabolic pathway. A series of enzyme-catalyzed reactions by which one substance is transformed into another.

Metabolic syndrome. A set of symptoms related to obesity, including inulin resistance, atherosclerosis, and hypertension.

Metabolically irreversible reaction. A reaction whose value of ΔG is large and negative so that the reaction cannot proceed in reverse.

Metabolism. The total of all degradative and biosynthetic cellular reactions.

Metabolite. A reactant, intermediate, or product of a metabolic reaction.

Metabolome. The complete set of metabolites produced by a cell or tissue.

Metabolomics. The study of all the metabolites produced by a cell or tissue.

Metagenomics. The analysis of all the DNA present in a sample in order to identify the genomes of the different species present.

Metal ion catalysis. A catalytic mechanism that requires the presence of a metal ion to lower the free energy of a reaction's transition state.

Metamorphic protein. A protein that has more than one stable tertiary structure and can easily switch between them.

Metastasis. The spread of cancerous cells from the site of a primary tumor to a secondary site.

Micelle. A globular aggregate of amphiphilic molecules in aqueous solution that are oriented such that polar segments form the surface of the aggregate and the nonpolar segments form a core that is out of contact with the solvent.

Michaelis constant (K_M). For an enzyme that follows the Michaelis–Menten model, $K_M = (k_{-1} + k_2)/k_1$; K_M is equal to the substrate concentration at which the reaction velocity is half-maximal.

Michaelis–Menten equation. A mathematical expression that describes the activity of an enzyme in terms of the substrate concentration ([S]), the enzyme's maximal velocity (V_{max}), and its Michaelis constant (K_M): $v_0 = V_{max}[S]/(K_M + [S])$.

Micro RNA (miRNA).   A 20- to 25-nucleotide double-stranded RNA that binds to and inactivates a number of complementary mRNA molecules in RNA interference.

Microarray.   A collection of DNA sequences that hybridize with RNA molecules and that can therefore be used to identify active genes. Also called a DNA chip.

Microbiome.   The collective genetic material belonging to the microbiota.

Microbiota.   The collection of microorganisms that live in or on the human body.

Microenvironment.   A group's immediate neighbors, whose chemical and physical properties may affect the group.

Microfilament.   See actin filament.

Microtubule.   A 240-Å-diameter cytoskeletal element consisting of a hollow tube of polymerized tubulin subunits.

Microvesicle.   See exosome.

Minor groove.   The narrower of the two grooves on a DNA double helix.

(–) end.   The end of a polymeric filament where growth is slower. See also (+) end.

miRNA.   See micro RNA.

Mismatch repair.   A DNA repair pathway that removes and replaces mispaired nucleotides on a newly synthesized DNA strand.

Mitochondrial matrix.   The gel-like solution of enzymes, substrates, cofactors, and ions in the interior of the mitochondrion.

Mitochondrion (pl. mitochondria).   The double-membrane-enveloped eukaryotic organelle in which aerobic metabolic reactions occur, including those of the citric acid cycle, fatty acid oxidation, and oxidative phosphorylation.

Mitosis.   The process of allocating equivalent sets of chromosomes to daughter cells during eukaryotic cell division.

Mixed inhibition.   A form of enzyme inhibition in which an inhibitor binds to the enzyme such that it causes the apparent V_{max} to decrease and the apparent K_M to increase or decrease.

Mobilization.   The process in which polysaccharides, triacylglycerols, and proteins are degraded to make metabolic fuels available.

Moderately repetitive DNA.   See interspersed repetitive DNA.

Molecular chaperone.   A protein that binds to unfolded or misfolded proteins in order to promote their normal folding.

Monoclonal antibody.   One of the identical immunoglobulins produced from a single B lymphocyte or from clones (copies) of that cell.

Monogenetic disease.   A disease linked to a defect in a single gene.

Monomer.   A structural unit from which a polymer is built up.

Monomorphic protein.   A protein with a single stable tertiary structure.

Monosaccharide.   A carbohydrate consisting of a single sugar molecule.

Motor protein.   An intracellular protein that couples the free energy of ATP hydrolysis to molecular movement relative to another protein that often acts as a track for the linear movement of the motor protein.

mRNA.   See messenger RNA.

Multienzyme complex.   A group of noncovalently associated enzymes that catalyze two or more sequential steps in a metabolic pathway.

Multifunctional enzyme.   A protein that carries out more than one chemical reaction.

Mutagen.   An agent that induces a mutation in an organism.

Mutation.   A heritable alteration in an organism's genetic material.

Myelin sheath.   The multilayer coating of sphingomyelin-rich membranes that insulates a mammalian neuron.

N

N-linked oligosaccharide.   An oligosaccharide linked to the amide group of a protein Asn residue.

N-terminus.   The end of a polypeptide that has a free amino group.

Nanopore sequencing.   A method that identifies each residue in a polynucleotide strand by monitoring how the current changes as the strand passes through a protein pore in a polarized membrane.

Native structure.   The fully folded conformation of a macromolecule.

Natural selection.   The evolutionary process by which the continued existence of a replicating entity depends on its ability to survive and reproduce under the existing conditions.

ncRNA.   See noncoding RNA.

Near-equilibrium reaction.   A reaction whose ΔG value is close to zero, so that it can operate in either direction depending on the substrate and product concentrations.

Negative effector.   A substance that diminishes an enzyme's activity through allosteric inhibition.

Nernst equation.   An expression of the relationship between the actual ($\mathcal{E}$) and standard reduction potential ($\mathcal{E}°'$) of a substance A: $\mathcal{E} = \mathcal{E}°' - (RT/n\mathcal{F}) \ln([A_{reduced}]/[A_{oxidized}])$.

Neurotransmitter.   A substance released by a nerve cell to alter the activity of a target cell.

Neutral solution.   A solution whose pH is equal to 7.0 ($[H^+] = 10^{-7}$ M).

Nick.   A single-strand break in a double-stranded nucleic acid.

Nick translation.   The progressive movement of a single-strand break (nick) in DNA through the actions of an exonuclease that removes residues followed by a polymerase that replaces them.

Nitrification.   The conversion of ammonia (NH_3) to nitrate (NO_3^-).

Nitrogen cycle.   A set of reactions, including nitrogen fixation, nitrification, and denitrification, for the interconversion of different forms of nitrogen.

Nitrogen fixation.   The process by which atmospheric N_2 is converted to a biologically useful form such as NH_3.

NMR spectroscopy.   See nuclear magnetic resonance spectroscopy.

Noncoding RNA (ncRNA).   An RNA molecule that is not translated into protein.

Noncoding strand.   The DNA strand that has a sequence complementary (except for the replacement of U with T) to the transcribed RNA; it is the template strand. Also called the antisense strand.

Noncompetitive inhibition.   A form of enzyme inhibition in which an inhibitor binds to an enzyme such that the apparent V_{max} decreases but K_M is not affected.

Noncyclic electron flow.   See linear electron flow.

Nonessential amino acid.   An amino acid that an organism can synthesize from common intermediates.

Nonhomologous end-joining.   A ligation process that repairs a double-stranded break in DNA.

Nonreducing sugar.   A saccharide with an anomeric carbon that has formed a glycosidic bond and cannot therefore act as a reducing agent.

Nonspontaneous process.   A thermodynamic process that has a net increase in free energy ($\Delta G > 0$) and can occur only with the input of free energy from outside the system. See also endergonic reaction.

Nuclear magnetic resonance (NMR) spectroscopy.   A spectroscopic method in which the signals emitted by atomic nuclei in a magnetic field can be used to determine the three-dimensional structure of a molecule.

Nuclear pore complex.   A large assembly of nucleoporin proteins that form a three-ring structure spanning the inner and outer nuclear membranes, with a gel-like pore that permits transport factors to escort proteins and RNA between the nucleus and cytosol.

Nuclease.   An enzyme that degrades nucleic acids.

Nucleic acid.   A polymer of nucleotide residues. The major nucleic acids are deoxyribonucleic acid (DNA) and ribonucleic acid (RNA). Also known as a polynucleotide.

Nucleolus.   A dense membraneless organelle inside nucleus, where ribosomal RNA genes are transcribed and ribosomal proteins are assembled with rRNA to make the large and small subunits of the ribosome.

Nucleophile.   A compound containing an electron-rich group. A nucleophile (nucleus-lover) reacts with an electrophile (electron-lover).

Nucleoside.   A compound consisting of a nitrogenous base linked to a five-carbon sugar (ribose or deoxyribose).

Nucleosome.   The disk-shaped complex of a histone octamer and DNA that represents the fundamental unit of DNA organization in eukaryotes.

Nucleotide.   A compound consisting of a nucleoside esterified to one or more phosphate groups. Nucleotides are the monomeric units of nucleic acids.

Nucleotide excision repair.   A DNA repair pathway in which a damaged single-stranded segment of DNA is removed and replaced with normal DNA.

O

O-linked oligosaccharide.   An oligosaccharide linked to the hydroxyl group of a protein Ser or Thr side chain.

Odorant.   The ligand for an olfactory (smell) receptor.

Okazaki fragments.   The short segments of DNA formed in the discontinuous lagging-strand synthesis of DNA.

Olfaction.   The sense of smell, mediated by olfactory receptors in sensory neurons in the nose.

Oligonucleotide.   A polynucleotide consisting of a few nucleotide residues.

Oligopeptide.   A polypeptide consisting of a few amino acid residues.

Oligosaccharide.   A polymeric carbohydrate containing a few monosaccharide residues. In glycoproteins, the groups are known as N-linked and O-linked oligosaccharides.

Omega-3 fatty acid.   A fatty acid with a double bond starting at the third carbon from the methyl (omega) end of the molecule.

Oncogene. A mutant gene that interferes with the normal regulation of cell growth and contributes to cancer.

Open reading frame (ORF). A portion of the genome that potentially codes for a protein.

Operon. A prokaryotic genetic unit that consists of several genes with related functions that are transcribed as a single mRNA molecule.

Opioid. A derivative or analog of morphine, which blocks pain perception but may also depress respiration.

Ordered mechanism. A multisubstrate reaction with a compulsory order of substrate binding to the enzyme.

ORF. See open reading frame.

Organelle. A membrane-enclosed compartment, with a specialized function, inside a eukaryotic cell.

Orientation effects. See proximity and orientation effects.

Osmosis. The movement of solvent from a region of low solute concentration to a region of high solute concentration.

Osteogenesis imperfecta. A disease caused by mutations in collagen genes and characterized by bone fragility and deformation.

Oxidant. See oxidizing agent.

Oxidation. A reaction in which a substance loses electrons.

Oxidative phosphorylation. The process by which the free energy obtained from the oxidation of metabolic fuels is used to generate ATP from $ADP + P_i$.

Oxidizing agent. A substance that can accept electrons, thereby becoming reduced. Also called an oxidant.

Oxyanion hole. A cavity in the active site of a serine protease that accommodates the reactants during the transition state and thereby lowers its energy.

Oxyhemoglobin. Hemoglobin that contains bound oxygen or is in the oxygen-binding conformation.

P

P site. The ribosomal binding site that accommodates a peptidyl–tRNA.

Palindrome. A segment of DNA that has the same sequence on each strand when read in the $5' \rightarrow 3'$ direction.

Parallel β sheet. See β sheet.

Partial oxygen pressure (pO_2). The concentration of gaseous O_2 in units of torr.

Passive transport. The thermodynamically spontaneous protein-mediated transmembrane movement of a substance from high to low concentration.

Pasteur effect. The greatly increased sugar consumption of yeast grown under anaerobic conditions compared to that of yeast grown under aerobic conditions.

Pathogen. An infectious agent, such as a virus, bacterium, or microscopic eukaryote, that causes illness.

PCR. See polymerase chain reaction.

Pentose. A five-carbon sugar.

Pentose phosphate pathway. A pathway for glucose degradation that yields ribose-5-phosphate and NADPH.

Peptide. A short polypeptide.

Peptide bond. An amide linkage between the α-amino group of one amino acid and the α-carboxylate group of another. Peptide bonds link the amino acid residues in a polypeptide.

Peptidoglycan. The cross-linked polysaccharides and polypeptides that form bacterial cell walls.

Peripheral membrane protein. A protein that is weakly associated with the surface of a biological membrane. Also called an extrinsic protein.

Peroxisome. A eukaryotic organelle with specialized oxidative functions, including fatty acid degradation.

p_{50}. The ligand concentration (or pressure for a gaseous ligand) at which a binding protein such as hemoglobin is half-saturated with ligand.

pH. A quantity used to express the acidity of a solution, equivalent to $-\log[H^+]$.

Phage. See bacteriophage.

Phagophore. The membrane-enclosed compartment that forms a cuplike shape around materials to be degraded by autophagy.

Phagosome. See autophagosome.

Pharmacokinetics. The behavior of a drug in the body, including its metabolism and excretion.

φ (phi). The torsion angle describing rotation around the N—Cα bond in a peptide group.

Phosphatase. An enzyme that hydrolyzes phosphoryl ester groups.

Phosphodiester bond. The linkage in which a phosphate group is esterified to two alcohol groups (e.g., two ribose units that join the adjacent nucleotide residues in a polynucleotide).

Phosphoinositide signaling system. A signal transduction pathway in which hormone binding to a cell-surface receptor induces phospholipase C to catalyze the hydrolysis of phosphatidylinositol bisphosphate to yield the second messengers inositol trisphosphate and diacylglycerol.

Phospholipase. An enzyme that hydrolyzes one or more bonds in a glycerophospholipid.

Phosphorolysis. The cleavage of a chemical bond by the substitution of a phosphate group rather than water.

Photoautotroph. An organism that obtains its building materials from inorganic compounds and its free energy from sunlight.

Photon. A packet of light energy.

Photooxidation. A mode of decay of an excited molecule, in which oxidation occurs through the transfer of an electron to an acceptor molecule.

Photophosphorylation. The synthesis of ATP from $ADP + P_i$ coupled to the dissipation of a proton gradient that has been generated through light-driven electron transport.

Photoreceptor. A light-absorbing molecule, or pigment.

Photorespiration. The consumption of O_2 and evolution of CO_2 by plants (a dissipation of the products of photosynthesis), a consequence of the competition between O_2 and CO_2 for ribulose bisphosphate carboxylase.

Photosynthesis. The light-driven incorporation of CO_2 into organic compounds.

pI. See isoelectric point.

P_i. Inorganic phosphate or a phosphoryl group: HPO_4^- or PO_4^{3-}.

Pigment. See photoreceptor.

Ping pong mechanism. An enzymatic reaction in which one or more products are released before all the substrates have bound to the enzyme.

Pinocytosis. Endocytosis of small amounts of extracellular fluid and solutes.

pK. A quantity used to express the tendency for an acid to donate a proton (dissociate); equal to $-\log K$, where K is the dissociation constant.

Planck's law. An expression for the energy (E) of a photon: $E = hc/\lambda$, where c is the speed of light, λ is its wavelength, and h is Planck's constant (6.626×10^{-34} J·s).

Plasmid. A small circular DNA molecule that autonomously replicates and may be used as a vector for recombinant DNA.

(+) end. The end of a polymeric filament where growth is faster. See also (−) end.

P:O ratio. The ratio of the number of molecules of ATP synthesized from $ADP + P_i$ to the number of atoms of oxygen reduced.

Point mutation. The substitution of one base for another in DNA, arising from mispairing during DNA replication or from chemical alterations of existing bases.

Polarity. Having an uneven distribution of charge.

Poly(A) tail. The sequence of adenylate residues that is post-transcriptionally added to the 3′ end of eukaryotic mRNAs.

Polygenic disease. A disease linked to variations in more than one gene.

Polymer. A molecule consisting of numerous smaller units that are linked together in an organized manner.

Polymerase. An enzyme that catalyzes the addition of nucleotide residues to a polynucleotide. DNA is synthesized by DNA polymerase, and RNA is synthesized by RNA polymerase.

Polymerase chain reaction (PCR). A procedure for amplifying a segment of DNA by repeated rounds of replication centered between primers that hybridize with the two ends of the DNA segment of interest. See also quantitative PCR (qPCR) and reverse transcriptase quantitative PCR (RT-qPCR).

Polynucleotide. See nucleic acid.

Polypeptide. A polymer consisting of amino acid residues linked in linear fashion by peptide bonds.

Polyprotein. A polypeptide that undergoes proteolysis after its synthesis to yield several separate protein molecules.

Polyprotic acid. A substance that has more than one acidic proton and therefore has multiple ionization states.

Polysaccharide. A polymeric carbohydrate containing multiple monosaccharide residues. Also called a glycan.

Polysome. An mRNA transcript bearing multiple ribosomes in the process of translating the mRNA.

Porin. A β barrel protein in the outer membrane of bacteria, mitochondria, or chloroplasts that forms a weakly solute-selective pore.

Positive effector. A substance that boosts an enzyme's activity through allosteric activation.

Post-translational processing. The removal or derivatization of amino acid residues following their incorporation into a polypeptide.

pO_2. See partial oxygen pressure.

Primary structure. The sequence of residues in a polymer.

Primase. The enzyme that synthesizes a segment of RNA to be extended by DNA polymerase during DNA replication.

Primer. An oligonucleotide that base pairs with a template polynucleotide strand and is extended through template-directed polymerization.

Prion. An infectious protein that causes its cellular counterparts to misfold and aggregate, thereby leading to the development of a disease such as transmissible spongiform encephalopathy.

Probe. A labeled single-stranded DNA or RNA segment that can hybridize with a DNA or RNA of interest in a screening procedure.

Processing. See RNA processing and posttranslational processing.

Processivity. A property of a motor protein or other enzyme that undergoes many reaction cycles before dissociating from its track or substrate.

Product inhibition. A form of enzyme inhibition in which the reaction product acts as a competitive inhibitor.

Prokaryote. A unicellular organism that lacks a membrane-bounded nucleus. All bacteria and archaea are prokaryotes.

Proliferation. Uncontrolled multiplication of the cells.

Promoter. The DNA sequence at which RNA polymerase binds to initiate transcription.

Proofreading. An additional catalytic activity of an enzyme, which acts to correct errors made by the primary enzymatic activity.

Prosthetic group. An organic group (such as a coenzyme) that is permanently associated with a protein.

Protease. An enzyme that catalyzes the hydrolysis of peptide bonds.

Protease inhibitor. An agent, often a protein, that reacts incompletely with a protease so as to inhibit further proteolytic activity.

Proteasome. A multiprotein complex with a hollow cylindrical core in which cellular proteins are degraded to peptides in an ATP-dependent process.

Protein. A macromolecule that consists of one or more polypeptide chains.

Proteoglycan. An extracellular aggregate of protein and glycosaminoglycans.

Proteome. The complete set of proteins synthesized by a cell.

Proteomics. The study of all the proteins synthesized by a cell.

Protofilament. One of the linear polymers of tubulin subunits that forms a microtubule.

Proton jumping. The rapid movement of a proton among hydrogen-bonded water molecules.

Proton wire. A series of hydrogen-bonded water molecules and protein groups that can relay protons from one site to another.

Protonmotive force. The free energy of the electrochemical proton gradient that forms during electron transport.

Proximity and orientation effects. A catalytic mechanism in which reacting groups are brought close together in an enzyme active site to accelerate the reaction.

Pseudogene. A nonfunctional genomic copy of a true gene, which has lost its function through evolution.

ψ (psi). The torsion angle describing rotation around the Cα—C bond in a peptide group.

Purine. A derivative of the compound purine, such as the nucleotide base adenine or guanine.

Purinosome. A complex of enzymes that catalyze the reactions of purine nucleotide synthesis.

Pyrimidine. A derivative of the compound pyrimidine, such as the nucleotide base cytosine, uracil, or thymine.

Q

Q cycle. The cyclic flow of electrons involving a semiquinone intermediate in Complex III of mitochondrial electron transport and in photosynthetic electron transport.

qPCR. See quantitative PCR.

Quantitative PCR (qPCR). A variation of the polymerase chain reaction in which the amount of amplified DNA is monitored as it is generated. Also known as real-time PCR.

Quantum yield. The ratio of carbon atoms fixed or oxygen molecules produced to the number of photons absorbed by the photosynthetic machinery.

Quaternary structure. The spatial arrangement of a macromolecule's individual subunits.

Quorum sensing. The ability of cells to monitor population density by detecting the concentrations of extracellular substances.

R

R. See gas constant.

R group. A symbol for a variable portion of a molecule, such as the side chain of an amino acid.

R state. One of two conformations of an allosteric protein; the other is the T state.

Raft. An area of a lipid bilayer with a distinct lipid composition and near-crystalline consistency.

Ramachandran diagram. This diagram visually illustrates the permissible or allowed combinations of ϕ and ψ angles in the peptide backbone, as well of those which are disallowed or unfavourable due to steric constraints.

Random mechanism. A multisubstrate reaction without a compulsory order of substrate binding to the enzyme.

Rate constant (k). The proportionality constant between the velocity of a chemical reaction and the concentration(s) of the reactant(s).

Rate-determining reaction. The slowest step in a multistep sequence, such as a metabolic pathway, whose rate determines the rate of the entire sequence.

Rate equation. A mathematical expression for the time-dependent progress of a reaction as a function of reactant concentration.

Rational drug design. The synthesis of more effective drugs based on detailed knowledge of the target molecule's structure and function.

Reactant. One of the starting materials for a chemical reaction.

Reaction center. A chlorophyll-containing protein where photooxidation takes place.

Reaction coordinate. A line representing the progress of a reaction, part of a graphical presentation of free energy changes during a reaction.

Reaction specificity. The ability of an enzyme to discriminate between possible substrates and to catalyze a single type of chemical reaction.

Reactive oxygen species. Free radicals and their reaction products, such as $\cdot O_2^-$, H_2O_2, and $\cdot OH$, derived enzymatically and nonenzymatically from O_2.

Reading frame. The grouping of nucleotides in sets of three whose sequence corresponds to a polypeptide sequence.

Real-time PCR. See quantitative PCR.

Receptor. A binding protein that is specific for its ligand and elicits a discrete biochemical effect when its ligand is bound.

Receptor-mediated endocytosis. Endocytosis of an extracellular component as a result of its specific binding to a cell surface receptor.

Receptor tyrosine kinase. A cell-surface receptor whose intracellular domains become active as Tyr-specific kinases as a result of extracellular ligand binding.

Recombinant DNA. A DNA molecule containing DNA segments from different sources.

Recombination. The exchange of polynucleotide strands between separate DNA segments; recombination is one mechanism for repairing damaged DNA by allowing a homologous segment to serve as a template for replacement of the damaged bases.

Redox center. A group that can undergo an oxidation–reduction reaction.

Redox reaction. A chemical reaction in which one substance is reduced and another substance is oxidized.

Reducing agent. A substance that can donate electrons, thereby becoming oxidized. Also called a reductant.

Reducing sugar. A saccharide with an anomeric carbon that has not formed a glycosidic bond and can therefore act as a reducing agent.

Reductant. See reducing agent.

Reduction. A reaction in which a substance gains electrons.

Reduction potential ($\mathcal{E}$). A measure of the tendency of a substance to gain electrons.

Regular secondary structure. A segment of a polymer in which the backbone adopts a regularly repeating conformation; the opposite of irregular secondary structure.

Release factor (RF). A protein that recognizes a stop codon and causes a ribosome to terminate polypeptide synthesis.

Renaturation. The refolding of a denatured macromolecule so as to regain its native conformation.

Replication. The process of making an identical copy of a DNA molecule. During DNA replication, the parental polynucleotide strands separate so that each can direct the synthesis of a complementary daughter strand, resulting in two complete DNA double helices.

Replication fork. The point in a replicating DNA molecule where the two parental strands separate in order to serve as templates for the synthesis of new strands.

Replisome. The complex of proteins that unwind the DNA helix at the replication fork and synthesize complementary DNA strands using both parental strands as templates.

Repressor. A protein that binds at or near a gene so as to prevent its transcription.

Residue. A term for a monomeric unit after it has been incorporated into a polymer.

Resonance stabilization. The effect of delocalization of electrons in a molecule that cannot be depicted by a single structural diagram.

Respirasome. A respiratory supercomplex containing Complexes I, III, and IV of the mitochondrial electron transport chain, which carries out the entire process of electron transfer from NADH to O_2.

Respiration. See cellular respiration.

Respiratory acidosis. A low blood pH caused by insufficient elimination of CO_2 (carbonic acid) by the lungs.

Respiratory alkalosis. A high blood pH caused by the excessive loss of CO_2 (carbonic acid) from the lungs.

Restriction digest. The generation of a set of DNA fragments by the action of a restriction endonuclease.

Restriction endonuclease. A bacterial enzyme that cleaves a specific DNA sequence.

Reverse transcriptase. A DNA polymerase that uses RNA as its template.

Reverse transcriptase quantitative PCR (RT-qPCR). A variation of the polymerase chain reaction in which cellular messenger RNA is first reverse-transcribed to DNA before being amplified and quantified by PCR; this method provides information about expressed DNA sequences.

RF.   See release factor.

Ribonucleic acid.   See RNA.

Ribosomal RNA (rRNA).   The RNA molecules that comprise most of the mass of the ribosome and catalyze peptide bond formation.

Ribosome.   The RNA-and-protein particle that synthesizes polypeptides under the direction of mRNA.

Ribosome profiling.   A technique based on sequencing and identifying mRNA segments undergoing translation, which are protected from ribonuclease digestion by the presence of an active ribosome.

Ribosome recycling factor (RRF).   A protein that binds to a ribosome after protein synthesis to prepare it for another round of translation.

Riboswitch.   A segment of mRNA that changes conformation in response to binding a ligand in order to expose a sequence that can affect transcription termination (in prokaryotes), splicing (in eukaryotes), or some other aspect of gene expression, such as translation initiation.

Ribozyme.   An RNA molecule that has catalytic activity.

RNA (Ribonucleic acid).   A polymer of ribonucleotides, such as messenger RNA (mRNA), transfer RNA (tRNA), and ribosomal RNA (rRNA).

RNA interference (RNAi).   A phenomenon in which short RNA segments direct the degradation of complementary mRNA, thereby inhibiting gene expression.

RNA polymerase.   See polymerase.

RNA processing.   The addition, removal, or modification of nucleotides in an RNA molecule that is necessary to produce a fully functional RNA.

RNA world.   A hypothetical time before the evolution of DNA or protein, when RNA stored genetic information and functioned as a catalyst.

RRF.   See ribosome recycling factor.

rRNA.   See ribosomal RNA.

RT-qPCR.   See reverse transcriptase quantitative PCR.

S

S.   See entropy.

Saccharide.   See carbohydrate.

Salvage pathway.   A pathway that reincorporates an intermediate of nucleotide degradation into a new nucleotide, thereby minimizing the need for the nucleotide biosynthetic pathways.

Saturated fatty acid.   A fatty acid that does not contain any double bonds in its hydrocarbon chain.

Saturation.   The state in which all of a macromolecule's ligand-binding sites are occupied by ligands.

Schiff base.   An imine that forms between an amine and an aldehyde or ketone.

Scissile bond.   The bond that is to be cleaved during a proteolytic reaction.

Scramblase.   A translocase that catalyzes the equilibration of membrane lipids between the two bilayer leaflets.

SDS-PAGE.   A form of polyacrylamide gel electrophoresis in which denatured polypeptides are separated by size in the presence of the detergent sodium dodecyl sulfate. See electrophoresis.

Second messenger.   An intracellular ion or molecule that acts as a signal for an extracellular event such as ligand binding to a cell-surface receptor.

Second-order reaction.   A reaction whose rate is proportional to the square of the concentration of one reactant or to the product of the concentrations of two reactants.

Secondary active transport.   Transmembrane transport of one substance that is driven by the free energy of an existing gradient of a second substance.

Secondary structure.   The local spatial arrangement of a polymer's backbone atoms without regard to the conformations of its substituent side chains.

Selection.   A technique for distinguishing cells that contain a particular feature, such as resistance to an antibiotic.

Semiconservative replication.   The mechanism of DNA duplication in which each new molecule contains one strand from the parent molecule and one newly synthesized strand.

Sense strand.   See coding strand.

Serine protease.   A peptide-hydrolyzing enzyme that has a reactive Ser residue in its active site.

Set-point.   A body weight that is maintained through regulation of fuel metabolism and that resists change when an individual attempts to alter fuel consumption or expenditure.

Signal peptide.   A short sequence in a membrane or secretory protein that binds to the signal recognition particle in order to direct the translocation of the protein across a membrane.

Signal recognition particle (SRP).   A complex of protein and RNA that recognizes membrane and secretory proteins and mediates their binding to a membrane for translocation.

Signal transduction.   The process by which an extracellular signal transmits information to the cell interior by binding to a cell-surface receptor such that binding triggers a series of intracellular events.

Silencer.   A DNA sequence some distance from the transcription start site, where a repressor of transcription may bind.

Single-nucleotide polymorphism (SNP).   A nucleotide sequence variation in the genomes of two individuals from the same species.

siRNA.   See small interfering RNA.

Sister chromatids.   The two identical molecules that are the products of eukaryotic DNA replication and that remain together until cell division.

Size-exclusion chromatography.   A procedure in which macromolecules are separated on the basis of their size and shape. Also called gel filtration chromatography.

Small interfering RNA (siRNA).   A 20- to 25-nucleotide double-stranded RNA that targets for destruction a fully complementary mRNA molecule in RNA interference.

Small nuclear RNA (snRNA).   Highly conserved RNAs that participate in eukaryotic mRNA splicing.

Small nucleolar RNA (snoRNA).   RNA molecules that direct the sequence-specific methylation of eukaryotic rRNA transcripts.

snoRNA.   See small nucleolar RNA.

SNP.   See single-nucleotide polymorphism.

snRNA.   See small nuclear RNA.

Solute.   The substance that is dissolved in water or another solvent to make a solution.

Solvation.   The state of being surrounded by solvent molecules.

Specificity pocket.   A cavity on the surface of a serine protease, whose chemical characteristics determine the identity of the substrate residue on the N-terminal side of the bond to be cleaved.

Sphingolipid.   An amphipathic lipid containing an acyl group, a palmitate derivative, and a polar head group attached to a serine backbone. In sphingomyelins, the head group is a phosphate derivative.

Sphingomyelin.   See sphingolipid.

Spliceosome.   A complex of protein and snRNA that carries out the splicing of immature mRNA molecules.

Splicing.   The process by which introns are removed and exons are joined to produce a mature RNA transcript.

Spontaneous process.   A thermodynamic process that has a net decrease in free energy ($\Delta G < 0$) and occurs without the input of free energy from outside the system. See also exergonic reaction.

SRP.   See signal recognition particle.

Stacking interactions.   The stabilizing van der Waals interactions between successive (stacked) bases in a polynucleotide.

Staggered conformation.   Where the torsion angle is 180°.

Standard conditions.   A set of conditions including a temperature of 25°C, a pressure of 1 atm, and reactant concentrations of 1 M. Biochemical standard conditions include a pH of 7.0 and a water concentration of 55.5 M.

Standard free energy change ($\Delta G°'$).   The force that drives reactants to reach their equilibrium values when the system is in its biochemical standard state.

Standard reduction potential ($\mathcal{E}°'$).   A measure of the tendency of a substance to gain electrons (to be reduced) under standard conditions.

Steady state.   A set of conditions under which the formation and degradation of individual components are balanced such that the system does not change over time.

Stereocilia.   Microfilament-stiffened cell processes on the surface of cells in the inner ear, which are deflected in response to sound waves.

Steroid hormone.   A subset of bioactive lipids, occupy a crucial position in the intricate web of physiological signaling and regulation within organisms.

Sticky ends.   Single-stranded extensions of DNA that are complementary, often because they have been generated by the action of the same restriction endonuclease.

Stroma.   The gel-like solution of enzymes and small molecules in the interior of the chloroplast; the site of carbohydrate synthesis.

Substrate.   A reactant in an enzymatic reaction.

Substrate-level phosphorylation.   The transfer of a phosphoryl group to ADP that is directly coupled to another chemical reaction.

Subunit.   One of several polypeptide chains that make up a protein.

Sugar–phosphate backbone.   The chain of (deoxy)ribose groups linked by phosphodiester bonds in a polynucleotide chain.

Suicide substrate.   A molecule that chemically inactivates an enzyme only after undergoing part of the normal catalytic reaction.

Supercoiling.   A topological state of DNA in which the helix is underwound or overwound so that the molecule tends to writhe or coil up on itself.

Supercomplex.   An assembly of mitochondrial electron-transport proteins that may contain variable numbers of Complexes I, III, and IV.

Symport. Transport that involves the simultaneous transmembrane movement of two molecules in the same direction.

Synaptic vesicle. A vesicle loaded with neurotransmitters to be released from the end of an axon.

T

T lymphocyte. A type of white blood cell that responds to the presence of an antigen in part by releasing signals that help regulate immunoglobulin production in B lymphocytes.

T state. One of two conformations of an allosteric protein; the other is the R state.

Tandemly repeated DNA. Clusters of short DNA sequences that are repeated side-by-side and are present at millions of copies in the human genome. Formerly known as highly repetitive DNA.

Tastant. The ligand for a receptor that detects sweet, bitter, or umami substances.

TATA box. A eukaryotic promoter element with an AT-rich sequence located upstream from the transcription start site.

Tautomer. One of a set of isomers that differ only in the positions of their hydrogen atoms.

Telomerase. An enzyme that uses an RNA template to polymerize deoxynucleotides and thereby extend the 3′-ending strand of a eukaryotic chromosome.

Telomere. The end of a linear eukaryotic chromosome, which consists of tandem repeats of a short G-rich sequence on the 3′-ending strand and its complementary sequence on the 5′-ending strand.

Terpenoid. See isoprenoid.

Tertiary structure. The entire three-dimensional structure of a single-chain polymer, including the conformations of its side chains.

Tetramer. An assembly consisting of four monomeric units.

Tetrose. A four-carbon sugar.

Thalassemia. A hereditary disease caused by insufficient synthesis of hemoglobin, which results in anemia.

Thermogenesis. The process of generating heat by muscular contraction or by metabolic reactions.

Thick filament. A muscle cell structural element that is composed of several hundred myosin molecules.

Thin filament. A muscle cell structural element that consists primarily of an actin filament.

Thioester. A compound containing an ester linkage to a sulfur rather than an oxygen atom.

Thioester bond. An ester linkage to a sulfur rather than an oxygen atom.

3′ end. The terminus of a polynucleotide whose C3′ is not esterified to another nucleotide residue.

Thylakoid. The membranous structure in the interior of a chloroplast that is the site of the light reactions of photosynthesis.

T_m. See melting temperature.

Topoisomerase. An enzyme that alters DNA supercoiling by breaking and resealing one or both strands.

Torsion angle. The angle described by successive bonds in a polymeric chain. The torsion angles φ and ψ describe the conformation of a polypeptide backbone.

Trace element. An element that is present in small quantities in a living organism.

Transamination. The transfer of an amino group from an amino acid to an α-keto acid to yield a new α-keto acid and a new amino acid.

Transcription. The process by which RNA is synthesized using a DNA template, thereby transferring genetic information from the DNA to the RNA.

Transcription factor. A protein that promotes the transcription of a gene by binding to DNA sequences at or near the gene or by interacting with other proteins that do so. See also general transcription factor.

Transcriptome. The set of all the RNA molecules produced by a cell.

Transcriptomics. The study of the genes that are transcribed in a certain cell type or at a certain time.

Transfer RNA (tRNA). The small L-shaped RNAs that deliver specific amino acids to ribosomes according to the sequence of a bound mRNA.

Transgenic organism. An organism that stably expresses a foreign gene.

Transition mutation. A point mutation in which one purine (or pyrimidine) is replaced by another purine (or pyrimidine).

Transition state. The point of highest free energy, or the structure that corresponds to that point, in the reaction coordinate diagram of a chemical reaction.

Transition state analog. A stable substance that geometrically and electronically resembles the transition state of a reaction and that therefore may inhibit an enzyme that catalyzes the reaction.

Translation. The process of transforming the information contained in the nucleotide sequence of an RNA to the corresponding amino acid sequence of a polypeptide as specified by the genetic code.

Translocase. An enzyme that catalyzes the movement of a substance from one side of a membrane to the other.

Translocation. The movement of tRNA and mRNA, relative to the ribosome, that occurs following formation of a peptide bond and that allows the next mRNA codon to be translated.

Translocon. The complex of membrane proteins that mediates the transmembrane movement of a polypeptide.

Transmissible spongiform encephalopathy (TSE). A fatal neurodegenerative disease caused by infection with a prion.

Transpeptidation. The ribosomal process in which the peptidyl group attached to a tRNA is transferred to the aminoacyl group of another tRNA, forming a new peptide bond and lengthening the polypeptide by one residue at its C-terminus.

Transposable element. A segment of DNA, sometimes including genes, that can move (be copied) from one position to another in a genome.

Transverse diffusion. The movement of a membrane component from one leaflet of a bilayer to the other. Also called flip-flop.

Transversion mutation. A point mutation in which a purine is replaced by a pyrimidine or vice versa.

Treadmilling. The addition of monomeric units to one end of a polymer and their removal from the opposite end such that the length of the polymer remains unchanged.

Triacylglycerol. A lipid in which three fatty acids are esterified to a glycerol backbone. Also called a triglyceride.

Triglyceride. See triacylglycerol.

Trimer. An assembly consisting of three monomeric units.

Triose. A three-carbon sugar.

Triple helix. The right-handed helical structure formed by three left-handed helical polypeptide chains in collagen.

Trisaccharide. A carbohydrate consisting of three monosaccharides.

tRNA. See transfer RNA.

TSE. See transmissible spongiform encephalopathy.

Tumor suppressor gene. A gene whose loss or mutation may lead to cancer.

Turnover number. See catalytic constant.

U

Uncompetitive inhibition. A form of enzyme inhibition in which an inhibitor binds to an enzyme–substrate complex such that the apparent V_{max} and K_M are both decreased to the same extent.

Uncoupler. A substance that allows the proton gradient across a membrane to dissipate without ATP synthesis so that electron transport proceeds without oxidative phosphorylation.

Unimolecular reaction. A reaction involving one molecule.

Uniport. Transport that involves transmembrane movement of a single molecule.

Unsaturated fatty acid. A fatty acid that contains at least one double bond in its hydrocarbon chain.

Urea cycle. A cyclic metabolic pathway in which amino groups are converted to urea for disposal.

Usher syndrome. A genetic disease characterized by profound deafness and retinitis pigmentosa that leads to blindness, caused in some cases by a defective myosin protein.

V

v. Velocity (rate) of a reaction.

Vaccination. See immunization.

van der Waals interaction. A weak noncovalent association between molecules that arises from the attractive forces between polar groups (dipole–dipole interactions) or between nonpolar groups whose fluctuating electron distribution gives rise to temporary dipoles (London dispersion forces).

van der Waals radius. The distance from an atom's nucleus to its effective electronic surface.

Variable domain. A structural domain of an immunoglobulin heavy or light chain, whose highly variable surface loops define the antigen-binding site.

Variable residue. A position in a polypeptide that is occupied by different residues in evolutionarily related proteins; its substitution has little or no effect on protein function.

Vector. A DNA molecule, such as a plasmid, that can accommodate a segment of foreign DNA.

Vesicle. A fluid-filled sac enclosed by a lipid-bilayer membrane.

Virion. A single virus particle.

Virus. A nonliving infectious particle consisting of genetic information (DNA or RNA) inside a protein capsule that may be surrounded by a lipid membrane. A virus requires a host cell in order to replicate.

Vitamin. A metabolically required substance that cannot be synthesized by an animal and must therefore be obtained from the diet.

V_{max}. Maximal velocity of an enzymatic reaction.

v_0. Initial velocity of an enzymatic reaction.

W

Warburg effect. The increased rate of glycolysis observed in cancerous tissues.

Wobble hypothesis. An explanation for the nonstandard base pairing between tRNA and mRNA at the third codon position, which allows a tRNA to recognize more than one codon.

X

X-Ray crystallography. A method for determining three-dimensional molecular structures from the diffraction pattern produced by exposing a crystal of a molecule to a beam of X-rays.

Y

Y. See fractional saturation.

Z

Z. The net charge of an ion.

Z-scheme. A Z-shaped diagram indicating the electron carriers and their reduction potentials in the photosynthetic electron transport system of plants and cyanobacteria.

Zinc finger. A protein structural motif consisting of 20–60 residues, including Cys and His residues to which one or two Zn^{2+} ions are tetrahedrally coordinated.

Zymogen. The inactive precursor (proenzyme) of a proteolytic enzyme.

Page references followed by T indicate tables. Page references followed by F indicate figures.

Nucleic Acid Bases, Nucleosides, and Nucleotides

Base Formula	Base (X = H)	Nucleoside (X = ribose or deoxyribose)	Nucleotide (X = ribose phosphate or deoxyribose phosphate)
	Adenine (A)	Adenosine	Adenosine monophosphate (AMP)
	Guanine (G)	Guanosine	Guanosine monophosphate (GMP)
	Cytosine (C)	Cytidine	Cytidine monophosphate (CMP)
	Thymine (T)	Thymidine	Thymidine monophosphate (TMP)
	Uracil (U)	Uridine	Uridine monophosphate (UMP)

The Standard Genetic Code

First Position (5′ end)	Second Position								Third Position (3′ end)
	U		C		A		G		
U	UUU	Phe	UCU	Ser	UAU	Tyr	UGU	Cys	U
	UUC	Phe	UCC	Ser	UAC	Tyr	UGC	Cys	C
	UUA	Leu	UCA	Ser	UAA	Stop	UGA	Stop	A
	UUG	Leu	UCG	Ser	UAG	Stop	UGG	Trp	G
C	CUU	Leu	CCU	Pro	CAU	His	CGU	Arg	U
	CUC	Leu	CCC	Pro	CAC	His	CGC	Arg	C
	CUA	Leu	CCA	Pro	CAA	Gln	CGA	Arg	A
	CUG	Leu	CCG	Pro	CAG	Gln	CGG	Arg	G
A	AUU	Ile	ACU	Thr	AAU	Asn	AGU	Ser	U
	AUC	Ile	ACC	Thr	AAC	Asn	AGC	Ser	C
	AUA	Ile	ACA	Thr	AAA	Lys	AGA	Arg	A
	AUG	Met	ACG	Thr	AAG	Lys	AGG	Arg	G
G	GUU	Val	GCU	Ala	GAU	Asp	GGU	Gly	U
	GUC	Val	GCC	Ala	GAC	Asp	GGC	Gly	C
	GUA	Val	GCA	Ala	GAA	Glu	GGA	Gly	A
	GUG	Val	GCG	Ala	GAG	Glu	GGG	Gly	G

Common Functional Groups and Linkages in Biochemistry

Compound name	Structure[a]	Functional group
Amine[b]	RNH_2 or RNH_3^+ R_2NH or $R_2NH_2^+$ R_3N or R_3NH^+	$-N\langle$ or $\overset{+}{-N}-$ (amino group)
Alcohol	ROH	—OH (hydroxyl group)
Thiol	RSH	—SH (sulfhydryl group)
Ether	ROR	—O— (ether linkage)
Aldehyde	$R-\overset{O}{\overset{\|}{C}}-H$	$-\overset{O}{\overset{\|}{C}}-$ (carbonyl group), $R-\overset{O}{\overset{\|}{C}}-$ (acyl group)
Ketone	$R-\overset{O}{\overset{\|}{C}}-R$	$-\overset{O}{\overset{\|}{C}}-$ (carbonyl group), $R-\overset{O}{\overset{\|}{C}}-$ (acyl group)
Carboxylic acid[b] (Carboxylate)	$R-\overset{O}{\overset{\|}{C}}-OH$ or $R-\overset{O}{\overset{\|}{C}}-O^-$	$-\overset{O}{\overset{\|}{C}}-OH$ (carboxyl group) or $-\overset{O}{\overset{\|}{C}}-O^-$ (carboxylate group)
Ester	$R-\overset{O}{\overset{\|}{C}}-OR$	$-\overset{O}{\overset{\|}{C}}-O-$ (ester linkage)
Thioester	$R-\overset{S}{\overset{\|}{C}}-OR$	$-\overset{S}{\overset{\|}{C}}-O-$ (thioester linkage)
Amide	$R-\overset{O}{\overset{\|}{C}}-NH_2$ $R-\overset{O}{\overset{\|}{C}}-NHR$ $R-\overset{O}{\overset{\|}{C}}-NR_2$	$-\overset{O}{\overset{\|}{C}}-N\langle$ (amido group)
Imine[b]	$R{=}NH$ or $R{=}NH_2^+$ $R{=}NR$ or $R{=}NHR^+$	$\rangle C{=}N-$ or $\rangle C{=}\overset{+}{N}\langle^H$ (imino group)
Phosphoric acid ester[b, c]	$R-O-\overset{O}{\overset{\|}{P}}-OH$ or $\quad\quad\overset{\|}{OH}$ $R-O-\overset{O}{\overset{\|}{P}}-O^-$ $\quad\quad\overset{\|}{O^-}$	$-O-\overset{O}{\overset{\|}{P}}-O-$ (phosphoester linkage) $\quad\quad\overset{\|}{OH}$ $-\overset{O}{\overset{\|}{P}}-OH$ or $-\overset{O}{\overset{\|}{P}}-O^-$ (phosphoryl group) $\overset{\|}{OH}\quad\quad\overset{\|}{O^-}$
Diphosphoric acid ester[b, d]	$R-O-\overset{O}{\overset{\|}{P}}-O-\overset{O}{\overset{\|}{P}}-OH$ or $\quad\quad\overset{\|}{OH}\quad\quad\overset{\|}{OH}$ $R-O-\overset{O}{\overset{\|}{P}}-O-\overset{O}{\overset{\|}{P}}-O^-$ $\quad\quad\overset{\|}{O^-}\quad\quad\overset{\|}{O^-}$	$-O-\overset{O}{\overset{\|}{P}}-O-\overset{O}{\overset{\|}{P}}-O-$ (phosphoanhydride linkage) $\quad\quad\overset{\|}{OH}\quad\quad\overset{\|}{OH}$ $-\overset{O}{\overset{\|}{P}}-O-\overset{O}{\overset{\|}{P}}-OH$ or $-\overset{O}{\overset{\|}{P}}-O-\overset{O}{\overset{\|}{P}}-O^-$ $\overset{\|}{OH}\quad\quad\overset{\|}{OH}\quad\quad\overset{\|}{O^-}\quad\quad\overset{\|}{O^-}$ (diphosphoryl group, pyrophosphoryl group)

[a] R represents any carbon-containing group. In a molecule with more than one R group, the groups may be the same or different.

[b] Under physiological conditions, these groups are ionized and hence bear a positive or negative charge.

[c] When R = H, the molecule is inorganic phosphate (abbreviated P_i), usually $H_2PO_4^-$ or HPO_4^{2-}.

[d] When R = H, the molecule is pyrophosphate (abbreviated PP_i).

Useful Constants

Avogadro's number	6.02×10^{23} molecules $\cdot$ mol^{-1}
Gas constant (R)	8.314 J $\cdot$ K^{-1} $\cdot$ mol^{-1}
Faraday ($\mathcal{F}$)	$96{,}485$ J $\cdot$ V^{-1} $\cdot$ mol^{-1}
Kelvin (K)	$°C + 273$

Key Equations

Henderson–Hasselbalch equation

$$pH = pK + \log \frac{[A^-]}{[HA]}$$

Michaelis–Menten equation

$$v_0 = \frac{V_{max}[S]}{K_M + [S]}$$

Lineweaver–Burk equation

$$\frac{1}{v_0} = \left(\frac{K_M}{V_{max}}\right)\frac{1}{[S]} + \frac{1}{V_{max}}$$

Nernst equation

$$\mathcal{E} = \mathcal{E}^{°\prime} - \frac{RT}{n\mathcal{F}} \ln \frac{[A_{reduced}]}{[A_{oxidized}]} \quad \text{or} \quad \mathcal{E} = \mathcal{E}^{°\prime} - \frac{0.026\,\text{V}}{n} \ln \frac{[A_{reduced}]}{[A_{oxidized}]}$$

Thermodynamics equations

$$\Delta G = \Delta H - T\Delta S$$
$$\Delta G^{°\prime} = -RT \ln K_{eq}$$
$$\Delta G = \Delta G^{°\prime} + RT \ln \frac{[C][D]}{[A][B]}$$
$$\Delta G^{°\prime} = -n\mathcal{F}\Delta \mathcal{E}^{°\prime}$$